Handbook of Experimental Pharmacology

Volume 193

Constance N. Wilson
Endacea, Inc.,
2 Davis Drive P.O. Box 12076
Research Triangle Park NC
27709–2076
USA
cwilson@endacea.nctda.org

S. Jamal Mustafa
West Virginia University
Health Sciences Center
Department of Basic Pharmaceutical Sciences
2267 Health Sciences S.
P.O. Box 9104
Morgantown WV 26506–9104
USA
smustafa@hsc.wvu.edu

ISSN 0171-2004 e-ISSN 1865-0325
ISBN 978-3-540-89614-2 e-ISBN 978-3-540-89615-9
DOI 10.1007/978-3-540-89615-9
Springer Dordrecht Heidelberg London New York

Library of Congress Control Number: 2009921799

Cover design: SPi Publisher Services

Printed on acid-free paper

Springer is part of Springer Science+Business Media (www.springer.com)

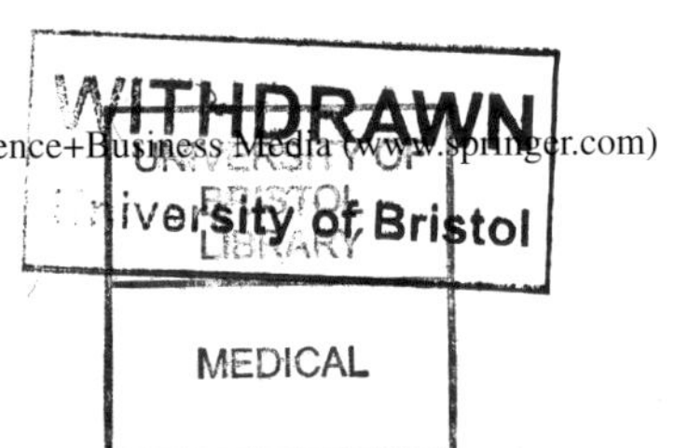

Constance N. Wilson • S. Jamal Mustafa
Editors

Adenosine Receptors in Health and Disease

Contributors

M.P. Abbracchio, S. Bar-Yehuda, P.G. Baraldi, A. Bauer, L. Belardinelli, I. Biaggioni, M.R. Blackburn, P.A. Borea, R. Brown, A.R. Carta, S. Ceruti, J.W. Chisholm, J.P. Clancy, G. Com, G. Cristalli, B.N. Cronstein, A.K. Dhalla, E. Elzein, I. Feoktistov, P. Fishman, S. Gessi, J.P. Headrick, K. Ishiwata, A.A. Ivanov, K.A. Jacobson, P. Jenner, R.V. Kalla, W.F. Kiesman, K.N. Klotz, A.M. Klutz, R.D. Lasley, M. Morelli, R.R. Morrison, E. Morschl, C.E. Müller, S.J. Mustafa, A. Nadeem, H. Osswald, C.P. Page, A. Pelleg, J.D. Powell, D. Preti, G.M. Reaven, J.A. Ribeiro, A.M. Sebastião, D. Spina, T.W. Stone, M. Synowitz, M.A. Tabrizi, B. Teng, D.K. Tosh, V. Vallon, C.O. Vance, R. Volpini, C.N. Wilson, J. Zablocki

Preface

Since the first description of adenosine receptors 30 years ago, based on the valuable scientific discoveries and contributions by individuals working in the field of adenosine receptor (AR) research around the world elucidating AR molecular structure, pharmacology, and function, and the intensive efforts in chemistry to identify selective ligands for ARs, molecules that target all four AR subtypes, A_1, A_{2A}, A_{2B}, and A_3ARs have advanced to clinical trials with a recent FDA approval and an NDA submission. As contributing authors to this volume of the Handbook of Experimental Pharmacology (HEP), "Adenosine Receptors in Health and Disease", these scientists describe the impact of their discoveries and contributions, as well as those by others, on defining the role of ARs in a number of different diseases and the advancement of this field of science and medicine. Since the inception of this area of basic science research, it has truly been an incredible experience for all of us in academia and the pharmaceutical industry to participate in and observe this captivating and fast-moving field advance from the bench to the clinic.

In the A_1AR area, A_1AR agonists have been tested in humans for the following conditions: atrial arrhythmias (Tecadenoson, Selodenoson and PJ-875); Type II diabetes (GR79236, ARA, and CVT-3619); and angina (BAY-68–4986). New partial A_1AR agonists are in development, including CVT-3619, that have the potential to provide enhanced insulin sensitivity without cardiovascular side effects and tachyphylaxis. Based on the diuretic/natriuretic and renoprotective effects of A_1ARs in the kidney, A_1AR antagonists are currently in late-stage clinical development, including KW3902 (rolofylline, Phase III), BG9928 (Adentri®, Phase III), and SLV320 (Phase II), for acute decompensated heart failure (ADHF) with renal impairment. All three have high affinity for the human A_1AR subtype and demonstrate diuretic and renal protective effects in humans with ADHF with renal impairment. Moreover, to date, two PET ligands have been successfully tested in humans for the visualization of A_1ARs in the brain, [^{18}F]CPFPX and [^{11}C]MPDX. The use of these PET imaging agents may provide valuable insights into sleep disorders and neurodegenerative disorders, e.g. Alzheimer's Disease (AD).

In the A_{2A}AR area, A_{2A}AR agonists are currently in clinical trials, with one recent FDA approval and one NDA submission for the following indications: myocardial perfusion imaging (recently FDA approved Lexiscan^{TM}, regadenoson, CVT-3146; CorVue, binodenoson, MRE-0470, WRC-0470, NDA submission; apadenoson, ATL-146e), and wound healing (sonedenoson, MRE 0094). A_{2A}AR antagonists have been tested in clinical trials for Parkinson's Disease (PD), including istradefylline, KW 6002; BIIB014, V2006; and SCH 58261. Moreover, two A_{2A}AR PET ligands have been successfully tested in humans for the visualization of A_{2A}ARs in the brain, $[^{11}C]$TMSX and $[^{11}C]$KW-6002. The use of these PET imaging agents may provide valuable insights into PD, psychiatric diseases, and perhaps drug addiction.

In the A_{2B}AR area, a mixed A_{2B}/A_3AR antagonist, QAF 805, was tested in humans with asthma and an A_{2B}AR antagonist, CVT 6883, is in clinical development for asthma and currently is in Phase I clinical trials.

In the A_3AR area, A_3AR agonists are in clinical trials for the following indications: rheumatoid arthritis, dry eye syndrome, psoriasis (CF 101), and liver cancer, hepatitis, and liver regeneration (CF 102).

A number of other molecules that target AR subtypes and that are at various stages of preclinical development appear to be promising drug candidates for asthma, inflammation, sepsis, ischemia-reperfusion organ injury, fibrosis, ADHF with renal impairment, PD, AD, cancer, diabetes, obesity, glaucoma, and as coronary vasodilators for myocardial imaging. Moreover, based on the growing scientific evidence supporting the role of ARs in other neurodegenerative diseases and drug abuse and addiction, it is expected that AR-based drug candidates will enter clinical trials to target these diseases. We look forward with anticipation to the advancement of these promising drug candidates towards the clinic and their approval. We expect they will significantly alter the life styles and outcomes of patients with these diseases.

It has been our pleasure to work closely with the world-renowned AR scientists who contributed to this volume of the HEP. We are extremely grateful for their invaluable contributions to this area of science and medicine, which will be realized for generations to come. In this volume of the HEP, all of us have tried to present chapters with up-to-date information about the role of ARs in health and disease and the importance of ARs as drug targets for a number of different diseases. It was our intention to present this information in such a way that those who are not as closely associated with this area of science and medicine and with different interests and backgrounds can understand and appreciate its significance. We are especially indebted to Springer for providing us the opportunity to contribute this volume of the HEP and to Susanne Dathe for her support and successfully managing this project.

Research Triangle Park, NC, *Constance N. Wilson*
Morgantown, WV, *S. Jamal Mustafa*

Contents

Contributors

M.P. Abbracchio Laboratory of Molecular and Cellular Pharmacology of Purinergic Transmission, Department of Pharmacological Sciences, University of Milan, via Balzaretti, 9 Milan, Italy, mariapia.abbracchio@unimi.it

S. Bar-Yehuda Can-Fite BioPharma, 10 Bareket St., Kiryat Matalon, Petach Tikva, 49170, Israel, sara@canfite.co.il

Pier Giovanni Baraldi Dipartimento di Scienze Farmaceutiche, Università di Ferrara, Via Fossato di Mortara 17-19, 44100 Ferrara, Italy, baraldi@unife.it

Andreas Bauer Institute of Neuroscience and Biophysics (INB-3), Research Center Jülich, 52425 Jülich, Germany, an.bauer@fz-juelich.de

Luiz Belardinelli Department of Pharmacological Sciences, CV Therapeutics, 3172 Porter Drive, Palo Alto, CA 94304, USA, luiz.belardinelli@cvt.com

Italo Biaggioni Division of Clinical Pharmacology, 556 RRB, Vanderbilt University, 2220 Pierce Ave, Nashville, TN 37232, USA, italo.biaggioni@vanderbilt.edu

Michael R. Blackburn Department of Biochemistry and Molecular Biology, The University of Texas–Houston Medical School, 6431 Fannin, Houston, TX 77030, USA, michael.r.blackburn@uth.tmc.edu

P.A. Borea University of Ferrara, Department of Clinical and Experimental Medicine, Pharmacology Unit Via Fossato di Mortara 17-19, 44100 Ferrara, Italy, bpa@dns.unife.it

Rachel Brown Sackler Institute of Pulmonary Pharmacology, Division of Pharmaceutical Science, School of Biomedical and Health Science, King's College London, London SE1 1UL UK, rachel.2.brown@kcl.ac.uk

Anna R. Carta Department of Toxicology and Center of Excellence for Neurobiology of Addiction, University of Cagliari, via Ospedale 72, 09124 Cagliari, Italy, acarta@unica.it

S. Ceruti Laboratory of Molecular and Cellular Pharmacology of Purinergic Transmission, Department of Pharmacological Sciences, University of Milan, via Balzaretti, 9 Milan, Italy, stefania.ceruti@unimi.it

Jeffrey W. Chisholm Department of Pharmacological Sciences, CV Therapeutics, 3172 Porter Drive, Palo Alto, CA 94304, USA, jeff.chisholm@cvt.com

J.P. Clancy Department of Pediatrics, University of Alabama, 620 ACC, 1600 7th Ave South, Birmingham, AL 35233, UK, jpclancy@peds.uab.edu

Gulnur Com University of Arkansas Medical Sciences, Arkansas Children's Hospital, 800 Marshall Street, slot 512-17, Little Rock, AR 72202-3591, USA, comgulnur@uams.edu

Gloria Cristalli Dipartimento di Scienze Chimiche, Università di Camerino, via S. Agostino 1, 62032 Camerino (MC), Italy, gloria.cristalli@unicam.it

Bruce N. Cronstein Division of Clinical Pharmacology, Department of Medicine, NYU School of Medicine, 550 First Ave., NBV16N1, New York, NY 10016, USA, cronsb01@med.nyu.edu

Arvinder K. Dhalla Department of Pharmacological Sciences, CV Therapeutics, 3172 Porter Drive, Palo Alto, CA 94304, USA, arvinder.dhalla@cvt.com

Elfatih Elzein CV Therapeutics Inc., 3172 Porter Drive, Palo Alto, CA 94304, USA, elfatih.elzein@cvt.com

Igor Feoktistov Division of Cardiovascular Medicine, 360 PRB, Vanderbilt University, 2220 Pierce Ave, Nashville, TN 37232-6300, USA, igor.feoktistov@vanderbilt.edu

P. Fishman Can-Fite BioPharma, 10 Bareket St., Kiryat Matalon, Petach Tikva, 49170, Israel, pnina@canfite.co.il

S. Gessi University of Ferrara, Department of Clinical and Experimental Medicine, Pharmacology Unit Via Fossato di Mortara 17–19, 44100 Ferrara, Italy, gss@dns.unife.it

John P. Headrick Heart Foundation Research Centre, School of Medical Science, Griffith University, Southport, QLD 4217, Australia, j.headrick@griffith.edu.au

Kiichi Ishiwata Positron Medical Center, Tokyo Metropolitan Institute of Gerontology, 1-1, Nakacho, Itabashi, Tokyo 173-0011, Japan, ishiwata@pet.tmig.or.jp

Andrei A. Ivanov Bldg. 8A, Rm. B1A-23, NIH, NIDDK, LBC, Bethesda, MD 20892-0810, USA, Ivanovan@niddk.nih.gov

Kenneth A. Jacobson Molecular Recognition Section, Laboratory of Bio-organic Chemistry, National Institute of Diabetes and Digestive and Kidney Diseases, National Institutes of Health, Bldg. 8A, Rm. B1A-19, Bethesda, MD 20892-0810, USA, kajacobs@helix.nih.gov

Peter Jenner Neurodegenerative Diseases Research Centre, School of Health and Biomedical Sciences, King's College, London, SE1 1UL, UK, peter.jenner@kcl.ac.uk

Rao V Kalla Department of Bioorganic Chemistry, CV Therapeutics Inc., 3172 Porter Drive, Palo Alto, CA 94304, USA, rao.kalla@cvt.com

William F. Kiesman Biogen Idec, 14 Cambridge Center, Cambridge, MA 02142, USA, william.kiesman@biogenidec.com

K.N. Klotz Universität Würzburg, Institut für Pharmakologie und Toxikologie, Versbacher Str. 9, 97078 Würzburg, Germany, klotz@toxi.uni-wuerzburg.de

Athena M. Klutz Bldg. 8A, Rm. B1A-23, NIH, NIDDK, LBC, Bethesda, MD 20892-0810, KlutzA@niddk.nih.gov

Robert D. Lasley Department of Physiology, Wayne State University School of Medicine, Detroit, MI 48201, USA, rlasley@med.wayne.edu

Micaela Morelli Department of Toxicology and Center of Excellence for Neurobiology of Addiction, University of Cagliari, via Ospedale 72, 09124 Cagliari, Italy, morelli@unica.it and CNR Institute of Neuroscience, Cagliari, Italy

R. Ray Morrison Division of Critical Care Medicine, St. Jude Children's Research Hospital, Memphis, TN, USA

Eva Morschl Department of Biochemistry and Molecular Biology, The University of Texas–Houston Medical School, 6431 Fannin, Houston, TX 77030, USA, eva.morschl@uth.tmc.edu

Christa E. Müller University of Bonn, Pharmaceutical Institute, Pharmaceutical Chemistry I, An der Immenburg 4, 53121 Bonn, Germany, christa.mueller@uni-bonn.de

S. Jamal Mustafa Department of Physiology and Pharmacology, School of Medicine, West Virginia University, Morgantown, WV 26505-9229, USA, smustafa@hsc.wvu.edu

Ahmed Nadeem Department of Physiology and Pharmacology, School of Medicine, West Virginia University, Morgantown, WV 26505-9229, USA, anadeem@hsc.wvu.edu

Hartmut Osswald Department of Pharmacology, Medical Faculty, University of Tübingen, Wilhelmstrasse 56, 72074 Tübingen, Federal Republic of Germany, hartmut.osswald@uni-tuebingen.de

Clive P. Page Sackler Institute of Pulmonary Pharmacology, Division of Pharmaceutical Science, School of Biomedical and Health Science, King's College London, London SE1 1UL UK, clive.page@kcl.ac.uk

Amir Pelleg Department of Medicine, College of Medicine, Drexel University, Philadelphia, PA, USA, ap33@drexel.edu

J.D. Powell The Sidney Kimmel Comprehensive Cancer Center, Johns Hopkins University School of Medicine, CRB I Building, Room 443, 1650 Orleans Street, Baltimore, MD 21231, USA, poweljo@jhmi.edu

Delia Preti Dipartimento di Scienze Farmaceutiche, Università di Ferrara, Via Fossato di Mortara 17-19, 44100 Ferrara, Italy, delia.preti@unife.it

Gerald M. Reaven Division of Cardiovascular Medicine, Stanford University School of Medicine, 300 Pasteur Dr. CVRC MC: 5406, Stanford, CA 94305, USA, gReaven@cvmed.stanford.edu

Joaquim A. Ribeiro Institute of Pharmacology and Neurosciences, Institute of Molecular Medicine, University of Lisbon, 1649-028 Lisbon, Portugal, jaribeiro@fm.ul.pt

Ana M. Sebastião Institute of Pharmacology and Neurosciences, Institute of Molecular Medicine, University of Lisbon, 1649-028 Lisbon, Portugal, anaseb@fm.ul.pt

Domenico Spina Sackler Institute of Pulmonary Pharmacology, Division of Pharmaceutical Science, School of Biomedical and Health Science, King's College London, London SE1 1UL UK, domenico.spina@kcl.ac.uk

T.W. Stone Institute of Biomedical and Life Sciences, University of Glasgow, Glasgow G12 8QQ, UK, T.W.Stone@bio.gla.ac.uk

M. Synowitz Department of Neurosurgery, Charité-Universitätsmedizin Berlin, Augustenburger Platz 1, 13353 Berlin, Germany, Michael.Synowitz@charite.de

Mojgan Aghazadeh Tabrizi Dipartimento di Scienze Farmaceutiche, Università di Ferrara, Via Fossato di Mortara 17–19, 44100 Ferrara, Italy, tbj@unife.it

Bunyen Teng Department of Physiology and Pharmacology, School of Medicine, West Virginia University, Morgantown, WV 26505-9229, USA

Dilip K. Tosh Bldg. 8A, Rm. B1A-15, NIH, NIDDK, LBC, Bethesda, MD 20892-0810, USA, ToshD@niddk.nih.gov

Volker Vallon Departments of Medicine and Pharmacology, University of California San Diego and VA San Diego Healthcare System, 3350 La Jolla Village Dr (9151), San Diego, CA 92161, USA, vvallon@ucsd.edu

Constance O. Vance Endacea, Inc. 2 Davis Drive, P.O Box 12076, Research Triangle Park, NC 27709-2076, USA, cvance@endacea.nctda.org

Rosaria Volpini Dipartimento di Scienze Chimiche, Università di Camerino, via S. Agostino 1, 62032 Camerino (MC), Italy, rosaria.volpini@unicam.it

Constance N. Wilson Endacea, Inc., P.O. Box 12076 (Mail), 2 Davis Drive (Courier), Research Triangle Park, NC 27709-2076, USA, cwilson@endacea.nctda.org

Jeff Zablocki Department of Bioorganic Chemistry, CV Therapeutics Inc., 3172 Porter Drive, Palo Alto, CA 94304, USA, jeff.zablocki@cvt.com

Introduction to Adenosine Receptors as Therapeutic Targets

Kenneth A. Jacobson

Contents

Abstract Adenosine acts as a cytoprotective modulator in response to stress to an organ or tissue. Although short-lived in the circulation, it can activate four subtypes of G protein-coupled adenosine receptors (ARs): A_1, A_{2A}, A_{2B}, and A_3. The alkylxanthines caffeine and theophylline are the prototypical antagonists of ARs, and their stimulant actions occur primarily through this mechanism. For each of the four AR subtypes, selective agonists and antagonists have been introduced and used to develop new therapeutic drug concepts. ARs are notable among the GPCR family in the number and variety of agonist therapeutic candidates that have been proposed. The selective and potent synthetic AR agonists, which are typically much longer lasting in the body than adenosine, have potential therapeutic applications based on their anti-inflammatory (A_{2A} and A_3), cardioprotective (preconditioning by A_1

K.A. Jacobson (✉)
Molecular Recognition Section, Laboratory of Biooorganic Chemistry, National Institute of Diabetes and Digestive and Kidney Diseases, National Institutes of Health, Bldg. 8A, Rm. B1A-19, Bethesda, MD 20892-0810, USA
kajacobs@helix.nih.gov

C.N. Wilson and S.J. Mustafa (eds.), *Adenosine Receptors in Health and Disease*, Handbook of Experimental Pharmacology 193, DOI 10.1007/978-3-540-89615-9_1,

and A_3 and postconditioning by A_{2B}), cerebroprotective (A_1 and A_3), and antinociceptive (A_1) properties. Potent and selective AR antagonists display therapeutic potential as kidney protective (A_1), antifibrotic (A_{2A}), neuroprotective (A_{2A}), and antiglaucoma (A_3) agents. AR agonists for cardiac imaging and positron-emitting AR antagonists are in development for diagnostic applications. Allosteric modulators of A_1 and A_3 ARs have been described. In addition to the use of selective agonists/antagonists as pharmacological tools, mouse strains in which an AR has been genetically deleted have aided in developing novel drug concepts based on the modulation of ARs.

Keywords Adenosine receptors · G protein-coupled receptors · Purines · Nucleosides · Imaging · Allosteric modulation · Agonists · Antagonists

Abbreviations

ADHF	Acute decompensated heart failure
ADP	Adenosine diphosphate
AMP	Adenosine 5′-monophosphate
AMP579	[1*S*-[1α, 2β, 3β, 4α(*S**)]]-4-[7-[[1-[(3-Chlorothien-2-yl)methyl] propyl]amino]-3*H*-imidazo[4,5-*b*]pyrid-3-yl]-*N*-ethyl 2,3-dihydroxycyclopentanecarboxamide
AR	Adenosine receptor
ATP	Adenosine triphosphate
BAY 60–6583	2-[6-Amino-3,5-dicyano-4-[4-(cyclopropylmethoxy)phenyl] pyridin-2-ylsulfanyl]acetamide
BAY 68–4986	6-Amino-2-(2-(4-chlorophenyl)thiazol-4-ylthio)-4-(4-(2-hydroxyethoxy)phenyl)-5-isocyanonicotinonitrile
BG9719	1,3-Dipropyl-8-(2-(5,6-epoxy)norbornyl)xanthine
BG9928	3-[4-(2,6-Dioxo-1,3-dipropyl-2,3,6,7-tetrahydro-1*H*-purin-8-yl)-bicyclo[2.2.2]oct-1-yl]-propionic acid
BIIB014	3-(4-Amino-3-methylbenzyl)-7-(2-furyl)-3*H*- [1,2,3]triazolo [4,5-*d*]pyrimidine-5-amine (V2006)
CD39	Apyrase
CD73	Ecto-5′-nucleotidase
CF101	N^6-(3-Iodobenzyl)-5′-*N*-methylcarboxamidoadenosine (IB-MECA)
CF102	2-Chloro-N^6-(3-iodobenzyl)-5′-*N*-methylcarboxamidoadenosine (Cl-IB-MECA)
CP-608,039	(2*S*, 3*S*, 4*R*, 5*R*)-3-Amino-5-{6-[5-chloro-2-(3-methylisoxazol-5-ylmethoxy)benzylamino]purin-9-yl-l-4-hydroxytetrahydrofuran-2-carboxylic acid methylamide
CP-532,903	(2*S*, 3*S*, 4*R*, 5*R*)-3-Amino-5-{6-[2,5-dichlorobenzylamino]purin-9-yl-l-4-hydroxytetrahydrofuran-2-carboxylic acid methylamide

CPFPX	8-Cyclopentyl-1-propyl-3-(3-fluoropropyl)-xanthine
CVT-3146	1-[6-Amino-9-[(2*R*, 3*R*, 4*S*, 5*R*)-3,4-dihydroxy-5-(hydroxymethyl)oxolan-2-yl]purin-2-yl]-*N*-methylpyrazole-4-carboxamide
CVT-6883	3-Ethyl-1-propyl-8-[1-(3-trifluoromethylbenzyl)-1*H*-pyrazol-4-yl]-3,7-dihydropurine-2,6-dione
EL	Extracellular loop
ENT	Equilibrative nucleoside transporter
E-NTPDase	Ectonucleoside triphosphate diphosphohydrolase
ERK	Extracellular receptor signal-induced kinase
FK 453	(+)-(*R*)-(1-(*E*)-3-(2-Phenylpyrazolo(1,5-*a*)pyridin-3-yl)acryl)-2-piperidine ethanol
FR194921	2-(1-Methyl-4-piperidinyl)-6-(2-phenylpyrazolo[1,5-*a*]pyridin-3-yl)-3(2*H*)-pyridazinone
GPCRs	G protein-coupled receptors
GR79236	N^6-[(1*S*, 2*S*)-2-Hydroxycyclopentyl]adenosine
GRKs	G-protein-coupled receptor kinases
IL	Intracellular loop
KW3902	8-(Noradamantan-3-yl)-1,3-dipropylxanthine
KW6002	8-[(*E*)-2-(3,4-Dimethoxyphenyl)vinyl]-1,3-diethyl-7-methylpurine-2,6-dione
L-97-1	3-[2-(4-Aminophenyl)-ethyl]-8-benzyl-7-{2-ethyl-(2-hydroxy-ethyl)-amino]-ethyl}-1-propyl-3,7-dihydro-purine-2,6-dione
MAP	Mitogen-activated protein
MAPK	Mitogen-activated protein kinases
MRE0094	2-[2-(4-Chlorophenyl)ethoxy]adenosine
MRE-0470	2-[{Cyclohexylmethylene}hydrazino]adenosine (WRC-0470, binodenoson)
MRS5147	(1′*R*, 2′*R*, 3′*S*, 4′*R*, 5′*S*)-4′-[2-Chloro-6-(3-bromobenzylamino)-purine]-2′, 3′-*O*-dihydroxybicyclo-[3.1.0]hexane
N-0861	(±)-N^6-Endonorbornan-2-yl-9-methyladenine
NNC-21-0136	2-Chloro-N^6-[(*R*)-[(2-benzothiazolyl)thio]-2-propyl]-adenosine
OT-7999	5-*N*-Butyl-8-(4-trifluoromethylphenyl)-3*H*-[1,2,4]triazolo-[5,1-*i*]purine
PET	Positron emission tomography
PI3K	Phosphoinositide-3 kinase
T-62	(2-Amino-4,5,6,7-tetrahydrobenzo[*b*]thiophen-3-yl)-(4-chlorophenyl)-methanone
SLV-320	4-[(2-Phenyl-7*H*-pyrrolo[2,3-*d*]pyrimidin-4-yl)amino]-*trans*-cyclohexanol
SDZ WAG94	N^6-Cyclohexyl-2′-*O*-methyl-adenosine
TM	Transmembrane helix
VER6947	2-Amino-*N*-benzyl-6-(furan-2-yl)-9*H*-purine-9-carboxamide
VER7835	2-Amino-6-(furan-2-yl)-*N*-(thiophen-2-ylmethyl)-9*H*-purine-9-carboxamide

V2006	see BIIB014
WRC-0571	8-(*N*-Methylisopropyl)amino-N^6-(5′-endohydroxy-endonorbornan-2-yl-9-methyladenine
ZM241385	4-2-[7-Amino-2-(2-furyl)-1,2,4-triazolo[1,5-*a*][1,3,5]triazin-5-yl-amino]ethylphenol

1 Introduction

Extracellular adenosine acts as a cytoprotective modulator, under both physiological and pathophysiological conditions, in response to stress to an organ or tissue (Fredholm et al. 2001; Haskó et al. 2008; Jacobson and Gao 2006). This protective response might take the form of increased blood supply (vasodilation or angiogenesis) (Ryzhov et al. 2008), ischemic preconditioning (in the heart, brain, or skeletal muscle) (Akaiwa et al. 2006; Cohen and Downey 2008; Liang and Jacobson 1998; Zheng et al. 2007), and/or suppression of inflammation (activation and infiltration of inflammatory cells, production of cytokines and free radicals) (Chen et al. 2006b; Martin et al. 2006; Ohta and Sitkovsky 2001). Adenosine acts on cell surface receptors that are coupled to intracellular signaling cascades. There are four subtypes of G-protein-coupled receptors (GPCRs); i.e., four distinct sequences of adenosine receptors (ARs) termed A_1, A_{2A}, A_{2B}, and A_3 (Fig. 1). The second messengers associated with the ARs are historically defined with respect to the adenylate cyclase system (Fredholm and Jacobson 2009). The A_1 and A_3 receptors inhibit the production of cyclic AMP through coupling to G_i. The A_{2A} and A_{2B} subtypes are coupled to G_s or G_o to stimulate adenylate cyclase. Furthermore, the A_{2B} subtype, which has the lowest affinity ($K_i > 1\,\mu M$) of all the subtypes for native adenosine, is also coupled to G_q (Ryzhov et al. 2006). Adenosine has the highest affinity at the A_1 and A_{2A} ARs (K_i values in binding of 10–30 nM at the high affinity sites), and the affinity of adenosine at the A_3AR is intermediate (ca. 1 μM at the rat A_3AR) (Jacobson et al. 1995).

Effector mechanisms other than the adenylate cyclase and phospholipase C are associated with the stimulation of ARs. For example, adenosine action can activate phosphoinositide 3-kinase (PI3K), mitogen-activated protein kinases (MAPKs), and extracellular receptor signal-induced kinase (ERK) (Schulte and Fredholm 2003). The indirect regulation by adenosine of MAPKs can have effects on differentiation, proliferation, and apoptosis (Che et al. 2007; Fredholm et al. 2001; Jacobson and Gao 2006; Schulte and Fredholm 2003). Thus, the A_3AR activates Akt to inhibit apoptosis. These actions may be initiated through the β, γ subunits of the G proteins, which can also lead to the coupling of ARs to ion channels. The influx of calcium ions or the efflux of potassium ions can be induced by the activation of the A_1AR. The arrestin pathway, which has the dual role of signal transmission and downregulation of the receptor, is also activated by ARs (Klaasse et al. 2008; Penn et al. 2001). The A_{2A}AR forms a tight complex with G_s by a process described as "restricted collision coupling" (Zezula and Freissmuth 2008). The A_{2A}AR also

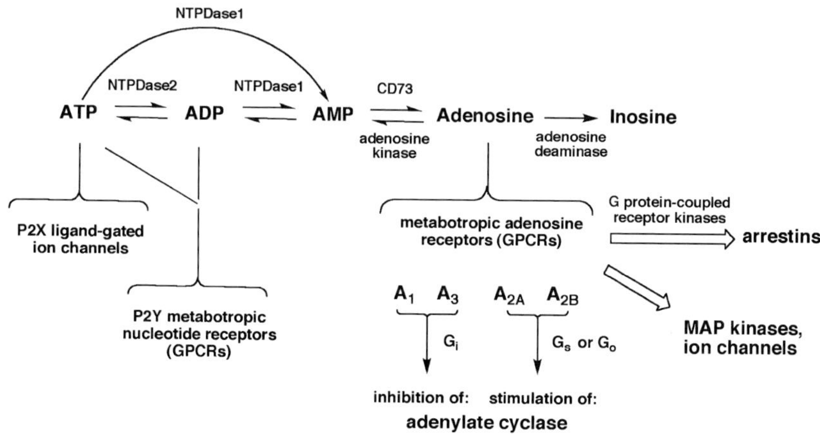

Fig. 1 Interconversion of extracellular adenine nucleotides and adenosine and their associated signaling pathways. These molecules may originate from intracellular sources. For example, adenosine may cross the plasma membrane through an equilibrative nucleoside transporter (ENT)1. The four subtypes of adenosine receptors (ARs) are grouped according to effects on adenylate cyclase. Inosine at micromolar concentrations also activates the A_3AR. Various extracellular nucleotides activate seven subtypes of P2X receptors and eight subtypes of P2Y, which are not specified here. The ARs and P2Y receptors are G-protein-coupled receptors (GPCRs), while the P2X receptors are ionotropic receptors. The ectonucleoside triphosphate diphosphohydrolases NTPDase1 and NTPDase2 are also known as CD39 (apyrase) and CD39L1, respectively. NTPDases3 and 8 (not shown) are also involved in breakdown of extracellular nucleotides

binds to additional "accessory" proteins, such as alpha-actinin, ARNO, USP4 and translin-associated protein-X (Zezula and Freissmuth 2008).

Adenosine suppresses various cytotoxic processes, such as cytokine-induced apoptosis. In the brain, both neuronal and glial cell functions are regulated by adenosine (Björklund et al. 2008; Fredholm et al. 2005). Adenosine acts as a local modulator of the action of various other neurotransmitters, including biogenic amines and excitatory amino acids. Adenosine attenuates the release of many stimulatory neurotransmitters and can counteract the excitotoxicity associated with excessive glutamate release in the brain. Adenosine can also modulate the interaction of neurotransmitters, such as dopamine, with their own receptors. In the periphery, adenosine has been shown to attenuate excessive inflammation, to promote wound healing, and to protect tissue against ischemic damage (Chen et al. 2006a; Haskó et al. 2008). In the cardiovascular system, adenosine promotes vasodilation, vascular integrity, and angiogenesis, and also counteracts the lethal effects of prolonged ischemia on cardiac myocytes and skeletal muscle (Cohen and Downey 2008; Zheng et al. 2007).

Therapeutic applications, both in the central nervous system and in the periphery, are being explored for selective AR agonists and antagonists. A large body of medicinal chemistry has been created around the four AR subtypes, such that selective agonists and antagonists are now available for each. These ligands have been used as pharmacological probes to introduce many new drug concepts. Mouse strains in which an AR has been genetically deleted (each of the subtypes has now been deleted) have also been useful in developing novel drug concepts based on the modulation of ARs (Fredholm et al. 2005).

Adenosine itself is short-lived in the circulation, which has allowed its clinical use in the treatment of paroxysmal supraventricular tachycardia and in radionuclide myocardial perfusion imaging (Cerqueira 2006). The many selective and potent synthetic AR agonists, which are typically much longer lasting in the body than adenosine, have been slower to enter a clinical pathway than adenosine. Recently, the first such synthetic adenosine agonist, Lexiscan (regadenoson, CV Therapeutics, Palo Alto, CA, USA), an A_{2A}AR agonist, was approved for diagnostic use (Lieu et al. 2007).

Synthetic adenosine agonists have potential therapeutic applications based on their anti-inflammatory (A_{2A} and A_3) (Haskó et al. 2008; Ohta and Sitkovsky 2001), cardioprotective (preconditioning of the ischemic heart muscle by activation of the A_1 and A_3 ARs and its postconditioning by A_{2B}AR activation) (Cohen and Downey 2008), cerebroprotective (A_1 and A_3) (Chen et al. 2006a; Knutsen et al. 1999; von Lubitz et al. 1994), and antinociceptive (A_1) (Johansson et al. 2001) properties. Potent and selective AR antagonists display therapeutic potential as kidney protective (A_1) (Gottlieb et al. 2002), antifibrotic (A_{2A}) (Che et al. 2007), neuroprotective (A_{2A}) (Yu et al. 2004), antiasthmatic (A_{2B}) (Holgate 2005), and antiglaucoma (A_3) (Yang et al. 2005) agents. A_3AR agonists have been proposed for the treatment of a wide range of autoimmune inflammatory conditions, such as rheumatoid arthritis, inflammatory bowel diseases, psoriasis, etc. (Guzman et al. 2006; Kolachala et al. 2008; Madi et al. 2007), and also for cardiac and brain ischemia. A_1AR agonists are useful in preclinical models of cardiac arrythmia and ischemia and in pain. Adenosine agonists are also of interest for the treatment of sleep disorders (Porkka-Heiskanen et al. 1997). Activation of the A_{2B}AR protects against vascular injury (Yang et al. 2008).

The alkylxanthines caffeine and theophylline are the prototypical antagonists of ARs, and their stimulant actions are produced primarily through blocking the depressant actions of adenosine through the A_1 and A_{2A} ARs (Fredholm and Jacobson 2009). Prior to the work of Rall, Daly, and other pioneers in the field, the stimulant actions of the alkylxanthines were thought to occur as a result of inhibition of phosphodiesterases. It is true that caffeine inhibits phosphodiesterases and has other actions, such as stimulation of calcium release, but these non-AR-mediated actions require higher concentrations of caffeine than are typically ingested in the human diet (Fredholm and Jacobson 2009).

The nonselective AR antagonist theophylline has been in use as an antiasthmatic drug (Holgate 2005), although its use is now limited as a result of side effects on the central nervous system and the renal system. Adenosine antagonists of various selectivities remain of interest as potential drugs for treating asthma (Wilson 2008). A large number of synthetic AR antagonists that are much more potent and selective than the prototypical alkylxanthines have been introduced, although none have yet been approved for clinical use. For example, AR antagonists have been proposed for neurodegenerative diseases (such as Parkinson's disease and Alzheimer's disease) (Schwarzschild et al. 2006), although a well-advanced A_{2A}AR antagonist

KW6002 (Istradefylline) (8-[(*E*)-2-(3,4-dimethoxyphenyl)vinyl]-1,3-diethyl-7-methylpurine-2,6-dione, Kyowa Hakko Kirin Co. Ltd, Tokyo, Japan) was recently denied FDA approval for the treatment of Parkinson's disease (LeWitt et al. 2008).

2 Sources and Fate of Extracellular Adenosine

Adenosine is not a classical neurotransmitter because it is not principally produced and released vesicularly in response to neuronal firing. Most tissues in the body and cells in culture release adenosine to the extracellular medium, from where it can feed back and act as an autocoid on the ARs present locally. The basal levels of extracellular adenosine have been estimated as roughly 100 nM in the heart and 20 nM in the brain, which would only partially activate the ARs present (Fredholm et al. 2005). In the case of severe ischemic stress, the levels can rapidly rise to the micromolar range, which would cause a more intense and generalized activation of the four subtypes of ARs. Nevertheless, it is thought that the exogenous administration of highly potent and selective AR agonists in such cases of severe ischemic challenge might still provide additional benefit beyond that offered by the endogenous adenosine generated (Jacobson and Gao 2006; Yan et al. 2003).

Extracellular adenosine may arise from intracellular adenosine or from the breakdown of the adenine nucleotides, such as adenosine triphosphate (ATP), outside the cell (Fig. 1). Adenosine, which is present in a higher concentration inside than outside the cell, does not freely diffuse across the cell membrane. There are nucleoside transporters, such as the equilibrative nucleoside transporter (ENT), ENT1, which bring it to the extracellular space. Extracellular nucleotides activate their own receptors, known as P2Y metabotropic and P2X inotropic receptors (Burnstock 2008). Extracellular nucleotides may also originate from cytosolic sources, including by vesicular release exocytosis, passage through channels, and cell lysis. Ectonucleotidases break down the adenine nucleotides in stages to produce free extracellular adenosine at the terminal step (Zimmermann 2000). For example, the extracellular enzyme ectonucleoside triphosphate diphosphohydrolase 1 (E-NTPDase1) converts ATP and adenosine diphosphate (ADP) to adenosine monophosphate (AMP). A related ectonucleotidase, E-NTPDase2, primarily hydrolyzes 5′-triphosphates to 5′-diphosphates. The final and critical step, with respect to AR activation, of conversion of AMP to adenosine is carried out by ecto-5′-nucleotidase, also known as CD73. Overexpression of CD73 has been proposed to protect organs under stress by the formation of cytoprotective adenosine (Beldi et al. 2008). The adenosine produced extracellularly is also subject to metabolic breakdown by adenosine deaminase to produce inosine or (re)phosphorylation by adenosine kinase to produce AMP. Therefore, when an organ is under stress there is a highly complex and time-dependent interplay of the activation of many receptors in the same vicinity. In addition to the direct activation of ARs by selective agonists or their blockade by selective antagonists, inhibition of the metabolic or transport pathways surrounding adenosine is also being explored for therapeutic purposes (McGaraughty et al. 2005).

3 Adenosine Receptor Structure

The ARs, as GPCRs, share the structural motif of a single polypeptide chain forming seven transmembrane helices (TMs), with the N-terminus being extracellular and the C-terminus being cytosolic (Costanzi et al. 2007). These helices, consisting of 25–30 amino acid residues each, are connected by six loops, i.e., three intracellular (IL) and three extracellular (EL) loops. The extracellular regions contain sites for posttranslational modifications, such as glycosylation. The A_1 and A_3 ARs also contain sites for palmitoylation in the C-terminal domain. The A_{2A}AR has a long C-terminal segment of more than 120 amino acid residues, which is not required for coupling to G_s, but can serve as a binding site for "accessory" proteins (Zezula and Freissmuth 2008). The sequence identity between the human A_1 and A_3 ARs is 49%, and the human A_{2A} and A_{2B} ARs are 59% identical. Particular conserved residues point to specific functions. For example, there are two characteristic His residues in TMs 6 and 7 of the A_1, A_{2A}, and A_{2B} ARs. In the A_3AR, the His residue in TM6 is lacking but another His residue has appeared at a new location in TM3. All of these His residues have been indicated by mutagenesis to be important in the recognition and/or activation function of the receptor (Costanzi et al. 2007; Kim et al. 2003).

Recently, the human A_{2A}AR joined the shortlist of GPCRs for which an X-ray crystallographic structure has been determined (Jaakola et al. 2008). The reported structure (Fig. 2) contained a bound high-affinity antagonist ligand, ZM241385 (4-2-[7-amino-2-(2-furyl)-1,2,4-triazolo[1,5-*a*][1,3,5]triazin-5-yl-amino]ethylphenol), which is moderately selective for the A_{2A}AR. Prior to this dramatic step in bringing ARs into the age of structural biology, homology modeling of the ARs, based on a rhodopsin template, was the principal means of AR structural prediction and was useful in interpreting mutagenesis data. The modeling has defined two subregions within the putative agonist binding site (Costanzi et al. 2007; Kim et al. 2003). This putative binding site is located within the barrel or cleft created by five of the seven TMs (excluding TM1 and TM2), approximately one-third of the distance across the membrane from the exofacial side. The ribose moiety of adenosine binds in a hydrophilic region defined by TMs 3 and 7, and the adenine moiety binds in a largely hydrophobic region surrounded by TMs 5 and 6. Thus, the region of adenosine in the binding site is approximately the same as the position of the retinal in rhodopsin. Even the importance of the Lys residue in TM7 of rhodopsin that forms the covalent association (Schiff base) with retinal is conserved by analogy in the ARs, i.e., with a His residue that occurs at the same position (7.43) in all of the ARs. The His residue is predicted by molecular modeling to associate with the ribose moiety of adenosine. Features of the putative binding site of adenosine have been reviewed recently (Costanzi et al. 2007). Different labs have not been in agreement on the precise placement of the adenosine moiety when docked in the receptor. However, the major modeling publications in this area have zeroed in on the same limited region of the receptor structure for coordination of adenosine. One can consider the modeling approach to provide insights that are subject to refinement over time, as more is learned from mutagenesis studies and the modeling

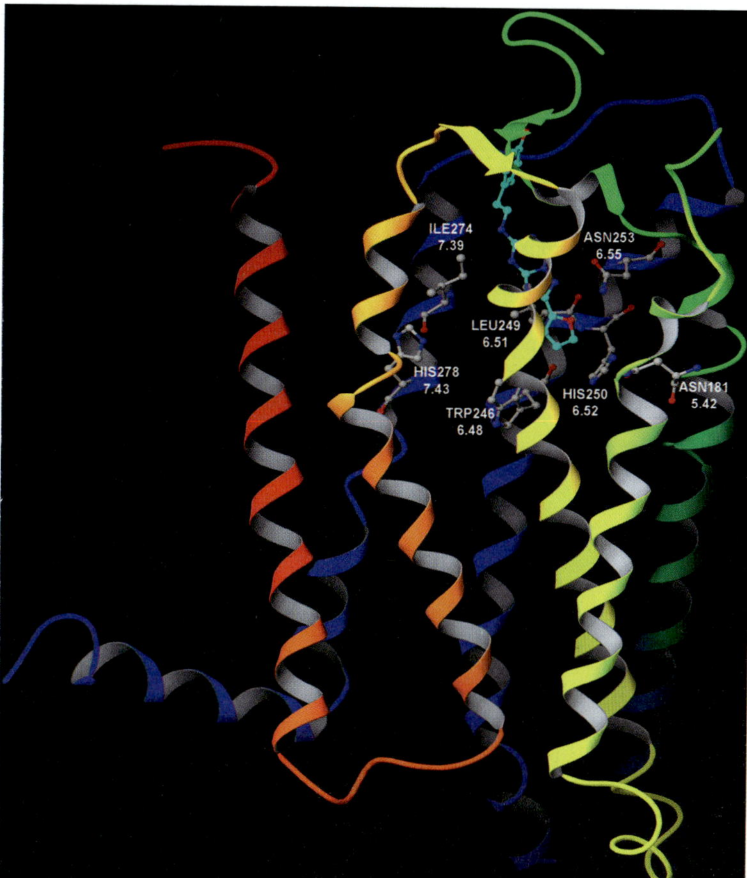

Fig. 2 X-ray crystallographic structure of the human A_{2A} adenosine receptor (AR), showing the bound antagonist ZM241385 (Jaakola et al. 2008). The structure of the A_{2A}AR is colored by region: N-terminus and transmembrane helical (TM) domain 1 in *orange*, TM2 in *ochre*, TM3 in *yellow*, TM4 in *green*, TM5 in *cyan*, TM6 in *blue*, TM7 and C-terminus in *purple*. The *p*-hydroxyphenylethyl moiety of the antagonist ligand points toward the exofacial side of the receptor

templates and computational methods are refined (Ivanov et al. 2009). Many amino acid residues predicted by molecular modeling to be involved in the coordination of antagonists by the A_{2A}AR were indeed in proximity to the bound ZM241385 in the X-ray structure, although the molecule was somewhat rotated from the orientation predicted in various docking models. These residues include Asn253 in TM6, which hydrogen bonds to the exocyclic NH of agonists and various antagonists in the AR models. The same residue was found to form a hydrogen bond with the exocyclic NH of ZM241385.

Dimerization has been proposed to occur between ARs, leading to homo- or heterodimers (Franco et al. 2006). Dimerization between ARs and other receptors has also been proposed; for example, A_1AR/D_1 dopamine receptor dimers and $A_{2A}AR/D_2$ dopamine receptor dimers (Franco et al. 2006). Heterodimers of the A_1AR with either $P2Y_1$ or $P2Y_2$ nucleotide receptors or with metabotropic glutamate receptors have been detected (Prinster et al. 2005). The pharmacological properties of these heterodimers may differ dramatically from the properties of each monomer alone. For example, the $A_1AR/P2Y_1$ dimers have been characterized pharmacologically and were found to be inhibited by known nucleotide antagonists but not activated by known nucleotide agonists of the $P2Y_1$ receptor (Nakata et al. 2005). Dimers of A_{2A} adenosine/D_2 dopamine receptors are present in striatum and display a modified pharmacology relative to each of the individual subtypes. These receptor dimers are drug development targets for Parkinson's disease (Schwarzschild et al. 2006).

4 Regulation of Adenosine Receptors

Similar to the function and regulation of other GPCRs, both activation and desensitization of the ARs occur after agonist binding. Interaction of the activated ARs with the G proteins leads to second messenger generation and classical physiological responses. Interaction of the activated ARs with G protein-coupled receptor kinases (GRKs) leads to their phosphorylation. Downregulation of ARs should be considered in both the basic pharmacological studies and with respect to the possible therapeutic application of agonists. AR responses desensitize rapidly, and this phenomenon is associated with receptor downregulation, internalization and degradation. The internalization and desensitization of ARs has been reviewed recently (Klaasse et al. 2008). Mutagenesis has been applied to analyze the molecular basis for the differences in the kinetics of the desensitization response displayed by various AR subtypes. The most rapid downregulation among the AR subtypes is generally seen with the A_3AR, due to phosphorylation by GRKs. The $A_{2A}AR$ is only slowly desensitized and internalized as a result of agonist activation.

5 Adenosine Receptor Agonists and Antagonists in Preclinical and Clinical Trials

Potent and selective AR agonists and antagonists have been synthesized for all four AR subtypes, with selective $A_{2B}AR$ agonists being the most recently reported (Baraldi et al. 2009). Some of these ligands are selective for a single AR subtype, and others have mixed selectivity for several subtypes. Thus, numerous pharmacological tools for studying the ARs are available, and some of these compounds have advanced to clinical studies (Baraldi et al. 2008; Elzein and Zablocki 2008; Giorgi and Nieri 2008; Moro et al. 2006).

A general caveat in the design of selective agonists and antagonists is the frequent observation of a variation of affinity for a given compound at the same subtype in different species. There are many examples of marked species dependence of ligand affinity at the ARs (Jacobson and Gao 2006; Yang et al. 2005). Therefore, caution must be used in generalizing the selectivity of a given compound from one species to another. In general, one must be cognizant of potential species differences for both AR agonists and antagonists.

5.1 Adenosine Receptor Agonists

Nearly all AR agonists reported are adenosine derivatives. A noteworthy exception is the class of pyridine-3,5-dicarbonitrile derivatives that fully activate ARs and that display varied selectivity at the AR subtypes (Beukers et al. 2004). One such compound is the A_{2B}AR-selective agonist BAY 60–6583 (2-[6-amino-3,5-dicyano-4-[4-(cyclopropylmethoxy)phenyl]pyridin-2-ylsulfanyl]acetamide) (Cohen and Downey 2008; Eckle et al. 2007). Another AR agonist of nonnucleoside structure is BAY 68–4986 (Capadenoson), which is a selective A_1AR agonist in clinical trials for the oral treatment of stable angina pectoris (Mittendorf and Wuppertal 2008). The structure–activity relationships (SARs) of adenosine derivatives as agonists of the ARs have been thoroughly probed (Jacobson and Gao 2006; Yan et al. 2003), and representative agonists are shown in Fig. 3. In general, substitution at the N6 position with certain alkyl, cycloalkyl, and arylalkyl groups increases selectivity for the A_1AR. Substitution with an N^6-benzyl group or substituted benzyl group increases selectivity for the A_3AR. Substitution at the 2 position, especially with ethers, secondary amines, and alkynes, often results in high selectivity for the A_{2A}AR.

All of the A_1AR agonists shown in Fig. 3 contain a characteristic N6 modification. The singly substituted A_1AR agonists NNC-21-0136 (2-chloro-N^6-[(*R*)-[(2-benzothiazolyl)thio]-2-propyl]-adenosine) and GR79236 (N^6-[(1*S*, 2*S*)-2-hydroxycyclopentyl]adenosine) (Merkel et al. 1995) and the doubly substituted selodenoson have been clinical candidates. NNC-21-0136 was the result of a program to develop CNS-selective AR agonists for use in treating stroke and other neurodegenerative conditions (Knutsen et al. 1999). A_1AR agonists are of interest for use in treating cardiac arrhythmias [for which adenosine itself, under the name Adenocard (Astellas Pharma, Inc., Tokyo, Japan), is in widespread use]. The A_1AR agonist SDZ WAG94 (2′-*O*-methyl-N^6-cyclohexyladenosine) was under consideration for treatment of diabetes (Ishikawa et al. 1998). The AR agonist of mixed selectivity AMP579 ([1*S*-[1α, 2β, 3β, 4α(S*)]]-4-[7-[[1-[(3-chlorothien-2-yl)methyl]propyl]amino]-3*H*-imidazo[4,5-b]pyrid-3-yl] *N*-ethyl-2,3-dihydroxycyclopentanecarboxamide) has cardioprotective properties (Cohen and Downey 2008). The 2-substituted A_{2A}AR agonists ATL-146e (4-{3-[6-amino-9-(5-ethylcarbamoyl-3,4-dihydroxy-tetrahydro-furan-2-yl)-9*H*-purin-2-yl]-prop-2-ynyl}-cyclohexanecarboxylic acid methyl ester), binodenoson (2-[{cyclohexylmethylene}hydrazino]adenosine, MRE-0470 or WRC-0470), and MRE0094 (2-[2-(4-chlorophenyl)ethoxy]adenosine) have been cardiovascular clinical candidates (Awad et al. 2006; Desai et al. 2005; Udelson et al. 2004). Several of the A_{2A}AR agonists shown in Fig. 3 contain the 5′-uronamide

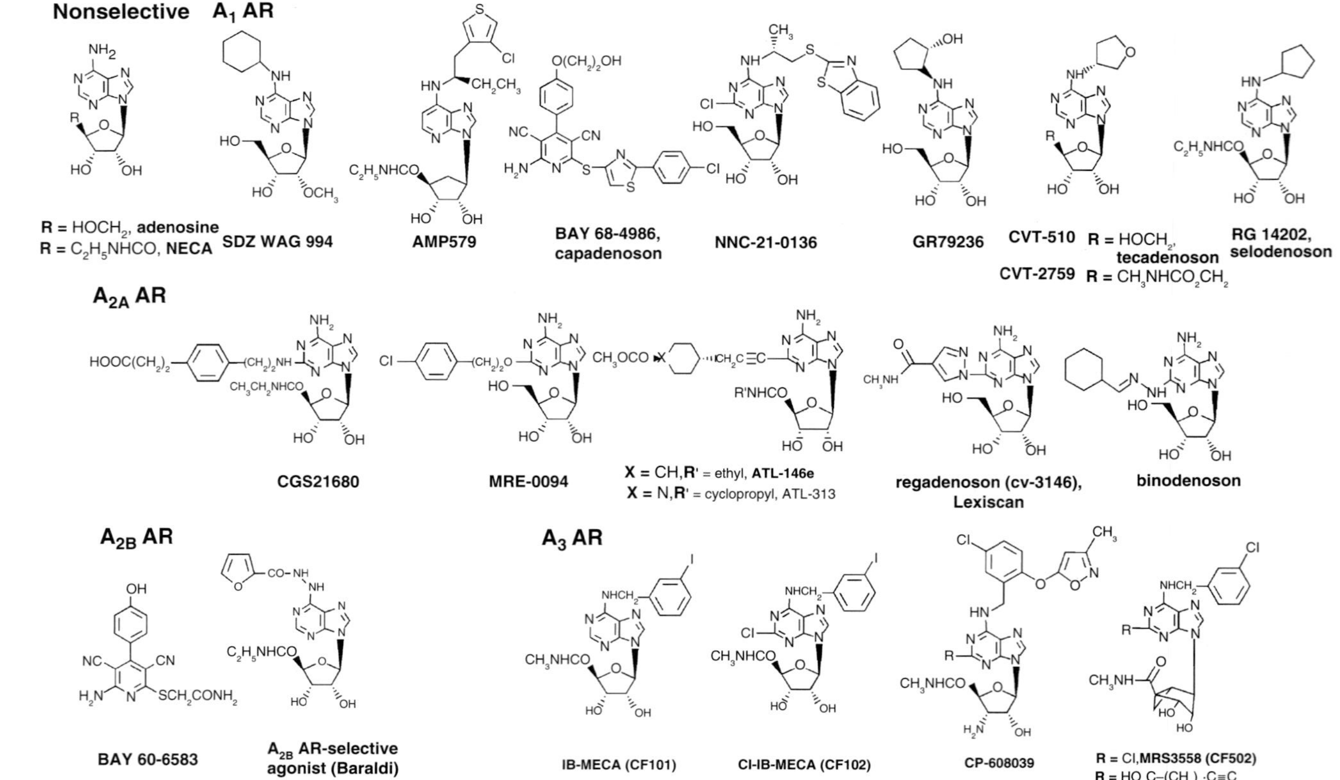

Fig. 3 Structures of selected adenosine receptor (AR) agonists. K_i values in binding are available in references (Baraldi et al. 2008; Jacobson and Gao 2006; Yan et al. 2003)

modification, characteristic of NECA; others have the adenosine-like CH_2OH group. Such agonists are of interest for use as vasodilatory agents in cardiac imaging [adenosine itself, under the name Adenoscan (Astellas Pharma, Inc., Tokyo, Japan), is in use for this purpose] and in suppressing inflammation (Cerqueira 2006). CVT-3146 (1-[6-amino-9-[(2*R*, 3*R*, 4*S*, 5*R*)-3,4-dihydroxy-5-(hydroxymethyl)oxolan-2-yl]purin-2-yl]-*N*-methylpyrazole-4-carboxamide, Lexiscan, regadenoson) is already approved for diagnostic imaging (Lieu et al. 2007).

All of the A_3AR agonists shown in Fig. 3 contain the NECA-like 5′-uronamide modification and have nanomolar affinity at the receptor. CP-608,039 ((2*S*, 3*S*, 4*R*, 5*R*)-3-amino-5-{6-[5-chloro-2-(3-methylisoxazol-5-ylmethoxy)benzylamino]purin-9-yl-l-4-hydroxytetrahydrofuran-2-carboxylic acid methylamide) and its N^6-(2,5-dichlorobenzyl) analog CP-532,903 ((2*S*, 3*S*, 4*R*, 5*R*)-3-amino-5-{6-[2, 5-dichlorobenzylamino]purin-9-yl-l-4-hydroxytetrahydrofuran-2-carboxylic acid methylamide) (Wan et al. 2008) (not shown) are selective A_3 agonists that were developed for cardioprotection. CF101 (N^6-(3-iodobenzyl)-5′-*N*-methylcarboxamidoadenosine, IB-MECA) is being studied by Can-Fite Biopharma (Petah-Tikva, Israel) for the treatment of rheumatoid arthritis (Phase IIb), dry eye syndrome (Phase II) and psoriasis (Phase II) (http://clinicaltrials.gov). Can-Fite Biopharma is also developing the A_3AR agonist CF102 (2-chloro-N^6-(3-iodobenzyl)-5′-*N*-methylcarboxamidoadenosine, Cl-IB-MECA) for the treatment of liver conditions, including liver cancer, hepatitis infections and liver tissue regeneration (Bar-Yehuda et al. 2008; Madi et al. 2004). The North conformation of the ribose ring was found to be the preferred conformation at the A_3AR, which accounts for the high potency and selectivity of the rigid analog MRS3558 ((1′S, 2′R, 3′S, 4′R, 5′S)-4′-{2-chloro-6-[(3-chlorophenylmethyl)amino]purin-9-yl}-1-(methylaminocarbonyl)bicyclo[3.1.0]hexane-2,3-diol) at the human and rat A_3ARs (Ochaion et al. 2008). The bicyclic ring constrains the ribose-like moiety in the desired conformation. The recent generation agonist in the same chemical series MRS5151 ((1′S, 2′R, 3′S, 4′R, 5′S)-4′-[6-(3-chlorobenzylamino)-2-(5-hydroxycarbonyl-1-pentynyl)-9-yl]-2′, 3′-dihydroxybicyclo[3.1.0]hexane-1′-carboxylic acid *N*-methylamide) is designed to be A_3AR selective in at least three different species, including mouse (Melman et al. 2008a).

Recently, macromolecular conjugates (e.g., dendrimers) of chemically functionalized AR agonists were introduced as potent polyvalent activators of the receptors that are qualitatively different in pharmacological characteristics in comparison to the monomeric agonists (Kim et al. 2008; Klutz et al. 2008). The feasibility of using dendrimer conjugates to bind to AR dimers was studied using a molecular modeling approach (Ivanov and Jacobson 2008).

5.2 *Adenosine Receptor Antagonists*

The newer and most selective AR antagonists are more chemically diverse than the classical 1,3-dialkylxanthines, which have been used pharmacologically as antagonists of the A_1 and A_2 ARs. A range of AR antagonists and their synthetic methods were recently reviewed (Baraldi et al. 2008; Moro et al. 2006).

Purine AR antagonists, including both xanthine and adenine derivatives, have provided a wide range of receptor subtype selectivity, depending on the substitution (Fig. 4). In general, modifications of the xanthine scaffold at the 8 position with aryl or cycloalkyl groups has led to high affinity and selectivity for the A_1AR. Highly selective xanthine antagonists of the A_1AR (e.g., the epoxide derivative BG 9719 (1,3-dipropyl-8-(2-(5,6-epoxy)norbornyl)xanthine) and the more water soluble BG9928 (3-[4-(2,6-dioxo-1,3-dipropyl-2,3,6,7-tetrahydro-1*H*-purin-8-yl)

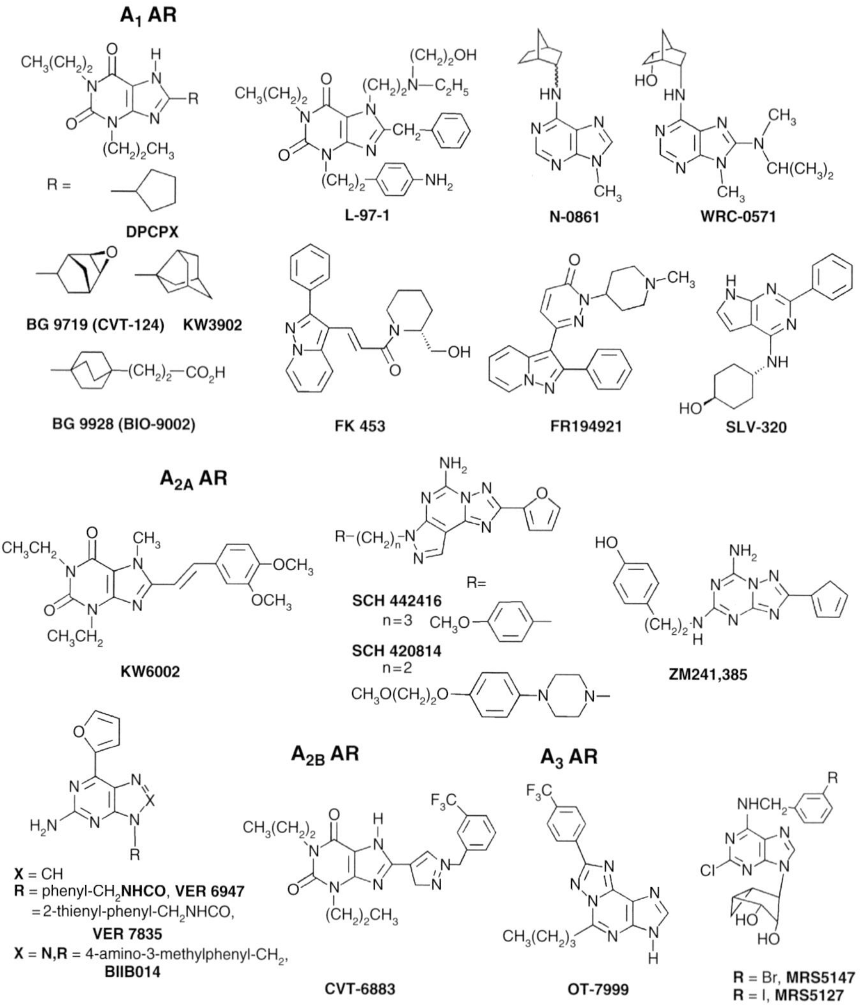

Fig. 4 Structures of selected adenosine receptor (AR) antagonists. K_i values in binding are available in references (Baraldi et al. 2008; Jacobson and Gao 2006)

-bicyclo[2.2.2]oct-1-yl]-propionic acid, Biogen Idec, Cambridge, MA, USA), as well as KW3902 (8-(noradamantan-3-yl)-1,3-dipropylxanthine, Merck and Co., Inc., Whitehouse Station, NJ, USA) have been (BG 9719) (Gottlieb et al. 2002) or are currently (BG9928 and KW3902) (Cotter et al. 2008; Dittrich et al. 2007; Givertz et al. 2007; Greenberg et al. 2007) in clinical trials for treatment of acute decompensated heart failure (ADHF) with renal impairment. In dogs, both BG9719 and BG9928 have high affinity for both the A_1AR and A_{2B}AR (Auchampach et al. 2004) with A_{2B}/A_1 ratios of 21 and 24, respectively (Doggrell 2005). The selectivity of BG 9928 for the human A_1AR compared to the human A_{2B}AR is 12 (Kiesman et al. 2006). The 8-cyclopentyl derivative DPCPX (8-cyclopentyl-1,3-dipropylxanthine), also known as CPX, which is selective for the A_1AR in the rat with nanomolar affinity but less selective at the human AR subtypes, has been in clinical trials for cystic fibrosis through a non-AR-related mechanism (Arispe et al. 1998). The highly selective A_1AR antagonist L-97-1 (3-[2-(4-aminophenyl)-ethyl]-8-benzyl-7-{2-ethyl - (2-hydroxy-ethyl)-amino]-ethyl}-1-propyl-3,7-dihydro-purine-2,6-dione, Endacea Inc., Research Triangle Park, NC, USA) is water soluble and in late preclinical development for the treatment of asthma (Wilson 2008). As in the cases of DPCPX, BG 9719, N-0861 (($\pm$)-N^6-endonorbornan-2-yl-9-methyladenine), and others, a persistent problem in the development of A_1AR antagonists is low aqueous solubility, e.g., high lipophilicity, corresponding low water solubility, and low bioavailability (Hess 2001); thus, A_1AR antagonists, e.g., BG 9928 and L-97-1, with good water solubility are preferable clinical candidates. Moreover, a persistent problem in the use of xanthine derivatives as AR antagonists is their interaction at the A_{2B}AR. Modification of xanthines at the 8 position with certain aryl groups has given rise to preclinical candidates that are selective for the A_{2B}AR (e.g., CVT-6883, 3-ethyl-1-propyl-8-[1-(3-trifluoromethylbenzyl)-1H-pyrazol-4-yl]-3,7-dihydropurine-2,6-dione, CV Therapeutics, Palo Alto, CA, USA) (Mustafa et al. 2007). Use of the adenine derivatives WRC-0571 (8-(N-methylisopropyl)amino-N^6-(5′-endohydroxy-endonorbornan-2-yl-9-methyladenine) as an inverse agonist at the A_1AR provides A_1AR selective antagonism without blocking the A_{2B}AR (Martin et al. 1996). Nonxanthine antagonists of the A_1AR have also been shown to have high receptor subtype selectivity, e.g., FK453 (Terai et al. 1995) and SLV 320 (4-[(2-phenyl-7H-pyrrolo[2,3-d]pyrimidin-4-yl)amino]-trans-cyclohexanol, Solvay Pharmaceuticals SA, Brussels, Belgium) (Hocher et al. 2008). Moreover, various nonxanthine A_1AR antagonists have been or are currently being explored for clinical applications (Jacobson and Gao 2006). For example, SLV 320 is in clinical trials as an intravenous treatment for ADHF with renal impairment (http://clinicaltrials.gov).

Modification of xanthines at the 8 position with alkenes (specifically styryl groups) has led to selectivity for the A_{2A}AR. Such derivatives include the A_{2A}AR antagonist KW6002 (istradefylline), which has been in clinical trials. Some 8-styrylxanthine derivatives, such as CSC (8-(3-chlorostyryl)caffeine), have been discovered to inhibit monoamine oxidase-B, as well as the A_{2A}AR (Vlok et al. 2006). The triazolotriazine ZM241385 and the pyrazolotriazolopyrimidine

SCH 442416 (5-amino-7-(3-(4-methoxy)phenylpropyl)-2-(2-furyl)pyrazolo[4,3-*e*]-1,2,4-triazolo[1,5-*c*]pyrimidine) are highly potent A_{2A}AR antagonists (Moresco et al. 2005; Palmer et al. 1996). ZM241385 also binds to the human A_{2B}AR with moderate affinity, and has been used as a radioligand at that subtype (Ji and Jacobson 1999). SCH 442416 displays $> 23,000$-fold selectivity for the human A_{2A}AR (K_i 0.048 nM) in comparison to human A_1AR and an $IC_{50} > 10\,\mu M$ at the A_{2B} and A_3 ARs. A_{2A}AR antagonists, such as the xanthine KW6002 and the nonxanthines SCH 442416, VER 6947 (2-amino-*N*-benzyl-6-(furan-2-yl)-9*H*-purine-9-carboxamide), and VER 7835 (2-amino-6-(furan-2-yl)-*N*-(thiophen-2-ylmethyl)-9*H*-purine-9-carboxamide), are of interest for use in treating Parkinson's disease (Gillespie et al. 2008; LeWitt et al. 2008; Schwarzschild et al. 2006). The A_{2A}AR antagonist BIIB014 (V2006) has begun Phase II clinical trials (Biogen Idec, Cambridge, MA, USA, in partnership with Vernalis, Cambridge, UK) for Parkinson's disease (Jordan 2008).

Cyclized derivatives of xanthines, such as PSB-11 (8-ethyl-4-methyl-2-phenyl-(8*R*)-4,5,7,8-tetrahydro-1*H*imidazo[2.1-*i*]purin-5-one), are A_3AR-selective, and similar compounds have been explored by Kyowa Hakko. Selective A_3AR antagonists, such as the heterocyclic derivatives OT-7999 (5-*n*-butyl-8-(4-trifluoromethylphenyl)-3*H*-[1,2,4]triazolo-[5,1-*i*]purine), are being studied for the treatment of glaucoma (Okamura et al. 2004), and other such antagonists are under consideration for treatment of cancer, stroke, and inflammation (Gessi et al. 2008; Jacobson and Gao 2006). MRS5147 (($1'R, 2'R, 3'S, 4'R, 5'S$)-4′-[2-chloro-6-(3-bromobenzylamino)-purine]-2′, 3′-*O*-dihydroxybicyclo-[3.1.0]hexane) and its 3-iodo analog MRS5127 are highly selective A_3AR antagonists in both human and rat, based on a conformationally constrained ribose-like ring that is truncated at the 5′ position (Melman et al. 2008b). No selective A_3AR antagonists have yet reached human trials. However, an antagonist of mixed A_{2B}/A_3AR selectivity in the class of 5-heterocycle-substituted aminothiazoles from Novartis (Horsham, UK), QAF 805 (Press et al. 2005), was in a Phase Ib clinical trial for the treatment of asthma. This antagonist failed to decrease sensitivity to the bronchoconstrictive effects of AMP in asthmatics (Pascoe et al. 2007).

5.3 Radioligands for In Vivo Imaging

With the established relevance of ARs to human disease states, it has been deemed useful to develop high-affinity imaging ligands for these receptors, for eventual diagnostic use in the CNS and in the periphery. Ligands for in vivo positron emission tomographic (PET) imaging of A_1, A_{2A}, and A_3 ARs have been developed. For example, the xanthine [^{18}F]CPFPX (8-cyclopentyl-1-propyl-3-(3-fluoropropyl)-xanthine, similar in structure to DPCPX) and the nonxanthine [^{11}C]FR194921 (2-(1-methyl-4-piperidinyl)-6-(2-phenylpyrazolo[1,5-*a*]pyridin-3-yl)-3(2*H*)-pyridazinone) have been developed as centrally-active PET tracers for imaging of the A_1AR in the brain (Bauer et al. 2005). The first PET ligand for the $A_{2A}AR$ was [7-methyl-^{11}C]-(E)-8-(3,4,5-trimethoxystyryl)-1,3,7-trimethylxanthine ([^{11}C]TMSX) (Ishiwata et al. 2000). This is a caffeine analog related to the series of KW6002, introduced by the Kyowa Hakko. 5-Amino-7-(3-(4-[^{11}C]methoxy)phenylpropyl)-2-(2-furyl)pyrazolo[4,3-*e*]-1,2,4-triazolo[1,5-*c*]pyrimidine ([^{11}C]SCH442416) has recently been explored as a PET agent in the noninvasive in vivo imaging of the human $A_{2A}AR$ (Moresco et al. 2005). [^{11}C]SCH442416 displays an extremely high affinity at the human $A_{2A}AR$ (K_i 0.048 nM). Recently, an A_3AR PET ligand, [F-18]FE@SUPPY (5-(2-fluoroethyl) 2,4-diethyl-3-(ethylsulfanylcarbonyl)-6-phenylpyridine-5-carboxylate), based on a series of pyridine A_3AR antagonists, was introduced (Wadsak et al. 2008). Several nucleoside derivatives that bind with nanomolar affinity at the A_3AR and that contain ^{76}Br for PET imaging were recently reported, including the antagonist MRS5147 (Kiesewetter et al. 2008).

6 Allosteric Modulation of Adenosine Receptors

In addition to directly acting AR agonists and antagonists, allosteric modulators of A_1 and A_3 ARs have been introduced (Gao et al. 2005). Allosteric modulators have advantages over the directly acting (orthosteric) receptor ligands in that they would magnify the effect of the native adenosine released in response to stress at a specific site or tissue and, in theory, would not induce a biological effect in the absence of an agonist. Various allosteric enhancers of the activation of ARs by agonists are under consideration as clinical candidates. The benzoylthiophenes, represented by PD-81,723 (Fig. 5), were the first AR allosteric modulators to be identified. A structurally related benzoylthiophene derivative known as T-62 ((2-amino-4,5,6,7-tetrahydrobenzo[*b*]thiophen-3-yl)-(4-chlorophenyl)-methanone), which acts as a selective positive enhancer of the A_1AR, like PD-81,723 (2-amino-4,5-dimethyl-3-thienyl-[3-trifluoromethylphenyl]methanone), had progressed toward clinical trials for neuropathic pain (Li et al. 2004). LUF6000 (N-(3,4-dichlorophenyl)-2-cyclohexyl-1H-imidazo[4,5-*c*]quinolin-4-amine) is a selective positive enhancer of the human A_3AR (Gao et al. 2008).

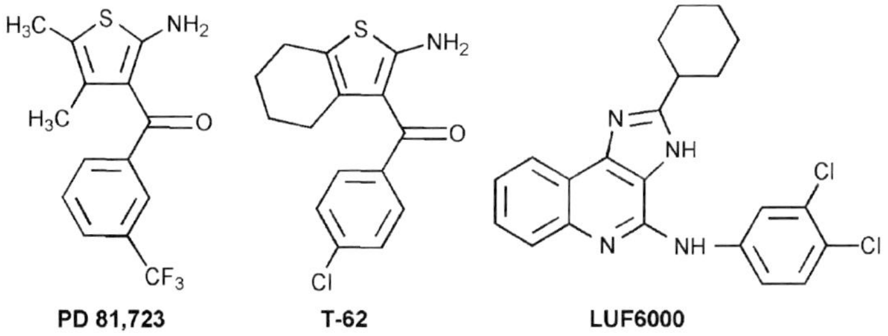

Fig. 5 Allosteric modulators of adenosine receptors (ARs)

7 Genetic Deletion of Adenosine Receptors

Deletion of each of the four AR subtypes has been carried out, and the resulting single-AR knockout (KO) mice are viable and not highly impaired in function (Fredholm et al. 2005; Yang et al. 2008). The pharmacological profile indicates that the analgesic effect of adenosine is mediated by the A_1AR, and analgesia is lost in mice in which the A_1AR has been genetically eliminated. Genetic KO of the A_1AR in mice removes the discriminative-stimulus effects but not the arousal effect of caffeine and increases anxiety and hyperalgesia. Study of A_{2A}AR KO mice reveals functional interaction between the spinal opioid receptors and peripheral ARs. A_1AR KO mice demonstrate a decreased thermal pain threshold, whereas A_{2A}AR null mice demonstrate an increased threshold to noxious heat stimulation, supporting an A_1AR-mediated inhibitory and an A_{2A}AR-mediated excitatory effect on pain transduction pathways. KO of the A_{2A}AR eliminates the arousal effect of caffeine. Genetic KO of the A_{2A}AR also suggests a link to increased anxiety and protected against damaging effects of ischemia and the striatal toxin 3-nitropropionic acid. Genetic KO of the A_3AR leads to increased neuronal damage in a model of carbon monoxide-induced brain injury. Neutrophils lacking A_3ARs show correct directionality but diminished speed of chemotaxis (Chen et al. 2006b). Although studies on A_{2B}AR KO mice have been reported (Yang et al. 2008), the importance of A_{2B}AR in the brain still awaits future investigation.

8 Conclusions

In conclusion, adenosine is released in response to organ stress or tissue damage and displays cytoprotective effects, in general, both in the brain and in the periphery. When excessive activity occurs in a given organ, adenosine acts as an endogenous quieting substance, to either reduce the energy demand or increase the energy supply to that organ. Nearly every cell type in the body expresses one or more of the AR subtypes, which indicates the central role of this feedback system in protecting organs and tissues and in tissue regeneration. Thus, a common theme to the therapeutic applications proposed for agonists is that adenosine acts as a cytoprotective modulator in response to stress to an organ or tissue.

Selective agonists and antagonists have been introduced and used to develop new therapeutic drug concepts. ARs are notable among the GPCR family in terms of the number and variety of agonist drug candidates that have been proposed. Thus, this has led to new experimental agents based on anti-inflammatory (A_{2A} and A_3), cardioprotective (preconditioning by A_1 and A_3 and postconditioning by A_{2B}), cerebroprotective (A_1 and A_3), and antinociceptive (A_1) effects. Potent and selective AR antagonists display therapeutic potential as kidney-protective (A_1), antifibrotic (A_{2A}), neuroprotective (A_{2A}), and antiglaucoma (A_3) agents. Adenosine agonists for cardiac imaging and positron-emitting adenosine antagonists are in development for diagnostic use. Allosteric modulation of A_1 and A_3 ARs has been demonstrated.

In addition to selective agonists/antagonists, mouse strains in which an AR has been genetically deleted have been useful in developing novel drug concepts based on modulation of ARs.

Acknowledgements Support from the Intramural Research Program of the NIH, National Institute of Diabetes and Digestive and Kidney Diseases is gratefully acknowledged. Dr. Andrei A. Ivanov, NIDDK, prepared the image shown in Fig. 2. Dr. Zhang-Guo Gao and Dr. Dale Kiesewetter are acknowledged for helpful discussions.

References

Akaiwa K, Akashi H, Harada H, Sakashita H, Hiromatsu S, Kano T, Aoyagi S (2006) Moderate cerebral venous congestion induces rapid cerebral protection via adenosine A_1 receptor activation. Brain Res 1122:47–55

Arispe N, Ma J, Jacobson KA, Pollard HB (1998) Direct activation of cystic fibrosis transmembrane conductance regulator (CFTR) channels by CPX and DAX. J Biol Chem 273:5727–5734

Auchampach JA, Jin X, Moore J, Wan TC, Kreckler LM, Ge ZD, Narayanan J, Whalley E, Kiesman W, Ticho B, Smits G, Gross GJ (2004) Comparison of three different A_1 adenosine receptor antagonists on infarct size and multiple cycle ischemic preconditioning in anesthetized dogs. J Pharmacol Exp Ther 308:846–856

Awad AS, Huang L, Ye H, Duong ET, Bolton WK, Linden J, Okusa MD (2006) Adenosine A_{2A} receptor activation attenuates inflammation and injury in diabetic nephropathy. Am J Physiol Renal Physiol 290: F828–F837

Baraldi PG, Tabrizi MA, Gessi S, Borea PA (2008) Adenosine receptor antagonists: translating medicinal chemistry and pharmacology into clinical utility. Chem Rev 108:238–263

Baraldi PG, Tabrizi MA, Fruttarolo F, Romagnoli R, Preti D (2009) Recent improvements in the development of A_{2B} adenosine receptor agonists. Purinergic Signal 4(4):287–303

Bar-Yehuda S, Stemmer SM, Madi L, Castel D, Ochaion A, Cohen S, Barer F, Zabutti A, Perez-Liz G, Del Valle L, Fishman P (2008) The A_3 adenosine receptor agonist CF102 induces apoptosis of hepatocellular carcinoma via de-regulation of the Wnt and NF-kappaB signal transduction pathways. Int J Oncol 33:287–295

Bauer A, Langen KJ, Bidmon H, Holschbach MH, Weber S, Olsson RA, Coenen HH, Zilles K (2005) 18F-CPFPX PET identifies changes in cerebral A_1 adenosine receptor density caused by glioma invasion. J Nucl Med 46:450–454

Beldi G, Wu Y, Sun X, Imai M, Enjyoji K, Csizmadia E, Candinas D, Erb L, Robson SC (2008) Regulated catalysis of extracellular nucleotides by vascular CD39/ENTPD1 is required for liver regeneration. Gastroenterology 135:1751–1760

Beukers MW, Chang LC, von Frijtag Drabbe Künzel JK, Mulder-Krieger T, Spanjersberg RF, Brussee J, IJzerman AP (2004) New, non-adenosine, high-potency agonists for the human adenosine A_{2B} receptor with an improved selectivity profile compared to the reference agonist N-ethylcarboxamidoadenosine. J Med Chem 47:3707–3709

Björklund O, Shang M, Tonazzini I, Daré E, Fredholm BB (2008) Adenosine A_1 and A_3 receptors protect astrocytes from hypoxic damage. Eur J Pharmacol 596:6–13

Burnstock G (2008) Purinergic signalling and disorders of the central nervous system. Nat Rev Drug Discov 7:575–590

Cerqueira MD (2006) Advances in pharmacologic agents in imaging: new A_{2A} receptor agonists. Curr Cardiol Rep 8:119–122

Che J, Chan ES, Cronstein BN (2007) Adenosine A_{2A} receptor occupancy stimulates collagen expression by hepatic stellate cells via pathways involving protein kinase A, Src, and extracellular signal-regulated kinases 1/2 signaling cascade or p38 mitogen-activated protein kinase signaling pathway. Mol Pharmacol 72:1626–1636

Chen GJ, Harvey BK, Shen H, Chou J, Victor A, Wang Y (2006a) Activation of adenosine A_3 receptors reduces ischemic brain injury in rodents. J Neurosci Res 84:1848–1855

Chen Y, Corriden R, Inoue Y, Yip L, Hashiguchi N, Zinkernagel A, Nizet V, Insel PA, Junger WG (2006b) ATP release guides neutrophil chemotaxis via $P2Y_2$ and A_3 receptors. Science 314:1792–1795

Cohen MV, Downey JM (2008) Adenosine: trigger and mediator of cardioprotection. Basic Res Cardiol 103:203–215

Costanzi S, Ivanov AA, Tikhonova IG, Jacobson KA (2007) Structure and function of G protein-coupled receptors studied using sequence analysis, molecular modeling, and receptor engineering: adenosine receptors. Front Drug Design Disc 3:63–79

Cotter G, Dittrich HC, Weatherley BD, Bloomfield DM, O'Connor CM, Metra M, Massie BM, PROTECT Steering Committee, Investigators, and Coordinators (2008) The PROTECT pilot study: a randomized, placebo-controlled, dose-finding study of the adenosine A_1 receptor antagonist rolofylline in patients with acute heart failure and renal impairment. J Cardiac Fail 14:631–640

Desai A, Victor-Vega C, Gadangi S, Montesinos MC, Chu CC, Cronstein B (2005) Adenosine A_{2A} receptor stimulation increases angiogenesis by down-regulating production of the antiangiogenic matrix protein thrombospondin 1. Mol Pharmacol 67:1406–1413

Dittrich HC, Gupta DK, Hack TC, Dowling T, Callahan J, Thomson S (2007) The effect of KW-3902, an adenosine A_1 receptor antagonist, on renal function and renal plasma flow in ambulatory patients with heart failure and renal impairment. J Card Failure 13:609–617

Doggrell SA (2005) BG-9928 (Biogen Idec). Curr Opin Investig Drugs 6:962–968

Eckle T, Krahn T, Grenz A, Köhler D, Mittelbronn M, Ledent C, Jacobson MA, Osswald H, Thompson LF, Unertl K, Eltzschig HK (2007) Cardioprotection by ecto-5′-nucleotidase (CD73) and A_{2B} adenosine receptors. Circulation 115:1581–1590

Elzein E, Zablocki J (2008) A_1 adenosine receptor agonists and their potential therapeutic applications. Expert Opin Investig Drugs 17:1901–1910

Franco R, Casadó V, Mallol J, Ferrada C, Ferré S, Fuxe K, Cortés A, Ciruela F, Lluis C, Canela EI (2006) The two-state dimer receptor model: a general model for receptor dimers. Mol Pharmacol 69:1905–1912

Fredholm BB, Jacobson KA (2009) John W. Daly and the early characterization of adenosine receptors. Heterocycles 79:73–83

Fredholm BB, IJzerman AP, Jacobson KA, Klotz KN, Linden J (2001) International Union of Pharmacology. XXV. Nomenclature and classification of adenosine receptors. Pharmacol Rev 53:527–552

Fredholm BB, Chen JF, Masino SA, Vaugeois JM (2005) Actions of adenosine at its receptors in the CNS: insights from knockouts and drugs. Annu Rev Pharmacol Toxicol 45:385–412

Gao ZG, Kim SK, IJzerman AP, Jacobson KA (2005) Allosteric modulation of the adenosine family of receptors. Mini Rev Med Chem 5:545–553

Gao ZG, Ye K, Göblyös A, IJzerman AP, Jacobson KA (2008) Flexible modulation of agonist efficacy at the human A_3 adenosine receptor by an imidazoquinoline allosteric enhancer LUF6000 and its analogues. BMC Pharmacol 8:20

Gessi S, Merighi S, Varani K, Leung E, Mac Lennan S, Borea PA (2008) The A_3 adenosine receptor: an enigmatic player in cell biology. Pharmacol Ther 117:123–140

Gillespie RJ, Cliffe IA, Dawson CE, Dourish CT, Gaur S, Jordan AM, Knight AR, Lerpiniere J, Misra A, Pratt RM, Roffey J, Stratton GC, Upton R, Weiss SM, Williamson DS (2008) Antagonists of the human adenosine A_{2A} receptor. Part 3: Design and synthesis of pyrazolo[3,4-*d*]pyrimidines, pyrrolo[2,3-*d*]pyrimidines and 6-arylpurines. Bioorg Med Chem 18:2924–2929

Giorgi I, Nieri P (2008) Therapeutic potential of A_1 adenosine receptor ligands: a survey of recent patent literature. Expert Opin Ther Patents 18:677–691

Givertz MM, Massie BM, Fields TK, Pearson LL, Dittrich HC (2007) The effect of KW-3902, an adenosine A_1-receptor antagonist, on diuresis and renal function in patients with acute decompensated heart failure and renal impairment or diuretic resistance. J Am Coll Cardiol 50:1551–1560

Gottlieb SS, Brater DC, Thomas I, Havranek E, Bourge R, Goldman S, Dyer F, Gomez M, Bennett D, Ticho B, Beckman E, Abraham WT (2002) BG9719 (CVT-124), an A_1 adenosine receptor antagonist, protects against the decline in renal function observed with diuretic therapy. Circulation 105:1348–1353

Greenberg B, Ignatius T, Banish D, Goldman S, Havranek E, Massie BM, Zhu Y, Ticho B, Abraham WT (2007). Effects of multiple oral doses of an A_1 adenosine receptor antagonist, BG 9928, in patients with heart failure. J Am Coll Cardiol 50:600–606

Guzman J, Yu JG, Suntres Z, Bozarov A, Cooke H, Javed N, Auer H, Palatini J, Hassanain HH, Cardounel AJ, Javed A, Grants I, Wunderlich JE, Christofi FL (2006) ADOA3R as a therapeutic target in experimental colitis: proof by validated high-density oligonucleotide microarray analysis. Inflamm Bowel Dis 12:766–789

Haskó G, Linden J, Cronstein B, Pacher P (2008) Adenosine receptors: therapeutic aspects for inflammatory and immune diseases. Nat Rev Drug Discov 7:759–770

Hess S (2001) Recent advances in adenosine receptor antagonist research. Expert Opin Ther Patents 11:1533–1561

Hocher B, Fischer Y, Witte K, Ziegler D (2008) Use of adenosine A_1 antagonists in radiocontrast media induced nephropathy. US Patent Appl 20080027082

Holgate ST (2005) The identification of the adenosine A_{2B} receptor as a novel therapeutic target in asthma. Br J Pharmacol 145:1009–1015

Ishikawa J, Mitani H, Bandoh T, Kimura M, Totsuka T, Hayashi S (1998) Hypoglycemic and hypotensive effects of 6-cyclohexyl-2′-*O*-methyl-adenosine, an adenosine A_1 receptor agonist, in spontaneously hypertensive rat complicated with hyperglycemia. Diab Res Clin Pract 39:3–9

Ishiwata K, Noguchi J, Wakabayashi S, Shimada J, Ogi N, Nariai T, Tanaka A, Endo K, Suzuki F, Senda M (2000) ^{11}C-labeled KF18446: a potential central nervous system adenosine A_{2A} receptor ligand. J Nucl Med 41:345–354

Ivanov AA, Jacobson KA (2008) Molecular modeling of a PAMAM-CGS21680 dendrimer bound to an A_{2A} adenosine receptor homodimer. Bioorg Med Chem Lett 18:4312–4315

Ivanov AA, Baak D, Jacobson KA (2009) Evaluation of homology modeling of G protein-coupled receptors in light of the A_{2A} adenosine receptor crystallographic structure. J Med Chem, doi: 10.1021/jm801533x

Jaakola VP, Griffith MT, Hanson MA, Cherezov V, Chien EYT, Lane JR, IJzerman JR, Stevens RC (2008) The 2.6 Angstrom crystal structure of a human A_{2A} adenosine receptor bound to an antagonist. Science 322(5905):1211–1217

Jacobson KA, Gao ZG (2006) Adenosine receptors as therapeutic targets. Nat Rev Drug Disc 5:247–264

Jacobson KA, Kim HO, Siddiqi SM, Olah ME, Stiles GL, von Lubitz DKJE (1995) A_3 adenosine receptors: design of selective ligands and therapeutic prospects. Drugs Future 20:689–699

Ji XD, Jacobson KA (1999) Use of the triazolotriazine [^{3}H]ZM 241385 as a radioligand at recombinant human A_{2B} adenosine receptors. Drug Des Discov 16:217–226

Johansson B, Halldner L, Dunwiddie TV, Masino SA, Poelchen W, Giménez-Llort L, Escorihuela RM, Fernández-Teruel A, Wiesenfeld-Hallin Z, Xu XJ, Hårdemark A, Betsholtz C, Herlenius E, Fredholm BB (2001) Hyperalgesia, anxiety, and decreased hypoxic neuroprotection in mice lacking the adenosine A_1 receptor. Proc Natl Acad Sci USA 98:9407–9412

Jordan AM (2008) Science and serendipity: discovery of novel, orally bioavailable adenosine A_{2A} antagonists for the treatment of Parkinson's disease. Abstract MEDI-015, 236th ACS National Meeting, Philadelphia, PA, 17–21 Aug 2008

Kiesewetter DO, Lang L, Ma Y, Bhattacharjee AK, Gao ZG, Joshi BV, Melman A, Castro S, Jacobson KA (2008) Synthesis and characterization of [^{76}Br]-labeled high affinity A_3 adenosine receptor ligands for positron emission tomography. Nucl Med Biol 36:3–10

Kiesman WF, Zhao J, Conlon PR, Dowling JE, Petter RC, Lutterodt F, Jin X, Smits G, Fure M, Jayaraj A, Kim J, Sullivan GW, Linden J (2006) Potent and orally bioavailable 8-bicyclo[2.2.2]octylxanthines as adenosine A_1 receptor antagonists. J Med Chem 49:7119–7131

Kim SK, Gao, ZG, Van Rompaey P, Gross AS, Chen A, Van Calenbergh S, Jacobson KA (2003) Modeling the adenosine receptors: comparison of binding domains of A_{2A} agonist and antagonist. J Med Chem 46:4847–4859

Kim Y, Hechler B, Klutz A, Gachet C, Jacobson KA (2008) Toward multivalent signaling across G protein-coupled receptors from poly(amidoamine) dendrimers. Bioconjugate Chem 19: 406–411

Klaasse EC, IJzerman AP, de Grip WJ, Beukers MW (2008) Internalization and desensitization of adenosine receptors. Purinergic Signal 4:21–37

Klutz AM, Gao ZG, Lloyd J, Shainberg A, Jacobson KA (2008) Enhanced A_3 adenosine receptor selectivity of multivalent nucleoside-dendrimer conjugates. J Nanobiotechnol 6:12

Knutsen LJ, Lau J, Petersen H, Thomsen C, Weis JU, Shalmi M, Judge ME, Hansen AJ, Sheardown MJ (1999) *N*-Substituted adenosines as novel neuroprotective A_1 agonists with diminished hypotensive effects. J Med Chem 42:3463–3477

Kolachala VL, Bajaj R, Chalasani M, Sitaraman SV (2008) Purinergic receptors in gastrointestinal inflammation. Am J Physiol Gastrointest Liver Physiol 294:G401–G410

LeWitt PA, Guttman M, Tetrud JW, Tuite PJ, Mori A, Chaikin P, Sussman NM (2008) Adenosine A_{2A} receptor antagonist istradefylline (KW-6002) reduces "off" time in Parkinson's disease: a double-blind, randomized, multicenter clinical trial (6002-US-005). Ann Neurol 63:295–302

Li X, Bantel C, Conklin D, Childers SR, Eisenach JC (2004) Repeated dosing with oral allosteric modulator of adenosine A_1 receptor produces tolerance in rats with neuropathic pain. Anesthesiology 100:956–961

Liang BT, Jacobson KA (1998) A physiological role of the adenosine A_3 receptor: sustained cardioprotection. Proc Natl Acad Sci USA 95:6995–6999

Lieu HD, Shryock JC, von Mering GO, Gordi T, Blackburn B, Olmsted AW, Belardinelli L, Kerensky RA (2007) Regadenoson, a selective A_{2A} adenosine receptor agonist, causes dose-dependent increases in coronary blood flow velocity in humans. J Nucl Cardiol 14:514–520

Madi L, Ochaion A, Rath-Wolfson L, Bar-Yehuda S, Erlanger A, Ohana G, Harish A, Merimski O, Barer F, Fishman P (2004) The A_3 adenosine receptor is highly expressed in tumor versus normal cells: potential target for tumor growth inhibition. Clin Cancer Res 10:4472–4479

Madi L, Cohen S, Ochayin A, Bar-Yehuda S, Barer F, and Fishman P (2007) Overexpression of A_3 adenosine receptor in peripheral blood mononuclear cells in rheumatoid arthritis: involvement of nuclear factor-kappa B in mediating receptor level. J Rheumatol 34:20–26

Martin PL, Wysocki RJ Jr, Barrett RJ, May JM, Linden J (1996) Characterization of 8-(*N*-methylisopropyl)amino-N^6-(5′-endohydroxy-endonorbornyl)-9-methyladenine (WRC-0571), a highly potent and selective, non-xanthine antagonist of A_1 adenosine receptors. J Pharmacol Exp Ther 276:490–499

Martin L, Pingle SC, Hallam DM, Rybak LP, Ramkumar V (2006) Activation of the adenosine A_3 receptor in RAW 264.7 cells inhibits lipopolysaccharide-stimulated tumor necrosis factor-alpha release by reducing calcium-dependent activation of nuclear factor-kappaB and extracellular signal-regulated kinase 1/2. J Pharmacol Exp Ther 316:71–78

McGaraughty S, Cowart M, Jarvis MF, Berman RF (2005) Anticonvulsant and antinociceptive actions of novel adenosine kinase inhibitors. Curr Top Med Chem 5:43–58

Melman A, Gao ZG, Kumar D, Wan TC, Gizewski E, Auchampach JA, Jacobson KA (2008a) Design of (*N*)-methanocarba adenosine 5′-uronamides as species-independent A_3 receptor-selective agonists. Bioorg Med Chem Lett 18:2813–2819

Melman A, Wang B, Joshi BV, Gao ZG, de Castro S, Heller CL, Kim SK, Jeong LS, Jacobson KA (2008b) Selective A_3 adenosine receptor antagonists derived from nucleosides containing a bicyclo[3.1.0]hexane ring system. Bioorg Med Chem 16:8546–8556

Merkel LA, Hawkins ED, Colussi DJ, Greenland BD, Smits GJ, Perrone MH, Cox BF (1995) Cardiovascular and antilipolytic effects of the adenosine agonist GR 79236. Pharmacology 51:224–236

Mittendorf J, Wuppertal D (2008) BAY 68–4986 (Capadenoson): the first non-purinergic adenosine A_1 agonist for the oral treatment of stable angina pectoris. Fachgruppe Medizinische Chemie Annual Meeting, Regensburg, Germany, 2–5 March 2008, doi: 10.1002/cmdc.200800114

Moresco RM, Todde S, Belloli S, Simonelli P, Panzacchi A, Rigamonti M, Galli-Kienle M, Fazio F (2005) In vivo imaging of adenosine A_{2A} receptors in rat and primate brain using [^{11}C]SCH442416. Eur J Nucl Med Mol Imag 32:405–413

Moro S, Gao ZG, Jacobson KA, Spalluto G (2006) Progress in pursuit of therapeutic adenosine receptor antagonists. Med Res Rev 26:131–159

Mustafa SJ, Nadeem A, Fan M, Zhong H, Belardinelli L, Zeng D (2007) Effect of a specific and selective A_{2B} adenosine receptor antagonist on adenosine agonist AMP and allergen-induced airway responsiveness and cellular influx in a mouse model of asthma. J Pharmacol Exp Ther 320:1246–1251

Nakata H, Yoshioka K, Kamiya T, Tsuga H, Oyanagi K (2005) Functions of heteromeric association between adenosine and P2Y receptors. J Mol Neurosci 26:233–238

Ochaion A, Bar-Yehuda S, Cohen S, Amital H, Jacobson KA, Joshi BV, Gao ZG, Barer F, Zabutti A, Del Valle L, Perez-Liz G, Fishman P (2008) The A_3 adenosine receptor agonist CF502 inhibits the PI3K, PKB/Akt and NF-κB signaling pathways in synoviocytes from rheumatoid arthritis patients and in adjuvant induced arthritis. Biochem Pharmacol 76:482–494

Ohta A, Sitkovsky M (2001) Role of G-protein-coupled adenosine receptors in downregulation of inflammation and protection from tissue damage. Nature 41:916–920

Okamura T, Kurogi Y, Hashimoto K, Sato S, Nishikawa H, Kiryu K, Nagao Y (2004) Structure–activity relationships of adenosine A_3 receptor ligands: new potential therapy for the treatment of glaucoma. Bioorg Med Chem Lett 14:3775–3779

Palmer TM, Poucher SM, Jacobson KA, Stiles GL (1996) 125I-4-(2-[7-Amino-2-{furyl}{1,2,4} triazolo{2,3-*a*}{1,3,5}triazin-5-ylaminoethyl)phenol (^{125}I-ZM241385), a high affinity antagonist radioligand selective for the A_{2A} adenosine receptor. Mol Pharmacol 48:970–974

Pascoe SJ, Knight H, Woessner R (2007) QAF805, an A_{2b}/A_3 adenosine receptor antagonist does not attenuate AMP challenge in subjects with asthma. Am J Resp Crit Care Med 175:A682

Penn RB, Pascual RM, Kim YM, Mundell SJ, Krymskaya VP, Panettieri RA Jr, Benovic JL (2001) Arrestin specificity for G protein-coupled receptors in human airway smooth muscle. J Biol Chem 276:32648–32656

Porkka-Heiskanen T, Strecker RE, Thakkar M, Bjorkum AA, Greene RW, McCarley RW (1997) Adenosine: a mediator of the sleep-inducing effects of prolonged wakefulness. Science 276:1265–1268

Press NJ, Taylor RJ, Fullerton JD, Tranter P, McCarthy C, Keller TH, Brown L, Cheung R, Christie J, Haberthuer S, Hatto JD, Keenan M, Mercer MK, Press NE, Sahri H, Tuffnell AR, Tweed M, Fozard JR (2005) A new orally bioavailable dual adenosine A_{2B}/A_3 receptor antagonist with therapeutic potential. Bioorg Med Chem Lett 15:3081–3085

Prinster SC, Hague C, Hall RA (2005) Heterodimerization of G protein-coupled receptors: specificity and functional significance. Pharmacol Rev 57:289–298

Ryzhov S, Goldstein AE, Biaggioni I, Feoktistov I (2006) Cross-talk between G(s)- and G(q)-coupled pathways in regulation of interleukin-4 by A_{2B} adenosine receptors in human mast cells. Mol Pharmacol 70:727–735

Ryzhov S, Novitskiy SV, Zaynagetdinov R, Goldstein AE, Carbone DP, Biaggioni I, Dikov MM, Feoktistov I (2008) Host A_{2B} adenosine receptors promote carcinoma growth. Neoplasia 10:987–995

Schulte G, Fredholm BB (2003) Signalling from adenosine receptors to mitogen-activated protein kinases. Cell Signal 15:813–827

Schwarzschild MA, Agnati L, Fuxe K, Chen JF, Morelli M (2006) Targeting adenosine A_{2A} receptors in Parkinson's disease. Trends Neurosci 29:647–54

Terai T, Kita Y, Kusunoki T, Shimazaki T, Ando T, Horiai H, Akahane A, Shiokawa Y, Yoshida K (1995) A novel non-xanthine adenosine A_1 receptor antagonist. Eur J Pharmacol 279:217–225

Udelson JE, Heller GV, Wackers FJ, Chai A, Hinchman D, Coleman PS, Dilsizian V, DiCarli M, Hachamovitch R, Johnson JR, Barrett RJ, Gibbons RJ (2004) Randomized, controlled dose-ranging study of the selective adenosine A_{2a} receptor agonist binodenoson for pharmacological stress as an adjunct to myocardial perfusion imaging. Circulation 109:457–464

Vlok N, Malan SF, Castagnoli N Jr, Bergh JJ, Petzer JP (2006) Inhibition of monoamine oxidase B by analogues of the adenosine A_{2A} receptor antagonist (*E*)-8-(3-chlorostyryl)caffeine (CSC). Bioorg Med Chem 14:3512–3521

von Lubitz DKJE, Lin RC, Popik P, Carter MF, Jacobson KA (1994) Adenosine A_3 receptor stimulation and cerebral ischemia. Eur J Pharmacol 263:59–67

Wadsak W, Mien LK, Shanab K, Ettlinger DE, Haeusler D, Sindelar K, Lanzenberger RR, Spreitzer H, Viernstein H, Keppler BK, Dudczak R, Kletter K, Mitterhauser M (2008) Preparation and first evaluation of [^{18}F]FE@SUPPY: a new PET tracer for the adenosine A_3 receptor. Nucl Med Biol 35:61–66

Wan TC, Ge ZD, Tampo A, Mio Y, Bienengraeber MW, Tracey WR, Gross GJ, Kwok WM, Auchampach JA (2008) The A_3 adenosine receptor agonist CP-532,903 [N^6-(2,5-dichlorobenzyl)-3′-aminoadenosine-5′-*N*-methylcarboxamide] protects against myocardial ischemia/reperfusion injury via the sarcolemmal ATP-sensitive potassium channel. J Pharmacol Exp Ther 324:234–243

Wilson CN (2008) Adenosine receptors and asthma in humans. Br J Pharmacol 155:475–486

Yan L Burbiel JC Maass A, Müller CE (2003) Adenosine receptor agonists: from basic medicinal chemistry to clinical development. Expert Opin Emerg Drugs 8:537–576

Yang H, Avila MY, Peterson-Yantorno K, Coca-Prados M, Stone RA, Jacobson KA, Civan MM (2005) The cross-species A_3 adenosine-receptor antagonist MRS 1292 inhibits adenosine-triggered human nonpigmented ciliary epithelial cell fluid release and reduces mouse intraocular pressure. Curr Eye Res 30:747–754

Yang D, Koupenova M, McCrann DJ, Kopeikina KJ, Kagan HM, Schreiber BM, Ravid K (2008) The A_{2b} adenosine receptor protects against vascular injury. Proc Natl Acad Sci USA 105: 792–796

Yu L, Huang Z, Mariani J, Wang Y, Moskowitz M, Chen JF (2004) Selective inactivation or reconstitution of adenosine A_{2A} receptors in bone marrow cells reveals their significant contribution to the development of ischemic brain injury. Nat Med 10:1081–1087

Zezula J, Freissmuth M (2008) The A_{2A}-adenosine receptor: a GPCR with unique features? Br J Pharmacol 153(Suppl 1):S184–S190

Zheng J, Wang R, Zambraski E, Wu D, Jacobson KA, Liang BT (2007) A novel protective action of adenosine A_3 receptors: attenuation of skeletal muscle ischemia and reperfusion injury. Am J Physiol Heart Circ Physiol 293:3685–3691

Zimmermann H (2000) Extracellular metabolism of ATP and other nucleotides. Naunyn–Schmiedeberg's Arch Pharmacol 362:299–309

A_1 Adenosine Receptor Antagonists, Agonists, and Allosteric Enhancers

William F. Kiesman, Elfatih Elzein, and Jeff Zablocki

Contents

Abstract Intense efforts of many pharmaceutical companies and academicians in the A_1 adenosine receptor (AR) field have led to the discovery of clinical candidates that are antagonists, agonists, and allosteric enhancers. The A_1AR antagonists currently in clinical development are KW3902, BG9928, and SLV320. All three have high affinity for the human (h) A_1AR subtype (hA_1 $K_i < 10\,nM$), > 200-fold selectivity over the hA_{2A} subtype, and demonstrate renal protective effects in multiple animal models of disease and pharmacologic effects in human subjects. In the A_1AR agonist area, clinical candidates have been discovered for the following conditions: atrial arrhythmias (tecadenoson, selodenoson and PJ-875); Type II diabetes and insulin sensitizing agents (GR79236, ARA, RPR-749, and CVT-3619); and angina

W.F. Kiesman (✉)
14 Cambridge Center, Cambridge, MA 02142, USA
william.kiesman@biogenidec.com

C.N. Wilson and S.J. Mustafa (eds.), *Adenosine Receptors in Health and Disease*, Handbook of Experimental Pharmacology 193, DOI 10.1007/978-3-540-89615-9_2,

(BAY 68–4986). The challenges associated with the development of any A_1AR agonist are to obtain tissue-specific effects but avoid off-target tissue side effects and A_1AR desensitization leading to tachyphylaxis. For the IV antiarrhythmic agents that act as ventricular rate control agents, a selective response can be accomplished by careful IV dosing paradigms. The treatment of type II diabetes using A_1AR agonists in the clinic has met with limited success due to cardiovascular side effects and a well-defined desensitization of full agonists in human trials (GR79236, ARA, and RPR 749). However, new partial A_1AR agonists are in development, including CVT-3619 (hA_1 AR K_i = 55 nM, hA_{2A}:hA_{2B}:hA_3 > 200:1,000:20, CV Therapeutics), which have the potential to provide enhanced insulin sensitivity without cardiovascular side effects and tachyphylaxis. The nonnucleosidic A_1AR agonist BAY 68–4986 (capadenoson) represents a novel approach to angina wherein both animal studies and early human studies are promising. T-62 is an A_1AR allosteric enhancer that is currently being evaluated in clinical trials as a potential treatment for neuropathic pain. The challenges associated with developing A_1AR antagonists, agonists, or allosteric enhancers for therapeutic intervention are now well defined in humans. Significant progress has been made in identifying A_1AR antagonists for the treatment of edema associated with congestive heart failure (CHF), A_1AR agonists for the treatment of atrial arrhythmias, type II diabetes and angina, and A_1AR allosteric enhancers for the treatment of neuropathic pain.

Keywords A1 adenosine receptor agonists · A1 adenosine receptor antagonists · Acutely decompensated heart failure · Adentri · Cardiorenal syndrome · Congestive heart failure · Anti-arrhythmic agents · Tecadenoson · Selodenoson · Insulin sensitizing agents · CVT-3619 · Angina · Capadenoson · Allosteric enhancers · Neuropathic pain · Type II diabetes · BG9928 · KW3902 · SLV320

Abbreviations

ACE	Angiotensin-converting enzyme
ADHF	Acutely decompensated heart failure
ALT	Alanine aminotransferase
APD	Action potential duration
AR	Adenosine receptor
ARA	(1*S*, 2*R*, 3*R*)-3-((trifluoromethoxy)methyl)-5-(6-(1-(5-(trifluromethyl)pyridine-2-yl)pyrrolidin-3-ylamino)-9*H*-purin-9-yl) cyclopentane-1,2-diol
ARB	Angiotensin II receptor blocker
AST	Aspartate aminotransferase
AUC	Area under the curve
(A–V) node	Atrioventricular
BG9719	(8-(3-Oxa-tricyclo[3.2.1.$0^{2,4}$]oct-6-yl)-1,3-dipropyl-3, 7-dihydropurine-2,6-dione)

BG9928	(3-(4-(2,6-Dioxo-1,3-dipropyl-2,3,6,7-tetrahydro-1*H*-purin-8-yl) bicyclo[2.2.2]octan-1-yl)propanoic acid)
cAMP	Cyclic AMP
Capadenoson	(BAY 68–4986) (2-amino-6-((2-(4-chlorophenyl)thiazol-4-yl) methylthio)-4-(4-(2-hydroxyethoxy)phenyl)pyridine-3, 5-dicarbonitrile)
CHA	N^6-Cyclohexyl adenosine
CHF	Congestive heart failure
CHO	Chinese hamster ovary
CK	Creatinine kinase
CL	Total body clearance
C_{max}	Maximal plasma concentration
CPA	N^6-Cyclopentyl adenosine
CrCl	Creatinine clearance
CV	Cardiovascular
CVT-3619	(2*S*, 3*S*, 4*R*)-2-((2-fluorophenylthio)methyl)-5-(6-((1*R*, 2*R*)-2-hydroxycyclopentylamino)-9*H*-purin-9-yl)tetrahydrofuran-3,4-diol
DPCPX	1,3-Dipropyl-8 cyclopentylxanthine
ED_{50}	50% Efficient dose
F (%)	% Oral bioavailability
GFR	Glomerular filtration rate
GR79236	(3*R*, 4*S*, 5*R*)-2-(6-((1*S*, 2*S*)-2-hydroxycyclopentylamino)-9*H*-purin-9-yl)-5-(hydroxymethyl)tetrahydrofuran-3,4-diol
HF	Heart failure
HSL	Hormone sensitive lipase
IP	Intraperitoneal
IPC	Ischemic preconditioning
IV	Intravenous
KW3902	(3-Noradamantyl-1,3-dipropylxanthine)
L-(NAME)	*N*-o-Nitro-L-arginine methyl ester
MRT	Mean residence time
NEFA	Nonesterified fatty acids
PVST	Paroxysmal supraventricular tachycardia
RCM	Radiocontrast media
RPF	Renal plasma flow
SAR	Structure–activity relationship
Selodenoson	(2*S*, 3*S*, 4*R*)-5-(6-(Cyclopentylamino)-9*H*-purin-9-yl)-*N*-ethyl-3,4-dihydroxytetrahydrofuran-2-carboxamide)
(S–A)	Sinoatrial
SLV320	(4-(2-Phenyl-7*H*-pyrrolo[2,3-*d*]pyrimidin-4-ylamino) cyclohexanol)
$t_{1/2}$	Half-life
T-62	(2-Amino-4,5,6,7-tetrahydrobenzo[*b*]thiophen-3-yl) (4-chlorophenyl)methanone
Tecadenoson	(2*R*, 3*S*, 4*R*)-2-(Hydroxymethyl)-5-(6-((*R*)-tetrahydrofuran-3-ylamino)-9*H*-purin-9-yl)tetrahydrofuran-3,4-diol)

TG's	Triglycerides
T2D	Type II diabetes
UNaV	Urinary sodium excretion
UV	Urine volume
V_{ds}	Volume of distribution

1 Introduction

The A_1 adenosine receptor (AR), a member of the P_1 family of seven-transmembrane adenosine receptors, couples to G_i to decrease the secondary messenger cAMP. The P1 family of adenosine receptors consists of the members A_1, A_{2A}, A_{2B}, and A_3, which have high sequence homology with many conserved residues at the active sites; however, sufficient differences are found for each active site such that selective agonists and antagonists have been generated for each receptor subtype (Akhari et al. 2006; Dhalla et al., 2003; Fredholm et al. 2001; Jacobson and Gao 2006). The major goal of this review is twofold: to highlight the structure–affinity relationships (SAR) of A_1AR antagonists, agonists, and allosteric enhancers, and to give an overview of the A_1AR antagonists, agonists, and allosteric enhancers currently under development for various indications.

2 A_1 Adenosine Receptor Antagonists

From the earliest reports on the physiologic effects of theophylline and adenosine to the three active clinical programs today (see Fig. 1), the study of AR ligands has a long and rich history (Baraldi et al. 2008; Jacobson and Gao 2006). There are a number of excellent reviews on A_1AR antagonists (Baraldi et al. 2008; Hess 2001; Moro et al. 2006; van Galen et al. 1992; Yuzlenko and Kieć-Kononowicz 2006). Because these reviews discuss the historical development of this class of molecules, their structure–activity relationships (SARs), pharmacology, and therapeutic applications, a comprehensive review of A_1AR antagonists will not be presented here. It is important to note that a number of A_1AR antagonists have entered clinical trials; however, problems with high lipophilicity and corresponding low water solubility and bioavailability have limited their clinical development (Hess 2001). This section of our review will present brief overviews of the most advanced A_1 adenosine receptor (A_1AR) antagonists that are promising drug candidates currently in clinical trials. The discussion will address SARs around the lead molecules, highlights of pharmacology in healthy animals and disease models, and top-line human clinical trial results.

The 1,3-dialkylxanthine core has been the mainstay of A_1AR antagonists since the isolation of theophylline in 1886 (Kossel 1888) (Fig. 1). One hundred years passed before replacement of the methyl substituents with *n*-propyl chains and

theophylline adenosine KW3902 BG9928 SLV320

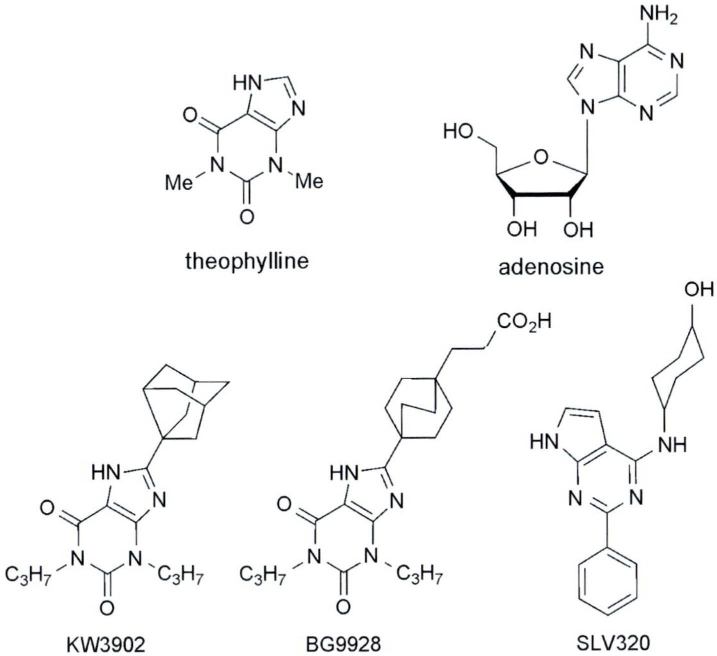

Fig. 1 Dialkylxanthine and adenine-based adenosine A_1 receptor ligands

the installation of cyclopentane at the C8 position led to the discovery of 1,3-dipropyl-8-cyclopentyl xanthine (DPCPX), which has been used as a radioligand and pharmacologic probe for the in vivo effects of A_1AR antagonism in living systems (Shamim et al. 1988). Since then, significant effort has been directed toward garnering improvements in activity and selectivity on the well-optimized 1,3-dipropylxanthine, and has led to the discovery of two molecules that are in active clinical development programs: KW3902 (3-noradamantyl-1,3-dipropylxanthine) (Suzuki et al. 1992) and BG9928 (3-(4-(2,6-dioxo-1,3-dipropyl-2,3,6,7-tetrahydro-1*H*-purin-8-yl)bicyclo[2.2.2]octan-1-yl)propanoic acid) (Kiesman et al. 2006a). A structurally distinct nonxanthine series based upon the adenine substructure of adenosine itself has also been developed and is represented by clinical candidate SLV320 (4-(2-phenyl-7*H*-pyrrolo[2,3-*d*]pyrimidin-4-ylamino)cyclohexanol) (Kalk et al. 2007). All three of these A_1AR antagonists display high affinity for the A_1AR and significant selectivity over the A_{2A}AR (Table 1).

2.1 KW3902

To further characterize the hydrophobic interactions between the 8 position in the xanthine series and the A_1AR binding site, Shimada et al. (1991, 1992) investigated substitutions of the 8-cyclopentyl ring in DPCPX **1** (Table 2). Clipping the cyclopentyl ring into two ethyl groups **2** led to a loss of guinea pig (gp)A_1 affinity but

Table 1 Binding affinities for selected A_1 adenosine receptor antagonists

	K_i (nM)[a]					
	rA_1	rA_{2A}	hA_1	hA_{2A}	hA_{2B}	hA_3
KW3902	0.2[b]	170[b]	8	673	296	4,390
BG9928	1.3	2,440	7	6,410	90	>10,000
SLV320	2.5[c]	501[c]	1[c]	398[c]	3980[c]	200[c]

[a]Receptor binding experiments using cloned human receptors in CHO (hA_1, 0.3 nM ^{125}I-aminobenzyladenosine (^{125}IABA)) or HEK293-derived cell membranes (hA_{2A}, 0.7 nM ^{125}I-ZM241385; hA_{2B}, 0.5 nM ^{125}I-3-(4-aminobenzyl)-8-phenyloxyacetate-1-propyl-xanthine; and hA_3, 0.6 nM ^{125}I-ABA); rA_1 binding measured as inhibition of [^{3}H]-DPCPX to rat forebrain membranes; rA_{2A} binding measured as inhibition of [^{3}H] ZM241385 in rat striatal membranes (Kiesman et al. 2006a)

[b]rA_1 binding measured as inhibition of [^{3}H]-CHA to rat forebrain membranes; rA_{2A} binding measured as inhibition of [^{3}H] CGS21680 in rat striatal membranes (Suzuki et al. 1992)

[c]Receptor binding experiments using cloned human receptors in CHO (hA_1: [^{3}H]-DPCPX or HEK293-derived cell membranes (hA_{2A}: [^{3}H]-CGS21680; hA_{2B}: [^{3}H]-DPCPX; hA_3: [^{3}H]-AB-MECA or rat cerebral cortex (rA_1: [^{3}H]-CCPA, or striatal membranes (rA_{2A}: [^{3}H]-CGS21680 (Kalk et al. 2007)

had no effect on rat (r)A_{2A} binding. When compared to **1**, dicyclopropyl substitution **3** showed enhanced potency and selectivity versus gpA_1 and rA_{2A} receptors, respectively, while the addition of *gem*-dimethyl substitution **4** led to diminished gpA_1 affinity but a remarkable decrease in rA_{2A} binding (>170-fold decrease). Bicyclo- and tricycloalkane systems **5, 7, 8** were then examined to determine if restrictions in conformational flexibility around the cyclopentyl ring in **1** effected the antagonist activity. Interestingly, the 3-noradamantyl system **8** stood out, with increases in gpA_1 affinity ($K_i = 1.3$ nM and high selectivity over the rA_{2A} receptor (890-fold; Nonanka et al. 1996). Introduction of a methylene linker between the bulky polycyclic alkane in the 8 position gave compound **9** significantly reduced gpA_1 affinity (>65-fold).

Animal Studies

The diuretic activity of KW3902 was examined in saline-loaded, conscious Wistar rats (Suzuki et al. 1992). The antagonist was orally administered in a saline suspension and the urine collected and analyzed for sodium content. Urine volume (UV) and urinary sodium excretion (UNaV) both increased in a dose-dependent manner, with maximal effects observed between 0.1 and 0.4 mg kg^{-1} (Fig. 2).

During the development of intravenously (IV) injectable formulations for KW3902, which has a solubility in water of <1 μg mL^{-1}, Hosokawa et al. (2002) investigated the effects of a lipid emulsion and liposome formulation on the pharmacokinetics of KW3902 and its metabolite (M1-KW3902) in comparison to a 1 N NaOH–DMSO-containing formulation (Fig. 3). They reported no

Table 2 Structure–activity relationships for 8-substituted 1,3-dipropylxanthines

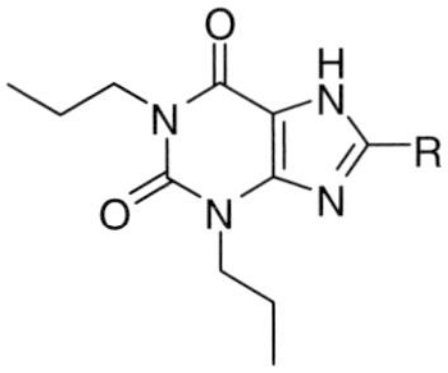

Cmpd	R	K_i (nM)[a] gpA$_1$	rA$_{2A}$	rA$_{2A}$/gpA$_1$
1		6.4 ± 0.35	590 ± 48	92
		$(0.49 \pm 0.06)^b$ rat	–	–
		$(17 \pm 6.5)^c$ dog	–	–
2	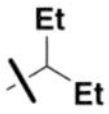 Et, Et	19 ± 1.0	570 ± 44	30
3		3.0 ± 0.21	430 ± 5.8	140
		$(0.919 \pm 0.04)^b$ rat	–	–
4		22	>100,000	>4,500
5		3.83 ± 0.32	440 ± 42	120
6	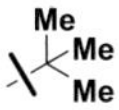 Me, Me, Me	31	2,300	74
7	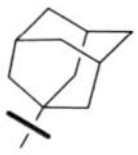	13 ± 2.8	$5,100 \pm 1,100$	390
8		1.3 ± 0.12	380 ± 30	290
		$(0.19 \pm 0.042)^b$, rat	$(170 \pm 16)^c$ (rat)	890 (rat)
		10 ± 2.6^c dog	–	–
9	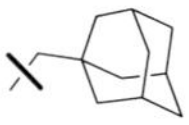	880	>100,000	>110

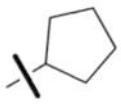

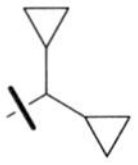

[a]gpA$_1$ binding was carried out with N^6-[^{3}H]cyclohexyladenosine ([^{3}H]-CHA)in guinea pig (gp) forebrain membranes and rA$_{2A}$, binding was carried out with N-[^{3}H]ethyladenosine-5′-uronamide in the presence of 50 nM cyclopentyladenosine in rat (r) striatal membranes (Shimada et al. 1991)
[b]rA$_1$ binding measured as inhibition of [^{3}H]-CHA to rat forebrain membranes (Shimada et al. 1992)
[c]rA$_{2A}$ binding measured as inhibition of [^{3}H] CGS21680 in rat striatal membranes and dA$_1$ (dog) measured with [^{3}H]-CHA in dog forebrain membranes (Nonanka et al. 1996). All K_i measurements are given as mean ± SEM for 3–5 determinations

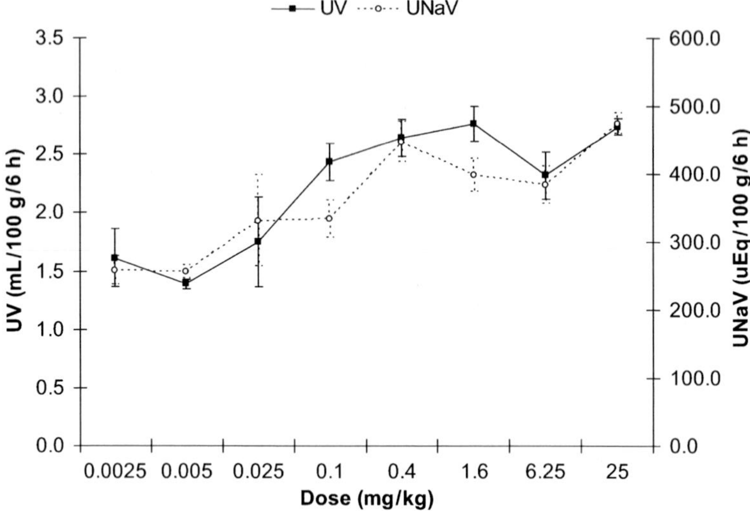

Fig. 2 Dose–response for urine volume (UV) in mL $100\,g^{-1}6\,h^{-1}$ (mean ± SEM) and urinary sodium excretion (UNaV) in μEq $100\,g^{-1}$ $6\,h^{-1}$ (mean ± SEM) over 6 h following oral doses of KW3902 ranging from 0.0025 to 25 mg kg^{-1} in male rats

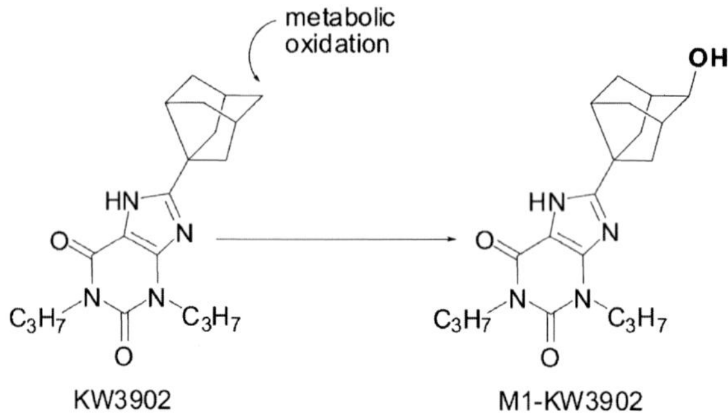

Fig. 3 Chemical structure of the metabolite of KW3902, M1-KW3902

significant differences in elimination half-life ($t_{1/2}$), area under the curve (AUC), total body clearance (CL), and mean residence time (MRT) for all of the formulations investigated. Table 3 summarizes the pharmacokinetic parameters measured for KW3902 and M1-KW3902 in the 1 N NaOH–DMSO formulation. The lipid formulation, however, did prevent the precipitation of KW3902 after IV injection, and it was suggested that this formulation may be used in further clinical studies.

The renal protective activity of KW3902 was investigated in a rat model of glycerol-induced acute renal failure (Suzuki et al. 1992). The antagonist was administered intraperitoneally (IP), and after 30 min glycerol (50% v/v in sterile saline; 0.8 mL $100\,g^{-1}$) was injected subcutaneously. After a subsequent 24-h hold time, blood was collected and serum creatinine and urea nitrogen were determined

Table 3 Pharmacokinetic parameters of KW3902 and metabolite M1-KW3902 after IV administration of KW3902 to rats

	Free KW3902		Metabolite M1-KW3902	
	0.1 mg kg^{-1}	1 mg kg^{-1}	0.1 mg kg^{-1}	1 mg kg^{-1}
$t_{1/2}$ (h)	1.0 ± 1.1	1.6 ± 0.6	4.7 ± 1.2	7.9 ± 3.9
$AUC_{0-\infty}$ (ng h^{-1} mL^{-1})	72 ± 50	561 ± 18	486 ± 113	$4,916 \pm 2,457$
CL (L h^{-1} kg^{-1})	1.87 ± 1.11	1.79 ± 0.06	–	–
MRT (h)	1.5 ± 1.9	1.5 ± 0.3	7.3 ± 1.7	11.3 ± 5.6
V_{ds} (L kg^{-1})	1.66 ± 1.10	2.72 ± 0.51	–	–

Values represent the means ± standard deviation of three experiments.

$t_{1/2}$, Half-life; *AUC*, area under the curve; *CL*, total body clearance; *MRT*, mean residence time; V_{ds}, volume of distribution

Table 4 Renal protective activity of KW3902 in rats[a]

Dose (mg kg^{-1}, IP)	Serum creatinine Concentration (mg dL^{-1})			Serum urea nitrogen (mg dL^{-1})		
	Vehicle	Treated	% Inhibition	Vehicle	Treated	% Inhibition
0.01	4.03 ± 0.23	1.83 ± 0.08[b]	55	130.5 ± 7.0	45.2 ± 3.9[b]	65
0.1	2.89 ± 0.18	1.40 ± 0.14[c]	52	138.4 ± 9.8	48.1 ± 7.0[c]	65
1	3.75 ± 0.43	2.00 ± 0.14[c]	47	123.2 ± 14.2	66.4 ± 6.3[c]	46

[a] All values are the means ± SEM; significant difference from vehicle-treated group

[b] $P < 0.01$

[c] $P < 0.001$

(Table 4). Typically after glycerol injection, serum creatinine and urea nitrogen increase seven- to tenfold in rats. Pretreatment with KW3902 significantly reduced (50–60%) the negative renal effects of glycerol-induced acute renal failure (Suzuki et al. 1992).

The mechanism of the protective effects of A_1AR antagonism on two additional nephrotoxic models of acute renal failure in vivo, including renal accumulation of gentamicin (Yao 2000) and radiocontrast media (RCM) (Yao et al. 2001), were examined. From these studies, it was suggested that KW3902 inhibited the action of endogenous adenosine and increased renal blood flow, which led to suppression of intrarenal accumulation of gentamicin, whereas in RCM-induced nephropathy, it prevented the drop in glomerular filtration rate (GFR); a marker for kidney function) that occurs in *N*-o-nitro-L-arginine methyl ester (*L*-NAME) hypertensive rats. In the RCM model, it is unclear whether the action of the antagonist reduces RCM uptake and cellular toxicity by inhibiting sodium transport in the proximal tubule or by another mechanism.

Adenosine has been found, through its actions on A_2ARs (presumably A_{2A}ARs), to play a protective role in the ischemic preconditioning of multiple organs, including the liver (Peralta et al. 1999), heart (Forman et al. 1993), and lung (Khimenko et al. 1995). Magata (Magata et al. 2007) recently reported that blockade of the A_1AR with KW3902 attenuated hepatic ischemia-reperfusion injury in dogs. Two groups of female beagles ($n = 6$) underwent a 2 h total hepatic vascular exclusion; one group served as the control and the other received 1 μg kg^{-1} min^{-1} of KW3902 via continuous intraportal infusion for 60 min prior to ischemia. It was noted that although peripheral IV infusion of KW3902 was effective in earlier studies (Nonaka et al. 1996), no beneficial effects were seen in the hepatic ischemia-reperfusion model by this route of administration. Thus, treatment by KW3902 via the portal vein proved beneficial in a number of outcomes in the study. The two-week survival of the control group was 16.7% versus 83.3% ($P < 0.05$) for the treated group. Serum alanine aminotransferase (ALT) levels were significantly inhibited by the A_1AR antagonist; the control group rose rapidly to 12,625 ± 1,010 U L^{-1} after 6 h of reperfusion, while the treated animals peaked at 24 h at 2,352 ± 452 U L^{-1}. In addition, the number of infiltrating neutrophils in the hepatic tissue of the KW3902 group (41.17 ± 11.01 at 60 min) was significantly lower than that of the control group (63.6 ± 23.6 at 60 min) (Fig. 4).

Treatment with KW3902 in the liver ischemia-reperfusion model prevented the bradycardia seen in the control animals just after reperfusion, significantly increased adenine nucleotide levels in the ischemic tissues, and regulated the microcirculatory disturbances, resulting in greater hepatic tissue blood flow. It was concluded that for adenosine to protect the liver from ischemia-reperfusion injury it is necessary to block A_1AR activation.

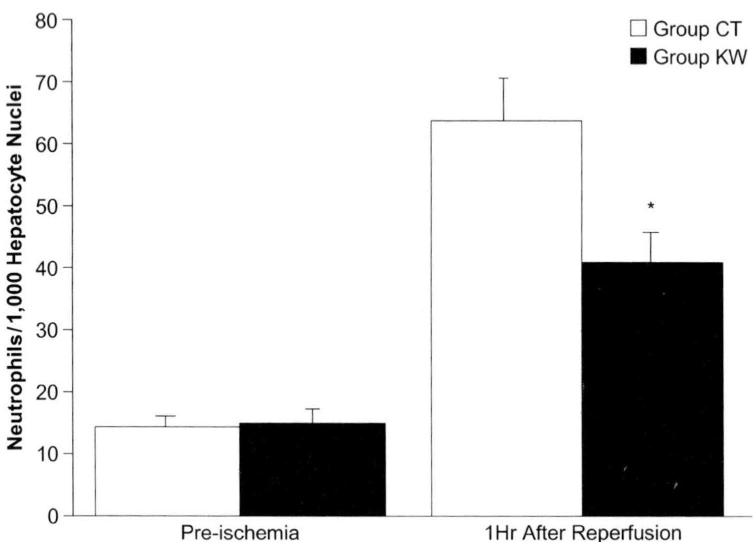

Fig. 4 The number of infiltrated neutrophils in liver tissue. Data are expressed as mean ± SEM. Group CT, negative control ($n = 6$); group KW, treated KW3902 ($n = 6$). $^*P < 0.05$ versus group CT. Reprinted with permission from Magata et al. (2007)

Clinical Studies

Building upon the earlier studies of the renal effects of another A_1AR antagonist, BG9719 (8-(3-oxa-tricyclo[3.2.1.$0^{2,4}$]oct-6-yl)-1,3-dipropyl-3,7-dihydropurine-2,6-dione) (Gotttlieb et al. 2002), in patients with congestive heart failure (CHF), Dittrich and colleagues (Dittrich et al. 2007) examined the renal vasodilatory effects of 30 mg doses of KW3902 in patients with ambulatory heart failure (HF). The two-way crossover study followed patients with CHF with mild renal impairment (median GFR 50 mL min^{-1}) and compared IV administration of placebo or KW3902 oil emulsion followed by IV furosemide (80 mg). GFR and renal plasma flow were assessed by iothalamate and *para*-aminohippurate clearances over 8 h. After a three- to eight-day washout period, subjects crossed over to either active treatment or placebo, again followed by furosemide. Renal plasma flow (Fig. 5) and GFR increased by 48% ($P < 0.05$ vs. placebo) and 32% ($P < 0.05$ vs. placebo), respectively, over baseline for 8 h postKW3902 administration, supporting the conclusion that blockade of the A_1AR leads to vasodilation and increases in filtration rates in patients with HF with reduced kidney function. There also appeared to be a persistent positive effect on GFR (approx. 10 mL min^{-1} increase in GFR over previous baseline) seen in the crossover patients who received KW3902 in the first dose. The pharmacokinetics of the parent compound or its metabolites could not account for the change in the baseline GFR values.

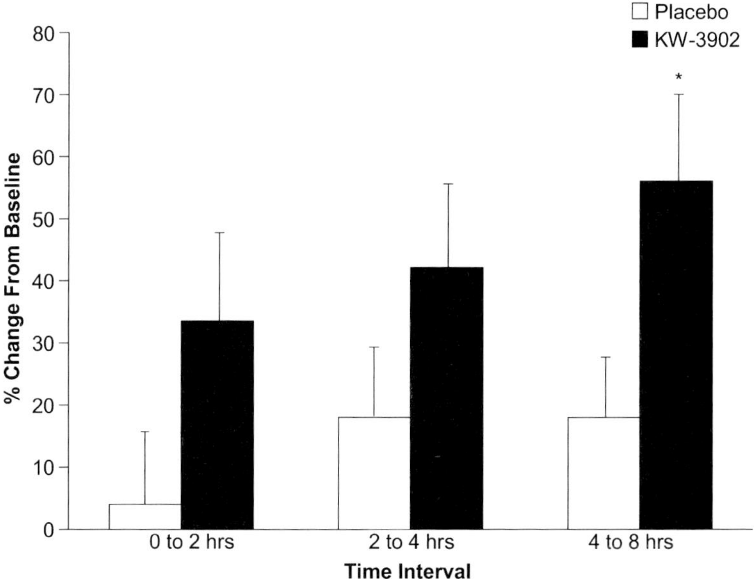

Fig. 5 Renal plasma flow (RPF). The percent change in RPF from baseline for KW-3902 and placebo in the presence of furosemide ($n = 23$). Values shown are for all subjects mean ± SEM. P values reflect analysis of RPF percent change between KW-3902 and placebo using log RPF values ($^*P < 0.05$). Reprinted with permission from Dittrich et al. (2007)

In a related clinical investigation, Givertz et al. (2007) examined the dose-dependent effects of KW3902 on diuresis and renal function in two subsets of patients with acutely decompensated heart failure (ADHF) with either renal impairment or diuretic resistance. In the first protocol, patients with volume overload and creatinine clearance (CrCl) of 20–80 mL min^{-1} received either placebo or one of four doses (2.5, 15, 30, or 60 mg) of KW3902 as a 2 h IV infusion for up to three days of treatment. All four doses increased urine output during the 6 h period following administration (Fig. 6). There were no significant differences in systolic blood pressure or heart rate in any of the treatment groups. A transient decrease in serum creatinine was noted on day 2 of treatment for all dose levels (-0.03 to -0.08 mg dL^{-1} for KW3902 arms vs. $+0.04$ mg dL^{-1} for placebo). This effect was maintained on day 4 (or the day of discharge), except for the 60 mg dose level.

In the diuretic-resistant population, single infusions of KW3902 (10, 30, or 60 mg) were given to patients with an average baseline CrCl of 34 mL min^{-1}. Urine output increased for all dose levels (ranging from $+22$ to $+24$ mL h^{-1}), whereas the placebo arm saw a decrease in urine output (-29 mL h^{-1}). The CrCl data for this subset of patients was complex. In general, the placebo arm had decreases in CrCl over 24 h with similar trending from the 60 mg dose treatment group. However, the 10 and 30 mg doses showed increases in CrCl over the 24 h period. The inverted relationship between KW3902 dose and renal function (CrCl) at high doses bears some resemblance to the dose–response relationship seen at higher doses of the A_1AR antagonist BG9719 (Gottlieb et al. 2002). Whether these similarities are due

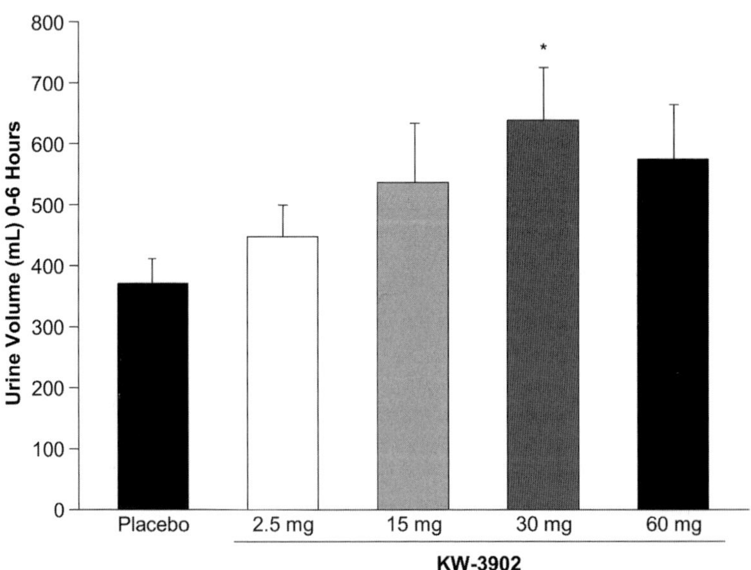

Fig. 6 Urine output in first 6 h after administration of KW3902. Cumulative urine volume (mean ± SEM) 6 h after initiation of placebo or KW-3902 in patients with acutely decompensated heart failure (ADHF) with renal impairment ($^{*}P = 0.02$ vs. placebo). Reprinted with permission from Givertz et al. (2007)

to cross-activity against the other adenosine receptors in the kidney at high doses or in other tissues is unclear and may require further study. Phase III clinical trials of KW3902 in patients with ADHF are currently underway, and limited releases of the data have recently appeared (Novacardia Press Release 2007).

2.2 BG9928

At the same time that formulation development began on KW3902, another selective A_1AR antagonist, BG9719, which possessed adequate pharmacologic activity, was used to demonstrate proof of concept for A_1AR antagonism in animals (Pfister et al. 1997) and humans (Gottlieb et al. 2000, 2002). However, the poor pharmaceutical properties of this molecule (low aqueous solubility and a tendency to rearrange to inactive products in both acidic and basic media) led Kiesman et al. (Kiesman et al. 2006a, b) to design more pharmaceutically acceptable antagonists by exploring the placement of polar substituents on linearly substituted 8-cycloalkyl 1,3-dipropylxanthines. For structurally related imidazolines, see Vu et al. (2006). The binding affinities of selected 8-cyclohexyl and 8-bicyclo[2.2.2]octyl xanthines are listed in Table 5.

The bicyclo[2.2.2]octyl analogs **11, 13, 15** had better A_1AR binding affinities than the related cyclohexyl variants **10, 12, 14**, and maintained similar A_{2A}AR activity. A significant improvement in receptor selectivity ($hA_{2A}/hA_1 = 161$ vs. 22) came with the replacement of the dimethylamino functional group in **15** with the carboxylic acid in **16**. Further optimization of the bridgehead chain led to the propionic acid **18**, BG9928, and single-digit nanomolar ($rA_1 = 1.3\,nM$ and $hA_1 = 7.4\,nM$) activity and high receptor selectivities ($rA_{2A}/rA_1 = 1{,}880$; $hA_{2A}/hA_1 = 915$).

The functional antagonist activity of BG9928 (Kiesman et al. 2006a) was confirmed by examining the blockade or increasing doses of the compound on the inhibitory effects of N^6-cyclopentyl adenosine (CPA) on the beat rates of isolated rat atria. Administration of the antagonist restored the atrial beat rates to their maxima and effectively blocked the negative chronotropic activity of CPA ($EC_{50} = 16.1 \pm 7.7\,nM$). In a separate set of isolated rat atria experiments, BG9928 was found to have a pA_2 (antagonist potency) of 9.8.

Animal Studies

Single oral doses of BG9928 administered to male Sprague–Dawley rats (Fig. 7) led to increases in urine volume (UV) and sodium excretion (UNaV) with a 50% efficient dose (ED_{50}) of approximately $15\,\mu g\,kg^{-1}$ (Kiesman et al. 2006a; Ticho et al. 2003). The increases in urinary potassium excretion were proportional to volume increases, confirming the potassium-neutral diuresis commonly observed with A_1AR antagonists. The dose–response relationships are similar to those seen with KW3902 (Fig. 2); however, the magnitude of the pharmacodynamic effect is smaller

Table 5 Structure–activity relationships for 1,4-linearly-substituted 8-cycloalkylxanthines

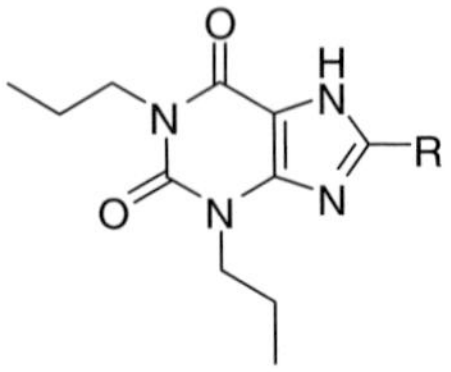

Cmpd	R	K_i (nM)[a] or % of specific radioligand binding[b]				
		hA_1	hA_{2A}	hA_{2B}	hA_3	hA_{2A}/hA_1
10	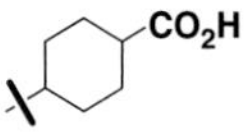 CO_2H	(31%)	(75%)	(69%)	(88%)	–
11	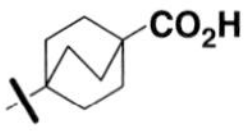 CO_2H	33	1,070	(48%)	(100%)	32
12	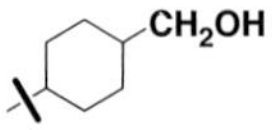 CH_2OH	41	313	(18%)	(77%)	8
13	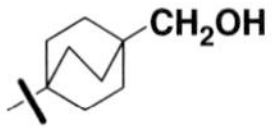 CH_2OH	16	414	(27%)	(73%)	26
14	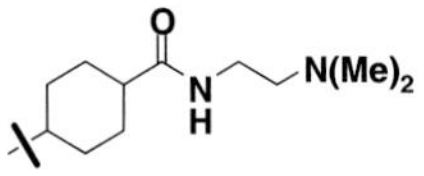 O, NH, $N(Me)_2$	12	168	(16%)	(91%)	14
15	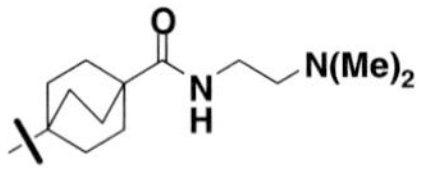 O, NH, $N(Me)_2$	6	132	(3%)	(79%)	22
16	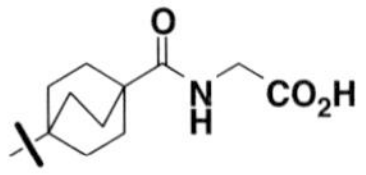 O, NH, CO_2H	49	7,880	(53%)	(70%)	161
17	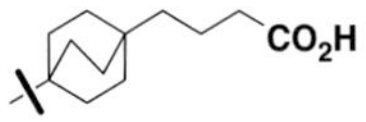 CO_2H	29	–	127	(26%)	–
		(4.0) rat	(50%) rat	–	–	(∼250) rat
18	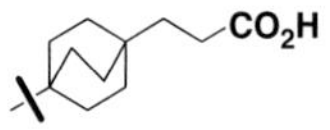 CO_2H	7.4	6,410	90	>10,000	915
		(1.3) rat	(2,440) rat	–	–	(1,880) rat
19	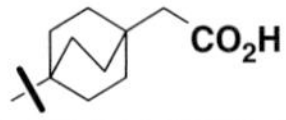 CO_2H	(22.5) rat	(8,960) rat	–	–	(398) rat

[a]Receptor binding experiments using cloned human receptors in CHO (hA_1, 0.3 nM ^{125}I-aminobenzyladenosine (^{125}IABA) or HEK293-derived cell membranes (hA_{2A}, 0.7 nM ^{125}I-ZM241385; hA_{2B}, 0.5 nM ^{125}I-3-(4-aminobenzyl)-8-phenyloxyacetate-1-propyl-xanthine; and hA_3, 0.6 nM ^{125}I-ABA); rA_1 binding measured as inhibition of [^{3}H]-DPCPX to rat forebrain membranes; rA_{2A} binding measured as inhibition of [^{3}H] ZM241385 in rat striatal membranes. All K_i values were calculated from binding curves generated from the mean of four determinations per concentration (seven antagonist concentrations), with variation in individual values of <15% (Kiesman et al. 2006a)

[b]Data are presented as percent of radioligand bound in the presence of target compound relative to control

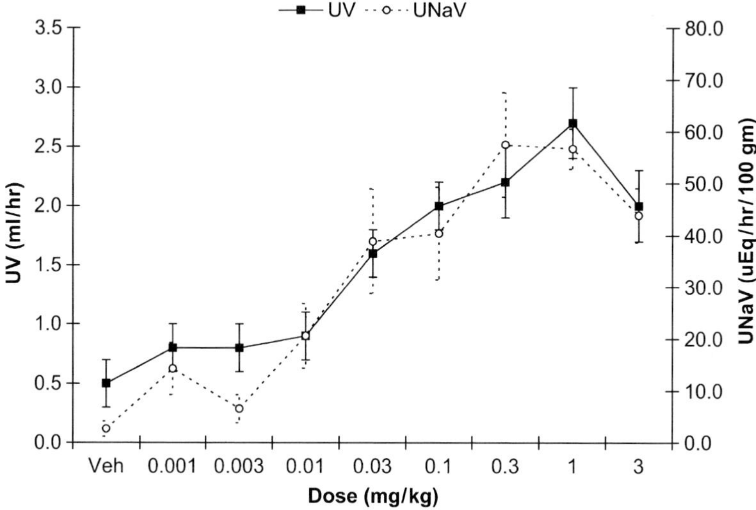

Fig. 7 Dose–response for urine volume (UV) in mL h^{-1} (mean ± SEM) and urinary sodium excretion (UNaV) in μEq h^{-1} 100 g^{-1} (mean ± SEM) over 4 h following single oral doses of vehicle (0.5% carboxymethyl cellulose suspension, n = 3) or BG9928 ranging from 0.001 to 3 mg kg^{-1} in rats (0.001 mg kg^{-1}, 0.003 mg kg^{-1}, 0.01 mg kg^{-1}, each $n = 4$; 0.03 mg kg^{-1}, 0.1 mg kg^{-1}, 0.3 mg kg^{-1}, each $n = 5$; 1.0 mg kg^{-1}, 3.0 mg kg^{-1}, each $n = 3$). Adapted with permission from Kiesman et al. (2006a)

Table 6 Pharmacokinetic parameters of BG9928 following single oral dose administration

	F (%)	$t_{1/2}$ (h)	CL (mL min^{-1} kg^{-1})	V_{ds} (L kg^{-1})
Rat	99	3.14 ± 0.14	1.56 ± 0.26	0.32 ± 0.02
Dog	78	6.40 ± 4.0	11.8 ± 0.6	2.64 ± 1.29
Monkey	94	11.1 ± 4.2	5.82 ± 0.45	4.25 ± 0.70

$n = 3$ male rats, 4 male dogs, and 4 male cynomolgus monkeys
F(%), Percent bioavailability; $t_{1/2}$, half-life; *CL*, total body clearance; V_{ds}, volume of distribution

in the BG9928 study than in the KW3902 study because the rats in the BG9928 model were not saline-loaded prior to treatment.

The single oral-dose pharmacokinetic profile of 1 mg kg^{-1} BG9928 was assessed in the rat, dog, and cynomolgus monkey (Table 6). Bioavailability was nearly complete in the rat and cynomolgus monkey, slightly lower in the dog, and followed the clearance difference amongst the species. Elimination half-lives were similar in the rat and dog (3–6 h), and longer in the cynomolgus monkey (11 h).

A rat model was designed to mimic the sodium-retentive state of patients with CHF (Ticho et al. 2003). Rats were given an oral dose of 100 mg kg^{-1} furosemide and placed on a low-sodium diet for up to six days. One group ($n = 8$) then received a single IV bolus of 30 mg kg^{-1} furosemide, while the other group ($n = 9$) received IV furosemide and 1 mg kg^{-1} of BG9928. The results are depicted in Fig. 8.

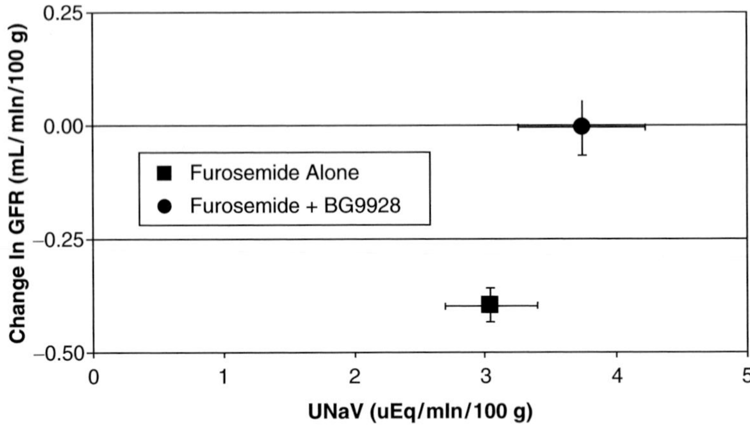

Fig. 8 Renal protective (mean ± SEM change in glomerular filtration rate [GFR]) and natriuretic (mean ± SEM) urinary sodium excretion (UNaV) effect of 1 mg kg^{-1} BG9928 administered IV in combination with furosemide (30 mg kg^{-1} IV) ($n = 9$, *circles*) compared with furosemide alone (30 mg kg^{-1}IV) ($n = 8$, *squares*) in low-sodium-retention rats. Reprinted with permission from Ticho et al. (2003)

Furosemide increased natriuresis and reduced GFR by approximately 50% over baseline. The addition of BG9928 not only further increased the natriuresis (+0.71 μEq min^{-1} 100 gm^{-1}) but also maintained the GFR, therefore preserving renal function. Similar data were presented for a Phase II proof-of-concept clinical trial for BG9719, a precursor compound to BG9928 (Gottlieb et al. 2002).

The interplay of A_1AR antagonism and ischemic preconditioning (IPC), specifically the effects of DPCPX, BG9719, and BG9928, in an in vivo dog model of myocardial infarction was examined (Auchampach et al. 2004). The study was composed of three arms in which the dogs ($n = 6$–12) were subjected to 60 min of left anterior descending coronary artery occlusion, followed by 3 h of reperfusion. Infarct size was assessed by triphenyltetrazolium chloride staining. In protocol 1, the dogs received vehicle or 1 mg kg^{-1} of the antagonist as a pretreatment, followed by continuous infusion at 10 μg kg^{-1} min^{-1} over the occlusion time. In protocol 2, the dogs received the same pretreatment as before but also received four 5-min occlusion/reperfusion preconditioning cycles. In protocol 3, the antagonists were not administered until 10 min prior to release of the occlusion and continued for 1 h into the reperfusion. Figure 9 summarizes the infarct size measurements across all three protocols.

Pretreatment with DPCPX or BG9928 reduced the myocardial infarct size by 51% and 49%, respectively, and none of the three antagonists blocked the protection of the myocardium afforded by the brief multiple-cycle IPC. In the most challenging experiment, it was found that treatment with either DPCPX or BG9928 just prior to reperfusion, a situation that more closely resembles clinical intervention in the course of treatment for myocardial infarction, reduced the infarct sizes by 43%

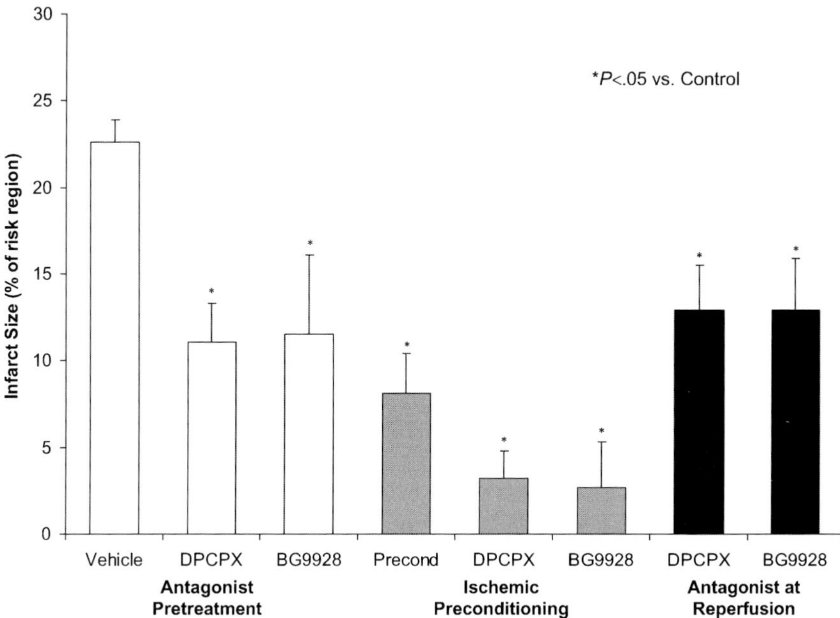

Fig. 9 Myocardial infarct size data (infarct size expressed as a percent of the area at risk) from antagonist pretreatment (protocol 1), ischemic preconditioning (protocol 2), and antagonist at reperfusion (protocol 3). $P < 0.05$ vs. control Adapted with permission from Auchampach et al. (2004)

and 45%, respectively. The study concluded that treatment with BG9928 provided cardioprotective effects that reduced infarct size and did not interfere with the protective effects of multiple-cycle IPC.

Clinical Studies

Greenberg et al. (2007) described the results of a placebo-controlled dose-escalation study designed to assess the pharmacokinetics and clinical effects of oral BG9928 in patients with HF. The study was conducted in 50 patients with HF, an ejection fraction of $\leqslant 40\%$ documented in the past 12 months, and who were on standard therapy including angiotensin-converting enzyme (ACE) inhibitors or angiotensin II receptor blocker (ARB) therapy and diuretics.

The pharmacokinetics of oral BG9928 in humans compared favorably to data from the earlier animal studies (Kiesman et al. 2006a) (Table 6). BG9928 was rapidly absorbed, with a $t_{\max}$ of 1.5–3.1 h and similar $C_{\max}$, $t_{1/2}$, and clearances for days 1, 6, and 10. Steady-state AUC was reached by day 6, and the elimination half-life (14–25 h) was consistent with once-daily dosing. Patients received BG9928 (3, 15, 75, or 225 mg) or placebo orally for ten days and were evaluated for changes in sodium excretion (primary endpoint), potassium excretion, creatinine clearance,

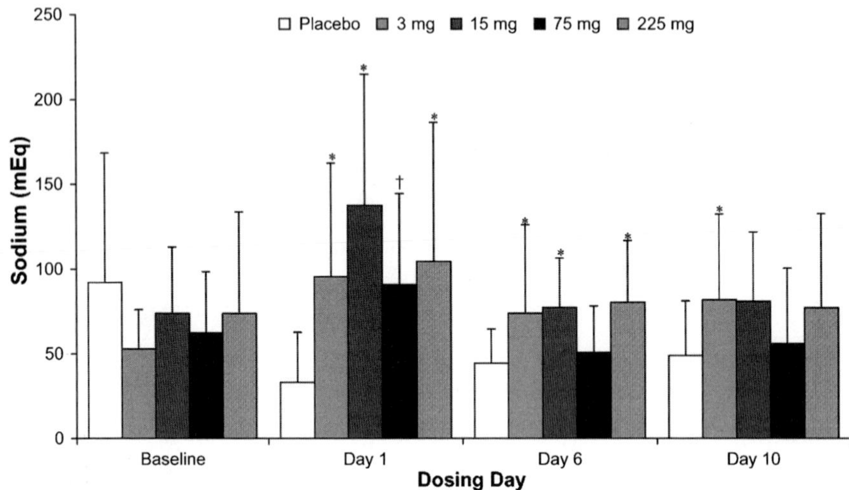

Fig. 10 Cumulative urinary sodium excretion (mEq) over the interval of 0–8 h at baseline and after placebo or BG9928 administration on days 1, 6, and 10. $^{*}P < 0.05$ vs. placebo; $^{\dagger}P < 0.055$ vs. placebo; $n = 10$ per group except for 15 and 75 mg dose groups on day 10 ($n = 9$ per group). Reproduced with permission from Greenberg et al. (2007)

and body weight. In humans, BG9928 increased sodium excretion compared with placebo and maintained the natriuresis over the ten-day study period (Fig. 10) with little kaliuresis. These data followed the same trends seen in the previous rat and monkey studies (Kiesman et al. 2006a; Ticho et al. 2003) (Fig. 7).

Use of ACE inhibitors, ARBs, and diuretics in patients with HF can adversely affect renal function and depress GFR. Despite the significant increases in natriuresis, adjusted CrCl was unchanged over the study period for all of the treatment groups (Fig. 11), suggesting that BG9928 may have had a protective effect on renal function.

Patients who received daily doses of greater than 3 mg had a reduction in body weight (−0.6, −0.7, −0.5 kg) versus a net weight gain of +0.3 kg for the placebo group at the end of the study (Fig. 12).

Patients receiving BG9928 also showed favorable directional trends in other measures of clinical status, including New York Heart Association functional class (five BG9928-treated patients improved by one level); Cody edema score (mean change from day 1 to day 11: 0 for placebo and up to −0.6 for the treated groups; a negative number indicates an improvement in HF signs); and physician's global assessment. Thus, positive effects were observed for all treated groups. However, there were no significant differences in clinical status for the short duration of the trial. This is the first clinical assessment of chronic oral dosing of an A_1 AR antagonist in humans. Future studies are planned to examine clinical status and renal preservation with both oral and parenteral BG9928 in patients with HF.

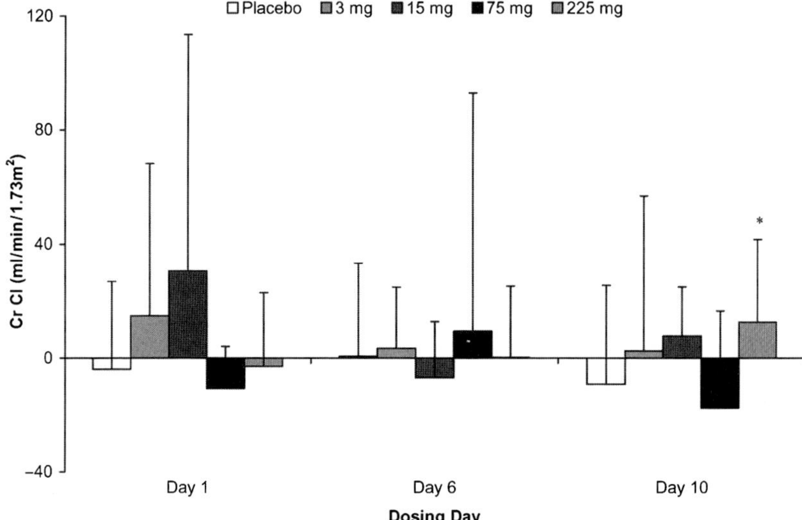

Fig. 11 Absolute change from baseline in adjusted creatinine clearance (CrCl) (ml min^{-1} 1.73 m^{-2}) for the interval of 2–24 h after placebo or BG9928 administration on days 1, 6, and 10. $^*P < 0.05$ vs. placebo; $n = 7$–10 per group. Reproduced with permission from Greenberg et al. (2007)

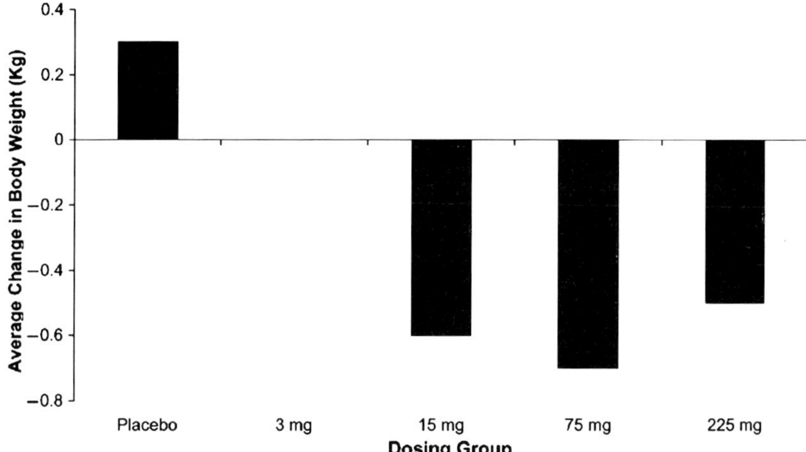

Fig. 12 Average change in body weight (kg) from baseline to day 11 after placebo or BG9928 administration. $n = 10$ per group. Reproduced with permission from Greenberg et al. (2007)

2.3 SLV320

Unlike the two xanthine-based A$_1$AR antagonists described in the preceding sections, SLV320 (see Fig. 1) contains an N^6-substituted-7-deazapurine core (specifically a 2-phenyl-pyrrolopyrimidine) with an N^6-*trans*-cyclohexanol side chain. The

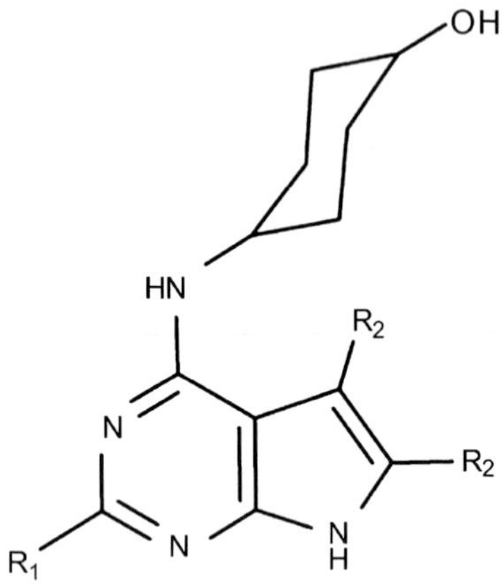

Fig. 13 Structure of pyrrolopyrimidine series

Table 7 Structure–activity relationships for pyrrolopyrimidines

			K_i (nM) or % of specific radioligand binding[a]				
Compd	R_1	R_2	hA_1	hA_{2A}	hA_{2B}	hA_3	hA_{2A}/hA_1
20	Ph	Me	~22	933	138	22	42
21	4-Pyridyl	H	8.0	131	1,031	(54%)	16
22	3-Chlorophenyl	H	8.0	531	530	419	66
23	3-Fluorophenyl	H	1.8	206	802	270	114
24	Ph	H	3.7	630	2,307	630	170

[a]Receptor binding experiments performed using membranes from yeast cells containing cloned human receptors (hA_1, [^{3}H]-DPCPX) or HEK293-derived cell membranes (hA_{2A}, [^{3}H]-CGS21680; hA_{2B}, [^{3}H]-DPCPX; and hA_3, 0.6 nM ^{125}I-AB-MECA) (Castelhano et al. 2005)

published SAR data around the pyrrolopyrimidine series (Fig. 13, Table 7) are only described within a series of patent filings (Castelhano et al. 2005). Dimethyl analog **20** had equipotent A_1 and A_3 AR affinities, a result that contrasts strikingly with those for the earlier xanthine systems (Tables 2 and 2.5). Removal of the two methyl groups ($R_2 = Me$) from compound **20** led to a tenfold increase in affinity for the A_1AR **24** and a significant reduction in A_3AR binding (22–630 nM) (Table 7). Substitution on the phenyl ring has small effects upon A_1AR activity and, in general, decreased selectivity versus the A_{2A}AR.

The functional antagonist activity of SLV320 was confirmed in experiments involving transient A_1AR-mediated bradycardia in anesthetized rats. Bolus injections of adenosine (100 μg kg^{-1}) lowered the heart rate in rats, and subsequent pretreatment both IV and orally with the antagonist caused a dose-dependent increase back to baseline in heart rate, with ED_{50} values of 0.25 and 0.49 mg kg^{-1}, respectively. Similar to the results for BG9928 (Greenberg et al. 2007; Ticho et al. 2003), no significant hemodynamic effects were seen in anesthetized rats (heart rate, systolic arterial pressure, or diastolic arterial pressure) with single IV bolus doses of between 0.1 and 5 mg kg^{-1}.

In a model of chronic renal failure and myocardial fibrosis, rats were subjected to either sham operations or removal of 5/6 of their kidneys (5/6 NX animals) (Kalk

Table 8 Plasma analytes, GFR, and albumin excretion at the end of the 5/6 NX study

Parameter	Sham	Sham + SLV320	5/6NX	5/6 NX +SLV320
CK ($U\,L^{-1}$)	481 ± 159	255 ± 31	$1,267 \pm 324$	196 ± 64^d
AST ($U\,L^{-1}$)	41 ± 2	41 ± 1	60 ± 5^a	35 ± 1^d
ALT ($U\,L^{-1}$)	36 ± 1	31 ± 1	43 ± 1^a	32 ± 1^d
Creatinine ($mg\,L^{-1}$)	4.7 ± 0.3	4.3 ± 0.2	7.3 ± 0.3^b	7.6 ± 0.42^b
GFR ($mL\ min^{-1}\ 100\,g^{-1}$)	0.45 ± 0.05	0.52 ± 0.02	0.32 ± 0.02^a	0.31 ± 0.01^a
Urinary albumin excretion ($mg\ 24\,h^{-1}$)	0.06 ± 0.02	0.07 ± 0.02	2.31 ± 0.39^b	$1.08 \pm 0.19^{b,c}$

Values given as mean ± SEM
$^a P < 0.05$ versus sham
$^b P < 0.001$ versus sham
$^c P < 0.05$ versus sham
$^d P < 0.001$ versus 5/6 NX
CK, Creatinine kinase; *AST*, aspartate aminotransferase; *ALT*, serum alanine aminotransferase; *GFR*, glomerular filtration rate

et al. 2007). The effects of SLV320 on markers for cardiomyopathy and clinical chemistry were examined. Treatment with the A_1AR antagonist completely abolished higher creatinine kinase (CK) plasma levels as well as elevated ALT and aspartate aminotransferase (AST) in the nephrectomized animals (Table 8). The creatinine levels and GFR measurements of the 5/6 NX animals showed diminished renal function, as expected, when compared to the sham group. Treatment with SLV320 did not significantly lower creatinine or increase GFR. In addition, although albuminuria was higher in the 5/6 NX group, treatment with SLV320 led to a 50% reduction in albumin excretion and exerted beneficial effects on renal disease progression.

No significant differences in cardiac histology were seen between the arms of the study; however, immunohistochemistry uncovered a significant increase in collagen I and III in the untreated 5/6 NX group compared to the SLV320-treated group (Fig. 14). This study was the first to demonstrate that an A_1AR antagonist inhibited markers of myocardial fibrosis without changes in blood pressure. The experiments also agree with two previous studies (Amann et al. 1998a, b) that concluded that uremia promotes cardiac fibrosis independently of hypertension.

It was recently reported that the clinical development of the oral form of SLV320 was suspended and little information is available on the results of human clinical trials with the intravenous product.

3 A_1 Adenosine Receptor Agonists

Agonism at A_1ARs may provide benefit for the following disease states: paroxysmal supraventricular tachycardia (PSVT)—break the atrial arrhythmia to return to sinus rhythm (Belardinelli and Lerman 1990; Belardinelli et al. 1995; DiMarco

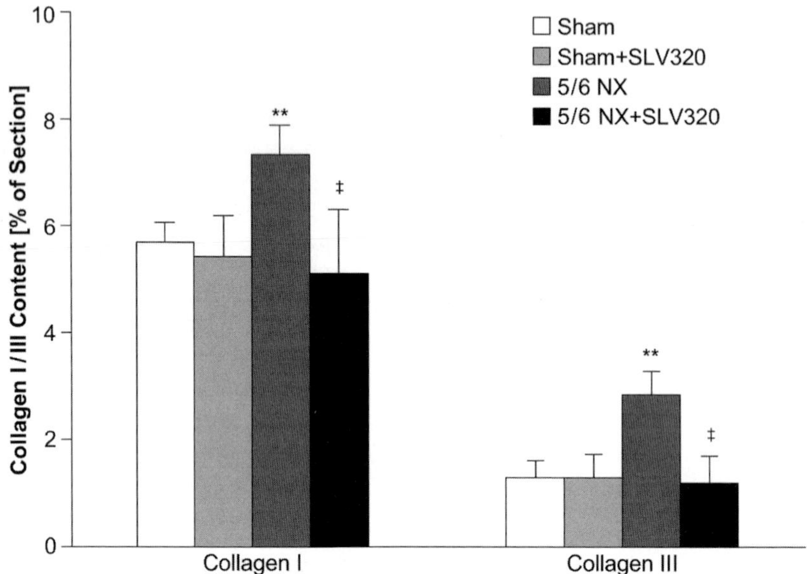

Fig. 14 Collagen I and III in rat hearts from nephrectomized rats versus normal controls. Values are given as mean ± SEM. Unpaired t-test was applied to detect significant differences between the study groups. $^{**}P < 0.001$ vs. sham; $^{\ddagger}P < 0.001$ vs. 5/6 NX. Reprinted with permission from Kalk et al. (2007)

et al. 1985; Lerman and Belardinelli 1991), atrial fibrillation (AF)—provide ventricular rate control (Wang et al. 1996; Zablocki et al. 2004), type II diabetes (T2D)—lower nonesterified fatty acid (NEFA) levels and triglycerides (TG's), as well as enhancing insulin sensitivity (Fatholahi et al. 2006; Gardner et al. 1994; Hoffman et al. 1986; Roden et al. 1996), and angina (Liu et al. 1991; Miura and Tsuchida 1999; Mizumura et al. 1996). The A_1AR is found in the A–V and S–A nodes, where stimulation by an A_1AR agonist results in negative dromotropic and chronotropic effects, respectively (Belardinelli et al. 1995; Wang et al. 1996). These cardiovascular (CV) effects are often side effects of A_1AR agonists that are being pursued for the other indications. Multiple full A_1AR agonists, tecadenoson (2*R*, 3*S*, 4*R*)-2-(hydroxymethyl)-5-(6-((*R*)-tetrahydrofuran-3-ylamino)-9*H*-purin-9-yl)tetrahydrofuran-3,4-diol), selodenoson (2*S*, 3*S*, 4*R*)-5-(6-(cyclopentylamino)-9*H*-purin-9-yl)-*N*-ethyl-3,4-dihydroxytetrahydrofuran-2-carboxamide) and PJ-875 are being pursued as intravenous clinical agents for the treatment of atrial arrhythmias, and the progress of these compounds will be highlighted below. In addition to the full A_1AR agonists, CV Therapeutics has reported on orally bioavailable partial A_1AR agonists that slow AV nodal conduction without causing third-degree AV block. For T2D, although both full and partial agonists will lower NEFA levels, a partial A_1AR agonist has the potential to do so with fewer side effects (Song et al. 2002; Srinivas et al. 1997; Stephenson 1997; Wu et al. 2001). Partial A_1AR agonists have the potential to provide for a selective targeted response, avoiding CV side effects (Wu et al. 2001). Plus, partial A_1AR agonists may be able to avoid receptor desensitization due to overstimulation that can lead to tachyphylaxis. In T2D, overstimulation of hormone-sensitive lipase (HSL) due to enhanced beta-adrenergic agonism on adipocytes leads to elevated NEFA levels within T2D patients from the 0.4–0.5 mM range up to the 0.8–1.2 mM range. Elevated NEFA levels have been shown to decrease skeletal muscle uptake of glucose, lower insulin release from the pancreas, and increase glucose production in the liver (Boden et al. 2005; Dhalla et al. 2007a; Ferrannini et al. 1983; Green 1987; Itani et al. 2002; Sako and Grill 1990). Decreasing NEFA levels through A_1AR agonism has an insulin-sensitizing effect in animal models. Several full A_1AR agonists (GR79236, ARA, and RPR749) have been evaluated in animal models and in clinical trials for the treatment of T2D, and the progress and challenges of these compounds will be described. CVT-3619 (2*S*, 3*S*, 4*R*)-2-((2-fluorophenylthio)methyl)-5-(6-((1*R*, 2*R*)-2-hydroxycyclopentyl-amino)-9*H*-purin-9-yl)tetrahydrofuran-3,4-diol), a partial agonist, is in preclinical development. The SAR leading up to the discovery of this partial agonist, as well as its efficacy in animal models of T2D, will be highlighted. The last indication for A_1AR agonists that we describe in this review is the treatment of angina. One A_1AR agonist (BAY 68–4986) is currently under clinical evaluation, and the progress of this compound will be described.

3.1 Intravenous Antiarrhythmic Agents: Tecadenoson, Selodenoson, Phenylsulfide, Phenylethers, PJ-875

The first A_1AR agonists to enter clinical development with the exception of adenosine were the IV antiarrhythmic agents, tecadenoson (Ellenbogen et al. 2005; Peterman and Sanoski 2005; Prystowsky et al. 2003) and selodenoson (Bayes et al. 2003; ClinicalTrials.gov 2005) (Fig. 15). The N6 lipophilic substituents are the key structure features that impart high affinity and selectivity for the A_1AR: N^6-(*R*)-3-tetrahydrofuranyl for tecadenoson (A_1AR $K_i = 3$ nM) and N^6-cyclopentyl for selodenoson (A_1AR $K_i = 6$ nM). Both compounds are similar in structure to N^6-cyclopentyl adenosine (CPA), but key structural differences are found in each molecule relative to CPA that impart beneficial pharmacological and pharmaceutical properties. For tecadenoson, the furan oxygen is favorable for imparting enhanced binding selectivity and additional solubility. For selodenoson, the 5′-*N*-ethyl carboxamide is favorable for both A_1AR and A_3 AR affinities, and it enhances oral activity. In Phase I clinical trials, both compounds were found to be safe and well tolerated at their specified IV bolus or infusion doses. With regards to efficacy, tecadenoson demonstrated favorable conversion (90%) of acute PSVT at the 300–600 μg bolus dose without significant adverse events. The A_1AR, which is highly expressed in atrial and AV nodal tissues, exerts its effects in the heart through lowering cAMP and direct activation of the inward rectifying potassium current, IK(Ado) (Belardinelli et al. 2005). In addition, A_1AR activation in the heart inhibits the

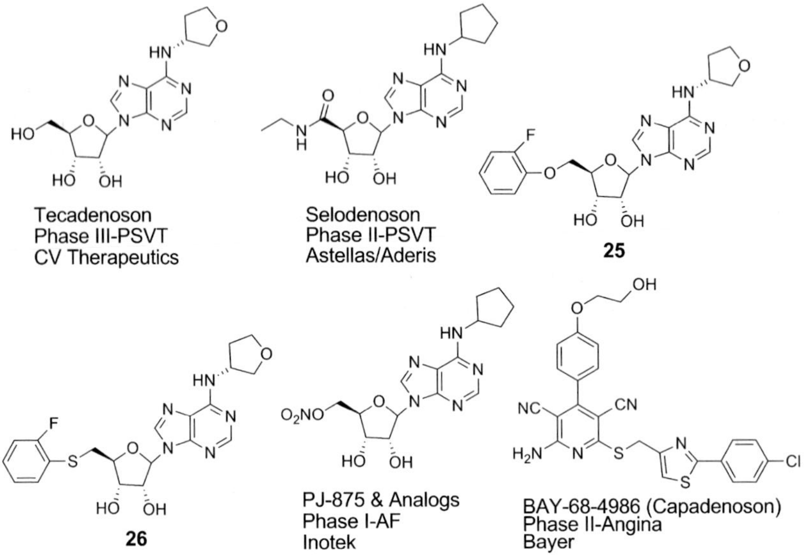

Tecadenoson
Phase III-PSVT
CV Therapeutics

Selodenoson
Phase II-PSVT
Astellas/Aderis

25

26

PJ-875 & Analogs
Phase I-AF
Inotek

BAY-68-4986 (Capadenoson)
Phase II-Angina
Bayer

Fig. 15 Antiarrhythmic and antianginal A_1AR agonists

catecholamine-stimulated ion currents such as pacemaker current and L-type calcium currents (Belardinelli et al. 2005). The result of A_1AR activation in the heart is prolongation of the AV nodal refractory period, reducing sinoatrial pacemaker rate and shortening the action potential (Belardinelli et al. 2005). Because of the shortening of the atrial action potential duration (APD), it is not unexpected to have some atrial fibrillation (AF) after PSVT conversion, and this was found to be extremely low with tecadenoson (<1%), but the incidence of AF with adenosine following IV bolus is reported to be 11% and 15% in different studies (Ellenbogen et al. 2005).

Selodenoson was evaluated for the treatment of AF in a dose-ranging infusion study where it was infused for 15 min at doses of 2, 4, 6, 8, 10, and 12 $\mu g\ kg^{-1}$, where it provided for effective ventricular rate control in a dose-dependent manner with minimal side effects (Bayes et al. 2003). CV Therapeutics' scientists have described a number of partial A_1AR agonists as potential oral antiarrhythmic agents that do not cause high-degree AV block at high concentrations (Morrison et al. 2004). These partial agonists were obtained by incorporating aromatic ethers and sulfides at the 5′ position of the full agonist tecadenoson, a strategy that is known to decrease intrinsic activity with respect to GTP shift and induction of $[^{35}S]GTP\gamma S$ binding to G-protein (Yan et al. 2003). The 5′ substitution caused a significant drop in affinity for the A_1AR when compared to tecadenoson, and the 5′-aromatic ethers had greater affinity and potency for the A_1AR than the corresponding 5′-sulfides. Comparing the two lead molecules from both series (**25** and **26**) (Fig. 15), the 2-fluorophenyl ether analog **25** displayed higher affinity for the A_1AR ($K_i = 12$ nM) and sixfold greater potency ($EC_{50} = 200$ nM) in slowing AV nodal conduction than the 2-fluorophenyl sulfide **26** without causing third-degree AV block. In addition, compound **25** exhibited greater oral bioavailability (81%) relative to **26**. To our knowledge, compound **25** is the most potent partial A_1 AR agonist known to date; however, after oral administration, a small amount of the extremely potent full A_1AR agonist tecadenoson was generated; thus, compound 25 was unacceptable for further clinical development as an oral partial A_1AR agonist for chronic use. PJ-875 is a third A_1AR agonist in clinical development for AF from Inotek (DailyDrugNews.com 2008). The structure of PJ-875 has not been publicly disclosed; however, Inotek's patent application focuses on a 5′-nitrate ester of CPA with high A_1AR affinity (A_1AR K_i = 1 nM) and a 5′-nitrate ester of tecadenoson (A_1 AR K_i = 10 nM) (Jagtap et al. 2005). In Phase I clinical trials, PJ-875 did not have serious side effects, and Phase II clinical trials are planned (DailyDrugNews.com 2008).

The initial clinical trials with full A_1AR agonists in a controlled IV setting demonstrate that it may be possible to obtain antiarrhythmic properties with minimal CNS side effects. In addition, CV Therapeutics has discovered that the antiarrhythmic properties of a full A_1AR agonist, tecadenoson, can be augmented by coadminstration of a subtherapeutic dose of a short-acting beta-blocker, esmolol, to achieve pronounced ventricular rate control effects in animal models (Belardinelli and Dhalla 2003). This combination approach of beta-blocker and A_1AR agonist will be interesting to watch in the clinic. Due to the pronounced CV effects at low doses of full A_1AR agonists, it is clear that a partial A_1AR agonist may be required to achieve tissue selectivity for other indications such as T2D.

3.2 Insulin-Sensitizing Agents: GR79236, ARA, CVT-3619

The therapeutic use of A_1 AR agonists as antilipolytic agents has been tried in the clinic; however, the CV effects mediated by the A_1AR agonists are a potential obstacle to the successful use of A_1AR agonists for this indication. A second challenge associated with the use of A_1AR agonists as antilipolytic agents is the development of acute tolerance to the antilipolytic effects due to receptor desensitization (Dhalla et al. 2007a; IJzerman et al. 1995). One potential solution is to discover a partial A_1AR agonist that is capable of eliciting a greater effect in the adipocytes than in the heart (i.e., tissue selectivity). By definition, a partial agonist is a low-efficacy ligand that, in contrast to the full agonist, elicits only a submaximal biological response, and is hence less prone to receptor desensitization.

GR79236 ((3*R*, 4*S*, 5*R*)-2-(6-((1*S*, 2*S*)-2-hydroxycyclopentylamino)-9*H*-purin -9-yl)-5-(hydroxymethyl)tetrahydrofuran-3,4-diol) and ARA ((1*S*,2*R*,3*R*)-3-((trifluoromethoxy)methyl)-5-(6-(1-(5-(trifluromethyl)pyridine-2-yl)pyrrolidin-3-ylamino)-9*H*-purin-9-yl)cyclopentane-1,2-diol) (Fig. 16), the two full A_1AR agonists, have demonstrated that A_1AR agonism can have a pronounced effect on NEFA and TG levels in both acute and chronic animal models, thus establishing the potential of this approach for the treatment of T2D (Bigot et al. 2004; Merkel et al. 1995).

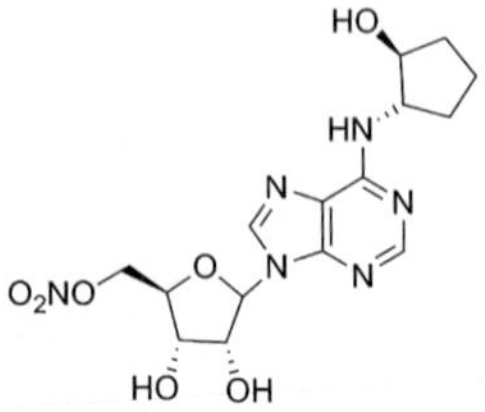

GR-79236
Phase I-Type II Diabetes
GSK

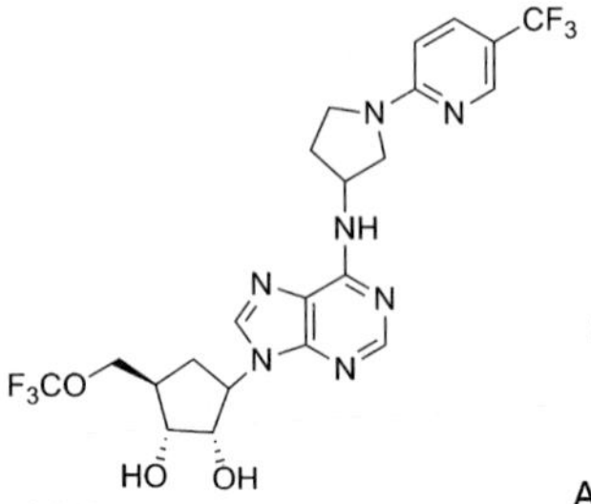

ARA
Phase I-Type I Diabetes
Aventis

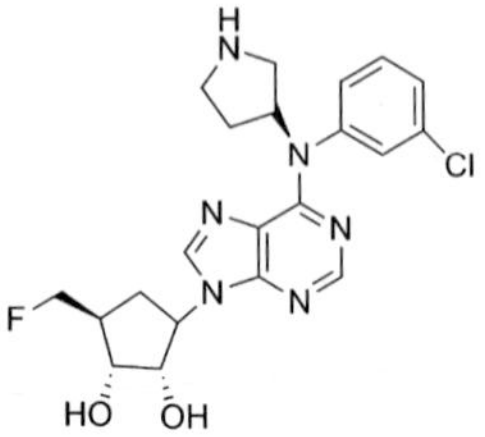

ARA Analog **27**
Phase I-Type I Diabetes
Aventis

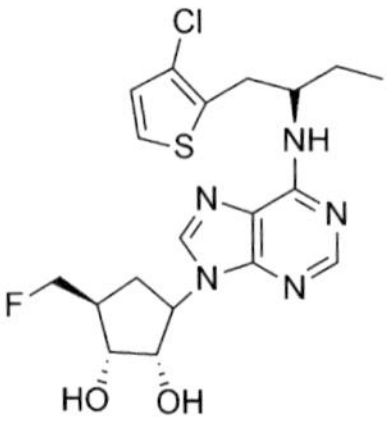

ARA Analog **28**
Phase I-Type I Diabetes
Aventis

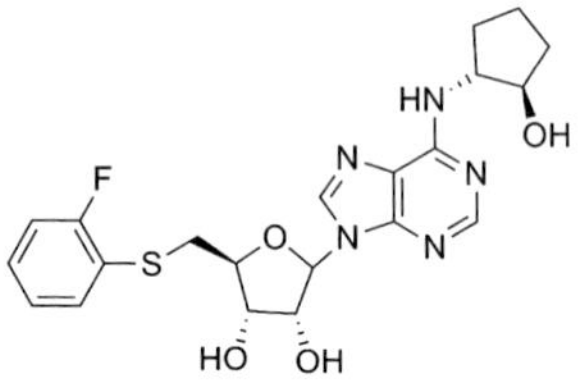

CVT-3619
Preclinical-Type II Diabetes
CV Therapeutics

Fig. 16 Antidiabetic and insulin-sensitizing A_1AR agonists

Plus, the use of these full A_1AR agonists in clinical trials has resulted in a better understanding of the desensitization of the A_1AR and some potential limitations of using a full A_1AR agonist in a chronic setting. In early in vitro studies, GR79236 demonstrated that it inhibited catecholamine-induced lipolysis in adipocytes at low concentrations (Green et al. 1990; Qu et al. 1997; Webster et al. 1996). In addition, GR79236 was demonstrated to reduce NEFA levels by 50% in normal fasted rats (Qu et al. 1997). However, in a fructose-fed rat model of noninsulin-dependent diabetes, GR79236 (1 mg kg^{-1} per day for eight days oral administration) did not enhance insulin sensitivity, but it did significantly lower NEFA and TGs (Webster et al. 1996). ARA, a full A_1AR agonist, has both animal data and clinical trial data supporting its effects on NEFA (Bigot et al. 2004). ARA is a C-sugar wherein the ribose oxygen is replaced by a carbon, and the ribose 5′-hydroxyl group was replaced by fluoro (as in compounds **27** and **28**) or a trifluoromethoxy group, as in ARA (Fig. 16). Compound **28** has lower affinity for the A_1AR with its unusual disubstituted N6 substituent containing an anilino moiety and a 3-pyrrolidinyl group. This is expected, since in most models of A_1AR agonist binding to the receptor, the N–H on N6 is involved in a key hydrogen-bonding interaction to the asparagine 254 side chain (IJzerman et al. 1995). ARA exhibited high affinity and selectivity for the A_1AR agonist (K_i = 1.7 nM and 4.5 nM in rat brain and rat adipocytes, respectively) (Zannikos et al. 2001). ARA demonstrated some tissue selectivity, being less potent (100- to 200-fold) in inducing A_1AR-mediated bradycardia than in inducing A_1AR inhibition of lipolysis in rat and human adipocytes. This selective effect is most likely due to the high density and/or efficiency of A_1 AR coupling in adipocytes. Although ARA was effective at lowering plasma FFA when administered intravenously to fasted healthy volunteers in a Phase I clinical study, the rapid appearance of tolerance to its FFA-lowering ability was clearly evident (Zannikos et al. 2001).

These clinical findings support the need for a partial A_1AR agonist.

Partial A_1AR agonists were considered as an alternative to full agonists to avoid receptor desensitization. CVT-3619 ((2*S*, 3*S*, 4*R*)-2-((2-fluorophenylthio)methyl)-5-(6-((1*R*, 2*R*)-2-hydroxycyclopentylamino)-9*H*-purin-9-yl) tetrahydrofuran-3,4-diol) (Fig. 16), a selective partial A_1AR agonist devoid of CV effects, is being developed by CV Therapeutics as an antilipolytic agent (Dhalla et al. 2007a; Fatholai et al. 2006). This clinical candidate was obtained by further optimization of the 5′-phenylsulfide derivatives of tecadenoson (described earlier). The binding affinity of CVT-3619 for rat epididymal adipocytes was 14 nM (K_i, high affinity). CVT-3619 reduced forskolin-induced cAMP accumulation in both epididymal and inguinal adipocytes, with EC_{50} values of 5.9 nM and 44 nM, respectively. The maximal effect of CVT-3619 at reducing cAMP levels in adipocytes was similar to that of CPA, suggesting that CVT-3619 is a full agonist with respect to reduction of cAMP. Plus, CVT-3619 reduced the forskolin-stimulated release of NEFA from both epididymal and inguinal adipocytes, with EC_{50} values of 47 nM and 170 nM, respectively. However, CVT-3619 was found to be a partial agonist with respect to forskolin (1 μM)-stimulated NEFA release from epididymal and inguinal adipocytes, with only 42 and 58%, respectively, of CPA's effect. Most likely, the presence of a large receptor reserve and/or a higher efficacy of coupling of A_1AR in the adipocytes can explain the fact that CVT-3619 reduced the cAMP content of epididymal adipocytes with an EC_{50} value that was lower than the K_i value from the binding assay, and the EC_{50} value to reduce the release of NEFA was also much lower than the K_i. Furthermore, the high A_1AR receptor reserve in the adipocyte relative to the heart can explain the 1,000-fold functional selectivity of CVT-3619 to decrease epididymal adipose tissue lipolysis in the rat ($EC_{15} = 30$ nM) relative to the atrial rate (both A_1AR-mediated effects) (Fatholai et al. 2006). With respect to CV side effects, CVT -3619 (10 nM–30 μM) caused only a small increase in S–H interval (6 ms) without causing second- or higher-degree AV block; however, CPA significantly prolonged the S–H interval (38 ms) and caused second- or higher-degree AV block at concentrations >30 nM. In normal, overnight-fasted awake rats, at doses of 2.5, 5 and 10 mg kg^{-1}, CVT-3619 lowered FFA by 31%, 47% and 57% from baseline, respectively (Dhalla et al. 2007b). In addition, CVT-3619 significantly reduced serum TG levels and increased insulin sensitivity in rats (Dhalla et al. 2007b). The ED_{50} of insulin to inhibit lipolysis was potentiated fourfold by a single dose (0.5 mg kg^{-1}) of CVT-3619, suggesting that CVT-3619 increases insulin sensitivity in adipose tissue. The antilipolytic effects of CVT-3619 in rats (given twice daily) were well maintained for up to six weeks of treatment, and no tachyphylaxis or receptor desensitization were observed. Based on the above data, CVT-3619 is in preclinical development by CV Therapeutics as a partial A_1AR agonist for the potential treatment of T2D in order to avoid CV effects and receptor desensitization. For more information on the effects of partial A_1AR agonists in diabetes and obesity, the reader is referred to Chap. 9 of this volume, "A_1 Adenosine Receptor: Role in Diabetes and Obesity" (Dhalla et al.).

3.3 Angina Agents: Capadenoson (Nonnucleoside: BAY 68–4986)

Bayer chemists were first to make a key discovery that a heterocyclic class of compounds devoid of a ribose moiety can function as agonists at the adenosine receptor, although the first compounds were nonselective (Erguden et al. 2007). IJzerman and colleagues followed this with a further elaboration of the heterocyclic class of agonists in order to introduce some receptor selectivity (Chang et al. 2005). The Bayer chemists then reported the development of a compound from this very novel class of compounds. The oral A_1AR agonist capadenoson (BAY 68–4986), 2-amino-6-((2-(4-chlorophenyl)thiazol-4-yl)methylthio)-4-(4-(2-hydroxyethoxy)phenyl)pyridine-3,5-dicarbonitrile (Fig. 15), was evaluated in a Phase II double-blinded, placebo-controlled multicenter study in patients with stable angina and coronary heart disease studying doses of 1, 2.5, 5, 10 and 20 mg. A 10 mg dose of capadenoson significantly reduced heart rate at peak exercise compared to placebo. Capadenoson is currently under going further studies and is anticipated to finish Phase III clinical trials by 2009 (Bays et al. 2007).

4 Allosteric Enhancers

4.1 Neuropathic Pain: T-62

A different approach to A_1AR agonism is to use the endogenous adenosine levels to activate the receptor coupled with an allosteric enhancer of the A_1AR. This approach has the theoretical advantage of fewer side effects, since it relies on adenosine being produced at the target tissue. In some disease states, adenosine release is a natural compensatory process to help the tissue restore balance. The A_1AR allosteric enhancer will take advantage of this local adenosine release and provide activation of a local A_1AR. The SAR of A_1AR allosteric enhancers has evolved over many years, with major contributions from IJzerman and Baraldi (Baraldi et al. 2007; Van der Klein et al. 1999). The common structural theme that has emerged is a 2-amino-3-acyl-thiophenyl core as exemplified by the lead compound in the area, T-62 (2-amino-4,5,6,7-tetrahydrobenzo[*b*]thiophen-3-yl)(4-chlorophenyl)methanone; (Fig. 17), a compound discovered by Baraldi et al. and developed by King Pharmaceuticals for neuropathic pain (Baraldi et al. 2007; Obata et al. 2003; Pan et al. 2001). T-62 demonstrated efficacy for reducing pain

Fig. 17 Allosteric A_1AR enhancer

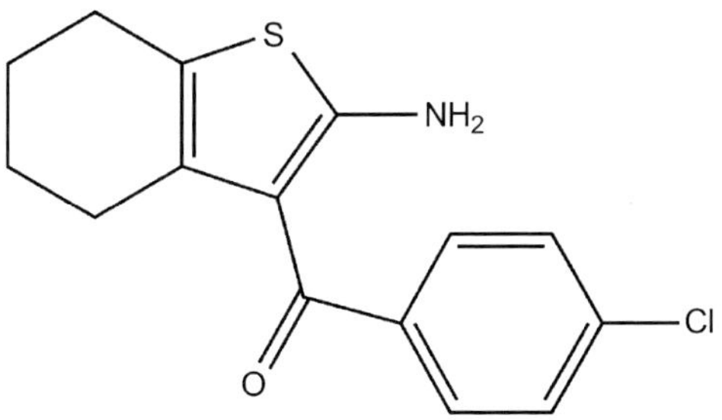

hypersensitivity in a plantar surgical injury rat model (0.3–1 mcg intrathecal administration) in a dose-dependent manner. The dose of T-62 required for an antihyperalgesic effect was reduced by half when clonidine was coadministered, and this effect was 40% of the maximum possible effect. T-62 is under clinical evaluation in patients with postherpetic neuralgia experiencing pain. It will be interesting to see how the lead compound T-62 does in clinical trials of neuropathic pain, since it may drive further research in the area of A_1AR allosteric enhancers.

5 Conclusion

A considerable body of research over the past 20 years in the A_1AR field has resulted in the identification of clinical candidates for A_1AR antagonism, agonism, and allosteric modification. From a pharmacological perspective, the developmental path for A_1AR antagonists should theoretically be easier due to the challenges associated with developing A_1AR agonists, such as receptor desensitization and the risk of pronounced CV and CNS side effects. With two of the three active A_1AR antagonist clinical programs (KW3902 and BG9928) in Phase III human clinical trials, there is optimism in the cardiology community that an A_1AR antagonist will be available for patient use in the coming years (Dohadwala and Givertz 2008). Partial A_1AR agonism with CVT-3619, for example, may represent a way to avoid both CV and CNS side effects, which makes CVT-3619 an interesting compound to watch as it proceeds to the clinic. BAY 68–4986 opens up the A_1AR agonist field with the advent of nonribose partial agonists that possess a longer half-life for chronic agents that are no longer limited by the high polarity of the ribose ring. The A_1AR allosteric enhancer T-62 has demonstrated promising results in animal models of neuropathic pain, and is currently undergoing clinical evaluation. Based on these important scientific and clinical advances, therapeutics that target the A_1AR (A_1AR antagonists, A_1AR agonists, and allosteric enhancers) may show long-awaited clinical success in the near future.

References

Akhari R, Burbiel JC, Hockemeyer J, Muller CE (2006) Recent progress in the development of adenosine receptor ligands as anti-inflammatory drugs. Curr Top Med Chem 6:1375–1399

Amann K, Breitbach M, Ritz E, Mall G (1998a) Myocyte/capillary mismatch in the heart of uremic patients. J Am Soc Nephrol 9:1018–1022

Amann K, Kronenberg G, Gehlen F, Wessels S, Orth S, Munter K (1998b) Cardiac remodeling in experimental renal failure—an immunohistochemical study. Nephrol Dial Transplant 13:1958–1966

Auchampach JA, Jin X, Moore J, Wan TC, Kreckler LM, Ge ZD, Narayanan J, Whalley E, Kiesman W, Ticho B, Smits G, Gross GJ (2004) Comparison of three different A_1 adenosine receptor antagonists on infarct size and multiple cycle ischemic preconditioning in anesthetized dogs. J Pharmacol Exp Ther 308:846–856

Baraldi PG, Iaconinoto MA, Moorman AR, Carrion MD, Cara CL, Preti D, Lopez OC, Fruttarolo F, Tabrizi MA, Romagnoli R (2007) Allosteric enhancers for A_1 adenosine receptor. Minirev Med Chem 7:559–569

Baraldi PG, Tabrizi MA, Gessi S, Borea PA (2008) Adenosine receptor antagonists: translating medicinal chemistry and pharmacology into clinical utility. Chem Rev 108:238–263

Bayes M, Rabasseda X, Prous JR (2003) Gateways to clinical trials. Methods Find Exp Clin Pharmacol 25:831–855

Bays M, Rabasseda X, Prous JR (2007) Gateways to clinical trials. Method Find Exp Clin Pharmacol 29(5):359–373

Belardinelli L, Dhalla A (2003) Method of treating arrhythmias comprising administration of an A_1 adenosine agonist with a beta blocker, calcium channel blocker, or a cardiac glycoside. US Patent WO03088978

Belardinelli L, Lerman BB (1990) Electrophysiological basis for use of adenosine in the diagnosis and treatment of cardiac arrhythmias. Br Heart J 63:3–34

Belardinelli L, Shryock JC, Song Y, Wang D, Srinivas M (1995) Ionic basis of the electrophysiological actions of adenosine on cardiomyocytes. FASEB J 5:359–365

Belardinelli L, Shryock JC, Wu L, Song Y (2005) Use of preclinical assays to predict risk of drug-induced torsades de pointes. Heart Rhythm 2(2 Suppl):S16–22

Bigot A, Stengelin S, Johne G, Herling A, Muller G, Hock FJ, Myers MR (2004) Novel adenosine analogues and their use as pharmaceutical agents. US Patent WO04003002

Boden G, She P Mozzoli M, Cheung P, Gumireddy K, Reddy P, Xiang X, Luo Z, Ruderman N (2005) Free fatty acids produce insulin resistance and activate the proinflammatory nuclear factor-{Kappa}B pathway in rat liver. Diabetes 54:3458–3465

Castelhano Al, McKibben B, Witter DJ (2005) Compounds specific to adenosine A_1 receptor and uses thereof. US Patent 6,878,716

Chang LC, Von Kuenzel JK, Mulder-Krieger T, Spanjersberg RF, Roerink F, Van Den Hout G, Beukers MW, Brussee J, IJzerman AP (2005) A series of ligands displaying a remarkable agonistic–antagonistic profile at the adenosine A_1 receptor. J Med Chem 48(6):2045–2053

ClinicalTrials.gov (2005) Aderis Pharmaceuticals: safety and efficacy study of an A1-adenosine receptor agonist to slow heart rate atrial fibrillation. http://www.clinicaltrials.gov/ct/show/NCT00040001?order=1

DailyDrugNews.com (2008) Inotek begins a Phase I trial of INO-8875 for the treatment of glaucoma (Daily Essentials, 20 June 2008). http://www.dailydrugnews.com

Dhalla AK, Shryock JC, Shreeniwas R, Beraldinelli L (2003) Pharmacology and therapeutic application of A_1 adenosine receptor ligands. Curr Top Med Chem 3:369–385

Dhalla AK, Wong MY, Voshol PJ, Belardinelli L, Reaven GM (2007a) A_1-adenosine receptor partial agonist lowers plasma FFA and improves insulin resistance induced by high-fat diet in rodents. Am J Physiol Endocrinol Metab 292:E1358–E1363

Dhalla AK, Melissa S, Smith M, Wong M, Shryock JC, Beraldinelli L (2007b) Antilipolytic activity of a novel partial A_1 adenosine receptor agonist devoid of cardiovascular effects: comparison with nicotinic acid. J Pharmacol Exp Therp 321:327–333

DiMarco JP, Sellers TD, Lerman BB Greenberg ML, Berne RM, Belardinelli L (1985) Diagnostic and therapeutic use of adenosine in patients with supraventricular tachycardias. J Am Coll Cardiol 6:417–425

Dittrich HC, Gupta D, Hack TC, Dowling T, Callahan J, Thomson S (2007) The effect of KW-3902, an adenosine A_1 receptor antagonist, on renal function and renal plasma flow in ambulatory patients with heart failure and renal impairment. J Cardiac Fail 13:609–617

Dohadwala MM, Givertz MM (2008) Role of adenosine antagonism in cardiorenal syndrome. Cardiovasc Ther 26:276–286

Ellenbogen KA, O'Neill G, Prystowsky EN, Camm JA, Meng L, Lieu HD, Jerling M, Shreeniwas R, Beraldinelli L, Wolff AA (2005) Trial to evaluate the management of paroxysmal supraventricular tachycardia during an electrophysiology study with Tecadenoson. Circulation 111:3202–3208

Erguden JK, Karig G, Rosentretter U, Albrecht B, Henninger K, Hutter J, Diedrichs N, Nell P, Arndt S, Hubsch W, Knorr A, Schlemmer KH, Brohm P (2007) Substituted phenylaminothiazoles and use thereof. US Patent WO0773855

Fatholahi M, Xiang Y, Wu Y, Li Y, Wu L, Dhalla AK, Belardinelli L, Shryock JC (2006) A novel partial agonist of the A_1-adenosine receptor and evidence of receptor homogeneity in adipocytes. J Pharmacol Exp Ther 317:676–684

Ferrannini E, Barrett EJ, Bevilacqua S, DeFronzo RA (1983) Effect of fatty acids on glucose production and utilization in man. J Clin Invest 72:1737–1747

Forman M, Velasco CE, Jackson EK (1993) Adenosine attenuates reperfusion injury following regional myocardial ischemia. Cardiovasc Res 27:9–17

Fredholm BB, IJzerman AP, Jacobson KA, Klotz, KN Linden J (2001) International Union of Pharmacology. XXXV: nomenclature and classification of adenosine receptors. Pharmacol Rev 53:527–552

Gardner CJ, Twissell DJ, Coates J, Strong P (1994) The effects of GR79236 on plasma fatty acid concentrations, heart rate and blood pressure in the conscious rats. Eur J Pharmacol 257:117–121

Givertz MM, Massie BM, Fields TK, Pearson LL, Dittrich HC, on behalf of the CKI-201 and CKI-202 investigators (2007) The effects of KW-3902, an adenosine A_1-receptor antagonist, on diuresis and renal function in patients with acute decompensated heart failure and renal impairment or diuretic resistance. J Am Coll Cardiol 50:1551–1560

Gottlieb SS, Skettino SL, Wolff A, Beckman E, Fisher ML, Freudenberger R, Gladwell T, Marshall J, Cines M, Bennett, DLiitschwager EB (2000) Effects of BG9719 (CVT-124), an A_1-adenosine receptor antagonist, and furosemide on glomerular filtration rate and natriuresis in patients with congestive heart failure. J Am Coll Cardiol 35:56–59

Gottlieb SS, Brater C, Thomas I, Havranek E, Bourge R, Goldman S, Dyer F, Gomez M, Bennett D, Ticho B, Beckman E, Abraham WT (2002) BG9719 (CVT-124), an A_1 adenosine receptor antagonist, protects against the decline in renal function observed with diuretic therapy. Circulation 105:1348–1353

Green A (1987) Adenosine receptor down-regulation and insulin resistance following prolonged incubation of adipocytes with an A_1 adenosine receptor agonist. J Biol Chem 262:15702–15707

Green A, Johnson JL, Milligan G (1990) Down-regulation of G_i subtypes by prolonged incubation of adipocytes with an A_1 adenosine receptor agonist. J Biol Chem 265:5206–5210

Greenberg B, Thomas I, Banish D, Goldman S, Havranek E, Massie B, Zhu Y, Ticho B, Abraham WT (2007) Effects of multiple oral doses of an A_1 adenosine antagonist, BG9928, in patients with heart failure. J Am Coll Cardiol 50:600–606

Hess S (2001) Recent advances in adenosine receptor antagonist research. Expert Opin Ther Patents 11:1533–1561

Hoffman BB, Chang H, Dall'Agilo E, Reaven GM (1986) Desensitization of adenosine receptor-mediated inhibition of lipolysis. The mechanism involves the development of enhanced cyclic adenosine monophosphate accumulation in tolerant adipocytes. J Clin Invest 78:185–190

Hosokawa T, Yamauchi M, Yamamoto Y, Iwata K, Mochizuki H, Kato Y (2002) Role of the lipid emulsion on an injectable formulation of lipophilic KW-3902, a newly synthesized adenosine A1-receptor antagonist. Biol Pharm Bull 25:492–498

IJzerman AP, van der Wenden NM, van Galen PJM, Jacobson KA (1995) Molecular modeling of adenosine A_1 and A_{2a} receptors. In: Belardinelli L, Pelleg A (eds) Adenosine and adenine nucleotides: from molecular biology to integrative physiology. Kluwer, Boston, MA, pp 27–37

Itani SI, Ruderman NB, Schmieder F, Boden G (2002) Lipid-induced insulin resistance in human muscle is associated with changes in diacylglycerol protein kinase C and IKappaB-alpha. Diabetes 51:2005–2011

Jacobson KA, Gao ZG (2006) Adenosine receptors as therapeutic targets. Nat Rev Drug Disc 5:247–264

Jagtap P, Szabo C, Salzman AL (2005) Purine derivatives as adenosine A_1 receptor agonists and methods of use thereof. WO Patent Appl 2005/117910

Kalk P, Eggert B, Relle K, Godes M, Heiden S, Sharkovska Y, Fischer Y, Ziegler D, Bielenberg GW, Hocher B (2007) The adenosine A1 receptor antagonist SLV320 reduces myocardial fibrosis in rats with 5/6 nephrectomy without affecting blood pressure. Br J Pharmacol 151:1025–1032

Khimenko PL, Moore TM, Hill LW, Wilson PS, Coleman S, Rizzo A, Taylor AE (1995) Adenosine A_2 receptors reverse ischemia-reperfusion lung injury independent of beta-receptors. J Appl Physiol 78:990–996

Kiesman WF, Zhao J, Conlon PR, Dowling JE, Petter RC, Lutterodt F, Jin X, Smits G, Fure M, Jayaraj A, Kim J, Sullivan GW, Linden J (2006a) Potent and orally bioavailable 8-bicyclo[2.2.2]octylxanthines as adenosine A_1 receptor antagonists. J Med Chem 49:7119–7131

Kiesman WF, Zhao J, Conlon PR, Petter RC, Jin X, Smits G, Lutterodt F, Sullivan GW, Linden J (2006b) Norbornyllactone-substituted xanthines as adenosine A_1 receptor antagonists. Bioorg Med Chem 14:3654–3661

Kossel A (1888) Über eine neue Base aus dem Pflanzenreich. Chem Ber 21:2164–2167

Lerman BB, Belardinelli L (1991) Cardiac electrophysiology of adenosine. Basis and clinical concepts. Circulation 83:1499–1509

Liu GS, Thornton J, Van Winkle DM, Stanely AW, Olsson RA, Downey JM (1991) Protection against infarction afforded by preconditioning is mediated by A_1 adenosine receptors in rabbit heart. Circulation 84:350–356

Magata S, Taniguchi M, Suzuki T, Shimamura T, Fukai M, Furukawa H, Fujita M, Todo S (2007) The effect of antagonism of adenosine A_1 receptor against ischemia and reperfusion injury of the liver. J Surg Res 139:7–14

Merkel LA, Hawkins ED, Colussi DJ, Greenland BD, Smits GJ, Perrone MH, Cox BF (1995) Cardiovascular and antilipolytic effects of the adenosine agonist GR79236. Pharmacology 51:224–36

Mizumura T, Auchampach JA, Linden J, Burns RF, Gross GJ (1996) PD 81,723 an allosteric enhancer of the A_1 adenosine receptor, lowers the threshold for ischemic preconditioning in dogs. Circ Res 79:415–423

Miura T, Tsuchida A (1999) Adenosine and preconditioning revisited. Clin Exp Pharm Physiol 26:92–99

Moro S, Gao Z.-G, Jacobson KA, Spalluto G (2006) Progress in the pursuit of therapeutic adenosine receptor antagonists. Med Res Rev 26:131–159

Morrison CF, Elzein E, Jiang B, Ibrahim P, Maa T, Wu L, Zeng D, Fong I, Lustig D, Leung K, Zablocki J (2004) Structure–affinity relationships of 5′-aromatic ethers and 5′-aromatic sulfides as partial A_1 adenosine agonists, potential supraventricular anti-arrhythmic agents. Bioorg Med Lett 14:3793–3797

Nonaka H, Ichimura M, Takeda M, Kanda T, Shimada J, Suzuki F, Kase H (1996) KW-3902, a selective high affinity antagonist for adenosine A_1 receptors. Br J Pharmacol 117:1645–1652

Novacardia, Inc. (2007) Pilot phase 3 results of Novacardia's KW-3902 for acute congestive heart failure. In: Heart Failure Congress 2007, Hamburg, Germany, 9–12 June 2007 (see http://www.medicalnewstoday.com/articles/73876.php)

Obata H, Li X, Eisenach JC (2003) A synergistic effect of intrathecal administration of T62 and clonidine in the rat postoperative pain model. In: 33rd Annu Meet Soc Neurosci Proc, New Orleans, LA, 8–12 Nov 2003, Abstract 589.7

Pan H.-L, Xu Z, Leung E, Eisenach JC (2001) Allosteric adenosine modulation to reduce allodynia. Anesthesiology 95:416

Peralta C, Hotter G, Closa D, Prats N, Xaus C, Gelpi E, Rosello-Catafau J (1999) The protective role of adenosine in inducing nitric oxide synthesis in rat liver ischemia preconditioning is mediated by activation of adenosine A_2 receptors. Hepatology 29:126–132

Peterman C, Sanoski CA (2005) Tecadenoson: a novel, selective A_1 adenosine receptor agonist. Cardiol Rev 13:315–321

Pfister JR, Bellardinelli L, Lee G, Lum RT, Milner P, Stanley WC, Linden J, Baker SP, Schreiner G (1997) Synthesis and biological evaluation of the enantiomers of the potent and selective A_1-adenosine antagonist 1,3-dipropyl-8-[2-(5,6-epoxynorbonyl)]-xanthine. J Med Chem 40:1773–1778

Prystowsky EN, Niazi I, Curtis AB, Wilber DJ, Bahnson T, Ellenbogen K, Dhalla A, Bloomfield DM, Gold M, Kadish A, Fogel RI, Gonzalez MD, Belardinelli L, Shreeniwas R, Wolff AA (2003) Termination of paroxysmal supraventricular tachycardia by tecadenoson (CVT-510), a novel A_1-adenosine receptor agonist. J Am Coll Cardiol 42:1098–1102

Qu X, Cooney G, Donnelly R (1997) Short-term metabolic and haemodynamic effects of GR79236 in normal and fructose-fed rats. Eur J Pharmacol 338:269–276

Roden, M, Price, TB, Perseghin, G, Petersen, KF, Rothman, DL, Cline, GW, Shulman, GI (1996) Mechanism of free fatty acid-induced insulin resistance in humans. J Clin Investig 97: 2859–2865

Sako Y, Grill VE (1990) A 48-hours lipid infusion in the rat time-dependently inhibits glucose-induced insulin secretion and B cell oxidation through a process likely coupled to fatty acid oxidation. Endocrinology 127:1580–1589

Shamim MT, Ukena D, Padgett WL, Hong O, Daly JW (1988) 8-Aryl-and 8-cycloalkyl-1,3-dipropylxanthines: further potent and selective antagonists for A_1-adenosine receptors. J Med Chem 31:613–617

Shimada J, Suzuki F, Nonaka H, Karasawa A, Mizumoto H, Ohno T, Kubo K, Ishii A (1991) 8-(Dicyclopropylmethyl)-1,3 dipropylxanthine: a potent and selective adenosine A_1, antagonist with renal protective and diuretic activities. J Med Chem 34:466–469

Shimada J, Suzuki F, Nonaka H, Ishii A (1992) 8-Polycycloalkyl-1,3-dipropylxanthines as potent and selective antagonists for A_1-adenosine receptors. J Med Chem 35:924–930

Song Y, Wu L, Shryock JC, Beralinelli, L (2002) Selective attenuation of isoproterenol-stimulated arrhythmic activity by a partial agonist of adenosine A_1 receptor. Circulation 105:118–123

Srinivas M, shryock JC, Dennis DM, Baker SP, Beraldinelli L (1997) Differential A_1 adenosine receptor reserve for two actions of adenosine on guinea pig atrial myocytes. Mol Pharmacol 52:683–691

Stephenson RP (1997) A modification of receptor theory. Br J Pharmacol 120:106–120

Suzuki F, Shimada J, Mizumoto H, Karasawa A, Kubo K, Nonaka H, Ishii A, Kawakita T (1992) Adenosine A_1 antagonists. 2. Structure–activity relationships on diuretic activities and protective effects against acute renal failure. J Med Chem 35:3066–3075

Ticho B, Whalley E, Gill A, Lutterodt F, Jin X, Auchampach J, Smits G (2003) Renal effects of BG9928, an A_1 adenosine receptor antagonist, in rats and nonhuman primates. Drug Dev Res 58:486–492

Van der Klein PAM, Kourounakis AP, IJzerman AP (1999) Allosteric modulation of the adenosine A_1 receptor. Synthesis and biological evaluation of novel 2-amino-3-benzoylthiophenes as allosteric enhancers of agonist binding. J Med Chem 42:3629–3635 van Galen PJM, Stiles GL, Michaels G, Jacobson KA (1992) Adenosine A_1 and A_2 receptors: structure–function relationships. Med Res Rev 12:423–471

Vu CB, Kiesman WF, Conlon PR, et al (2006) Tricyclic imidazoline derivatives as potent and selective adenosine A_1 receptor antagonists. J Med Chem 49:7132–7139

Wang D, Shryock JC, Belardinelli L (1996) Cellular basis for the negative dromotropic effect of adenosine on rabbit single atrioventricular nodal cell. Circ Res 78:697–706

Webster JM, Heseltine L, Taylor R (1996) In vitro effect of adenosine agonist GR79236 on insulin sensitivity of glucose utilization in rat soleus and human rectus abdominus muscle. Biochim Biophys Acta 1316:109–113

Wu L, Belardinelli L, Zablocki JA, Palle V, Shryock JC (2001) A partial agonist of the A_1-adenosine receptor selectively slows AV conduction in guinea pig hearts. Am J Physiol Heart Cir 280:H334–H343

Yan L, Burbiel JC, Maass A, Muller CE (2003) Adenosine receptor agonists from basic medicinal chemistry to clinical development. Expert Opin Emerg Drugs 8:537–576

Yao K (2000) Effect of KW-3902, a selective adenosine A_1-receptor antagonist, on accumulation of gentamicin in the proximal renal tubules in rats. Yakugaku Zasshi 120:801–805

Yao K, Heyne N, Erley CM, Risler T, Osswald H (2001) The selective adenosine A1 receptor antagonist KW-3902 prevents radiocontrast media-induced nephropathy in rats with chronic nitric oxide deficiency. Eur J Pharmacol 414:99–104

Yuzlenko O, Kieć-Kononowicz K (2006) Potent adenosine A_1 and A_{2A} receptors antagonists: recent developments. Curr Med Chem 13:3609–3625

Zablocki JA, Wu L, Shryock J, Belardinelli L (2004) Partial A(1) adenosine receptor agonists from a molecular perspective and their potential use as chronic ventricular rate control agents during atrial fibrillation (AF). Curr Top Med Chem 4:839–54

Zannikos PN, Rohatagi S, Jensen BK (2001) Pharmacokinetic-pharmacodynamic modeling of the antilipolytic effects of an adenosine receptor agonist in healthy volunteers. J Clin Pharmacol 41:61–69

Recent Developments in Adenosine A_{2A} Receptor Ligands

Gloria Cristalli, Christa E. Müller, and Rosaria Volpini

Contents

Abstract The development of potent and selective agonists and antagonists of adenosine receptors (ARs) has been a target of medicinal chemistry research for several decades, and recently the US Food and Drug Administration has approved Lexiscan^{TM}, an adenosine derivative substituted at the 2 position, for use as a pharmacologic stress agent in radionuclide myocardial perfusion imaging. Currently, some other adenosine A_{2A} receptor (A_{2A}AR) agonists and antagonists are undergoing preclinical testing and clinical trials. While agonists are potent antiinflammatory agents also showing hypotensive effects, antagonists are being developed for the treatment of Parkinson's disease.

However, since there are still major problems in this field, including side effects, low brain penetration (for the targeting of CNS diseases), short half-life, or lack of in vivo effects, the design and development of new AR ligands is a hot research topic.

G. Cristalli (✉)
Dipartimento di Scienze Chimiche, Università di Camerino, via S. Agostino 1, 62032 Camerino (MC), Italy
gloria.cristalli@unicam.it

C.N. Wilson and S.J. Mustafa (eds.), *Adenosine Receptors in Health and Disease*, Handbook of Experimental Pharmacology 193, DOI 10.1007/978-3-540-89615-9_3,

This review presents an update on the medicinal chemistry of A_{2A}AR agonists and antagonists, and stresses the strong need for more selective ligands at the human A_{2A}AR subtype, in particular in the case of agonists.

Keywords Adenosine receptor · Adenosine A_{2A} receptor · A_{2A} agonists · A_{2A} antagonists · Nucleosides · Xanthines · Adenines · Nitrogen (poly)heterocyclic compounds

Abbreviations

ADA	Adenosine deaminase
Ado	Adenosine
AK	Adenosine kinase
AR	Adenosine receptor
CCPA	2-Chloro-N^6-cyclopentyladenosine
CHA	N^6-Cyclohexyladenosine
CHO	Chinese hamster ovarian
CNS	Central nervous system
CPA	N^6-Cyclopentyladenosine
HEAdo	2-(Hexyn-1-yl)adenosine
HENECA	2-Hexynyl-NECA
MECA	*N*-Methylcarboxamidoadenosine
NECA	*N*-Ethylcarboxamidoadenosine
PEAdo	2-Phenylethynyladenosine
PENECA	2-PhenylethynylNECA
PHPAdo	2-Phenylhydroxypropynyladenosine
PHPNECA	2-PhenyhydroxypropynylNECA
PIA	N^6-(2-Phenylisopropyl)adenosine
QSAR	Quantitative structure–activity relationships

1 Adenosine A_{2A} Receptor Agonists

1.1 Adenosine

The clinical utility of adenosine (Ado, **1**, Fig. 1) was recognized late in the 1980s by Belardinelli and Pelleg, and it soon became clear that the unmodified molecule is of restricted interest as a tool for the study of adenosine receptors due to its susceptibility to extensive metabolism by a number of enzymes (Klotz 2000). In fact, the observation that the activity of exogenous Ado on the mammalian cardiovascular system is of short duration because of the rapid uptake of Ado into red blood cells and tissues (Pfleger et al. 1969), its phosphorylation by adenosine kinase

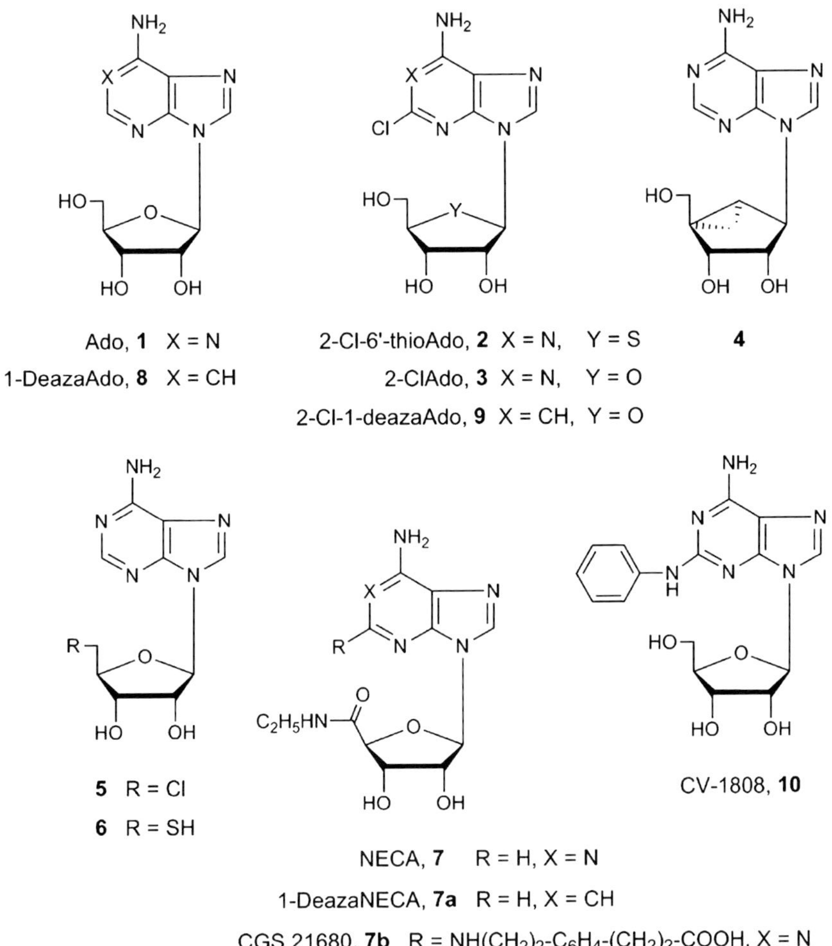

Fig. 1 A$_{2A}$AR agonists

(AK), and its conversion to inosine by adenosine deaminase (ADA) (Cristalli et al. 2001) led many labs to carry out several modifications of the Ado structure in order to find stable and selective ligands for the four adenosine receptor subtypes.

Almost all AR agonists known so far are derivatives of the physiological agonist Ado (Table 1). One exception is a set of substituted pyridines recently found to be agonists for human adenosine A$_{2B}$ receptor (A$_{2B}$AR) (Beukers et al. 2004). Many attempts to modify the Ado structure led to the conclusion that the Ado scaffold must be conserved, although three positions in the molecule may be modified to increase affinity to specific receptor subtypes without destroying the agonistic efficacy: the 5′ position of the ribose and the 2 and N^6 positions of the purine (Cristalli et al. 2003). It must be underlined that any of these modifications render the agonists metabolically stable.

Table 1 Affinities of AR agonists in radioligand binding assays at A_1AR, $A_{2A}AR$, and A_3AR, and effects on adenylate cyclase activity at the $A_{2B}AR$

Cpd	K_i (A_1AR)[a]	K_i ($A_{2A}AR$)[a]	EC_{50} ($A_{2B}AR$)[b]	K_i (A_3AR)[a]
2	300 r	20 r	–	1,090 r
3	9.3 r	63 r	24,000 r	1,890 r
7	10 r	7.8 r	–	113 r
	63 r	16 r	3,100 h	10 h
	14 h	20 h	2,400 h	6.2 h
7a	51 r	580 r	16,000 h	703 r
7b	1,400 r	19 r		584 r
	290 h	27 h	88,800 h	67 h
8	115 r	2,900 r	–	–
9	226 r	163 r	–	2,480 r
10	400 r	100 r	–	–
11	977 r	68 r	–	–
	530 h	62 h	–	310 h
12	130 r	17 r	–	–
	221 h	9.3 h	3,490 h	54 h
13	48,000 h	270 h	>100,000 h	900 h
14	11,700 r	22 r	–	–
15	2,800 r	22 r	–	–
	1,730 h	92 h	–	83 h
16	701 r	109 r	–	–
	806 r[c]	246 r[c]	–	28 r[c]
	395 h	363 h	>100,000 h	16 h
17	98 r	2.2 r	–	–
	111 r[c]	5.2 r[c]	–	24 r[c]
	18 h	5.7 h	≈100,000 h	4.7 h
18	3.4 r	1.9 r	–	–
	3.9 r[c]	5.3 r[c]	–	0.98 r[c]
	0.67 h	7.0 h	2,400 h	3.3 h
19a	1.5 r[c]	1.0 r[c]	–	0.40 r[c]
	0.44 h	29 h	6,200 h	5.0 h
19b	2.5 r[c]	1.6 r[c]	–	24 r[c]
	0.67 h	1.8 h	920 h	1.4 h
20	332 r	14 r	–	–
21	618 r	757 r	–	–
23	0.6 r	462 r	–	–
26	356 h	1.0 h	2,780 h[d]	100 h
27	473 r	9.7 r	–	–
28	130 r	2.2 r	–	26 r
	160 r[c]	1.0 r[c]	–	18 r[c]
	60 h	6.4 h	6,100 h	2.4 h
29	2.5 r	0.9 r	–	–
	3.8 r	2.7 r	–	7.7 r
	2.7 h	3.1 h	1,100 h	0.42 h

(continued)

Table 1 (continued)

Cpd	K_i (A_1AR)[a]	K_i (A_{2A}AR)[a]	EC_{50} (A_{2B}AR)[b]	K_i (A_3AR)[a]
29a	5.9 r	2.6 r	–	–
	2.7 r[c]	16 r[c]	–	0.46 r[c]
	1.9 h	39 h	2,400 h	5.5 h
29b	4.0 r	0.5 r	–	–
	5.5 r[c]	1.8 r[c]	–	2.6 r[c]
	2.1 h	2.0 h	220 h	0.75 h
30	698 r	120 r	–	–
	1,000 r[c]	267 r[c]	–	768 r[c]
	560 h	620 h	>100,000 h	6.2 h
31	28 r	5.5 r	–	–
	77 h	0.2 h	–	45 h
32a	251 r	1.6 r	–	–
32b	951 r	70 r	–	–
33	>10,000 p	85 p	–	–
	380 r[c]	15 r[c]	–	46 r[c]
	189 h	24 h	>100,000 h	86 h
34	1,100 r	330 r	45,000 h	6.4 h
35	63 r	12 r	5,300 h	108 h
36	403 h	49 h	>100,000 h	16 h
37	1,700 h	720 h	>100,000 h	246 h
38	32% r	115 r	–	5,640 r
39	48% r	82 r	–	3,160 r
40	–	1,122 p	–	–
41	5,836 h	2,895 h	–	–

[a]Binding data from different species: rat (*r*), human (*h*) or pig (*p*) A_1AR, A_{2A}AR, and A_3AR, expressed as K_i (nM)
[b]Measurement of receptor-stimulated adenylate cyclase activity at rat (*r*) or human (*h*) A_{2B}AR, expressed as EC_{50} (nM)
[c]Unpublished data
[d]Binding data

1.2 Ribose-Modified Adenosine Derivatives

A variety of modifications of the Ado ribose ring in several positions were carried out in order to get information on the essential points of agonist activity, and possibly to obtain more active and stable compounds (Yan et al. 2003; Akkari et al. 2006). Most alterations of either the structure or the stereochemistry of the ribose resulted in a loss of receptor binding potency and possibly intrinsic activity (Siddiqi et al. 1995).

Compounds in which the furanose ring was modified have been synthesized in order to improve stability, since the glycosidic bonds of adenine riboside derivatives are subject to scission in vivo. Results have shown that the sugar moiety must be maintained as a ribose ring, but that in some cases the endocyclic oxygen ring atom can be replaced with a sulfur atom (**2**, Fig. 1) (Siddiqi et al. 1995) or a methylene group (carbonucleoside). Comparison of 2-ClAdo (**3**) and the thio–ribosyl analog **2** showed a 3.2-fold higher affinity of the latter at the A_{2A}AR, whereas its adenosine A_1 receptor (A_1AR) affinity was reduced by 32-fold. In contrast, compounds **2** and **3** were of similar potency at the adenosine A_3 receptor (A_3AR) (Siddiqi et al. 1995). Carbonucleosides showed generally weak A_{2A}AR selectivity and low affinity for A_3AR. Carbocyclic modification of the agonists ribose resulted in nonglycosidic compounds that are potentially more biologically stable. The synthesis of a variety of methanocarba analogs of Ado was reported (**4**, Fig. 1) (Jacobson et al. 2000). These compounds contain a fused cyclopropane ring that constrains the pseudosugar ring in either a North (*N*) or South (*S*) conformation, with the aim of defining the role of sugar puckering in stabilizing the AR-bound conformation. Such modifications lead to compounds endowed with very low A_{2A}AR affinity and high A_1AR and A_3AR selectivity.

The 2′- and 3′-hydroxy groups of the ribose moiety appear to be essential for full agonist activity (Mathot et al. 1995; Siddiqi et al. 1995; van der Wenden et al. 1995; Vittori et al. 2000), whereas the substitution of the 5′-hydroxyl group of Ado is better tolerated, although the removal of this group results in a decrease in potency (van der Wenden et al. 1995). Moreover, 5′-modified Ados are also less expected to be incorporated into DNA due to their resistance to phosphorylation by AK (IJzerman and van der Wenden 1997).

Substitution of the 5′-hydroxyl group with a chlorine or a thiol group (**5** and **6**, Fig. 1) has been observed to increase affinity for ARs (Taylor et al. 1986; van der Wenden et al. 1998). However, it has been observed that the 5′-chloro-5′-deoxy modification of N^6-substituted Ados can increase A_1AR selectivity by reducing A_2 receptor potency (Taylor et al. 1986). A number of changes have been made to the riboses of a range of Ado analogs (Siddiqi et al. 1995). Most of the compounds with modified ribose in these studies were not substrates for ADA, and hence all were resistant to metabolism.

The introduction of an *N*-alkylcarboxamido group in position 5′ was well tolerated by all AR subtypes, and produced the most active compounds, such as NECA (**7**, Fig. 1) (Prasad et al. 1980), a nonselective AR agonist. On the other hand, *N*-ethylthiocarboxamidoAdo showed a decrease in affinity compared with NECA at all AR subtypes (de Zwart et al. 1999a). In particular, the 5′-*N*-ethyluronamide group enhances receptor affinity for all AR subtypes and it leads to a further increase in the agonist activity and/or selectivity, especially if other substituents are simultaneously present at position 2 of the Ado (Prasad et al. 1980; Hutchison et al. 1990; Cristalli et al. 1995; Baraldi et al. 1998a; de Zwart et al. 1999a). Structure–activity relationships showed that the 5′-*N*-ethyl-, 5′-*N*-methyl- and 5′-*N*-cyclopropylcarboxamido substitutions give the most potent agonists (Prasad et al. 1980).

1.3 Purine-Modified Adenosine Derivatives

In general, modification of the purine scaffold results in compounds with reduced receptor binding affinity compared with the corresponding Ado analogs

(Müller and Scior 1993; IJzerman et al. 1994). In particular, 1-deazaAdo (**8**) and its N^6-substituted derivatives are A_1AR selective, while the nitrogen atoms in the 3 and 7 positions are required for high affinity of Ado analogs at all subtypes (Bruns 1980; Cristalli et al. 1985; Siddiqi et al. 1995; de Zwart et al. 1998). On the other hand, 2-chloro-1-deazaAdo (**9**) showed an A_{2A}AR and A_3AR affinity similar to that of compound **3** (which is slightly A_1AR selective), and a reduced A_1AR activity, thus being slightly selective for the A_{2A}AR (Cristalli et al. 1988). Furthermore, **8** was reported to possess ADA inhibitory activity (Cristalli et al. 2001).

1.3.1 2- or N^6-Substituted Adenosine Derivatives

In the last 35 years, a significant number of C2-substituted Ado derivatives were synthesized and tested for their affinities at A_1AR and A_{2A}AR, and the first Ado derivative found to have some A_{2A}AR selectivity was CV-1808 (**10**, Fig. 1) (Bruns et al. 1986). A number of substitutions were made with amine (Francis et al. 1991), hydrazine (Niiya et al. 1992a, b; Viziano et al. 1995), alkoxyl (Daly et al. 1993; Matova et al. 1997), alkythio (Hasan et al. 1994; Cristalli 2000; Volpini et al. 2004), and alkynyl groups (Abiru et al. 1992, 1995; Cristalli et al. 1992; Matsuda et al. 1992; Volpini et al. 2002; Ohno et al. 2004), and the compounds with a phenylethyl (or cyclohexylethyl) group directly linked to the heteroatom (**11–15**, Fig. 2) or a triple bond (**16–18**) showed the highest A_{2A}AR affinities (Cristalli et al. 2007).

Substitutions with hydrazine led to 2-(N'-alkylidenehydrazino) and 2-(N'-aralkylidenehydrazino)Ado derivatives (Niiya et al. 1992a, b). Among these molecules, we should mention WRC-0470 (2-cyclohexylmethylidenehydrazinoAdo, also known as MRE-0470 or SHA-174 or Binodenoson, **13**) discovered at Nelson/Whitby Research and developed at Discovery Therapeutics, and now in clinical trial for myocardial perfusion imaging.

The alkynyl derivatives 2-phenylethynylAdo (PEAdo, **16**), 2-(hexyn-1-yl)Ado (HEAdo, **17**), (R, S)-2-phenylhydroxypropynylAdo ((R, S)-PHPAdo, **18**), and the corresponding diastereomers **19a** and **19b** were tested in binding studies on rat membrane A_1AR, A_{2A}AR (Cristalli et al. 1992), and A_3AR (Cristalli et al., unpublished results) and on the four human recombinant receptor subtypes, stably transfected into Chinese hamster ovarian (CHO) cells (the potency at the A_{2B}AR was measured with adenylate cyclase activity assays) (Volpini et al. 2002). All the compounds showed A_{2A}AR affinity in the low nanomolar range, and HEAdo was also shown to be slightly A_{2A}AR selective in rat membrane (A_1AR/A_{2A}AR $\approx$ 20 and A_3AR/A_{2A}AR $\approx$ 5). The phenylhydroxypropynyl derivatives are generally very potent, but are not selective at both rat and human AR subtypes. Partial and full reduction of the HEAdo triple bond led to E- and Z-alkenyl isomers **20** and **21** and 2-hexylAdo, respectively, among which the *trans* isomer **20** showed good A_{2A}AR affinity and modest selectivity (A_1AR/A_{2A}AR $\approx$ 24), while 2-hexylAdo proved to be inactive at both A_1AR and A_{2A}AR subtypes (Vittori et al. 1996). More recently, broad screening was carried out with the aim of characterizing the affinity and selectivity of 2-alkoxyAdo derivatives at A_3AR subtypes.

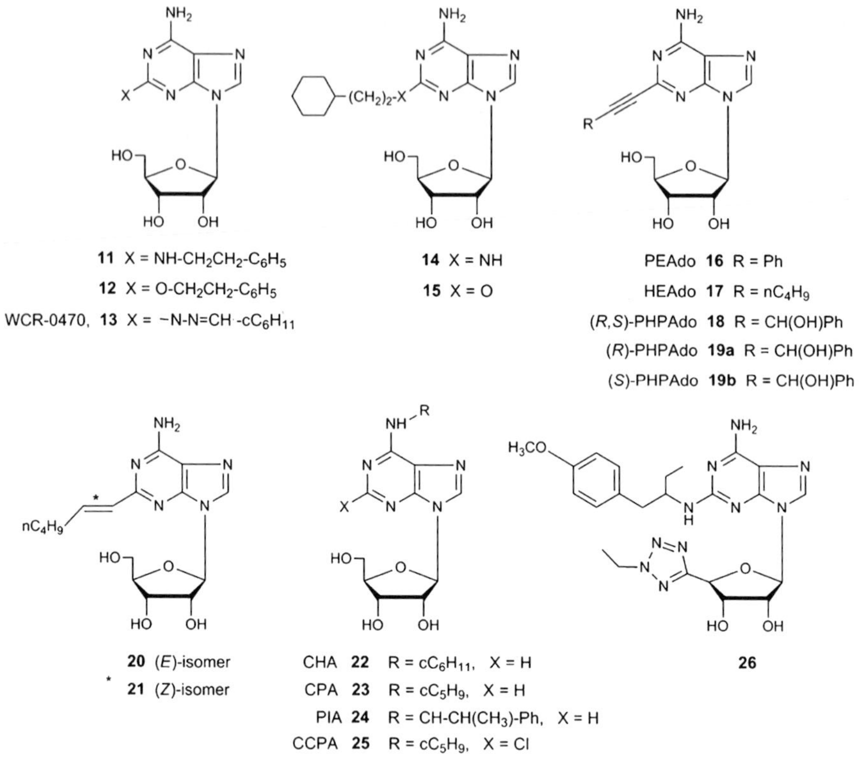

11 X = NH-CH_2CH_2-C_6H_5
12 X = O-CH_2CH_2-C_6H_5
WCR-0470, **13** X = –N-N=CH -cC_6H_{11}

14 X = NH
15 X = O

PEAdo **16** R = Ph
HEAdo **17** R = nC_4H_9
(*R,S*)-PHPAdo **18** R = CH(OH)Ph
(*R*)-PHPAdo **19a** R = CH(OH)Ph
(*S*)-PHPAdo **19b** R = CH(OH)Ph

20 (*E*)-isomer
21 (*Z*)-isomer

CHA **22** R = cC_6H_{11}, X = H
CPA **23** R = cC_5H_9, X = H
PIA **24** R = CH-CH(CH_3)-Ph, X = H
CCPA **25** R = cC_5H_9, X = Cl

26

Fig. 2 A_{2A}AR agonists: various Ado derivatives

These single substitutions at the 2 position, previously found to contribute to the affinity for the rat A_{2A}AR, were also proven to be important for affinity and selectivity at the human A_{2A}AR ortholog (Gao et al. 2004).

In general, substitution of Ado at the N^6 position (and in particular disubstitution with bulky substituents at the C2 and N^6 positions) is detrimental to A_{2A}AR affinity (Müller and Scior 1993). In fact, the first known subtype-selective Ado derivatives modified at the N^6 position, such as N^6-cyclohexylAdo (CHA, **22**), N^6-cyclopentylAdo (CPA, **23**), and N^6-(2-phenylisopropyl)Ado (PIA, **24**) showed A_1AR selectivity (Daly 1982). Furthermore, substituents in this position were more recently also shown to enhance A_3AR affinity and selectivity (Knutsen et al. 1999; Volpini et al. 2002).

In a series of 1-deaza analogs of Ados, it turned out that 2-chloro substitution in addition to an N^6-cyclopentyl increases A_1AR selectivity (Cristalli et al. 1988). The respective modification in Ado led to the development of 2-chloro-N^6-cyclopentylAdo (CCPA, **25**) as the most potent and selective A_1AR ligand characterized in rat brain (Lohse et al. 1988; Klotz et al. 1989).

1.4 *Ribose- and Purine-Modified Adenosine Derivatives*

The majority of A_{2A}AR-selective agonists are 2-substituted Ado derivatives bearing an *N*-alkylcarboxamido modification at the ribose 5′ position, as in NECA (Hutchison et al. 1990; Cristalli et al. 1992, 1994b, 1995, 1996, 2003, 2007; Homma et al. 1992; Vittori et al. 1996; de Zwart et al. 1998; Müller 2000a). Also, Ado derivatives bearing bulky substituents in the C2 position and NECA derivatives with bulky substituents in the N^6 position are not selective versus A_1AR and A_3AR. N^6 and C2 substitution are helpful to improve A_3AR agonist activity, even if substitution at both N^6 and C2 with large substituents led to a large drop in affinity when combined (Baraldi et al. 1998a). This effect at A_{2A}AR had been observed in a series of Ado derivatives developed as A_{2A}AR agonists (Müller and Scior 1993). QSAR (quantitative structure–activity relationship) studies on different N^6-arylcarbamoyl, 2-arylalkynyl-N^6-arylcarbamoyl, and N^6-carboxamide Ado derivatives showed that the main determinants of the affinity at A_{2A}ARs were the bulkiness of the substituents attached at the 2 and 5′ positions and the stereoselectivity of the Ado derivatives (Gonzalez et al. 2005). Moreover, the synthesis and potential human A_{2A}AR agonistic activity of Ado derivatives containing an ethyl-substituted tetrazole moiety at the 4′ position of the ribose and an amino alcohol at the 2 position of the adenine core were reported (Bosch et al. 2004). The activities of these compounds were tested in radioligand binding assays using the four cloned human ARs. The compounds have also been profiled in cAMP assays using human receptors expressed on transfected CHO cells, and in functional assays using rat aorta, guinea pig aorta, and guinea pig tracheal rings. Results of these experiments show that substitution at the *para* position of the phenyl ring at the 2 side-chain by different groups greatly increases the affinity for A_{2A}AR. At the same time, the tested substituted derivatives have reduced affinity for A_1AR and A_3AR, thus greatly improving the A_1AR/A_{2A}AR and A_3AR/A_{2A}AR selectivity. Among the tested Ado derivatives, compound **26**, lacking the hydroxyl group in the side chain, was the most potent and selective in binding studies.

1.4.1 2-Substituted NECA Derivatives

The 4′-uronic acid ethyl ester analog of Ado, NECA, was reported in the early 1980s to be a potent coronary vasodilator and hypotensive (Prasad et al. 1980), and a good inhibitor of platelet aggregation induced by ADP (Cusack and Hourani 1981). However, NECA showed little or no A_2 selectivity in either functional or binding studies (Cristalli et al. 1994a, b; Klotz et al. 1999).

A series of 2-(arylalkylamino)-NECA derivatives were synthesized and evaluated for their A_1AR and A_{2A}AR binding profiles in rat brain membranes soon after the first Ado derivative with some A_{2A}AR selectivity, CV-1808 (**10**, Fig. 1), was reported. As in the case of arylalkylaminoAdos, the phenylethylamino analog of NECA **27** (Fig. 3) showed the highest rat A_{2A}AR affinity in the series and a greater than 2,000-fold separation between A_2 (coronary vasodilation) and A_1AR

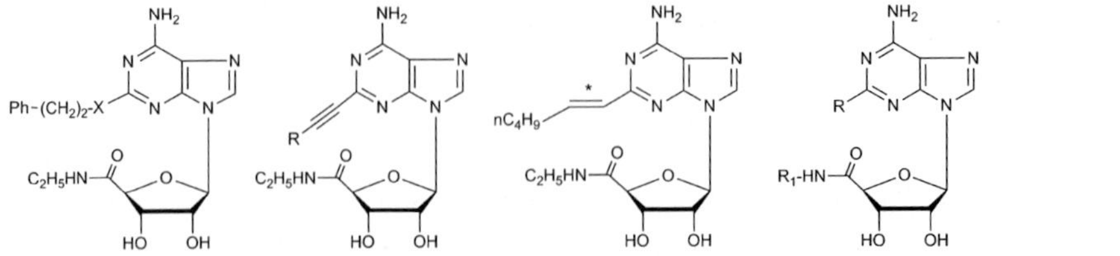

27 X = NH
33 X = S

HENECA **28** R = n-C_4H_9
(*R,S*)-PHPNECA **29** R = CH(OH)Ph
(*R*)-PHPNECA **29a** R = CH(OH)Ph
(*S*)-PHPNECA **29b** R = CH(OH)Ph
PENECA **30** R = Ph
ATL-146e, **31** R = cC_6H_{10}-4-$COOCH_3$

32a (*E*)-isomer
32b (*Z*)-isomer

MECA **34** R = H, $R_1 = CH_3$
NCPCA **35** R = H, $R_1 = cC_3H_5$
36 R = 1-hexynyl, $R_1 = cC_5H_9$
37 R = 1-hexynyl, $R_1 = CH_2$-Ph

Fig. 3 A_{2A}AR agonists: NECA derivatives

(negative chronotropic effect) receptor-mediated events. Among these compounds, CGS 21680 (**7b**, Fig. 1) proved to be an A_{2A}AR-selective agonist that was 140-fold selective vs. A_1AR in a rat model (Hutchison et al. 1989). This molecule was selected for extensive biological evaluation (Hutchison et al. 1989) and tritiation for use as an A_{2A}AR-selective ligand for receptor binding (Jarvis et al. 1989). However, due to a similar affinity of CGS 21680 for A_3AR and the remarkable species variation observed for the A_1AR, with an over tenfold higher affinity of this compound for the human subtype (Klotz et al. 1998), it can no longer be considered an A_{2A}AR-selective agonist. In any case, it has been the ligand of choice to distinguish A_{2A}AR- and A_{2B}AR-mediated effects so far.

The synthesis and evaluation of 2-alkynyl derivatives of NECA, bearing from five to eight linear carbon atom chains, was driven by the same observations that led to the synthesis and testing of 2-alkynylAdos (Cristalli et al. 1992). Affinities for A_1AR and A_{2A}AR were determined in rat membranes using radioligand competition assays. All compounds showed good A_1AR and A_{2A}AR affinities (K_i in the nanomolar range) and moderate A_{2A}AR selectivity (Cristalli et al. 1992). Among this series of 2-substituted compounds tested at rat receptors, 2-hexynyl-NECA (HENECA, **28**, Fig. 3) exhibited 60-fold A_{2A}AR selectivity compared to the A_1AR subtype. The pharmacological profile of this compound was characterized by studies carried out by Monopoli and coworkers, using in vitro and in vivo models (Monopoli et al. 1994). In addition to the binding studies on both rat and bovine brain, which confirmed the moderate A_{2A}AR versus A_1AR selectivity, HENECA was administered intraperitoneally in conscious spontaneously hypertensive rats, and it caused a dose-dependent reduction in systolic blood pressure with minimal reflex tachycardia. It also appeared to penetrate the central nervous system, as shown by its protection against pentylenetetrazole-induced convulsions in rats (Monopoli et al. 1994). In another work, administration of HENECA i.p. induced Fos-like immunoreactivity in the rat nucleus accumbens shell, lateral septal nucleus, and dorso–medial striatum, similar to that induced by atypical neuroleptics (Pinna et al. 1997).

The therapeutic potential of HENECA for the treatment of cardiovascular and psychotic diseases led to the synthesis of a series of 2-alkynyl, 2-cycloalkynyl, 2-aralkynyl, and 2-heteroaralkynyl derivatives of NECA that were tested in binding and functional assays to evaluate their potency for the A_{2A}AR compared to A_1AR (Cristalli et al. 1994b; Cristalli et al. 1995). Results showed that good A_{2A}AR affinities of the compounds were obtained with large 2-substituents containing a relatively rigid spacer, but that the affinity was reduced by introducing the bulkier naphthyl ring at the 2 position.

High agonist potency was found by introducing an α-hydroxy group into the alkynyl chain of NECA derivatives and obtaining compounds like 2-phenylhydroxypropynylNECA ((*R*, *S*)-PHPNECA, **29**), which was endowed with subnanomolar affinity in binding studies (K_i A_1AR = 2.5 nM and K_i A_{2A}AR = 0.9 nM) and was 16-fold more potent than NECA (**7**) as a platelet aggregation inhibitor. The problem with these analogs is that they also possess good A_1AR affinity, resulting in low A_{2A}AR selectivity. The diastereoisomer separation of a PHPNECA racemic mixture was accomplished obtaining compounds **29a** and **29b**. Binding tests in rat membranes showed that the (*S*)-diastereomer **29b** is about fivefold more potent and selective than the (*R*)-diastereomer **29a** as an agonist of the A_{2A}AR receptor subtype (**29b**, K_i A_{2A}AR = 0.5 nM; **29a**, K_i A_{2A}AR = 2.6 nM, Table 1) (Camaioni et al. 1997).

Things changed in the late 1990s after the cloning of the four human AR subtypes and their stable transfection into CHO cells. In fact, it was then possible to carry out comparative studies in a similar cellular background, utilizing binding studies (A_1AR, A_{2A}AR, A_3AR) or adenylate cyclase activity assays (A_{2B}AR) (Klotz et al. 1998). Transfected CHO cells were employed to screen for some nucleosides previously considered A_{2A}AR selective, and following this screening none of the prototypical AR agonists exhibited high affinity and selectivity for the human A_{2A}AR subtype. Both NECA and CGS 21680, which were available as radioligands for this subtype, demonstrated reduced affinity at the human as compared to the rat receptor, whereas HENECA (**28**) also showed high affinity at human A_{2A}AR and A_3AR, with tenfold and 25-fold selectivity versus the A_1AR subtype, respectively (K_i A_1AR = 60 nM, K_i A_{2A}AR = 6.4 nM, and K_i A_3AR = 2.4 nM). Interestingly, the potency for A_{2B}AR receptor is comparable with that of **7** (**28**: EC_{50} A_{2B} = 6.1 μM against **7** EC_{50} A_{2B} = 2.4 μM) (Cristalli et al. 1998), and it was also confirmed that **29** is a highly potent, nonselective agonist at A_1AR, A_{2A}AR, and A_3AR subtypes with a K_i in the low nanomolar range at the three subtypes. In the A_{2B}AR functional test, it was found that **29** (EC_{50} A_{2B} = 1.1 μM) is twofold more potent than **7**, and the (*S*)-diastereomer **29b** showed an EC_{50} A_{2B} in the nanomolar range (EC_{50} = 220 nM). It must be underlined that this was the first case of a NECA derivative substituted in the 2 position with a bulky group and showing good potency at the human A_{2B}AR subtype (Klotz et al. 1999; Lambertucci et al. 2003; Vittori et al. 2004). On the other hand, CGS 21680 was about 100-fold weaker than (*R*, *S*)-PHPNECA at the same subtype, with EC_{50} A_{2B} = 89 μM (Cristalli et al. 1998). The substituent linked to the triple bond allowed modulation of selectivity at the A_3AR, and the presence of a phenyl ring conjugated to the triple bond was detrimental for all the subtypes with the exception of the A_3AR; for example, PENECA (**30**) showed high potency and good selectivity for the A_3AR subtype (Klotz et al. 1999; Vittori et al. 2005). Anyway, the introduction of an alkyl spacer group restored high A_{2A}AR affinity and selectivity, as in phenylpentynyl–NECA.

Another A_{2A}AR agonist, apadenoson (ATL-146e, **31**, Fig. 3), was prepared following the literature activity on alkynyl derivatives. In fact, this molecule is a NECA derivative bearing in the 2 position a propynyl–cyclohexanecarboxylic acid methyl ester group, and binding assays are reported in which the affinity to recombinant human A_{2A}AR is measured as high- and low-affinity K_i values (0.2 and 67.9 nM, respectively) (Murphree et al. 2002).

Other developments include 2-(aralkenyl)-substituted Ado and NECA derivatives (Vittori et al. 1996), and (*E*)-isomers (**32a**, Fig. 3) were 15- to 50-fold more potent at A_{2A}AR than the corresponding (*Z*)-isomers (**32b**). Alkenyl–NECA derivatives, such as (*E*)-2-hexenyl-NECA (**32a**), displayed similar potency as A_{2A}AR agonists to the corresponding alkynyl derivatives, but showed higher selectivity versus A_1AR (Vittori et al. 1996). In this series, the *N*-ethylcarboxamido modification of the ribose was critical to increasing A_{2A}AR affinity. In addition, some 2-arylalkylthio analogs of NECA were synthesized and tested in radioligand binding studies, and the 2-phenylethylthio derivative (**33**) proved to be the most potent and selective agonist at the pig and rat A_{2A}AR (Volpini et al. 2004).

In conclusion, the affinities at the human and rat A_{2A}AR are ranked as follows: PHPNECA $\geqslant$ HENECA $>$ NECA $>$ CGS 21680 $>$ PENECA, even though none of these compounds are selective towards both A_1AR and A_3AR subtypes at the same time. Thus, so far, no satisfactory A_{2A}AR-selective agonists are available. In 2001, four new derivatives that are structurally similar to the 2-alkynyl derivatives of NECA that were previously reported (Cristalli et al. 2003) were evaluated by competitive binding assays employing the A_{2A}AR in rat striatal membranes and A_1AR of rat cortex. Hence, the A_{2A}AR against A_1AR selectivity was evaluated, but no A_{2A}AR against A_3AR selectivity was reported (Rieger et al. 2001). As some 2-alkynyl derivatives of NECA had been previously reported to behave as potent A_3AR agonists, affinity at this receptor should be measured before claiming selectivity for the reported compounds.

1.4.2 Ribose- and Purine-Modified NECA Derivatives

A few modifications of the ribose moiety of NECA have been reported (Jacobson et al. 1995; Volpini et al. 1998, 1999; de Zwart et al. 1999a). The ethyl group of the *N*-alkylcarboxamido function was substituted by a methyl or a cyclopropyl group, and this modification seems to be the only one that is well tolerated by the rat A_{2A}AR (see compounds **34** (MECA) and **35** in Fig. 3 and Table 1, K_i $A_{2A}AR = 330$ and 12 nM, respectively) (de Zwart et al. 1999a). On the other hand, replacing the same ethyl substituent in the 5′ position of **28** with a cyclopentyl or benzyl group brought about a significant decrease in affinity at all of the receptor subtypes (see compounds **36** and **37** in Table 2, K_i $A_{2A}AR = 49$ and 720 nM, respectively)

(Volpini et al. 1999). Some deoxy and dideoxy derivatives of **34** have been described, and the general effect of these modifications is a reduced affinity at all receptor subtypes (Jacobson et al. 1995; Volpini et al. 1998). However, the removal of the 3′-hydroxy group seems to be better tolerated by the A_{2A}AR than the removal of the corresponding group in the 2′ position (Cristalli et al., unpublished results).

The only purine-modified analog of NECA that has been synthesized and tested so far is 1-deazaNECA (**7a**, Fig. 1) (Cristalli et al. 1988; Siddiqi et al. 1995). As in the case of the other 1-deazaAdo analogs, the affinity of 1-deazaNECA at all ARs is reduced in comparison to that of the parent compound NECA (**7**)—in fact it is about tenfold less active than NECA—but 1-deazaNECA is clearly more active than the parent compound 1-deazaAdo (**8**) as an inhibitor of platelet aggregation and as a stimulator of cyclic AMP accumulation. However, in contrast to 2-chloro-1-deazaAdo (**9**), which was the only 1-deaza analog showing slight A_{2A}AR-selectivity, the potency of 1-deazaNECA at A_1AR, A_{2B}AR, and A_3AR is diminished by a factor of about 5, whereas that at the A_{2A}AR subtype is about 60-fold lower than that of NECA. Hence, 1-deazaNECA proved to be a moderate A_{2A}AR agonist.

1.5 Agonist Radioligands

[^{3}H]NECA was introduced as a ligand for the A_2 receptor (K_d values of between 31 and 46 nM), but further studies demonstrated that it is a prototypical nonselective ligand (Gessi et al. 2000). It labels A_1AR, A_{2A}AR, and A_3AR with similar affinities, with a slight preference for the A_3AR subtype (Bruns et al. 1986). CGS 21680 was introduced as an A_2-selective agonist and it was also developed as a tritiated ligand (Jarvis et al. 1989), but (as reported above) this molecule is not an ideal tool for the characterization of A_{2A}ARs, particularly if differentiation from A_3AR is required. The tritiated compound displays a K_d value of 32 nM at the human A_{2A}AR and therefore shows a comparable potency to [^{3}H]NECA (Wan et al. 1990).

1.6 Partial Agonists

Recently, a series of 2,8-disubstituted Ado derivatives were synthesized and tested. Most of these compounds appeared to have A_{2A}AR affinities in the low micromolar or nanomolar range, and also displayed reduced intrinsic activities compared to the reference agonist CGS 21680 (**7b**); hence, they behaved as partial agonists (van Tilburg et al. 2003).

The introduction of 8-alkylamino substituents led to a reduction in A_{2A}AR affinity but also to an increase in selectivity versus the A_3AR subtype. In particular, the 8-methylamino and 8-propylamino derivatives of **17** (**38** and **39**, respectively, Fig. 4)

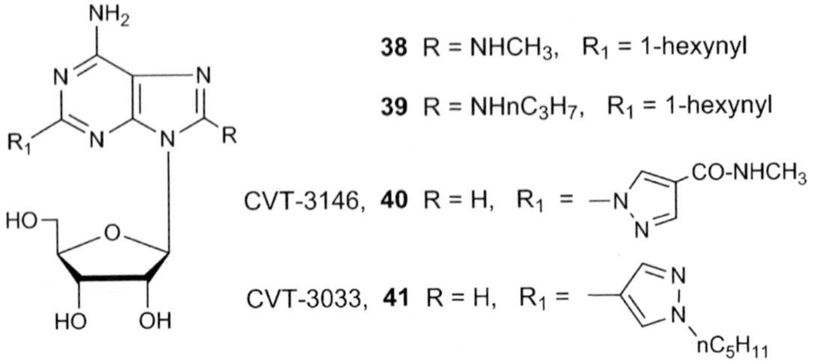

38 R = $NHCH_3$, R_1 = 1-hexynyl

39 R = $NHnC_3H_7$, R_1 = 1-hexynyl

CVT-3146, **40** R = H, R_1 = (1-pyrazolyl, 4-CO-$NHCH_3$)

CVT-3033, **41** R = H, R_1 = (4-pyrazolyl, 1-nC_5H_{11})

Fig. 4 A_{2A}AR partial agonists

showed K_i A_{2A}AR affinity values of 115 and 82 nM, respectively, and 49- and 26-fold selectivities for the A_{2A}AR versus the A_3AR.

Other Ado derivatives that were substituted at the 2 position with 1-pyrazolyl (Lexiscan, regadenoson, CVT-3146, **40**) or 4-pyrazolyl (CVT-3033, **41**) rings were found to be short-acting functionally selective coronary vasodilators with good potency, but they possessed low affinity for A_{2A}AR ($K_i = 1{,}122$ and 2,895 nM, respectively) (Zablocki et al. 2001). One of these, Lexiscan, appears to be a weak partial agonist in stimulating cAMP accumulation in PC12 cells but a full and potent agonist in inducing coronary vasodilation, a response that has a very large A_{2A}AR reserve (Gao et al. 2001; Eggbrecht and Gossl 2006; Gordi 2006).

Very recently, the US Food and Drug Administration (FDA) has approved injected Lexiscan for use as a pharmacologic stress agent in radionuclide myocardial perfusion imaging (MPI) (CVT 2008).

2 Adenosine A_{2A} Receptor Antagonists

In the last few years, A_{2A}AR antagonists have become attractive pharmacological tools due to their potential as novel drugs for the treatment of Parkinson's disease (PD) and restless legs syndrome, Alzheimer's disease, and their antidepressive and neuroprotective activities (Impagnatiello et al. 2000; Cacciari et al. 2003; Xu et al. 2005; Jacobson and Gao 2006; Moro et al. 2006; Schapira et al. 2006; Schwarzschild et al. 2006; Cristalli et al. 2007; Dall'Igna et al. 2007; Fuxe et al. 2007; Yu et al. 2008; Salamone et al. 2008). In addition, A_{2A}AR antagonists seem to protect against cellular death induced by ischemia, and may also be active as cognition enhancers, antiallergic agents, analgesics, positive inotropics, and even for the treatment of alcoholism and alcohol and cannabis abuse (Ledent et al. 1997; Richardson et al. 1997; Monopoli et al. 1998; Brambilla et al. 2003; Pedata et al. 2005; Melani et al. 2006; Ferré et al. 2007; Thorsell et al. 2007; Bilkei-Gorzo et al. 2008; Takahashi et al. 2008). A_{2A}ARs are expressed in high density in restricted areas of

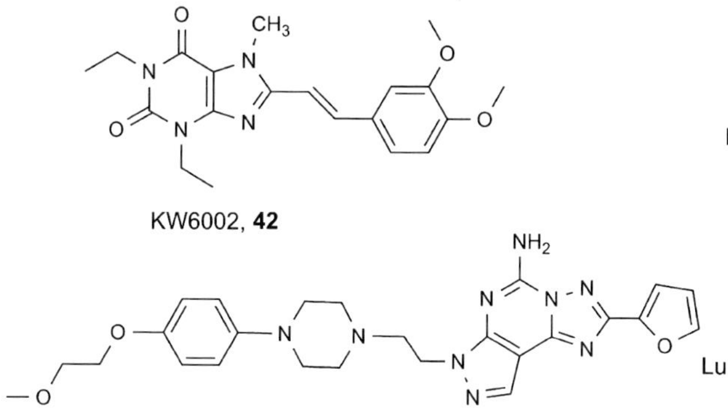

KW6002, **42**

BIIB014, **44**

SCH 420814 **43**

Lu AA47040, **45**

Fig. 5 A_{2A}AR antagonists in clinical trials

the brain, namely in the caudate-putamen (striatum), and there they are coexpressed with dopamine D_2 and cannabinoid CB_1 receptors (Carriba et al. 2007; Ferré et al. 2008). The restricted expression as well as the promising pharmacological potential of A_{2A}AR antagonists has led to extensive efforts to develop potent and selective A_{2A}AR antagonists (Yuzlenko and Kiec-Kononowicz 2006; Müller and Ferré 2007; Baraldi et al. 2008). Four different A_{2A}AR antagonists are currently being studied in clinical trials, istradefylline (KW-6002, **42**), preladenant (SCH-420814, **43**), BIIB014 (V2006, **44**), and Lu AA47040 (**45**). The structures of the latter two compounds have not been disclosed (Fig. 5).

Several heterocyclic classes of compounds have been studied as A_{2A}AR antagonists; these can generally be divided into xanthine and non-xanthine derivatives. The xanthine analogs represent the prototypical group of antagonists, and modifications of the xanthine scaffold resulted in a comprehensive collection of derivatives, among which several compounds showed distinct subtype selectivity. A second class of heterocyclic compounds can be envisaged as adenine-derived structures (Cacciari et al. 2003; Vu 2005; Moro et al. 2006; Müller and Ferré 2007). Very recently, other heterocyclic structures related to neither xanthine nor adenine derivatives have been described. These are based on lead structures identified by the screening of large compound libraries (Müller and Ferré 2007). The present review focuses on antagonists published in scientific articles. Thorough reviews on the patent literature have recently been published (Vu 2005; Müller and Ferré 2007).

2.1 *Xanthine Derivatives*

Years ago it was reported that caffeine was the "most widely consumed behaviorally active substance in the world" (Fredholm et al. 1999). In fact, the vast majority of people on our planet have enjoyed the CNS effects of the AR antagonist caffeine long before the physiological effects of Ado were discovered. Naturally occurring xanthines like caffeine or theophylline generally have affinities at the micromolar level, with the highest affinity being at the $A_{2A}AR$, and this receptor subtype appears to be relevant to the activation caused by caffeine (Ledent et al. 1997; Svenningsson et al. 1997). Hence, the xanthine scaffold represented an important starting point for the development of antagonists of this family of receptors (Daly et al. 1991).

A large number of modifications at the 1, 3, 7 and 8 positions have been performed with the aim of obtaining potent and selective $A_{2A}AR$ antagonists. The first xanthine derivative considered an $A_{2A}AR$ antagonist was 3,7-dimethyl-1-propargylxanthine (DMPX, **46**, Fig. 6, Table 2), even though this compound proved to be poorly active (K_i rA_{2A} and hA_{2A} = 16 and 2 μM, respectively) and moderately selective against the A_1AR and $A_{2B}AR$ subtypes (Daly et al. 1986, 1991). Nevertheless, this compound has been widely used in in vivo studies because of its good water solubility and bioavailability (Daly et al. 1986; Seale et al. 1988; Thorsell et al. 2007). Further studies on DMPX derivatives led to the 2-*O*-methyl-1-propargylxanthine derivative **47**, endowed with an affinity in the high nanomolar range (K_i $A_{2A}AR$ = 105 nM) at the $A_{2A}AR$ subtype and significant selectivity in comparison to the A_1AR (45-fold) (Müller and Stein 1996; Müller et al. 1998a).

Starting from these observations, a program to screen various 1,3,8-substituted xanthines led to the discovery of the first very potent and selective $A_{2A}AR$ antagonists (Erickson et al. 1991; Jacobson et al. 1993a; Nonaka et al. 1994a; Müller and Stein 1996; Müller 2000b). In particular, 3-chlorostyrylcaffeine (CSC, **48**) showed

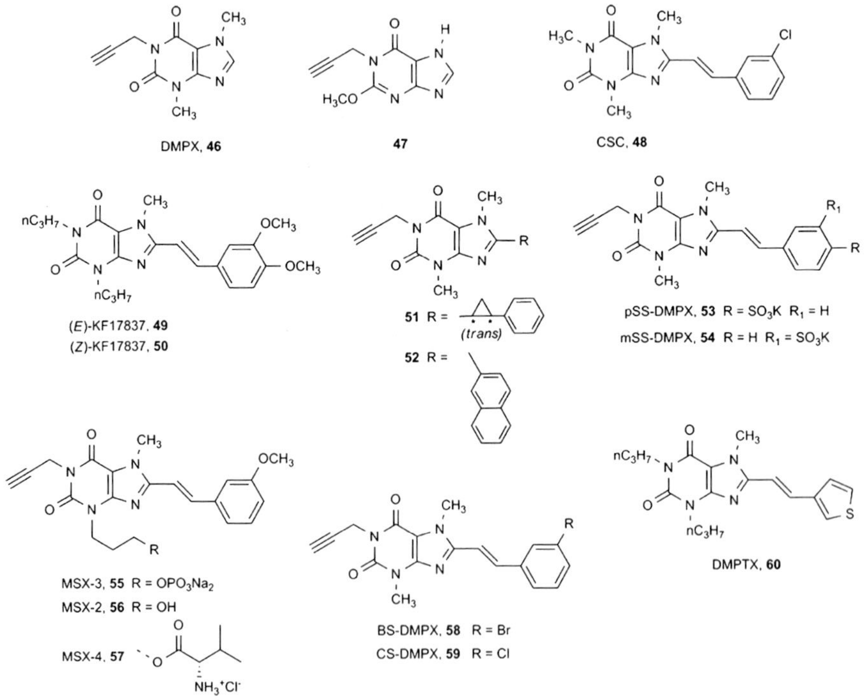

DMPX, **46** · **47** · CSC, **48** · (*E*)-KF17837, **49** · (*Z*)-KF17837, **50** · **51** R = (*trans*) · **52** R = · pSS-DMPX, **53** R = SO_3K R_1 = H · mSS-DMPX, **54** R = H R_1 = SO_3K · MSX-3, **55** R = OPO_3Na_2 · MSX-2, **56** R = OH · MSX-4, **57** · BS-DMPX, **58** R = Br · CS-DMPX, **59** R = Cl · DMPTX, **60**

Fig. 6 $A_{2A}AR$ antagonists: xanthines

Table 2 Affinities of AR antagonists in radioligand binding assays at A_1AR, A_{2A}AR and A_3AR. For A_{2B}AR, radioligand binding assays values are reported where available; for some compounds, values are related to the effects on adenylate cyclase activity

Cpd	K_i (A_1AR)[a]	K_i (A_{2A}AR)[a]	K_i (A_{2B}AR)[a]	K_i (A_3AR)[a]
42	580 r	13 r	–	–
	2,830 h	36 h	1,800 h	>3,000 h
43	–	2.5 r	–	–
	>1000 h	1.1 h	>1,700 h	>1,000 h
46	45,000 r	16,000 r	2,500 m	–
	–	2,000 h	4,130 h	>10,000 h
47	4,700 r	105 r	–	–
48	28,000 r	54 r	–	–
49	62 r	1 r	–	–
50	>10,000 r	860 r	–	–
51	4,600 r	1,700 r	–	–
52	980 r	380 r	–	–
53	4,900 r	240 r	–	–
	–	–	–	>100,000 h
54	8,900r	300 r	–	–
	–	–	–	>100,000 h
56	900r	8 r	–	–
	2,500 h	5.0 h	–	>10,000 h
58	1,200 r	8.2 r	–	–
59	1,300 r	13 r	–	–
60	561 r	19 r	–	–
61	21 r	3.3 r	–	–
	4.4 h	0.43 h	25 h	85 h
62	270 r	21 r	–	–
63	3.3 r	1.2 r	–	–
64	121 r	2.3 r	–	–
	549 h	1.1 h	>10,000 h	>10,000 h
65	504 r	2.4 r	–	–
	350 h	1.2 h	>10,000 h	>10,000 h
66	444 r	1.7 r	–	–
	–	–	–	>10,000 h
67	741 r	0.94 r	–	–
	1,111 h	1.5 h	–	>10,000 h
68	1,815 r	0.048 r	–	–
	1,111 h	0.5 h	>10,000 h	>10,000 h
69	4,927 h	4.63 h	>10,000 h	>10,000 h
70	139 h	140 h	>10,000 h	>10,000 h
71	2,160 h	0.22 h	>10,000 h	>10,000 h
72	369 h	3.8 h	>10,000 h	>10,000 h
73	15 b	6.5 b	–	–
	–	–	–	>10,000 h
74	83 h	0.8 h	–	–
75	257 r	1.8 r	–	–
	774 h	1.6 h	28 h	743 h

(continued)

Table 2 (continued)

Cpd	K_i (A_1AR)[a]	K_i ($A_{2A}AR$)[a]	K_i ($A_{2B}AR$)[a]	K_i (A_3AR)[a]
76	1,270 p	14 p	–	–
77	320 r	1 r	–	–
78	100 r	1.1 r	–	–
79	208 h	1.4 h	865 h	476 h
80	12.5 r	1.2 r	–	–
	7.9 m	1.6 m	–	–
	9.0 h	1.8 h	>557 h	–
81	17 h	1.1 h	112 h	1,472 h
82	170 h	1.7 h	141 h	1,931h
83	24 h	3.7 h	380 h[b]	4,700 h
84	2,400 h	46 h	>30,000 h[b]	21,000 h
85	150 h	19 h	690 h[b]	3,100 h
86	23 h	1.7 h	569 h[b]	1,090 h
87	9.4 h	3.8 h	780 h[b]	17.6 h
88	5.8 h	2.2 h	521 h[b]	16 h
89	71.8 h	6.6 h	352 h	>10,000 h
90	–	26 r	–	–
	266 h	2.7 h	–	–
91	–	9.4 r	–	–
	60 h	0.4 h	–	–
94	–	90 r	–	–
	1,380 h	20 h	–	–

[a]Binding data from different species: rat (*r*), human (*h*), pig (*p*), bovine (*b*) or mouse (*m*) A_1AR, $A_{2A}AR$, $A_{2B}AR$, and A_3AR, expressed as K_i (nM)
[b]Effects on adenylate cyclase activity at the human (*h*) $A_{2B}AR$ expressed as K_i (nM)

high affinity at the $A_{2A}AR$ (54 nM) and high selectivity in comparison to the A_1AR subtype (560-fold) (Jacobson et al. 1993a). In addition, it is a relatively potent monoaminoxidase type B (MAO-B) inhibitor, which may contribute to its pharmacological effects in models of Parkinson's disease (Petzer et al. 2003; van den Berg et al. 2007). Another compound, (*E*)-1,3-dipropyl-8-(3,4-dimethoxystyryl)-7-methylxanthine ((*E*)-KF17837, **49**), proved to be potent in the nanomolar range at the $A_{2A}AR$ subtype (1 nM) and significantly selective in comparison to the A_1AR (62-fold) (Nonaka et al. 1994a). However, several problems have initially limited the use of these xanthine derivatives as pharmacological tools for studying the $A_{2A}AR$ subtype, in particular their low water solubility (Jackson et al. 1993) and the rapid photoisomerization that they undergo when exposed to daylight in dilute solution (Nonaka et al. 1993; Müller et al. 1998a). It should be noted that this isomerization process does not occur when styrylxanthines are administered orally as solid substances, but the phenomenon happens very rapidly during binding studies performed in buffer solution and in the presence of light (Müller et al. 1998a). In particular, after photoisomerization, (*E*)-KF17837 becomes a stable mixture of ca. 18% (*E*) and ca. 82% (*Z*, **50**) isomers, and the binding data change (K_i $A_{2A}AR$ = 7.9 nM, K_i A_1AR = 390 nM) (Nonaka et al. 1993). Another problem associated with 8-styrylxanthine derivatives is their tendency to undergo light-induced dimerization ([2 + 2]-cycloaddition reaction) in the solid state, yielding weakly active cyclobutane derivatives (Hockemeyer et al. 2004).

To overcome this photoisomerization, the styryl moiety has been replaced with different functional groups (e.g., triple bond, cyclopropyl ring, **51**, a 2-naphthyl residue, **52**) (Müller et al. 1997c), or a tricyclic constrained structure (Kiec-Kononowicz et al. 2001; Drabczynska et al. 2003, 2004, 2006, 2007). In many cases, a significant loss of affinity was observed by such modifications. Substitution of the ethenyl group with an azo structure has also been performed. The compounds obtained retained selectivity but showed only moderate affinity (Müller et al. 1997b).

Different approaches have been utilized to improve the water solubility of styrylxanthines, such as the introduction of polar groups on the phenyl ring and the preparation of phosphate or amino-acid prodrugs. The introduction of a sulfonate group on the phenyl ring of the styryl moiety at the *para*- (**53**) or *meta*- (**54**) position led to water-soluble derivatives endowed with only high nanomolar affinity at the A_{2A}AR but retaining selectivity (Müller et al. 1998b). Tricyclic styryl-substituted imidazo[2,1-*i*]purin-5-one derivatives showed enhanced water-solubility but reduced A_{2A}AR affinity and selectivity (Müller et al. 2002). The prodrug approach has been much more successful. In fact, MSX-3 (**55**), which is the phosphate prodrug of MSX-2 (3-(3-hydroxypropyl)-8-(*m*-methoxystyryl)-1-propargylxanthine, **56**), is stable and highly soluble (15 mM) in aqueous solution but readily cleaved by phosphatases to liberate MSX-2, which showed a very high affinity (rat and human A_{2A}AR $K_i = 8$ and 5 nM, respectively) and selectivity for the A_{2A}AR (Sauer et al. 2000; Hockemeyer et al. 2004). Recently, an L-valine ester prodrug of MSX-2 has been described, named MSX-4 (**57**), which shows good water solubility as a hydrochloride as well as high stability in artificial gastric fluid and at physiological pH values, but is readily cleaved by esterases (Vollmann et al. 2008). It is expected that the L-amino acid ester prodrug can be absorbed via an active transport mechanism by L-amino acid carrier proteins.

All of these studies strongly suggest that the xanthine family should be reconsidered as A_{2A}AR antagonists. In fact, the antagonist KW-6002 (istradefylline: 1,3-diethyl-8-(3,4-dimethoxystyryl)-7-methylxanthine, **42**; human A_{2A}AR $K_i = 36$ nM) is already in Phase III clinical trials for the treatment of basal ganglia disorders such as Parkinson's disease (Knutsen and Weiss 2001; Weiss et al. 2003; Kalda et al. 2006). This compound showed a $(E)/(Z)$ stable equilibrium ratio of 19:81 with good affinity and selectivity but most importantly a very high anticataleptic activity (0.03 mg kg^{-1}, p.o.) in a mouse haloperidol model (Shimada et al. 1997).

Further modifications of all the positions of the xanthine nucleus were introduced and investigated. For example, the bioisosteric replacement of one of the alkenyl CH groups of the 8-styryl residue with nitrogen led to more potent and selective antagonists for the A_{2A}ARs, but the compounds were highly unstable in aqueous solution because of their imine (Schiff base) structure (Müller et al. 1997b). The introduction of a propargyl or an *n*-propyl residue at the 1 position in combination with the 8-styryl group seems to increase affinity at the A_{2A}AR subtypes while retaining the selectivity. These studies led to the discovery of two compounds, named BS-DMPX (3,7-dimethyl-1-propargyl-8-(3-bromostyryl)xanthine, **58**) and CS-DMPX (3,7-dimethyl-1-propargyl-8-(3-chlorostyryl)xanthine, **59**), which could be considered lead compounds of this series (Müller et al. 1997a). Methyl substitution at the 3 and 7 positions appears to be desirable for achieving both affinity and selectivity at A_{2A}AR subtypes (Shamim et al. 1989; Erickson et al. 1991; Del Giudice et al. 1996). However, large substituents are also tolerated at the 3 position (Massip et al. 2006). The bioisosteric replacement of the phenyl ring with a thienyl moiety led to DPMTX ((*E*)-1,3-dipropyl-7-methyl-8-[2-(3-thienyl)ethenyl]xanthine, **60**) which showed high affinity and selectivity (Del Giudice et al. 1996). Regarding the substitutions at the 8 position, it has been demonstrated that an aromatic ring attached to an ethenyl group is essential for both affinity and selectivity at the A_{2A}AR (Erickson et al. 1991; Jacobson et al. 1993b; Del Giudice et al. 1996). 8-Styryl-9-deazaxanthine derivatives were nearly as potent as the corresponding xanthine derivatives at A_{2A}ARs (Grahner et al. 1994).

2.2 Adenine Derivatives and Related Heterocyclic Compounds

Due to the initial problems with xanthine derivatives, such as poor water solubility and photoisomerization, many scientists searched for alternative heterocyclic derivatives for use as lead compounds. The first promising A_{2A}AR antagonists with a non-xanthine structure were CGS 15943 (9-chloro-2-(2-furanyl)-[1,2,4]triazolo[1,5-*c*]quinazolin-5-amine, **61**, Fig. 7) (Williams et al. 1987; Francis et al. 1988; Kim et al. 1996; Baraldi et al. 2000) and CP-66713 (4-amino-8-chloro-1-phenyl-[1,2,4]triazolo[4,3-*a*]quinoxaline, **62**) (Sarges et al. 1990), compounds that were not very A_{2A}AR selective but were important as starting points for developing new non-xanthine structures as A_{2A}AR antagonists. All of these structures are reminiscent of the nucleobase adenine, a partial structure of Ado.

A few years later, the synthesis of 8FB-PTP (5-amino-8-(4-fluorobenzyl)-2-(2-furyl)pyrazolo[4,3-*e*]-1,2,4-triazolo[1,5-*c*]pyrimidine, **63**), a bioisoster of **61**, was reported (Gatta et al. 1993; Dionisotti et al. 1994). Here, the phenyl ring was replaced by a substituted pyrazole nucleus; this compound showed good affinity but no selectivity for A_{2A}ARs. Structure–activity relationship studies on the pyrazolo-triazolo-pyrimidine nucleus were carried out with the aim of determining the important features for high A_{2A}AR potency and selectivity, focusing on the presence of a free amino group at the 5 position and a furan ring at the triazole ring. The role of the substituents on the pyrazole ring was explored. Results showed that the substituents at the 7 and 8 positions were influential. In particular, substitutions at the 7 position gave selective compounds, whereas the same substitution at the 8 position resulted in potent but nonselective derivatives (Baraldi et al. 1994, 1996a, 2001). Furthermore, replacement of the pyrazole ring with a triazole led to affinity retention but also a complete loss of selectivity (Baraldi et al. 1996b). Recently, the pyrazole was replaced by an imidazole ring with great success (Silverman et al. 2007).

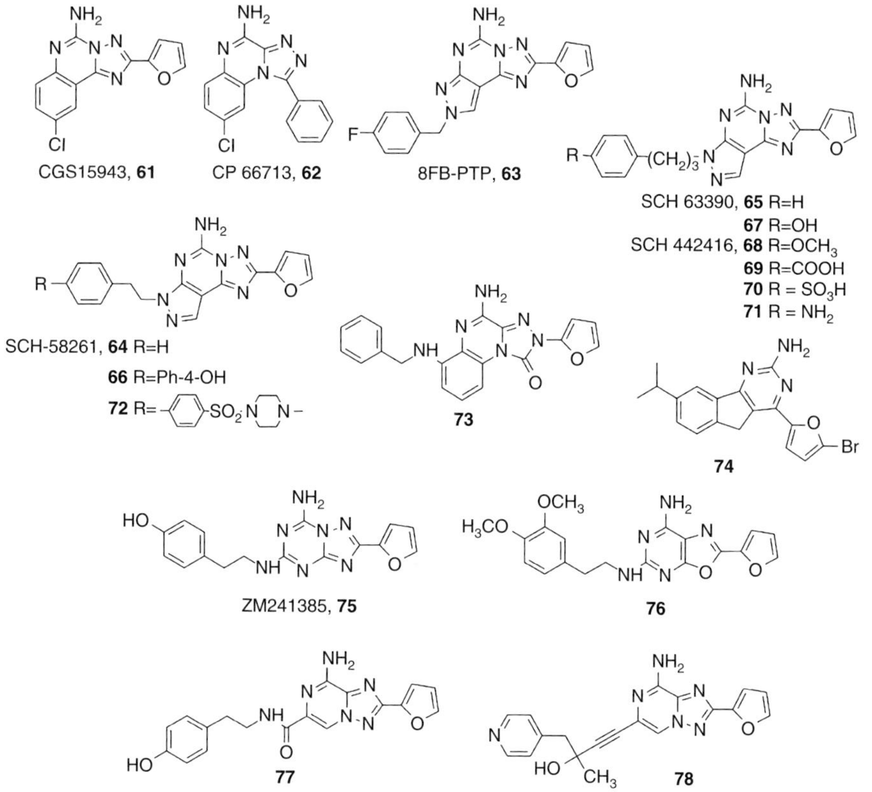

Fig. 7 A_{2A}AR antagonists: nonxanthine derivatives

Two selected compounds named SCH-58261 (5-amino-7-(2-phenylethyl)-2-(2-furyl)pyrazolo[4,3-*e*]1,2,4-triazolo[1,5-*c*]pyrimidine, **64**, Fig. 7) and SCH-63390 (5-amino-7-(3-phenylpropyl)-2-(2-furyl)pyrazolo[4,3-*e*]1,2,4-triazolo[1,5-*c*]pyrimidine, **65**) proved to be very potent and selective A_{2A}AR antagonists at both rat and human receptors (Baraldi et al. 1996a, b, 1998b; Zocchi et al. 1996a).

Problems with low water solubility affected even these non-xanthine compounds, and the poor bioavailability limited their use as pharmacological tools. To improve the hydrophilicities of these derivatives, polar functions were introduced on the phenyl ring located on the side chain of the pyrazole nucleus. The presence of a hydroxyl group at the phenyl ring in the *para* positions of compounds **64** and **65** led to compounds **66** (5-amino-7-[4-(4-hydroxyphenyl)ethyl]-2-(2-furyl)pyrazolo[4,3-*e*]1,2,4-triazolo[1,5-*c*]pyrimidine) and **67** (5-amino-7-[3-(4-hydroxyphenyl)propyl]-2-(2-furyl)pyrazolo[4,3-*e*]1,2,4-triazolo[1,5-*c*]pyrimidine), which showed slightly enhanced hydrophilicity and also a significant increase in both affinity and selectivity (Baraldi et al. 1998b). To understand the nature of the hydrogen bond, the phenolic hydroxy group was substituted with a methoxy group (thus reducing compound hydrophilicity), leading to

SCH-442416 (5-amino-7-[3-(4-methoxyphenyl)propyl]-2-(2-furyl)pyrazolo[4,3-*e*]-1,2,4-triazolo[1,5-*c*]pyrimidine, **68**). This derivative showed an increased potency and remarkable selectivity for the A_{2A}AR, and so it has been used as a tool for PET studies in its ^{11}C-labeled form (Todde et al. 2000). The introduction of oxygen-containing groups on the phenyl ring did not confer sufficient water solubility on the derivative, so it appeared necessary to introduce different functionalities to address this problem. Several polar functions such as carboxylic (**69**) and sulfonic acid (**70**) functions were introduced for this purpose and, as expected, an increased solubility was observed, especially in the case of the sulfonate. Unfortunately, a great loss of affinity and selectivity was observed at the same time. The introduction of an amino group at the *para* position of the phenyl ring gave compound **71** ($K_i = 0.22$ nM, $hA_1AR/hA_{2A}AR = 9820$), which yielded the best results in terms of affinity and selectivity, without improving the water solubility. Sulfonamido derivatives seem to exhibit a good balance between solubility and affinity (**72**) (Baraldi et al. 2002). Structure–activity relationships for this group of compounds indicated that the tricyclic structure of the pyrazolo-triazolo-pyrimidine, the presence of the furan ring, the exocyclic 5-amino group, and the arylalkyl substituent on the nitrogen at the 7 position are probably crucial to their affinities and selectivities for the A_{2A}AR subtype.

A recent series of pyrazolo-triazolo-pyrimidine derivatives was obtained by modifying the phenylethyl substituent of **64** with substituted phenylpiperazinethyl groups (Neustadt et al. 2007). Introduction of fluorine atoms into the phenyl ring enhanced the affinity to subnanomolar values and the compounds displayed potent peroral activity, but their solubility still remained poor. Further introduction of ether substituents led to derivatives with high affinities and selectivities for A_{2A}ARs and improved water solubilities. In particular, one of these compounds (SCH-420814, preladenant, **43**) exhibited high affinities for both rat and human A_{2A}ARs, with K_i values of 2.5 and 1.1 nM, respectively. In addition, the compound is very selective for human A_{2A}ARs over A_1AR, A_{2B}AR, and A_3AR. Interestingly, the compound did not show significant binding against a panel of 59 unrelated receptors, enzymes, and ion channels. preladenant is now in Phase II clinical trials for dyskinesia in Parkinson's disease (Neustadt et al. 2007). Recently, the pyrazole moiety in these tricyclic derivatives was replaced by an imidazole ring, yielding 3*H*-[1,2,4]triazolo[5,1-*i*]purin-5-amine derivatives. The isomer of SCH-420814 displayed promising in vitro and in vivo profiles (Silverman et al. 2007).

The triazoloquinoxaline (Colotta et al. 1999, 2000, 2003) and the indenopyrimidine (Matasi et al. 2005) series possess promising features as A_{2A}AR antagonists. The triazoloquinoxaline nucleus seems to be very sensitive to any kind of variation and modification: alkylation of the amino group, replacement of the amino group by a carbonyl function, and substitution on the phenyl ring all reduced A_{2A}AR affinity. In this class, only compound **73** (4-amino-6-benzylamino-1,2-dihydro-2-1,2,4-triazolo[4,3-*a*]quinoxalin-1-one) showed a favorable binding profile (Colotta et al. 1999, 2000, 2003). In contrast, the indenopyrimidine derivatives are very promising, and the derivative **74** shows affinity in the nanomolar range and good selectivity against the A_1AR subtype. It must be underlined that binding data at A_{2B}AR and A_3AR are lacking, so it is not possible to fully assess this compound with regard to potentially being an ideal A_{2A}AR antagonist (Matasi et al. 2005). Anyway, these structures showed several problems, such as poor water solubility and (most importantly) complex and difficult synthetic accessibility.

Therefore, researchers focused their attention on simplified analogs like bicyclic systems, and the Zeneca group reported on a compound named ZM241385 (4-[2-[[7-amino-2-(2-furyl)[1,2,4]-triazolo[2,3-*a*][1,3,5]triazin-5-yl]amino]ethyl]phenol, **75**), which proved to be one of the most potent A_{2A}AR antagonists ever reported, and which had a favorable water solubility (Caulkett et al. 1995; Poucher et al. 1995; de Zwart et al. 1999b; Weiss et al. 2003; Moro et al. 2006). This compound also binds with high affinity to human A_{2B}AR, and its tritiated form is actually used in radioligand binding studies for this receptor subtype (Ji and Jacobson 1999).

In the last few years, Biogen Idec Inc. has developed a large series of triazolotriazine and triazolopyrimidine analogs bearing various substituents, and a few compounds have shown high potency and selectivity for the A_{2A}AR as compared with the A_1AR (Peng et al. 2004; Vu et al. 2004a, b, c, 2005; Yang et al. 2007). However, the lack of binding data for the A_{2B}AR and A_3AR prevents any comparison of the derivatives with other fully characterized compounds. Interestingly, some of these derivatives showed good oral efficacy in a rodent catalepsy model of Parkinson's disease (Peng et al. 2004; Vu et al. 2004a, b, c, 2005).

Among synthesized isosters of the triazolotriazine nucleus, some oxazolopyrimidines (**76**) (Holschbach et al. 2006) and triazolopyrazines (**77**, **78**) should be mentioned (Dowling et al. 2005; Yao et al. 2005). All of these compounds showed good A_{2A}AR potency and selectivity against the A_1AR, but full characterization at the four AR subtypes has not been completed. Some pyrazolopyrimidines have also been reported (Chebib et al. 2000), but in all cases the affinities and/or selectivities were only moderate.

A thieno[3,2-*d*]pyrimidine, VER-6623 (**79**, Fig. 8), showed a high affinity for A_{2A}AR ($K_i = 1.4$ nM), but it also had low or poor oral bioavailability (Weiss et al. 2003; Yang et al. 2007). Very recently, a potent A_1AR and A_{2A}AR dual antagonist, 5-[5-amino-3-(4-fluorophenyl)pyrazin-2-yl]-1-isopropylpyridine-2(1*H*)-one (ASP5854, **80**), was synthesized and tested in models of Parkinson's disease and cognition (Mihara et al. 2007). The binding affinities of **80** for human A_1AR and A_{2A}AR were 9.0 and 1.8 nM, respectively. This compound also showed antagonistic action on A_1AR and A_{2A}AR agonist-induced increases in intracellular Ca^{2+} concentration, and in vivo tests showed that this molecule improves motor impairment, is neuroprotective via A_{2A}AR antagonism, and also enhances cognitive function through A_1AR antagonism.

The development of A_{2A}AR antagonists also made use of non-xanthine imidazopyrimidine (purine)-type structures, and some of these derivatives (recently reported by several groups) seem to be very promising. Some compounds, like VER-6947 (**81**) and VER-7835 (**82**), show human A_{2A}AR K_i values of around 1 nM (Weiss et al. 2003), while some 6-(2-furanyl)-9*H*-purin-2-amino derivatives

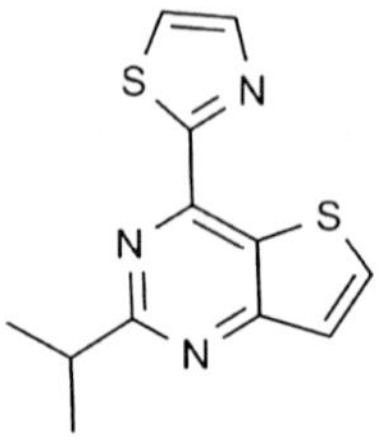

VER-6623, **79**

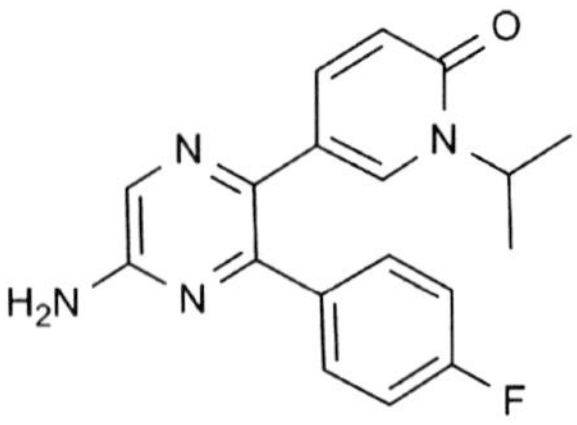

ASP5854, **80**

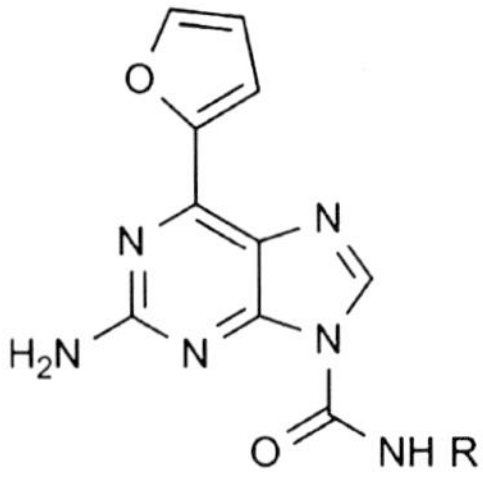

VER-6497, **81** R = CH_2-Ph
VER-7835, **82** R = CH_2-Furyl

Fig. 8 A_{2A}AR antagonists: nonxanthine derivatives (2)

are endowed with A_{2A}AR affinities in the low nanomolar range and a good level of selectivity against the other receptor subtypes (Kiselgof et al. 2005).

In the late 1990s, Cristalli and coworkers reported the synthesis of a number of 9-ethylpurines bearing various substituents in the 2, 6 or 8 positions (Camaioni et al. 1998). 9-Ethyladenine showed micromolar affinities at the human A_1AR and A_{2A}AR subtypes, but the introduction of a bromine atom in the 8 position led to an enhancement of the binding affinity at all AR subtypes. Recently, rat model studies on the derivatives ANR-152 (9-ethyl-8-furyl-adenine, **83**, Fig. 9) and ANR-94 (8-ethoxy-9-ethyl-adenine, **84**) were reported. It should be noted that **83** was more potent at A_{2A}AR than at A_1AR, with poor selectivity against A_1AR, while the replacement of furan ring with an ethoxy function (**84**) (Klotz et al. 2003) led to a decrease in affinity but a significant increase in selectivity. Study results showed that both of these derivatives are able to ameliorate motor deficits in rat models of Parkinson's disease (Pinna et al. 2005).

The 2 and 8 positions of adenine were further explored through the introduction of alkynyl chains, and while the 2-alkynyl derivatives possessed good affinity and were slightly selective for the human A_{2A}AR, the affinities of the 8-alkynyl derivatives at the human A_1AR, A_{2A}AR, and A_{2B}AR proved to be lower than those of the corresponding 2-alkynyl derivatives, with improved binding data for the human A_3AR subtype (Volpini et al. 2005). The observation that the introduction at the 2 position of phenylethylamino or phenethoxy groups resulted in compounds with increased A_{2A}AR affinity (Camaioni et al. 1998) led to the synthesis of 9-ethyladenine derivatives substituted at the 2 position with phenylalkylamino

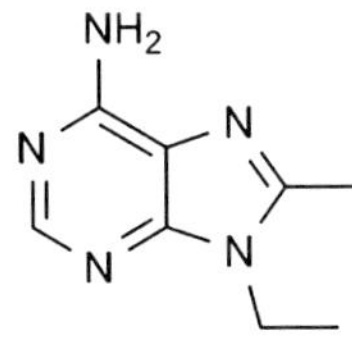

ANR 152, **83**

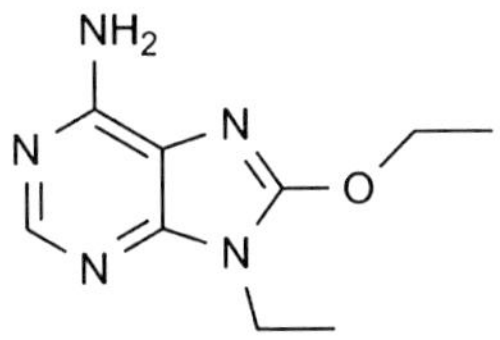

ANR 94, **84**

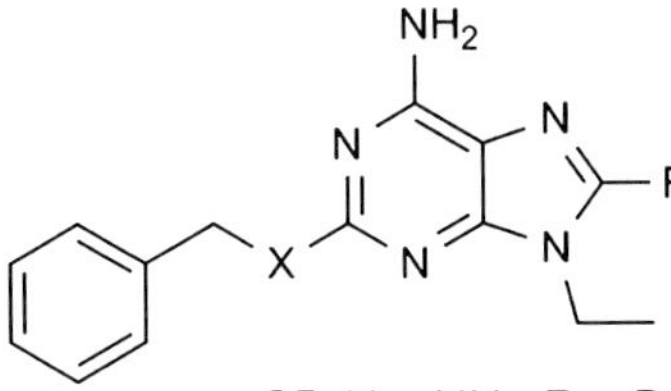

85 X = NH, R = Br
86 X = O, R = Br
87 X = NH, R = Furyl
88 X = O, R = Furyl

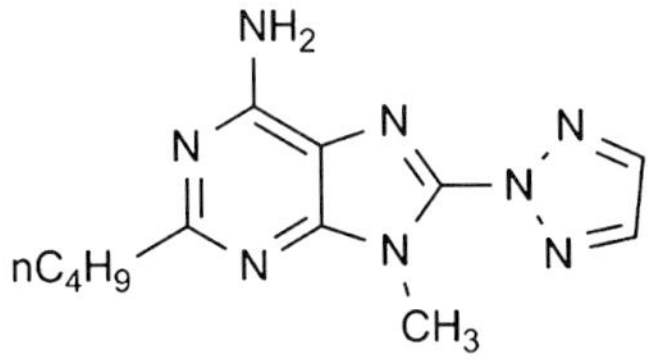

ST1535, **89**

Fig. 9 $A_{2A}AR$ antagonists: adenine derivatives

and phenylalkoxy groups and bearing a bromine atom in the 8 position (**85** and **86**, respectively) (Lambertucci et al. 2007b). This series was synthesized and tested in binding affinity assays at human ARs, and the new compounds showed good affinity and selectivity at $A_{2A}AR$. In particular, the introduction of a bromine atom at the 8 position increased the affinity of these compounds, leading to ligands with K_i values in the nanomolar range. Further substitution of the bromine atom of **85** and **86** with a 2-furyl group led to compounds **87** and **88** respectively, which maintained the $A_{2A}AR$ affinity at low nanomolar levels, but with reduced selectivity versus A_1AR and A_3AR (Cristalli et al., unpublished results).

A new series of 2,6-substituted 9-propyladenines has been recently synthesized and reported (Lambertucci et al. 2007a). Results show that the introduction of bulky chains at the N^6 position of 9-propyladenine significantly increases binding affinities at the human A_1AR and A_3AR, while the presence of a chlorine atom at the 2 position results in unequivocal effects depending on the receptor subtype and/or on the substituent present in the N^6 position. In any case, the presence in the 2 position of a chlorine atom favors the interaction with the $A_{2A}AR$ subtype. Among other adenine derivatives reported as $A_{2A}AR$ antagonists, ST1535 (2-*n*-butyl-9-methyl-8-[1,2,3]triazol-2-yl-9*H*-purin-6-ylamine, **89**, Fig. 9) (Minetti et al. 2005) proved to be quite potent but barely selective against A_1AR. Nevertheless, this compound was selected for in vivo studies and was shown to induce a dose-related increase in locomotor activity.

Slee and colleagues developed a series of aminopyrimidine derivatives that were acylated at the amino group (2-amino-*N*-pyrimidin-4-yl acetamides) and showed high water solubility (Slee et al. 2008c). The lead compound **90** was optimized with regard to replacement of the metabolically problematic furan ring (Slee et al. 2008a), reducing its effects on hERG channels (Slee et al. 2008b); it showed high affinity at

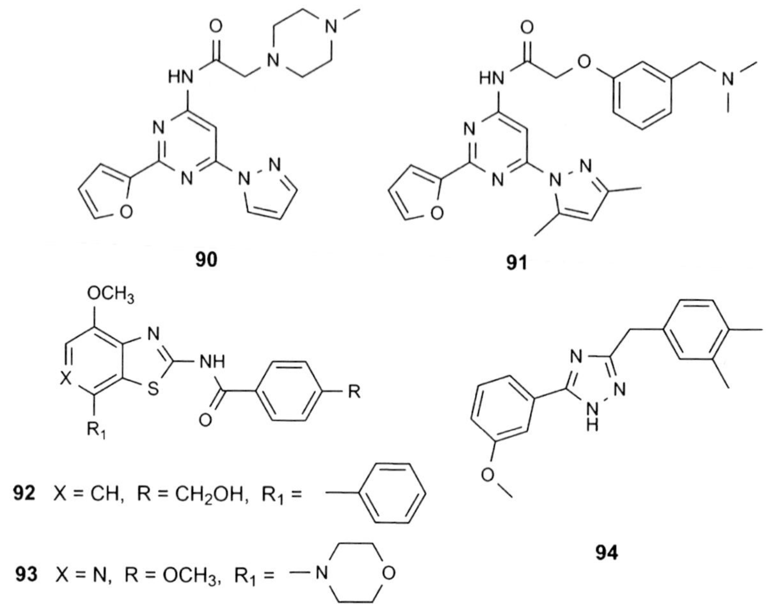

92 X = CH, R = CH_2OH, R_1 =

93 X = N, R = OCH_3, R_1 =

Fig. 10 Various heterocyclic compounds

both human and rodent $A_{2A}ARs$, as well as $A_{2A}AR$ selectivity (Zhang et al. 2008) and efficacy in rodent catalepsy models after peroral application, yielding **91** as a new lead structure (Fig. 10).

2.3 *Heterocyclic Compounds Unrelated to Adenine or Xanthine*

Simplified heterocyclic compounds, such as benzothiazole (Flhor and Riemer 2006) and 1,2,4-triazole (Alanine et al. 2004) derivatives (**92–94**), have been reported by the Roche group. These derivatives have been identified by high-throughput screening of compound libraries and are structurally related to neither xanthine nor to adenine derivatives. These compounds appear to be promising new lead compounds for the development of $A_{2A}AR$ antagonists for therapeutic applications (Müller and Ferré 2007).

2.4 *Antagonist Radioligands*

A number of $A_{2A}AR$ antagonist radioligands have been developed, and again they can be divided into xanthine and non-xanthine derivatives. Among the xanthine derivatives, three biotin conjugates of 1,3-dipropyl-8-phenylxanthine were reported in 1985 as being able to bind competitively to the ARs, but only in the absence of avidin. Results were interpreted in terms of the possible reorientation of the ligands at the receptor binding site (Jacobson et al. 1985). A few years later, a study on a radiolabeled amine-functionalized derivative of 1,3-dipropyl-8-phenylxanthine ([^{3}H]XAC) as an A_2 antagonist at human platelets was published. This molecule exhibited a K_d value at the nanomolar level, and it was reported as the first antagonist radioligand with high affinity at A_2ARs (Jacobson et al. 1986; Ukena et al. 1986). In the mid 1990s, the tritiated derivative of KF17837S (the equilibrium mixture of (*E*)- and (*Z*)-KF17837 isomers) was shown to bind to rat striatal membranes in a saturable and reversible way, with K_d values at low nanomolar concentrations (Nonaka et al. 1994b). In another study, ^{11}C-labeled (*E*)-KF17837 was synthesized and tested, and it was proposed as a potential positron emission tomography (PET) radioligand for mapping the A_{2A}ARs in the heart and the brain (Ishiwata et al. 1996, 1997). Further studies on radiolabeled xanthine derivatives as A_{2A}AR radioligands were carried out by preparing and testing an ^{11}C-labeled selective A_{2A}AR antagonist, (*E*)-8-(3-chlorostyryl)-1,3-dimethyl-7-[^{11}C]methylxanthine [^{11}C]CSC). This molecule was shown to accumulate in the striatum, and PET studies on rabbits showed a fast brain uptake of [^{11}C]CSC, reaching a maximum in less than 2 min (Marian et al. 1999). Few years later, iodinated and brominated styrylxanthine derivatives labeled with ^{11}C were tested as in vivo probes (Ishiwata et al. 2000c). [7-Methyl-^{11}C]-(*E*)-3,7-dimethyl-8-(3-iodostyryl)-1-propargylxanthine ([^{11}C]IS-DMPX) and [7-methyl-^{11}C]-(*E*)-8-(3-bromostyryl)-3,7-dimethyl-1-propargylxanthine ([^{11}C] BS-DMPX) showed K_i affinities of 8.9 and 7.7 nM respectively, and high A_{2A}AR/A_1AR selectivity values. Unfortunately, biological studies proved that the two ligands were only slightly concentrated in the striatum, and that the two compounds were not suitable as in vivo ligands because of low selectivity for the striatal A_{2A}ARs and a high nonspecific binding (Ishiwata et al. 2000c). Another A_{2A}AR antagonist radioligand was prepared, [^{3}H]3-(3-hydroxypropyl)-7-methyl-8-(*m*-methoxystyryl)-1-propargylxanthine ([^{3}H]MSX-2). This molecule showed high affinity (K_d = 8.0 nM) for A_{2A}AR, with saturable and reversible binding, and also a selectivity of at least two orders of magnitude versus all other AR subtypes (Müller et al. 2000). A very interesting xanthine derivative that acts as A_{2A}AR radioligand was found in [^{11}C]KF18446 ([7-methyl-^{11}C]-(*E*)-8-(3,4,5-trimethoxystyryl)-1,3,7-trimethylxanthine, also named (^{11}C)TMSX) (Ishiwata et al. 2000a, b, 2002, 2003a, b). Ex vivo autoradiography for this molecule showed a high striatal uptake and a high uptake ratio of the striatum in comparison to other brain regions; [^{11}C]KF18446 was therefore proposed as a suitable radioligand for mapping A_{2A}ARs of the brain by PET (Mishina et al. 2007). In 2001, the synthesis and the testing of [^{11}C]KW-6002 as a PET ligand was reported. This molecule showed high retention in the striatum, but it also bound to extrastriatal regions, so its potential as a PET ligand appeared to require further investigation (Hirani et al. 2001).

Among nonxanthine derivatives, in 1995 the synthesis of [^{125}I]-4-(2-[[7-amino-(2-furyl)[1,2,4]-triazolo[2,3-*a*][1,3,5]triazin-5-yl]amino]ethyl)phenol ([^{125}I]ZM241

385) and its characterization as a radioligand in A_{2A}AR-expressing membranes was reported (Palmer et al. 1995). This molecule proved to be a highly selective antagonist radioligand for studying A_{2A}ARs within some species. [^{3}H]ZM241385 showed A_{2A}AR affinity at subnanomolar levels (Alexander and Millns 2001; DeMet and Chicz-DeMet 2002; Kelly et al. 2004; Uustare et al. 2005) and, as reported above, it later also proved to be a high-affinity ligand for A_{2B}AR receptors, and is actually used in radioligand binding studies of this receptor subtype (Ji and Jacobson 1999). Another important A_{2A}AR antagonist radioligand was obtained with [^{3}H]SCH-58261, which showed a K_d value of about 1 nM (Zocchi et al. 1996b). Biological results showed that this compound directly labels striatal A_{2A}ARs in vivo, and it could be an excellent tool for studying A_{2A}AR brain distribution and its occupancy of various antagonists. Additional studies suggested that [^{3}H]SCH-58261 is a useful tool for autoradiography studies, and indicated that it was the first available radioligand for the characterization of the A_{2A}AR subtype in platelets (Dionisotti et al. 1996, 1997; Zocchi et al. 1996b; Fredholm et al. 1998; El Yacoubi et al. 2001).

References

Abiru T, Miyashita T, Watanabe Y, Yamaguchi T, Machida H, Matsuda A (1992) Nucleosides and nucleotides. 107. 2-(cycloalkylalkynyl)adenosines: adenosine A_2 receptor agonists with potent antihypertensive effects. J Med Chem 35:2253–2260

Abiru T, Endo K, Machida H (1995) Differential vasodilatory action of 2-octynyladenosine (YT-146), an adenosine A_2 receptor agonist, in the isolated rat femoral artery and vein. Eur J Pharmacol 281:9–15

Akkari R, Burbiel JC, Hockemeyer J, Müller CE (2006) Recent progress in the development of adenosine receptor ligands as antiinflammatory drugs. Curr Top Med Chem 6:1375–1399

Alanine A, Anselm L, Steward L, Thomi S, Vifian W, Groaning MD (2004) Synthesis and SAR evaluation of 1,2,4-triazoles as A_{2A} receptor antagonists. Bioorg Med Chem Lett 14:817–821

Alexander SP, Millns PJ (2001) [^{3}H]ZM241385—an antagonist radioligand for adenosine A_{2A} receptors in rat brain. Eur J Pharmacol 411:205–210

Baraldi PG, Manfredini S, Simoni D, Zappaterra L, Zocchi C, Dionisotti S, Ongini E (1994) Synthesis of new pyrazolo[4,3-e]1,2,4-triazolo[1,5-c] pyrimidine and 1,2,3-triazolo[4, 5-e]1,2,4-triazolo[1,5-c] pyrimidine displaying potent and selective activity as A_{2A} adenosine receptor antagonists. Bioorg Med Chem Lett 4:2539–2544

Baraldi PG, Cacciari B, Spalluto G, Pineda de las Infantas y Villatoro MJ, Zocchi C, Dionisotti S, Ongini E (1996a) Pyrazolo[4,3-*e*]-1,2,4-triazolo[1,5-*c*]pyrimidine derivatives: potent and selective A_{2A} adenosine antagonists. J Med Chem 39:1164–1171

Baraldi PG, Cacciari B, Spalluto G, Pineda de las Infants MJ, Zocchi C, Ferrara S, Dionisotti S, Ferra DS (1996b) 1,2,3-Triazolo[5,4-*e*]1,2,4-triazolo[1,5-*c*]pyrimidine derivatives: a new class of A_{2A} adenosine receptor antagonists. Farmaco 51:297–300

Baraldi PG, Cacciari B, Pineda de Las Infantas MJ, Romagnoli R, Spalluto G, Volpini R, Costanzi S, Vittori S, Cristalli G, Melman N, Park KS, Ji XD, Jacobson KA (1998a) Synthesis and biological activity of a new series of N^6-arylcarbamoyl, 2-(Ar)alkynyl-N^6-arylcarbamoyl, and N^6-carboxamido derivatives of adenosine-5′-N-ethyluronamide as A_1 and A_3 adenosine receptor agonists. J Med Chem 41:3174–3185

Baraldi PG, Cacciari B, Spalluto G, Bergonzoni M, Dionisotti S, Ongini E, Varani K, Borea PA (1998b) Design, synthesis, and biological evaluation of a second generation of pyrazolo[4,

3-*e*]-1,2,4-triazolo[1,5-*c*]pyrimidines as potent and selective A_{2A} adenosine receptor antagonists. J Med Chem 41:2126–2133

Baraldi PG, Cacciari B, Romagnoli R, Spalluto G, Moro S, Klotz KN, Leung E, Varani K, Gessi S, Merighi S, Borea PA (2000) Pyrazolo[4,3-*e*]1,2,4-triazolo[1,5-*c*]pyrimidine derivatives as highly potent and selective human A_3 adenosine receptor antagonists: influence of the chain at the *N*(8) pyrazole nitrogen. J Med Chem 43:4768–4780

Baraldi PG, Cacciari B, Romagnoli R, Klotz KN, Spalluto G, Varani K, Gessi S, Merighi S, Borea PA (2001) Pyrazolo[4,3-*e*]1,2,4-triazolo[1,5-*c*]pyrimidine derivatives as adenosine receptor ligands: a starting point for searching for A_{2B} adenosine receptor antagonists. Drug Dev Res 53:225–235

Baraldi PG, Cacciari B, Romagnoli R, Spalluto G, Monopoli A, Ongini E, Varani K, Borea PA (2002) 7-Substituted 5-amino-2-(2-furyl)pyrazolo[4,3-*e*]-1,2,4-triazolo[1,5-*c*]pyrimidines as A_{2A} adenosine receptor antagonists: a study on the importance of modifications at the side chain on the activity and solubility. J Med Chem 45:115–126

Baraldi PG, Tabrizi MA, Gessi S, Borea PA (2008) Adenosine receptor antagonists: translating medicinal chemistry and pharmacology into clinical utility. Chem Rev 108:238–263

Beukers MW, Chang LC, von Frijtag Drabbe Kunzel JK, Mulder-Krieger T, Spanjersberg RF, Brussee J, IJzerman AP (2004) New, non-adenosine, high-potency agonists for the human adenosine A_{2B} receptor with an improved selectivity profile compared to the reference agonist *N*-ethylcarboxamidoadenosine. J Med Chem 47:3707–3709

Bilkei-Gorzo A, Abo-Salem OM, Hayallah AM, Michel K, Müller CE, Zimmer A (2008) Adenosine receptor subtype-selective antagonists in inflammation and hyperalgesia. Naunyn Schmiedebergs Arch Pharmacol 377:65–76

Bosch MP, Campos F, Niubo I, Rosell G, Diaz JL, Brea J, Loza MI, Guerrero A (2004) Synthesis and biological activity of new potential agonists for the human adenosine A_{2A} receptor. J Med Chem 47:4041–4053

Brambilla R, Cottini L, Fumagalli M, Ceruti S, Abbracchio MP (2003) Blockade of A_{2A} adenosine receptors prevents basic fibroblast growth factor-induced reactive astrogliosis in rat striatal primary astrocytes. Glia 43:190–194

Bruns RF (1980) Adenosine receptor activation in human fibroblasts: nucleoside agonists and antagonists. Can J Physiol Pharmacol 58:673–691

Bruns RF, Lu GH, Pugsley TA (1986) Characterization of the A_2 adenosine receptor labeled by [^{3}H]NECA in rat striatal membranes. Mol Pharmacol 29:331–346

Cacciari B, Pastorin G, Spalluto G (2003) Medicinal chemistry of A_{2A} adenosine receptor antagonists. Curr Top Med Chem 3:403–411

Camaioni E, Di Francesco E, Vittori S, Volpini R, Cristalli G (1997) Adenosine receptor agonists: synthesis and biological evaluation of the diastereoisomers of 2-(3-hydroxy-3-phenyl-1-propyn-1-yl)NECA. Bioorg Med Chem 5:2267–2275

Camaioni E, Costanzi S, Vittori S, Volpini R, Klotz KN, Cristalli G (1998) New substituted 9-alkylpurines as adenosine receptor ligands. Bioorg Med Chem 6:523–533

Carriba P, Ortiz O, Patkar K, Justinova Z, Stroik J, Themann A, Müller C, Woods AS, Hope BT, Ciruela F, Casado V, Canela EI, Lluis C, Goldberg SR, Moratalla R, Franco R, Ferré S (2007) Striatal adenosine A_{2A} and cannabinoid CB_1 receptors form functional heteromeric complexes that mediate the motor effects of cannabinoids. Neuropsychopharmacology 32:2249–2259

Caulkett PWR, Jones G, McPartlin M, Renshaw ND, Stewart SK, Wright B (1995) Adenine isosteres with bridgehead nitrogen. Part 1. Two independent syntheses of the [1,2,4]triazolo[1,5-*a*][1,3,5]triazine ring system leading to a range of substituents in the 2, 5 and 7 positions. J Chem Soc Perkin Trans 1:801

Chebib M, McKeveney D, Quinn RJ (2000) 1-Phenylpyrazolo[3,4-d]pyrimidines; structure–activity relationships for C6 substituents at A_1 and A_{2A} adenosine receptors. Bioorg Med Chem 8:2581–2590

Colotta V, Catarzi D, Varano F, Cecchi L, Filacchioni G, Martini C, Trincavelli L, Lucacchini A (1999) 4-Amino-6-benzylamino-1,2-dihydro-2-phenyl-1,2,4-triazolo [4,3-alpha]-quinoxalin-

1-one: a new A_{2A} adenosine receptor antagonist with high selectivity versus A_1 receptors. Arch Pharm 332:39–41

Colotta V, Catarzi D, Varano F, Cecchi L, Filacchioni G, Martini C, Trincavelli L, Lucacchini A (2000) 1,2,4-Triazolo[4,3-*a*]quinoxalin-1-one: a versatile tool for the synthesis of potent and selective adenosine receptor antagonists. J Med Chem 43:1158–1164

Colotta V, Catarzi D, Varano F, Filacchioni G, Martini C, Trincavelli L, Lucacchini A (2003) Synthesis of 4-amino-6-(hetero)arylalkylamino-1,2,4-triazolo[4,3-*a*]quinoxalin-1-one derivatives as potent A_{2A} adenosine receptor antagonists. Bioorg Med Chem 11:5509–5518

Cristalli G (2000) Patent no WO2001062768

Cristalli G, Grifantini M, Vittori S, Balduini W, Cattabeni F (1985) Adenosine and 2-chloadenosine deaza analogues as adenosine receptor agonists. Nucleosides Nucleotides 4:625–639

Cristalli G, Franchetti P, Grifantini M, Vittori S, Klotz KN, Lohse MJ (1988) Adenosine receptor agonists: synthesis and biological evaluation of 1-deaza analogues of adenosine derivatives. J Med Chem 31:1179–1183

Cristalli G, Eleuteri A, Vittori S, Volpini R, Lohse MJ, Klotz KN (1992) 2-Alkynyl derivatives of adenosine and adenosine-5′-*N*-ethyluronamide as selective agonists at A_2 adenosine receptors. J Med Chem 35:2363–2368

Cristalli G, Vittori S, Thompson RD, Padgett WL, Shi D, Daly JW, Olsson RA (1994a) Inhibition of platelet aggregation by adenosine receptor agonists. Naunyn–Schmiedeberg's Arch Pharmacol 349:644–650

Cristalli G, Volpini R, Vittori S, Camaioni E, Monopoli A, Conti A, Dionisotti S, Zocchi C, Ongini E (1994b) 2-Alkynyl derivatives of adenosine-5′-*N*-ethyluronamide: selective A_2 adenosine receptor agonists with potent inhibitory activity on platelet aggregation. J Med Chem 37:1720–1726

Cristalli G, Camaioni E, Vittori S, Volpini R, Borea PA, Conti A, Dionisotti S, Ongini E, Monopoli A (1995) 2-Aralkynyl and 2-heteroalkynyl derivatives of adenosine-5′-*N*-ethyluronamide as selective A_{2A} adenosine receptor agonists. J Med Chem 38:1462–1472

Cristalli G, Camaioni E, Di Francesco E, Volpini R, Vittori S (1996) Chemical and pharmacological profile of selective adenosine receptor agonists. In: Giardinà D, Piergentili S, Pigini M (eds) Perspectives in receptor research. Pharmaco Chemistry Library, Elsevier, Amsterdam, pp 165–180

Cristalli G, Camaioni E, Costanzi S, Vittori S, Volpini R, Klotz KN (1998) Characterization of potent ligands at human recombinant adenosine receptors. Drug Dev Res 45:176–181

Cristalli G, Costanzi S, Lambertucci C, Lupidi G, Vittori S, Volpini R, Camaioni E (2001) Adenosine deaminase: functional implications and different classes of inhibitors. Med Res Rev 21:105–128

Cristalli G, Lambertucci C, Taffi S, Vittori S, Volpini R (2003) Medicinal chemistry of adenosine A_{2A} receptor agonists. Curr Top Med Chem 3:387–401

Cristalli G, Cacciari B, Dal Ben D, Lambertucci C, Moro S, Spalluto G, Volpini R (2007) Highlights on the development of A_{2A} adenosine receptor agonists and antagonists. ChemMedChem 2:260–281

Cusack NJ, Hourani SM (1981) 5′-*N*-Ethylcarboxamidoadenosine: a potent inhibitor of human platelet aggregation. Br J Pharmacol 72:443–447

CVT (2008) CV Therapeutics and Astellas announce FDA approval for Lexiscan(TM) (regadenoson) injection. http://www.cvt.com/PressRelease.aspx?releaseID=1128317

Dall'Igna OP, Fett P, Gomes MW, Souza DO, Cunha RA, Lara DR (2007) Caffeine and adenosine A_{2A} receptor antagonists prevent beta-amyloid (25–35)-induced cognitive deficits in mice. Exp Neurol 203:241–245

Daly JW (1982) Adenosine receptors: targets for future drugs. J Med Chem 25:197–207

Daly JW, Padgett WL, Shamim MT (1986) Analogues of caffeine and theophylline: effect of structural alterations on affinity at adenosine receptors. J Med Chem 29:1305–1308

Daly JW, Hide I, Müller CE, Shamim M (1991) Caffeine analogs: structure–activity relationships at adenosine receptors. Pharmacology 42:309–321

Daly JW, Padgett WL, Secunda SI, Thompson RD, Olsson RA (1993) Structure-activity relationships for 2-substituted adenosines at A_1 and A_2 adenosine receptors. Pharmacology 46:91–100

de Zwart M, Link R, von Frijtag Drabbe Kunzel JK, Cristalli G, Jacobson KA, Townsend-Nicholson A, IJzerman AP (1998) A functional screening of adenosine analogues at the adenosine A_{2B} receptor: a search for potent agonists. Nucleosides Nucleotides 17:969–985

de Zwart M, Kourounakis A, Kooijman H, Spek AL, Link R, von Frijtag Drabbe Kunzel JK, IJzerman AP (1999a) 5′-*N*-substituted carboxamidoadenosines as agonists for adenosine receptors. J Med Chem 42:1384–1392

de Zwart M, Vollinga RC, Beukers MW, Sleegers DF, Von Frijtag Drabbe Künzel J, de Groote M, IJzerman AP (1999b) Potent antagonists for the human adenosine A_{2B} receptor. Derivatives of the triazolotriazine adenosine receptor antagonist ZM241385 with high affinity. Drug Dev Res 48:95–103

Del Giudice MR, Borioni A, Mustazza C, Gatta F, Dionisotti S, Zocchi C, Ongini E (1996) (*E*)-1-(Heterocyclyl or cyclohexyl)-2-[1,3,7-trisubstituted(xanthin-8-yl)]ethenes as adenosine A_{2A} receptors antagonists. Eur J Med Chem 31:59–63

DeMet EM, Chicz-DeMet A (2002) Localization of adenosine A_{2A}-receptors in rat brain with [^{3}H]ZM-241385. Naunyn–Schmiedeberg's Arch Pharmacol 366:478–481

Dionisotti S, Conti A, Sandoli D, Zocchi C, Gatta F, Ongini E (1994) Effects of the new A_2 adenosine receptor antagonist 8FB-PTP, an 8-substituted pyrazolo-triazolo-pyrimidine, on in vitro functional models. Br J Pharmacol 112:659–665

Dionisotti S, Ferrara S, Molta C, Zocchi C, Ongini E (1996) Labeling of A_{2A} adenosine receptors in human platelets by use of the new nonxanthine antagonist radioligand [^{3}H]SCH 58261. J Pharmacol Exp Ther 278:1209–1214

Dionisotti S, Ongini E, Zocchi C, Kull B, Arslan G, Fredholm BB (1997) Characterization of human A_{2A} adenosine receptors with the antagonist radioligand [^{3}H]-SCH 58261. Br J Pharmacol 121:353–360

Dowling JE, Vessels JT, Haque S, Chang HX, van Vloten K, Kumaravel G, Engber T, Jin X, Phadke D, Wang J, Ayyub E, Petter RC (2005) Synthesis of [1,2,4]triazolo[1,5-*a*]pyrazines as adenosine A_{2A} receptor antagonists. Bioorg Med Chem Lett 15:4809–4813

Drabczynska A, Schumacher B, Müller CE, Karolak-Wojciechowska J, Michalak B, Pekala E, Kiec-Kononowicz K (2003) Impact of the aryl substituent kind and distance from pyrimido[2, 1-*f*]purindiones on the adenosine receptor selectivity and antagonistic properties. Eur J Med Chem 38:397–402

Drabczynska A, Müller CE, Schumacher B, Hinz S, Karolak-Wojciechowska J, Michalak B, Pekala E, Kiec-Kononowicz K (2004) Tricyclic oxazolo[2,3-f]purinediones: potency as adenosine receptor ligands and anticonvulsants. Bioorg Med Chem 12:4895–4908

Drabczynska A, Müller CE, Lacher SK, Schumacher B, Karolak-Wojciechowska J, Nasal A, Kawczak P, Yuzlenko O, Pekala E, Kiec-Kononowicz K (2006) Synthesis and biological activity of tricyclic aryloimidazo-, pyrimido-, and diazepinopurinediones. Bioorg Med Chem 14:7258–7281

Drabczynska A, Müller CE, Schiedel A, Schumacher B, Karolak-Wojciechowska J, Fruzinski A, Zobnina W, Yuzlenko O, Kiec-Kononowicz K (2007) Phenylethyl-substituted pyrimido[2, 1-*f*]purinediones and related compounds: structure–activity relationships as adenosine A_1 and A_{2A} receptor ligands. Bioorg Med Chem 15:6956–6974

Eggbrecht H, Gossl M (2006) Regadenoson (CV Therapeutics/Astellas). Curr Opin Investig Drugs 7:264–271

El Yacoubi M, Ledent C, Parmentier M, Ongini E, Costentin J, Vaugeois JM (2001) In vivo labelling of the adenosine A_{2A} receptor in mouse brain using the selective antagonist [^{3}H]SCH 58261. Eur J Neurosci 14:1567–1570

Erickson RH, Hiner RN, Feeney SW, Blake PR, Rzeszotarski WJ, Hicks RP, Costello DG, Abreu ME (1991) 1,3,8-trisubstituted xanthines. Effects of substitution pattern upon adenosine receptor A_1/A_2 affinity. J Med Chem 34:1431–1435

Ferré S, Diamond I, Goldberg SR, Yao L, Hourani SM, Huang ZL, Urade Y, Kitchen I (2007) Adenosine A_{2A} receptors in ventral striatum, hypothalamus and nociceptive circuitry implications for drug addiction, sleep and pain. Prog Neurobiol 83:332–347

Ferré S, Ciruela F, Borycz J, Solinas M, Quarta D, Antoniou K, Quiroz C, Justinova Z, Lluis C, Franco R, Goldberg SR (2008) Adenosine A_1–A_{2A} receptor heteromers: new targets for caffeine in the brain. Front Biosci 13:2391–2399

Flhor A, Riemer C (2006) Patent no. WO 200507592

Francis JE, Cash WD, Psychoyos S, Ghai G, Wenk P, Friedmann RC, Atkins C, Warren V, Furness P, Hyun JL, Stone GA, Desai M, Williams M (1988) Structure–activity profile of a series of novel triazoloquinazoline adenosine antagonists. J Med Chem 31:1014–1020

Francis JE, Webb RL, Ghai GR, Hutchison AJ, Moskal MA, deJesus R, Yokoyama R, Rovinski SL, Contardo N, Dotson R, Barclay B, Stone GA, Jarvis MF (1991) Highly selective adenosine A_2 receptor agonists in a series of *N*-alkylated 2-aminoadenosines. J Med Chem 34:2570–2579

Fredholm BB, Lindstrom K, Dionisotti S, Ongini E (1998) [^{3}H]SCH 58261, a selective adenosine A_{2A} receptor antagonist, is a useful ligand in autoradiographic studies. J Neurochem 70:1210–1216

Fredholm BB, Bättig K, Holmén J, Nehlig A, Zvartau EE (1999) Actions of caffeine in the brain with special reference to factors that contribute to its widespread use. Pharmacol Rev 51:83–133

Fuxe K, Marcellino D, Genedani S, Agnati L (2007) Adenosine A_{2A} receptors, dopamine D_2 receptors and their interactions in Parkinson's disease. Mov Disord 22:1990–2017

Gao Z, Li Z, Baker SP, Lasley RD, Meyer S, Elzein E, Palle V, Zablocki JA, Blackburn B, Belardinelli L (2001) Novel short-acting A_{2A} adenosine receptor agonists for coronary vasodilation: inverse relationship between affinity and duration of action of A_{2A} agonists. J Pharmacol Exp Ther 298:209–218

Gao ZG, Mamedova LK, Chen P, Jacobson KA (2004) 2-Substituted adenosine derivatives: affinity and efficacy at four subtypes of human adenosine receptors. Biochem Pharmacol 68:1985–1993

Gatta F, Del Giudice MR, Borioni A, Borea PA, Dionisotti S, Ongini E (1993) Synthesis of imidazo[1,2-*c*]pyrazolo[4,3-e]pyrimidines, pyrazolo[4,3-*e*]1,2,4-triazolo[1,5-*c*]pyrimidines and 1,2,4-triazolo[5,1-*i*]purines: new potent adenosine A_2 receptor antagonists. Eur J Med Chem 28:569–577

Gessi S, Varani K, Merighi S, Ongini E, Borea PA (2000) A_{2A} adenosine receptors in human peripheral blood cells. Br J Pharmacol 129:2–11

Gonzalez MP, Teran C, Teijeira M, Gonzalez-Moa MJ (2005) GETAWAY descriptors to predicting A_{2A} adenosine receptors agonists. Eur J Med Chem 40:1080–1086

Gordi T (2006) Patent no. WO2006044856

Grahner B, Winiwarter S, Lanzner W, Müller CE (1994) Synthesis and structure–activity relationships of deazaxanthines: analogs of potent A_1- and A_2-adenosine receptor antagonists. J Med Chem 37:1526–1534

Hasan A, Hussain T, Mustafa SJ, Srivastava PC (1994) 2-Substituted thioadenine nucleoside and nucleotide analogues: synthesis and receptor subtype binding affinities (1). Bioconjug Chem 5:364–369

Hirani E, Gillies J, Karasawa A, Shimada J, Kase H, Opacka-Juffry J, Osman S, Luthra SK, Hume SP, Brooks DJ (2001) Evaluation of [4-O-methyl-^{11}C]KW-6002 as a potential PET ligand for mapping central adenosine A_{2A} receptors in rats. Synapse 42:164–176

Hockemeyer J, Burbiel JC, Müller CE (2004) Multigram-scale syntheses, stability, and photoreactions of A_{2A} adenosine receptor antagonists with 8-styrylxanthine structure: potential drugs for Parkinson's disease. J Org Chem 69:3308–3318

Holschbach MH, Bier D, Stusgen S, Wutz W, Sihver W, Coenen HH, Olsson RA (2006) Synthesis and evaluation of 7-amino-2-(2(3)-furyl)-5-phenylethylamino-oxazolo[5,4-d]pyrimidines as potential A_{2A} adenosine receptor antagonists for positron emission tomography (PET). Eur J Med Chem 41:7–15

Homma H, Watanabe Y, Abiru T, Murayama T, Nomura Y, Matsuda A (1992) Nucleosides and nucleotides. 112. 2-(1-Hexyn-1-yl)adenosine-5′-uronamides: a new entry of selective A_2 adenosine receptor agonists with potent antihypertensive activity. J Med Chem 35:2881–2890

Hutchison AJ, Webb RL, Oei HH, Ghai GR, Zimmerman MB, Williams M (1989) CGS 21680C, an A_2 selective adenosine receptor agonist with preferential hypotensive activity. J Pharmacol Exp Ther 251:47–55

Hutchison AJ, Williams M, de Jesus R, Yokoyama R, Oei HH, Ghai GR, Webb RL, Zoganas HC, Stone GA, Jarvis MF (1990) 2-(Arylalkylamino)adenosin-5′-uronamides: a new class of highly selective adenosine A_2 receptor ligands. J Med Chem 33:1919–1924

IJzerman AP, Von Frijtag Drabbe Kuenzel JK, Vittori S, Cristalli G (1994) Purine-substituted adenosine derivatives with small N^6-substituents as adenosine receptor agonists. Nucleosides Nucleotides 13:2267–2281

IJzerman AP, van der Wenden NM (1997) Modulators of adenosine uptake, release, and inactivation. In: Jacobson KA, Jarvis MF (eds) Purinergic approaches in experimental therapeutics. Wiley-Liss, New York, pp 129–148

Impagnatiello F, Bastia E, Ongini E, Monopoli A (2000) Adenosine receptors in neurological disorders. Emerg Ther Targets 4:635–663

Ishiwata K, Noguchi J, Toyama H, Sakiyama Y, Koike N, Ishii S, Oda K, Endo K, Suzuki F, Senda M (1996) Synthesis and preliminary evaluation of [^{11}C]KF17837, a selective adenosine A_{2A} antagonist. Appl Radiat Isot 47:507–511

Ishiwata K, Sakiyama Y, Sakiyama T, Shimada J, Toyama H, Oda K, Suzuki F, Senda M (1997) Myocardial adenosine A_{2A} receptor imaging of rabbit by PET with [^{11}C]KF17837. Ann Nucl Med 11:219–225

Ishiwata K, Noguchi J, Wakabayashi S, Shimada J, Ogi N, Nariai T, Tanaka A, Endo K, Suzuki F, Senda M (2000a) ^{11}C-labeled KF18446: a potential central nervous system adenosine A_{2A} receptor ligand. J Nucl Med 41:345–354

Ishiwata K, Ogi N, Shimada J, Nonaka H, Tanaka A, Suzuki F, Senda M (2000b) Further characterization of a CNS adenosine A_{2A} receptor ligand [^{11}C]KF18446 with in vitro autoradiography and in vivo tissue uptake. Ann Nucl Med 14:81–89

Ishiwata K, Shimada J, Wang WF, Harakawa H, Ishii S, Kiyosawa M, Suzuki F, Senda M (2000c) Evaluation of iodinated and brominated [^{11}C]styrylxanthine derivatives as in vivo radioligands mapping adenosine A_{2A} receptor in the central nervous system. Ann Nucl Med 14:247–253

Ishiwata K, Ogi N, Hayakawa N, Oda K, Nagaoka T, Toyama H, Suzuki F, Endo K, Tanaka A, Senda M (2002) Adenosine A_{2A} receptor imaging with [^{11}C]KF18446 PET in the rat brain after quinolinic acid lesion: comparison with the dopamine receptor imaging. Ann Nucl Med 16:467–475

Ishiwata K, Kawamura K, Kimura Y, Oda K, Ishii K (2003a) Potential of an adenosine A_{2A} receptor antagonist [^{11}C]TMSX for myocardial imaging by positron emission tomography: a first human study. Ann Nucl Med 17:457–462

Ishiwata K, Wang WF, Kimura Y, Kawamura K, Ishii K (2003b) Preclinical studies on [^{11}C]TMSX for mapping adenosine A_{2A} receptors by positron emission tomography. Ann Nucl Med 17:205–211

Jackson EK, Herzer WA, Suzuki F (1993) KF17837 is an A_2 adenosine receptor antagonist in vivo. J Pharmacol Exp Ther 267:1304–1310

Jacobson KA, Gao ZG (2006) Adenosine receptors as therapeutic targets. Nat Rev Drug Discov 5:247–264

Jacobson KA, Kirk KL, Padgett W, Daly JW (1985) Probing the adenosine receptor with adenosine and xanthine biotin conjugates. FEBS Lett 184:30–35

Jacobson KA, Ukena D, Kirk KL, Daly JW (1986) [^{3}H]Xanthine amine congener of 1,3-dipropyl-8-phenylxanthine: an antagonist radioligand for adenosine receptors. Proc Natl Acad Sci USA 83:4089–4093

Jacobson KA, Gallo-Rodriguez C, Melman N, Fischer B, Maillard M, van Bergen A, van Galen PJ, Karton Y (1993a) Structure–activity relationships of 8-styrylxanthines as A_2-selective adenosine antagonists. J Med Chem 36:1333–1342

Jacobson KA, Shi D, Gallo-Rodriguez C, Manning M, Jr., Müller C, Daly JW, Neumeyer JL, Kiriasis L, Pfleiderer W (1993b) Effect of trifluoromethyl and other substituents on activity of xanthines at adenosine receptors. J Med Chem 36:2639–2644

Jacobson KA, Siddiqi SM, Olah ME, Ji XD, Melman N, Bellamkonda K, Meshulam Y, Stiles GL, Kim HO (1995) Structure–activity relationships of 9-alkyladenine and ribose-modified adenosine derivatives at rat A_3 adenosine receptors. J Med Chem 38:1720–1735

Jacobson KA, Ji X, Li AH, Melman N, Siddiqui MA, Shin KJ, Marquez VE, Ravi RG (2000) Methanocarba analogues of purine nucleosides as potent and selective adenosine receptor agonists. J Med Chem 43:2196–2203

Jarvis MF, Schulz R, Hutchison AJ, Do UH, Sills MA, Williams M (1989) [^{3}H]CGS 21680, a selective A_2 adenosine receptor agonist directly labels A_2 receptors in rat brain. J Pharmacol Exp Ther 251:888–893

Ji XD, Jacobson KA (1999) Use of the triazolotriazine [^{3}H]ZM 241385 as a radioligand at recombinant human A_{2B} adenosine receptors. Drug Des Discov 16:217–226

Kalda A, Yu L, Oztas E, Chen JF (2006) Novel neuroprotection by caffeine and adenosine A_{2A} receptor antagonists in animal models of Parkinson's disease. J Neurol Sci 248:9–15

Kelly M, Bailey A, Ledent C, Kitchen I, Hourani S (2004) Characterization of [^{3}H]ZM 241385 binding in wild-type and adenosine A_{2A} receptor knockout mice. Eur J Pharmacol 504:55–59

Kiec-Kononowicz K, Drabczynska A, Pekala E, Michalak B, Müller CE, Schumacher B, Karolak-Wojciechowska J, Duddeck H, Rockitt S, Wartchow R (2001) New developments in A_1 and A_2 adenosine receptor antagonists. Pure Appl Chem 73:1411–1420

Kim YC, Ji XD, Jacobson KA (1996) Derivatives of the triazoloquinazoline adenosine antagonist (CGS15943) are selective for the human A_3 receptor subtype. J Med Chem 39:4142–4148

Kiselgof E, Tulshian DB, Arik L, Zhang H, Fawzi A (2005) 6-(2-Furanyl)-9*H*-purin-2-amine derivatives as A_{2A} adenosine antagonists. Bioorg Med Chem Lett 15:2119–2122

Klotz KN (2000) Adenosine receptors and their ligands. Naunyn–Schmiedeberg's Arch Pharmacol 362:382–391

Klotz KN, Lohse MJ, Schwabe U, Cristalli G, Vittori S, Grifantini M (1989) 2-Chloro-N^6-[^{3}H]cyclopentyladenosine ([^{3}H]CCPA)—a high affinity agonist radioligand for A_1 adenosine receptors. Naunyn–Schmiedeberg's Arch Pharmacol 340:679–683

Klotz KN, Hessling J, Hegler J, Owman C, Kull B, Fredholm BB, Lohse MJ (1998) Comparative pharmacology of human adenosine receptor subtypes—characterization of stably transfected receptors in CHO cells. Naunyn-Schmiedeberg's Arch Pharmacol 357:1–9

Klotz KN, Camaioni E, Volpini R, Kachler S, Vittori S, Cristalli G (1999) 2-Substituted *N*-ethylcarboxamidoadenosine derivatives as high-affinity agonists at human A_3 adenosine receptors. Naunyn–Schmiedeberg's Arch Pharmacol 360:103–108

Klotz KN, Kachler S, Lambertucci C, Vittori S, Volpini R, Cristalli G (2003) 9-Ethyladenine derivatives as adenosine receptor antagonists: 2- and 8-substitution results in distinct selectivities. Naunyn–Schmiedeberg's Arch Pharmacol 367:629–634

Knutsen LJ, Weiss SM (2001) KW-6002 (Kyowa Hakko Kogyo). Curr Opin Investig Drugs 2:668–673

Knutsen LJ, Lau J, Petersen H, Thomsen C, Weis JU, Shalmi M, Judge ME, Hansen AJ, Sheardown MJ (1999) *N*-substituted adenosines as novel neuroprotective A_1 agonists with diminished hypotensive effects. J Med Chem 42:3463–3477

Lambertucci C, Volpini R, Costanzi S, Taffi S, Vittori S, Cristalli G (2003) 2-Phenylhydroxypropynyladenosine derivatives as high potent agonists at A_{2B} adenosine receptor subtype. Nucleosides Nucleotides Nucleic Acids 22:809–812

Lambertucci C, Cristalli G, Dal Ben D, Kachare DD, Bolcato C, Klotz KN, Spalluto G, Volpini R (2007a) New 2,6,9-trisubstituted adenines as adenosine receptor antagonists: a preliminary SAR profile. Purinergic Signal 3:339–346

Lambertucci C, Vittori S, Mishra RC, Dal Ben D, Klotz KN, Volpini R, Cristalli G (2007b) Synthesis and biological activity of trisubstituted adenines as A_{2A} adenosine receptor antagonists. Nucleosides Nucleotides Nucleic Acids 26:1443–1446

Ledent C, Vaugeois JM, Schiffmann SN, Pedrazzini T, El Yacoubi M, Vanderhaeghen JJ, Costentin J, Heath JK, Vassart G, Parmentier M (1997) Aggressiveness, hypoalgesia and high blood pressure in mice lacking the adenosine A_{2A} receptor. Nature 388:674–678

Lohse MJ, Klotz KN, Schwabe U, Cristalli G, Vittori S, Grifantini M (1988) 2-Chloro-N^6-cyclopentyladenosine: a highly selective agonist at A_1 adenosine receptors. Naunyn–Schmiedeberg's Arch Pharmacol 337:687–689

Marian T, Boros I, Lengyel Z, Balkay L, Horvath G, Emri M, Sarkadi E, Szentmiklosi AJ, Fekete I, Tron L (1999) Preparation and primary evaluation of [^{11}C]CSC as a possible tracer for mapping adenosine A_{2A} receptors by PET. Appl Radiat Isot 50:887–893

Massip S, Guillon J, Bertarelli D, Bosc JJ, Leger JM, Lacher S, Bontemps C, Dupont T, Müller CE, Jarry C (2006) Synthesis and preliminary evaluation of new 1- and 3-[1-(2-hydroxy-3-phenoxypropyl)]xanthines from 2-amino-2-oxazolines as potential A_1 and A_{2A} adenosine receptor antagonists. Bioorg Med Chem 14:2697–2719

Matasi JJ, Caldwell JP, Hao J, Neustadt B, Arik L, Foster CJ, Lachowicz J, Tulshian DB (2005) The discovery and synthesis of novel adenosine receptor (A_{2A}) antagonists. Bioorg Med Chem Lett 15:1333–1336

Mathot RA, Van der Wenden EM, Soudijn W, IJzerman AP, Danhof M (1995) Deoxyribose analogues of N^6-cyclopentyladenosine (CPA): partial agonists at the adenosine A_1 receptor in vivo. Br J Pharmacol 116:1957–1964

Matova MM, Nacheva RN, Boicheva SV (1997) QSAR analysis of 2-alkyloxy and 2-aralkyloxy adenosine A_1- and A_2-agonists. Eur J Med Chem 32:505–513

Matsuda A, Shinozaki M, Yamaguchi T, Homma H, Nomoto R, Miyasaka T, Watanabe Y, Abiru T (1992) Nucleosides and nucleotides. 103. 2-Alkynyladenosines: a novel class of selective adenosine A_2 receptor agonists with potent antihypertensive effects. J Med Chem 35:241–252

Melani A, Gianfriddo M, Vannucchi MG, Cipriani S, Baraldi PG, Giovannini MG, Pedata F (2006) The selective A_{2A} receptor antagonist SCH 58261 protects from neurological deficit, brain damage and activation of p38 MAPK in rat focal cerebral ischemia. Brain Res 1073–1074:470–480

Mihara T, Mihara K, Yarimizu J, Mitani Y, Matsuda R, Yamamoto H, Aoki S, Akahane A, Iwashita A, Matsuoka N (2007) Pharmacological characterization of a novel, potent adenosine A_1 and A_{2A} receptor dual antagonist, 5-[5-amino-3-(4-fluorophenyl)pyrazin-2-yl]-1-isopropylpyridine-2(1H)-one (ASP5854), in models of Parkinson's disease and cognition. J Pharmacol Exp Ther 323:708–719

Minetti P, Tinti MO, Carminati P, Castorina M, Di Cesare MA, Di Serio S, Gallo G, Ghirardi O, Giorgi F, Giorgi L, Piersanti G, Bartoccini F, Tarzia G (2005) 2-*n*-Butyl-9-methyl-8-[1,2,3]triazol-2-yl-9H-purin-6-ylamine and analogues as A_{2A} adenosine receptor antagonists. Design, synthesis, and pharmacological characterization. J Med Chem 48:6887–6896

Mishina M, Ishiwata K, Kimura Y, Naganawa M, Oda K, Kobayashi S, Katayama Y, Ishii K (2007) Evaluation of distribution of adenosine A_{2A} receptors in normal human brain measured with [^{11}C]TMSX PET. Synapse 61:778–784

Monopoli A, Conti A, Zocchi C, Casati C, Volpini R, Cristalli G, Ongini E (1994) Pharmacology of the new selective A_{2A} adenosine receptor agonist 2-hexynyl-5′-*N*-ethylcarboxamidoadenosine. Arzneimittelforschung 44:1296–1304

Monopoli A, Lozza G, Forlani A, Mattavelli A, Ongini E (1998) Blockade of adenosine A_{2A} receptors by SCH 58261 results in neuroprotective effects in cerebral ischaemia in rats. Neuroreport 9:3955–3959

Moro S, Gao ZG, Jacobson KA, Spalluto G (2006) Progress in the pursuit of therapeutic adenosine receptor antagonists. Med Res Rev 26:131–159

Müller CE (2000a) Adenosine receptor ligands—recent developments, part I. Agonists. Curr Med Chem 7:1269–1288

Müller CE (2000b) A_{2A} Adenosine receptor antagonists—future drugs for Parkinson's disease? Drugs Fut 25:1043

Müller CE, Ferré S (2007) Blocking striatal adenosine A_{2A} receptors: a new strategy for basal ganglia disorders. Recent Patents CNS Drug Discov 2:1–21

Müller CE, Scior T (1993) Adenosine receptors and their modulators. Pharm Acta Helv 68:77–111

Müller CE, Stein B (1996) Adenosine receptor antagonists: structures and potential therapeutic applications. Curr Pharm Des 2:501–530

Müller CE, Geis U, Hipp J, Schobert U, Frobenius W, Pawlowski M, Suzuki F, Sandoval-Ramirez J (1997a) Synthesis and structure–activity relationships of 3,7-dimethyl-1-propargylxanthine derivatives, A_{2A}-selective adenosine receptor antagonists. J Med Chem 40:4396–4405

Müller CE, Sauer R, Geis U, Frobenius W, Talik P, Pawlowski M (1997b) Aza-analogs of 8-styrylxanthines as A_{2A}-adenosine receptor antagonists. Arch Pharm 330:181–189

Müller CE, Schobert U, Hipp J, Geis U, Frobenius W, Pawlowski M (1997c) Configurationally stable analogs of styrylxanthines as A_{2A} adenosine receptor antagonist. Eur J Med Chem 32:709–719

Müller CE, Deters D, Dominik A, Pawlowski M (1998a) Syntheses of paraxanthine and isoparaxanthine analogs (1,7- and 1,9-substituted xanthine derivatives). Synthesis 93:1428–1436

Müller CE, Sandoval-Ramirez J, Schobert U, Geis U, Frobenius W, Klotz KN (1998b) 8-(Sulfostyryl)xanthines: water-soluble A_{2A}-selective adenosine receptor antagonists. Bioorg Med Chem 6:707–719

Müller CE, Maurinsh J, Sauer R (2000) Binding of [^{3}H]MSX-2 (3-(3-hydroxypropyl)-7-methyl-8-(*m*-methoxystyryl)-1-propargylxanthine) to rat striatal membranes—a new, selective antagonist radioligand for A_{2A} adenosine receptors. Eur J Pharm Sci 10:259–265

Müller CE, Thorand M, Qurishi R, Diekmann M, Jacobson KA, Padgett WL, Daly JW (2002) Imidazo[2,1-*i*]purin-5-ones and related tricyclic water-soluble purine derivatives: potent A_{2A}- and A_3-adenosine receptor antagonists. J Med Chem 45:3440–3450

Murphree LJ, Marshall MA, Rieger JM, MacDonald TL, Linden J (2002) Human A_{2A} adenosine receptors: high-affinity agonist binding to receptor-G protein complexes containing Gbeta(4). Mol Pharmacol 61:455–462

Neustadt BR, Hao J, Lindo N, Greenlee WJ, Stamford AW, Tulshian D, Ongini E, Hunter J, Monopoli A, Bertorelli R, Foster C, Arik L, Lachowicz J, Ng K, Feng KI (2007) Potent, selective, and orally active adenosine A_{2A} receptor antagonists: arylpiperazine derivatives of pyrazolo[4,3-*e*]-1,2,4-triazolo[1,5-*c*]pyrimidines. Bioorg Med Chem Lett 17:1376–1380

Niiya K, Olsson RA, Thompson RD, Silvia SK, Ueeda M (1992a) 2-(N'-alkylidenehydrazino)adenosines: potent and selective coronary vasodilators. J Med Chem 35:4557–4561

Niiya K, Thompson RD, Silvia SK, Olsson RA (1992b) 2-(N'-aralkylidenehydrazino)adenosines: potent and selective coronary vasodilators. J Med Chem 35:4562–4566

Nonaka Y, Shimada J, Nonaka H, Koike N, Aoki N, Kobayashi H, Kase H, Yamaguchi K, Suzuki F (1993) Photoisomerization of a potent and selective adenosine A_2 antagonist, (*E*)-1,3-dipropyl-8-(3,4-dimethoxystyryl)-7-methylxanthine. J Med Chem 36:3731–3733

Nonaka H, Ichimura M, Takeda M, Nonaka Y, Shimada J, Suzuki F, Yamaguchi K, Kase H (1994a) KF17837 ((*E*)-8-(3,4-dimethoxystyryl)-1,3-dipropyl-7-methylxanthine), a potent and selective adenosine A_2 receptor antagonist. Eur J Pharmacol 267:335–341

Nonaka H, Mori A, Ichimura M, Shindou T, Yanagawa K, Shimada J, Kase H (1994b) Binding of [^{3}H]KF17837S, a selective adenosine A_2 receptor antagonist, to rat brain membranes. Mol Pharmacol 46:817–822

Ohno M, Gao ZG, Van Rompaey P, Tchilibon S, Kim SK, Harris BA, Gross AS, Duong HT, Van Calenbergh S, Jacobson KA (2004) Modulation of adenosine receptor affinity and intrinsic efficacy in adenine nucleosides substituted at the 2-position. Bioorg Med Chem 12:2995–3007

Palmer TM, Poucher SM, Jacobson KA, Stiles GL (1995) ^{125}I-4-(2-[7-Amino-2-[2-furyl][1,2,4]triazolo[2,3-*a*][1,3,5] triazin-5-yl-amino]ethyl)phenol, a high affinity antagonist radioligand selective for the A_{2A} adenosine receptor. Mol Pharmacol 48:970–974

Pedata F, Gianfriddo M, Turchi D, Melani A (2005) The protective effect of adenosine A_{2A} receptor antagonism in cerebral ischemia. Neurol Res 27:169–174

Peng H, Kumaravel G, Yao G, Sha L, Wang J, Van Vlijmen H, Bohnert T, Huang C, Vu CB, Ensinger CL, Chang H, Engber TM, Whalley ET, Petter RC (2004) Novel bicyclic piperazine derivatives of triazolotriazine and triazolopyrimidines as highly potent and selective adenosine A_{2A} receptor antagonists. J Med Chem 47:6218–6229

Petzer JP, Steyn S, Castagnoli KP, Chen JF, Schwarzschild MA, Van der Schyf CJ, Castagnoli N (2003) Inhibition of monoamine oxidase B by selective adenosine A_{2A} receptor antagonists. Bioorg Med Chem 11:1299–1310

Pfleger K, Seifen E, Schondorf H (1969) Inosine potentiation of the effect of adenosine on the heart. Biochem Pharmacol 18:43–51

Pinna A, Wardas J, Cristalli G, Morelli M (1997) Adenosine A_{2A} receptor agonists increase Fos-like immunoreactivity in mesolimbic areas. Brain Res 759:41–49

Pinna A, Volpini R, Cristalli G, Morelli M (2005) New adenosine A_{2A} receptor antagonists: actions on Parkinson's disease models. Eur J Pharmacol 512:157–164

Poucher SM, Keddie JR, Singh P, Stoggall SM, Caulkett PW, Jones G, Coll MG (1995) The in vitro pharmacology of ZM 241385, a potent, non-xanthine A_{2A} selective adenosine receptor antagonist. Br J Pharmacol 115:1096–1102

Prasad RN, Bariana DS, Fung A, Savic M, Tietje K, Stein HH, Brondyk H, Egan RS (1980) Modification of the 5′ position of purine nucleosides. 2. Synthesis and some cardiovascular properties of adenosine-5′-(*N*-substituted)carboxamides. J Med Chem 23:313–319

Richardson PJ, Kase H, Jenner PG (1997) Adenosine A_{2A} receptor antagonists as new agents for the treatment of Parkinson's disease. Trends Pharmacol Sci 18:338–344

Rieger JM, Brown ML, Sullivan GW, Linden J, Macdonald TL (2001) Design, synthesis, and evaluation of novel A_{2A} adenosine receptor agonists. J Med Chem 44:531–539

Salamone JD, Betz AJ, Ishiwari K, Felsted J, Madson L, Mirante B, Clark K, Font L, Korbey S, Sager TN, Hockemeyer J, Müller CE (2008) Tremorolytic effects of adenosine A2A antagonists: implications for parkinsonism. Front Biosci 13:3594–3605

Sarges R, Howard HR, Browne RG, Lebel LA, Seymour PA, Koe BK (1990) 4-Amino [1,2,4]triazolo[4,3-*a*]quinoxalines. A novel class of potent adenosine receptor antagonists and potential rapid-onset antidepressants. J Med Chem 33:2240–2254

Sauer R, Maurinsh J, Reith U, Fülle F, Klotz KN, Müller CE (2000) Water-soluble phosphate prodrugs of 1-propargyl-8-styrylxanthine derivatives, A_{2A}-selective adenosine receptor antagonists. J Med Chem 43:440–448

Schapira AH, Bezard E, Brotchie J, Calon F, Collingridge GL, Ferger B, Hengerer B, Hirsch E, Jenner P, Le Novere N, Obeso JA, Schwarzschild MA, Spampinato U, Davidai G (2006) Novel pharmacological targets for the treatment of Parkinson's disease. Nat Rev Drug Discov 5:845–854

Schwarzschild MA, Agnati L, Fuxe K, Chen JF, Morelli M (2006) Targeting adenosine A_{2A} receptors in Parkinson's disease. Trends Neurosci 29:647–654

Seale TW, Abla KA, Shamim MT, Carney JM, Daly JW (1988) 3,7-Dimethyl-1-propargylxanthine: a potent and selective in vivo antagonist of adenosine analogs. Life Sci 43:1671–1684

Shamim MT, Ukena D, Padgett WL, Daly JW (1989) Effects of 8-phenyl and 8-cycloalkyl substituents on the activity of mono-, di-, and trisubstituted alkylxanthines with substitution at the 1-, 3-, and 7-positions. J Med Chem 32:1231–1237

Shimada J, Koike N, Nonaka H, Shiozaki S, Yanagawa K, Kanda T, Kobayashi H, Ichimura M, Nakamura J, Kase H, Suzuki F (1997) Adenosine A_{2A} antagonists with potent anti-cataleptic activity. Bioorg Med Chem Lett 7:2349–2352

Siddiqi SM, Jacobson KA, Esker JL, Olah ME, Ji XD, Melman N, Tiwari KN, Secrist JA 3rd, Schneller SW, Cristalli G, et al. (1995) Search for new purine- and ribose-modified adenosine analogues as selective agonists and antagonists at adenosine receptors. J Med Chem 38:1174–1188

Silverman LS, Caldwell JP, Greenlee WJ, Kiselgof E, Matasi JJ, Tulshian DB, Arik L, Foster C, Bertorelli R, Monopoli A, Ongini E (2007) 3*H*-[1,2,4]-Triazolo[5,1-*i*]purin-5-amine derivatives as adenosine A_{2A} antagonists. Bioorg Med Chem Lett 17:1659–1662

Slee DH, Chen Y, Zhang X, Moorjani M, Lanier MC, Lin E, Rueter JK, Williams JP, Lechner SM, Markison S, Malany S, Santos M, Gross RS, Jalali K, Sai Y, Zuo Z, Yang C, Castro-Palomino

JC, Crespo MI, Prat M, Gual S, Diaz JL, Saunders J (2008a) 2-Amino-*N*-pyrimidin-4-ylacetamides as A_{2A} receptor antagonists: 1. Structure–activity relationships and optimization of heterocyclic substituents. J Med Chem 51:1719–1729

Slee DH, Moorjani M, Zhang X, Lin E, Lanier MC, Chen Y, Rueter JK, Lechner SM, Markison S, Malany S, Joswig T, Santos M, Gross RS, Williams JP, Castro-Palomino JC, Crespo MI, Prat M, Gual S, Diaz JL, Jalali K, Sai Y, Zuo Z, Yang C, Wen J, O'Brien Z, Petroski R, Saunders J (2008b) 2-Amino-*N*-pyrimidin-4-ylacetamides as A_{2A} receptor antagonists: 2. Reduction of hERG activity, observed species selectivity, and structure–activity relationships. J Med Chem 51:1730–1739

Slee DH, Zhang X, Moorjani M, Lin E, Lanier MC, Chen Y, Rueter JK, Lechner SM, Markison S, Malany S, Joswig T, Santos M, Gross RS, Williams JP, Castro-Palomino JC, Crespo MI, Prat M, Gual S, Diaz JL, Wen J, O'Brien Z, Saunders J (2008c) Identification of novel, water-soluble, 2-amino-*N*-pyrimidin-4-yl acetamides as A_{2A} receptor antagonists with in vivo efficacy. J Med Chem 51:400–406

Svenningsson P, Nomikos GG, Ongini E, Fredholm BB (1997) Antagonism of adenosine A_{2A} receptors underlies the behavioural activating effect of caffeine and is associated with reduced expression of messenger RNA for NGFI-A and NGFI-B in caudate-putamen and nucleus accumbens. Neuroscience 79:753–764

Takahashi RN, Pamplona FA, Prediger RD (2008) Adenosine receptor antagonists for cognitive dysfunction: a review of animal studies. Front Biosci 13:2614–2632

Taylor MD, Moos WH, Hamilton HW, Szotek DS, Patt WC, Badger EW, Bristol JA, Bruns RF, Heffner TG, Mertz TE (1986) Ribose-modified adenosine analogues as adenosine receptor agonists. J Med Chem 29:346–353

Thorsell A, Johnson J, Heilig M (2007) Effect of the adenosine A_{2A} receptor antagonist 3,7-dimethyl-propargylxanthine on anxiety-like and depression-like behavior and alcohol consumption in Wistar rats. Alcohol Clin Exp Res 31:1302–1307

Todde S, Moresco RM, Simonelli P, Baraldi PG, Cacciari B, Spalluto G, Varani K, Monopoli A, Matarrese M, Carpinelli A, Magni F, Kienle MG, Fazio F (2000) Design, radiosynthesis, and biodistribution of a new potent and selective ligand for in vivo imaging of the adenosine A_{2A} receptor system using positron emission tomography. J Med Chem 43:4359–4362

Ukena D, Jacobson KA, Kirk KL, Daly JW (1986) A [3H]amine congener of 1,3-dipropyl-8-phenylxanthine. A new radioligand for A_2 adenosine receptors of human platelets. FEBS Lett 199:269–274

Uustare A, Vonk A, Terasmaa A, Fuxe K, Rinken A (2005) Kinetic and functional properties of [^{3}H]ZM241385, a high affinity antagonist for adenosine A_{2A} receptors. Life Sci 76:1513–1526

van den Berg D, Zoellner KR, Ogunrombi MO, Malan SF, Terre'Blanche G, Castagnoli N Jr, Bergh JJ, Petzer JP (2007) Inhibition of monoamine oxidase B by selected benzimidazole and caffeine analogues. Bioorg Med Chem 15:3692–3702

van der Wenden EM, von Frijtag Drabbe Kunzel JK, Mathot RA, Danhof M, IJzerman AP, Soudijn W (1995) Ribose-modified adenosine analogues as potential partial agonists for the adenosine receptor. J Med Chem 38:4000–4006

van der Wenden EM, Carnielli M, Roelen HC, Lorenzen A, von Frijtag Drabbe Kunzel JK, IJzerman AP (1998) 5′-Substituted adenosine analogs as new high-affinity partial agonists for the adenosine A_1 receptor. J Med Chem 41:102–108

van Tilburg EW, Gremmen M, von Frijtag Drabbe Kunzel J, de Groote M, IJzerman AP (2003) 2,8-Disubstituted adenosine derivatives as partial agonists for the adenosine A_{2A} receptor. Bioorg Med Chem 11:2183–2192

Vittori S, Camaioni E, Di Francesco E, Volpini R, Monopoli A, Dionisotti S, Ongini E, Cristalli G (1996) 2-Alkenyl and 2-alkyl derivatives of adenosine and adenosine-5′-*N*-ethyluronamide: different affinity and selectivity of *E*- and *Z*-diastereomers at A_{2A} adenosine receptors. J Med Chem 39:4211–4217

Vittori S, Lorenzen A, Stannek C, Costanzi S, Volpini R, IJzerman AP, Kunzel JK, Cristalli G (2000) *N*-Cycloalkyl derivatives of adenosine and 1-deazaadenosine as agonists and partial agonists of the A_1 adenosine receptor. J Med Chem 43:250–260

Vittori S, Costanzi S, Lambertucci C, Portino FR, Taffi S, Volpini R, Klotz KN, Cristalli G (2004) A_{2B} adenosine receptor agonists: synthesis and biological evaluation of 2-phenylhydroxypropynyl adenosine and NECA derivatives. Nucleosides Nucleotides Nucl Acids 23:471–481

Vittori S, Volpini R, Lambertucci C, Taffi S, Klotz KN, Cristalli G (2005) 2-Substituted 5′-*N*-methylcarboxamidoadenosine (MECA) derivatives as A_3 adenosine receptor ligands. Nucleosides Nucleotides Nucl Acids 24:935–938

Viziano M, Ongini E, Conti A, Zocchi C, Seminati M, Pocar D (1995) 2-[*N*′-(3-Arylallylidene)hydrazino]adenosines showing A_{2A} adenosine agonist properties and vasodilation activity. J Med Chem 38:3581–3585

Vollmann K, Qurishi R, Hockemeyer J, Müller CE (2008) Synthesis and properties of a new water-soluble prodrug of the adenosine A_{2A} receptor antagonist MSX-2. Molecules 13:348–359

Volpini R, Camaioni E, Vittori S, Barboni L, Lambertucci C, Cristalli G (1998) Synthesis of new nucleosides by coupling of chloropurines with 2- and 3-deoxy derivatives of N-methyl-D-ribofuranuronamide. Helv Chim Acta 81:145–152

Volpini R, Camaioni E, Costanzi S, Vittori S, Klotz KN, Cristalli G (1999) Synthesis of di- and tri-substituted adenosine derivatives and their affinities at human adenosine receptor subtypes. Nucleosides Nucleotides 18:2511–2520

Volpini R, Costanzi S, Lambertucci C, Taffi S, Vittori S, Klotz KN, Cristalli G (2002) N^6-Alkyl-2-alkynyl derivatives of adenosine as potent and selective agonists at the human adenosine A_3 receptor and a starting point for searching A_{2B} ligands. J Med Chem 45:3271–3279

Volpini R, Costanzi S, Lambertucci C, Portino FR, Taffi S, Vittori S, Klotz KN, Cristalli G (2004) Adenosine receptor agonists: synthesis and binding affinity of 2-(aryl)alkylthioadenosine derivatives. ARKIVOC 301–311

Volpini R, Costanzi S, Lambertucci C, Vittori S, Martini C, Trincavelli ML, Klotz KN, Cristalli G (2005) 2- and 8-Alkynyl-9-ethyladenines: synthesis and biological activity at human and rat adenosine receptors. Purinergic Signal 1:173–181

Vu CB (2005) Recent advances in the design and optimization of adenosine A_{2A} receptor antagonists. Curr Opin Drug Discov Dev 8:458–468

Vu CB, Pan D, Peng B, Sha L, Kumaravel G, Jin X, Phadke D, Engber T, Huang C, Reilly J, Tam S, Petter RC (2004a) Studies on adenosine A_{2A} receptor antagonists: comparison of three core heterocycles. Bioorg Med Chem Lett 14:4831–4834

Vu CB, Peng B, Kumaravel G, Smits G, Jin X, Phadke D, Engber T, Huang C, Reilly J, Tam S, Grant D, Hetu G, Chen L, Zhang J, Petter RC (2004b) Piperazine derivatives of [1,2,4]triazolo[1,5-a][1,3,5]triazine as potent and selective adenosine A_{2A} receptor antagonists. J Med Chem 47:4291–4299

Vu CB, Shields P, Peng B, Kumaravel G, Jin X, Phadke D, Wang J, Engber T, Ayyub E, Petter RC (2004c) Triamino derivatives of triazolotriazine and triazolopyrimidine as adenosine A_{2A} receptor antagonists. Bioorg Med Chem Lett 14:4835–4838

Vu CB, Pan D, Peng B, Kumaravel G, Smits G, Jin X, Phadke D, Engber T, Huang C, Reilly J, Tam S, Grant D, Hetu G, Petter RC (2005) Novel diamino derivatives of [1,2,4]triazolo[1,5-a][1,3,5]triazine as potent and selective adenosine A_{2A} receptor antagonists. J Med Chem 48:2009–2018

Wan W, Sutherland GR, Geiger JD (1990) Binding of the adenosine A_2 receptor ligand [^{3}H]CGS 21680 to human and rat brain: evidence for multiple affinity sites. J Neurochem 55:1763–1771

Weiss SM, Benwell K, Cliffe IA, Gillespie RJ, Knight AR, Lerpiniere J, Misra A, Pratt RM, Revell D, Upton R, Dourish CT (2003) Discovery of nonxanthine adenosine A_{2A} receptor antagonists for the treatment of Parkinson's disease. Neurology 61:S101–S106

Williams M, Francis J, Ghai G, Braunwalder A, Psychoyos S, Stone GA, Cash WD (1987) Biochemical characterization of the triazoloquinazoline, CGS 15943, a novel, non-xanthine adenosine antagonist. J Pharmacol Exp Ther 241:415–420

Xu K, Bastia E, Schwarzschild M (2005) Therapeutic potential of adenosine A_{2A} receptor antagonists in Parkinson's disease. Pharmacol Ther 105:267–310

Yan L, Burbiel JC, Maass A, Müller CE (2003) Adenosine receptor agonists: from basic medicinal chemistry to clinical development. Expert Opin Emerg Drugs 8:537–576

Yang M, Soohoo D, Soelaiman S, Kalla R, Zablocki J, Chu N, Leung K, Yao L, Diamond I, Belardinelli L, Shryock JC (2007) Characterization of the potency, selectivity, and pharmacokinetic profile for six adenosine A_{2A} receptor antagonists. Naunyn–Schmiedeberg's Arch Pharmacol 375:133–144

Yao G, Haque S, Sha L, Kumaravel G, Wang J, Engber TM, Whalley ET, Conlon PR, Chang H, Kiesman WF, Petter RC (2005) Synthesis of alkyne derivatives of a novel triazolopyrazine as A_{2A} adenosine receptor antagonists. Bioorg Med Chem Lett 15:511–515

Yu L, Shen HY, Coelho JE, Araujo IM, Huang QY, Day YJ, Rebola N, Canas PM, Rapp EK, Ferrara J, Taylor D, Müller CE, Linden J, Cunha RA, Chen JF (2008) Adenosine A_{2A} receptor antagonists exert motor and neuroprotective effects by distinct cellular mechanisms. Ann Neurol 63:338–346

Yuzlenko O, Kiec-Kononowicz K (2006) Potent adenosine A_1 and A_{2A} receptors antagonists: recent developments. Curr Med Chem 13:3609–3625

Zablocki J, Palle V, Blackburn B, Elzein E, Nudelman G, Gothe S, Gao Z, Li Z, Meyer S, Belardinelli L (2001) 2-Substituted pi system derivatives of adenosine that are coronary vasodilators acting via the A_{2A} adenosine receptor. Nucleosides Nucleotides Nucl Acids 20:343–360

Zhang X, Rueter JK, Chen Y, Moorjani M, Lanier MC, Lin E, Gross RS, Tellew JE, Williams JP, Lechner SM, Markison S, Joswig T, Malany S, Santos M, Castro-Palomino JC, Crespo MI, Prat M, Gual S, Diaz JL, Saunders J, Slee DH (2008) Synthesis of N-pyrimidinyl-2-phenoxyacetamides as adenosine A_{2A} receptor antagonists. Bioorg Med Chem Lett 18: 1778–1783

Zocchi C, Ongini E, Conti A, Monopoli A, Negretti A, Baraldi PG, Dionisotti S (1996a) The non-xanthine heterocyclic compound SCH 58261 is a new potent and selective A_{2A} adenosine receptor antagonist. J Pharmacol Exp Ther 276:398–404

Zocchi C, Ongini E, Ferrara S, Baraldi PG, Dionisotti S (1996b) Binding of the radioligand [^{3}H]-SCH 58261, a new non-xanthine A_{2A} adenosine receptor antagonist, to rat striatal membranes. Br J Pharmacol 117:1381–1386

Recent Developments in A_{2B} Adenosine Receptor Ligands

Rao V. Kalla, Jeff Zablocki, Mojgan Aghazadeh Tabrizi, and Pier Giovanni Baraldi

Contents

Abstract A selective, high-affinity A_{2B} adenosine receptor (AR) antagonist will be useful as a pharmacological tool to help determine the role of the A_{2B}AR in inflammatory diseases and angiogenic diseases. Based on early A_{2B}AR-selective ligands with nonoptimal pharmaceutical properties, such as **15** (MRS 1754: $K_i(hA_{2B}) = 2\,nM$; $K_i(hA_1) = 403\,nM$; $K_i(hA_{2A}) = 503\,nM$, and $K_i(hA_3) = 570\,nM$), several groups have discovered second-generation A_{2B}AR ligands that are suitable for development. Scientists at CV Therapeutics have discovered the selective, high-affinity A_{2B}AR antagonist **22**, a 8-(4-pyrazolyl)-xanthine derivative, (CVT-6883, $K_i(hA_{2B}) = 22\,nM$; $K_i(hA_1) = 1{,}940\,nM$; $K_i(hA_{2A}) = 3{,}280$; and $K_i(hA_3) = 1{,}070\,nM$). Compound **22** has demonstrated favorable pharmacokinetic (PK) properties ($T_{1/2} = 4\,h$ and $F > 35\%$ rat), and it is a functional antagonist at

R.V. Kalla (✉)
Department of Medicinal Chemistry, CV Therapeutics Inc., 3172 Porter Drive, Palo Alto, CA 94304, USA
rao.kalla@cvt.com

C.N. Wilson and S.J. Mustafa (eds.), *Adenosine Receptors in Health and Disease*, Handbook of Experimental Pharmacology 193, DOI 10.1007/978-3-540-89615-9_4,

the $A_{2B}AR$ ($K_B = 6$ nM). In a mouse model of asthma, compound **22** demonstrated a dose-dependent efficacy supporting the role of the $A_{2B}AR$ in asthma. In two Phase I clinical trails, **22** (CVT-6883) was found to be safe, well tolerated, and suitable for once-daily dosing. Baraldi et al. have independently discovered a selective, high-affinity $A_{2B}AR$ antagonist, **30** (MRE2029F20), 8-(5-pyrazolyl)-xanthine ($K_i(hA_{2B}) = 5.5$ nM; $K_i(hA_1) = 200$ nM; $K_i(hA_{2A}, A_3) > 1{,}000$), that has been selected for development in conjunction with King Pharmaceuticals. Compound **30** has been demonstrated to be a functional antagonist of the $A_{2B}AR$, and it has been radiolabeled for use in pharmacological studies. A third compound, **58** (LAS-38096), is a 2-aminopyrimidine derivative (discovered by the Almirall group) that has high $A_{2B}AR$ affinity and selectivity ($K_i(hA_{2B}) = 17$ nM; $K_i(hA_1) > 1{,}000$ nM; $K_i(hA_{2A}) > 2{,}500$; and $K_i(hA_3) > 1{,}000$ nM), and **58** has been moved into preclinical safety testing. A fourth selective, high-affinity $A_{2B}AR$ antagonist, **54** (OSIP339391 $K_i(hA_{2B}) = 0.5$ nM; $K_i(hA_1) = 37$ nM; $K_i(hA_{2A}) = 328$; and $K_i(hA_3) = 450$ nM) was discovered by the OSI group. The three highly selective, high-affinity $A_{2B}AR$ antagonists that have been selected for development should prove useful in subsequent clinical trials that will establish the role of the $A_{2B}ARs$ in various disease states.

Keywords Adenosine receptor antagonist · Asthma · CVT-6883 · MRE2029F20 · LAS-38096

Abbreviations

AR	Adenosine receptor
BALF	Bronchoalveolar lavage fluid
BSMCs	Bronchial smooth muscle cells
cAMP	Cyclic adenosine monophosphate
CGS-21680	2-[*p*-(2-Carboxyethyl)phenylethylamino]-5′-*N*-ethylcarboxamidoad enosine
CPA	N^6-Cyclopentyladenosine
DAG	Diacylglycerol
HBECs	Human bronchial epithelial cells
HRECs	Human retinal endothelial cells
IL	Interleukin
IP3	(1,4,5)Inositol triphosphate
MCP-1	Monocyte chemotactic protein-1
NECA	5′-*N*-Ethylcarboxamidoadenosine
NIDDM	Noninsulin-dependent diabetesmellitus
NSAIDs	Nonsteroidal antiinflammatory drugs
PK	Pharmacokinetic
SAR	Structure–activity relationship
VEGF	Vascular endothelial growth factor

1 Introduction

The A_{2B} adenosine receptor (AR) is a member of the P_1 family of seven-transmembrane ARs, and it couples to G_s to increase cAMP and G_{q11} to increase (1,4,5)inositol triphosphate (IP_3)/diacylglycerol (DAG) (Fozard and McCarth 2002). The A_{2B}AR has been found to be located in smooth muscle cells of the vascular, intestinal, and bronchial tissue, chromaffin tissue, mast cells, and the brain. The goal of this review is to highlight the structure–affinity relationships (SAR) of A_{2B}AR antagonists that started with early lead compounds with nonoptimized pharmaceutical properties, which served as the genesis for second-generation compounds with high selectivity and affinity that have become development candidates. We will only briefly mention the major indications to demonstrate the potential utility of A_{2B}AR antagonists, since the potential indications of A_{2B}AR have been described in detail elsewhere (Feoktistov et al. 1998; Holgate ST 2005; Kurukulasuriya et al. 2003; Harada et al. 2001a; Hayallah et al. 2002). Although the A_{2B}AR antagonists described in this review are currently under investigation in order to fully define the role of the A_{2B}AR in disease states, early in vitro and in vivo experiments suggest that A_{2B}AR antagonists may be beneficial for the following diseases: asthma—A_{2B}AR mediates inflammatory cytokine release (Holgate ST 2005; Zhong et al. 2004, 2005, 2006); diabetes—A_{2B}AR mediates gluconeogenesis (Harada et al. 2001a, b; Kurukulasuriya et al. 2003); diabetic retinopathy (Feoktistov et al. 2004) and cancer (Zeng et al. 2003)—A_{2B}AR mediates angiogenesis (Belardinelli and Grant 2001; Feoktistov et al. 2004).

A number of studies have suggested that activation of the A_{2B}AR may play an important role in asthma. Activation of A_{2B}ARs on human bronchial smooth muscle cells (BSMCs) has been shown to induce the release of the inflammatory cytokines interleukin (IL)-6 and monocytic chemotactic protein-1 (MCP-1) (Zhong et al. 2004), on human lung fibroblasts the release of IL-6 and differentiation of fibroblasts into myofibroblasts (Zhong et al. 2005), and on human bronchial epithelial cells (HBECs) the release of IL-19, which in turn activates human monocytes to induce the release of TNF-α, which in turn upregulates A_{2B}AR expression on HBECs (Zhong et al. 2006). Adenosine levels are elevated in the bronchoalveolar lavage fluid (BALF) of asthmatics relative to healthy volunteers (Driver et al. 1993). Moreover, when AMP is administered to asthmatics and healthy normal individuals, it provides a source of adenosine that leads to bronchoconstriction in asthmatics but not normals (Cushley et al. 1984). Furthermore, an adenosine uptake blocker, dipyridamole, can precipitate asthma (Fozard and McCarth 2002). Therefore, the above evidence supports the notion that adenosine plays a role in asthma, and that its effects may be, at least in part, mediated through the A_{2B}AR. For more information on the role of A_{2B}ARs in asthma, the reader is referred to Chap. 11 of this volume, "Adenosine Receptors and Asthma" (Wilson et al.).

Scientists at Eisai have provided evidence that the A_{2B}AR antagonists and/or mixed A_{2B}/A_1AR antagonists may be useful in the treatment of diabetes. First, they demonstrated that the adenosine agonist analogs NECA [5′-*N*-ethylcarboxamidoadenosine] (nonselective), CPA [N^6-cyclopentyladenosine] (A_1-selective) and

CGS-21680 [2-[*p*-(2-carboxyethyl)phenylethylamino]-5′-*N*-ethylcarboxamidoadenosine] (A_{2A}-selective) stimulate glucose production from rat hepatocytes, with NECA having the most pronounced effect (Harada et al. 2001a). Then, the Eisai researchers found that their high-affinity A_{2B}AR antagonists that possess low selectivity over A_1AR and A_{2A}AR block NECA-induced glucose production in rat hepatocytes. Eisai found in a separate study that the inhibition of glucose production was best correlated with the A_{2B}AR affinity of the compounds used in a diabetes model (Harada et al. 2001b). Specifically, a nonselective high-affinity A_{2B}AR antagonist (**52**) was found to lower plasma glucose following oral dosing (10 and 30 mg kg^{-1} bodyweight) in a mouse model of noninsulin-dependent diabetes mellitus (NIDDM), KK-A^y mice (Harada et al. 2001b). It is clear from the above studies that adenosine likely plays a role in glucose production; however, the studies should be repeated with the highly selective, high-affinity A_{2B}AR antagonists described in this review.

Angiogenesis plays a major role in diabetic retinopathy and certain cancers. In proliferative diabetic retinopathy, it has been shown that activation of the A_{2B}AR on human retinal endothelial cells (HRECs) leads to new vessel formation that has uncontrolled growth, resulting in an increase in the permeability of the vasculature (Feoktistov et al. 2004). Support for a role of the A_{2B}AR in angiogenesis in HRECs was demonstrated when NECA caused a concentration-dependent increase in vascular endothelial growth factor (VEGF) mRNA in HRECs which was blocked by antisense oligonucleotides for the A_{2B}AR. Moreover, scientists at CV Therapeutics have suggested that labeled antibodies directed against the A_{2B}AR are potentially useful tools in detecting and possibly preventing the angiogenesis associated with gliomas, colon cancer, and solid tumors (Belardinelli and Grant 2001).

Finally, since the nonspecific AR antagonist caffeine is known to have intrinsic antinociceptive properties when used in combination with nonsteroidal anti-inflammatory drugs (NSAIDs) or opiates, Müller and coworkers investigated and discovered that A_{2B}AR antagonists possess antinociceptive effects in a hot plate test (Abo-Salem et al. 2004). In the same study, Müller et al. found that an A_{2B}AR antagonist was found to synergize with morphine for an enhanced antinociceptive effect in the same manner as caffeine.

2 A_{2B} Adenosine Receptor Antagonists

2.1 Xanthine-Based Antagonists

The naturally occurring alkylxanthines theophylline (**1**) and caffeine (**2**) are considered classical antagonists for the ARs. They exhibit weak affinity towards the A_{2B}AR, with no selectivity against the other ARs (Jacobson et al. 1999). Substitution of the dipropyl groups at the 1,3-methyl position of theophylline, as in 1,3-dipropyl xanthine, resulted in a 15-fold enhancement of A_{2B}AR affinity. The introduction of hydrophobic substitution at the 8 position of the xanthine core increases affinity towards the ARs. For example, 8-cyclopentyl-1,3-dipropyl xanthine (**3**, DPCPX), a known A_1AR antagonist, displays good affinity for both A_1 ($K_i = 0.9\,\text{nM}$) and A_{2B} ($K_i = 56\,\text{nM}$)ARs. Introduction of aryl substitution at the 8 position of theophylline, as in 1,3-dimethyl-8-phenyl xanthine (**4**), led to good affinity for the A_{2B}AR ($K_i = 415\,\text{nM}$), which is a 22-fold enhancement in affinity compared to theophylline. Further substitution of uncharged electron-donating groups like a methoxy group (**5**) or a hydroxyl group (**6**) at the *para* position of the 8-phenyl group enhances the A_{2B}AR affinity. The replacement of the 1,3-methyl groups of **4** with *n*-propyl groups, as in **7**, increased the A_{2B}AR affinity (Jacobson et al. 1999).

In an effort to develop potent and selective adenosine receptor antagonists, Jacobson et al. selected the 1,3-dipropyl-8-(*p*-hydoxyphenyl) xanthine as a lead compound to explore the effect of functionalized congeners (Kim et al. 1999). Initially, the effects of carboxylic acids, amine derivatives and amino acid conjugates at the *para* position of the phenyl group were explored, and it was observed that there was no clear preference for these groups with respect to the A_{2B}AR affinity. The neutral biotin conjugates of various chain lengths were considerably less potent than the parent amine, whereas the L-thienyl alanine derivative displayed good affinity ($K_i = 6.9\,\text{nM}$) at human A_{2B}AR. The high-affinity compounds observed in the series of 8-phenylxanthine functionalized congeners were XCC [8-{4-[(carboxymethyl)oxy]phenyl}-1,3-dipropylxanthine, **8**], its hydrazide derivative (**9**), and another substituted amide derivative (**10**) (Jacobson et al. 1999) (Table 1).

The hydrazide was reacted with various mono- and dicarboxylic acids, and this structural change further enhanced selectivity, as exemplified by the dimethylmalamide derivative (**11**, MRS1595). Further exploration of the amide derivatives

Table 1 A_1, A_{2A}, A_{2B} and A_3AR binding affinities of xanthine derivatives at human A_{2B}ARs expressed in HEK-293 cells

Compound	R	hK_i nM				A_{2B} Selectivity		
		A_{2B}	A_1	A_{2A}	A_3	A_1/A_{2B}	A_{2A}/A_{2B}	A_3/A_{2B}
8	OH	13	58	2,200	–	4	169	–
9	$NHNH_2$	14	323	21	217	23	1	16
10	$-NC(O)CH_2CH_2C(O)-$	10	153	127	227	15	13	23
11 (MRS1595)	NHNdimethylmaloyl	27	3,030	1,970	670	110	74	25
12	NH–Ph(4-CF_3)	2.14	61	238	213	29	110	100
13 (MRS1706)	NH–Ph(4-COCH3)	1.4	157	112	230	110	81	170
14	NH–Ph(4-I)	2	293	5,140	1,270	140	2,400	600
15 (MRS1754)	NH–Ph(4-CN)	1.97	403	503	570	205	255	290

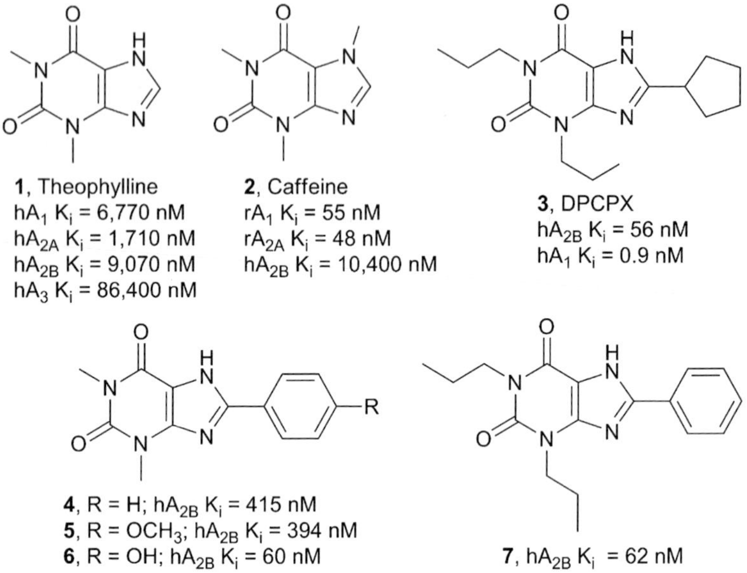

1, Theophylline
hA_1 K_i = 6,770 nM
hA_{2A} K_i = 1,710 nM
hA_{2B} K_i = 9,070 nM
hA_3 K_i = 86,400 nM

2, Caffeine
rA_1 K_i = 55 nM
rA_{2A} K_i = 48 nM
hA_{2B} K_i = 10,400 nM

3, DPCPX
hA_{2B} K_i = 56 nM
hA_1 K_i = 0.9 nM

4, R = H; hA_{2B} K_i = 415 nM
5, R = OCH_3; hA_{2B} K_i = 394 nM
6, R = OH; hA_{2B} K_i = 60 nM

7, hA_{2B} K_i = 62 nM

Fig. 1 1,3,7,8-Substituted xanthines

by condensing the carboxylic acid group of **8** with substituted phenylamines led to the discovery of several high-affinity and selective compounds. The anilides substituted at the *para* position with an electron-withdrawing group (**12**) showed good affinity and selectivity. Also, the *para*-acetophenone analog **13** displayed good affinity and selectivity. The *para*-halogen-substituted derivatives exhibited high affinity for the A_{2B}AR, with the *para*-iodo derivative (**14**) demonstrating the highest selectivity compared to the other halo derivatives (Kim et al. 2000). The *para*-cyano anilide derivative (**15**, MRS 1754) displayed high affinity for the A_{2B}AR (K_i = 1.97 nM) and 205-, 255-, and 290-fold selectivity versus human $A_1/A_{2A}/A_3$ARs, respectively (Kim et al. 2002) (Table 1).

The 1-alkyl-8-phenyl(cyclopentyl) xanthine derivatives were found to exhibit high affinity for A_{2B}ARs (Hayallah et al. 2002). In this study, the 1,8-disubstituted xanthine derivatives were shown to be equipotent to or more potent than 1,3,8-trisubstituted xanthines at A_{2B}ARs, but generally less potent at A_1 and A_{2A}, and much less potent at A_3AR subtypes. 1-Propyl-8-*p*-sulfophenylxanthine (**16**, PSB1115) was the most selective compound of this family, exhibiting a K_i value of 53 nM at human A_{2B}AR. This compound is highly water soluble due to its sulfonate functional group. The 4-nitrophenylester of PSB1115 is also reported to be a potential prodrug despite its significant binding affinity for the A_1AR subtype ($rA_{2B} = 5.4$ nM, $rA_1 = 3.6$ nM) (Hayallah et al. 2002). 1-Butyl-8-*p*-carboxyphenylxanthine (**17**), another polar analog bearing a carboxylate functional group, exhibited a K_i value of 24 nM for A_{2B}ARs, 49-fold selectivity versus human and 20-fold selectivity versus rat A_1AR subtype, and greater than 150-fold selectivity versus human A_{2A} and A_3ARs (Yan and Müller 2004) (Fig. 2).

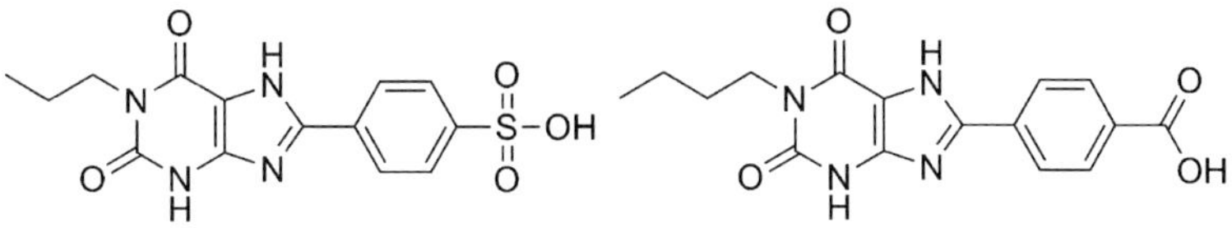

16, PSB-1115
rA_1 K_i = 2200 nM,
rA_{2A} K_i = 24,000 nM,
hA_{2B} K_i = 53.4 nM,
hA_3 K_i = 14% displ. at 10 mM

17
rA_1 K_i = 481 nM
rA_{2A} K_i = 3,800 nM
hA_{2B} K_i = 24.0 nM
hA_3 K_i = 4,622

Fig. 2 1,8-Substituted xanthines

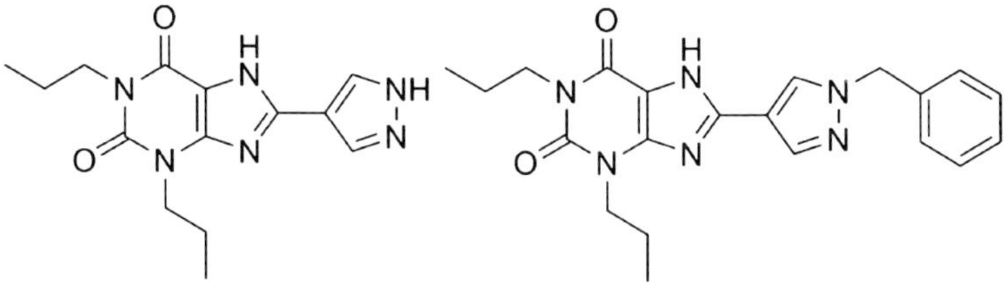

18
hA_{2B} = 9 nM
hA_1 = 20 nM
hA_{2A} = 230 nM
hA_3 = 23 nM

19
hA_{2B} = 11 nM
hA_1 = 76 nM
hA_{2A} = 290 nM
hA_3 = 170 nM

Fig. 3 8-(Pyrazol-4-yl)xanthines as A_{2B}AR antagonists

Chemists at CV Therapeutics have explored various heterocycles as bioisosteric replacements for the 8-phenyl group of 1,3-dipropyl-8-phenyl xanthines, and they observed that 1,3-dipropyl-8-(4-pyrazolyl)xanthine **18** displayed a high affinity for the A_{2B}AR (Kalla et al. 2004). Substitution of the N-1 pyrazole with a benzyl group, as in **19**, retained the A_{2B}AR affinity compared to the phenethyl and phenpropyl groups (Fig. 3). Further substitution of the phenyl ring with electron-withdrawing groups, for example CF_3 at the *meta*-position (**20**), increased the A_{2B}AR selectivity. Replacing the 1,3-dipropyl groups of the xanthine core with 1,3-dimethyl groups resulted in **21**, which has both high affinity and selectivity for the A_{2B}AR (Kalla et al. 2006). Exploration of differential substitution at the N-1 and N-3 positions of the xanthine core led to compound **22** (CVT-6883), which has good affinity for the A_{2B}AR and displayed good selectivity (Elzein et al. 2008). The introduction of monosubstitution at the N-1 position of the xanthine core, as in compound **23**, enhanced the selectivity compared to the disubstituted derivative **20** (Kalla et al. 2008). The N-3 monosubstituted derivative **24** lost the A_{2B}AR affinity, and this is in agreement with Hayallah et al.'s observation for the 8-phenyl series that 1,8-disubstituted xanthine derivatives display better A_{2B}AR affinities and selectivities than the 3,8-disubstituted xanthine derivatives (Hayallah et al. 2002). Replacing the phenyl group of **20** with different heterocycles, including 3-phenyl-

1,2,4-oxadiazoles, 5-phenyl-1,2,4-oxadiazoles and 3-phenyl-isoxazoles, resulted in compounds that display high affinity and good selectivity regardless of the substitution at the N-1 and N-3 positions on the xanthine core (**25** and **26**) (Elzein et al. 2006). The N-1 monosubstituted oxadiazole **27** and isoxazole **28** displayed high affinity and selectivity for the A_{2B}AR (Kalla et al. 2008). Compound **22** (6883) antagonized the NECA-induced cAMP accumulation in HEK-A_{2B} cells and NIH 3T3 cells (Sun et al. 2006), and compound **26** (6694) completely abolished the NECA-induced cAMP accumulation in BSMCs (Zhong et al. 2004), confirming that these compounds are antagonists for the hA_{2B}AR. In a mouse model of asthma, compound **22** demonstrated a dose-dependent blocking effect on NECA-induced increases in airway reactivity (Mustafa et al. 2007). Also, in this mouse model of allergic asthma, compound **22** significantly reduced the late allergic airway response and inflammatory cells in BALF, supporting the role of the A_{2B}ARs in asthma (Table 2).

Several heterocycles, such as pyrazole, isoxazole, pyridine and pyridazine, linked by different spacers (substituted acetamido, oxyacetamido and urea moieties) at the 8 position of the xanthine nucleus were investigated (Baraldi et al. 2004a). The synthesized compounds showed A_{2B}AR affinities in the nanomolar range and good levels of selectivity, as evaluated in radioligand binding assays at human A_1, A_{2A}, A_{2B}, and A_3ARs. This study allowed the identification of the derivatives 2-(3,4-dimethoxy-phenyl)-*N*-[5-(2,6-dioxo-1,3-dipropyl-2,3,6,7-tetrahydro-1*H*-purin-8-yl)-1-methyl-1*H*-pyrazol-3-yl]-acetamide (**29**, MRE2028-F20) [$K_i(hA_{2B}) = 38$ nM, K_i (hA_1, hA_{2A}, hA_3) $> 1{,}000$ nM], *N*-benzo[1,3]dioxol-5-yl-2-[5-(2,6-dioxo-1,3-dipropyl-2,3,6,7-tetrahydro-1*H*-purin-8-yl)-1-methyl-1*H*-pyrazol-3-yloxy]-acetamide (**30**, MRE2029F20) [K_i $(hA_{2B}) = 5.5$ nM, K_i $(hA_1) = 200$ nM, (hA_{2A}, hA_3) $> 1{,}000$], and *N*-(3,4-dimethoxy-phenyl)-2-[5-(2,6-dioxo-1,3-dipropyl-2,3,6,7-tetrahydro-1*H*-purin-8-yl)-1-methyl-1*H*-pyrazol-3-yloxy]acetamide (**31**, MRE2030F20) [K_i ($hA_{2B} = 12$ nM, K_i (hA_1, hA_{2A}, hA_3) $> 1{,}000$ nM] (Fig. 4), which showed high affinity at the A_{2B}AR subtype and very good selectivity versus the other AR subtypes. The derivatives with higher affinity at human A_{2B}AR proved to be antagonists in the cyclic AMP assay, capable of inhibiting the stimulatory effect of NECA (100 nM) with IC_{50} values in the nanomolar range and a trend similar to that observed in the binding assay.

Compounds **32, 33** (Fig. 4) bearing the isoxazole nucleus at the 8 position showed lower affinities at the A_{2B}AR than the corresponding 8-pyrazole derivatives. However, replacing the pyrazole ring with an isoxazole enhanced selectivity versus the A_1AR. Consequently, the radiolabeled analog of compound *N*-benzo[1,3]dioxol-5-yl-2-[5-(1,3-diallyl-2,6-dioxo-2,3,6,7-tetrahydro-1*H*-purin-8-yl)-1-methyl-1*H*-pyrazol-3-yloxy]-acetamide (**30**) was prepared in the tritium-labeled form [^{3}H] MRE2029F20, which displayed a K_d value of 1.65 ± 0.10 nM). This compound was found to be a selective, high-affinity radioligand that is useful for characterizing recombinant human A_{2B}ARs (Baraldi et al. 2004b). Very recently, the same authors also described a series of 1,3-dipropyl-8-(1-phenylacetamide-1*H*-pyrazol-3-yl)-xanthines as selective A_{2B}AR antagonists (Aghazadeh Tabrizi et al. 2008). The 4-chlorophenyl derivative **34** (Fig. 4) was found to be the most

Table 2 A_1, A_{2A}, A_{2B} and A_3AR binding affinities of 8-(4-pyrazolyl)xanthine analogs

Compound	R^1	R^3	R	hK_i nM				A_{2B} Selectivity		
				A_{2B}	A_1	A_{2A}	A_3	A_1/A_{2B}	A_{2A}/A_{2B}	A_3/A_{2B}
20	Propyl	Propyl	CF_3	14	160	400	140	12	27	10
21	Methyl	Methyl	CF_3	1	990	690	1,000	990	690	1,000
22 (CVT-6883)	Propyl	Ethyl	CF_3	22	1,940	3,280	1,070	88	149	48
23	Propyl	H	CF_3	8	>6,000	>5,000	700	>750	>620	80
24	H	Propyl	H	>6,000	nd	nd	nd	nd	nd	nd
25 (X = N)	Propyl	Propyl	CF_3	21	>6,000	>5,000	1,300	>290	>240	60
26 (X = N) (CVT-6694)	Propyl	H	Cl	7	>6,000	>5,000	>9,000	>850	>700	>1,280
27 (X = N)	Cyclopropyl methyl	H	CF_3	13	>6,000	>5,000	>9,000	>460	>380	>700
28 (X = C)	Cyclopropyl methyl	H	CF_3	15	>6,000	>5,000	>9,000	>400	>333	>600

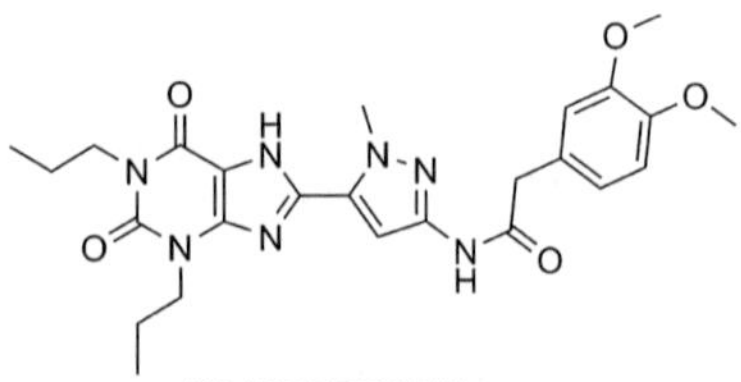

29, MRE2028F20
hA_{2B} K_i = 38 nM
hA_1, hA_{2A}, hA_3 > 1000 nM

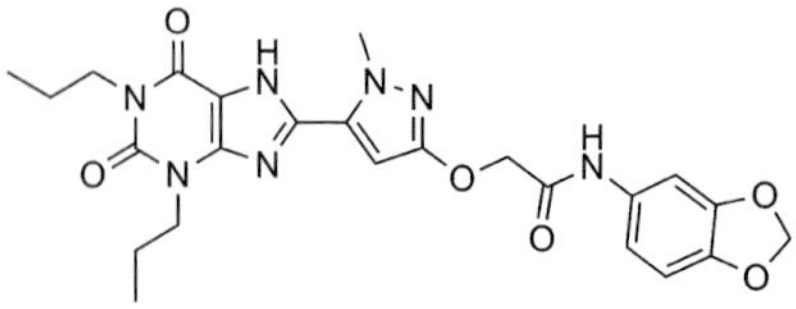

30, MRE2029F20
hA_{2B} K_i = 5.5 nM
hA_1 K_i = 200 nM
hA_{2A}, hA_3 > 1000nM

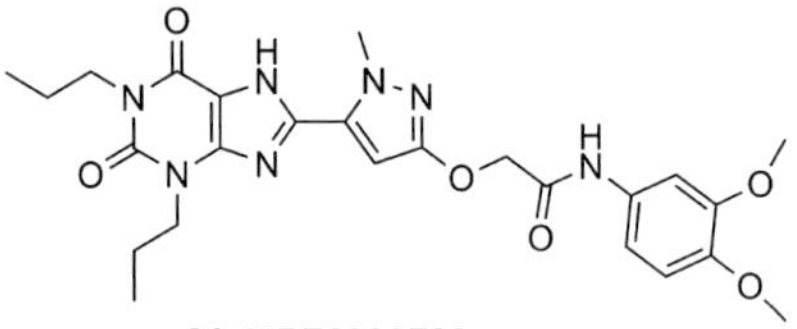

31, MRE2030F20
hA_{2B} K_i = 12 nM
hA_1, hA_{2A}, hA_3 > 1000 nM

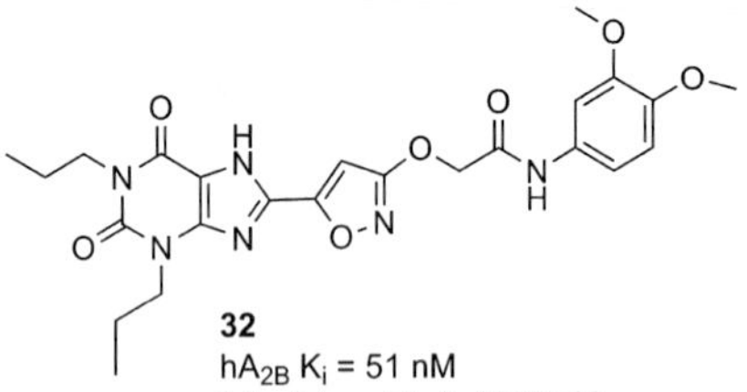

32
hA_{2B} K_i = 51 nM
hA_1, hA_{2A}, hA_3 > 1000 nM

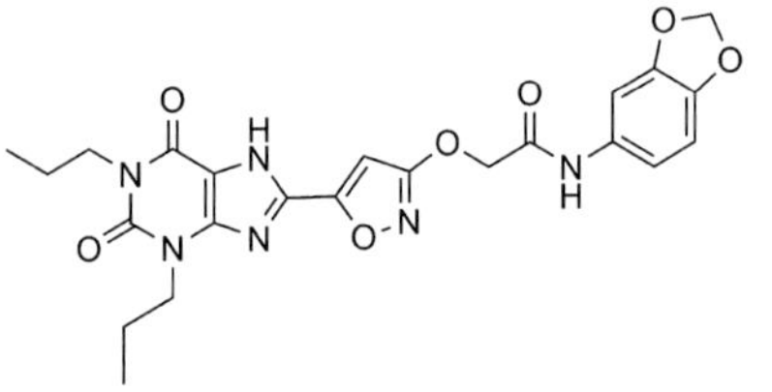

33
hA_{2B} K_i = 47 nM
hA_1, hA_{2A}, hA_3 > 1000 nM

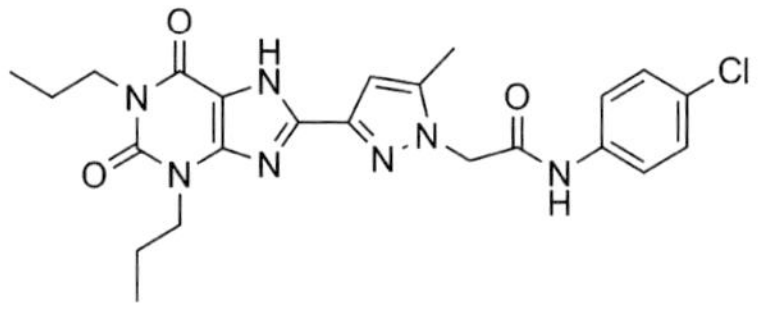

34
hA_{2B} K_i = 7.0 nM
hA_1, hA_{2A}, hA_3 > 1000 nM

Fig. 4 1,3,8-Substituted xanthines

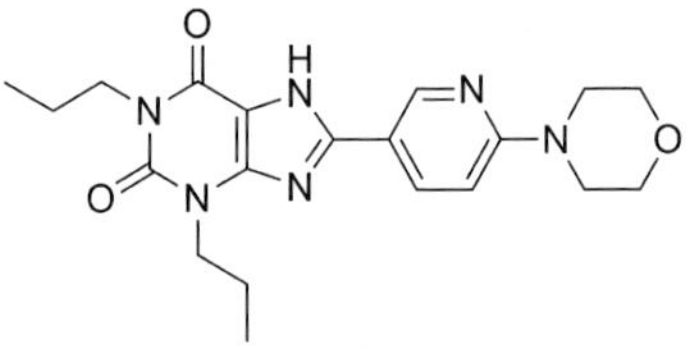

35 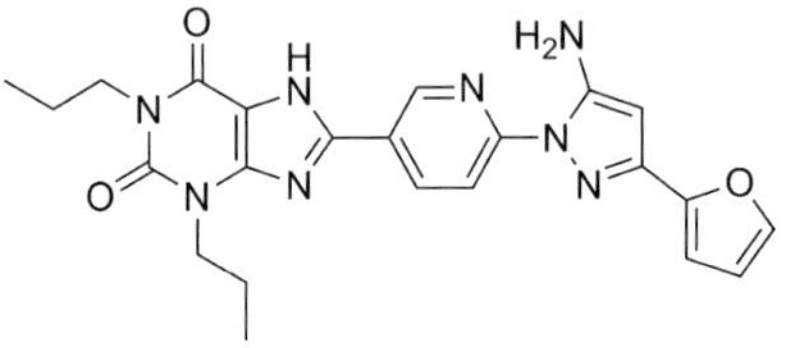 **36**

Fig. 5 Substituted 8-pyridyl xanthines as A_{2B}AR antagonists

potent ($K_i A_{2B} = 7.0$ nM) and selective compound within the series (A_1, A_{2A}, A_3/ $A_{2B} > 140$).

Scientists at Adenosine Therapeutics describe a series of 8-(3-pyridyl)xanthines that have high affinity for the A_{2B}AR. The morpholino derivative **35** and the pyrazolyl derivative **36** possess better than 100 nM affinity for the A_{2B}AR based upon the published patent application (Wang et al. 2006). The compounds demonstrate that the six-membered pyridyl ring can serve as a favorable linker, providing high affinity for the A_{2B}AR in a similar manner to the 4-pyrazolyl and 5-pyrazolyl linkers described above (Fig. 5).

2.2 Deazaxanthine-Based Antagonists

In the xanthine family, the 8-substituted-9-deaza-xanthines are reported to be antagonists with nanomolar affinities for the A_{2B}AR (Carotti et al. 2006; Stefanachi et al. 2008). The 1-, 3-, 8-, and 9-substituted-deazaxanthines of general structure **37** (Fig. 6) were prepared and evaluated for their binding affinities at the recombinant human ARs, in particular at the hA_{2B} and hA_{2A}AR subtypes. 1,3-Dimethyl-8-phenoxy-(*N*-*p*-halo-phenyl)-acetamido-9-deazaxanthine derivatives appeared to be the most interesting leads, with some of them, such as the compound (*N*-(4-bromo-phenyl)-2-[4-(1,3-dimethyl-2,4-dioxo-2,3,4,5-tetrahydro-1*H*-pyrrolo[3,2-d]pyrimidin-6-yl)-phenoxy]-acetamide (**38**), showing high hA_{2B}AR affinities and selectivity over hA_{2A} and hA_3ARs, but low selectivity over hA_1AR. Structure–affinity relationships suggested that the binding potency at the hA_{2B}AR was mainly modulated by the steric (lipophilic) properties of the substituents at positions 1 and 3 and by the electronic and lipophilic characteristics of the substituents at position 8. Electron-withdrawing groups in the *para*-position of the anilide phenyl ring increased the activity. Regarding the hA_{2B}/hA_{2A}AR selectivity,

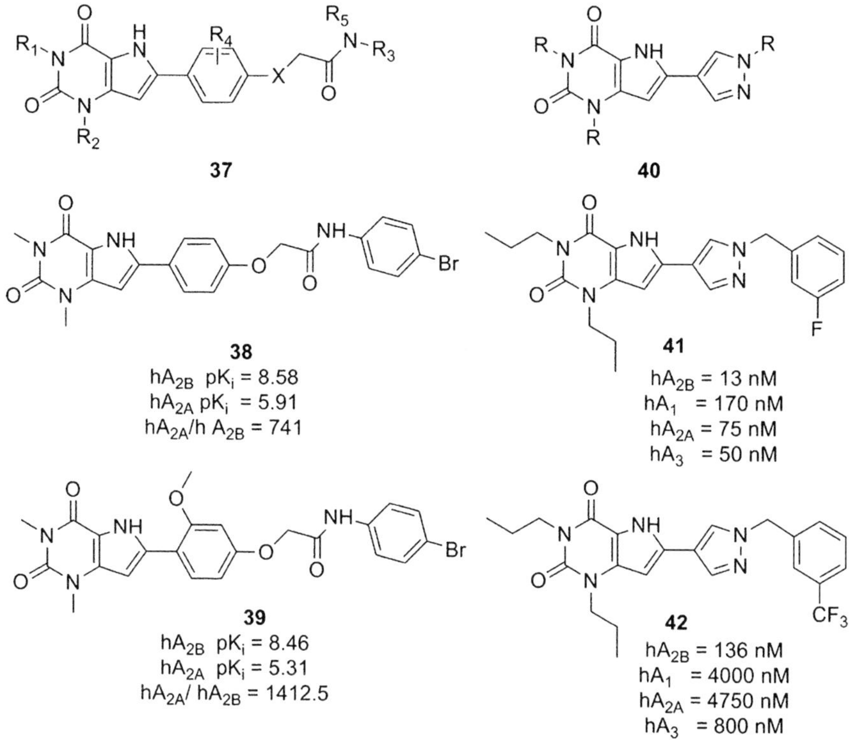

37

40

38
hA_{2B} pK_i = 8.58
hA_{2A} pK_i = 5.91
hA_{2A}/h A_{2B} = 741

41
hA_{2B} = 13 nM
hA_1 = 170 nM
hA_{2A} = 75 nM
hA_3 = 50 nM

39
hA_{2B} pK_i = 8.46
hA_{2A} pK_i = 5.31
hA_{2A}/ hA_{2B} = 1412.5

42
hA_{2B} = 136 nM
hA_1 = 4000 nM
hA_{2A} = 4750 nM
hA_3 = 800 nM

Fig. 6 General structures and binding affinities of deazaxanthine derivatives

the most interesting result came from the introduction of a methoxy substituent in the *ortho* position of the 8-phenyl ring, which led to an enhancement in selectivity for compound **39**, making it 1,412-fold more selective for the $hA_{2B}AR$ over the $hA_{2A}AR$.

The 9-deaza analog **40** of the 8-(4-pyrazolyl)xanthine class was synthesized to compare its $A_{2B}AR$ affinity and selectivity (Kalla et al. 2005). Compound **41**, a 9-deaza derivative with a *meta*-fluoro substitution on the pyrazole ring, has the same affinity as the direct xanthine analog. Compound **42**, a *meta*-CF_3 derivative and a direct analog of compound **20**, displayed a lower affinity for the $A_{2B}AR$ but good selectivity.

A new series of 4-(1,3-dialkyl-2,4-dioxo-2,3,4,5-tetrahydro-1*H*-pyrrolo[3, 2-*d*] pyrimidin-6-yl)benzenesulfonamides (**43–46**, Table 3) are also reported to be potent $A_{2B}AR$ antagonists (Esteve et al. 2006). In this series, the 6-(4-{[4-(4-bromobenzyl)piperazin-1-yl]sulfonyl}phenyl)-1,3-dimethyl-1*H*-pyrrolo[3,2-d] pyrimidine-2,4(3*H*, 5*H*)-dione (**45**) showed a high affinity for the $A_{2B}AR$ (IC_{50} = 1 nM) and selectivity. The presence of metabolically stable benzenesulfonamide in this novel class of compounds improved their physiochemical properties, resulting in increased oral bioavailability.

Table 3 Deazaxanthine-benzenesulfonamides as $A_{2B}AR$ antagonists

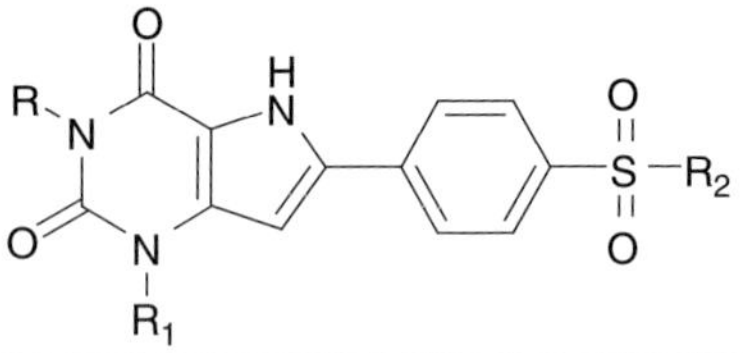

Compound	R	R_1	R_2	hA_{2B}	hA_1	hA_3
				IC_{50} (nM)		
43	CH_3	CH_3	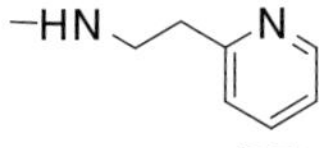 –HN–CH₂CH₂–(2-pyridyl)	14	150	2,085
44	CH_3	CH_3	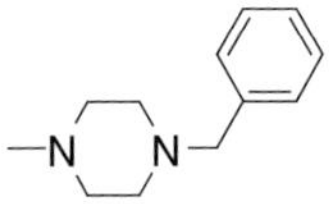 –N(piperazine)N–CH₂–phenyl	16	415	3,169
45	CH_3	CH_3	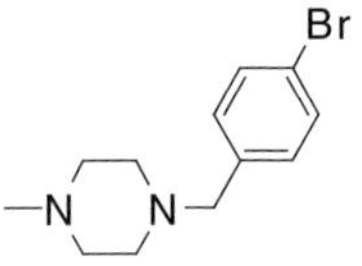 –N(piperazine)N–CH₂–(4-Br-phenyl)	1	183	12,260
46	nC_3H_7	CH_3	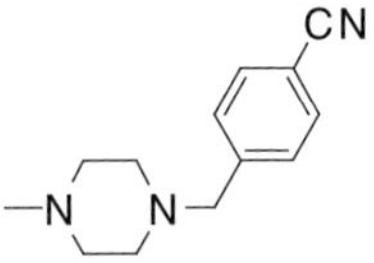 –N(piperazine)N–CH₂–(4-CN-phenyl)	6	370	950

2.3 *Adenine-Based Antagonists*

Adenine has proven to be a useful core for the development of AR antagonists. Modifications of adenine led to various 2- and 8-substituted derivatives with moderate affinities at all four AR subtypes (Campioni et al. 1998). Generally, it appears that 2-substitution led to nonselective antagonists or to antagonists with high affinities at both A_1 and A_{2A}ARs. In the series of 8-bromo-9-alkyl-adenines **47–50** (Table 4), the presence of a propyl group at the 9 position and a bromine at the 8 position, such as in compound **49** (hA_{2B} K_i = 200 nM), increased the affinity and selectivity for the A_{2B} receptor in comparison to the parent 9-ethyladenine, with a K_i value of 0.84 μM. The experimental data show that different substituents in the 8 position result in compounds with quite different pharmacological features. The 8-phenethylamino, 8-phenethoxy, and 8-(ar)alkynyl compounds generally showed lower potency at all receptors than compound **48**. Replacement of the N9 ethyl with a methyl group retained A_{2B}AR affinity and decreased affinity at all other AR subtypes, while the N9 bulky groups led to derivatives with higher selectivity versus A_{2B}AR (Lambertucci et al. 2000; Volpini et al. 2003).

The 2-alkynyl-8-aryl-9-methyladenine derivatives were also synthesized as candidate hypoglycemic agents (Harada et al. 2001a). These analogs were eval-

Table 4 Adenine derivatives as A_{2B}AR antagonists

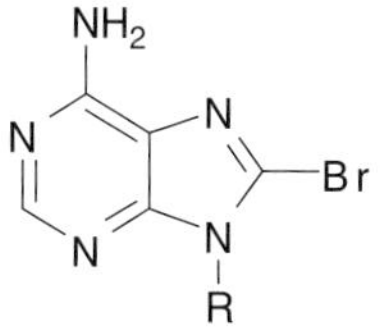

	R	hA_{2B}	hA_1	hA_{2A}	hA_3
		K_i (nM)		K_i (nM)	
47	CH_3	720	570	120	>100,000
48	C_2H_5	840	280	52	28,000
49	C_3H_7	200	1,100	300	>100,000
50	cC_5H_9	860	1,000	1,900	4,200
51	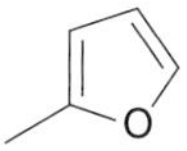	14	25	11	–
52	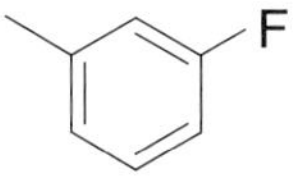	23	14	16	540

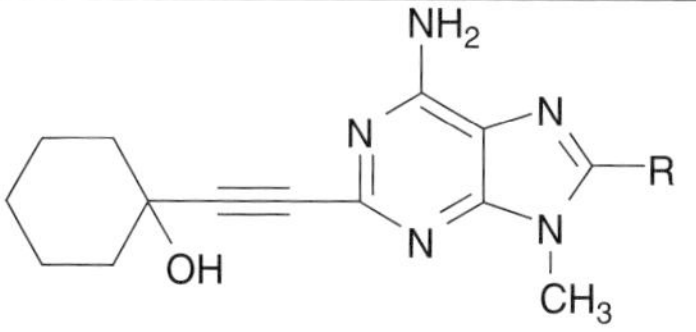

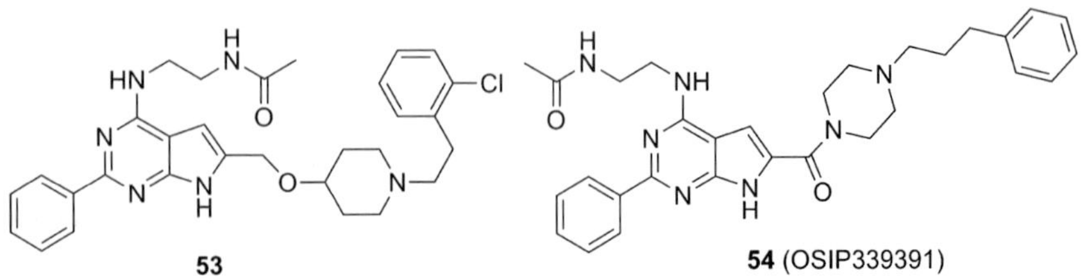

53
hA_{2B} = 5 nM
hA_1 ~ 125 nM
hA_{2A} ~ 125 nM
hA_3 ~ 125 nM

54 (OSIP339391)
hA_{2B} = 0.5 nM
hA_1 = 37 nM
hA_{2A} = 328 nM
hA_3 = 450 nM

Fig. 7 Deazapurines as A_{2B}AR antagonists

uated for inhibitory activity on *N*-ethylcarboxamidoadenosine (NECA)-induced glucose production in primary cultured rat hepatocytes. The introduction of various heteroaromatic rings and (substituted) phenyl rings at the 8 position of 9-methyladenine, and of other alkynyl groups at the 2 position, was investigated. The aromatic groups in the 8 position significantly increased the potency, and the preferred substituents at the 8 position of adenine were the 2-furyl and 3-fluorophenyl groups (**51** and **52** respectively, Table 4). Another modification at the alkynyl side chain, changes in ring size, cleavage of the ring, and removal of the hydroxyl group were all well tolerated. Compound **52** is a nonspecific adenosine antagonist, but it was hypothesized that its inhibition of hepatic glucose production via the A_{2B}AR could be at least one of the mechanisms associated with its in vivo activity.

Scientists at OSI Pharmaceuticals have discovered several very high affinity A_{2B}AR antagonists, **53** and **54** (OSIP339391), that are based on a deaza-adenine scaffold (Castelhano et al. 2003) (Fig. 7). Compound **54** has extremely high affinity for the A_{2B}AR, making it suitable for use as a radiolabeled ligand for competitive binding assays, and it possesses moderate selectivity over hA_1, hA_{2A} and hA_3ARs (>70-fold). The tritium-labeled **54** was prepared from the acetylene precursor, affording a compound with radiolabeled OSIP339391 that had a selectivity of greater than 70-fold for A_{2B}ARs over other human AR subtypes (Stewart et al. 2004). The radiolabel was introduced by hydrogenation of the acetylenic precursor, affording a compound with high specific activity. These compounds should possess good water solubilities, since they incorporate a basic piperidine or piperazine moiety that should be protonated at physiological pH. It would be interesting to know whether these compounds are metabolically stable and whether they have favorable PK properties, but this information has not been reported.

2.4 2-Aminopyridine-Based Antagonists

2-Aminopyridine derivatives that are selective A_{2B}AR antagonists have been reported (Harada et al. 2004). The core structure of this class of compounds is

Table 5 2-Aminopyridine derivatives as $A_{2B}AR$ antagonists

[Structure: 2-aminopyridine scaffold with H_2N, N, R_2, NC, R_1]

	R_1	R_2	hA_1 K_i(nM)	hA_{2A} K_i(nM)	$hA_{2B}IC_{50}$ (nM)
55	4-Pyridine	2-Furyl	990	23	2.7
56	4-Pyridine	3-Fluoro-phenyl	66	22	3.7
57	1-Ethyl-6-oxo-3-pyridine	2-Furyl	400	7	6.5

a 2-aminopyridine that presents a cyano group at the 3 position together with substituents at the 4 and 5 positions (**55–57**). Few data were presented, but analog **55** (6-amino-2-(2-furyl)-3,4′-bipyridine-5-carbonitrile; Table 5) showed at least ninefold selectivity and good affinity at $A_{2B}AR$. Apparently, the introduction of a furyl group at the 2 position and a pyridine group at the 4 position (**55**) introduced some selectivity with respect to the A_1AR and $A_{2A}AR$, respectively.

2.5 *Bipyrimidine-Based Antagonists*

2-Amino-substituted pyrimidines have been identified as suitable templates for the construction of adenosine $A_{2B}AR$ antagonists (Vidal et al. 2007c). Several compounds of this novel series of *N*-heteroaryl-4′-(furyl)-4,5′-bipyrimidin-2′-amines (Table 6) were very selective over other ARs and had a low nanomolar affinity at the $A_{2B}AR$. The introduction of unsubstituted nitrogen-containing heterocycles in R_1, such as pyridine, pyrimidine or pyrazine, yielded compounds **58–67**. Among these, the 3-pyridyl derivative **58** (LAS-38096) was found to show high potency and selectivity. Substitution by a methoxy group (**61** and **62**) led to a drop in potency while retaining good selectivity. Compounds **63** and **64** were investigated to evaluate the effect of substitution of the nitrogen atoms of the 3-pyridyl or the 3-pyrimidinyl rings with a hydrogen bond acceptor. The cyano derivative **63** showed lower affinity but an increase in selectivity versus $A_{2A}AR$. The *N*-oxide **64** had a twofold decrease in affinity compared to that of the corresponding reduced analog **58**. The introduction of a pyridone moiety yielded compound **65**, which was found to be one of the most potent and selective compounds within the series. Alkylation at the pyridine nitrogen had a slightly detrimental effect on $A_{2B}AR$ potency for the more lipophilic analog **67**. LAS-38096, which represents the lead for this series, was capable of inhibiting $A_{2B}AR$-mediated NECA-dependent increases in intracellular cAMP, with IC_{50} values of 321 nM and 349 nM in cells expressing human and mouse ARs, respectively; it also displayed a favorable PK profile in preclinical species. The efficacy of compound **58** was evaluated in vivo in an allergic mouse model, and the mice showed significantly less bronchial

Table 6 2-Amino substituted pyrimidines as $A_{2B}AR$ antagonists

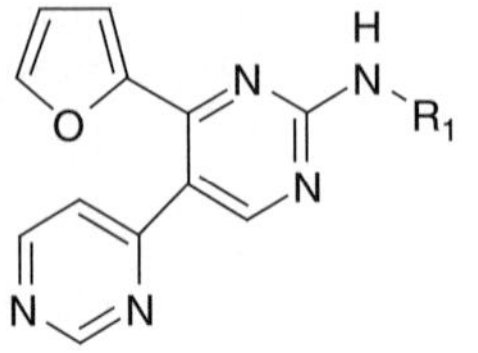

	R_1	hA_{2B}	hA_{2A}	hA_1	hA_3
		K_i (nM)			
58 (LAS-38096)	3-Pyridyl	17	>2, 500	>1, 000	>1, 000
59	3-Pyrimidinyl	24	>2, 500	>10, 000	>1, 000
60	2-Pyrazinyl	116	>2, 500	>10, 000	>1, 000
61	6-Methoxypyridin-3-yl	115	>2, 500	>10, 000	>1, 000
62	2-Methoxypyrimidin-5-yl	39	>2, 500	>10, 000	>1, 000
63	5-Cyano-pyridin-3-yl	69	>2, 500	>10, 000	>1, 000
64	1-Oxido-pyridin-3-yl	34	>2, 500	>10, 000	>1, 000
65	6-Oxo-1,6-dihydropyridin-3-yl	16	>2, 500	>10, 000	>1, 000
66	1-Methyl-6-oxo-1,6-dihydropyridin-3-yl	28	>2, 500	>10, 000	>1, 000
67	1-Cyanopropylmethyl-6-oxo-1,6- dihydropyridin-3-yl	119	>2, 500	>10, 000	>1, 000

hyperresponsiveness, mucus production, and a slight decrease in eosinophil infiltration and Th2 cytokine levels (Aparici et al. 2006).

2.6 *Pyrimidone-Based Antagonists*

A series of compounds with a pyrimidine-4-(3H)-one core structure has been reported as antagonists for the $A_{2B}AR$ (Harada et al. 2003). However, few data were reported, and data on the A_3AR are lacking. The most representative compound, 2-amino-6-(2-furyl)-3-methyl-5-(4-pyridyl)pyrimidin-4-(3H)-one (**69**), derived by introducing a methyl group at the 3 position of the unsubstituted analog **68**, gave K_i values of 966 nM and 493 nM, respectively, against A_1 and $A_{2A}ARs$ in binding assays. It also inhibited NECA-stimulated cAMP production in A_{2B}-transfected CHO-K1 cells, with an IC_{50} value of 71 nM (Table 7).

2.7 *Imidazopyridine-Based Antagonists*

The imidazopyridine nucleus was recently identified in a patent as a core structure in a new series of $A_{2B}AR$ antagonists (Vidal et al. 2005). This patent presented few data, and data on the A_1, A_{2A} and A_3ARs are lacking. Several compounds (**70–75**) of this new class had low nanomolar (<10 nM) affinities for the $A_{2B}AR$ (Table 8).

Table 7 Pyrimidone derivatives as $A_{2B}AR$ antagonists

	R	hA_{2B} IC_{50} (nM)	hA_1	hA_{2A}
			K_i (nM)	
68	H	256	1108	345
69	CH3	71	966	493

Table 8 Imidazopyridines as $A_{2B}AR$ antagonists

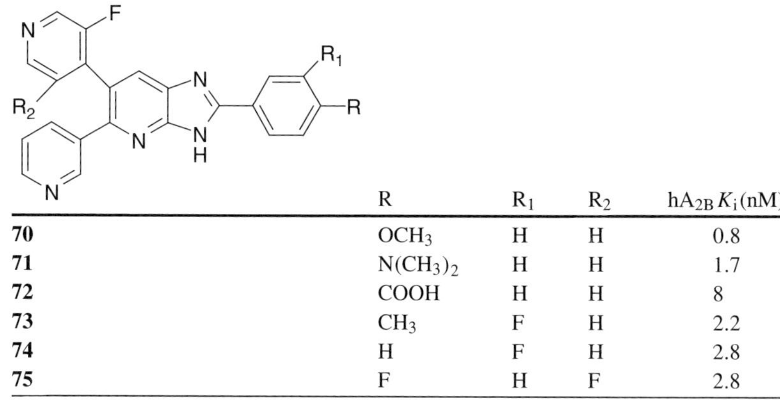

	R	R_1	R_2	hA_{2B} K_i (nM)
70	OCH_3	H	H	0.8
71	$N(CH_3)_2$	H	H	1.7
72	COOH	H	H	8
73	CH_3	F	H	2.2
74	H	F	H	2.8
75	F	H	F	2.8

2.8 Pyrazine-Based Antagonists

Scientists at Almirall Pharmaceuticals have found that pyrazine derivatives are novel potent antagonists of $A_{2B}ARs$ (Vidal et al. 2007a, b). Table 9 shows the binding activities for human $A_{2B}AR$ of some of these compounds (**76–83**). In this patent, affinity data for other AR subtypes are lacking. Generally, these 2-aminopyrazines present the pyridine nucleus at the 5 and 6 positions. Apparently, the introduction of a furyl group at the 6 position of the pyrazine ring (**79**) was tolerated by $A_{2B}ARs$. The imidazopyrazine **80** also showed high affinity at $A_{2B}AR$. The lead compound of this new series is 2-(3-fluoropyridin-4-yl)-3,6-di(3-pyridyl)-5*H*-pyrrolo[2,3-*b*]pyrazine, **83** ($K_i = 0.9$ nM).

Table 9 Pyrazine derivatives as A_{2B}AR antagonists

Compound	Chemical structure	$hA_{2B}\ K_i$ (nM)	Compound	Chemical structure	$hA_{2B}\ K_i$ (nM)
76	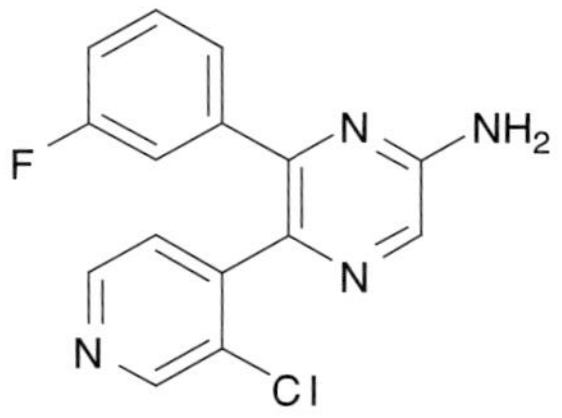 F, N, NH_2, N, N, Cl	4	**80**	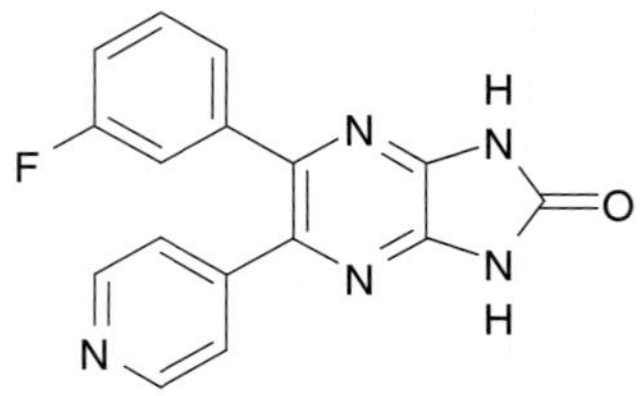 F, N, H, N, O, N, N, H, N	4
77	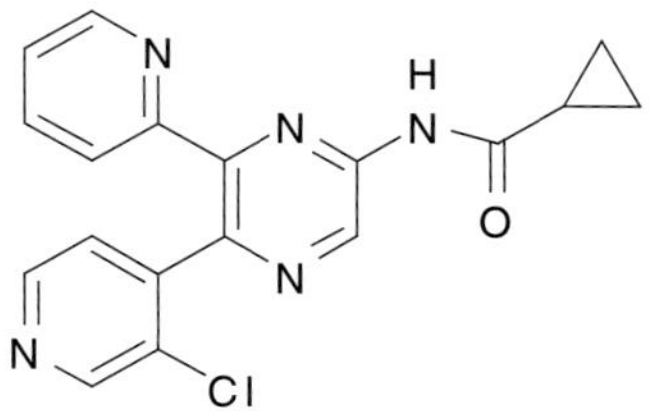 N, N, H, N, O, N, N, Cl	16	**81**	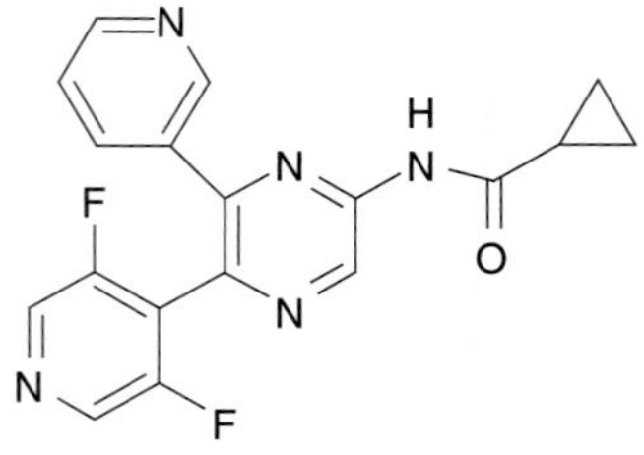 N, F, N, H, N, O, N, N, F	3

78	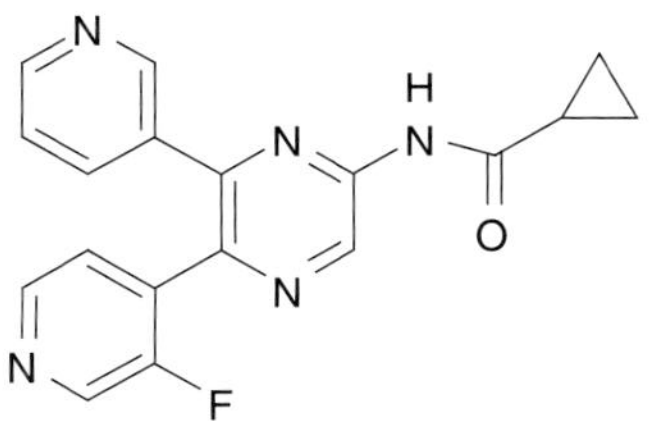	19	**82**	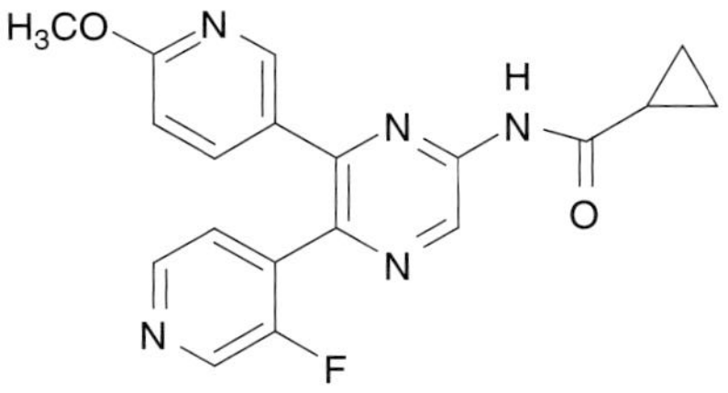	26
79	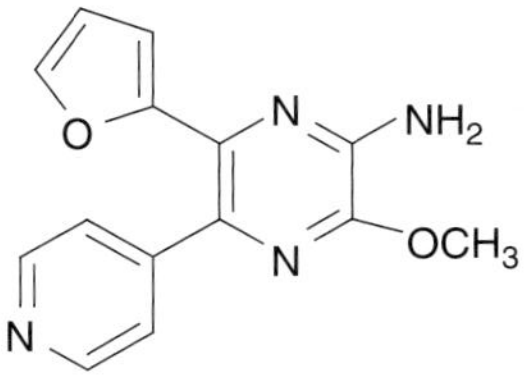	9	**83**	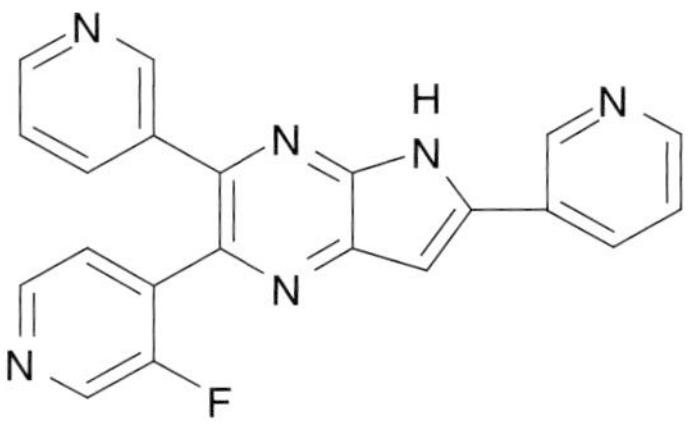	0.9

2.9 Pyrazolo-Triazolo-Pyrimidine-Based Antagonists

Pyrazolo[4,3-*e*]1,2,4-triazolo-[1,5-*c*]pyrimidine derivatives were initially investigated for the development of selective A_3AR antagonists (Baraldi et al. 2000, 2002). From preliminary studies, it has been demonstrated that the N^5-unsubstituted derivatives show different binding profiles according to the substitution position (N7 or N8) on the pyrazole nucleus. The N7 derivatives showed high affinity for the human A_{2A}ARs but did not bind to the human A_{2B}AR subtype (range $> 1\,\mu$M). The N^8-substituted derivatives (derivatives with phenethyl or *iso*-pentyl groups, such as in compounds **84** and **85**, Table 10) displayed affinity in the nanomolar range to human A_{2B}AR, but no selectivity versus the A_1 and A_{2A}AR subtypes was observed. In parallel studies on human A_3AR antagonists, it was observed that the introduction of a phenylacetyl group at the N5 position (**86**) produces an increase in affinity at the A_3AR. In fact, a combination of an aryl acetyl moiety at the N5 position and a phenylethyl or phenylpropyl group at the N8 position led to compounds **87** and **88**, which were found to be nonselective AR antagonists. The introduction of an alkylcarbamoyl moiety at the N5 position yielded **89**, which is also a nonselective AR antagonist (Baraldi et al. 2001). The introduction of a α-naphthoyl chain at the N5 position instead of a phenyl group (**90**) was tolerated by the A_{2B}AR but not the other AR subtypes, resulting in the most selective A_{2B}AR antagonist of this series (Pastorin et al. 2003).

Table 10 Pyrazolo[4,3-*e*]1,2,4-triazolo-[1,5-*c*]pyrimidines as A_{2B}AR antagonists

	R	R_1	hA_1	hA_{2B}	hA_{2A}	hA_3
			K_i (nM)			
84	H	$(CH_2)_2Ph$	1	5	0.31	2, 030
85	H	Iso-pentyl	2	9	0.8	700
86	$COCH_2Ph$	Methyl	702	165	423	0.81
87	$COCH_2Ph$	$(CH_2)_2Ph$	120	35	60	45
88	$COCH_2Ph$	$(CH_2)_3Ph$	75	40	60	121
89	$CO(CH_2)_3NH_3$	$(CH_2)_2Ph$	1.6	27	54	65
90	α-Naphthoyl	$(CH_2)_3Ph$	1, 100	20	800	300

3 Conclusion

The challenge to obtain second-generation, selective, high-affinity A_{2B}AR antagonists has been met, as exemplified by the xanthines 8-(4-(*N*-1-benzyl-pyrazolyl))xanthine **22** (A_{2B}AR $K_i = 1$ nM, CV Therapeutics) and 8-(5-pyrazolyl) xanthine **30** (A_{2B}AR $K_i = 5.5$ nM, Baraldi et al. and King Pharmaceuticals) and the pyrimidine 2-(amino-5-pyrimidinyl)pyrimidine **58** (A_{2B}AR $K_i = 17$ nM, Almiral Prodesfarma). These compounds should prove useful as tools to define the role of the A_{2B}AR in various disease states, including asthma, diabetes, cancer, and management of inflammatory pain in clinical trials in the near future.

References

Abo-Salem OM, Hayallah AM, Bilkei-Gorzo A, Filipek B, Zimmer A, Müller CE (2004) Antinociceptive effects of novel A_{2B} adenosine receptor antagonists. J Pharmacol Exp Ther 308:358–366

Aghazadeh Tabrizi M, Baraldi PG, Preti D, Romagnoli R, Saponaro G, Baraldi S, Moorman Allan R, Zaid AN, Varani, Borea PA (2008) 1,3-Dipropyl-8-(1-phenylacetamide-1*H*-pyrazol-3-yl)-xanthine derivatives as highly potent and selective human A_{2B} adenosine receptor antagonists. Bioorg Med Chem 16:2419–2430

Aparici M, Nueda A, Beleta J, Prats N, Fernandez R, Miralpeix M (2006) A potent adenosine A_{2B} receptor antagonist attenuates methacholine-induced bronchial hyperresponsiveness, mucus production and IgE levels in an allergic mouse model (Poster 162). In: CIA Symp, Malta, 5–10 May 2006

Baraldi PG, Cacciari B, Romagnoli R, Merighi S, Varani K, Borea PA, Spalluto G (2000) A_3 adenosine receptor ligands: history and perspectives. Med Res Rev 20:103–128

Baraldi PG, Cacciari B, Romagnoli R, Spalluto G, Varani K, Gessi S, Merighi S, Borea PA (2001) Pyrazolo[4,3-*e*]1,2,4-triazolo[1,5-c]pyrimidine derivatives: a new pharmacological tool for the characterization of the human A_3 adenosine receptor. Drug Dev Res 52:406–415

Baraldi PG, Cacciari B, Moro S, Spalluto G, Pastorin G, Da Ros T, Klotz KN, Varani K, Gessi S, Borea PA (2002) Synthesis, biological activity, and molecular modeling investigation of new pyrazolo[4,3-*e*]-1,2,4-triazolo[1,5-*c*]pyrimidine derivatives as human A_3 adenosine receptor antagonists. J Med Chem 45:770–780

Baraldi PG, Aghazadeh Tabrizi M, Preti D, Bovero A, Romagnoli R, Fruttarolo F, Zaid AN, Moorman Allan R, Varani K, Gessi S, Merighi S, Borea PA (2004a) Design, synthesis, and biological evaluation of new 8-heterocyclic xanthine derivatives as highly potent and selective human A_{2B} adenosine receptor antagonists. J Med Chem 47: 1434–1447

Baraldi PG, Aghazadeh Tabrizi M, Preti D, Bovero A, Romagnoli R, Fruttarolo F, Moorman Allan R, Varani K, Borea PA (2004b) [^{3}H]-MRE 2029-F20, a selective antagonist radioligand for the human A_{2B} adenosine receptors. Bioorg Med Chem Lett 14:3607–3610

Belardinelli L, Grant MB (2001) Method for identifying and using A_{2B} adenosine receptor antagonists to mediate mammalian cell proliferation. WO Patent 01060350

Campioni E, Costanzi S, Vittori S, Volpini R, Klotz KN, Cristalli G (1998) New substituted 9-alkylpurines as adenosine receptor ligands. Bioorg Med Chem 6:523–533

Carotti A, Cadavid MI, Centeno NB, Esteve C, Loza MI, Martinez A, Nieto R, Ravina E, Sanz F, Segarra V, Sotelo E, Stefanachi A, Vidal B (2006) Design, synthesis, and structure–activity relationships of 1-,3-,8-, and 9-substituted-9-deazaxanthines at the human A_{2B} adenosine receptor. J Med Chem 49:282–299

Castelhano AL, McKibben B, Steinig AG (2003) Pyrrololopyrimidine A_{2B} selective antagonist compound, their synthesis and use. WO Patent 2003053361

Cushley MJ, Tattersfield AE, Holgate ST (1984) Adenosine-induced bronchoconstriction in asthma: antagonism by inhaled theophylline. Am Rev Respir Dis 129:380–384

Driver AG, Kukoly CA, Ali S, Mustafa SJ (1993) Adenosine in bronchoalveolar lavage fluid in asthma. Am Rev Respir Dis 148:91–97

Elzein E, Kalla R, Li X, Perry T, Parkhill E, Palle V, Varkhedkar V, Gimbel A, Zeng D, Lustig D, Leung K, Zablocki J (2006) Novel 1,3-dipropyl-8-(1-heteroarylmethyl-1*H*-pyrazol-4-yl)-xanthine derivatives as high affinity and selective A_{2B} adenosine receptor antagonists. Bioorg Med Chem Lett 16:302–306

Elzein E, Kalla RV, Li X, Perry T, Gimbel A, Zeng D, Lustig D, Leung K, Zablocki J (2008) Discovery of a novel A_{2B} adenosine receptor antagonist as a clinical candidate for chronic inflammatory airway diseases. J Med Chem 51:2267–2278

Esteve C, Nueda A, Díaz JL, Beleta J, Cárdenas A, Lozoya E, Cadavid MI, Loza MI, Ryder H, Vidal B (2006) New pyrrolopyrimidin-6-yl benzenesulfonamides: potent A_{2B} adenosine receptor antagonists. Bioorg Med Chem Lett 16:3642–3645

Feoktistov I, Wells JN, Biaggioni I (1998) Adenosine A_{2B} receptors as therapeutic targets. Drug Dev Res 45:198–206

Feoktistov I, Ryzhov S, Zhong H, Goldstein AE, Matafonov A, Zeng D, Biaggioni I (2004) Hypoxia modulates adenosine receptors in human endothelial and smooth muscle cells toward an A_{2B} angiogenic phenotype. Hypertension 44:649–654

Fozard JR, McCarth C (2002) Adenosine receptor ligands as potential therapeutics in asthma. Curr Opin Investig Drugs 3:69–77

Fredholm BB, IJzerman AP, Jacobson KA, Kloz K-N, Linden J (2001) International Union of Pharmacology. XXV. Nomenclature and classification of adenosine receptors. Pharmacol Rev 53:527–552

Harada H, Asano O, Hoshino Y, Yoshikawa S, Matsukura M, Kabasawa Y, Niijima J, Kotake Y, Watanabe N, Kawata T, Inoue T, Horizoe T, Yasuda N, Minami H, Nagata K, Murakami M, Nagaoka J, Kobayashi S, Tanaka I, Abe S (2001a) 2-Alkynyl-8-aryl-9-methyladenines as novel adenosine receptor antagonists: Their synthesis and structure-activity relationships toward hepatic glucose production induced via agonism of the A_{2B} receptor. J Med Chem 44:170–179

Harada H, Asano O, Kawata T, Inoue T, Horizoe T, Yasuda N, Nagata K, Murakami M, Nagaoka J, Kobayashi S, Tanaka I, Abe S (2001b) 2-Alkynyl-8-aryladenines possessing an amide moiety: their synthesis and structure-activity relationships of effects on hepatic glucose production induced via agonism of the A_{2B} adesnosine receptor. Bioorg Med Chem 9:2709–2726

Harada H, Asano O, Miyazawa S, Yasuda N, Kabasawa Y, Ueda M, Yasuda M, Kotake Y (2003) Pyrimidone compounds and pharmaceutical compositions containing the same. WO 2003035640

Harada H, Asano O, Yasuda N, Ueda M. Yasuda M, Miyazawa S (2004) 2-Aminopyridine compounds and thereof as drugs. US Patent 6750232

Hayallah AM, Sandoval-Ramırez JS, Reith U, Schobert U, Preiss B, Schumacher B, Daly JW, Müller CE (2002) 1,8-Disubstituted xanthine derivatives: synthesis of potent A_{2B}-selective adenosine receptor antagonists. J Med Chem 45:1500–1510

Holgate ST (2005) The Quintiles Prize Lecture 2004: the identification of the adenosine A_{2B} receptor as a novel therapeutic target in asthma. Br J Pharmacol 145:1009–1015

Jacobson KA, IJzerman AP, Linden J (1999) 1,3-Dialkylxanthie derivatives having high potency as antagonists at human A_{2B} adenosine receptors. Drug Dev Res 47:45–53

Kalla R, Perry T, Elzein E, Varkhedkar V, Li X, Ibrahim P, Palle V, Xiao D, Zablocki J (2004) A_{2B} adenosine receptor antagonists. US Patent 6,825,349

Kalla R, Elzein E, Marquart T, Perry T, Li X, Zablocki J (2005) A_{2B} adenosine receptor antagonists. WO Patent 2005042534

Kalla R, Elzein E, Perry T, Li X, Palle V, Varkhedkar V, Gimbel A, Maa T, Zeng D, Zablocki J (2006) Novel 1,3-disubstituted 8-(1-benzyl-1*H*-pyrazol-4-yl)xanthines: High affinity and selective A_{2B} sdenosine receptor antagonists. J Med Chem 49:3682–3692

Kalla R, Elzein E, Perry T, Li X, Gimbel A, Yang M, Zeng D, Zablocki J (2008) Selective, high affinity A_{2B} adenosine receptor antagonists: N-1 monosubstituted 8-(pyrazol-4-yl)xanthines. Bioorg Med Chem Lett 18:1397–1401

Kim Y-C, Karton Y, Ji X-D, Melman N, Linden J, Jacobson KA (1999) Acyl hydrazide derivatives of a xanthine carboxylic congener (XCC) as selective antagonists at human A_{2B} adenosine receptor. Drug Dev Res 47:178–188

Kim Y-C, Ji X-D, Melman N, Linden J, Jacobson KA (2000) Anilide derivatives of a 8-phenylxanthine carboxylic congener are highly potent and selective antagonists at human A_{2B} adenosine receptors. J Med Chem 43:1165–1172

Kim S-A, Marshall MA, Melman N, Kim HS, Müller CE, Linden J, Jacobson KA (2002) Structure–activity relationships at human and rat A_{2B} adenosine receptors of xanthine derivatives substituted at the 1-, 3-, 7-, and 8-positions. J Med Chem 45:2131–2138

Kurukulasuriya R, Link JT, Madar DJ, Pei Z, Richards SJ, Rhode JJ, Souers AJ, Szczepakieicz BG (2003) Potential drug targets and progress towards pharmacologic inhibition of hepatic glucose production. Curr Med Chem 10:123–153

Lambertucci C, Campioni E, Costanzi G, Kachler S, Klotz KN, Volpini R, Cristalli G, Vittori S (2000) A bromine atom in C-8 position of 9-substituted adenines enhnces A_{2A}. Drug Dev Res 50:67–74

Mustafa SJ, Nadeem A, Fan M, Zhong H, Belardinelli, Zeng D (2007) Effect of a specific and selective A_{2B} adenosine receptor antagonist on adenosine agonist AMP and allergen-induced airway responsiveness and cellular influx in a mouse model of asthma. J Pharmacol Exp Ther 320:1246–1251

Pastorin G, Da Ros T, Spalluto G, Deflorian F, Moro S, Cacciari B, Baraldi PG, Gessi S, Varani K, Borea PA (2003) Pyrazolo[4,3-*e*]-1,2,4-triazolo[1,5-*c*]pyrimidine derivatives as adenosine receptor antagonists. Influence of the N5 substituent on the affinity at the human A_3 and A_{2B} adenosine receptor subtypes: a molecular modelling investigation. J Med Chem 46:4287–4296

Stefanachi A, Nicolotti O, Leonetti F, Cellamare S, Campagna F, Loza MI, Brea JM, Mazza F, Gavuzzo E and Carotti A (2008) 1,3-Dialkyl-8-(hetero)aryl-9-OH-9-deazaxanthines as potent A_{2B} adenosine receptor antagonists: design, synthesis, structure–affinity and structure–selectivity relationships. Bioorg Med Chem 16:9780–9789

Stewart M, Steinig AG, Ma C, Song J-P, McKibben B, Castelhano AL, MacLennan SJ (2004) [^{3}H]OSIP339391, a selective, novel, and high affinity antagonist radioligand for adenosine A_{2B} receptors. Biochem Pharmacol 68:305–312

Sun C-X, Zhong H, Mohsenin A, Morschi E, Chunn JL, Molina JG, Belardinelli L, Zeng D, Blackburn M (2006) Role of A_{2B} adenosine receptor signaling in adenosine-dependent pulmonary inflammation and injury. J Clin Invest 116:2173–2182

Vidal JB, Esteve TC (2005) Pyrimidin-2-amine derivates and their use as A_{2B} adenosine receptor antagonists. WO Patent 2005040155

Vidal JB, Esteve TC, Soca PL, Eastwood PR (2007a) Pyrazine derivatives useful as adenosine receptor antagonists. WO Patent 2007017096

Vidal JB, Fonquerna PS, Eastwood PR, Aiguade BJ, Cardus FA, Carranco MI, Gonzalez RJ, Paredes AS (2007b) Imidazopyridine derivatives as A_{2B} adenosine receptor antagonists. WO Patent 2007039297

Vidal JB, NA, Esteve TC, Domenech T, Benito S, Reinoso RF, Pont M, Calbet M, Lopez R, Cadavid MI, Loza MI, Cardenas A, Godessart N, Beleta J, Warrellow G, Ryder H (2007c) Discovery and characterization of 4′-(2-furyl)-*N*-pyridin-3-yl-4,5′-bipyrimidin-2′-amine (LAS38096), a potent, selective, and efficacious A_{2B} adenosine receptor antagonist. J Med Chem 50:2732–2736

Volpini R, Costanzi S, Vittori S, Cristalli G, Klotz KN (2003) Medicinal chemistry and pharmacology of A_{2B} adenosine receptors. Curr Top Med Chem 3:427–443

Wang G, Rieger JM, Thompson RD (2006) Pyridyl substituted xanthines. WO Patent 2006091896

Yan L, Müller CE (2004) Preparation, properties, reactions, and adenosine receptor affinities of sulfophenylxanthine nitrophenyl esters: toward the development of sulfonic acid prodrugs with peroral bioavailability. J Med Chem 47:10311043

Zeng D, Maa T, Wang U, Feoktistov I, Biaggioni I, Belardinelli L (2003) Expression and function of A_{2B} adenosine receptors in the U87MG tumor cells. Drug Dev Res 58:405–411

Zhong H, Belardinelli L, Maa T, Feoktistov I, Biaggioni I, Zeng D (2004) A_{2B} adenosine receptors increase cytokine release by bronchial smooth muscle cells. Am J Respir Cell Mol Biol 30:118–125

Zhong H, Belardinelli L, Maa T, Zeng D (2005) Synergy between A2B adenosine receptors and-hypoxia in activating human lung fibroblasts. Am J Respir Cell Mol Biol 32:2–8

Zhong H, Wu Y, Belardinelli L, Zeng D (2006) A2B adenosine receptors induce IL-19 from bronchial epithelial cells and results in TNF-alpha increase. Am J Respir Cell Mol Biol 35:587–592

Medicinal Chemistry of the A_3 Adenosine Receptor: Agonists, Antagonists, and Receptor Engineering

Kenneth A. Jacobson, Athena M. Klutz, Dilip K. Tosh, Andrei A. Ivanov, Delia Preti, and Pier Giovanni Baraldi

Contents

Abstract A_3 adenosine receptor (A_3AR) ligands have been modified to optimize their interaction with the A_3AR. Most of these modifications have been made to the N^6 and C2 positions of adenine as well as the ribose moiety, and using a combination of these substitutions leads to the most efficacious, selective, and potent ligands. A_3AR agonists such as IB-MECA and Cl-IB-MECA are now advancing into Phase II clinical trials for treatments targeting diseases such as cancer, arthritis, and psoriasis. Also, a wide number of compounds exerting high potency and selectivity in antagonizing the human (h)A_3AR have been discovered. These molecules are generally characterized by a notable structural diversity, taking into account that aromatic nitrogen-containing monocyclic (thiazoles and thiadiazoles), bicyclic

---

K.A. Jacobson (✉)
Molecular Recognition Section, Laboratory of Bioorganic Chemistry, National Institute of Diabetes, Digestive and Kidney Diseases, National Institutes of Health, Bldg. 8A, Rm. B1A-19, Bethesda, MD 20892-0810, USA
kajacobs@helix.nih.gov

C.N. Wilson and S.J. Mustafa (eds.), *Adenosine Receptors in Health and Disease,* Handbook of Experimental Pharmacology 193, DOI 10.1007/978-3-540-89615-9_5,

(isoquinoline, quinozalines, (aza)adenines), tricyclic systems (pyrazoloquinolines, triazoloquinoxalines, pyrazolotriazolopyrimidines, triazolopurines, tricyclic xanthines) and nucleoside derivatives have been identified as potent and selective A_3AR antagonists. Probably due to the "enigmatic" physiological role of A_3AR, whose activation may produce opposite effects (for example, concerning tissue protection in inflammatory and cancer cells) and may produce effects that are species dependent, only a few molecules have reached preclinical investigation. Indeed, the most advanced A_3AR antagonists remain in preclinical testing. Among the antagonists described above, compound OT-7999 is expected to enter clinical trials for the treatment of glaucoma, while several thiazole derivatives are in development as antiallergic, antiasthmatic and/or antiinflammatory drugs.

Keywords A_3 adenosine receptor · A_3 adenosine receptor agonist · A_3 adenosine receptor antagonist · Purines · Structure activity relationship · Nucleoside · G protein-coupled receptor · Neoceptor

Abbreviations

ADME	Absorption, distribution, metabolism, and excretion
AR	Adenosine receptor
b	Bovine
cAMP	Cyclic adenosine monophosphate
CHO cells	Chinese hamster ovary cells
Cl–IB–MECA	2-Chloro-N^6-(3-iodobenzyl)-5′-*N*-methylcarboxamidoadenosine
CoMFA	Comparative molecular field analysis
CVT-3146	1-{9-[(4*S*, 2*R*, 3*R*, 5*R*)-3,4-Dihydroxy-5-(hydroxymethyl) oxolan-2-yl]-6-aminopurin-2-yl}pyrazol-4-yl)-*N*-methylcarboxamide
DBXRM	7-β-D-Ribofuronamide
DHP	1,4-Dihydropyridine
Et	Ethyl
GPCR	G-protein-coupled receptor
h	Human
HEK293 cells	Human embryonic kidney 293 cells
I–AB–MECA	N^6-(4-Amino-3-iodobenzyl)-5′-*N*-methylcabroxamidoadenosine
IB–MECA	N^6-(3-Iodobenzyl)-5′-*N*-methylcarboxamidoadenosine
KF-26777	2-(4-Bromophenyl)-7,8-dihydro-4-propyl-1*H*-imidazo[2,1-*i*]purin-5(4*H*)-one
LJ-529	2-Chloro-N^6-(3-iodobenzyl)-4′-thioadenosine-5′-methyluronamide
LJ-1251	(2*R*, 3*R*, 4*S*)-2-(2-Chloro-6-(3-iodobenzylamino)-9*H*-purin-9-yl)tetrahydrothiophene-3,4-diol

LJ-1416	(2*R*, 3*R*, 4*S*)-2-(2-Chloro-6-(3-chlorobenzylamino)-9*H*-purin-9-yl)tetrahydrothiophene-3,4-diol
LUF6000	*N*-(3,4-Dichloro-phenyl)-2-cyclohexyl-1*H*-imidazo[4,5-*c*] quinolin-4-amine
Me	Methyl
MRE-3005-F20	5-*N*-(4-Methoxyphenylcarbamoyl)amino-8-ethyl-2-(2-furyl) pyrazolo[4,3-*e*]-1,2,4-triazolo[1,5-*c*]pyrimidine
MRE-3008-F20	5-*N*-(4-Methoxyphenylcarbamoyl)amino-8-propyl-2-(2-furyl) pyrazolo[4,3-*e*]-1,2,4-triazolo[1,5-*c*]pyrimidine
MRS1191	1,4-Dihydro-2-methyl-6-phenyl-4-(phenylethynyl)-3, 5-pyridinedicarboxylic acid, 3-ethyl 5-(phenylmethyl) ester
MRS1220	*N*-[9-Chloro-2-(2-furanyl)[1,2,4]triazolo[1,5-*c*]quinazolin-5-yl]benzeneacetamide
MRS1292	(2*R*, 3*R*, 4*S*, 5*S*)-2-[N^6-3-Iodobenzyl)adenos-9′-yl]-7-aza-1-oxa-6-oxospiro[4.4]-nonan-4,5-diol
MRS1523	5-Propyl-2-ethyl-4-propyl-3-(ethylsulfanylcarbonyl)-6-phenylpyridine-5-carboxylate
MRS3558	(1′*S*, 2′*R*, 3′*S*, 4′*R*, 5′*S*)-4-{2-Chloro-6-[(3-iodophenylmethyl) amino]purin-9-yl}-1-(methylaminocarbonyl)bicyclo-[3.1.0]-hexane-2,3-diol
MRS3777	2-(Phenyloxy)-N^6-cyclohexyladenine
MRS5127	(1′*R*, 2′*R*, 3′*S*, 4′*R*, 5′*S*)-4′-[2-chloro-6-(3-iodobenzylamino)-purine]-2′, 3′-*O*-dihydroxybicyclo-[3.1.0]hexane
MRS5147	(1′*R*, 2′*R*, 3′*S*, 4′*R*, 5′*S*)-4′-[2-chloro-6-(3-bromobenzylamino)-purine]-2′, 3′-*O*-dihydroxybicyclo-[3.1.0]hexane
MRS5151	(1′*S*, 2′*R*, 3′*S*, 4′*S*, 5′*S*)-4′-[6-(3-chlorobenzylamino)-2-(5-hydroxycarbonyl-1-pentynyl)-9-yl]-2′, 3′-dihydroxybicyclo[3.1.0]hexane-1′-carboxylic acid *N*-methylamide
NECA	adenosine 5′-*N*-ethyluronamide
OT-7999	5-*n*-Butyl-8-(4-trifluoromethylphenyl)-3*H*-[1,2,4]triazolo-[5,1-*i*]purine
Pr	Propyl
PSB-10	8-Ethyl-1,4,7,8-tetrahydro-4-methyl-2-(2,3,5-trichlorophenyl)-5*H*-imidazo[2,1-*i*]purin-5-one
PSB-11	(*R*)-4-Methyl-8-ethyl-2-phenyl-4,5,7,8-tetrahydro-1*H*-imidazo[2,1-*i*]purin-5-one
QSAR	Quantitative structure–activity relationships
r	Rat
SARs	Structure–activity relationships
TM	Transmembrane domain
VUF 5574	*N*-(2-Methoxyphenyl)-*N*′-(2-(3-pyridyl)quinazolin-4-yl)urea
VUF 8504	4-Methoxy-*N*-(3-(2-pyridinyl)-1-isoquinolinyl)benzamide

1 Introduction

The four subtypes of adenosine receptors (ARs), designated A_1, A_{2A}, A_{2B}, and A_3, are all seven-transmembrane spanning (7TM) receptors that couple to G proteins. The A_3AR inhibits adenylate cyclase through coupling to G_i. A_3AR activation may lead to an activation of the phospholipase C pathway through the β, γ subunit. The A_3AR is found at a high receptor density in the lungs, liver, and in immune cells such as neutrophils and macrophages, as well as at lower densities in the heart and brain (Fredholm et al. 2001) The A_3AR is expressed in neurons in the brain (Lopes et al. 2003; Yaar et al. 2002).

ARs in general, and the A_3AR in particular, are involved in many of the body's cytoprotective functions. Recently, agents that act at the A_3AR have been targeted for pharmaceutical development based on their anti-inflammatory, anticancer, and cardioprotective effects. For example, activation of the cardiac A_3AR preconditions cardiac myocytes against ischemic damage (Strickler et al. 1996; Tracey et al. 2003) and protects against apoptosis. Selective A_3AR agonists have been shown to protect cardiac muscle in various ischemic models and are protective against the cardiotoxic effects of the anticancer drug doxorubicin (Shneyvais et al. 2001). A_3AR antagonists are of interest as potential antiglaucoma agents (Yang et al. 2005) and as anticancer agents (Gessi et al. 2008).

Agonist ligands for the ARs, including the A_3AR, are almost exclusively nucleoside derivatives. The search for antagonists of the A_3AR in the early 1990s initially encountered an unanticipated difficulty: the lack of an obvious lead structure. Previously, efforts to develop antagonist ligands for the A_1 and A_{2A} ARs focused on xanthine derivatives. However, at the A_3AR, the prototypical AR antagonists (i.e., the xanthines) are typically much weaker in binding than at the other AR subtypes. This observation stimulated the screening of structurally diverse heterocyclic molecules as potential antagonists (Moro et al. 2006). Chemically diverse leads were discovered in this process that were subsequently optimized to achieve high antagonist selectivity for the A_3AR.

While the A_3AR may be activated by orthosteric agonists that are competitive with adenosine, the action of nucleosides at this receptor may also be enhanced by allosteric modulators. Several heterocyclic classes of positive allosteric modulators of the A_3AR, including 1*H*-imidazo-[4,5-*c*]quinolines such as LUF6000 (*N*-(3,4-dichloro-phenyl)-2-cyclohexyl-1*H*-imidazo[4,5-*c*]quinolin-4-amine) and pyridinylisoquinolines, have been reported (Gao et al. 2005; Göblyös et al. 2006).

The structure–activity relationships (SARs) of nucleoside derivatives in binding to the A_3AR and other ARs have been extensively studied, leading to the development of both selective agonists and, more recently, antagonists. Most of the useful modifications of adenosine **1** (Fig. 1) to achieve high A_3AR affinity and selectivity have been made at the N^6 or C2 positions of adenine or on the ribose group of adenosine. The systematic probing of SAR of both adenosine derivatives and nonpurine antagonists is frequently guided by molecular modeling (Kim et al. 2003), in which the receptor protein is modeled based on structural homology to the light receptor, rhodopsin. The effects of substitution at various

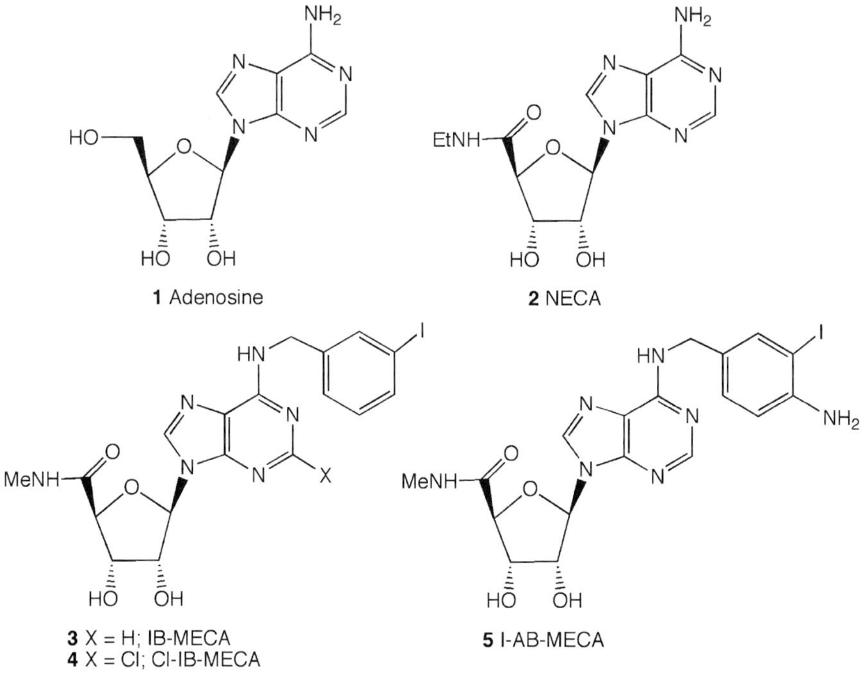

1 Adenosine

2 NECA

3 X = H; IB-MECA
4 X = Cl; Cl-IB-MECA

5 I-AB-MECA

Fig. 1 Structures of adenosine and widely used agonist probes of the A_3AR

sites (i.e., on the nucleobase and ribose moiety) on both the affinity and relative efficacy of nucleoside derivatives at the A_3AR have been extensively probed (Gao et al. 2003, 2004). The approach initially taken to identify agonists for the newly cloned A_3AR was to screen known AR ligands in binding assays. The agonist NECA **2** (adenosine 5′-*N*-ethyluronamide) was found to be highly potent but nonselective for this receptor (Zhou et al. 1992). The structural features that promoted A_3AR potency were combined, leading to the first selective A_3 agonist, IB–MECA (N^6-(3-iodobenzyl)-5′-*N*-methylcarboxamidoadenosine), developed in 1993 at the National Institutes of Health (Jacobson et al. 1993). This potent A_3AR agonist IB–MECA **3** and its more selective 2-chloro analog, Cl–IB–MECA **4** (2-chloro-N^6-(3-iodobenzyl)-5′-*N*-methylcarboxamidoadenosine), are used widely as pharmacological tools. A related derivative **5** is widely used as an iodinated radioligand for the A_3AR. IB–MECA and Cl–IB–MECA have entered clinical trials for the treatment of rheumatoid arthritis and cancer (Baharav et al. 2005; Ohana et al. 2001).

One problem encountered in refining selective A_3AR ligands into pharmaceutically useful agents has been the species dependence of binding. This difference in affinity reflects the difference in sequence between the rodent and the human receptors, with only a 74% sequence identity between the rat (r) and human (h) A_3ARs (Fredholm et al. 2001). The species-dependence of A_3AR affinity is particularly pronounced for agonists that contain small alkyl N^6 substituents and for various heterocyclic antagonists, both of which are more potent in binding to the human than to the rat A_3AR (Yang et al. 2005). The first report of a cloned receptor sequence to be later identified as an A_3AR was that of the rat (Meyerhof et al. 1991; Zhou et al. 1992), and this species was initially used for screening purposes. Nevertheless, many of the nucleoside analogs that were shown to be rat A_3AR agonists, including Cl–IB–MECA and IB–MECA, were later found to be moderately selective for the human A_3AR after it was cloned (Jacobson and Gao, 2006; Salvatore et al. 1993).

The ligand recognition within the putative binding site of the ARs has also been probed through extensive mutagenesis to confirm the predictions concerning ligand recognition made using molecular modeling (Kim et al. 2003). The hydrophobic environment surrounding the purine ring of AR agonists, as found in the putative $A_{2A}AR$ model, is defined mainly by residues of TM5 and TM6 (Kim et al. 2003). This region is very similar to the putative binding region of hydrophobic heterocyclic (e.g., triazolopyrimidine) antagonists. An exocyclic amino group is common to both adenosine agonists and to typical heterocyclic antagonists, and this amine is generally required to donate a hydrogen bond to the receptor protein. Amino acid residues involved in the ligand recognition in the putative A_{2A} and A_3 AR binding sites have been reviewed (Kim et al. 2003).

2 A_3AR Agonists

The subtype selectivity of adenosine derivatives as AR agonists has been probed extensively, principally through modification of the N^6-amine moiety (where large hydrophobic groups tend to produce A_1AR and A_3AR selectivity, Table 1) and the C2 position (where large hydrophobic groups tend to produce $A_{2A}AR$ selectivity, but have also been shown to enhance A_3AR selectivity, Table 2). The ribose moiety is less amenable than the adenine moiety to the addition of steric bulk, although substitution of the $5'$-CH_2OH moiety with certain amides, ethers, or other hydrophilic groups has resulted in enhancement of A_3AR selectivity.

The binding of a nucleoside to the A_3AR and its activation of the receptor are separate processes that appear to have distinct structural requirements. There is no general correlation between the affinity of a given nucleoside derivative in binding to the A_3AR and its ability to fully vs. partially activate the receptor (Table 1). Specific functionality on the nucleoside structure that lowers efficacy relative to that of a full agonist (e.g., NECA) has been identified. For example, N^6-benzyl and certain 2-position substituents on the adenine moiety reduce the relative efficacy at the A_3AR. 2-Chloro alone does not reduce A_3AR efficacy, but, in combination with a substituted N^6-benzyl moiety, it leads to a further reduction (Gao et al. 2002). Other N^6 substitutions have been studied using the same criteria. For example, the relative efficacy of N^6-(2-phenylethyl) derivatives is extremely sensitive to substitution of the phenyl ring and the β-methylene carbon (Tchilibon et al. 2004).

Table 1 Binding affinities of monosubstituted adenosine derivatives (N^6-substituted) at the human A_3AR expressed in CHO cells and at A_1 and A_{2A} ARs, and maximal A_3AR agonist effect

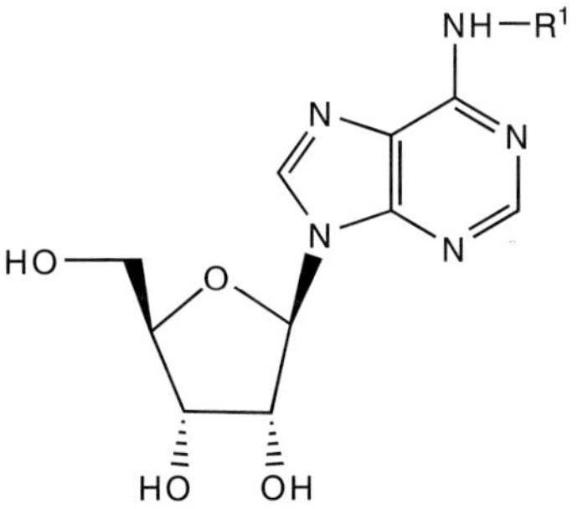

Compound	Substitution R^1 =	pK_i at A_1AR^a	pK_i at $A_{2A}AR^a$	pK_i at A_3AR^a	%Activation, A_3AR^a
6	CH_3	7.22[b]	<5[b]	8.03	96
7	CH_2CH_3	8.31[b]	5.05[b]	8.34	102
8	OCH_3	6.65[b]	<5[b]	7.55	107
9	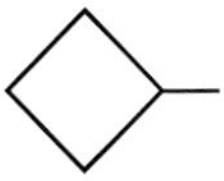	9.15[b]	5.76[b]	8.19	100
10	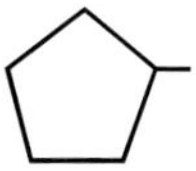	9.35[b]	6.34[b]	7.14	97
11	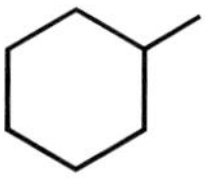	9.05[b]	6.28[b]	7.14	76
12	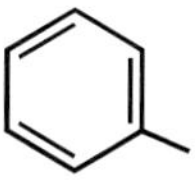	8.48[b]	6.18[b]	7.83	102
13	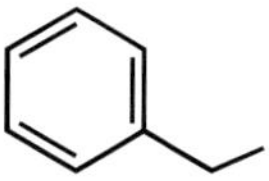	6.76[b]	6.55[b]	7.38	55
14	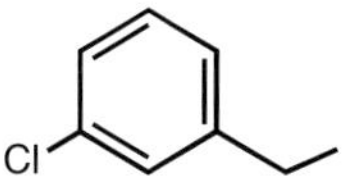 Cl	7.35	<5[b]	8.36	80
15	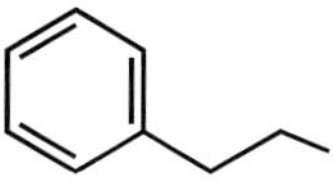	7.89, 7.89[b]	7.17, 7.17[b]	8.68, 6.62[b]	84
16	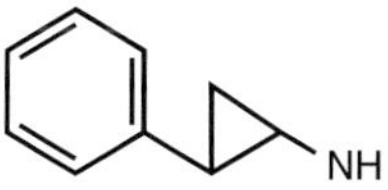 NH	6.91	5.60	9.06	101
17	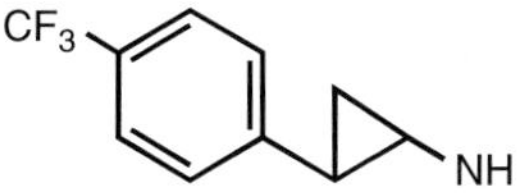 CF_3 NH	6.98	5.60	8.72	101

(continued)

Table 1 (continued)

Compound	Substitution R^1 =	pK_i at A_1AR^a	pK_i at $A_{2A}AR^a$	pK_i at A_3AR^a	%Activation, A_3AR^a
18	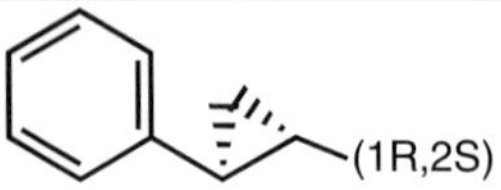 (1R,2S)	7.82[b]	5.52[b]	9.20	100
19	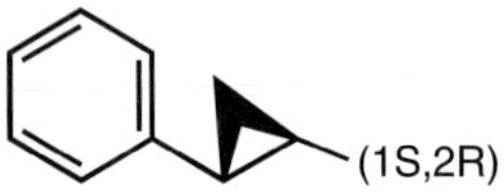 (1S,2R)	7.93[b]	5.60[b]	7.62	87
20	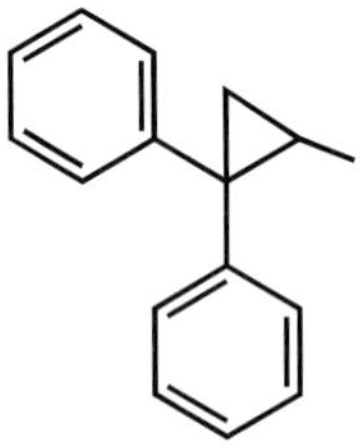	8.15	6.24	7.04	100
21	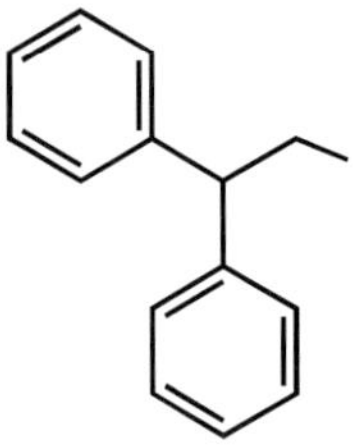	7.30	6.29	8.41	0
22	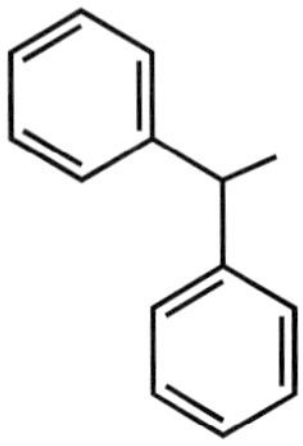	6.31	<5	5.48	87
23	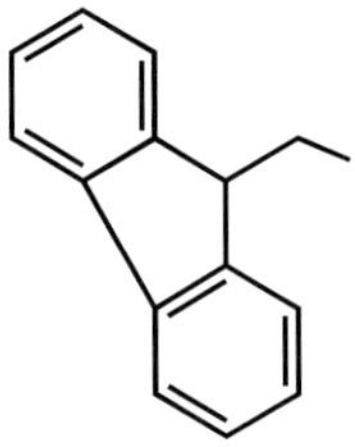	7.85	6.84	9.04	99

[a] A_3AR binding experiments were performed with membranes prepared from adherent CHO cells stably transfected with cDNA encoding the human A_3AR, using as radioligand $[^{125}I]N^6$-(4-amino-3-iodobenzyl)adenosine-5′-*N*-methyluronamide ($[^{125}I]$I–AB–MECA; 2000 Ci/mmol) at a final concentration of 0.5 nM, in Tris·HCl buffer (50 mM, pH 8.0) containing 10 mM $MgCl_2$, 1 mM EDTA. Nonspecific binding was determined using 10 μM Cl–IB–MECA. The mixtures were incubated at 25°C for 60 min. Maximal A_3AR agonist effect is the inhibition of forskolin-stimulated adenylate cyclase at 10 μM using a reference value for Cl–IB–MECA of 100%. (Gao et al. 2003; Tchilibon et al. 2004).

[b] In rat brain (Gao et al. 2003; Tchilibon et al. 2004).

Table 2 Binding affinities of monosubstituted adenosine derivatives (2-ether-substituted) at the human A_3AR expressed in CHO cells and at A_1 and A_{2A} ARs, and maximal A_3AR agonist effect

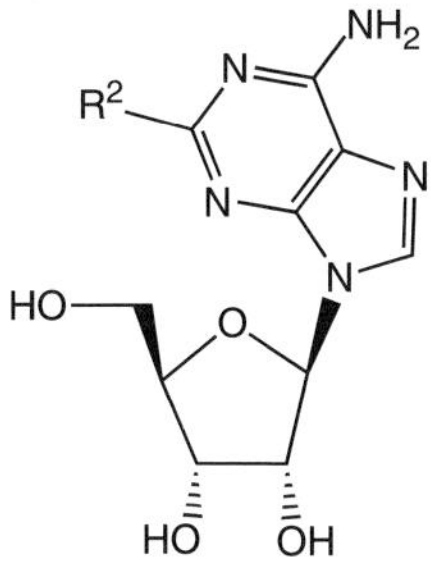

Compound	Substitution R^1 =	pK_i at A_1AR^a	pK_i at $A_{2A}AR^a$	pK_i at A_3AR^a	%Activation, A_3AR^a
29	Cl	8.12	6.20	7.06	100
30	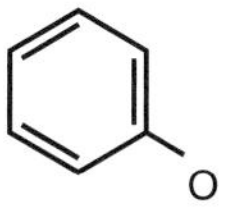 O	5.29	7.36	6.44	32
31	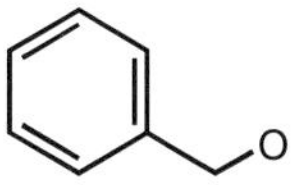 O	6.19	6.23	6.93	17
32	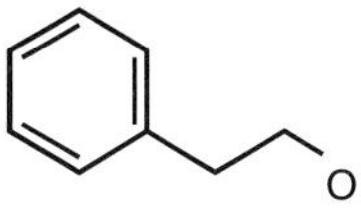 O	6.66	8.03	7.27	71
33	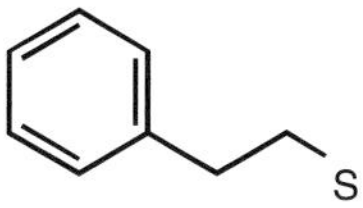 S	5.43	6.23	5.71	ND
34	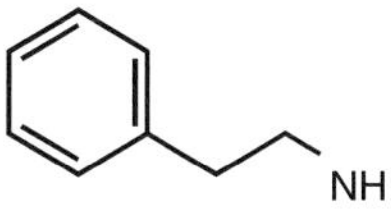 NH	6.28	7.21	6.51	72
35	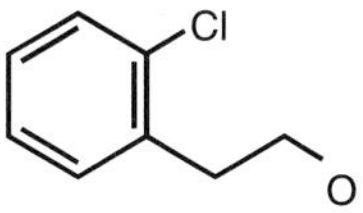 Cl, O	6.66	7.75	6.85	1
36	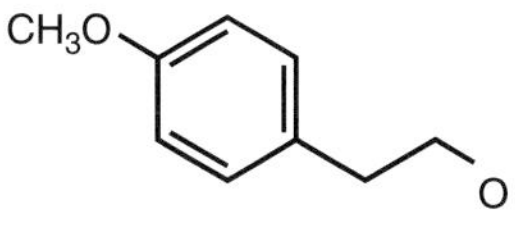 CH_3O, O	6.54	7.19	6.98	91
37	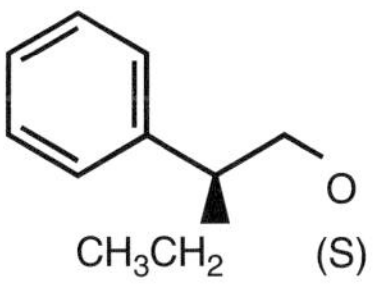 O, CH_3CH_2 (S)	5.32	7.57	6.76	0

(continued)

Table 2 (continued)

Compound	Substitution R^1 =	pK_i at A_1AR^a	pK_i at $A_{2A}AR^a$	pK_i at A_3AR^a	%Activation, A_3AR^a
38	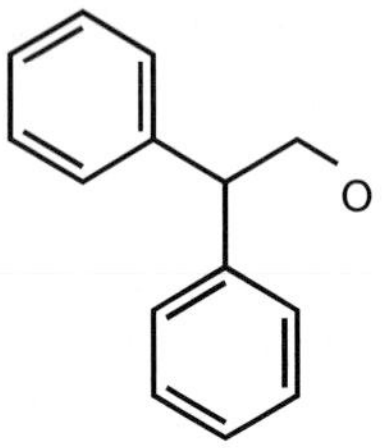	<5	6.51	7.27	0

[a] A_3AR binding experiments were performed with membranes prepared from adherent CHO cells stably transfected with cDNA encoding the human A_3AR, using as radioligand [^{125}I]N^6-(4-amino-3-iodobenzyl)adenosine-5′-*N*-methyluronamide ([^{125}I]I–AB–MECA; 2000 Ci/mmol) at a final concentration of 0.5 nM, in Tris·HCl buffer (50 mM, pH 8.0) containing 10 mM $MgCl_2$, 1 mM EDTA. Nonspecific binding was determined using 10 μM Cl–IB–MECA. The mixtures were incubated at 25°C for 60 min. ND, not determined. Maximal A_3AR agonist effect is inhibition of forskolin-stimulated adenylate cyclase at 10 μM using a reference value for Cl–IB–MECA of 100% (Gao et al. 2004)

Modifications of the ribose moiety have also been explored for effects on both A_3AR binding affinity and efficacy (Gao et al. 2004; van Tilburg et al. 2002). SAR studies also indicate that flexibility in the ribose 5′ region is a prerequisite for A_3AR activation, in concert with a proposed required rotation of TM6 (Kim et al. 2006). Thus, with proper manipulation of groups at the N^6 and/or ribose moieties, a high-affinity agonist may be converted into a selective A_3AR antagonist. Conversely, agonist function may be maintained fully with proper derivatization of the ribose moiety. A flexible 5′-uronamide moiety is particularly well suited to maintaining efficacy, and even overcomes the reduction of efficacy induced by various adenine substituents at the N^6 and C2 positions.

2.1 *Substitution of the Adenine Moiety of Adenine Nucleosides*

2.1.1 N^6 Position

Multiple studies have been undertaken to optimize the N^6 position of adenosine in order to design selective A_3AR agonists (Table 1, **6–23**). Addition of small groups such as methyl (**6**) and oxymethyl (**8**) to the N^6 amine gave at least a tenfold increase in potency over adenosine and increased the selectivity of the ligand for the human A_3AR over other human ARs (Volpini et al. 2007). However, increasing the alkyl chain length to an ethyl group (**7**) increased the affinity of the ligand for both the A_3 and A_1 ARs, thus, decreasing the selectivity (Gao et al. 2003). Larger alkyl chains were not well tolerated at the N^6 position, and increased branching of the chain caused a decrease in A_3AR affinity and efficacy. Various cycloalkyl groups were also appended to the N^6-amino group. Adding an N^6-cyclobutyl (**9**) or cyclopentyl (**10**) ring resulted in agonists that had greater affinity to the A_1AR than the A_3AR (Gao et al. 2003) but were full agonists at the A_3AR. Analogs bearing larger N^6-cycloalkyl rings such as **11** were only partially efficacious as A_3AR agonists. When a benzyl ring was attached to the N^6 amine (**13**), the compound was three- to four-fold selective in binding to the human A_3AR in comparison to the A_1 and A_{2A} ARs, but only displayed a 55% relative efficacy at the A_3AR. N^6-Phenyladenosine (**12**) was fully efficacious as an A_3AR agonist. N^6-(2-Phenylethyl)adenosine (**15**) was the most potent in binding to the A_3AR among a series of arylalkyl-substituted homologs. However, the N^6-benzyl and the N^6-phenyl substituents provided greater selectivity than 2-phenylethyl for the A_3AR. Generally, halogen substitution at the 3 position of the N^6-benzyl ring caused an increase in A_3AR affinity and selectivity. For example, N^6-(3-chlorobenzyl)adenosine (**14**) showed a tenfold selectivity for the A_3AR and a nanomolar affinity. Halogen substitution at other positions of the ring frequently decreased the A_3AR affinity.

Addition of certain larger N^6 substituents also increased the potency and affinity of the ligands at the A_3AR. For instance, N^6-(*trans*-2-phenylcycloprpropyl)adeno sine (**16**) was a full agonist with high selectivity and a subnanomolar potency (Gao et al. 2003). Further variations on this substituent were prepared, and the importance of conformational factors in the relative efficacy was demonstrated. The addition of one bond to bridge the phenyl rings could change an antagonist into an agonist. Thus, while N^6-(2,2-diphenylethyl)adenosine (**21**) was an antagonist at the A_3AR, adding a bond between the phenyl groups to create N^6-(9-fluorenylmethyl)adenosine (**23**) restored the efficacy. This compound also had a subnanomolar A_3AR affinity but was less selective than N^6-(*trans*-2-phenylcyclopropyl)adenosine (Tchilibon et al. 2004). The most selective compound of the series was N^6-(*trans*-2-(3-trifluoromethyl)phenyl)-1-cyclopropyl adenosine (**17**), which had a 100-fold selectivity at the A_3AR in comparison to the A_1AR.

Various labs have combined sterically bulky N^6 groups with a 5′-uronamide moiety on the ribose group to make potent, selective A_3AR agonists. The first A_3AR-selective compounds combined a 5′-*N*-alkyluronamide with an N^6-benzyl group (Gallo-Rodriguez et al. 1994; Jacobson et al. 1993; van Galen et al. 1994). One of the most common A_3 agonists, Cl–IB–MECA (Fig. 1), has an N^6-iodobenzyl group, a 2-chloro group, and a 5′-methylcarboxamido group. This compound has a K_i of 0.33 nM at the rat A_3AR, but K_i values of only 2,500 and 1,400 nM at rat A_1 and A_{2A} ARs, respectively (Kim et al. 1994a). At the human ARs, the binding affinities of Cl–IB–MECA are (nM): A_1AR 220, A_{2A}AR 5400, and A_3AR 1.4. Thus, Cl–IB–MECA is more selective for the rat A_3AR than the human A_3AR (Melman et al. 2008a). The ^{125}I form of I–AB–MECA (N^6-(4-amino-3-iodobenzyl)-5′-*N*-methylcarboxamidoadenosine, **5**, Fig. 1) is commonly used as a high-affinity radioligand for characterizing binding to the A_3AR of various species.

Baraldi et al. (1998) prepared a series of N^6-substituted-aminosulfonylphenyl derivatives of NECA (e.g., compound **24**). Among these compounds, the most favorable substituents of the sulfonamido group for increasing affinity at the A_3AR were small alkyl groups, such as ethyl or allyl moieties, and disubstitution of the sulfonamido group. The A_3AR selectivity was increased by the addition of a saturated heterocyclic ring, such as piperidine or morpholine, to the sulfonamido moiety.

Finally, A_3 selective fluorescent probes have also been made by attaching 7-nitrobenzofurazan fluorophores to NECA derivatives using an alkyl spacer (e.g., compound **25**). These compounds displayed 500-fold selectivity at the A_3AR and bound in the low nanomolar range (Cordeaux et al. 2008).

2.1.2 Adenine 2 Position

Many modifications at the 2 position of adenosine (Table 2, **29–38**) tend to increase $A_{2A}AR$ potency, but some additions have been found to contribute to A_3AR selectivity. Adding a simple 2-chloro group (**29**) increased the A_3AR affinity in comparison to adenosine, but it also significantly increased the potency at the A_1AR (Gao et al. 2004). Generally, 2-ether modifications decreased A_3AR affinity, with certain exceptions. For example, adding a 2-*i*-pentyloxy moiety increased A_3AR affinity threefold, and the compound was slightly selective. 2-Benzyloxy substitution (**31**) decreased the efficacy to 17% of the full agonist Cl–IB–MECA. 2-Phenylethyloxy substitution (**32**) often increased affinity at both the A_3 and A_{2A} ARs, but many such analogs displayed a decreased efficacy as A_3AR agonists. Other 2-ethers, such as 2-(2,2-diphenylethyloxy)adenosine (**38**), were A_3 AR antagonists, in curious parallel to the effect of the same group when placed at the N^6 position (Tchilibon et al. 2004).

Many other substitutions at the 2 position of adenosine were combined with previously introduced substitutions at the N^6 position of adenosine. For instance, adding a 2-cyano group to N^6-(3-iodobenzyl)adenosine created an A_3AR antagonist, but when the 2-cyano group was added to N^6-methyladenosine, the compound was a full agonist that was 30-fold selective for the human A_3AR in comparison to the A_1AR (Ohno et al. 2004). However, when other small modifications were made at the 2 position of N^6-methyladenosine, such as an amino or a trifluoromethyl group, there was a decrease in selectivity and affinity toward the A_3AR. Elzein et al. (2004) synthesized a series of 2-pyrazolyl-N^6-substituted adenosine derivatives that were very potent and selective for the A_3AR. Cosyn et al. (2006b) found that several 2-triazol-1-yl substitutions of N^6-methyladenosine increased affinity at the A_3AR. However, in order to maintain efficacy, a 5′-ethyluronamide was necessary. 9-(5-Ethylcarbamoyl-β-D-ribofuranosyl)-N^6-methyl-2-(4-pyridin-2-yl-1,2,3-triazol-1-yl)adenine **26** (LC257, Fig. 2) was a full agonist with a K_i of 1.8 nM at the A_3AR and a minimum of 900-fold selectivity over other ARs.

Additions at the 2 position of NECA often increased potency and/or selectivity. For instance, 2-(3-hydroxy-3-phenyl)propyn-1-yl-NECA **27** (PHPNECA) (Fig. 2) exhibited a subnanomolar affinity at the A_3 receptor (Volpini et al. 2002). Also, Zhu et al. (2006) made a series of N^6-ethyl-2-alkynyl-NECA derivatives which had sub- to low nanomolar affinities and were very selective in comparison to the A_{2A} and A_{2B} ARs, with some selectivity over the A_1AR. The most potent compound in that series (**28**) had a (*p*-(methoxy)phenyl)alkynyl substituent at the 2 position.

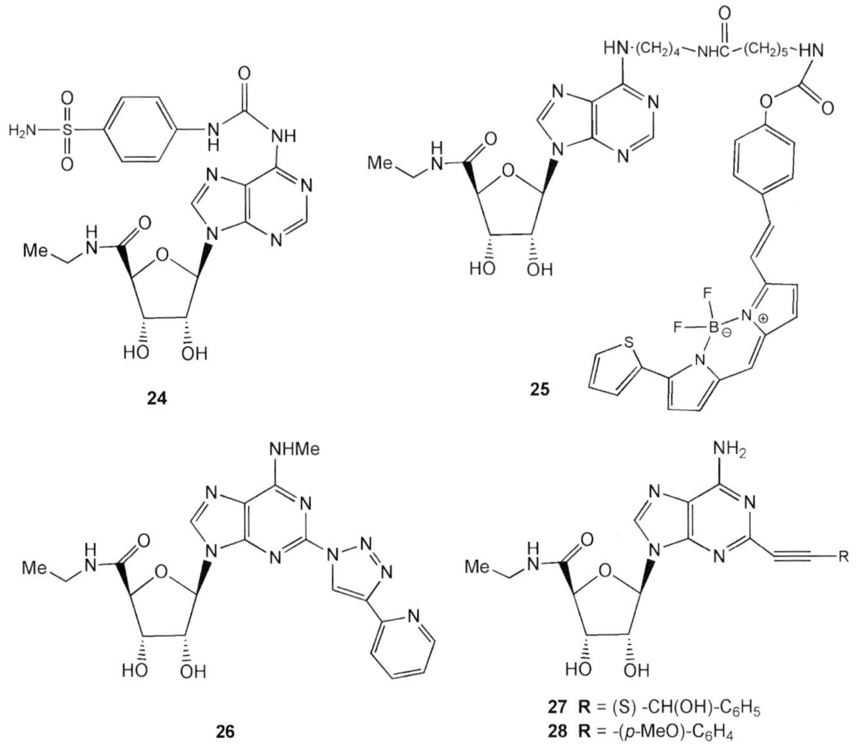

27 **R** = (S) -CH(OH)-C_6H_5
28 **R** = -(*p*-MeO)-C_6H_4

Fig. 2 Structures of a novel, multiply substituted A_3AR agonists

2.2 *Ribose Modifications*

2.2.1 Modification of Ribose Hydroxyl Groups

Many modifications have been made to the ribose ring. As mentioned above, the 5′-*N*-alkyluronamide modification has been particularly fruitful. Gallo-Rodriguez et al. (1994) initially found that adding a 5′-*N*-methyluronamide group to N^6-benzyl derivatives increased the binding affinity at all three ARs examined and resulted in several of the compounds gaining selectivity for the A_3AR. They also found that adding a 5′-*N*-ethyluronamide more than doubled the potency of several N^6-benzyl derivatives of adenosine. Other modifications at the ribose 5′ position, such as alkylthioethers (van Tilburg et al. 2002) have been found to modulate affinity and efficacy at the A_3AR.

Both the 2′- and the 3′- hydroxyl groups contribute to the binding process, since replacing either of these groups in Cl–IB–MECA with a fluoro group caused a significant drop in both affinity and efficacy (Gao et al. 2004). A less drastic decrease in binding and efficacy was seen when the 3′-hydroxyl of the adenosine analogs was replaced with an amino group (DeNinno et al. 2003). When a methylene spacer

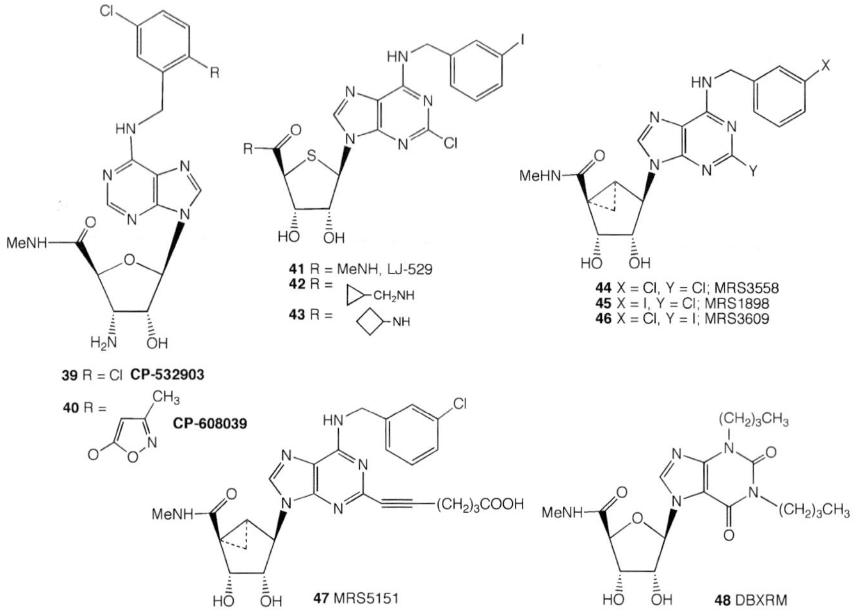

Fig. 3 Structures of ribose ring-modified selective A_3AR agonist probes

was added between the 3′-amino and ribose groups, there was a total loss of affinity (Van Rompaey et al. 2005). Also, 3′-deoxy-3′-acetylamino analogs were weak at the A_3AR. However, DeNinno et al. (2003, 2006) found that the 3′-amino substitution was tolerated and gave high selectivity when the 5′ and N^6 positions of adenosine were also appropriately modified in compounds **39** and **40** (Fig. 3). Replacement of the 3′-hydroxyl with an azido group generally abolished A_3AR activation. The 2′-hydroxyl group appeared to be more important than the 3′-hydroxyl group, because when it was replaced with the fluoro group there was no binding or activation of the A_3AR (Gao et al. 2004).

2.2.2 Modification of the Pentose Ring

4′-Thio derivatives were usually equipotent or slightly more potent at ARs than their oxygen equivalents (Jeong et al. 2006a). Many 4′-thio derivatives of adenosine have been found to be full agonists. For example, LJ-529 **41** (2-chloro-N^6-(3-iodobenzyl)-4′-thioadenosine-5′-methyluronamide) (Fig. 3) is a highly potent ligand (K_i = 0.38 nM against [^{125}I]–AB–MECA binding to the human A_3AR expressed in CHO cells). In the same 4′-thio-modified series, a wide variety of ribose 5′-alkyluronamides have shown that there is tolerance for groups larger than N-ethyl (Jeong et al. 2006a). For example, compounds **42** and **43** were full agonists with K_i values of 3.6 and 18 nM at the hA_3AR, respectively. The nature of the N-alkyl or N-arylalkyl group can modulate affinity and efficacy at the A_3AR (Jeong et al. 2008).

However, when the thio modification was combined with shifting the adenine moiety of Cl–IB–MECA from the 1′ to the 4′ position of the ribose ring, the compound was curiously transformed into a potent antagonist (Gao et al. 2004).

Ring-constrained nucleosides have been used to define conformational preferences at the A_3AR. Medicinal chemists frequently utilize the approach of conformationally constraining otherwise flexible molecules to probe the "active" conformation(s) and to increase ligand affinity by overcoming the energy barriers needed to attain this preferred conformation. Nucleoside analogs containing novel rigid ring systems in place of the ribose ring have been explored as ligands for the ARs. The focus on conformational factors of the ribose or ribose-like moiety allows the introduction of general modifications that lead to enhanced potency and selectivity at certain subtypes of these receptors. One ring system selected for this purpose is the methanocarba (bicyclo[3.1.0]hexane) ring system, which has been incorporated in either of two isomeric forms that adopt either a North (N) or South (S) envelope conformation (Jacobson et al. 2000; Marquez et al. 1996). These ribose modifications were combined with known enhancing modifications at other positions on the molecule to explore the resulting SARs. (N)-Methanocarba-adenosine was favored in binding at the A_3AR by 150-fold over the (S) conformation and by 2.5-fold over adenosine. Doubly modified nucleoside derivatives containing the (N)-methanocarba ring system have confirmed that this conformation of the pseudoribose ring is highly preferred over the (S) conformation for agonists at the A_3AR in general.

Introducing an (N)-methanocarba modification to adenosine 5′-ethyluronamide increased the human A_3AR binding affinity by sixfold. This modification also demonstrated that the ring oxygen is not required for binding or activation of the receptor (Lee et al. 2001).

Highly selective ring-constrained agonists of the A_3AR have been designed and synthesized based on the (N)-methanocarba ring system (Fig. 3). This led to the introduction of MRS3558 **44** ((1′*R*,2′*R*,3′*S*,4′*R*,5′*S*)-4-{2-chloro-6-[(3-iodophenylmethyl)amino]purin-9-yl}-1-(methylaminocarbonyl)bicyclo-[3.1.0]-hexane-2,3-diol) as a full agonist with subnanomolar potency at the A_3AR and its congeners (e.g., **45** and **46**) as full agonists with nanomolar potency at the A_3AR (Tchilibon et al. 2005). The SAR of MRS3558 and related congeners as A_3AR agonists (Melman et al. 2008a) was recently explored in detail. The utility of MRS3558 in treating lung injury was shown in a model of ischemia reperfusion lung injury (Matot et al. 2006). In this series of (N)-methanocarba nucleosides, a 5′-uronamide moiety is needed in order to achieve full efficacy at the A_3AR. The corresponding 5′-alcohol is an antagonist of the A_3AR. The 5′-uronamide moiety overcomes the loss of efficacy associated with substitution of the N^6 and ribose ring moieties. Thus, in the (N)-methanocarba series, as in the ribose series, a freely rotating 5′-uronamide that is able to make and break multiple hydrogen bonds provides a necessary degree of flexibility during the receptor activation step.

2.3 Nonadenine Nucleosides and Nonnucleosides as A_3AR Agonists

Occasionally, nonadenine nucleotides are also found to activate the A_3AR. For instance, xanthines such as caffeine are generally found to act as antagonists, but *N*-methyl-1,3-dibutylxanthine 7-β-D-ribofuronamide **48** (DBXRM) acted as a moderately selective A_3AR agonist (Kim et al. 1994b).

A series of atypical, nonnucleoside agonist ligands that activated various ARs were reported (Chang et al. 2005). In addition to compounds in this family of pyridine-3,5-dicarbonitriles that were selective agonists of the A_1AR, various members of this series substantially activated the A_3AR.

2.4 Further Optimization of A_3AR Agonists Using Multiple Modifications

Interestingly, certain modifications (such as a 5′-alkylamide or an N^6-methyl group) can restore efficacy to previously modified compounds. For instance, adding a 2-chloro group to N^6-cyclopentyladenosine creates an A_3AR antagonist (Gao et al. 2002), but activation is restored by the 5′-methylcarboxamide and 4-thio substitutions. This is particularly interesting since 4′-thioadenosine is also an A_3AR antagonist, and 2-chloro-4-thioadenosine is only a partial agonist (Jeong et al. 2006b).

A series of (N)-methanocarba-2,N^6-disubstituted adenine nucleosides were made by Tchilibon et al. (2004), who found that adding the (N)-methanocarba, 2-chloro, and 5′-methyluronamido groups significantly improved the selectivity and efficacy of several compounds. For instance, N^6-(2,2-diphenylethyl)adenosine was an A_3AR antagonist with 12-fold and 130-fold selectivity over A_1 and A_{2A} ARs, respectively. However, by adding the above substitutions, the compound became a full agonist with a K_i of 0.69 nM and a selectivity of close to 2,000-fold over A_1 and A_{2A} AR (Tchilibon et al. 2004). Also, the 2-cyano derivative of N^6-methyl adenosine was a full agonist whereas the 2-cyano derivative of N^6-(2-phenylcyclopropyl) adenosine was an A_3AR antagonist (Ohno et al. 2004).

Adding several substitutions may also improve selectivity for the A_3AR. Adding an N^6-methyl group and 2-chloro group to 4′-thioadenosine-5′-methyluronamide created a compound with a K_i of 0.28 nM and a nearly 5,000-fold selectivity for the A_3AR (Jeong et al. 2006a). A series of these compounds was made by varying the N^6 and 5′ groups. While none of these derivatives could match the potency and selectivity of the original compound, it was found that 4′-thioadenosine derivatives were often more potent than their oxy counterparts. The most potent compound was 9-(3-amino-3-deoxy-5-methylcarbamoyl-β-D-ribofuranosyl)-2-amino-N^6-methylpurine. Another highly substituted yet extremely potent N^6-methyl derivative is 2-chloro-N^6-methyl-4-thioadenosine-5-methyluronamide, which has a K_i of 0.28 nM (Jeong et al. 2006a). N^6-Methylation also seems to improve human A_3AR selectivity, as N^6-methyl-2-(2-phenylethyl)-adenosine is much more selective than 2-(2-phenylethyl)-NECA (Volpini et al. 2002). While large 2-position substitutions are not always tolerated, (2*R*,3*S*,4*R*)-tetra hydro-2-(hydroxymethyl)-5-(6-(methylamino)-2-(4-pyridin-2-yl)-1*H*-pyrazol-1-yl) -9*H*-purin-9-yl) furan-3,4-diol had a K_i of 2 nM and was extremely selective (Van Rompaey et al. 2005).

Recently, new potent and A_3-selective N^6,2-disubstituted adenosine derivatives have been reported. Volpini et al. (2007) made a series of N^6-methoxy-2-alkyladenosine derivatives, of which N^6-methoxy-2-*p*-acetylphenylethylMECA was the most potent and selective. This compound had a K_i of 2.5 nM at the human A_3AR and selectivities of 21,000 and 4,200 against A_1 and A_{2A} ARs, respectively. Recently, a series of water-soluble A_3AR agonists were synthesized (DeNinno et al. 2006). Of these compounds, (2*S*,3*S*,4*R*,5*R*)-3-amino-5-{6-[5-chloro-2-(2-oxo-2-piperazin-1-yl-ethoxy)-benzylamino]-purin-9-yl}-4-hydroxy-tetrahydro-furan-2-carboxylic acid methylamide was the most potent/selective derivative, with a K_i of 10 nM. Van Rompaey et al. (2005) found that adding additional substitutions to 3-amino-3-deoxyadenosine increased the potency, but these compounds were only partial agonists. 9-[3-Amino-3-deoxy-5-(methylcarbamoyl)-β-D-ribofuranosyl]-N^6-(5-chloro2-methoxybenzyl)adenine had a K_i of 27 nM and was extremely selective for the A_3AR, but had an efficacy of only 51%. Cosyn et al. (2006a) made a series of 3′-amino-3′-deoxy congeners that were highly selective for the A_3AR.

The selectivity at the mouse A_3AR of analogs containing the (N)-methanocarba ring system was reduced due to an increased tolerance of this ring system at the mouse A_1AR (Melman et al. 2008a). Substitution of the 2-chloro atom with iodo or hydrophobic alkynyl groups tended to increase the A_3AR selectivity (up to 430-fold) in mouse and preserve it in human. Extended and chemically functionalized alkynyl chains attached at the C2 position of the purine moiety preserved A_3AR selectivity more effectively than similar chains attached at the 3 position of the N^6-benzyl group. For example, the carboxylic acid congener MRS5151 **47** (Fig. 3) is a highly potent agonist (K_i 2.38 nM at hA_3AR) and is selective in binding at human (6,260-fold) and mouse (431-fold) A_3ARs in comparison to A_1ARs in the same species.

3 A_3AR Antagonists

Initial attempts at obtaining potent and highly selective A_3AR antagonists focused on wide pharmacological screening of different heterocyclic compounds (Jacobson et al. 1995; Ji et al. 1996; Siddiqi et al. 1995). One of the first nonxanthine heterocyclic derivatives (Fig. 4) found to be selective for the human A_3AR (K_i 0.65 nM) was MRS1220 (*N*-[9-chloro-2-(2-furanyl)[1,2,4]triazolo[1,5-*c*]quinazolin-5-yl]benzeneacetamide) **49**, which was based on appropriate acylation of the exocyclic amino group of this class of known AR antagonists (Kim et al.

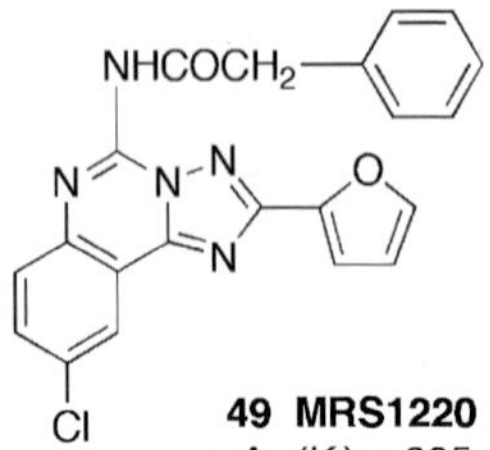

49 MRS1220
rA_1 (K_i) = 305 nM
rA_{2A} (K_i) = 52 nM
hA_3 (K_i) = 0.65 nM

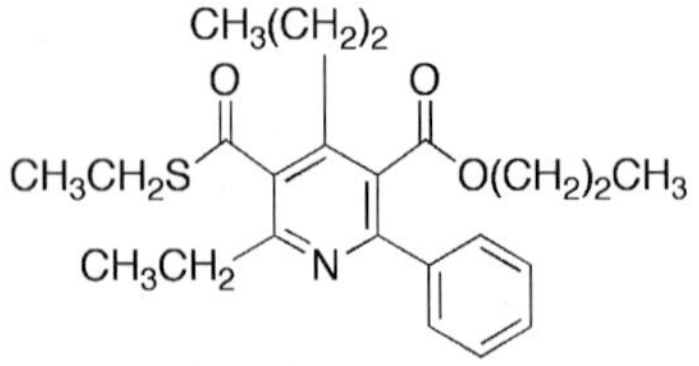

50 MRS1523
rA_1 (K_i) = 15,600 nM
rA_{2A} (K_i) = 2050 nM
hA_3 (K_i) = 18.9 nM

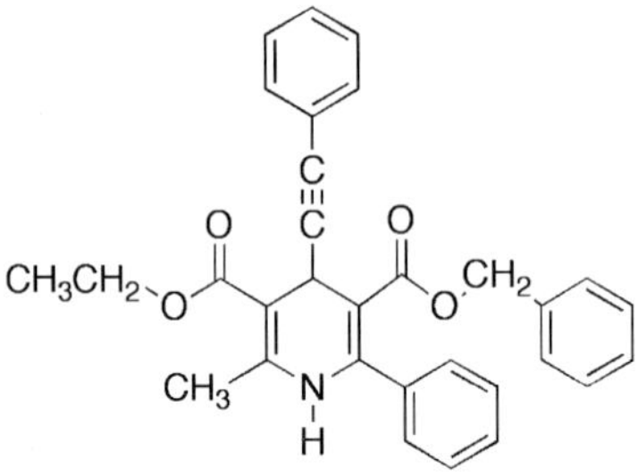

51 MRS1191
rA_1 (K_i) = 40,100 nM
rA_{2A} (K_i) = >100,000 nM
hA_3 (K_i) = 31.4 nM

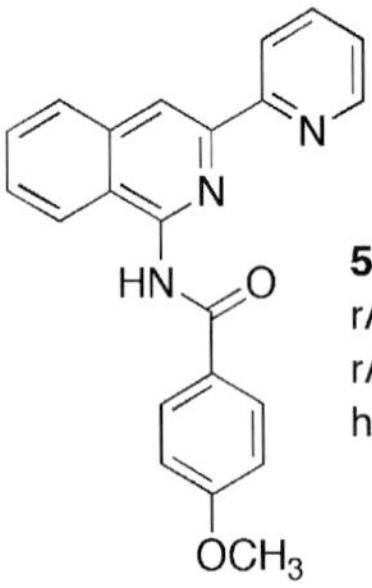

52 VUF 8504
rA_1 = 14 % (10 μM)
rA_{2A} = 0 % (10 μM)
hA_3 (K_i) = 17 nM

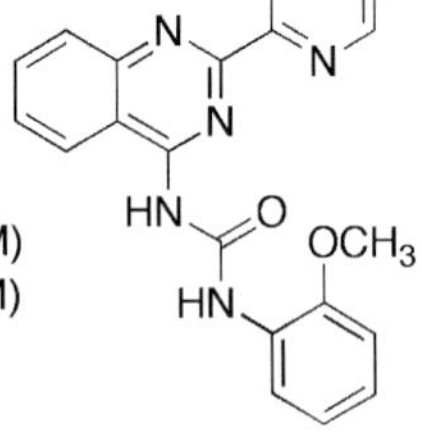

53 VUF 5574
rA_1 = 52 % (10 μM)
rA_{2A} = 43 % (10 μM)
hA_3 (K_i) = 4 nM

Fig. 4 Structures of heterocyclic derivatives that are widely used as selective human A_3AR antagonists

1996). During subsequent evaluations, different classes of nonxanthine nitrogen-containing molecules were identified as potent A_3AR antagonists: flavonoids, 1,4-dihydropyridines and pyridines, triazoloquinazolines, isoquinolines and quinazolines (Baraldi et al. 2003a; Müller et al. 2003). The 1,4-dihydropyridine (DHP) derivative MRS1191 (1,4-dihydro-2-methyl-6-phenyl-4-(phenylethynyl)-3,5-pyridinedicarboxylic acid, 3-ethyl 5-(phenylmethyl) ester) **51** was structurally optimized for binding to the A_3AR (K_i 31 nM) from library screening that identified various DHP calcium channel blockers as weak A_3AR antagonists (Jacobson et al. 1997). The pyridine derivative MRS1523 (5-propyl-2-ethyl-4-propyl-3-(ethylsulfanylcarbonyl)-6-phenylpyridine-5-carboxylate) **50** was the first heterocyclic A_3AR antagonist to display considerable potency and selectivity for the rat A_3AR (K_i 113 nM), as well as the human (18.9 nM) and mouse A_3AR (Li et al. 1998). In this section, the recent advancements in this field have been summarized, with particular attention paid to the most important reports of the last five years.

3.1 Recent Developments in Nonpurine Heterocycles

3.1.1 Thiazole and Thiadiazole

IJzerman and coworkers investigated a series of 3-(2-pyridinyl)-isoquinoline derivatives for their affinity at the A_3AR (Van Muijlwijk-Koezen et al. 2000). The effect of an additional nitrogen atom was valued by synthesizing bioisosteric quinazoline derivatives. The compounds VUF 8504 (4-methoxy-*N*-(3-(2-pyridinyl)-1-isoquinolinyl)benzamide, **52**) and VUF 5574, (*N*-(2-methoxyphenyl)-*N*′-(2-(3-pyridyl)quinazolin-4-yl)urea, **53**) (Fig. 4) display considerable A_3AR affinity and appreciable selectivity versus A_1 and A_{2A} AR subtypes.

The bicyclic system of isoquinoline and quinazoline has been replaced by several monocyclic rings (Van Muijlwijk-Koezen et al. 2001). Some thiazole and thiadiazole derivatives were shown to be most promising candidates for the identification of new A_3AR ligands.

The derivative *N*-[3-(4-methoxy-phenyl)-[1,2,4]thiadiazol-5-yl]-acetamide (**54**, Fig. 5) has been claimed to be the most potent A_3AR antagonist of the series, exhibiting a K_i value of 0.79 nM at hA_3AR and antagonistic properties in a cAMP functional assay (Jung et al. 2004). A series of potent and selective A_3AR antagonists have been obtained via an optimization study of compound **55** that revealed that a 5-(pyridine-4-yl) moiety on the 2-aminothiazole ring was optimal for enhanced receptor potency and selectivity (Press et al. 2004). Of particular note, *N*-[4-(3,4,5-trimethoxyphenyl)-5-pyridin-4-ylthiazol-2-yl]-acetamide **56** showed subnanomolar affinity at the human A_3AR as a competitive antagonist of $[^{125}I]$I–AB–MECA, binding with 1,000-fold selectivity versus the other ARs.

Binding affinity data on thiazole and thiadiazole derivatives at the hA_3AR have been subjected to QSAR analysis (Bhattacharya et al. 2005). This study disclosed the importance of the molecular electrostatic potential surface (Wang–Ford charges)

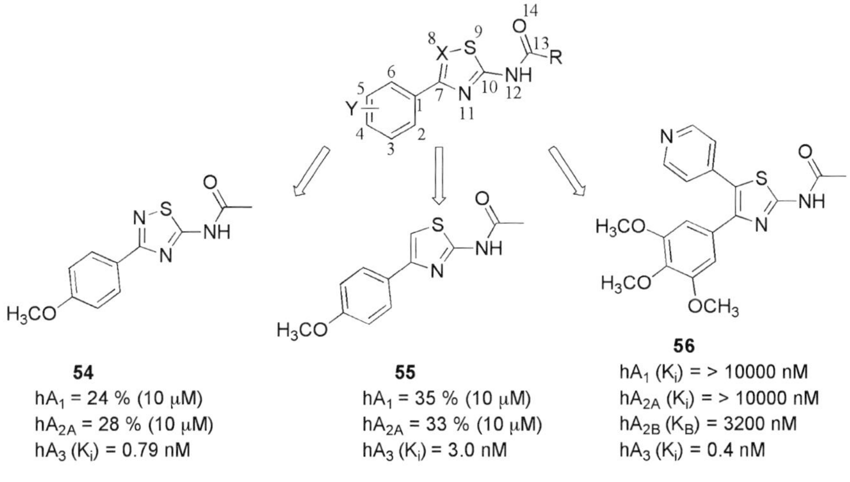

54
hA_1 = 24 % (10 μM)
hA_{2A} = 28 % (10 μM)
hA_3 (K_i) = 0.79 nM

55
hA_1 = 35 % (10 μM)
hA_{2A} = 33 % (10 μM)
hA_3 (K_i) = 3.0 nM

56
hA_1 (K_i) = > 10000 nM
hA_{2A} (K_i) = > 10000 nM
hA_{2B} (K_B) = 3200 nM
hA_3 (K_i) = 0.4 nM

Fig. 5 Thiazole and thiadiazole derivatives as human A_3AR antagonists

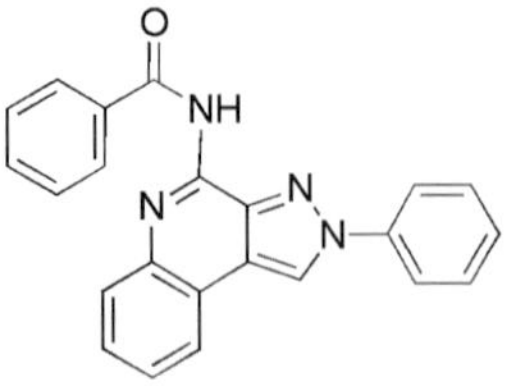

57
bA_1 = 42 % (20 μM)
hA_1 = 0 % (10 μM)
bA_{2A} = 3 % (20 μM)
hA_{2A} = 9 % (10 μM)
hA_3 (K_i) = 2.1 nM

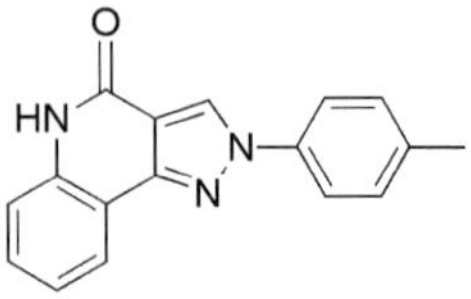

58
hA_1 (K_i) > 1 μM
hA_{2A} (K_i) > 1 μM
hA_{2B} (IC_{50}) > 1 μM
hA_3 (K_i) = 9 nM

Fig. 6 Pyrazoloquinoline derivatives as human A_3AR antagonists

in relation to atoms C2, C5, C7, X8 and S9 (Fig. 5), the last two playing the most important roles. Furthermore, the A_3AR binding affinity increases with decreasing lipophilicity of the compounds and in the presence of short alkyl chains—methyl (Me) or ethyl (Et)—at the R position.

3.1.2 Pyrazoloquinolines

The binding affinities at bovine A_1 and A_{2A} ARs and at human cloned A_3ARs of some 2-arylpyrazolo[3,4-*c*]quinolin-4-ones along with their corresponding 4-amines and 4-substituted-amino derivatives were reported by Colotta et al. (2000). The 4-benzoylamido derivative **57** (Fig. 6) displayed one of the best binding profiles of the series of A_3AR antagonists. The same group recently reported an extension of the SAR study of this class of compounds (Colotta et al. 2007) which highlighted that bulky and lipophilic acyl–amino groups at the 4 position seemed able to promote hA_3AR potency and selectivity. Selected compounds of these series were tested in an in vitro rat model of cerebral ischemia and prevented the irreversible failure of synaptic activity induced by oxygen and glucose deficiency in the hippocampus, thus confirming that potent and selective A_3AR antagonists may substantially increase the tissue resistance to ischemic damage.

The synthesis and the affinity profile at ARs of a series of 2-phenyl-2,5-dihydro-pyrazolo[4,3-*c*]quinolin-4-ones, conceived as structural isomers of the parent 2-arylpyrazolo[3,4-*c*]quinoline derivatives, have also been reported (Baraldi et al. 2005a). Some of the synthesized compounds showed A_3AR affinities in the nanomolar range and good selectivities, as evaluated in radioligand binding assays at hARs. In particular, substitution at the 4 position of the 2-phenyl ring with methyl, methoxy, or chlorine and the presence of a 4-oxo functionality gave good activity and selectivity (**58**).

3.1.3 Triazoloquinoxalines

Triazolo[4,3-*a*]quinoxalines

Interesting studies performed in the last decade by Colotta and coworkers highlighted that the 1,2,4-triazolo[4,3-*a*]quinoxalin-1-one moiety is an attractive

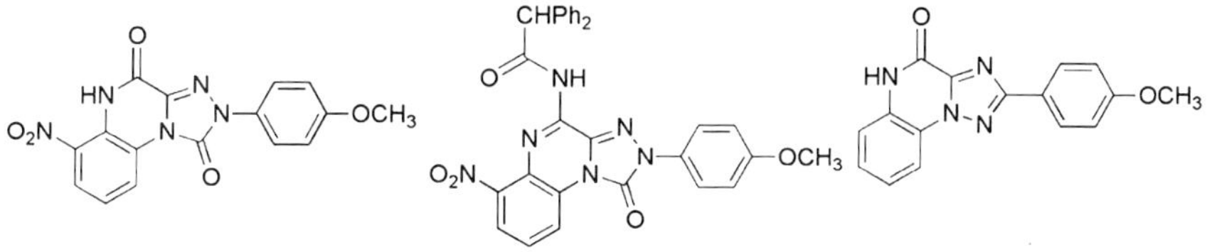

59
hA_1 = 0% (10 μM)
bA_{2A} = 14 % (20 μM)
hA_3 (K_i) = 4.7 nM

60
hA_1 = 11% (10 μM)
hA_{2A} = 2 % (10 μM)
hA_3 (K_i) = 0.8 nM

61
bA_1 = 0 % (34 μM)
bA_{2A} = 8.7 % (34 μM)
hA_3 (K_i) = 2.1 nM

Fig. 7 Triazoloquinoxaline derivatives as A_3AR antagonists

scaffold for obtaining potent and selective hA_3AR antagonists (Colotta et al. 2004; Lenzi et al. 2006). Intensive efforts in the chemical synthesis of compounds based on the systematic substitution of the 2, 4 and 6 positions of the tricyclic template, along with molecular modeling investigations performed to rationalize the experimental SAR findings, led to the identification of optimal structural requirements for A_3AR affinity and selectivity. In particular, the introduction into the triazoloquinoxaline moiety of a 4-oxo (**59**) or a 4-*N*-amido (**60**, Fig. 7) function affords selective and/or potent A_3AR antagonists, indicating that a $C=O$ group (either extranuclear or nuclear) is necessary for A_3AR affinity. This suggested that the probable engagement of this site of the molecule is a hydrogen bond with the A_3AR binding site. Hindering and lipophilic acyl–amino moieties at the 4 position showed enhanced A_3AR affinity (**60**). Substitution of the 4 position of the 2-phenyl ring with a methoxy or a nitro group and 6-nitro substitution, as well as the combination of these substituents, afforded nanomolar A_3AR affinity and better A_3AR selectivity. 1-Oxo, 6-nitro, and 4-amino groups have been proposed to be involved in hydrogen bonds that anchor the antagonists to the binding site.

Triazolo[1,5-*a*]quinoxalines

Some 2-aryl-8-chloro-1,2,4-triazolo[1,5-*a*]quinoxaline derivatives have been synthesized and tested in radioligand binding assays at bovine (b) A_1 and bA_{2A}ARs and at hA_1 and hA_3ARs (Catarzi et al. 2005a, b). The SAR of these compounds are in agreement with those of previously reported for 2-aryl-1,2,4-triazolo[4,3-*a*]quinoxalines and 2-arylpyrazolo[3,4/4,3-c]quinolines, thus suggesting a similar AR-binding mode. These studies provided some interesting compounds; among them, 2-(4-methoxyphenyl)-1,2,4-triazolo[1,5-*a*]quinoxalin-4-one (**61**, Fig. 7) is the most potent and selective hA_3AR antagonist of this series.

3.1.4 Pyrazolo[4,3-*e*]-1,2,4-triazolo[1,5-*c*]pyrimidines

The first example of an AR antagonist containing the pyrazolo-triazolo-pyrimidine scaffold (Cacciari et al. 2007) was reported by Gatta and coworkers (Gatta et al. 1993).

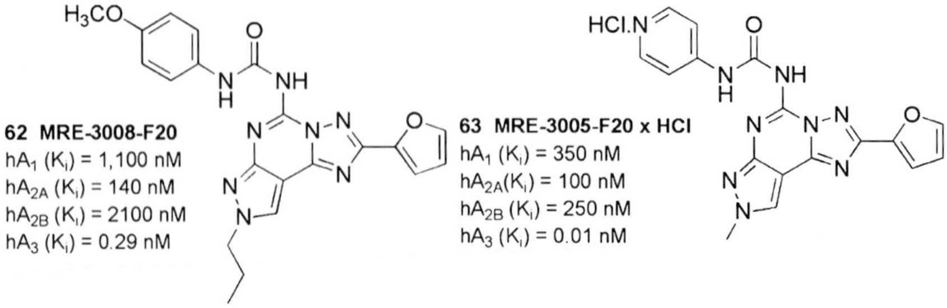

62 MRE-3008-F20
hA_1 (K_i) = 1,100 nM
hA_{2A} (K_i) = 140 nM
hA_{2B} (K_i) = 2100 nM
hA_3 (K_i) = 0.29 nM

63 MRE-3005-F20 x HCl
hA_1 (K_i) = 350 nM
hA_{2A}(K_i) = 100 nM
hA_{2B} (K_i) = 250 nM
hA_3 (K_i) = 0.01 nM

Fig. 8 A_3AR antagonists based on a pyrazolo-triazolo-pyrimidine scaffold

A wide number of compounds (MRE series) originated from the structure–activity optimization work based on systematic substitution at the C2, C5, C9, N7, and N8 positions (Baraldi et al. 2002a; 2003b; 2006). The N^7-substituted derivatives were found to bind principally to the hA_{2A}AR (Baraldi et al. 2002b), while the most potent and selective hA_3AR antagonists in this series were derived from the combination of a small alkyl chain at the N^8-pyrazole position with a (substituted)phenylcarbamoyl chain at the N5 position (Baraldi et al. 2003a). The compound designated MRE-3008-F20, (5-*N*-(4-methoxyphenylcarbamoyl)amino-8-propyl-2-(2-furyl)pyrazolo[4,3-*e*]-1,2,4-triazolo[1,5-*c*]pyrimidine, **62**) (Fig. 8), one of several high-affinity antagonists of this series, is a highly potent A_3AR ligand ($K_i = 0.29$ nM against [^{125}I]I–AB–MECA binding to human AR receptors expressed in HEK293 cells) with good selectivity over the other hARs. It showed antagonist activity in a functional assay blocking the effect of IB–MECA on cAMP production in CHO cells with an IC_{50} value of 4.5 nM. [^{3}H]MRE 3008-F20 shows a K_d value of 0.82 ± 0.08 nM and a B_{max} value of 297 ± 28 fmol mg^{-1} protein (Varani et al. 2000).

An important problem with the pyrazolo-triazolo-pyrimidine series was the low water solubilities typically observed, which could limit their use as pharmacological and diagnostic ligands. The bioisosteric replacement of the phenyl ring of the 5-phenylcarbamoyl moiety with a 4-pyridyl moiety (Maconi et al. 2002) provided high water solubility while enhancing hA_3AR affinity. Compound MRE-3005-F20, (5-*N*-(4-methoxyphenylcarbamoyl)amino-8-ethyl-2-(2-furyl)pyrazolo[4,3-*e*]-1,2,4-triazolo[1,5-*c*]pyrimidine, **63**) and the corresponding HCl salt, which showed very high affinities and good selectivities at the hA_3 receptor subtype, with K_i values in the picomolar range (40 and 10 pM, respectively), can be considered ideal candidates for pharmacological and clinical investigations of the hA_3AR subtype. Receptor modeling ascribed this increase in affinity, compared to neutral arylcarbamate derivatives, to strong electrostatic interactions between the pyridinium moiety and the side chain carbonyl oxygen atoms of Asn274 and Asn278, both located on TM7. Additional studies suggested that involvement of the residue Tyr254 in a hydrogen bond with the pyridyl ring was responsible for both enhanced receptor affinity and selectivity (Tafi et al. 2006). The replacement of the N^5-pyridine moiety with several N^5-heteroaryl rings produced a general loss of affinity and selectivity at the hA_3AR (Pastorin et al. 2006).

In order to rationally design and synthesize hA_3AR antagonists with improved binding and/or absorption, distribution, metabolism, and excretion (ADME) profiles, and as suitable clinical candidates, different molecular modeling investigations have been carried out in the last years. Particular attention has been paid to the pyrazolo[4,3-*e*]-1,2,4-triazolo[1,5-*c*]pyrimidine family, the most potent class of A_3AR antagonists ever reported (Tafi et al. 2006). A combined target-based (high-throughput molecular docking) and ligand-based (CoMFA) (comparative molecular field analysis) drug design approach has recently been performed by Moro and coworkers (Moro et al. 2005), which defined a novel "Y-shaped" binding motif for pyrazolo-triazolo-pyrimidines and rationally delineated some key ligand–receptor interactions for this class of molecules as follows: (1) steric control around the 3 and 4 positions of the N^5-phenyl ring justifies the decrease in affinity of 3- or 4-substituted-phenyl derivatives; (2) an important $\pi-\pi$ interaction takes place between the 2-furyl ring and two phenylalanine residues of the binding site; (3) a hydrophobic pocket, bordered by two hydrophilic amino acids, surrounds the N8 interaction area; and (4) strong hydrogen bonding is possible between a residue of Asn and the N4 of the triazolo ring.

3.1.5 Various Heterocycles

In the last few years, other classes of heterocyclic compounds have been identified as A_3AR antagonists, but large structural dissimilarities meant that none of these could be classified into particular family groups. The quinoxaline derivative **64** (Fig. 9) deserves to be mentioned here, not only because of its good binding profile as an A_3AR antagonist, but also (especially) due to the novelty of the strategy applied to its design, which was based on a 3D database-searching approach (Novellino et al. 2005). There is increasing evidence of the importance of 2D/3D database searching as a valuable tool to discover novel lead compounds for the A_3AR and for other G-protein-coupled receptors (GPCRs) (Costanzi et al. 2008).

The structural manipulation of a series of phenyltriazolobenzotriazindiones, previously described as ligands at the central benzodiazepine receptor, led Da Settimo

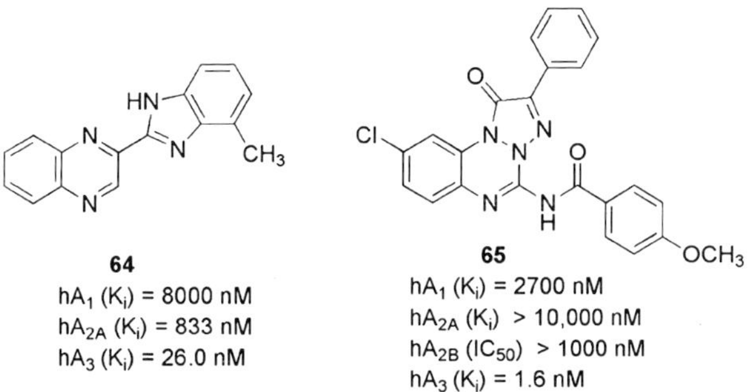

Fig. 9 A_3AR antagonists based on quinoxaline and triazolobenzotriazinone scaffolds

and coworkers to the identification of a series of aminophenyltriazolobenzotriazinones. Among these, compound **65**, a result of a systematic lead optimization, stands out for its remarkable potency and selectivity at the A_3AR (K_i values at the A_1, A_{2A}, A_3ARs of 2,700 > 10,000, 1.6 nM, respectively, and IC_{50} value from cAMP assay at the A_{2B} > 1,000 nM) (Da Settimo et al. 2007). Interestingly, the triazolobenzotriazinone nucleus is isomeric with that of the triazoloquinoxalinone series described above (compounds **59–61**, Fig. 7).

3.2 Purine Derivatives

3.2.1 Adenines

The first class of A_3AR-selective antagonists with a bicyclic structure strictly correlated to the adenine nucleus was claimed in 2005 by Biagi and coworkers (Biagi et al. 2005). The authors described the synthesis of a series of N^6-ureidosubstituted-2-phenyl-9-benzyl-8-azaadenines whose adenine-like structure was responsible for the antagonist activity and whose phenylcarbamoyl group ensures selectivity at the A_3AR. The structure–activity relationship studies were performed based on the systematic optimization of substituents at the 2, 6 and 9 positions of the bicyclic scaffold, and led to the desired enhancement of A_1/A_3 selectivity (compound **66**, Fig. 10).

Basing on the finding that the known differentiation agent "reversine" (2-(4-morpholinoanilino)-N^6-cyclohexyladenine) exerted a moderate antagonist activity at the hA_3AR (K_i value of 0.66 μM), Jacobson and coworkers developed a series of reversine analogs, focusing their attention on the substitution pattern at the 2 and N^6 positions of the adenine scaffold (Perreira et al. 2005). One of most interesting compounds in terms of hA_3AR affinity and selectivity, MRS3777, (2-(phenyloxy)-N^6-cyclohexyladenine, **67**), combines the N^6-cyclohexyl moiety of reversine with a 2-phenyloxy group. A few derivatives tested in binding assays to the rat A_3AR seemed to reflect the species dependence of the affinity typical of most known nonnucleoside A_3AR antagonists, and were shown to be inactive at 10 μM in this species.

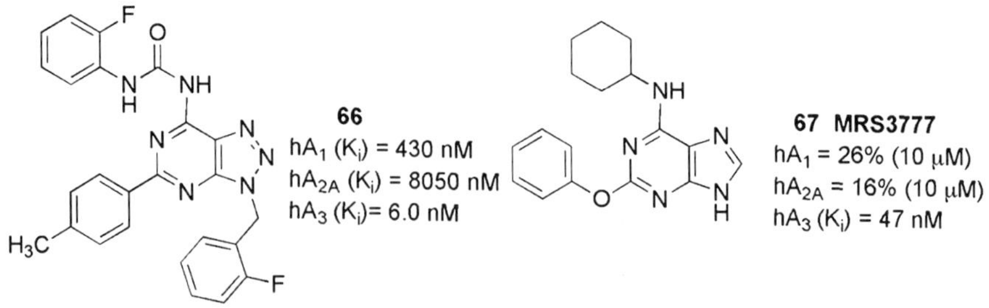

66
hA_1 (K_i) = 430 nM
hA_{2A} (K_i) = 8050 nM
hA_3 (K_i)= 6.0 nM

67 MRS3777
hA_1 = 26% (10 μM)
hA_{2A} = 16% (10 μM)
hA_3 (K_i) = 47 nM

Fig. 10 A_3AR antagonists based on nonnucleoside adenine scaffolds

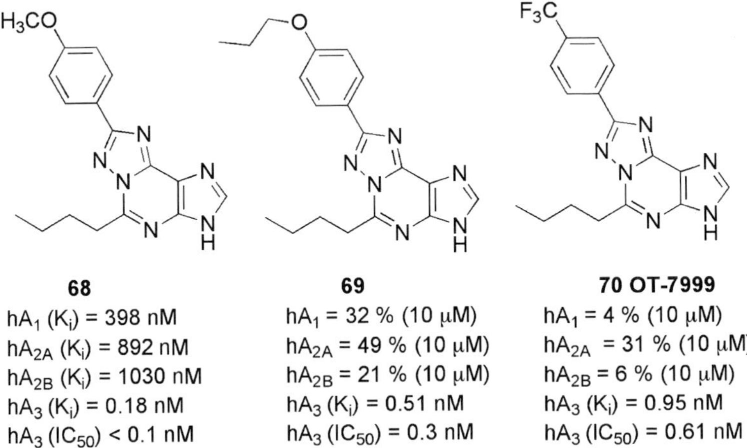

Fig. 11 A_3AR antagonists based on triazolopurine scaffolds

3.2.2 Triazolopurines

Okamura et al. (Okamura et al. 2002, 2004a) recently reported the study of a new series of 1,2,4-triazolo[5,1-*i*]purines. This research group highlighted the structural similarity between the new class of compounds and the triazoloquinazoline derivatives and consequently evaluated the corresponding A_3AR affinities. These investigations led to potent and selective hA_3AR ligands, the most potent of which are reported in Fig. 11 (**68,69**). In particular, 5-*n*-butyl-8-(4-*n*-propoxyphenyl)-3*H*-[1,2,4]triazolo[5,1-*i*]purine (**69**) exhibited the best selectivity profile of this series (affinity ratios vs. other AR subtypes > 19,600). Compound (**70**), 5-*n*-butyl-8-(4-trifluoromethylphenyl)-3*H*-[1,2,4]triazolo-[5,1-*i*]purine (OT-7999), significantly reduced intraocular pressure in cynomolgus monkeys at 2–4 h following topical application (500 mcg) (Okamura et al. 2004b).

3.2.3 Tricyclic Xanthines

Natural antagonists for ARs such as caffeine and theophylline show, in general, low affinity for the A_3AR subtype (Baraldi et al. 2003a; van Galen et al. 1994). In a recent study, the approach based on the ring annelation of xanthine derivatives for the development of AR antagonists was considered in depth (Drabczyńska et al. 2003).

Some pyrido[2,1-*f*]purine-2,4-dione derivatives, which could be considered tricyclic xanthine derivatives, have been reported to exert subnanomolar affinity at the hA_3AR (Priego et al. 2002). The most potent compound of this recent series is the 1-benzyl-3-propyl-1*H*, 3*H*-pyrido[2,1-*f*]purine-2,4-dione derivative (**71**, Fig. 12), which presents a K_i value of 4.0 ± 0.3 nM at hA_3AR. The replacement of the benzyl nucleus at the 1 position with a methyl moiety caused dramatic losses of

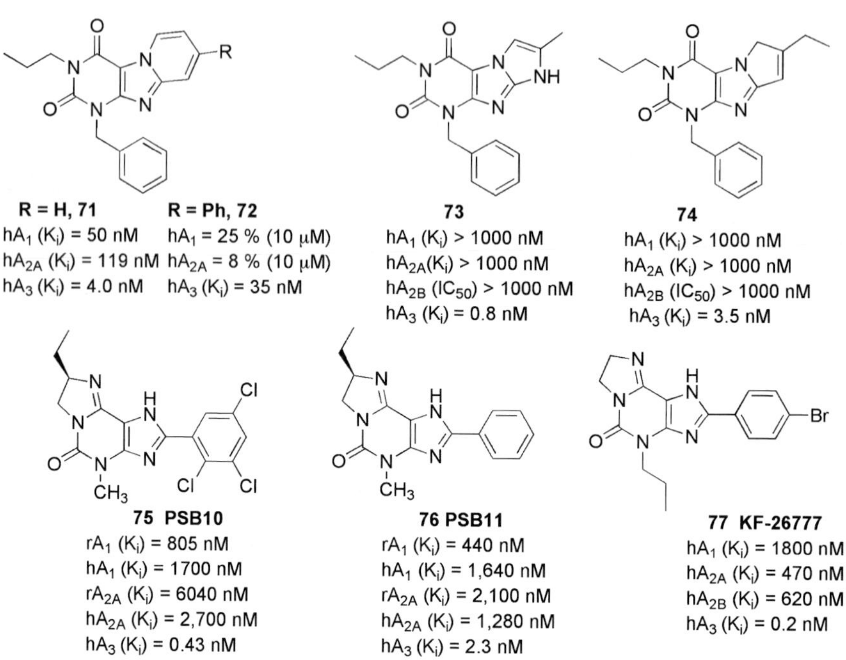

Fig. 12 A_3AR antagonists based on tricyclic xanthine scaffolds

both affinity and selectivity. The effect of the replacement of the pyridine ring of the pyrido[2,1-*f*]purine-2,4-dione core with different five-membered heterocycles was examined. In particular, the synthesis and the SAR profile at the ARs of a series of 1-benzyl-3-propyl-7-aryl/alkyl-1*H*,6*H*-pyrrolo[2,1-*f*]purine-2,4-dione and 1-benzyl-3-propyl-7-aryl/alkyl-1*H*,8*H*-imidazo[2,1-*f*]purine-2,4-dione derivatives were recently reported (Baraldi et al. 2005b). Among the examined tricycles, the imidazo[2,1-*f*]purine-2,4-dione derivatives were two- to tenfold more potent than the corresponding pyrrolo[2,1-*f*]purine-2,4-dione derivatives. The best results were obtained with the introduction of small alkyl chains at the 7 position (1-benzyl-7-methyl-3-propyl-1*H*,8*H*-imidazo[2,1-*f*]purine-2,4-dione **73**, 1-benzyl-7-ethyl-3-propyl-1*H*,6*H*-pyrrolo[2,1-*f*]purine-2,4-dione **74**, Fig. 12). Compound **73** shows a subnanomolar affinity towards the target A_3AR, with noteworthy selectivity with respect to the other AR subtypes (K_i (hA_3) $= 0.8$ nM, K_i (hA_1/hA_3) $= 3,163$, K_i (hA_{2A}/hA_3) $> 6{,}250$, IC_{50} (hA_{2B})/K_i (hA_3) $= 2{,}570$).

The synthesis and biological evaluation of a series of fused xanthine derivatives was investigated by Müller and coworkers (Müller et al. 2002a). In particular, the (*R*)-4-methyl-8-ethyl-2-phenyl-4,5,7,8-tetrahydro-1*H*-imidazo[2,1-*i*]purin-5-one (PSB-11 (**76**), Fig. 12) exhibited a K_i value of 2.3 nM for the A_3AR and good selectivity vs. all other AR subtypes. The radiolabeled derivative of this compound ([^{3}H]PSB-11) exhibited a K_d value of 4.9 nM and a B_{max} value of 3,500 fmol mg^{-1} of protein in human A_3AR binding in transfected CHO cells (Müller et al. 2002b). An important innovation of such a series, in comparison with xanthines, is a significant increase in water solubility due to the introduction of a basic nitrogen atom, which can be protonated in physiological conditions. Compound PSB-10, bearing a 2,3,5-trichlorophenyl moiety at the 2 position, showed inverse agonist activity in binding studies in CHO cells expressing recombinant hA_3ARs ($IC_{50} = 4$ nM) (Ozola et al. 2003). The 2-(4-bromophenyl)-derivative named KF-26777 (**77**) with subnanomolar affinity at the hA_3AR ($K_i = 0.2$ nM) and high selectivity over A_1, A_{2A} and A_{2B} ARs (9,000-, 23,500-, 31,000-fold, respectively) was considered a potential lead molecule for development for the treatment of brain ischemia and inflammatory diseases such as asthma (Saki et al. 2002).

3.3 *Nucleoside-Derived A_3AR Antagonists*

Based on the observation that the relative efficacy of purine nucleosides depends on structural features (see Sect. 2), new subtype-selective nucleoside antagonists of the A_3AR have been designed. One of the first such antagonists was the rigid spirolactam MRS1292 (**78**) (Fig. 13, (2*R*,3*R*,4*S*,5*S*)-2-[N^6-3-iodobenzyl)adenos-9′-yl]-7-aza-1-oxa-6-oxospiro[4.4]-nonan-4,5-diol) (Gao et al. 2002), which binds potently and selectively to the rat and human A_3ARs but does not activate these receptors, and thus acts as an antagonist.

Modeling/mutagenesis of ARs has focused on distinct residues related to ligand binding and the relative efficacy of adenosine derivatives, and on a conserved Trp residue (6.48) which is involved in the activation process (termed a "rotamer switch," Shi et al. 2002). Docking studies of agonists suggest that the activation pathway of the A_3AR involves a characteristic anticlockwise rotation of this residue, as viewed from the exofacial side (Kim et al. 2006). The docking of MRS1292 (**78**) to the A_3AR model is not accompanied by rotation of this residue, as occurs with nucleoside agonists, consistent with its action as an antagonist (Kim et al. 2006). Moreover, the affinity and selectivity of MRS1292 occurs across species, unlike most other heterocyclic antagonists for the A_3AR reported. This allows its use in nonprimate (e.g., murine) experimental animals used as clinical models. For example, MRS1292 applied directly to the eye in mouse has been shown to be effective in reducing intraocular pressure, which may be predictive of its utility as an antiglaucoma agent (Yang et al. 2005).

The removal of the ability of the 5′-*N*-alkyluronamide to donate a hydrogen bond was found to convert agonists into selective antagonists (Gao et al. 2006a). In both the 4′-oxo and the 4′-thio series, *N*-methylation of an *N*-methylamide (i.e., to form a dimethylamide) resulted in potent and selective A_3AR antagonists. Recently, nucleosides that are truncated at the 4′ position were found to act as A_3AR antagonists. For example, (2*R*,3*R*,4*S*)-2-(2-chloro-6-(3-chlorobenzylamino)-9*H*-purin-9-yl)tetrahydrothiophene-3,4-diol (LJ-1416, **80**) and (2*R*,3*R*,4*S*)-2- (2-chloro-6-(3-iodobenzylamino)-9*H*-purin-9-yl)tetrahydrothiophene-3,4-diol (LJ-1251, **81**) (Fig. 13) (Jeong et al. 2007) displayed K_i values of 1.66 and 4.16 nM, respectively, at the human A_3AR,

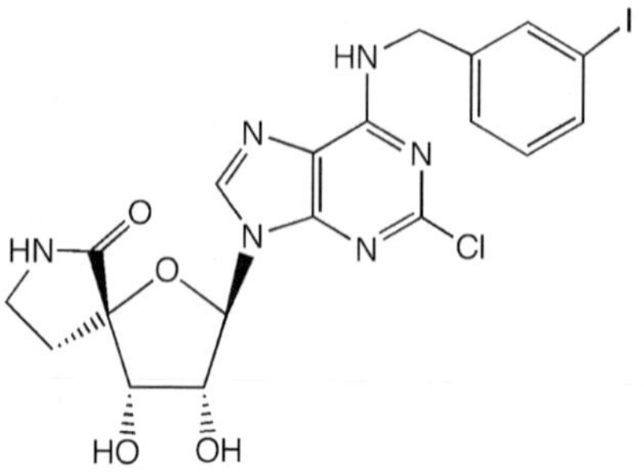

78 MRS1292

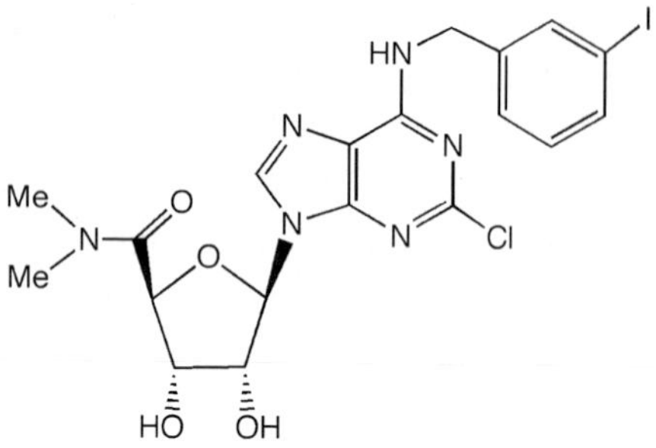

79 MRS3771

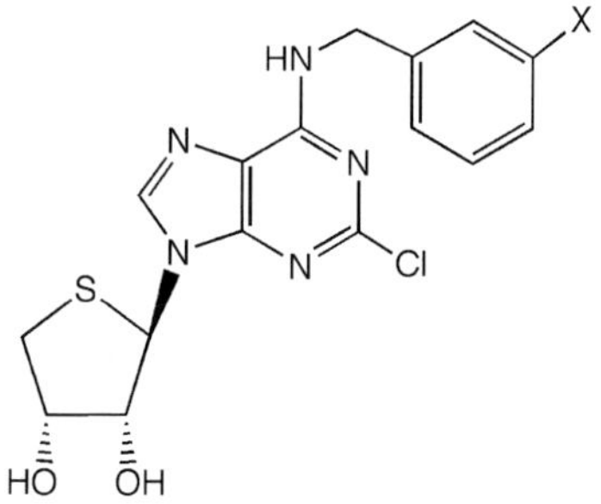

80 X = Cl; LJ-1416
81 X = I; LJ-1251

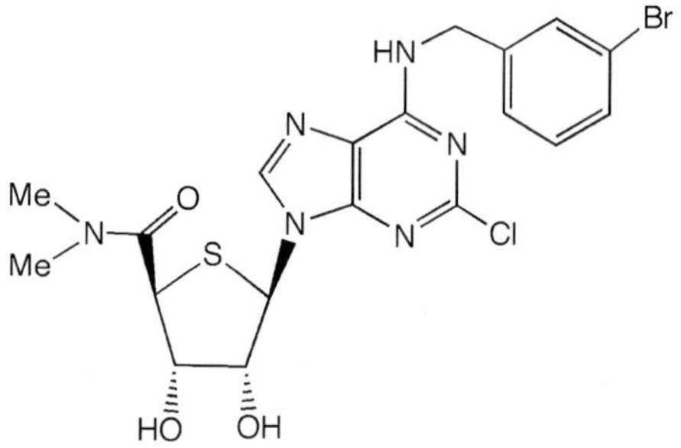

82 LJ-1351

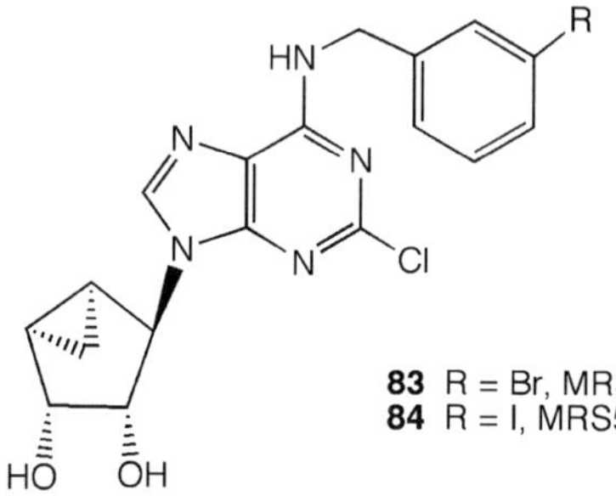

83 R = Br, MRS5147
84 R = I, MRS5127

Fig. 13 A_3AR antagonists based on nucleoside scaffolds

with >600-fold selectivity in comparison to the A_1AR. LJ-1251 was shown to have neuroprotective properties in an ischemia model in the rat hippocampus (Pugliese et al. 2007). Truncation at the 4′ position of A_3AR agonist in the (N)-methanocarba series produces potent and selective A_3AR antagonists (Melman et al. 2008b), such as the 3-bromo derivative $1'R, 2'R, 3'S, 4'R, 5'S$)-4′-[2-chloro-6-(3-bromobenzylamino)-purine]-2′, 3′-*O*-dihydroxybicyclo-[3.1.0]hexane (MRS5147) (**83**, Fig. 13) (2,900-fold selective for hA_3 vs. hA_1AR) or its 3-iodo analog, MRS5127 (**84**) (2,400-fold selective for hA_3 vs. hA_1AR). MRS5127 (**84**) displayed a K_B (Schild constant) value of 8.9 nM as an antagonist of the human A_3AR in a functional assay.

4 Engineering of the A_3AR to Avoid Side Effects of Conventional Synthetic Agonists

Although selective agonists of several of the ARs have been known for years, their use as pharmaceutical agents has been impeded by undesirable side effects of exogenously administered adenosine derivatives. In spite of the clinically useful protective properties of adenosine agonists observed in experimental animals, such as protection against ischemic damage and suppression of excessive inflammation, none of the selective synthetic agonists have yet been approved for human therapeutic use. The A_{2A}AR-selective agonist Lexiscan (regadenoson, CV Therapeutics, Palo Alto, CA, USA) (CVT-3146, 1-{9-[(4*S*, 2*R*, 3*R*, 5*R*)-3,4-dihydroxy-5-(hydroxymethyl)oxolan-2-yl]-6-aminopurin-2-yl}pyrazol-4-yl)-*N*-methylcarboxamide) was recently approved for cardiac imaging in patients. The only other adenosine agonist currently in clinical use is adenosine itself, for the treatment of supraventricular tachycardia and as an aid in cardiac imaging.

Since ARs are widespread in the body, in order to overcome inherent nonselectivity of activating the native ARs using synthetic agonists, we have introduced the concept of neoceptors, by which the putative ligand binding site of a 7TM receptor is re-engineered for activation by synthetic agonists (neoligands) that are built to have a structural complementarity. This is a molecular modeling approach to receptor engineering by which a mutant receptor (neoceptor) is designed for selective activation by a novel synthetic ligand (neoligand) at concentrations that do not activate the native receptor. An amino acid residue of the receptor and a functional group of the ligand moiety thought to be in close proximity can be modified in a complementary fashion so that the two groups exhibit a novel mode of interaction (e.g., reversing the polarity in a salt bridge or introducing unique hydrogen-bonding sites). If a stabilizing interaction exists between these two groups, an increase in affinity is expected at the mutant receptor relative to the native receptor. This strategy is intended for eventual use in gene therapy and may also be useful in mechanistic elucidation, using neoceptor–neoligand pairs that are pharmacologically orthogonal with respect to the native species. Neoceptors have so far been applied successfully to A_{2A} and A_3 ARs (Gao et al. 2006b; Jacobson et al. 2001, 2005). Compounds **85–87** (Fig. 14) were found to interact selectively with the H272E mutant hA_3AR. All three compounds activated this neoceptor.

5 Conclusions

A_3AR ligands have been modified to optimize their interaction with the A_3AR. Most of these modifications have been made to the N^6 and 2 positions of adenine as well as the ribose moiety, and using a combination of these substitutions leads to the most efficacious, selective, and potent ligands. A_3AR agonists such as IB–MECA and Cl–IB–MECA are now advancing into Phase II clinical trials for treatments targeting diseases such as cancer, arthritis, and psoriasis.

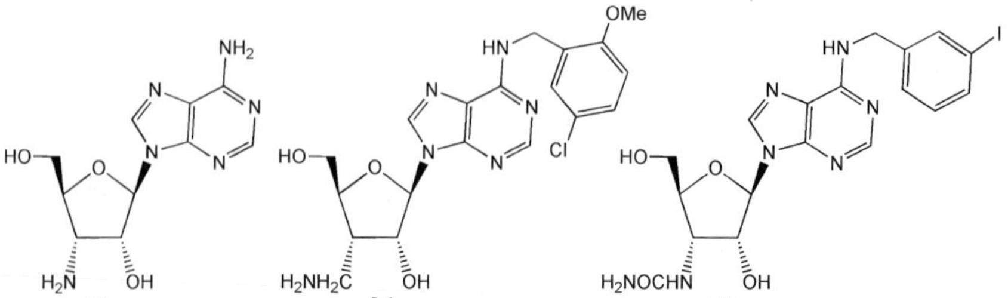

Fig. 14 Compounds that interact selectively with the H272E mutant hA_3AR neoceptor

Also, a wide number of compounds exerting high potency and selectivity in antagonizing the hA_3AR have been discovered. These molecules are generally characterized by a notable structural diversity, taking into account that aromatic nitrogen-containing monocyclic (thiazoles and thiadiazoles), bicyclic (isoquinoline, quinozalines, (aza)adenines), tricyclic systems (pyrazoloquinolines, triazoloquinoxalines, pyrazolotriazolopyrimidines, triazolopurines, tricyclic xanthines) and nucleoside derivatives have been identified as potent and selective A_3AR antagonists. Probably due to the "enigmatic" physiological role of A_3AR, whose activation may produce opposite effects (for example, concerning tissue protection in inflammatory and cancer cells) and may produce effects that are species dependent, only a few molecules have reached preclinical investigation. Indeed, the most advanced A_3AR antagonists remain in preclinical biological testing. Among the antagonists described above, compound OT-7999 is expected to enter clinical trials for the treatment of glaucoma, while several thiazole derivatives are in development as antiallergic, antiasthmatic and/or anti-inflammatory drugs.

Acknowledgements This research was supported in part by the Intramural Research Program of the National Institute of Diabetes and Digestive and Kidney Diseases, National Institutes of Health, Bethesda, MD, USA.

References

Baharav E, Bar-Yehuda S, Madi L, Silberman D, Rath-Wolfson L, Halpren M, Ochaion A, Weinberger A, Fishman P (2005) Antiinflammatory effect of A_3 adenosine receptor agonists in murine autoimmune arthritis models. J Rheumatol 32:469–476

Baraldi PG, Cacciari B, Pineda de las Infantas MJ, Romagnoli R, Spalluto G, Volpini R, Costanzi S, Vittori S, Cristalli G, Melman N, Park K-S, Ji X-d, Jacobson KA (1998) Synthesis and biological activity of a new series of N^6-arylcarbamoyl-, 2-(ar)alkynyl-N^6-arylcarbamoyl, and N^6-carboxamido- derivatives of adenosine-5′-N-ethyluronamide (NECA) as A_1 and A_3 adenosine receptor agonists. J Med Chem 41:3174–3185

Baraldi PG, Cacciari B, Borea PA, Varani K, Pastorin G, Da Ros T, Spalluto G (2002a) Pyrazolo-triazolo-pyrimidine derivatives as adenosine receptor antagonists: a possible template for adenosine receptor subtypes? Curr Pharm Design 8:99–110

Baraldi PG, Cacciari B, Romagnoli R, Spalluto G, Monopoli A, Ongini E, Varani K, Borea PA (2002b) 7-Substituted 5-amino-2-(2-furyl)pyrazolo[4,3-*e*]-1,2,4-triazolo[1,5-*c*]pyrimidines as A_{2A} adenosine receptor antagonists: a study on the importance of modifications at the side chain on the activity and solubility. J Med Chem 45:115–126

Baraldi PG, Tabrizi MA, Fruttarolo F, Bovero A, Avitabile B, Preti D, Romagnoli R, Merighi S, Gessi S, Varani K, Borea PA (2003a) Recent developments in the field of A_3 adenosine receptor antagonists. Drug Dev Res 58:315–329

Baraldi PG, Fruttarolo F, Tabrizi MA, Preti D, Romagnoli R, El-Kashef H, Moorman A, Varani K, Gessi S, Merighi S, Borea PA (2003b) Design, synthesis, and biological evaluation of C9- and C2-substituted pyrazolo[4,3-*e*]-1,2,4-triazolo[1,5-*c*]pyrimidines as new A_{2A} and A_3 adenosine receptors antagonists. J Med Chem 46:1229–1241

Baraldi PG, Tabrizi MA, Preti D, Bovero A, Fruttarolo F, Romagnoli R, Zaid NA, Moorman AR, Varani K, Borea PA (2005a) New 2-arylpyrazolo[4,3-*c*]quinoline derivatives as potent and selective human A_3 adenosine receptor antagonists. J Med Chem 48:5001–5008

Baraldi PG, Preti D, Tabrizi MA, Fruttarolo F, Romagnoli R, Zaid NA, Moorman AR, Merighi S, Varani K, Borea PA (2005b) New pyrrolo[2,1-*f*]purine-2,4-dione and imidazo[2,1-*f*]purine-2,4-dione derivatives as potent and selective human A_3 adenosine receptor antagonists. J Med Chem 48:4697–4701

Baraldi PG, Tabrizi MA, Romagnoli R, El-Kashef H, Preti D, Bovero A, Fruttarolo F, Gordaliza M, Borea PA (2006) Pyrazolo[4,3-*e*][1,2,4]triazolo[1,5-*c*]pyrimidine template: organic and medicinal chemistry approach. Curr Org Chem 10:259–275

Bhattacharya P, Leonard JT, Roy K (2005) Exploring QSAR of thiazole and thiadiazole derivatives as potent and selective human adenosine A_3 receptor antagonists using FA and GFA techniques. Bioorg Med Chem 13:1159–1165

Biagi G, Bianucci AM, Coi A, Costa B, Fabbrini L, Giorgi I, Livi O, Micco I, Pacchini F, Santini E, Leonardi M, Nofal FA, Salernid OL, Scartonia V (2005) 2,9-Disubstituted-N^6-(arylcarbamoyl)-8-azaadenines as new selective A_3 adenosine receptor antagonists: synthesis, biochemical and molecular modelling studies. Bioorg Med Chem 13:4679–4693

Cacciari B, Bolcato C, Spalluto G, Klotz KN, Bacilieri M, Deflorian F, Moro S (2007) Pyrazolo-triazolo-pyrimidines as adenosine receptor antagonists: a complete structure–activity profile. Purinergic Signal 3:183–193

Catarzi D, Colotta V, Varano F, Calabri FR, Lenzi O, Filacchioni G, Trincavelli L, Martini C, Tralli A, Christian M, Moro S (2005a) 2-Aryl-8-chloro-1,2,4-triazolo[1,5-*a*]quinoxalin-4-amines as highly potent A_1 and A_3 adenosine receptor antagonists. Bioorg Med Chem 13:705–715

Catarzi D, Colotta V, Varano F, Lenzi O, Filacchioni G, Trincavelli L, Martini C, Montopoli C, Moro S (2005b) 1,2,4-Triazolo[1,5-*a*]quinoxaline as a versatile tool for the design of selective human A_3 adenosine receptor antagonists: synthesis, biological evaluation, and molecular modeling studies of 2-(hetero)aryl- and 2-carboxy-substitued derivatives. J Med Chem 48:7932–7945

Chang LC, von Frijtag Drabbe Künzel JK, Mulder-Krieger T, Spanjersberg RF, Roerink SF, van den Hout G, Beukers MW, Brussee J, IJzerman AP (2005) A series of ligands displaying a remarkable agonistic–antagonistic profile at the adenosine A_1 receptor. J Med Chem 48: 2045–2053

Colotta V, Catarzi D, Varano F, Cecchi L, Filacchioni G, Martini C, Trincavelli L, Lucacchini A (2000) Synthesis and structure–activity relationships of a new set of 2-arylpyrazolo [3,4-*c*]quinoline derivatives as adenosine receptor antagonists. J Med Chem 43:3118–3124

Colotta V, Catarzi D, Varano F, Calabri FR, Lenzi O, Filacchioni G, Martini C, Trincavelli L, Deflorian F, Moro S (2004) 1,2,4-Triazolo[4,3-*a*]quinoxalin-1-one moiety as an attractive scaffold to develop new potent and selective human A_3 adenosine receptor antagonists: synthesis, pharmacological, and ligand–receptor modeling studies. J Med Chem 47:3580–3590

Colotta V, Catarzi D, Varano F, Capelli F, Lenzi O, Filacchioni G, Martini C, Trincavelli L, Ciampi O, Pugliese AM, Pedata F, Schiesaro A, Morizzo E, Moro S (2007) New 2-arylpyrazolo[3,4-*c*]quinoline derivatives as potent and selective human A_3 adenosine receptor antagonists.

Synthesis, pharmacological evaluation, and ligand–receptor modeling studies. J Med Chem 50:4061–4074

Cordeaux Y, Briddon SJ, Alexander SP, Kellam B, Hill SJ (2008) Agonist-occupied A_3 adenosine receptors exist within heterogeneous complexes in membrane microdomains of individual living cells. FASEB J 22:850–860

Costanzi S, Tikhonova IG, Harden TK, Jacobson KA (2008) Ligand and structure-based methodologies for the prediction of the activity of G protein-coupled receptor ligands. J Comput Aided Mol Des doi:10.1007/s10822-008-9218-3

Cosyn L, Gao ZG, Van Rompaey P, Lu C, Jacobson KA, Van Calenbergh S (2006a) Synthesis of hypermodified adenosine derivatives as selective adenosine A_3 receptor ligands. Bioorg Med Chem 14:1403–1412

Cosyn L, Palaniappan KK, Kim SK, Duong HT, Gao ZG, Jacobson KA, Van Calenbergh S (2006b) 2-Triazole-substituted adenosines: a new class of selective A_3 adenosine receptor agonists, partial agonists, and antagonists. J Med Chem 49:7373–7383

Da Settimo F, Primofiore G, Taliani S, Marini AM, La Motta C, Simorini F, Salerno S, Sergianni V, Tuccinardi T, Martinelli A, Cosimelli B, Greco G, Novellino E, Ciampi O, Trincavalle ML, Martini C (2007) 5-Amino-2-phenyl[1,2,3]triazolo[1,2-*a*][1,2,4]benzotriazin-1-one: a versatile scaffold to obtain potent and selective A_3 adenosine receptor antagonists. J Med Chem 50:5676–5684

DeNinno MP, Masamune H, Chenard LK, DiRico KJ, Eller C, Etienne JB, Tickner JE, Kennedy SP, Knight DR, Kong J, Oleynek JJ, Tracey WR, Hill RJ (2003) 3′-Aminoadenosine-5′-uronamides: discovery of the first highly selective agonist at the human adenosine A_3 receptor. J Med Chem 46:353–355

DeNinno MP, Masamune H, Chenard LK, DiRico KJ, Eller C, Etienne JB, Tickner JE, Kennedy SP, Knight DR, Kong J, Oleynek JJ, Tracey WR, Hill RJ (2006) The synthesis of highly potent, selective, and water-soluble agonists at the human adenosine A_3 receptor. Bioorg Med Chem Lett 16:2525–2527

Drabczyńska A, Schumacher B, Müller CE, Karolak-Wojciechowska J, Michalak B, Pękala E, Kieć-Kononowicz K (2003) Impact of the aryl substituent kind and distance from pyrimido[2,1-*f*]purindiones on the adenosine receptor selectivity and antagonistic properties. Eur J Med Chem 38:397–402

Elzein E, Palle V, Wu Y, Maa T, Zeng D, Zablocki J (2004) 2-Pyrazolyl-N^6-substituted adenosine derivatives as high affinity and selective adenosine A_3 receptor agonists. J Med Chem 47: 4766–4773

Fredholm BB, IJzerman AP, Jacobson KA, Klotz KN, Linden J (2001) International Union of Pharmacology. XXV. Nomenclature and classification of adenosine receptors. Pharmacol Rev 53:527–552

Gallo-Rodriguez C, Ji X-D, Melman N, Siegman BD, Sanders LH, Orlina J, Fischer B, Pu Q-L, Olah ME, van Galen PJM, Stiles GL, Jacobson KA (1994) Structure–activity relationships of N^6-benzyladenosine-5′-uronamides as A_3-selective adenosine agonists. J Med Chem 37: 636–646

Gao ZG, Kim SK, Biadatti T, Chen W, Lee K, Barak D, Kim SG, Johnson CR, Jacobson KA (2002) Structural determinants of A_3 adenosine receptor activation: nucleoside ligands at the agonist/antagonist boundary. J Med Chem 45:4471–4484

Gao ZG, Blaustein J, Gross AS, Melman N, Jacobson KA (2003) N^6-Substituted adenosine derivatives: selectivity, efficacy, and species differences at A_3 adenosine receptors. Biochem Pharmacol 65: 1675–1684

Gao ZG, Mamedova L, Chen P, Jacobson KA (2004) 2-Substituted adenosine derivatives: affinity and efficacy at four subtypes of human adenosine receptors. Biochem Pharmacol 68: 1985–1993

Gao ZG, Kim SK, IJzerman AP, Jacobson KA (2005) Allosteric modulation of the adenosine family of receptor. Mini Rev Med Chem 5:545–553

Gao ZG, Joshi BV, Klutz A, Kim SK, Lee HW, Kim HO, Jeong LS, Jacobson KA (2006a) Conversion of A_3 adenosine receptor agonists into selective antagonists by modification of the 5′-ribofuran-uronamide moiety. Bioorg Med Chem Lett 16:596–601

Gao ZG, Duong HT, Sonina T, Lim SK, Van Rompaey P, Van Calenbergh S, Mamedova L, Kim HO, Kim MJ, Kim AY, Liang BT, Jeong LS, Jacobson KA (2006b) Orthogonal activation of the reengineered A_3 adenosine receptor (neoceptor) using tailored nucleoside agonists. J Med Chem 49:2689–2702

Gatta F, Del Giudice MR, Borioni A, Borea PA, Dionisotti S, Ongini E (1993) Synthesis of imidazo[1,2-*c*]pyrazolo[4,3-*e*]pyrimidines, pyrazolo[4,3-*e*]1,2,4-triazolo[1,5-*c*]pyrimidines and 1,2,4-triazolo[5,1-*i*]purines: new potent adenosine A_2 receptor antagonists. Eur J Med Chem 28:569–576

Gessi S, Merighi S, Varani K, Leung E, Mac Lennan S, Borea PA (2008) The A_3 adenosine receptor: an enigmatic player in cell biology. Pharmacol Ther 117:123–140

Göblyös A, Gao ZG, Brussee J, Connestari R, Neves Santiago S, Ye K, IJzerman AP, Jacobson KA (2006) Structure–activity relationships of 1*H*-imidazo[4,5-*c*]quinolin-4-amine derivatives new as allosteric enhancers of the A_3 adenosine receptor. J Med Chem 49:3354–3361

Jacobson KA, Gao ZG (2006) Adenosine receptors as therapeutic targets. Nat Rev Drug Discov 5:247–264

Jacobson KA, Nikodijevic O, Shi D, Gallo-Rodriguez C, Olah ME, Stiles GL, Daly JW (1993) A role for central A_3-adenosine receptors: mediation of behavioral depressant effects. FEBS Lett 336:57–60

Jacobson KA, Siddiqi SM, Olah ME, Ji XD, Melman N, Bellamkonda K, Meshulmam Y, Stiles GL, Kim HO (1995) Structure–activity relationships of 9-alkyladenine and ribose modified adenosine derivatives at rat A_3 adenosine receptors. J Med Chem 38:1720–1735

Jacobson KA, Park KS, Jiang J-L, Kim YC, Olah ME, Stiles GL, Ji X-D (1997) Pharmacological characterization of novel A_3 adenosine receptor-selective antagonists. Neuropharmacology 36:1157–1165

Jacobson KA, Ji X-d, Li AH, Melman N, Siddiqui MA, Shin KJ, Marquez VE, Ravi RG (2000) Methanocarba analogues of purine nucleosides as potent and selective adenosine receptor agonists. J Med Chem 43:2196–2203

Jacobson KA, Gao ZG, Chen A, Barak D, Kim SA, Lee K, Link A, Van Rompaey P, Van Calenbergh S, Liang BT (2001) Neoceptor concept based on molecular complementarity in GPCRs: a mutant adenosine A_3 receptor with selectively enhanced affinity for amine-modified nucleosides. J Med Chem 44:4125–4136

Jacobson KA, Ohno M, Duong HT, Kim SK, Tchilibon S, Cesnek M, Holy A, Gao ZG (2005) A neoceptor approach to unraveling microscopic interactions between the human A_{2A} adenosine receptor and its agonists. Chem Biol 12:237–247

Jeong LS, Lee HW, Jacobson KA, Kim HO, Shin DH, Lee JA, Gao ZG, Lu C, Duong HT, Gunaga P, Lee SK, Jin DZ, Chun MW, Moon HR (2006a) Structure–activity relationships of 2-chloro-N^6-substituted-4′-thioadenosine-5′-uronamides as highly potent and selective agonists at the human A_3 adenosine receptor. J Med Chem 49:273–281

Jeong LS, Lee HW, Kim HO, Jung JY, Gao ZG, Duong HT, Rao S, Jacobson KA, Shin DH, Lee JA, Gunaga P, Lee SK, Jin DZ, Chun MW (2006b) Design, synthesis, and biological activity of N^6-substituted-4′-thioadenosines at the human A_3 adenosine receptor. Bioorg Med Chem 14:4718–4730

Jeong LS, Choe SA, Gunaga P, Kim HO, Lee HW, Lee SK, Tosh DK, Patel A, Palaniappan KK, Gao ZG, Jacobson KA, Moon HR (2007) Discovery of a new nucleoside template for human A_3 adenosine receptor ligands: D-4′-thioadenosine derivatives without 4′-hydroxymethyl group as highly potent and selective antagonists. J Med Chem 50:3159–3162

Jeong LS, Lee HW, Kim HO, Tosh D, Pal S, Choi WJ, Gao ZG, Patel AR, Williams W, Jacobson KA, Kim HD (2008) Structure–activity relationships of 2-chloro-N^6-substituted-4′-thioadenosine-5′-*N*, *N*-dialkyluronamides as human A_3 adenosine receptor antagonists. Bioorg Med Chem 18:1612–1616

Ji Xd, Melman N, Jacobson KA (1996) Interactions of flavonoids and other phytochemicals with adenosine receptors. J Med Chem 39:781–788

Jung K-Y, Kim S-K, Gao Z-G, Gross AS, Melman N, Jacobson KA, Kim YC (2004) Structure–activity relationships of thiazole and thiadiazole derivatives as potent and selective human adenosine A_3 receptor antagonists. Bioorg Med Chem 12:613–623

Kim HO, Ji X-D, Siddiqi SM, Olah ME, Stiles GL, Jacobson KA (1994a) 2-Substitution of N^6-benzyladenosine-5′-uronamides enhances selectivity for A_3-adenosine receptors. J Med Chem 37:3614–3621

Kim HO, Ji X-D, Melman N, Olah ME, Stiles GL, Jacobson KA (1994b) Selective ligands for rat A_3-adenosine receptors: structure–activity relationships of 1,3-dialkylxanthine-7-riboside derivatives. J Med Chem 37:4020–4030

Kim YC, Ji X-D, Jacobson KA (1996) Derivatives of the triazoloquinazoline adenosine antagonist (CGS15943) are selective for the human A_3 receptor subtype. J Med Chem 39:4142–4148

Kim SK, Gao Z-G, Van Rompaey P, Gross AS, Chen A, Van Calenbergh S, Jacobson KA (2003) Modeling the adenosine receptors: comparison of binding domains of A_{2A} agonist and antagonist J Med Chem 46:4847–4859

Kim SK, Gao ZG, Jeong LS, Jacobson KA (2006) Docking studies of agonists and antagonists suggest an activation pathway of the A_3 adenosine receptor. J Mol Graph Model 25:562–577

Lee K, Ravi RG, Ji X-D, Marquez VE, Jacobson KA (2001) Ring-constrained (N)methanocarba-nucleosides as adenosine receptor agonists: independent 5′-uronamide and 2′-deoxy modifications. Bioorg Med Chem Lett 11:1333–1337

Lenzi O, Colotta V, Catarzi D, Varano F, Filacchioni G, Martini C, Trincavelli L, Ciampi O, Varani K, Marighetti F, Morizzo E, Moro S (2006) 4-Amido-2-aryl-1,2,4-triazolo[4,3-*a*]quinoxalin-1-ones as new potent and selective human A_3 adenosine receptor antagonists. Synthesis, pharmacological evaluation, and ligand–receptor modeling studies. J Med Chem 49:3916–3925

Li AH, Moro S, Melman N, Ji XD, Jacobson KA (1998) Structure–activity relationships and molecular modeling of 3,5-diacyl-2,4-dialkylpyridine derivatives as selective A_3 adenosine receptor antagonists. J Med Chem 41:3186–3201

Lopes LV, Rebola N, Pinheiro PC, Richardson PJ, Oliveira CR, Cunha RA (2003) Adenosine A_3 receptors are located in neurons of the rat hippocampus. Neuroreport 14:1645–1648

Maconi A, Pastorin G, Da Ros T, Spalluto G, Gao ZG, Jacobson KA, Baraldi PG, Cacciari B, Varani K, Moro S, Borea PA (2002) Synthesis, biological properties, and molecular modeling investigation of the first potent, selective, and water-soluble human A_3 adenosine receptor antagonist. J Med Chem 45:3579–82

Marquez VE, Siddiqui MA, Ezzitouni A, Russ P, Wang J, Wagner RW, Matteucci MD (1996) Nucleosides with a twist. Can fixed forms of sugar ring pucker influence biological activity in nucleosides and oligonucleotides. J Med Chem 39:3739–3747

Matot I, Weininger CF, Zeira E Galun E, Joshi BV, Jacobson KA (2006) A_3 Adenosine receptors and mitogen activated protein kinases in lung injury following in-vivo reperfusion. Crit Care 10:R65, doi:10.1186/cc4893

Melman, A, Gao, ZG, Kumar, D, Wan, TC, Gizewski, E, Auchampach, JA, Jacobson, KA (2008a) Design of (N)-methanocarba adenosine 5′-uronamides as species-independent A_3 receptor-selective agonists. Bioorg Med Chem Lett 18:2813–2819

Melman A, Wang B, Joshi BV, Gao ZG, de Castro S, Heller CL, Kim SK, Jeong LS, Jacobson KA (2008b) Selective A_3 adenosine receptor antagonists derived from nucleosides containing a bicyclo[3.1.0]hexane ring system. Bioorg Med Chem 16:8546–8556

Meyerhof W, Müller-Brechlin R, Richter D (1991) Molecular cloning of a novel putative G-protein coupled receptor expressed during rat spermiogenesis. FEBS Lett 284:155–160

Moro S, Braiuca P, Deflorian F, Ferrari C, Pastorin G, Cacciari B, Baraldi PG, Varani K, Borea PA, Spalluto G (2005) Combined target-based and ligand-based drug design approach as a tool to define a novel 3D-pharmacophore model of human A_3 adenosine receptor antagonists: pyrazolo[4,3-*e*]1,2,4-triazolo[1,5-*c*]pyrimidine derivatives as a key study. J Med Chem 48:152–162

Moro S, Spalluto G, Gao ZG, Jacobson KA (2006) Progress in pursuit of therapeutic adenosine receptor antagonists. Med Res Rev 26:131–159

Müller CE, Thorand M, Qurishi R, Diekmann M, Jacobson KA, Padgett WL, Daly JW (2002a) Imidazo[2,1-*i*]purin-5-ones and related tricyclic water-soluble purine derivatives: potent A_{2A}- and A_3-adenosine receptor antagonists. J Med Chem 45:3440–3450

Müller CE, Diekmann M, Thorand M, Ozola V (2002b) [^{3}H]8-Ethyl-4-methyl-2-phenyl-(8R)-4,5,7,8-tetrahydro-1*H*-imidazo[2,1-*i*]-purin-5-one ([^{3}H]PSB-11), a novel high affinity antagonist radioligand for human A_3 adenosine receptors. Bioorg Med Chem Lett 12:501–503

Müller CE (2003) Medicinal chemistry of adenosine A_3 receptor ligands. Curr Top Med Chem 3:445–462

Novellino E, Barbara Cosimelli, Marina Ehlardo, Giovanni Greco, Manuela Iadanza, Antonio Lavecchia, Rimoli MG, Sala A, Da Settimo A, Primofiore G, Da Settimo F, Taliani S, La Motta C, Klotz KN, Tuscano D, Trincavelli ML, Martini C (2005) 2-(Benzimidazol-2-yl)quinoxalines: a novel class of selective antagonists at human A_1 and A_3 adenosine receptors designed by 3D database searching. J Med Chem 48:8253–8260

Ohana G, Bar-Yehuda S, Barer F, Fishman P (2001) Differential effect of adenosine on tumor and normal cell growth: focus on the A_3 adenosine receptor. J Cell Physiol 186:19–23

Ohno M, Gao ZG, Van Rompaey P, Tchilibon S, Kim SK, Harris BA, Blaustein J, Gross AS, Duong HT, Van Calenbergh S, Jacobson KA (2004) Modulation of adenosine receptor affinity and intrinsic efficacy in nucleosides substituted at the 2-position. Bioorg Med Chem 12: 2995–3007

Okamura K, Kurogi Y, Nishikawa H, Hashimoto K, Fujiwara H, Nagao Y (2002) 1,2,4-triazolo[5,1-*i*]purine derivatives as highly potent and selective human adenosine A_3 receptor ligands. J Med Chem 45:3703–3708

Okamura T, Kurogi Y, Hashimoto K, Nagao Y (2004a) Facile synthesis of fused 1,2,4-triazolo[1,5-*c*]-pyrimidine derivatives as human adenosine A_3 receptor ligands. Bioorg Med Chem Lett 14:2443–2446

Okamura T, Kurogi Y, Hashimoto K, Sato S, Nishikawa H, Kiryu K, Nagao Y (2004b) Structure-activity relationships of adenosine A_3 receptor ligands: new potential therapy for the treatment of glaucoma. Bioorg Med Chem Lett 14:3775–3779

Ozola V, Thorand M, Diekmann M, Qurishi R, Schumacher B, Jacobson KA, Müller CE (2003) 2-Phenylimidazo[2,1-*i*]purin-5-ones: structure–activity relationships and characterization of potent and selective inverse agonists at human A_3 adenosine receptors. Bioorg Med Chem 11:347–356

Pastorin G, Da Ros T, Bolcato C, Montopoli C, Moro S, Cacciari B, Baraldi PG, Varani K, Borea PA, Spalluto G (2006) Synthesis and biological studies of a new series of 5 heteroarylcarbamoylaminopyrazolo[4,3-*e*]1,2,4-triazolo[1,5-*c*]pyrimidines as human A_3 adenosine receptor antagonists. Influence of the heteroaryl substituent on binding affinity and molecular modeling investigations. J Med Chem 49:1720–1729

Perreira M, Jiang J, Klutz AM, Gao ZG, Shainberg A, Lu C, Thomas CJ, Jacobson KA (2005) Reversine and its 2-substituted adenine derivatives as potent and selective A_3 adenosine receptor antagonists. J Med Chem 48:4910–4918

Press NJ, Keller TH, Tranter P, Beer D, Jones K, Faessler A, Heng R, Lewis C, Howe T, Gedeck P, Mazzoni L, Fozard JR (2004) New highly potent and selective adenosine A_3 receptor antagonists. Curr Top Med Chem 4:863–870

Priego EM, von Frijtag Drabbe Kuenzel J, IJzerman AP, Camarasa MJ, Pérez-Pérez MJ (2002) Pyrido[2,1-*f*]purine-2,4-dione derivatives as a novel class of highly potent human A_3 adenosine receptor antagonists. J Med Chem 45:3337–3344

Pugliese AM, Coppi E, Volpini R, Cristalli G, Corradetti R, Jeong LS, Jacobson KA, Pedata F (2007) Role of adenosine A_3 receptors on CA1 hippocampal neurotransmission during oxygen-glucose deprivation episodes of different duration. Biochem Pharmacol 74:768–779

Saki M, Tsumuki H, Nonaka H, Shimada J, Ichimura M (2002) KF26777 (2-(4-bromophenyl)-7,8-dihydro-4-propyl-1*H*-imidazo[2,1-*i*]purin-5(4*H*)-one dihydrochloride), a new potent and selective adenosine A_3 receptor antagonist. Eur J Pharmacol 444:133–144

Salvatore CA, Jacobson MA, Taylor HE, Linden J, Johnson, RG (1993) Molecular cloning and characterization of the human A_3 adenosine receptor. Proc Natl Acad Sci 90:10365–10369

Shi L, Liapakis G, Xu R, Guarnieri F, Ballesteros JA, Javitch JA (2002) Beta2 adrenergic receptor activation. Modulation of the proline kink in transmembrane 6 by a rotamer toggle switch. J Biol Chem 277:40989–40996

Shneyvais V, Mamedova L, Zinman T, Jacobson KA, Shainberg A (2001) Activation of A_3 adenosine receptor protects against doxorubicin-induced cardiotoxicity. J Mol Cell Cardiol 33:1249–1261

Siddiqi SM, Jacobson KA, Esker JL, Olah ME, Ji XD, Melman N, Tiwari KN, Secrist JA III, Schneller SW, Cristalli G, Stiles GL, Johnson Cr, IJzerman AP (1995) Search for new purine- and ribose-modified adenosine analogues as selective agonists and antagonists at adenosine receptors. J Med Chem 38:1174–1188

Strickler J, Jacobson KA, Liang BT (1996) Direct preconditioning of cultured chick ventricular myocytes: novel functions of cardiac adenosine A_{2A} and A_3 receptors. J Clin Invest 98: 1773–1779

Tafi A, Bernardini C, Botta M, Corelli F, Andreini M, Martinelli A, Ortore G, Baraldi PG, Fruttarolo F, Borea PA, Tuccinardi T (2006) Pharmacophore based receptor modeling: the case of adenosine A_3 receptor antagonists. An approach to the optimization of protein models. J Med Chem 49:4085–4097

Tchilibon S, Kim SK, Gao ZG, Harris BA, Blaustein J, Gross AS, Melman N, Jacobson KA (2004) Exploring distal regions of the A_3 adenosine receptor binding site: sterically-constrained N^6-(2-phenylethyl)adenosine derivatives as potent ligands. Bioorg Med Chem 12: 2021–2034

Tchilibon S, Joshi BV, Kim SK, Duong HT, Gao ZG, Jacobson KA (2005) (N)-Methanocarba 2,N^6-disubstituted adenine nucleosides as highly potent and selective A_3 adenosine receptor agonists. J Med Chem 48:1745–1758

Tracey WR, Magee WP, Oleynek JJ, Hill RJ, Smith AH, Flynn DM, Knight DR (2003) Novel N^6-substituted adenosine $5'$-N-methyluronamides with high selectivity for human adenosine A_3 receptors reduce ischemic myocardial injury. Am J Physiol Heart Circ Physiol 285: H2780–H2787

van Galen PJM, van Bergen AH, Gallo-Rodriguez C, Melman N, Olah ME, IJzerman AP, Stiles GL, Jacobson KA (1994) A binding site model and structure–activity relationships for the rat A_3 adenosine receptor. Mol Pharmacol 45:1101–1111

van Muijlwijk-Koezen JE, Timmerman H, Link R, von der Goot H, Menge WMPB, von Frijtag von Drabbe Künzel JK, de Groote M, IJzerman AP (2000) Isoquinoline and quinazoline urea analogues as antagonists for the human adenosine A_3 receptor. J Med Chem 43:2227–2238

van Muijlwijk-Koezen JE, Timmerman H, Vollinga RC, von Drabbe Künzel JF, de Groote M, Visser S, IJzerman AP (2001) Thiazole and thiazole analogues as novel class of adenosine receptor antagonists. J Med Chem 44:749–762

Van Rompaey P, Jacobson KA, Gross AS, Gao ZG, Van Calenbergh S (2005) Exploring human adenosine A_3 receptor complementarity and activity for adenosine analogues modified in the ribose and purine moiety. Bioorg Med Chem 13:973–983

van Tilburg EW, von Frijtag Drabbe Kunzel J, de Groote M, IJzerman AP (2002) 2, 5′-Disubstituted adenosine derivatives: evaluation of selectivity and efficacy for the adenosine A_1, A_{2A}, and A_3 receptor. J Med Chem 45:420–429

Varani K, Merighi S, Gessi S, Klotz KN, Leung E, Baraldi PG, Cacciari B, Romagnoli R, Spalluto G, Borea PA (2000) [^{3}H]MRE 3008F20: a novel antagonist radioligand for the pharmacological and biochemical characterization of human A_3 adenosine receptors. Mol Pharmacol 57: 968–975

Volpini R, Costanzi S, Lambertucci C, Taffi S, Vittori S, Klotz KN, Cristalli G (2002) N^6-Alkyl-2-alkynyl derivatives of adenosine as potent and selective agonists at the human adenosine A_3 receptor and a starting point for searching A_{2B} ligands. J Med Chem 45:3271–3279

Volpini R, Dal Ben D, Lambertucci C, Taffi S, Vittori S, Klotz KN, Cristalli GJ (2007) N^6-Methoxy-2-alkynyladenosine derivatives as highly potent and selective ligands at the human A_3 adenosine receptor. J Med Chem 50:1222–1230

Yaar R, Lamperti ED, Toselli PA, Ravid K (2002) Activity of the A_3 adenosine receptor gene promoter in transgenic mice: characterization of previously unidentified sites of expression. FEBS Lett 532:267–272

Yang H, Avila MY, Peterson-Yantorno K, Coca-Prados M, Stone RA, Jacobson KA, Civan MM (2005) The cross-species A_3 adenosine-receptor antagonist MRS1292 inhibits adenosine-triggered human nonpigmented ciliary epithelial cell fluid release and reduces mouse intraocular pressure. Curr Eye Res 30: 747–754

Zhou QY, Li C, Olah ME, Johnson RA, Stiles GL, Civelli O (1992) Molecular cloning and characterization of an adenosine receptor: the A_3 adenosine receptor. Proc Natl Acad Sci USA 89:7432–7436

Zhu R, Frazier CR, Linden J, Macdonald TL (2006) N^6-Ethyl-2-alkynyl NECAs, selective human A_3 adenosine receptor agonists. Bioorg Med Chem Lett 16:2416–2418

Adenosine Receptors and the Heart: Role in Regulation of Coronary Blood Flow and Cardiac Electrophysiology

S. Jamal Mustafa, R. Ray Morrison, Bunyen Teng, and Amir Pelleg

Contents

Abstract Adenosine is an autacoid that plays a critical role in regulating cardiac function, including heart rate, contractility, and coronary flow. In this chapter, current knowledge of the functions and mechanisms of action of coronary flow regulation and electrophysiology will be discussed. Currently, there are four known

S.J. Mustafa (✉)
Department of Physiology and Pharmacology, School of Medicine, West Virginia University, Morgantown, WV 26505-9229, USA
smustafa@hsc.wvu.edu

C.N. Wilson and S.J. Mustafa (eds.), *Adenosine Receptors in Health and Disease*, Handbook of Experimental Pharmacology 193, DOI 10.1007/978-3-540-89615-9_6,

adenosine receptor (AR) subtypes, namely A_1, A_{2A}, A_{2B}, and A_3. All four subtypes are known to regulate coronary flow. In general, A_{2A}AR is the predominant receptor subtype responsible for coronary blood flow regulation, which dilates coronary arteries in both an endothelial-dependent and -independent manner. The roles of other ARs and their mechanisms of action will also be discussed. The increasing popularity of gene-modified models with targeted deletion or overexpression of a single AR subtype has helped to elucidate the roles of each receptor subtype. Combining pharmacologic tools with targeted gene deletion of individual AR subtypes has proven invaluable for discriminating the vascular effects unique to the activation of each AR subtype.

Adenosine exerts its cardiac electrophysiologic effects mainly through the activation of A_1AR. This receptor mediates direct as well as indirect effects of adenosine (i.e., anti-β-adrenergic effects). In supraventricular tissues (atrial myocytes, sinuatrial node and atriovetricular node), adenosine exerts both direct and indirect effects, while it exerts only indirect effects in the ventricle. Adenosine exerts a negative chronotropic effect by suppressing the automaticity of cardiac pacemakers, and a negative dromotropic effect through inhibition of AV-nodal conduction. These effects of adenosine constitute the rationale for its use as a diagnostic and therapeutic agent. In recent years, efforts have been made to develop A_1R-selective agonists as drug candidates that do not induce vasodilation, which is considered an undesirable effect in the clinical setting.

Keywords A_1 adenosine receptor · A_{2A} adenosine receptor · A_{2B} adenosine receptor · A_3 adenosine receptor · Endothelium · Coronary artery · Smooth muscle · Adenosine receptor knockout · Phospholipase C · MAPK · Adenosine receptor agonist · Adenosine receptor antagonist · Sinus node · AV node · Cardiac electrophysiology · PSVT · Anti-beta adrenergic action

Abbreviations

AC	Adenylate cyclase
AH	Atrial to His bundle activation time (representative of AV-nodal conduction time)
AR	Adenosine receptor
ATP	Adenosine 5′-triphosphate
AV	Atrioventricular
AVN	AV-nodal
CCPA	2-Chloro-N^6-cyclopentyl-adenosine
CF	Coronary flow
CGS-21680	2-[*p*-(2-carboxyethyl)]-phenylethyl-amino-5′-*N*-ethylcarboxamidoadenosine
CGS-22492	2-[(2-Cyclohexylethyl)amino]-adenosine
Cox-I	Cyclooxygenase I

CPA	N^6-Cyclopentyladenosine
DAD	Delayed afterdepolarizations
DPCPX	1,3-Dipropyl-8-cyclopentylxanthine
DPMA	N^6-[2-(3,5-Dimethoxyphenyl)-2-(2-methoxyphenyl] ethyl adenosine
ECG	Electrocardiogram
ERK	Extracellular regulated kinase
HV	His bundle to ventricular activation time
HUT	Head-up tilt table test
I_{Ca}	Inward calcium current
I_{CaL}	Inward L-type Ca^{2+} current
I_{Cl}	Chloride current
I_{f}	Hyperpolarization-activated current (“funny” current)
$I_{\mathrm{KAdo,Ach}}$	Outward potassium current
$I_{\mathrm{K,ATP}}$	ATP-dependent outward potassium current
I_{Ti}	Transient inward current
JNK	Jun N-terminal kinase
KO	Knockout
L-NMA	N^{G}-Methyl-L-arginine
LAD	Left anterior descending artery
LQTS	Long QT interval syndrome
MAPK	Mitogen-activated protein kinase
NECA	Adenosine-5′-N-ethylcarboxamide
NO	Nitric oxide
PDBu	Phorbol 12,13-dibutyrate
PI_3-kinase	Phosphatidylinositol 3-kinase
PLC	Phospholipase C
PKA	Protein kinase A
PKB (Akt)	Protein kinase B
PKC	Protein kinase C
PR	P wave to R wave interval on the ECG
PSVT	Paroxysmal supraventricular tachycardia
QT	Q wave–T wave interval in the ECG
QTc	Corrected QT interval
RR	R wave–R wave interval in the ECG
SN	Sinus node
SR	Sarcoplasmic reticulum
SSS	Sick sinus syndrome
SVT	Supraventricular tachycardia
VF	Ventricular fibrillation
VT	Ventricular tachycardia

1 General Background: The Adenosine Hypothesis

The heart is an astounding organ, capable of pumping over 8,000 liters of blood through the efficient operation of ~100,000 heartbeats per day. To place this in perspective, the total volume of blood ejected by the heart in a single day weighs over nine tons, and over one's lifetime the volume of blood pumped by the heart could fill the Empire State Building! The energy required to perform work of this magnitude is almost exclusively derived from aerobic oxidation of various substrates in the form of adenosine 5′-triphosphate (ATP) (Ingwall 2007; Knaapen et al. 2007). Cycling through up to 30 times its own weight in ATP per day (Ingwall and Weiss 2004; Neubauer 2007), the heart consumes more oxygen than any other organ (Taegtmeyer et al. 2005). Yet, because the myocardial ATP content is relatively small (4–6 μmol g^{-1}) compared to the rapid basal rate of ATP expenditure (30 μmol g^{-1} per minute), it is absolutely crucial that ATP production, and therefore oxygen supply, is closely matched across a broad range of cardiac work loads (Deussen et al. 2006). This inextricable link between myocardial function and metabolic demand is the basis of the "adenosine hypothesis" for the metabolic regulation of coronary flow (Berne 1963; Gerlach and Deuticke 1966).

Using an anesthetized open-chest working dog heart model, Berne demonstrated that myocardial hypoxia results in coronary venous efflux of adenine nucleotides, and that adenosine induces coronary dilation (Berne 1963). Together, these findings led to the following hypothesis:

> *Reduction in myocardial oxygen tension by hypoxemia, decreased coronary blood flow, or increased oxygen utilization by the myocardial cell leads to the breakdown of adenine nucleotides to adenosine. The adenosine diffuses out of the cell and reaches the coronary arterioles via the interstitial fluid and produces arteriolar dilation. The resultant increase in coronary blood flow elevates myocardial oxygen tension, thereby reducing the rate of degradation of adenine nucleotides, and decreases the interstitial fluid concentration of adenosine by washout and enzymatic destruction. This feedback mechanism serves to adjust coronary blood flow to meet the new metabolic requirements and a new steady state is achieved.* (Berne 1963)

Soon afterward, it was demonstrated that adenosine levels increase almost threefold within as little as 5 s of myocardial ischemia in vivo (Olsson 1970), and the incremental increase in coronary flow correlates highly with this rapid release of endogenous adenosine (Rubio et al. 1974). In the 45 years since the adenosine hypothesis was proposed, extensive investigation has established that adenosine serves as both a "sensor" of imbalances in energetic supply and demand and as a local metabolic regulator of coronary flow (Berne 1980; Deussen et al. 2006; Hori and Kitakaze 1991; Morrison et al. 2007; Tune et al. 2004). Although other effectors confer built-in redundancy for control of the coronary circulation (nitric oxide, ATP-sensitive potassium channels, acidosis, carbon dioxide, pO2, etc.; Deussen et al. 2006), it is clear that under conditions of impaired oxygen supply-to-demand ratio, rapid local production of adenosine leads to marked coronary dilation.

2 Adenosine and Coronary Regulation

Adenosine is an autacoid that plays a critical role in regulating coronary circulation. Adenosine is produced by the action of ecto-5′-nucleotidase on extracellular ATP released from the parenchymal tissue (including endothelium). Extracellular adenosine interacts with specific cell-surface receptors located on the smooth muscle and endothelial cells of the coronary artery to produce relaxation. Currently, there are four known adenosine receptor (AR) subtypes, namely A_1, A_{2A}, A_{2B}, and A_3. Although all four AR subtypes are found in coronary smooth muscle cells, only A_{2A}AR and A_{2B}AR have been shown to be present on coronary endothelial cells (Olanrewaju et al. 2002, 2000). Recently, A_3AR has been localized on endothelial cells in mouse aorta, leading to contraction of smooth muscle through cyclooxygenase I (Cox-I) (Ansari et al. 2007).

To date, pharmacological interventions using adenosine or its analogs are mostly directed toward adenosine-mediated effects on the cardiovascular system, such as the treatment of supraventricular arrhythmia, pharmacological stress myocardial perfusion imaging, congestive heart failure, controlling blood pressure, attenuating reperfusion injury following regional myocardial infarction, reducing infarct size, reducing incidence of arrhythmias, and improving postischemic cardiac function (Geraets and Kienzle 1992; Neubauer 2007; Peart and Headrick 2007; Smits et al. 1998). In the coronary circulation, A_{2A}AR plays a pivotal role in controlling vasodilation, while other receptors play a lesser role (Frobert et al. 2006; Hodgson et al. 2007; Morrison et al. 2002; Talukder et al. 2003). For instance, A_{2B}AR also mediates coronary vasodilation, while both A_1AR and A_3AR have been found to negatively modulate coronary vasodilation induced by A_{2A}AR and/or A_{2B}AR activation (Morrison et al. 2002; Talukder et al. 2002a; Tawfik et al. 2006). However, the significance of A_1, A_{2B}, and A_3ARs in coronary flow regulation remains to be fully elucidated.

The distribution of ARs along the branches of coronary arteries also varies. In the porcine heart, expression of A_1 and A_{2A}AR proteins has been documented in the left anterior descending artery (LAD), but only A_{2A}ARs are expressed coronary arterioles (Hein et al. 2001). Another study found that A_1, A_{2A}, and A_{2B}ARs are also expressed in coronary arterioles and venules (Wang et al. 2005). Functional studies in A_{2A}AR knockout (KO) mice suggested that A_{2B}AR may be more important in regulating larger coronary arteries (e.g., the LAD) than previously thought (Teng et al. 2008).

3 Endothelium-Dependent and Endothelium-Independent Regulation

It has been suggested that both A_{2A}AR and A_{2B}AR mediate hyperpolarization of smooth muscle and nitric oxide (NO) release from coronary artery endothelium (Hasan et al. 2000; Olanrewaju et al. 2002; Watts et al. 1998). Cell culture studies have demonstrated the involvement of A_{2A}AR- and A_{2B}AR-mediated NO release in porcine and human coronary endothelial cells (Li et al. 1998; Olanrewaju and Mustafa 2000). However, very few functional studies demonstrated that NO release is responsible for A_{2A}AR- or A_{2B}AR-mediated coronary vasodilation. Inhibition of NO synthase has been found to limit basal coronary flow (CF) in various species (Flood et al. 2002; Zatta and Headrick 2005). It has been shown in porcine coronary arterial rings that N^G-methyl-L-arginine (L-NMA, 30 μM), an NO synthase inhibitor, attenuated the relaxations of endothelium-intact but not endothelium-denuded rings induced by adenosine-5′-*N*-ethylcarboxamide (NECA), a nonselective adenosine agonist, and 2-[*p*-(2-carboxyethyl)]phenylethyl-amino-5′-*N*-ethylcarboxamidoadenosine (CGS-21680), a selective A_{2A}AR agonist (Abebe et al. 1995). It has been speculated that endogenously released adenosine and prostanoids induce NO- and/or K_{ATP} channel-dependent vasodilation and thereby modulate basal coronary tone (Flood et al. 2002; Hein et al. 2001; Talukder et al. 2002b; Zatta and Headrick 2005). Using two different NO synthase inhibitors in isolated hearts from wild-type and A_{2A}AR KO mice, it was found that A_{2A}AR plays a significant role in background NO release, thus affecting basal coronary tone (Teng et al. 2008). The role of A_{2B}AR in NO release remains to be determined, however.

The ARs responsible for endothelial-independent relaxation of coronary artery smooth muscle have not been conclusively determined; however, both A_{2A}AR and A_{2B}AR have been implicated (Morrison et al. 2002; Talukder et al. 2003, 2002b). A study with denuded porcine coronary arteries clearly demonstrated that A_{2A}AR plays a predominant role in endothelial-independent vasodilation, while A_{2B}AR may play a minor role (Teng et al. 2005).

4 Baseline Coronary Flow Control

It has been shown that both A_{2A}AR and A_{2B}AR mediate endogenous and exogenous adenosine-induced dilation of mouse coronary arteries (Morrison et al. 2002; Talukder et al. 2003). A_{2A}AR activation also contributes significantly to basal NO release and basal tone in coronary circulation (Flood et al. 2002; Teng et al. 2008; Zatta and Headrick 2005).

The cardiovascular effects of A_{2B}AR activation are similar to those mediated by A_{2A}AR; however, the affinity of adenosine to the latter is lower (Feoktistov and Biaggioni 1997; Hack and Christie 2003; Schulte and Fredholm 2003). The role of A_{2A}AR in basal vascular tone remains to be determined. However, it has been speculated that under pathological conditions such as ischemia, A_{2B}AR may be upregulated to compensate for the downregulation of A_{2A}AR-mediated responses. Indeed, an upregulation of A_{2B}AR gene expression has been found in ischemic mouse hearts (Ashton et al. 2003; Morrison et al. 2007). A more recent study has also demonstrated upregulation of A_{2B}AR in coronary arteries of A_{2A}AR gene KO mice, suggesting that A_{2B}AR provides a supportive role to the predominantly A_{2A}-mediated control of the coronary circulation (Teng et al. 2008).

5 Second-Messenger Systems

It has been well recognized that A_1 and A_3 ARs are coupled to $G_i/G_o/G_q$ proteins and inhibit the activity of adenylate cyclase (AC), while A_{2A} and A_{2B}ARs are coupled to G_s and activate AC, leading to cyclic adenosine 5′-monophosphate (cAMP) accumulation and subsequent activation of protein kinase A (PKA) (Fredholm et al. 2000). Indeed, in coronary arteries, where A_{2A}AR is predominant, A_{2A}AR-induced vasodilation is mediated mainly by the cAMP-dependent pathway (Hussain and Mustafa 1993; Rekik and Mustafa 2003). However, other second-messenger systems, such as phosphatidylinositol 3-kinase, tyrosine kinase and phospholipase C (PLC), may also be activated by ARs (Ansari et al. 2008; Peart and Headrick 2007; Tawfik et al. 2005), but their roles in mediating the effects of adenosine on the coronary vasculature have not been clearly defined. In addition, crosstalk between the cAMP/PKA pathway and the PLC/PKC pathway has also been reported (Germack and Dickenson 2004). Currently, the tangled web of these two second-messenger systems has garnered the most attention in studies of AR mechanism of action in cardiovascular tissue.

5.1 *cAMP–MAPK*

Following the activation of G_s protein by A_{2A}AR and A_{2B}AR, various second messenger signaling pathways including mitogen-activated protein kinases (MAPK) are initiated. The signal transduction pathway from G-protein-coupled receptors to MAPK is not fully understood, and may vary in different cell types (Fredholm et al. 2000). There are three well-characterized MAPKs: extracellular regulated kinase (ERK), or p42/44, p38, and jun *N*-terminal kinase (JNK). They seem to play a role in ischemic preconditioning, postconditioning (Morrison et al. 2007), smooth muscle cell growth, vascular smooth muscle migration, and vascular contraction (Haq et al. 1998; Kalyankrishna and Malik 2003; Wilden et al. 1998). Adenosine is reported to stimulate all MAPKs in the perfused rat heart (Haq et al. 1998). Agonist binding to A_{2A}AR can result in both activation and inhibition of ERK phosphorylation, depending on the type of cell expressing these receptors, and so can the second messenger pathway controlled by A_{2A}AR (Fredholm et al. 2000). A_{2B}AR is the only subtype capable of activating all three types of MAPKs (ERK1/2, p38, and JNK). It has also been shown that the same concentration of NECA and adenosine induces ERK1/2 phosphorylation to a greater extent than cAMP production (Fredholm et al. 2000). The involvement of p38 MAPK in adenosine-induced vasodilation has been recently reported (Teng et al. 2005); however, the role of MAPKs in the regulation of vascular tone requires more complete characterization.

There are a few reports linking ARs to p38 MAPK that provide clues as to which mediators are involved in the activation of p38 MAPK. A recent report demonstrated that cAMP inhibits p38 MAPK activation in endothelial cells derived from human umbilical vein (Rahman et al. 2004). In contrast, PKA was found to activate p38

MAPK in macrophages (Chio et al. 2004). Furthermore, the signaling pathways both up- and downstream of the p38 MAPK pathway are diverse, which may explain why p38 can be activated and create crosstalk among various stimuli (Eckle et al. 2007; Ono and Han 2000). For instance, it has been reported that p38 MAPK plays a significant role in angiotensin II-induced contraction (Meloche et al. 2000; Watts et al. 1998), while others have found that p38 MAPK is involved in adenosine-induced vasodilation (Teng et al. 2005). It is also possible that different p38 MAPK subtypes ($p38_{\alpha}$, $p38_{\beta}$, and $p38_{\gamma}$) are responsible for signaling via different pathways. Further investigation is needed to clarify the relationship between ARs and MAPKs vis-à-vis coronary regulation.

5.2 *PLC–PKC*

By virtue of differential coupling to either G_s (A_{2A} and A_{2B}ARs) or G_i proteins (A_1AR and A_3AR), along with variable tissue distribution of AR subtypes, adenosine elicits both relaxation (A_{2A}- and A_{2B}-mediated) and constriction (A_1- and A_3-mediated) in the peripheral and coronary vasculature. While this is discussed in further detail below with regard to coronary regulation, recent evidence supports a role for the phospholipase C (PLC)–protein kinase C (PKC) system in A_1AR-mediated contraction of aortic vascular smooth muscle (Tawfik et al. 2005). Specifically, isolated aortic rings from wild-type and A_1AR-KO mice were treated with adenosine, NECA, a nonselective AR agonist or 2-chloro-N^6-cyclopentyl-adenosine (CCPA), an A_1AR selective agonist, demonstrating uniform contractile responses in the 100 nM to 1 μM range in wild-type aortas only. Adenosine-induced vasoconstriction was not observed in aortas from A_1AR knockout mice with either nonselective (adenosine, NECA) or A_1-selective (CCPA) agonists, and the contractile response in wild-type aortas was eliminated by an A_1AR-selective antagonist, 1,3-dipropyl-8-cyclopentylxanthine (DPCPX). CCPA-mediated contraction in wild-type aortic rings was also eliminated by the PLC inhibitor U-73122, indicating a role for the PLC–PKC pathway in adenosine-mediated vasoconstriction (Tawfik et al. 2005). Other studies have shown that A_1AR enhances PKC expression in porcine coronary arteries (Marala and Mustafa 1995a, b, c). Moreover, a PKC inhibitor, GO-6893, was able to inhibit ENBA-induced contraction in mouse aorta (Ansari et al. 2008). Taken together, these findings suggest that the PLC–PKC pathway has a major role in A_1AR-mediated vascular tone (i.e., contraction of coronary arteries and the aorta).

5.3 *Other Second Messengers*

Phosphatidylinositol 3-kinase (PI_3 kinase) activates protein kinase B (PKB, also known as Akt), which phosphorylates and activates a cyclic nucleotide phosphodiesterase, 3B. Increases in cyclic nucleotide concentrations inhibit agonist-induced contraction of vascular smooth muscle. A PI_3-kinase inhibitor, LY 294002, has been shown to inhibit KCl, phorbol 12,13-dibutyrate (PDBu), and serotonin-induced contraction in bovine carotid artery smooth muscle strips, suggesting that the PI_3-kinase pathway plays a role in vascular smooth muscle tone (Komalavilas et al. 2001). A recent study provided the first evidence that A_{2B}AR-mediated cAMP formation activates ERK1/2 via a pathway dependent on PI_3 kinase, tyrosine kinases and Rap1 in CHO cells (Schulte and Fredholm 2002).

Tyrosine kinase and PKC are also found to trigger the MAPK system (Fredholm et al. 2000; Lowes et al. 2002; Robinson and Dickenson 2001; Yang et al. 2000; Zhao et al. 2001). Studies in A_1AR-induced delayed preconditioning in rabbits have suggested an important role for tyrosine kinase and PKC. These studies also speculate that the p38 MAPK/Hsp27 pathway may be a distal effector of this protection (Dana et al. 2000).

5.4 K^+ Channels

All four major types of K^+ channels (K_{ATP}, K_v, K_{IR} and K_{Ca}) are present in both coronary endothelial and smooth muscle cells (Frieden et al. 1999; Glavind-Kristensen et al. 2004; Kim et al. 2003; Li et al. 1999; Liu et al. 2001; Quayle et al. 1997; Rogers et al. 2007; Sun Park et al. 2006). However, the involvement of K^+ channels in adenosine-induced responses remains unclear. Activation of A_{2B}AR activates K_{ATP} channels in human and guinea pig coronary arteries independent of NO (Kemp and Cocks 1999; Mutafova-Yambolieva and Keef 1997; Niiya et al. 1994). K_{ATP} channels have been shown to mediate A_{2A}AR-induced vasodilation in systemic artery circulation and afferent arterioles of rat kidney and porcine coronary arterioles (Bryan and Marshall 1999; Hein et al. 2001; Tang et al. 1999). A study of cultured porcine coronary endothelial cells demonstrated the involvement of both K_{ATP} and K_{ca} in A_{2A} and A_{2B}AR-mediated hyperpolarization (Olanrewaju et al. 2002), which also leads to NO release (Olanrewaju and Mustafa 2000). Another study also suggested that activation of A_1AR on endothelial cells leads to K_{ATP} channel opening and subsequent Ca^{2+} influx, and an increase in $[Ca^{2+}]_{i,}$, which may lead to direct activation of eNOS via the Ca^{2+}–calmodulin pathway and NO release (Ray and Marshall 2006). In guinea pig coronary artery smooth muscle cells, the A_{2A}AR selective agonists CGS-21680 and N^6-[2-(3,5-dimethoxyphenyl)-2-(2-methoxyphenyl] ethyl adenosine (DPMA) failed to induce hyperpolarization, while the nonselective agonist NECA induced glibenclamide-sensitive hyperpolarization, suggesting that A_{2B}AR may be the only AR subtype involved in K_{ATP}-induced hyperpolarization in coronary smooth muscle (Mutafova-Yambolieva and Keef 1997). Further studies using patch clamp techniques will be valuable in clarifying the role of K^+ channels in adenosine-mediated vasoregulation.

6 Insight from Adenosine Receptor Gene-Modified Models

Most of what is known about adenosine-mediated coronary regulation is derived from pharmacologic studies using a broad spectrum of experimental models (Abebe et al. 1994; Belardinelli et al. 1998; Berne 1980; Deussen et al. 2006; Flood et al. 2002; Hasan et al. 2000; Hori and Kitakaze 1991; Makujina et al. 1992; Mustafa and Abebe 1996). Applying an ever-expanding collection of highly selective AR analogs (both agonists and antagonists) in such a variety of models has confirmed that the A_{2A}AR is the predominant subtype mediating adenosine-induced coronary vasodilation (Belardinelli et al. 1998; Shryock et al. 1998). However, the pharmacologic approach is limited by the selectivity of the ligands and/or potency, and frequently results in only indirect evidence that activation of other AR subtypes modifies adenosine-mediated coronary responses (Kemp and Cocks 1999; Makujina et al. 1992; Talukder et al. 2002b). The adventage of gene-modified models with targeted deletion or overexpression of a single AR subtype has allowed a more complete evaluation of adenosine-mediated responses than previously possible through agonist/antagonist studies alone. The cardiovascular phenotypes of several AR KO/overexpression models are reviewed by Ashton et al. (2007); Table 1 summarizes available data on the vascular phenotypes of A_1, A_{2A}, A_{2B}, and A_3AR KO models.

6.1 A_{2A}AR KO Mouse

The A_{2A}AR KO mouse model was developed and characterized by Ledent et al. 1997. Although the model was developed primarily as a tool for Parkinson's disease research, A_{2A} knockout mice were noted to be hypertensive, suggesting a direct vascular/cardiovascular effect of this targeted deletion. Subsequent studies combined the specificity of A_{2A}AR deletion with the traditional pharmacologic approach to demonstrate that although basal coronary flow was unchanged by A_{2A}AR deletion (Morrison et al. 2002, 2007; Talukder et al. 2003), adenosine-induced coronary dilation was significantly impaired in isolated hearts (Morrison et al. 2002; Talukder et al. 2003) and in isolated coronary arteries (Teng et al. 2008). During recovery from global ischemia, isolated A_{2A}AR KO hearts also demonstrated reduced coronary flow compared to wild-type littermate controls (Morrison et al. 2007). Together, these studies indicate that A_{2A}AR plays a primary role in murine coronary regulation. Importantly, the observation that a nonselective adenosine analog, NECA, induced coronary dilation in hearts lacking A_{2A}ARs documented for the first time that other AR subtypes modulate adenosine-induced coronary regulation (Morrison et al. 2002). This NECA-induced coronary dilation in A_{2A} knockout hearts was attenuated by alloxazine, a putatively selective A_{2B}AR antagonist, indicating that A_{2B}ARs act in concert with A_{2A}ARs to elicit murine coronary dilation (Morrison et al. 2002; Talukder et al. 2003).

Table 1 Vascular phenotype of AR KO mice

Gene deletion	Experimental model (refs.)	Vascular phenotype (refs.)
A_1AR KO	• Isolated aortic rings (Tawfik et al. 2005) • Isolated hearts (Morrison et al. 2006; Reichelt et al. 2005; Salloum et al. 2007; Tawfik et al. 2006) • Postischemic isolated hearts (Morrison et al. 2006; Salloum et al. 2007)	• Enhanced basal coronary flow (Morrison et al. 2006; Tawfik et al. 2006) • Unchanged basal coronary flow (Reichelt et al. 2005; Salloum et al. 2007) • Enhanced adenosinergic dilation (Tawfik et al. 2005; Tawfik et al. 2006) • Unchanged adenosinergic dilation (Reichelt et al. 2005) • Reduced postischemic coronary flow (Morrison et al. 2006) • Unchanged postischemic coronary flow (Salloum et al. 2007)
A_{2A}AR KO	• In vivo tail cuff pressure (Ledent et al. 1997) • Isolated hearts (Morrison et al. 2002, 2007; Talukder et al. 2003) • Isolated coronary arteries (Teng et al. 2008)	• Hypertension (Ledent et al. 1997) • Unchanged basal coronary flow (Morrison et al. 2002, 2007; Talukder et al. 2003) • Impaired adenosinergic coronary dilation (Morrison et al. 2002, 2007; Talukder et al. 2003; Teng et al. 2008) • Reduced postischemic coronary flow (Morrison et al. 2007)
A_{2B}AR KO	• In vivo tail-cuff pressure (Yang et al. 2006; Hua et al. 2007) • Isolated hearts (unpublished observations from Mustafa's group)	• Normal basal blood pressure (Yang et al. 2006; Hua et al. 2007) • Unchanged basal coronary flow (unpublished observations from Mustafa's group) • Unchanged postischemic coronary flow (unpublished observations from Morrison's group)
A_3AR KO	• In vivo tail-cuff pressure (Ge et al. 2006; Zhao et al. 2000) • Isolated hearts (Cerniway et al. 2001; Ge et al. 2006; Harrison et al. 2002; Talukder et al. 2002a) • Postischemic isolated hearts (Cerniway et al. 2001; Harrison et al. 2002) • Isolated aortic rings (Ansari et al. 2007)	• Normal basal blood pressure (Ge et al. 2006; Zhao et al. 2000) • Enhanced adenosinergic hypotension (Zhao et al. 2000) • Enhanced adenosinergic coronary dilation (Talukder et al. 2002a) • Unchanged basal coronary flow (Cerniway et al. 2001; Ge et al. 2006; Harrison et al. 2002; Talukder et al. 2002a) • Unchanged postischemic coronary flow (Cerniway et al. 2001; Harrison et al. 2002) • Reduced A_3-mediated vasoconstriction (Ansari et al. 2007)

6.2 A_{2B} AR KO Mouse

Limited data exist regarding the vascular phenotype of A_{2B} KO mice, as this is the latest of the AR KO models to be developed. Most recent reports on studies with A_{2B} KO mice demonstrate a critical role for the A_{2B}AR in protecting against excessive vascular adhesion and injury (Yang et al. 2008, 2006), hypoxia-induced vascular leak (Eckle et al. 2008), and infarct size associated with regional ischemia-reperfusion (Eckle et al. 2007). In two distinct A_{2B}AR KO in vivo models, tail-cuff measurements showed no differences in resting blood pressure (Hua et al. 2007; Yang et al. 2006). The effects of A_{2B}AR deletion on coronary flow are not yet reported. However, using the A_{2B} KO model recently characterized by Hua et al. (2007), preliminary data have indicated that targeted deletion of A_{2B}ARs has no effect on either basal or postischemic coronary flow compared to wild-type hearts (Morrison et al., unpublished observations). While prior studies have shown that A_{2B}ARs plays a role in adenosine-induced coronary dilation (Morrison et al. 2002), it is not clear whether the absence of A_{2B}ARs would result in the attenuation of this effect.

6.3 A_1AR and A_3AR KO Mouse

Based on data obtained in earlier pharmacologic studies, along with more recent data from studies using A_{2A} and A_{2B}AR knockout mice, it can be concluded that adenosine-induced coronary dilation is predominantly mediated by A_{2A}AR and to a smaller extent by A_{2B}AR. As noted above, the vasodilatory effect of both A_{2A} and A_{2B}AR activation is largely due to their coupling to G_s proteins and the resultant activation of AC and production of cAMP (Hussain and Mustafa 1993; Rekik and Mustafa 2003). Since both A_1AR and A_3AR are G_i coupled, and their activation attenuates cAMP production by AC inhibition, it is not unreasonable to expect that A_1AR and/or A_3AR activation would negatively modulate A_{2A}AR- and A_{2B}AR-mediated vasodilation. Early evidence of A_1AR-mediated coronary vasoconstriction is derived from pharmacologic studies (Hussain and Mustafa 1995), but more recently this phenomenon has been confirmed in both coronary (Tawfik et al. 2006) and aortic (Tawfik et al. 2005) vasculature using A_1AR KO mouse. Similarly, early attempts to clarify the relative roles of each AR subtype in aortic and coronary vasoregulation using pharmacologic agents (Talukder et al. 2002b) have been followed by direct and convincing evidence from A_3AR KO mice demonstrating that A_3 activation causes vasoconstriction in both the coronary (Talukder et al. 2003) and aortic vasculatures (Ansari et al. 2007). Thus, while activation of A_{2A} and A_{2B} ARs leads to coronary dilation, A_1 and A_3AR activation negatively modulates this effect through vasoconstriction.

Using functional studies with even more selective and potent pharmacologic ligands, targeted gene deletion of individual AR subtypes has proven invaluable for discriminating the vascular effects unique to the activation of each AR subtype.

Coming full circle, it seems fitting that information gained from these models is now being harnessed to improve our pharmacologic approach to both diagnostic and therapeutic interventions in the clinical management of heart disease.

7 Clinical Application of Selective A_{2A}AR Agonists for the Detection of Coronary Artery Disease

Adenosine (Adenoscan) has been used as a pharmacological stress agent in conjunction with radionuclide myocardial perfusion imaging (MPI) in patients unable to undergo adequate exercise stress. Dipyridamole, an adenosine uptake blocker, was also used for this purpose for several years prior to the approval of Adenoscan by the US Food and Drug Administration (Cerqueira 2006). Due to frequent side effects (e.g., bronchospasm, AV nodal conduction block) of these two agents, there was a need for better selective drugs for myocardial stress testing. In the late 1980s to early 1990s (Abebe et al. 1994; Mustafa and Askar 1985), it was discovered that adenosine-induced vasodilation of coronary arteries of several species, including humans, was mediated predominately by A_{2A}AR (Ramagopal et al. 1988; Shryock et al. 1998). This discovery led to the development of more selective agonists for the A_{2A}AR subtype, including the Ciba–Geigy (Novartis) compound CGS-21680 (Francis et al. 1991; Hutchison et al. 1989). CGS-21680 was later discovered to be a very selective agonist for the A_{2A}AR subtype in a number of species, including humans (Abebe et al. 1994; Makujina et al. 1992). It was shown that CGS-21680 and another A_{2A}-selective compound from Ciba–Geigy (2-[(2-cyclohexylethyl)amino]-adenosine, CGS-22492) produced significant relaxation in isolated human coronary arteries from organ donors (Makujina et al. 1992). However, it was also discovered that the nonselective analog NECA produced greater relaxation than the A_{2A}-selective CGS-21680 and CGS-22492, suggesting that there was another AR subtype (possibly A_{2B}AR) causing this additional relaxation (Makujina et al. 1992).

Since ARs were not cloned at the time of the study described above, and little was known about A_{2B}ARs, the possibility was left open that another AR subtype contributes to relaxation of human coronary arteries. It was not until the availability of A_{2A}AR KO mouse that an unequivocal demonstration of the role that the A_{2B}AR plays in the regulation of coronary flow became possible (Morrison et al. 2002). Using A_{2A}AR KO mice, Mustafa and his coworkers showed that NECA increased coronary flow, whereas CGS-21680 did not have an effect (Morrison et al. 2002). Moreover, it has recently been reported that there is a compensatory upregulation of the A_{2B}AR receptor in A_{2A}AR KO mice (Teng et al. 2008), further lending support to the theory that most likely the A_{2B}AR is responsible for NECA-induced vasorelaxation of coronary arteries in the A_{2A}AR KO mouse model and perhaps in human coronary arteries. Validation of the role of A_{2B}ARs in functional responses in coronary arteries will be determined in A_{2B}AR KO mouse hearts, as the A_{2B}AR KO mouse model has just recently become available. This is an area of active investigation in Mustafa's group.

These data strongly suggest that, in addition to the A_{2A}AR, the A_{2B}AR also plays a role in the regulation of coronary flow in humans and animals. Therefore, complete dilation of the coronary vascular bed to determine coronary reserve in patients with suspected coronary artery disease may require the use of a combination of A_{2A}AR and A_{2B}AR agonists. These observations are supported by a recent report (Nitenberg et al. 2007) showing that intracoronary infusion of adenosine (60 μg) elicits a lower hyperemic response than postocclusion hyperemia (30 s). These authors concluded that the use of an adenosine infusion represents a potential source of error in determination of coronary reserve, and may result in an underestimation of the physiological significance of coronary stenosis. It is true that adenosine, being the natural endogenous nonselective AR agonist, will activate all four AR subtypes, including A_{2A} and A_{2B}. However, adenosine is a less potent and nonselective agonist that will also activate A_1 and A_3 ARs, causing a reduction in coronary flow (Talukder et al. 2002a; Tawfik et al. 2006), which serves to counter the increase in flow due to A_{2A} and A_{2B} AR activation.

If A_{2B}ARs play an important role in human coronary vasodilation, as suggested by earlier studies (Makujina et al. 1992), then re-evaluating the sole use of A_{2A}AR-selective agonists in myocardial perfusion stress testing may be warranted. It is possible that adjunctive use of selective A_{2B}AR agonists in concert with currently employed highly selective A_{2A} agonists may lead to a more complete evaluation of both coronary artery disease and coronary reserve. This becomes important in light of the fact that two A_{2A}-selective AR agonists are in Phase III clinical trials (binodenoson, MRE-0470/WRC-0470, Aderis Pharmaceuticals; apadenoson, ATL-146e, Adenosine Therapeutics) and another, Lexiscan™ (regadenoson, CVT-3146, CV Therapeutics), has recently received FDA approval for use in pharmacological stress myocardial perfusion imaging. Newly available and highly selective A_{2B}AR analogs are beginning to advance our understanding of the role of A_{2B}ARs in the heart, and it is plausible to envision their adjunctive use for coronary dilation in this clinical setting.

8 Cardiac Electrophysiology of Adenosine: Recent Developments

8.1 Introduction

This section focuses on several aspects of the cardiac electrophysiology of adenosine and gives an update on clinical applications of second-generation AR ligands. For a broader discussion of the cardiac electrophysiology of adenosine, the reader is referred to several previously published reviews (Belardinelli et al. 1995; Dhalla et al. 2003; Pelleg and Belardinelli 1993; Pelleg et al. 2002; Shen and Kurachi 1995; Zablocki et al. 2004).

Adenosine is a ubiquitous adenine nucleoside found in every cell of the human body; it is released into the extracellular space under physiologic and pathophysiologic conditions. The actions of extracellular adenosine are mediated by four subtypes of AR coupled to G proteins: A_1, A_{2A}, A_{2B} and A_3. In the heart, the electrophysiologic effects of adenosine are mediated mainly by A_1AR. The latter receptor mediates the direct effects as well as the indirect effects; i.e., the anti-β-adrenergic effects of adenosine (Dobson et al. 1987; Schrader et al. 1977).

8.2 Basic Aspects

8.2.1 Negative Chronotropic Action

Adenosine suppresses the activity of cardiac pacemakers including the sinus node (SN), atrio–ventricular (AV) junction, and His–Purkinje system; an inverse relationship between pacemaker hierarchy and sensitivity to adenosine was observed. Specifically, the following sensitivity cascade has been observed: Purkinje fibers > His bundle > AV junction > SN (for references, see Pelleg et al. 1990a). This action is mediated by A_1AR and the activation of a potassium outward current ($I_{KAdo,Ach}$), as well as the suppression of inward calcium current (I_{Ca}) and the hyperpolarization-activated current ("funny" current) (I_f) (Belardinelli et al. 1988; Zaza et al. 1996). Since norepinephrine shifts the activation curve of I_f to the right (DiFrancesco and Borer 2007) and enhances I_{Ca}, the antiadrenergic action of adenosine can also play an important role in its negative chronotropic effects. Data obtained in vitro were interpreted to suggest that the suppression of I_f is more relevant than the activation of $I_{KAdoAch}$ to the modulation of SN automaticity by adenosine (Zaza et al. 1996). However, data obtained in vivo suggest that I_f plays a larger role in the pacemaker activity of His–Purkinje fibers vs. SN (Pelleg et al. 1990a). Specifically, in dogs with complete AV nodal conduction block where SN and ventricular pacemakers were operating concurrently but independently, adenosine suppressed the activities of both pacemakers in a dose-dependent manner through the activation of A_1AR; however, in the presence of isoproterenol, the dose–response to adenosine in the SN and in the ventricular pacemaker shifted to the left and right, respectively (Pelleg et al. 1990a). Thus, the accentuation of the adenosine's action in the SN seemed to be the result of its suppression of isoproterenol-enhanced I_f and I_{Ca} (indirect, anti-β-adrenergic action), which was added to its induced $I_{KAdoAch}$ (i.e., direct action), while in the ventricular pacemakers, the action of adenosine was mediated mainly by its suppression of I_f, an action which was attenuated in the presence of isoproterenol due the rightward shift of the I_f activation curve induced by the catecholamine (DiFrancesco and Borer 2007). This interpretation agrees well with the maximal diastolic potentials of approximately -65 mV and -90 mV in the SN and His–Purkinje pacemaker cells, respectively, as well as the I_f activation curve, which indicates fractional activation (i.e., activated channel probability of 0.33) and full activation (i.e., activated channel probability of 1.0) of this current at membrane potentials of −65 mV and −90 mV, respectively (DiFrancesco and Borer 2007). The limited yet significant role of I_f in the pacemaking mechanism in the SN is indicated by the fact that CsCl and ZD7288, which are known blockers of I_f, slowed but did not arrest spontaneous pacemaking in SN cells (Denyer and Brown 1990; Sanders et al. 2006). Further support for this interpretation was given by the fact that in this canine model, quinidine, which suppresses acetylcholine-induced $I_{KAdoAch}$, and probably also adenosine-induced $I_{KAdoAch}$, attenuated the negative chronotropic action of adenosine in the SN but not in ventricular pacemakers (Pelleg et al. 1990a). Thus, data obtained in isolated single cells in vitro should be extrapolated to the in vivo setting with great caution; the lack of electrotonic interactions and constitutive neural input (among other factors) in commonly used in vitro models may affect this process.

Overexpression of A_1AR was associated with (i) a 20-fold increase in the potency of 2-chloroadenosine in slowing heart rate and a 35% reduction in maximal heart rate induced by β-adrenoceptor stimulation (Headrick et al. 2000), (ii) a reduced positive chronotropic response to exercise, and (iii) little effect on the resting heart rate (Kirchhof et al. 2003). Interestingly, overexpression of A_3AR was associated with depressed heart rate preferentially at rest (Fabritz et al. 2004). These data give further support to the notion that A_1AR mediates the negative chronotropic action and anti-β-adrenergic effects of adenosine. The role of A_3AR, if any, in the cardiac electrophysiology of adenosine remains to be determined.

8.2.2 Negative Dromotropic Action

The negative dromotropic action of adenosine is manifested in the prolongation of the PR and AH intervals as well as complete AV nodal (AVN) conduction block. Adenosine does not alter the HV interval; therefore, its dromotropic action is mainly due to its effects on the AVN. The seminal work of Belardinelli et al. (see (Belardinelli et al. 1987) elucidated the mechanisms of action of adenosine on the AVN; their major findings were: (i) adenosine mediates hypoxia/ischemia-induced AVN conduction block; (ii) adenosine hyperpolarizes cell membrane potential, shortens action potential duration, slows the recovery of I_{Ca}, and prolongs postrepolarization refractoriness in isolated single AVN cells; (iii) these actions of adenosine are mediated by A_1AR, and; (iv) the degree of amplification of A_1AR occupancy as determined by the negative dromotropic response to adenosine is relatively minimal, indicating "tight" coupling between receptor occupancy and its physiologic outcome (Belardinelli et al. 1981; Clemo and Belardinelli 1986; Clemo et al. 1987; Dennis et al. 1992). A subsequent study confirmed that A_1AR and a pertussis toxin-sensitive G protein mediate the AVN conduction block associated with global myocardial ischemia in vivo (Xu et al. 1993).

8.2.3 Adenosine's Effects on Atrial and Ventricular Myocardium

In the atria, adenosine exerts direct and indirect anti-β-adrenergic effects. The activation of $I_{KAdoAch}$ in atrial myocytes, which is mediated by A_1AR and pertussis toxin-sensitive G protein, results in shortened action potential duration and refractoriness (Pelleg et al. 1996), thereby facilitating re-entry. Indeed, a common side effect of adenosine is the induction of transient atrial fibrillation (Pelleg et al. 2002). Recently, Hove-Madsen et al. (Hove-Madsen et al. 2006) have demonstrated that $A_{2A}AR$ is expressed in the human right atrium and distributed in a banded pattern along the Z lines, overlapping with the ryanodine receptor. In this study, an $A_{2A}AR$-selective agonist did not affect the L-type inward Ca^{2+} current (I_{CaL}) amplitude, but it did increase spontaneous calcium release from the sarcoplasmic reticulum (SR) and reduce the fast time constant for I_{Ca} inactivation (Hove-Madsen et al. 2006). These data were interpreted to suggest that activation of the $A_{2A}AR$ stimulates the ryanodine receptor itself (Hove-Madsen et al. 2006).

In general, adenosine does not directly affect ventricular myocytes; although direct activation by adenosine of the ATP-dependent potassium outward current ($I_{K,\,ATP}$) in isolated rat ventricular myocytes has been proposed (Kirsch et al. 1990), subsequent studies in vitro and in vivo failed to support this hypothesis (Song et al. 2002; Xu et al. 1994). Adenosine exerts pronounced anti-β-adrenergic effects in the ventricular myocardium, which are mediated by A_1AR and reduced intracellular levels of cAMP (Belardinelli and Isenberg 1983). Adenosine attenuates the catecholamine-dependent increase in inward L-type Ca^{2+} current (I_{CaL}), the delayed rectifier potassium current and chloride current (I_{Cl}). In addition, adenosine attenuates I_{CaL}- and transient inward current (I_{Ti})-dependent afterdepolarizations and triggered activity (Song et al. 1992). Interestingly, adenosine terminated episodes of ventricular tachycardia (VT) and abolished the delayed afterdepolarizations (DAD) associated with digoxin toxicity in the perfused guinea-pig heart in vitro and guinea-pig and canine hearts in vivo (Fogaça and Leal-Cardoso 1985; Xu et al. 1995). Because catecholamines play a mechanistic role in digoxin-induced DAD and triggered activity, it was concluded that this antiarrhythmic effect of adenosine was mediated by its anti-β-adrenergic action (Xu et al. 1995). Indeed, several studies have indicated that adenosine can exert an antiarrhythmic effect in the setting of other catecholamine/cAMP-dependent ventricular tachycardias (see below).

8.3 *Clinical Aspects*

8.3.1 Supraventricular Tachycardias

The seminal work of Belardinelli et al. in the late 1970s and early 1980s led in 1989 to the introduction of adenosine as an effective and safe antiarrhythmic drug for the acute termination of paroxysmal supraventricular tachycardia (PSVT) involving the AVN (Adenocard) (DiMarco et al. 1983; for reviews, see Pelleg and

Kutalek 1997; Pelleg et al. 2002). The rationale for the use of adenosine as an antiarrhythmic drug in this setting is derived from its potent suppression of AVN conduction; the latter breaks or slows down re-entrant circuits involving the AVN. However, it has also led to several "off label" uses of adenosine as a diagnostic drug, including the differential diagnosis of broad QRS complex tachycardia (i.e., SVT with aberrant ventricular conduction vs. VT), and assessment of accessory AV pathway ablation (Conti et al. 1995; Keim et al. 1992).

In recent years, a second generation of adenosine receptor-related drug candidates has been developed (Hutchinson and Scammells 2004). For example, tecadenoson (CVT-510; CV Therapeutics, Inc.) is a novel selective A_1AR agonist that is being evaluated as a drug candidate for the acute suppression of PSVT (Cheung and Lerman 2003; Peterman and Sanoski 2005). Clinical trials have shown that the drug effectively terminates PSVT without the side effects caused by the activation of ARs other than the A_1AR, which is associated with the use of adenosine in this setting.

Focal atrial tachycardias are a group of SVTs characterized by the concentric spread of a wave of depolarization from a specific localized source, the mechanism of which includes abnormal automaticity, triggered activity and microreentry (Lindsay 2007). The response to programmed atrial stimulation as well as several pharmacologic agents including adenosine has been used to differentiate these mechanisms. Regarding adenosine, data obtained in recent years support the hypothesis that adenosine-induced suppression or termination of a focal atrial tachycardia is indicative of a microreentry mechanism rather than abnormal automaticity or triggered activity (Iwai et al. 2002; Markowitz et al. 2007, 1999).

8.3.2 Ventricular Tachycardia/Fibrillation

Due to its anti-β-adrenergic actions in the ventricular myocardium, adenosine can affect catecholamine-dependent ventricular arrhythmias. In an early study, adenosine terminated sustained, exercise-triggered VT in four patients with structurally normal hearts (Lerman et al. 1986). Observations in this study have led to the hypothesis that the mechanism of the adenosine-sensitive VT is cAMP-mediated triggered activity (Lerman et al. 1986). Data obtained in subsequent studies in similar patients have supported this hypothesis and indicated that the action of adenosine is mediated by A_1AR (Lerman 1993). Idiopathic repetitive nonsustained monomorphic VT, which is characterized by frequent ectopic beats and salvos of VT, and typically occurs at rest, can also be sensitive to adenosine (Lerman et al. 1995). Among the idiopathic VT, the right outflow tract VT, which is the most common form (and presents as repetitive monomorphic VT or exercise-induced VT) and the left outflow tract VT (Nogami 2002) are both adenosine sensitive (Iwai et al. 2006; Lerman et al. 1997). Thus, the responsiveness to adenosine suggests that the mechanism of these tachycardias is probably cAMP-mediated triggered activity (Lerman et al. 2000).

In a swine model of prolonged ventricular fibrillation (VF), a selective A_1AR antagonist accelerated the deterioration in the VF waveform; this finding was interpreted to suggest that endogenous adenosine exerts cardioprotective effects during sudden cardiac arrest associated with VF (Mader et al. 2006). However, data obtained in human subjects raise doubts regarding the use of an A_1AR antagonist in this setting. Specifically, because endogenous adenosine (which accumulates during hypoxia and ischemia) may perpetuate asystole, the use of aminophylline, a nonselective AR antagonist, in the setting of cardiac arrest has been proposed as an acute pharmacologic intervention to improve resuscitation outcome (Viskin et al. 1993). However, subsequent studies have shown that aminophylline offers no benefits in this situation (Hayward et al. 2007).

8.4 Adenosine as a Diagnostic Tool

Several diagnostic applications of adenosine, in addition to the diagnosis of broad QRS complex tachycardia mention above, have been proposed. Viskin et al. (Viskin et al. 2006) have shown that by provoking transient bradycardia followed by sinus tachycardia, adenosine challenge induces changes in QT interval that could be useful in distinguishing patients with long QT syndrome (LQTS) from healthy subjects. Specifically, adenosine challenge resulted in dissimilar responses in patients with LQTS and healthy subjects; the largest difference was recorded during maximal bradycardia, where the difference between the mean QT and QTc values of the two groups was 121 ms (vs. a 59 ms difference at baseline) and 125 ms (vs. a 55 ms difference at baseline), respectively (Viskin et al. 2006). These observations by Viskin et al. (2006) explain the several cases of adenosine-induced polymorphic ventricular tachycardia (i.e., Torsade-de-Pointe; see review by Pelleg et al. 2002).

Several studies have indicated that adenosine can also identify patients with sick sinus syndrome (SSS). For example, Fragakis et al. (2007) found that, when a cut-off value of 525 ms for sinus recovery time (i.e., the time elapsed from sinus arrest until the emergence of the first sinus beat) was used as an indicator of sinus node dysfunction, sinus node recovery time (corrected for baseline rate) had 74% and a specificity of 100% for diagnosis of SSS, while the recovery time following adenosine had a sensitivity of 94 and a specificity of 84%, respectively (Fragakis et al. 2007). Earlier studies reported similar values; i.e., 80% sensitivity and 97% specificity (Burnett et al. 1999), and 67% and 100%, respectively (Resh et al. 1992).

Adenosine has been used in the diagnosis of patients with neurally mediated syncope; i.e., vasovagal syncope and syncope of unknown origin. Based on its sympathomimetic action (direct via activation of chemoreceptors and indirect via the baroreflex), adenosine has been proposed as an adjuvant provocative agent in the protocol of head-up tilt table test (HUT) (Mittal et al. 2004; Shen et al. 1996). The fact that adenosine plasma levels (Carrega et al. 2007; Saadjian et al. 2002) and the number of A_{2A}AR, which were upregulated (Carrega et al. 2007), were higher in patients with a positive HUT was interpreted to suggest that endogenous adenosine mediates syncope in a specific cohort of syncopal patients. However, prolonged adenosine induced AV block in conjunction with HUT in patients with unexplained syncope failed to predict recurrent syncopal episodes (Cheung et al. 2004).

Unfortunately, many studies as well as reviews of those mentioned above (and other similar studies) have treated the cardiovascular effects of adenosine and ATP as being identical. While ATP mimics adenosine due to its rapid degradation to the nucleoside by ectoenzymes, the reverse does not hold; specifically, before its degradation, ATP triggers a cardio-cardiac central vagal reflex mediated by the activation of $P2X_{2/3}$ receptors localized on vagal sensory nerve terminals in the left ventricle; adenosine is devoid of this action (Xu et al. 2005). Thus, the negative chronotropic and dromotropic actions of ATP are mediated by adenosine and the vagus nerve (Pelleg et al. 1997); a mechanism found in cat, dog and man, but not in rodents (Pelleg et al. 1990b). Brignole et al. (Brignole et al. 2003) and Flammang et al. (2006) used bolus intravenous injections of ATP as a diagnostic tool in patients with syncope of unknown cause (for references, see Parry et al. 2006). The former group has used a maximal RR interval >6 s while the latter group has used complete AVN block duration >10 s as an indication of a positive test. In view of the fact the both ATP and adenosine suppress ventricular escape rhythms (Lerman et al. 1988; Pelleg et al. 1986), it is difficult to interpret the RR interval data as the rate of ATP degradation, and hemodynamic factors can directly affect this parameter. Using the RR interval >10 s criterion, it seems that ATP can identify a cohort of elderly patients in whom the mechanism of syncope is bradycardia and who may benefit from pacemaker therapy (see Flammang and AMS Investigators 2006; Flammang et al. 2005; Parry et al. 2006). In these patients, the bradycardia can be due to SN dysfunction, AVN dysfunction, abnormal vagal input to the heart, and any combination of these three causes.

8.5 *Future Prospects*

In view of the current efforts by the pharmaceutical industry, one can expect the introduction of AR ligands as new drugs for the treatment and diagnosis of cardiac arrhythmias in the near future. This would constitute a quantum step forward in the harnessing of adenosine signal transduction for the benefit of patients.

Acknowledgement We thank would like to Luiz Belardinelli, M.D. for critical reading of the manuscript and his helpful comments. Also, we would like to acknowledge the support of NIH (HL-027339-SJM; HL-074001-RRM).

References

Abebe W, Makujina SR, Mustafa SJ (1994) Adenosine receptor-mediated relaxation of porcine coronary artery in presence and absence of endothelium. Am J Physiol 266:H2018–H2025

Abebe W, Hussain T, Olanrewaju H, Mustafa SJ (1995) Role of nitric oxide in adenosine receptor-mediated relaxation of porcine coronary artery. Am J Physiol 269:H1672–H1678

Ansari HR, Nadeem A, Tilley SL, Mustafa SJ (2007) Involvement of COX-1 in A_3 adenosine receptor-mediated contraction through endothelium in mice aorta. Am J Physiol Heart Circ Physiol 293:H3448–H3455

Ansari H, Teng B, Nadeem A, Schnermann J, Mustafa S (2008) A_1 adenosine receptor-activated protein kinase C signaling in A_1 knock-out mice coronary artery smooth muscle cells. FASEB J 22:: 1152.11

Ashton KJ, Nilsson U, Willems L, Holmgren K, Headrick JP (2003) Effects of aging and ischemia on adenosine receptor transcription in mouse myocardium. Biochem Biophys Res Commun 312:367–372

Ashton KJ, Peart JN, Morrison RR, Matherne GP, Blackburn MR, and Headrick JP (2007) Genetic modulation of adenosine receptor function and adenosine handling in murine hearts: insights and issues. J Mol Cell Cardiol 42:693–705

Belardinelli L, Isenberg G (1983) Actions of adenosine and isoproterenol on isolated mammalian ventricular myocytes. Circ Res 53:287–297

Belardinelli L, Mattos EC, Berne RM (1981) Evidence for adenosine mediation of atrioventricular block in the ischemic canine myocardium. J Clin Invest 68:195–205

Belardinelli L, West GA, Clemo SHF (1987). Regulation of atrioventricular node function by adenosine. In: Gerlach E, Becker B (eds) Topics and perspectives of adenosine research. Springer, Berlin, pp 344–355

Belardinelli L, Giles WR, West A (1988) Ionic mechanisms of adenosine actions in pacemaker cells from rabbit heart. J Physiol 405:615–633

Belardinelli L, Shryock JC, Song Y, Wang D, Srinivas M (1995) Ionic basis of the electrophysiological actions of adenosine on cardiomyocytes. FASEB J 9:359–365

Belardinelli L, Shryock JC, Snowdy S, Zhang Y, Monopoli A, Lozza G, Ongini E, Olsson RA, Dennis DM (1998) The A_{2A} adenosine receptor mediates coronary vasodilation. J Pharmacol Exp Ther 284:1066–1073

Berne RM (1963) Cardiac nucleotides in hypoxia: possible role in regulation of coronary blood flow. Am J Physiol 204:317–322

Berne RM (1980) The role of adenosine in the regulation of coronary blood flow. Circ Res 47: 807–813

Brignole M, Donateo P, Menozzi C (2003) The diagnostic value of ATP testing in patients with unexplained syncope. Europace 5:425–428

Bryan PT, Marshall JM (1999) Cellular mechanisms by which adenosine induces vasodilatation in rat skeletal muscle: significance for systemic hypoxia. J Physiol 514 (1): 163–175

Burnett D, Abi-Samra F, Vacek JL (1999) Use of intravenous adenosine as a noninvasive diagnostic test for sick sinus syndrome. Am Heart J 137:435–438

Carrega L, Saadjian AY, Mercier L, Zouher I, Berge-Lefranc JL, Gerolami V, Giaime P, Sbragia P, Paganelli F, Fenouillet E, Levy S, Guieu RP (2007) Increased expression of adenosine A_{2A} receptors in patients with spontaneous and head-up-tilt-induced syncope. Heart Rhythm 4:870–876

Cerniway RJ, Yang Z, Jacobson MA, Linden J, Matherne GP (2001) Targeted deletion of A(3) adenosine receptors improves tolerance to ischemia-reperfusion injury in mouse myocardium. Am J Physiol Heart Circ Physiol 281:H1751–H1758

Cerqueira MD (2006) Advances in pharmacologic agents in imaging: new A_{2A} receptor agonists. Curr Cardiol Rep 8:119–122

Cheung JW, Lerman BB (2003) CVT-510: a selective A_1 adenosine receptor agonist. Cardiovasc Drug Rev 21:277–292

Cheung JW, Stein KM, Markowitz SM, Iwai S, Guttigoli AB, Shah BK, Yarlagadda RK, Lerman BB, Mittal S (2004) Significance of adenosine-induced atrioventricular block in patients with unexplained syncope. Heart Rhythm 1:664–668

Chio CC, Chang YH, Hsu YW, Chi KH, Lin WW (2004) PKA-dependent activation of PKC, p38 MAPK and IKK in macrophage: implication in the induction of inducible nitric oxide synthase and interleukin-6 by dibutyryl cAMP. Cell Signal 16:565–575

Clemo HF, Belardinelli L (1986) Effect of adenosine on atrioventricular conduction. I: Site and characterization of adenosine action in the guinea pig atrioventricular node. Circ Res 59: 427–436

Clemo HF, Bourassa A, Linden J, Belardinelli L (1987) Antagonism of the effects of adenosine and hypoxia on atrioventricular conduction time by two novel alkylxanthines: correlation with binding to adenosine A_1 receptors. J Pharmacol Exp Ther 242:478–484

Conti JB, Belardinelli L, Curtis AB (1995) Usefulness of adenosine in diagnosis of tachyarrhythmias. Am J Cardiol 75:952–955

Dana A, Skarli M, Papakrivopoulou J, Yellon DM (2000) Adenosine A(1) receptor induced delayed preconditioning in rabbits: induction of p38 mitogen-activated protein kinase activation and Hsp27 phosphorylation via a tyrosine kinase- and protein kinase C-dependent mechanism. Circ Res 86:989–997

Dennis D, Jacobson K, Belardinelli L (1992) Evidence of spare A_1-adenosine receptors in guinea pig atrioventricular node. Am J Physiol 262:H661–H671

Denyer JC, Brown HF (1990) Pacemaking in rabbit isolated sino-atrial node cells during Cs+ block of the hyperpolarization-activated current if. J Physiol 429:401–409

Deussen A, Brand M, Pexa A, Weichsel J (2006) Metabolic coronary flow regulation—current concepts. Basic Res Cardiol 101:453–464

Dhalla AK, Shryock JC, Shreeniwas R, Belardinelli L (2003) Pharmacology and therapeutic applications of A_1 adenosine receptor ligands. Curr Top Med Chem 3:369–385

DiFrancesco D, Borer JS (2007) The funny current: cellular basis for the control of heart rate. Drugs 67(Suppl 2):15–24

DiMarco JP, Sellers TD, Berne RM, West GA, Belardinelli L (1983) Adenosine: electrophysiologic effects and therapeutic use for terminating paroxysmal supraventricular tachycardia. Circulation 68:1254–1263

Dobson JG, Jr., Fenton RA, Romano FD (1987) The cardiac anti-adrenergic effect of adenosine. Prog Clin Biol Res 230:331–343

Eckle T, Krahn T, Grenz A, Kohler D, Mittelbronn M, Ledent C, Jacobson MA, Osswald H, Thompson LF, Unertl K, Eltzschig HK (2007) Cardioprotection by ecto-5′-nucleotidase (CD73) and A_{2B} adenosine receptors. Circulation 115:1581–1590

Eckle T, Faigle M, Grenz A, Laucher S, Thompson LF, Eltzschig HK (2008) A_{2B} adenosine receptor dampens hypoxia-induced vascular leak. Blood 111:2024–2035

Fabritz L, Kirchhof P, Fortmuller L, Auchampach JA, Baba HA, Breithardt G, Neumann J, Boknik P, Schmitz W (2004) Gene dose-dependent atrial arrhythmias, heart block, and brady-cardiomyopathy in mice overexpressing A(3) adenosine receptors. Cardiovasc Res 62:: 500–508

Feoktistov I, Biaggioni I (1997) Adenosine A_{2B} receptors. Pharmacol Rev 49:381–402

Flammang D, Pelleg A, Benditt DG (2005) The adenosine triphospate (ATP) test for evaluation of syncope of unknown origin. J Cardiovasc Electrophysiol 16:1388–1389

Flammang D, AMS Investigators (2006) The ATP test: a useful test for identifying the cardioinhibitory mechanism of syncope of unknown origin and for directing the therapy. Europace 8(Suppl 1):55

Flood AJ, Willems L, Headrick JP (2002) Coronary function and adenosine receptor-mediated responses in ischemic-reperfused mouse heart. Cardiovasc Res 55:161–170

Fogaça RTH, Leal-Cardoso JH (1985) Effects of adenosine on digitalis induced arrhythmias. Braz J Med Biol Res 18:663A

Fragakis N, Iliadis I, Sidopoulos E, Lambrou A, Tsaritsaniotis E, Katsaris G (2007) The value of adenosine test in the diagnosis of sick sinus syndrome: susceptibility of sinus and atrioventricular node to adenosine in patients with sick sinus syndrome and unexplained syncope. Europace 9:559–562

Francis JE, Webb RL, Ghai GR, Hutchison AJ, Moskal MA, deJesus R, Yokoyama R, Rovinski SL, Contardo N, Dotson R, Barclay B, Stone GA, Jarvis MF (1991) Highly selective adenosine A_2 receptor agonists in a series of N-alkylated 2-aminoadenosines. J Med Chem 34:2570–2579

Fredholm BB, Arslan G, Halldner L, Kull B, Schulte G, Wasserman W (2000) Structure and function of adenosine receptors and their genes. Naunyn–Schmiedeberg's Arch Pharmacol 362:364–374

Frieden M, Sollini M, Beny J (1999) Substance P and bradykinin activate different types of KCa currents to hyperpolarize cultured porcine coronary artery endothelial cells. J Physiol 519(2):361–371

Frobert O, Haink G, Simonsen U, Gravholt CH, Levin M, Deussen A (2006) Adenosine concentration in the porcine coronary artery wall and A_{2A} receptor involvement in hypoxia-induced vasodilatation. J Physiol 570:375–384

Ge ZD, Peart JN, Kreckler LM, Wan TC, Jacobson MA, Gross GJ, Auchampach JA (2006) Cl-IB-MECA [2-chloro-N^6-(3-iodobenzyl)adenosine-5′-N-methylcarboxamide] reduces ischemia/reperfusion injury in mice by activating the A_3 adenosine receptor. J Pharmacol Exp Ther 319:1200–1210

Geraets D, Kienzle M (1992) Clinical use of adenosine. Iowa Med 82:25–28

Gerlach E, Deuticke B (1966) Comparative studies on the formation of adenosine in the myocardium of different animal species in oxygen deficiency. Klin Wochenschr 44:1307–1310

Germack R, Dickenson JM (2004) Characterization of ERK1/2 signalling pathways induced by adenosine receptor subtypes in newborn rat cardiomyocytes. Br J Pharmacol 141:329–339

Glavind-Kristensen M, Matchkov V, Hansen VB, Forman A, Nilsson H, Aalkjaer C (2004) KATP-channel-induced vasodilation is modulated by the Na,K-pump activity in rabbit coronary small arteries. Br J Pharmacol 143:872–880

Hack SP, Christie MJ (2003) Adaptations in adenosine signaling in drug dependence: therapeutic implications. Crit Rev Neurobiol 15:235–274

Haq SE, Clerk A, Sugden PH (1998) Activation of mitogen-activated protein kinases (p38-MAPKs, SAPKs/JNKs and ERKs) by adenosine in the perfused rat heart. FEBS Lett 434:305–308

Harrison GJ, Cerniway RJ, Peart J, Berr SS, Ashton K, Regan S, Paul Matherne G, Headrick JP (2002) Effects of A(3) adenosine receptor activation and gene knock-out in ischemic-reperfused mouse heart. Cardiovasc Res 53:147–155

Hasan AZ, Abebe W, Mustafa SJ (2000) Antagonism of coronary artery relaxation by adenosine A_{2A}-receptor antagonist ZM241385. J Cardiovasc Pharmacol 35:322–325

Hayward E, Showler L, Soar J (2007) Aminophylline in bradyasystolic cardiac arrest. Emerg Med J 24:582–583

Headrick JP, Gauthier NS, Morrison RR, Matherne GP (2000) Chronotropic and vasodilatory responses to adenosine and isoproterenol in mouse heart: effects of adenosine A_1 receptor overexpression. Clin Exp Pharmacol Physiol 27:185–190

Hein TW, Wang W, Zoghi B, Muthuchamy M, Kuo L (2001) Functional and molecular characterization of receptor subtypes mediating coronary microvascular dilation to adenosine. J Mol Cell Cardiol 33:271–282

Hodgson JM, Dib N, Kern MJ, Bach RG, Barrett RJ (2007) Coronary circulation responses to binodenoson, a selective adenosine A_{2A} receptor agonist. Am J Cardiol 99:1507–1512

Hori M, Kitakaze M (1991) Adenosine, the heart, and coronary circulation. Hypertension 18: 565–574

Hove-Madsen L, Prat-Vidal C, Llach A, Ciruela F, Casado V, Lluis C, Bayes-Genis A, Cinca J, Franco R (2006) Adenosine A_{2A} receptors are expressed in human atrial myocytes and modulate spontaneous sarcoplasmic reticulum calcium release. Cardiovasc Res 72:292–302

Hua X, Kovarova M, Chason KD, Nguyen M, Koller BH, Tilley SL (2007) Enhanced mast cell activation in mice deficient in the A_{2b} adenosine receptor. J Exp Med 204:117–128

Hussain T, Mustafa SJ (1993) Regulation of adenosine receptor system in coronary artery: functional studies and cAMP. Am J Physiol 264:H441–H447

Hussain T, Mustafa SJ (1995) Binding of A_1 adenosine receptor ligand [3H]8-cyclopentyl-1,3-dipropylxanthine in coronary smooth muscle. Circ Res 77:194–198

Hutchinson SA, Scammells PJ (2004) A(1) adenosine receptor agonists: medicinal chemistry and therapeutic potential. Curr Pharm Des 10:2021–2039

Hutchison AJ, Webb RL, Oei HH, Ghai GR, Zimmerman MB, Williams M (1989) CGS 21680C, an A_2 selective adenosine receptor agonist with preferential hypotensive activity. J Pharmacol Exp Ther 251:47–55

Ingwall JS (2007) On substrate selection for ATP synthesis in the failing human myocardium. Am J Physiol Heart Circ Physiol 293:H3225–H3226

Ingwall JS, Weiss RG (2004) Is the failing heart energy starved? On using chemical energy to support cardiac function. Circ Res 95:135–145

Iwai S, Markowitz SM, Stein KM, Mittal S, Slotwiner DJ, Das MK, Cohen JD, Hao SC, Lerman BB (2002) Response to adenosine differentiates focal from macroreentrant atrial tachycardia: validation using three-dimensional electroanatomic mapping. Circulation 106:2793–2799

Iwai S, Cantillon DJ, Kim RJ, Markowitz SM, Mittal S, Stein KM, Shah BK, Yarlagadda RK, Cheung JW, Tan VR, Lerman BB (2006) Right and left ventricular outflow tract tachycardias: evidence for a common electrophysiologic mechanism. J Cardiovasc Electrophysiol 17: 1052–1058

Kalyankrishna S, Malik KU (2003) Norepinephrine-induced stimulation of p38 mitogen-activated protein kinase is mediated by arachidonic acid metabolites generated by activation of cytosolic phospholipase A(2) in vascular smooth muscle cells. J Pharmacol Exp Ther 304:761–772

Keim S, Curtis AB, Belardinelli L, Epstein ML, Staples ED, Lerman BB (1992) Adenosine-induced atrioventricular block: a rapid and reliable method to assess surgical and radiofrequency catheter ablation of accessory atrioventricular pathways. J Am Coll Cardiol 19: 1005–1012

Kemp BK, Cocks TM (1999) Adenosine mediates relaxation of human small resistance-like coronary arteries via A_{2B} receptors. Br J Pharmacol 126:1796–1800

Kim N, Chung J, Kim E, Han J (2003) Changes in the Ca^{2+}-activated K^+ channels of the coronary artery during left ventricular hypertrophy. Circ Res 93:541–547

Kirchhof P, Fabritz L, Fortmuller L, Matherne GP, Lankford A, Baba HA, Schmitz W, Breithardt G, Neumann J, Boknik P (2003) Altered sinus nodal and atrioventricular nodal function in freely moving mice overexpressing the A_1 adenosine receptor. Am J Physiol Heart Circ Physiol 285:H145–H153

Kirsch GE, Codina J, Birnbaumer L, Brown AM (1990) Coupling of ATP-sensitive K^+ channels to A_1 receptors by G proteins in rat ventricular myocytes. Am J Physiol 259:H820–H826

Knaapen P, Germans T, Knuuti J, Paulus WJ, Dijkmans PA, Allaart CP, Lammertsma AA, Visser FC (2007) Myocardial energetics and efficiency: current status of the noninvasive approach. Circulation 115:918–927

Komalavilas P, Mehta S, Wingard CJ, Dransfield DT, Bhalla J, Woodrum JE, Molinaro JR, Brophy CM (2001) PI3-kinase/Akt modulates vascular smooth muscle tone via cAMP signaling pathways. J Appl Physiol 91:1819–1827

Ledent C, Vaugeois JM, Schiffmann SN, Pedrazzini T, El Yacoubi M, Vanderhaeghen JJ, Costentin J, Heath JK, Vassart G, Parmentier M (1997) Aggressiveness, hypoalgesia and high blood pressure in mice lacking the adenosine A_{2a} receptor. Nature 388:674–678

Lerman BB (1993) Response of nonreentrant catecholamine-mediated ventricular tachycardia to endogenous adenosine and acetylcholine. Evidence for myocardial receptor-mediated effects. Circulation 87:382–390

Lerman BB, Belardinelli L, West GA, Berne RM, DiMarco JP (1986) Adenosine-sensitive ventricular tachycardia: evidence suggesting cyclic AMP-mediated triggered activity. Circulation 74:270–280

Lerman BB, Wesley RC, Jr., DiMarco JP, Haines DE, Belardinelli L (1988) Antiadrenergic effects of adenosine on His-Purkinje automaticity. Evidence for accentuated antagonism. J Clin Invest 82:2127–2135

Lerman BB, Stein K, Engelstein ED, Battleman DS, Lippman N, Bei D, Catanzaro D (1995) Mechanism of repetitive monomorphic ventricular tachycardia. Circulation 92:421–429

Lerman BB, Stein KM, Markowitz SM (1997) Mechanisms of idiopathic left ventricular tachycardia. J Cardiovasc Electrophysiol 8:571–583

Lerman BB, Stein KM, Markowitz SM, Mittal S, Slotwiner DJ (2000) Ventricular arrhythmias in normal hearts. Cardiol Clin 18:265–291

Li J, Fenton RA, Wheeler HB, Powell CC, Peyton BD, Cutler BS, Dobson JG, Jr. (1998) Adenosine A_{2a} receptors increase arterial endothelial cell nitric oxide. J Surg Res 80:357–364

Li PL, Zhang DX, Zou AP, Campbell WB (1999) Effect of ceramide on KCa channel activity and vascular tone in coronary arteries. Hypertension 33:1441–1446

Lindsay BD (2007) Focal and macroreentrant atrial tachycardia: from bench to bedside and back to the bench again. Heart Rhythm 4:1361–1363

Liu Y, Terata K, Rusch NJ, Gutterman DD (2001) High glucose impairs voltage-gated K(+) channel current in rat small coronary arteries. Circ Res 89:146–152

Lowes VL, Ip NY, Wong YH (2002) Integration of signals from receptor tyrosine kinases and G protein-coupled receptors. Neurosignals 11:5–19

Mader TJ, Menegazzi JJ, Betz AE, Logue ES, Callaway CW, Sherman LD (2006) Adenosine A_1 receptor antagonism hastens the decay in ventricular fibrillation waveform morphology during porcine cardiac arrest. Resuscitation 71:254–259

Makujina SR, Sabouni MH, Bhatia S, Douglas FL, Mustafa SJ (1992) Vasodilatory effects of adenosine A_2 receptor agonists CGS 21680 and CGS 22492 in human vasculature. Eur J Pharmacol 221:243–247.

Marala RB, Mustafa SJ (1995a) Adenosine A_1 receptor-induced upregulation of protein kinase C: role of pertussis toxin-sensitive G protein(s). Am J Physiol 269:H1619–H1624

Marala RB, Mustafa SJ (1995b) Adenosine analogues prevent phorbol ester-induced PKC depletion in porcine coronary artery via A_1 receptor. Am J Physiol 268:H271–H277

Marala RB, Mustafa SJ (1995c) Modulation of protein kinase C by adenosine: involvement of adenosine A_1 receptor-pertussis toxin sensitive nucleotide binding protein system. Mol Cell Biochem 149–150:51–58

Markowitz SM, Nemirovksy D, Stein KM, Mittal S, Iwai S, Shah BK, Dobesh DP, Lerman BB (2007) Adenosine-insensitive focal atrial tachycardia: evidence for de novo micro-re-entry in the human atrium. J Am Coll Cardiol 49:1324–1333

Markowitz SM, Stein KM, Mittal S, Slotwiner DJ, Lerman BB (1999) Differential effects of adenosine on focal and macroreentrant atrial tachycardia. J Cardiovasc Electrophysiol 10:489–502

Meloche S, Landry J, Huot J, Houle F, Marceau F, Giasson E (2000) p38 MAP kinase pathway regulates angiotensin II-induced contraction of rat vascular smooth muscle. Am J Physiol Heart Circ Physiol 279:H741–H751

Mittal S, Stein KM, Markowitz SM, Iwai S, Guttigoli A, Lerman BB (2004) Single-stage adenosine tilt testing in patients with unexplained syncope. J Cardiovasc Electrophysiol 15:637–640

Morrison RR, Talukder MA, Ledent C, Mustafa SJ (2002) Cardiac effects of adenosine in A(2A) receptor knockout hearts: uncovering A(2B) receptors. Am J Physiol Heart Circ Physiol 282:H437–H444

Morrison RR, Teng B, Oldenburg PJ, Katwa LC, Schnermann JB, Mustafa SJ (2006) Effects of targeted deletion of A_1 adenosine receptors on postischemic cardiac function and expression of adenosine receptor subtypes. Am J Physiol Heart Circ Physiol 291:H1875–H1882

Morrison RR, Tan XL, Ledent C, Mustafa SJ, Hofmann PA (2007) Targeted deletion of A_{2A} adenosine receptors attenuates the protective effects of myocardial postconditioning. Am J Physiol Heart Circ Physiol 293:H2523–H2529

Mustafa SJ, Abebe W (1996) Coronary vasodilation by adenosine: receptor subtypes and mechanisms of action. Drug Dev Res 39:308–313

Mustafa SJ, Askar AO (1985) Evidence suggesting an Ra-type adenosine receptor in bovine coronary arteries. J Pharmacol Exp Ther 232:49–56

Mutafova-Yambolieva VN, Keef KD (1997) Adenosine-induced hyperpolarization in guinea pig coronary artery involves A_{2b} receptors and KATP channels. Am J Physiol 273:H2687–H2695

Neubauer S (2007) The failing heart—an engine out of fuel. N Engl J Med 356:1140–1151

Niiya K, Uchida S, Tsuji T, Olsson RA (1994) Glibenclamide reduces the coronary vasoactivity of adenosine receptor agonists. J Pharmacol Exp Ther 271:14–19

Nitenberg A, Durand E, Delatour B, Sdiri W, Raha S, Lafont A (2007) Postocclusion hyperemia provides a better estimate of coronary reserve than intracoronary adenosine in patients with coronary artery stenosis. J Invasive Cardiol 19:390–394

Nogami A (2002) Idiopathic left ventricular tachycardia: assessment and treatment. Card Electrophysiol Rev 6:448–457

Olanrewaju HA, Mustafa SJ (2000) Adenosine A(2A) and A(2B) receptors mediated nitric oxide production in coronary artery endothelial cells. Gen Pharmacol 35:171–177

Olanrewaju HA, Qin W, Feoktistov I, Scemama JL, Mustafa SJ (2000) Adenosine A(2A) and A(2B) receptors in cultured human and porcine coronary artery endothelial cells. Am J Physiol Heart Circ Physiol 279:H650–H656

Olanrewaju HA, Gafurov BS, Lieberman EM (2002) Involvement of K^+ channels in adenosine A_{2A} and A_{2B} receptor-mediated hyperpolarization of porcine coronary artery endothelial cells. J Cardiovasc Pharmacol 40:43–49

Olsson RA (1970) Changes in content of purine nucleoside in canine myocardium during coronary occlusion. Circ Res 26:301–306

Ono K, Han J (2000) The p38 signal transduction pathway: activation and function. Cell Signal 12:1–13

Parry SW, Nath S, Bourke JP, Bexton RS, Kenny RA (2006) Adenosine test in the diagnosis of unexplained syncope: marker of conducting tissue disease or neurally mediated syncope? Eur Heart J 27:1396–1400

Peart JN, Headrick JP (2007) Adenosinergic cardioprotection: multiple receptors, multiple pathways. Pharmacol Ther 114:208–221

Pelleg A, Belardinelli L (1993) Cardiac electrophysiology and pharmacology of adenosine: basic and clinical aspects. Cardiovasc Res 27:54–61

Pelleg A, Kutalek SP (1997) Adenosine in the mammalian heart: nothing to get excited about. Trends Pharmacol Sci 18:236–238

Peterman C, Sanoski CA (2005) Tecadenoson: a novel, selective A_1 adenosine receptor agonist. Cardiol Rev 13:315–321

Pelleg A, Mitamura H, Mitsuoka T, Michelson EL, Dreifus LS (1986) Effects of adenosine and adenosine 5′-triphosphate on ventricular escape rhythm in the canine heart. J Am Coll Cardiol 8:1145–1151

Pelleg A, Hurt C, Miyagawa A, Michelson EL, Dreifus LS (1990a) Differential sensitivity of cardiac pacemakers to exogenous adenosine in vivo. Am J Physiol 258:H1815–H1822

Pelleg A, Hurt CM, Michelson EL (1990b) Cardiac effects of adenosine and ATP. Ann N Y Acad Sci 603:19–30

Pelleg A, Hurt CM, Hewlett EL (1996) ATP shortens atrial action potential duration in the dog: role of adenosine, the vagus nerve, and G protein. Can J Physiol Pharmacol 74:15–22

Pelleg A, Katchanov G, Xu J (1997) Autonomic neural control of cardiac function: modulation by adenosine and adenosine 5′-triphosphate. Am J Cardiol 79:11–14

Pelleg A, Pennock RS, Kutalek SP (2002) Proarrhythmic effects of adenosine: one decade of clinical data. Am J Ther 9:141–147

Quayle JM, Nelson MT, Standen NB (1997) ATP-sensitive and inwardly rectifying potassium channels in smooth muscle. Physiol Rev 77:1165–1232

Rahman A, Anwar KN, Minhajuddin M, Bijli KM, Javaid K, True AL, Malik AB (2004) cAMP Targeting of p38 MAP kinase Inhibits thrombin-induced NF-{kappa}B activation and ICAM-1 expression in endothelial cells. Am J Physiol Lung Cell Mol Physiol 287:L1017–L1024

Ramagopal MV, Chitwood RW Jr, Mustafa SJ (1988) Evidence for an A_2 adenosine receptor in human coronary arteries. Eur J Pharmacol 151:483–486

Ray CJ, Marshall JM (2006) The cellular mechanisms by which adenosine evokes release of nitric oxide from rat aortic endothelium. J Physiol 570:85–96

Reichelt ME, Willems L, Molina JG, Sun CX, Noble JC, Ashton KJ, Schnermann J, Blackburn MR, Headrick JP (2005) Genetic deletion of the A_1 adenosine receptor limits myocardial ischemic tolerance. Circ Res 96:363–367

Rekik M, Mustafa JS (2003) Modulation of A_{2A} adenosine receptors and associated Galphas proteins by ZM 241385 treatment of porcine coronary artery. J Cardiovasc Pharmacol 42:736–744

Resh W, Feuer J, Wesley RC, Jr. (1992) Intravenous adenosine: a noninvasive diagnostic test for sick sinus syndrome. Pacing Clin Electrophysiol 15:2068–2073

Robinson AJ, Dickenson JM (2001) Regulation of p42/p44 MAPK and p38 MAPK by the adenosine A(1) receptor in DDT(1)MF-2 cells. Eur J Pharmacol 413:151–161

Rogers PA, Chilian WM, Bratz IN, Bryan RM, Jr., Dick GM (2007) H_2O_2 activates redox- and 4-aminopyridine-sensitive Kv channels in coronary vascular smooth muscle. Am J Physiol Heart Circ Physiol 292:H1404–H1411

Rubio R, Wiedmeier VT, Berne RM (1974) Relationship between coronary flow and adenosine production and release. J Mol Cell Cardiol 6:561–566

Saadjian AY, Levy S, Franceschi F, Zouher I, Paganelli F, Guieu RP (2002) Role of endogenous adenosine as a modulator of syncope induced during tilt testing. Circulation 106:569–574

Salloum FN, Das A, Thomas CS, Yin C, Kukreja RC (2007) Adenosine A(1) receptor mediates delayed cardioprotective effect of sildenafil in mouse. J Mol Cell Cardiol 43:545–551

Sanders L, Rakovic S, Lowe M, Mattick PA, Terrar DA (2006) Fundamental importance of Na^{+}-Ca^{2+} exchange for the pacemaking mechanism in guinea-pig sino-atrial node. J Physiol 571:639–649

Schrader J, Baumann G, Gerlach E (1977) Adenosine as inhibitor of myocardial effects of catecholamines. Pflugers Arch 372:29–35

Schulte G, Fredholm BB (2002) Adenosine A_{2B} receptors activate extracellular signal-regulated kinase ERK 1/2 and stress-activated protein kinase p38. In: XIVth World Congress of Pharmacology, San Francisco, CA, 7–12 July 2002, 44(2):A262

Schulte G, Fredholm BB (2003) Signalling from adenosine receptors to mitogen-activated protein kinases. Cell Signal 15:813–827

Shen WK, Kurachi Y (1995) Mechanisms of adenosine-mediated actions on cellular and clinical cardiac electrophysiology. Mayo Clin Proc 70:274–291

Shen WK, Hammill SC, Munger TM, Stanton MS, Packer DL, Osborn MJ, Wood DL, Bailey KR, Low PA, Gersh BJ (1996) Adenosine: potential modulator for vasovagal syncope. J Am Coll Cardiol 28:146–154

Shryock JC, Snowdy S, Baraldi PG, Cacciari B, Spalluto G, Monopoli A, Ongini E, Baker SP, Belardinelli L (1998) A_{2A}-adenosine receptor reserve for coronary vasodilation. Circulation 98:711–718

Smits GJ, McVey M, Cox BF, Perrone MH, Clark KL (1998) Cardioprotective effects of the novel adenosine A_1/A_2 receptor agonist AMP 579 in a porcine model of myocardial infarction. J Pharmacol Exp Ther 286:611–618

Song Y, Thedford S, Lerman BB, Belardinelli L (1992) Adenosine-sensitive afterdepolarizations and triggered activity in guinea pig ventricular myocytes. Circ Res 70:743–753

Song Y, Wu L, Shryock JC, Belardinelli L (2002) Selective attenuation of isoproterenol-stimulated arrhythmic activity by a partial agonist of adenosine A_1 receptor. Circulation 105:118–123

Sun Park W, Kyoung Son Y, Kim N, Boum Youm J, Joo H, Warda M, Ko JH, Earm YE, Han J (2006) The protein kinase A inhibitor, H-89, directly inhibits KATP and Kir channels in rabbit coronary arterial smooth muscle cells. Biochem Biophys Res Commun 340:1104–1110

Taegtmeyer H, Wilson CR, Razeghi P, Sharma S (2005) Metabolic energetics and genetics in the heart. Ann N Y Acad Sci 1047:208–218

Talukder MA, Morrison RR, Jacobson MA, Jacobson KA, Ledent C, Mustafa SJ (2002a) Targeted deletion of adenosine A(3) receptors augments adenosine-induced coronary flow in isolated mouse heart. Am J Physiol Heart Circ Physiol 282:H2183–H2189

Talukder MA, Morrison RR, Mustafa SJ (2002b) Comparison of the vascular effects of adenosine in isolated mouse heart and aorta. Am J Physiol Heart Circ Physiol 282:H49–H57

Talukder MA, Morrison RR, Ledent C, Mustafa SJ (2003) Endogenous adenosine increases coronary flow by activation of both A_{2A} and A_{2B} receptors in mice. J Cardiovasc Pharmacol 41:562–570

Tang L, Parker M, Fei Q, Loutzenhiser R (1999) Afferent arteriolar adenosine A_{2a} receptors are coupled to KATP in in vitro perfused hydronephrotic rat kidney. Am J Physiol 277:F926–F933

Tawfik HE, Schnermann J, Oldenburg PJ, Mustafa SJ (2005) Role of A_1 adenosine receptors in regulation of vascular tone. Am J Physiol Heart Circ Physiol 288:H1411–H1416

Tawfik HE, Teng B, Morrison RR, Schnermann J, Mustafa SJ (2006) Role of A_1 adenosine receptor in the regulation of coronary flow. Am J Physiol Heart Circ Physiol 291:H467–H472

Teng B, Qin W, Ansari HR, Mustafa SJ (2005) Involvement of p38-mitogen-activated protein kinase in adenosine receptor-mediated relaxation of coronary artery. Am J Physiol Heart Circ Physiol 288:H2574–H2580

Teng B, Ledent C, Mustafa SJ (2008) Up-regulation of A(2B) adenosine receptor in A(2A) adenosine receptor knockout mouse coronary artery. J Mol Cell Cardiol 44:905–914

Tune JD, Gorman MW, Feigl EO (2004) Matching coronary blood flow to myocardial oxygen consumption. J Appl Physiol 97:404–415

Viskin S, Belhassen B, Roth A, Reicher M, Averbuch M, Sheps D, Shalabye E, Laniado S (1993) Aminophylline for bradyasystolic cardiac arrest refractory to atropine and epinephrine. Ann Intern Med 118:279–281

Viskin S, Rosso R, Rogowski O, Belhassen B, Levitas A, Wagshal A, Katz A, Fourey D, Zeltser D, Oliva A, Pollevick GD, Antzelevitch C, Rozovski U (2006) Provocation of sudden heart rate oscillation with adenosine exposes abnormal QT responses in patients with long QT syndrome: a bedside test for diagnosing long QT syndrome. Eur Heart J 27:469–475

Wang J, Whitt SP, Rubin LJ, Huxley VH (2005) Differential coronary microvascular exchange responses to adenosine: roles of receptor and microvessel subtypes. Microcirculation 12: 313–326

Watts SW, Florian JA, Monroe KM (1998) Dissociation of angiotensin II-stimulated activation of mitogen-activated protein kinase kinase from vascular contraction. J Pharmacol Exp Ther 286:1431–1438

Wilden PA, Agazie YM, Kaufman R, Halenda SP (1998) ATP-stimulated smooth muscle cell proliferation requires independent ERK and PI3K signaling pathways. Am J Physiol 275: H1209–H1215

Xu J, Tong H, Wang L, Hurt CM, Pelleg A (1993) Endogenous adenosine, A_1 adenosine receptor, and pertussis toxin sensitive guanine nucleotide binding protein mediate hypoxia induced AV nodal conduction block in guinea pig heart in vivo. Cardiovasc Res 27:134–140

Xu J, Wang L, Hurt CM, Pelleg A (1994) Endogenous adenosine does not activate ATP-sensitive potassium channels in the hypoxic guinea pig ventricle in vivo. Circulation 89:1209–1216

Xu J, Hurt CM, Pelleg A (1995) Digoxin-induced ventricular arrhythmias in the guinea pig heart in vivo: evidence for a role of endogenous catecholamines in the genesis of delayed afterdepolarizations and triggered activity. Heart Vessels 10:119–127

Xu J, Kussmaul W, Kurnik PB, Al-Ahdav M, Pelleg A (2005) Electrophysiological-anatomic correlates of ATP-triggered vagal reflex in the dog. V. Role of purinergic receptors. Am J Physiol Regul Integr Comp Physiol 288:R651–R655

Yang ZW, Wang J, Zheng T, Altura BT, Altura BM (2000) Low $[Mg^{2+}]_o$ induces contraction of cerebral arteries: roles of tyrosine and mitogen-activated protein kinases. Am J Physiol Heart Circ Physiol 279:H185–H194

Yang D, Zhang Y, Nguyen HG, Koupenova M, Chauhan AK, Makitalo M, Jones MR, St Hilaire C, Seldin DC, Toselli P, Lamperti E, Schreiber BM, Gavras H, Wagner DD, Ravid K (2006) The A_{2B} adenosine receptor protects against inflammation and excessive vascular adhesion. J Clin Invest 116:1913–1923

Yang D, Koupenova M, McCrann DJ, Kopeikina KJ, Kagan HM, Schreiber BM, Ravid K (2008) The A_{2b} adenosine receptor protects against vascular injury. Proc Natl Acad Sci USA 105: 792–796

Zablocki JA, Wu L, Shryock J, Belardinelli L (2004) Partial A(1) adenosine receptor agonists from a molecular perspective and their potential use as chronic ventricular rate control agents during atrial fibrillation (AF). Curr Top Med Chem 4:839–854

Zatta AJ, Headrick JP (2005) Mediators of coronary reactive hyperaemia in isolated mouse heart. Br J Pharmacol 144:576–587

Zaza A, Rocchetti M, DiFrancesco D (1996) Modulation of the hyperpolarization-activated current ($I(f)$) by adenosine in rabbit sinoatrial myocytes. Circulation 94:734–741

Zhao Z, Makaritsis K, Francis CE, Gavras H, Ravid K (2000) A role for the A_3 adenosine receptor in determining tissue levels of cAMP and blood pressure: studies in knock-out mice. Biochim Biophys Acta 1500:280–290

Zhao TC, Hines DS, Kukreja RC (2001) Adenosine-induced late preconditioning in mouse hearts: role of p38 MAP kinase and mitochondrial K(ATP) channels. Am J Physiol Heart Circ Physiol 280:H1278–H1285

Adenosine Receptors and Reperfusion Injury of the Heart

John P. Headrick and Robert D. Lasley

Contents

Abstract Adenosine, a catabolite of ATP, exerts numerous effects in the heart, including modulation of the cardiac response to stress, such as that which occurs during myocardial ischemia and reperfusion. Over the past 20 years, substantial evidence has accumulated that adenosine, administered either prior to ischemia or during reperfusion, reduces both reversible and irreversible myocardial injury. The latter effect results in a reduction of both necrosis or myocardial infarction (MI) and apoptosis. These effects appear to be mediated via the activation of one or more G-protein-coupled receptors (GPCRs), referred to as A_1, A_{2A}, A_{2B} and A_3 adenosine receptor (AR) subtypes. Experimental studies in different species and models suggest that activation of the A_1 or A_3ARs prior to ischemia is cardioprotective. Further experimental studies reveal that the administration of A_{2A}AR agonists during reperfusion can also reduce MI, and recent reports suggest that A_{2B}ARs may also play an important role in modulating myocardial reperfusion injury. Despite convincing

J.P. Headrick (✉)
Heart Foundation Research Centre, School of Medical Science, Griffith University, Southport, Queensland, 4217, Australia
j.headrick@griffith.edu.au

C.N. Wilson and S.J. Mustafa (eds.), *Adenosine Receptors in Health and Disease*, Handbook of Experimental Pharmacology 193, DOI 10.1007/978-3-540-89615-9_7,

experimental evidence for AR-mediated cardioprotection, there have been only a limited number of clinical trials examining the beneficial effects of adenosine or adenosine-based therapeutics in humans, and the results of these studies have been equivocal. This review summarizes our current knowledge of AR-mediated cardioprotection, and the roles of the four known ARs in experimental models of ischemia-reperfusion. The chapter concludes with an examination of the clinical trials to date assessing the safety and efficacy of adenosine as a cardioprotective agent during coronary thrombolysis in humans.

Keywords Adenosine receptor subtypes · Cardioprotection · Ischemia · Myocardial infarction · Reperfusion · Signaling

Abbreviations

AR	Adenosine receptor
CCPA	2 Chloro-N^6-cyclopentyladenosine
CHF	Congestive heart failure
CSC	8-(13-Chlorostyryl) caffeine
FMLP	Formyl–Met–Leu–Phe
GPCR	G-protein-coupled receptor
I/R	Ischemia-reperfusion
iNOS	Inducible nitric oxide synthase
IPC	Ischemic preconditioning
KO	Knockout
MI	Myocardial infarction
mito K_{ATP}	Mitochondrial ATP-sensitive K^+ channels
MPTP	Mitochondrial permeability transition pore
NECA	5′-*N*-Ethyl-carboxamidoadenosine
p38-MAPK	p38 Mitogen-activated protein kinase
PC	Preconditioning
PIA	N^6-1-(Phenyl-2*R*-isopropyl)adenosine
PKC	Protein kinase C
PTCA	Percutaneous transluminal coronary angioplasty
ROS	Reactive O_2 species
SPECT	Single-photon emission computed tomography
STEMI	ST-segment elevation myocardial infarction
SR	Sarcoplasmic reticulum

1 Introduction

Since extent of myocardial cell death is the primary determinant of outcome from planned or unplanned cardiac ischemia, protective strategies to limit this damage during ischemia-reperfusion (I/R) are highly sought after. It is now clear that a number of GPCR families can activate cytoprotective responses. These receptors, including the adenosine, opioid and bradykinin families, may act not only as acute "retaliatory" systems mediating immediate responses to injurious stimuli, but function as sensors of low-level stress to initiate a signaling cascade culminating in the expression of more prolonged protected phenotypes. These adaptive or hormesis responses predate mammals, and offer potential as targets for therapeutic cardioprotection.

The AR family, composed of A_1, A_{2A}, A_{2B}, and A_3 subtypes, has been implicated in both acute protection and adaptive preconditioning (PC) responses. Not only does preischemic activation of ARs generate potent protection, but significant evidence indicates that this receptor class also mediates powerful cardioprotection when targeted during the reperfusion phase. This brief review focuses on temporal properties of AR-mediated cardioprotection (prior to, during, after ischemia), their contributions to PC responses, and their relevance to the protection of human myocardium.

2 Cardioprotection with Tonic A_1AR Agonism: A_1AR Overexpression

Given early evidence of cardioprotection in response to adenosine and (subsequently) selective A_1AR agonism, the A_1AR subtype seemed an obvious target for manipulating myocardial ischemic tolerance. To test the hypothesis that A_1AR density (rather than endogenous [ligand]) limits the resistance of the heart to I/R, Matherne and colleagues developed a cardiac-specific A_1AR overexpression model. The model employed a construct containing the rat A_1AR gene under the control of a mutated α-myosin heavy chain promoter (Matherne et al. 1997), with extent of A_1AR expression varying across the lines generated (with up to 100-fold overexpression of coupled A_1ARs). The resulting phenotype was characterized by modest bradycardia, conduction disturbances, and a small increase in heart mass in some lines (Matherne et al. 1997; Gauthier et al. 1998; Kirchhof et al. 2003). Initial studies of I/R revealed profound reductions in cell death and contractile dysfunction compared with wild-type hearts (Matherne et al. 1997; Headrick et al. 1998; Morrison et al. 2000). Tolerance to hypoxic challenge (Cerniway et al. 2002), and long-term cold storage of hearts (Crawford et al. 2005) were also improved. Cardioprotection was evident in isolated tissue preparations (Matherne et al. 1997; Headrick et al. 1998) and in vivo (Yang et al. 2002). These outcomes were consistent with protective effects of artificially enhanced A_1 (and A_3) expression in isolated myocytes (Dougherty et al. 1998). Differing components of cardiac protection were apparent, with reduced necrosis and infarction (Matherne et al. 1997; Morrison et al. 2000; Yang et al. 2002), inhibition of apoptosis (Regan et al. 2003; Crawford et al. 2005), enhancement of bioenergetic state during ischemia (Headrick et al. 1998), and selective modulation of contractile injury: A_1AR overexpression consistently reduces diastolic (and not systolic) dysfunction during I/R (Matherne et al. 1997; Reichelt et al. 2007). The latter suggests that A_1ARs selectively target processes underlying diastolic contracture (e.g., Ca^{2+} handling, myofibrillar function).

While the signaling basis of cardioprotection with A_1AR overexpression remains to be established, analysis to date implicates players common to protective signaling in wild-type tissue, including mitochondrial ATP-sensitive K^+ (mito K_{ATP}) channels and inducible nitric oxide synthase (iNOS) (Headrick et al. 2000; Nayeem et al. 2003). Curiously, mito K_{ATP} channels (or 5-hydroxydecanoate-sensitive targets) were not implicated in protection against hypoxia (Cerniway et al. 2002). Other work supports a role for p38 mitogen-activated protein kinase (p38-MAPK)-dependent signaling, though this remains to be more fully tested (Jones et al. 1999). Sarcoplasmic reticulum (SR) Ca^{2+} handling is impaired (Zucchi et al. 2002), which could contribute to specific aspects of associated cardioprotection. Another interesting outcome with A_1AR overexpression is restoration of ischemic resistance in aged hearts: aging may limit the capacity of hearts to withstand damage during I/R (Willems et al. 2005), and this effect was reversed by A_1AR overexpression in mice (Headrick et al. 2003b), in parallel with restoration of adenosine responsiveness.

In terms of PC responses, overexpression of A_1ARs mimics the benefit with this stimulus, actually surpassing the degree of protection with ischemic PC (IPC) (Morrison et al. 2000). Protection with A_1AR overexpression is also nonadditive with IPC, suggesting a commonality of signaling/end-effectors and/or maximally effective protection with A_1AR overexpression. However, the latter is inconsistent with reports that acute application of adenosine (Peart et al. 2002) or A_1AR agonist (Nayeem et al. 2003) can augment the protection with A_1AR overexpression.

Overexpression of A_1ARs in cardiac cells did confirm the hypothesis that normal levels of A_1AR expression in wild-type hearts do appear to limit the extent of cardioprotection possible, and thus the heart's intrinsic resistance to I/R (Matherne et al. 1997). Nonetheless, pharmacologically activating A_1ARs does provide benefit in wild-type hearts (see Sect. 2.1.1 below), demonstrating that normally expressed A_1ARs can be targeted to achieve further cardioprotection. This may reflect additional effects of transient AR agonism (and induction of a short-lived PC state), as opposed to the longer-lived effects of tonic A_1AR activity in transgenic hearts.

3 Cardioprotection via Preischemic AR Activation: A Role in PC Responses

Since its discovery by Murry and colleagues (Murry et al. 1986), the molecular basis of IPC has been the subject of intense investigation. An ultimate goal is translation to the clinical setting, enabling activation of similar protection in cardiac patients. Through a simplified scheme, we can examine the roles of ARs in PC responses from the viewpoint of the initial "trigger" phase and the subsequent "mediation" phase.

The initial and rather crude ischemic trigger of PC is now known to involve the release and actions of several GPCR ligands (including opioids, bradykinin, and adenosine). A "threshold" model for triggering PC has evolved, in which summation of multiple GPCR stimuli is required to activate delayed protection (Goto et al. 1995; Baba et al. 2005). The response may involve not only summation of GPCR triggers but also downstream kinase signaling (Vahlhaus et al. 1998). The kinase cascades involved in PC have been elaborated over recent years, and are currently thought to converge on modulation of mitochondrial effectors, including K_{ATP} channels and the mitochondrial permeability transition pore (MPTP) (Murphy 2004; Hausenloy and Yellon 2007; Liem et al. 2008). Nonetheless, there remains considerable disagreement regarding the roles of different signaling components, and putative end-effectors, in AR-mediated protection and PC. As the focus of this review is on AR involvement in cardioprotection, and since the signaling basis of PC responses has been very well addressed in recent reviews (Murphy 2004; Downey et al. 2007; Hausenloy and Yellon 2007), interested readers are directed to these for further details.

3.1 *Adenosine as a Preischemic Trigger of PC*

It should be clarified that true PC describes a delayed protective state persisting in the absence of the initial stimulus. Many studies refer to “preconditioning” effects when assessing preischemic receptor or pathway activation. However, application of receptor agonists up to induction of ischemia (with no intervening washout) will modify the same targets during ischemia and possibly early reperfusion. This is an inherent limitation to in vivo studies, since exogenously applied AR agonists (or antagonists) may be slowly removed and thus exert potentially long-lasting effects beyond the desired “window.” Thus, while discussion of the effects of preischemic AR activation (or antagonism) can be informative in terms of roles of ARs in PC responses, these experimental scenarios do not simulate PC per se.

In seeking a released factor capable of transducing protection with PC, adenosine seemed a likely candidate: adenosine release increases rapidly in response to different conditions of stress (Headrick et al. 2003a); the interstitial concentrations achieved are sufficient to activate one or more AR subtypes (Van Wylen 1994; Lasley et al. 1995a; Headrick 1996; Harrison et al. 1998); rapid transport and catabolism ensures a brief extracellular half-life and localized signaling; and exogenous AR agonists appear to induce similar protective states.

3.1.1 AR-Triggered Pharmacological PC

In early work Liu et al. showed that preischemic treatment with adenosine or N^6-1-(phenyl-2R-isopropyl) adenosine (PIA) mimicked the protective effects of PC in rabbit myocardium (Liu et al. 1991). Subsequent studies confirmed protection via preischemic A_1AR agonism in different models and species (Lasley and Mentzer 1992; Thornton et al. 1992; Liu and Downey 1992; Tsuchida et al. 1993; Strickler et al. 1996; Carr et al. 1997; Liang and Jacobson 1998; de Jonge and de Jong

1999; de Jonge et al. 2002; Germack et al. 2004; Germack and Dickenson 2005). Toombs and colleagues not only showed that preischemic adenosine limited infarct size (Toombs et al. 1992), but further showed that activation of 8-ρ-sulfophenyltheophylline-sensitive ARs (likely A_1 and/or A_2ARs) during the ischemic period itself was required for protection.

Preischemic activation of the A_3AR subtype can also generate cardiac protection. Strickler et al. (1996) presented some of the first evidence that A_3AR activation prior to ischemia could confer protection against ischemia-like insult in myocytes (of avian origin), while Tracey and colleagues acquired evidence for A_3AR-triggered protection in rabbit hearts (Tracey et al. 1997). Other groups confirmed A_3AR-mediated protection in multiple models (Strickler et al. 1996; Carr et al. 1997; Liang and Jacobson 1998; de Jonge et al. 2002; Maddock et al. 2002; Germack et al. 2004; Germack and Dickenson 2005; Wan et al. 2008). Indeed, Liang and Jacobson (1998) found that the A_3AR induced a more sustained state of protection than the A_1AR when activated prior to ischemia.

In contrast to PC-like effects of A_1AR or A_3AR agonism, preischemic activation of A_{2A}ARs or A_{2B}ARs is generally ineffective in limiting myocardial injury during subsequent I/R (Thornton et al. 1992; Lasley and Mentzer 1992; Maddock et al. 2002; Germack and Dickenson 2005). Studies with the natural agonist adenosine yield mixed results, likely due to rapid uptake and catabolism of extracellular adenosine, complications of potent hemodynamic actions of the endogenous agonist, and the impact of mixed AR activation on different cell types.

3.1.2 ARs as Intrinsic Triggers of IPC

Studies demonstrating PC-like responses to preischemic AR activation provided support for AR involvement in IPC. To more directly test for a role of AR activation in triggering nonpharmacological forms of PC, AR antagonists or adenosine deaminase have been added, often in both trigger and mediation phases, to limit any contributions from ARs. A number of these studies independently provided no evidence for essential roles for ARs in PC (Liu and Downey 1992; Lasley et al. 1993; Hendrikx et al. 1993; Bugge and Ytrehus 1995; Lasley et al. 1995b), leading to premature elimination of this class of GPCRs as contributing to PC (Cave et al. 1993; Li and Kloner 1993). In the context of protective thresholds and contributions of multiple stimuli, a more accurate conclusion may be that the roles of ARs in triggering/mediating PC are redundant, with other concomitant stimuli (e.g., endogenous opioids and bradykinin) being able to compensate and surpass the signaling threshold required for protection.

On the other hand, considerable evidence supporting essential AR involvement in PC has been reported. Studies employing different AR antagonists or adenosine deaminase supported roles in rabbit (Liu et al. 1991; Tsuchida et al. 1992; Thornton et al. 1993; Urabe et al. 1993; Weinbrenner et al. 1997) rat (Headrick 1996; de Jonge and de Jong 1999; de Jonge et al. 2001; Tani et al. 1998), dog (Auchampach and Gross 1993; Hoshida et al. 1994), and pig (Schulz et al. 1995;

Vogt et al. 1996; Louttit et al. 1999). Early studies of PC responses in human myocardium also supported involvement of endogenous adenosine, likely via A_1ARs (Walker et al. 1995; Tomai et al. 1996).

Reasons for differing outcomes with AR blockade in varied models of PC are not clear. Evidence has been presented for substantial species differences in adenosine handling and receptor activation (Headrick 1996), which might dictate differing roles for adenosine and certainly contribute to differing abilities of competitive AR antagonists to limit these responses. Moreover, the affinity and selectivity of AR ligands varies across species, and in the event of poor solubility, bioavailability may limit the effects of a ligand. Furthermore, the relative contributions of adenosine and ARs in triggering PC may be species dependent, with a greater and essential contribution in rodent myocardium. Nonetheless, evidence for essential AR involvement has been reported in large animal models (Auchampach and Gross 1993; Hoshida et al. 1994; Schulz et al. 1995; Vogt et al. 1996; Louttit et al. 1999) and in human tissue (Walker et al. 1995; Tomai et al. 1996; Ikonomidis et al. 1997). Responses may be model specific, in part, since some aspects of I/R injury are dependent upon blood components and activation of pathways for inflammation, while others are intrinsic to the myocardial cells themselves (and these cell-dependent responses may also vary across species). Thus, injury and counteracting protective processes may differ between ex vivo or blood-free models and the in situ myocardium. Finally, differences reported with the use of AR antagonists in PC studies may be related to the nature and duration of the PC stimulus (see below), which may influence the contribution of ARs to protection.

In terms of the identity of the ARs implicated in triggering PC, initial work supported the involvement of A_1ARs (Liu et al. 1991; Tsuchida et al. 1992; Auchampach and Gross 1993). However, subsequent studies (Armstrong and Ganote 1994, 1995; Liu et al. 1994; Wang et al. 1997) demonstrated that partially selective A_3AR antagonism also impaired the protective efficacy of PC. Liang and colleagues documented A_1AR and A_3AR involvement in PC responses in chick cardiomyocytes (Strickler et al. 1996; Liang and Jacobson 1998), while Wang et al. (1997) reported additive contributions from A_1AR and A_3ARs to optimize PC in rabbit myocytes. Although other studies initially supported A_3AR involvement in IPC in intact rabbit myocardium (Tracey et al. 1997), this group subsequently presented evidence of a quantitatively more critical role for A_1AR vs. A_3AR (Hill et al. 1998). More recent studies confirm that endogenous adenosine contributes to IPC via A_1AR and/or A_3AR activation, though the contribution of ARs may be dependent upon the nature and duration of the PC stimulus, being less important with shorter periods of triggering ischemia (Liem et al. 2001, 2008). This is consistent with earlier observations of Schulz et al. in pigs (1995).

Ultimately, preservation of AR-dependent protection in human myocardial tissue is of key importance. Walker and colleagues provided some of the first support for mediation of PC by ARs in human myocardium (Walker et al. 1995). Cleveland et al. (1996, 1997) subsequently confirmed AR-mediated PC responses in human myocardial tissue. Carr et al. (1997) further established that A_1ARs and A_3ARs trigger PC in human atrial muscle, while Ikonomidis et al. (1997) demonstrated AR dependence of PC in human pediatric myocytes. Thus, AR-mediation of PC is relevant to human myocardium. Indeed, an early study by Tomai et al. (1996) supported A_1AR-dependent PC in patients undergoing coronary angioplasty. Furthermore, the importance of ARs in determining resistance to myocardial ischemia is supported by associations between AR polymorphisms, specifically for A_1 and A_3ARs, and infarct size in patients with ischemic cardiomyopathy (Tang et al. 2007).

3.1.3 Evidence from Gene-Modified Models

Essential contributions of ARs to PC are borne out by recent gene manipulation studies. Analysis of A_3AR gene knockout (KO) in mice revealed no impact on induction of IPC (Guo et al. 2001), apparently negating an essential role for this AR subtype. However, A_1AR KO eliminates protection with both IPC (Lankford et al. 2006) and remote PC triggered by cerebral ischemia (Schulte et al. 2004). Moreover, ecto-5′-nucleotidase deletion also eliminates protection with IPC, supporting an essential role for endogenous adenosine generated at the cell surface (Eckle et al. 2007). This latter study also confirmed an essential role for ARs in IPC, although their data differed in implicating only the A_{2B}AR. The basis of this discrepancy is not clear, but may, in part, be model related (in vivo vs. in vitro). This latter observation is, however, consistent with recent data from the laboratory of Downey and colleagues, who reported evidence for protein kinase C (PKC) dependent sensitization of A_{2B}ARs during the trigger or ischemic phases and their role in protection during the subsequent reperfusion phase (Kuno et al. 2007).

Of course, a limitation inherent to gene deletion (or overexpression) is an inability to distinguish events temporally. Since gene deletion eliminates the actions of targeted ARs at all time points, it is unclear from such work when the receptors are involved. For example, A_1ARs or A_3ARs may trigger protection with IPC prior to or during ischemia, while recent evidence implicates a role for A_{2B}AR in mediating the protection with PC during the reperfusion phase (Kuno et al. 2007). This A_{2B}AR-mediated protection during reperfusion could depend to some extent upon A_1AR and/or A_3AR activation of PKC prior to or during ischemia. Such complex responses are not amenable to interrogation by gene manipulation.

3.2 AR Activity During Ischemia

Cardioprotective effects of PC and preischemic GPCR activation were initially thought to manifest primarily during ischemia itself (Cohen et al. 2000). Preischemic AR agonism (or A_1AR overexpression) modifies substrate and energy metabolism, H^+ and Ca^{2+} accumulation, and contracture development during the ischemic episode (Lasley et al. 1990; Fralix et al. 1993; Lasley and Mentzer 1993; Headrick 1996). Similarly, there is evidence of specific protective actions of adenosine and A_1ARs during ischemia versus reperfusion (Peart and Headrick 2000;

Peart et al. 2003). IPC also modifies ischemic events relevant to tissue protection (de Jonge and de Jong 1999), reducing purine moiety accumulation and washout (Van Wylen 1994; Lasley et al. 1995a; Harrison et al. 1998; de Jonge et al. 2002) and ionic perturbations (Fralix et al. 1993). Such observations are consistent with the idea that modulation of injury during ischemia itself contributes to overall protection and improved postischemic outcome. This is supported by early work of Thornton et al. (1993), who showed that protection with IPC is mediated, at least in part, via intrinsic activation of A_1ARs during the subsequent ischemic insult. Studies such as that of Stambaugh et al. (1997) also show that AR activation throughout the period of ischemia/hypoxia is beneficial.

While a majority of studies across differing species support beneficial actions of either exogenously or intrinsically activated ARs during myocardial ischemia, there are a small number of reports of improved outcomes with AR antagonists applied prior to ischemia in vivo (and thus reflecting possible blockade of ARs prior to, during, or following ischemia). Neely et al. (1996) initially documented infarct limitation with three different A_1AR antagonists, DPCPX (1,3 dipropyl-8-cyclopentylxanthine), XAC (xanthine amine congener) and bamiphylline, in a feline regional myocardial infarct model. To rule out that the possibility that these A_1AR antagonists were producing their effects via a nonspecific intracellular action (i.e., inhibition of intracellular enzymes, e.g., phosphodiesterases), Forman and colleagues (2000) reported that another (albeit poorly selective) A_1AR antagonist, DPSPX (1,3-dipropyl-8-*p*-sulfophenylxanthine), which is negatively charged and thus does not accumulate in intracellular spaces because of its high water solubility, also reduced infarct size in dogs. Because DPSPX significantly reduced FMLP (formyl–Met–Leu–Phe)-induced chemoattraction of human neutrophils, the authors of this study suggested that this A_1AR antagonist produced sustained myocardial protection in dogs by reducing inflammation. However, DPSPX is also known to interact with the A_{2B}AR (Feoktistov and Biaggioni 1997), and at the doses applied in this study, to block A_2-dependent coronary dilation (Forman et al. 2000). A later detailed study by Auchampach et al. (2004) described the effect of three different A_1AR antagonists, DPCPX, BG 9928 (1,3-dipropyl-8-[1-(4-propionate)-bicyclo-[2,2,2]octyl)]xanthine) and BG 9719 (1,3-dipropyl-8-[2-(5,6-epoxynorbornyl) xanthine), of varying specificities in a regional myocardial infarct model in vivo in dogs. A_1AR antagonists could limit infarct size in dog hearts, though only with those agents (DPCPX and BG 9928) that also antagonized A_{2A}AR-mediated coronary dilation and possessed appropriate affinities for A_{2B}ARs, raising the possibility of actions at multiple AR subtypes. An alternative explanation by the authors of this study was that differences in the pharmacokinetic and pharmacodynamic properties of BG 9719 may have limited the in vivo potency of this A_1AR antagonist in these studies. They additionally showed that the A_1AR antagonists DPCPX and BG 9928 were equally protective when applied just prior to reperfusion or throughout ischemia-reperfusion, suggesting a primarily postischemic mode of action.

The basis of these mixed observations remains to be determined, though they do raise the possibility of opposing effects of ARs through cell-specific responses. For example, A_1AR activity may augment chemotaxis and neutrophil-dependent injury, whereas the same receptor limits injury in cardiomyocytes. A number of studies confirm a lack of any infarct-sparing effects of nonselective or subtype-specific AR antagonists in vivo in multiple species (Toombs et al. 1992; Tsuchida et al. 1992; Auchampach and Gross 1993; Thornton et al. 1993; Zhao et al. 1993; Hoshida et al. 1994; Baba et al. 2005; Kin et al. 2005; Lasley et al. 2007). However, with the exception of the study by Zhao et al. (1993), the antagonists used in these studies were administered as single doses and not as continuous infusions or multiple doses to achieve a steady state plasma concentration of the AR antagonist, as was done by Neely et al., Forman et al., and Auchampach et al.. Moreover, problems with the selectivity of AR antagonists for specific AR subtypes, particularly during in vivo studies, limit their interpretation with respect to the definitive roles of the four AR subtypes in the setting of acute myocardial ischemia-reperfusion injury.

4 Reperfusion Injury and ARs in Experimental Studies

Although reperfusion is necessary to salvage ischemic myocardium, the process of restoring blood flow also contributes to the total injury observed in ischemic-reperfused myocardium. Reperfusion injury is caused by intracellular calcium overload and oxidative stress induced by the formation of reactive O_2 species (ROS) in the presence of decreased cellular redox state. Reperfusion injury in intact animals and in humans following myocardial ischemia durations of >15 min produces irreversible injury that is also associated with a general inflammatory process including the release of numerous cytokines, adhesion and infiltration of neutrophils across the damaged coronary endothelium, platelet aggregation, and activation of the complement cascade (Ambrosio and Tritto 1999; Park and Lucchesi 1999; Verma et al. 2002).

Similar to the beneficial protective effects of AR agonists discussed in the first sections of this chapter, there is now convincing evidence that the activation of ARs during reperfusion is cardioprotective in animal models. However, in contrast to reports nearly 20 years old documenting the cardioprotective effects of adenosine treatment prior to ischemia, initial studies on the effects of treatment with adenosine after reperfusion were much more controversial. Two initial reports in canine models indicated that intracoronary and intravenous adenosine infusions for the first 1–2.5 h of reperfusion after 90 min coronary occlusions significantly reduced infarct size after 24 and 72 h reperfusion, respectively (Olafsson et al. 1987; Pitarys et al. 1991). In both of these studies, the ischemic myocardium from animals treated with adenosine exhibited significantly less neutrophil accumulation and erythrocyte plugging of capillaries. These observations are consistent with adenosine's ability to inhibit both neutrophil adherence to endothelium (Cronstein et al. 1992) and platelet aggregation (Söderbäck et al. 1991). Several subsequent reports were, however, unable to reproduce these positive findings (Homeister et al. 1990; Goto et al. 1991; Vander Heide and Reimer 1996). Negative results with adenosine treatment following reperfusion may be due to the use of inadequate doses, which must be high enough to overcome its rapid uptake and metabolism by red blood cells and endothelial cells.

However, high concentrations of adenosine can be associated with severe hypotension, reflex tachycardia, and coronary steal. These side effects will likely limit the use of adenosine as a cardioprotective agent in humans.

4.1 Effects of the A_{2A}AR During Reperfusion

Despite the contradictory reports regarding the beneficial effects of adenosine as a reperfusion treatment, there have been an increasing number of reports that reperfusion treatments with infusions of certain AR agonists are cardioprotective. Such studies support the hypothesis that the cardioprotective effects of adenosine are mediated primarily via activation of one or more AR subtypes. The majority of such studies indicate that the infusion of adenosine A_{2A}AR agonists during reperfusion reduces myocardial infarct size. It appears that the first such study was conducted by Norton et al. (1992), who reported that the A_{2A}AR agonist CGS21680 (4-[2-[[6-Amino-9-(*N*-ethyl-*b*-D-ribofuranuronamidosyl)-9*H*-purin-2-yl]amino]ethyl]benzenepropanoic acid), infused during reperfusion in vivo, significantly reduced myocardial infarct size measured after 48 h of reperfusion in rabbits in the absence of hypotension. Subsequent studies have reproduced similar infarct size-reducing effects of reperfusion A_{2A}AR stimulation in dogs, pigs, rats, and mice (Schlack et al. 1993; Zhao et al. 1996; Jordan et al. 1997; Budde et al. 2000; Lasley et al. 2001; Boucher et al. 2005; Yang et al. 2005, 2006).

Although there is a significant expression of A_{2A}ARs on vascular cells (vascular smooth muscle and endothelial cells), and activation of this receptor is associated with coronary vasodilatation, the beneficial effects of reperfusion A_{2A}AR agonists are independent of increased coronary blood flow and can be achieved without systemic hypotension. The prevailing current hypothesis for the beneficial A_{2A}AR effects during reperfusion are related to its anti-inflammatory properties, such as inhibition of neutrophil production of ROS and adherence to endothelium (Visser et al. 2000; Sullivan et al. 2001). Recent studies in mice further suggest that this A_{2A}AR-mediated reperfusion protection is due to effects on bone marrow-derived cells, more specifically to $CD4^{+}$ T-helper lymphocytes (Toufektsian et al. 2006).

However, two additional studies conducted in intact animal models of myocardial stunning indicate that reperfusion treatment with A_{2A}AR agonists can exert beneficial effects in the absence of severe inflammation and myocardial necrosis. In porcine regionally stunned myocardium, an intracoronary infusion of the A_{2A}AR agonist CGS21680, initiated after 2 h reperfusion following 15 min coronary occlusion, significantly increased regional preload-recruitable stroke work and stroke work area, both of which are load-insensitive parameters of cardiac contractility. This effect, which appeared to be independent of increased coronary blood flow, occurred in stunned (i.e., no infarction was detected), but not normal, myocardium (Lasley et al. 2001). The fact that the A_{2A}AR agonist exerted its beneficial effects 2 h after reperfusion suggests that the improvement in regional contractility is likely to have been independent of a reduction in myocardial reperfusion injury, but rather may have been a true positive inotropic effect. Using another myocardial stunning model in dogs, Glover et al. (2007) observed that the A_{2A}AR agonist ATL-146e, given just prior and during reperfusion following multiple brief (5 min) coronary occlusions, improved reperfusion wall thickening in the absence of any increase in coronary blood flow. Infusion of ATL-146e had no effect on regional function in normally perfused myocardium. Whether these beneficial effects of reperfusion A_{2A}AR stimulation in the absence of necrosis are due to a direct effect on the myocardium remains to be determined.

Although the evidence implicating the anti-inflammatory effects of postischemic A_{2A}AR activation in the setting of myocardial infarction is compelling, the above two studies in stunned myocardium indicate that A_{2A}AR activation may also protect the reperfused heart via mechanisms independent of neutrophils and inflammatory processes, as well as increased coronary blood flow. There are several reports that A_{2A}ARs are expressed in porcine, human, and rat ventricular myocytes (Marala and Mustafa 1998; Kilpatrick et al. 2002), which raises the possibility that the beneficial effect of A_{2A}AR agonists during reperfusion may also be due to direct effects on the cardiac myocyte. There have been numerous studies over the past 15 years investigating the effects of A_{2A}AR agonists on cardiac myocyte physiology, but these reports have yielded conflicting findings (Shryock et al. 1993; Stein et al. 1994; Xu et al. 1996, 2005; Boknik et al. 1997; Woodiwiss et al. 1999; Hleihel et al. 2006; Hove-Madsen et al. 2006). The majority of these reports indicate that A_{2A}AR activation alone exerts little, if any, direct effects on normal cardiac ventricular myocytes. However, it is possible that during myocardial ischemia, when endogenous adenosine levels increase and multiple AR subtypes are activated, cardiomyocyte A_{2A}AR may modulate the cardioprotective effects of adenosine.

There remain several interesting and incomplete aspects to our understanding of the cardioprotective effects of reperfusion AR agonist treatment. Although A_{2A}AR agonists administered during reperfusion have been shown to be cardioprotective in intact animals, the administration of A_{2A}AR antagonists does not exacerbate myocardial injury or infarct size in normal animals (Kin et al. 2005; Reid et al. 2005; Lasley et al. 2007). However, there is evidence that the A_{2A}AR does participate in the cardioprotective effect of ischemic postconditioning. Ischemic postconditioning is the phenomenon by which brief interruptions in coronary flow during the initial minutes of reperfusion following a prolonged occlusion reduce myocardial infarct size. This phenomenon is thus somewhat analogous to ischemic preconditioning, which was described earlier. The AR antagonist ZM241385 (4-(2-[7-amino-2-(2-furyl)[1,2,4]triazolo[2,3-*a*][1,3,5]triazin-5-ylamino]ethyl)phenol), which exhibits some selectivity for the A_{2A}AR subtype, has been shown to block ischemic postconditioning in vivo in rat hearts and in isolated perfused mouse hearts (Kin et al. 2005). A more recent report indicated that ischemic postconditioning could not be induced in mouse hearts from A_{2A}AR KO mice (Morrison et al. 2007). These findings indicate that stimulation of A_{2A}ARs plays a pivotal role in reducing myocardial reperfusion injury. Observations in isolated buffer perfused hearts in these latter two reports further support the hypothesis that this protective effect is mediated, at least in part, by the cardiomyocyte A_{2A}AR.

As described above, there are now numerous reports indicating that the infusion of $A_{2A}AR$ agonists during reperfusion is cardioprotective. Although the administration of $A_{2A}AR$ agonists prior to ischemia does not reduce myocardial ischemia-reperfusion injury, there is increasing evidence that $A_{2A}ARs$ may modulate the protective effects of A_1AR stimulation. Reid et al. (2005) and Lasley et al. (2007) reported that the $A_{2A}AR$ antagonist ZM241385 blocked the infarct reducing effects of preischemic treatments with three different AR agonists—AMP579 ($1S$-[$1a, 2b, 3b, 4a(S^*)$]-4-[7-[[2-(3-chloro-2-thienyl)-1-methylpropyl]amino]-3H-imidazo[4,5-b]pyridyl-3-yl]cyclopentane carboxamide), 2 chloro-N^6-cyclopentyladenosine (CCPA), $5'$-N-ethyl-carboxamidoadenosine (NECA)—in two different studies. The $A_{2A}AR$ antagonist did not alter the A_1AR-induced bradycardia with these agonists, indicating that the A_1AR was not blocked; however, the ability of ZM241385 to block the protection by these AR agonists was comparable to that achieved with the A_1AR antagonist DPCPX. Preliminary observations in one of these studies suggested that the $A_{2A}AR$ antagonist partially blunted the effects of AMP579 on preischemic mitogen-activated protein kinase (MAPK) signaling (Reid et al. 2005). These findings regarding the effects of $A_{2A}AR$ antagonists on A_1AR cardioprotection are supported by an increasing number of reports of interactions between AR subtypes, including the formation of heterodimers (Karcz-Kubicha et al. 2003; O'Kane and Stone 1998; Lopes et al. 1999, 2002; Nakata et al. 2005).

There is also evidence that the beneficial effects of reperfusion AR agonist treatments may involve interactions among AR subtypes. In the isolated perfused rabbit heart, a reperfusion infusion (500 nM) of the AR agonist AMP579, which has a high affinity for both A_1 and $A_{2A}ARs$ (Smits et al. 1998), reduced infarct size—an effect that was blocked by 8-(13-chlorostyryl) caffeine (CSC), which exhibits some selectivity for $A_{2A}ARs$, but not by the A_1AR antagonist DPCPX (Xu et al. 2001). The beneficial effect of AMP579 was mimicked by the nonselective agonist NECA at a dose (100 nM) activating both A_1 and $A_{2A}ARs$, but not by the $A_{2A}AR$ agonist CGS21680 (50 nM). Kis et al. (2003) reported similar findings in the intact rabbit, where an infusion of AMP579 during reperfusion reduced infarct size, and this effect was blocked by the $A_{2A}AR$ antagonist ZM241385 but not mimicked by the same dose of the $A_{2A}AR$ agonist CGS21680. It is not clear why these studies did not observe protection with the $A_{2A}AR$ agonist alone, when numerous other studies have reported such protection; however, these findings support a role for the $A_{2A}AR$ in reduction of myocardial injury. Since ZM241385 has some affinity for $A_{2B}ARs$, it is also possible that the effects of this agent could be due to antagonism of this receptor subtype (Hasan et al. 2000).

4.2 *Effects of A_1 and A_3ARs During Reperfusion*

To date, the primary emphasis on AR reduction of reperfusion injury has focused on the role of the $A_{2A}AR$. However, given that there are four AR subtypes, all of which appear to be expressed in the heart, it is possible that one or more of these other AR subtypes may modulate reperfusion injury. The one exception to this hypothesis is the A_1AR. Although, as described in the first section of this chapter, there is significant evidence that A_1AR agonists administered prior to ischemia are protective, it is clear that A_1AR agonists administered during reperfusion are not protective (Thornton et al. 1992; Baxter et al. 2000). There is evidence that A_3AR activation during reperfusion may be cardioprotective, as studies in isolated hearts and intact animals indicate that the A_3AR agonists IBMECA (1-deoxy-1-[6-[[(3-iodophenyl)methyl]amino]-9*H*-purin-9-yl]-*N*-methyl-*b*-D-ribofuranuronamide) and Cl-IBMECA (1-[2-chloro-6-[[(3-iodophenyl)methyl]amino]-9*H*-purin-9-yl]-1-deoxy-*N*-methyl-*b*-D-ribofuranuronamide), administered during reperfusion, reduce myocardial infarct size (Maddock et al. 2002; Auchampach et al. 2003; Park et al. 2006). In two of these studies, the effects of the A_3AR agonists were blocked by A_3AR antagonists (Maddock et al. 2002; Park et al. 2006). Interestingly, in the former study (Maddock et al. 2002) the reperfusion A_3AR agonist protection was also blocked by the A_{2A}AR antagonist CSC. Finally, Kin et al. (2005) observed that postconditioning could be blocked by an A_3AR antagonist. Thus, in contrast to the A_1AR, activation of the A_3AR either prior to ischemia or during reperfusion appears to be cardioprotective.

4.3 Emerging Roles for the A_{2B}AR During Reperfusion

With respect to the fourth AR subtype, only now are a limited number of studies supporting a role for the A_{2B}AR in modulating myocardial reperfusion injury appearing. Investigations of this receptor in the heart have been hindered by the fact that there are no radioligand binding studies defining A_{2B}AR receptor density or affinity in mammalian myocardium or cardiomyocytes. The role of this receptor has also been hindered by the lack of studies with well-characterized, selective A_{2B}AR agonists and antagonists. To date there are four pharmacological studies providing some evidence for the involvement of A_{2B}ARs, although the results are conflicting. Auchampach et al. (2004) reported that reperfusion treatments with DPCPX and BG 9928, but not BG 9719, all of which are selective A_1AR antagonists, reduced infarct size in dogs by ~40%. These effects were compared to radioligand binding studies performed with recombinant canine ARs expressed in HEK cells, and blockade of canine A_1 (heart rate) and A_{2A}AR (coronary conductance) effects. Based on these observations, the authors concluded that DPCPX and BG 9928 may exert their infarct-reducing effects by blocking A_{2B}ARs; however, they could not discount the possibility that DPCPX and BG 9928 reduced infarct size by blocking A_1ARs.

Three additional studies in rabbit heart models of ischemia/reperfusion concluded that A_{2B}AR activation, rather than inhibition, contributes to reperfusion cardioprotection (Solenkova et al. 2006; Phillip et al. 2006; Kuno et al. 2007). In the first of these studies, the infarct-reducing effect of IPC was blocked by the A_{2B}AR antagonist, MRS1754 (*N*-(4-cyanophenyl)-2-[4-(2,3,6,7-tetrahydro-2,6-dioxo-1,

3-dipropyl-1*H*-purin-8-yl)phenoxy]-acetamide), but not an A_{2A}AR antagonist, CSC, administered at the onset of reperfusion. Subsequently, Phillip et al. (2006) reported that the cardioprotective effect of NECA administration at reperfusion (i.e., pharmacological postconditioning) in intact rabbits was blocked by MRS1754. Interestingly, a previous report from this same laboratory concluded that the reperfusion protection induced by NECA was due to A_{2A}AR activation (Xu et al. 2001). More recently, Kuno et al. (2007) demonstrated that a novel A_{2B}AR agonist, BAY 60–6583, administered during reperfusion, is protective. Given the apparent expression of multiple AR subtypes in the heart and their possible interactions, as well as the lack of selectivity for many of the commonly used AR agonists and antagonists, studies in AR KO mice will likely be needed to address the question of the A_{2B}AR, as well as the definitive roles of other AR subtypes. Interestingly, the results of a recent study by Eckle et al. (2007) indicated that in vivo IPC was ablated in A_{2B}AR KO mice, but not in mice lacking A_1, A_{2A} or A_3 receptors.

5 Reperfusion Injury and ARs in Human Myocardium

Despite all of the experimental evidence to date indicating the cardioprotective effects of adenosine and AR agonists, there have been very few studies examining the beneficial effects of these agents in humans in the setting of myocardial ischemia-reperfusion and thrombolysis. The initial such report was the acute myocardial infarction study of adenosine (AMISTAD) trial conducted between December 1994 and July 1997, the results of which were published in 1999 (Mahaffey et al. 1999). This was an open-label, placebo-controlled, randomized study to determine the safety and efficacy of adenosine as an adjunct to thrombolytic therapy in the treatment of acute myocardial infarction (MI). The effect of an intravenous infusion of adenosine ($70\,\mu g\,kg^{-1}\,min^{-1}$) for 3 h was compared to a placebo infusion in patients treated with thrombolysis within 6 h of the onset of an MI. After modification for slow enrollment, 197 patients were included, with the primary end-point being myocardial infarct size, as determined by Tc-99m sestamibi single-photon emission computed tomography (SPECT) imaging 5–7 days after enrollment. The results indicated that there was a 33% relative reduction in infarct size in patients that received adenosine ($p = 0.03$). Patients with an anterior MI exhibited a 67% relative reduction in infarct size, whereas there was no beneficial effect in patients with a nonanterior MI. Patients receiving adenosine, particularly those with nonanterior MI, experienced more bradycardia, heart block, hypotension and ventricular arrhythmias (Mahaffey et al. 1999).

There is a significant amount of preclinical data on the efficacy of AR agonists in reducing myocardial reperfusion injury, and these studies are clearly more consistently positive than the often contradictory findings with adenosine. Despite this wealth of information, today there remains only one documented clinical trial examining the effects of an AR agonist in the setting of clinical myocardial ischemia-reperfusion injury, the ADMIRE (AMP579 Delivery for Myocardial Infarction

REduction) study. This was a double-blind, multicenter, placebo-controlled trial of 311 patients undergoing primary percutaneous transluminal coronary angioplasty (PTCA) after acute ST-segment elevation MI (Kopecky et al. 2003). Patients were randomly assigned to placebo or to one of three different doses of AMP579 (15, 30 or 60 μg kg^{-1}) continuously infused over 6 h. This AR agonist, which has a high affinity for both A_1 and A_{2A}ARs, has been shown to reduce experimental myocardial ischemia-reperfusion in multiple species when administered both prior to ischemia or during reperfusion (Merkel et al. 1998; McVey et al. 1999; Meng et al. 2000; Xu et al. 2001; Kis et al. 2003; Kristo et al. 2004). The primary end-point was final myocardial infarct size measured by technetium Tc-99m sestamibi scanning at 120–216 h after PTCA. Secondary end-points included myocardial salvage and salvage index at the same time interval (in a subset of patients), left ventricular ejection fraction, duration of hospitalization, heart failure at 4–6 weeks, and cardiac events at four weeks and six months. Results indicated that there was no difference in final infarct size or in any of the secondary end-points. There was a trend towards increased myocardial salvage in patients with anterior MI. The authors of this study concluded that, based on the pharmacokinetic data, the maximal dose used in this trial was comparable to the lowest dose proven effective in animal studies.

The promising results of AMISTAD I led to a second trial (AMISTAD II) to determine the effects of adenosine infusion on clinical outcomes and infarct size in ST-segment elevation myocardial infarction (STEMI) patients undergoing reperfusion therapy (Ross et al. 2005). A total of 2,118 patients receiving thrombolysis or primary angioplasty were randomized to a 3 h infusion of either adenosine (50 or 70 μg kg^{-1} min^{-1}) or placebo. The primary end-point was new congestive heart failure (CHF) beginning $>$24 h after randomization, or the first rehospitalization for CHF, or death from any cause within six months. Infarct size was measured in a subset of 243 patients by Tc-99m sestamibi tomography. There was no effect of either adenosine dose on primary end-points, although patients receiving the higher dose (70 μg kg^{-1} min^{-1}) exhibited a median infarct size (11%) that was significantly lower ($p = 0.023$) than that of the placebo group (median infarct size 23%). It was concluded that a larger clinical trial was warranted to determine whether the decreased infarct size observed with adenosine was associated with enhanced long-term outcome. A post hoc subanalysis of these data indicated that patients receiving the adenosine infusion within 3 h of the onset of symptoms exhibited significantly reduced mortality at one and six months, and event-free survival was enhanced compared to patients treated with placebo (Kloner et al 2006).

Given all of the experimental evidence supporting the cardioprotective effects of AR agonists administered either prior to ischemia or during reperfusion, there clearly needs to more research and development into the synthesis, screening, and testing of potent, selective AR agonists. Basic scientists must also utilize consistent experimental models to determine the specific contributions of the multiple AR subtypes and their mechanisms of action. Because animal efficacy studies do not always translate to human efficacy, preclinical models with high relevance to humans and that closely simulate the human condition should be designed. Finally, clinical trials must be better designed along the lines of the information learned from the multitude of preclinical studies and clinical studies performed to date.

6 Impact of Age and Disease

Ischemic heart disease occurs predominantly in the elderly population (affecting up to 50% of those over 65), and can be associated with multiple underlying disease states, including atherosclerosis, hyperlipidemia, hypertension, and diabetes. From a clinical perspective, it is thus essential that protective strategies derived from research into PC or other protective modalities are effective across age groups and in diseased hearts. Unfortunately, aging limits or even abrogates protection with PC (Abete et al. 1996; Fenton et al. 2000; Schulman et al. 2001), AR activation (Gao et al. 2000; Schulman et al. 2001; Headrick et al. 2003b; Willems et al. 2005), and other GPCR stimuli (Peart et al. 2007). Newly discovered postconditioning is also impaired (Przyklenk et al. 2008). These age-dependent failures may stem from ineffective activation of key components of downstream signaling cascades (Peart et al. 2007; Przyklenk et al. 2008). On the other hand, age-related failure of AR-dependent protection is not universally observed. For example, Kristo et al. (2005) found no age-related changes in functional AR sensitivity, and augmentation of the infarct-sparing actions of adenosine. Thus, adenosine's role in aged hearts as well as the efficacy of cardioprotection in these hearts by targeting ARs with adenosine or AR agonists are questions that remain open.

Disease states underlying or contributing to ischemic disorders (when intrinsic protective responses such as PC are more important) can also impair these responses. For example, Ghosh et al. (2001) showed failure of PC in diabetic human myocardium, which may also reflect abnormalities in distal signaling cascades. In terms of AR responses, Donato et al. (2007) showed not only involvement of A_1ARs (and the mito K_{ATP} channel) in ischemic PC in normal hearts, but confirmed the ability of this stimulus to limit ischemic injury in hypercholesterolemic hearts. Moreover, A_1 and A_3 AR-triggered PC responses appear to be preserved in hypertrophic myocardium (Hochhauser et al. 2007). Thus, the few studies to date do support the preservation of AR-mediated protection in animal models of some relevant disease states. Whether this extends to patients suffering from chronic forms of cardiovascular disease remains to be established. It is worth considering that combined effects of age and disease may well underlie the rather modest benefit obtained with adenosine in clinical trials (AMISTAD I and II) versus the profound protective responses observed in the laboratory.

References

Abete P, Ferrara N, Cioppa A, Ferrara P, Bianco S, Calabrese C, Cacciatore F, Longobardi G, Rengo F (1996) Preconditioning does not prevent postischemic dysfunction in aging heart. J Am Coll Cardiol 27:1777–1786

Ambrosio G, Tritto I (1999) Reperfusion injury: experimental evidence and clinical implications. Am Heart J 138:S69–S75

Armstrong S, Ganote CE (1994) Adenosine receptor specificity in preconditioning of isolated rabbit cardiomyocytes: evidence of A_3 receptor involvement. Cardiovasc Res 28:1049–1056

Armstrong S, Ganote CE (1995) In vitro ischaemic preconditioning of isolated rabbit cardiomyocytes: effects of selective adenosine receptor blockade and calphostin C. Cardiovasc Res 29:647–652

Auchampach JA, Gross GJ (1993) Adenosine A_1 receptors, K_{ATP} channels, and ischemic preconditioning in dogs. Am J Physiol 264:H1327–H1336

Auchampach JA, Ge ZD, Wan TC, Moore J, Gross GJ (2003) A_3 adenosine receptor agonist IB-MECA reduces myocardial ischemia-reperfusion injury in dogs. Am J Physiol Heart Circ Physiol 285:H607–H613

Auchampach JA, Jin X, Moore J, Wan TC, Kreckler LM, Ge ZD, Narayanan J, Whalley E, Kiesman W, Ticho B, Smits G, Gross GJ (2004) Comparison of three different A1 adenosine receptor antagonists on infarct size and multiple cycle ischemic preconditioning in anesthetized dogs. J Pharmacol Exp Ther 308:846–856

Baba K, Minatoguchi S, Zhang C, Kariya T, Uno Y, Kawai T, Takahashi M, Takemura G, Fujiwara H (2005) α1-Receptor or adenosine A_1-receptor dependent pathway alone is not sufficient but summation of these pathways is required to achieve an ischaemic preconditioning effect in rabbits. Clin Exp Pharmacol Physiol 32:263–268

Baxter GF, Hale SL, Miki T, Kloner RA, Cohen MV, Downey JM, Yellon DM (2000) Adenosine A_1 agonist at reperfusion trial (AART): results of a three-center, blinded, randomized, controlled experimental infarct study. Cardiovasc Drugs Ther 14:607–614

Bokník P, Neumann J, Schmitz W, Scholz H, Wenzlaff H (1997) Characterization of biochemical effects of CGS 21680C, an A_2-adenosine receptor agonist, in the mammalian ventricle. J Cardiovasc Pharmacol 30: 750–758

Boucher M, Wann BP, Kaloustian S, Massé R, Schampaert E, Cardinal R, Rousseau G (2005) Sustained cardioprotection afforded by A_{2A} adenosine receptor stimulation after 72 h of myocardial reperfusion. J Cardiovasc Pharmacol 45:439–446

Budde JM, Velez DA, Zhao Z-Q, Clark KL, Morris CD, Muraki S, Guyton RA, Vinten-Johansen J (2000) Comparative study of AMP579 and adenosine in inhibition of neutrophil-mediated vascular and myocardial injury during 24 h of reperfusion. Cardiovasc Res 47:294–305

Bugge E, Ytrehus K (1995) Ischaemic preconditioning is protein kinase C dependent but not through stimulation of alpha adrenergic or adenosine receptors in the isolated rat heart. Cardiovasc Res 29:401–406

Carr CS, Hill RJ, Masamune H, Kennedy SP, Knight DR, Tracey WR, Yellon DM (1997) Evidence for a role for both the adenosine A_1 and A_3 receptors in protection of isolated human atrial muscle against simulated ischaemia. Cardiovasc Res 36:52–59

Cave AC, Collis CS, Downey JM, Hearse DJ (1993) Improved functional recovery by ischaemic preconditioning is not mediated by adenosine in the globally ischaemic isolated rat heart. Cardiovasc Res 27:663–668

Cerniway RJ, Morrison RR, Byford AM, Lankford AR, Headrick JP, Van Wylen DG, Matherne GP (2002) A_1 adenosine receptor overexpression decreases stunning from anoxia-reoxygenation: role of the mitochondrial K_{ATP} channel. Basic Res Cardiol 97:232–238

Cleveland JC Jr, Wollmering MM, Meldrum DR, Rowland RT, Rehring TF, Sheridan BC, Harken AH, Banerjee A (1996) Ischemic preconditioning in human and rat ventricle. Am J Physiol 271:H1786–H1794

Cleveland JC Jr, Meldrum DR, Rowland RT, Banerjee A, Harken AH (1997) Adenosine preconditioning of human myocardium is dependent upon the ATP-sensitive K^+ channel. J Mol Cell Cardiol 29: 175–182

Cohen MV, Baines CP, Downey JM (2000) Ischemic preconditioning: from adenosine receptor to K_{ATP} channel. Annu Rev Physiol 62:79–109

Crawford M, Ford S, Henry M, Matherne GP, Lankford A (2005) Myocardial function following cold ischemic storage is improved by cardiac-specific overexpression of A_1-adenosine receptors. Can J Physiol Pharmacol 83:493–498

Cronstein BN, Levin RI, Philips M, Hirschhorn R, Abramson SB, Weissmann G (1992) Neutrophil adherence to endothelium is enhanced via adenosine A_1 receptors and inhibited via adenosine A_2 receptors. J Immunol 148:2201–2206

de Jonge R, de Jong JW (1999) Ischemic preconditioning and glucose metabolism during low-flow ischemia: role of the adenosine A_1 receptor. Cardiovasc Res 43: 909–918

de Jonge R, de Jong JW, Giacometti D, Bradamante S (2001) Role of adenosine and glycogen in ischemic preconditioning of rat hearts. Eur J Pharmacol 414:55–62

de Jonge R, Out M, Maas WJ, de Jong JW (2002) Preconditioning of rat hearts by adenosine A_1 or A_3 receptor activation. Eur J Pharmacol 441:165–172

Donato M, D'Annunzio V, Berg G, Gonzalez G, Schreier L, Morales C, Wikinski RL, Gelpi RJ (2007) Ischemic postconditioning reduces infarct size by activation of A_1 receptors and $K^+{}_{ATP}$ channels in both normal and hypercholesterolemic rabbits. J Cardiovasc Pharmacol 49:287–292

Dougherty C, Barucha J, Schofield PR, Jacobson KA, Liang BT (1998) Cardiac myocytes rendered ischemia resistant by expressing the human adenosine A_1 or A_3 receptor. FASEB J 12: 1785–1792

Downey JM, Davis AM, Cohen MV (2007) Signaling pathways in ischemic preconditioning. Heart Fail Rev 12:181–188

Eckle T, Krahn T, Grenz A, Köhler D, Mittelbronn M, Ledent C, Jacobson MA, Osswald H, Thompson LF, Unertl K, Eltzschig HK (2007) Cardioprotection by ecto-5′-nucleotidase (CD73) and A_{2B} adenosine receptors. Circulation 115:1581–1590

Fenton RA, Dickson EW, Meyer TE, Dobson JG Jr (2000) Aging reduces the cardioprotective effect of ischemic preconditioning in the rat heart. J Mol Cell Cardiol 32:1371–1375

Feoktistov I, Biaggioni I (1997) Adenosine A_{2B} receptors. Pharmacol Rev 49:381–402

Forman MB, Vitola JV, Velasco CE, Murray JJ, Raghvendra KD, Jackson EK (2000) Sustained reduction in myocardial reperfusion injury with an adenosine receptor antagonist: possible role of the neutrophil chemoattractant response. J Pharmacol Exp Ther 292:929–938

Fralix TA, Murphy E, London RE, Steenbergen C (1993) Protective effects of adenosine in the perfused rat heart: changes in metabolism and intracellular ion homeostasis. Am J Physiol 264: C986–C994

Gao F, Christopher TA, Lopez BL, Friedman E, Cai G, Ma XL (2000) Mechanism of decreased adenosine protection in reperfusion injury of aging rats. Am J Physiol Heart Circ Physiol 279:H329–H338

Gauthier NS, Morrison RR, Byford AM, Jones R, Headrick JP, Matherne GP (1998) Functional genomics of transgenic overexpression of A_1 adenosine receptors in the heart. Drug Dev Res 45:402–409

Germack R, Dickenson JM (2005) Adenosine triggers preconditioning through MEK/ERK1/2 signalling pathway during hypoxia/reoxygenation in neonatal rat cardiomyocytes. J Mol Cell Cardiol 39:429–442

Germack R, Griffin M, Dickenson JM (2004) Activation of protein kinase B by adenosine A_1 and A_3 receptors in newborn rat cardiomyocytes. J Mol Cell Cardiol 37:989–999

Ghosh S, Standen NB, Galiñianes M (2001) Failure to precondition pathological human myocardium. J Am Coll Cardiol 37:711–718

Glover DK, Ruiz M, Takehana K, Petruzella FD, Rieger JM, Macdonald TL, Watson DD, Linden J, Beller GA (2007) Cardioprotection by adenosine A_{2A} agonists in a canine model of myocardial stunning produced by multiple episodes of transient ischemia. Am J Physiol Heart Circ Physiol 292:H3164–H3171

Goto M, Miura T, Iliodoromitis EK, O'Leary EL, Ishimoto R, Yellon DM, Iimura O (1991) Adenosine infusion during early reperfusion failed to limit myocardial infarct size in a collateral deficient species. Cardiovasc Res 25:943–949

Goto M, Liu Y, Yang XM, Ardell JL, Cohen MV, Downey JM (1995) Role of bradykinin in protection of ischemic preconditioning in rabbit hearts. Circ Res 77:611–621

Guo Y, Bolli R, Bao W, Wu WJ, Black RG Jr, Murphree SS, Salvatore CA, Jacobson MA, Auchampach JA (2001) Targeted deletion of the A_3 adenosine receptor confers resistance to myocardial ischemic injury and does not prevent early preconditioning. J Mol Cell Cardiol 33:825–830

Harrison GJ, Willis RJ, Headrick JP (1998) Extracellular adenosine levels and cellular energy metabolism in ischemically preconditioned rat heart. Cardiovasc Res 40:74–87

Hasan AZMA, Abebe W, Mustafa SJ (2000) Antagonism of porcine coronary artery relaxation by adenosine A_{2A} receptor antagonist ZM 241385. J Cardiovasc Pharmacol 35:322–325

Hausenloy DJ, Yellon DM (2007) Preconditioning and postconditioning: united at reperfusion. Pharmacol Ther 116:173–191

Headrick JP (1996) Ischemic preconditioning: bioenergetic and metabolic changes and the role of endogenous adenosine. J Mol Cell Cardiol 28:1227–1240

Headrick JP, Gauthier NS, Berr SS, Morrison RR, Matherne GP (1998) Transgenic A_1 adenosine receptor overexpression markedly improves myocardial energy state during ischemia-reperfusion. J Mol Cell Cardiol 30:1059–1064

Headrick JP, Gauthier NS, Morrison R, Matherne GP (2000) Cardioprotection by K_{ATP} channels in wild-type hearts and hearts overexpressing A_1-adenosine receptors. Am J Physiol Heart Circ Physiol 279:H1690–H1697

Headrick JP, Hack B, Ashton KJ (2003a) Acute adenosinergic cardioprotection in ischemic-reperfused hearts. Am J Physiol Heart Circ Physiol 285:H1797–H1818

Headrick JP, Willems L, Ashton KJ, Holmgren K, Peart J, Matherne GP (2003b) Ischaemic tolerance in aged mouse myocardium: the role of adenosine and effects of A_1 adenosine receptor overexpression. J Physiol 549:823–833

Hendrikx M, Toshima Y, Mubagwa K, Flameng W (1993) Improved functional recovery after ischemic preconditioning in the globally ischemic rabbit heart is not mediated by adenosine A_1 receptor activation. Basic Res Cardiol 88:576–593

Hill RJ, Oleynek JJ, Magee W, Knight DR, Tracey WR (1998) Relative importance of adenosine A_1 and A_3 receptors in mediating physiological or pharmacological protection from ischemic myocardial injury in the rabbit heart. J Mol Cell Cardiol 30:579–585

Hleihel W, Lafoux A, Ouaini N, Divet A, Huchet-Cadiou C (2006) Adenosine affects the release of Ca^{2+} from the sarcoplasmic reticulum via A_{2A} receptors in ferret skinned cardiac fibres. Exp Physiol 91:681–691

Hochhauser E, Leshem D, Kaminski O, Cheporko Y, Vidne BA, Shainberg A (2007) The protective effect of prior ischemia reperfusion adenosine A_1 or A_3 receptor activation in the normal and hypertrophied heart. Interact Cardiovasc Thorac Surg 6:363–368

Homeister JW, Hoff PT, Fletcher DD, Lucchesi BR (1990) Combined adenosine and lidocaine administration limits myocardial reperfusion injury. Circulation 82:595–608

Hoshida S, Kuzuya T, Nishida M, Yamashita N, Oe H, Hori M, Kamada T, Tada M (1994) Adenosine blockade during reperfusion reverses the infarct limiting effect in preconditioned canine hearts. Cardiovasc Res 28:1083–1088

Hove-Madsen L, Prat-Vidal C, Llach A, Ciruela F, Casadó V, Lluis C, Bayes-Genis A, Cinca J, Franco R (2006) Adenosine A_{2A} receptors are expressed in human atrial myocytes and modulate spontaneous sarcoplasmic reticulum calcium release. Cardiovasc Res 72:292–302

Ikonomidis JS, Shirai T, Weisel RD, Derylo B, Rao V, Whiteside CI, Mickle DA, Li RK (1997) Preconditioning cultured human pediatric myocytes requires adenosine and protein kinase C. Am J Physiol Heart Circ Physiol 272:H1220–H1230

Jones R, Lankford AR, Byford AM, Matherne GP (1999) Inhibition of p38 MAPK in hearts overexpressing A_1 adenosine receptors. Circulation 100:I563

Jordan JE, Zhao ZQ, Sato H, Taft S, Vinten-Johansen J (1997) Adenosine A_2 receptor activation attenuates reperfusion injury by inhibiting neutrophil accumulation, superoxide generation and coronary endothelial adherence. J Pharmacol Exp Ther 280:301–309

Karcz-Kubicha M, Quarta D, Hope BT, Antoniou K, Muller CE, Morales M, Schindler CW, Goldberg SR, Ferre S (2003) Enabling role of adenosine A_1 receptors in adenosine A_{2A} receptor-mediated striatal expression of c-Fos. Eur J Neurosci 18:296–302

Kilpatrick EL, Narayan P, Mentzer RM, Lasley RD (2002) Rat ventricular myocyte adenosine A_{2a} receptor activation fails to alter cAMP or contractility: role of receptor localization. Am J Physiol Heart Circ Physiol 282:H1035–H1040

Kin H, Zatta AJ, Lofye MT, Amerson BS, Halkos ME, Kerendi F, Zhao ZQ, Guyton RA, Headrick JP, Vinten-Johansen J (2005) Postconditioning reduces infarct size via adenosine receptor activation by endogenous adenosine. Cardiovasc Res 67:124–133

Kirchhof P, Fabritz L, Fortmuller L, Matherne GP, Lankford A, Baba HA, Schmitz W, Breithardt G, Neumann J, Boknik P (2003) Altered sinus nodal and atrioventricular nodal function in freely moving mice overexpressing the A_1 adenosine receptor. Am J Physiol Heart Circ Physiol 285:H145–H153

Kis A, Baxter GF, Yellon DM (2003) Limitation of myocardial reperfusion injury by AMP579, an adenosine A_1/A_{2A} receptor agonist: role of A_{2A} receptor and Erk1/2. Cardiovasc Drugs Ther 17:415–425

Kloner RA, Forman MB, Gibbons RJ, Ross AM, Alexander RW, Stone GW (2006) Impact of time to therapy and reperfusion modality on the efficacy of adenosine in acute myocardial infarction: the AMISTAD-2 trial. Eur Heart J 27:2400–2405

Kopecky SL, Aviles RJ, Bell MR, Lobl JK, Tipping D, Frommell G, Ramsey K, Holland AE, Midei M, Jain A, Kellett M, Gibbons RJ (2003) AmP579 Delivery for Myocardial Infarction REduction study. A randomized, double-blinded, placebo-controlled, dose-ranging study measuring the effect of an adenosine agonist on infarct size reduction in patients undergoing primary percutaneous transluminal coronary angioplasty: the ADMIRE (AmP579 Delivery for Myocardial Infarction REduction) study. Am Heart J 146:146–152

Kristo G, Yoshimura Y, Keith BJ, Stevens RM, Jahania SA, Mentzer Jr RM, Lasley RD (2004) The adenosine A_1/A_{2A} receptor agonist AMP579 induces both acute and delayed preconditioning against in vivo myocardial stunning. Am J Physiol Heart Circ Physiol 287:H2746–H2753

Kristo G, Yoshimura Y, Keith BJ, Mentzer RM Jr, Lasley RD (2005) Aged rat myocardium exhibits normal adenosine receptor-mediated bradycardia and coronary vasodilation but increased adenosine agonist-mediated cardioprotection. J Gerontol A Biol Med Sci 60:1399–1404

Kuno A, Critz SD, Cui L, Solodushko V, Yang XM, Krahn T, Albrecht B, Philipp S, Cohen MV, Downey JM (2007) Protein kinase C protects preconditioned rabbit hearts by increasing sensitivity of adenosine A_{2b}-dependent signaling during early reperfusion. J Mol Cell Cardiol 43:262–271

Lankford AR, Yang JN, Rose'Meyer R, French BA, Matherne GP, Fredholm BB, Yang Z (2006) Effect of modulating cardiac A_1 adenosine receptor expression on protection with ischemic preconditioning. Am J Physiol Heart Circ Physiol 290:H1469–H1473

Lasley RD, Mentzer RM Jr (1992) Adenosine improves recovery of postischemic myocardial function via an adenosine A_1 receptor mechanism. Am J Physiol 263:H1460–H1465

Lasley RD, Mentzer RM Jr (1993) Adenosine increases lactate release and delays onset of contracture during global low flow ischaemia. Cardiovasc Res 27: 96–101

Lasley RD, Rhee JW, Van Wylen DG, Mentzer RM Jr (1990) Adenosine A_1 receptor mediated protection of the globally ischemic isolated rat heart. J Mol Cell Cardiol 22:39–47

Lasley RD, Anderson GM, Mentzer RM Jr (1993) Ischaemic and hypoxic preconditioning enhance postischaemic recovery of function in the rat heart. Cardiovasc Res 27:565–570

Lasley RD, Konyn PJ, Hegge JO, Mentzer RM Jr (1995a) Effects of ischemic and adenosine preconditioning on interstitial fluid adenosine and myocardial infarct size. Am J Physiol Heart Circ Physiol 269:H1460–H1466

Lasley RD, Noble MA, Konyn PJ, Mentzer RM Jr (1995b) Different effects of an adenosine A_1 analogue and ischemic preconditioning in isolated rabbit hearts. Ann Thorac Surg 60: 1698–1703

Lasley RD, Jahania MS, Mentzer RM Jr (2001) Beneficial effects of adenosine A_{2a} agonist CGS-21680 in infarcted and stunned porcine myocardium. Am J Physiol Heart Circ Physiol 280:H1660–H1666

Lasley RD, Kristo G, Keith BJ, Mentzer RM Jr (2007) The A_{2a}/A_{2b} receptor antagonist ZM-241385 blocks the cardioprotective effect of adenosine agonist pretreatment in in vivo rat myocardium. Am J Physiol Heart Circ Physiol 292:H426–H431

Li Y, Kloner RA (1993) The cardioprotective effects of ischemic 'preconditioning' are not mediated by adenosine receptors in rat hearts. Circulation 87:1642–1648

Liang BT, Jacobson KA (1998) A physiological role of the adenosine A_3 receptor: sustained cardioprotection. Proc Natl Acad Sci USA 95:6995–6999

Liem DA, van den Doel MA, de Zeeuw S, Verdouw PD, Duncker DJ (2001) Role of adenosine in ischemic preconditioning in rats depends critically on the duration of the stimulus and involves both A_1 and A_3 receptors. Cardiovasc Res 51:701–708

Liem DA, Manintveld OC, Schoonderwoerd K, McFalls EO, Heinen A, Verdouw PD, Sluiter W, Duncker DJ (2008) Ischemic preconditioning modulates mitochondrial respiration, irrespective of the employed signal transduction pathway. Transl Res 151:17–26

Liu Y, Downey JM (1992) Ischemic preconditioning protects against infarction in rat heart. Am J Physiol 263:H1107–H1112

Liu GS, Thornton J, Van Winkle DM, Stanley AWH, Olsson RA, Downey JM (1991) Protection against infarction afforded by preconditioning is mediated by A_1 adenosine receptors in rabbit heart. Circulation 84:350–356

Liu GS, Richards SC, Olsson RA, Mullane K, Walsh RS, Downey JM (1994) Evidence that the adenosine A_3 receptor may mediate the protection afforded by preconditioning in the isolated rabbit heart. Cardiovasc Res 28:1057–1061

Lopes LV, Cunha RA, Ribeiro JA (1999) ZM 241385, an adenosine A_{2A} receptor antagonist, inhibits hippocampal A_1 receptor responses. Eur J Pharmacol 383:395–398

Lopes LV, Cunha RA, Kull B, Fredholm BB, Ribeiro JA (2002) Adenosine A_{2A} receptor facilitation of hippocampal synaptic transmission is dependent on tonic A_1 receptor inhibition. Neuroscience 112:319–329

Louttit JB, Hunt AA, Maxwell MP, Drew GM (1999) The time course of cardioprotection induced by GR79236, a selective adenosine A_1-receptor agonist, in myocardial ischaemia-reperfusion injury in the pig. J Cardiovasc Pharmacol 33:285–291

Maddock HL, Mocanu MM, Yellon DM (2002) Adenosine A_3 receptor activation protects the myocardium from reperfusion/reoxygenation injury. Am J Physiol Heart Circ Physiol 283:H1307–H1313

Mahaffey KW, Puma JA, Barbagelata NA, DiCarli MF, Leesar MA, Browne KF, Eisenberg PR, Bolli R, Casas AC, Molina-Viamonte V, Orlandi C, Blevins R, Gibbons RJ, Califf RM, Granger CB (1999) Adenosine as an adjunct to thrombolytic therapy for acute myocardial infarction: results of a multicenter, randomized, placebo-controlled trial: the Acute Myocardial Infarction STudy of ADenosine (AMISTAD) trial. J Am Coll Cardiol 34:1711–1720

Marala RB, Mustafa SJ (1998) Immunological characterization of adenosine A_{2A} receptors in human and porcine cardiovascular tissues. J Pharmacol Exp Ther 286:1051–1057

Matherne GP, Linden J, Byford AM, Gauthier NS, Headrick JP (1997) Transgenic A_1 adenosine receptor overexpression increases myocardial resistance to ischemia. Proc Natl Acad Sci USA 94:6541–6546

McVey MJ, Smits GJ, Cox BF, Kitzen JM, Clark KL, Perrone MH (1999) Cardiovascular pharmacology of the adenosine A_1/A_2-receptor agonist AMP579: coronary hemodynamic and cardioprotective effects in the canine myocardium. J Cardiovasc Pharmacol 33:703–710

Meng H, McVey M, Perrone MH, Clark KL (2000) Intravenous AMP 579, a novel adenosine A_1/A_{2a} receptor agonist, induces a delayed protection against myocardial infarction in minipig. Eur J Pharmacol 387:101–105

Merkel L, Rojas CJ, Jarvis MF, Cox BF, Fink C, Smits GJ, Spada AP, Perrone MH, Clark KL (1998) Pharmacological characterization of AMP 579, a novel adenosine A_1/A_2 receptor agonist and cardioprotective. Drug Dev Res 45:30–43

Morrison RR, Jones R, Byford AM, Stell AR, Peart J, Headrick JP, Matherne GP (2000) Transgenic overexpression of cardiac A_1 adenosine receptors mimics ischemic preconditioning. Am J Physiol Heart Circ Physiol 279:H1071–H1078

Morrison RR, Tan XL, Ledent C, Mustafa SJ, Hofmann PA (2007) Targeted deletion of A_{2A} adenosine receptors attenuates the protective effects of myocardial postconditioning. Am J Physiol Heart Circ Physiol 293:H2523–H2529

Murphy E (2004) Primary and secondary signaling pathways in early preconditioning that converge on the mitochondria to produce cardioprotection. Circ Res 94:7–16

Murry CE, Jennings RB, Reimer KA (1986) Preconditioning with ischemia: a delay of lethal cell injury in ischemic myocardium. Circulation 74:1124–1136

Nakata H, Yoshioka K, Kamiya T, Tsuga H, Oyanagi K (2005) Functions of heteromeric association between adenosine and P2Y receptors. J Mol Neurosci 26:233–238

Nayeem MA, Matherne GP, Mustafa SJ (2003) Ischemic and pharmacological preconditioning induces further delayed protection in transgenic mouse cardiac myocytes over-expressing adenosine A_1 receptors (A_1AR): role of A_1AR, iNOS and K_{ATP} channels. Naunyn–Schmiedeberg's Arch Pharmacol 367:219–226

Neely CR, DiPierro FV, Kong M, Greelish JP, Gardner TJ (1996) A_1 adenosine receptor antagonists block ischemia-reperfusion injury of the heart. Circulation 94(Suppl 2):II376–II380

Norton ED, Jackson EK, Turner MB, Virmani R, Forman MB (1992) The effects of intravenous infusions of selective adenosine A_1-receptor and A_2-receptor agonists on myocardial reperfusion injury. Am Heart J 123:332–338

O'Kane EM, Stone TW (1998) Interaction between adenosine A_1 and A_2 receptor-mediated responses in the rat hippocampus in vitro. Eur J Pharmacol 362:17–25

Olafsson B, Forman MB, Puett DW, Pou A, Cates CU, Friesinger GC, Virmani R (1987) Reduction of reperfusion injury in the canine preparation by intracoronary adenosine: importance of the endothelium and the no-reflow phenomenon. Circulation 76:1135–1145

Park JL, Lucchesi BR (1999) Mechanisms of myocardial reperfusion injury. Ann Thorac Surg 68:1905–1912

Park SS, Zhao H, Jang Y, Mueller RA, Xu Z (2006) N^6-(3-Iodobenzyl)-adenosine-5′-N-methylcarboxamide confers cardioprotection at reperfusion by inhibiting mitochondrial permeability transition pore opening via glycogen synthase kinase 3 beta. J Pharmacol Exp Ther 318:124–131

Peart J, Headrick JP (2000) Intrinsic A_1 adenosine receptor activation during ischemia or reperfusion improves recovery in mouse hearts. Am J Physiol Heart Circ Physiol 279:H2166–H2175

Peart J, Flood A, Linden J, Matherne GP, Headrick JP (2002) Adenosine-mediated cardioprotection in ischemic-reperfused mouse heart. J Cardiovasc Pharmacol 39:117–129

Peart J, Willems L, Headrick JP (2003) Receptor and non-receptor-dependent mechanisms of cardioprotection with adenosine. Am J Physiol Heart Circ Physiol 284:H519–H527

Peart JN, Gross ER, Headrick JP, Gross GJ (2007) Impaired p38 MAPK/HSP27 signaling underlies aging-related failure in opioid-mediated cardioprotection. J Mol Cell Cardiol 42:972–980

Philipp S, Yang XM, Cui L, Davis AM, Downey JM, Cohen MV (2006) Postconditioning protects rabbit hearts through a protein kinase C-adenosine A_{2b} receptor cascade. Cardiovasc Res 70:308–314

Pitarys CJ II, Virmani R, Vildibill HD Jr, Jackson EK, Forman MB (1991) Reduction of myocardial reperfusion injury by intravenous adenosine administered during the early reperfusion period. Circulation 83:237–247

Przyklenk K, Maynard M, Darling CE, Whittaker P (2008) Aging mouse hearts are refractory to infarct size reduction with post-conditioning. J Am Coll Cardiol 51:1393–1398

Regan SE, Broad M, Byford AM, Lankford AR, Cerniway RJ, Mayo MW, Matherne GP (2003) A_1 adenosine receptor overexpression attenuates ischemia-reperfusion-induced apoptosis and caspase 3 activity. Am J Physiol Heart Circ Physiol 284:H859–H866

Reichelt ME, Willems L, Peart JN, Ashton KJ, Matherne GP, Blackburn MR, Headrick JP (2007) Modulation of ischaemic contracture in mouse hearts: a 'supraphysiological' response to adenosine. Exp Physiol 92:175–185

Reid EA, Kristo G, Yoshimura Y, Ballard-Croft C, Keith BJ, Mentzer RM Jr, Lasley RD (2005) In vivo adenosine receptor preconditioning reduces myocardial infarct size via subcellular ERK signaling. Am J Physiol Heart Circ Physiol 288:H2253–H2259

Ross AM, Gibbons RJ, Stone GW, Kloner RA, Alexander RW; AMISTAD-II Investigators (2005) A randomized, double-blinded, placebo-controlled multicenter trial of adenosine as an adjunct to reperfusion in the treatment of acute myocardial infarction (AMISTAD-II). J Am Coll Cardiol 45:1775–1780

Schlack W, Schafer M, Uebing A, Schafer S, Borchard U, Thamer V (1993) Adenosine A_2-receptor activation at reperfusion reduces infarct size and improves myocardial wall function in dog heart. J Cardiovasc Pharmacol 22:89–96

Schulman D, Latchman DS, Yellon DM (2001) Effect of aging on the ability of preconditioning to protect rat hearts from ischemia-reperfusion injury. Am J Physiol Heart Circ Physiol 281:H1630–H1636

Schulte G, Sommerschild H, Yang J, Tokuno S, Goiny M, Lövdahl C, Johansson B, Fredholm BB, Valen G (2004) Adenosine A_1 receptors are necessary for protection of the murine heart by remote, delayed adaptation to ischaemia. Acta Physiol Scand 182:133–143

Schulz R, Rose J, Post H, Heusch G (1995) Involvement of endogenous adenosine in ischaemic preconditioning in swine. Pflugers Arch 430:273–282

Shryock, J, Song Y, Wang D, Baker SP, Olsson RA, Belardinelli L (1993) Selective A_2-adenosine receptor agonists do not alter action potential duration, twitch shortening, or cAMP accumulation in guinea pig, or rabbit isolated ventricular myocytes. Circ Res 72:194–205

Smits GJ, McVey M, Cox BF, Perrone MH, Clark KL (1998) Cardioprotective effects of the novel A_1/A_2 receptor agonist AMP 579 in a porcine model of myocardial infarction. J Pharmacol Exp Ther 286:611–618

Söderbäck U, Sollevi A, Wallen NH, Larsson PT, Hjemdahl P (1991) Anti-aggregatory effects of physiological concentrations of adenosine in human whole blood as assessed by filtragometry. Clin Sci 81:691–694

Solenkova NV, Solodushko V, Cohen MV, Downey JM (2006) Endogenous adenosine protects preconditioned heart during early minutes of reperfusion by activating Akt. Am J Physiol Heart Circ Physiol 290:H441–H449

Stambaugh K, Jacobson KA, Jiang JL, Liang BT (1997) A novel cardioprotective function of adenosine A_1 and A_3 receptors during prolonged simulated ischemia. Am J Physiol 273: H501–H505

Stein B, Schmitz W, Scholz H, Seeland C (1994) Pharmacological characterization of A_2-adenosine receptors in guinea-pig ventricular cardiomyocytes. J Mol Cell Cardiol 26:403–414

Strickler J, Jacobson KA, Liang BT (1996) Direct preconditioning of cultured chick ventricular myocytes. Novel functions of cardiac adenosine A_{2a} and A_3 receptors. J Clin Invest 98: 1773–1779

Sullivan GW, Rieger JM, Scheld WM, Macdonald TL, Linden J (2001) Cyclic AMP–dependent inhibition of human neutrophil oxidative activity by substituted 2-propynylcyclohexyl adenosine A_{2A} receptor agonists. Br J Pharmacol 132:1017–1026

Tang Z, Diamond MA, Chen JM, Holly TA, Bonow RO, Dasgupta A, Hyslop T, Purzycki A, Wagner J, McNamara DM, Kukulski T, Wos S, Velazquez EJ, Ardlie K, Feldman AM (2007) Polymorphisms in adenosine receptor genes are associated with infarct size in patients with ischemic cardiomyopathy. Clin Pharmacol Ther 82:435–440

Tani M, Suganuma Y, Takayama M, Hasegawa H, Shinmura K, Ebihara Y, Tamaki K (1998) Low concentrations of adenosine receptor blocker decrease protection by hypoxic preconditioning in ischemic rat hearts. J Mol Cell Cardiol 30:617–626

Thornton JD, Liu GS, Olsson RA, Downey JM (1992) Intravenous pre-treatment with A_1-selective adenosine analogues protects the heart against infarction. Circulation 85:659–665

Thornton JD, Thornton CS, Downey JM (1993) Effect of adenosine receptor blockade: preventing protective preconditioning depends on time of initiation. Am J Physiol 265:H504–H508

Tomai F, Crea F, Gaspardone A, Versaci F, De Paulis R, Polisca P, Chiariello L, Gioffrè PA (1996) Effects of A_1 adenosine receptor blockade by bamiphylline on ischaemic preconditioning during coronary angioplasty. Eur Heart J 17:846–853

Toombs CF, McGee S, Johnston WE, Vinten-Johansen J (1992) Myocardial protective effects of adenosine. Infarct size reduction with pretreatment and continued receptor stimulation during ischemia. Circulation 86:986–894

Toufektsian MC, Yang Z, Prasad KM, Overbergh L, Ramos SI, Mathieu C, Linden J, French BA (2006) Stimulation of A_{2A}-adenosine receptors after myocardial infarction suppresses inflammatory activation and attenuates contractile dysfunction in the remote left ventricle. Am J Physiol Heart Circ Physiol 290:H1410–H1418

Tracey WR, Magee W, Masamune H, Kennedy SP, Knight DR, Buchholz RA, Hill RJ (1997) Selective adenosine A_3 receptor stimulation reduces ischemic myocardial injury in the rabbit heart. Cardiovasc Res 33:410–415

Tsuchida A, Miura T, Miki T, Shimamoto K, Iimura O (1992) Role of adenosine receptor activation in myocardial infarct size limitation by ischaemic preconditioning. Cardiovasc Res 26:456–461

Tsuchida A, Liu GS, Wilborn WH, Downey JM (1993) Pretreatment with the adenosine A_1 selective agonist, 2-chloro-N^6-cyclopentyladenosine (CCPA), causes a sustained limitation of infarct size in rabbits. Cardiovasc Res 27:652–656

Urabe K, Miura T, Iwamoto T, Ogawa T, Goto M, Sakamoto J, Iimura O (1993) Preconditioning enhances myocardial resistance to postischaemic myocardial stunning via adenosine receptor activation. Cardiovasc Res 27:657–662

Vahlhaus C, Schulz R, Post H, Rose J, Heusch G (1998) Prevention of ischemic preconditioning only by combined inhibition of protein kinase C and protein tyrosine kinase in pigs. J Mol Cell Cardiol 30:197–209

Van Wylen DG (1994). Effect of ischemic preconditioning on interstitial purine metabolite and lactate accumulation during myocardial ischemia. Circulation 89:2283–2289

Vander Heide RS, Reimer KA (1996) Effect of adenosine therapy at reperfusion on myocardial infarct size in dogs. Cardiovasc Res 31:711–718

Verma S, Fedak PW, Weisel RD, Butany J, Rao V, Maitland A, Li RK, Dhillon B, Yau TM (2002) Fundamentals of reperfusion injury for the clinical cardiologist. Circulation 105:2332–2336

Visser SS, Theron AJ, Ramafi G, Ker JA, Anderson R (2000) Apparent involvement of the A_{2A} subtype adenosine receptor in the anti-inflammatory interactions of CGS 21680, cyclopentyladenosine, and IB-MECA with human neutrophils. Biochem Pharmacol 60:993–999

Vogt AM, Htun P, Arras M, Podzuweit T, Schaper W (1996) Intramyocardial infusion of tool drugs for the study of molecular mechanisms in ischemic preconditioning. Basic Res Cardiol 91:389–400

Walker DM, Walker JM, Pugsley WB, Pattison CW, Yellon DM (1995) Preconditioning in isolated superfused human muscle. J Mol Cell Cardiol 27:1349–1357

Wan TC, Ge ZD, Tampo A, Mio Y, Bienengraeber MW, Tracey WR, Gross GJ, Kwok WM, Auchampach JA (2008) The A_3 adenosine receptor agonist CP-532,903 (N^6-(2,5-dichlorobenzyl)-3′-aminoadenosine-5′-N-methylcarboxamide) protects against myocardial ischemia/reperfusion injury via the sarcolemmal ATP-sensitive potassium channel. J Pharmacol Exp Ther 324:234–243

Wang J, Drake L, Sajjadi F, Firestein GS, Mullane KM, Bullough DA (1997) Dual activation of adenosine A_1 and A_3 receptors mediates preconditioning of isolated cardiac myocytes. Eur J Pharmacol 320:241–248

Weinbrenner C, Liu GS, Cohen MV, Downey JM (1997) Phosphorylation of tyrosine 182 of p38 mitogen-activated protein kinase correlates with the protection of preconditioning in the rabbit heart. J Mol Cell Cardiol 29:2383–2391

Willems L, Ashton KJ, Headrick JP (2005) Adenosine-mediated cardioprotection in the aging myocardium. Cardiovasc Res 66:245–255

Woodiwiss AJ, Honeyman TW, Fenton RA, Dobson JG Jr (1999) Adenosine A_{2a}-receptor activation enhances cardiomyocyte shortening via Ca^{2+}-independent and -dependent mechanisms. Am J Physiol 276:H1434–H1441

Xu H, Stein B, Liang B (1996) Characterization of a stimulatory adenosine A_{2a} receptor in adult rat ventricular myocyte. Am J Physiol 270:H1655–H1661

Xu Z, Downey JM, Cohen MV (2001) AMP 579 reduces contracture and limits infarction in rabbit heart by activating adenosine A_2 receptors. J Cardiovasc Pharmacol 38:474–481

Xu Z, Park SS, Mueller RA, Bagnell RC, Patterson C, Boysen PG (2005) Adenosine produces nitric oxide and prevents mitochondrial oxidant damage in rat cardiomyocytes. Cardiovasc Res 65:803–812

Yang Z, Cerniway RJ, Byford AM, Berr SS, French BA, Matherne GP (2002) Cardiac overexpression of A_1-adenosine receptor protects intact mice against myocardial infarction. Am J Physiol Heart Circ Physiol 282:H949–H955

Yang Z, Day YJ, Toufektsian MC, Ramos SI, Marshall M, Wang XQ, French BA, Linden J (2005) Infarct-sparing effect of A_{2A}-adenosine receptor activation is due primarily to its action on lymphocytes. Circulation 111:2190–2197

Yang Z, Day YJ, Toufektsian MC, Xu Y, Ramos SI, Marshall MA, French BA, Linden J (2006). Myocardial infarct-sparing effect of adenosine A_{2A} receptor activation is due to its action on $CD4^{+}T$ lymphocytes. Circulation 114:2056–2064

Zhao Z-Q, McGee S, Nakanishi K, Toombs CF, Johnston WE, Ashar MS, Vinten-Johansen J (1993) Receptor-mediated cardioprotective effects of endogenous adenosine are exerted primarily during reperfusion after coronary occlusion in the rabbit. Circulation 88:709–719

Zhao Z-Q, Sato H, Williams MW, Fernandez AZ, Vinten-Johansen J (1996) Adenosine A_2-receptor activation inhibits neutrophil-mediated injury to coronary endothelium. Am J Physiol (Heart Circ Physiol) 271:H1456–H1464

Zucchi R, Cerniway RJ, Ronca-Testoni S, Morrison RR, Ronca G, Matherne GP (2002) Effect of cardiac A_1 adenosine receptor overexpression on sarcoplasmic reticulum Cardiovasc Res 53:326–333

Adenosine Receptors and Inflammation

Michael R. Blackburn, Constance O. Vance, Eva Morschl, and Constance N. Wilson

Contents

Abstract Extracellular adenosine is produced in a coordinated manner from cells following cellular challenge or tissue injury. Once produced, it serves as an autocrine- and paracrine-signaling molecule through its interactions with seven-membrane-spanning G-protein-coupled adenosine receptors. These signaling pathways have widespread physiological and pathophysiological functions. Immune

M.R. Blackburn (✉)
Department of Biochemistry and Molecular Biology,
The University of Texas–Houston Medical School, 6431 Fannin, Houston, TX 77030, USA
michael.r.blackburn@uth.tmc.edu

C.N. Wilson and S.J. Mustafa (eds.), *Adenosine Receptors in Health and Disease*,
Handbook of Experimental Pharmacology 193, DOI 10.1007/978-3-540-89615-9_8,

cells express adenosine receptors and respond to adenosine or adenosine agonists in diverse manners. Extensive in vitro and in vivo studies have identified potent anti-inflammatory functions for all of the adenosine receptors on many different inflammatory cells and in various inflammatory disease processes. In addition, specific proinflammatory functions have also been ascribed to adenosine receptor activation. The potent effects of adenosine signaling on the regulation of inflammation suggest that targeting specific adenosine receptor activation or inactivation using selective agonists and antagonists could have important therapeutic implications in numerous diseases. This review is designed to summarize the current status of adenosine receptor signaling in various inflammatory cells and in models of inflammation, with an emphasis on the advancement of adenosine-based therapeutics to treat inflammatory disorders.

Keywords Adenosine · Adenosine Receptor · G-Protein Coupled Receptor · Inflammation

Abbreviations

A_1AR	Adenosine A_1 receptor
$A_{2A}AR$	Adenosine A_{2A} receptor
$A_{2B}AR$	Adenosine A_{2B} receptor
A_3AR	Adenosine A_3 receptor
AC	Adenylate cyclase
ADA	Adenosine deaminase
AR	Adenosine receptor
CD26	Dipeptidyl pepsidase
CD39	Ectonucleoside triphosphate diphosphohydrolase
CD73	Ecto-5′-nucleotidase
CLP	Cecal ligation and puncture
CNTs	Concentrative nucleoside transporters
COPD	Chronic obstructive pulmonary disease
DC	Dendritic cell
ERK	Extracellular signal-related protein kinase
fMLP	Formyl methionyl-leucyl-phenylalanine
GRK	G-protein-coupled receptor kinase
HIF	Hypoxia-inducible factor
HMVECs	Human microvascular endothelial cells
HPAECs	Human pulmonary artery endothelial cells
HSP	Heat shock protein
HUVEC	Human umbilical vein endothelial cell
ICAM-1	Intracellular adhesion molecule-1
IFN	Interferon
IL	Interleukin

imDC	Immature dendritic cell
KO	Knockout
LPS	Lipopolysaccharide
MAPK	Mitogen-activated protein kinase
MHC	Major histocompatibility complex
MIP	Macrophage inflammatory protein
NF-κB	Nuclear factor kappa B
NK	Natural killer
NR	Normothermic recirculation
PAF	Platelet-activating factor
PBMCs	Peripheral blood mononuclear cells
PDCs	Plasmacytoid dendritic cells
PI3K	Phosphoinositide 3-kinase
PKC	Protein kinase C
PLC	Phospholipase C
PTX	Pertussis toxin
TCR	T-cell receptor
TNF-α	Tumor necrosis factor alpha
Tregs	Regulatory T cells
VCAM-1	Vascular cell adhesion molecule-1
VEGF	Vascular endothelial growth factor

1 Introduction

Adenosine is an endogenous signaling molecule that engages cell surface adenosine receptors to regulate numerous physiological and pathological processes (Fredholm et al. 2001). Extracellular adenosine is produced in excess in response to cellular stress, largely from the breakdown of released adenine nucleotides. Substantial evidence demonstrates that adenosine is an important signaling molecule and adenosine receptors are important molecular targets in the pathophysiology of inflammation. All inflammatory cells express adenosine receptors, and research into the consequences of adenosine receptor activation has identified numerous avenues for adenosine-based therapeutic intervention. Indeed, adenosine-based approaches are currently being developed for the treatment of various disorders where inflammatory modulation is a key component (reviewed in Jacobson and Gao 2006). This chapter was designed to review the contribution of adenosine and adenosine receptors to the regulation of key inflammatory and immune responses.

1.1 Adenosine Production and Metabolism

Regulation of extracellular adenosine levels is orchestrated by the actions of proteins that regulate adenosine production, metabolism and transport across the plasma membrane. The release and catabolism of adenine nucleotides to adenosine is believed to be the major route of adenosine production following cellular stress or injury. Possible routes of ATP release include the constitutive release of ATP through vesicle fusion with the plasma membrane, and programmed release through membrane channels such as the ATP binding cassette family of membrane transporters, including the cystic fibrosis transmembrane conductance regulator (Reisin et al. 1994) and multiple drug resistance channels (Roman et al. 2001), connexin hemichannels (Cotrina et al. 1998), maxi-ion channels (Bell et al. 2003), stretch-activated channels (Braunstein et al. 2001) and voltage-dependent anion channels (Okada et al. 2004). A number of different cell types are sources of adenine nucleotides, including platelets, neurons, and endothelial cells. In addition, inflammatory cells such as mast cells (Marquardt et al. 1984), neutrophils (Madara et al. 1993) and eosinophils (Resnick et al. 1993) are able to release adenine nucleotides and adenosine into the local environment.

Extracellular ATP is rapidly dephosphorylated by ectonucleoside triphosphate diphosphohydrolases such as CD39 to form ADP and AMP (Kaczmarek et al. 1996), and extracellular AMP is dephosphorylated to adenosine by the 5′-nucleotidase CD73 (Resta et al. 1998). CD39 and CD73 are widely expressed on the surface of cells and are essential for the production of adenosine following cellular stress or injury (Thompson et al. 2004; Volmer et al. 2006). Recent findings demonstrate that CD73 and CD39 are novel markers on regulatory T cells (Tregs), where they serve to convert extracellular adenine nucleotides to adenosine, which in turn promotes immunosuppressive activities (Kobie et al. 2006; Deaglio et al. 2007). This process is an example of the concerted role of extracellular adenosine production and signaling in the regulation of inflammatory processes. Adenosine is also generated inside cells by either the dephosphorylation of AMP by cytosolic nucleotidases (Sala-Newby et al. 1999) or the hydrolysis of *S*-adenosylhomocysteine (Hermes et al. 2005). Alterations in cellular metabolic load or methylation reactions that utilize *S*-adenosylmethonine as a methyl donor can lead to increased intracellular adenosine levels and subsequent release.

Adenosine is transported across the plasma membrane by both facilitated and cotransport mechanisms. The facilitated nucleoside transporters, known as the equilibrative nucleoside transporters, are bidirectional transporters (Baldwin et al. 2004). They are widely distributed in mammalian tissues, and play a major role in transporting adenosine in and out of the cell. Adenosine transported across the cell membrane also occurs through concentrative nucleoside transporters (CNTs), which are Na^+-dependent concentrative transporters (Gray et al. 2004). The tissue distributions of the CNTs vary, with CNT1 localized primarily to epithelial cells, while CNT2 and CNT3 are more widely distributed.

Finally, adenosine is metabolized by one of two pathways. It can be phosphorylated to form AMP intracellularly by the enzyme adenosine kinase (Spychala et al. 1996), or it can be deaminated to inosine by adenosine deaminase (ADA) (Blackburn and Kellems 1996). ADA is a predominantly cytosolic enzyme. However, it is also found outside the cell as a component of plasma. In humans,

ADA can complex with the cell surface protein CD26 (Hashikawa et al. 2004). This interaction may play an important role in localizing adenosine metabolism to certain regions of the cell surface to impact adenosine signaling. These enzymes, together with rapid cellular uptake, serve to regulate the levels of intra- and extracellular adenosine. In homeostatic situations, adenosine levels range from 10 to 200 nM, whereas extracellular adenosine levels can be elevated to 10–100 μM in hypoxic or stressed tissue environments (Fredholm 2007). The concerted production and metabolism of adenosine is an important mechanism that contributes to the ability of this signaling molecule to regulate aspects of immunobiology and tissue homeostasis.

1.2 Adenosine Receptors

Adenosine exerts its effects by interacting with receptors located on the cell surface. Four adenosine receptor subtypes, A_1, A_{2A}, A_{2B} and A_3, have been defined by pharmacological and molecular biological approaches (Fredholm et al. 2001). These receptors belong to the superfamily of G-protein-coupled receptors and are characterized by seven-transmembrane-spanning α-helical domains with an extracellular amine terminus and a cytoplasmic carboxy terminus. Receptor subtypes are distinguished based on their affinity for adenosine, pharmacological profiles, G-protein coupling and signaling pathways, and genetic sequence. The physiological effects of adenosine are mediated by intracellular signaling processes that are specific to the receptor subtype and the type of cell. The adenosine A_1 receptor (A_1AR) is coupled to the pertussis toxin (PTX)-sensitive inhibitory G proteins (G_i) or G_o. Activation of the A_1AR can lead to the activation a number of effector systems, including adenylate cyclase (AC), phospholipase A_2, phospholipase C (PLC), potassium channels, calcium channels, and guanylate cyclase (Akbar et al. 1994; Olah and Stiles 2000; Fredholm et al. 2001). The primary changes in second messengers associated with A_1AR activation are decreased production of cAMP or increased Ca^{2+}, depending on the effector system. Like the A_1AR, the adenosine A_3 receptor (A_3AR) is coupled to the PTX-sensitive G_i protein and also to G_q (Fredholm et al. 2001). Activation of the A_3AR results in an inhibition of AC (leading to decreased cAMP) or stimulation of PLC and phospholipase D (Gessi et al. 2008). The adenosine A_{2A} receptor (A_{2A}AR) and adenosine A_{2B} receptor (A_{2B}AR) share a relatively high homology and are coupled to Gs (Fredholm et al. 2001), leading to increased levels of cAMP. In addition, the A_{2B}AR has been shown to couple to G_q (Feoktistov et al. 2002), thereby regulating intracellular Ca^{2+} levels. In general, the A_1AR, A_{2A}AR and A_3AR subtypes have high affinity for adenosine, while the A_{2B}AR has a lower affinity (Fredholm 2007).

2 Adenosine Receptors on Immune Cells

2.1 Neutrophils

Neutrophils are the most abundant leukocyte and represent the body's first line of defense in response to a pathogenic challenge; they are the predominant leukocyte involved in acute inflammation (Burg and Pillinger 2001; Edwards 1994; Witko-Sarsat et al. 2000). All four adenosine receptor subtypes are expressed on neutrophils (Bours et al. 2006; Marone et al. 1992; Fortin et al. 2006; Fredholm 2007). At submicromolar adenosine concentrations, A_1AR activation on human neutrophils produces a proinflammatory response by promoting chemotaxis and adherence to the endothelium (Bours et al. 2006; Cronstein et al. 1990, 1992; Forman et al. 2000; Rose et al. 1988). A_1AR-mediated chemotaxis in neutrophils is disrupted by PTX, an agent that inhibits the function of G_i-linked receptors, and requires an intact microtubule system (Cronstein et al. 1990, 1992).

Activation of A_{2A}AR and A_{2B}ARs on neutrophils is anti-inflammatory. High concentrations (micromolar) of adenosine inhibit neutrophil adhesion to endothelial cells by activating A_{2A}AR and A_{2B}ARs on neutrophils (Bours et al. 2006; Eltzschig et al. 2004; Sullivan et al. 2004; Thiel et al. 1996; Wakai et al. 2001). In human neutrophils, A_{2A}AR activation inhibits the formation of reactive oxygen species (Cronstein et al. 1983, 1990; Salmon and Cronstein 1990). In addition, A_{2A}AR activation inhibits the adherence of *N*-formyl methionyl-leucyl-phenylalanine (fMLP)-activated neutrophils to endothelium (Cronstein et al. 1992) and downregulates Mac-1 (Wollner et al. 1993), β_2-integrin (Thiel et al. 1996; Zalavary and Bengtsson 1998), and L-selectin (Thiel et al. 1996). Activation of the A_{2A}AR also downregulates the activity of other endothelial cell surface proteins, including vascular cell adhesion molecule-1 (VCAM-1), intracellular adhesion molecule-1 (ICAM-1) (McPherson et al. 2001), alpha 4/beta 1 integrin VLA4 (Sullivan et al. 2004), and platelet cell adhesion molecule (Cassada et al. 2002). Activation of A_{2A}ARs on activated human neutrophils produces an anti-inflammatory effect by decreasing the formation of the proinflammatory cytokine tumor necrosis factor alpha (TNF-α) (Harada et al. 2000, Thiel and Chouker 1995), chemokines such as macrophage inflammatory protein (MIP)-1α/CCL3, MIP-1β/CCL4, MIP-2α/CXCL2, and MIP-3α/CCL20 (McColl et al. 2006), and leukotriene LTB4 (Flamand et al. 2000, 2002; Grenier et al. 2003; Krump et al. 1996, 1997; Krump and Borgeat 1999; Surette et al. 1999), and platelet activating factor (PAF) (Flamand et al. 2006). Other important immunoregulatory effects mediated by the A_{2A}AR include the inhibition of Fc gamma (Fcγ) receptor-mediated neutrophil phagocytosis and inhibition of degranulation (Bours et al. 2006; Cronstein et al. 1983; Harada et al. 2000; Salmon and Cronstein 1990; Sullivan et al. 1999; Visser et al. 2000; Zalavary et al. 1994, Zalavary and Bengtsson 1998). Activation of the A_{2B}AR inhibits neutrophil extravasation across human umbilical vein endothelial cell (HUVEC) monolayers and inhibits the release of vascular endothelial growth factor (VEGF) (Wakai et al. 2001).

Conflicting reports suggest that activation of A_3ARs on neutrophils may produce proinflammatory or anti-inflammatory effects. Studies with A_3AR knockout mice suggest that the A_3AR promotes recruitment of neutrophils to lungs during sepsis (Inoue et al. 2008). Moreover, A_3ARs play an important role in the migration of human neutrophils in response to chemoattractant molecules released by microbes (Chen et al. 2006). In isolated human neutrophils, extracellular adenosine (1–1,000 nM) induces a redistribution of A_3ARs to the neutrophil's leading edge, the portion of the membrane closest to the chemoattractant stimulus (Chen et al. 2006). In addition, selective A_3AR antagonists inhibit fMLP-mediated chemotaxis in human neutrophils (Chen et al. 2006). In other studies, activation of A_3ARs on human neutrophils has been shown to counteract inflammation by inhibiting degranulation and oxidative burst (Bouma et al. 1997; Fishman and Bar-Yehuda 2003; Gessi et al. 2002).

2.2 Monocytes and Macrophages

Monocytes and macrophages are a heterogenous group of mononuclear cells that present an early line of innate immune defense. They represent a primary source of inflammatory modulators and are highly adaptable with a phenotype that can change rapidly in response to the local environment of the inflamed tissue (Hasko et al. 2007; Rutherford et al. 1993). Macrophages also serve an important role in terminating the inflammatory process, which is critical for preventing excessive tissue injury (Duffield 2003; Gilroy et al. 2004; Hasko et al. 2007; Wells et al. 2005; Willoughby et al. 2000). All four adenosine receptors are expressed on monocytes and macrophages, although expression levels differ markedly throughout the maturation and differentiation process (Eppell et al. 1989; Thiele et al. 2004). In quiescent monocytes, adenosine receptor expression is low and is increased following activation by inflammatory stimuli. It is hypothesized that the temporal changes in the expression of adenosine receptor subtypes play an important role in the resolution of inflammation. In human monocytes, A_1AR activation produces a proinflammatory effect whereas A_{2A}AR activation produces an anti-inflammatory effect. A key function of the A_1AR is a rapid enhancement of the activity of the Fcγ receptor (Salmon et al. 1993). Activation of A_{2A}ARs limits inflammatory reactions by inhibiting phagocytosis in monocytes (Salmon et al. 1993) and macrophages (Eppell et al. 1989), decreasing the production of reactive oxygen species (Thiele et al. 2004), and altering cytokine release. In addition, A_3AR activation inhibits fMLP-triggered respiratory burst in human monocytes (Broussas et al. 1999).

Monocytes and macrophages are a primary source of TNF-α, a proinflammatory cytokine involved in the pathophysiology of a number of chronic inflammatory diseases. Early studies suggested that activation of the A_{2A}AR suppresses production of TNF-α in human monocytes activated by bacterial lipopolysaccharide (LPS) (Le Vraux et al. 1993). In primary cultures of human monocytes activated by LPS (Zhang et al. 2005) and LPS-stimulated mouse macrophages (Ezeamuzie and Khan

2007), activation of the $A_{2A}AR$ attenuated the release of TNF-α, whereas activation of the A_1AR and A_3AR subtypes had no effect on the formation of TNF-α (Zhang et al. 2005). Similar results were obtained in studies with primary cultures of mouse peritoneal macrophages, in which activation of the $A_{2A}AR$ inhibited LPS-induced TNF-α release, while activation of the A_3AR had no effect (Kreckler et al. 2006). In other studies, activation of the A_3AR was shown to inhibit LPS-induced TNF-α release in vitro in the RAW 264.7 murine leukemia macrophage line (Haskó et al. 1996; Martin et al. 2006), U937 human leukemic macrophage cell line (Sajjadi et al. 1996), murine J774.1 macrophages (Bowlin et al. 1997; McWhinney et al. 1996) and in vivo in endotoxemic mice (Hasko et al. 1996). In the RAW 264.7 macrophage line, the inhibitory effects of A_3ARs were mediated by a mechanism involving Ca^{2+}-dependent activation of nuclear factor-kappa B (NF-κB) (Martin et al. 2006).

Interleukin (IL)-12. IL-12 is a proinflammatory cytokine that is produced in response to certain bacterial and parasitic infections. IL-12 activates naïve T lymphocytes to mount a T helper 1 response. The production of IL-12 is modulated by adenosine and ARs (Hasko et al. 1998, 2000, 2007; Le Vraux et al. 1993; Link et al. 2000). Pharmacological studies (Hasko et al. 2000; Le Vraux et al. 1993; Link et al. 2000) and studies with $A_{2A}AR$ knockout (KO) mice (Hasko et al. 2000) have demonstrated that $A_{2A}AR$ activation downregulates IL-12 production, thereby producing an anti-inflammatory response. In human peripheral monocytes, $A_{2A}AR$ activation decreases IL-12 and IL-12p40 (Link et al. 2000). The effects of the $A_{2A}AR$ on IL-12 production are strongly influenced by the presence of proinflammatory cytokines (Khoa et al. 2001). In THP-1 monocytic cells, TNF-α and IL-1 enhanced $A_{2A}AR$-mediated inhibition of IL-12 production, whereas interferon (IFN)-γ attenuated $A_{2A}AR$-mediated inhibition of IL-12 production (Khoa et al. 2001). The effects of TNF-α and IL-1 were associated with an upregulation of $A_{2A}ARs$, while IFN-γ effects were associated with downregulation of $A_{2A}ARs$.

Activation of the A_3AR negatively regulates the synthesis of IL-12 in murine RAW 264.7 macrophages (Szabo et al. 1998), human monocytes (la Sala et al. 2005), and mice treated with LPS (Hasko et al. 1998). The A_3AR-mediated effects appear to be mediated through the phosphatidyl inositol 3-kinase signaling pathway (la Sala et al. 2005). Taken together, these studies suggest an anti-inflammatory role for the A_3AR via negative regulation of IL-12.

IL-10. IL-10 is an anti-inflammatory cytokine (Kotenko 2002; Moore et al. 1993, Mosmann 1994; Hasko et al. 2007) that functions by inhibiting the secretion of proinflammatory cytokines, including TNF-α and IL-12 (Moore et al. 2001). IL-10 is produced by T helper 2 cells, monocytes, and macrophages (Moore et al. 2001). Following the induction of proinflammatory cytokines, IL-10 regulates the termination of inflammatory processes. Both the $A_{2A}AR$ and $A_{2B}AR$ subtypes have been implicated in the stimulation of IL-10 production in monocytes and macrophages (Haskó et al. 1996, 2000, 2007; Khoa et al. 2001; Link et al. 2000; Nemeth et al. 2005).

Other cytokines, chemokines, and adhesion molecules. Treatment of peripheral blood mononuclear cells (PBMCs) with IL-18, a proinflammatory cytokine released by T cells and dendritic cells, results in increased TNF-α, IL-12, IFN-γ release, and increased expression of ICAM-1 (Takahashi et al. 2003). In PBMCs, adenosine inhibited the IL-18-induced release of TNF-α, IL-12, and IFN-γ, and expression of ICAM-1. This inhibitory effect was mimicked by an A_{2A}AR agonist and blocked by A_{2A}AR antagonism (Takahashi et al. 2007a). Moreover, the A_{2A}AR-mediated anti-inflammatory effects on the IL-18-induced production of TNF-α, IL-12, IFN-γ, and ICAM-1 were reversed by an A_1AR agonist and an A_3AR agonist. The results of these studies suggest that the anti-inflammatory effect of adenosine on human PBMCs activated by IL-18 occurs by activation of the A_{2A}AR; however, an A_1AR proinflammatory effect predominates when the A_{2A}AR is saturated with agonist. Thus, the net effect of adenosine on PBMCs activated by IL-18 is a function of the activation of multiple adenosine receptor subtypes, including an anti-inflammatory effect via A_{2A}ARs and proinflammatory effects via A_1AR and A_3ARs (Takahashi et al. 2007a).

With respect to activation of A_{2B}AR and A_3ARs on monocytes and macrophages, in both in vivo and in vitro studies, activation of the A_{2B}AR induces the release of the proinflammatory cytokine IL-6 from macrophages (Ryzhov et al. 2008a), and activation of the A_3AR inhibits the production of MIP-α in LPS-stimulated RAW 264.7 macrophages (Szabo et al. 1998) and inhibits tissue factor expression in LPS-stimulated human macrophages (Broussas et al. 2002). In human monocytes, A_1AR activation induces the release of VEGF (Clark et al. 2007).

2.3 Dendritic Cells

Dendritic cells (DCs) are highly specialized antigen-presenting cells that play an important role in the initiation and regulation of immune responses by migrating to sites of injury and infection, processing antigens, and activating naive T cells (Banchereau and Steinman 1998; Macagno et al. 2007). Immature DCs (imDCs) undergo a maturation process following exposure to proinflammatory signals, including pathogens, LPS, TNF-α, IL-1, and IL-6 (Banchereau and Steinman 1998). The maturation process results in decreased phagocytic activity and increased expression of membrane major histocompatibility complex (MHC), CD54, CD80, CD83, and CD86. Mature DCs release a number of cytokines, including TNF-α, IL-12 and IL-10. IL-12 is a major contributor to the differentiation of Th1 cells. In human blood, DCs are classified as the CD1c$^+$ DCs and the CD123$^+$ DCs (Shortman and Liu 2002). CD123$^+$ DCs, also known as plasmacytoid DCs (PDCs), are located in blood and secrete IFN-γ (Siegal et al. 1999). In addition, PDCs are powerful regulators of T-cell responses (Gilliet and Liu 2002; Kadowaki et al. 2000).

Adenosine receptors are differentially expressed on human DCs (Fossetta et al. 2003; Hofer et al. 2003; Panther et al. 2001, 2003; Schnurr et al. 2004). Immature, undifferentiated human DCs express mRNAs for the A_1AR, A_{2A}AR and A_3AR but not for the A_{2B}AR (Fossetta et al. 2003; Hofer et al. 2003; Panther et al. 2001; Schnurr et al. 2004). Activation of the A_1AR and A_3AR subtypes in undifferentiated DCs induces chemotaxis and mobilization of intracellular Ca^{2+}, while activation of the $A_{2A}AR$ subtype has no effect (Panther et al. 2001; Fossetta et al. 2003). Activation of the $A_{2A}AR$, but not A_1AR and A_3ARs, in imDCs is linked to increased cell surface expression of CD80, CD86, human leukocyte antigen-DR, and MHC-I (Panther et al. 2003). Activation of A_1ARs in resting DCs suppresses vesicular MHC class I cross-presentation by a G_i-mediated pathway (Chen et al. 2008).

Following treatment with LPS to induce differentiation and maturation, human DCs primarily express the $A_{2A}AR$ (Fossetta et al. 2003; Panther et al. 2001). Activation of the $A_{2A}AR$ increases AC activity and inhibits production of the proinflammatory cytokine IL-12, thereby reducing the ability of the DC to promote the differentiation of T cells to the Th-1 phenotype, and stimulates the production of the anti-inflammatory cytokine IL-10 (Banchereau and Steinman 1998; Panther et al. 2001, 2003).

In immature PDCs, adenosine acting via the A_1AR promotes the migration of PDCs to the site of infection. As PDCs differentiate and mature, the expression of the A_1AR is downregulated, corresponding to a decrease in migratory capability. In mature PDCs, the $A_{2A}AR$ is the predominant subtype and $A_{2A}AR$ activation decreases the production of IL-6, IL-12 and IFN-α (Schnurr et al. 2004). Moreover, IL-3-induced maturation of human PDCs results in a downregulation of A_1ARs and an upregulation of $A_{2A}ARs$ (Schnurr et al. 2004). The mouse DC line XS-106 expresses functional adenosine $A_{2A}AR$ and A_3ARs (Dickenson et al. 2003). $A_{2A}AR$ activation increases cAMP levels and p42/p44 mitogen-activated protein kinase (MAPK) phosphorylation, whereas activation of the A_3AR inhibits cAMP accumulation and increases in p42/p44 MAPK phosphorylation. Functionally, the activation of both subtypes produces a partial inhibition of LPS-induced release of TNF-α.

2.4 Lymphocytes

Lymphocytes are critically involved in adaptive immunity (Alam and Gorska 2003; Larosa and Orange 2008). Adenosine regulates multiple physiologic processes and inflammatory actions on lymphocytes (Bours et al. 2006; Marone et al. 1986, 1992; Priebe et al. 1988, 1990a, b, c). In early studies, it was demonstrated in mixed human lymphocytes that R-PIA (N^6-*R*-phenylisopropyladenosine) and low concentrations of adenosine (1–100 nM) inhibit cAMP accumulation in human lymphocytes via an A_1AR mechanism, while high concentrations of adenosine (100 nM–100 μM) stimulate cAMP via an $A_{2A}AR$ mechanism (Marone et al. 1986, 1992).

$CD4^+$ and $CD8^+$ T lymphocytes express $A_{2A}AR$, $A_{2B}AR$, and A_3ARs (Gessi et al. 2004, 2005; Huang et al. 1997; Hoskin et al. 2008; Koshiba et al. 1997, 1999; Mirabet et al. 1997). In activated human $CD4^+$ and $CD8^+$ T lymphocytes, $A_{2B}AR$ expression is increased and $A_{2B}AR$ activation is linked to decreased IL-2 production (Mirabet et al. 1999). Activation of human $CD4^+$ T lymphocytes with phytohemagglutinin results in increases in A_3AR mRNA and protein levels that are accompanied by increased agonist potency (Gessi et al. 2004).

A number of studies suggest that $A_{2A}AR$ engagement on $CD4^+$ T lymphocytes results in anti-inflammatory effects. In mouse $CD4^+$ T lymphocytes, $A_{2A}AR$ engagement inhibits T-cell receptor (TCR)-mediated production of IFN-γ (Lappas et al. 2005). TCR activation results in $A_{2A}AR$ mRNA upregulation, which functions as an anti-inflammatory mechanism for limiting T-cell activation and subsequent macrophage activation in inflamed tissues (Lappas et al. 2005). In vitro and in vivo studies suggest that the $A_{2A}ARs$ selectively inhibit TCR-activated T cells, thereby inhibiting lymphocyte inflammatory activity (Apasov et al. 1995, 2000; Erdmann et al. 2005; Huang et al. 1997; Lappas et al. 2005). Activation of the $A_{2A}AR$ on $CD4^+$ T lymphocytes prevents myocardial ischemia-reperfusion injury by inhibiting the accumulation and activation of $CD4^+$ T cells in the reperfused heart (Yang et al. 2006b). Moreover, an anti-inflammatory role in chronic inflammation was demonstrated for the $A_{2A}AR$ in an in vivo murine model of inflammatory bowel disease, where activation of the $A_{2A}AR$ attenuated the production of IFN-γ, TNF-α, and IL-4 in mesenteric T lymphocytes in a rabbit model of colitis (Odashima et al. 2005).

In a mixed lymphocyte reaction of human PBMCs and lymphocytes, adenosine-in duced inhibition of IL-18-induced increases in IL-12, IFN-γ, ICAM-1, and lymphocyte proliferation was blocked by an $A_{2A}AR$ antagonist, ZM-241385 (4-(2-[7-amino-2-(2-furyl)[1,2,4]triazolo[2,3-*a*][1,3,5]triazin-5-ylamino]ethyl)phenol), was enhanced by an A_1AR antagonist, DPCPX (8-cyclopentyl-1,3-dipropylxanthine), and an A_3AR antagonist, MRS1220 (*N*-[9-chloro-2-(2-furanyl)[1,2,4]-triazolo [1,5-*c*]quinazolin-5-yl]benzene acetamide), and was not affected by an $A_{2B}AR$ antagonist (Takahashi et al. 2007b). Moreover, the anti-inflammatory effect of an $A_{2A}AR$ agonist, CGS 21680 (2-(*p*-(2-carnonylethyl) phenylethylamino)-5-*N*-ethylcarboxamido adenosine) on IL-18-induced increases in IL-12, IFN-γ, ICAM-1, and lymphocyte proliferation were reversed by A_1AR and A_3AR agonists. These results suggest that the anti-inflammatory effects of adenosine in a mixed lymphocyte reaction are mediated by $A_{2A}ARs$; however, an A_1AR proinflammatory effect predominates when the $A_{2A}AR$ is saturated with agonist. As such, the net effect of adenosine on a mixed lymphocyte reaction activated by IL-18 is a function of activation of multiple adenosine receptor subtypes, including an anti-inflammatory effect via $A_{2A}ARs$ and proinflammatory effects via A_1AR and A_3ARs (Takahashi et al. 2007b).

In primary cultures of B lymphocytes, activation of B-cell antigen receptors results in the activation of NF-κB pathways (Minguet et al. 2005). Adenosine inhibits the NF-κB pathway by a mechanism related to increased cAMP levels and activation of protein kinase A. This study suggests that adenosine-mediated signals represent an important step in mediating the activation of B lymphocytes.

In activated human and mouse natural killer (NK) cells, adenosine inhibited the production of cytokines and chemokines (Raskovalova et al. 2005, 2006). In in vitro studies with lymphocytes derived from mouse spleen, A_1AR activation increased NK cell activity while A_2AR activation decreased NK cell activity (Priebe et al. 1990a). In mouse LAK cells, the adenosine agonist CADO (2-chloroadenosine) inhibited the cytotoxic activity and attenuated the production of IFN-γ, granulocyte macrophage colony-stimulating factor, TNF-α, and MIP-1α (Lokshin et al. 2006). Taken together, these results suggest that elevated adenosine levels in tumors may inhibit the tumoricidal effects of activated NK cells (Raskovalova et al. 2005; Lokshin et al. 2006). In addition, recent studies have shown that adenosine exhibits anti-inflammatory activities by engaging A_{2A}ARs on regulatory cells (Deaglio et al. 2007).

2.5 *Mast Cells*

Mast cells are important effector cells of allergic diseases such as asthma (Shimizu and Schwartz 1997). They can be stimulated to release mediators that have both immediate and chronic effects on airway constriction and inflammation. Adenosine can impact both the degranulation of mast cells and the production of inflammatory mediators. Rodent and human mast cells express the A_{2A}AR, A_{2B}AR and A_3AR (Feoktistov and Biaggioni 1995; Salvatore et al. 2000; Zhong et al. 2003b; Ryzhov et al. 2008b). Engagement of the A_3AR on rodent mast cells mediates degranulation in a manner that involves phosphoinositide 3-kinase (PI3K) activation and increase in intracellular Ca^{2+} (Salvatore et al. 2000; Zhong et al. 2003b). With regards to humans, it is not clear which adenosine receptor mediates the degranulation of mast cells, particularly in the airways. However, emerging evidence suggests that the A_{2B}AR mediates the production and release of proinflammatory mediators such as IL-8, IL-4 and IL-13 from both mouse and human mast cells (Feoktistov and Biaggioni 1995; Ryzhov et al. 2004, 2008b). The role of A_3AR and A_{2B}AR contributions to mast cell degranulation and the production of mediators are areas of active research that will aid in the development of adenosine-based therapeutics for diseases such as asthma, where mast cells play an important role.

2.6 *Eosinophils*

Eosinophils are involved in the pathophysiology of allergic diseases, including asthma (Frigas and Gleich 1986; Frigas et al. 1991; Gleich et al. 1983). During airway inflammation, eosinophils infiltrate tissues and release inflammatory mediators, including leukotrienes, reactive oxygen species, and granular proteins such as major basic protein. Activation of the A_1AR on human eosinophils enhances O_2^- release (Ezeamuzie and Philips 1999), whereas activation of A_3ARs on human eosinophils elevates intracellular Ca^{2+} (Kohno et al. 1996), inhibits PAF-induced chemotaxis (Knight et al. 1997; Walker et al. 1997), inhibits C5a-induced degranulation (Ezeamuzie and Philips 1999, 2001), and inhibits C5a-induced O_2^- release (Ezeamuzie et al. 1999). Eosinophils isolated from the lungs of patients with airway inflammation have higher levels of A_3AR mRNA compared to controls (Walker et al. 1997). In contrast to these findings, adenosine and A_3AR engagement has been shown to have proinflammatory effects on mouse (Young et al. 2004) and guinea pig eosinophils (Walker 1996). Together, these studies have led to the suggestion that selective A_3AR ligands may be useful therapies for the treatment of eosinophil-dependent inflammatory disorders such as asthma.

2.7 Endothelial Cells

Under normal physiological conditions, the endothelium provides several important regulatory and protective functions by serving as a physical barrier with both anticoagulant and anti-inflammatory properties (Hordijk 2006; Mehta and Malik 2006; Sands and Palmer 2005). An initiating event in inflammation is the recruitment and adhesion of leukocytes to the vascular endothelium and changes in endothelial permeability that permit the passage of leukocytes out of the vasculature and into the site of infection or tissue damage.

Adenosine receptors are expressed heterogeneously on endothelial cells, with the predominant subtypes generally being $A_{2A}AR$ and $A_{2B}AR$ (Deguchi et al. 1998; Feoktistov et al. 2002, 2004; Iwamoto et al. 1994; Khoa et al. 2003; Lennon et al. 1998; Olanrewaju et al. 2000; Sexl et al. 1997). In cell culture, endothelial cells derived from different sources have unique expression patterns of adenosine receptor subtypes (Feoktistov et al. 2002, 2004; Khoa et al. 2003). For example, mRNA levels of the $A_{2A}AR$ are approximately tenfold greater than mRNA levels for the $A_{2B}AR$ in HUVECs, whereas mRNA expression of the $A_{2B}AR$ is approximately fourfold greater than $A_{2A}AR$ mRNA expression levels in human microvascular endothelial cells (HMVECs) (Feoktistov et al. 2002). In endothelial cells, activation of the $A_{2A}AR$ inhibits the expression of VCAM-1 (Zernecke et al. 2006), E-selectin (Bouma et al. 1996; Hasko and Cronstein 2004), and tissue factor (Deguchi et al. 1998). Furthermore, activation of the $A_{2A}AR$ (Sullivan et al. 1999) and $A_{2B}AR$ (Eltzschig et al. 2003; Lennon et al. 1998; Yang et al. 2006a) is associated with decreased permeability of the vascular endothelium. These studies suggest that the $A_{2A}AR$ and $A_{2B}AR$ on endothelial cells play an important role in the prevention and mitigation of the inflammatory process. As opposed to these anti-inflammatory effects, activation of A_1ARs on human pulmonary artery endothelial cells (HPAECs) induces the release of thromboxane A_2 and IL-6, substances that are cytotoxic to endothelial cells and increase endothelial permeability (Wilson and Batra 2002). A_1AR antagonists prevented endothelial adhesion and digestion of the endothelial plasmalemma of alveolar capillaries by granulocytes, as well as the diapedesis of neutrophils toward the alveolar lumen in endotoxin-induced acute lung injury (Neely et al. 1997). In addition, activation of A_1ARs and A_3ARs on stimulated HUVECs results in an upregulation and downregulation of tissue factor expression, respectively, representing a potential mechanism for regulating the procoagulant activity of vascular endothelial cells in vivo by adenosine receptors (Deguchi et al. 1998).

3 Regulation of Adenosine Receptor Expression in Inflammatory Environments

Adenosine receptor expression is under dynamic regulation during various forms of physiological and pathophysiological stress, including hypoxia/ischemia and inflammation. For example, a distinct time-dependent alteration in adenosine receptor levels was observed in primary HMVECs subjected to hypoxic culture conditions (Eltzschig et al. 2003). After 12 h, hypoxia induced a selective upregulation of the A_{2B}AR, while at later time points (18 and 24 h), expression levels of the A_1AR and A_{2A}AR were downregulated and expression levels of the A_3AR were unchanged. This example demonstrates how a single stimulus can lead to complex alterations in adenosine receptor subtype expression.

Expression of the A_1AR is upregulated under conditions of stress. Numerous studies have demonstrated that stress-induced upregulation of the A_1AR involves increased transcriptional regulation by NF-κB, including in vitro oxidative stress (Nie et al. 1998), in vivo oxidative stress (Ford et al. 1997), in vivo cerebral ischemia (Lai et al. 2005), in vitro hyperosmotic stress (Pingle et al. 2004), in vivo exposure to LPS (Jhaveri et al. 2007), cardiac dysfunction induced by TNF-α overexpression (Funakoshi et al. 2007), and sleep deprivation stress (Basheer et al. 2007). Studies with genetically modified mice lacking the p50 subunit of NF-κB underscore the role of NF-κB in regulating A_1AR expression under basal conditions and pathogenic conditions (Jhaveri et al. 2007). LPS, an activator of NF-κB, increases A_1AR expression levels in the cortices of wild-type but not NF-κB p50 KO mice. In addition, expression of the A_1AR is upregulated in the bronchial epithelium and bronchial smooth muscle of asthmatics (Brown et al. 2008).

In a number of different cell types, the expression of the A_{2A}AR increases following exposure to proinflammatory conditions (Thiel et al. 2003). Following exposure to proinflammatory cytokines, including TNF-α and IL-1β, the expression and functional activity of the A_{2A}AR increases in cultured human monocytic THP-1 cells (Khoa et al. 2001), HMVECs (Nguyen et al. 2003), isolated human neutrophils (Fortin et al. 2006), and A549 human lung epithelial cells (Morello et al. 2006). In A549 cells, the upregulation of A_{2A}AR expression is regulated by NF-κB (Morello et al. 2006). Conversely, IFN-γ downregulates A_{2A}AR expression in THP-1 cells (Khoa et al. 2001) and HMVECs (Nguyen et al. 2003). Following exposure to LPS, mRNA levels for the A_{2B}AR and A_3AR were slightly upregulated in primary mouse intraperitoneal macrophage and WEHI-3 cells (Murphree et al. 2005); however, A_{2A}AR mRNA levels increased dramatically. The increased transcription of A_{2A}AR mRNA in the mouse intraperitoneal macrophages occurred via an NF-κB pathway. In WEHI-3 cells, the LPS-induced upregulation of A_{2A}AR mRNA was accompanied by an increase in cell surface A_{2A}AR expression and increased A_{2A}AR agonist-mediated cAMP production. Functionally, A_{2A}AR agonists inhibited TNF-α production with greater potency in the LPS-treated mouse intraperitoneal macrophages as compared to untreated control cells. These findings demonstrate the role of inflammatory stimuli in the upregulation of A_{2A}AR signaling.

In models of inflammatory bowel disease, characterized by altered levels of proinflammatory cytokines and local tissue hypoxia (Taylor and Colgan 2007), the expression of A_1AR, A_3AR (Sundaram et al. 2003) and A_{2B}ARs (Kolachala et al. 2005a) are altered. In a rabbit model of chronic ileitis, transcription of the A_1AR and A_3AR is upregulated in the ileum (Sundaram et al. 2003). The A_{2B}AR is upregulated in intestinal epithelia of a mouse model of colitis and in human intestinal epithelial mucosa during active colitis (Kolachala et al. 2005a). In T84 human colonic mucosal epithelial cells, TNF-α increases A_{2B}AR mRNA and protein levels (Kolachala et al. 2005a). IFN-γ inhibits A_{2B}AR-mediated effects without changing protein expression or A_{2B}R membrane recruitment (Kolachala et al. 2005b). Thus, in inflammatory conditions of the bowel, the regulation of the low-affinity A_{2B}AR occurs via direct effects on receptor expression and indirect effects on signal transduction pathways (Kolachala et al. 2005a, b).

Expression of the A_{2B}AR is upregulated during hypoxic and ischemic conditions (Linden 2001; Eltzschig et al. 2003; Zhong et al. 2005). In primary cultures of human lung fibroblasts, hypoxia induces an increase in A_{2B}AR expression levels (Zhong et al. 2005). Moreover, activation of the A_{2B}AR acts synergistically with hypoxia to increase the release of IL-6 from fibroblasts and promotes differentiation to myofibroblasts, suggesting that the upregulation of the A_{2B}AR may be relevant to chronic lung inflammatory diseases such as asthma and chronic obstructive pulmonary disease (COPD) (for more information on the role of the A_{2B}AR in asthma, see Chap. 11 of this volume, "Adenosine Receptors and Asthma," by Wilson et al.).

Hypoxia also selectively increases the expression of the A_{2B}R in HMVECs and T84 cells (Kong et al. 2006). The A_{2B}AR promoter contains a functional binding site for hypoxia-inducible factor (HIF)-1α, a transcriptional regulator that is important for adaptive responses to hypoxia. Disruption of this element blocks hypoxia-induced A_{2B}AR upregulation, and hypoxia-induced A_{2B}AR expression is directly proportional to HIF-1α activity (Kong et al. 2006). In an in vivo mouse model of colitis, a disorder characterized by increased HIF-1α, A_{2B}AR expression in colon endothelial tissue was increased (Kong et al. 2006). Moreover, HIF-1α KO mice have decreased A_{2B}AR levels in intestinal epithelia. Thus, HIF-1α directly regulates the expression of the A_{2B}AR by modulating gene transcription.

In an in vivo mouse model of LPS-induced peritonitis, a distinct time course for the differential expression of adenosine receptor subtypes is observed (Rogachev et al. 2006). In mouse mesothelial cells, the early stages of peritonitis are characterized by an induction of the A_1AR, with a peak in receptor protein at 12 h and a return to baseline by 24 h. During this phase of peritonitis, activation of the A_1AR is proinflammatory and results in the recruitment and extravasation of leukocytes, with the peak in A_1AR expression correlating with peak leukocyte counts. The A_{2A}AR protein reached a plateau between 12 and 24 h, and the expression of the A_{2B}AR reached a peak after 48 h in mesothelial cells. Functionally, the A_{2A}AR reduced TNF-α and IL-6 levels and decreased leukocyte accumulation. A similar adenosine receptor upregulation profile and time course was observed for the A_{2A}AR and A_{2B}ARs in mouse peritoneal neutrophils (Rogachev et al. 2006). In addition, the effect of proinflammatory cytokines on adenosine receptors was evaluated in human primary peritoneal mesothelial cells (Rogachev et al. 2006). Following exposure to IL-1 and TNF-α, early proinflammatory cytokines, mRNA and protein expression levels of $A_{2A}AR$ and $A_{2B}ARs$ were upregulated. IFN-γ, secreted later during the course of peritonitis, decreased $A_{2A}AR$ levels but increased $A_{2B}AR$ expression levels in human primary peritoneal mesothelial cells. These results suggest that the acute phase of the peritoneal infection involves a proinflammatory A_1AR response and increased release of proinflammatory cytokines, which upregulate the expression of the anti-inflammatory $A_{2A}AR$ and $A_{2B}ARs$ (Rogachev et al. 2006). These findings further demonstrate the intricate regulation of adenosine receptor expression in specific inflammatory environments (Rogachev et al. 2006).

3.1 *Adenosine Receptor Desensitization*

Desensitization, a mechanism by which a cell attenuates its response to prolonged agonist stimulation, has been studied for all four adenosine receptor subtypes and is driven by a number of factors, including receptor subtype, compartmentalization and scaffolding, and the complement of intracellular proteins involved in the desensitization and signaling process (Klaasse et al. 2008). While both the A_1AR and A_3AR are G_i-coupled receptors, their desensitization responses to agonist stimulation are very different. The cloned rat A_3AR desensitizes rapidly via the phosphorylation of serine and threonine residues on the intracellular carboxy terminus (Palmer et al. 1995a, 1996; Palmer and Stiles 2000). In contrast, the cloned human A_1AR is not phosphorylated in response to agonist and only becomes desensitized after prolonged agonist exposure (Ferguson et al. 2000, 2002).

A number of in vitro studies with cultured cells have demonstrated agonist-mediated $A_{2A}AR$ desensitization, including Chinese hamster ovary cells (Palmer et al. 1994), rat pheochromocytoma PC-12 cells (Chang et al. 1997), NG108-15 mouse neuroblastoma × rat glioma hybrid cells (Mundell and Kelly 1998; Mundell et al. 1998), rat aortic vascular smooth muscle cells (Anand-Srivastava et al. 1989), and bovine aortic endothelial cells (Luty et al. 1989). In addition, $A_{2A}AR$ desensitization has been demonstrated in native tissue, including rat brain (Barraco et al. 1996) and porcine coronary artery (Makujina and Mustafa 1993). Desensitization of both the $A_{2A}AR$ and $A_{2B}AR$ is mediated by the G-protein-coupled receptor kinase (GRK) 2 isozyme (Mundell et al. 1998). In human astroglial cells, chronic treatment with TNF-α increases the functional responsiveness of $A_{2B}ARs$ (Trincavelli et al. 2004); however, short-term treatment with TNF-α causes $A_{2B}AR$ phosphorylation, impaired $A_{2B}AR$–G protein coupling, and reduced cAMP production (Trincavelli et al. 2008).

In addition to direct effects on expression level, adenosine receptor signaling can be modified by indirect changes in intracellular signal transduction components. In HMVECs treated with TNF-α and IL-1β, there is an increase in $A_{2A}AR$ activity related to receptor upregulation and increased levels of the G protein β4 isoform (Nguyen et al. 2003). Moreover, TNF-α prevents $A_{2A}AR$ desensitization in human monocytoid THP-1 cells by blocking the translocation of GRK2 and β-arrestin to the cell membrane, which together with TNF-α stimulation results in upregulation of the A_{2A}AR (Khoa et al. 2001) and enhanced A_{2A}AR activity (Khoa et al. 2006). In T84 human colonic mucosal epithelial cells treated with IFN-γ, a reduction in A_{2B}AR signaling occurs in response to a downregulation of AC isoforms 5 and 7 without affecting A_{2B}AR expression levels or membrane recruitment (Kolachala et al. 2005b). In human astrocytoma ADF cells, TNF-α increased A_{2B}AR functional responses and receptor G-protein coupling without altering expression levels. This increased functional response was mediated by attenuating agonist-mediated phosphorylation and desensitization of the A_{2B}AR (Trincavelli et al. 2004).

Thus, desensitization is an important phenomenon that contributes to the net effect of adenosine signaling on specific cell types involved in inflammation and to the development of agonists as therapeutic agents, since the potential for tolerance/tachyphylaxis as an unwanted effect could limit their efficacy with chronic use.

4 Adenosine Receptor Contributions to the Regulation of Inflammation

4.1 A_1AR and Inflammatory Responses

4.1.1 Historical Perspective

A seminal study published in 1983 by Cronstein and colleagues demonstrated that the A_1AR mediates proinflammatory events and the A_2AR mediates antiinflammatory effects in isolated human neutrophils (Cronstein et al. 1983). In recent years, the role of the A_1AR in inflammation has been extensively studied using a number of approaches, including selective agonists and antagonists, monoclonal antibodies, selective antisense molecules, and genetically modified animals (Bours et al. 2006; Hasko and Cronstein 2004; Salmon et al. 1993; Sun et al. 2005). These studies have contributed to the delineation of the role of A_1AR in inflammation.

4.1.2 Proinflammatory Effects

Activation of the A_1ARs produces proinflammatory effects in a number of different tissues and cell types. On human neutrophils, A_1AR activation induces neutrophil chemotaxis, adherence to endothelial cells, and Fcγ receptor-mediated phagocytosis and O_2^- generation (Cronstein et al. 1990, 1992; Forman et al. 2000; Salmon and Cronstein 1990). In cultured human monocytes, the A_1AR enhances Fcγ receptor-mediated phagocytosis (Salmon et al. 1993; Salmon and Cronstein 1990), and promotes multinucleated giant cell formation on synovial fluid mononuclear phagocytes of patients with rheumatoid arthritis (Salmon et al. 1993; Merrill et al.

1997). Furthermore, A_1AR activation induces VEGF release from human monocytes (Clark et al. 2007). In human PBMCs, A_1AR antagonist enhanced and A_1AR agonist reversed the anti-inflammatory effects of adenosine mediated by $A_{2A}ARs$ on expression of ICAM-1 and production of IFN-γ, IL-12 and TNF-α in the presence of IL-18 (Takahashi et al. 2007a). In addition, A_1AR antagonism enhanced and A_1AR agonism reversed the anti-inflammatory effects of adenosine mediated by $A_{2A}ARs$ on expression of ICAM-1 and production of IL-12 and IFN-γ and lymphocyte proliferation during a human mixed lymphocyte reaction (Takahashi et al. 2007b). These findings suggest that activation of the A_1AR on a number of different inflammatory cells results in proinflammatory effects.

A proinflammatory role for the A_1AR has also been demonstrated in in vivo studies in a number of different species and disease states, including a rat model of pancreatitis (Satoh et al. 2000), ischemia-reperfusion injury of the lung (Neely and Keith 1995), ischemia-reperfusion injury of the heart in cats (Neely et al. 1996), dogs (Auchampach et al. 2004; Forman et al. 2000), and rats (Katori et al. 1999), ischemia-reperfusion injury of the liver in dogs (Magata et al. 2007) and pigs (Net et al. 2005), and endotoxin-induced lung injury in cats (Neely et al. 1997). Moreover, in an allergic model of asthma, L-97-1 (3-[2-(4-aminophenyl)-ethyl]-8-benzyl-7-2-ethyl-(2-hydroxy-ethyl)-amino]-ethyl-1-propyl-3,7-dihydro-purine-2,6-dione), a selective A_1AR antagonist reduced airway inflammation following allergen challenge, specifically reducing the number of eosinophils, neutrophils and lymphocytes in the airways (Nadeem et al. 2006). Recently, Ponnoth et al. have shown a proinflammatory role for A_1AR in vascular inflammation using a mouse model of allergic asthma (Ponnoth et al. 2008).

In rat models of acute pancreatitis induced with cerulein or taurocholate, the pancreas showed morphological changes that included interstitial edema and leukocyte infiltration (Satoh et al. 2000). Intraperitoneal administration of CCPA (2-chloro-N^6-cyclopentyladenosine), a selective A_1AR agonist, produced similar dose- and time-dependent effects on leukocyte infiltration and interstitial edema in pancreatic tissue, while $A_{2A}AR$ and A_3AR agonists had no effect (Satoh et al. 2000). The proinflammatory histopathological effects produced by CCPA in this model were attenuated by FK-838 (6-oxo-3-(2-phenylpyrazolo[1,5-α]pyridin-3-yl)-1(6*H*)-pyridazinebutanoic acid), an A_1AR-selective antagonist. These results suggest that activation of the A_1AR may play an important role in the tissue damage observed in acute pancreatitis.

Whole animal studies in models of ischemia-reperfusion in the lungs (Neely and Keith 1995), heart (Auchampach et al. 2004; Forman et al. 2000; Neely et al. 1996), and liver (Magata et al. 2007; Net et al. 2005) demonstrated that the A_1AR plays a proinflammatory role in these systems. In a feline model of ischemia-reperfusion injury of the lung, infusion of the A_1AR antagonists XAC (xanthine amine congener) or DPCPX reduced the percentage of injured alveoli (Neely and Keith 1995). In addition, DPCPX prevented endothelial damage, as well as margination and adhesion of neutrophils to pulmonary endothelial cells. Moreover, in a feline regional cardiac infarct model, pretreatment with the A_1AR antagonists DPCPX, bamiphylline, and XAC prevented ischemia-reperfusion injury; in other words, reduced infarct size (Neely et al. 1996). Similarly, in the canine model of myocardial ischemia-reperfusion, pre- and posttreatment with the A_1AR antagonist DPSPX (1,3-dipropyl-8-*p*-sulfophenylxanthine) decreased the area of cardiac necrosis and improved regional ventricular function (Forman et al. 2000). Based on studies with isolated human neutrophils demonstrating that DPSPX and DPCPX blocked fMLP-induced chemoattraction, it was hypothesized that the cardioprotective effect of the A_1AR antagonist DPSPX in the canine model was due to inhibition of neutrophil chemoattraction (Forman et al. 2000). For more information on A_1ARs and ischemia-reperfusion injury of the heart, please refer to Chap. 7 of this volume, "Adenosine Receptors and Reperfusion Injury of the Heart," by Headrick and Lasley.

The role of the A_1AR in hepatic ischemia-reperfusion injury was studied in a model using total hepatic vascular exclusion in beagles (Magata et al. 2007) and in a normothermic recirculation (NR) model of liver transplantation in pigs (Net et al. 2005). In the canine model, pretreatment with KW3902 (8-(noradamantan-3-yl)-1,3-dipropylxanthine), an A_1AR antagonist, significantly increased survival following hepatic ischemia-reperfusion (Magata et al. 2007). Moreover, histopathological examination of liver tissue revealed that pretreatment with KW3902 preserved hepatic architecture and decreased the infiltration of neutrophils into hepatic tissue. In the porcine model, NR following warm ischemia reversed the injury associated with liver transplantation and increased five-day survival. This protective effect of NR was simulated by the preadministration of adenine (Net et al. 2005). Blockade of the A_1AR with DPCPX during NR further protected the liver. Taken together, these studies suggest that the A_1AR plays a proinflammatory role in hepatic ischemia-reperfusion injury.

In an in vivo feline model of LPS-induced lung injury, blockade of the A_1AR prevents acute lung injury (Neely et al. 1997). In this model, an intralobar arterial infusion of LPS produced dose-dependent lung injury characterized by perivascular and peribronchial edema and hemorrhage, margination of neutrophils along the venular endothelium, thickened alveolar septae, alveolar infiltration of neutrophils and macrophages, alveolar edema, and alveolar hemorrhagic necrosis. In this study, lungs from animals treated with the A_1AR antagonists DPCPX or bamiphylline could not be distinguished from controls, suggesting that LPS-induced pulmonary injury involves activation of the A_1AR (Neely et al. 1997). To further evaluate A_1AR function in LPS-induced lung injury, LPS from various Gram-negative bacterial sources were evaluated using cultured cells derived from HPAECs (Wilson and Batra 2002). LPSs from *Escherichia coli*, *Salmonella typhimurium*, *Klebsiella pneumoniae*, and *Pseudomonas aeruginosa* bind directly to the A_1AR. Additional studies with HPAECs demonstrated that the CCPA- and LPS-induced release of IL-6 and thromboxane A_2, cytotoxic substances that increase permeability of the endothelium, was blocked by DPCPX. Together, these studies suggest that activation of the A_1AR on pulmonary artery endothelial cells by LPS during sepsis directly contributes to the pathology of acute lung injury (Neely et al. 1997; Wilson and Batra 2002).

To demonstrate the efficacy of an A_1AR antagonist as an antiendotoxin, antisepsis adjunctive therapy in combination with antibiotics in sepsis, the A_1AR antagonist

L-97-1 has been tested in a model of polymicrobial sepsis and endotoxemia (rat cecal ligation and puncture, CLP). Administration of L-97-1 as an intravenous therapy postCLP improved the seven-day survival in a dose-dependent manner (30–40% survival) as compared to untreated CLP controls (17% survival) or antibiotics alone (23% survival) (Wilson et al. 2006). In combination with antibiotics, L-97-1 increased survival to 50–70% in a dose-dependent manner. Improvement in seven-day survival was statistically significant for L-97-1 versus CLP, as well as for L-97-1 versus antibiotics. Moreover, L-97-1 plus antibiotics had a significant trend towards increased survival time based on the dose of L-97-1. Furthermore, efficacy for the A_1AR antagonist L-97-1 has been demonstrated in a bioterrorism animal model of pneumonic plague (Wilson, Endacea, Inc., unpublished data). In this model, rats are infected via intratracheal administration with *Yersinia pestis*, a Gram-negative bacterium that releases endotoxin, a major virulence factor for *Y. pestis*. In these studies, L-97-1 plus antibiotics (ciprofloxacin) improves six-day survival and lung injury scores versus antibiotics alone in 72 h delay treatment groups. During sepsis, the expression of A_1ARs is upregulated (Rogachev et al. 2006), and furthermore, LPS upregulates A_1AR expression (Jhaveri et al. 2007). These studies, taken together with the findings that A_1AR antagonists block LPS-induced acute lung injury and improve survival in both a CLP and a Gram-negative sepsis model induced by *Y. pestis*, suggest that the A_1AR is an important target in sepsis, and that A_1AR antagonists may represent an attractive class of compounds for development as antisepsis drugs.

Collectively, in vitro studies in inflammatory cells and in vivo studies in animal models suggest that the A_1AR is an important target in inflammation, and that A_1AR antagonists may be efficacious as anti-inflammatory drugs. Several biotechnology, biopharmaceutical, and pharmaceutical companies have engaged in developing A_1AR antagonists for different medical conditions. Phase I/II/III clinical trials demonstrate that A_1AR antagonists as a class of drugs appear to be safe in humans. For example, Aderis Pharmaceuticals (formerly Discovery Therapeutics) developed an A_1AR antagonist, N-0861 (N^6-endonorboran-2-yl-9-methyladenine), for the treatment of bradyarrhythmias (Bertolet et al. 1996). This compound was in Phase I/IIa clinical trials until it was put on clinical hold due to solubility problems. The high volume of diluent required to administer the drug intravenously, not the safety of N-0861, prevented further clinical development of this molecule. CV Therapeutics (Palo Alto, CA, USA) licensed CVT 124 to Biogen as BG-9719 (ENX) (1,3-dipropyl-8-[2-(5,6-epoxynorbornyl)]xanthine) for the treatment of congestive heart failure with renal impairment (Gottlieb et al. 2002). In clinical trials, both N-0861 and BG-9719 were well tolerated. However, problems with solubility, bioavailability, and formulation prevented the further clinical development of these A_1AR antagonists (Bertolet et al. 1996; Gottlieb et al. 2002; Doggrell 2005). Biogen (Biogen Idec, Cambridge, MA, USA) is developing another A_1AR antagonist, Adentri (BG 9928; 1,3-dipropyl-8-[1-(4-propionate)-bicyclo-[2,2,2]octyl]xanthine) for chronic congestive heart failure and renal impairment. This molecule is safe and is in Phase III clinical trials (Doggrell 2005; Greenberg et al. 2007; Press release Biogen, August 21, 2008). Two other A_1AR antagonists, SLV320 (Solvay) and KW-3902 (rolofylline) (previously NovaCardia, now Merck) are in Phase II and Phase III clinical trials for congestive heart failure with renal impairment, respectively, and are well tolerated (Givertz et al. 2007; Dittrich et al. 2007; http://www.clinicaltrials.gov).

4.1.3 Anti-inflammatory Effects

Studies with genetically modified mice suggest that the A_1AR also has context-dependent anti-inflammatory functions (Sun et al. 2005; Joo et al. 2007; Lee and Emala 2000; Lee et al. 2004a, b, 2007). A_1AR function was evaluated in adenosine deaminase (ADA) knockout mice by generating ADA/A_1AR double-knockout mice. ADA knockout mice exhibit increased levels of adenosine and increased levels of the A_1AR transcript, which was most predominant in activated alveolar macrophages (Sun et al. 2005). These animals developed pulmonary inflammation, characterized by an increase in macrophages, eosinophils, fibrosis, and airway hyperreactivity. Pulmonary inflammation was exacerbated in mice lacking both ADA and the A_1AR (Sun et al. 2005). The lungs of ADA/A_1AR double-knockout mice were characterized by higher levels of cytokines, including IL-4 and IL-13, and chemokines such as eotaxin 2 and thymus- and activation-regulated chemokine. Interestingly, lung adenosine levels in ADA/A_1AR double-knockout mice were approximately 200% higher than those found in ADA-deficient mice (Sun et al. 2005). Furthermore, an anti-inflammatory role of the A_1AR was demonstrated in A_1AR knockout mice with experimental allergic encephalomyelitis, an in vivo model of multiple sclerosis (Tsutsui et al. 2004). A_1AR knockout mice exhibited severe demyelination and axonal injury, enhanced activation of macrophages and microglial cells, increased transcription of proinflammatory cytokines, and decreased transcription of anti-inflammatory cytokines.

The anti-inflammatory role associated with the A_1AR has been extensively studied in models of renal ischemia-reperfusion in mice and rats and in cultured renal tubule cells (Joo et al. 2007; Lee and Emala 2000; Lee et al. 2004a, 2007). Initially, the anti-inflammatory role for the A_1AR was described in a rat model of renal ischemia-reperfusion, where preconditioning, adenosine, and an A_1AR agonist, R PIA, produced a protective effect improving renal function and morphology (Lee and Emala 2000). Interestingly, an A_3AR agonist, IB-MECA (1-deoxy-1-(6-((3-iodophenyl)methyl)amino-9*H*-purin-9-yl)-*N*-methyl-D-ribofuranuronamide), worsened and an A_3AR antagonist, MRS 1191 (3-ethyl-5-benzyl-2-methyl-4-phenylethynyl-6-phenyl-1,4-dihydropyridine-3,5-dicarboxylate), improved renal function in this model. The protective effect of the A_3AR antagonist was greater than that of adenosine. In these studies, DPCPX blocked the protective effect of adenosine but not that of preconditioning (Lee and Emala 2000). Subsequently, a protective effect for CCPA, an A_1AR agonist, was demonstrated in a mouse model of renal ischemia-reperfusion (Lee et al. 2004b). In these studies, DPCPX worsened renal function and increased expression of inflammatory markers, necrosis and apoptosis, and blocked the protective effect of CCPA. Next, the protective effect of the A_1AR in renal ischemia-reperfusion was studied in A_1AR knockout mice (Lee et al. 2004a). In these studies, A_1AR knockout mice showed worsened renal function and histology compared to the wild-type controls. Moreover, DPCPX increased markers of renal inflammation while CCPA reduced markers of renal inflammation. Interestingly, the A_3AR antagonist MRS 1191 improved renal function in A_1AR knockout mice with an efficacy similar to that produced by CCPA in wild-type mice (Lee et al. 2004a). Collectively, these studies suggest that both A_1AR and A_3ARs play an important role in ischemia-reperfusion injury in the kidney of rats and mice.

In mice, the mechanism of renal protection was found to consist of an acute and a delayed phase. Renal protection involved A_1AR-mediated phosphorylation of ERK MAPK and Akt, which are involved in the upregulation of cytoprotective genes (Joo et al. 2007). Activation of the A_1AR also resulted in increased phosphorylation of heat shock protein (HSP) 27 (Joo et al. 2007; Lee et al. 2007), a molecular chaperone involved in the cytoprotection of cellular proteins through the prevention of denaturation and aggregation under conditions of oxidative stress (Joo et al. 2007). In contrast, A_1AR knockout mice had decreased levels of basal HSP27 (Lee et al. 2007). Specific inhibitors of HSP synthesis blocked the A_1AR-mediated renal protection in A_1AR wild-type mice. Inhibition of G_i proteins with PTX blocked both the early phase and the late phase protective effects mediated by the A_1AR. The early phase of the A_1AR-mediated antiinflammatory effect was blocked with chelerythrine, a protein kinase C (PKC) inhibitor. The early and delayed phases of renal protection were blocked by deletion of PI3K gamma and inhibition of Akt, but not inhibition of ERK.

The role of A_1AR activation has also been studied in an immortalized porcine renal tubule cell line (LLC–PK1 cells) overexpressing the human A_1AR, and in primary cultures of renal proximal tubule cells from A_1AR knockout mice (Lee et al. 2007). In the LLC–PK1 cells, overexpression of the A_1AR was associated with increased basal expressions of total and phosphorylated HSP27, reportedly due to A_1AR-mediated stimulation of p38 and MAPK. Renal epithelial cells overexpressing the A_1AR showed decreased peroxide-induced necrosis and TNF-α-induced apoptosis, which was blocked by selective blockade of the A_1AR. In contrast, primary cultures of proximal tubule cells from A_1AR knockout mice showed increased levels of necrosis and apoptosis. Taken together, these studies suggest that A_1AR activation exerts a protective preconditioning effect in renal ischemia-reperfusion by modulating the inflammatory response and tissue necrosis, and that this process may involve HSP27.

Furthermore, studies with A_1AR knockout and wild-type mice suggest that activation of the A_1AR protects against sepsis (Gallos et al. 2005). Following CLP, mortality was increased in both the A_1AR knockout mice and in wild-type mice treated with DPCPX to antagonize A_1ARs. In addition, A_1AR knockout mice had increased levels of TNF-α, suggesting that the A_1AR modulates TNF-α production during sepsis. Finally, renal tissue in the A_1AR knockout mice exhibited increased levels of neutrophils, ICAM-1, and proinflammatory cytokines, indicating a higher degree of renal dysfunction induced by sepsis. These results suggest that the A_1AR attenuates the inflammatory response and diminishes the hyperacute inflammatory response characteristic of sepsis.

The differences in the studies suggesting both a proinflammatory role for the A_1AR in the pancreas, lung, heart, and liver as well as an anti-inflammatory role in the lung and kidney may be due to differences in the models (i.e., genetically modified mice and cell lines overexpressing adenosine receptors versus other models). For example, in the A_1AR knockout models and cell lines, other proteins such as the $A_{2B}AR$ and A_3ARs may be responsible for the protective effects of adenosine described. There is substantial evidence suggesting that the protective effect of preconditioning is mediated by PTX-sensitive G_i-coupled proteins, including the A_1AR and A_3ARs. The studies that described a protective effect of both A_1AR and A_3ARs in the kidney are consistent with what is reported in the literature for other species. Therefore, the protective effects of adenosine and the selective A_1AR agonists R-PIA and CCPA in studies of renal ischemia-reperfusion are not surprising. The protective effect of overexpression of the A_1AR in LLC–PKC1 cells is also not surprising, for the same reasons.

The deleterious effects of DPCPX on the renal function and histology of the kidney in the rat and mouse is surprising in light of the protective effect of a number of A_1AR antagonists in different models of inflammation. DPCPX is at best tenfold selective for A_1ARs versus $A_{2B}ARs$ (Fredholm et al. 2001). In the in vivo renal ischemia-reperfusion studies and sepsis studies in mice, it is possible that DPCPX may be blocking the anti-inflammatory effects of the G_s-coupled $A_{2B}AR$ (Yang et al. 2006a). Although another highly selective A_1AR antagonist, FSCPX 8-cyclopentyl-3-[3-[[4-(fluorosulfonyl)benzoyl]oxy]propyl]-1-propylxanthine), reversed the resistance to cell death in LLC–PK1 cells produced by overexpressing the A_1AR, FSCPX was used in a high concentration (20 μM), and a dose–response relationship for the blocking effects of FSCPX was not demonstrated. In other models of inflammation, anti-inflammatory effects for highly selective A_1AR antagonists, including BG-9928 in ischemia-reperfusion injury of the heart in dogs (Auchampach et al. 2004), FK-838 in pancreatitis in the rat (Satoh et al. 2000), FK-352 ([*R*]-1-[(*E*)-3-(2-phenylpyrazolo[1,5-*a*]pyridin-3-yl) acryloyl]piperidin-2-yl acetic acid) in ischemia-reperfusion injury of the heart in rats (Katori et al. 1999), and L-97-1 in allergic asthma in the rabbit (Nadeem et al. 2006), were demonstrated.

Differences in the studies described above suggesting both a proinflammatory and an anti-inflammatory role for the A_1AR may be due to differences in (i) the A_1AR-activated signaling pathway that results in tissue injury (i.e., proinflammatory pathway) versus that for protection (i.e., anti-inflammatory pathway), and which pathway predominates as a function of species and the stage/progression of injury; (ii) predominant inflammatory cell type as a function of species; in other words, the role of the macrophage where the transcript for the A_1AR was high in mice (Sun et al. 2005) versus the predominant role of the neutrophil in acute inflammation in other species; (iii) expression of A_1ARs on endothelial cells and different inflammatory cells, possibly a function of species differences; (iv) intracellular signaling and desensitization mechanisms as a function of species or cell/tissue/organ; and (v) density of homo- and heterodimers of adenosine receptors and their functional properties as a function of the cell type, organ, and species. For example, there is evidence from in vitro studies with native neural tissue as well as in vivo studies to suggest that adenosine receptors, including the A_1AR, can form homo- and heterodimers that have unique pharmacological profiles and functional effects, including altered ligand affinity, G-protein coupling, and desensitization characteristics (Ciruela et al. 2006; Ferre et al. 2008; Franco et al. 2006; Fuxe et al. 1998; Nakata et al. 2004). To date, adenosine receptor dimerization has not been studied in cells from the immune system, but it is likely that this phenomenon is relevant in inflammatory processes. These considerations are of particular significance given that inflammatory cells undergo significant phenotype changes (e.g., adenosine receptor expression levels, altered cytokine profiles, altered cell surface protein levels) that are unique to various physiological and pathophysiological challenges. Finally, differences in the phenotypes of genetically modified animals and cells are very complex and are not yet completely understood. It is possible that genetically manipulated animals and cells exhibit compensatory expression or functions of other proteins that alter the phenotype of cells and organs in a manner that is not fully appreciated at this time.

4.2 A_{2A}AR and Inflammatory Responses

4.2.1 Historical Perspective

Numerous investigations in cellular and animal model systems have provided evidence that A_{2A}AR signaling pathways are active in limiting inflammation and tissue injury (Hasko and Cronstein 2004; Linden 2005; Sitkovsky and Ohta 2005; Hasko and Pacher 2008). Some of the earliest observations that A_{2A}AR signaling is anti-inflammatory came from Cronstein and colleagues, who demonstrated that engagement of the A_{2A}AR could inhibit elicited superoxide formation from neutrophils (Cronstein et al. 1983). Expression of the A_{2A}AR has subsequently been found on most inflammatory cells (Sitkovsky et al. 2004), where it has numerous anti-inflammatory properties, including inhibiting T-cell activation (Huang et al. 1997; Erdmann et al. 2005) and limiting the production of inflammatory mediators such as IL-12, TNF-α and INFγ (Hasko et al. 2000; Pinhal-Enfield et al. 2003; Lappas et al. 2005). Ohta and colleagues performed a series of studies in vivo using A_{2A}AR knockout mice to demonstrate that this receptor plays an important role in limiting the degree of inflammatory mediator production and tissue injury in response to challenges with concanavalin A or endotoxin (Ohta and Sitkovsky 2001). Subthreshold doses of these agents that caused minimal responses in wild-type mice led to extensive inflammatory mediator production, tissue damage and death in A_{2A}AR knockout mice. Thus, adenosine signaling through the A_{2A}AR appears to serve as a critical endogenous regulator of tissue inflammation and damage. Given that hypoxia and subsequent adenosine generation is likely an acute response to numerous injuries, this pathway is likely to have important and widespread implications in dictating the balance of tissue injury and repair.

4.2.2 Anti-inflammatory Effects

Substantial lines of evidence suggest that the $A_{2A}AR$ is the major adenosine receptor mediating the anti-inflammatory properties of adenosine (Hasko and Pacher 2008). The ability of $A_{2A}AR$ activation to suppress Th1 cytokine and chemokine expression by immune cells is likely the dominant mechanism involved. For example, $A_{2A}AR$ activation can attenuate IL-12, INFγ and TNF-α production from important immunomodulatory cells such as monocytes (Hasko et al. 2000), dendritic cells (Panther et al. 2003) and T cells (Lappas et al. 2005). The ability to diminish the production of such cardinal inflammatory molecules likely contributes to the decreased inflammation and tissue damage due to effector cell activation that is often seen with $A_{2A}AR$ activation. However, there is also evidence that $A_{2A}AR$ activation can prevent effector cell activities such as neutrophil migration (Cronstein et al. 1992) and oxidative burst (Cronstein et al. 1983). Collectively, these anti-inflammatory properties of the $A_{2A}AR$ represent a sensitive and widespread mechanism for the immunoregulation of tissue injury and repair.

Findings in disease-relevant animal models suggest that $A_{2A}AR$ activation on immune cells is beneficial in environments associated with acute inflammation and hypoxia. $A_{2A}AR$ agonists have remarkable anti-inflammatory and tissue-protective effects in models of ischemic liver damage (Day et al. 2004,) myocardial injury (Lasley et al. 2001; Glover et al. 2005), spinal cord injury (Reece et al. 2004), renal injury (Day et al. 2003), inflammatory bowel disease (Naganuma et al. 2006), and lung transplantation (Ross et al. 1999). Many of these models involve postischemic environments and suggest that $A_{2A}AR$ activation (on various immune cells) limits or inhibits the degree of inflammation and subsequent tissue damage. Activation of the $A_{2A}AR$ has also been shown to play an important role in the promotion of wound healing and angiogenesis (Montesinos et al. 2002), and the $A_{2A}AR$ and A_3AR are responsible for the anti-inflammatory actions of methotrexate in the treatment of inflammatory arthritis (Montesinos et al. 2003). Collectively, these studies suggest that activation of the $A_{2A}AR$ has a significant impact on stemming inflammation and tissue damage in a number of disease-relevant models, suggesting that there may be numerous clinical benefits from the use of $A_{2A}AR$-activating compounds.

Recent studies have utilized bone marrow transplantation approaches together with gene knockout and selective $A_{2A}AR$ agonist treatments to identify populations of immune cells that contribute to the anti-inflammatory properties of this receptor in disease models. In a model of ischemia-reperfusion liver injury, activation of the $A_{2A}AR$ with the selective agonist ATL146e (4-(3-[6-amino-9-(5-ethylcarbamoyl-3,4-dihydroxy-tetrahydro-furan-2-yl)-9*H*-purin-2-yl]-prop-2-ynyl)- cyclohexanecarboxylic acid methyl ester) was associated with decreased inflammation, and the liver was protected from damage brought about by reperfusion following ischemia (Day et al. 2004). When $A_{2A}AR$ knockout mice were subjected to the same insult, the effectiveness of ATL146e was lost. Moreover, $A_{2A}AR$ knockout mice exhibited increased liver damage, suggesting that endogenous adenosine is involved in the tissue protection seen. Subsequent studies using bone marrow transplantation approaches and $A_{2A}AR$ knockout mice suggested that it was $A_{2A}AR$ on bone marrow-derived cells that conferred $A_{2A}AR$ agonist protection. Subsequent studies identified CD1d-activated NK T cells as being the critical cells mediating the protective effects of $A_{2A}AR$ agonist treatment in this model, where $A_{2A}AR$ engagement reduced the production of IFN-γ from NK T-cells in association with blocking liver reperfusion injury (Lappas et al. 2006). Bone marrow transplantation studies using $A_{2A}AR$ knockout mice were also used to demonstrate that the protective effect of $A_{2A}AR$ agonist in a model of renal ischemia-reperfusion injury was due to $A_{2A}AR$ activation on marrow-derived cells (Day et al. 2003). Although the exact cell type has not been identified, there is evidence to suggest that it is a cell type other than macrophages, which have been shown to be important in mediating the protective effects of $A_{2A}AR$ agonism in a model of diabetic nephropathy (Awad et al. 2006). Similar approaches demonstrate that $A_{2A}AR$ expression on bone marrow-derived cells is responsible for $A_{2A}AR$ agonist anti-inflammatory and tissue-protective effects in models of myocardial infarction (Yang et al. 2006b), acute lung injury (Reutershan et al. 2007), and spinal cord compression injury (Li et al. 2006).

Inflammatory bowel diseases such as Crohn's disease and ulcerative colitis are associated with severe tissue inflammation and damage. $A_{2A}AR$ activation has anti-inflammatory and tissue protective properties in several studies investigating inflammation in the gastrointestinal tract (Odashima et al. 2005; Cavalcante et al. 2006; Naganuma et al. 2006). $A_{2A}AR$ knockout mice are more sensitive to experimental colitis, and treatment with the $A_{2A}AR$ agonist ATL 146e is associated with decreased leukocyte infiltration, inflammatory mediator production and necrosis in a model of inflammatory bowel disease (Odashima et al. 2005). CD25+ CD4+ Tregs play an important role in regulating inflammatory responses, including those associated with inflammatory bowel disease (Izcue et al. 2006). Recent studies have identified $A_{2A}ARs$ on Tregs as playing an important role in regulating inflammation in inflammatory bowel disease (Naganuma et al. 2006). Tregs isolated from wild-type mice and transferred to immunodeficient mice together with colitis-inducing CD4+ T cells were able to confer protection from the development of colitis, whereas Tregs isolated from $A_{2A}AR$ knockout mice were not. These studies highlight the importance of $A_{2A}AR$ signaling as an anti-inflammatory pathway in inflammatory bowel disease.

Given that endogenous adenosine acting through the $A_{2A}AR$ appears to be a potent regulator of inflammation and tissue injury, it stands to reason that mechanisms must exist to tightly regulate adenosine's actions during the natural course of the inflammatory response. This could occur at multiple levels, including the regulation of adenosine production and the availability of effective receptor signaling pathways. A recent study by Deaglio and colleagues provided new and interesting information on the mechanisms of adenosine generation and immunoregulation by Tregs (Deaglio et al. 2007). Extracellular adenosine is generated from the dephosphorylation of extracellular nucleotides (Zimmermann 2000). ATP and ADP are converted to AMP by the ectonucleoside triphosphate diphosphohydrolase CD39. AMP is dephosphorylated to adenosine by the ectonucleotidase CD73. Both of these enzymes play critical roles in producing extracellular adenosine (Deaglio et al. 2007). A newly recognized feature of Tregs is that they express a unique combination of both CD39 and CD73 together with the forkhead transcription factor Foxp-3 (Deaglio et al. 2007). These findings provide an important new signature for defining Tregs, but more importantly they demonstrate that the production of adenosine through this cascade on the surface of Tregs is important to the A_{2A}AR-mediated immunosuppressive effects of these cells (Deaglio et al. 2007). These findings provide an elegant example of how the coordinate regulation of adenosine production and signaling can impact the immune response.

Anti-inflammatory properties of A_{2A}AR signaling have also been noted in animal models of inflammatory lung disease. In a model of LPS-induced lung injury, treatment with the A_{2A}AR agonist ATL202 was associated with decreased recruitment of neutrophils to the lung, together with reduced cytokine levels and pulmonary edema (Reutershan et al. 2007). There was enhanced neutrophil recruitment in A_{2A}AR knockout mice treated with LPS and bone marrow transplantation, as well as tissue-specific A_{2A}AR deletion studies suggested that A_{2A}AR expression on leukocytes was important in the anti-inflammatory effects seen. In addition to models of acute lung injury, anti-inflammatory effects of A_{2A}AR signaling have been noted in allergic lung inflammation. Treatment of allergic rats with the A_{2A}AR agonist resulted in diminished pulmonary inflammation (Fozard et al. 2002). Similar findings were seen in a mouse model of allergic lung inflammation (Bonneau et al. 2006). A_{2A}AR knockout allergic mice have also been shown to have higher lung inflammation as compared to A_{2A}AR wild-type mice upon allergen challenge (Nadeem et al. 2007). Finally, in a recent study on the ADA knockout model of adenosine-dependent lung inflammation and damage, genetic removal of the A_{2A}AR led to enhanced pulmonary inflammation, mucus production and alveolar airway destruction (Mohsenin et al. 2007), further implicating A_{2A}AR signaling pathways as important anti-inflammatory networks in the lung. These findings suggest that A_{2A}AR agonism may be beneficial in the treatment of diseases such as asthma; however, recent investigations into this in humans have been inconclusive (Luijk et al. 2008).

4.2.3 Detrimental Aspects of A_{2A}AR Engagement

It is becoming increasingly clear that the suppression of various T-cell functions is a major mechanism by which A_{2A}AR signaling limits tissue inflammation and damage in response to acute injury. Whereas this has obvious benefits in protecting tissues and promoting repair, there are examples where this paradigm is detrimental. A recent study by Ohta and colleagues demonstrated that A_{2A}AR-mediated immunosuppressive activities serve to protect cancer cells from the activities of antitumor T cells, and thus promote the survival and growth of tumors (Ohta et al. 2006). It was shown that the hypoxic environment of certain tumors promotes adenosine formation, and that treatment of wild-type mice with A_{2A}AR antagonists can decrease tumor size. Moreover, injection of A_{2A}AR knockout mice with cancer cells was associated with remarkable decreases in tumor size and animal survival relative to what was seen in wild-type mice injected with the same cancer cells. This effect appears to be mediated largely by CD8+ T cells and the production of INFγ and TNF-α. These findings suggest that the well-characterized anti-inflammatory properties of $A_{2A}AR$ signaling may actually serve to protect certain tumors from the body's attempt to eliminate them. This raises several attractive avenues for novel cancer therapies. The use of $A_{2A}AR$ antagonists or strategies to lower adenosine levels in tumors may prove beneficial in allowing the immune system to attack cancer cells. In addition, dampening the anti-inflammatory effects of $A_{2A}AR$ signaling in certain tumors with the development of targeted antitumor T cells with $A_{2A}AR$ gene deletion may improve strategies for cancer immunotherapy.

Another area where $A_{2A}AR$ receptor signaling has received substantial attention as a potential target for therapeutic intervention is neurodegenerative disorders. Adenosine levels markedly increase in the brain in response to hypoxic, traumatic, and inflammatory insults (Pedata et al. 2001). Interestingly, engagement of the $A_{2A}AR$ in brain injury models appears to have both protective and detrimental effects. Akin to what is seen in other organ systems (see above), activation of the $A_{2A}AR$ has been shown to reduce brain damage in kainate-induced hippocampal injury and hemorrhagic brain injury (Jones et al. 1998; Mayne et al. 2001). In contrast, $A_{2A}AR$ antagonists have been shown to attenuate ischemic brain injury (Monopoli et al. 1998) and neurotoxicity induced by kainate and quinolinate (Jones et al. 1998; Popoli et al. 2002). In addition, $A_{2A}AR$ antagonists are protective in neurotoxic models of Parkinson's disease (Ikeda et al. 2002). The contribution of $A_{2A}AR$ signaling to the promotion of neuronal injury has been validated genetically. $A_{2A}AR$ knockout mice were found to have smaller infarct volumes and neural behavioral deficit scores following ischemic brain injury than wild-type mice (Yu et al. 2004). The $A_{2A}AR$ is expressed on many cell types in the brain, including neuronal components such as striatal neurons and glial cells (Svenningsson et al. 1999), endothelial cells and various bone marrow-derived cells such as neutrophils, macrophages and dendritic cells (Hasko and Cronstein 2004). Engagement of $A_{2A}ARs$ on different cells during different types or even stages of injury may be responsible for the destructive or protective effects of this receptor in the injured brain. A series of experiments using bone marrow transplantation of cells from $A_{2A}AR$ knockout mice and tissue-specific knockout of the $A_{2A}AR$ in neuronal cells has recently provided evidence that it is the expression of the $A_{2A}AR$ on bone marrow-derived cells that is responsible for the detrimental effects of $A_{2A}AR$ signaling in ischemic brain injury and 3-nitropropionic acid-induced striatal damage (Yu et al. 2004; Huang et al. 2006). The mechanisms underlying potential protective effects of $A_{2A}AR$ are not clear. However, the observation that both protective and detrimental effects have been noted in different types of brain injury suggests that multiple sites of action for injury-induced adenosine production and $A_{2A}AR$ signaling must be understood before $A_{2A}AR$ agonists and antagonists can be effectively utilized in the treatment of neurodegenerative disorders. For more information on $A_{2A}ARs$ in neuroprotection and neurodegenerative diseases, please refer to other chapters in this volume, including Chap. 16, "Adenosine Receptors and the Central Nervous System" (by Sebastião and Ribeiro), Chap. 17, "Adenosine Receptors and Neurological

Disease: Neuroprotection and Neurodegeneration" (by Stone et al.), and Chap. 18, "Adenosine A_{2A} Receptors and Parkinson's Disease" (by Morelli et al.).

Thus, studies in cellular models and preclinical investigations in animal models suggest that A_{2A}AR agonists will be useful in the treatment of many diseases where inflammation is a detrimental component, while A_{2A}AR antagonists may be beneficial in the treatment of neurological disorders such as Parkinson's disease. As clinical trials advance, it will become evident whether these preclinical observations translate into beneficial effects in humans (Schwarzschild et al. 2006; Gao and Jacobson 2007).

4.3 A_{2B}AR and Inflammatory Responses

4.3.1 Historical Perspective

Initial identification of the A_2ARs was based on the ability of this class of receptor to activate AC (Londos et al. 1980). Both high-affinity and low-affinity A_2AR subtypes were described (Bruns et al. 1986), and it was not until the successful molecular cloning of these receptors in the early 1990s that it became clear that the high-affinity A_2AR was the A_{2A}AR and the low-affinity A_2AR was the A_{2B}AR (Pierce et al. 1992; Rivkees and Reppert 1992). Subsequently, studies in HMC-1 cells have shown that the A_{2B}AR can not only couple AC through G_s (Feoktistov and Biaggioni 1997), but that it also interacts with G_q to activate PLC (Feoktistov et al. 1999). However, the majority of in vitro studies in normal human cell lines, including human airway epithelial and bronchial smooth muscle cells as well as lung fibroblasts, and in vivo studies in mice suggest that the primary signaling pathway for the A_{2B}AR is via AC, to produce an increase in intracellular cAMP (see below). The A_{2B}AR is expressed on most inflammatory cells, and its expression is induced in hypoxic and inflammatory environments (Xaus et al. 1999a; Eltzschig et al. 2003). Furthermore, as with the other adenosine receptors, both anti- and proinflammatory activities have been associated with A_{2B}AR activation. However, the observation that the A_{2B}AR has a relatively low affinity for adenosine has led to the notion that A_{2B}AR activation may serve important functions in pathological situations where adenosine production is increased (Fredholm 2007). Deciphering the contributions of A_{2B}AR signaling in models of tissue inflammation and tissue injury is an active area of research with exciting possibilities for novel adenosine-based therapeutics.

4.3.2 Anti-inflammatory Effects

Perhaps the best-characterized anti-inflammatory actions associated with the A_{2B}AR are its ability to inhibit monocyte and macrophage functions. Along these lines, IFN-γ can upregulate the $A_{2B}AR$ expression on macrophages as part of a proposed mechanism for macrophage deactivation (Xaus et al. 1999a). Consistent with this, $A_{2B}AR$ activation can inhibit the production or release of proinflammatory cytokines such as TNF-α and IL-1β from macrophages or monocytes, and it can inhibit macrophage proliferation following inflammatory stimulation (Xaus et al. 1999b; Sipka et al. 2005; Kreckler et al. 2006). In addition, $A_{2B}AR$ activation can increase the production of IL-10 from macrophages (Nemeth et al. 2005), a process that can be considered anti-inflammatory. As with the $A_{2A}AR$, these anti-inflammatory effects on macrophages likely stem from the ability of the $A_{2B}AR$ to couple to adenylate cyclase to increase cAMP levels.

Recent observations in $A_{2B}AR$ knockout mice provide compelling evidence that this receptor is associated with anti-inflammatory events in vivo. Yang and colleagues published on the initial characterization of $A_{2B}AR$ knockout mice and demonstrated that these mice show evidence of increased inflammation at baseline, in that levels of cytokines such as TNF-α and IL-6 were elevated in naïve $A_{2B}AR$ knockout mice, while IL-10 levels were elevated (Yang et al. 2006a). Exposure of $A_{2B}AR$ knockout mice to LPS resulted in the exaggerated production of TNF-α and IL-6, further suggesting an anti-inflammatory role for the $A_{2B}AR$ in vivo. These findings were surprising given data suggesting that $A_{2B}AR$ signaling is proinflammatory in vitro and in vivo (see below). Furthermore, these studies suggest a role for physiological $A_{2B}AR$ activation, which is paradoxical considering the notion that $A_{2B}AR$ signaling is only activated during pathological situations when adenosine levels are high, and that it is a low-affinity receptor. Interestingly, in $A_{2B}AR$ knockout mice there was evidence of increased expression of vascular adhesion molecules that mediate inflammation (Yang et al. 2006a), a finding that was associated with the $A_{2B}AR$-dependent production of cytokines from macrophages. These findings are consistent with the anti-inflammatory properties previously attributed to this receptor on macrophages.

A second manuscript by Hua and colleagues using independently generated $A_{2B}AR$ knockout mice demonstrated enhanced mast-cell activation in the absence of the $A_{2B}AR$ (Hua et al. 2007). These studies showed a reduction in basal levels of cAMP in mast cells isolated from $A_{2B}AR$ knockout mice, suggesting that $A_{2B}AR$ engagement may play a role in regulating mast-cell activation at baseline. Furthermore, experiments in this study demonstrated that mice lacking the $A_{2B}AR$ exhibit increased sensitivity to IgE-mediated anaphylaxis, suggesting that this receptor may limit the magnitude of antigen-driven responses on mast cells. These findings are somewhat paridoxical given the touted proinflammatory functions of the $A_{2B}AR$ receptor on HMC-1 cells (Feoktistov et al. 1998). A recent manuscript by Ryzhov and colleagues probed this further by examining adenosine-dependent effects on mast cells isolated from $A_{2B}AR$ knockout mice (Ryzhov et al. 2008b). These studies confirmed the anti-inflammatory effects of $A_{2B}AR$ signaling on mast cells and demonstrated that the $A_{2B}AR$ is necessary for antigen-induced proinflammatory cytokine production from these cells. Thus, the $A_{2B}AR$ seems to play both anti- and proinflammatory functions on mast cells.

A recent study by Eckle and colleagues demonstrated increased vascular leakage in $A_{2B}AR$ knockout mice exposed to hypoxia (Eckle et al. 2008). In addition, there was an increase in neutrophil influx into tissues of $A_{2B}AR$ knockout mice exposed to hypoxia. Bone marrow transplantation studies suggested that this enhanced neutrophilia was, in part, due to $A_{2B}AR$ expression on bone marrow-derived cells. The ability of the $A_{2B}AR$ to regulate vascular leak represents a potentially major anti-inflammatory role for this receptor. Moreover, the anti-inflammatory effects of $A_{2B}ARs$ are not surprising for a G_s-coupled receptor that increases intracellular cAMP, and are similar to those produced by other agents that increase cAMP, in other words $A_{2A}AR$ agonists and phosphodiesterase IV inhibitors.

4.3.3 Proinflammatory Effects

Numerous studies have demonstrated a proinflammatory function for the $A_{2B}AR$, largely through its regulation of proinflammatory cytokine and chemokine production. A prominent example of this is the promotion of IL-6 release from a number of cells, including intestinal (Sitaraman et al. 2001) and airway epithelial cells (Sun et al. 2008), macrophages (Ritchie et al. 1997; Ryzhov et al. 2008a), pulmonary fibroblasts (Zhong et al. 2005), bronchial smooth muscle cells (Zhong et al. 2003a), astrocytes (Schwaninger et al. 1997) and cardiomyocytes (Wagner et al. 1999). In addition, $A_{2B}AR$ activation can promote the release of IL-8 from HMC-1 cells (Feoktistov and Biaggioni 1995), IL-4 and IL-13 (Ryzhov et al. 2004, 2008b) from HMC-1 cells and murine bone marrow-derived mast cells, and the release of IL-19 from airway epithelial cells (Zhong et al. 2006) and MCP-1 from bronchial smooth muscle cells (Zhong et al. 2003a). In addition, $A_{2B}AR$ activation can stimulate the production of VEGF (Feoktistov et al. 2003; Ryzhov et al. 2008b), which can be considered proinflammatory in certain disease states. The $A_{2B}AR$ driven production of these proinflammatory molecules has been attributed to both G_s and G_q pathways (Feoktistov et al. 1999; Sitaraman et al. 2001).

$A_{2B}AR$ expression has been shown to be increased in the gastrointestinal track during inflammatory bowel disease and colitis (Hosokawa et al. 1999). Evidence in animal models of inflammatory bowel disease demonstrates that $A_{2B}AR$ activation can stimulate the release of IL-6, an important proinflammatory cytokine in inflammatory bowel disease, from the apical surface of the colonic epithelium (Sitaraman et al. 2001). This increased IL-6 secretion is proposed to promote the degranulation of neutrophils and contribute to disease progression. In support of this model, emerging work from the Sitaraman laboratory suggests that treatment of animal models of inflammatory bowel disease with the $A_{2B}AR$ antagonist ATL 801 inhibits IL-6 production and is associated with an improvement in clinical and histological signs in these models (Kolachala et al. 2008a, b). Collectively, these findings suggest that $A_{2B}AR$ signaling may play a role in the progression of inflammatory bowel disease, and so an $A_{2B}AR$ antagonists may have therapeutic benefit in related conditions.

As mentioned earlier, engagement of the $A_{2B}AR$ on HMC-1 cells appears to have predominantly proinflammatory functions (Feoktistov et al. 1998); however, recent studies in $A_{2B}AR$ knockout mice have revealed that this receptor plays an anti-inflammatory function in bone marrow-derived mast cells (Hua et al. 2007). Understanding the complexity of $A_{2B}AR$ signaling in this cell type may relate to the distinct inflammatory functions of this cell type. Following antigen priming and stimulation, mast cells undergo a degranulation process where preformed mediators are released (Shimizu and Schwartz 1997). This process is part of an acute response to antigen that mediates important processes such as bronchoconstriction in airways. In addition to this acute response, mast cells are stimulated to produce and release cytokines and chemokines. This inflammatory response is part of a more chronic or late-stage response that can promote additional tissue inflammation and injury. The recent work from Ryzhov and colleagues demonstrates that engagement of the $A_{2B}AR$ on mouse mast cells does not contribute to adenosine's ability to promote mast cell degranulation; however, the $A_{2B}AR$ does contribute to the production of IL-13 and VEGF (Ryzhov et al. 2008b). Thus, adenosine-mediated degranulation is likely mediated by the A_3AR in rodents, while the $A_{2B}AR$ regulates mediator production. It is important to note that significant species differences have been noted in relation to adenosine's effects on mast cells, and relatively little is known about the contribution of the $A_{2B}AR$ in human mast cells. Continued efforts to define the functions of adenosine receptors on mast cells in specific disease environments such as asthma will be critical to the development of adenosine-based therapeutics targeting mast-cell effector activities.

The apparent low affinity of adenosine for the $A_{2B}AR$ suggests that this receptor may have important roles in pathological environments where adenosine levels are elevated (Fredholm 2007). Consistent with this, work in the ADA-deficient model of adenosine-dependent lung inflammation and injury has demonstrated proinflammatory features of $A_{2B}AR$ signaling (Sun et al. 2006). In this model, mice that lack ADA exhibit progressive increases in lung adenosine concentrations in association with progressive pulmonary inflammation and tissue remodeling (Blackburn et al. 2000; Chunn et al. 2005). Noted features include the accumulation of activated alveolar macrophages that produce numerous inflammatory mediators, including IL-6, CXCL1, TGF-β1 and osteopontin (Sun et al. 2006). Production of these mediators is associated with alveolar airway destruction, mucus cell metaplasia and pulmonary fibrosis. $A_{2B}AR$ expression is elevated in the lungs of ADA-deficient mice, and treatment of these mice with the selective $A_{2B}AR$ antagonist CVT-6883 resulted in decreased production of proinflammatory mediators from macrophages, which was associated with decreased alveolar airway enlargement and pulmonary fibrosis (Sun et al. 2006). Similarly, a recent study using a mouse model of ragweed sensitization and challenge in the lung revealed that $A_{2B}AR$ antagonism with CVT-6883 (3-ethyl-1-propyl-8-[1-(3-trifluoromethylbenzyl)-1*H*-pyrazol-4-yl]-3,7-dihydropurine-2,6-dione) was associated with decreased airway inflammation and airway hyperreactivity (Mustafa et al. 2007). These studies demonstrate that $A_{2B}AR$ signaling plays a proinflammatory role in the lung, and suggest that $A_{2B}AR$ antagonists may prove beneficial in the treatment of lung disorders such as asthma, COPD, and pulmonary fibrosis.

4.4 A_3AR and Inflammatory Responses

4.4.1 Historical Perspective

The A_3AR was first identified through molecular cloning from a rat testis cDNA library based on 40% sequence homology with the canine A_1AR and $A_{2A}AR$ (Meyerhof et al. 1991). A_1AR, $A_{2A}AR$ and $A_{2B}AR$ are antagonized by methylxanthines, such as caffeine, theophylline and enprofylline, while the A_3AR is relatively xanthine insensitive, which may have been a reason for its relatively late discovery. In 1992, Zhou et al. cloned the A_3AR from rat striatum, expressed the protein in Chinese hamster ovary cells, and showed that A_3AR engagement leads to inhibition of AC (Zhou et al. 1992). Later it was shown that inhibition of AC is achieved through activation of the pertussis toxin-sensitive Gia2,3 protein (Palmer et al. 1995b). Ligand binding can also result in activation of PLC through $G_{q/11}$ or the $\beta\gamma$ subunits, leading to increased release of Ca^{2+} (Abbracchio et al. 1995). More recent studies revealed several additional intracellular pathways that can be accessed by the A_3AR in different cell types to promote tissue-specific functions (for a review, see Gessi et al. 2008). Homologs of the A_3AR gene have been cloned from several species, and only 74% sequence homology was found between the genes from rat and human or sheep, while there is 85% homology between human and sheep. The recently cloned and characterized equine A_3AR gene shows a high degree of sequence homology to the human and sheep genes, but has a different pharmacological profile (Brandon et al. 2006). These species differences make it possible to design highly selective ligands for the human A_3AR, but the disadvantage is that these ligands cannot be adequately tested in rodent models. In an effort to circumvent this problem, Yamano and his colleagues created an A_3AR-humanized mouse by replacing the mouse A_3AR gene with the human gene (Yamano et al. 2006). When bone marrow-derived mast cells from the A_3AR-humanized mice were treated with an A_3AR agonist, an elevation of intracellular Ca^{2+} concentration was observed, and this increase could be completely antagonized by a human-selective A_3AR antagonist. However, the A_3AR agonist did not potentiate antigen-dependent degranulation, probably because the agonist-stimulated human A_3AR could not activate the phosphorylation of either ERK 1/2 or protein kinase B due to uncoupling of the receptor from G proteins (Yamano et al. 2006). To overcome the uncoupling, the group generated A_3AR functionally humanized mice by replacing the mouse A_3AR gene with a chimeric human/mouse sequence in which the intracellular regions of the human receptor were substituted for the corresponding regions of the mouse A_3AR. Activation of the chimeric A_3AR led to intracellular Ca^{2+} elevation and activation of the PI3Kγ-signaling pathway, which are equivalent to the actions induced by the A_3AR in wild-type mice. The human A_3AR antagonist had the same binding affinities for this chimeric receptor as for the human A_3AR, and completely antagonized the potentiation of antigen-dependent mast-cell degranulation. These studies provided the first direct evidence that the uncoupling of mouse G protein(s) to the human A_3AR is due to a sequence difference in the intracellular regions of the receptor protein critical for G-protein recognition/coupling. It is expected that the A_3AR functionally humanized mice can be employed for pharmacological evaluations of the human A_3AR antagonists (Yamano et al. 2006).

4.4.2 Anti-inflammatory Effects

The A_3AR has been shown to suppress LPS-induced TNF-α production in vitro from various macrophage cell lines (Le Vraux et al. 1993; McWhinney et al. 1996; Sajjadi et al. 1996; Martin et al. 2006) and microglial cells (Lee et al. 2006b), where, depending on the cell type, different signal transduction pathways are responsible for the inhibition. This inhibitory effect was also assessed in vivo by treating wild-type and A_3AR knockout mice with the A_3AR agonist 2-Cl–IB–MECA (2-chloro-N^6-(3-iodobenzyl)-adenosine-5′-N-methyluronamide) following LPS challenge, resulting in decreased TNF-α production that was more pronounced in wild-type mice (Salvatore et al. 2000). In peritoneal macrophages isolated from A_3AR knockout or wild-type mice, treatment with IB–MECA reduced TNF-α release to the same extent (Kreckler et al. 2006). Both of these in vivo studies demonstrate that $A_{2A}AR$ activation inhibits the production of TNF-α regardless of the presence of A_3AR. Recent studies with human monocytes implicate both the $A_{2A}AR$ and $A_{2B}AR$ in the regulation of LPS-induced TNF-α production (Zhang et al. 2005; Hasko et al. 2007). In human neutrophils, both the A_3AR and the $A_{2A}AR$ are involved in the reduction of O_2^- generation (Bouma et al. 1997; Gessi et al. 2002), and the A_3AR also promotes neutrophil migration (Chen et al. 2006), thus performing both anti- and proinflammatory actions, respectively.

4.4.3 A_3AR in Disease Progression and Potential Agonist Therapies

A_3AR agonists can exert significant protective effects in animal models of arthritis. In a collagen-induced arthritis model, A_3AR activation inhibited CCL3 (MIP-1α) production (Szabo et al. 1998), while in autoimmune arthritis models, suppression of TNF-α production was found (Baharav et al. 2005). Moreover, inhibition of proinflammatory cytokine production was achieved by inhibiting the PI3K NF-κβ signaling pathway in adjuvant-induced arthritis (Madi et al. 2007). Methotrexate, a therapeutic agent that is widely used to treat arthritis, exerts its anti-inflammatory effect through adenosine, and was shown to upregulate the expression of A_3AR on peripheral blood mononuclear cells both in rats with adjuvant-induced arthritis and in patients with rheumatoid arthritis (Ochaion et al. 2006). Concomitant treatment with IB–MECA and methotrexate resulted in additive anti-inflammatory effects in the adjuvant-induced arthritis animal model. IB–MECA (CF-101) has been tested in Phase I and Phase IIa clinical trials, where it was found to be safe, well tolerated, and shows evidence of an anti-inflammatory effects in patients with rheumatoid arthritis (Silverman et al. 2008). IB–MECA was also found to be protective in other mouse models of inflammatory diseases. In endotoxemic mice, pretreatment with IB–MECA decreased mortality by reducing IL-12 and IFN-γ production independently of IL-10 production (Hasko et al. 1998). A similar effect was observed in different mouse models of colititis (Mabley et al. 2003) and septic peritonitis (Lee et al. 2006a), where IB–MECA treatment decreased the expression of proinflammatory cytokines, mainly TNF-α, while in A_3AR knockout mice inflammation was heightened. The A_3AR has been reported to have a protective role in vivo in lung injury following ischemia and reperfusion (Matot et al. 2006). Recently, in the bleomycin-induced lung injury model, significantly enhanced inflammatory cell recruitment was observed in the lungs of A_3AR knockout mice due to elevated expression of the chemokines CXCL-1, CCL11 (eotaxin-1) and GM-CSF (Morschl et al. 2008). These observations suggest that A_3AR agonists may represent a new family of orally bioavailable drugs in the treatment of inflammatory diseases (Bar-Yehuda et al. 2007). For more information on A_3ARs and inflammation, please refer to Chap. 10 of this volume, "A_3 Adenosine Receptor: Pharmacology and Role in Disease," by Borea et al.

4.4.4 Proinflammatory Effects

Early studies in mast cells indicated that A_3AR activation leads to increased inflammation by inducing the release of mediators and the potentiation of antigen-dependent degranulation (Ramkumar et al. 1993; Fozard et al. 1996; Reeves et al. 1997). A_3AR activation can enhance the degranulation of mast cells isolated from mouse lung through elevations of intracellular Ca^{2+} mediated by the coupling of G_i to PI3K (Zhong et al. 2003b). In contrast, in canine mast cells (Auchampach et al. 1997), degranulation was mediated by the $A_{2B}AR$ instead of the A_3AR. The specific adenosine receptors involved in the degranulation of human mast cells are not known.

A_3AR activation in sensitized guinea pigs resulted in increased inflammatory cell recruitment to the lung (Spruntulis and Broadley 2001). In rat mast cells, activation of the A_3AR inhibited apoptosis through protein kinase B phosphorylation (Gao et al. 2001), and this enhanced survival may contribute to inflammatory cell expansion in inflamed tissues.

A_3AR mRNA expression is elevated in transbronchial biopsy samples from asthma and COPD patients, where expression is localized to infiltrating eosinophils rather than mast cells (Walker et al. 1997). In ADA-deficient mice, which exhibit adenosine-mediated lung disease, genetic removal of the A_3AR or treatment with an A_3AR antagonist, MRS-1523 (3-propyl-6-ethyl-5-[(ethylthio)carbonyl]-2phenyl-4-propyl-3-pyridine carboxylate), prevented airway eosinophilia and decreased mucus production, suggesting that A_3AR signaling contributes to the regulation of features of chronic lung disease (Young et al. 2004). Additional ex vivo studies with mouse eosinophils confirmed the results of in vitro observations with human eosinophils, where A_3AR activation suppressed eosinophil chemotaxis (Ezeamuzie and Philips 1999). These observations suggest that the diminished eosinophilia in the ADA/A_3AR double-knockout mice or in the antagonist-treated ADA-deficient mice in Young's study is not a direct effect on the eosinophils. Indirect regulators may be cytokines and chemokines that are known to be involved in eosinophil recruitment, but these were not affected by removal or inhibition of the A_3AR in ADA-deficient mice, suggesting that other mediators, such as proteases, extracellular matrix proteins and cell adhesion molecules, may be responsible (Young et al. 2004).

In addition to regulating chemotaxis, the A_3AR is also important for eosinophil activation and degranulation, although results are contradictory. For example, A_3AR activation decreased degranulation and O_2^- production in human eosinophils isolated from blood (Ezeamuzie and Philips 1999; Ezeamuzie 2001), while there was no degranulation in the absence of the A_3AR assessed by eosinophil peroxidase release in the bronchoalveolar lavage fluid in bleomycin-challenged mice (Morschl et al. 2008). Moreover, the treatment of human eosinophils with Cl−IB−MECA (a selective A_3AR agonist) elevated intracellular Ca^{2+} levels, suggesting the presence of PLC-coupled A_3AR and supporting the role of A_3AR in eosinophil degranulation and chemotaxis, which are both Ca^{2+}-driven events (Kohno et al. 1996).

4.4.5 Potential Use of Antagonist in the Treatment of Inflammation

Early observations showing an A_3AR-mediated enhancement of antigen-dependent degranulation of mast cells in mice and bone marrow-derived cell lines (Reeves et al. 1997; Salvatore et al. 2000) suggested that selective A_3AR antagonists may have therapeutic potential as antiasthmatic agents. A compound with dual antagonist properties for both the $A_{2B}AR$ and the A_3AR, QAF 805 (Novartis), is under development as an antiasthma drug (Press et al. 2005). However, this mixed A_{2B}/A_3 AR antagonist has now entered human clinical trials and has failed to increase the PC_{20} for AMP versus placebo in 24 AMP-sensitive asthmatics in a placebo-controlled, double-blind, randomized, two-way crossover Phase Ib clinical trial (Pascoe et al. 2007). Moreover, researchers at GlaxoSmithKline developed a compound with dual $A_{2A}AR$ agonist and A_3AR antagonist effects that was able to inhibit both the production of reactive oxygen species and degranulation from human eosinophils and neutrophils in vitro, but provided very little clinical benefit when used in a clinical study for the treatment of allergic rhinitis (Bevan et al. 2007; Rimmer et al. 2007).

Although several potent and selective antagonists of the human A_3AR have been identified, they show extremely low binding affinity for the rodent A_3AR (typically 1,000 times lower), and since rodent models are used for the pharmacological evaluation of new therapeutic agents, this poses a serious drawback. The humanized A_3AR chimera mice (Yamano et al. 2006) may prove useful in overcoming this problem; however, further in vivo studies are needed to confirm that these mice can be utilized to test human A_3AR-selective compounds.

The A_3AR exerts both pro- and anti-inflammatory effects on different cell types and cell functions, but how these cells interact and influence each other in their microenvironment is still not known. Methods to examine A_3AR protein expression in situ are not sensitive, and it is possible that cells with low surface expression of A_3AR may be important in the regulation of inflammation in a manner that has not yet been appreciated.

Most cells express multiple types of adenosine receptors, and their actions can be overlapping or opposing, which may be an important mechanism to keep cell function in balance. It might prove beneficial to design analogs with dual or multiple affinities towards different types of adenosine receptors in order to influence various inflammatory actions at the same time. There are some reports of ligands that have $A_{2B}AR$ and A_3AR antagonist or $A_{2A}AR$ agonist and A_3AR antagonist properties that were designed to treat allergic airway diseases and that show promise in cell culture experiments, but further studies need to be performed with animal models of inflammation and in humans.

5 Conclusions

Orchestrated responses of cells to injury are essential for survival. As part of the body's ability to respond and recover from infection and injury, inflammatory processes help to limit infection and promote pathways for wound healing and the establishment of homeostasis. During various injurious situations, cells are placed under stress and must adapt to survive during the resolution of injury. Adenosine production and signaling has emerged as a major mechanism whereby cells respond to injury and regulate inflammation. There are precise mechanisms for regulating the production of extracellular adenosine at the cell surface, and there are now numerous studies demonstrating that this process helps to set in place pathways that can limit detrimental inflammatory processes while promoting beneficial inflammatory processes and promoting wound healing. The extensive research into the contributions of individual adenosine receptors on various immune cells, which has been extensively reviewed here, suggests that selective adenosine receptor agonists and antagonists may prove useful in regulating the immune response and hence the treatment of various injuries or diseases states. However, work from genetically modified mice and the use of selective adenosine receptor ligands in vivo have shown us that the path forward for the use of adenosine-based therapeutics will present many challenges. The engagement of all of the adenosine receptors has potent anti-inflammatory and tissue protective features in many situations. However, demonstrated proinflammatory and tissue destructive properties can also be ascribed to each of the adenosine receptors. Though this seems paradoxical, it may, in general terms, highlight the importance of adenosine signaling in regulating the balance between tissue injury and repair. For example, stimulation of anti-inflammatory adenosine receptor pathways will likely serve to stem inflammatory processes associated with numerous infections and challenges, as well as to promote wound-healing features such as angiogenesis and matrix deposition; however, overstimulation of such wound healing processes may actually promote disease. In addition, activation of anti-inflammatory pathways, such as the downregulation IL-12 and upregulation of IL-10, may bias the tissue environment toward Th2-like inflammation, which may present exacerbations of inflammation in certain environments. Thus, therapeutic approaches must take into account numerous factors, including the stage of the disease, the immunological and pathological processes involved, and the duration of treatment. Finally, the numerous observations that adenosine receptor expression increases in inflammatory environments suggest that we must learn more about receptor number or availability on the cell surface during specific inflammatory insults. Despite these challenges, it is clear that selective adenosine receptor engagement can regulate many of the features of inflammation, and with time and continued research, adequate approaches will be developed for the treatment of human disease with adenosine-based approaches.

In this regard, it is clear that adenosine receptors are important molecular targets for adenosine-based therapeutics for the entire spectrum from inflammation to immune suppression. Approaches utilizing adenosine receptor-based therapeutics will be dependent on the role of the adenosine receptors in mechanisms of disease in humans, the timing of treatment with respect to the therapeutic window and the stage/progression of injury, and the duration and monitoring of treatment for both beneficial effects and adverse events. A number of adenosine receptor-based ligands with good safety profiles and high selectivity are available for testing in humans. Preclinical efficacy in animal models does not always translate into human efficacy. The development of preclinical model systems with relevance to the human condition of inflammation is essential for successful drug discovery. Only by testing adenosine receptor-based ligands that are safe and selective for the adenosine receptor subtypes in humans will we understand the role of these receptors in human conditions of inflammation, which will allow for the successful development of human therapeutics towards these important molecular targets.

References

Abbracchio MP, Brambilla R, Ceruti S, Kim HO, von Lubitz DK, Jacobson KA, Cattabeni F (1995) G protein-dependent activation of phospholipase C by adenosine A_3 receptors in rat brain. Mol Pharmacol 48:1038–1045

Akbar M, Okajima F, Tomura H, Shimegi S, Kondo Y (1994) A single species of A_1 adenosine receptor expressed in Chinese hamster ovary cells not only inhibits cAMP accumulation but also stimulates phospholipase C and arachidonate release. Mol Pharmacol 45:1036–1042

Alam R, Gorska M (2003) Lymphocytes. J Allergy Clin Immunol 111:S476–S485

Anand-Srivastava MB, Cantin M, Ballak M, Picard S (1989) Desensitization of the stimulatory A_2 adenosine receptor-adenylate cyclase system in vascular smooth muscle cells from rat aorta. Mol Cell Endocrinol 62:273–279

Apasov S, Koshiba M, Redegeld F, Sitkovsky MV (1995) Role of extracellular ATP and P1 and P2 classes of purinergic receptors in T-cell development and cytotoxic T lymphocyte effector functions. Immunol Rev 146:5–19

Apasov S, Chen JF, Smith P, Sitkovsky M (2000) A(2A) receptor dependent and A(2A) receptor independent effects of extracellular adenosine on murine thymocytes in conditions of adenosine deaminase deficiency. Blood 95:3859–3867

Auchampach JA, Jin X, Wan TC, Caughey GH, Linden J (1997) Canine mast cell adenosine receptors: cloning and expression of the A_3 receptor and evidence that degranulation is mediated by the A_{2B} receptor. Mol Pharmacol 52:846–860

Auchampach JA, Jin X, Moore, J, Wan TC, Kreckler LM, Ge ZD, Narayanan J, Whalley E, Kiesman W, Ticho B, Smits G, Gross GJ (2004) Comparison of three different A_1 adenosine receptor antagonists on infarct size and multiple cycle ischemic preconditioning in anesthetized dogs. J Pharmacol Exp Ther 308:846–856

Awad AS, Huang L, Ye H, Duong ET, Bolton WK, Linden J, Okusa MD (2006) Adenosine A_{2A} receptor activation attenuates inflammation and injury in diabetic nephropathy. Am J Physiol Renal Physiol 290:F828–F837

Baharav E, Bar-Yehuda S, Madi L, Silberman D, Rath-Wolfson L, Halpren M, Ochaion A, Weinberger A, Fishman P (2005) Antiinflammatory effect of A_3 adenosine receptor agonists in murine autoimmune arthritis models. J Rheumatol 32:469–476

Baldwin SA, Beal PR, Yao SY, King AE, Cass CE, Young JD (2004) The equilibrative nucleoside transporter family, SLC29. Pflugers Arch 447:735–743

Banchereau J, Steinman RM (1998) Dendritic cells and the control of immunity. Nature 392:245–252

Bar-Yehuda S, Silverman MH, Kerns WD, Ochaion A, Cohen S, Fishman P (2007) The anti-inflammatory effect of A_3 adenosine receptor agonists: a novel targeted therapy for rheumatoid arthritis. Expert Opin Investig Drugs 16:1601–1613

Barraco RA, Helfman CC, Anderson GF (1996) Augmented release of serotonin by adenosine A_{2a} receptor activation and desensitization by CGS 21680 in the rat nucleus tractus solitarius. Brain Res 733:155–161

Basheer R, Bauer A, Elmenhorst D, Ramesh V, McCarley RW (2007) Sleep deprivation upregulates A_1 adenosine receptors in the rat basal forebrain. Neuroreport 18:1895–1899

Bell PD, Lapointe JY, Sabirov R, Hayashi S, Peti-Peterdi J, Manabe K, Kovacs G, Okada Y (2003) Macula densa cell signaling involves ATP release through a maxi anion channel. Proc Natl Acad Sci USA 100:4322–4327

Bertolet BD, Belardinelli L, Franco EA, Nichols WW, Kerensky RA, Hill JA (1996) Selective attenuation by N-0861 (N^6-endonorboran-2-yl-9-methyladenine) of cardiac A_1 adenosine receptor-mediated effects in humans. Circulation 93:1871–1876

Bevan N, Butchers PR, Cousins R, Coates J, Edgar EV, Morrison V, Sheehan MJ, Reeves J, Wilson DJ (2007) Pharmacological characterisation and inhibitory effects of (2*R*, 3*R*, 4*S*, 5*R*)-2-(6-amino-2-{[(1*S*)-2-hydroxy-1-(phenylmethyl)ethyl]amino}-9*H*-purin-9-yl)-5-(2-ethyl-2*H*-tetrazol-5-yl)tetrahydro-3,4-furandiol, a novel ligand that demonstrates both adenosine A(2A) receptor agonist and adenosine A(3) receptor antagonist activity. Eur J Pharmacol 564:219–225

Blackburn MR, Kellems RE (1996) Regulation and function of adenosine deaminase in mice. Prog Nucl Acid Res Mol Biol 55:195–226

Blackburn MR, Volmer JB, Thrasher JL, Zhong H, Crosby JR, Lee JJ, Kellems RE (2000) Metabolic consequences of adenosine deaminase deficiency in mice are associated with defects in alveogenesis, pulmonary inflammation, and airway obstruction. J Exp Med 192:159–170

Bonneau O, Wyss D, Ferretti S, Blaydon C, Stevenson CS, Trifilieff A (2006) Effect of adenosine A_{2A} receptor activation in murine models of respiratory disorders. Am J Physiol Lung Cell Mol Physiol 290:L1036–L1043

Bouma MG, van den Wildenberg FA, Buurman WA (1996) Adenosine inhibits cytokine release and expression of adhesion molecules by activated human endothelial cells. Am J Physiol 270:C522–529

Bouma MG, Jeunhomme TM, Boyle DL, Dentener MA, Voitenok NN, van den Wildenberg FA, Buurman WA (1997) Adenosine inhibits neutrophil degranulation in activated human whole blood: involvement of adenosine A_2 and A_3 receptors. J Immunol 158:5400–5408

Bours MJL, Swennen ELR, Virgilio FD, Cronstein BN, Dagnelie PC (2006) Adenosine 5′-phosphate and adenosine as endogenous sigaling molecules in immunity and inflammation. Pharmcol Ther 112:358–404

Bowlin TL, Borcherding DR, Edwards CK 3rd, McWhinney CD (1997) Adenosine A_3 receptor agonists inhibit murine macrophage tumor necrosis factor-alpha production in vitro and in vivo. Cell Mol Biol 43:345–349

Brandon CI, Vandenplas M, Dookwah H, Murray TF (2006) Cloning and pharmacological characterization of the equine adenosine A_3 receptor. J Vet Pharmacol Ther 29:255–263

Braunstein GM, Roman RM, Clancy JP, Kudlow BA, Taylor AL, Shylonsky VG, Jovov B, Peter K, Jilling T, Ismailov, II, Benos DJ, Schwiebert LM, Fitz JG, Schwiebert EM (2001) Cystic fibrosis transmembrane conductance regulator facilitates ATP release by stimulating a separate ATP release channel for autocrine control of cell volume regulation. J Biol Chem 276:6621–6630

Broussas M, Cornillet-Lefèbvre P, Potron G, Nguyen P (1999) Inhibition of fMLP-triggered respiratory burst of human monocytes by adenosine: involvement of A_3 adenosine receptor. J Leukoc Biol 66:495–501

Broussas M, Cornillet-Lefèbvre P, Potron G, Nguyên P (2002) Adenosine inhibits tissue factor expression by LPS-stimulated human monocytes: involvement of the A_3 adenosine receptor. Thromb Haemost 88:123–130

Brown RA, Clarke GW, Ledbetter CL, Hurle MJ, Denyer JC, Simcock DE, Coote JE, Savage TJ, Murdoch RD, Page CP, Spina D, O'Connor BJ (2008) Elevated expression of adenosine A_1 receptor in bronchial biopsy specimens from asthmatic subjects. Eur Respir J 31:311–319

Bruns RF, Lu GH, Pugsley TA (1986) Characterization of the A_2 adenosine receptor labeled by [3H]NECA in rat striatal membranes. Mol Pharmacol 29:331–346

Burg ND, Pillinger MH (2001) The neutrophil: function and regulation in innate and humoral immunity. Clin Immunol 99:7–17

Cassada DC, Tribble CG, Long SM, Laubach VE, Kaza AK, Linden J, Nguyen BN, Rieger JM, Fiser SM, Kron IL, Kern JA (2002) Adenosine A_{2A} analogue ATL-146e reduces systemic tumor necrosing factor-alpha and spinal cord capillary platelet–endothelial cell adhesion molecule-1 expression after spinal cord ischemia. J Vasc Surg 35:994–998

Cavalcante IC, Castro MV, Barreto AR, Sullivan GW, Vale M, Almeida PR, Linden J, Rieger JM, Cunha FQ, Guerrant RL, Ribeiro RA, Brito GA (2006) Effect of novel A_{2A} adenosine receptor agonist ATL 313 on *Clostridium difficile* toxin A-induced murine ileal enteritis. Infect Immun 74:2606–2612

Chang YH, Conti M, Lee YC, Lai HL, Ching YH, Chern Y (1997) Activation of phosphodiesterase IV during desensitization of the A_{2A} adenosine receptor-mediated cyclic AMP response in rat pheochromocytoma (PC12) cells. J Neurochem 69:1300–1309

Chen L, Fredholm BB, Jondal M (2008) Adenosine, through the A(1) receptor, inhibits vesicular MHC class I cross-presentation by resting DC. Mol Immunol 45:2247–2254

Chen Y, Corriden R, Inoue Y, Yip L, Hashiguchi N, Zinkernagel A, Nizet V, Insel PA, Junger WG (2006) ATP release guides neutrophil chemotaxis via $P2Y_2$ and A_3 receptors. Science 314:1792–1795

Chunn JL, Molina JG, Mi T, Xia Y, Kellems RE, Blackburn MR (2005) Adenosine-dependent pulmonary fibrosis in adenosine deaminase-deficient mice. J Immunol 175:1937–1946

Ciruela F, Casadó V, Rodrigues RJ, Luján R, Burgueño J, Canals M, Borycz J, Rebola N, Goldberg SR, Mallol J, Cortés A, Canela EI, López-Giménez JF, Milligan G, Lluis C, Cunha RA, Ferré S, Franco R (2006) Presynaptic control of striatal glutamatergic neurotransmission by adenosine A_1–A_{2A} receptor heteromers. J Neurosci 26:2080–2087

Clark AN, Youkey R, Liu X, Jia L, Blatt R, Day YJ, Sullivan GW, Linden J, Tucker AL (2007) A_1 adenosine receptor activation promotes angiogenesis and release of VEGF from monocytes. Circ Res 101:1130–1138

Cotrina ML, Lin JH, Nedergaard M (1998) Cytoskeletal assembly and ATP release regulate astrocytic calcium signaling. J Neurosci 18:8794–8804

Cronstein BN, Kramer SB, Weissmann G, Hirschhorn R (1983) Adenosine: a physiological modulator of superoxide anion generation by human neutrophils. J Exp Med 158:1160–1177

Cronstein BN, Daguma L, Nichols D, Hutchison AJ, Williams M (1990) The adenosine/neutrophil paradox resolved: human neutrophils possess both A_1 and A_2 receptors that promote chemotaxis and inhibit O_2 generation, respectively. J Clin Invest 85:1150–1157

Cronstein BN, Levin RI, Philips M, Hirschhorn R, Abramson SB, Weissmann G (1992) Neutrophil adherence to endothelium is enhanced via adenosine A_1 receptors and inhibited via adenosine A_2 receptors. J Immunol 148:2201–2206

Day YJ, Huang L, McDuffie MJ, Rosin DL, Ye H, Chen JF, Schwarzschild MA, Fink JS, Linden J, Okusa MD (2003) Renal protection from ischemia mediated by A_{2A} adenosine receptors on bone marrow-derived cells. J Clin Invest 112:883–891

Day YJ, Marshall MA, Huang L, McDuffie MJ, Okusa MD, Linden J (2004) Protection from ischemic liver injury by activation of A_{2A} adenosine receptors during reperfusion: inhibition of chemokine induction. Am J Physiol Gastrointest Liver Physiol 286:G285–G293

Deaglio S, Dwyer KM, Gao W, Friedman D, Usheva A, Erat A, Chen JF, Enjyoji K, Linden J, Oukka M, Kuchroo VK, Strom TB, Robson SC (2007) Adenosine generation catalyzed by CD39 and CD73 expressed on regulatory T cells mediates immune suppression. J Exp Med 204:1257–1265

Deguchi, H; Tekeya, H, Urano, H; Gabazza, EC; Zhou H; Suzuki, K (1998) Adenosine regulates tissue factor expression on endothelial cells. Throm Res 91:57–64

Dickenson JM, Reeder S, Rees B, Alexander S, Kendall D (2003) Functional expression of adenosine A_{2A} and A_3 receptors in the mouse dendritic cell line XS-106. Eur J Pharmacol 474:43–51

Dittrich HC, Gupta DK, Hack TC, Dowling T, Callahan J, Thomson S (2007) The effect of KW-3902, an adenosine A_1 receptor antagonist, on renal function and renal plasma flow in ambulatory patients with heart failure and renal impairment. J Card Failure 13:609–617

Doggrell SA (2005) BG-9928 Biogen Idec. Curr Opin Investig Drugs 6:962–968

Duffield JS (2003) The inflammatory macrophage: a story of Jekyll and Hyde. Clin Sci 104:27–38

Eckle T, Faigle M, Grenz A, Laucher S, Thompson LF, Eltzschig HK (2008) A_{2B} adenosine receptor dampens hypoxia-induced vascular leak. Blood 111:2024–2035

Edwards SW (1994) Biochemistry and physiology of the neutrophil. Cambridge Univeristy Press, New York

Eltzschig HK, Ibla JC, Furuta GT, Leonard MO, Jacobson KA, Enjyoji K, Robson SC, Colgan SP (2003) Coordinated adenine nucleotide phosphohydrolysis and nucleoside signaling in posthypoxic endothelium: role of ectonucleotidases and adenosine A_{2B} receptors. J Exp Med 198:783–796

Eltzschig HK, Thompson LF, Karhausen J, Cotta RJ, Ibla JC, Robson SC, Colgan SP (2004) Endogenous adenosine produced during hypoxia attenuates neutrophil accumulation: coordination by extracellular nucleotide metabolism. Blood 104:3986–3992

Eppell BA, Newell AM, Brown EJ (1989) Adenosine receptors are expressed during differentiation of monocytes to macrophages invitro. J Immunol 143:4141–4145

Erdmann AA, Gao ZG, Jung U, Foley J, Borenstein T, Jacobson KA, Fowler DH (2005) Activation of Th1 and Tc1 cell adenosine A_{2A} receptors directly inhibits IL-2 secretion in vitro and IL-2-driven expansion in vivo. Blood 105:4707–4714

Ezeamuzie CI, Philips E (1999) Adenosine A_3 receptors on human eosinophils mediate inhibition of degranulation and superoxide anion release. Br J Pharmacol 127:188–194

Ezeamuzie CI (2001) Involvement of A(3) receptors in the potentiation by adenosine of the inhibitory effect of theophylline on human eosinophil degranulation: possible novel mechanism of the anti-inflammatory action of theophylline. Biochem Pharmacol 61:1551–1559

Ezeamuzie CI, Khan I (2007) The role of adenosine A_2 receptors in the regulation of TNF-α production and PGE2 release in mouse peritoneal macrophages. Int Immunopharm 7:483–490

Feoktistov I, Biaggioni I (1995) Adenosine A_{2b} receptors evoke interleukin-8 secretion in human mast cells. An enprofylline-sensitive mechanism with implications for asthma. J Clin Invest 96:1979–1986

Feoktistov I, Biaggioni I (1997) Adenosine A_{2B} receptors. Pharmacol Rev 49:381–402

Feoktistov I, Polosa R, Holgate ST, Biaggioni I (1998) Adenosine A_{2B} receptors: a novel therapeutic target in asthma? Trends Pharmacol Sci 19:148–153

Feoktistov I, Goldstein AE, Biaggioni I (1999) Role of p38 mitogen-activated protein kinase and extracellular signal-regulated protein kinase kinase in adenosine A_{2B} receptor-mediated interleukin-8 production in human mast cells. Mol Pharmacol 55:726–734

Feoktistov I, Goldstein AE, Ryzhov S, Zeng D, Belardinelli L, Voyno-Yasenetskaya T, Biaggioni I (2002) Differential expression of adenosine receptors in human endothelial cells: role of A_{2B} receptors in angiogenic factor regulation. Circ Res 90:531–538

Feoktistov I, Ryzhov S, Goldstein AE, Biaggioni I (2003) Mast cell-mediated stimulation of angiogenesis: cooperative interaction between A_{2B} and A_3 adenosine receptors. Circ Res 92:485–492

Feoktistov I, Ryzhov S, Zhong H, Goldstein AE, Matafonov A, Zeng D (2004) Hypoxia modulates adenosine receptors in human endothelial and smooth muscle cells towards an A_{2B} angiogenic phenotype. Hypertension 44:649–654

Ferguson G, Watterson KR, Palmer TM (2000) Subtype-specific kinetics of inhibitory adenosine receptor internalization are determined by sensitivity to phosphorylation by G protein-coupled receptor kinases. Mol Pharmacol 57:546–552

Ferguson G, Watterson KR, Palmer TM (2002) Subtype-specific regulation of receptor internalization and recycling by the carboxyl-terminal domains of the human A_1 and rat A_3 adenosine receptors: consequences for agonist-stimulated translocation of arrestin 3. Biochemistry 41:14748–14761

Ferre S, Ciruela F, Borycz J, Solinas M, Quarta D, Antoniou K, Quiroz C, Justinova Z, Lluis C, Franco R, Goldberg SR (2008) Adenosine A_1–A_{2A} receptor heteromers: new targets for caffeine in the brain. Front Biosci 13:2391–2399

Fishman P, Bar-Yehuda S (2003) Pharmacology and therapeutic applications of A_3 receptor subtype. Curr Top Med Chem 3:463–469

Flamand N, Boudreault S, Picard S, Austin M, Surette ME, Plante H, Krump E, Vallée MJ, Gilbert C, Naccache P, Laviolette M, Borgeat P (2000) Adenosine, a potent natural suppressor of arachidonic acid release and leukotriene biosynthesis in human neutrophils. Am J Respir Crit Care Med 161:S88–94

Flamand N, Surette ME, Picard S, Bourgoin S, Borgeat P (2002) Cyclic AMP-mediated inhibition of 5-lipoxygenase translocation and leukotriene biosynthesis in human neutrophils. Mol Pharmacol 62:250–256

Flamand N, Lefebvre J, Lapointe G, Picard S, Lemieux L, Bourgoin SG, Borgeat P (2006) Inhibition of platelet-activating factor biosynthesis by adenosine and histamine in human neutrophils: involvement of cPLA2alpha and reversal by lyso-PAF. J Leukoc Biol 79:1043–1051

Ford MS, Nie Z, Whitworth C, Rybak LP, Ramkumar V (1997) Up-regulation of adenosine receptors in the cochlea by cisplatin. Hear Res 111:143–152

Forman MB, Vitola JV, Velasco CE, Murray JJ, Dubey RK, Jackson EK (2000) Sustained reduction in myocardial reperfusion injury with an adenosine receptor antagonist: possible role of the neutrophil chemoattractant response. J Pharmacol Exp Ther 292:929–938

Fortin A, Harbour D, Fernandes M, Borgeat P, Bourgoin S (2006) Differential expression of adenosine receptors in human neutrophils: up-regulation by specific Th1 cytokines and lipopolysaccharide. J Leukoc Biol 79:574–585

Fossetta J, Jackson J, Deno G, Fan X, Du XK, Bober L, Soudé-Bermejo A, de Bouteiller O, Caux C, Lunn C, Lundell D, Palmer RK (2003) Pharmacological analysis of calcium responses mediated by the human A_3 adenosine receptor in monocyte-derived dendritic cells and recombinant cells. Mol Pharmacol 63:342–350

Fozard JR, Pfannkuche HJ, Schuurman HJ (1996) Mast cell degranulation following adenosine A3 receptor activation in rats. Eur J Pharmacol 298:293–297

Fozard JR, Ellis KM, Villela Dantas MF, Tigani B, Mazzoni L (2002) Effects of CGS 21680, a selective adenosine A_{2A} receptor agonist, on allergic airways inflammation in the rat. Eur J Pharmacol 438:183–188

Franco R, Casadó V, Mallol J, Ferrada C, Ferré S, Fuxe K, Cortés A, Ciruela F, Lluis C, Canela EI (2006) The two-state dimer receptor model: a general model for receptor dimers. Mol Pharmacol 69:1905–1912

Fredholm BB (2007) Adenosine, an endogenous distress signal, modulates tissue damage and repair. Cell Death Differ 14:1315–1323

Fredholm BB, AP IJ, Jacobson KA, Klotz KN, Linden J (2001) International Union of Pharmacology. XXV. Nomenclature and classification of adenosine receptors. Pharmacol Rev 53:527–552

Frigas E, Gleich GJ (1986) The eosinophil and the pathophysiology of asthma. J Allergy Clin Immunol 77:527–537

Frigas E, Motojima S, Gleich GJ (1991) The eosinophilic injury to the mucosa of the airways in the pathogenesis of bronchial asthma. Eur Respir J 13:123–135

Funakoshi H, Zacharia LC, Tang Z, Zhang J, Lee LL, Good JC, DE, Higuchi Y, Koch WJ, Jackson EK, Chan TO, Feldman AM (2007) A_1 adenosine receptor upregulation accompanies decreasing myocardial adenosine levels in mice with left ventricular dysfunction. Circulation 115:2307–2315

Fuxe K, Ferré S, Zoli M, Agnati LF (1998) Integrated events in central dopamine transmission as analyzed at multiple levels. Evidence for intramembrane adenosine A_{2A}/dopamine D_2 and adenosine A_1/dopamine D_1 receptor interactions in the basal ganglia. Brain Res Brain Res Rev 26:258–273

Gallos G, Ruyle TD, Emala CW, Lee HT (2005) A_1 adenosine receptor knockout mice exhibit increased mortality, renal dysfunction, and hepatic injury in murine septic peritonitis. Am J Physiol Renal Physiol 289:F369–F376

Gao Z, Li BS, Day YJ, Linden J (2001) A_3 adenosine receptor activation triggers phosphorylation of protein kinase B and protects rat basophilic leukemia 2H3 mast cells from apoptosis. Mol Pharmacol 59:76–82

Gao ZG, Jacobson KA (2007) Emerging adenosine receptor agonists. Expert Opin Emerg Drugs 124:79–492

Gessi S, Varani K, Merighi S, Cattabriga E, Iannotta V, Leung E, Baraldi PG, Borea PA (2002) A(3) adenosine receptors in human neutrophils and promyelocytic HL60 cells: a pharmacological and biochemical study. Mol Pharmacol 61:415–424

Gessi S, Varani K, Merighi S, Cattabriga E, Avitabile A, Gavioli R, Fortini C, Leung E, Mac Lennan S, Borea PA (2004) Expression of A_3 adenosine receptors in human lymphocytes: up-regulation in T cell activation. Mol Pharm 65:711–719

Gessi S, Varani K, Merighi S, Cattabriga E, Pancaldi C, Szabadkai Y, Rizzuto R, Klotz KN, Leung E, Mac Lennan S, Baraldi PG, Borea PA (2005) Expression, pharmacological profile, and functional coupling of A_{2B} receptors in a recombinant system and in peripheral blood cells using a novel selective antagonist radioligand, [3H]MRE 2029-F20. Mol Pharmacol 67:2137–2147

Gessi S, Merighi S, Varani K, Leung E, Mac Lennan S, Borea PA (2008) The A_3 adenosine receptor: an enigmatic player in cell biology. Pharmacol Ther 117:123–140

Gilliet M, Liu YJ (2002) Generation of human CD8 T regulatory cells by CD40 ligand-activated plasmacytoid dendritic cells. J Exp Med 195:695–704

Gilroy DW, Newson J, Sawmynaden P, Willoughby DA, Croxtall JD (2004) A novel role for phospholipase A_2 isoforms in the checkpoint control of acute inflammation. FASEB J 18:489–498

Givertz MM, Massie BM, Fields TK, Pearson LL, Dittrich HC (2007) The effect of KW-3902, an adenosine A_1-receptor antagonist, on diuresis and renal function in patients with acute decompensated heart failure and renal impairment or diuretic resistance. J Am Coll Cardiol 50:1551–1560

Gleich GJ, Loegering DA, Frigas E, Filley WV (1983) The eosinophil granule major basic protein: biological activities and relationship to bronchial asthma. Monogr Allergy 18:277–283

Glover DK, Riou LM, Ruiz M, Sullivan GW, Linden J, Rieger JM, Macdonald TL, Watson DD, Beller GA (2005) Reduction of infarct size and postischemic inflammation from ATL-146e, a highly selective adenosine A_{2A} receptor agonist, in reperfused canine myocardium. Am J Physiol Heart Circ Physiol 288:H1851–H1858

Gottlieb SS, Brater C, Thomas I, Havranek E, Bourge R, Goldman S, Dyer F, Gomez M, Bennett D, Ticho B, Beckman E, Abraham WT (2002) BG9719 (CVT-124), an A_1 adenosine receptor antagonist, protects against the decline in renal function observed with diuretic therapy. Circulation 105:1348–1353

Gray JH, Owen RP, Giacomini KM (2004) The concentrative nucleoside transporter family, SLC28. Pflugers Arch 447:728–734

Greenberg B, Ignatius T, Banish D, Goldman S, Havranek E, Massie BM, Zhu Y, Ticho B, Abraham WT (2007) Effects of multiple oral doses of an A_1 adenosine receptor antagonist, BG 9928, in patients with heart failure. J Am Coll Cardiol 50:600–606

Grenier S, Flamand N, Pelletier J, Naccache PH, Borgeat P, Bourgoin SG (2003) Arachidonic acid activates phospholipase D in human neutrophils; essential role of endogenous leukotriene B4 and inhibition by adenosine A_{2A} receptor engagement. J Leukoc Biol 73:530–539

Harada N, Okajima K, Murakami K, Usune S, Sato C, Ohshima K, Katsuragi T (2000) Adenosine and selective A(2A) receptor agonists reduce ischemia/reperfusion injury of rat liver mainly by inhibiting leukocyte activation. J Pharmacol Exp Ther 294:1034–1042

Hashikawa T, Hooker SW, Maj JG, Knott-Craig CJ, Takedachi M, Murakami S, Thompson LF (2004) Regulation of adenosine receptor engagement by ecto-adenosine deaminase. FASEB J 18:131–133

Hasko G, Cronstein BN (2004) Adenosine: an endogenous regulator of innate immunity. Trends Immunol 25:33–39

Hasko G, Pacher P (2008) A_{2A} receptors in inflammation and injury: lessons learned from transgenic animals. J Leukoc Biol 83:447–455

Haskó G, Szabó C, Németh ZH, Kvetan V, Pastores SM, Vizi ES (1996) Adenosine receptor agonists differentially regulate IL-10, TNF-alpha, and nitric oxide production in RAW 264.7 macrophages and in endotoxemic mice. J Immunol 10:4634–4640

Hasko G, Nemeth ZH, Vizi ES, Salzman AL, Szabo C (1998) An agonist of adenosine A_3 receptors decreases interleukin-12 and interferon-gamma production and prevents lethality in endotoxemic mice. Eur J Pharmacol 358:261–268

Hasko G, Kuhel DG, Chen JF, Schwarzschild MA, Deitch EA, Mabley JG, Marton A, Szabo C (2000) Adenosine inhibits IL-12 and TNF-[alpha] production via adenosine A_{2a} receptor-dependent and independent mechanisms. FASEB J 14:2065–2074

Hasko G, Pacher P, Deitch EA, Vizi ES (2007) Shaping of monocyte and macrophage function by adenosine receptors. Pharmacol Ther 113:264–275

Hermes M, von Hippel S, Osswald H, Kloor D (2005) S-Adenosylhomocysteine metabolism in different cell lines: effect of hypoxia and cell density. Cell Physiol Biochem 15:233–244

Hofer S, Ivarsson L, Stoitzner P, Auffinger M, Rainer C, Romani N, Heufler C (2003) Adenosine slows migration of dendritic cells but does not affect other aspects of dendritic cell maturation. J Invest Dermatol 121:300–307

Hordijk, PL (2006) Endothelial signalling events during leukocyte transmigration. FEBS J 273:4408–441

Hoskin DW, Mader JS, Furlong SJ, Conrad DM, Blay J (2008) Inhibition of T cell and natural killer cell function by adenosine and its contribution to immune evasion by tumor cells. Int J Oncol 32:527–535

Hosokawa T, Kusugami K, Ina K, Ando T, Shinoda M, Imada A, Ohsuga M, Sakai T, Matsuura T, Ito K, Kaneshiro K (1999) Interleukin-6 and soluble interleukin-6 receptor in the colonic mucosa of inflammatory bowel disease. J Gastroenterol Hepatol 14:987–996

Hua X, Kovarova M, Chason KD, Nguyen M, Koller BH, Tilley SL (2007) Enhanced mast cell activation in mice deficient in the A_{2b} adenosine receptor. J Exp Med 204:117–128

Huang QY, Wei C, Yu L, Coelho JE, Shen HY, Kalda A, Linden J, Chen JF (2006) Adenosine A_{2A} receptors in bone marrow-derived cells but not in forebrain neurons are important contributors to 3-nitropropionic acid-induced striatal damage as revealed by cell-type-selective inactivation. J Neurosci 26:11371–11378

Huang S, Apasov S, Koshiba M, Sitkovsky M (1997) Role of A_{2a} extracellular adenosine receptor-mediated signaling in adenosine-mediated inhibition of T-cell activation and expansion. Blood 90:1600–1610

Ikeda K, Kurokawa M, Aoyama S, Kuwana Y (2002) Neuroprotection by adenosine A_{2A} receptor blockade in experimental models of Parkinson's disease. J Neurochem 80:262–270

Inoue Y, Chen Y, Hirsh MI, Yip L, Junger WG (2008) A_3 and $P2Y_2$ receptors control their recruitment of neutrophils to the lungs in a model of sepsis. Shock 30:173–177

Iwamoto T, Umemura S, Toya Y, Uchibori T, Kogi K, Tagaki N, Ishii M (1994) Identification of adenosine A_2 receptor–cAMP system in human aortic endothelial cells. Biochem Biophys Res Commun 199:905–910

Izcue A, Coombes JL, Powrie F (2006) Regulatory T cells suppress systemic and mucosal immune activation to control intestinal inflammation. Immunol Rev 212:256–271

Jacobson KA, Gao ZG (2006) Adenosine receptors as therapeutic targets. Nat Rev Drug Discov 5:247–264

Jhaveri KA, Reichensperger J, Toth LA, Sekino Y, Ramkumar V (2007) Reduced basal and lipopolysaccharide-stimulated adenosine A_1 receptor expression in the brain of nuclear factor-kappaB p50–/– mice. Neuroscience 146:415–426

Jones PA, Smith RA, Stone TW (1998) Protection against hippocampal kainate excitotoxicity by intracerebral administration of an adenosine A_{2A} receptor antagonist. Brain Res 800:328–335

Joo JD, Kim M, Horst P, Kim J, D'Agati VD, Emala CW Sr, Lee HT (2007) Acute and delayed renal protection against renal ischemia and reperfusion injury with A_1 adenosine receptors. Am J Physiol Renal Physiol 293:F1847–F1857

Kaczmarek E, Koziak K, Sevigny J, Siegel JB, Anrather J, Beaudoin AR, Bach FH, Robson SC (1996) Identification and characterization of CD39/vascular ATP diphosphohydrolase. J Biol Chem 271:33116–33122

Kadowaki N, Antonenko S, Lau JY, Liu YJ (2000) Natural interferon alpha/beta-producing cells link innate and adaptive immunity. J Exp Med 192:219–226

Katori M, Tamaki T, Tanaka M, Konoeda Y, Yokota N, Hayashi T, Uchida Y, Hui Y, Takahashi Y, Kakita A, Kawamura A (1999) Cytoprotective role of antioxidant stress protein induced by adenosine A_1 receptor antagonist in rat heart ischemic injury. Transplant Proc 31:1016–1017

Khoa ND, Montesinos MC, Reiss AB, Delano D, Awadallah N, Cronstein BN (2001) Inflammatory cytokines regulate function and expression of adenosine A(2A) receptors in human monocytic THP-1 cells. J Immunol 167:4026–4032

Khoa, ND; Montesinos MC, Williams, AJ, Kelly M, Cronstein, BN (2003) Th1 cytokines regulate adenosine receptors and their downstream signaling elements in human microvascular endothelial cells. J Immunol 171:3991–3998

Khoa ND, Postow M, Danielsson J, Cronstein BN (2006) Tumor necrosis factor-alpha prevents desensitization of $G_{\alpha s}$-coupled receptors by regulating GRK2 association with the plasma membrane. Mol Pharmacol 69:1311–1319

Klaasse EC, Ijzerman AP, de Grip WJ, Beukers MW (2008) Internalization and desensitization of adenosine receptors. Purinergic Signal 4:21–37

Knight K, Zheng X, Rocchini C, Jacobson M, Bai T, Walker B (1997) Adenosine A_3 receptor stimulation inhibits migration of human eosinophils. J Leukoc Biol 62:465–468

Kobie JJ, Shah PR, Yang L, Rebhahn JA, Fowell DJ, Mosmann TR (2006) T regulatory and primed uncommitted CD4 T cells express CD73, which suppresses effector CD4 T cells by converting 5′-adenosine monophosphate to adenosine. J Immunol 177:6780–6786

Kohno Y, Ji X, Mawhorter SD, Koshiba M, Jacobson KA (1996) Activation of A_3 adenosine receptors on human eosinophils elevates intracellular calcium. Blood 88:3569–3574

Kolachala V, Asamoah V, Wang L, Obertone TS, Ziegler TR, Merlin D, Sitaraman SV (2005a) TNF-alpha upregulates adenosine 2b (A_{2b}) receptor expression and signaling in intestinal epithelial cells: a basis for A_{2b}R overexpression in colitis. Cell Mol Life Sci 62:2647–2657

Kolachala V, Asamoah V, Wang L, Srinivasan S, Merlin D, Sitaraman SV (2005b) Interferon-gamma down-regulates adenosine 2b receptor-mediated signaling and short circuit current in the intestinal epithelia by inhibiting the expression of adenylate cyclase. J Biol Chem 280:4048–4057

Kolachala VL, Bajaj R, Chalasani M, Sitaraman SV (2008a) Purinergic receptors in gastrointestinal inflammation. Am J Physiol Gastrointest Liver Physiol 294:G401–G410

Kolachala VL, Ruble BK, Vijay-Kumar M, Wang L, Mwangi S, Figler HE, Figler RA, Srinivasan S, Gewirtz AT, Linden J, Merlin D, Sitaraman SV (2008b) Blockade of adenosine A_{2B} receptors ameliorates murine colitis. Br J Pharmacol 155:127–137

Kong T, Westerman KA, Faigle M, Eltzschig HK, Colgan SP (2006) HIF-dependent induction of adenosine A_{2B} receptor in hypoxia. FASEB J 20:2242–2250

Koshiba M, Kojima H, Huang S, Apasov S, Sitkovsky MV (1997) Memory of extracellular adenosine A_{2A} purinergic receptor-mediated signaling in murine T cells. J Biol Chem 272:25881–25889

Koshiba M, Rosin DL, Hayashi N, Linden J, Sitkovsky MV (1999) Patterns of A_{2A} extracellular adenosine receptor expression in different functional subsets of human peripheral T cells. Flow cytometry studies with anti-A_{2A} receptor monoclonal antibodies. Mol Pharmacol 55:614–624

Kotenko SV (2002) The family of IL-10-related cytokines and their receptors: related, but to what extent? Cytokine Growth Factor Rev 13:223–240

Kreckler LM, Wan TC, Ge ZD, Auchampach JA (2006) Adenosine inhibits tumor necrosis factor-alpha release from mouse peritoneal macrophages via A_{2A} and A_{2B} but not the A_3 adenosine receptor. J Pharmacol Exp Ther 317:172–180

Krump E, Borgeat P (1999) Adenosine. An endogenous inhibitor of arachidonic acid release and leukotriene biosynthesis in human neutrophils Adv Exp Med Biol 447:107–115

Krump, E, Lemay G, Borgeat P (1996) Adenosine A_2 receptor-induced inhibition of leukotriene B4 synthesis in whole blood ex vivo. Br J Pharmacol 117:1639–1644

Krump E, Picard S, Mancini J, Borgeat P (1997) Suppression of leukotriene B4 biosynthesis by endogenous adenosine in ligand-activated human neutrophils. J Exp Med 186:1401–1406

La Sala A, Gadina M, Kelsall BL (2005) G(i)-protein-dependent inhibition of IL-12 production is mediated by activation of the phosphatidylinositol 3-kinase-protein 3 kinase B/Akt pathway and JNK. J Immunol 175:2994–2999

Lai DM, Tu YK, Liu IM, Cheng JT (2005) Increase of adenosine A_1 receptor gene expression in cerebral ischemia of Wistar rats. Neurosci Lett 387:59–61

Lappas CM, Rieger JM, Linden J (2005) A_{2A} adenosine receptor induction inhibits IFN-gamma production in murine CD4+ T cells. J Immunol 174:1073–1080

Lappas CM, Day YJ, Marshall MA, Engelhard VH, Linden J (2006) Adenosine A_{2A} receptor activation reduces hepatic ischemia reperfusion injury by inhibiting CD1d-dependent NKT cell activation. J Exp Med 203:2639–2648

Larosa DF, Orange JS (2008) Lymphocytes. J Allergy Clin Immunol 121:S364–S369

Lasley RD, Jahania MS, Mentzer RM, Jr (2001) Beneficial effects of adenosine A(2a) agonist CGS-21680 in infarcted and stunned porcine myocardium. Am J Physiol Heart Circ Physiol 280:H1660–1666

Le Vraux V, Chen YL, Masson I, De Sousa M, Giroud JP, Florentin I, Chauvelot-Moachon L (1993) Inhibition of human monocyte TNF production by adenosine receptor agonists. Life Sci 52:1917–1924

Lee HT, Emala CW (2000) Protective effects of renal ischemic preconditioning and adenosine pretreatment: role of A(1) and A(3) receptors. Am J Physiol Renal Physiol 278:F380–F387

Lee HT, Xu H, Nasr SH, Schnermann J, Emala CW (2004a) A_1 adenosine receptor knockout mice exhibit increased renal injury following ischemia and reperfusion. Am J Physiol Renal Physiol 286:F298–F306

Lee HT, Gallos G, Nasr SH, Emala CW (2004b) A_1 adenosine receptor activation inhibits inflammation, necrosis, and apoptosis after renal ischemia-reperfusion injury in mice. J Am Soc Nephrol 15:102–111

Lee HT, Kim M, Jan M, Penn RB, Emala CW (2007) Renal tubule necrosis and apoptosis modulation by A_1 adenosine receptor expression. Kidney Int 71:1249–1261

Lee HT, Kim M, Joo JD, Gallos G, Chen JF, Emala CW (2006a) A_3 adenosine receptor activation decreases mortality and renal and hepatic injury in murine septic peritonitis. Am J Physiol Regul Integr Comp Physiol 291:R959–R969

Lee JY, Jhun BS, Oh YT, Lee JH, Choe W, Baik HH, Ha J, Yoon KS, Kim SS, Kang I (2006b) Activation of adenosine A_3 receptor suppresses lipopolysaccharide-induced TNF-alpha production through inhibition of PI 3-kinase/Akt and NF-kappaB activation in murine BV2 microglial cells. Neurosci Lett 396:1–6

Lennon PF, Taylor CT, Stahl, GL, Colgan, SP (1998) Neutrophil-derived 5′-adenosine monophosphate promotes endothelial barrier function via CD-73-mediated conversion to adenosine and endothelial A_{2B} receptor activation. J Exp Med 188:1433–1443

Li Y, Oskouian RJ, Day YJ, Rieger JM, Liu L, Kern JA, Linden J (2006) Mouse spinal cord compression injury is reduced by either activation of the adenosine A_{2A} receptor on bone marrow-derived cells or deletion of the A_{2A} receptor on non-bone marrow-derived cells. Neuroscience 141:2029–2039

Linden J (2001) Molecular approach to adenosine receptors: receptor-mediated mechanisms of tissue protection. Annu Rev Pharmacol Toxicol 41:775–787

Linden J (2005) Adenosine in tissue protection and tissue regeneration. Mol Pharmacol 67:1385–1387

Link AA, Kino T, Worth JA, McGuire JL, Crane ML, Chrousos GP, Wilder RL, Elenkov IJ (2000) Ligand-activation of the adenosine A_{2a} receptors inhibits IL-12 production by human monocytes. J Immunol 164:436–442

Lokshin A, Raskovalova T, Huang X, Zacharia LC, Jackson EK, Gorelik E (2006) Adenosine-mediated inhibition of the cytotoxic activity and cytokine production by activated natural killer cells. Cancer Res 66:7758–7765

Londos C, Cooper DM, Wolff J (1980) Subclasses of external adenosine receptors. Proc Natl Acad Sci USA 77:2551–2554

Luijk B, van den Berge M, Kerstjens HA, Postma DS, Cass L, Sabin A, Lammers JW (2008) Effect of an inhaled adenosine A_{2A} agonist on the allergen-induced late asthmatic response. Allergy 63:75–80

Luty J, Hunt JA, Nobbs PK, Kelly E, Keen M, MacDermot J (1989) Expression and desensitisation of A_2 purinoceptors on cultured bovine aortic endothelial cells. Cardiovasc Res 23:303–307

Mabley J, Soriano F, Pacher P, Hasko G, Marton A, Wallace R, Salzman A, Szabo C (2003) The adenosine A_3 receptor agonist, N^6-(3-iodobenzyl)-adenosine-5′-N-methyluronamide, is protective in two murine models of colitis. Eur J Pharmacol 466:323–329

Macagno A, Napolitani G, Lanzavecchia A, Sallusto F (2007) Duration, combination and timing: the signal integration model of dendritic cell activation. Trends Immunol 28:227–233

Madara JL, Patapoff TW, Gillece-Castro B, Colgan SP, Parkos CA, Delp C, Mrsny RJ (1993) 5′-Adenosine monophosphate is the neutrophil-derived paracrine factor that elicits chloride secretion from T84 intestinal epithelial cell monolayers. J Clin Invest 91:2320–2325

Madi L, Cohen S, Ochayin A, Bar-Yehuda S, Barer F, Fishman P (2007) Overexpression of A_3 adenosine receptor in peripheral blood mononuclear cells in rheumatoid arthritis: involvement of nuclear factor-kappaB in mediating receptor level. J Rheumatol 34:20–26

Magata S, Taniguchi M, Suzuki T, Shimamura T, Fukai M, Furukawa H, Fujita M, Todo S (2007) The effect of antagonism of adenosine A_1 receptor against ischemia and reperfusion injury of the liver. J Surg Res 139:7–14

Makujina SR, Mustafa SJ (1993) Adenosine-5′-uronamides rapidly desensitize the adenosine A_2 receptor in coronary artery. J Cardiovasc Pharmacol 22:506–509

Marone G, Vigorita S, Triggiani M, Condorelli M (1986) Adenosine receptors on human lymphocytes. Adv Exp Med Biol 195:7–14

Marone G, Petracca R, Vigorita S, Genovese A, Casolaro V (1992) Adenosine receptors on human leukocytes. IV. Characterization of an A_1/Ri receptor. Int J Clin Lab Res 22:235–242

Marquardt DL, Gruber HE, Wasserman SI (1984) Adenosine release from stimulated mast cells. Proc Natl Acad Sci USA 81:6192–6196

Martin L, Pingle SC, Hallam DM, Rybak LP, Ramkumar V (2006) Activation of the adenosine A_3 receptor in RAW 264.7 cells inhibits lipopolysaccharide-stimulated tumor necrosis factor-alpha release by reducing calcium-dependent activation of nuclear factor-kappaB and extracellular signal-regulated kinase 1/2. J Pharmacol Exp Ther 316:71–78

Matot I, Weiniger CF, Zeira E, Galun E, Joshi BV, Jacobson KA (2006) A_3 adenosine receptors and mitogen-activated protein kinases in lung injury following in vivo reperfusion. Crit Care 10:R65

Mayne M, Fotheringham J, Yan HJ, Power C, Del Bigio MR, Peeling J, Geiger JD (2001) Adenosine A_{2A} receptor activation reduces proinflammatory events and decreases cell death following intracerebral hemorrhage. Ann Neurol 49:727–735

McColl SR, St-Onge M, Dussault AA, Laflamme C, Bouchard L, Boulanger J, Pouliot M (2006) Immunomodulatory impact of the A_{2A} adenosine receptor on the profile of chemokines produced by neutrophils. FASEB J 1:187–189

McPherson JA, Barringhaus KG, Bishop GG, Sanders JM, Rieger JM, Hesselbacher SE, Gimple LW, Powers ER, Macdonald T, Sullivan G, Linden J, Sarembock IJ (2001) Adenosine A(2A) receptor stimulation reduces inflammation and neointimal growth in a murine carotid ligation model. Arterioscler Thromb Vasc Biol 21:791–796

McWhinney CD, Dudley MW, Bowlin TL, Peet NP, Schook L, Bradshaw M, De M, Borcherding DR, Edwards CK (1996) Activation of adenosine A_3 receptors on macrophages inhibits tumor necrosis factor-alpha. Eur J Pharmacol 310:209–216

Mehta D, Malik AB (2006) Signaling mechanisms regulating endothelial permeability. Physiol Rev 86:279–367

Merrill JT, Shen C, Schreibman D, Coffey D, Zakharenko O, Fisher R, Lahita RG, Salmon J, Cronstein BN (1997) Adenosine A_1 receptor promotion of multinucleated giant cell formation by human monocytes: a mechanism for methotrexate-induced nodulosis in rheumatoid arthritis. Arthritis Rheum 40:1308–1315

Meyerhof W, Paust HJ, Schonrock C, Richter D (1991) Cloning of a cDNA encoding a novel putative G-protein-coupled receptor expressed in specific rat brain regions. DNA Cell Biol 10:689–694

Minguet S, Huber M, Rosenkranz L, Schamel WW, Reth M, Brummer T (2005) Adenosine and cAMP are potent inhibitors of the NF-kappa B pathway downstream of immunoreceptors. Eur J Immunol 35:31–41

Mirabet M, Mallol J, Lluis C, Franco R (1997) Calcium mobilization in Jurkat cells via A_{2b} adenosine receptors. Br J Pharmacol 122:1075–1082

Mirabet M, Herrera C, Cordero OJ, Mallol J, Lluis C, Franco R (1999) Expression of A_{2B} adenosine receptors in human lymphocytes: their role in T cell activation. J Cell Sci 112:491–502

Mohsenin A, Mi T, Xia Y, Kellems RE, Chen JF, Blackburn MR (2007) Genetic removal of the A_{2A} adenosine receptor enhances pulmonary inflammation, mucin production, and angiogenesis in adenosine deaminase-deficient mice. Am J Physiol Lung Cell Mol Physiol 293:L753–L761

Monopoli A, Lozza G, Forlani A, Mattavelli A, Ongini E (1998) Blockade of adenosine A_{2A} receptors by SCH 58261 results in neuroprotective effects in cerebral ischaemia in rats. Neuroreport 9:3955–3959

Montesinos MC, Desai A, Chen JF, Yee H, Schwarzschild MA, Fink JS, Cronstein BN (2002) Adenosine promotes wound healing and mediates angiogenesis in response to tissue injury via occupancy of A(2A) receptors. Am J Pathol 160:2009–2018

Montesinos MC, Desai A, Delano D, Chen JF, Fink JS, Jacobson MA, Cronstein BN (2003) Adenosine A_{2A} or A_3 receptors are required for inhibition of inflammation by methotrexate and its analog MX-68. Arthritis Rheum 48:240–247

Moore KW, O'Garra A, de Waal Malefyt R, Vieira P, Mosmann TR (1993) Interleukin-10. Annu Rev Immunol 11:165–190

Moore KW, de Waal Malefyt R, Coffman RL, O'Garra A (2001) Interleukin-10 and the interleukin-10 receptor. Annu Rev Immunol 19:683–765

Morello S, Ito K, Yamamura S, Lee KY, Jazrawi E, Desouza P, Barnes P, Cicala C, Adcock IM (2006) IL-1 beta and TNF-alpha regulation of the adenosine receptor (A_{2A}) expression: differential requirement for NF-kappa B binding to the proximal promoter. J Immunol 177:7173–7183

Morschl E, Molina JG, Volmer J, Mohsenin A, Pero RS, Hong JS, Kheradmand F, Lee JJ, Blackburn MR (2008) A_3 adenosine receptor signaling influences pulmonary inflammation and fibrosis. Am J Respir Cell Mol Biol 39(6):697–705

Mosmann TR (1994) Properties and functions of interleukin-10. Adv Immunol 56:1–26

Mundell SJ, Kelly E (1998) Evidence for co-expression and desensitization of A_{2a} and A_{2b} adenosine receptors in NG108–15 cells. Biochem Pharmacol 55:595–603

Mundell SJ, Luty JS, Willets J, Benovic JL, Kelly E (1998) Enhanced expression of G protein-coupled receptor kinase 2 selectively increases the sensitivity of A_{2A} adenosine receptors to agonist-induced desensitization. Br J Pharmacol 125:347–356

Murphree LJ, Sullivan GW, Marshall MA, Linden J (2005) Lipopolysaccharide rapidly modifies adenosine receptor transcripts in murine and human macrophages: role of NF-kappaB in A(2A) adenosine receptor induction. Biochem J 391:575–580

Mustafa SJ, Nadeem A, Fan M, Zhong H, Belardinelli L, Zeng D (2007) Effect of a specific and selective A(2B) adenosine receptor antagonist on adenosine agonist AMP and allergen-induced airway responsiveness and cellular influx in a mouse model of asthma. J Pharmacol Exp Ther 320:1246–1251

Nadeem A, Obiefuna PC, Wilson CN, Mustafa SJ (2006) Adenosine A_1 receptor antagonist versus montelukast on airway reactivity and inflammation. Eur J Pharmacol 551:116–124

Nadeem A, Fan A, Ansari HR, Ledent C, Mustafa SJ (2007) Airway inflammation and hyperreponsiveness in A_{2A} adenosine receptor deficient allergic mice. Am J Physiol Lung Cell Mol Physiol 292:1335–1344

Naganuma M, Wiznerowicz EB, Lappas CM, Linden J, Worthington MT, Ernst PB (2006) Cutting edge: critical role for A_{2A} adenosine receptors in the T cell-mediated regulation of colitis. J Immunol 177:2765–2769

Nakata H, Yoshioka K, Kamiya T (2004) Purinergic-receptor oligomerization: implications for neural functions in the central nervous system. Neurotox Res 6:291–297

Neely CF, Keith IM (1995) A_1 adenosine receptor antagonists block ischemia-reperfusion injury of the lung. Am J Physiol 268:L1036–L1046

Neely CF, Jin J, Keith IM (1997) A_1-adenosine receptor antagonists block endotoxin-induced lung injury. Am J Physiol 272:L353–L361

Neely CF, DiPierro FV, Kong M, Greelish JP, Gardner TJ (1996) A_1 adenosine receptor antagonists block ischemia-reperfusion injury of the heart. Circulation 94:II376–I1380

Nemeth ZH, Lutz CS, Csoka B, Deitch EA, Leibovich SJ, Gause WC, Tone M, Pacher P, Vizi ES, Hasko G (2005) Adenosine augments IL-10 production by macrophages through an A2B receptor-mediated posttranscriptional mechanism. J Immunol 175:8260–8270

Net M, Valero R, Almenara R, Barros P, Capdevila L, López-Boado MA, Ruiz A, Sánchez-Crivaro F, Miquel R, Deulofeu R, Taurá P, Manyalich M, García-Valdecasas JC (2005) The effect of normothermic recirculation is mediated by ischemic preconditioning in NHBD liver transplantation. Am J Transplant 5:2385–2392

Nguyen DK, Montesinos MC, Williams AJ, Kelly M, Cronstein BN (2003) Th1 cytokines regulate adenosine receptors and their downstream signaling elements in human microvascular endothelial cells. J Immunol 171:3991–3998

Nie Z, Mei Y, Ford M, Rybak L, Marcuzzi A, Ren H, Stiles GL, Ramkumar V (1998) Oxidative stress increases A_1 adenosine receptor expression by activating nuclear factor kappa B. Mol Pharmacol 53:663–669

Ochaion A, Bar-Yehuda S, Cohn S, Del Valle L, Perez-Liz G, Madi L, Barer F, Farbstein M, Fishman-Furman S, Reitblat T, Reitblat A, Amital H, Levi Y, Molad Y, Mader R, Tishler M, Langevitz P, Zabutti A, Fishman P (2006) Methotrexate enhances the anti-inflammatory effect of CF101 via up-regulation of the A_3 adenosine receptor expression. Arthritis Res Ther 8:R169

Odashima M, Bamias G, Rivera-Nieves J, Linden J, Nast CC, Moskaluk CA, Marini M, Sugawara K, Kozaiwa K, Otaka M, Watanabe S, Cominelli F (2005) Activation of A_{2A} adenosine receptor attenuates intestinal inflammation in animal models of inflammatory bowel disease. Gastroenterology 129:26–33

Ohta A, Sitkovsky M (2001) Role of G-protein-coupled adenosine receptors in downregulation of inflammation and protection from tissue damage. Nature 414:916–920

Ohta A, Gorelik E, Prasad SJ, Ronchese F, Lukashev D, Wong MK, Huang X, Caldwell S, Liu K, Smith P, Chen JF, Jackson EK, Apasov S, Abrams S, Sitkovsky M (2006) A_{2A} adenosine receptor protects tumors from antitumor T cells. Proc Natl Acad Sci USA 103:13132–13137

Okada SF, O'Neal WK, Huang P, Nicholas RA, Ostrowski LE, Craigen WJ, Lazarowski ER, Boucher RC (2004) Voltage-dependent anion channel-1 (VDAC-1) contributes to ATP release and cell volume regulation in murine cells. J Gen Physiol 124:513–526

Olah ME, Stiles GL (2000) The role of receptor structure in determining adenosine receptor activity. Pharmacol Ther 85:55–75

Olanrewaju HA, Qin W, Feoktistove I, Scemama JL, Mustafa SJ (2000) Adenosine A_{2a} and A_{2B} receptors in cultured human and porcine cornary artery endothelial cells. Am J Physiol 279:H650–H656

Palmer TM, Stiles GL (2000) Identification of threonine residues controlling the agonist-dependent phosphorylation and desensitization of the rat A(3) adenosine receptor. Mol Pharmacol 57:539–545

Palmer TM, Gettys TW, Jacobson KA, Stiles GL (1994) Desensitization of the canine A_{2a} adenosine receptor: delineation of multiple processes. Mol Pharmacol 45:1082–1094

Palmer TM, Benovic JL, Stiles GL (1995a) Agonist-dependent phosphorylation and desensitization of the rat A_3 adenosine receptor. Evidence for a G-protein-coupled receptor kinase-mediated mechanism. J Biol Chem 270:29607–29613

Palmer TM, Gettys TW, Stiles GL (1995b) Differential interaction with and regulation of multiple G-proteins by the rat A_3 adenosine receptor. J Biol Chem 270:16895–16902

Palmer TM, Benovic JL, Stiles GL (1996) Molecular basis for subtype-specific desensitization of inhibitory adenosine receptors. Analysis of a chimeric A_1–A_3 adenosine receptor. J Biol Chem 271:15272–15278

Panther E, Idzko M, Herouy Y et al. (2001) Expression and function of adenosine receptors in human dendritic cells. FASEB J 15:1963–1970

Panther E, Corinti S, Idzko M, Herouy Y, Napp M, la Sala A, Girolomoni G, Norgauer J (2003) Adenosine affects expression of membrane molecules, cytokine and chemokine release, and the T-cell stimulatory capacity of human dendritic cells. Blood 101:3985–3990

Pascoe SJ, Knight H, Woessner R (2007) QAF805, an A_{2b}/A_3 adenosine receptor antagonist does not attenuate AMP challenge in subjects with asthma. Am J Resp Crit Care Med 175:A682

Pedata F, Corsi C, Melani A, Bordoni F, Latini S (2001) Adenosine extracellular brain concentrations and role of A_{2A} receptors in ischemia. Ann N Y Acad Sci 939:74–84

Pierce KD, Furlong TJ, Selbie LA, Shine J (1992) Molecular cloning and expression of an adenosine A_{2b} receptor from human brain. Biochem Biophys Res Commun 187:86–93

Pingle SC, Mishra S, Marcuzzi A, Bhat SG, Sekino Y, Rybak LP, Ramkumar V (2004) Osmotic diuretics induce adenosine A_1 receptor expression and protect renal proximal tubular epithelial cells against cisplatin-mediated apoptosis. J Biol Chem 279:43157–43167

Pinhal-Enfield G, Ramanathan M, Hasko G, Vogel SN, Salzman AL, Boons GJ, Leibovich SJ (2003) An angiogenic switch in macrophages involving synergy between Toll-like receptors 2, 4, 7, and 9 and adenosine A(2A) receptors. Am J Pathol 163:711–721

Ponnoth DS, Nadeem A, Mustafa SJ (2008) Adenosine-mediated alteration of vascular reactivity and inflammation in a murine model of asthma. Am J Physiol Heart Circ Physiol 294(5):H2158–H2165

Popoli P, Pintor A, Domenici MR, Frank C, Tebano MT, Pezzola A, Scarchilli L, Quarta D, Reggio R, Malchiodi-Albedi F, Falchi M, Massotti M (2002) Blockade of striatal adenosine A_{2A} receptor reduces, through a presynaptic mechanism, quinolinic acid-induced excitotoxicity: possible relevance to neuroprotective interventions in neurodegenerative diseases of the striatum. J Neurosci 22:1967–1975

Press NJ, Taylor RJ, Fullerton JD, Tranter P, McCarthy C, Keller TH, Brown L, Cheung R, Christie J, Haberthuer S, Hatto JD, Keenan M, Mercer MK, Press NE, Sahri H, Tuffnell AR, Tweed M, Fozard JR (2005) A new orally bioavailable dual adenosine A_{2B}/A_3 receptor antagonist with therapeutic potential. Bioorg Med Chem Lett 15:3081–3085

Priebe T, Kandil O, Nakic M, Pan BF, Nelson JA (1988) Selective modulation of antibody response and natural killer cell activity by purine nucleoside analogues. Cancer Res 48:4799–4803

Priebe T, Platsoucas CD, Nelson JA (1990a) Adenosine receptors and modulation of natural killer cell activity by purine nucleosides. Cancer Res 50:4328–4331

Priebe T, Platsoucas CD, Seki H, Fox FE, Nelson JA (1990b) Purine nucleoside modulation of functions of human lymphocytes. Cell Immunol 129:321–328

Priebe T, Ruiz L, Nelson JA (1990c) Role of natural killer cells in the modulation of primary antibody production by purine nucleosides and their analogs. Cell Immunol 130:513–519

Ramkumar V, Stiles GL, Beaven MA, Ali H (1993) The A_3 adenosine receptor is the unique adenosine receptor which facilitates release of allergic mediators in mast cells. J Biol Chem 268:16887–16890

Raskovalova T, Huang X, Sitkovsky M, Zacharia LC, Jackson EK, Gorelik E (2005) Gs protein-coupled adenosine receptor signaling and lytic function of activated NK cells. J Immunol 175:4383–4391

Raskovalova T, Lokshin A, Huang X, Jackson EK, Gorelik E (2006) Adenosine-mediated inhibition of cytotoxic activity and cytokine production by IL-2/NKp46-activated NK cells: involvement of protein kinase A isozyme I (PKA I). Immunol Res 36:91–99

Reece TB, Okonkwo DO, Ellman PI, Warren PS, Smith RL, Hawkins AS, Linden J, Kron IL, Tribble CG, Kern JA (2004) The evolution of ischemic spinal cord injury in function, cytoarchitecture, and inflammation and the effects of adenosine A_{2A} receptor activation. J Thorac Cardiovasc Surg 128:925–932

Reeves JJ, Jones CA, Sheehan MJ, Vardey CJ, Whelan CJ (1997) Adenosine A_3 receptors promote degranulation of rat mast cells both in vitro and in vivo. Inflamm Res 46:180–184

Reisin IL, Prat AG, Abraham EH, Amara JF, Gregory RJ, Ausiello DA, Cantiello HF (1994) The cystic fibrosis transmembrane conductance regulator is a dual ATP and chloride channel. J Biol Chem 269:20584–20591

Resnick MB, Colgan SP, Patapoff TW, Mrsny RJ, Awtrey CS, Delp-Archer C, Weller PF, Madara JL (1993) Activated eosinophils evoke chloride secretion in model intestinal epithelia primarily via regulated release of 5′-AMP. J Immunol 151:5716–5723

Resta R, Yamashita Y, Thompson LF (1998) Ecto-enzyme and signaling functions of lymphocyte CD73. Immunol Rev 161:95–109

Reutershan J, Cagnina RE, Chang D, Linden J, Ley K (2007) Therapeutic anti-inflammatory effects of myeloid cell adenosine receptor A_{2a} stimulation in lipopolysaccharide-induced lung injury. J Immunol 179:1254–1263

Rimmer J, Peake HL, Santos CM, Lean M, Bardin P, Robson R, Haumann B, Loehrer F, Handel ML (2007) Targeting adenosine receptors in the treatment of allergic rhinitis: a randomized, double-blind, placebo-controlled study. Clin Exp Allergy 37:8–14

Ritchie PK, Spangelo BL, Krzymowski DK, Rossiter TB, Kurth E, Judd AM (1997) Adenosine increases interleukin 6 release and decreases tumour necrosis factor release from rat adrenal zona glomerulosa cells, ovarian cells, anterior pituitary cells, and peritoneal macrophages. Cytokine 9:187–198

Rivkees SA, Reppert SM (1992) RFL9 encodes an A_{2b}-adenosine receptor. Mol Endocrinol 6:1598–1604

Rogachev B, Ziv NY, Mazar J, Nakav S, Chaimovitz C, Zlotnik M, Douvdevani A (2006) Adenosine is upregulated during peritonitis and is involved in downregulation of inflammation. Kidney Int 70:675–681

Roman RM, Lomri N, Braunstein G, Feranchak AP, Simeoni LA, Davison AK, Mechetner E, Schwiebert EM, Fitz JG (2001) Evidence for multidrug resistance-1 P-glycoprotein-dependent regulation of cellular ATP permeability. J Membr Biol 183:165–173

Rose FR, Hirschhorn R, Weissmann G, Cronstein BN (1988) Adenosine promotes neutrophil chemotaxis. J Exp Med 167:1186–94

Ross SD, Tribble CG, Linden J, Gangemi JJ, Lanpher BC, Wang AY, Kron IL (1999) Selective adenosine-A_{2A} activation reduces lung reperfusion injury following transplantation. J Heart Lung Transplant 18:994–1002

Rutherford MS, Witsell A, Schook LB (1993) Mechanisms generating functionally heterogeneous macrophages: chaos revisited. J Leukoc Biol 53:602–618

Ryzhov S, Goldstein AE, Matafonov A, Zeng D, Biaggioni I, Feoktistov I (2004) Adenosine-activated mast cells induce IgE synthesis by B lymphocytes: an A_{2B}-mediated process involving Th2 cytokines IL-4 and IL-13 with implications for asthma. J Immunol 172:7726–7733

Ryzhov SV, Zaynagetdinov R, Goldstein AE, Novitskiy SV, Blackburn MR, Biaggioni I, Feoktistov I (2008a) Effect of A_{2B} adenosine receptor gene ablation on adenosine-dependent regulation of proinflammatory cytokines. J Exp Pharmacol Ther 324:694–700

Ryzhov S, Zaynagetdinov R, Goldstein AE, Novitskiy SV, Dikov MM, Blackburn MR, Biaggioni I, Feoktistov I (2008b) Effect of A_{2B} adenosine receptor gene ablation on proinflammatory adenosine signaling in mast cells. J Immunol 180:7212–7220

Sajjadi FG, Takabayashi K, Foster AC, Domingo RC, Firestein GS (1996) Inhibition of TNF-alpha expression by adenosine: role of A_3 adenosine receptors. J Immunol 156:3435–3442

Sala-Newby GB, Skladanowski AC, Newby AC (1999) The mechanism of adenosine formation in cells. Cloning of cytosolic 5′-nucleotidase-I. J Biol Chem 274:17789–17793

Salmon JE, Cronstein BN (1990) Fc gamma receptor-mediated functions in neutrophils are modulated by adenosine receptor occupancy. A_1 receptors are stimulatory and A_2 receptors are inhibitory. J Immunol 145:2235–2240

Salmon JE, Brogle N, Brownlie C, Edberg JC, Kimberly RP, Chen BX, Erlanger BF (1993) Human mononuclear phagocytes express adenosine A_1 receptors. A novel mechanism for differential regulation of Fc gamma receptor function. J Immunol 151:2775–2785

Salvatore CA, Tilley SL, Latour AM, Fletcher DS, Koller BH, Jacobson MA (2000) Disruption of the A(3) adenosine receptor gene in mice and its effect on stimulated inflammatory cells. J Biol Chem 275:4429–4434

Sands WA, Palmer TM (2005) Adenosine receptors and the control of endothelial cell function in inflammatory disease. Immunol Lett 101:1–11

Satoh A, Shimosegawa T, Satoh K, Ito H, Kohno Y, Masamune A, Fujita M, Toyota T (2000) Activation of adenosine A_1-receptor pathway induces edema formation in the pancreas of rats. Gastroenterology 119:829–836

Schnurr M, Toy T, Shin A, Hartmann G, Rothenfusser S, Soellner J, Davis ID, Cebon J, Maraskovsky E (2004) Role of adenosine receptors in regulating chemotaxis and cytokine production of plasmacytoid dendritic cells. Blood 103:1391–1397

Schwaninger M, Neher M, Viegas E, Schneider A, Spranger M (1997) Stimulation of interleukin-6 secretion and gene transcription in primary astrocytes by adenosine. J Neurochem 69:1145–1150

Schwarzschild MA, Agnati L, Fuxe K, Chen JF, Morelli M (2006) Targeting adenosine A_{2A} receptors in Parkinson's disease. Trends Neurosci 29:647–654

Sexl V, Mancusi G, Holler C, Gloria-Maercker E, Schutz Freissmuth M (1997) Stimulation of mitogen-activated protein kinase via the A2a receptor in human primary endothelial cells. J Biol Chem 272:5792–5799

Shortman K, Liu YJ (2002) Mouse and human dendritic cell subtypes. Nat Rev Immunol 2:151–161

Shimizu Y, Schwartz LB (1997) Mast cell involvement in asthma. Lippincot-Raven, Philadelphia

Siegal FP, Kadowaki N, Shodell M, Fitzgerald-Bocarsly PA, Shah K, Ho S, Antonenko S, Liu, YJ (1999) The nature of the principal type 1 interferon-producing cells in human blood. Science 284:1835–1837

Silverman MH, Strand V, Markovits D, Nahir M, Reitblat T, Molad Y, Rosner I, Rozenbaum M, Mader R, Adawi M, Caspi D, Tishler M, Langevitz P, Rubinow A, Friedman J, Green L, Tanay A, Ochaion A, Cohen S, Kerns WD, Cohn I, Fishman-Furman S, Farbstein M, Yehuda SB, Fishman P (2008) Clinical evidence for utilization of the A_3 adenosine receptor as a target to treat rheumatoid arthritis: data from a phase II clinical trial. J Rheumatol 35:41–48

Sipka S, Kovacs I, Szanto S, Szegedi G, Brugos L, Bruckner G, Jozsef Szentmiklosi A (2005) Adenosine inhibits the release of interleukin-1beta in activated human peripheral mononuclear cells. Cytokine 31:258–263

Sitaraman SV, Merlin D, Wang L, Wong M, Gewirtz AT, Si-Tahar M, Madara JL (2001) Neutrophil–epithelial crosstalk at the intestinal lumenal surface mediated by reciprocal secretion of adenosine and IL-6. J Clin Invest 107:861–869

Sitkovsky MV, Ohta A (2005) The 'danger' sensors that STOP the immune response: the A_2 adenosine receptors? Trends Immunol 26:299–304

Sitkovsky MV, Lukashev D, Apasov S, Kojima H, Koshiba M, Caldwell C, Ohta A, Thiel M (2004) Physiological control of immune response and inflammatory tissue damage by hypoxia-inducible factors and adenosine A_{2A} receptors. Annu Rev Immunol 22:657–682

Spruntulis LM, Broadley KJ (2001) A_3 receptors mediate rapid inflammatory cell influx into the lungs of sensitized guinea-pigs. Clin Exp Allergy 31:943–951

Spychala J, Datta NS, Takabayashi K, Datta M, Fox IH, Gribbin T, Mitchell BS (1996) Cloning of human adenosine kinase cDNA: sequence similarity to microbial ribokinases and fructokinases. Proc Natl Acad Sci USA 93:1232–1237

Sullivan GW, Linden J, Buster BL, Scheld WM (1999) Neutrophil A_{2A} adenosine receptor inhibits inflammation in a rat model of meningitis: synergy with the type IV phosphodiesterase inhibitor, rolipram. J Infect Dis 180:1550–1560

Sullivan GW, Lee DD, Ross WG, DiVietro JA, Lappas CM, Lawrence MB, Linden J (2004) Activation of A_{2A} adenosine receptors inhibits expression of alpha 4/beta 1 integrin (very late antigen-4) on stimulated human neutrophils. J Leukoc Biol 75:127–134

Sun CX, Young HW, Molina JG, Volmer JB, Schnermann J, Blackburn MR. (2005) A protective role for the A_1 adenosine receptor in adenosine-dependent pulmonary injury. J Clin Invest 115:35–43

Sun CX, Zhong H, Mohsenin A, Morschl E, Chunn JL, Molina JG, Belardinelli L, Zeng D, Blackburn MR (2006) Role of A_{2B} receptor signaling in adenosine-dependent pulmonary inflammation and injury. J Clin Invest 116:1–10

Sun Y, Wu F, Sun F, Huang P (2008) Adenosine promotes IL-6 release in airway epithelia. J Immunol 180:4173–4181

Sundaram U, Hassanain H, Suntres Z, Yu JG, Cooke HJ, Guzman J, Christofi FL (2003) Rabbit chronic ileitis leads to up-regulation of adenosine A_1/A_3 gene products, oxidative stress, and immune modulation. Biochem Pharmacol 65:1529–1538

Surette ME, Krump E, Picard S, Borgeat P (1999) Activation of leukotriene synthesis in human neutrophils by exogenous arachidonic acid: inhibition by adenosine A(2a) receptor agonists and crucial role of autocrine activation by leukotriene B(4). Mol Pharmacol 56:1055–1062

Svenningsson P, Le Moine C, Fisone G, Fredholm BB (1999) Distribution, biochemistry and function of striatal adenosine A_{2A} receptors. Prog Neurobiol 59:355–396

Szabo C, Scott GS, Virag L, Egnaczyk G, Salzman AL, Shanley TP, Hasko G (1998) Suppression of macrophage inflammatory protein (MIP)-1alpha production and collagen-induced arthritis by adenosine receptor agonists. Br J Pharmacol 125:379–387

Takahashi HK, Iwagaki H, Tamura R, Xue D, Sano M, Mori S, Yoshino T, Tanaka N, Nishibori M (2003) Unique regulation profile of prostaglandin E1 on adhesion molecule expression and cytokine production in human peripheral blood mononuclear cells. J Pharmacol Exp Ther 307:1188–1195

Takahashi HK, Iwagaki H, Hamano R, Wake H, Kanke T, Liu K, Yoshino T, Tanaka N, Nishibori M (2007a) Effects of adenosine on adhesion molecule expression and cytokine production in human PBMC depend on the receptor subtype activated. Br J Pharmacol 150:816–822

Takahashi HK, Iwagaki H, Hamano R, Kanke T, Liu K, Sadamori H, Yagi T, Yoshino T, Sendo T, Tanaka N, Nishibori M (2007b) Effect of adenosine receptor subtypes stimulation on mixed lymphocyte reaction. Eur J Pharmacol 564:204–210

Taylor CT, Colgan SP (2007) Hypoxia and gastrointestinal disease. J Mol Med 85:1295–1300

Thiele A, Kronstein, R. Wetzel A, Gerth A, Nieber K, Hauschildt S (2004) Regulation of adenosine receptor subtypes during cultivation of human monocytes: role of receptors in preventing lipopolysaccharide-triggered respiratory burst. Infect Immun 72:1349–1357

Thiel M, Chouker A (1995) Acting via A_2 receptors, adenosine inhibits the production of tumor necrosis factor-alpha of endotoxin-stimulated human polymorphonuclear leukocytes. J Lab Clin Med 126:275–282

Thiel M, Chambers JD, Chouker A, Fischer S, Zourelidis C, Bardenheuer HJ, Arfors KE, Peter K (1996) Effect of adenosine on the expression of beta(2) integrins and L-selectin of human polymorphonuclear leukocytes in vitro. J Leukoc Biol 59:671–682

Thiel M, Caldwell CC, Sitkovsky MV (2003) The critical role of adenosine A_{2A} receptors in downregulation of inflammation and immunity in the pathogenesis of infectious diseases. Microb Infect 5:515–526

Thompson LF, Eltzschig HK, Ibla JC, Van De Wiele CJ, Resta R, Morote-Garcia JC, Colgan SP (2004) Crucial role for ecto-5′-nucleotidase (CD73) in vascular leakage during hypoxia. J Exp Med 200:1395–1405

Trincavelli ML, Marroni M, Tuscano D, Ceruti S, Mazzola A, Mitro N, Abbracchio MP, Martini C (2004) Regulation of A_{2B} adenosine receptor functioning by tumour necrosis factor a in human astroglial cells. J Neurochem 91:1180–1190

Trincavelli ML, Tonazzini I, Montali M, Abbracchio MP, Martini C (2008) Short-term TNF-Alpha treatment induced A2B adenosine receptor desensitization in human astroglial cells. J Cell Biochem 104:150–161

Tsutsui S, Schnermann J, Noorbakksh F, Henry S, Yong VW, Winston BW, Warren K, Power C (2004) A_1 adenosine receptor upregulation and activation attenuates neuroinflammation and demyelination in a model of multiple sclerosis. J Neurosci 24:1521–1529

Visser SS, Theron AJ, Ramafi G, Ker JA, Anderson R (2000) Apparent involvement of the A(2A) subtype adenosine receptor in the anti-inflammatory interactions of CGS 21680, cyclopentyladenosine, and IB-MECA with human neutrophils. Biochem Pharmacol 60:993–999

Volmer JB, Thompson LF, Blackburn MR (2006) Ecto-5′-nucleotidase (CD73)-mediated adenosine production is tissue protective in a model of bleomycin-induced lung injury. J Immunol 176:4449–4458

Wagner DR, Kubota T, Sanders VJ, McTiernan CF, Feldman AM (1999) Differential regulation of cardiac expression of IL-6 and TNF-alpha by A_2- and A_3-adenosine receptors. Am J Physiol 276:H2141–H2147

Wakai A, Wang JH, Winter DC, Street JT, O'Sullivan RG, Redmond HP (2001) Adenosine inhibits neutrophil vascular endothelial growth factor release and transendothelial migration via A_{2B} receptor activation. Shock 15:297–301

Walker BA (1996) Effects of adenosine on guinea pig pulmonary eosinophils. Inflammation 20:11–21

Walker BA, Jacobson MA, Knight DA, Salvatore CA, Weir T, Zhou D, Bai TR (1997) Adenosine A_3 receptor expression and function in eosinophils. Am J Respir Cell Mol Biol 16:531–537

Wells CA, Ravasi T, Hume DA (2005) Inflammation suppressor genes: please switch out all the lights. J Leukoc Biol 78:9–13

Willoughby DA, Moore AR, Colville-Nash PR, Gilroy D (2000) Resolution of inflammation. Int J Immunopharmacol 22:1131–1135

Wilson CN, Batra VK (2002) Lipopolysaccharide binds to and activates A(1) adenosine receptors on human pulmonary artery endothelial cells. J Endotoxin Res 8:263–271

Wilson CN, Vance CO, Goto M (2006) A_1 adenosine receptor antagonist, L-97–1, improves survival and acute lung injury following cecal ligation and perforation. J Leukoc Biol Suppl 62

Witko-Sarsat V, Rieu P, Descamps-Latscha B, Lesavre P, Halbwachs-Mecarelli L (2000) Neutrophils: molecules, functions and pathophysiological aspects. Lab Invest 80:617–653

Wollner A, Wollner S, Smith JB (1993) Acting via A_2 receptors, adenosine inhibits the upregulation of Mac-1 (Cd11b/CD18) expression on FMLP-stimulated neutrophils. Am J Respir Cell Mol Biol 9:179–185

Xaus J, Mirabet M, Lloberas J, Soler C, Lluis C, Franco R, Celada A (1999a) IFN-gamma upregulates the A_{2B} adenosine receptor expression in macrophages: a mechanism of macrophage deactivation. J Immunol 162:3607–3614

Xaus J, Valledor AF, Cardo M, Marques L, Beleta J, Palacios JM, Celada A (1999b) Adenosine inhibits macrophage colony-stimulating factor-dependent proliferation of macrophages through the induction of p27kip-1 expression. J Immunol 163:4140–4149

Yamano K, Inoue M, Masaki S, Saki M, Ichimura M, Satoh M (2006) Generation of adenosine A_3 receptor functionally humanized mice for the evaluation of the human antagonists. Biochem Pharmacol 71:294–306

Yang D, Zhang Y, Nguyen HG, Koupenova M, Chauhan AK, Makitalo M, Jones MR, St Hilaire C, Seldin DC, Toselli P, Lamperti E, Schreiber BM, Gavras H, Wagner DD, Ravid K (2006a) The A_{2B} adenosine receptor protects against inflammation and excessive vascular adhesion. J Clin Invest 116:1913–1923

Yang Z, Day YJ, Toufektsian MC, Xu Y, Ramos SI, Marshall MA, French BA, Linden J (2006b) Myocardial infarct-sparing effect of adenosine A_{2A} receptor activation is due to its action on CD4+ T lymphocytes. Circulation 114:2056–2064

Young HW, Molina JG, Dimina D, Zhong H, Jacobson M, Chan L-NL, Chan T-S, Lee JJ, Blackburn MR (2004) A_3 adenosine receptor signaling contributes to airway inflammation and mucus production in adenosine deaminase-deficient mice. J Immunol 173:1380–1389

Yu L, Huang Z, Mariani J, Wang Y, Moskowitz M, Chen JF (2004) Selective inactivation or reconstitution of adenosine A_{2A} receptors in bone marrow cells reveals their significant contribution to the development of ischemic brain injury. Nat Med 10:1081–1087

Zalavary S, Stendahl O, Bengtsson T (1994) The role of cyclic AMP, calcium and filamentous actin in adenosine modulation of Fc receptor-mediated phagocytosis in human neutrophils. Biochim Biophys Acta 1222:249–256

Zalavary S, Bengtsson T (1998) Adenosine inhibits actin dynamics in human neutrophils: evidence for the involvement of cAMP. Eur J Cell Biol 75:128–139

Zernecke A, Bidzhekov K, Ozüyaman B, Fraemohs L, Liehn EA, Lüscher-Firzlaff JM, Lüscher B, Schrader J, Weber C (2006) CD73/ecto-5′-nucleotidase protects against vascular inflammation and neointima formation. Circulation 113:2120–2127

Zhang JG, Hepburn L, Cruz G, Borman RA, Clark KL (2005) The role of adenosine A_{2A} and A_{2B} receptors in the regulation of TNF-alpha production by human monocytes. Biochem Pharmacol 69:883–889

Zhong H, Belardinelli L, Maa T, Feoktistov I, Biaggioni I, Zeng D (2003a) A_{2B} adenosine receptors increase cytokine release by bronchial smooth muscle cells. Am J Respir Cell Mol Biol

Zhong H, Sergiy S, Molina JG, Sanborn BM, Jacobson MA, Tilley SL, Blackburn MR (2003b) Activation of murine lung mast cells by the adenosine A_3 receptor. J Immunol 170:338–345

Zhong H, Belardinelli L, Maa T, Zeng D (2005) Synergy between A_{2B} adenosine receptors and hypoxia in activating human lung fibroblasts. Am J Respir Cell Mol Biol 32:2–8

Zhong H, Wu Y, Belardinelli L, Zeng D (2006) A_{2B} adenosine receptors induce IL-19 from bronchial epithelial cells, resulting in TNF-alpha increase. Am J Respir Cell Mol Biol 35:587–592

Zhou QY, Li C, Olah ME, Johnson RA, Stiles GL, Civelli O (1992) Molecular cloning and characterization of an adenosine receptor: the A_3 adenosine receptor. Proc Natl Acad Sci USA 89:7432–7436

Zimmermann H (2000) Extracellular metabolism of ATP and other nucleotides. Naunyn–Schmiedeberg's Arch Pharmacol 362:299–309

A_1 Adenosine Receptor: Role in Diabetes and Obesity

Arvinder K. Dhalla, Jeffrey W. Chisholm, Gerald M. Reaven, and Luiz Belardinelli

Contents

Abstract Adenosine mediates its diverse effects via four subtypes (A_1, A_{2A}, A_{2B} and A_3) of G-protein-coupled receptors. The A_1 adenosine receptor (A_1AR) subtype is the most extensively studied and is well characterized in various organ systems. The A_1ARs are highly expressed in adipose tissue, and endogenous adenosine has been shown to tonically activate adipose tissue A_1ARs. Activation of the A_1ARs in adipocytes reduces adenylate cyclase and cAMP content and causes inhibition of lipolysis. The role of A_1ARs in lipolysis has been well characterized by using several selective A_1AR agonists as well as A_1AR knockout mice. However, the contribution of A_1ARs to the regulation of lipolysis in pathological conditions like insulin resistance, diabetes and dyslipidemia, where free fatty acids (FFA) play

A.K. Dhalla (✉)
Department of Pharmacological Sciences, CV Therapeutics Inc., 3172 Porter Drive, Palo Alto, CA 94304, USA
arvinder.dhalla@cvt.com

C.N. Wilson and S.J. Mustafa (eds.), *Adenosine Receptors in Health and Disease*, Handbook of Experimental Pharmacology 193, DOI 10.1007/978-3-540-89615-9_9,

an important role, has not been well characterized. Pharmacological agents that reduce the release of FFA from adipose tissue and thus the availability of circulating FFA have the potential to be useful for insulin resistance and hyperlipidemia. Toward this goal, several selective and efficacious agonists of the A_1ARs are now available, and some have entered early-phase clinical trials; however, none have received regulatory approval yet. Here we review the existing knowledge on the role of A_1ARs in insulin resistance, diabetes and obesity, and the progress made in the development of A_1AR agonists as antilipolytic agents, including the challenges associated with this approach.

Keywords A_1 Adenosine Receptor · Antilipolytic · Insulin Resistance · Diabetes · Obesity

Abbreviations

A_1AR	A_1 Adenosine receptor
A_{2A}AR	A_{2A} Adenosine receptor
A_{2B}AR	A_{2B} Adenosine receptor
AR	Adenosine receptor
ATGL	Adipose triglyceride lipase
BMI	Body mass index
cAMP	Cyclic adenosine monophosphate
CCPA	2-Chloro-N^6-cyclopentyladenosine
FFA	Free fatty acids
GPCR	G-protein-coupled receptor
HPIA	Hydroxyphenylisopropyl adenosine
HSL	Hormone-sensitive lipase
KO	Knockout
NFkB	Nuclear factor kappa beta
PIA	Phenylisopropyladenosine
PI3	Phosphoinositide 3
PKA	Protein kinase A
PKC	Protein kinase C
R-PIA	N^6-(R)-Phenylisopropyladenosine
SPA	N^6-(p-Sulfophenyl)adenosine
TG	Triglycerides
VLDL-TG	Very low density lipoprotein triglyceride

1 Introduction

In 1961, Dole demonstrated that adenosine and some adenosine metabolites inhibited the breakdown of triglycerides (TG) to FFA in isolated rat epididymal fat pads (Dole 1961). Subsequent studies indicated that the initial event in hormone-induced lipolysis was a rapid rise in intracellular cyclic adenosine monophosphate (cAMP) content (Butcher et al. 1965; Fain and Malbon 1979; Fain et al. 1972). Interestingly, lipolysis remained elevated for some time, and then began to decrease despite the continued presence of excess stimulatory agents (Butcher et al. 1965). Investigations into this phenomenon revealed that lipolysis could be restored by the addition of fresh incubation medium, suggesting that a component of the medium was limiting lipolysis, or that an inhibitor of lipolysis was being generated by the cells (Ho and Sutherland 1971; Schwabe et al. 1973). Schwabe et al. demonstrated that the inhibitory effect on lipolysis could be minimized by the addition of adenosine deaminase, as well as by reducing the number of incubated cells (Schwabe et al. 1973; Schwabe and Ebert 1974). Finally, Fain et al. showed that adenosine and adenosine analogs inhibited adenylate cyclase and antagonized the stimulation of cAMP by catecholamines (Fain et al. 1972), thereby inhibiting catecholamine-induced lipolysis in adipocytes. Later work established that inhibition of lipolysis by adenosine is mediated by activation of the A_1 adenosine receptor (A_1AR), and that this receptor was potentially an important regulator of lipolysis, fatty acid storage and tissue partitioning of fat (Fain et al. 1972; Johansson et al. 2007b; Schwabe et al. 1974). The focus of this review is on the role of A_1AR in the regulation of lipolysis in adipose tissue and its consequences for insulin resistance, diabetes and dyslipidemia. The potential of A_1AR agonists as therapeutically useful antilipolytic agents is also discussed.

Adenosine, an endogenous nucleoside, mediates its pharmacological actions via four distinct G-protein-coupled receptors (GPCR), classified as A_1, A_{2A}, A_{2B} and A_3 adenosine receptors (ARs). The role of A_1ARs in mammalian physiology and pharmacology has been well established (Dhalla et al. 2003; Fredholm and Sollevi 1986; Fredholm et al. 2001; Jacobson et al. 1996). The primary actions of adenosine mediated via A_1ARs include decreases in; heart rate and atrial contractility, release of neurotransmitters, lipolysis and renal function (for reviews see Belardinelli et al. 1989; Fredholm and Sollevi 1986; Gao and Jacobson 2007; Linden 1991; Press et al. 2007). A_1ARs mediate the inhibitory effects of adenosine and are primarily coupled to pertussis toxin-sensitive inhibitory guanine nucleotide-binding (G_i and G_O) proteins (Munshi et al. 1991). These G proteins regulate adenylate cyclases, kinases and ion channels (e.g., potassium) that play crucial roles in various cellular functions (Belardinelli et al. 1989; Brechler et al. 1990; Okajima et al. 1989). As depicted in Fig. 1, activation of A_1ARs in adipocytes causes inhibition of adenylate cyclase activity, reduction of cyclic AMP formation, inhibition of protein kinase A (PKA), and inhibition of lipolysis (Dobson 1978; Fain and Malbon 1979; Fain et al. 1972; Londos and Wolff 1977; Schrader et al. 1977). Although a direct inhibition of hormone-sensitive lipase (HSL) by A_1AR agonists has not been demonstrated, because of the well-established role of HSL and more recently adipose triglyceride lipase (ATGL) in lipolysis, it is assumed that inhibition of lipolysis by adenosine and its analogs is due to the activation of A_1ARs, resulting in the inhibition of HSL and/or ATGL.

The means by which activation of A_1ARs causes inhibition of adenylate cyclase and reduction in cAMP have been well characterized. A_1AR agonists inhibit

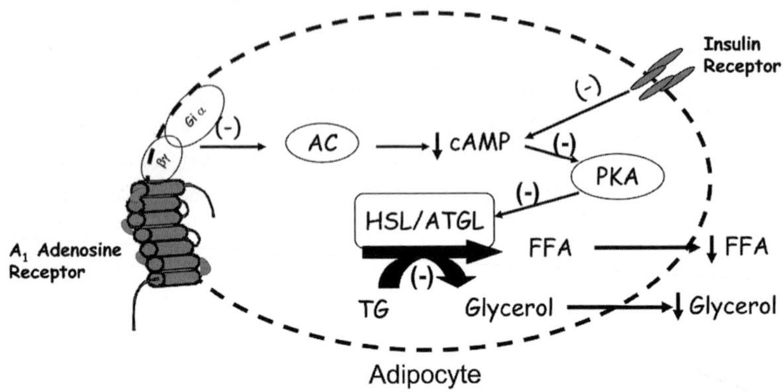

Fig. 1 Schematic representation of mechanisms by which A_1 adenosine receptors (A_1ARs) mediate antilipolytic effects in adipocytes. Activation of A_1ARs causes inhibition of adenylate cyclase (AC) activity via G_i (inhibitory GTP-binding protein), reduction of cyclic AMP (cAMP) formation, and inhibition of protein kinase A (PKA), leading to a reduction of hormone-sensitive lipase (HSL) and/or adipose triglyceride lipase (ATGL) activity which results in inhibition of the breakdown of triglycerides (TG) to free fatty acids (FFA)

the activation of adenylate cyclase activity caused by stimulatory agents (e.g., catecholamines), but have minimal effect in the absence of such agents (Dobson 1983; Dobson et al. 1986; Schrader et al. 1977). The affinity of adenosine for the adipocyte A_1ARs is estimated to be in the low micromolar range and is similar to that seen in cardiac myocytes (Liang et al. 2002; Srinivas et al. 1997). Uncoupling of the receptor from G protein with pertussis toxin (which causes ADP ribosylation of the α-subunit of G_i) attenuates the inhibition by adenosine of adenylate cyclase activity (Moreno et al. 1983), as well as the antilipolytic effects mediated by A_1AR activation. Hence, it has been proposed that A_1ARs and G_i proteins are intimately coupled in adipocytes, and that one is not likely to be affected independently of each other, suggesting that inactivation of G proteins cannot be overcome by activating a greater number of adenosine receptors (Liang et al. 2002). Functional uncoupling of A_1ARs from its effectors leads to de novo synthesis of A_1ARs by nuclear factor kappa beta (NF-kB) and protein kinase C (PKC) activation (Jajoo et al. 2006). Each A_1AR appears to activate a certain number of G protein molecules (more than one), and amplification of G protein activation appears to be independent of total number of receptors (Baker et al. 2000). Thus, it has been suggested that the A_1AR–G protein activation ratio may be a better measure of cell responsiveness to agonists than the independent quantification of receptors and G proteins (Baker et al. 2000).

In addition to direct antilipolytic effects, adenosine and its analogs have been shown to modulate insulin action and insulin sensitivity in muscle and adipose tissue (Budohoski et al. 1984; Rolband et al. 1990) which is suggested to be mediated via A_1ARs. For instance, a partial A_1 agonist, CVT-3619 (2-{6-[((1*R*,2*R*)-2-hydroxycyclopentyl)amino]purin-9-yl}(4*S*,5*S*,2*R*,3*R*)-5-[(2-fluorophenylthio)methyl]oxolane-3,4-diol) decreased the EC_{50} for insulin to inhibit lipolysis in vivo by fourfold, suggesting that CVT-3619 increases insulin sensitivity

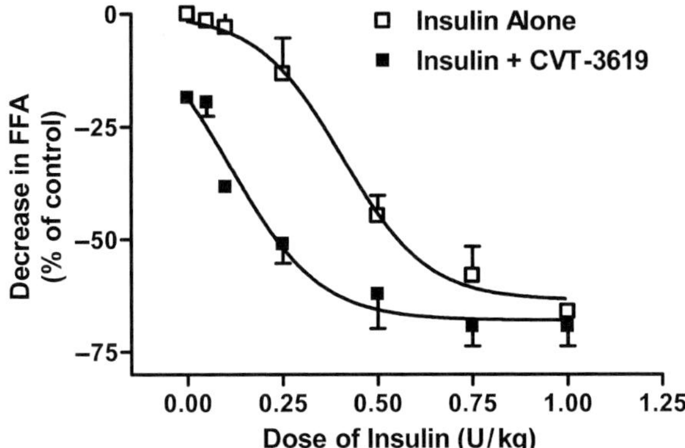

Fig. 2 CVT-3619 potentiates the effect of insulin to reduce FFA levels. Shown are the dose–response curves for the effect of insulin in reducing FFA in the absence and presence of CVT-3619 (0.5 mg kg^{-1}) in awake rats. Both insulin and CVT-3619 were given via i.p. injection. Each data point is the mean ± SEM of the maximal (peak effect) percent decrease in FFA levels from baseline for 3–5 rats. The doses of insulin that cause a 50% decrease (ED_{50}) in FFA level in the absence and presence of CVT-3619 were 0.4 (0.3916–0.4208, 95% CI) and 0.1 (0.0935–0.133) U kg^{-1}, respectively

in adipose tissue (Fig. 2). Phenylisopropyladenosine (PIA), a full A_1AR agonist, potentiates the insulin-induced activation of phosphoinositide 3-kinase (PI3K), a second messenger for insulin actions, in rat adipocytes (Takasuga et al. 1999). In summary, the role of adenosine in the regulation of lipolysis and insulin homeostasis is well established, and these effects are mediated by the A_1AR.

2 A_1AR Expression in Adipose Tissue

A_1ARs are highly expressed in adipose tissue (Dhalla et al. 2003; Trost and Schwabe 1981; Ukena et al. 1984b), and are sensitive to eliciting a functional response with low affinity ligands due to highly efficient coupling (Liang et al. 2002). A_1AR density (Bmax) is reported to be comparable in adipose tissue from rat (690 fm mg^{-1} protein) and humans (360–1,800 fm mg^{-1} protein) (Green et al. 1989; Liang et al. 2002), although lower numbers (72–95 fm mg^{-1}) for human adipose tissue have also been reported (Larrouy et al. 1991). It should be noted that the results of the binding studies depend on the ligand used to determine receptor density and, in general (as expected), agonist radioligands give lower numbers than that obtained by using antagonist radioligands (Kollias-Baker et al. 1997; Larrouy et al. 1991; Leung et al. 1990). Gene expression and functional studies using rat epididymal adipocytes and mouse Ob17 cells demonstrated that A_1ARs are expressed and functionally active in differentiated adipocytes (Borglum et al. 1996;Vassaux et al. 1993). In contrast, A_2AR expression and functional activity is generally decreased following differentiation (Borglum et al. 1996; Vassaux et al. 1993). Human and rat A_1ARs are 90% homologous (Tatsis-Kotsidis and Erlanger 1999). In studies using human primary adipocyte membranes, the human A_1AR protein was found to have a slightly higher molecular weight than rat and a lower affinity for hydroxyphenylisopropyl adenosine (HPIA) (Green et al. 1989). While multiple A_1AR mRNA (messenger ribonucleic acids) transcripts have been identified, the common variant in human adipocytes is identical to the one characterized in human brain (Tatsis-Kotsidis and Erlanger 1999). Taken together, the available data suggests that human and rodent A_1ARs are structurally and functionally similar. The discrepancies regarding the A_1AR density in human adipose tissue remain to be resolved.

There is some evidence that the number of A_1ARs differs between fat depots. White adipose tissue from rats has a much higher affinity and binding capacity for the A_1AR agonist PIA than brown adipose tissue (Saggerson and Jamal 1990). Functionally, these differences result in white adipose tissue lipolysis being threefold more sensitive to inhibition by PIA than brown adipose tissue (Saggerson and Jamal 1990). Adenosine receptor gene expression and A_1AR-mediated inhibition of lipolysis in epididymal and inguinal adipose tissue of normal rats were not different (Fatholahi et al. 2006). A_1AR protein expression was fourfold higher and Bmax was twofold higher in subcutaneous adipose tissue than visceral adipose tissue isolated from women with body mass index (BMI) $> 42\,\mathrm{kg\,m^{-2}}$ (Barakat et al. 2006), even though mRNA expression was similar in both fat depots. Higher receptor numbers in subcutaneous tissue could mean that the A_1AR plays a larger role in regulating subcutaneous adipose fat storage than visceral fat storage; however, functional data to support this possibility is lacking. Finally, because it appears that measurement of gene expression may not accurately reflect the A_1AR number, determination of receptor number using either radiological binding studies or protein expression may be necessary to quantitate the number of receptors in various fat depots.

3 Adipocyte A_1AR Function and Regulation in Disease Models

A_1AR level (protein expression or receptor number) and activity in adipose tissue in models of insulin resistance, diabetes and obesity has not been systematically characterized. There are relatively few studies describing changes at the A_1AR expression and function in insulin resistance (Vannucci et al. 1989) and obesity (Kaartinen et al. 1994). More specifically, it has been reported that A_1AR signaling is more active in adipocytes from obese Zucker rats (Berkich et al. 1995), resulting in increased sensitivity to inhibition by A_1AR agonists (Vannucci et al. 1989). A_1AR- mediated inhibition of adenylate cyclase was increased in adipocytes from Zucker fatty rats (Vannucci et al. 1989). Tonic activity of A_1ARs on lipolysis has also been shown to be higher in obese rats (LaNoue and Martin 1994). Kaartinen et al., using adipocytes (large and small) from the same fat depot, showed that the large fat cells were more responsive to the inhibition of forskolin-stimulated adenylate cyclase activity by PIA than the small cells (Kaartinen et al. 1991). It has been suggested that the changes in the receptor–effector complex on adipocytes could influence the effectiveness and tissue selectivity of adenosine and its analogs. This notion is supported by data from in vivo studies showing that the A_1AR agonist, ARA, ([1*S*,2*R*,3*R*,5*R*]-3-methoxymethyl-5-[6-(1-[5-trifluoromethyl-pyridin-2-yl] pyrrolidin-3-[*S*]-ylamino)-purin-9-yl]cyclopentane-1,2-diol) is more potent in inhibiting tissue lipolysis in Zucker fatty rats (which are not diabetic) as compared to Wistar rats, and is equally efficacious/effective in lowering plasma FFA concentrations (Schoelch et al. 2004). Recently, it was reported that A_1AR gene expression was similar in normal (SD) and diabetic ZDF rats (although cell surface A_1AR number was not measured), and that inhibition of lipolysis by a partial A_1AR agonist is not different in SD and ZDF rats (Dhalla et al. 2008).

On the other hand, it has been reported that concentration of inhibitory protein (G_{i1}) is lower in fat cells isolated from obese subjects as compared to lean subjects, and that the decrease is negatively correlated with BMI (Kaartinen et al. 1994). Furthermore, it has been shown that N^6-(p-sulfophenyl) adenosine (SPA), a selective A_1AR agonist, is less potent in diabetic ZDF rats as compared to control rats, resulting in a short-lasting antilipolytic effect (van Schaick et al. 1998b). Because no difference in the pharmacokinetics of SPA between control and diabetic rats was observed, it was suggested that metabolic alterations in diabetic ZDF rats might be associated with an altered sensitivity to A_1AR agonists.

Overall, the available data suggest that A_1AR-mediated responses may be more sensitive to ligands in adipose tissue isolated from animal models of obesity. Thus, it has been proposed that that inhibition of lipolysis due to excessive activity of A_1AR may lead to obesity (Barakat et al. 2006); however, there are not enough data to support this hypothesis. On the contrary, data from studies with in vivo pharmacological modulators of A_1AR function and adipose tissue specific A_1AR transgenic mice show that activation of A_1ARs in models of insulin resistance and obesity results in improvement in insulin sensitivity with no significant effect on weight gain (Dhalla et al. 2007b; Dong et al. 2001; Schoelch et al. 2004). Overexpression of A_1ARs in adipose tissue protects mice from diet-induced insulin resistance (Dong et al. 2001), whereas the number of A_1ARs in adipocytes from obese individuals has been reported to be decreased (Kaartinen et al. 1991). Similarly, HSL-deficient mice are also resistant to weight gain on a high-fat diet (Harada et al. 2003; Osuga et al. 2000). No significant changes in weight gain were observed up to four months in A_1AR knockout (KO) mice as compared to wild-type mice (Johansson et al. 2007a), suggesting that weight gain is not associated with A_1AR activity. In summary, the preponderance of the published data suggests that the inhibition of adipocyte lipolysis may simply result in a redistribution of fat to adipose tissue rather than an accumulation of fat and weight gain.

A simple comparison of A_1AR expression in obese vs. lean adipocytes from human and animal models does not provide insight into how A_1ARs modulate circulating FFA concentrations, insulin sensitivity or type 2 diabetes. To understand the regulation of lipolysis through the A_1ARs, studies correlating A_1AR expression and function to disease end-points, like circulating FFA, insulin and glucose homeostasis, body-weight changes and whole body fat distribution, are needed. This can be achieved by using selective ligands (agonists and antagonists) of the A_1AR, as discussed below.

4 Inhibition of Lipolysis: A Therapeutic Approach

Under normal conditions, FFAs released from the adipose tissue due to lipolysis are an important source of fuel for many tissues. However, when the effect of insulin in inhibiting FFA release from adipose tissue is reduced (adipose tissue insulin resistance), chronic increases in circulating FFA concentrations occur (Reaven 1995). In addition to aggravating muscle insulin resistance, increases in plasma FFA concentrations have other adverse metabolic effects that play an important role in the pathogenesis of type 2 diabetes (Boden 2001; Fruhbeck and Gomez-Ambrosi 2002; Fruhbeck et al. 2001;Wyne 2003). Included in the consequences of chronically elevated FFA concentrations are increased deposition of TG in tissues such as skeletal muscle, liver, pancreas and heart; contributing to the defects in insulin stimulation of muscle glucose uptake and glucose-stimulated insulin secretion that characterize patients with type 2 diabetes (Boden et al. 2005; Itani et al. 2002; Roden et al. 1996; Sako and Grill 1990).

Thus, pharmacological inhibition of lipolysis to lower plasma FFA would seem to be an attractive therapeutic approach for the management of insulin resistance and diabetes (Bays et al. 2004; Boden 2001, 2002; Jensen 2006; Langin 2006; Large and Arner 1998; Reaven 1995). Because FFAs are continually being mobilized from adipose tissue via lipolysis, inhibition of release of FFA can be expected to affect various metabolic processes. Despite overwhelming evidence of a role of elevated FFA in insulin resistance and diabetes, very few inhibitors of lipolysis are available for either experimental or clinical use. Nicotinic acid and its analog acipimox are the only well-characterized antilipolytic agents that are currently used for treatment of dyslipidemia (Carlson 2005; Vega et al. 2005). Their therapeutic usefulness is limited because the initial decrease in plasma FFA levels is followed by a rebound effect that leads to transient increases in FFA and insulin resistance (Poynten et al. 2003). In addition, nicotinic acid has an unfavorable side-effect profile, and it has been suggested that it may not be an appropriate drug to use in the treatment of diabetic patients as it can increase plasma glucose levels (Garg and Grundy 1990; Grundy et al. 2002; McKenny et al. 1994; Poynten et al. 2003). Lowering circulating FFA levels by inhibiting adipose tissue lipolysis by A_1AR agonists can potentially fulfill an unmet need for novel antilipolytic agents in the treatment of pathological conditions where FFA are elevated. In this context, activation of A_1ARs has been shown to result in lowering plasma glucose levels in streptozotocin-induced diabetic rats (Cheng et al. 2000; Nemeth et al. 2007; Reaven et al. 1988). Dipyridamole, which increases endogenous adenosine levels, also lowers plasma glucose levels in a dose-dependent manner (Cheng et al. 2000). In addition, because FFA flux to the liver is an important modulator of very low density lipoprotein–triglyceride (VLDL–TG) synthesis and secretion, lowering plasma FFA concentrations by inhibiting lipolysis reduces the supply of FFA to the liver, thereby decreasing hepatic VLDL–TG production and circulating plasma TG concentrations. Therefore, antilipolytic effects of A_1AR agonists may be beneficial in a variety of conditions wherein plasma FFA and TG concentrations are elevated, including insulin resistance, diabetes and dyslipidemia.

4.1 A_1AR Agonists

A large number of selective and potent A_1AR agonists and antagonists have been synthesized over the last three decades for research and therapeutic purposes (Ashton et al. 2008; Cappellacci et al. 2008; Klotz et al. 1989; Klotz 2000; Morrison et al. 2004; Muller 2001; Palle et al. 2004). Ligands with high (several hundredfold) selectivity for the A_1AR versus one or two of the other subtypes of ARs have been synthesized and characterized; however, ligands with high selectivity for the A_1AR versus all three other subtypes of ARs have not been described to date (Klotz 2000). Several agonists of the A_1AR have been developed as potential antilipolytic agents, and a few of them have entered early-phase clinical trials (Cox et al. 1997; Dhalla et al. 2007a; Fraser et al. 2003; Fredholm et al. 2001; Hoffman et al. 1986b; Ishikawa et al. 1998; Jacobson et al. 1992; Klotz 2000; Leblanc and Soucy 1994; Press et al. 2007; Schoelch et al. 2004; Shah et al. 2004; Strong et al. 1993; Van der Graaf et al. 1999; van Schaick et al. 1998b; Wagner et al. 1995; Zannikos et al. 2001). However, none has received regulatory approval. In the following section we summarize the salient features of the few selected A_1AR ligands that have been reported to have antilipolytic effects.

SDZ WAG-994 (*N*-cyclohexyl-2′-*O*-methyladenosine) is an orally bioavailable, selective A_1AR agonist and is one of the first compounds developed as a potential therapeutic agent based on A_1AR agonism. In isolated adipocytes, SDZ WAG-994 inhibited lipolysis and increased insulin-dependent glucose uptake with an $EC_{50} = 8$ nM (Foley et al. 1997; Wagner et al. 1995). Single doses of SDZ WAG-994 given to normal rats caused dose-dependent decreases in serum FFA, glucose and insulin levels, with maximal reductions of up to 68% (Ishikawa et al. 1998). The hypoglycemic effect of SDZ WAG-994 was abolished when FFA levels were kept constant by infusion of lipids, suggesting that glucose lowering by SDZ WAG-994 was due to FFA reduction (Foley et al. 1997). Although SDZ WAG-994 has beneficial metabolic effects in various animal models, the antilipolytic effects could not be separated from cardiovascular effects. For example, a 66% reduction in FFA with a dose of 3 mg kg^{-1} was accompanied by a 73% reduction in heart rate and 50% reduction in mean arterial pressure in rats (Cox et al. 1997).

GR79236 (*N*-[(1*S*,2*S*)-2-hydroxycyclopentyl]-adenosine) is a selective and very potent A_1AR agonist (Strong et al. 1993). GR79236 inhibited lipolysis in isolated adipocytes and was shown to reduce plasma FFA and glucose levels in normal rats (Gardner et al. 1994; Merkel et al. 1995; Strong et al. 1993). Rats fed a high fructose diet also showed improved glucose tolerance; however, a significant reduction of blood pressure following GR79236 treatment was observed (Qu et al. 1997). In a rat model of diabetic ketoacidosis, GR79236 reduced plasma FFA but did not affect blood glucose levels. The lack of effect on glucose was proposed to be a result of stimulation of gastric emptying and enhanced absorption of stomach contents (Thompson et al. 1994). However, in addition to its potent antilipolytic effects, GR79236 also caused hypotension and bradycardia in conscious rats (Merkel et al. 1995).

ARA is an AR agonist with high affinity and selectivity for A_1ARs (Schoelch et al. 2004). ARA treatment significantly lowered dialysate glycerol levels in subcutaneous and visceral adipose tissue and gastrocnemius muscle, as measured by the tissue microdialysis technique in Wistar and Zucker fatty rats. ARA treatment caused significant reductions in plasma FFA, glycerol and TGs and an improvement in insulin sensitivity in Zucker fatty rats. The effects of ARA on heart rate and blood pressure were not reported in this study (Schoelch et al. 2004).

CVT-3619 is a selective and partial agonist of the A_1AR (Fatholahi et al. 2006). The binding affinity of CVT-3619 for A_1AR is 113 nM. The K_i values for CVT-3619 to bind to A_{2A}, A_{2B} and A_3 ARs are >5,000 nM, suggesting that CVT-3619 has very low affinity for these receptor subtypes (Fatholahi et al. 2006). The antilipolytic effects of CVT-3619, in vitro and in vivo, have been well characterized (Dhalla et al. 2007b; Fatholahi et al. 2006). CVT-3619 reduces forskolin-induced increase in cAMP and FFA levels in epididymal adipocytes. The IC_{50} values for CVT-3619 for reducing cAMP levels and FFA release are 6 and 47 nM, respectively (Fatholahi et al. 2006). CVT-3619 lowers circulating FFA and triglyceride (TG) levels (20–60%) in a dose-dependent manner in awake rats at doses (1–10 mg kg^{-1}) that do not have any significant effect on heart rate or blood pressure. In a two-week high fat diet-induced model of insulin resistance in rats, pretreatment with CVT-3619 prevented the development of insulin resistance (Dhalla et al. 2007a).

Tecadenoson (6-(N-3′-(R)-tetrahydrofuranyl)-amino-purine riboside), a potent and selective A_1AR agonist, has been shown to lower FFA levels in rats in a dose-dependent manner (Fraser et al. 2003). Although tecadenoson was not designed to be an antilipolytic agent, infusion of tecadenoson also reversibly reduced elevated FFA levels in a pilot Phase I study in patients (data on file at CV Therapeutics). Interestingly, the antilipolytic effect of tecadenoson was observed at doses that did not affect heart rate or P–R interval (AV nodal conduction). This result is consistent with the presence of a high receptor density and greater receptor reserve in adipocytes as compared to the heart (Liang et al. 2002) (see Sect. 6.1).

Pharmacokinetics and pharmacodynamics of another AR agonist, referred to as ARA, which has high affinity for A_1ARs and A_2 adenosine receptors (A_{2A}ARs), were determined in a Phase I clinical study with two parallel groups of 13 healthy males following administration of a single 6 h intravenous infusion of ARA or placebo (Zannikos et al. 2001). ARA was found to have high clearance (Cl: 0.79 L h^{-1} kg^{-1}), with a modest volume of distribution (Vss: 0.91 L kg^{-1}) and short half-life ($t_{1/2}$: approximately 1 h). The reduction in circulating levels of FFA by ARA was related to its plasma concentrations using a modified Emax (maximal effective concentration)-based tolerance model, and the EC_{50} value was 17.0 ng mL^{-1}. The results of this study led to the suggestion that the use of A_1AR agonists as antilipolytic drugs may be limited due to the potential development of tolerance or desensitization, and that a period free from the agonist may be required before the response of FFA returns to baseline conditions. It should be noted that this agonist, which was given as a continuous IV infusion for 6 h, is not selective for A_1ARs and also has high affinity for A_{2A}ARs (Zannikos et al. 2001). Activation of A_{2A}ARs would increase FFA levels indirectly by causing sympathetic stimulation (Dhalla et al. 2006), and therefore will counteract the effect mediated by A_1ARs.

RPR749 and its methylated metabolite RPR772 are reported to be orally active and selective A_1AR agonists that inhibit lipolysis and lower plasma FFA and TG levels in various animal models (Shah et al. 2004); however, preclinical reports on these compounds have not been described. The pharmacokinetics and pharmacodynamics (effect on FFA) of RPR749 were evaluated in humans in a double-blind, placebo-controlled, parallel-group, randomized Phase I study with a single oral dose of up to 200 mg (Shah et al. 2004). Six parallel groups of eight healthy men (six active and two placebo/group) were enrolled in the study. Plasma samples were collected for up to 72 h postdose. RPR749 was safe and well tolerated as a single oral dose up to 200 mg. Plasma concentrations of RPR749 were approximately 30-fold higher than the mean RPR772 plasma concentrations. The mean terminal half-lives of RPR749 and RPR772 were similar (approximately 16.4 h). Serum FFA concentrations decreased (between 25 and 70%) in all treatment groups, with the maximal decrease in the 200 mg dose group. However, significant decreases in FFA concentrations were also observed in the placebo group. RPR749 seems to have pharmacological properties that may be beneficial in treating insulin resistance and hyperlipidemia; however, further development of this compound has been discontinued for reasons that are not publicly disclosed.

In summary, data in the literature suggests that A_1AR agonists are a viable approach to lowering FFA by inhibiting adipose tissue lipolysis; however, a number of hurdles need to be overcome (as described in Sect. 5) before this class of molecules can be successfully developed as antilipolytic drugs.

4.2 A_1AR Antagonists

Interestingly, an antagonist of the A_1AR, BW-1433 (1,3-dipropyl-8-[*p*-(carboxyethynyl) phenyl]xanthine), has also been reported to improve glucose tolerance in Zucker rats after a six-week treatment (Xu et al. 1998). This is contrast to the many studies showing that glucose tolerance improves in response to treatment with A_1AR agonists (Dhalla et al. 2007a; Hoffman et al. 1986b; Schoelch et al. 2004; van Schaick et al. 1998b). Given that agonists of A_1AR inhibit lipolysis and lower circulating FFA, one would expect an increase in FFA levels with administration of an A_1AR antagonist, which will lead to a worsening of insulin resistance. The paradoxical findings of Xu et al. 1998 can be explained based on the observation that treatment with BW-1433 actually resulted in a very small increase in FFA levels. Furthermore, the increase in FFA was transient and disappeared after seven days. On the other hand, one-week treatment with BW-1433 resulted in a selective increase in the number of A_1ARs in the adipose tissue. It has been reported that endogenous adenosine tonically inhibits adipose tissue A_1ARs (Liang et al. 2002). Therefore, improvement in glucose tolerance with BW-1433 may be due to the inhibitory effects of endogenous adenosine acting on a higher background of A_1ARs upregulated by chronic treatment with BW-1433 (Xu et al. 1998). In this regard, overexpression of A_1ARs in mice has also been shown to improve glucose tolerance in a model of diet-induced insulin resistance and obesity (Dong et al. 2001). Furthermore, BW-1433 is a nonselective antagonist of the AR subtypes, and has been reported to have higher affinity for A_{2B} adenosine receptors (A_{2B}ARs) (US patent no. 6060481). Thus, the antidiabetic effects of BW-1433 may not solely be due to A_1AR antagonism, as suggested (Xu et al. 1998).

5 Challenges for the Development of A_1AR Agonists as Therapeutic Agents

In general, the challenges for the development of A_1AR agonists as therapeutic agents are similar to those for other GPCR agonists. They include at least (a) selectivity, (b) receptor downregulation, and (c) receptor desensitization. The development of A_1AR agonists as antilipolytic agents has been further limited by the concurrent side effects induced by the activation of A_1ARs in other organs such as heart, kidney and brain (Belardinelli et al. 1989; Wu et al. 2001). The following section describes the specific challenges and hurdles that need to be overcome for the successful development of A_1AR ligands as antilipolytic agents.

5.1 Receptor Density and Distribution

A_1ARs are widely expressed in both the central nervous system and peripheral tissues, as shown by radioligand binding (Gould et al. 1997; Kollias-Baker et al. 1995; Ukena et al. 1984a), in situ hybridization (Reppert et al. 1991; Schiffmann et al. 1990), immunohistochemistry (Rivkees 1995), mRNA expression (Dixon et al. 1996; Tatsis-Kotsidis and Erlanger 1999; Shen et al. 2005) and PET (positron emission tomography) scanning (Meyer et al. 2003). Under normal physiological conditions, A_1ARs are found at their highest density in the brain (cortex, cerebellum and hippocampus), followed by adipose tissue. Moderate to high densities of A_1ARs appear to be present on specialized cells of the thyroid, spinal cord, eye, adrenal gland, kidney, and sinoatrial and atrioventricular nodal tissues of the heart. In most other tissues, including the cardiac ventricles, lung, pancreas, liver and GI tract, the expression of A_1ARs is low (Lohse et al. 1984; Dixon et al. 1996). The expression of A_1ARs is increased under conditions of oxidative stress, ischemia, inflammation and diabetes (Funakoshi et al. 2007; Grden et al. 2007; Lai et al. 2005; Liu et al. 2003; Nie et al. 1998; Pawelczyk et al. 2005; Rogachev et al. 2006).

Receptor density and coupling efficiency of receptor activation to response (i.e., receptor reserve) influence and determine the responses to receptor agonists in various organs in the intact animal; in general, an agonist is more potent or efficacious where the receptor number is high (Strange 2008). Thus, although adipose tissue has the second highest receptor density of A_1ARs, ubiquitous expression of A_1ARs is one of the biggest challenges to the development of antilipolytic agents without eliciting unwanted side effects. However, the high density of receptors efficiently coupled to a functional response enables the adipocytes to respond with high sensitivity even to low-affinity ligands such as adenosine, which can inhibit lipolysis at concentrations as low as 1–2 nM (Liang et al. 2002). Thus, differences in receptor reserve and coupling efficiency in various organs can be exploited to achieve functional selectivity, as described in Sect. 6.1.

5.2 *Receptor Desensitization*

Desensitization of the response or tachyphylaxis is a potential problem when considering a receptor agonist for long-term therapeutic use. It has been suggested that the A_1AR desensitizes slowly (hours to days) (Baker et al. 2000; Green et al. 1990; Hoffman et al. 1986a; Parsons and Stiles 1987), which was attributed to the finding that A_1AR does not undergo phosphorylation upon short-term agonist treatment (Gao et al. 1999; Palmer et al. 1996). However, rapid acute tolerance to the FFA-lowering effect of the AR agonist, ARA, has been reported following a continuous intravenous infusion of the agonist to healthy volunteers (Zannikos et al. 2001). Whether this was due to desensitization of the FFA-lowering effect of the agonist on A_1ARs or due to stimulation of A_{2A} receptors because of lack of selectivity of this agonist is not known.

Mechanism(s) underlying desensitization of A_1ARs include changes in the following: (a) receptor number and/or affinity; (b) receptor to G-protein-coupling efficiency, and; (c) quantity of G proteins (Kenakin 1984). Desensitization of adipocyte responses to AR agonists is accompanied by downregulation of A_1ARs, loss of high-affinity receptor sites, and decrease in alpha subunit of G_i protein (Green 1987; Green et al. 1990; Jajoo et al. 2006; Parsons and Stiles 1987). The changes in G proteins (an increase in $G_{\alpha\sigma}$; decreases in $G_{\alpha\iota 1}$ and $G_{\alpha\iota 2}$) have been suggested to be at the protein level, as no differences were detected in their respective transcripts (Longabaugh et al. 1989). Using an A_1–A_3 chimeric receptor, it was demonstrated that A_1ARs lack the necessary molecular determinants for desensitization in response to short-term agonist treatment (Palmer et al. 1996). Chronic exposure to agonists has been shown to induce phosphorylation, clustering and desensitization of A_1ARs (Ciruela et al. 1997). Recent findings indicate that A_1ARs aggregate with adenosine deaminase on the cell surface and then translocate either into or out of caveolae upon treatment with an agonist (Gines et al. 2001; Lasley et al. 2000), suggesting another mechanism for regulation of A_1AR signaling. Regardless of the underlying mechanism, receptor desensitization is a potential problem for the development of A_1AR agonists as therapeutic agents for chronic use. Desensitization of the antilipolytic effect of A_1ARs has been shown to occur with prolonged and continuous exposure to high concentrations of full A_1AR agonists in in vitro and in vivo studies (Green 1987; Hoffman et al. 1986a; Zannikos et al. 2001). Prolonged incubation of isolated adipocytes with the A_1AR agonist

PIA resulted in downregulation of A_1ARs (Green 1987). Adipocytes isolated from epididymal fat pads of rats treated with PIA for six days were insensitive to the acute inhibitory effects of PIA on lipolysis (Hoffman et al. 1986a).

It should be noted that the abovementioned studies on receptor desensitization have been conducted using full agonists (e.g., N^6-(R)-phenylisopropyladenosine (R-PIA)), which are more likely to cause desensitization. It has been suggested that partial agonists of G-protein-coupled receptors may cause less receptor desensitization than full agonists (Kovoor et al. 1998; Vachon et al. 1987). Consistent with this notion was the finding that both the magnitude and the duration of the FFA-lowering effect of the partial A_1AR agonist CVT-3619 were similar for at least three consecutive (1 h apart) injections, suggesting that the effect of this agonist does not undergo acute desensitization; i.e., tachyphylaxis (Dhalla et al. 2007a). However, to our knowledge, desensitization upon continuous infusion of partial agonists has not been investigated.

On the other hand, A_1ARs can be upregulated by chronic exposure to AR antagonists (Murray 1982; Szot et al. 1987; Xu et al. 1998). Upregulation of the receptors is also accompanied by an increase in the functional response to the agonist (Szot et al. 1987; Wu et al. 1989). Chronic exposure to the AR antagonists theophylline and caffeine resulted in an increase in the total number of A_1ARs, an enhanced coupling of the receptors to G_i protein, and an increased quantity of $G_{\alpha i}$ protein (Ramkumar et al. 1988; Wu et al. 1989). Treatment with BW-1433, an antagonist of A_1ARs, in insulin-resistant rats led to an increase in A_1AR number and an improvement in glucose tolerance (Xu et al. 1998), suggesting that an increase in A_1AR number in adipose tissue is beneficial; perhaps by increasing the tonic effects of adenosine on lipolysis due to the increase in receptor density (Liang et al. 2002). Transgenic mice overexpressing A_1ARs have been shown to be protected from high fat diet-induced obesity and glucose intolerance (Dong et al. 2001). The proposed beneficial effects of coffee consumption on insulin resistance may also be a result of an increase in A_1ARs by caffeine, an unselective and nonspecific AR antagonist.

6 Possible Solutions to the Challenges Involved in Developing A_1AR Agonists as Antilipolytic Agents

The unwanted effects of A_1AR agonists can theoretically be minimized, and functional selectivity achieved with these compounds, as a consequence of differential receptor reserve for the A_1AR-mediated responses in nonadipose (e.g., cardiac) and adipose tissue (Fraser et al. 2003; Liang et al. 2002; Wu et al. 2001). Furthermore, by using partial agonists (low-efficacy agonists), the differences in receptor reserve between various tissues (e.g., cardiac vs. adipose) are sufficient to achieve functional selectivity of drug action (Dhalla et al. 2007a; van Schaick et al. 1998a), as discussed below.

6.1 Receptor Reserve

Receptor reserve is a term used to indicate that maximal response to a given agonist in a given cell or tissue can be obtained by activation of less than 100% of the receptors. It is thus an indicator of both the number of receptors that mediate a response and the efficiency (amplification) of receptor-to-response coupling. Because receptor reserve is dependent on the intrinsic activity of the agonist, each value of receptor reserve must specify the agonist, response and tissue for which the measurement is pertinent. When receptor reserve is present, the agonist concentration–response curve lies to the left of the concentration–receptor occupancy curve, and hence the K_A value (concentration of agonist that occupies 50% of receptors) is greater than the EC_{50} value (concentration of agonist that causes a half-maximal response). That is, the potency of the agonist is greater than its affinity when receptor reserve is present, whereas the potency and affinity of an agonist are equal when receptor reserve is absent. The ratio between K_A and EC_{50} is referred to as the pharmacological shift ratio (PSR), and is used to quantify the extent of receptor reserve (Ruffolo 1982). Receptor reserves for A_1AR-mediated responses in cardiac and adipose tissues have been measured by using a specific, selective and irreversible A_1AR antagonist (Baker et al. 2000; Liang et al. 2002; Morey et al. 1998; Srinivas et al. 1997). As stated above, the extent of A_1AR reserve is unique for each agonist, response, and tissue such that the receptor reserve for the A_1AR agonist 2 chloro-N^6-cyclopentyladenosine (CCPA) is higher than that for R-PIA for reducing the cAMP content of adipocytes (Liang et al. 2002). The EC_{50} for adenosine to reduce lipolysis in human adipocytes has been reported to be 6 nM, suggesting that receptor reserve of A_1ARs is also present in human adipocytes (Kather 1988). Results of these studies provide much of the rationale for the use of partial agonists (discussed below) for antilipolytic effects. In summary, significant differences in agonist receptor reserve among tissues (and/or responses) can thus be exploited to selectively elicit responses in tissues with the highest receptor reserves. Greater selectivity is more likely to be achieved with partial agonists (discussed below) than with full agonists (Dhalla et al. 2007a; van Schaick et al. 1998a), because the partial agonist is expected to elicit a robust response only in tissues with relatively high receptor reserve, whereas the full agonist will elicit a robust response wherever A_1ARs are present.

6.2 Partial Agonists

Partial agonists are useful for achieving drug selectivity for the target tissue and minimizing toxicity caused by the activation of the same class of receptors in nontarget tissues (de Ligt and IJzerman 2002). In contrast to a full agonist, a partial agonist is a low-efficacy receptor ligand that elicits a submaximal response, even when all ($\geqslant$95%) available receptors are occupied (Stephenson 1997; Wu et al. 2001). Therefore, a partial agonist is less effective than a full agonist in evoking a response in a tissue(s) with low amplification of the signal transduction pathway (Kenakin et al. 1992), and will elicit fewer responses in the intact animal than a full agonist. Partial agonists have pharmacological features that offer several advantages over a full agonist for therapeutic purposes, as described below.

Studies of structure–activity relationships for adenosine derivatives have led to the discovery of a few selective A_1 partial agonists (Cristalli et al. 1988; Lorenzen et al. 1997; Mathot et al. 1995; Muller 2001; Palle et al. 2004; van Tilburg et al. 2001), which have been synthesized by substitution at the 2′-, 3′- and 5′-hydroxyl groups of the ribose moiety of adenosine (de Ligt and IJzerman 2002; Klotz 2000; Morrison et al. 2004; Poulsen and Quinn 1998; van der Wenden et al. 1995). A few studies reporting the antilipolytic and cardiac effects of partial A_1AR agonists have appeared (Dhalla et al. 2007a; Fatholahi et al. 2006; Srinivas et al. 1997; van Schaick et al. 1998a; Wu et al. 2001). However, few partial agonists with demonstrated high affinity and selectivity for the A_1AR have been described in the literature. Therefore, for the discussion of partial agonists in the following section, we will focus on the antilipolytic effects of CVT-3619, a recently discovered novel partial A_1AR agonist (Dhalla et al. 2007a; Fatholahi et al. 2006).

6.2.1 Organ and Response Selectivity

In adipose tissue, a partial agonist can provoke a maximal functional response at concentrations that do not affect the heart rate because receptor density is high and receptor reserve is large in adipocytes, whereas it is low in the heart. For example, CVT-3619 causes minimal or no A_1AR-mediated slowing of heart rate (Fatholahi et al. 2006), indicating that it is a partial agonist for this cardiac response. In contrast, CVT-3619 is a full agonist to decrease cAMP content in rat adipocytes and decrease plasma levels of FFA (Dhalla et al. 2007a; Fatholahi et al. 2006). As shown in Fig. 3, while the full agonist CPA decreases lipolysis and markedly reduces

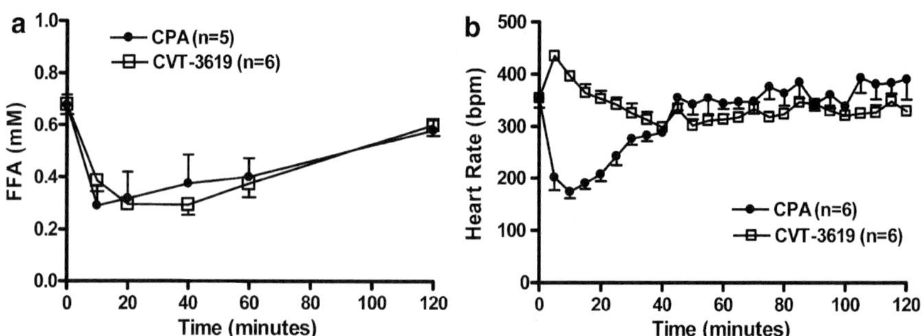

Fig. 3 a–b Comparison of the **a** antilipolytic effect and **b** cardiac effect of CVT-3619 to that of full agonist CPA. For FFA measurements, animals were fasted overnight and were treated with CVT-3619 (10 mg kg^{-1}, p.o.) or CPA (20 μg kg^{-1}, i.p.). Effect of CVT-3619 and CPA on heart rate was determined by telemetry in awake rats at the same doses. Data are presented as the mean ± SEM derived from the number of animals indicated in the parentheses after each group

heart rate, the partial agonist CVT-3619 reduces plasma FFA concentration to similar extent without causing bradycardia. Administration of another partial A_1 agonist, CVT-2759 ({[(5-{6-[(3*R*)oxolan-3-yl]amino}purin-9-yl)(3*S*,2*R*,4*R*)-3,4-dihydroxyoxolan-2-yl]-methoxy}-*N*-methylcarboxamide), to the intact rat caused a greater response in adipocytes (decrease of lipolysis) than in the heart (decrease of heart rate) (Liang et al. 2002).

6.2.2 Less Receptor Desensitization than Full Agonist

It has been suggested that partial agonists of G-protein-coupled receptors cause less receptor desensitization than full agonists (Kovoor et al. 1998; Vachon et al. 1987). The FFA lowering effect of the A_1AR partial agonist CVT-3619 was found to be devoid of acute desensitization (Dhalla et al. 2007a). The acute antilipolytic effect of CVT-3619, given twice daily, was maintained up to six weeks of treatment (unpublished data). However, it remains to be determined whether the antilipolytic effect of CVT-3619 is sustained over long-term use (months) or continuous infusion. Thus, additional work with new partial agonists of the A_1AR is needed to confirm the hypothesis that partial agonists are less prone to cause desensitization and the unwanted therapeutic effect of tolerance/tachyphylaxis.

6.2.3 Functions as an Antagonist of a Full Agonist

Partial agonists may also be considered ligands that display both agonistic and antagonistic effects. When both a full agonist and partial agonist are present, the partial agonist acts as a competitive antagonist, competing with the full agonist for receptor occupancy and producing a net decrease in the receptor activation observed with the full agonist alone. A partial agonist will displace the concentration–response curve of a full agonist to the left, just as an antagonist would (Stephenson 1997). This has also been shown to be the case for the A_1AR agonists. For example, CVT-2759 was shown to cause a leftward shift of the adenosine concentration–response curve to slow AV nodal conduction in guinea pig isolated heart (Wu et al. 2001). Interestingly, it is unlikely that a partial A_1AR will reduce diuresis, an effect expected from a full A_1AR agonist, because it will antagonize the antidiuretic effect of endogenous adenosine (a full agonist).

7 Conclusions

A_1AR agonists are potent inhibitors of adipose tissue lipolysis and have the potential to be developed as clinically useful antilipolytic agents. Because of the ubiquitous presence of these receptors, achieving organ and response selectivity is most important for developing A_1AR agonists as successful antilipolytic agents, and this may be possible with partial agonists. Additional studies are needed to clearly understand the role of A_1AR in the regulation of lipolysis in various pathological conditions where lipolysis may be dysfunctional and the expression of A_1ARs is changed (e.g., during oxidative stress and inflammation).

References

Ashton TD, Baker SP, Hutchinson SA, Scammells PJ (2008) N^6-Substituted C5′-modified adenosines as A_1 adenosine receptor agonists. Bioorg Med Chem 16:1861–1873

Baker SP, Scammells PJ, Belardinelli L (2000) Differential A(1)-adenosine receptor reserve for inhibition of cyclic AMP accumulation and G-protein activation in DDT(1) MF-2 cells. Br J Pharmacol 130:1156–1164

Barakat H, Davis J, Lang D, Mustafa SJ, McConnaughey MM (2006) Differences in the Expression of the adenosine A_1 receptor in adipose tissue of obese black and white women. J Clin Endocrinol Met 91:1882–1886

Bays H, Mandarino L, DeFronzo RA (2004) Role of the adipocyte, free fatty acids, and ectopic fat in pathogenesis of type 2 diabetes mellitus: peroxisomal proliferator-activated receptor agonists provide a rational therapeutic approach. J Clin Endocrinol Met 89:463–478

Belardinelli L, Linden J, Berne RM (1989) The cardiac effects of adenosine. Prog Cardiovasc Dis 32:73–97

Berkich DA, Luthin DR, Woodard RL, Vannucci SJ, Linden J, LaNoue KF (1995) Evidence for regulated coupling of A_1 adenosine receptors by phosphorylation in Zucker rats. Am J Physiol 268: E693–E704

Boden G (2001) Free fatty acids—the link between obesity and insulin resistance. Endocr Pract 7:44–51

Boden G (2002) Obesity and diabetes mellitus—how are they linked? West Indian Med J 51(Suppl 1):51–54

Boden G, She P, Mozzoli M, Cheung P, Gumireddy K, Reddy P, Xiang X, Luo Z, Ruderman N (2005) Free fatty acids produce insulin resistance and activate the proinflammatory nuclear factor-{kappa}b pathway in rat liver. Diabetes 54:3458–3465

Borglum JD, Vassaux G, Richelsen B, Gaillard D, Darimont C, Ailhaud G, Negrel R (1996) Changes in adenosine A_1- and A_2-receptor expression during adipose cell differentiation. Mol Cell Endocrinol 117:17–25

Brechler V, Pavoine C, Lotersztajn S, Garbarz E, Pecker F (1990) Activation of Na^+/Ca^+ exchange by adenosine in ewe heart sarcolemma is mediated by a pertussis toxin-sensitive G protein. J Biol Chem 265(28):16851–16855

Budohoski L, Challiss RA, McManus B, Newsholme EA (1984) Effects of analogues of adenosine and methyl xanthines on insulin sensitivity in soleus muscle of the rat. FEBS Lett 167:1–4

Butcher RW, Ho RJ, Meng HC, Sutherland EW (1965) Adenosine 3′,5′-monophosphate in biological materials. II. The measurement of adenosine 3′, 5′-monophosphate in tissues and the role of the cyclic nucleotide in the lipolytic response of fat to epinephrine. J Biol Chem 240:4515–4523

Cappellacci L, Franchetti P, Vita P, Petrelli R, Lavecchia A, Costa B, Spinetti F, Martini C, Klotz KN, Grifantini M (2008) 5′-Carbamoyl derivatives of 2′-*C*-methyl-purine nucleosides as selective A_1 adenosine receptor agonists: affinity, efficacy, and selectivity for A_1 receptor from different species. Bioorg Med Chem 16:336–353

Carlson LA (2005) Nicotinic acid: the broad-spectrum lipid drug. A 50th anniversary review. J Intern Med 258:94–114

Cheng JT, Chi TC, Liu IM (2000) Activation of adenosine A_1 receptors by drugs to lower plasma glucose in streptozotocin-induced diabetic rats. Auton Neurosci 83:127–133

Ciruela F, Saura C, Canela EI, Mallol J, Lluis C, Franco R (1997) Ligand-induced phosphorylation, clustering, and desensitization of A_1 adenosine receptors. Mol Pharmacol 52:788–797

Cox BF, Clark KL, Perrone MH, Welzel GE, Greenland BD, Colussi DJ, Merkel LA (1997) Cardiovascular and metabolic effects of adenosine A_1-receptor agonists in streptozotocin-treated rats. J Cardiovasc Pharmacol 29:417–426

Cristalli G, Franchetti P, Grifantini M, Vittori S, Klotz KN, Lohse MJ (1988) Adenosine receptor agonists: synthesis and biological evaluation of 1-deaza analogues of adenosine derivatives. J Med Chem 31:1179–1183

de Ligt RA, IJzerman AP (2002) Intrinsic activity at adenosine A_1 receptors: partial and inverse agonism. Curr Pharm Des 8:2333–2344

Dhalla AK, Shryock JC, Shreeniwas R, Belardinelli L (2003) Pharmacology and therapeutic applications of A_1 adenosine receptor ligands. Curr Top Med Chem 3:369–385

Dhalla AK, Wong MY, Wang WQ, Biaggioni I, Belardinelli L (2006) Tachycardia caused by A_{2A} adenosine receptor agonists is mediated by direct sympathoexcitation in awake rats. JPET 316:695–702

Dhalla A, Santikul M, Smith M, Wong MY, Shryock J, Belardinelli L (2007a) Anti-lipolytic activity of a novel partial A_1 adenosine receptor agonist devoid of cardiovascular effects: Comparison with nicotinic acid. JPET 321:327–333

Dhalla AK, Wong MY, Voshol PJ, Belardinelli L, Reaven GM (2007b) A_1 adenosine receptor partial agonist lowers plasma FFA and improves insulin resistance induced by high-fat diet in rodents. Am J Physiol Endocrinol Metab 292:E1358–E1363

Dhalla AK, Santikul M, Chisholm JW, Belardinelli L, Reaven GM (2008) Comparison of the antilipolytic effects of an A_1 adenosine receptor partial agonist in normal and diabetic rats. Diabetes Obes Metab 11:95–101

Dixon AK, Gubitz AK, Sirinathsinghji DJS, Richardson PJ, Freeman TC (1996) Tissue distribution of adenosine receptor mRNAs in the rat. Br J Pharmacol 118:1461–1468

Dobson JG Jr (1978) Reduction by adenosine of the isoproterenol-induced increase in cyclic adenosine $3', 5'$-monophosphate formation and glycogen phosphorylase activity in rat heart muscle. Circ Res 43:785–792

Dobson JG Jr (1983) Interaction between adenosine and inotropic interventions in guinea pig atria. Am J Physiol 245: H475–H480

Dobson JG Jr, Ordway RW, Fenton RA (1986) Endogenous adenosine inhibits catecholamine contractile responses in normoxic hearts. Am J Physiol 251:H455–H462

Dole VP (1961) Effect of nucleic acid metabolites on lipolysis in adipose tissue. J Biol Chem 236:3125–3130

Dong Q, Ginsberg HN, Erlanger BF (2001) Overexpression of the A_1 adenosine receptor in adipose tissue protects mice from obesity-related insulin resistance. Diabetes Obes Metab 3:360–366

Fain JN, Malbon CC (1979) Regulation of adenylate cyclase by adenosine. Mol Cell Biochem 25:143–169

Fain JN, Pointer RH, Ward WF (1972) Effects of adenosine nucleosides on adenylate cyclase, phosphodiesterase, cyclic adenosine monophosphate accumulation, and lipolysis in fat cells. J Biol Chem 247:6866–6872

Fatholahi M, Xiang Y, Wu Y, Li Y, Wu L, Dhalla AK, Belardinelli L, Shryock JC (2006) A novel partial agonist of the A_1-adenosine receptor and evidence of receptor homogeneity in adipocytes. JPET 317:676–684

Foley JE, Anderson RC, Bell PA, Burkey BF, Deems RO, de Souza C, Dunning BE (1997) Pharmacological strategies for reduction of lipid availability. Ann N Y Acad Sci 827:231–245

Fraser H, Gao Z, Ozeck MJ, Belardinelli L (2003) *N*-[3-(*R*)-Tetrahydrofuranyl]-6-aminopurine riboside, an A_1 adenosine receptor agonist, antagonizes catecholamine-induced lipolysis without cardiovascular effects in awake rats. J Pharmacol Exp Ther 305:225–231

Fredholm BB, Sollevi A (1986) Cardiovascular effects of adenosine. Clin Physiol 6:1–21

Fredholm BB, IJzerman AP, Jacobson KA, Klotz KN, Linden J (2001) International Union of Pharmacology. XXV. Nomenclature and classification of adenosine receptors. Pharmacol Rev 53:527–552

Fruhbeck G, Gomez-Ambrosi J (2002) Depot-specific differences in the lipolytic effect of leptin on isolated white adipocytes. Med Sci Monit 8:BR47–BR55

Fruhbeck G, Gomez-Ambrosi J, Salvador J (2001) Leptin-induced lipolysis opposes the tonic inhibition of endogenous adenosine in white adipocytes. FASEB J 15:333–340

Funakoshi H, Zacharia LC, Tang Z, Zhang J, Lee LL, Good JC, Herrmann DE, Higuchi Y, Koch WJ, Jackson EK, Chan TO, Feldman AM (2007) A_1 adenosine receptor upregulation accompanies decreasing myocardial adenosine levels in mice with left ventricular dysfunction. Circulation 115:2307–2315

Gao ZG, Jacobson KA (2007) Emerging adenosine receptor agonists. Expert Opin Emerg Drugs 12:479–492

Gao Z, Robeva AS, Linden J (1999) Purification of A_1 adenosine receptor-G-protein complexes: effects of receptor down-regulation and phosphorylation on coupling. Biochem J 338 (Pt 3):729–736

Gardner CJ, Twissell DJ, Coates J, Strong P (1994) The effects of GR79236 on plasma fatty acid concentrations, heart rate and blood pressure in the conscious rat. Eur J Pharmacol 257:117–121

Garg A, Grundy SM (1990) Nicotinic acid as therapy for dyslipidemia in non-insulin-dependent diabetes mellitus. JAMA 264:723–726

Gines S, Ciruela F, Burgueno J, Casado V, Canela EI, Mallol J, Lluis C, Franco R (2001) Involvement of caveolin in ligand-induced recruitment and internalization of A(1) adenosine receptor and adenosine deaminase in an epithelial cell line. Mol Pharmacol 59:1314–1323

Gould J, Morton MJ, Sivaprasadarao A, Bowmer CJ, Yates MS (1997) Renal adenosine A_1 receptor binding characteristics and mRNA levels during the development of acute renal failure in the rat. Br J Pharmacol 120:947–953

Grden M, Podgorska M, Szutowicz A, Pawelczyk T (2007) Diabetes-induced alterations of adenosine receptors expression level in rat liver. Exp Mol Pathol 83:392–398

Green A (1987) Adenosine receptor down-regulation and insulin resistance following prolonged incubation of adipocytes with an A_1 adenosine receptor agonist. J Biol Chem 262:15702–15707

Green A, Swenson S, Johnson JL, Partin M (1989) Characterization of human adipocyte adenosine receptors. Biochem Biophys Res Commun 163:137–142

Green A, Johnson JL, Milligan G (1990) Down-regulation of G_i sub-types by prolonged incubation of adipocytes with an A_1 adenosine receptor agonist. J Biol Chem 265:5206–5210

Grundy SM, Vega GL, McGovern ME, Tulloch BR, Kendall DM, Fitz-Patrick D, Ganda OP, Rosenson RS, Buse JB, Robertson DD, Sheehan JP, for the Diabetes Multicenter Research Group (2002) Efficacy, safety, and tolerability of once-daily niacin for the treatment of dyslipidemia associated with type 2 diabetes: results of the assessment of diabetes control and evaluation of the efficacy of Niaspan trial. Arch Intern Med 162:1568–1576

Harada K, Shen WJ, Patel S, Natu V, Wang J, Osuga J, Ishibashi S, Kraemer FB (2003) Resistance to high-fat diet-induced obesity and altered expression of adipose-specific genes in HSL-deficient mice. Am J Physiol Endocrinol Metab 285: E1182–E1195

Ho RJ, Sutherland EW (1971) Formation and release of a hormone antagonist by rat adipocytes. J Biol Chem. 246:6822–6827

Hoffman BB, Chang H, Dall'Aglio E, Reaven GM (1986a) Desensitization of adenosine receptor-mediated inhibition of lipolysis. The mechanism involves the development of enhanced cyclic adenosine monophosphate accumulation in tolerant adipocytes. J Clin Invest 78:185–190

Hoffman BB, Dall'Aglio E, Hollenbeck C, Chang H, Reaven GM (1986b) Suppression of free fatty acids and triglycerides in normal and hypertriglyceridemic rats by the adenosine receptor agonist phenylisopropyladenosine. JPET 239:715–718

Ishikawa J, Mitani H, Bandoh T, Kimura M, Totsuka T, Hayashi S (1998) Hypoglycemic and hypotensive effects of 6-cyclohexyl-2′-O-methyl-adenosine, an adenosine A_1 receptor agonist, in spontaneously hypertensive rat complicated with hyperglycemia. Diabetes Res Clin Pract 39:3–9

Itani SI, Ruderman NB, Schmieder F, Boden G (2002) Lipid-induced insulin resistance in human muscle is associated with changes in diacylglycerol, protein kinase C, and IkappaB-alpha. Diabetes 51:2005–2011

Jacobson KA, van Galen PJ, Williams M (1992) Adenosine receptors: pharmacology, structure–activity relationships, and therapeutic potential. J Med Chem 35:407–422

Jacobson KA, Von Lubitz DKJE, Daly JW, Fredholm BB (1996) Adenosine receptor ligands: differences with acute versus chronic treatment. TIPS 17:108–113

Jajoo S, Mukherjea D, Pingle S, Sekino Y, Ramkumar V (2006) Induction of adenosine A_1 receptor expression by pertussis toxin via an adenosine 5′-diphosphate ribosylation-independent pathway. JPET 317:1–10

Jensen MD (2006) Adipose tissue as an endocrine organ: implications of its distribution on free fatty acid metabolism. Eur Heart J 8: B13–B19

Johansson SM, Salehi A, Sandstrom ME, Westerblad H, Lundquist I, Carlsson PO, Fredholm BB, Katz A (2007a) A_1 receptor deficiency causes increased insulin and glucagon secretion in mice. Biochem Pharmacol 74:1628–1635

Johansson SM, Yang JN, Lindgren E, Fredholm BB (2007b) Eliminating the antilipolytic adenosine A_1 receptor does not lead to compensatory changes in the antilipolytic actions of PGE2 and nicotinic acid. Acta Physiol 190:87–96

Kaartinen JM, Hreniuk SP, Martin LF, Ranta S, LaNoue KF, Ohisalo JJ (1991) Attenuated adenosine-sensitivity and decreased adenosine-receptor number in adipocyte plasma membranes in human obesity. Biochem J 279 (Pt 1):17–22

Kaartinen JM, LaNoue KF, Ohisalo JJ (1994) Quantitation of inhibitory G-proteins in fat cells of obese and normal-weight human subjects. Biochim Biophys Acta 1201:69–75

Kather H (1988) Purine accumulation in human fat cell suspensions. Evidence that human adipocytes release inosine and hypoxanthine rather than adenosine. J Biol Chem. 263:8803–8809

Kenakin TP (1984) The classification of drugs and drug receptors in isolated tissues. Pharmacol Rev 36:165–222

Kenakin TP, Bond RA, Bonner TI (1992) Definition of pharmacological receptors. Pharmacol Rev 44:351–362

Klotz KN (2000) Adenosine receptors and their ligands. Naunyn–Schmiedeberg's Arch Pharmacol 362:382–391

Klotz K-N, Lohse MJ, Schwabe U, Cristalli G, Vittori S, Grifantini M (1989) 2-Chloro-N^6-[3H]cyclopentyladenosine ([3H]CCPA): a high 1 affinity agonist radioligand for A_1 adenosine receptors. Naunyn–Schmiedeberg's Arch Pharmacol 340:679–683

Kollias-Baker C, Shryock JC, Belardinelli L (1995) Myocardial adenosine receptors. In: Belardinelli L, Pelleg A (eds) Adenosine and adenine nucleotides: from molecular biology to integrative physiology. Kluwer, Boston, pp 221–228

Kollias-Baker CA, Ruble J, Jacobson M, Harrison JK, Ozeck M, Shryock JC, Belardinelli L (1997) Agonist-independent effect of an allosteric enhancer of the A_1 adenosine receptor in CHO cells stably expressing the recombinant human A_1 receptor. JPET 281:761–768

Kovoor A, Celver JP, Wu A, Chavkin C (1998) Agonist induced homologous desensitization of mu-opioid receptors mediated by G protein-coupled receptor kinases is dependent on agonist efficacy. Mol Pharmacol 54:704–711

Lai DM, Tu YK, Liu IM, Cheng JT (2005) Increase of adenosine A_1 receptor gene expression in cerebral ischemia of Wistar rats. Neurosci Lett 387:59–61

Langin D (2006) Adipose tissue lipolysis as a metabolic pathway to define pharmacological strategies against obesity and the metabolic syndrome. Pharmacol Res 53:482–491

LaNoue KF, Martin LF (1994) Abnormal A_1 adenosine receptor function in genetic obesity. FASEB J 8:72–80

Large V, Arner P (1998) Regulation of lipolysis in humans. Pathophysiological modulation in obesity, diabetes, and hyperlipidaemia. Diabet Metab 24:409–418

Larrouy D, Galitzky J, Lafontan M (1991) A_1 adenosine receptors in the human fat cell: tissue distribution and regulation of radioligand binding. Eur J Pharmacol 206:139–147

Lasley RD, Narayan P, Uittenbogaard A, Smart EJ (2000) Activated cardiac adenosine A_1 receptors translocate out of caveolae. J Biol Chem 275:4417–4421

Leblanc J, Soucy J (1994) Hormonal dose–response to an adenosine receptor agonist. Can J Physiol Pharmacol 72:113–116

Leung E, Jacobson KA, Green RD (1990) Analysis of agonist–antagonist interactions at A_1 adenosine receptors. Mol Parmacol 38:72–83

Liang HX, Belardinelli L, Ozeck MJ, Shryock JC (2002) Tonic activity of the rat adipocyte A_1-adenosine receptor. Br J Pharmacol 135:1457–1466

Linden J (1991) Structure and function of A_1 adenosine receptors. FASEB J 5:2668–2676

Liu IM, Tzeng TF, Tsai CC, Lai TY, Chang CT, Cheng JT (2003) Increase in adenosine A_1 receptor gene expression in the liver of streptozotocin-induced diabetic rats. Diabet Metab Res Rev 19:209–215

Lohse MJ, Ukena D, Schwabe U (1984) Adenosine receptors on heart muscle. Lancet 2:355

Londos C, Wolff J (1977) Two distinct adenosine-sensitive sites on adenylate cyclase. Proc Natl Acad Sci USA 74:5482–5486

Longabaugh JP, Didsbury J, Spiegel A, Stiles GL (1989) Modification of the rat adipocyte A_1 adenosine receptor-adenylate cyclase system during chronic exposure to an A_1 adenosine receptor agonist: alterations in the quantity of Gs alpha and Gi alpha are not associated with changes in their mRNAs. Mol Pharmacol 36:681–688

Lorenzen A, Sebastiao AM, Sellink A, Vogt H, Schwabe U, Ribeiro JA, IJzerman AP (1997) Biological activities of N^6, C8-disubstituted adenosine derivatives as partial agonists at rat brain adenosine A_1 receptors. Eur J Pharmacol 334:299–307

Mathot RAA, van der Wenden EM, Soudijn W, IJzerman AP, Danhof M (1995) Deoxyribose analogues of N^6-cyclopentyladenosine (CPA): partial agonists at the adenosine A_1 receptor in vivo. Br J Pharmacol 116:1957–1964

McKenny JM, Proctor JD, Harris S, Chinchili VM (1994) A comparison of the efficacy and toxic effects of sustained vs immediate-release niacin in hypercholesterolemic patients. JAMA 271:672–710

Merkel LA, Hawkins ED, Colussi DJ, Greenland BD, Smits GJ, Perrone MH, Cox BF (1995) Cardiovascular and antilipolytic effects of the adenosine agonist GR79236. Pharmacology 51:224–236

Meyer P, Bier D, Holschbach M, Cremer M, Tellmann L, Bauer A (2003) In vivo imaging of rat brain A_1 adenosine receptor occupancy by caffeine. Eur J Nucl Med Mol Imag 30:1440

Moreno FJ, Mills I, Garcia-Sainz JA, Fain JN (1983) Effects of pertussis toxin treatment on the metabolism of rat adipocytes. J Biol Chem 258(18):10938–10943

Morey TE, Belardinelli L, Dennis DM (1998) Validation of Furchgott's method to determine agonist-dependent A_1-adenosine receptor reserve in guinea-pig atrium. Br J Pharmacol 123:1425–1433

Morrison CF, Elzein E, Jiang B, Ibrahim PN, Marquart T, Palle V, Shenk KD, Varkhedkar V, Maa T, Wu L, Wu Y, Zeng D, Fong I, Lustig D, Leung K, Zablocki JA (2004) Structure–affinity relationships of 5′-aromatic ethers and 5′-aromatic sulfides as partial A_1 adenosine agonists, potential supraventricular anti-arrhythmic agents. Bioorg Med Chem Lett 14:3793–3797

Muller CE (2001) A_1 adenosine receptors and their ligands: overview and recent developments. Il Farmaco 56:77–80

Munshi R, Pang IH, Sternweis PC, Linden J (1991) A_1 adenosine receptors of bovine brain couple to guanine nucleotide-binding proteins Gi1, Gi2, and Go. J Biol Chem 266:22285–22289

Murray TF (1982) Up-regulation of rat cortical adenosine receptors following chronic administration of theophylline. Eur J Pharmacol 82:113–114

Nemeth ZH, Bleich D, Csoka B, Pacher P, Mabley JG, Himer L, Vizi ES, Deitch EA, Szabo C, Cronstein BN, Hasko G (2007) Adenosine receptor activation ameliorates type 1 diabetes. FASEB J 21:2379–2388

Nie Z, Mei Y, Ford M, Rybak L, Marcuzzi A, Ren H, Stiles GL, Ramkumar V (1998) Oxidative stress increases A_1 adenosine receptor expression by activating nuclear factor kappa B. Mol Pharmacol 53:663–669

Okajima F, Sato K, Sho K, Kondo Y (1989) Stimulation of adenosine receptor enhances A_1-adrenergic receptor-mediated activation of phospholipase C and Ca^{2+} mobilization in a pertussis toxin-sensitive manner in FRTL-5 thyroid cells. FEBS Lett 248:145–149

Osuga J, Ishibashi S, Oka T, Yagyu H, Tozawa R, Fujimoto A, Shionoiri F, Yahagi N, Kraemer FB, Tsutsumi O, Yamada N (2000) Targeted disruption of hormone-sensitive lipase results in male sterility and adipocyte hypertrophy, but not in obesity. Proc Natl Acad Sci USA 97:787–792

Palle VP, Varkhedkar V, Ibrahim P, Ahmed H, Li Z, Gao Z, Ozeck M, Wu Y, Zeng D, Wu L, Leung K, Chu N, Zablocki JA (2004) Affinity and intrinsic efficacy (IE) of 5′-carbamoyl adenosine analogues for the A_1 adenosine receptor–efforts towards the discovery of a chronic ventricular rate control agent for the treatment of atrial fibrillation (AF). Bioorg Med Chem Lett 14:535–539

Palmer TM, Benovic JL, Stiles GL (1996) Molecular basis for subtype-specific desensitization of inhibitory adenosine receptors. Analysis of a chimeric A_1–A_3 adenosine receptor. J Biol Chem 271:15272–15278

Parsons WJ, Stiles GL (1987) Heterologous desensitization of the inhibitory A_1 adenosine receptor–adenylate cyclase system in rat adipocytes. Regulation of both Ns and Ni. J Biol Chem 262:841–847

Pawelczyk T, Grden M, Rzepko R, Sakowicz M, Szutowicz A (2005) Region-specific alterations of adenosine receptors expression level in kidney of diabetic rat. Am J Pathol 167:315–325

Poulsen S, Quinn RJ (1998) Adenosine receptors: new opportunities for future drugs. Bioorg Med Chem 6:619–641

Poynten AM, Gan SK, Kriketos AD, O'Sullivan A, Kelly JJ, Ellis BA, Chisholm DJ, Campbell LV (2003) Nicotinic acid-induced insulin resistance is related to increased circulating fatty acids and fat oxidation but not muscle lipid content. Metabolism 52:699–704

Press NJ, Gessi S, Borea PA, Polosa R (2007) Therapeutic potential of adenosine receptor antagonists and agonists. Expert Opin Ther Patents 17:979–991

Qu X, Cooney G, Donnelly R (1997) Short-term metabolic and haemodynamic effects of GR79236 in normal and fructose-fed rats. Eur J Pharmacol 338:269–276

Ramkumar V, Bumgarner JR, Jacobson KA, Stiles GL (1988) Multiple components of the A_1 adenosine receptor–adenylate cyclase system are regulated in rat cerebral cortex by chronic caffeine ingestion. J Clin Invest 82:242–247

Reaven GM (1995) The fourth musketeer: from Alexandre Dumas to Claude Bernard. Diabetologia 38:3–13

Reaven GM, Chang H, Ho H, Jeng CY, Hoffman BB (1988) Lowering of plasma glucose in diabetic rats by antilipolytic agents. Am J Physiol 254:E23–E30

Reppert SM, Weaver DR, Stehle JH, Rivkees SA (1991) Molecular cloning and characterization of a rat A_1-adenosine receptor that is widely expressed in brain and spinal cord. Mol Endocrin 5:1037–1048

Rivkees SA (1995) The ontogeny of cardiac and neural A_1 adenosine receptor expression in rats. Dev Brain Res 89:202–213

Roden M, Price TB, Perseghin G, Petersen KF, Rothman DL, Cline GW, Shulman GI (1996) Mechanism of free fatty acid-induced insulin resistance in humans. J Clin Invest 97:2859–2865

Rogachev B, Ziv NY, Mazar J, Nakav S, Chaimovitz C, Zlotnik M, Douvdevani A (2006) Adenosine is upregulated during peritonitis and is involved in downregulation of inflammation. Kidney Int 70:675–681

Rolband GC, Furth ED, Staddon JM, Rogus EM, Goldberg AP (1990) Effects of age and adenosine in the modulation of insulin action on rat adipocyte metabolism. J Gerontol 45:B174–B178

Ruffolo RR (1982) Important concepts of receptor theory. J Auton Pharmacol 2:277–295

Saggerson ED, Jamal Z (1990) Differences in the properties of A_1-type adenosine receptors in rat white and brown adipocytes. Biochem J 269:157–161

Sako Y, Grill VE (1990) A 48-h lipid infusion in the rat time-dependently inhibits glucose-induced insulin secretion and B cell oxidation through a process likely coupled to fatty acid oxidation. Endocrinology 127:1580–1589

Schiffmann SN, Libert F, Vassart G, Dumont JE, Vanderhaeghen JJ (1990) A cloned G protein-coupled protein with a distribution restricted to striatal medium-sized neurons. Possible relationship with D_1 dopamine receptor. Brain Res 519:333–337

Schoelch C, Kuhlmann J, Gossel M, Mueller G, Neumann-Haefelin C, Belz U, Kalisch J, Biemer-Daub G, Kramer W, Juretschke HP, Herling AW (2004) Characterization of adenosine-A_1 receptor-mediated antilipolysis in rats by tissue microdialysis, 1H-spectroscopy, and glucose clamp studies. Diabetes 53:1920–1926

Schrader J, Baumann G, Gerlach E (1977) Adenosine as inhibitor of myocardial effects of catecholamines. Pflugers Arch 372:29–35

Schwabe U, Ebert R (1974) Stimulation of cyclic adenosine $3', 5'$-monophosphate accumulation and lipolysis in fat cells by adenosine deaminase. Naunyn–Schmiedeberg's Arch Pharmacol 282:33–44

Schwabe U, Ebert R, Erbler C (1973) Adenosine release from isolated fat cells and its significance for the effects of hormones on cyclic $3'$, $5'$-AMP levels and lipolysis. Naunyn–Schmiedeberg's Arch Pharmacol 276:133–148

Schwabe U, Schonhofer PS, Ebert R (1974) Facilitation by adenosine of the action of insulin on the accumulation of adenosine $3', 5'$-monophosphate, lipolysis, and glucose oxidation in isolated fat cells. Eur J Biochem 46:537–545

Shah B, Rohatagi S, Natarajan C, Kirkesseli S, Baybutt R, Jensen BK (2004) Pharmacokinetics, pharmacodynamics, and safety of a lipid-lowering adenosine A_1 agonist, RPR749, in healthy subjects. Am J Ther 11:175–189

Shen J, Halenda SP, Sturek M, Wilden P (2005) Novel mitogenic effect of adenosine on coronary artery smooth muscle cells. Circ Res 96:982–990

Srinivas M, Shryock JC, Dennis DM, Baker SP, Belardinelli L (1997) Differential A_1 adenosine receptor reserve for two actions of adenosine on guinea pig atrial myocytes. Mol Pharmacol 52:683–691

Stephenson RP (1997) A modification of receptor theory. 1956. Br J Pharmacol 120:106–120

Strange PG (2008) Agonist binding, agonist affinity and agonist efficacy at G protein-coupled receptors. Br J Pharmacol 153:1353–1363

Strong P, Anderson R, Coates J, Ellis F, Evans B, Gurden MF, Johnstone J, Kennedy I, Martin DP (1993) Suppression of non-esterified fatty acids and triacylglycerol in experimental animals by the adenosine analogue GR79236. Clin Sci 84:663–669

Szot P, Sanders RC, Murray TF (1987) Theophylline-induced upregulation of A_1-adenosine receptors associated with reduced sensitivity to convulsants. Neuropharmacology 26:1173–1180

Takasuga S, Katada T, Ui M, Hazeki O (1999) Enhancement by adenosine of insulin-induced activation of phosphoinositide 3-kinase and protein kinase B in rat adipocytes. J Biol Chem 274:19545–19550

Tatsis-Kotsidis I, Erlanger BF (1999) Initiation of a process of differentiation by stable transfection of ob17 preadipocytes with the cDNA of human A_1 adenosine receptor. Biochem Pharmacol 58:167–170

Thompson CS, Strong P, Mikhailidis DP (1994) Interactions between insulin and the antilipolytic agent GR79236 in ketoacidotic diabetic rats. J Drug Dev 6:183–186

Trost T, Schwabe U (1981) Adenosine receptors in fat cells. Identification by (−)-N^6-[3H]phenylisopropyladenosine binding. Mol Pharmacol 19:228–235

Ukena D, Furler R, Lohse MJ, Engel G, Schwabe U (1984a) Labelling of Ri adenosine receptors in rat fat cell membranes with (−)-[^{125}iodo]N^6-hydroxyphenylisopropyladenosine. Naunyn–Schmiedeberg's Arch Pharmacol 326:233–240

Ukena D, Poeschla E, Schwabe U (1984b) Guanine nucleotide and cation regulation of radioligand binding to Ri adenosine receptors of rat fat cells. Naunyn–Schmiedeberg's Arch Pharmacol 326:241–247

Vachon L, Costa T, Herz A (1987) Opioid receptor desensitization in NG 108–15 cells. Differential effects of a full and a partial agonist on the opioid-dependent GTPase. Biochem Pharmacol 36:2889–2897

Van der Graaf PH, van Schaick EA, Visser SAG, De Greef HJMM, IJzerman AP, Danhof M (1999) Mechanism-based pharmacokinetic–pharmacodynamic modeling of antilipolytic effects of adenosine A_1 receptor agonists in rats: prediction of tissue-dependent efficacy in vivo. J Pharmacol Exp Ther 290:702–709

van der Wenden EM, von Frijtag Drabbe Kunzel JK, Mathot RA, Danhof M, IJzerman AP, Soudijn W (1995) Ribose-modified adenosine analogues as potential partial agonists for the adenosine receptor. J Med Chem 38:4000–4006

van Schaick EA, Tukker HE, Roelen HCPF, IJzerman AP, Danhof M (1998a) Selectivity of action of 8-alkylamino analogues of N^6-cyclopentyladenosine in vivo: haemodynamic versus anti-lipolytic responses in rats. Br J Pharmacol 124:607–618

van Schaick EA, Zuideveld KP, Tukker HE, Langemeijer MW, IJzerman AP, Danhof M (1998b) Metabolic and cardiovascular effects of the adenosine A_1 receptor agonist N^6-(*p*-sulfophenyl)adenosine in diabetic Zucker rats: influence of the disease on the selectivity of action. JPET 287:21–30

van Tilburg EW, van der Klein PA, Frijtag Drabbe KJ, de Groote M, Stannek C, Lorenzen A, IJzerman AP (2001) 5′-*O*-Alkyl ethers of *N*,2-substituted adenosine derivatives: partial agonists for the adenosine A_1 and A_3 receptors. J Med Chem 44:2966–2975

Vannucci SJ, Klim CM, Martin LF, LaNoue KF (1989) A_1-adenosine receptor-mediated inhibition of adipocyte adenylate cyclase and lipolysis in Zucker rats. Am J Physiol 257: E871-E878

Vassaux G, Gaillard D, Mari B, Ailhaud G, Negrel R (1993) Differential expression of adenosine A_1 and A_2 receptors in preadipocytes and adipocytes. Biochem Biophys Res Commun 193:1123–1130

Vega GL, Cater NB, Meguro S, Grundy SM (2005) Influence of extended-release nicotinic acid on nonesterified fatty acid flux in the metabolic syndrome with atherogenic dyslipidemia. Am J Cardiol 95:1309–1313

Wagner H, Milavec-Krizman M, Gadient F, Menninger K, Schoeffter P, Tapparelli C, Pfannkuche H-J, Fozard JR (1995) General pharmacology of SDZ WAG 994, a potent selective and orally active adenosine A_1 receptor agonist. Drug Dev Res 34:276–288

Wu SN, Linden J, Visentin S, Boykin M, Belardinelli L (1989) Enhanced sensitivity of heart cells to adenosine and up-regulation of receptor number after treatment of guinea pigs with theophylline. Circ Res 65:1066–1077

Wu L, Belardinelli L, Zablocki JA, Palle V, Shryock JC (2001) A partial agonist of the A(1)-adenosine receptor selectively slows AV conduction in guinea pig hearts. Am J Physiol Heart Circ Physiol 280:H334–H343

Wyne KL (2003) Free fatty acids and type 2 diabetes mellitus. Am J Med 115(Suppl 8A):29S–36S

Xu B, Berkich DA, Crist GH, LaNoue KF (1998) A_1 adenosine receptor antagonism improves glucose tolerance in Zucker rats. Am J Physiol 274:E271–E279

Zannikos PN, Rohatagi S, Jensen BK (2001) Pharmacokinetic–pharmacodynamic modeling of the antilipolytic effects of an adenosine receptor agonist in healthy volunteers. J Clin Pharmacol 41:61–69

A_3 Adenosine Receptor: Pharmacology and Role in Disease

P.A. Borea, S. Gessi, S. Bar-Yehuda, and P. Fishman

Contents

Abstract The study of the A_3 adenosine receptor (A_3AR) represents a rapidly growing and intense area of research in the adenosine field. The present chapter will provide an overview of the expression patterns, molecular pharmacology and functional role of this A_3AR subtype under pathophysiological conditions. Through studies utilizing selective A_3AR agonists and antagonists, or A_3AR knockout mice, it is now clear that this receptor plays a critical role in the modulation of ischemic diseases as well as in inflammatory and autoimmune pathologies. Therefore, the potential therapeutic use of agonists and antagonists will also be described. The discussion will principally address the use of such compounds in the treatment of brain and heart ischemia, asthma, sepsis and glaucoma. The final part concentrates on the molecular basis of A_3ARs in autoimmune diseases such as rheumatoid arthritis, and includes a description of clinical trials with the selective agonist CF101. Based on this chapter, it is evident that continued research to discover agonists and antagonists for the A_3AR subtype is warranted.

P.A. Borea (✉)
Chair of Pharmacology, Faculty of Medicine, University of Ferrara,
Department of Clinical and Experimental Medicine, Pharmacology Unit Via Fossato di Mortara 17–19, 44100 Ferrara, Italy
bpa@dns.unife.it

C.N. Wilson and S.J. Mustafa (eds.), *Adenosine Receptors in Health and Disease*,
Handbook of Experimental Pharmacology 193, DOI 10.1007/978-3-540-89615-9_10,

Keywords A3 Adenosine receptor · Gene and tissue localization · Ischemic conditions · Inflammation · Autoimmune diseases

Abbreviations

$\Delta\psi$	Mitocondrial membrane potential
A_1 $AR A_1$	Adenosine receptor
A_{2A} $AR A_{2A}$	Adenosine receptor
A_{2B} $AR A_{2B}$	Adenosine receptor
$A_3AR^{-/-}$	Functional deletions of the A_3AR
A_3 $AR A_3$	Adenosine receptor
AC	Adenylyl cyclase
ACR	American College of Rheumatology
ADA	Adenosine deaminase
$ADA^{-/-}$	Adenosine deaminase deficient
AIA	Adjuvant-induced arthritis
AICAR	Aminoimidazole carboxamide ribonucleotide
AR	Adenosine receptor
Ca^{2+}	Calcium
cAMP	Cyclic adenosine monophosphate
$CHO - hA_3$	Chinese hamster ovary cells transfected with human A_3AR
Cl–IB–MECA	2-Chloro-N^6-(3-iodobenzyl)-N-methyl-5′-carbamoyladenosine
CNS	Central nervous system
ConA	Concanavalin A
COPD	Chronic obstructive pulmonary disease
COX	Cyclooxygenase
CP-532,903	N^6-(2,5-Dichlorobenzyl)-3-aminoadenosine-5-N -methylcarboxamide
DAG	1,2-Diacylglycerol
DMARDs	Disease-modifying antirheumatic drugs
DPCPX	8-Cyclopentyl-1,3-dipropylxanthine
ERK1/2	Extracellular signal-regulated kinases
GPCR	G-protein-coupled receptor
GRKs	G-protein-coupled receptor kinases
GSK-3β	Glycogen synthase kinase
HIF-1α	Hypoxia-inducible factor 1α
IB–MECA	N^6-(3-Iodobenzyl)-adenosine-5′-N-methylcarboxamide
IκB	Inhibitor of κB
IKK	IκB kinase
IL	Interleukin
IPC	Ischemic preconditioning
IP_3	Inositol triphosphate
JNK	Jun N-terminal kinase

K_{ATP}	ATP-sensitive potassium
KO	Knockout
LPS	Lipopolysaccharide
MAPK	Mitogen-activated protein kinase
MBP	Myelin basic protein
MEK	Mitogen-activated protein kinase kinase
Mito	Mitochondrial
MTX	Methotrexate
mPTP	Mitochondrial permeability transition pore
MRS 1191	3-Ethyl-5-benzyl-2-methyl-4-phenylethynyl-6-phenyl-1,4-(±) -dihydropyridine-3,5-dicarboxylate
MRS 1523	5-Propyl-2-ethyl-4-propyl-3-(ethylsulfanylcarbonyl)-6 -phenylpyridine-5-carboxylate
NF-kB	Nuclear factor kappa B
NOS	Nitric oxide synthase
OGD	Oxygen and glucose deprivation
p38	Stress-activated protein kinase with molecular weight 38 kDa
PBMC	Peripheral blood mononuclear cells
PI3K	Phosphoinositide 3-kinase
PKA	Protein kinase A
PKB/Akt	Protein kinase B
PKC	Protein kinase C
PLC	Phospholipase C
PLD	Phospholipase D
PTX	Pertussis toxin
RA	Rheumatoid arthritis
RBL	Rat basophilic leukemia
STAT3	Signal transducer and activator of transcription 3
$t_{1/2}$	Half-life
TNF-α	Tumor necrosis factor alpha

1 Cloning, Distribution and Gene Structure of the A_3 Adenosine Receptor (A_3AR)

The A_3AR is the last member of the adenosine family of G-protein-coupled receptors (GPCR) to have been cloned. It was originally isolated as an orphan receptor from rat testis and designated tgpcr1, and it has 40% sequence homology with canine A_1 and A_{2A} adenosine receptor (AR) subtypes (Meyerhof et al. 1991). Subsequently, an identical clone was obtained from rat striatum, initially named R226, and functionally expressed in Chinese hamster ovary (CHO) cells (Zhou et al. 1992). Homologs of the rat striatal A_3AR have been identified and cloned from sheep hypophysial pars tuberalis and from human striatum and heart (Linden et al. 1993; Sajjadi and Firestein 1993; Salvatore et al. 1993). Recently, an equine A_3AR

was cloned and pharmacologically characterized, and revealed a high degree of sequence similarity with that of human and sheep A_3AR transcripts (Brandon et al. 2006). Whilst the cDNA sequences of A_1, A_{2A} and A_{2B} ARs are highly conserved between rat and human, interspecies differences in A_3AR structure are large, with the rat A_3AR showing only 72% sequence homology with that of sheep and human. This led to different pharmacological profiles for the species homologs, especially in relation to antagonist binding (Jacobson and Gao 2006).

Interspecies differences have also been found in the peripheral expression of A_3AR mRNA. In particular, the tissue distribution of the human A_3AR transcript was found to be more similar to the sheep than to the rat homolog. In the rat, it has a very narrow distribution, being expressed mainly in the testes, lung, kidneys, heart and brain; in the sheep, the A_3AR transcript is expressed in lung, spleen, pars tuberalis and pineal gland; in the human it is highly expressed in lung and liver and at a moderate level in heart, kidney, placenta and brain (Dixon et al. 1996; Linden 1994; Rivkees 1994; Salvatore et al. 1993; Zhou et al. 1992). The presence of A_3AR protein has been evaluated through radioligand binding and immunoassays in primary cells, tissues and cell lines of differing origins. The first cell line that was demonstrated to have high levels of endogenous A_3AR was the rat mast cell line RBL-2H3, where binding experiments detected a density of about 1 pmol mg^{-1} of protein (Olah et al. 1994; Ramkumar et al. 1993). Low levels of A_3AR binding sites have been observed in the mouse, rat, gerbil and rabbit brain (Jacobson et al. 1993; Ji et al. 1994). No direct evidence of the presence of A_3AR has been obtained in cardiomyocytes (Peart and Headrick 2007), even though several functional studies reported that it was responsible for cardioprotection (Cross et al. 2002; Headrick and Peart 2005; Shneyvays et al. 1998, 2001; Tracey et al. 1997; Xu et al. 2006). Recently, functional A_3ARs have been detected on mice aorta, mediating contraction through a cyclooxygenase (COX)-dependent mechanism (Ansari et al. 2007). Importantly, the A_3AR was found at high levels in a variety of primary cells involved in inflammatory responses, including human eosinophils (Khono et al. 1996a), neutrophils (Bouma et al. 1997; Chen et al. 2006a; Gessi et al. 2002), monocytes (Broussas et al. 1999, 2002; Thiele et al. 2004), macrophages (McWhinney et al. 1996; Szabo et al. 1998), dendritic cells (Dickenson et al. 2003; Fossetta et al. 2003; Hofer et al. 2003; Panther et al. 2001) and lymphocytes (Gessi et al. 2004a). Finally, very high expression of A_3AR protein was observed in a variety of cancer cell lines (Gessi et al. 2001, 2007; Merighi et al. 2001, 2003; Suh et al. 2001) and in cancer tissues, suggesting a role for this subtype as a tumor marker (Gessi et al. 2004b; Madi et al. 2004).

The A_3AR receptor coding region was found to be divided into two exons separated by a single intron of about 2.2 kb. The upstream sequence does not contain a TATA-like motif, but it has a CCAAT sequence and consensus binding sites for SP1, NF-IL6, GATA1 and GATA3 transcription factors (Murrison et al. 1996). The involvement of the latter factors in transcriptional control of this gene would be consistent with a role of the receptor in immune function. Bioinformatics studies revealed the presence of nuclear factor kappa B (NF-κB) in the A_3AR promoter, demonstrating the role of this transcription factor in determining A_3AR expression level (Bar-Yehuda et al. 2007).

The A_3AR has been mapped on human chromosome 1 (Atkinson et al. 1997) and consists of 318 amino acid residues. The A_3AR subtype is a GPCR characterized by its C-terminal portion facing the intracellular compartment and seven transmembrane-spanning domains. Differently to other adenosine receptors, the C-terminal region presents multiple serine and threonine residues, which may serve as potential sites of phosphorylation that are important for receptor desensitization upon agonist application (Palmer et al. 1995a, b).

2 A_3 Adenosine Receptor (A_3AR) Signal Transduction

The first second-messenger systems found to be associated with A_3AR activation were adenylyl cyclase (AC) activity, which is inhibited, and phospholipase C (PLC), which is stimulated, through G_i and Gq protein coupling, respectively (Abbracchio et al. 1995; Ramkumar et al. 1993). Activation of PLC is responsible for inositol triphosphate (IP_3) and intracellular calcium (Ca^{2+}) elevation in a variety of cellular models. Initially, A_3AR agonist-induced effects on Ca^{2+} mobilization were observed in HL-60 cells, in human eosinophils, and in cardiomyocytes, where a high micromolar EC_{50} value was shown by 2-chloro-N^6-(3-iodobenzyl)-N-methyl-5′-carbamoyladenosine (Cl–IB–MECA), making it difficult to reconcile this functional effect with its high affinity in binding and cyclic adenosine monophosphate (cAMP) inhibition assays (Kohno et al. 1996a, b; Shneyvays et al. 2000). Other studies reporting similar results followed (Gessi et al. 2001, 2002; Merighi et al. 2001; Reshkin et al. 2000; Shneyvays et al. 2004), suggesting the possibility that a GPCR might have different potencies in different signaling pathways in the same cellular system (Schulte and Fredholm 2000; Fredholm et al. 2000). Recently, through a transgenic mammalian animal model that expresses apoaequorin, allowing intracellular Ca^{2+} concentrations to be measured in living organisms, functional expression of A_3AR in pancreatic cells was observed and Cl–IB–MECA was effective in increasing calcium at the micromolar level (Yamano et al. 2007).

On the other hand, there have also been studies showing that this A_3AR agonist has nanomolar affinity in calcium mobilization studies (Fossetta et al. 2003), suggesting that this pathway may be differentially activated by A_3ARs depending on the cellular system investigated. Recently, a role for A_3AR activation in the reduction of calcium increase induced by $P2X_7$ receptors in retinal ganglion cells has also been reported, shifting the balance of purinergic action from that of death to the preservation of life (Zhang et al. 2006).

In addition, other intracellular pathways have been described as being linked with A_3AR activation. Starting from the pioneering work by Schulte and Fredholm, who reported the coupling of all adenosine receptors with mitogen-activated protein kinases (MAPKs), a plethora of studies has followed showing the modulation of these kinases by A_3AR in different recombinant and native cell lines (Schulte and Fredholm 2000). A_3AR signaling in CHO cells expressing human A_3AR (CHO – hA_3) triggers the stimulation of extracellular signal-regulated kinases (ERK1/2) through βγ release from pertussis toxin (PTX)-sensitive G proteins, phosphoinositide 3-kinase (PI3K), Ras and mitogen-activated protein kinase kinase (MEK) (Schulte and Fredholm 2002). Functional A_3AR activating ERK1/2 has also been described in microglia cells, where a biphasic, partly G_i-protein-dependent influence on the phosphorylation of the ERK1/2 has been found (Hammarberg et al. 2003). In colon cancer cells, after adenosine deaminase (ADA) treatment, A_3AR activation stimulates cell proliferation through ERK1/2 activation (Gessi et al. 2007), whilst in melanoma cells it stimulates PI3K-dependent phosphorylation of protein kinase B (PKB/Akt), leading to the reduction of basal levels of ERK1/2 phosphorylation (Merighi et al. 2005a). MAPK kinase activation is also responsible for adenosine-mediated hypoxia-inducible factor-α (HIF1-α) stimulation in melanoma, colon carcinoma and glioblastoma cells (Merighi et al. 2005b, 2006, 2007a). An active MAPK signaling pathway appears to be essential for A_3AR phosphorylation, desensitization and internalization (Trincavelli et al. 2002a). ERK1/2 are also involved in cardiac hypertrophy and can play a protective role in ischemic myocardium. Interestingly, A_3AR activation in rat cardiomyocytes has been demonstrated to increase ERK1/2 phosphorylation by involving $G_{i/o}$ proteins, protein kinase C (PKC), and tyrosine kinase-dependent pathways (Germack and Dickenson 2004). An ERK-dependent signal has been also reported in the protective effects of A_3AR activation in lung injury following *in vivo* reperfusion (Matot et al. 2006). Another important pathway triggered by adenosine via A_3AR is that of PI3K/Akt. There is evidence that A_3AR activation mediates phosphorylation of PKB/Akt, protecting rat basophilic leukemia (RBL)-2H3 mast cells from apoptosis through the βγ subunits of G_i proteins and PI3K-β (Gao et al. 2001). In contrast, it has been reported that A_3AR activation is able to decrease the levels of PKA, a downstream effector of cAMP, and of the phosphorylated form of PKB/Akt in melanoma cells. It is well known that protein kinase A (PKA) and PKB/Akt phosphorylate and inactivate glycogen synthase kinase 3β (GSK-3β), a serine/threonine kinase that acts as a key element in the Wnt signaling pathway. In its active form, GSK-3β suppresses mammalian cell proliferation (Fishman et al. 2002a, b, 2004). This implies the deregulation of the Wnt signaling pathway, which is generally active during embryogenesis and tumorigenesis in order to increase cell cycle progression and cell proliferation (Fishman et al. 2002b). Support for this mechanistic pathway comes from the work of Chung et al. (2006), who demonstrated the deregulation of the Wnt pathway in LJ-529 breast cancer cells. A central role of PI3K has been demonstrated for A_3AR-induced p38 and ERK1/2 stimulation in CHO-hA_3 and in immortalized N13 microglial cells (Hammarberg et al. 2003). Furthermore, it was found that serine 727 phosphorylation of signal transducer and activator of transcription 3 (STAT3) is a possible downstream target of A_3AR-mediated ERK1/2 activation (Hammarberg et al. 2004). Modulation of these pathways is relevant, as it may represent the molecular basis for the apoptotic-modulating effects of the A_3AR. Activation of PI3K-Akt-phospho-BAD by A_3AR has been observed recently in glioblastoma cells, leading to cell survival in hypoxic conditions (Merighi et al. 2007b). Contrasting results have been obtained on PI3K modulation related to cytokine production by A_3ARs. For example, in BV2 mouse microglial cells, A_3AR stimulation inhibited LPS-induced PI3K/Akt activation, leading to the inhibition of tumor necrosis factor alpha (TNF-α) and the NF-kB pathway (Lee et al. 2006b). However, in human monocytes, N^6-(3-iodobenzyl)-adenosine-5′-N-methylcarboxamide (IB–MECA) activated the PI3K/Akt signaling pathway and induced the phosphorylation of the MAPK p38, ERK, and Jun N-terminal kinase (JNK), thus leading to a reduction in interleukin (IL)-12 production (Haskó et al. 1998; la Sala et al. 2005). The PI-3K/PKB and MEK1/ERK1/2 pathways, which are involved in cell survival, have been linked with preconditioning effects induced by A_3AR activation in cardiomyocytes from newborn rats during hypoxia/reoxygenation (Germack and Dickenson 2005). Cardioprotection at reperfusion has been observed after A_3AR activation of the PI3K/Akt pathway, leading to a reduction of GSK-3β and mitochondrial permeability transition pore opening (mPTP) (Park et al. 2006). In the heart, A_3AR mediates cardioprotective effects through ATP-sensitive potassium (K_{ATP}) channel activation. Moreover, it is coupled to the activation of RhoA and the subsequent stimulation of phospholipase D (PLD), which in turn mediates the protection of cardiac myocytes from ischemia (Lee et al. 2001; Mozzicato et al. 2004; Parsons et al. 2000).

2.1 A_3 Adenosine Receptor (A_3AR) Desensitization

The initial characterization of the A_3AR expressed in RBL-2H3 rat mast cells demonstrated that agonist-stimulated calcium mobilization is subject to a rapid, homologous desensitization that is apparent after only a few minutes of agonist exposure (Ali et al. 1990). This phenomenon in GPCR-coupled receptors is typically triggered by receptor phosphorylation induced by either second-messenger-activated kinases or G-protein-coupled receptor kinases (GRKs). In the case of A_3AR, it has been demonstrated that desensitization of the rat subtype after 10 min of agonist exposure is associated with rapid phosphorylation on serine and threonine residues by a GRK2 kinase (Palmer et al. 1995a). This was related to a reduction of 30–40% in the number of high-affinity binding sites and to a functional receptor desensitization, as manifested by an eightfold increase in the IC_{50} value of IB–MECA-mediated inhibition of cAMP levels. It has been reported that under conditions in which the A_3AR undergoes agonist-dependent phosphorylation and desensitization, the A_1 adenosine receptor (A_1AR) was not affected. Indeed, the A_3R contains multiple serine and threonine residues in the region of the C-terminal tail that are important for phosphorylation by GRK2. Therefore, a chimeric A_1–A_3AR obtained by introducing the extreme C-terminal 14-amino acid segment of the A_3AR into the A_1AR, expressed in CHO cells, undergoes rapid desensitization, suggesting that the C-terminal domain of the A_3AR is the site for phosphorylation by GRK2, 3 and 5-kinases, with the last being less important (Palmer et al. 1996). It has been also demonstrated that in response to short-term agonist exposure, A_3ARs internalize profoundly and rapidly ($t_{1/2} = 10$ min) over a time frame that follows the onset of receptor phosphorylation, in contrast to the A_1AR, which internalized quite slowly ($t_{1/2} = 90$ min). A nonphosphorylated A_3AR mutant failed to internalize over a 60 min time course, suggesting that receptor phosphorylation was essential for rapid A_3AR internalization to occur. In addition, fusion onto the A_1AR of the A_3AR C-terminal domain containing the sites for phosphorylation by GRKs conferred rapid agonist-induced internalization kinetics ($t_{1/2} = 10$ min) on the resulting chimeric AR; this suggests that GRK-stimulated phosphorylation of threonine residues within the C-terminal domain of the A_3AR is obligatory to observe rapid agonist-mediated internalization of the receptor (Ferguson et al. 2000). In particular, the amino acid residues in the C-terminus responsible for rapid desensitization were Thr(307), Thr(318), and Thr(319). Individually mutating each residue demonstrated that Thr(318) and Thr(319) are the major sites of phosphorylation. Phosphorylation at Thr(318) appeared to be necessary to observe phosphorylation at Thr(319), but not vice versa. In addition, the mutation of two predicted palmitoylation-site cysteine residues proximal to the regulatory domain resulted in the appearance of an agonist-independent basal phosphorylation. Therefore, GRK-mediated phosphorylation of the C-terminal tail of the A_3AR in situ appears to follow a sequential mechanism, perhaps involving receptor depalmitoylation, with phosphorylation at Thr(318) being particularly important (Palmer et al. 2000).

The agonist-induced internalization of the human A_3ARs in CHO-hA_3 cells and the relationship between internalization, desensitization and resensitization have been investigated. Agonist-induced internalization of A_3 adenosine receptors was directly demonstrated by immunogold electron microscopy, which revealed the localization of these receptors in plasma membranes and intracellular vesicles. Moreover, short-term exposure of these cells to the agonist caused rapid desensitization, as tested in AC assays. Subsequent removal of the agonist led to restoration of the receptor function and recycling of the receptors to the cell surface. Blockade of internalization and recycling demonstrated that internalization did not affect signal desensitization, whereas recycling of internalized receptors was implicated in the signal resensitization (Trincavelli et al. 2000). These mechanisms have been also evaluated on native A_3ARs in human astrocytoma cells. Short-term exposure to the agonist Cl–IB–MECA caused rapid receptor desensitization, within 15 min. Agonist-induced desensitization was accompanied by receptor internalization. A_3AR internalized with rapid kinetics, within 30 min, after cell exposure to Cl–IB–MECA. After desensitization, the removal of agonist led to the restoration of A_3AR functioning through receptor recycling to the cell surface within 120 min (Trincavelli et al. 2002a). The involvement of ERK 1 and 2 in A_3AR phosphorylation has been demonstrated. A_3AR mediated the activation of ERK 1/2 with typical transient monophasic kinetics within 5 min. The activation was not affected by hypertonic sucrose cell pretreatment, suggesting that this effect occurred independently of receptor internalization. The exposure of cells to the MEK inhibitor PD98059 prevented MAPK activation and inhibited homologous A_3AR desensitization and internalization, impairing agonist-mediated receptor phosphorylation. PD98059 inhibited the membrane translocation of GRK2, which is involved in A_3AR homologous phosphorylation, suggesting that the MAPK cascade is involved in A_3AR regulation by a feedback mechanism which controls GRK2 activity and probably involves direct receptor phosphorylation (Trincavelli et al. 2002b). Receptor activation, internalization and recycling events have also been described in B16F10 murine melanoma cells, where they play an important role in turning on/off receptor-mediated signal transduction pathways. It has been observed that melanoma cells highly express A_3AR on the cell surface, which is rapidly internalized to the cytosol and "sorted" to the endosomes for recycling and to the lysosomes for degradation. Receptor distribution in the lysosomes was consistent with the downregulation of receptor protein expression and was followed by mRNA and protein resynthesis. Receptor binding experiments reveal a reduction in receptor density after 15 and 60 min, and a full recovery after 24 h (Madi et al. 2003). In an *in vivo* prostate cancer model, chronic treatment of the tumor-bearing rats with IB–MECA resulted in receptor downregulation shortly after treatment. Interestingly, full recovery of the A_3AR was noted after 24 h, demonstrating the continuing presence of the receptor upon chronic agonist treatment (Fishman et al. 2003).

As for the effect of prolonged agonist exposure of CHO cells expressing a recombinant rat A_3AR, it has been shown that this induces a desensitization of receptor function that is associated with the downregulation of specific G protein subunits (Palmer et al. 1995b). Given the structural and pharmacological differences displayed by rat and human A_3ARs, it has been reported that in CHOhA_3 cells the prolonged agonist exposure results in not only a receptor density decrease and functional desensitization but that it also induces a sensitization of the stimulatory pathway of AC by increasing its activity by 1.5- to 2.5-fold. This sensitization was not a consequence of the downregulation of G_i proteins induced by agonist treatment, and was not associated with sustained or transient increases in the expression of Gs. Moreover, it was not due to the synthesis of new proteins, because cycloheximide treatment failed to inhibit sensitization. Instead, the inability of the sensitization process to alter the forskolin-stimulated AC activity in the presence of manganese chloride, which uncouples AC from G-protein regulation, suggested that prolonged A_3AR activation increased the coupling efficiency between Gs and AC catalytic units (Palmer et al. 1997). This phenomenon might provide a molecular basis for the observation that for many of the effects mediated by adenosine receptors, acute and chronic agonist treatment often produce opposite effects. A marked downregulation of A_3ARs following prolonged agonist exposure (1–24 h) has been observed also by Trincavelli and collegues (2002a). After downregulation, the recovery of receptor functioning was slow (24 h) and associated with the restoration of receptor levels close to control values.

3 A_3 Adenosine Receptor (A_3AR) and Ischemic Brain Disease

Despite low levels of A_3AR message in the central nervous system (CNS), one of the first effects observed following intraperitoneal injection of an A_3AR agonist (IB–MECA) was depression of locomoter activity in mice (Jacobson et al. 1993). However, contrasting results on how A_3AR activation might influence neuronal activity in rat brain in both normoxic and hypoxic conditions have been reported, making it difficult to understand whether an A_3AR agonist or antagonist would be better to treat cerebral ischemia.

It has been reported that in the CA1 region of the rat hippocampus, A_3AR had no direct effect on synaptically evoked excitatory responses, whilst it induced heterologous desensitization of A_1ARs, thus limiting adenosine-mediated cerebroprotection (Dunwiddie et al. 1997). Moreover, in the CA3 area of immature rat hippocampal slices, it has been observed that Cl–IB–MECA facilitates epileptiform discharges, suggesting that activation of A_3ARs following a rise in endogenous adenosine facilitates excitation, thus again limiting the known inhibitory and neuroprotective effects of adenosine in immature brain (Laudadio and Psarropoulou 2004).

Other studies suggested that A_3AR activation in cortical neurons mediated a depression of synaptic transmission by inhibiting glutamate release additionally to and independently from the A_1ARs, thus providing neuroprotection (Brand et al. 2001; Lewerenz et al. 2003; Lopes et al. 2003). These contrasting actions may lead to both protective or deleterious effects during ischemia, when adenosine concentrations rise to levels that activate the A_3AR subtype. In particular, it has been reported that chronic preischemic treatment with the agonist IB–MECA before forebrain ischemia in gerbils induces a significant protection of neurons and a reduction in the subsequent mortality, whilst acute administration of the drug results in a pronounced worsening of neuronal damage and postischemic mortality (von Lubitz et al. 1994). Accordingly, for the acute effect, cell death induction was also observed in cell cultures of rat cerebellar granule neurons where high concentrations of Cl–IB–MECA induced lactate dehydrogenase release, neuronal cell death and glutamate-mediated neurotoxicity (Sei et al. 1997).

Destructive and protective actions of A_3AR stimulation have also been demonstrated in astroglial cells, where Cl–IB–MECA at nanomolar doses was responsible for "trophic effects" related to the reorganization of the actin cytoskeleton, whilst it was a mediator of apoptosis in the micromolar range (Abbracchio et al. 1997, 2001; Appel et al. 2001; Di Iorio et al. 2002). It has been suggested that astrocyte death induced during severe metabolic stress by A_3AR activation may isolate the worst-affected tissue by physically excising cells from sites of irreversible injury (von Lubitz 1999). This may help to shift energetic resources to less severely injured tissue (the penumbra) and increase the chance of survival for the penumbra (von Lubitz 1999). It was later demonstrated that desensitization/downregulation of the A_3AR may be the basis of cytoprotection, suggesting a role for this receptor in induction of cell death (Trincavelli et al. 2002a). The effect of A_3AR activation has been investigated during "preconditioning," a phenomenon consisting of exposure to brief periods of hypoxia or ischemia that result in increased tolerance to severe ischemic insults (Dawson and Dawson 2000; Ishida et al. 1997). Pugliese et al. (2003) demonstrated that blocking A_3AR during preconditioning episodes improved the recovery of field excitatory postsynaptic potential, and this suggests that the stimulation of A_3ARs by endogenous adenosine may be one of the negative stimuli that eventually leads to the irreversible loss of neurotransmission during prolonged ischemic episodes. Later, this group reported that A_3AR antagonism prevented or delayed the appearance of anoxic depolarisation induced by prolonged (7 min) oxygen and glucose deprivation (OGD) episodes, and exerted a protective effect on neurotransmission, supporting the evidence that A_3AR stimulation is deleterious during prolonged ischemia (Pugliese et al. 2006). Interestingly, the same results in terms of protection were obtained after the extended application of an A_3AR agonist, suggesting that A_3ARs undergo rapid desensitization following exposure to exogenous agonist (Pugliese et al. 2007). In contrast, when the A_3AR agonist was applied for a short time, so that it was unable to cause receptor desensitization, A_3AR activation was responsible for a depression of synaptic activity, like that obtained after A_1AR activation. This is in agreement with previous data concerning A_3AR-mediated inhibition of excitatory neurotransmission in rat cortical neurons (Hentschel et al. 2003). Therefore, it has been suggested that prolonged ischemic conditions could be crucial in switching the effects of A_3AR stimulation from protective to dangerous by counteracting the inhibitory effects of A_1AR on excitatory neurotransmission or potentiation of excitotoxic glutamate effects. On this basis, it has been speculated that A_3AR stimulation by adenosine released during brief periods of ischemia might exert A_1AR-like protective effects on neurotransmission. Prolonged periods of ischemia are able to change the A_3AR-mediated effects from protective to dangerous (Pugliese et al. 2007).

The potential neuroprotective actions of the A_3AR have been further demonstrated using mice with functional deletions of the A_3AR ($A_3AR^{-/-}$). The $A_3AR^{-/-}$ mice reveal a number of CNS functions where the A_3ARs play a role, including nociception, locomotion, behavioral depression and neuroprotection. Pharmacologic or genetic suppression of A_3AR function enhances some aspects of motor function and suppresses pain processing at supraspinal levels. In response to repeated episodes of hypoxia, $A_3AR^{-/-}$ mice show an increase in neurodegeneration, suggesting the possible use of A_3AR agonists in the treatment of ischemic, degenerative conditions of the CNS (Fedorova et al. 2003). Other authors found that the purine inosine exerted protective effects in stroke animals, in terms of reduced bradykinesia and cerebral infarction induced by middle cerebral artery occlusion, and suggest that they were mediated by A_3AR activation (Shen et al. 2005). Accordingly, in cortical culture, Cl–IB–MECA pretreatment antagonized the hypoxia-mediated decrease in cell viability. Animals subjected to focal cerebral ischemia and treated with Cl–IB–MECA showed increased locomotor activity and decreased cerebral infarction. In these animals, Cl–IB–MECA also reduced the density of TUNEL labeling in the lesioned cortex. Furthermore, in $A_3AR^{-/-}$ mice, an increase in cerebral infarction was found compared with the A_3AR wild-type controls, suggesting that A_3ARs are tonically activated during ischemia. Additionally, intracerebroventricular pretreatment with Cl–IB–MECA decreased the size of infarction in the wild-type controls, but not in the $A_3AR^{-/-}$ animals, suggesting that Cl–IB–MECA induced protection through the A_3ARs (Chen et al. 2006b).

Different evidence suggests that some of the neuroprotection induced by A_3AR derives from its modulation of the brain immune system (Daré et al. 2007; Haskó et al. 2005). A_3AR stimulation induces the synthesis of neuroprotective chemokine ligand 2 (Wittendorp et al. 2004). Moreover, in lipopolysaccharide (LPS)-treated

BV2 microglial cells, A_3AR activation suppresses TNF-α production by inhibiting PI3K/Akt and NF-kB activation, suggesting that selective ligands of this receptor may have therapeutic potential for the modulation and possible treatment of brain inflammation (Lee et al. 2006b).

It has long been known that adenosine plays an important role in ischemia, and abundant evidence indicates that it is an endogenous neuroprotective agent. Apart from the well-established protective role exerted by adenosine through A_1AR activation, a lot of work has been carried out to shed light on the effects exerted through A_3AR stimulation. Even though the role of this AR subtype in neuroprotection has been enigmatic for a long time, new data from *in vitro* and *in vivo* $A_3AR^{-/-}$ mice models suggest a neuroprotective role. It can also be speculated that apparently contrasting results concerning protective effects induced through A_3AR block may be explained by the very fast internalization and desensitization of the A_3AR, making agonist exposure therapeutically equivalent to antagonist occupancy of the receptor.

4 A_3 Adenosine Receptor (A_3AR) and Ischemic Heart Disease

One of the most important topics in the area of A_3AR-targeted therapy is the protective role of this adenosine receptor subtype in cardiac ischemia. To date, several studies have pointed to the evidence that the A_3AR is a key player in adenosine-induced cardioprotection during and following ischemia-reperfusion (Headrick and Peart 2005). Following the discovery of ischemic preconditioning (IPC) as a mechanism to reduce infarct size (Murry et al. 1986), and the identification of adenosine as one of the mediators of this phenomenon, a lot of work has been done that attributes A_1AR with a major role in adenosine-mediated effects. Liu et al. (1994) found that the A_1AR antagonist 8-cyclopentyl-1,3-dipropylxanthine (DPCPX) was not able to abolish the anti-infarct effect induced by IPC in rabbit, thus suggesting the possible involvement of another adenosine subtype that they pharmacologically identified as the A_3AR (Liu et al. 1994). Furthermore, it was demonstrated in rabbit that IB–MECA reproduced IPC, suggesting the involvement of A_3AR subtype modulation, and there was also a lack of efficacy in reducing IPC-induced cardioprotection by A_1AR-selective antagonists in dog models (Auchampach et al. 1997a, 2004). In terms of the timing of cardioprotection, some reports have indicated that preischemic A_3AR agonism is effective and necessary, while others suggested that protection occurs postischemia, and still others have found that A_3AR agonism is able to trigger an anti-infarct response with either pre- or postischemic treatment (Auchampach et al. 2003). Pretreatment with an A_3AR agonist is responsible for cardioprotection, and may be classified into classic or early preconditioning, in which adenosine treatment occurs for 5 min, before exposure to ischemia (Armstrong and Ganote 1994; Tracey et al. 1997, 1998; Wang et al. 1997), and delayed or late preconditioning, in which adenosine treatment occurs 24 h before the induction of ischemia (Takano et al. 2001; Zhao et al. 2002). The mechanism involved in these effects (shared with the A_1AR subtype) was shown to be the activation of PKC and the regulation of mitochondrial K_{ATP} channels (Auchampach et al. 1997a; Thourani et al. 1999). However, in avian cells it has been reported that the signaling pathways activated by the A_1ARs and A_3ARs are distinct and involve selective coupling to PLC and PLD/RhoA, respectively (Lee et al. 2001; Mozzicato et al. 2004). Recently, by studying the cardioprotective profile of the A_3AR agonist N^6-(2,5-dichlorobenzyl)-3-aminoadenosine-5-N-methylcarboxamide (CP-532,903) in an isolated mouse heart model of global ischemia and reperfusion and an *in vivo* mouse model of infarction, it has been found that A_3AR activation provides ischemic protection by facilitating the opening of the sarcolemmal isoform of the K_{ATP} channel (Wan et al. 2008). In addition, roles for MAPK and Akt/PI3 kinase have been documented for early preconditioning (Germack and Dickenson 2004, 2005), whilst for late preconditioning the involvement of NF-kB, synthesis of inducible nitric oxide synthase (NOS) and mitochondrial K_{ATP} channels has been suggested (Zhao et al. 2002). This was not recognized by Takano et al., who reported that an NOS-dependent pathway was implicated in the effect mediated through A_1AR, but not in A_3AR activation (Takano et al. 2001). In any case, late preconditioning is more relevant than early preconditioning due to its sustained duration and the possibility of maintaining patients in a protracted, preconditioned, defensive state.

The cardioprotective effects of A_3ARs were also detected in A_3AR-over expressing mice, where infarct size was lower than in wild-type mice after *in vivo* regional ischemia and reperfusion (Black et al. 2002). In these animals, A_3ARs overexpression decreased basal heart rate and contractility, preserved ischemic ATP, and decreased postischemic dysfunction (Cross et al. 2002). Recent evidence obtained by using pharmacological agents and genetic methods suggest that Cl–IB–MECA protects against myocardial ischemia/reperfusion injury in mice via A_3AR activation. These conclusions were suggested by experiments with a selective A_3AR antagonist and by evaluating the A_3AR agonist effects on A_3AR knockout (KO) mice. Interestingly, in this paper, by using congenic (C57BL/6) A_3AR KO mice, the deletion of the A_3AR gene itself has no effect on ischemic tolerance, suggesting that the previous contradictory results from the same and other groups (Cerniway et al. 2001; Guo et al. 2001; Harrison et al. 2002) can probably be explained by differences in the genetic backgrounds of the mice rather than specific deletion of the A_3AR gene. Interestingly, additional studies using wild-type mice treated with compound 48/80 (a condensation product of p-methoxyphenethyl methylamine with formaldehyde) to deplete mast cell contents exclude the possibility that Cl–IB–MECA exerts a cardioprotective effect by releasing mediators from mast cells (Ge et al. 2006) and support the idea that therapeutic strategies focusing on the A_3AR subtype are a novel and useful approach to protecting the ischemic myocardium. However, an important question arises from these data. Preconditioning obtained through adenosine receptor modulation may have clinical relevance (for example in cardiac surgery), but pretreatment is rarely permitted during acute myocardial infarction. For this reason, it would be more useful to achieve a protective effect from ischemia-reperfusion injury when the drug is administered postischemia or during reperfusion. Literature data indicate that A_3AR agonism is able to protect the heart when given after the onset of ischemia or during reperfusion, suggesting its role in the treatment of acute myocardial infarction. In particular, Vinten-Johansen's group has reported that A_3AR agonist administration at reperfusion protects isolated rabbit hearts by reducing neutrophil activation (Jordan et al. 1999). After that, other studies also demonstrated a cardioprotective effect after A_3AR activation upon reperfusion in rat (Maddock et al. 2002), guinea pig (Maddock et al. 2003), and dog (Auchampach et al. 2003) hearts. As for the molecular mechanism involved in this effect, it has been reported that the opening of mPTP plays a crucial role in myocardial ischemia/reperfusion injury and that blockade of the pore opening is cardioprotective (Suleiman et al. 2001; Weiss et al. 2003). Interestingly, the inhibition of mPTP opening through the activation of PI3K/Akt and the consequent inhibition of glycogen synthase kinase after the activation of A_3AR have been reported (Park et al. 2006).

Despite the fact that the bulk of literature has reported the efficacy of adenosine in triggering cardioprotection, clinical trials to test adenosine as an adjunct to reperfusion therapy in patients with acute myocardial infarction have revealed controversial results (Mahaffey et al. 1999; Ross et al. 2005).

It has been commented that these discrepancies may be due to the age differences between animals used for experimental work and patients tested in clinical trials. In particular, experimental work has been done in healthy young adult animals, whilst heart disease is a typical pathology of the elderly population (Cohen and Downey 2008; Peart and Headrick 2007). Following this proposal, Ashton et al. (2003) reported reduced A_3AR and increased A_{2B} adenosine receptor ($A_{2B}AR$) mRNA levels with aging, similar to what happens during ischemia in young hearts (Jenner and Rose'meyer 2006). Additionally, a reduction in A_1AR has been observed during ischemia in aged hearts. Although it is just a hypothesis, decreased A_1AR and A_3AR expression might be responsible for the puzzling results mentioned above. Therefore, it is possible that differences in the modulation of adenosine receptor subtypes occur during aging and, due to the differences and simultaneous involvement of all AR subtypes in cardioprotection (Philipp et al. 2006; Solenkova et al. 2006), it is possible that a better understanding of their interplay and age dependence will provide insights into the treatment of ischemic injuries in the myocardium.

5 A_3 Adenosine Receptor (A_3AR) and Inflammatory Diseases

The role of A_3AR in inflammatory diseases is currently controversial, and both anti- and proinflammatory effects have been attributed to its activation. One of the first therapeutic applications that was hypothesized for A_3AR antagonists was the treatment of asthma. In fact, it was reported that in rodents, A_3AR activation was responsible for mast cell degranulation (Fozard et al. 1996; Ramkumar et al. 1993; Shepherd et al. 1996). This was confirmed by Salvatore et al., who showed that the potentiating effect of Cl–IB–MECA on antigen-dependent mast cell degranulation disappeared in A_3AR KO mice, and that the inhibition of LPS-induced

TNF-α production was lower in mice lacking the A_3AR subtype (Salvatore et al. 2000). The involvement of A_3ARs in mast cell degranulation was further confirmed in murine lung mast cells, where it was dependent on intracellular Ca^{2+} elevations through G_i and PI3K coupling (Zhong et al. 2003). In addition, it has been reported that A_3AR mRNA was higher in lung tissue of patients with airway inflammation, and that A_3AR activation mediates rapid inflammatory cell influx into the lungs of sensitized guinea pigs (Spruntulis and Broadley 2001; Walker et al. 1997). Furthermore A_3AR activation in RBL-2H3 mast cells was found to inhibit the apoptosis of inflammatory cells expressing A_3ARs in inflamed tissues, thus allowing inflammatory cell expansion (Gao et al. 2001). However, in contrast with these findings, it has been demonstrated that human and canine mast cell degranulation is mediated by A_{2B}ARs instead of A_3ARs (Auchampach et al. 1997b; Feoktistov and Biaggioni 1995; Ryzhov et al. 2004). This discrepancy reflects the low overall coidentity of human and rat at the aminoacid level of A_3AR, and questions the role of the A_3AR as a target for asthma therapy. Another discrepant result that questions the involvement of A_3ARs in asthma is the recent finding that in the lung parenchymal strip from brown Norway rats, where contraction in response to adenosine is mast cell-mediated, the receptor involved shows similarities to the A_3AR, but Cl–IB–MECA is a high-affinity antagonist and 5-propyl-2-ethyl-4-propyl-3-(ethylsulfanylcarbonyl)-6-phenylpyridine-5-carboxylate (MRS 1523) and 3-ethyl-5-benzyl-2-methyl-4-phenylethynyl-6-phenyl-1,4-(±)-dihydropyridine-3,5-dicarboxylate (MRS 1191) are inactive at concentrations that substantially exceed their affinities for the rat A_3AR, suggesting a puzzling A_3AR pharmacological picture in brown Norway rats (Wolber and Fozard 2005). However the idea of using an A_3AR antagonist in the treatment of asthma seems to be sustained by data obtained in other cells involved in this pathology. In fact, a high expression of A_3AR transcript levels has been found in eosinophilic infiltrates of the lungs of patients affected by asthma and chronic obstructive pulmonary disease (COPD) (Walker et al. 1997). Interestingly, similar findings were seen in the lungs of adenosine deaminase-deficient ($ADA^{-/-}$) mice that showed adenosine-mediated lung disease. Treatment of $ADA^{-/-}$ mice with MRS 1523, a selective A_3AR antagonist, prevented airway eosinophilia and mucus production. Similar results were obtained in the lungs of ADA/A_3AR double KO mice, suggesting that A_3AR signaling plays an important role in regulating chronic lung disease, and that A_3AR antagonism may be useful for reducing eosinophilia (Young et al. 2004). These results are in contrast with experiments performed in human eosinophils ex vivo, where chemotaxis, degranulation and superoxide anion production were reduced by A_3AR activation (Ezeamuzie and Philips 1999). This discrepancy was later attributed to the ex vivo nature of the chemotaxis experiments performed, suggesting that the diminished airway eosinophilia seen in the lungs of $ADA^{-/-}$ mice following the disruption of A_3AR is not a cell-autonomous effect of eosinophils, but may be due to the modulation of key regulatory molecules from other cells that express A_3ARs and that affect eosinophil migration (Young et al. 2004). However the clinical relevance of the A_3AR subtype in the pathogenesis of asthma remains a conundrum, and differences in pharmacology between the A_3AR subtypes from different species make it difficult to determine whether an A_3AR agonist or antagonist could provide a better treatment for asthma. A novel A_{2A} adenosine receptor ($A_{2A}AR$) agonist/A_3AR antagonist used in a randomized, double-blind, placebo-controlled study for the treatment of allergic rhinitis demonstrated limited clinical benefits in both the early- and late-phase responses to intranasal allergen challenge, even though it reduced the release of some mediators after allergen challenge (Rimmer et al. 2007). However, as correctly pointed out by the authors, the study had a variety of defects. As an example, the dose of the drug was limited by the narrow therapeutic index, due to side effects like tachycardia, raising the possibility that higher doses of new compounds with fewer side effects might be more efficacious. Therefore, it is possible that future studies targeting a different receptor (perhaps the $A_{2B}AR$) or using dual $A_3AR/A_{2B}AR$ antagonists will be more successful (Press et al. 2005).

Recently, it has been demonstrated that A_3AR activation decreases mortality and renal and hepatic injury in murine septic peritonitis (Lee et al. 2006a). Higher levels of endogenous TNF-α were observed in A_3AR KO mice after sepsis induction in comparison to wild-type animals, and IB–MECA significantly reduced mortality in mice lacking the A_1AR or $A_{2A}AR$ but not the A_3AR, demonstrating the specificity of the A_3AR agonist in activating A_3AR subtype and mediating protection against sepsis-induced mortality (Lee et al. 2006a). A similar mortality reduction associated with a decrease in interleukin (IL)-12 and interferon-gamma production induced by A_3AR activation was previously observed in endotoxemic mice (Haskó et al. 1998). In addition, other investigators reported reduced inflammation and increased survival following A_3AR activation in two murine models of colitis (Mabley et al. 2003). Furthermore, a protective role for A_3AR in lung injury following *in vivo* reperfusion has been reported. Rivo et al. demonstrated that in a spontaneously breathing cat model, IB–MECA given both before ischemia-reperfusion or during reperfusion conferred powerful protection against reperfusion lung injury, which was associated with decreased apoptosis (Rivo et al. 2004a). This effect was found to be mediated by a NOS-independent pathway and involved the opening of K_{ATP} channels (Rivo et al. 2004b). The signaling pathway linked to this effect was further identified in the A_3AR-induced upregulation of phosphorylated ERK (Matot et al. 2006). Furthermore, a reduction in the recruitment of neutrophils to the lungs after sepsis was found to be mediated by A_3AR and $P2Y_2$ receptors, suggesting that targeting these receptors might be useful to control acute lung tissue injury in sepsis (Inoue et al. 2008). Recently, a role for A_3AR activation has also been reported in the protection of skeletal muscle from ischemia and reperfusion injury. Because the use of an A_3AR agonist is not associated with cardiac or hemodynamic depression, the A_3AR represents a potential therapeutic target because of its ability to ameliorate skeletal muscle injury (Zheng et al. 2007). In contrast, it has been demonstrated that A_3AR activation exacerbates renal dysfunction, and mice lacking A_3ARs show better renal function following renal ischemia reperfusion injury (Lee and Emala 2000; Lee et al. 2003). Expression of A_3ARs is upregulated in ocular ischemic diseases and in conditions associated with oxidative stress. The A_3AR-selective agonist IB–MECA did not affect intraocular pressure in $A_3AR^{-/-}$ mice, but raised it in $A_3AR^{+/+}$ mice (Avila et al. 2002). The use of a cross-species A_3AR antagonist in the mouse reduced intraocular pressure (Yang et al. 2005). Activation of A_3AR leads to the regulation of chloride channels in nonpigmented ciliary epithelial cells, suggesting that A_3AR agonists would increase aqueous humor secretion and thereby intraocular pressure *in vivo*, whilst A_3AR antagonists may represent a specific approach for treating ocular hypertension (Mitchell et al. 1999; Okamura et al. 2004; Schlotzer-Schrehardt et al. 2005). Unfortunately, there are currently no A_3AR antagonists in clinical phases. However, in light of the plethora of biological effects attributed to A_3ARs, substantial efforts in medicinal chemistry have been directed towards developing antagonists for the A_3AR subtype. As a result, a number of molecules are in biological testing as therapeutic agents for asthma and COPD, glaucoma, stroke, cardiac hypoxia and cerebral ischemia (Baraldi et al. 2008; Press et al. 2007).

5.1 A_3 Adenosine Receptor (A_3AR) and Autoimmune Inflammatory Diseases

During the last decade, new immunotherapy approaches have been introduced for the treatment of autoimmune diseases. Anti TNF-α monoclonal antibody drugs are now widely used since they are one of the most effective classes of biological drugs. This treatment has remarkable effects on several autoimmune diseases, including rheumatoid arthritis (RA), Crohn's disease, psoriasis, ankylosing spondylitis, and others (Cordoro and Feldman 2007; Danese et al. 2007; McLeod et al. 2007; Rigopoulos et al. 2008; Tilg et al. 2007; Toussirot et al. 2007; Valesini et al. 2007). Anti TNF-α drugs are considered disease-modifying antirheumatic drugs (DMARDs) which modulate the pathophysiology of autoimmune diseases, but at the same time these drugs may interfere with host defense and disease pathology, resulting in severe adverse events (Desai et al. 2006; Hansen et al. 2007; Mader and Keystone 2007).

Recent findings indicate that the inhibition of TNF-α by adenosine is mediated via the A_3AR (Lee et al. 2006b; Levy et al. 2006). Selective agonists to A_3AR such as IB–MECA, Cl–IB–MECA and MRS3558 inhibit TNF-α production both *in vitro* and *in vivo* (Baharav et al. 2005; Fishman et al. 2006; Hasko et al. 1996; Lee et al. 2006a, b; Martin et al. 2006; Ochaion et al. 2008; Rath-Wolfson et al. 2006). It was further shown that A_3AR is overexpressed in inflammatory tissues derived from adjuvant-induced arthritis (AIA) experimental models. Interestingly, A_3AR overexpression was also found in the peripheral blood mononuclear cells (PBMCs) of the arthritic animals, reflecting receptor status in the remote inflammatory organs (Bar Yehuda et al. 2007; Fishman et al. 2006; Ochaion et al. 2006; Rath-Wolfson et al. 2006). These findings are in agreement with data obtained from patients with colorectal cancer, who demonstrated that elevated expression of A_3ARs in this cancer is reflected in PBMCs (Gessi et al. 2004a).

The ability of A_3AR agonists to inhibit TNF-α and the upregulation of the receptor in inflammatory cells led to the development of the concept that A_3AR may be a specific target to combat inflammation.

In this section, data from *in vivo* experiments demonstrating the anti-inflammatory effect of A_3AR agonists and the molecular mechanisms involved will be presented. In addition, results from a human clinical study in RA patients showing the ability of IB–MECA to improve signs and symptoms of arthritis, as well as the safety of the drug will be presented. The utilization of A_3AR as a biological predictive marker to be analyzed prior to treatment with the agonist will be discussed.

IB–MECA, Cl–IB–MECA and MRS3558 act as potent anti-inflammatory agents in experimental animal models of various inflammatory diseases. IB–MECA was tested in three experimental models that imitate Crohn's disease, including a rat chronic model of 2,4,6-trinitrobenzene sulfonic acid-induced colitis, dextran sodium sulfate-induced colitis and spontaneous colitis found in IL-10 gene-deficient mice. Treatment with IB–MECA (1.5 mg kg^{-1} b.i.d., 1 or 3 mg kg^{-1} per day, accordingly) protected against colitis (Guzman et al. 2006; Mabley et al. 2003).

Studies were performed to explore the mechanisms by which A_3AR agonists produce their anti-inflammatory effect. The effects of IB–MECA, MRS3558 and in some experiments Cl–IB–MECA on the development of arthritis in experimental animal models were extensively studied. The agonists suppressed the clinical and pathological manifestations of arthritis in the mouse collagen-induced arthritis model, in the rat AIA model, and in the rat tropomyosin-induced arthritis model (Baharav et al. 2005; Fishman et al. 2006; Ochaion et al. 2008; Rath-Wolfson et al. 2006). The mechanism of action entailed direct effects of the A_3AR agonists on cells from synovial tissue and paw, which included deregulation of the NF-κB signaling pathway manifested by downregulation of PI3K, PKB/Akt, IκB kinase (IKK) and inhibitor of κB (IκB), resulting in decreased expression levels of TNF-α and apoptosis of inflammatory cells (Baharav et al. 2005; Fishman et al. 2006; Ochaion et al. 2008; Rath-Wolfson et al. 2006). IB–MECA affected T-cell-mediated responses by inhibiting T regulatory cell proliferation and adoptive transfer in AIA rats upon treatment of the donors with the agonist (Bar Yehuda et al. 2007). MRS3558 was also able to induce a dose-dependent inhibitive effect on the proliferation of fibroblast-like synoviocytes cultured from synovial fluids from RA patients via the same mechanism (Ochaion et al. 2008).

An important finding that supported the selection of IB–MECA as a drug candidate for the treatment of RA was its efficacy in enhancing the anti-inflammatory effect of methotrexate (MTX)(Ochaion et al. 2006). The latter is the most widely used DMARD, and it is the "gold standard" therapy, which other systemic medications are compared (Weinblatt et al. 1985). It was suggested by Montesinos et al. (2003) that the anti-inflammatory effect of MTX is mediated by adenosine, produced in the cells upon metabolism of MTX. When MTX is taken up by cells, it is converted to long-lived polyglutamates known to inhibit the activity of aminoimidazole carboxamide ribonucleotide (AICAR) transformylase, thereby leading to an increase in the cellular level of AICAR. AICAR inhibits adenosine degradation, resulting in its accumulation in the extracellular fluid (Baggott et al. 1999; Chan et al. 2002; Laghi Pasini et al. 1997). Adenosine has been reported to exhibit a number of anti-inflammatory effects. It was further shown that the anti-inflammatory effect of MTX is mediated via A_{2A}ARs and the A_3ARs (Cronstein et al. 1994; Montesinos et al. 2003). Thus, it seems that the enhanced anti-inflammatory effects of IB–MECA and MTX are mediated via the A_3AR. This hypothesis was confirmed in a study where the combined treatment of AIA rats with IB–MECA and MTX resulted in an additive anti-inflammatory effect. Mechanistic studies revealed that MTX induced upregulation of the A_3AR in inflammatory cells from AIA rats, making the cells more susceptible to treatment with IB–MECA. It was further found that A_3AR is overexpressed in PBMCs of RA patients treated with MTX, suggesting that combined treatment with A_3AR and MTX in RA patients may be beneficial (Ochaion et al. 2006).

The above preclinical data, demonstrating the marked anti-inflammatory effect of IB–MECA (designated as CF101), prompted the initiation of a clinical development program to look at the safety and efficacy of CF101 for the treatment of RA. In a Phase I study conducted in healthy subjects, CF101 was given orally, and its plasma half-life ($t_{1/2}$) was 8 h. Single oral doses of CF101 of 1 and 5 mg were well tolerated, whereas at 10 mg, CF101 was associated with adverse events including asymptomatic sinus tachycardia and mild elevations of systolic blood pressure. These events are presumed to represent effects on cardiovascular ARs, which are most likely to be A_{2A}AR-mediated at the plasma concentrations for the 10 mg dose. In a subsequent trial of twice-daily repeat-dose testing of CF101, 4 mg every 12 h, the schedule was found to be well tolerated in male volunteers, with an adverse event profile comparable to placebo (van Troostenburg et al. 2004).

The safety and efficacy of CF101 was studied in a Phase IIa study in RA patients. The trial was a multicenter (ten sites), randomized, double-blind, parallel-group study and included 74 patients with active RA who failed between one and four DMARDs, excluding the biologic drugs. CF101 was administered at doses of 0.1, 1.0 and 4.0 mg twice-daily, orally for 12 weeks. The primary efficacy end-point was American College of Rheumatology (ACR) 20 (ACR response outcome score for 20% improvement in a number of different measurements) at week 12. CF101 reduced disease activity, showing maximal response at 1 mg, with somewhat lower responses at 0.1 and 4 mg. At week 12, there were 60, 36, and 12% of the patients receiving CF101 1 mg who achieved ACR 20, 50, and 70 responses, respectively. The respective mean percentage reduction in the number of tender and swollen joints was ~80% in all dose groups. CF101 was well tolerated with no dose-limiting side effects. During this study, blood was withdrawn from patients at baseline (prior to drug administration) and at week 12. A statistically significant direct correlation was found between A_3AR overexpression at baseline and ACR 50 at week 12, demonstrating that A_3AR may be a predictive biomarker. Overall it was concluded that CF101 showed a clinical response in this Phase IIa study without dose-limiting side effects in patients with active RA. A_3AR levels may be a predictive surrogate marker of response to this therapy (Silverman et al. 2008). More clinical studies are underway to explore the effect of A_3AR agonists in RA and additional autoimmune inflammatory diseases.

6 Conclusion

Knowledge of the structure and function of the A_3AR has evolved dramatically in the last decade, and now this subtype, which originally appeared to be quite enigmatic in terms of its effects, has started to reveal its secrets. A synopsis of A_3AR-regulated pathways and functions is provided in Fig. 1. It appears evident that a plethora of biological functions have been attributed to the A_3AR in ischemic and inflammatory pathologies, and substantial efforts in medicinal chemistry have been directed at developing agonists and antagonists that target this AR subtype.

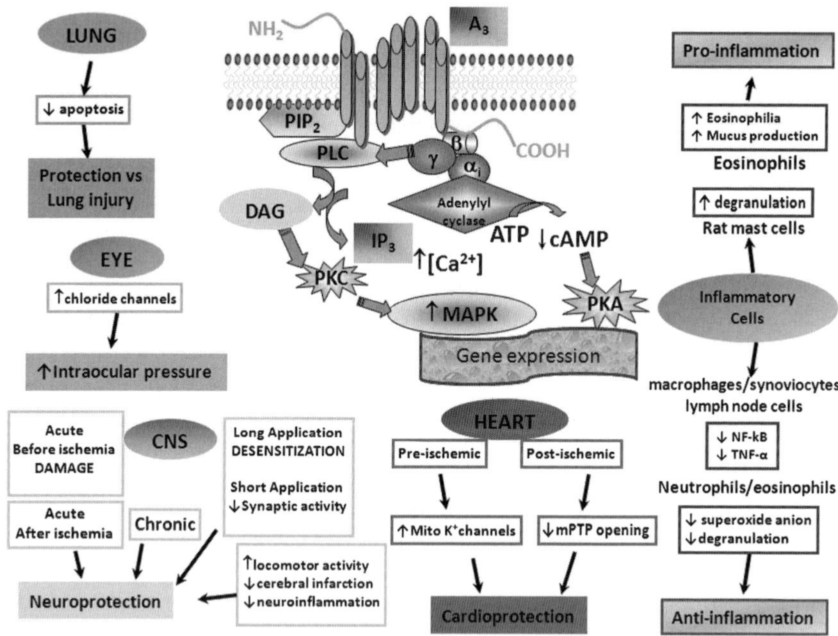

Fig. 1 Synopsis of the A_3 adenosine receptor (A_3AR)-regulated pathways and functions. Activation of the A_3AR results in the modulation of different intracellular pathways. Classically, this adenosine receptor subtype is coupled to the inhibition of adenylyl cyclase, leading to a reduction in cAMP levels. In addition, it may activate phospholipase C (PLC), through $G\beta\gamma$ subunits, inducing an increase in intracellular calcium and activation of protein kinase C (PKC). Recently, it has also been demonstrated that it is coupled to mitogen-activated protein kinases (MAPKs), suggesting its involvement in cell growth, survival, death and differentiation. Activation of the A_3AR subtype induces protective effects in the CNS, heart and lung, and both pro- and antiinflammatory effects in peripheral blood cells

As a result, there are currently A_3AR agonists in clinical phases for several autoimmune diseases, such as RA. Unfortunately, there are no A_3AR antagonists currently in clinical development, but a number of molecules are in biological testing as therapeutic agents for asthma and COPD, glaucoma, and stroke, which are waiting to enter the clinical arena. This is only the starting point of more expensive and challenging work, and it is likely that, with the availability of both selective ligands and animal models, several roles of the A_3AR that are currently ambiguous will be clearer in the near future. This will allow the chemistry and pharmacology of the A_3AR to be utilized clinically with the development of selective molecules for this important target that may improve the outcomes of patients with a number of diseases.

References

Abbracchio MP, Brambilla R, Ceruti S, Kim HO, von Lubitz DK, Jacobson KA, Cattabeni F (1995) G protein-dependent activation of phospholipase C by adenosine A_3 receptors in rat brain. Mol Pharmacol 48:1038–1045

Abbracchio MP, Rainaldi G, Giammarioli AM, Ceruti S, Brambilla R, Cattabeni F, Barbieri D, Franceschi C, Jacobson KA, Malorni W (1997) The A_3 adenosine receptor mediates cell spreading, reorganization of actin cytoskeleton, and distribution of Bcl-XL: studies in human astroglioma cells. Biochem Biophys Res Commun 241:297–304

Abbracchio MP, Cimurri A, Ceruti S, Cattabeni F, Falzano L, Giammarioli AM, Jacobson KA, Trincavelli L, Martini C, Malorni W, Fiorentini C (2001) The A_3 adenosine receptor induces cytoskeleton rearrangement in human astrocytoma cells via a specific action on Rho proteins. Ann New York Acad Sci 939:63–73

Ali H, Cunha-Melo JR, Saul WF, Beaven MA (1990) Activation of phospholipase C via adenosine receptors provides synergistic signals for secretion in antigen-stimulated RBL-2H3 cells. Evidence for a novel adenosine receptor. J Biol Chem 265:745–753

Ansari HR, Nadeem A, Tilley SL, Mustafa SJ (2007) Involvement of COX-1 in A_3 adenosine receptor-mediated contraction through endothelium in mice aorta. Am J Physiol Heart Circ Physiol 293:H3448–H3455

Appel E, Kazimirsky G, Ashkenazi E, Kim SG, Jacobson KA, Brodie C (2001) Roles of BCL-2 and caspase 3 in the adenosine A_3 receptor-induced apoptosis. J Mol Neurosci 17:285–292

Armstrong S, Ganote CE (1994) Adenosine receptor specificity in preconditioning of isolated rabbit cardiomyocytes: evidence of A_3 receptor involvement. Cardiovasc Res 28:1049–1056

Ashton KJ, Nilsson U, Willems L, Holmgren K, Headrick JP (2003) Effects of aging and ischemia on adenosine receptor transcription in mouse myocardium. Biochem Biophys Res Commun 312:367–372

Atkinson MR, Townsend-Nicholson A, Nicholl JK, Sutherland GR, Schofield PR (1997) Cloning, characterisation and chromosomal assignment of the human adenosine A_3 receptor (ADORA3) gene. Neurosci Res 29:73–79

Auchampach JA, Rizvi A, Qiu Y, Tang XL, Maldonado C, Teschner S, Bolli R (1997a) Selective activation of A_3 adenosine receptors with N^6-(3-iodobenzyl)adenosine-5′-N-methyluronamide protects against myocardial stunning and infarction without hemodynamic changes in conscious rabbits. Circ Res 80:800–809

Auchampach JA, Jin X, Wan TC, Caughey GH, Linden J (1997b) Canine mast cell adenosine receptors: cloning and expression of the A_3 receptor and evidence that degranulation is mediated by the A_{2B} receptor. Mol Pharmacol 52:846–860

Auchampach JA, Ge Z-D, Wan TC, Moore J, Gross GJ (2003) A_3 adenosine receptor agonist IB-MECA reduces myocardial ischemia-reperfusion injury in dogs. Am J Physiol Heart Circ Physiol 285:H607–H613

Auchampach JA, Jin X, Moore J, Wan TC, Kreckler LM, Ge ZD, Narayanan J, Whalley E, Kiesman W, Ticho B, Smits G, Gross GJ (2004) Comparison of three different A_1 adenosine receptor antagonists on infarct size and multiple cycle ischemic preconditioning in anesthetized dogs. J Pharmacol Exp Ther 308:846–856

Avila MY, Stone RA, Civan MM (2002) Knockout of A_3 adenosine receptors reduces mouse intraocular pressure. Invest Ophthalmol Vis Sci 43:3021–3026

Baggott JE, Morgan SL, Sams WM, Linden J (1999) Urinary adenosine and aminoimidazolecarboxamide excretion in methotrexate-treated patients with psoriasis. Arch Dermatol 135:813–817

Baharav E, Bar-Yehuda S, Madi L, Silberman D, Rath-Wolfson L, Halpren M, Ochaion A, Weinberger A, Fishman P (2005) Antiinflammatory effect of A_3 adenosine receptor agonists in murine autoimmune arthritis models. J Rheumatol 32:469–476

Baraldi PG, Tabrizi MA, Gessi S, Borea PA (2008) Adenosine receptor antagonists: translating medicinal chemistry and pharmacology into clinical utility. Chem Rev 108:238–263

Bar-Yehuda S, Silverman MH, Kerns WD, Ochaion A, Cohen S, Fishman P (2007) The anti-inflammatory effect of A3 adenosine receptor agonists: a novel targeted therapy for rheumatoid arthritis. Expert Opin Investig Drugs 16:1601–1613

Black RG, Guo Y, Ge Z-D, Murphree SS, Prabhu SD, Jones WK, Bolli R, Auchampach JA (2002) Gene dosage-dependent effects of cardiac-specific overexpression of the A_3 adenosine receptor. Circ Res 91:165–172

Bouma MG, Jeuhomme TMMA, Boyle DL, Dentener MA,Voitenok NN, van den Wildenberg FAJM, Buurman WA (1997) Adenosine inhibits neutrophil degranulation in activated human whole blood. J Immunol 158:5400–5408

Brand A, Vissiennon Z, Eschke D, Nieber K (2001) Adenosine A_1 and A_3 receptors mediate inhibition of synaptic transmission in rat cortical neurons. Neuropharmacology 40:85–95

Brandon CI, Vandenplas M, Dookwah H, Murray TF (2006) Cloning and pharmacological characterization of the equine adenosine A_3 receptor. J Vet Pharmacol Ther 29:255–263

Broussas M, Cornillet-Lefebvre P, Potron G, Nguyen P (1999) Inhibition of fMLP-triggered respiratory burst of human monocytes by adenosine: involvement of A_3 adenosine receptor. J Leukoc Biol 66:495–501

Broussas M, Cornillet-Lefebvre P, Potron G, Nguyen P (2002) Adenosine inhibits tissue factor expression by LPS-stimulated human monocytes: involvement of the A_3 adenosine receptor. Thromb Haemost 88:123–130

Cerniway RJ, Yang Z, Jacobson MA, Linden J, Matherne GP (2001) Targeted deletion of A_3 adenosine receptors improves tolerance to ischemia-reperfusion injury in mouse myocardium. Am J Physiol Heart Circ Physiol 281:H1751–H1758

Chan ES, Cronstein BN (2002) Molecular action of methotrexate in inflammatory diseases. Arthritis Res 4:266–273

Chen Y, Corriden R, Inoue Y, Yip L, Hashiguchi N, Zinkernagel A, Nizet V, Insel PA, Junger WG (2006a) ATP release guides neutrophil chemotaxis via P2Y2 and A_3 receptors. Science 314:1792–1795

Chen GJ, Harvey BK, Shen H, Chou J, Victor A, Wang Y (2006b) Activation of adenosine A_3 receptors reduces ischemic brain injury in rodents. J Neurosci Res 84:1848–1855

Chung H, Jung JY, Cho SD, Hong KA, Kim HJ, Shin DH, Kim H, Kim HO, Shin DH, Lee HW, Jeong LS, Kong G (2006) The antitumor effect of LJ-529, a novel agonist to A_3 adenosine receptor, in both estrogen receptor-positive and estrogen receptor-negative human breast cancers. Mol Cancer Ther 5:685–692

Cohen MV, Downey JM (2008) Adenosine: trigger and mediator of cardioprotection. Basic Res Cardiol 103(3):203–215

Cordoro KM, Feldman SR (2007) TNF-alpha inhibitors in dermatology. Skin Ther Lett 12:4–6

Cronstein BN (1994) Adenosine, an endogenous anti-inflammatory agent. J Appl Physiol 76:5–13

Cross HR, Murphy E, Black RG, Auchampach J, Steenbergen C (2002) Overexpression of A_3 adenosine receptors decreases heart rate, preserves energetics, and protects ischemic hearts. Am J Physiol Heart Circ Physiol 283:H1562–H1568

Danese S, Pagano N, Angelucci E, Stefanelli T, Repici A, Omodei P, Daperno M, Malesci A (2007) Tumor necrosis factor-alpha monoclonal antibodies for Crohn's disease: tipping the balance. Curr Med Chem 14:1489–1497

Daré E, Schulte G, Karovic O, Hammarberg C, Fredholm BB (2007) Modulation of glial cell functions by adenosine receptors. Physiol Behav 92(1–2):15–20

Dawson VL, Dawson TM (2000) Neuronal ischaemic preconditioning. Trends Pharmacol Sci 21:423–424

Desai SB, Furst DE (2006) Problems encountered during anti-tumour necrosis factor therapy. Best Pract Res Clin Rheumatol 20:757–590

Dickenson JM, Reeder S, Rees B, Alexander S, Kendall D (2003) Functional expression of adenosine A_{2A} and A_3 receptors in the mouse dendritic cell line XS-106. Eur J Pharmacol 474:43–51

Di Iorio P, Kleywegt S, Ciccarelli R, Traversa U, Andrew CM, Crocker CE, Werstiuk ES, Rathbone MP (2002) Mechanisms of apoptosis induced by purine nucleosides in astrocytes. Glia 38:179–190

Dixon AK, Gubitz AK, Sirinathsinghji DJS, Richardson PJ, Freeman TC (1996) Tissue distribution of adenosine receptor mRNAs in the rat. Br J Pharmacol 118:1461–1468

Dunwiddie TV, Diao L, Kim HO, Jiang J-L, Jacobson KA (1997) Activation of hippocampal adenosine A_3 receptors produces a desensitization of A_1 receptor-mediated responses in rat hippocampus. J Neurosci 17:607–614

Ezeamuzie CI, Philips E (1999) Adenosine A_3 receptors on human eosinophils mediate inhibition of degranulation and superoxide anion release. Br J Pharmacol 127:188–194

Fedorova IM, Jacobson MA, Basile A, Jacobson KA (2003) Behavioral characterization of mice lacking the A_3 adenosine receptor: sensitivity to hypoxic neurodegeneration. Cell Mol Neurobiol 23:431–447

Feoktistov I, Biaggioni I (1995) Adenosine A_{2B} receptors evoke interleukin-8 secretion in human mast cells. An enprofylline-sensitive mechanism with implications for asthma. J Clin Invest 96:1979–1986

Ferguson G, Watterson KR, Palmer TM (2000) Subtype-specific kinetics of inhibitory adenosine receptor internalization are determined by sensitivity to phosphorylation by G protein-coupled receptor kinases. Mol Pharmacol 57:546–552

Fishman P, Bar-Yehuda S, Madi L, Cohn I (2002a) A_3 adenosine receptor as a target for cancer therapy. Anticancer Drug 13:1–8

Fishman P, Madi L, Bar-Yehuda S, Barer F, Del Valle, Khalili K (2002b) Evidence for involvement of Wnt signalling pathway in IB-MECA mediated suppression of melanoma cells. Oncogene 21:4060–4064

Fishman P, Bar-Yehuda S, Ardon E, Rath-Wolfson L, Barrer F, Ochaion A, Madi L (2003) Targeting the A3 adenosine receptor for cancer therapy: inhibition of prostate carcinoma cell growth by A3AR agonist. Anticancer Res 23:2077–2083

Fishman P, Bar-Yehuda S, Ohana G, Barer F, Ochaion A, Erlanger A, Madi L (2004) An agonist to the A2 adenosine receptor inhibits colon carcinoma growth in mice via modulation of GSK-3β and NF-kB. Oncogene 23:2465–2471

Fishman P, Bar-Yehuda S, Madi L, Rath-Wolfson L, Ochaion A, Cohen S, Baharav E (2006) The PI3K-NF-kappaB signal transduction pathway is involved in mediating the anti-inflammatory effect of IB-MECA in adjuvant-induced arthritis. Arthritis Res Ther 8:R33

Fossetta J, Jackson J, Deno G, Fan X, Du XK, Bober L, Soudé-Bermejo A, de Bouteiller O, Caux C, Lunn C, Lundell D, Palmer RK (2003) Pharmacological analysis of calcium responses mediated by the human A_3 adenosine receptor in monocyte-derived dendritic cells and recombinant cells. Mol Pharmacol 63:342–350

Fozard JR, Pfannkuche H-J, Schuurman H-J (1996) Mast cell degranulation following adenosine A_3 receptor activation in rats. Eur J Pharmacol 298:293–297

Fredholm BB, Arslan G, Halldner L, Kull B, Schulte G, Wasserman W (2000) Structure and function of adenosine receptors and their genes. Naunyn–Schmiedeberg's Arch Pharmacol 362:364–374

Gao Z, Li BS, Day YJ, Linden J (2001) A_3 adenosine receptor activation triggers phosphorylation of protein kinase B and protects rat basophilic leukemia 2H3 mast cells from apoptosis. Mol Pharmacol 59:76–82

Ge Z-D, Peart JN, Kreckler LM, Wan TC, Jacobson MA, Gross GJ, Auchampach JA (2006) Cl-IB-MECA [2-Chloro-N^6-(3-iodobenzyl)adenosine-5′-N-methylcarboxamide] reduces ischemia/reperfusion injury in mice by activating the A_3 adenosine receptor. J Pharm Exp Ther 319:1200–1210

Germack R, Dickenson JM (2004) Characterization of ERK 1/2 signalling pathways induced by adenosine receptor subtype in newborn rat cardiomyocytes. Br J Pharmacol 141:329–339

Germack R, Dickenson JM (2005) Adenosine triggers preconditioning through MEK/ERK1/2 signalling pathway during hypoxia/reoxygenation in neonatal rat cardiomyocytes. J Mol Cell Cardiol 39:429–442

Gessi S, Varani K, Merighi S, Morelli A, Ferrari D, Leung E, Baraldi PG, Spalluto G, Borea PA (2001) Pharmacological and biochemical characterization of A_3 adenosine receptors in Jurkat T cells. Br J Pharmacol 134:116–126

Gessi S, Varani K, Merighi S, Cattabriga E, Iannotta V, Leung E, Baraldi PG, Borea PA (2002) A_3 adenosine receptors in human neutrophils and promyelocytic HL60 cells: a pharmacological and biochemical study. Mol Pharmacol 61:415–424

Gessi S, Cattabriga E, Avitabile A, Gafa' R, Lanza G, Cavazzini L, Bianchi N, Gambari R, Feo C, Liboni A, Gullini S, Leung E, Mac-Lennan S, Borea PA (2004a) Elevated expression of A_3 adenosine receptors in human colorectal cancer is reflected in peripheral blood cells. Clin Cancer Res 10:5895–5901

Gessi S, Varani K, Merighi S, Cattabriga E, Avitabile A, Gavioli R, Fortini C, Klotz KN, Leung E, MacLennan S, Borea PA (2004b) Expression of A_3 adenosine receptors in human lymphocytes: up-regulation in T cell activation. Mol Pharmacol 65:711–719

Gessi S, Merighi S, Varani K, Cattabriga E, Benini A, Mirandola P, Leung E, Mac Lennan S, Feo C, Baraldi S, Borea PA (2007) Adenosine receptors in colon carcinoma tissues and colon tumoral cell lines: focus on the A_3 adenosine subtype. J Cell Physiol 211:826–836

Guo Y, Bolli R, Bao W, Wu WJ, Black RG, Murphree SS, Salvatore CA, Jacobson MA, Auchampach JA (2001) Targeted deletion of the A_3 adenosine receptor confers resistance to myocardial ischemic injury and does not prevent early preconditioning. J Mol Cell Cardiol 33:825–830

Guzman J, Yu JG, Suntres Z, Bozarov A, Cooke H, Javed N, Auer H, Palatini J, Hassanain HH, Cardounel AJ, Javed A, Grants I, Wunderlich JE, Christofi FL (2006) ADOA3R as a therapeutic target in experimental colitis: proof by validated high-density oligonucleotide microarray analysis. Inflamm Bowel Dis 12:766–789

Hammarberg C, Schulte G, Fredholm BB (2003) Evidence for functional adenosine A_3 receptors in microglia cells. J Neurochem 86:1051–1054

Hammarberg C, Fredholm BB, Schulte G (2004) Adenosine A_3 receptor-mediated regulation of p38 and extracellular-regulated kinase ERK1/2 via phosphatidylinositol-3′-kinase. Biochem Pharmacol 67:129–134

Hansen RA, Gartlehner G, Powell GE, Sandler RS (2007) Serious adverse events with infliximab: analysis of spontaneously reported adverse events. Clin Gastroenterol Hepatol 5:729–735

Harrison GJ, Cerniway RJ, Peart J, Berr SS, Ashton K, Regan S, Matherne GP, Headrick JP (2002) Effects of A_3 adenosine receptor activation and gene knock-out in ischemic-reperfused mouse heart. Cardiovasc Res 53:147–155

Haskó G, Szabó C, Németh ZH, Kvetan V, Pastores SM, Vizi ES (1996) Adenosine receptor agonists differentially regulate IL-10, TNF-alpha, and nitric oxide production in RAW 264.7 macrophages and in endotoxemic mice. J Immunol 157:4634–4640

Haskó G, Nemeth ZH, Vizi ES, Salzman AL, Szabo C (1998) An agonist of adenosine A_3 receptors decreases interleukin-12 and interferon-gamma production and prevents lethality in endotoxemic mice. Eur J Pharmacol 358:261–268

Haskó G, Pacher P, Vizi ES, Illes P (2005) Adenosine receptor signalling in the brain immune system. Trends Pharmacol Sci 26:511–516

Headrick JP, Peart J (2005) A_3 adenosine receptor-mediated protection of the ischemic heart. Vasc Pharmacol 42:271–279

Hentschel S, Lewerenz A, Nieber K (2003) Activation of A_3 receptors by endogenous adenosine inhibits synaptic transmission during hypoxia in rat cortical neurons. Restor Neurol Neurosci 21:55–63

Hofer S, Ivarsson L, Stoitzner P, Auffinger M, Rainer C, Romani N, Heufler C (2003) Adenosine slows migration of dendritic cells but does not affect other aspects of dendritic cell maturation. J Invest Dermatol 121:300–307

Inoue Y, Chen Y, Hirsh MI, Yip L, Junger WG (2008) A_3 and P2Y2 receptors control the recruitment of neutrophils to the lungs in a mouse model of sepsis. Shock 30:173–177

Ishida T, Yarimizu K, Gute DC, Korthuis RJ (1997) Mechanisms of ischemic preconditioning. Shock 8:86–94

Jacobson KA, Gao Z-G (2006) Adenosine receptors as therapeutic targets. Nat Rev Drug Discov 5:247–264

Jacobson KA, Nikodijevic O, Shi D, Gallo-Rodriguez C, Olah ME, Stiles GL, Daly JW (1993) A role for central A_3-adenosine receptors: mediation of behavioural depressant effects. FEBS Lett 336:57–60

Jenner TL, Rose'meyer RB (2006) Adenosine A3 receptor mediated coronary vasodilation in the rat heart: changes that occur with maturation. Mech Ageing Dev 127:264–273

Ji X-D, von Lubitz D, Olah ME, Stiles GL, Jacobson KA (1994) Species differences in ligand affinity at central A_3-adenosine receptors. Drug Dev Res 33:51–59

Jordan JE, Thourani VH, Auchampach JA, Robinson JA, Wang N-P, Vinten-Johansen J (1999) A_3 adenosine receptor activation attenuates neutrophil function and neutrophil-mediated reperfusion injury. Am J Physiol 277:H1895–H1905

Kohno Y, Ji X, Mawhorter SD, Koshiba M, Jacobson KA (1996a) Activation of A_3 adenosine receptors on human eosinophils elevates intracellular calcium. Blood 88:3569–3574

Kohno Y, Sei Y, Koshiba M, Kim HO, Jacobson KA (1996b) Induction of apoptosis in HL-60 human promyelocytic leukemia cells by adenosine A_3 receptor agonists. Biochem Biophys Res Comm 219:904–910

Laghi Pasini F, Capecchi PL, Di Perri T (1997) Adenosine plasma levels after low dose methotrexate administration. J Rheumatol 24:2492–2493

la Sala A, Gadina M, Kelsall BL (2005) G(i)-protein-dependent inhibition of IL-12 production is mediated by activation of the phosphatidylinositol 3-kinase–protein 3 kinase B/Akt pathway and JNK. J Immunol 175:2994–2999

Laudadio MA, Psarropoulou C (2004) The A_3 adenosine receptor agonist 2-CL-IB-MECA facilitates epileptiform discharges in the CA3 area of immature rat hippocampal slices. Epilepsy Res 59:83–94

Lee HT, Emala CW (2000) Protective effects of renal ischemic preconditioning and adenosine pre-treatment: role of A_1 and A_3 receptors. Am J Physiol Renal Physiol 278:F380–F387

Lee JE, Bokoch G, Liang B (2001) A novel cardioprotective role of RhoA: new signaling mechanism for adenosine. FASEB J 15:1886–1894

Lee HT, Ota-Setlik A, Xu H, D'Agati VD, Jacobson MA, Emala CW (2003) A_3 adenosine receptor knockout mice are protected against ischemia- and myoglobinuria-induced renal failure. Am J Physiol Renal Physiol 284:F267–F273

Lee HT, Kim M, Joo JD, Gallos G, Chen JF, Emala CW (2006a) A3 adenosine receptor activation decreases mortality and renal and hepatic injury in murine septic peritonitis. Am J Physiol Regul Integr Comp Physiol 291:R959–R969

Lee JY, Jhun BS, Oh YT, Lee JH, Choe W, Baik HH, Ha J, Yoon K-S, Kim SS, Kang I (2006b) Activation of adenosine A_3 receptor suppresses lipopolysaccaride-induced TNF-α production through inhibition of PI-3-kinase/Akt and NF-kB activation in murine BV2 microglial cells. Neurosci Lett 396:1–6

Levy O, Coughlin M, Cronstein BN, Roy RM, Desai A, Wessels MR (2006) The adenosine system selectively inhibits TLR-mediated TNF-alpha production in the human newborn. J Immunol 177:1956–1966

Lewerenz A, Hentschel S, Vissiennon Z, Michael S, Nieber K (2003) A_3 receptors in cortical neurons: pharmacological aspects and neuroprotection during hypoxia. Drug Dev Res 58:420–427

Linden J (1994) Cloned adenosine A_3 receptors: pharmacological properties, species differences and receptor functions. Trends Pharmacol Sci 15:298–306

Linden J, Taylor HE, Robeva AS, Tucker AL, Stehle JH, Rivkees SA, Fink JS, Reppert SM (1993) Molecular cloning and functional expression of a sheep A_3 adenosine receptor with widespread tissue distribution. Mol Pharmacol 44:524–532

Liu GS, Richards SC, Olsson RA, Mullane K, Walsh RS, Downey JM (1994) Evidence that the adenosine A3 receptor may mediate the protection afforded by preconditioning in the isolated rabbit heart. Cardiovasc Res 28:1057–1061

Lopes LV, Rebola N, Costenla AR, Halldner L, Jacobson MA, Oliveira CR, Richardson PJ, Fredholm BB, Ribeiro JA, Cunha RA (2003) Adenosine A_3 receptors in the rat hippocampus: lack of interaction with A_1 receptors. Drug Dev Res 58:428–438

Mabley J, Soriano F, Pacher P, Haskó G, Marton A, Wallace R, Salzman A, Szabo C (2003) The adenosine A_3 receptor agonist, N^6-(3-iodobenzyl)-adenosine-5′-N-methyluronamide, is protective in two murine models of colitis. Eur J Pharmacol 466:323–329

Maddock HL, Mocanu MM, Yellon DM (2002) Adenosine A_3 receptor activation protects the myocardium from reperfusion/reoxygenation injury. Am J Physiol Heart Circ Physiol 283:H1307–H1313

Maddock HL, Gardner NM, Khandoudi N, Bril A, Broadley KJ (2003) Protection from myocardial stunning by ischaemia and hypoxia with the adenosine A_3 receptor agonist, IB-MECA. Eur J Pharmacol 477:235–245

Mader R, Keystone E (2007) Optimizing treatment with biologics. J Rheumatol 80:16–24

Madi L, Bar-Yehuda S, Barer F, Ardon E, Ochaion A, Fishman P (2003) A_3 adenosine receptor activation in melanoma cells: association between receptor fate and tumor growth inhibition. J Biol Chem 278:42121–42130

Madi L, Ochaion A, Rath-Wolfson L, Bar-Yehuda S, Erlanfer A, Ohana G, Harish A, Merimski O, Barer F, Fishman P (2004) The A_3 adenosine receptor is highly expressed in tumor versus normal cells: potential target for tumor growth inhibition. Clin Cancer Res 10:4472–4479

Mahaffey KW, Puma JA, Barbagelata NA, DiCarli MF, Leesar MA, Browne KF, Eisenberg PR, Bolli R, Casas AC, Molina-Viamonte V, Orlandi C, Blevins R, Gibbons RJ, Califf RM, Granger CB (1999) Adenosine as an adjunct to thrombolytic therapy for acute myocardial infarction: results of a multicenter, randomized, placebo-controlled trial: the Acute Myocardial Infarction Study of ADenosine (AMISTAD) trial. J Am Coll Cardiol 34(6):1711–1720

Martin L, Pingle SC, Hallam DM, Rybak LP, Ramkumar V (2006) Activation of the adenosine A_3 receptor in RAW 264.7 cells inhibits lipopolysaccharide-stimulated tumor necrosis factor-α release by reducing calcium-dependent activation of nuclear factor-kB and extracellular signal-regulated kinase 1/2. J Pharmacol Exp Ther 316:71–78

Matot I, Weiniger CF, Zeira E, Galun E, Joshi BV, Jacobson KA (2006) A_3 adenosine receptors and mitogen-activated protein kinases in lung injury following in vivo reperfusion. Crit Care 10:R65

McLeod C, Bagust A, Boland A, Dagenais P, Dickson R, Dundar Y, Hill RA, Jones A, Mujica Mota R, Walley T (2007) Adalimumab, etanercept and infliximab for the treatment of ankylosing spondylitis: a systematic review and economic evaluation. Health Technol Assess 11:1–158

McWhinney CD, Dudley MW, Bowlin TL, Peet NP, Schook L, Bradshaw M, De M, Borcheerding DR, Edwards CK (1996) Activation of adenosine A_3 receptors on macrophages inhibits tumor necrosis factor-α. Eur J Pharmacol 310:209–216

Merighi S, Varani K, Gessi S, Cattabriga E, Iannotta V, Ulouglu C, Leung E, Borea PA (2001) Pharmacological and biochemical characterization of adenosine receptors in the human malignant melanoma A375. Br J Pharmacol 134:1215–1226

Merighi S, Mirandola P, Varani K, Gessi S, Leung E, Baraldi PG, Tabrizi MA, Borea PA (2003) A glance at adenosine receptors: novel target for antitumor therapy. Pharmacol Ther 100:31–48

Merighi S, Benini A, Mirandola P, Gessi S, Varani K, Leung E, Maclennan S, Borea PA (2005a) A_3 adenosine receptor activation inhibits cell proliferation via phosphatidylinositol 3-kinase/Akt-dependent inhibition of the extracellular signal-regulated kinase 1/2 phosphorylation in A375 human melanoma cells. J Biol Chem 280:19516–19526

Merighi S, Benini A, Mirandola P, Gessi S, Varani K, Leung E, MacLennan S, Baraldi PG, Borea PA (2005b) A_3 adenosine receptors modulate hypoxia-inducible factor-1alpha expression in human A375 melanoma cells. Neoplasia 10:894–903

Merighi S, Benini A, Mirandola P, Gessi S, Varani K, Leung E, Maclennan S, Borea PA (2006) Adenosine modulates vascular endothelial growth factor expression via hypoxia-inducible factor-1 in human glioblastoma cells. Biochem Pharmacol 72:19–31

Merighi S, Benini A, Mirandola P, Gessi S, Varani K, Simioni C, Leung E, Maclennan S, Borea PA (2007a) Caffeine inhibits adenosine-induced accumulation of hypoxia-inducible factor-1α, vascular endothelial growth factor and interleukin-8 expression in hypoxic human colon cancer cells. Mol Pharmacol 72:395–406

Merighi S, Benini A, Mirandola P, Gessi S, Varani K, Leung E, Maclennan S, Baraldi PG, Borea, PA (2007b) Hypoxia inhibits paclitaxel-induced apoptosis through adenosine-mediated phosphorylation of BAD in glioblastoma cells. Mol Pharmacol 72:162–172

Meyerhof W, Müller-Brechlin R, Richter D (1991) Molecular cloning of a novel putative G-protein coupled receptor expressed during rat spermiogenesis. FEBS Lett 284:155–160

Mitchell CH, Peterson-Yantorno K, Carre DA, McGlinn AM, Coca-Prados M, Stone RA, Civan MM (1999) A_3 adenosine receptors regulate Cl^- channels of nonpigmented ciliary epithelial cell. Am J Physiol 276:C659–C666

Montesinos MC, Desai A, Delano D, Chen JF, Fink JS, Jacobson MA, Cronstein BN (2003) Adenosine A_{2A} or A_3 receptors are required for inhibition of inflammation by methotrexate and its analog MX-68. Arthritis Rheum 48:240–247

Mozzicato S, Joshi BV, Jacobson A, Liang BT (2004) Role of direct RhoA-phospholipase D1 interaction in mediating adenosine-induced protection from cardiac ischemia. FASEB J 18:406–408

Murrison EM, Goodson SJ, Edbrooke MR, Harris CA (1996) Cloning and characterisation of the human adenosine A_3 receptor gene. FEBS Lett 384:243–246

Murry CE, Jennings RB, Reimer KA (1986) Preconditioning with ischemia: a delay of lethal cell injury in ischemic myocardium. Circulation 74:1124–1136

Ochaion A, Bar-Yehuda S, Cohn S, Del Valle L, Perez-Liz G, Madi L, Barer F, Farbstein M, Fishman-Furman S, Reitblat T, Reitblat A, Amital H, Levi Y, Molad Y, Mader R, Tishler M, Langevitz P, Zabutti A, Fishman P (2006) Methotrexate enhances the anti-inflammatory effect of CF101 via up-regulation of the A_3 adenosine receptor expression. Arthritis Res Ther 8:R169

Ochaion A, Bar-Yehuda S, Cohen S, Amital H, Jacobson KA, Joshi BV, Gao ZG, Barer F, Patoka R, Del Valle L, Perez-Liz G, Fishman P (2008) The A3 adenosine receptor agonist CF502 inhibits the PI3K, PKB/Akt and NF-κB signaling pathway in synoviocytes from rheumatoid arthritis patients and in adjuvant-induced arthritis rats. Biochem Pharmacol 76:482–494.

Okamura T, Kurogi Y, Hashimoto K, Nishikawa H, Nagao Y (2004) Facile synthesis of fused 1,2,4-triazolo[1,5-*c*]pyrimidine derivatives as human adenosine A_3 receptor ligands. Bioorg Med Chem Lett 14:3775–3779

Olah ME, Gallo-Rodriguez C, Jacobson KA, Stiles GL (1994) [^{125}I]AB-MECA, a high affinity radioligand for the rat A_3 adenosine receptor. Mol Pharmacol 45:978–982

Palmer TM, Stiles GL (2000) Identification of threonine residues controlling the agonist-dependent phosphorylation and desensitization of the rat A_3 adenosine receptor. Mol Pharmacol 57:539–545

Palmer TM, Benovic JL, Stiles GL (1995a) Agonist-dependent phosphorylation and desensitization of the rat A_3 adenosine receptor. J Biol Chem 270:29607–29613

Palmer TM, Gettys TW, Stiles GL (1995b) Differential interaction with and regulation of multiple G-proteins by the rat A_3 adenosine receptor. J Biol Chem 270:16895–16902

Palmer TM, Benovic JL, Stiles GL (1996) Molecular basis for subtype-specific desensitization of inhibitory adenosine receptors. Analysis of a chimeric A_1–A_3 adenosine receptor. J Biol Chem 271:15272–15278

Palmer TM, Harris CA, Coote J, Stiles GL (1997) Induction of multiple effects on adenylyl cyclase regulation by chronic activation of the human A_3 adenosine receptor. Mol Pharmacol 52:632–640

Panther E, Idzko M, Herouy Y, Rheinen H, Geboicke-Haerter PJ, Mrowietz U, Dichmann S, Norgauer J (2001) Expression and function of adenosine receptors in human dendritic cells. FASEB J 15:1963–1970

Park S-S, Zhao H, Jang Y, Mueller RA, Xu Z (2006) N^6-(3-Iodobenzyl)-adenosine-5′-N-methylcarboxamide confers cardioprotection at reperfusion by inhibiting mitochondrial permeability transition pore opening via glycogen synthase kinase 3β. J Pharm Exp Ther 318:124–131

Parsons M, Young L, Lee JE, Jacobson KA, Liang BT (2000) Distinct cardioprotective effects of adenosine mediated by differential coupling of receptor subtypes to phospholipases C and D. FASEB J 14:1423–1431

Peart JN, Headrick JP (2007) Adenosinergic cardioprotection: multiple receptors, multiple pathways. Pharmacol Ther 114:208–221

Philipp S, Yang XM, Cui L, Davis AM, Downey JM, Cohen MV (2006) Postconditioning protects rabbit hearts through a protein kinase C-adenosine A_{2B} receptor cascade. Cardiovasc Res 70:308–314

Press NJ, Taylor RJ, Fullerton JD, Tranter P, McCarthy C, Keller TH, Brown L, Cheung R, Christie J, Haberthuer S, Hatto JD, Keenan M, Mercer MK, Press NE, Sahri H, Tuffnell AR, Tweed M, Fozard JR (2005) A new orally bioavailable dual adenosine A_{2B}/A_3 receptor antagonist with therapeutic potential. Bioorg Med Chem Lett 15:3081–3085

Press NJ, Gessi S, Borea PA, Polosa R (2007) Therapeutic potential of adenosine receptor antagonists and agonists. Expert Opin Ther Patents 17:979–991

Pugliese AM, Latini S, Corradetti R, Pedata F (2003) Brief, repeated, oxygen-glucose deprivation episodes protect neurotransmission from a longer ischemic episode in the in vitro hippocampus: role of adenosine receptors. Br J Pharmacol 140(2):305–314

Pugliese AM, Coppi E, Spalluto G, Corradetti R, Pedata F (2006) A_3 adenosine receptor antagonists delay irreversible synaptic failure caused by oxygen and glucose deprivation in the rat CA1 hippocampus in vitro. Br J Pharmacol 147(5):524–532

Pugliese AM, Coppi E, Volpini R, Cristalli G, Corradetti R, Jeong LS, Jacobson KA, Pedata F (2007) Role of adenosine A_3 receptors on CA1 hippocampal neurotransmission during oxygen-glucose deprivation episodes of different duration. Biochem Pharmacol 74:768–779

Ramkumar RV, Stiles GL, Beaven MA, Ali H (1993) The A_3 adenosine receptor is the unique adenosine receptor which facilitates release of allergic mediators in mast cells. J Biol Chem 268:16887–16890

Rath-Wolfson L, Bar-Yehuda S, Madi L, Ochaion A, Cohen S, Zabutti A, Fishman P (2006) IB-MECA, an A3 adenosine receptor agonist prevents bone resorption in rats with adjuvant induced arthritis. Clin Exp Rheumatol 24:400–406

Reshkin SJ, Guerra L, Bagorda A, Debellis L, Cardone R, Li AH, Jacobson KA, Casavola V (2000) Activation of A_3 adenosine receptor induces calcium entry and chloride secretion in A6 cells. J Membr Biol 178:103–113

Rigopoulos D, Korfitis C, Gregoriou S, Katsambas AD (2008) Infliximab in dermatological treatment: beyond psoriasis. Expert Opin Biol Ther 8:123–133

Rimmer J, Peake HL, Santos CMC, Lean M, Bardin P, Robson R, Haumann B, Loehrer F, Handel, ML (2007) Targeting adenosine receptors in the treatment of allergic rhinitis: a randomized, double-blind, placebo-controlled study. Clin Exp Allergy 37:8–14

Rivkees SA (1994) Localization and characterization of adenosine receptor expression in rat testis. Endocrinology 135:2307–2313

Rivo J, Zeira E, Galun E, Matot I (2004a) Activation of A_3 adenosine receptor provides lung protection against ischemia-reperfusion injury associated with reduction in apoptosis. Am J Transplant 4:1941–1948

Rivo J, Zeira E, Galun E, Matot I (2004b) Activation of A_3 adenosine receptors attenuates lung injury after in vivo reperfusion. Anesthesiology 101:1153–1159

Ross AM, Gibbons RJ, Stone GW, Kloner RA, Alexander RW, AMISTAD-II Investigators (2005) A randomized, double-blinded, placebo-controlled multicenter trial of adenosine as an adjunct to reperfusion in the treatment of acute myocardial infarction (AMISTAD-II). J Am Coll Cardiol 45(11):1775–1780

Ryzhov S, Goldstein AE, Matafonov A, Zeng D, Biaggioni I, Feoktistov I (2004) Adenosine-activated mast cells induce IgE synthesis by B lymphocytes: an A_{2B}-mediated process involving Th2 cytokines IL-4 and IL-13 with implications for asthma. J Immunol 172:7726–7733

Sajjadi FG, Firestein GS (1993) cDNA cloning and sequence analysis of the human A_3 adenosine receptor. Biochim Biophys Acta 1179:105–107

Salvatore CA, Jacobson MA, Taylor HE, Linden J, Johnson RG (1993) Molecular cloning and characterization of the human A_3 adenosine receptor. Proc Natl Acad Sci USA 90:10365–10369

Salvatore CA, Tilley SL, Latour AM, Fletcher DS, Koller BH, Jacobson MA (2000) Disruption of the A_3 adenosine receptor gene in mice and its effect on stimulated inflammatory cells. J Biol Chem 275:4429–4434

Schlotzer-Schrehardt U, Zenkel M, Decking U, Haubs D, Kruse FE, Junemann A, Coca-Prados M, Naumann GO (2005) Selective upregulation of the A_3 adenosine receptor in eyes with pseudoexfoliation syndrome and glaucoma. Invest Ophthalmol Vis Sci 46:2023–2034

Schulte G, Fredholm BB (2000) Human adenosine A_1, A_{2A}, A_{2B} and A_3 receptors expressed in Chinese hamster ovary cells all mediate the phosphorylation of extracellular-regulated kinase 1/2. Mol Pharmacol 58:477–482

Schulte G, Fredholm BB (2002) Signaling pathway from the human adenosine A_3 receptor expressed in Chinese hamster ovary cells to the extracellular signal-regulated kinase 1/2. Mol Pharmacol 62:1137–1146

Sei Y, Von Lubitz DKJE, Abbracchio MP, Ji X-D, Jacobson A (1997) Adenosine A_3 receptor agonist-induced neurotoxicity in rat cerebellar granule neurons. Drug Dev Res 40:267–273

Shen H, Chen GJ, Harvey BK, Bickford PC, Wang Y (2005) Inosine reduces ischemic brain injury in rats. Stroke 36:654–659

Shepherd RK, Linden J, Duling BR (1996) Adenosine-induced vasoconstriction in vivo. Role of the mast cell and A3 adenosine receptor. Circ Res 78:627–634

Shneyvays V, Nawrath H, Jacobson KA, Shainberg A (1998) Induction of apoptosis in cardiac myocytes by an A_3 adenosine receptor agonist. Exp Cell Res 243:383–397

Shneyvays V, Jacobson KA, Li AH, Nawrath H, Zinman T, Isaac A, Shainberg A (2000) Induction of apoptosis in rat cardiocytes by A_3 adenosine receptor activation and its suppression by isoproterenol. Exp Cell Res 257:111–126

Shneyvays V, Mamedova LK, Zinman T, Jacobson KA, Shainberg A (2001) Activation of A_3 adenosine receptor protects against doxorubicin-induced cardiotoxicity. J Mol Cell Cardiol 33:1249–1261

Shneyvays V, Zinman T, Shainberg A (2004) Analysis of calcium responses mediated by the A_3 adenosine receptor in cultured newborn rat cardiac myocytes. Cell Calcium 36:387–396

Silverman MH, Strand V, Markovits D, Nahir M, Reitblat T, Molad Y, Rosner I, Rozenbaum M, Mader R, Adawi M, Caspi D, Tishler M, Langevitz P, Rubinow A, Friedman J, Green L, Tanay A, Ochaion A, Cohen S, Kerns WD, Cohn I, Fishman-Furman S, Farbstein M, Yehuda SB, Fishman P (2008) Clinical evidence for utilization of the A3 adenosine receptor as a target to treat rheumatoid arthritis: data from a Phase II clinical trial. J Rheumatol 35:41–48

Solenkova NV, Solodushko V, Cohen MV, Downey JM (2006) Endogenous adenosine protects preconditioned heart during early minutes of reperfusion by activating Akt. Am J Physiol Heart Circ Physiol 290:H441–H449

Spruntulis LM, Broadley KJ (2001) A_3 receptors mediate rapid inflammatory cell influx into the lungs of sensitized guinea-pigs. Clin Exp Allergy 31:943–951

Suh BC, Kim TD, Lee JU, Seong JK, Kim KT (2001) Pharmacological characterization of adenosine receptors in PGT-beta mouse pineal gland tumour cells. Br J Pharmacol 134:132–142

Suleiman M-S, Halestrap AP, Griffiths EJ (2001) Mitochondria: a target for myocardial protection. Pharmacol Ther 89:29–46

Szabo C, Scott GS, Virag L, Egnaczyk G, Salzman AL, Shanley TP, Haskó G (1998) Suppression of macrophage inflammatory protein (MIP)-1a production and collagen-induced arthritis by adenosine receptor agonists. Br J Pharmacol 125:379–387

Takano H, Bolli R, Black RG, Modani E, Tang X-L, Yang Z, Bhattacharya S, Auchampach JA (2001) A_1 or A_3 adenosine receptors induce late preconditioning against infarction in conscious rabbits by different mechanisms. Circ Res 88:520–528

Thiele A, Kronstein R, Wetzel A, Gerth A, Nieber K, Hauschildt S (2004) Regulation of adenosine receptor subtypes during cultivation of human monocytes: role of receptors in preventing lipopolysaccharide-triggered respiratory burst. Infect Immun 72:1349–1357

Thourani VH, Nakamura M, Ronson RS, Jordan JE, Zhao Z-Q, Levy JH, Szlam F, Guyton R, Vinten-Johansen J (1999) Adenosine A_3 receptor stimulation attenuates postischemic dysfunction through K_{ATP} channels. Am J Physiol 46:H228–H235

Tilg H, Moschen A, Kaser A (2007) Mode of function of biological anti-TNF agents in the treatment of inflammatory bowel diseases. Expert Opin Biol Ther 7:1051–1059

Toussirot E, Wendling D (2007) The use of TNF-alpha blocking agents in rheumatoid arthritis: an update. Expert Opin Pharmacother 8:2089–2107

Tracey WR, Magee W, Masamune H, Kennedy SP, Knight DR, Bucholz RA, Hill RJ (1997) Selective adenosine A_3 receptor stimulation reduces ischemic myocardial injury in the rabbit heart. Cardiovasc Res 33:410–415

Tracey WR, Magee W, Masamune H, Oleynek JJ, Hill RJ (1998) Selective activation of adenosine A_3 receptors with N^6-(3-chlorobenzyl)-5′-N-methylcarboxamidoadenosine (Cl-IBMECA) provides cardioprotection via KATP channel activation. Cardiovasc Res 40:138–145

Trincavelli ML, Tuscano D, Cecchetti P, Falleni A, Benzi L, Klotz KN, Gremigni V, Cattabeni F, Lucacchini A, Martini C (2000) Agonist-induced internalization and recycling of the human A(3) adenosine receptors: role in receptor desensitization and resensitization. J Neurochem 75:1493–1501

Trincavelli ML, Tuscano D, Marroni M, Falleni A, Gremigni V, Ceruti S, Abbracchio MP, Jacobson KA, Cattabeni F, Martini C (2002a) A_3 adenosine receptors in human astrocytoma cells: agonist-mediated desensitization, internalization, and down-regulation. Mol Pharmacol 62:1373–1384

Trincavelli ML, Tuscano D, Marroni M, Klotz K-N, Lucacchini A, Martini C (2002b) Involvement of mitogen protein kinase cascade in agonist-mediated human A_3 adenosine receptor regulation. Biochim Biophys Acta 1591:55–62

Valesini G, Iannuccelli C, Marocchi E, Pascoli L, Scalzi V, Di Franco M (2007) Biological and clinical effects of anti-TNFalpha treatment. Autoimmune Rev 7:35–41

van Troostenburg AR, Clark EV, Carey WD, Warrington SJ, Kerns WD, Cohn I, Silverman MH, Bar-Yehuda S, Fong KL, Fishman P (2004) Tolerability, pharmacokinetics and concentration-dependent hemodynamic effects of oral CF101, an A3 adenosine receptor agonist, in healthy young men. Int J Clin Pharmacol Ther 42:534–542

von Lubitz DKJE (1999) Adenosine and cerebral ischemia: therapeutic future or death of a brave concept? Eur J Pharmacol 371:85–102

von Lubitz DKJE, Lin RC-S, Popik P, Carter MF, Jacobson KA (1994) Adenosine A_3 receptor stimulation and cerebral ischemia. Eur J Pharmacol 263:59–67

Walker BAM, Jacobson MA, Knight DA, Salvatore CA, Weir T, Zhou D, Bai TR (1997) Adenosine A_3 receptor expression and function in eosinophils. Am J Respir Cell Mol Biol 16:531–537

Wan TC, Ge ZD, Tampo A, Mio Y, Bienengraeber MW, Tracey WR, Gross GJ, Kwok WM, Auchampach JA (2008) The A_3 adenosine receptor agonist CP-532,903 [N^6-(2,5-dichlorobenzyl)-3′-aminoadenosine-5′-N-methylcarboxamide] protects against myocardial ischemia/reperfusion injury via the sarcolemmal ATP-sensitive potassium channel. J Pharmacol Exp Ther 324:234–243

Wang J, Drake L, Sajjadi F, Firestein GS, Mullane KM, Bullough DA (1997) Dual activation of adenosine A_1 and A_3 receptors mediates preconditioning of isolated cardiac myocytes. Eur J Pharmacol 320:241–248

Weinblatt ME, Coblyn JS, Fox DA, Fraser PA, Holdsworth DE, Glass DN, Trentham DE (1985) Efficacy of low-dose methotrexate in rheumatoid arthritis. N Engl J Med 312:818–822

Weiss JN, Korge P, Honda HM, ping P (2003) Role of the mitochondrial permeability transition in myocardial disease. Circ Res 93:292–301

Wittendorp MC, Boddeke HWGM, Biber K (2004) Adenosine A_3 receptor-induced CCl2 synthesis in cultured mouse astrocytes. Glia 46:410–418

Wolber C, Fozard JR (2005) The receptor mechanism mediating the contractile response to adenosine on lung parenchymal strips from actively sensitised, allergen-challenged Brown Norway rats. Naunyn–Schmiedeberg's Arch Pharmacol 371:158–168

Xu Z, Jang Y, Mueller RA, Norfleet EA (2006) IB-MECA and cardioprotection. Cardiovasc Drug Rev 24:227–238

Yamano K, Mori K, Nakano R, Kusunoki M, Inoue M, Satoh M (2007) Identification of the functional expression of adenosine A_3 receptor in pancreas using transgenic mice expressing jellyfish apoaequorin. Transgenic Res16:429–435

Yang H, Avila MY, Peterson-Yantorno K, Coca-Prados M, Stone RA, Jacobson KA, Civan MM (2005) The cross-species A_3 adenosine-receptor antagonist MRS 1292 inhibits adenosine-triggered human nonpigmented ciliary epithelial cell fluid release and reduces mouse intraocular pressure. Curr Eye Res 30:747–754

Young HWJ, Molina JG, Dimina D, Zhong H, Jacobson M, Chan L-NL, Chan T-S, Lee JJ, Blackburn MR (2004) A_3 adenosine receptor signalling contributes to airway inflammation and mucus production in adenosine deaminase-deficient mice. J Immunol 173:1380–1389

Zhang X, Zhang M, Laties AM, Mitchell CH (2006) Balance of purines may determine life or death of retinal ganglion cells as A_3 adenosine receptors prevent loss following P2X7 receptor stimulation. J Neurochem 98:566–575

Zhao TC, Kukreja RC (2002) Late preconditioning elicited by activation of adenosine A_3 receptor in heart: role of NF-kB, iNOS and mitochondrial K_{ATP} channel. J Mol Cell Cardiol 34:263–277

Zheng J, Wang R, Zambraski E, Wu D, Jacobson KA, Liang BT (2007) Protective roles of adenosine A_1, A_{2A}, and A_3 receptors in skeletal muscle ischemia and reperfusion injury. Am J Physiol Heart Circ Physiol 293:H3685–H3691

Zhong H, Shlykov SG, Molina JG, Sanborn BM, Jacobson MA, Tilley S L, Blackburn MR (2003) Activation of murine lung mast cells by the adenosine A_3 receptor. J Immunol 170:338–345

Zhou Q-Y, Li C, Olah ME, Johnson RA, Stiles GL, Civelli O (1992) Molecular cloning and characterization of an adenosine receptor: the A_3 adenosine receptor. Proc Natl Acad Sci USA 89:7432–7436

Adenosine Receptors and Asthma

Constance N. Wilson, Ahmed Nadeem, Domenico Spina, Rachel Brown, Clive P. Page, and S. Jamal Mustafa

Contents

Abstract The pathophysiological processes underlying respiratory diseases like asthma are complex, resulting in an overwhelming choice of potential targets for the novel treatment of this disease. Despite this complexity, asthmatic subjects are uniquely sensitive to a range of substances like adenosine, thought to act indirectly to evoke changes in respiratory mechanics and in the underlying pathology, and thereby to offer novel insights into the pathophysiology of this disease. Adenosine is of particular interest because this substance is produced endogenously by many cells during hypoxia, stress, allergic stimulation, and exercise. Extracellular adenosine can be measured in significant concentrations within the airways; can be shown to activate adenosine receptor (AR) subtypes on lung resident cells and migrating inflammatory cells, thereby altering their function, and could therefore play a significant role in this disease. Many preclinical in vitro and in vivo studies have documented the roles of the various AR subtypes in regulating cell function and

C.N. Wilson (✉)
Endacea, Inc., P.O. Box 12076 (Mail), 2 Davis Drive (Courier), Research Triangle Park, NC 27709-2076, USA
cwilson@endacea.nctda.org

C.N. Wilson and S.J. Mustafa (eds.), *Adenosine Receptors in Health and Disease*, Handbook of Experimental Pharmacology 193, DOI 10.1007/978-3-540-89615-9_11,

how they might have a beneficial impact in disease models. Agonists and antagonists of some of these receptor subtypes have been developed and have progressed to clinical studies in order to evaluate their potential as novel antiasthma drugs. In this chapter, we will highlight the roles of adenosine and AR subtypes in many of the characteristic features of asthma: airway obstruction, inflammation, bronchial hyperresponsiveness and remodeling. We will also discuss the merit of targeting each receptor subtype in the development of novel antiasthma drugs.

Keywords Adenosine · Adenosine receptors · Asthma · Bronchial hyperresponsiveness · Airway smooth muscle · Airway remodeling · Airway inflammation

Abbreviations

AC	Adenylate cyclase
ADA	Adenosine deaminase
ADP	Adenosine diphosphate
AK	Adenosine kinase
AMP	Adenosine monophosphate
AR	Adenosine receptor
ATP	Adenosine triphosphate
BAL	Bronchoalveolar lavage
BHR	Bronchial hyperresponsiveness
BMMC	Bone marrow-derived mast cell
CFTR	Cystic fibrosis transmembrane conductance regulator
CPA	Cyclopentyadenosine
CXCR4	Chemokine receptor 4
cyto-5′-NT	Cytosolic form of nucleotidase
DPCPX	1,3-Dipropyl-8-cyclopentylxanthine
EAR	Early asthmatic response
ecto-5′-NT	Ecto-5′-nucleotidase
FEV1	Forced expiratory volume in 1 s
fMLP	Formyl-Met–Leu–Phe
HBEC	Human bronchial epithelial cell
HPRT	Hypoxanthine phosphoribosyltransferase
ICS	Inhaled corticosteroids
IgE	Immunoglobulin E
IL	Interleukin
IMP	Inosine monophosphate
iNOS	Inducible nitric oxide synthase
LABA	Long-acting beta-adrenoceptor agonist
LAR	Late asthmatic response
LPS	Lipopolysaccharide
MCP-1	Monocyte chemotactic protein-1

NADPH	Nicotinamide adenine dinucleotide phosphate
NF-κB	Nuclear factor kappa B
NTPDase	Nucleoside triphosphate diphosphohydrolase
PC	Provocative concentration
PEFR	Peak expiratory flow rate
PDE	Phosphodiesterase
PG	Prostaglandin
PLC	Phospholipase C
PNC	Purine nucleotide cycle
PNP	Purine nucleoside phosphorylase
POC	Proof of concept
RASON	Respiratory antisense oligonucleotide
RT-PCR	Reverse transcriptase polymerase chain reaction
SAHH	*S*-Adenosylhomocysteine hydrolase
TNF	Tumor necrosis factor
VEGF	Vascular endothelium growth factor

1 Adenosine: An Important Signaling Molecule in Asthma

Asthma is a lung disease characterized by airway hyperresponsiveness and inflammation. The pathogenesis of asthma involves the release of a broad array of mediators such as cysteinyl leukotrienes, histamine and cytokines from various cell types, leading to bronchoconstriction, proinflammatory effects, chemoattraction of leukocytes, and airway remodeling (Busse and Lemanske 2001). A number of clinical features distinguish asthmatic subjects from other respiratory diseases and may be considered characteristic of this phenotype (Avital et al. 1995). These include an exacerbation of disease following exposure to beta-adrenoceptor antagonists (Bond et al. 2007), an impairment in the ability to bronchodilate following deep inspiration (Slats et al. 2007), and their bronchoconstrictor sensitivity to a wide range of innocuous stimuli (Cockcroft and Davis 2006; Van Schoor et al. 2002). Various mechanisms have been proposed to account for this bronchial hyperresponsiveness (BHR) phenomenon, and these include increased airway smooth muscle function (An et al. 2007; Gil and Lauzon 2007), altered airway epithelial cell function (Holgate 2007), and the recruitment and activation of numerous inflammatory cells, including dendritic cells, T lymphocytes and eosinophils (Beier et al. 2007; Hammad and Lambrecht 2007; Jacobsen et al. 2007; Kallinich et al. 2007; Lloyd and Robinson 2007; Rosenberg et al. 2007), whose cell-derived products trigger a cascade of events within the lung that lead to airway epithelial cell damage, increased bronchial smooth muscle contractility and airway remodeling.

Asthmatic subjects bronchoconstrict in response to a number of physiological stimuli, such as exercise, distilled water, cold air and hypertonic saline, to which healthy subjects are refractory. Similarly, acidification, pollutants like sulfur dioxide, and chemical substances including adenosine, bradykinin and neuropeptides evoke bronchoconstriction in asthmatics but have little if any effect in nondiseased individuals. These agents are commonly referred to as indirect-acting stimuli, since they do not appear to mediate bronchoconstriction by the direct activation of airway smooth muscle. They are thought to elicit bronchospasm by activating a number of different cell types, including mast cells, vascular smooth muscle cells, vascular endothelial cells, and/or airway nerves (Spina and Page 1996, 2002; Van Schoor et al. 2000). It is therefore of interest that asthmatic subjects are sensitive to such stimuli whilst healthy subjects are invariably unresponsive to these agents (Van Schoor et al. 2000). This suggests that the mechanisms by which these stimuli provoke bronchoconstriction are upregulated in asthma and are characteristic of this phenotype.

Furthermore, airway inflammation appears to be correlated better with BHR to indirect stimuli like adenosine (van den Berge et al. 2001), bradykinin (Polosa et al. 1998; Roisman et al. 1996) and hypertonic saline (Sont et al. 1993) than it is to more direct-acting stimuli like methacholine. Similarly, during an exacerbation of BHR following the deliberate exposure of an asthmatic subject to an environmental allergen (e.g., house dust mite), there is a preferential increase in BHR to an indirect-acting stimulus like bradykinin in contrast to methacholine (Berman et al. 1995). On the other hand, a number of pharmacological drugs used to treat asthma, including nedocromil sodium and ipratropium bromide, suppress airway responsiveness to these indirect-acting stimuli, suggesting the likely involvement of neural reflexes (Van Schoor et al. 2000). Furthermore, it is now recognized that glucocorticosteroids preferentially suppress BHR to adenosine (Ketchell et al. 2002; van den Berge et al. 2001) and bradykinin (Reynolds et al. 2002) compared with methacholine.

It is common for clinicians to use stimuli like methacholine and histamine as provocative inhalation challenge agents to induce bronchoconstriction because these agents are relatively convenient to use. However, whilst there is a separation in airway responsiveness to these agents between asthmatic subjects and healthy individuals, there is also a considerable degree of overlap, and it has been suggested that airway responses to these agents may not be sensitive indicators of the asthma phenotype (Avital et al. 1995; O'Connor et al. 1999). In contrast, asthmatic subjects invariably bronchoconstrict in response to the indirect-acting stimuli described earlier, which provoke little if any response in otherwise healthy individuals or in subjects with other respiratory diseases (Avital et al. 1995; Van Schoor et al. 2000).

A growing body of evidence has emerged in support of the purine nucleoside adenosine in the pathogenic mechanisms of asthma (Spicuzza et al. 2006). This body of evidence is supported by the following reported findings. (a) In asthmatics adenosine levels are elevated in bronchoalveolar lavage (BAL) fluid (Driver et al. 1993), in the circulation following allergen inhalation (Mann et al. 1986a), and in exhaled breath condensate in patients with asthma (Csoma et al. 2005). (b) Adenosine given by inhalation causes a dose-dependent bronchoconstriction in subjects with asthma (Cushley et al. 1983; Polosa 2002; Rorke and Holgate 2002). (c) Inhalational challenge with adenosine monophosphate (AMP), which is metabolized locally by the ectonucleotidase 5′-nucleotidase to adenosine, increases the release of leukotrienes and other bronchoconstrictive mediators in asthmatics (Bucchioni et al. 2004). (d) Adenosine enhances mast-cell allergen-dependent activation (Polosa et al. 1995);

(e) Treatment with dipyridamole, a blocker of adenosine reuptake, significantly enhances the bronchoconstrictor response to inhaled adenosine in subjects with asthma (Crimi et al. 1988). (f) The sensitivity of airways to adenosine and AMP more closely reflects an inflammatory process and the phenotype for allergic asthma than the sensitivity of airways to other known inhalational bronchoprovocative agents, such as methacholine and histamine (de Meer et al. 2002; Holgate 2002; Spicuzza et al. 2003; van den Berge et al. 2001).

1.1 Adenosine Metabolism

The physiological effects of adenosine in asthma via its stimulation of cell-surface adenosine receptors (ARs) and subsequent downstream signaling pathways are a function of the local concentration of adenosine. Adenosine concentrations in unstressed cells and tissue are below 1 μM (estimates 10–100 nM); however, in metabolically stressed inflamed or ischemic tissues, adenosine levels may rise to 100 μM (Fredholm 2007; Hasko and Cronstein 2004). Lower concentrations of adenosine (10–100 nM) activate the high-affinity A_1, A_{2A}, and A_3 ARs and high adenosine concentrations (10 μM) stimulate low-affinity A_{2B}ARs (Fredholm 2007). Factors that determine the net effect of adenosine on specific cell and tissue function are AR expression and coupling to intracellular signaling pathways, all of which are tightly regulated in different tissues and cells.

The local adenosine concentration at its receptor subtypes is determined by several processes, which include extracellular and intracellular adenosine generation, adenosine release from cells, cellular reuptake and metabolism (Fig. 1). These processes are closely intertwined and strictly regulated. For, example, under the hypoxic and inflammatory conditions encountered in asthmatic airways, the increased intracellular dephosphorylation of adenosine 5′-triphosphate (ATP) to adenosine by the cytosolic metabolic enzyme 5′-nucleotidase may be accompanied by a suppression of the activity of the salvage enzyme adenosine kinase, which prevents the rephosphorylation of adenosine to AMP (Deussen 2000). These processes lead to high adenosine concentrations inside the cell and the release of adenosine from the dephosphorylation of AMP into the extracellular space through nucleoside transporters (Hyde et al. 2001; Pastor-Anglada et al. 2001).

The other major pathway that contributes to high extracellular adenosine concentrations during metabolic stress is release of adenine nucleotides (ATP, adenosine diphosphate (ADP), and AMP) from inflammatory and injured cells. This is followed by extracellular degradation to adenosine by a cascade of ectonucleotidases, which include CD39 (nucleoside triphosphate diphosphohydrolase (NTPDase)) and CD73 (5′-ectonucleotidase) (Eltzschig et al. 2004; Kaczmarek et al. 1996; Resta et al. 1998; Thompson et al. 2004; Zimmermann 1999). Adenosine accumulation is limited by its catabolism to inosine by adenosine deaminase. Inosine is finally degraded to the stable end-product uric acid (Hasko et al. 2000, 2004). Mechanisms of nucleotide release and metabolism, or adenosine release and metabolism, as well

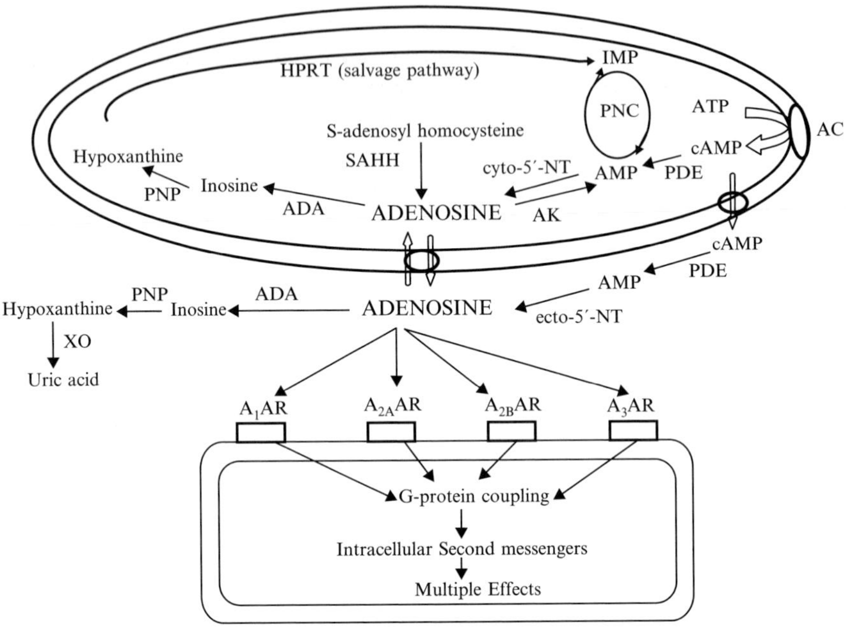

Fig. 1 Metabolism of adenosine. Adenosine is generated mainly by two enzymatic systems: intra/extracellularly localized nucleotidases and cytoplasmic *S*-adenosylhomocysteine hydrolase (SAHH). In response to hypoxia/cellular damage or other stressful/inflammatory stimuli, ATP is rapidly dephosphorylated by combined effects of adenylate cyclase (AC), phosphodiesterases (PDE) and nucleotidases to form intra/extracellular adenosine. Ecto-5′-nucleotidase (ecto-5′-NT) is one such enzyme that plays an important role in regulating local adenosine production for receptor signaling. Extracellular adenosine can interact with adenosine receptors (AR) that are coupled to heterotrimeric G proteins, which, in turn, couple AR activation to various effector molecules that can regulate second-messenger systems to influence cell and tissue function. Adenosine can also be deaminated to inosine by adenosine deaminase (ADA) that can exist intra- or extracellularly, or it can be transported into and out of the cells via membrane-associated nucleoside transporters. Intracellular adenosine is generated from the dephosphorylation of AMP by a cytosolic form of nucleotidase (cyto-5′-NT) or the hydrolysis of *S*-adenosylhomocysteine by SAHH. Adenosine can also be phosphorylated back to AMP by adenosine kinase (AK). AMP can also be directly deaminated to inosine monophosphate (IMP) by AMP deaminase. The reaction of phosphorylation predominates when adenosine occurs at a low physiological concentration ($<1\,\mu$M), whereas ADA is activated at higher concentrations of the substrate ($>10\,\mu$M). Hypoxanthine is formed after the removal of ribose from inosine by the actions of purine nucleoside phosphorylase (PNP) PNP has only negligible activity towards adenosine and degrades mainly inosine. Hypoxanthine can be salvaged back to IMP by hypoxanthine phosphoribosyltransferase (HPRT), which is again converted to AMP through the purine nucleotide cycle (PNC). Hypoxanthine can also enter the xanthine oxidase (XO) pathway to form xanthine and uric acid sequentially as byproducts

as transport mechanisms that account for the increased adenosine levels in exhaled breath condensate after exercise (Csoma et al. 2005), in the circulation following allergen inhalation (Mann et al. 1986a), and in BAL fluid (BAL adenosine concentration of $2.55 \pm 0.50\,\mu$M in asthmatics versus $0.72 \pm 0.16\,\mu$M in normals) (Driver et al. 1993) in human asthmatics, are yet to be determined.

There are several important cell types that are sources of extracellular adenosine. Neutrophils and endothelial cells release large amounts of adenosine at sites of metabolic distress, inflammation and infection (Cronstein et al. 1983; Gunther and Herring 1991; Madara et al. 1993; Rounds et al. 1994). Activated leukocytes are a major source of extracellular adenosine (Mann et al. 1986b). ADP released by platelets can be a significant source of adenosine after dephosphorylation (Marcus et al. 1995). Under conditions of stress including infection, activated macrophages can also serve as a major source of extracellular adenosine via ATP metabolism. Bacterial lipopolysaccharide (LPS) augments the release of ATP from macrophages (Sperlagh et al. 1998). Moreover, T-helper lymphocytes may be an important source of extracellular ATP. The presence of ecto-ATPase and antigen-triggered accumulation of extracellular ATP from T-helper cells has been reported (Apasov et al. 1995). In addition to inflammatory cells, airway epithelial cells and other structural cells in the lung may be important sources of high levels of adenosine in the airways of human asthmatics (Cohn et al. 2004).

1.2 Adenosine-Induced Bronchoconstriction, Airway Inflammation, and Airway Remodeling

In asthmatics, adenosine produces bronchoconstriction, inflammation, and airway plasma exudation, which lead to airway obstruction. Moreover, by acting on ARs, adenosine induces the release of inflammatory mediators that are important in the pathogenesis of airway remodeling in asthmatics. In both humans and animals, adenosine induces increases in BHR in asthmatics but not normal subjects, both in vivo following inhalation (Ali et al. 1994a; Cushley et al. 1983; Dahlen et al. 1983) and in vitro in small airways (Ali et al. 1994b; Bjorck et al. 1992). Adenosine produces bronchoconstriction in airways by directly acting on ARs in bronchial smooth muscle cells or indirectly by inducing the release of preformed and newly formed mediators from mast cells, and by acting on ARs on airway afferent sensory nerve endings (Hua et al. 2007a; Keir et al. 2006; Livingston et al. 2004; Polosa 2002). Multiple mechanisms may be involved in adenosine-induced bronchoconstriction; for example, the effects of adenosine in asthmatic subjects are sensitive to muscarinic receptor antagonists, suggesting that adenosine mediates obstruction indirectly (Crimi et al. 1992; Mann et al. 1985; Polosa et al. 1991), which would be consistent with the preclinical evidence that adenosine can activate afferent nerves in vivo (Hua et al. 2007a; Keir et al. 2006). However, since muscarinic antagonists do not completely abolish bronchoconstriction in response to adenosine, it is plausible to conclude that the "atropine-resistant" component of this response is mediated by direct activation of airway smooth muscle (Brown et al. 2008; Ethier and Madison 2006) and/or indirectly via mediators released from other cell types expressing these receptors.

Adenosine exposure through inhalation increases enhanced pause (Penh), a measure of airway resistance, in allergen-sensitized and -challenged mice (Fan and Mustafa 2002). This increase in enhanced pause due to adenosine was reversed by theophylline with methacholine-mediated enhanced pause being unaffected, suggesting the involvement of ARs (Fan and Mustafa 2002). This finding that adenosine-induced bronchoconstriction is mediated by ARs is supported by an earlier study in a rabbit model of allergic asthma, where adenosine-induced bronchoconstriction was blocked by theophylline (Ali et al. 1992). Following inhalation and its local metabolism to adenosine in the airway, AMP induced bronchoconstriction is attenuated by potent cyclooxygenase inhibitors, H_1 receptor and leukotriene receptor antagonists, suggesting that adenosine induces the release of prostaglandins, histamine and leukotrienes in the airways of asthmatics (Phillips and Holgate 1989; Rorke et al. 2002; Rutgers et al. 1999). Another study has shown that inhalation challenge with adenosine, but not methacholine, produces mild airway plasma exudation (Belda et al. 2005). Collectively, these effects of adenosine on airway nerves, contraction of bronchial smooth muscle, release of mast cell mediators, and airway edema produce airflow obstruction.

Adenosine produces inflammation in airways in allergic animals and humans. Animals with increased adenosine concentrations in the lung (adenosine deaminase (ADA)-deficient mice) develop severe pulmonary inflammation, with airway accumulation of eosinophils and activated macrophages, mast cell degranulation, and mucus metaplasia in the airways—features similar to that found in asthmatic bronchi (Blackburn et al. 2000; Chunn et al. 2001). Treatment of these mice with exogenous ADA to reduce adenosine concentrations results in the reversal of these asthmatic features (Chunn et al. 2001). In a mouse model of allergic asthma, inhalation of adenosine has also been shown to cause airway inflammation, as evidenced by an increased release of proinflammatory mediators from eosinophils and mast cells (Fan and Mustafa 2002, 2006; Oldenburg and Mustafa 2005; Tilley et al. 2003). Moreover, in human asthmatics, an inhalational challenge with AMP produced an increase in eosinophils and neutrophils in the sputum (Manrique et al. 2008; van den Berge et al. 2004).

Adenosine-mediated inflammation is not limited to the lung; it also reaches the systemic circulation. In a recent report in a mouse model of asthma activities of eosinophilic peroxidase, myeloperoxidase and beta-hexosaminidase were increased not only in the lung but also in the systemic circulation of allergic mice exposed to adenosine aerosol (Fan and Mustafa 2006). In human asthmatics, adenosine aerosol increases the release of neutrophil chemotactic factor in serum (Driver et al. 1991). Moreover, in a recent study it was demonstrated that adenosine-induced effects on urinary 9α, 11β-prostaglandin (PG) F_2 levels (a sensitive biomarker of mast cell degranulation) were enhanced during repeated low-dose allergen challenge in allergic asthmatics (Ihre et al. 2006). These earlier findings in asthmatics were confirmed by a recent study showing an increase in plasma 9α, 11β-PGF_2 levels after adenosine challenge in asthmatics (Bochenek et al. 2008). These studies suggest that following inhalation, adenosine enhances the release of systemic inflammatory mediators from sensitized inflammatory cells. Thus, following inhalation, adenosine not only produces inflammation in the airways of asthmatics but it also induces a systemic inflammatory response that would, in turn, amplify the inflammation locally in the airways of asthmatics.

Adenosine in the lung may also be involved in the airway remodeling process (Cohn et al. 2004). Pathogenic hallmarks of airway remodeling are mucous gland hyperplasia, subepithelial fibrosis, hypertrophy of bronchial smooth muscle, and angiogenesis (Cohn et al. 2004; Jarjour and Kelly 2002). In a recent report, substantial angiogenesis in the tracheas of ADA-deficient mice were seen in association with high levels of adenosine (Mohsenin et al. 2007). ADA replacement enzyme therapy in these mice resulted in a lowering of adenosine levels and reversal of tracheal angiogenesis. Moreover, in lung alveolar epithelial cells and lung fibroblasts, adenosine caused an induction of fibronectin (a matrix glycoprotein highly expressed in injured tissues that has been implicated in wound healing) mRNA and protein expression in a dose- and time-dependent manner (Roman et al. 2006). Furthermore, there appears to be a connection of IL-13 levels to high adenosine levels, ADA activity and airway remodeling (Blackburn et al. 2003). Studies in CC10 IL-13 Tg mice showed that IL-13 induced high levels of adenosine, inflammation, lung collagen content and subepithelial airway fibrosis and reduced ADA activity in the lung. ADA therapy administered to these mice decreased adenosine levels, inflammation, and subepithelial airway fibrosis (Blackburn et al. 2003). Moreover, in ADA-deficient mice, IL-13 was strongly induced. These findings suggest that Il-13 and adenosine stimulate one another to amplify the pathway that contributes to airway inflammation, fibrosis, and remodeling. Similar findings were also seen in the lungs of mice overexpressing the Th2 cytokine IL-4 (Ma et al. 2006).

2 Adenosine Receptors in Asthma

Collectively, the studies presented above suggest a strong role for adenosine not only in the bronchoconstriction of allergic airways but also in the progression and amplification of airway inflammation and airway remodeling. The effects of adenosine as an important signaling molecule in asthma may depend not only on the bioavailability of the nucleoside but also on the expression, density, and affinity of ARs, which are known to be finely modulated by physiological and/or pathological conditions, signaling mechanisms, the local metabolism of adenosine, and the predominant inflammatory cell types in the asthma model, which may be species specific (Chunn et al. 2001; Fan et al. 2003; Sun et al. 2005; Zhong et al. 2006).

Adenosine produces its effects in asthmatics by acting on membrane-bound extracellular ARs on target cells. Four subtypes of ARs (namely A_1, A_{2A}, A_{2B}, and A_3) have been cloned in humans, are expressed in the lung, and are all targets for drug development for human asthma (Polosa 2002; Rorke and Holgate 2002). These receptors are heptaspanning-transmembrane G-protein-coupled receptors. Three of the AR subtypes (A_1, A_{2A}, and A_{2B}) demonstrate 80–95% sequence homology across a wide evolutionary range of species (Fredholm et al. 2001). In contrast, the A_3ARs demonstrate significant species variation. Signal transduction by the ARs varies; not only among the subtypes but also for a particular subtype between different cell sources (Fredholm et al. 2001). A_1ARs were originally characterized

Table 1 Characteristics and pharmacology of adenosine receptors

	A_1AR	$A_{2A}AR$	$A_{2B}AR$	A_3AR
Agonists	CPA, CCPA, CHA, S-ENBA	CGS 21680, ATL146e, CV-1808, CVT-3146, MRE0740; MRE0094	BAY 60-6583	IB-MECA (CF 101), 2-Cl–IB–MECA (CF102), MRS3558 (CF502)
Antagonists	DPCPX, FSCPX, N-0861, BG-9719, BG-9928, WRC-0571; KW 3902, L-97-1, SLV320, EPI-2010	ZM 241385, KW 6002, SCH 58261, SCH 442416, CSC	IPDX, MRS1754, MRS1706, CVT-6883, CVT-5440	MRS1220, MRS1191, MRS1523, VUF 5574
Transduction mechanisms	$G_{i/o}$, ↓ cAMP; ↑K^+, ↓Ca^{2+} channels, PLA_2; $G_{\alpha 16}$ NF-κB, PLC, PKC, ↑IP3/DAG	$G_{s/olf}$, ↑cAMP	G_s/↑cAMP; G_q/PLC, ↑IP3/DAG	G_i/↓cAMP; ↑Ca^{2+}, ERK1/2, G_q/PLC, ↑IP3/DAG; PLD

References: Beukers et al. (2006); Baraldi et al. (2000); Fredholm (2007); Fredholm et al. (2001); Gao and Jacobson (2007); Gessi et al. (2008); Hess (2001); Liu and Wong (2004); Moro et al. (2006); Yuzlenko and Kiec-Kononowicz (2006)

as being coupled to pertussis-toxin-sensitive G_i-coupled signal transduction pathways, but in some cells they are directly associated with, and act through, ion channels. The A_2AR subtypes (A_{2A} and A_{2B}) are typically coupled to G_s-linked signal transduction pathways. In some cells, A_1AR receptor-mediated inhibition and $A_{2A}AR$-mediated stimulation of adenylate cyclase may coexist and their functions may be counterregulatory (Fredholm et al. 2001). A summary of the AR subtypes, their signal transduction mechanisms, and selective agonists and antagonists is presented in Table 1.

Adenosine receptors have been described on a number of different cell types that are important in the pathophysiology of asthma, including dendritic, antigen-presenting cells, human airway epithelial and bronchial smooth muscle cells, lymphocytes, mast cells, eosinophils, neutrophils, macrophages, fibroblasts and endothelial cells (Thiel et al. 2003; Young et al. 2006; Wilson 2008). Activation of ARs on these different cell types is responsible for inducing the release of mediators and cytokines, leading to BHR, inflammation, edema, and airway remodeling. Activation of ARs on afferent sensory airway nerves contributes to BHR in asthma (Hua et al. 2007a). The contributions of the different AR subtypes to the pathophysiology of asthma will be discussed in the following sections and are presented in Fig. 2. In this review, the pathophysiological role of each AR and its signaling in asthma is discussed. Furthermore, the targeting of ARs with selective agonists or antagonists as therapeutic strategies in the treatment of asthma is also discussed and is presented in Table 2.

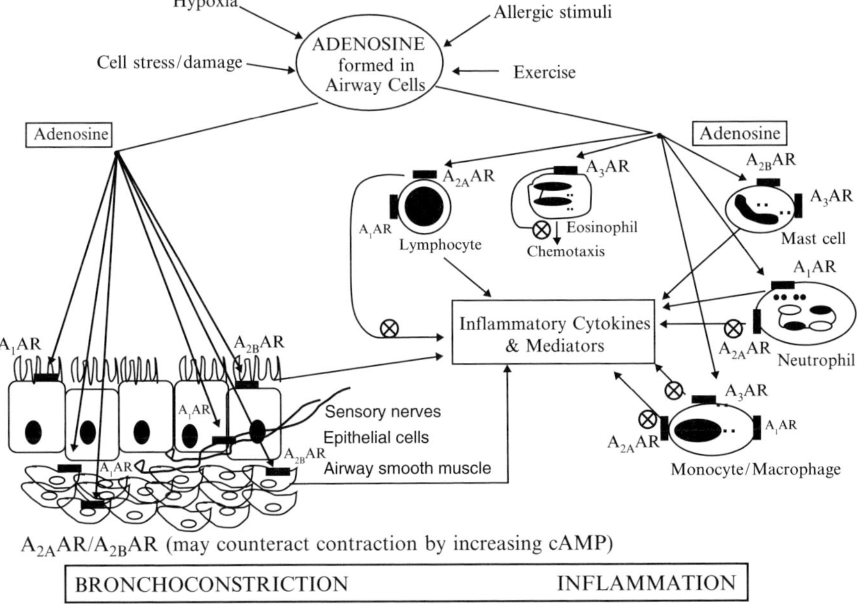

Fig. 2 Adenosine receptors and pathophysiology of asthma. By acting on adenosine receptors (ARs), A_1, A_{2A}, A_{2B}, and A_3 ARs, adenosine released under conditions of cellular stress as seen in asthmatic airways produces bronchoconstriction and inflammation. The net effect of adenosine on ARs will depend on the relative expression of these receptors on different cell types in asthmatic airways, and is concentration-dependent, as adenosine frequently exhibits opposing effects through the activation of AR subtypes expressed on the same cells coupled to different G proteins and signaling pathways. By acting on A_1ARs on bronchial smooth muscle cells and afferent sensory airway nerves, adenosine produces bronchoconstriction. By acting on A_1ARs on inflammatory leukocytes such as neutrophils, monocytes, macrophages, and lymphocytes, adenosine produces proinflammatory effects. Activation of A_{2A}ARs on the inflammatory cells suppresses the release of proinflammatory cytokines and mediators. Activation of A_{2A}ARs coupled to G_s and adenylate cyclase may also lead to bronchial smooth muscle relaxation via the cAMP–PKA (cyclic adenosine monophosphate–protein kinase A) pathway. Activation of A_{2B}ARs coupled to G_s and adenylate cyclase induce cytokine release from human bronchial epithelial and smooth muscle cells. Activation of A_{2B}ARs on murine bone marrow-derived mast cells (BMMCs) regulates the release of cytokines. The effect of adenosine on A_3ARs is species dependent. In mice, rats, and guinea pigs, activation of A_3ARs by adenosine produces bronchoconstriction, airway inflammation, mast cell degranulation, and mucus hyperplasia. In humans, activation of A_3ARs by adenosine produces anti-inflammatory effects, inhibition of chemotaxis and degranulation of eosinophils and cytokine release from monocytes. *Circled times* denote inhibition

2.1 A_1 Adenosine Receptors and Asthma

Until relatively recent times, the A_1AR received little attention as an important target in human asthma. However, a number of reports have demonstrated that expression of the A_1AR is upregulated in the airways of both animal models of allergic airway inflammation and human asthmatic subjects. Moreover, it is now appreciated

Table 2 Comparison of different potential therapeutic approaches targeting adenosine receptors in asthma

	A_1AR antagonists	A_{2A}AR agonists	A_{2B}AR antagonists	A_3AR agonists	A_3AR antagonists
Potential effects	Inhibition of bronchoconstriction, mucus hypersecretion, and inflammation	Bronchodilation and inhibition of inflammation	Inhibition of bronchoconstriction, inflammation, and airway remodeling	Inhibition of inflammation	Inhibit bronchoconstriction, inflammation, mucus hyperplasia
Disadvantages	No safety concerns reported to date for A_1AR antagonists in humans	CV side effects; tachyphylaxis; immune suppression	Reduce airway hydration; bronchoconstriction; inflammation	Tachyphylaxis; immune suppression	Inflammation
Latest developments in asthma	L-97–1 (Preclinical); EPI-2010 (Phase II; discontinued, no additional effect with ICSs)	GW328267X (Phase II; discontinued due to CV side effects)	CVT 6883 (Phase I); QAF 805 (Phase Ib)		QAF 805 (Phase Ib)
Pharmaceutical company involved in AR drug discovery	Epigenesis Pharmaceuticals; Endacea, Inc.; Biogen Idec; Merck; Solvay Pharmaceuticals; OSI Pharmaceuticals, Inc.	Glaxo Group Ltd; Pfizer; Novartis	CV Therapeutics; Novartis	Can-Fite Biopharma	Novartis

AR, Adenosine receptor; *CV*, cardiovascular; *ICSs*, inhaled corticosteroids

Adapted with permission from Wilson (2008)

that various functions relevant to asthma have also been associated with activation of the A_1AR, including bronchoconstriction, leukocyte activation and inflammation, BHR, and mucus secretion.

A pivotal study generated convincing evidence that A_1ARs could well play a significant role in the pathophysiology of asthma. The authors demonstrated that airway obstruction in response to aerosol administration of adenosine and allergen was inhibited in a rabbit model of allergic airway inflammation following treatment with antisense oligonucleotides as well as antagonist to this receptor (Ali et al. 1994a, b; Nyce and Metzger 1997). These data suggested that the A_1AR not only directly mediates bronchoconstriction following administration of exogenous adenosine, but that endogenous adenosine is an important component of the allergic response.

Although important species differences have been observed with regards to the expression and function(s) of the four AR subtypes, there is evidence that supports similar observations of an increased expression of the A_1AR and of A_1AR-induced bronchoconstriction in human asthmatic subjects. Firstly, it was demonstrated that adenosine-induced contraction of isolated bronchial tissue in vitro was greater in tissues obtained from asthmatic subjects than healthy subjects, and that this contraction could be significantly inhibited following preincubation with a selective A_1AR antagonist (Bjorck et al. 1992). Furthermore, it has been very recently demonstrated for the first time that expression of the A_1AR is increased in bronchial biopsies obtained from steroid-naïve mildly asthmatic subjects when compared with healthy subjects (Brown et al. 2008). This increased expression of the A_1AR appeared to be predominantly located in the airway epithelium and smooth muscle regions of the tissue, the latter observation thus correlating with the preclinical findings in the rabbit model of allergic asthma. In support of this, it has been demonstrated that activation of the A_1AR on human airway smooth muscle cells in vitro results in an increase in intracellular calcium mobilization, which could potentially mediate airway smooth muscle contraction (Ethier and Madison 2006). The finding of increased expression of A_1ARs in the airways and increased sensitivity of the airways to adenosine could well be of clinical significance. In asthmatics, the level of adenosine in plasma and exhaled breath condensate is increased following allergen or exercise challenge (Csoma et al. 2005; Mann et al. 1986a; Vizi et al. 2002) and therefore could lead to the activation of A_1ARs, thereby contributing toward airway obstruction during an acute exacerbation of asthma.

The report that the expression of A_1ARs is increased in bronchial biopsies of asthmatics is confirmed by the findings from another laboratory. In a preliminary study of a small number of human subjects, gene expression for A_1ARs is increased approximately 200% in bronchial tissue from small airways obtained from asthmatics ($n = 3$) versus normal subjects ($n = 3$) (Nadeem and Mustafa, unpublished data, West Virginia University). In these studies, expression of A_{2A}ARs is decreased while there is little to no change in the expression of A_{2B}ARs and A_3ARs in bronchial tissue from small airways in asthmatics versus normal subjects. The results of these studies were determined with the use of RT-PCR and confirmed with the use of western blots, with the exception of the A_{2B}AR, which was not tested in western blot studies.

A number of other studies using experimental animals have implicated a role for A_1ARs in mediating airway obstruction to adenosine. For example, the A_1AR agonist cyclopentyladenosine (CPA) selectively induces airway obstruction only in sensitized guinea pigs (Keir et al. 2006) and allergic rabbits (Ali et al. 1994a; el-Hashim et al. 1996). Further studies with the allergic rabbit model demonstrated that CPA also induced bronchoconstriction and stimulated IP_3 generation in airway smooth muscle (Abebe and Mustafa 1998). Allergic rabbits treated with the selective A_1AR antagonist L-97-1 ([3-(2-(4-aminophenyl)-ethyl]-8-benzyl-7-(2-ethyl-(2-hydroxy-ethyl)-amino]-ethyl)-1-propyl-3,7-dihydro-purine-2,6-dione]) provided bronchoprotection against inhaled adenosine (Obiefuna et al. 2005). However, atypical (Hannon et al. 2002) and adenosine A_1, A_{2B} and A_3 ARs (Fan et al. 2003; Hua et al. 2007a) have been suggested to mediate airway obstruction in response to adenosine in the brown Norway rat and mouse, respectively, underlying important species and strain differences.

Expression of the A_1AR has also been identified on a number of inflammatory cells. In general, these effects appear to be proinflammatory in nature. Activation of the A_1AR on human eosinophils, for example, promotes superoxide release (Ezeamuzie and Philips 1999). Furthermore, the A_1AR also mediates the respiratory burst in neutrophils (Salmon and Cronstein 1990), in addition to chemotaxis (Cronstein et al. 1990) and their adherence to endothelial cells (Cronstein et al. 1992). Furthermore, adenosine has been shown to promote monocyte phagocytosis (Salmon et al. 1993) and chemotaxis of immature dendritic cells (Panther et al. 2001), in addition to increasing the release of cytotoxic substances from endothelial cells that increase endothelial cell permeability (Wilson and Batra 2002) via the A_1AR.

The effects of adenosine upon inflammatory cells have been determined largely from in vitro experiments, and it should be noted that these effects are concentration dependent, as adenosine frequently exhibits opposing effects through the activation of other AR subtypes expressed on the same cells, since they are coupled to different G proteins. Thus, the relative expression of these receptors on inflammatory cells resident in asthmatic airways and the overall cellular effect of adenosine at the concentration present remain to be determined. It is likely, however, that the pattern of cellular expression for ARs changes following exposure to adenosine, since experimental evidence shows that an increased extracellular level of adenosine somewhat unusually appears to promote AR signaling. This was unequivocally demonstrated in mice partially deficient in ADA that consequently have high levels of adenosine in the lung (Chunn et al. 2001). Besides the severe pulmonary inflammation typical of this phenotype, these mice exhibited an increased transcript level for the A_1, A_{2B} and A_3 ARs.

In light of the many studies demonstrating the proinflammatory action attributed to activation of the A_1ARs, it is perhaps surprising that a preclinical study has purported to document an anti-inflammatory effect of A_1AR signaling (Sun et al. 2005). Adenosine deaminase is a ubiquitous enzyme responsible for the inactivation of adenosine, and mice deficient in this protein demonstrate profound pulmonary injury, the presence of elevated levels of macrophages, and increased mucus production. These indices of tissue damage were exacerbated in ADA double-knockout mice also deficient in the expression of A_1AR, thereby implicating the loss of an anti-inflammatory pathway mediated by this receptor (Sun et al. 2005). However, the relevance of this model to human asthma or chronic obstructive pulmonary disease is debatable, since two of the principal cell types observed in these diseases, namely eosinophils and neutrophils, respectively, are present in such small numbers (<1.7%). In contrast to these findings, the A_1AR antagonist L-97-1 inhibited the recruitment of eosinophils and neutrophils to the airways of allergic rabbits challenged with house dust mite antigen (Nadeem et al. 2006).

Very few studies have specifically addressed the question of whether activation of A_1ARs is important in the development of BHR. Animal models of allergic inflammation are characterized by increased sensitivity to inhaled histamine, and interference in A_1AR signaling following either treatment with an antisense against this receptor (Nyce and Metzger 1997) or the use of a selective antagonist (Nadeem et al. 2006; Obiefuna et al. 2005) provided some degree of protection against the development of BHR. One can only speculate as to the mechanism by which adenosine, released within the inflammatory milieu of the airways, causes BHR via an A_1AR-dependent mechanism. Activation of these receptors on inflammatory cells including mast cells, eosinophils, dendritic cells, and lymphocytes could stimulate the release of other inflammatory mediators that, in turn, increase the sensitivity of the airways. Alternatively, adenosine might stimulate C fibers, thereby lowering the threshold for the activation of afferent input into the nucleus tractus solitarius, and thus facilitating reflex activation of parasympathetic nerves (Chuaychoo et al. 2006; Hong et al. 1998).

The mechanism(s) by which adenosine mediates airway obstruction in vivo in animal models may constitute indirect components. For example, adenosine activates pulmonary C fibers in the rat (Hong et al. 1998) and in the guinea pig (Chuaychoo et al. 2006; Lee et al. 2004), and cholinergic neural pathways in conscious mice (Hua et al. 2007a) via an A_1AR-dependent mechanism. Moreover, the effect of activation of A_1ARs by a selective A_1AR agonist, CPA, was specific for nodose but not jugular ganglion-derived C fibers (Chuaychoo et al. 2006). The consequence of activating these nerves following the endogenous release of adenosine during an inflammatory response may be airway obstruction, a phenomenon that was abolished in guinea pigs chronically treated with capsaicin in order to chemically inactivate C fibers (Keir et al. 2006). Reflex activation of parasympathetic nerves was further implicated, since vagotomy or treatment with the muscarinic antagonist atropine attenuated bronchospasm induced by CPA (Keir et al. 2006). Moreover, in mice, an adenosine-induced increase in airway resistance was abolished in A_1AR knockout mice and following vagotomy in wild type mice, but not in A_{2A}, A_{2B}, or A_3 AR knockout mice (Hua et al. 2007a). In conscious mice, the adenosine-induced increase in airway resistance was significantly reduced by the selective A_1AR antagonist 1,3-dipropyl-8-cyclopentylxanthine (DPCPX) as well as atropine and bupivacaine, suggesting that the adenosine-induced bronchoconstriction was via the activation of A_1ARs on the cholinergic neural pathway. Similarly, the cholinergic-dependent reflex activation of tracheal smooth muscle in situ in response to CPA was mediated by the activation of A_1AR (Reynolds et al. 2008).

Finally, in addition to its effects on bronchoconstriction, leukocyte activation and inflammation, and BHR, the A_1AR may play an important role in mucus secretion and airway remodeling of human asthma. It has been shown that adenosine is able to induce mucus secretion via activation of the A_1AR in the canine trachea in vivo (Johnson and McNee 1985), which has now been confirmed in human bronchial epithelial cells in vitro, where activation of the A_1AR was shown to increase the expression of the *MUC2* mucin gene (McNamara et al. 2004). Thus, it could be speculated that the reported increased expression of A_1AR on asthmatic bronchial epithelium (Brown et al. 2008) promotes adenosine-induced mucin secretion, although the extent to which adenosine contributes to the overall mucus hypersecretion in asthma clearly remains to be determined. Further studies will hopefully precisely define the functional effects of the A_1AR expressed in human asthmatic epithelium. With respect to a potential role of A_1ARs in airway remodeling, recent reports, albeit not pertaining to the lung per se, suggest that activation of A_1ARs may play an important role in angiogenesis and fibrosis, cardinal features of airway remodeling in human asthma (Clark et al. 2007; Cohn et al. 2004; Kalk et al. 2007). For example, activation of A_1ARs on human monocytes induces the release of vascular endothelial growth factor (VEGF) (Clark et al. 2007), and an A_1AR antagonist with high affinity and high selectivity for the human A_1AR, SLV320, significantly reduced levels of collagen I and III in an animal model of myocardial fibrosis (Kalk et al. 2007).

Validation of the A_1AR as an important target for human asthma is supported by positive proof of concept (POC) results in patients with asthma for EPI-2010, an antisense ("knockout") compound that is a respiratory antisense oligonucleotide (RASON) for the human A_1AR, in a small clinical trial conducted by EpiGenesis Pharmaceuticals (Cranbury, NJ, USA). EpiGenesis reported that a single dose of EPI-2010 reduced the need for bronchodilator drugs to control asthma symptoms concomitant with a reduction in symptom scores, an effect that was statistically and clinically significant and lasted for one week following a single dose (Ball et al. 2003). However, disappointing results in a Phase II clinical trial with EPI-2010 administered to patients who were taking inhaled corticosteroids (ICSs) were reported (Langley et al. 2005). In this Phase II clinical trial, 146 patients with persistent airway obstruction (forced expiratory volume in 1 S (FEV_1) 74.5% predicted, $\geqslant$12% reversibility) and currently receiving ICSs were administered EPI 2010 (1, 3, or 9 mg) via nebulizer once or twice weekly for 29 days. In this clinical study there was no significant change in the FEV_1 after 29 days of treatment compared to baseline. It was concluded that EPI-2010 showed no additional therapeutic effect in patients currently receiving ICSs. Patients with a stable FEV_1 of 74.5% predicted have mild/moderate asthma, depending on the frequency of symptoms and magnitude of variability in the peak expiratory flow rate (PEFR). In patients with mild/moderate asthma treated with ICSs, the FEV_1 may be 90–100% of the predicted value when measured between exacerbations and without provocation. Thus, the FEV_1 is not a sensitive measure of asthma severity per se, vis-à-vis acute changes in airway function reflected by PEFR variability in ICS-treated patients with mild/moderate asthma. The lack of efficacy for EPI-2010 in this Phase II clinical trial (i.e., that EPI-2010 showed no additional therapeutic effect in patients taking ICSs) was not surprising.

Because of these effects of activation of A_1ARs on different cell types to produce bronchoconstriction, inflammation, mucous gland hyperplasia, angiogenesis, and fibrosis, all of which are important in the pathophysiology of human asthma, an A_1AR antagonist, L-97-1 (Endacea, Inc.), is in development as a once-daily, oral treatment for human asthma. L-97-1 is a water-soluble, small-molecule A_1AR antagonist with high affinity and high selectivity for the human A_1AR (Obiefuna et al. 2005). In an animal model of allergic asthma, L-97-1 blocks allergic airway responses, BHR to histamine, and airway inflammation (Nadeem et al. 2006; Obiefuna et al. 2005). A number of A_1AR antagonists have been or currently are in clinical trials for a number of different medical indications and, as a class, appear to be safe and well tolerated in humans (Barrett 1996; Bertolet et al. 1996; Dittrich et al. 2007; Doggrell 2005; Gaspardone et al. 1993; Givertz et al. 2007; Gottlieb et al. 2002; Greenberg et al. 2007).

2.2 A_{2A} Adenosine Receptors and Asthma

A_{2A}AR signaling in the pathophysiology of asthma may be critical considering the fact that A_{2A}ARs are present on most of the inflammatory cells (including neutrophils, mast cells, macrophages, eosinophils, platelets, and T cells; Lappas et al. 2005; Thiel et al. 2003). Activation of A_{2A}AR on these cell types is almost universally inhibitory, and therefore could modulate inflammatory events in the airways. The anti-inflammatory effects of activation of A_{2A}AR on these cell types include inhibition of chemotaxis, elastase release, phagocytosis, oxidative stress, adherence of neutrophils to endothelial cells, mast cell degranulation, and the release of proinflammatory cytokines (Lappas et al. 2005; Nadeem et al. 2007).

There are a multitude of mechanisms by which an agonist, acting through A_{2A}ARs, could suppress inflammation in asthmatic airways. In human neutrophils, stimulation of A_{2A}AR reduces neutrophil adherence to the endothelium, inhibits formyl-Met–Leu–Phe (fMLP)-induced oxidative burst, and inhibits superoxide anion generation (Visser et al. 2000). In monocytes and macrophages, activation of A_{2A}ARs inhibits LPS-induced tumor necrosis factor (TNF)-α expression (Bshesh et al. 2002). A_{2A}AR-deficient allergic mice have increased oxidative stress in the lung as well as the airway smooth muscle after ragweed/ovalbumin allergen challenge as compared to their wild type. This oxidative stress is caused by activation of inducible nitric oxide synthase (iNOS) and nicotinamide adenine dinucleotide phosphate (NADPH) oxidase signaling due to A_{2A}AR deficiency (Nadeem et al. 2007, 2008). Moreover, genetic removal of the A_{2A}AR from ADA-deficient mice results in enhanced inflammation (composed largely of macrophages and neutrophils, mucin production in the bronchial airways, and angiogenesis) relative to that seen in the lungs of ADA-deficient mice with the $A_{2A}AR$, suggesting a protective role of this receptor in pulmonary inflammation when adenosine levels are high (Mohsenin et al. 2007). $A_{2A}AR$-mediated suppression of inflammation is mainly thought to be mediated by activation of protein kinase A and cyclic AMP response element-binding protein (Allen-Gipson et al. 2005; Bshesh et al. 2002), and inhibition of nuclear factor kappa B (NF-κB) signaling (Bshesh et al. 2002; Lukashev et al. 2004; Nadeem et al. 2007).

Strong anti-inflammatory properties for $A_{2A}AR$ have been shown in an inflammatory disease model using $A_{2A}AR$ gene-deficient mice (Lukashev et al. 2004; Nadeem et al. 2007). Consistent with this, in rat and mouse animal models of allergic asthma, the selective $A_{2A}AR$ agonist CGS 21680 (2-*p*-(2-carboxyethyl) phenethylamino-5′-*N*-ethylcarboxoamido adenosine) significantly reduced the number of inflammatory cells in the BAL fluid during allergen-induced airway inflammation (Bonneau et al. 2006; Fozard et al. 2002). However, in these rodent animal models of allergic asthma, this selective $A_{2A}AR$ agonist reduced airway inflammation but not BHR. Moreover, in an $A_{2A}AR$-deficient allergic mouse model, not only was airway inflammation enhanced but BHR was too (Nadeem et al. 2007). The discrepancy in $A_{2A}AR$-deficient allergic mice on airway reactivity versus the earlier report wherein airway reactivity was not reduced with the $A_{2A}AR$ agonist, CGS 21680, in allergic mice is not apparent and may be due to differences in strains of mice.

Recently, the effects of a new $A_{2A}AR$ agonist, GW328267X, in human asthmatics was reported (Luijk et al. 2008). In this study, treatment with GW328267X delivered as an inhalational treatment did not protect against the late asthmatic response (LAR), expressed as the decline in FEV_1 after allergen challenge, or the accompanying increase in airway inflammation (Luijk et al. 2008). However, in an earlier study, GW328267X partially inhibited the early asthmatic response (EAR) and LAR after nasal allergen challenge in patients with allergic rhinitis (Rimmer et al. 2007). There may be several possible explanations for the observed discrepancies between these two human studies. First, this $A_{2A}AR$ agonist is not entirely selective for the $A_{2A}AR$; it also exhibits some inhibitory effect on A_3AR (Luijk et al. 2008). It is possible that inhibition of the A_3AR by GW328267X blocked the anti-inflammatory effects of A_3AR activation by adenosine, since it is reported that activation of the A_3AR in humans produces anti-inflammatory effects, including inhibition of migration of human eosinophils and inhibition of oxidative burst, degranulation and release of inflammatory cytokines in human neutrophils, monocytes, and macrophages (Fishman and Bar-Yehuda 2003). Thus, the inhibitory effect of GW328267X on A_3ARs may have counteracted possible beneficial effects of $A_{2A}AR$ activation. Secondly, it is possible that the dose of the GW328267X (inhaled dose, 25 μg twice daily) was subtherapeutic. It was previously determined that higher doses of GW328267X caused cardiovascular side effects (reduction in blood pressure and increase in heart rate) following inhalational delivery (Luijk et al. 2008).

As mentioned above, even following inhalational delivery in small doses, the cardiovascular side effects of A_{2A}AR agonists may limit their clinical development. Moreover, tachyphylaxis and immune suppression may limit the clinical efficacy and safety of A_{2A}AR agonists as antiasthma drugs. For example, with the chronic administration of A_{2A}AR agonists, tachyphylaxis to the bronchodilator and anti-inflammatory effects may occur via the desensitization of G_S-coupled intracellular signaling pathways (Sullivan 2003). This potential effect of A_{2A}AR agonists was evident with the chronic administration of CGS-21680 over a two-week period wherein tachyphylaxis to the blood pressure lowering effect was reported and prevented the development of this A_{2A}AR agonist as an antihypertensive agent (Webb et al. 1993). Furthermore, because A_{2A}AR agonists act via G_s to stimulate adenylate cyclase, they may be associated with an increased risk of sudden death in asthmatics in a similar fashion to that of long-acting β2-agonists (LABAs) (Salpeter et al. 2004). Moreover, activation of A_{2A}ARs produces neovascularization (angiogenesis) (Cronstein 2006; Montesinos et al. 1997; Montesinos et al. 2006). Because of this effect of A_{2A}ARs on neovascularization/angiogenesis, an A_{2A}AR agonist, MRE-0094, is in Phase II clinical trials as a treatment for wound healing in diabetic foot ulcers (Aderis Pharmaceuticals). However, angiogenesis is a cardinal feature of airway remodeling of human asthma, and despite the report that activation of A_{2A}ARs promotes wound healing in bronchial epithelial cells (Allen-Gipson et al. 2005), the effect of A_{2A}AR agonists on angiogenesis may limit their development as anti-asthma drugs.

Further to these clinical considerations for the development of A_{2A}AR agonists as antiasthma drugs, others include their potential to produce antitumor effects and immune suppression (Ohta et al. 2006; Sullivan 2003). Because A_{2A}AR agonists block oxidative and nonoxidative activity of neutrophils, cause functional repression and/or apoptosis of lymphocytes, and inhibit the release of (interleukin) IL-12, which promotes bacterial clearance in infection, these agents may cause immune suppression and predispose to infection (Sullivan 2003). Moreover, in an adenosine-rich tumor microenvironment, activation of A_{2A}ARs produces inhibition of antitumor T cells (Ohta et al. 2006). In mice, genetic deletion of the A_{2A}AR or the use of A_{2A}AR antagonists improved inhibition of tumor growth, destruction of metastasis and prevention of neovascularization by antitumor T cells. Despite what should be advantageous effects from the activation of A_{2A}ARs (i.e., bronchodilation and anti-inflammatory effects), the potential side effects of hypotension and tachycardia may limit the use of A_{2A}AR agonists as acute rescue antiasthma drugs, and the potential side effects of tachyphylaxis and immune suppression as well as the angiogenesis and antitumor effects produced by A_{2A}AR agonists may limit the use of these molecules as chronic maintenance antiasthma drugs.

2.3 *A_{2B} Adenosine Receptors and Asthma*

Studies in animals both in vitro and in vivo and human cell lines in vitro have suggested that A_{2B}ARs may play an important role in mediating airway reactivity, inflammation, and remodeling in asthma. A_{2B}ARs are coupled via both G_s and G_q proteins to intracellular signaling pathways, which results in the release of cytokines and other mediators that are important in the pathophysiology of human asthma (Feoktistov et al. 1999; Zhong et al. 2004). In a human mast cell line (HMC-1), by coupling primarily to G_q, activation of A_{2B}ARs by adenosine induces the release of inflammatory cytokines such as IL-4, IL-8 and IL-13 which, in turn, can induce immunoglobulin E (IgE) synthesis by B lymphocytes (Feoktistov et al. 1999; Ryzhov et al. 2004). Moreover, in these HMC-1 cells, the selective A_{2B}AR antagonists IPDX (3-isobutyl-8-pyrrolidinoxanthine) and MRS 1754 ([*N*-(4-cyanophenyl)-2-[4-(2,3,6,7-tetrahydro-2,6-dioxo-1,3-dipropyl-1*H*-purin-8-yl)-phenoxy]acetamide]) inhibited activation of HMC-1 cells induced by NECA (5′-*N*-ethylcarboxoamido adenosine), a stable analog of adenosine (Feoktistov et al. 2001). Since HMC-1 cells are derived from a highly malignant, undifferentiated human mastocytoma cancer, the relevance of these findings in this human mast cell line to that in IgE immunologically sensitized human mast cells in allergic asthma is unknown. In the allergic response, antigens bind and crosslink IgE molecules bound to the functional high-affinity receptor for IgE, FcεRI, on mast cells to induce degranulation and the release of a broad spectrum of proinflammatory mediators (Nilsson et al. 1994; Xiang et al. 2001). HMC-1 cells do not express FcεRI (Nilsson et al. 1994). For this reason the reference to HMC-1 cells as human mast cells in allergic conditions, including human asthma, is misleading.

The presence of A_{2B}ARs on IgE immunologically sensitized human mast cells has not been reported. However, bone marrow-derived mast cells (BMMCs) from mice do express FcεRI (Hua et al. 2007b). Moreover, as opposed to the HMC-1 cells, A_{2B}ARs on murine BMMCs (Hua et al. 2007b), as well as human bronchial epithelial and smooth muscle cells and fibroblasts (Zhong et al. 2003, 2004, 2005), are coupled to G_s and adenylate cyclase, as compared to HMC-1 cells, where they are primarily coupled to G_q and phospholipase C (PLC) (Feoktistov and Biaggioni 1995). Furthermore, as opposed to the HMC-1 cell line, mast cell activation is enhanced in mice deficient for the A_{2B}AR (Hua et al. 2007b). The authors of this study suggested that in mice lacking this G_s-coupled receptor, BMMCs expressing FcεRI have reduced levels of cyclic AMP and an excess of intracellular calcium via store-operated calcium channels following antigen activation, thereby increasing their sensitivity to antigen-mediated degranulation. In addition, these A_{2B}AR-deficient mice display an increased sensitivity to IgE-mediated tachyphylaxis. In a recent study, genetic ablation of the A_{2B}AR had no effect on A_3AR-dependent potentiation of antigen-induced degranulation in mouse BMMCs, but abrogated A_{2B}AR-induced release of IL-13 and VEGF. The authors of this study suggest that in the mouse the A_3AR regulates mast cell degranulation, whereas the A_{2B}AR regulates mediator release, e.g., IL-13 and VEGF (Ryzhov et al. 2008a)

As mentioned above, as opposed to that seen in HMC-1 cells, by coupling to G_s and adenylate cyclase, adenosine activation of A_{2B}AR increases the release of inflammatory cytokines from human bronchial epithelial cells (HBECs) (Zhong et al. 2006), human bronchial smooth muscle cells (Zhong et al. 2004) and human fibroblasts (Zhong et al. 2005). In human bronchial smooth muscle cells, activation of A_{2B}ARs induces the release of IL-6 and the chemokine monocyte chemotactic protein 1 (MCP-1) (Zhong et al. 2004). In HBECs, activation of A_{2B}ARs induces the release of IL-19, which in turn induces the release of TNF-α from monocytes, which in turn upregulates the expression of A_{2B}ARs on HBECs (Zhong et al. 2006). In human lung fibroblasts, activation of A_{2B}ARs induces the release of IL-6, which, in the presence of hypoxia, synergistically induced the differentiation of lung fibroblasts into myofibroblasts (Zhong et al. 2005). These effects of activation of A_{2B}ARs by NECA, a stable analog of adenosine, in human bronchial smooth muscle cells (Zhong et al. 2004) and HBECs (Zhong et al. 2006) are blocked by selective antagonists of the A_{2B}AR. Furthermore, in a recent study, genetic ablation of A_{2B}AR abrogated NECA-induced increases in IL-6 release from mouse peritoneal macrophages ex vivo and dramatically reduced the ability of NECA to increase IL-6 plasma levels in vivo (Ryzhov et al. 2008b). Moreover, stimulation of the A_{2B}AR on isolated mouse BMMCs can directly promote the production and secretion of IL-13 and VEGF (Ryzhov et al. 2008a). Taken together, these studies indicate that stimulation of A_{2B}AR is coupled to the release of proinflammatory cytokines, and may play an important role in airway remodeling of asthma.

Although the importance of these in vitro studies to support the role of the A_{2B}AR in vivo in humans with asthma remains to be determined, studies in animal models of allergic asthma support the role of this AR in asthma. In ragweed-sensitized allergic mice, airway challenge with adenosine increased bronchoconstrictor responses and amplified the pulmonary inflammatory response to an allergen challenge (Fan and Mustafa 2002, 2006). This increase in bronchoconstrictor responses and airway inflammation to adenosine was blocked by theophylline and attenuated by a specific antagonist of the A_{2B}AR, which suggests, in part, a role for the A_{2B}AR (Fan and Mustafa 2002; Fan et al. 2003; Mustafa et al. 2007). Moreover, in this allergic mouse model of asthma, adenosine-induced increases in β-hexosaminidase activity (a mast cell marker) were decreased by pretreatment with theophylline (Fan and Mustafa 2006). Furthermore, in another study involving the use of this allergic mouse model of asthma from this same group, aerosolized NECA- and AMP-elicited concentration-dependent increases in Penh were significantly attenuated by CVT-6883, an A_{2B}AR antagonist (Mustafa et al. 2007). In this study, an allergen challenge-induced increase in LAR was inhibited by CVT-6883, and the increase in the number of inflammatory cells in BAL fluid was also inhibited by CVT-6883 or theophylline.

These findings, that the A_{2B}AR antagonist CVT-6883 reduces inflammation in the lung in an animal model of allergic asthma, were demonstrated in another animal model of lung inflammation with a phenotype similar to allergic asthma, albeit not an allergic asthma animal model, ADA-deficient mice (Sun et al. 2006). As previously stated, ADA-deficient mice develop pulmonary inflammation, fibrosis, and enlargement of alveolar airspaces. In CVT-6883-treated ADA-deficient mice there was less pulmonary inflammation, fibrosis, and alveolar airspace enlargement (Sun et al. 2006). Moreover, in ADA-deficient mice, A_{2B}AR antagonism with CVT-6883 significantly reduced elevations in proinflammatory cytokines and chemokines as well as mediators of fibrosis and airway destruction (Sun et al. 2006). These findings in these animal models suggest that A_{2B}AR signaling influences pathways critical for airway reactivity and inflammation.

As opposed to these reports suggesting that activation of A_{2B}ARs play an important role in bronchoconstriction and airway inflammation in allergic asthma, recent reports suggest that activation of A_{2B}ARs may produce bronchorelaxant and anti-inflammatory effects. In a recent study in guinea pigs, NECA evoked relaxing responses of isolated tracheal preparations precontracted with histamine in normal and sensitized animals, and this effect was reversed by the A_{2B}AR antagonist MRS 1706 (Breschi et al. 2007). Moreover, in vitro desensitization with 100 μM NECA markedly reduced the relaxing effect of NECA, raising the possibility that higher adenosine levels in the lung might desensitize this receptor to cause bronchorelaxation (Breschi et al. 2007). Furthermore, activation of A_{2B}ARs may produce anti-inflammatory effects. In A_{2B}AR knockout/reporter gene-knockin mice, there was low-grade baseline inflammation, augmented release of proinflammatory cytokines (including TNF-α and IL-6), as well as leukocyte adhesion to the vasculature (Yang et al. 2006). This finding that TNF-α levels are increased in A_{2B}AR knockout mice was confirmed by a more recent report by the same group (Yang et al. 2008). In a femoral artery injury model that resembles restenosis following angioplasty, A_{2B}AR knockout mice had higher levels of TNF-α, an upregulator of chemokine receptor 4 (CXCR4), and proliferation of vascular smooth muscle cells (Yang et al. 2008).

It is possible that the bronchorelaxant and anti-inflammatory effects of A_{2B}ARs described above may be due to an increase in intracellular cyclic AMP levels following activation of A_{2B}ARs. It is well known that an increase in intracellular cyclic AMP produces relaxation of bronchial smooth muscle and bronchodilation, suppresses inflammation, and prevents changes in endothelial cells that lead to an increase in endothelial permeability. Given these effects of intracellular cyclic AMP, it is unclear why an approach to the treatment of asthma would be to block these salutary effects of a receptor coupled via G_s to adenylate cyclase (i.e., the A_{2B}AR). It is now reported that the use of A_{2B}AR antagonists may increase endothelial permeability (Lennon et al. 1998). Moreover, in human airway epithelial cells via coupling to G_s and adenylate cyclase, A_{2B}ARs play an important role in control of the cystic fibrosis transmembrane conductance regulator (CFTR)-operated Cl^- channel (Clancy et al. 1999; Huang et al. 2001). Because of the importance of this Cl^- channel in airway hydration, the use of A_{2B}AR antagonists may induce a cystic fibrosis-like phenotype associated with an increased viscosity of mucus in humans, and may therefore limit their development as antiasthma drugs. Thus, although it appears that A_{2B}ARs may play an important role in airway remodeling of human asthma, because of their effect on the CFTR-operated Cl^- channel in human airway epithelial cells and airway hydration, the safety of A_{2B}AR antagonists in human asthmatics remains to be determined. Moreover, the efficacy of A_{2B}AR antagonists may depend on the relative contribution of this G_s-coupled receptor to adenylate cyclase and increases in intracellular cyclic AMP to produce bronchodilation and anti-inflammatory effects.

With respect to the therapeutic approach to the $A_{2B}AR$ as a target in human asthma, based on the reports that activation of $A_{2B}ARs$ in HBECs (Zhong et al. 2006), human bronchial smooth muscle cells (Zhong et al. 2004) and human lung fibroblasts (Zhong et al. 2005) induces the release of mediators important in the pathophysiology of airway remodeling of human asthma, as well as the efficacy of the $A_{2B}AR$ antagonist CVT 6883 in an acceptable animal model of allergic asthma (Mustafa et al. 2007), CVT 6883 has entered Phase I clinical trials as an antiasthma drug (CV Therapeutics, Inc.). Moreover, a combined A_{2B}/A_3 AR antagonist, QAF 805 (Novartis), has been tested in humans as an antiasthma drug. This mixed A_{2B}/A_3 AR antagonist failed to increase the provocative concentration $(PC)_{20}$ for AMP (concentration of AMP required to reduce the FEV_1 by 20%) versus placebo in 24 AMP-sensitive asthmatics in a placebo-controlled, double-blind, randomized, two-way crossover Phase Ib clinical trial (Pascoe et al. 2007). The results of the clinical trials with CVT 6883 and other selective $A_{2B}AR$ antagonists should more clearly define the role of $A_{2B}ARs$ in human asthma, and are eagerly awaited.

2.4 A_3 Adenosine Receptors and Asthma

The functional relevance of the A_3AR in the pathogenesis of asthma is a matter of debate, primarily due to species differences. In humans, A_3ARs have been identified on eosinophils, neutrophils, and monocytes; however, they have not been identified on mast cells (Gessi et al. 2008; Walker et al. 1997). In rats and mice, A_3ARs play an important role in adenosine-induced mast cell degranulation, bronchoconstriction, eosinophilia, and mucus production; however, they exhibit poor sensitivity to methylxanthines (Fan et al. 2003; Ramkumar et al. 1993; Tilley et al. 2003; Young et al. 2006). In an A_3AR knockout mouse model, a selective A_3AR agonist, IB–MECA (N^6-(3-iodobenzyl)-adenosine-5′-N-methylcarboxamide) delivered via a nebulizer had no effect on lung mast cell degranulation compared to wild-type mice (Zhong et al. 2003). In murine primary lung mast cells, activation of A_3ARs induced mast cell histamine release in association with increases in intracellular calcium mediated through G_i and phosphoinositide 3-kinase signaling pathways (Zhong et al. 2003). Furthermore, in ADA-deficient mice, A_3ARs appear to be important in endogenous adenosine-induced lung mast cell degranulation in the absence of antigen stimulation (Zhong et al. 2003). Moreover, in ADA-deficient mice, an increase in eosinophils and mucus production were reversed by a selective A_3AR antagonist, suggesting an important role for A_3ARs in mediating the lung eosinophilia and mucus hyperplasia in this animal model (Young et al. 2004). Further to these studies, in a murine model of allergic asthma it appears that A_3ARs play an important role in adenosine-induced bronchoconstriction (Fan et al. 2003). Finally, in allergen-sensitized guinea pigs, an A_3AR antagonist, MRS-1220, significantly inhibited 5′-AMP-induced migration of eosinophils and macrophages into the airways (Spruntulis and Broadley 2001). Taken together, these studies suggest that A_3ARs play an important role in adenosine-induced mast cell degranulation as well as eosinophilia, mucus hyperplasia, and bronchoconstriction in mice and guinea pigs, and would support the approach to asthma with an A_3AR antagonist.

In humans, expression of A_3ARs is elevated in lung biopsies of patients with asthma, and is mostly localized on eosinophils where activation by adenosine via this receptor inhibits chemotaxis (Walker et al. 1997). This initial report describing this anti-inflammatory effect of activation of A_3ARs on human eosinophils was reproduced, and the studies were expanded by the same group to show that the activation of A_3ARs produced a dose-dependent inhibition in the chemotaxis of human eosinophils to platelet-activating factor, RANTES, and leukotriene B4, and this effect was completely reversed by selective A_3AR antagonists (Knight et al. 1997). Moreover, following these reports, another group reported that A_3ARs on human eosinophils mediate inhibition of both degranulation and superoxide anion release, and that therapeutic concentrations of theophylline inhibit the human eosinophil partly by acting as an A_3AR agonist, thus contributing to the mechanism of the anti-inflammatory action of this drug in vivo (Ezeamuzie 2001; Ezeamuzie and Philips 1999). However, another group studied IB–MECA-induced effects on free radical generation in eosinophils of asthmatics and reported that stimulation of A_3ARs does not appear to be a prime mechanism for free radical generation by human peripheral blood eosinophils (Reeves et al. 2000). Taken together, these studies in humans suggest that an A_3AR agonist should be considered as a therapeutic option for the treatment of human asthma, as opposed to the studies in mice, rats, and guinea pigs that suggest that the A_3AR target to treat asthma should be approached with an A_3AR antagonist.

Based on the reports in animals that activation of A_3ARs produce mast cell degranulation, bronchoconstriction, eosinophilia and mucus hyperplasia, and that activation of A_{2B}ARs on human mast cells may play an important role in human asthma, a combined A_{2B}/A_3 AR antagonist QAF 805 (Novartis) is under development as an antiasthma drug (Press et al. 2005; Pascoe et al. 2007). However, as mentioned above, this mixed A_{2B}/A_3 AR antagonist has now entered human clinical trials and it failed to increase the PC_{20} for AMP versus placebo in 24 AMP-sensitive asthmatics in a placebo-controlled, double-blind, randomized, two-way crossover study Phase Ib clinical trial (Pascoe et al. 2007). With respect to the use of A_3AR agonists as antiasthma drugs, the use of this new class of drugs for this therapeutic indication may be limited by hypotension, tolerance/tachyphylaxis, and immune suppression (Gessi et al. 2008). Because of the anti-inflammatory and specifically the anti-TNF-α effects of the activation of A_3ARs on human monocytes, an A_3AR agonist, CF-101, is in Phase IIb clinical trials for the treatment of rheumatoid arthritis (Can-Fite Biopharma). It is reported that CF-101 has an acceptable safety and tolerability profile in humans (van Troostenburg et al. 2004). In this report, bronchospasm was not reported as a side effect of CF-101; however, the patients in this study were taking another anti-inflammatory immune suppressant, methotrexate, and CF-101 has not been tested in humans with asthma. Thus, the safety and efficacy of A_3AR agonists as antiasthma drugs are yet to be determined in humans.

3 Conclusions and Future Directions

It is now well accepted that adenosine is an important signaling molecule in the pathogenesis of human asthma, and all AR subtypes are important targets for anti-asthma drug development for humans. A number of AR molecules with good safety profiles and selectivity are now available for testing in humans, in order to determine the role of ARs in human asthma and the therapeutic approach to these AR targets that will produce safe and effective antiasthma therapeutics. Because many patients with asthma are either not controlled or are noncompliant with current antiasthma therapies, a shift in focus towards new mechanisms and novel targets (e.g., adenosine signaling and AR targets) is necessary to discover new classes of drugs that are safe and effective that will not only control the symptoms of asthma, but interrupt the disease of airway remodeling and the progressive loss of lung function, and thus improve not only the quality of life, but the outcome for the patient with asthma.

Acknowledgment We thank would like to Michael R. Blackburn, Ph.D. (Professor, Department of Biochemistry and Molecular Biology, The University of Texas–Houston Medical School, Houston, TX, USA) for his critical review of this chapter and his helpful comments.

References

Abebe W, Mustafa SJ (1998) A_1 adenosine receptor-mediated Ins(1,4,5)P3 generation in allergic rabbit airway smooth muscle. Am J Physiol 275:L990–L997

Ali S, Mustafa SJ, Metzger WJ (1992) Adenosine-induced bronchoconstriction in an allergic rabbit model: antagonism by theophylline aerosol. Agents Actions 37:165–167

Ali S, Mustafa SJ, Metzger WJ (1994a) Adenosine receptor-mediated bronchoconstriction and bronchial hyper-responsiveness in allergic rabbit model. Am J Physiol 266(Lung Cell Mol Physiol 10):L271–L277

Ali S, Mustafa SJ, Metzger WJ (1994b) Adenosine-induced bronchoconstriction and contraction of airway smooth muscle from allergic rabbits with late-phase airway obstruction: evidence for an inducible adenosine A_1 receptor. J Pharmacol Exp Ther 268:1328–1334

Allen-Gipson DS, Wong J, Spurzem JR, Sisson JH, Wyatt TA (2005) Adenosine A_{2A} receptors promote adenosine-stimulated wound healing in bronchial epithelial cells. Am J Physiol Lung Cell Mol Physiol 290:L849–L855

An SS, Bai TR, Bates JH, Black JL, Brown RH, Brusasco V, Chitano P, Deng L, Dowell M, Eidelman DH, Fabry B, Fairbank NJ, Ford LE, Fredberg JJ, Gerthoffer WT, Gilbert SH, Gosens R, Gunst SJ, Halayko AJ, Ingram RH, Irvin CG, James AL, Janssen LJ, King GG, Knight DA, Lauzon AM, Lakser OJ, Ludwig MS, Lutchen KR, Maksym GN, Martin JG, Mauad T, McParland BE, Mijailovich SM, Mitchell HW, Mitchell RW, Mitzner W, Murphy TM, Pare PD, Pellegrino R, Sanderson MJ, Schellenberg RR, Seow CY, Silveira PS, Smith PG, Solway J, Stephens NL, Sterk PJ, Stewart AG, Tang DD, Tepper RS, Tran T, Wang L (2007) Airway smooth muscle dynamics: a common pathway of airway obstruction in asthma. Eur Respir J 29:834–860

Apasov S, Koshiba M, Redegeld F, Sitkovsky MV (1995) Role of extracellular ATP and P1 and P2 classes of purinergic receptors in T-cell development and cytotoxic T lymphocyte effector functions. Immunol Rev 146:5–19

Avital A, Springer C, Bar-Yishay E, Godfrey S (1995) Adenosine, methacholine, and exercise challenges in children with asthma or paediatric chronic obstructive pulmonary disease. Thorax 50:511–516

Ball HA, Sandrasagra A, Tang L, Scott MV, Wild J, Nyce JW (2003) Clinical potential of respirable antisense oligonucleotides (RASONs) in asthma. Am J Pharmacogenomics 3:97–106

Baraldi PG, Cacciari B, Romagnoli R, Merighi S, Varani K, Borea PA, Spalluto G (2000) A3 adenosine receptor ligands: history and perspectives. Med Res Rev 20:103–128

Barrett RJ (1996) Realizing the potential of adenosine-receptor-based therapeutics. Proc West Pharmacol Soc 39:61–66

Beier KC, Kallinich T, Hamelmann E (2007) T-cell co-stimulatory molecules: novel targets for the treatment of allergic airway disease. Eur Respir J 30:383–390

Belda J, Casan P, Martínez C, Margarit G, Giner J, Homs R, Granel C, Sanchis J (2005) Bronchial plasma exudation after adenosine monophosphate or methacholine challenge. J Asthma 42:885–890

Berman AR, Togias AG, Skloot G, Proud D (1995) Allergen-induced hyperresponsiveness to bradykinin is more pronounced than that to methacholine. J Appl Physiol 78:1844–1852

Bertolet BD, Belardinelli L, Franco EA, Nichols WW, Kerensky RA, Hill JA (1996) Selective attenuation by N-0861 (N^6-endonorboran-2-yl-9-methyladenine) of cardiac A_1 adenosine receptor-mediated effects in humans. Circulation 93:1871–1876

Beukers MW, Meurs I, Ijzerman AP (2006) Structure–affinity relationships of adenosine A2B receptor ligands. Med Res Rev 26:667–698

Bjorck T, Gustafsson LE, Dahlen S-E (1992) Isolated bronchi from asthmatics are hyperresponsive to adenosine, which apparently acts indirectly by liberation of leukotrienes and histamine. Am Rev Respir Dis 145:1087–1091

Blackburn MR, Volmer JB, Thrasher JL, Zhong H, Crosby JR, Lee JJ, Kellems RE (2000) Metabolic consequences of adenosine deaminase deficiency in mice are associated with defects in alveogenesis, pulmonary inflammation, and airway obstruction. J Exp Med 192:159–170

Blackburn MR, Lee CG, Young HWJ, Zhu Z, Chunn JL, Kang MJ, Banerjee SK, Elias JA (2003) Adenosine mediates IL-13-induced inflammation and remodeling in the lung and interacts in an IL-13-adenosine amplification pathway. J Clin Invest 112:332–344

Bochenek G, Nizankowska E, Gielicz A, Szczeklik A (2008) Mast cell activation after adenosine inhalation challenge in patients with bronchial asthma. Allergy 63:140–141

Bond RA, Spina D, Parra S, Page CP (2007) Getting to the heart of asthma: can "beta blockers" be useful to treat asthma? Pharmacol Ther 115:360–374

Bonneau O, Wyss D, Ferretti S, Blaydon C, Stevenson CS, Trifilieff A (2006) Effect of adenosine A_{2A} receptor activation in murine models of respiratory disorders. Am J Physiol Lung Cell Mol Physiol 290:L1036–L1043

Breschi MC, Blandizzi C, Fogli S, Martinelli C, Adinolfi B, Calderone V, Camici M, Martinotti E, Nieri P (2007) In vivo adenosine A(2B) receptor desensitization in guinea-pig airway smooth muscle: implications for asthma. Eur J Pharmacol 575:149–157

Brown RA, Clarke GW, Ledbetter CL, Hurle MJ, Denyer JC, Simcock DE, Coote JE, Savage TJ, Murdoch RD, Page CP, Spina D, O'Connor BJ (2008) Elevated expression of adenosine A_1 receptor in bronchial biopsy specimens from asthmatic subjects. Eur Respir J 31:311–319

Bshesh K, Zhao B, Spight D, Biaggioni I, Feokistov I, Denenberg A, Wong HR, Shanley TP (2002) The A_{2A} receptor mediates an endogenous regulatory pathway of cytokine expression in THP-1 cells. J Leukoc Biol 72:1027–1036

Bucchioni E, Csoma Z, Allegra L, Chung KF, Barnes PJ, Kharitonov SA (2004) Adenosine 5′-monophosphate increases levels of leukotrienes in breath condensate in asthma. Respir Med 98:651–655

Busse WW, Lemanske RF Jr (2001) Asthma. N Engl J Med 344:350–362

Chuaychoo B, Lee MG, Kollarik M, Pullmann R Jr, Undem BJ (2006) Evidence for both adenosine A_1 and A_{2A} receptors activating single vagal sensory C-fibres in guinea pig lungs. J Physiol 575:481–490

Chunn JL, Young HW, Banerjee SK, Colasurdo GN, Blackburn MR (2001) Adenosine-dependent airway inflammation and hyperresponsiveness in partially adenosine deaminase-deficient mice. J Immunol 167:4676–4685
Clancy JP, Ruiz FE, Sorscher EJ (1999) Adenosine and its nucleotides activate wild-type and R117H CFTR through an A2B receptor-coupled pathway. Am J Physiol 276(2 Pt 1): C361–C369
Clark AN, Youkey R, Liu X, Jia L, Blatt R, Day Y-J, Sullivan GW, Linden J, Tucker, AL (2007) A_1 adenosine receptor activation promotes angiogenesis and release of VEGF from monocytes. Circ Res 101:1–9
Cockcroft DW, Davis BE (2006) Mechanisms of airway hyperresponsiveness. J Allergy Clin Immunol 118:551–559
Cohn L, Elias JA, Chupp GL (2004) Asthma: mechanisms of disease persistence and progression. Annu Rev Immunol 22:789–815
Crimi N, Palermo F, Oliveri R, Maccarrone C, Palermo B, Vancheri C, Polosa R, Mistretta A (1988) Enhancing effect of dipyridamole inhalation on adenosine-induced bronchospasm in asthmatic patients. Allergy 43:179–183
Crimi N, Palermo F, Oliveri R, Palermo B, Polosa R, Mistretta A (1992) Protection of nedocromil sodium on bronchoconstriction induced by inhaled neurokinin A (NKA) in asthmatic patients. Clin Exp Allergy 22:75–81
Cronstein BN (2006) Adenosine receptors and wound healing, revised. Sci World J 6:984–991
Cronstein BN, Kramer SB, Weissmann G, Hirschhorn R (1983) Adenosine: a physiological modulator of superoxide anion generation by human neutrophils. J Exp Med 158:1160–1177
Cronstein BN, Daguma L, Nichols D, Hutchison AJ, Williams M (1990) The adenosine/neutrophil paradox resolved: human neutrophils possess both A_1 and A_2 receptors that promote chemotaxis and inhibit O_2 generation, respectively. J Clin Invest 85:1150–1157
Cronstein BN, Levin RI, Philips M, Hirschhorn R, Abramson SB, Weissmann G (1992) Neutrophil adherence to endothelium is enhanced via adenosine A_1 receptors and inhibited via adenosine A_2 receptors. J Immunol 148:2201–2206
Csoma Z, Huszár E, Vizi E, Vass G, Szabó Z, Herjavecz I, Kollai M, Horváth I (2005) Adenosine level in exhaled breath increases during exercise-induced bronchoconstriction. Eur Respir J 25:873–878
Cushley MJ, Tattersfield AE, Holgate ST (1983) Inhaled adenosine and guanosine on airway resistance in normal and asthmatic subjects. Br J Clin Pharmacol 15:161–165
Dahlen S.-E, Hansson G, Hedqvist P, Bjorck T, Granstrom E, Dahlen B (1983) Allergen challenge of lung tissue from asthmatics elicits bronchial contraction that correlates with the release of leukotrienes C4, D4, and E4. Proc Natl Acad Sci USA 80:1712–1716
de Meer G, Heederik D, Postma DS (2002) Bronchial responsiveness to adenosine 5′-monophosphate (AMP) and methacholine differ in their relationship with airway allergy and baseline FEV1. Am J Respir Crit Care Med 165:327–331
Deussen A (2000) Metabolic flux rates of adenosine in the heart. Naunyn–Schmiedeberg's Arch Pharmacol 362:351–363
Dittrich HC, Gupta DK, Hack TC, Dowling T, Callahan J, Thomson S (2007) The effect of KW-3902, an adenosine A_1 receptor antagonist, on renal function and renal plasma flow in ambulatory patients with heart failure and renal impairment. J Card Failure 13:609–617
Doggrell SA (2005) BG-9928 Biogen Idec. Curr Opin Investig Drugs 6: 962–968
Driver AG, Kukoly CA, Metzger WJ, Mustafa SJ (1991) Bronchial challenge with adenosine causes the release of serum neutrophil chemotactic factor in asthma. Am Rev Respir Dis 143:1002–1007
Driver AG, Kukoly CA, Ali S, Mustafa SJ (1993) Adenosine in bronchoalveolar lavage fluid in asthma. Am Rev Respir Dis 148:91–97
el-Hashim A, D'Agostino B, Matera MG, Page CP (1996) Characterization of adenosine receptors involved in adenosine-induced bronchoconstriction in allergic rabbits. Br J Pharmacol 119:1262–1268

Eltzschig HK, Thompson LF, Karhausen J, Cotta RJ, Ibla JC, Robson SC, Colgan SP (2004) Endogenous adenosine produced during hypoxia attenuates neutrophil accumulation: coordination by extracellular nucleotide metabolism. Blood 104:3986–3992

Ethier MF, Madison JM (2006) Adenosine A_1 receptors mediate mobilization of calcium in human bronchial smooth muscle cells. Am J Respir Cell Mol Biol 35:496–502

Ezeamuzie CI (2001) Involvement of A(3) receptors in the potentiation by adenosine of the inhibitory effect of theophylline on human eosinophil degranulation: possible novel mechanism of the anti-inflammatory action of theophylline. Biochem Pharmacol 61:1551–1559

Ezeamuzie CI, Philips E (1999) Adenosine A_3 receptors on human eosinophils mediate inhibition of degranulation and superoxide anion release. Br J Pharmacol 127:188–194

Fan M, Mustafa SJ (2002) Adenosine-mediated bronchoconstriction and lung inflammation in an allergic mouse model. Pul Pharmacol Ther 15:147–155

Fan M, Mustafa SJ (2006) Role of adenosine in airway inflammation in an allergic mouse model of asthma. Int Immunopharmacol 6:36–45

Fan M, Qin W, Mustafa SJ (2003) Characterization of adenosine receptor(s) involved in adenosine-induced bronchoconstriction in an allergic mouse model. Am J Physiol Lung Cell Mol Physiol 284:L1012–L1019

Feoktistov I, Biaggioni I (1995) Adenosine A_{2b} receptors evoke interleukin-8 secretion in human mast cells. An enprofylline-sensitive mechanism with implications for asthma. J Clin Invest 96:1979–1986

Feoktistov I, Goldstein AE, Biaggioni I (1999) Role of p38 mitogen-activated protein kinase and extracellular signal-regulated protein kinase in adenosine A_{2B} receptor-mediated interleukin-8 production in human mast cells. Mol Pharmacol 55:726–734

Feoktistov I, Garland EM, Goldstein AE, Zeng D, Belardinelli L, Wells JN, Biaggioni I (2001) Inhibition of human mast cell activation with the novel selective adenosine A(2B) receptor antagonist 3-isobutyl-8-pyrrolidinoxanthine (IPDX)(2). Biochem Pharmacol 62:1163–1173

Fishman P, Bar-Yehuda S (2003) Pharmacology and therapeutic applications of A3 receptor subtype. Curr Top Med Chem 3:463–469

Fozard JR, Ellis KM, Villela Dantas MF, Tigani B, Mazzoni L (2002) Effects of CGS 21680, a selective adenosine A_{2A} receptor agonist, on allergic airways inflammation in the rat. Eur J Pharmacol 438:183–188

Fredholm BB (2007) Adenosine, an endogenous distress signal, modulates tissue damage and repair. Cell Death Diff 14:1315–1323

Fredholm BB, IJzerman AP, Jacobson KA, Klotz KN, Linden J (2001) International Union of Pharmacology. XXV. Nomenclature and classification of adenosine receptors. Pharmacol Rev 53:527–552

Gao Z-G, Jacobson KA (2007) Emerging adenosine receptor agonists. Expert Opin Emerg Drugs 12:479–492

Gaspardone A, Crea F, Iamele M, Tomai F, Versaci F, Pellegrino A, Chiariello L, Gioffre PA (1993) Bamiphylline improves exercise-induced myocardial ischemia through a novel mechanism of action. Circulation 88:502–508

Gessi S, Merighi S, Varani K, Leung E, Lennan SM, Borea PA (2008) The A_3 adenosine receptor: an enigmatic player in cell biology. Pharmacol Ther 117:123–140

Gil FR, Lauzon AM (2007) Smooth muscle molecular mechanics in airway hyperresponsiveness and asthma. Can J Physiol Pharmacol 85:133–140

Givertz MM, Massie BM, Fields TK, Pearson LL, Dittrich HC (2007) The effect of KW-3902, an adenosine A_1-receptor antagonist, on diuresis and renal function in patients with acute decompensated heart failure and renal impairment or diuretic resistance. J Am Coll Cardiol 50:1551–1560

Gottlieb SS, Brater C, Thomas I, Havranek E, Bourge R, Goldman S, Dyer F, Gomez M, Bennett D, Ticho B, Beckman E, Abraham WT (2002) BG9719 (CVT-124), an A_1 adenosine receptor antagonist, protects against the decline in renal function observed with diuretic therapy. Circulation 105:1348–1353

Greenberg B, Ignatius T, Banish D, Goldman S, Havranek E, Massie BM, Zhu Y, Ticho B, Abraham WT (2007) Effects of multiple oral doses of an A_1 adenosine receptor antagonist, BG 9928, in patients with heart failure. J Am Coll Cardiol 50:600–606

Gunther GR, Herring MB (1991) Inhibition of neutrophil superoxide production by adenosine released from vascular endothelial cells. Ann Vasc Surg 5:325–330

Hammad H, Lambrecht BN (2007) Lung dendritic cell migration. Adv Immunol 93:265–278

Hannon JP, Tigani B, Wolber C, Williams I, Mazzoni L, Howes C, Fozard JR (2002) Evidence for an atypical receptor mediating the augmented bronchoconstrictor response to adenosine induced by allergen challenge in actively sensitized Brown Norway rats. Br J Pharmacol 135:685–696

Hasko G, Cronstein BN (2004) Adenosine: an endogenous regulator of innate immunity. Trends Immunol 25:33–39

Hasko G, Kuhel DG, Nemeth ZH, Mabley JG, Stachlewitz RF, Virag L, Lohinai Z, Southan GJ, Salzman AL, Szabo C (2000) Inosine inhibits inflammatory cytokine production by a posttranscriptional mechanism and protects against endotoxin-induced shock. J Immunol 164:1013–1019

Hasko G, Sitkovsky MV, Szabo C (2004) Immunomodulatory and neuroprotective effects of inosine. Trends Pharmacol Sci 25:152–157

Hess S (2001) Recent advances in adenosine receptor antagonist research. Expert Opin Ther Pat 11:1533–1561

Holgate ST (2002) Adenosine provocation: a new test for allergic type airway inflammation. Am J Respir Crit Care Med 165:317–319

Holgate ST (2007) The epithelium takes centre stage in asthma and atopic dermatitis. Trends Immunol 28:248–251

Hong JL, Ho CY, Kwong K, Lee LY (1998) Activation of pulmonary C fibres by adenosine in anaesthetized rats: role of adenosine A_1 receptors. J Physiol 508:109–118

Hua X, Erikson CJ, Chason KD, Rosebrock CN, Deshpande DA, Penn RB, Tilley SL (2007a) Involvement of A_1 adenosine receptors and neural pathways in adenosine-induced bronchoconstriction in mice. Am J Physiol Lung Cell Mol Physiol 293:L25–L32

Hua X, Kovarova M, Chason KD, Nguyen M, Koller BH, Tilley SL (2007b) Enhanced mast cell activation in mice deficient in the A_{2b} adenosine receptor. J Exp Med 204:117–128

Huang P, Lazarowski ER, Tarran R, Milgram SL, Boucher RC, Stutts MJ (2001) Compartmentalized autocrine signaling to cystic fibrosis transmembrane conductance regulator at the apical membrane of airway epithelial cells. Proc Natl Acad Sci USA 98:14120–14125

Hyde RJ, Cass CE, Young JD, Baldwin SA (2001) The ENT family of eukaryote nucleoside and nucleobase transporters: recent advances in the investigation of structure/function relationships and the identification of novel isoforms. Mol Membr Biol 18:53–63

Ihre E, Gyllfors P, Gustafsson LE, Kumlin M, Dahlén B (2006) Early rise in exhaled nitric oxide and mast cell activation in repeated low-dose allergen challenge. Eur Respir J 27:1152–1159

Jacobsen EA, Ochkur SI, Lee NA, Lee JJ (2007) Eosinophils and asthma. Curr Allergy Asthma Rep 7:18–26

Jarjour NN, Kelly EAB (2002) Pathogenesis of asthma. Med Clin North Am 86:925–936

Johnson HG, McNee ML (1985) Adenosine-induced secretion in the canine trachea: modification by methylxanthines and adenosine derivatives. Br J Pharmacol 86:63–67

Kaczmarek E, Koziak K, Sevigny J, Siegel JB, Anrather J, Beaudoin AR, Bach FH, Robson SC (1996) Identification and characterization of CD39/vascular ATP diphosphohydrolase. J Biol Chem 271:33116–33122

Kalk P, Eggert B, Relle K, Godes M, Heiden S, Sharkovska Y, Fischer Y, Ziegler D, Bielenberg G-W, Hocher B (2007) The adenosine A_1 receptor antagonist SLV320 reduces myocardial fibrosis in rats with 5/6 nephrectomy without affecting blood pressure. Br J Pharmacol 151:1025–1032

Kallinich T, Beier KC, Wahn U, Stock P, Hamelmann E (2007) T-cell co-stimulatory molecules: their role in allergic immune reactions. Eur Respir J 29:1246–1255

Keir S, Boswell-Smith V, Spina D, Page C (2006) Mechanism of adenosine-induced airways obstruction in allergic guinea pigs. Br J Pharmacol 47:720–728

Ketchell RI, Jensen MW, Lumley P, Wright AM, Allenby MI, O'Connor BJ (2002) Rapid effect of inhaled fluticasone propionate on airway responsiveness to adenosine 5′-monophosphate in mild asthma. J Allergy Clin Immunol 110:603–606

Knight D, Zheng X, Rocchini C, Jacobson M, Bai T, Walker B (1997) Adenosine A_3 receptor stimulation inhibits migration of human eosinophils. J Leukoc Biol 62:465–468

Langley SJ, Allen DJ, Houghton C, Woodcock A (2005) Efficacy of EPI-2010 (an inhaled respirable anti-sense oligonucleotide: RASON) in moderate/severe persistent asthma. Am J Resp Crit Care Med 171:A360

Lappas CM, Sullivan GW, Linden J (2005) Adenosine A_{2A} agonists in development for the treatment of inflammation. Expert Opin Investig Drugs 14:797–806

Lee MG, Kollarik M, Chuaychoo B, Undem BJ (2004) Ionotropic and metabotropic receptor mediated airway sensory nerve activation. Pulm Pharmacol Ther 17:355–360

Lennon PF, Taylor CT, Stahl GL, Colgan SP (1998) Neutrophil-derived 5′-adenosine monophosphate promotes endothelial barrier function via CD73-mediated conversion to adenosine and endothelial A_{2B} receptor activation. J Exp Med 188:1433–1443

Liu AMF, Wong YH (2004) G_{16}-mediated activation of nuclear factor κB by the adenosine A1 receptor involves c-Src, protein kinase C, and ERK signaling. J Biol Chem 279:53196–53204

Livingston M, Heaney LG, Ennis M (2004) Adenosine, inflammation, and asthma: a review. Inflamm Res 53:171–178

Lloyd CM, Robinson DS (2007) Allergen-induced airway remodelling. Eur Respir J 29:1020–1032

Luijk B, van den Berge M, Kerstjens HA, Postma DS, Cass L, Sabin A, Lammers JW (2008) Effect of an inhaled adenosine A_{2A} agonist on the allergen-induced late asthmatic response. Allergy 63:75–80

Lukashev D, Ohta A, Apasov S, Chen JF, Sitkovsky M (2004) Cutting edge: physiologic attenuation of proinflammatory transcription by the Gs protein-coupled A_{2A} adenosine receptor in vivo. J Immunol 173:21–24

Ma B, Blackburn MR, Lee CG, Homer RJ, Liu W, Flavell RA, Boyden L, Lifton RP, Sun C-X, Young HW, Elias JA (2006) Adenosine metabolism and murine strain-specific IL-4-induced inflammation, emphysema, and fibrosis. J Clin Invest 116:1274–1283

Madara JL, Patapoff TW, Gillece-Castro B, Colgan SP, Parkos CA, Delp C, Mrsny RJ (1993) 5′-Adenosine monophosphate is the neutrophil-derived paracrine factor that elicits chloride secretion from T84 intestinal epithelial cell monolayers. J Clin Invest 91:2320–2325

Mann JS, Cushley MJ, Holgate ST (1985) Adenosine-induced bronchoconstriction in asthma. Role of parasympathetic stimulation and adrenergic inhibition. Am Rev Respir Dis 132:1–6

Mann JS, Holgate ST, Renwick AG, Cushley MJ (1986a) Airway effects of purine nucleosides and nucleotides and release with bronchial provocation in asthma. J Appl Physiol 61:1667–1676

Mann JS, Renwick AG, Holgate ST (1986b) Release of adenosine and its metabolites from activated human leucocytes. Clin Sci 70:461–468

Manrique HA, Gómez FP, Muñoz PA, Peña AM, Barberà JA, Roca J, Rodríguez-Roisin R (2008) Adenosine 5′-monophosphate in asthma: gas exchange and sputum cellular responses. Eur Respir J 31:1205–1212

Marcus AJ, Safier LB, Broekman MJ, Islam N, Fliessbach JH, Hajjar KA, Kaminski WE, Jendraschak E, Silverstein RL, von Schacky C (1995) Thrombosis and inflammation as multicellular processes: significance of cell–cell interactions. Thromb Haemost 74: 213–217

McNamara N, Gallup M, Khong A, Sucher A, Maltseva I, Fahy J, Basbaum C (2004) Adenosine up-regulation of the mucin gene, MUC2, in asthma. FASEB J 18:1770–1772

Mohsenin A, Mi T, Xia Y, Kellems RE, Chen JF, Blackburn MR (2007) Genetic removal of the A_{2A} adenosine receptor enhances pulmonary inflammation, mucin production, and angiogenesis in adenosine deaminase-deficient mice. Am J Physiol Lung Cell Mol Physiol 293:L753–L761

Montesinos MC, Gadangi P, Longaker M, Sung J, Levine J, Nilsen D, Reibman J, Li M, Jiang C-K, Hirschhorn R, Recht PA, Ostad E, Levin RI, Cronstein BN (1997) Wound healing is accelerated by agonists of adenosine A2 (Gs-linked) receptors. J Exp Med 186:1615–1620

Montesinos MC, Shaw JP, Yee H, Shamamian P, Cronstein BN (2006) Adenosine A(2A) receptor activation promotes wound neovascularization by stimulating angiogenesis and vasculogenesis. Am J Pathol 164:1887–1892

Moro S, Gao Z-G, Jacobson KA, Spalluto G (2006) Progress in the pursuit of therapeutic adenosine receptor antagonists. Med Res Rev 26:131–159

Mustafa SJ, Nadeem A, Fan M, Zhong H, Belardinelli L, Zeng D (2007) Effect of a specific and selective A(2B) adenosine receptor antagonist on adenosine agonist AMP and allergen-induced airway responsiveness and cellular influx in a mouse model of asthma. J Pharmacol Exp Ther 320:1246–1251

Nadeem A, Obiefuna PC, Wilson CN, Mustafa SJ (2006) Adenosine A_1 receptor antagonist versus montelukast on airway reactivity and inflammation. Eur J Pharmacol 551:116–124

Nadeem A, Fan M, Ansari HR, Ledent C, Jamal Mustafa S (2007) Enhanced airway reactivity and inflammation in A_{2A} adenosine receptor-deficient allergic mice. Am J Physiol Lung Cell Mol Physiol 292:L1335–L1344

Nadeem A, Ponnoth DS, Batchelor T, Dey RD, Ledent C, Mustafa SJ (2008) Decreased tracheal relaxation via NADPH oxidase activation in A_{2A} AR deficent allergic mice. Am J Respir Crit Care Med 177:A487

Nilsson G, Blom T, Kusche-Gullberg M, Kjellen L, Butterfield JH, Sundstrom C, Nilsson K, Hellman L (1994) Phenotypic characterization of the human mast cell line HMC-1. Scand J Immunol 39:489–498

Nyce JW, Metzger WJ (1997) DNA antisense therapy for asthma in an animal model. Nature 385:721–725

Obiefuna PC, Batra VK, Nadeem A, Borron P, Wilson CN, Mustafa SJ (2005) A novel A_1 adenosine receptor antagonist, L-97-1 [3-[2-(4-aminophenyl)-ethyl]-8-benzyl-7-{2-ethyl-(2-hydroxy-ethyl)-amino]-ethyl}-1-propyl-3,7-dihydro-purine-2,6-dione], reduces allergic responses to house dust mite in an allergic rabbit model of asthma. J Pharmacol Exp Ther 315:329–336

O'Connor BJ, Crowther SD, Costello JF, Morley J (1999) Selective airway responsiveness in asthma. Trends Pharmacol Sci 20:9–11

Ohta A, Gorelik E, Prasad SJ, Ronchese F, Lukashev D, Wong MK, Huang X, Caldwell S, Liu K, Smith P, Chen JF, Jackson EK, Apasov S, Abrams S, Sitkovsky M (2006) A_{2A} adenosine receptor protects tumors from antitumor T cells. Proc Natl Acad Sci USA 103:13132–13137

Oldenburg PJ, Mustafa SJ (2005) Involvement of mast cells in adenosine-mediated bronchoconstriction and inflammation in an allergic mouse model. J Pharmacol Exp Ther 313:319–324

Panther E, Idzko M, Herouy Y, Rheinen H, Gebicke-Haerter PJ, Mrowietz U, Dichmann S, Norgauer J (2001) Expression and function of adenosine receptors in human dendritic cells. FASEB J 15:1963–1970

Pascoe SJ, Knight H, Woessner R (2007) QAF805, an A_{2b}/A_3 adenosine receptor antagonist does not attenuate AMP challenge in subjects with asthma. Am J Respir Crit Care Med 175:A682

Pastor-Anglada M, Casado FJ, Valdes R, Mata J, Garcia-Manteiga J, Molina M (2001) Complex regulation of nucleoside transporter expression in epithelial and immune system cells. Mol Membr Biol 18:81–85

Phillips GD, Holgate ST (1989) The effect of oral terfenadine alone and in combination with flurbiprofen on the bronchoconstrictor response to inhaled adenosine 5′-monophosphate in nonatopic asthma. Am Rev Respir Dis 139:463–469

Polosa R (2002) Adenosine-receptor subtypes: their relevance to adenosine-mediated responses in asthma and chronic obstructed pulmonary disease. Eur Respir J 20:488–496

Polosa R, Phillips GD, Rajakulasingam K, Holgate ST (1991) The effect of inhaled ipratropium bromide alone and in combination with oral terfenadine on bronchoconstriction provoked by adenosine 5′-monophosphate and histamine in asthma. J Allergy Clin Immunol 87:939–947

Polosa R, Ng WH, Crimi N, Vancheri C, Holgate ST, Church MK, Mistretta A (1995) Release of mast-cell-derived mediators after endobronchial adenosine challenge in asthma. Am J Respir Crit Care Med 151:624–629

Polosa R, Renaud L, Cacciola R, Prosperini G, Crimi N, Djukanovic R (1998) Sputum eosinophilia is more closely associated with airway responsiveness to bradykinin than methacholine in asthma. Eur Respir J 12:551–556

Press NJ, Taylor RJ, Fullerton JD, Tranter P, McCarthy C, Keller TH, Brown L, Cheung R, Christie J, Haberthuer S, Hatto JD, Keenan M, Mercer MK, Press NE, Sahri H, Tuffnell AR, Tweed M, Fozard JR (2005) A new orally bioavailable dual adenosine A_{2B}/A_3 receptor antagonist with therapeutic potential. Bioorg Med Chem Lett 15:3081–3085

Ramkumar V, Stiles GL, Beaven MA, Ali H (1993) The A_3 adenosine receptor is the unique adenosine receptor, which facilitates release of allergic mediators in mast cells. J Biol Chem 268:16887–16890

Reeves JJ, Harris CA, Hayes BP, Butchers PR, Sheehan MJ (2000) Studies on the effects of adenosine A_3 receptor stimulation on human eosinophils isolated from non-asthmatic or asthmatic donors. Inflamm Res 49:666–672

Resta R, Yamashita Y, Thompson LF (1998) Ecto-enzyme and signaling functions of lymphocyte CD73. Immunol Rev 161:95–109

Reynolds CJ, Togias A, Proud D (2002) Airways hyper-responsiveness to bradykinin and methacholine: effects of inhaled fluticasone. Clin Exp Allergy 32:1174–1179

Reynolds SM, Docherty R, Robbins J, Spina D, Page CP (2008) Adenosine induces a cholinergic tracheal reflex contraction in guinea pigs in vivo via an adenosine A_1 receptor-dependent mechanism. J Appl Physiol 105:187–196

Rimmer J, Peake HL, Santos CM, Lean M, Bardin P, Robson R, Haumann B, Loehrer F, Handel ML (2007) Targeting adenosine receptors in the treatment of allergic rhinitis: a randomized, double-blind, placebo-controlled study. Clin Exp Allergy 37:8–14

Roisman GL, Lacronique JG, Desmazes DN, Carre C, Le-Cae A, Dusser DJ (1996) Airway responsiveness to bradykinin is related to eosinophilic inflammation in asthma. Am J Respir Crit Care Med 153:381–390

Roman J, Rivera HN, Roser-Page S, Sitaraman SV, Ritzenthaler JD (2006) Adenosine induces fibronectin expression in lung epithelial cells: implications for airway remodeling. Am J Physiol Lung Cell Mol Physiol 290:L317–L325

Rorke S, Holgate ST (2002) Targeting adenosine receptors: novel therapeutic targets in asthma and chronic obstructive pulmonary disease. Am J Respir Med 1:99–105

Rorke S, Jennison S, Jeffs JA, Sampson AP, Holgate ST (2002) Role of cysteinyl leukotrienes in adenosine 5′-monophosphate induced bronchoconstriction in asthma. Thorax 57:323–327

Rosenberg HF, Phipps S, Foster PS (2007) Eosinophil trafficking in allergy and asthma. J Allergy Clin Immunol 119:1303–1310

Rounds S, Hsieh L, Agarwal KC (1994) Effects of endotoxin injury on endothelial cell adenosine metabolism. J Lab Clin Med 123:309–317

Rutgers SR, Koeter GH, Van Der Mark TW, Postma DS (1999) Protective effect of oral terfenadine and not inhaled ipratropium on adenosine 5′-monophosphate-induced bronchoconstriction in patients with COPD. Clin Exp Allergy 29:1287–1292

Ryzhov S, Goldstein AE, Matafonov A, Zeng D, Biaggioni I, Feoktistov I (2004) Adenosine-activated mast cells induce IgE synthesis by B lymphocytes: an A_{2B}-mediated process involving Th2 cytokines IL-4 and IL-13 with implications for asthma. J Immunol 172: 7726–7733

Ryzhov S, Zaynagetdinov R, Goldstein AE, Novitskiy SV, Dikov MM, Blackburn MR, Biaggioni I, Feoktistov I (2008a) Effect of A_{2B} adenosine receptor gene ablation on proinflammatory adenosine signaling in mast cells. J Immunol 180:7212–7220

Ryzhov S, Zaynagetdinov R, Goldstein AE, Novitskiy SV, Blackburn MR, Biaggioni I, Feoktistov I (2008b) Effect of A_{2B} adenosine receptor gene ablation on adenosine-dependent regulation of proinflammatory cytokines. J Pharmacol Exp Ther 324:694–700

Salmon JE and Cronstein BN (1990) Fc gamma receptor-mediated functions in neutrophils are modulated by adenosine receptor occupancy. A1 receptors are stimulatory and A2 receptors are inhibitory. J Immunol 145: 2235–2240

Salmon JE, Brogle N, Brownlie C, Edberg JC, Kimberly RP, Chen BX, Erlanger BF (1993) Human mononuclear phagocytes express adenosine A_1 receptors. A novel mechanism for differential regulation of Fc gamma receptor function. J Immunol 151:2775–2785

Salpeter SR, Ormiston TM, Salpeter EE (2004) Cardiovascular effects of beta-agonists in patients with asthma and COPD: a metanalysis. Chest 125:2309–2321

Slats AM, Janssen K, van SA, van der Plas DT, Schot R, van den Aardweg JG, de Jongste JC, Hiemstra PS, Mauad T, Rabe KF, Sterk PJ (2007) Bronchial inflammation and airway responses to deep inspiration in asthma and chronic obstructive pulmonary disease. Am J Respir Crit Care Med 176:121–128

Sont JK, Booms P, Bel EH, Vandenbroucke JP, Sterk PJ (1993) The determinants of airway hyperresponsiveness to hypertonic saline in atopic asthma in vivo. Relationship with sub-populations of peripheral blood leucocytes. Clin Exp Allergy 23:678–688

Sperlagh B, Hasko G, Nemeth Z, Vizi ES (1998) ATP released by LPS increases nitric oxide production in raw 264.7 macrophage cell line via P2Z/P2X7 receptors. Neurochem Int 33: 209–215

Spicuzza L, Bonfiglio C, Polosa R (2003) Research applications and implications of adenosine in diseased airways. Trends Pharmacol Sci 24:409–413

Spicuzza L, Di Maria G, Polosa R (2006) Adenosine in the airways: implications and applications. Eur J Pharmacol 533:77–88

Spina D, Page CP (1996) Airway sensory nerves in asthma: targets for therapy? Pulm Pharmacol 9:1–18

Spina D, Page CP (2002) Pharmacology of airway irritability. Curr Opin Pharmacol 2:264–272

Spruntulis LM, Broadley KJ (2001) A_3 receptors mediate rapid inflammatory cell influx into the lungs of sensitized guinea-pigs. Clin Exp Allergy 31:943–951

Sullivan GW (2003) Adenosine A_{2A} receptor agonists as anti-inflammatory agents. Curr Opin Investig Drugs 4:1313–1319

Sun CX, Young HW, Molina JG, Volmer JB, Schnermann J, Blackburn MR (2005) A protective role for the A1 adenosine receptor in adenosine-dependent pulmonary injury. J Clin Invest 115:35–43

Sun CX, Zhong H, Mohsenin A, Morschl E, Chunn JL, Molina JG, Belardinelli L, Zeng D, Blackburn MR (2006) Role of A_{2B} adenosine receptor signaling in adenosine-dependent pulmonary inflammation and injury. J Clin Invest 116:2173–2182

Thiel M, Caldwell CC, Sitkovsky MV (2003) The critical role of adenosine A_{2A} receptors in downregulation of inflammation and immunity in the pathogenesis of infectious diseases. Microb Infect 5:515–526

Thompson LF, Eltzschig HK, Ibla JC, Van De Wiele CJ, Resta R, Morote-Garcia JC, Colgan SP (2004) Crucial role for ecto-5′-nucleotidase (CD73) in vascular leakage during hypoxia. J Exp Med 200:1395–1405

Tilley SL, Tsai M, Williams CM, Wang ZS, Erikson CJ, Galli SJ, Koller BH (2003) Identification of A_3 receptor- and mast cell-dependent and -independent components of adenosine-mediated airway responsiveness in mice. J Immunol 171:331–337

van den Berge M, Kerstjens HA, Meijer RJ, de Reus DM, Koeter GH, Kauffman HF, Postma DS (2001) Corticosteroid-induced improvement in the PC20 of adenosine monophosphate is more closely associated with reduction in airway inflammation than improvement in the PC20 of methacholine. Am J Respir Crit Care Med 164:1127–1132

van den Berge M, Kerstjens HA, de Reus DM, Koeter GH, Kauffman HF, Postma DS (2004) Provocation with adenosine 5′-monophosphate, but not methacholine, induces sputum eosinophilia. Clin Exp Allergy 34:71–76

Van Schoor J, Joos GF, Pauwels RA (2000) Indirect bronchial hyperresponsiveness in asthma: mechanisms, pharmacology and implications for clinical research. Eur Respir J 16:514–533

Van Schoor J, Joos GF, Pauwels RA (2002) Effect of inhaled fluticasone on bronchial responsiveness to neurokinin A in asthma. Eur Respir J 19:997–1002

van Troostenburg A-R, Clark EV, Carey WDH, Warrington SJ, Kerns WD, Cohn I, Silverman MH, Bar-Yehuda S, Fong K-LL, Fishman P (2004) Tolerability, pharmacokinetics and concentration-dependent hemodynamic effects of oral CF 101, an A_3 adenosine receptor agonist, in healthy young men. Internat J Clin Pharmacol Ther 42:534–542

Visser SS, Theron AJ, Ramafi G, Ker JA, Anderson R (2000) Apparent involvement of the A_{2A} subtype adenosine receptor in the anti-inflammatory interactions of CGS 21680, cyclopentyladenosine, and IB-MECA with human neutrophils. Biochem Pharmacol 60:993–999

Vizi E, Huzzar E, Csoma Z, Boszormenyi-Nagy G, Barat E, Horvath I, Herjavecz I, Kollai M (2002) Plasma adenosine concentration increases during exercise: a possible contributing factor in exercise-induced bronchoconstriction in asthma. J Allergy Clin Immunol 109:446–448

Walker BAM, Jacobson MA, Knight DA, Salvatore CA, Weir T, Zhou D, Bai TR (1997) Adenosine A_3 receptor expression and function in eosinophils. Am J Respir Cell Mol Biol 16:531–537

Webb RL, Sills MA, Chovan JP, Peppard JV, Francis JE (1993) Development of tolerance to the antihypertensive effects of highly selective adenosine A_{2a} agonists upon chronic administration. J Pharmacol Exp Ther 267:287–295

Wilson CN, Batra VK (2002) Lipopolysaccharide binds to and activates A(1) adenosine receptors on human pulmonary artery endothelial cells. J Endotoxin Res 8:263–271

Wilson CN (2008) Adenosine receptors and asthma in humans. Br J Pharmacol 155:475–486

Xiang Z, Ahmed AA, Moller C, Nakayama K, Hatakeyama S, Nilsson G (2001) Essential role of the prosurvival bcl-2 homologue A1 in the mast cell survival after allergic activation. J Exp Med 194:1561–1569

Yang D, Zhang Y, Nguyen HG, Koupenova M, Chauhan AK, Makitalo M, Jones MR, Hilaire CS, Seldin DC, Toselli P, Lamperti E, Schreiber BM, Gavras H, Wagner DD, Ravid K (2006) The A(2B) adenosine receptor protects against inflammation and excessive vascular adhesion. J Clin Invest 116:1913–1923

Yang D, Koupenova M, McCrann DJ, Kopeikina KJ, Kagan HM, Schreiber BM, Ravid K (2008) The A_{2b} adenosine receptor protects against vascular injury. Proc Natl Acad Sci USA 105: 792–796

Young HW, Molina JG, Dimina D, Zhong H, Jacobson M, Chan LN, Chan TS, Lee JJ, Blackburn MR (2004) A_3 adenosine receptor signaling contributes to airway inflammation and mucus production in adenosine deaminase-deficient mice. J Immunol 173:1380–1389

Young HW, Sun CX, Evans CM, Dickey BF, Blackburn MR (2006) A_3 adenosine receptor signaling contributes to airway mucin secretion after allergen challenge. Am J Respir Cell Mol Biol 35:549–558

Yuzlenko O, Kiec-Kononowicz K (2006) Potent adenosine A_1 and A_{2A} receptor antagonists: recent developments. Curr Med Chem 13:3609–3625

Zhong H, Shlykov SG, Molina JG, Sanborn BM, Jacobson MA, Tilley SL, Blackburn MR (2003) Activation of murine lung mast cells by the adenosine A_3 receptor. J Immunol 171:338–345

Zhong H, Belardinelli L, Maa T, Feoktistov I, Biaggioni I, Zeng D (2004) A_{2B} adenosine receptors increase cytokine release by bronchial smooth muscle cells. Am J Respir Cell Mol Biol 30: 118–125

Zhong H, Belardinelli L, Maa T, Zeng D (2005) Synergy between A_{2B} adenosine receptors and hypoxia in activating human lung fibroblasts. Am J Respir Cell Mol Biol 32:2–8

Zhong H, Wu Y, Belardinelli L, Zeng D (2006) A2B adenosine receptors induce IL-19 from bronchial epithelial cells and results in TNF-alpha increase. Am J Respir Cell Mol Biol 35: 587–592

Zimmermann H (1999) Nucleotides and cd39: principal modulatory players in hemostasis and thrombosis. Nat Med 5:987–988

Adenosine Receptors, Cystic Fibrosis, and Airway Hydration

Gulnur Com and J.P. Clancy

Contents

Abstract Adenosine (Ado) regulates diverse cellular functions in the lung through its local production, release, metabolism, and subsequent stimulation of G-protein-coupled P1 purinergic receptors. The A_{2B} adenosine receptor ($A_{2B}AR$) is the predominant P1 purinergic receptor isoform expressed in surface airway epithelia, and Ado is an important regulator of airway surface liquid (ASL) volume through its activation of the cystic fibrosis transmembrane conductance regulator (CFTR). Through a delicate balance between sodium (Na^+) absorption and chloride (Cl^-) secretion, the ASL volume is optimized to promote ciliary activity and mucociliary clearance, effectively removing inhaled particulates. When CFTR is dysfunctional, the Ado/$A_{2B}AR$ regulatory system fails to optimize the ASL volume, leading to its depletion and interruption of mucociliary clearance. In cystic fibrosis (CF), loss of CFTR function and resultant mucus stasis leaves the lower airways susceptible to mucus obstruction, chronic bacterial infection, relentless inflammation, and eventually panbronchiectasis. Adenosine triphosphate (ATP) also regulates transepithelial Cl^- conductance, but through a separate system that relies on stimulation of $P2Y_2$ purinergic receptors, mobilization of intracellular calcium, and activation of calcium-activated chloride channels (CaCCs). These pathways remain functional in CF, and may serve a protective role in the disease. In this chapter, we will review

J.P. Clancy (✉)
Department of Pediatrics, University of Alabama, 620 ACC, 1600 7th Ave South, Birmingham, AL 35233, UK
jpclancy@peds.uab.edu

C.N. Wilson and S.J. Mustafa (eds.), *Adenosine Receptors in Health and Disease*, Handbook of Experimental Pharmacology 193, DOI 10.1007/978-3-540-89615-9_12,

our current understanding of how Ado and related nucleotides regulate CFTR and Cl^- conductance in the human airway, including the regulation of additional intracellular and extracellular signaling pathways that provide important links between ion transport and inflammation relevant to the disease.

Keywords Adenosine receptors · Cystic fibrosis · Adenosine · Airway hydration · P1 purinergic receptors

Abbreviations

Ach	Acetylcholine
ADA	Adenosine deaminase
Ado	Adenosine
AKAP	A kinase anchoring protein
AMP	Adenosine monophosphate
ASL	Airway surface liquid
ATP	Adenosine triphosphate
A_1AR	A_1 adenosine receptor
$A_{2A}AR$	A_2 adenosine receptor
$A_{2B}AR$	A_{2B} adenosine receptor
A_3AR	A_3 adenosine receptor
CaCC	Calcium-activated chloride channel
Ca^{++}	Calcium
cAMP	Cyclic adenosine monophosphate
CF	Cystic fibrosis
CFF	Cystic Fibrosis Foundation
CFTR	Cystic fibrosis transmembrane conductance regulator
Cl^-	Chloride
COPD	Chronic obstructive pulmonary disease
EBP50	Ezrin-binding protein 50
ENaC	Epithelial sodium channel
EP4	E4 prostaglandin receptor
FEV_1	Forced expiratory volume in one second
GI	Gastrointestinal
GPCR	G-protein-coupled receptor
GSH	Glutathione
HBEC	Human bronchial epithelial cell
HCO_3	Bicarbonate
ICAM-1	Intracellular adhesion molecule-1
IL-4	Interlukin-4
IL-5	Interlukin-5
IL-13	Interlukin-13
Iso	Isoproterenol

K^+	Potassium
MMP-9	Matrix metalloproteinase-9
MMP-12	Matrix metalloproteinase-12
MCC	Mucociliary clearance
MCP	Monocyte chemoattractant protein
Na^+	Sodium
NBD-1	Nucleotide-binding domain 1
NBD-2	Nucleotide-binding domain 2
NPD	Nasal potential difference
ORCC	Outwardly rectified chloride channel
PDZ	Postsynaptic density protein 95, disk-large tumor suppressor protein, *Zonula occludens* 1
PGE_2	Prostaglandin E_2
PKA	Protein kinase A
PS	Pancreatic sufficient
R domain	Regulatory domain
RT-PCR	Reverse transcriptase polymerase chain reaction
SCN^-	Thiocyanate
TIMP-1	Tissue inhibitor of metalloproteinase 1
TNF-α	Tumor necrosis factor α
UTP	Uridine triphosphate
VIP	Vasoactive intestinal peptide

1 Cystic Fibrosis and Airway Ion Transport

CF is a serious and life-threatening disease that affects approximately 30,000 US citizens and >70,000 people worldwide (CFF 2005; Pilewski and Frizzell 1999; Rowe et al. 2005). It is caused by autosomal recessive mutations in the gene encoding CFTR. The protein product is a membrane-localized traffic ATPase that functions as a Cl^- channel (Anderson et al. 1991), and it is also a regulator of many ion transport pathways (including nonCFTR Cl^- channels, Na^+, bicarbonate, ATP, glutathione (GSH), and potentially other small molecules (Hudson 2001; Moskwa et al. 2007; Park et al. 2002; Rowe et al. 2005; Schweibert et al. 1999). CFTR is expressed in many tissues, including the airways, the gastrointestinal (GI) tract and the hepatobiliary tree, the pancreatic ducts, the sweat glands, and in the male reproductive tract. In the airways, CFTR is expressed at high levels in submucosal gland ducts, and to a lesser extent at the apical cell membrane of the surface pseudostratified epithelium of medium and large airways, and in the distal epithelium of the small airways (Engelhardt et al. 1992, 1994).

Central to our understanding of CF pathophysiology is an appreciation of the role played by CFTR in regulating submucosal gland and airway surface liquid composition (Choi et al. 2007; Joo et al. 2002, 2004, 2006; Wine and Joo 2004; Wu et al. 2007). Much of the work describing defects in glandular function in CF have been recently reported by Wine and colleagues, examining gland activity in strips of airway tissue removed from CF patients undergoing lung transplantation. Submucosal gland secretory activity is regulated by neurogenic pathways [acetylcholine (Ach) and vasoactive intestinal peptide (VIP)] and also by local paracrine regulatory factors, which together hydrate the glandular secretions and promote their release upon gland stimulation. The released glandular contents provide a significant volume to the airway surface that is rich in a variety of antimicrobial peptides and factors. The glands therefore contribute to both the innate defense system through the release of antimicrobial molecules and through contributions to hydration of the airway. In CF, loss of CFTR function leads to thickening of glandular secretions and plugging of gland ducts. This can be seen pathologically as enlargement and hypertrophy of glands that are filled with retained secretions, demonstrating ductal dilation and hyperplasia of glandular acini. The glands fail to secrete in response to VIP, but retain Ach-stimulated secretion and release of abnormal glandular contents. In this model, glandular dysfunction is a primary cause of CF lung disease, initiating the obstructive, infectious, and inflammatory consequences of CFTR dysfunction (Choi et al. 2007; Wine and Joo 2004).

On the apical airway surface, epithelial cells are covered with numerous cilia that are bathed by a watery fluid compartment known as the airway surface liquid or ASL (Boucher 2007a; Matsui et al. 1998a; Rowe et al. 2005). The cilia beat rapidly within this compartment in a coordinated fashion across the epithelial sheet, and this arrangement allows swift and unimpeded ciliary activity that promotes cephalad movement of the overlying mucus blanket. This mucus or gel layer is constructed from hydrated mucins (MUC5AC and 5B) and also serves as a volume reservoir for the ASL. Trapped particulates are rapidly cleared from the airway surface, and this is accomplished prior to the activation of secondary host defense mechanisms. The end result is the continuous clearance of pathogens and the avoidance of unnecessary activation of host responses that could potentially injure the airway.

The composition of the ASL is determined in large part by the activity of epithelial ion transporters present in the airway cells, and these systems have been carefully characterized by researchers at the University of North Carolina using polarized mature airway epithelial monolayers isolated from human subjects. In the prevailing model of airway epithelial ion transport (see Fig. 1), a sodium (Na^+) absorptive phenotype is produced by the epithelial sodium channel (ENaC), which is a protease-regulated channel that absorbs Na^+ across the airway epithelium. There is complimentary passive Cl^- flow through CFTR (and potentially other CFTR-dependent Cl^- transport pathways) from the luminal compartment, with accompanying H_2O transport through the paracellular pathway and potentially aquaporins (Boucher 2007a, b; Donaldson et al. 2006). In addition to CFTR-dependent Cl^- transport, there is a CFTR-independent pathway that is regulated by cell calcium (Ca^{++}) that activates CaCCs (Boucher 2007a, b). These processes result in a "dry" luminal compartment, and provide a rational explanation for the ability of the large airways to avoid flooding in the face of the surface airway reduction that occurs as fluid moves from the smallest bronchioles to the mainstem bronchi. Loss of CFTR activity manifests as a loss of cAMP-dependent Cl^-

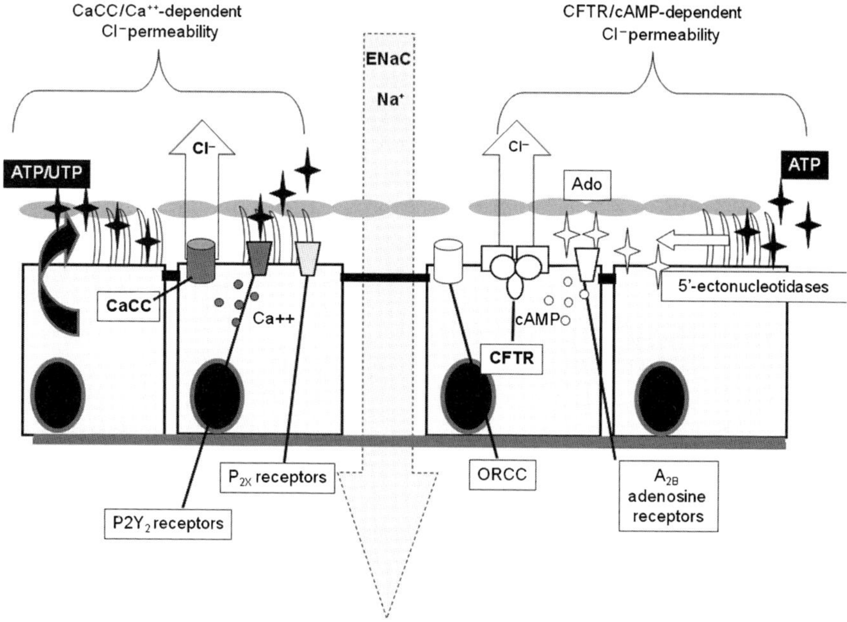

Fig. 1 Model of adenosine triphosphate (ATP) and adenosine (Ado)-regulated chloride (Cl^-) transport in airway cells. Na^+ absorption through epithelial sodium channels (ENaC) generally dominates transcellular ion flow, maintaining a dry luminal surface. On the *left*, ATP and uridine triphosphate (UTP) stimulate $P2Y_2$ receptors to mobilize intracellular calcium, and ATP stimulates P2X calcium receptor channels, inducing an influx of extracellular calcium. Increased intracellular Ca^{++} stimulates calcium-activated Cl^- channels (CaCCs), leading to transcellular Cl^- secretion. On the *right*, Ado nucleotides are dephosphorylated by 5′-ectonucleotidases on the cell surface, leading to the production of Ado. Ado can bind to A_{2B} adenosine receptors (A_{2B}ARs), stimulating local cAMP production and activation of CFTR. CFTR transports select anions such as Cl^- and bicarbonate (HCO_3^-) and thiocyanate (SCN^-). Activated CFTR also facilitates the release of ATP, which can positively couple to outwardly rectified Cl^- channels (ORCCs). The balance between Na^+ and Cl^- transport sets the airway surface liquid (ASL) volume, which is disturbed (reduced) in CF. For clarity, basolateral ion transport pathways and other apical Cl^- transport pathways are not included

transport across airway epithelia, and heightened Na^+ absorption through ENaC. The nature of the increased Na^+ absorption is increased ENaC activity, which is proposed to be secondary to the loss of direct negative regulation provided by intact CFTR, and protease activation of ENaC (Stutts et al. 1995; Tarran et al. 2006b). As Na^+ is absorbed across the airway, complimentary Cl^- transport occurs through a paracellular route, leading to a hyperpolarized surface epithelium and water depletion. Interestingly, the Ca^{++} regulated Cl^- transport pathway remains intact in the CF airway (and is upregulated relative to the nonCF phenotype). Despite this, the end result of CFTR dysfunction is a reduction in the ASL volume, which impedes normal ciliary beating (Boucher 2007a, b; Matsui et al. 1998a, b). In this model of CF lung disease, the interruption of mucus clearance by ASL volume depletion and ciliary collapse is the first in a series of cascading steps, including mucus obstruction of the airways, intermittent and subsequently chronic bacterial infection, and severe, sustained inflammation, that lead to irreversible bronchiectasis and eventual respiratory failure. Supportive care for CF continues to advance steadily, with the median life expectancy of persons with CF now exceeding 37 years (CFF 2006). Unfortunately, CF lung disease remains devastating for many patients, with the majority of patients who die from CF succumbing in early adulthood.

2 Regulation of Airway Cl^- Transport by Adenosine and Related Nucleotides

Figure 1 provides a simplified model of ion transport pathways in the surface airway epithelium, and of how Ado nucleotides on the epithelial surface regulate anion transport and mucociliary clearance. At least two pathways are operative; one dominated by Ca^{++} and CaCCs, and the other dominated by cAMP and CFTR. On the left, ATP-stimulated functions are highlighted, including (1) ATP regulation of ciliary beat frequency, (2) stimulation of $P2Y_2$ receptors, and (3) stimulation of P2X receptor channels (Alfahel et al. 1996; Bennett et al. 1996; Knowles et al. 1991; Korngreen and Priel 1996; Olivier et al. 1996; Zhang and Sanderson 2003; Zsembery et al. 2004). Recent reports by Tarran, Boucher and colleagues have led to a clearer view of how purinergic receptors regulate Cl^- transport, based on studies completed in their established primary human airway epithelial cultures (Button et al. 2007; Tarran et al. 2006a, b). In this model, shear stress produced by homeostatic processes such as the breathing cycle or coughing leads to the release of nucleotides from the airway epithelium. ATP can bind to $P2Y_2$ receptors on the epithelial surface, stimulating Ca^{++} release and Cl^- transport via CaCCs. New therapies in development that are designed to overcome defects in CF Cl^- transport exploit this signaling pathway (e.g., denufosol) and restore airway Cl^- conductance (Bye and Elkins 2007; Deterding et al. 2007). Early-phase studies have demonstrated that treated subjects exhibit improved lung function (forced expiratory volume in 1 s, or FEV_1) compared with placebo-treated controls following short-term exposure (28 days). Further studies will be needed to clarify that the short-term benefits of P2 receptor stimulation on lung function are durable over longer treatment periods, particularly since there is potential for off-target effects of nebulized nucleotides on other pulmonary P2Y receptors that may not be beneficial (Brunschweiger and Muller 2006). In addition to $P2Y_2$ receptors, extracellular ATP can stimulate P2X receptors on the luminal airway surface that function as ATP-regulated Ca^{++} entry channels, raising cell Ca^{++}, activating CaCCs and subsequent Cl^- conductance (Zsembery et al. 2003). Recent studies by Zsembery and colleagues indicate that zinc is an important cofactor to maximize P2X channel activity, which is further enhanced under low $[Na^+]$ conditions (Zsembery et al. 2004). Stimulation of Ca^{++}-dependent, CFTR-independent Cl^- transport typically produces a relatively short-lived spike in Cl^- conductance in vitro, likely due to the short half-life of ATP on the epithelial surface and the relatively short-lived effects of nucleotides on cellular Ca^{++} levels. As noted previously, Cl^- transport through this pathway is typically enhanced in CF patients and tissues relative to normal controls, which may be due to expansion of the ER and Ca^{++} storage, serving a protective function in the absence of CFTR activity (discussed in more detail below) (Knowles et al. 1991; Paradiso et al. 2001; Ribeiro et al. 2005a, b, 2006; Tarran et al. 2002).

On the right (Fig. 1), Ado-dependent Cl^- transport is highlighted, focusing on Ado and A_{2B}AR-regulation of CFTR. In current models of ion transport, released ATP can be dephosphorylated to Ado and stimulate P1 purinergic receptors (Boucher 2007a, b; Hirsh et al. 2007; Lazarowski et al. 2004). In the absence of airway infection or stress, baseline Ado levels on the airway surface are determined by a number of interrelated factors, including the production and release of Ado and related nucleotides by the surface epithelia, uptake by concentrative nucleoside transporters 2 and 3, and breakdown of Ado to inosine by ADA1 (Hirsh et al. 2007). The predominate P1 receptor found in airway cells is the A_{2B}AR, which couples to G_s and activates adenylyl cyclase, raising local cAMP concentrations and stimulating cAMP-dependent protein kinase A (PKA) (Cobb et al. 2002; Cobb and Clancy 2003; Hentchel-Franks et al. 2004; Huang et al. 2001; Li et al. 2006; Rollins et al. 2008; Tarran et al. 2001). PKA then phosphorylates the regulatory (R) domain of CFTR, activating the CFTR Cl^- channel (Cheng et al. 1991; Rowe et al. 2005). As a member of the ATP binding cassette protein family, CFTR also binds and hydrolyzes ATP through nucleotide binding domain 1 and 2 (NBD-1 and NBD-2) dimerization, an important step in the gating of Cl^- channel activity (Mense et al. 2006; Vergani et al. 2005a, b). Once activated, CFTR can positively couple to the outwardly rectified Cl^- channel (ORCC) through an ATP release process that is also sponsored by CFTR (Schwiebert et al. 1995; Schwiebert 1999). CFTR has also been implicated in the transport of other small molecules, including bicarbonate (HCO_3^-), GSH, and thiocyanate, each of which has described relationships to the pathogenesis of CF lung disease. The end result of CFTR activation is amplification of the CFTR Cl^- conductance signal and enhanced epithelial Cl^- secretion, promoting hydration and volume expansion of the ASL and luminal compartment. Work by Tarran and colleagues provide an elegant view of ASL volume regulation that is dependent on fluctuations in airway Ado concentrations. When the ASL volume falls, the relative Ado concentration rises (from its low baseline value to levels $>1\,\mu$M), promoting Ado binding to the low-affinity A_{2B}AR and activation of CFTR (Tarran et al. 2006b). When the ASL volume increases, Ado concentrations fall, reducing A_{2B}AR and CFTR activity and Cl^- transport. Thus, expansion and retraction of the ASL may be self-regulated by Ado concentrations that fluctuate, with secondary effects on A_{2B}AR and CFTR activity.

Figure 2 compares expression of the four P1 purinergic receptors (via real-time RT-PCR), demonstrating a relative expression profile of A_{2B}ARs $\gg A_{2A} > A_1 \sim A_3$ ARs in human bronchial epithelial cells (HBECs) [primary normal HBECs and immortalized CFBE41o– cells: ΔF508/ΔF508 genetic background]. The majority of published studies examining Ado regulation of CFTR and transepithelial Cl^- secretion implicate A_{2B}ARs as the primary P1 receptor responsible for CFTR

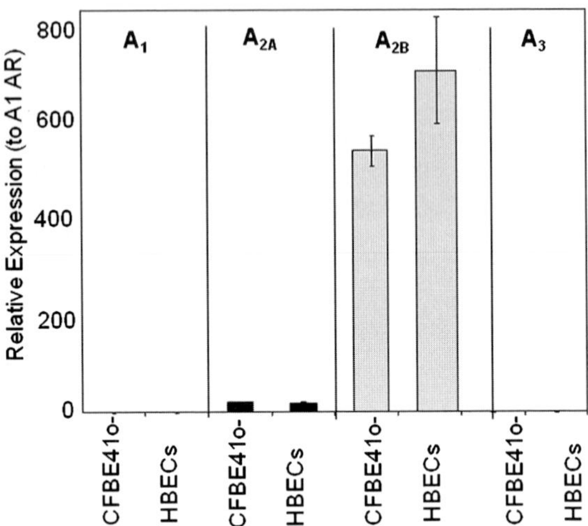

Fig. 2 P1 receptor expression in primary human bronchial epithelial cells (HBECs) and CFBE41o− cells. Isoform transcripts are compared to A_1 adenosine receptor (A_1AR) levels for the four known P1 purinergic receptors. In both cell types, A_{2B}AR expression dominates over A_1, A_{2A} and A_3 adenosine receptor (AR) transcripts ($*p < 0.01$). Methods: a TaqMan One Step RT-PCR protocol (Applied Biosystems, Foster City, CA, USA) was used to quantify P1 purinergic receptor mRNA transcripts using "Assays on Demand" Gene Expression Products, coupled with the ABI Prism 7500 sequence detection system (Applied Biosystems). Briefly, total RNA was isolated using the Qiagen RNeasy mini kit according to the manufacturer's instructions. To prevent possible DNA contamination, the samples were pretreated with RNase-free DNase (Qiagen, Valencia, CA, USA). Sequence-specific primers and probes for human P1 receptors and 18S rRNA were purchased from Assays on Demand (ABI, Foster City, CA, USA). TaqMan One Step PCR Master Mix Reagents Kit (ABI) was used for reverse transcriptase and PCR. The reaction volume was 25 μl, including 12.5 μl of 2 × Master Mix without UNG, 0.625 μl of 40 × MultiScribe and RNase Inhibitor Mix, 1.25 μl of 20 × target primer and probe, 5.625 μl of nuclease-free water (Ambion, Austin, TX, USA), and 5 μl of RNA sample. The reaction plates were covered with an optical cap and centrifuged briefly to remove bubbles. The thermocycler conditions were as follows: Stage 1: 48°C for 30 min; Stage 2: 95°C for 10 min; Stage 3: 95°C for 15 s, repeat 40 cycles, 60°C for 1 min. All experiments were run in triplicate on at least two separate days. Six experiments per condition were performed. The absolute value of the slope of log input amount vs. ΔCt was >0.1, implying that the efficiencies of AR isoform and 18S rRNA amplification were not equal. Therefore, the relative quantification of transcript levels (compared with endogenous 18S rRNA) was performed using the standard curve method (Li et al. 2006)

activation (Bebok et al. 2005; Clancy et al. 1999; Cobb et al. 2002; Cobb and Clancy 2003; Huang et al. 2001; Li et al. 2006; Tarran et al. 2005, 2006a, b). It is difficult to completely exclude participation of A_{2A}ARs (which also typically signal through G_s, adenylyl cyclase, cAMP and PKA), as low-level CFTR activation in monolayers can be seen at Ado concentrations that may be below the threshold to stimulate A_{2B}ARs. A recent report noted A_{2B}AR transcript levels that were approximately threefold higher than that of A_1 and A_{2A} receptor transcripts in laser-dissected specimens of human lower airway tissue (Rollins et al. 2008). The relative lack of A_{2B}AR-specific blockers also complicates the evaluation of P1 receptor regulation of CFTR.

Elegant studies completed by Huang and Stutts provided evidence for a tightly compartmentalized signaling complex at the airway epithelial surface comprising several proteins, including the A_{2B}AR, adenylate cyclase, A kinase anchoring proteins (AKAPs), PKAII, and CFTR (Huang et al. 2000, 2001). Their work indicated that all of the components necessary to activate CFTR through A_{2B}ARs were available within the boundary of a micropipette tip, and that signaling did not extend beyond this border to other transporters along the cell membrane. Their findings indicate that the coupling between CFTR and A_{2B}ARs is extremely efficient, as A_{2B}ARs can activate regionally localized CFTR without detectable effects on total cell cAMP. This regional regulation of CFTR by A_{2B}ARs underlies the importance of this signaling molecule in CFTR control and airway hydration in the upper and lower airways. This unique and highly compartmentalized signaling is also highlighted in Fig. 3, where cell cAMP levels produced by Ado are far lower than those produced by isoproterenol (Fig. 3a; Iso, a β_2 adrenergic receptor agonist), using concentrations that fully stimulate CFTR and Cl^- transport. Ado activates Cl^- transport across polarized airway epithelial cells when added to either the apical or basolateral compartment (Fig. 3b), which is sensitive to the relatively selective A_{2B}AR receptor blocker alloxazine. The results confirm that Ado stimulates Cl^- transport via A_{2B}ARs present on both the apical and basolateral membrane, and suggest that Ado may also directly regulate basolateral transporters (e.g., K^+ channels) that promote transepithelial Cl^- transport.

Using the nasal potential difference (NPD) measurement (an established bioelectric measure of Na^+ and CFTR-dependent Cl^- transport in the airway (Knowles et al. 1995; Rowe et al. 2007; Standaert et al. 2004)), our laboratory has confirmed that Ado activates CFTR-dependent Cl^- transport in the airways of human subjects, with robust responses seen in nonCF subjects that are absent in CF patients with severe, nonfunctional CFTR mutations (Hentchel-Franks et al. 2004). The response to 10 μM Ado exceeds that produced by 10 μM Iso (an agonist commonly used to detect CFTR activity in vivo), but the two agonists do not demonstrate additivity. Complementary NPD studies were recently completed in human subjects without CF and reported by Tarran and Rollins, in which Ado-stimulated Cl^- transport was inhibited by perfusion with caffeine, a known blocker of adenosine receptors (Rollins et al. 2008). These findings indicate that Ado activates predominately CFTR-dependent and not CFTR-independent Cl^- transport in vivo through stimulation of P1 receptors. Similar Cl^- transport results have also been demonstrated in NPDs completed in *Cftr*$^+$ mice, while Ado failed to activate Cl^- transport in *Cftr*$^{-/-}$ littermates (Cobb et al. 2002). We have also examined the capacity of Ado to stimulate Cl^- transport in patients with partial function CFTR mutations, mild disease with surface localized CFTR mutations, and/or nonclassic CF. Figure 4 summarizes NPD results obtained in six subjects with positive sweat tests (>60 mM) and clinical findings of CFTR dysfunction. Three subjects had two identified CFTR mutations, and three had one identified CFTR-causing mutation (demographic and relevant clinical information for these study subjects is summarized in Fig. 4a). Each subject underwent three standard NPD measurements using one of three agonist conditions in a random fashion (Fig. 4b; 10 μM adenosine

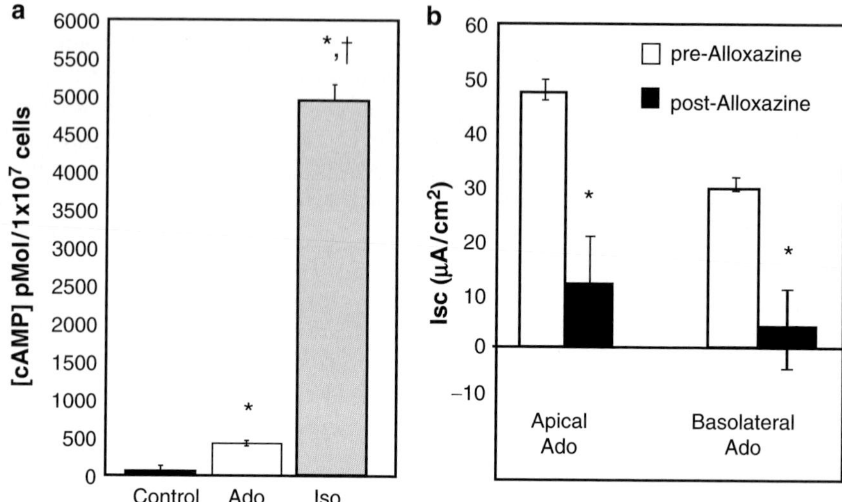

Fig. 3 **a–b** cAMP and Cl^- transport produced by Ado in airway cells. **a** Calu-3 cells were grown on impermeable supports and stimulated with agonists for 10 min prior to lysis and cAMP measurement by ELISA (Cayman Chemicals, Minneapolis, MN, USA). Cells were stimulated with 10 μM adenosine (Ado) or isoproterenol (Iso) (+100 μM papaverine, a nonspecific PDE inhibitor); $^*p < 0.001$ compared with control, $^\dagger p < 0.001$ compared with Ado alone ($n = 4$ experiments per condition). **b** Calu-3 cells grown as polarized monolayers were studied in Ussing chambers under voltage clamp conditions. Ado (10 μM) stimulated short circuit current (I_{sc}) when added to either the apical or basolateral compartment that was sensitive to pretreatment with the A_{2B}AR blocker alloxazine (20 μM). $n = 6$ experiments studied per condition. Methods: Calu-3 cells (immortalized human serous glandular cells) expressing wtCFTR were seeded on Costar 0.4 μm permeable supports (Bethesda, MD, USA; 5×10^5 cells per filter, 6.5 mm diameter) after coating with fibronectin. Cells were grown to confluence and then placed at an air/liquid interface (48 h) and mounted in modified Ussing chambers (Jim's Instruments, Iowa City, IA, USA), and initially bathed on both sides with identical Ringers solutions containing (in mM) 115 NaCl, 25 $NaHCO_3$, 2.4 KH_2PO_4, 1.24 K_2HPO_4, 1.2 $CaCl_2$, 1.2 $MgCl_2$, 10 D-glucose (pH 7.4). Bath solutions were vigorously stirred and gassed with $95\%O_2 : 5\% CO_2$. Solutions and chambers were maintained at 37°C. Short-circuit current (I_{sc}) measurements were obtained by using an epithelial voltage clamp (University of Iowa Bioengineering, Iowa City, IA, USA). A 3 mV pulse of duration 1 s was imposed every 100 s to monitor resistance, which was calculated using Ohm's law. To measure stimulated I_{sc}, the mucosal bathing solution was changed to a low-Cl^- solution containing (in mM) 1.2 NaCl, 115 Na gluconate, and all other components as above plus 100 μM amiloride. Ado (10 μM) in the presence or absence of alloxazine (20 μM) was added to the apical or basolateral solutions as indicated, and I_{sc} was measured (μA/cm^2; minimum 15 min of observation at each concentration) (Bebok et al. 2005)

(Ado), 10 μM isoproterenol (Iso), or 10 μM Ado and 10 μM Iso included in the fourth perfusate). As expected, the mean polarizing response to Iso (representative of CFTR-dependent Cl^- transport) was minimal, consistent with CFTR dysfunction (as predicted based on genotype, sweat Cl^- values, and/or clinical symptoms). In contrast, inclusion of Ado in the final perfusate produced a polarizing response distinct from that seen with Iso alone. These results, coupled with the previously

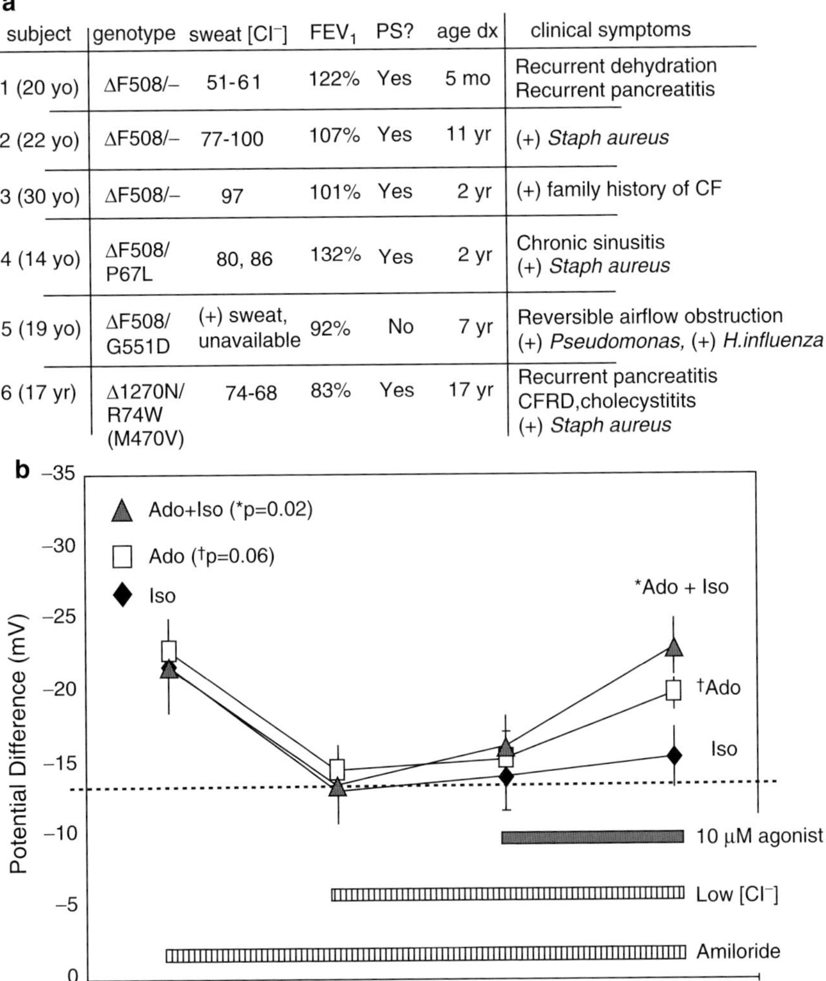

a

subject	genotype	sweat [Cl⁻]	FEV_1	PS?	age dx	clinical symptoms
1 (20 yo)	ΔF508/–	51-61	122%	Yes	5 mo	Recurrent dehydration Recurrent pancreatitis
2 (22 yo)	ΔF508/–	77-100	107%	Yes	11 yr	(+) *Staph aureus*
3 (30 yo)	ΔF508/–	97	101%	Yes	2 yr	(+) family history of CF
4 (14 yo)	ΔF508/ P67L	80, 86	132%	Yes	2 yr	Chronic sinusitis (+) *Staph aureus*
5 (19 yo)	ΔF508/ G551D	(+) sweat, unavailable	92%	No	7 yr	Reversible airflow obstruction (+) *Pseudomonas*, (+) *H.influenza*
6 (17 yr)	Δ1270N/ R74W (M470V)	74-68	83%	Yes	17 yr	Recurrent pancreatitis CFRD,cholecystitits (+) *Staph aureus*

Fig. 4 a–b Ado-stimulated Cl⁻ transport by nasal potential difference (NPD) in subjects with partial CFTR function and/or mild disease manifestations. In **a**, clinical information on the six subjects is summarized. All subjects had mild lung disease based on forced expiratory volume in 1 s (FEV_1), a relatively mild phenotype based on age at diagnosis, 5:6 were pancreatic sufficient (PS), and all demonstrated elevated sweat Cl⁻ values. In **b**, NPD values for the six subjects studied under the three conditions are compared. All subjects underwent three NPDs with varying CFTR agonists in the fourth perfusate (isoproterenol (Iso): 10 μM; adenosine (Ado): 10 μM; or Ado combined with Iso: 10 μM each). Inclusion of Ado in the perfusate produced enhanced Cl⁻ transport compared with Iso alone ($^{*}p < 0.01$, $^{\dagger}p = 0.06$). Methods: for these studies, we used a NPD protocol based on that published by Knowles and colleagues (Knowles et al. 1995), and previously described by our laboratory. Subjects underwent three NPD measurements over a two-week period, including 1 min perfusion with Ringers, 3 min perfusion with Ringers + amiloride (100 μM, to block ENaC), 3 min perfusion with low [Cl⁻] solution and amiloride (solution matching Ringers' except that [gluconate] = 115 mM and [Cl⁻] = 6 mM), and 3 min perfusion with low [Cl⁻] solution and amiloride and 10 μM isoproterenol (Iso), 10 μM Ado, or 10 μM Ado and 10 μM Iso. NPD values after the completion of the perfusion conditions are shown using methods as previously described (Hentchel-Franks et al. 2004)

published work from our laboratory and that by Stutts, confirm that CFTR activation by A_{2B}ARs is highly efficient and effective relative to β_2 adrenergic receptor stimulation. The results also raise questions regarding the sensitivity of current NPD protocols to detect partial function CFTR activity (Rowe et al. 2007), and how intracellular signaling pathways connect CFTR activation to A_{2B}ARs. The mechanism appears to be independent of protein:protein interactions mediated by the postsynaptic density protein 95, disk-large tumor suppressor protein, *Zonula occludens* 1 (PDZ) motif binding to ezrin binding protein 50 (EBP50), which has been shown to be operative in β_2 adrenergic receptor regulation of CFTR (Naren et al. 2003).

It is clear that an important function of Ado and related nucleotides in the airway is to regulate transepithelial ion transport and ASL volume, ciliary function, and coordinated mucociliary clearance. Indeed, recent work has demonstrated that Ado is a vital regulator of the minute-to-minute ASL depth, maintaining a volume that optimizes ciliary function via regulation of CFTR. Loss of CFTR function results in a decrease in the ASL volume secondary to unopposed/excessive Na^+ (and fluid) absorption via ENaC. Failure of this system is demonstrated in cystic fibrosis, where the absence of CFTR activity removes the positive contributions of Ado to airway anion and fluid homeostasis. A current model of cystic fibrosis offered by Boucher and colleagues implicates retention of transepithelial Cl^- and fluid flow via ATP, $P2Y_2$ and CaCC signaling that serves to maintain the ASL volume, MCC and cough clearance of lower airway secretions in the absence of CFTR activity (Boucher 2007a; Tarran et al. 2005). Unfortunately, this system eventually falters (possibly as a result of recurrent viral and/or bacterial infections) with the development of the hallmarks of CF airway disease.

3 Repercussions of Altered Adenosine Levels in the Airway

Ado and related nucleotides are ubiquitous signaling molecules that are regulators of a variety of pulmonary and airway processes, including ion transport, bronchial tone, mucus production, and inflammatory signaling. Ado and related nucleotides on the epithelial surface can have phosphate groups added to or removed from their 5′ end by surface kinases and 5′-ectonucleotidases (CD73), and the dominant effect is dependent upon the cell type, the balance of kinases/nucleotidases expressed, the nucleotide transport mechanisms at play, the activities of extracellular enzymes such as *ADA1* and *2*, and nucleoside transporters (concentrative and equilibrative) (Hirsh et al. 2007; Lazarowski et al. 2004). Paracellular transport of nucleosides does not appear to be of high importance under conditions where the airway epithelial integrity remains intact. In human airways, a dominant pathway of nucleotide metabolism includes the removal of phosphate groups in excess of phosphorylation, leading to the relative production of Ado on the airway surface (Hirsh et al. 2007). These processes generally maintain a short $t_{1/2}$ life of Ado in the ASL, and measurable levels are dependent on continuous and fluctuating production of Ado, nucleotide release and metabolism, and cellular uptake. Nucleotide release is a normal part of active homeostatic mechanisms that continuously function in the airway, including shear stress produced by the breathing cycle, cough, diffusion down chemical gradients from the intracellular to the extracellular compartment, and nucleotide release following stimulation of intracellular signaling pathways (Donaldson et al. 2000; Tarran et al. 2005, 2006a). Under conditions of stress, Ado and related nucleotides can accumulate to high levels on airway epithelia. Additionally, dead and apoptotic leukocytes, bacteria, and other microorganisms in the airway lumen are potential sources of surface nucleotides, contributing to levels that can be measured in excess of 50 μM in expectorated sputum (from CF patients) ex vivo (Li et al. 2006).

The four P1 purinergic receptors are all members of the G-protein-coupled receptor (GPCR) protein superfamily, but differ from one another in several defining features, including G-protein coupling, Ado affinity, and regulation of cell signaling (Cobb and Clancy 2003). For a more detailed discussion of P1 receptor structure, function and pharmacology, the reader is directed to excellent reviews within this publication. A_1ARs traditionally couple to G_i and inhibit cAMP production, while A_{2A} (high-affinity) and A_{2B} (low-affinity) ARs frequently couple to G_s and raise cell cAMP. The more recently identified A_3 adenosine receptor also inhibits cAMP production and has been linked to a variety of signaling pathways. Recent studies have demonstrated a more diverse signaling repertoire for A_1, A_{2A} and A_{2B} ARs, and the sum balance of Ado effects in vivo are at times difficult to predict (Caruso et al. 2006; Cobb and Clancy 2003; Polosa and Holgate 2006; Russo et al. 2006; Spicuzza et al. 2006). For example, A_1ARs have a role in Ado-induced bronchoconstriction and proinflammatory processes in animal models, and stimulation of MUC2 expression [a highly insoluble mucin that has been implicated in asthma pathology (McNamara et al. 2004)]. Stimulation of A_{2A}ARs expressed in leukocytes tends to activate a number of anti-inflammatory signaling pathways, while stimulation of A_{2B}ARs tends to favor proinflammatory mediator release, which is independent of its regulation of CFTR and Cl^- conductance (Spicuzza et al. 2006). A_3AR stimulation has been linked to several proinflammatory processes, including mast cell degranulation, airway responsiveness, and mucus production in animal models. In addition to airway epithelial cells, all P1 receptor isoforms are expressed at varying levels in resident macrophages and granulocytes of the lung, particularly in neutrophils, eosinophils and mast cells. Moreover, nebulized AMP has been shown to be a sensitive and specific agonist for the detection of allergic-based asthma in human subjects, producing reversible airway obstruction specifically in asthmatics but not in subjects with other chronic airway disorders, including chronic obstructive pulmonary disease (COPD) and cystic fibrosis (Caruso et al. 2006; Feoktistov et al. 1998; Polosa et al. 1995; Polosa and Holgate 1997, 2006; Polosa 2002; Russo et al. 2006; Spicuzza et al. 2006). For more detailed discussions regarding ARs, inflammation, and asthma, please refer to separate chapters in this volume, Chap. 8, "Adenosine Receptors and Inflammation" (by Blackburn et al.), and Chap. 11, "Adenosine Receptors and Asthma" (by Wilson et al.).

Ado has been reported to have both pro- and anti-inflammatory signaling properties, and is a significant paracrine regulator of inflammatory processes. An important series of studies recently reported by Blackburn and colleagues have demonstrated that mice deficient in adenosine deaminase (ADA) are prone to Ado accumulation in many organs, including the lungs (Blackburn et al. 2003; Blackburn 2003; Blackburn and Kellems 2005; Chunn et al. 2001). Elevations in pulmonary Ado levels produced a dramatic proinflammatory phenotype, with influx of neutrophils, and elevations of a variety of inflammatory chemokines, cytokines, and proteases including monocyte chemoattractant proteins (MCP-1, 2, 5), eotaxine, intracellular adhesion molecule 1 (ICAM-1), IL-4, IL-5, IL-13, TNF-α, matrix metalloproteases 9 and 12, and tissue inhibitor of metalloproteinase 1 (TIMP-1). In addition, extensive mucus production and bronchial plugging in response to Ado can further alter the airway structure. Subsequent studies implicated the A_{2B}AR as the primary receptor subtype responsible for the stimulation of many of these proinflammatory pathways, and highlighted the potential damaging effects of chronic Ado elevation in the lung (Sun et al. 2006). Treatment of ADA-deficient mice with A_3AR antagonists resulted in decrease in mucus production, suggesting that A_3AR signaling plays an important role in the development of Ado-stimulated mucus metaplasia (Young et al. 2004). In this animal model, the end result was chronic pulmonary injury and eventual fibrosis, with increased mortality seen in partially ADA-deficient mice by several months of age. These effects could be ameliorated by cotreatment with ADA, implicating Ado as the causative signaling molecule in this process.

Recently published work from our laboratory also implicates PLA_2 signaling following A_{2B}AR stimulation in airway epithelial cells, leading to the release of arachidonic acid and the accumulation of PGE_2 on the airway surface following stimulation with Ado or exposure to hypoxic conditions (Li et al. 2006). Both arachidonic acid and PGE_2 are effective stimuli of CFTR and transepithelial Cl^- transport in airway epithelia, and the EP4 receptor has recently been proposed to mediate CFTR-dependent halide transport produced by isoprostanes in Calu-3 cells (Cowley 2003; Joy and Cowley 2005, 2008; Li et al. 2006). Arachidonic acid is also a potent blocker of CFTR Cl^- channels from the cytoplasmic surface (Li et al. 2006; Linsdell 2000). These reports highlight the complex interrelationships between Ado signaling, ion transport and inflammation, and that Ado-stimulated mediators often serve several roles in the host response.

As evidence suggests that excessive Ado levels in the lung have a proinflammatory phenotype, one approach to reducing airway inflammation in conditions where Ado levels are elevated (asthma, COPD) could include inhibition of P1 receptor function. The A_{2B}AR is a logical target based on the work reported by Blackburn and others, and it is known to be expressed in granulocytes that contribute to airway pathology. A recent report by Rollins examined the impact of A_{2B}AR inhibition on ASL volume of human airway cells (Rollins et al. 2008). The results suggest that inhibition of A_{2B}AR activity in airway epithelial cells can lead to depletion of the ASL, producing a phenotype reminiscent of CF. Thus, the development of agents to target Ado-mediated inflammation in lung diseases may be complicated by off-target effects of receptor antagonists on airway hydration and innate defense.

4 Conclusions

ATP and Ado regulate distinct but interrelated Cl^- transport pathways that play important roles in the pathogenesis of cystic fibrosis. Results from several investigators support a model in which Ado is a central molecule in the disease, providing direct and indirect regulation to CFTR, the ASL volume, and inflammation. Evidence suggests that Ado serves a vital role in innate defense as a primary regulator of CFTR and thus a secondary regulator of mucocilary clearance. High levels of Ado appear to be causative of lower airway pathology, but targeting the $A_{2B}AR$ pathway to reduce inflammation is complicated by potential deleterious effects on mucociliary clearance. Continuing to clarify the relative roles of P1 receptor subtypes in airway homeostasis and pathologic responses should provide logical approaches to modulate these pathways to understand and potentially treat a variety of pulmonary diseases.

Acknowledgements The authors would like to thank Ms. Lijuan Fan, Dr. Yao Li, Ph.D., Dr. David Lozano, M.D. and Dr. Karen Hentschel-Franks, D.O. (UAB) for their help in completing the experiments included in this publication. Work in this publication was supported by NHLBI (HL67088), NIDDK P30 DK072482, and the Cystic Fibrosis Foundation (CFF R464-CR07, CLANCY02YO).

References

Alfahel E, Korngreen A, Parola AH, Priel Z (1996) Purinergically induced membrane fluidization in ciliary cells: characterization and control by calcium and membrane potential. Biophys J 70:1045–1053

Anderson MP, Gregory RJ, Thompson S, Souza DW, Paul S, Mulligan RC, Smith AE, Welsh MJ (1991) Demonstration that CFTR is a chloride channel by alteration of its anion selectivity. Science 253:202–205

Bebok Z, Collawn J, Wakefield J, Parker W, Varga K, Li Y, Sorscher EJ, Clancy JP (2005) Failure of cAMP agonists to activate rescued {Delta}F508 CFTR in CFBE41o– airway epithelial monolayers. J Physiol 569(Pt 2):601–615

Bennett WD, Olivier KN, Zeman KL, Hohneker KW, Boucher RC, Knowles MR (1996) Effect of uridine 5′-triphosphate plus amiloride on mucociliary clearance in adult cystic fibrosis. Am J Respir Crit Care Med 153:1796–1801

Blackburn MR (2003) Too much of a good thing: adenosine overload in adenosine-deaminase-deficient mice. Trends Pharmacol Sci 24:66–70

Blackburn MR, Kellems RE (2005) Adenosine deaminase deficiency: metabolic basis of immune deficiency and pulmonary inflammation. Adv Immunol 86:1–41

Blackburn MR, Lee CG, Young HW, Zhu Z, Chunn JL, Kang MJ, Banerjee SK, Elias JA (2003) Adenosine mediates IL-13-induced inflammation and remodeling in the lung and interacts in an IL-13-adenosine amplification pathway. J Clin Invest 112:332–44

Boucher RC (2007a) Airway surface dehydration in cystic fibrosis: pathogenesis and therapy. Annu Rev Med 58:157–70

Boucher RC (2007b) Evidence for airway surface dehydration as the initiating event in CF airway disease. J Intern Med 261:5–16

Brunschweiger A, Muller CE (2006) P2 receptors activated by uracil nucleotides: an update. Curr Med Chem 13:289–312

Button B, Picher M, Boucher RC (2007) Differential effects of cyclic and constant stress on ATP release and mucociliary transport by human airway epithelia. J Physiol 580:577–592

Bye PT, Elkins MR (2007) Other mucoactive agents for cystic fibrosis. Paediatr Respir Rev 8:30–39

Caruso M, Holgate ST, Polosa R (2006) Adenosine signalling in airways. Curr Opin Pharmacol 6:251–256

CFF (2005) www.portcf.org (Resources, Registry materials - 2005). Cystic Fibrosis Foundation, Bethesda, MD

CFF (2006) www.portcf.org (Resources, Registry materials - 2006). Cystic Fibrosis Foundation, Bethesda, MD

Cheng SH, Rich DP, Marshall J, Gregory RJ, Welsh MJ, Smith AE (1991) Phosphorylation of the R domain by cAMP-dependent protein kinase regulates the CFTR chloride channel. Cell 66:1027–1036

Choi JY, Joo NS, Krouse ME, Wu JV, Robbins RC, Ianowski JP, Hanrahan JW, Wine JJ (2007) Synergistic airway gland mucus secretion in response to vasoactive intestinal peptide and carbachol is lost in cystic fibrosis. J Clin Invest 117:3118–3127

Chunn JL, Young HW, Banerjee SK, Colasurdo GN, Blackburn MR (2001) Adenosine-dependent airway inflammation and hyperresponsiveness in partially adenosine deaminase-deficient mice. J Immunol 167:4676–4685

Clancy JP, Ruiz FE, Sorscher EJ (1999) Adenosine and its nucleotides activate wild-type and R117H CFTR through an A_{2B} receptor-coupled pathway. Am J Physiol 276:C361–C369

Cobb BR, Clancy JP (2003) Molecular and cell biology of adenosine receptors. In: Schwiebert EM (ed) Current topics in membranes, vol 54. Academic, New York

Cobb BR, Ruiz F, King CM, Fortenberry J, Greer H, Kovacs T, Sorscher EJ, Clancy JP (2002) A(2) adenosine receptors regulate CFTR through PKA and PLA(2). Am J Physiol Lung Cell Mol Physiol 282:L12–L25

Cowley EA (2003) Isoprostane-mediated secretion from human airway epithelial cells. Mol Pharmacol 64:298–307

Deterding RR, Lavange LM, Engels JM, Mathews DW, Coquillette SJ, Brody AS, Millard SP, Ramsey BW (2007) Phase 2 randomized safety and efficacy trial of nebulized denufosol tetrasodium in cystic fibrosis. Am J Respir Crit Care Med 176(4):362–369

Donaldson SH, Lazarowski ER, Picher M, Knowles MR, Stutts MJ, Boucher RC (2000) Basal nucleotide levels, release, and metabolism in normal and cystic fibrosis airways. Mol Med 6:969–982

Donaldson SH, Bennett WD, Zeman KL, Knowles MR, Tarran R, Boucher RC (2006) Mucus clearance and lung function in cystic fibrosis with hypertonic saline. N Engl J Med 354:241–250

Engelhardt JF, Yankaskas JR, Ernst SA, Yang Y, Marino CR, Boucher RC, Cohn JA, Wilson JM (1992) Submucosal glands are the predominant site of CFTR expression in the human bronchus. Nat Genet 2:240–248

Engelhardt JF, Zepeda M, Cohn JA, Yankaskas JR, Wilson JM (1994) Expression of the cystic fibrosis gene in adult human lung. J Clin Invest 93:737–749

Feoktistov I, Polosa R, Holgate ST, Biaggioni I (1998) Adenosine A_{2B} receptors: a novel therapeutic target in asthma? Trends Pharmacol Sci 19:148–153

Hentchel-Franks K, Lozano D, Eubanks-Tarn VL, Cobb BR, Fan L, Oster R, Sorscher EJ, Clancy JP (2004) Activation of airway Cl^- secretion in human subjects by adenosine. Am J Respir Cell Mol Biol 31:140–146

Hirsh AJ, Stonebraker JR, van Heusden CA, Lazarowski ER, Boucher RC, Picher M (2007) Adenosine deaminase 1 and concentrative nucleoside transporters 2 and 3 regulate adenosine on the apical surface of human airway epithelia: implications for inflammatory lung diseases. Biochemistry 46:10373–10383

Huang P, Trotter K, Boucher RC, Milgram SL, Stutts MJ (2000) PKA holoenzyme is functionally coupled to CFTR by AKAPs. Am J Physiol Cell Physiol 278:C417–C422

Huang P, Lazarowski ER, Tarran R, Milgram SL, Boucher RC, Stutts MJ (2001) Compartmentalized autocrine signaling to cystic fibrosis transmembrane conductance regulator at the apical membrane of airway epithelial cells. Proc Natl Acad Sci USA 98:14120–14125

Hudson VM (2001) Rethinking cystic fibrosis pathology: the critical role of abnormal reduced glutathione (GSH) transport caused by *CFTR* mutation. Free Radic Biol Med 30:1440–1461

Joo NS, Irokawa T, Wu JV, Robbins RC, Whyte RI, Wine JJ (2002) Absent secretion to vasoactive intestinal peptide in cystic fibrosis airway glands. J Biol Chem 277:50710–50715

Joo NS, Lee DJ, Winges KM, Rustagi A, Wine JJ (2004) Regulation of antiprotease and antimicrobial protein secretion by airway submucosal gland serous cells. J Biol Chem 279:38854–38860

Joo NS, Irokawa T, Robbins RC, Wine JJ (2006) Hyposecretion, not hyperabsorption, is the basic defect of cystic fibrosis airway glands. J Biol Chem 281:7392–7398

Joy A, Cowley EA (2005) Expression of prostenoid subtypes in Calu-3 cells. Ped PulmolS28

Joy AP, Cowley EA (2008) 8-iso-PGE2 stimulates anion efflux from airway epithelial cells via the EP4 prostanoid receptor. Am J Respir Cell Mol Biol 38:143–152

Knowles MR, Clarke LL, Boucher RC (1991) Activation by extracellular nucleotides of chloride secretion in the airway epithelia of patients with cystic fibrosis. N Engl J Med 325:533–538

Knowles MR, Paradiso AM, Boucher RC (1995) In vivo nasal potential difference: techniques and protocols for assessing efficacy of gene transfer in cystic fibrosis. Hum Gene Ther 6:445–455

Korngreen A, Priel Z (1996) Purinergic stimulation of rabbit ciliated airway epithelia: control by multiple calcium sources. J Physiol 497 (Pt 1):53–66

Lazarowski ER, Tarran R, Grubb BR, van Heusden CA, Okada S, Boucher RC (2004) Nucleotide release provides a mechanism for airway surface liquid homeostasis. J Biol Chem 279:36855–36864

Li Y, Wang W, Parker W, Clancy JP (2006) Adenosine regulation of cystic fibrosis transmembrane conductance regulator through prostenoids in airway epithelia. Am J Respir Cell Mol Biol 34:600–608

Linsdell P (2000) Inhibition of cystic fibrosis transmembrane conductance regulator chloride channel currents by arachidonic acid. Can J Physiol Pharmacol 78:490–499

Matsui H, Randell SH, Peretti SW, Davis CW, Boucher RC (1998a) Coordinated clearance of periciliary liquid and mucus from airway surfaces. J Clin Invest 102:1125–1131

Matsui H, Grubb BR, Tarran R, Randell SH, Gatzy JT, Davis CW, Boucher RC (1998b) Evidence for periciliary liquid layer depletion, not abnormal ion composition, in the pathogenesis of cystic fibrosis airways disease. Cell 95:1005–1015

McNamara N, Gallup M, Khong A, Sucher A, Maltseva I, Fahy J, Basbaum C (2004) Adenosine up-regulation of the mucin gene, MUC2, in asthma. FASEB J 18:1770–1772

Mense M, Vergani P, White DM, Altberg G, Nairn AC, Gadsby DC (2006) In vivo phosphorylation of CFTR promotes formation of a nucleotide-binding domain heterodimer. EMBO J 25:4728–4739

Moskwa P, Lorentzen D, Excoffon KJ, Zabner J, McCray PB, Jr., Nauseef WM, Dupuy C, Banfi B (2007) A novel host defense system of airways is defective in cystic fibrosis. Am J Respir Crit Care Med 175:174–183

Naren AP, Cobb B, Li C, Roy K, Nelson D, Heda GD, Liao J, Kirk KL, Sorscher EJ, Hanrahan J, Clancy JP (2003) A macromolecular complex of beta2-adrenergic receptor, CFTR, and ezrin/radixin/moesin-binding phosphoprotein 50 is regulated by PKA. Proc Natl Acad Sci USA 100:342–346

Olivier KN, Bennett WD, Hohneker KW, Zeman KL, Edwards LJ, Boucher RC, Knowles MR (1996) Acute safety and effects on mucociliary clearance of aerosolized uridine 5′-triphosphate $^{+/-}$ amiloride in normal human adults. Am J Respir Crit Care Med 154:217–223

Paradiso AM, Ribeiro CM, Boucher RC (2001) Polarized signaling via purinoceptors in normal and cystic fibrosis airway epithelia. J Gen Physiol 117:53–67

Park M, Ko SB, Choi JY, Muallem G, Thomas PJ, Pushkin A, Lee MS, Kim JY, Lee MG, Muallem S, Kurtz I (2002) The cystic fibrosis transmembrane conductance regulator interacts with and regulates the activity of the HCO_3^- salvage transporter human Na^+-HCO_3^- cotransport isoform 3. J Biol Chem 277:50503–50509

Pilewski JM, Frizzell RA (1999) Role of CFTR in airway disease. Physiol Rev 79:S215–S255

Polosa R (2002) Adenosine-receptor subtypes: their relevance to adenosine-mediated responses in asthma and chronic obstructive pulmonary disease. Eur Respir J 20:488–496

Polosa R, Holgate ST (1997) Adenosine bronchoprovocation: a promising marker of allergic inflammation in asthma? Thorax 52:919–923

Polosa R, Holgate ST (2006) Adenosine receptors as promising therapeutic targets for drug development in chronic airway inflammation. Curr Drug Targets 7:699–706

Polosa R, Ng WH, Crimi N, Vancheri C, Holgate ST, Church MK, Mistretta A (1995) Release of mast-cell-derived mediators after endobronchial adenosine challenge in asthma. Am J Respir Crit Care Med 151:624–629

Ribeiro CM (2006) The role of intracellular calcium signals in inflammatory responses of polarised cystic fibrosis human airway epithelia. Drugs R D 7:17–31

Ribeiro CM, Paradiso AM, Carew MA, Shears SB, Boucher RC (2005a) Cystic fibrosis airway epithelial Ca^{2+} i signaling: the mechanism for the larger agonist-mediated Ca^{2+} i signals in human cystic fibrosis airway epithelia. J Biol Chem 280:10202–10209

Ribeiro CM, Paradiso AM, Schwab U, Perez-Vilar J, Jones L, O'Neal W, Boucher RC (2005b) Chronic airway infection/inflammation induces a Ca^{2+}i-dependent hyperinflammatory response in human cystic fibrosis airway epithelia. J Biol Chem 280:17798–17806

Rollins BM, Burn M, Coakley RD, Chambers LA, Hirsh AJ, Clunes MT, Lethem MI, Donaldson SH, Tarran R (2008) A_{2B} adenosine receptors regulate the mucus clearance component of the lungs innate defense system. Am J Respir Cell Mol Biol 39:190–197

Rowe SM, Miller S, Sorscher EJ (2005) Cystic fibrosis. N Engl J Med 352:1992–2001

Rowe SM, Accurso F, Clancy JP (2007) Detection of cystic fibrosis transmembrane conductance regulator activity in early-phase clinical trials. Proc Am Thorac Soc 4:387–398

Russo C, Arcidiacono G, Polosa R (2006) Adenosine receptors: promising targets for the development of novel therapeutics and diagnostics for asthma. Fundam Clin Pharmacol 20:9–19

Schwiebert EM (1999) ABC transporter-facilitated ATP conductive transport. Am J Physiol 276:C1–C8

Schwiebert EM, Egan ME, Hwang TH, Fulmer SB, Allen SS, Cutting GR, Guggino WB (1995) *CFTR* regulates outwardly rectifying chloride channels through an autocrine mechanism involving ATP. Cell 81:1063–1073

Schweibert EM, Benos DJ, Egan ME, Stutts MJ, Guggino WB (1999) *CFTR* is a conductance regulator as well as a chloride channel. Physiol Rev 79:S145–S166

Spicuzza L, Di Maria G, Polosa R (2006) Adenosine in the airways: implications and applications. Eur J Pharmacol 533:77–88

Standaert TA, Boitano L, Emerson J, Milgram LJ, Konstan MW, Hunter J, Berclaz PY, Brass L, Zeitlin PL, Hammond K, Davies Z, Foy C, Noone PG, Knowles MR (2004) Standardized procedure for measurement of nasal potential difference: an outcome measure in multicenter cystic fibrosis clinical trials. Pediatr Pulmonol 37:385–392

Stutts MJ, Canessa CM, Olsen JC, Hamrick M, Cohn JA, Rossier BC, Boucher RC (1995) *CFTR* as a cAMP-dependent regulator of sodium channels. Science 269:847–850

Sun CX, Zhong H, Mohsenin A, Morschl E, Chunn JL, Molina JG, Belardinelli L, Zeng D, Blackburn MR (2006) Role of A_{2B} adenosine receptor signaling in adenosine-dependent pulmonary inflammation and injury. J Clin Invest 116:2173–2182

Tarran R, Lazarowski ER, Grubb BR, Matthews WG, Davis CW, Boucher RC (2001) Autoregulation of airway surface liquid (ASL) height involves *CFTR* and endogenous adenosine formation. Ped Pulmonol Suppl A71

Tarran R, Loewen ME, Paradiso AM, Olsen JC, Gray MA, Argent BE, Boucher RC, Gabriel SE (2002) Regulation of murine airway surface liquid volume by CFTR and Ca^{2+}-activated Cl^{-} conductances. J Gen Physiol 120:407–418

Tarran R, Button B, Picher M, Paradiso AM, Ribeiro CM, Lazarowski ER, Zhang L, Collins PL, Pickles RJ, Fredberg JJ, Boucher RC (2005) Normal and cystic fibrosis airway surface liquid homeostasis. The effects of phasic shear stress and viral infections. J Biol Chem 280:35751–35759

Tarran R, Button B, Boucher RC (2006a) Regulation of normal and cystic fibrosis airway surface liquid volume by phasic shear stress. Annu Rev Physiol 68:543–561

Tarran R, Trout L, Donaldson SH, Boucher RC (2006b) Soluble mediators, not cilia, determine airway surface liquid volume in normal and cystic fibrosis superficial airway epithelia. J Gen Physiol 127:591–604

Vergani P, Basso C, Mense M, Nairn AC, Gadsby DC (2005a) Control of the CFTR channel's gates. Biochem Soc Trans 33:1003–1007

Vergani P, Lockless SW, Nairn AC, Gadsby DC (2005b) CFTR channel opening by ATP-driven tight dimerization of its nucleotide-binding domains. Nature 433:876–880

Wine JJ, Joo NS (2004) Submucosal glands and airway defense. Proc Am Thorac Soc 1:47–53

Wu JV, Krouse ME, Wine JJ (2007) Acinar origin of CFTR-dependent airway submucosal gland fluid secretion. Am J Physiol Lung Cell Mol Physiol 292:L304–L311

Young HW, Molina JG, Dimina D, Zhong H, Jacobson M, Chan LN, Chan TS, Lee JJ, Blackburn MR (2004) A_3 adenosine receptor signaling contributes to airway inflammation and mucus production in adenosine deaminase-deficient mice. J Immunol 173:1380–1389

Zhang L, Sanderson MJ (2003) Oscillations in ciliary beat frequency and intracellular calcium concentration in rabbit tracheal epithelial cells induced by ATP. J Physiol 546:733–749

Zsembery A, Boyce AT, Liang L, Peti-Peterdi J, Bell PD, Schwiebert EM (2003) Sustained calcium entry through P2X nucleotide receptor channels in human airway epithelial cells. J Biol Chem 278:13398–13408

Zsembery A, Fortenberry JA, Liang L, Bebok Z, Tucker TA, Boyce AT, Braunstein GM, Welty E, Bell PD, Sorscher EJ, Clancy JP, Schwiebert EM (2004) Extracellular zinc and ATP restore chloride secretion across cystic fibrosis airway epithelia by triggering calcium entry. J Biol Chem 279:10720–10729

Adenosine Receptors in Wound Healing, Fibrosis and Angiogenesis

Igor Feoktistov, Italo Biaggioni, and Bruce N. Cronstein

Contents

Abstract Wound healing and tissue repair are critical processes, and adenosine, released from injured or ischemic tissues, plays an important role in promoting wound healing and tissue repair. Recent studies in genetically manipulated mice demonstrate that adenosine receptors are required for appropriate granulation tissue formation and in adequate wound healing. A_{2A} and A_{2B} adenosine receptors stimulate both of the critical functions in granulation tissue formation (i.e., new matrix production and angiogenesis), and the A_1 adenosine receptor (AR) may also contribute to new vessel formation. The effects of adenosine acting on these receptors is both direct and indirect, as AR activation suppresses antiangiogenic factor production by endothelial cells, promotes endothelial cell proliferation, and stimulates angiogenic factor production by endothelial cells and other cells present in

B.N. Cronstein (✉)
Paul R. Esserman Professor of Medicine, Division of Clinical Pharmacology, Department of Medicine, NYU School of Medicine, 550 First Ave., NBV16N1, New York, NY 10016, USA
cronsb01@med.nyu.edu

C.N. Wilson and S.J. Mustafa (eds.), *Adenosine Receptors in Health and Disease*, Handbook of Experimental Pharmacology 193, DOI 10.1007/978-3-540-89615-9_13,

the wound. Similarly, adenosine, acting at its receptors, stimulates collagen matrix formation directly. Like many other biological processes, AR-mediated promotion of tissue repair is critical for appropriate wound healing but may also contribute to pathogenic processes. Excessive tissue repair can lead to problems such as scarring and organ fibrosis and adenosine, and its receptors play a role in pathologic fibrosis as well. Here we review the evidence for the involvement of adenosine and its receptors in wound healing, tissue repair and fibrosis.

Keywords Adenosine receptors · Wound healing · Fibrosis · Angiogenesis · Neovascularization

Abbreviations

ADA	Adenosine deaminase
AR	Adenosine receptor
bFGF	Basic fibroblast growth factor
IFN	Interferon
IL	Interleukin
MMP	Matrix metalloproteinase
NECA	Adenosine 5′-*N*-ethyluronamide
TGF	Transforming growth factor
TNF-α	Tumor necrosis factor alpha
VEGF	Vascular endothelial growth factor

1 Introduction

Tissue repair is an essential homeostatic mechanism that involves a series of coordinated and overlapping phases: inflammation, neovascularization, new tissue generation, and tissue reorganization. In acute inflammation, tissue damage is followed by resolution, whereas in chronic inflammation, damage and repair continue concurrently. Inflammatory cells neutralize invading pathogens, remove waste and debris, and promote restoration of normal function, either through resolution or repair. Inflammation also promotes angiogenesis and vasculogenesis, the formation of new blood vessels, which in turn may enhance the recruitment of inflammatory cells and the subsequent laying down of extracellular matrix to repair tissue damage. Although usually beneficial to the organism, inflammation may lead to tissue damage, resulting in escalation of chronic inflammation. Furthermore, aberrant or inadequate repair can lead to excessive and poorly ordered matrix deposition and fibrosis, which affects normal tissue architecture and can ultimately disable the proper functioning of organs. Like matrix generation, overly exuberant vessel formation may lead to medical problems as well, and diabetic retinopathy and macular degeneration are examples of this phenomenon.

Extracellular accumulation of adenosine in response to tissue damage is an important event in the control of all aspects of tissue repair. The nature of adenosine's action depends on the magnitude of changes in extracellular adenosine concentrations as well as on the identity and expression levels of each adenosine receptor subtype on individual cell types. The role of adenosine in the regulation of inflammation is extensively covered in other chapters of this book. In this chapter, we will discuss the roles of specific adenosine receptors in the regulation of neovascularization and fibrosis in different organs and tissues.

2 Role of Adenosine in Neovascularization

Accumulating evidence indicates that adenosine is an important regulator of neovascularization, including angiogenesis and vasculogenesis. Stimulation of new blood vessel formation by adenosine was demonstrated in the chick chorioallantoic membrane and embryo (Adair et al. 1989; Dusseau et al. 1986; Dusseau and Hutchins 1988), the mouse retina (Afzal et al. 2003; Mino et al. 2001), and the optical tectum of *Xenopus leavis* tadpoles (Jen and Rovainen 1994). Adenosine reportedly modulates a number of steps involved in angiogenesis, including endothelial cell proliferation (Dubey et al. 2002; Ethier et al. 1993; Grant et al. 1999, 2001; Meininger et al. 1988; Meininger and Granger 1990; Van Daele et al. 1992), migration (Dubey et al. 2002; Grant et al. 2001; Lutty et al. 1998; Meininger et al. 1988; Teuscher and Weidlich 1985), and tube formation (Grant et al. 2001; Lutty et al. 1998). Adenosine has been also suggested to play an important role in adult vasculogenesis by directing the homing of endothelial progenitor cells to the site of tissue injury (Montesinos et al. 2004; Ryzhov et al. 2008b).

Adenosine has direct mitogenic effects on vascular cells that may contribute to angiogenesis (Ethier and Dobson Jr. 1997; Meininger et al. 1988; Sexl et al. 1995; Van Daele et al. 1992). However, the main proangiogenic actions of adenosine have been attributed to its ability to regulate the production of pro- and antiangiogenic substances. Adenosine modulates the release of angiogenic factors from various cells and tissues (Feoktistov et al. 2003, 2004; Gu et al. 1999, 2000; Hashimoto et al. 1994; Leibovich et al. 2002; Olah and Roudabush 2000; Pueyo et al. 1998; Takagi et al. 1996; Wakai et al. 2001; Zeng et al. 2003), thus regulating capillary growth in a paracrine fashion. In addition, adenosine can modulate release of angiogenic factors from endothelial cells (Desai et al. 2005; Feoktistov et al. 2002; Fischer et al. 1995, 1997; Grant et al. 1999; Khoa et al. 2003; Takagi et al. 1996), which may regulate capillary growth in an autocrine fashion.

All four adenosine receptor (AR) subtypes have been implicated in the regulation of neovascularization. In a similar manner to our early observation that the stimulation of A_1ARs on neutrophils increased their adherence to vascular endothelium (Cronstein et al. 1992), we have recently demonstrated that A_1ARs located on embryonic endothelial progenitor cells promote their adhesion to cardiac microvascular endothelial cells, suggesting an important role of this receptor subtype in vasculogenesis (Ryzhov et al. 2008b). A_1ARs have been also reported to upregulate vascular endothelial growth factor (VEGF) production from monocytes, thus promoting angiogenesis in an in vitro model (Clark et al. 2007). Among all of the AR subtypes, A_1ARs have the highest affinity to adenosine (Fredholm et al. 2001). It is possible, therefore, that engagement of the high-affinity A_1ARs is especially important for circulating cells moving toward a gradient of adenosine concentrations generated by tissue injury and/or hypoxia, whereas the lower-affinity A_2ARs are more important for the regulation of cells located in the vicinity of the injured or ischemic loci, where concentrations of adenosine are the highest.

Indeed, both A_2AR subtypes, A_{2A} and A_{2B}ARs, have been implicated in regulation of angiogenesis and vasculogenesis. Depending on tissue or cell studied, either one of these receptor subtypes can take the lead and play a dominant role in the regulation of angiogenic factors. For example, A_{2B}ARs upregulate the proangiogenic factors VEGF, basic fibroblast growth factor (bFGF), insulin-like factor 1, and IL-8 in human microvascular endothelial cells (Feoktistov et al. 2002; Grant et al. 1999). Conversely, A_{2A}ARs were reported to upregulate VEGF in macrophages (Leibovich et al. 2002; Pinhal-Enfield et al. 2003). However, A_{2B}ARs may also contribute to regulation of VEGF in these cells, since genetic deletion of A_{2B}ARs significantly decreased adenosine-dependent secretion of VEGF in mouse peritoneal macrophages (our unpublished observations). In addition, the stimulation of A_3ARs in mast cells and some tumors can result in the upregulation of certain proangiogenic factors, complementing the actions of adenosine mediated via A_{2B}ARs (Feoktistov et al. 2003; Merighi et al. 2005, 2007). Thus, the contribution of adenosine to the regulation of neovascularization can be dictated by the expression profile of AR subtypes and by the intracellular machinery to which they are coupled in specific cell types. Furthermore, the expressions of AR subtypes and their functions are subject to dynamic regulation by conditions present during inflammation, such as hypoxia and cytokine exposure (Bshesh et al. 2002; Eltzschig et al. 2003; Feoktistov et al. 2004; Khoa et al. 2003). Because the A_{2B}AR promoter contains a functional binding site for hypoxia-inducible factor (Kong et al. 2006), the onset of hypoxia strongly induces A_{2B}AR expression. Hypoxia-induced upregulation of A_{2B}ARs has been reported in human tumor cells (Zeng et al. 2003), rat hippocampus (Zhou et al. 2004), and human dermal microvascular endothelial cells (Eltzschig et al. 2003). This may have important functional implications for regulation of angiogenesis. For example, in human bronchial smooth muscle cells and human umbilical vein endothelial cells, adenosine does not stimulate VEGF secretion under normoxic conditions, but hypoxia increases expression of A_{2B}ARs, which are then able to stimulate VEGF release (Feoktistov et al. 2004). Similarly, treatment of human dermal microvascular endothelial cells with interferon (IFN)-γ increases A_{2B}AR expression but decreases A_{2A}AR levels. In contrast, other proinflammatory cytokines, such as interleukin (IL)-1 and tumor necrosis factor alpha (TNF-α) increase both A_{2A} and A_{2B} AR expression and function (Khoa et al. 2003). Because the expression and function of adenosine receptor subtypes may differ depending on the tissue and the nature of the tissue injury, we will next examine the role of AR subtypes in specific organs and pathological states.

2.1 Regulation of Neovascularization in the Skin

We have previously reported (Montesinos et al. 2002) that mice with genetically disrupted A_{2A}ARs form significantly fewer microvessels in healing wounds and in response to mechanical trauma by the formation of an air pouch (Montesinos et al. 2002). Furthermore, application of an A_{2A}AR agonist to wounds increases microvessel formation from both pre-existing endothelial cells and bone marrow-derived endothelial progenitors as compared to vehicle-treated mice, observations that provide the first in vivo evidence that A_{2A}AR occupancy promotes angiogenesis and vasculogenesis (Montesinos et al. 2002, 2004). Further studies indicate that the angiogenic effects of A_{2A}AR occupancy are mediated both directly on endothelial cells (increased endothelial cell migration and microvascular endothelial cell VEGF production; Khoa et al. 2003; Montesinos et al. 1997) and indirectly via promotion of VEGF production by macrophages (Leibovich et al. 2002). Desai and colleagues (Desai et al. 2005) have also reported evidence to indicate that A_{2A}AR occupancy suppresses the production of thrombospondin I, a potent inhibitor of angiogenesis, and this inhibition is responsible for enhanced vascular tube formation in vitro. Thus, there is growing evidence that A_{2A}ARs play an important role in skin neovascularization, and particularly during wound healing.

2.2 Regulation of Neovascularization in the Heart and Skeletal Muscles

Many studies have demonstrated that chronic elevation of tissue adenosine concentrations induced by the adenosine reuptake blocker dipyridamole (Adolfsson et al. 1981, 1982; Adolfsson 1986a, b; Belardinelli et al. 2001; Mall et al. 1987; Mattfeldt and Mall 1983; Symons et al. 1993; Tornling et al. 1978, 1980a, b; Tornling 1982a, b; Torry et al. 1992), or long-term administration of adenosine and its analogs (Hudlicka et al. 1986; Wothe et al. 2002; Ziada et al. 1984), promotes capillary proliferation in the heart and skeletal muscles. Antagonism of ARs with caffeine abrogated VEGF upregulation in skeletal muscles induced by local injection of adenosine 5′-*N*-ethyluronamide (NECA) into the mouse hind limb and produced a 46% reduction in neovascularization in a mouse ischemic hind limb model (Ryzhov et al. 2007). In the isolated heart model, adenosine but not selective A_{2A} or A_3 AR agonists increased retention of embryonic endothelial progenitors to microvascular endothelium, suggesting that A_1 and A_{2B}ARs may play an important role in the initial phase of vasculogenesis, promoting homing of endothelial progenitor cells to the site of ischemic injury (Ryzhov et al. 2008b). Indeed, endothelial progenitor cells and cardiac microvascular endothelial cells preferentially express functional A_1 and A_{2B}ARs, respectively, and both subtypes are involved in the regulation of the adhesion of endothelial progenitors to microvascular endothelial cells in the heart. Moreover, the interaction between P-selectin and its ligand PSGL-1 plays an important role in these process, and stimulation of A_{2B}ARs in cardiac microvascular endothelial cells induces rapid cell surface expression of P-selectin (Ryzhov et al. 2008b). These findings suggested a role for A_1 and A_{2B}ARs in myocardial vasculogenesis, and provided a rationale for the potential use of adenosine to stimulate engraftment in cell-based therapies.

2.3 *Regulation of Neovascularization in the Lung*

Angiogenesis is a feature of chronic lung diseases such as asthma and pulmonary fibrosis. Studies in adenosine deaminase (ADA)-deficient mice, characterized by elevated lung tissue levels of adenosine, strongly suggest a causal association between adenosine and an inflammatory phenotype (Blackburn et al. 2000; Blackburn 2003). These mice exhibit a lung phenotype with features of lung inflammation, bronchial hyperresponsiveness, enhanced mucus secretion, increased IgE synthesis, and elevated levels of proinflammatory cytokines and angiogenic factors that could be reversed by lowering adenosine levels with exogenous ADA (Blackburn et al. 2000). In particular, levels of the angiogenic chemokine CXCL1 (mouse functional homolog of human IL-8) are significantly elevated in an adenosine-dependent manner in the lungs of ADA-deficient mice, leading to substantial angiogenesis in the tracheas (Mohsenin et al. 2007a). The A_{2B}AR subtype appears to play an important role in this model, because pharmacological inhibition of A_{2B}ARs significantly reduced elevations in proinflammatory cytokines as well as mediators of airway remodeling induced by high adenosine levels in the lungs of ADA-deficient mice (Sun et al. 2006). In contrast, genetic removal of the A_{2A}AR enhances pulmonary inflammation, mucin production, and angiogenesis in ADA-deficient mice (Mohsenin et al. 2007b).

2.4 *Regulation of Neovascularization in Tumors*

Metabolically active solid tumors grow rapidly and routinely experience severe hypoxia and necrosis, which causes adenine nucleotide degradation and adenosine release. Expression of A_{2B}ARs was documented in various cancerous cells (Feoktistov and Biaggioni 1993, 1995; Panjehpour et al. 2005; Phelps et al. 2006; Rodrigues et al. 2007; Zeng et al. 2003), and analysis of gene expression in primary human tumors uncovered overexpression of A_{2B}ARs, suggesting their potential role in cancer biology (Li et al. 2005). Studies from different laboratories demonstrate that stimulation of A_{2B}ARs in cancer cell lines upregulates the production of angiogenic factors, suggesting that tumor A_{2B}ARs may promote neovascularization (Feoktistov et al. 2003; Merighi et al. 2007; Zeng et al. 2003). A_3ARs expressed in some tumor cell lines may also complement these A_{2B}AR-mediated effects by upregulating other proangiogenic factors (Feoktistov et al. 2003; Merighi et al. 2005,

2007). In addition, host tumor-infiltrating immune cells can also play an important role in tumor angiogenesis, since Lewis lung carcinoma isografts in A_{2B}AR knock-out mice contained lower VEGF levels and exhibited lower vessel density compared to tumors grafted in wild-type mice (Ryzhov et al. 2008a). Furthermore, treatment with A_{2A}/A_{2B}AR antagonists inhibited neovascularization of CL8-1 melanoma in mice (Ohta et al. 2006). Thus, there is growing evidence that adenosine acting via A_{2B} and possibly A_3 or A_{2A}ARs can promote tumor neovascularization. Involvement of different AR subtypes in the regulation of neovascularization is not surprising due to the multifaceted mechanism of blood vessel development.

3 Role of Adenosine in Fibrosis

3.1 A_{2A} Adenosine Receptor Agonists Promote Wound Healing

Recent reports indicate that topical application of an A_{2A}AR agonist increases the rate at which wounds close (Montesinos et al. 1997). That A_{2A}ARs were involved in this pharmacologic effect was demonstrated by the observation that a specific A_{2A}AR antagonist, but not antagonists at other ARs, reversed the effect of the selective A_{2A}AR agonist CGS21680 on wound healing. Treatment of wounds with this AR agonist promoted fibroblast migration in vitro, and in the AR agonist-treated mice there was an increase in matrix and fibroblast infiltration into the wounds (Montesinos et al. 1997). More recent studies demonstrate that a more highly selective A_{2A}AR agonist, sonedenoson, is a more potent promoter of wound healing than recombinant platelet derived growth factor (becaplermin) (Victor-Vega et al. 2002). The role of A_{2A}ARs in the promotion of wound healing was more fully confirmed by the observation that a selective A_{2A}AR agonist promotes wound healing in wild-type but not A_{2A}AR knockout mice (Montesinos et al. 2002; Victor-Vega et al. 2002). In these studies, there was a marked increase in the number of blood vessels in the healing wounds of wild-type mice treated with the A_{2A}AR agonist as compared to untreated controls. Absence of A_{2A}ARs was associated with disorganized granulation tissue although re-epithelialization was not delayed in the knockout mice. In contrast to this study, Sun and colleagues observed that N^6-cyclopentyladenosine, a relatively selective A_1AR agonist, promotes wound healing (Sun et al. 1999). In this study, there was no confirmation that the high concentrations of the agonist used were indeed selective for A_1ARs or whether the phenomenon could be mediated by A_{2A}ARs. These findings indicate that A_{2A}ARs stimulate wound healing by modulating inflammatory cell, endothelial cell and fibroblast functions that promote wound healing. A topical A_{2A}AR agonist, sonedenoson, is currently undergoing testing in Phase II clinical trials for the treatment of diabetic foot ulcers.

3.2 A_{2A} Adenosine Receptor Occupancy Stimulates Fibroblast Matrix Production

Replacement of the collagenous matrix of the skin and other tissues is an integral part of wound healing. Once the debris and destroyed matrix at the site of injury are eliminated, fibroblasts lay down a new matrix. This matrix may be remodeled over a longer period of time and the wound develops the characteristic appearance of a scar. A_{2A}AR occupancy stimulates fibroblasts to synthesize type I and III collagen at an increased level, similar to that induced by the growth factor transforming growth factor (TGF)-β, and downregulates matrix metalloproteinase (MMP) 9 but not MMP2 (Chan et al. 2006a).

The observation that adenosine, acting at A_{2A}ARs, stimulates the formation of matrix suggests the possibility that adenosine A_{2A}ARs play a role in fibrosing conditions and scarring, a hypothesis confirmed by in vivo experiments. Animals lacking A_{2A}ARs or treated with an A_{2A}AR antagonist were protected from developing diffuse dermal fibrosis in response to bleomycin (Chan et al. 2006a). The role of A_{2A}ARs in fibrosis in tissues outside of the skin is less clear. Prior studies have demonstrated that A_{2B}ARs regulate production of collagen in pulmonary and cardiac fibroblasts (Chen et al. 2004; Dubey et al. 2000, 2001), but other studies have demonstrated that A_{2A}ARs regulate collagen I and III production by hepatic stellate cells (Che et al. 2007), the fibroblasts of the liver, and A_{2A}AR knockout mice are protected from developing hepatic fibrosis following treatment with either CCl_4 or thioacetamide (Chan et al. 2006b). These observations help to explain the protection against death from liver disease provided by coffee drinking (Corrao et al. 1994, 2001; Gallus et al. 2002; Klatsky et al. 1993, 2006; Klatsky and Armstrong 1992; Ruhl et al. 2005; Sharp et al. 1999; Tverdal and Skurtveit 2003), since caffeine is a relatively weak and nonselective AR antagonist which offers some protection (although not complete) from the development of hepatic fibrosis in murine models (Chan et al. 2006b).

In a murine model of diffuse dermal fibrosis resembling scleroderma, we have also found that A_{2A}ARs play a central role in the development of fibrosis. A_{2A}ARs are present on human dermal fibroblasts and, when occupied, regulate collagen production by these cells (Chan et al. 2006a). Mice treated with subcutaneous bleomycin develop diffuse dermal fibrosis and we found that both A_{2A}AR knockout mice and mice treated with a selective A_{2A}AR antagonist were protected from the development of bleomycin-induced dermal fibrosis (Chan et al. 2006a). These results are consistent with the hypothesis that A_{2A}ARs play a role in organ and tissue fibrosis and that blockade or elimination of these receptors can prevent fibrosis.

Recently published indirect evidence provides further support for a role for adenosine and its receptors in dermal fibrosis. Imiquimod is an immune modulator that promotes a shift from Th2- to Th1-type immune responses (reviewed in Schon and Schon 2007) by mechanisms that have not been fully evaluated. Studies in inflammatory cells indicate that imiquimod, at pharmacologically relevant concentrations, is an A_{2A}AR antagonist, and that this may account for its immunological effects (Schon et al. 2006). Imiquimod, applied topically, has been used to treat morphea, a skin disease characterized by localized fibrosis, and its use has been advocated for the treatment of Dupuytren's contracture, another fibrosing disease (Dytoc et al. 2005; Man and Dytoc 2004; Namazi 2006; Schon et al. 2006). While intriguing (and supporting the clinical relevance of this work), we do realize the anecdotal nature of these reports.

3.3 A_{2B} Adenosine Receptor Occupancy Regulates Fibroblast Collagen Production and Fibrosis

As described above, a number of recent studies have demonstrated that cardiac and pulmonary fibroblasts express A_{2B}ARs that regulate their production of collagen (Chen et al. 2004; Dubey et al. 1997, 1998; Zhong et al. 2005). Stimulation of A_{2B}ARs in cardiac fibroblasts inhibited their proliferation, protein synthesis and collagen production (Chen et al. 2004; Dubey et al. 1997, 1998). Furthermore, it has been demonstrated in vivo that long-term stimulation of A_{2B}ARs after myocardial infarction prevents cardiac remodeling (Wakeno et al. 2006). In contrast, studies in ADA-deficient mice indicate that these animals develop pulmonary inflammation and pulmonary fibrosis that appear to be mediated by A_{2B}ARs (Sun et al. 2006), thus suggesting a role for A_{2B}ARs in pulmonary fibrosis. Based on these studies and the results described above, it is reasonable to conclude that adenosine can either inhibit (heart) or stimulate (skin, liver, lungs) fibrosis, and that adenosine-regulated fibrosis is mediated by different receptors depending on which organ is studied (skin and liver vs. heart and lungs).

3.4 A_1 Adenosine Receptors Play a Role in Cardiac and Vascular Fibrosis

Recently, Kalk and coworkers reported that SLV320 (Solvay Pharmaceuticals), a highly selective A_1AR antagonist, reduced myocardial fibrosis in a model of uremic cardiomyopathy (Kalk et al. 2007). In this model, partially (5/6) nephrectomized rats were treated with SLV320 or vehicle and myocardial fibrosis was markedly reduced, as was albuminuria, without any change in blood pressure or other factors that might have accounted for the change. Another problem associated with fibrosis and abnormal "wound" healing that may be mediated by A_1ARs is intimal hyperplasia and stenosis following stent placement, and recent studies suggest that an A_1AR antagonist diminishes both intimal hyperplasia and smooth muscle proliferation in a model of stent stenosis (Edwards et al. 2008) Thus, A_1ARs may also play a role in fibrosis, although their role seems to be confined to the cardiovascular system.

4 Conclusion

Adenosine and its receptors play important roles in both matrix production and neovascularization, processes that are critical for wound healing and tissue repair. Moreover, adenosine and its receptors play a direct role in stimulating fibrosis in the skin, lungs and liver, but inhibiting fibrosis in the heart. Adenosine and its receptors may also play an important role in physiologic and pathologic angiogenesis. Targeting of ARs to promote wound healing and neovascularization of ischemic tissues or to diminish pathologic fibrosis and angiogenesis is currently underway.

References

Adair TH, Montani JP, Strick DM, Guyton AC (1989) Vascular development in chick embryos: a possible role for adenosine. Am J Physiol 256:H240–H246

Adolfsson J (1986a) The time dependence of training-induced increase in skeletal muscle capillarization and the spatial capillary to fibre relationship in normal and neovascularized skeletal muscle of rats. Acta Physiol Scand 128:259–266

Adolfsson J (1986b) Time dependence of dipyridamole-induced increase in skeletal muscle capillarization. Arzneimittel-Forschung 36:1768–1769

Adolfsson J, Ljungqvist A, Tornling G, Unge G (1981) Capillary increase in the skeletal muscle of trained young and adult rats. J Physiol 310:529–532

Adolfsson J, Tornling G, Unge G, Ljungqvist A (1982) The prophylactic effect of dipyridamole on the size of myocardial infarction following coronary artery occlusion. Acta Pathol Microbiol Immunol Scand A Pathol 90:273–275

Afzal A, Shaw LC, Caballero S, Spoerri PE, Lewin AS, Zeng D, Belardinelli L, Grant MB (2003) Reduction in preretinal neovascularization by ribozymes that cleave the A_{2B} adenosine receptor mRNA. Circ Res 93:500–506

Belardinelli R, Belardinelli L, Shryock JC (2001) Effects of dipyridamole on coronary collateralization and myocardial perfusion in patients with ischaemic cardiomyopathy. Eur Heart J 22:1205–1213

Blackburn MR (2003) Too much of a good thing: adenosine overload in adenosine-deaminase-deficient mice. Trends Pharmacol Sci 24:66–70

Blackburn MR, Volmer JB, Thrasher JL, Zhong H, Crosby JR, Lee JJ, Kellems RE (2000) Metabolic consequences of adenosine deaminase deficiency in mice are associated with defects in alveogenesis, pulmonary inflammation, and airway obstruction. J Exp Med 192:159–170

Bshesh K, Zhao B, Spight D, Biaggioni I, Feoktistov I, Denenberg A, Wong HR, Shanley TP (2002) The A_{2A} receptor mediates an endogenous regulatory pathway of cytokine expression in THP-1 cells. J Leukoc Biol 72:1027–1036

Chan ES, Fernandez P, Merchant AA, Montesinos MC, Trzaska S, Desai A, Tung CF, Khoa DN, Pillinger MH, Reiss AB, Tomic-Canic M, Chen JF, Schwarzschild MA, Cronstein BN (2006a) Adenosine A_{2A} receptors in diffuse dermal fibrosis: pathogenic role in human dermal fibroblasts and in a murine model of scleroderma. Arthritis Rheum 54:2632–2642

Chan ES, Montesinos MC, Fernandez P, Desai A, Delano DL, Yee H, Reiss AB, Pillinger MH, Chen JF, Schwarzschild MA, Friedman SL, Cronstein BN (2006b) Adenosine A_{2A} receptors play a role in the pathogenesis of hepatic cirrhosis. Br J Pharmacol 148:1144–1155

Che J, Chan ES, Cronstein BN (2007) Adenosine A_{2A} receptor occupancy stimulates collagen expression by hepatic stellate cells via pathways involving protein kinase A, Src, and extracellular signal-regulated kinases 1/2 signaling cascade or p38 mitogen-activated protein kinase signaling pathway. Mol Pharmacol 72:1626–1636

Chen Y, Epperson S, Makhsudova L, Ito B, Suarez J, Dillmann W, Villarreal F (2004) Functional effects of enhancing or silencing adenosine A_{2B} receptors in cardiac fibroblasts. Am J Physiol 287:H2478–H2486

Clark AN, Youkey R, Liu X, Jia L, Blatt R, Day YJ, Sullivan GW, Linden J, Tucker AL (2007) A_1 adenosine receptor activation promotes angiogenesis and release of VEGF from monocytes. Circ Res 101:1130–1138

Corrao G, Lepore AR, Torchio P, Valenti M, Galatola G, D'Amicis A, Arico S, di Orio F (1994) The effect of drinking coffee and smoking cigarettes on the risk of cirrhosis associated with alcohol consumption. A case–control study. Provincial Group for the Study of Chronic Liver Disease. Eur J Epidemiol 10:657–664

Corrao G, Zambon A, Bagnardi V, D'Amicis A, Klatsky A, Collaborative SIDECIR Group. (2001) Coffee, caffeine, and the risk of liver cirrhosis. Ann Epidemiol 11:458–465

Cronstein BN, Levin RI, Philips M, Hirschhorn R, Abramson SB, Weissmann G (1992) Neutrophil adherence to endothelium is enhanced via adenosine A_1 receptors and inhibited via adenosine A_2 receptors. J Immunol 92:2201–2206

Desai A, Victor-Vega C, Gadangi S, Montesinos MC, Chu CC, Cronstein BN (2005) Adenosine A_{2A} receptor stimulation increases angiogenesis by down-regulating production of the antiangiogenic matrix protein thrombospondin 1. Mol Pharmacol 67:1406–1413

Dubey RK, Gillespie DG, Mi Z, Jackson EK (1997) Exogenous and endogenous adenosine inhibits fetal calf serum-induced growth of rat cardiac fibroblasts: role of A_{2B} receptors. Circulation 96:2656–2666

Dubey RK, Gillespie DG, Jackson EK (1998) Adenosine inhibits collagen and protein synthesis in cardiac fibroblasts: role of A_{2B} receptors. Hypertension 31:943–948

Dubey RK, Gillespie DG, Shue H, Jackson EK (2000) A_{2B} receptors mediate antimitogenesis in vascular smooth muscle cells. Hypertension 35:267–272

Dubey RK, Gillespie DG, Zacharia LC, Mi Z, Jackson EK (2001) A_{2B} receptors mediate the antimitogenic effects of adenosine in cardiac fibroblasts. Hypertension 37:716–721

Dubey RK, Gillespie DG, Jackson EK (2002) A_{2B} adenosine receptors stimulate growth of porcine and rat arterial endothelial cells. Hypertension 39:530–535

Dusseau JW, Hutchins PM (1988) Hypoxia-induced angiogenesis in chick chorioallantoic membranes: a role for adenosine. Respir Physiol 71:33–44

Dusseau JW, Hutchins PM, Malbasa DS (1986) Stimulation of angiogenesis by adenosine on the chick chorioallantoic membrane. Circ Res 59:163–170

Dytoc M, Ting PT, Man J, Sawyer D, Fiorillo L (2005) First case series on the use of imiquimod for morphoea. Br J Dermatol 153:815–820

Edwards JM, Alloosh MA, Long XL, Dick GM, Lloyd PG, Mokelke EA, Sturek M (2008) Adenosine A_1 receptors in neointimal hyperplasia and in-stent stenosis in Ossabaw miniature swine. Coronary Artery Dis 19:27–31

Eltzschig HK, Ibla JC, Furuta GT, Leonard MO, Jacobson KA, Enjyoji K, Robson SC, Colgan SP (2003) Coordinated adenine nucleotide phosphohydrolysis and nucleoside signaling in posthypoxic endothelium: role of ectonucleotidases and adenosine A_{2B} receptors. J Exp Med 198:783–796

Ethier MF, Dobson JG Jr (1997) Adenosine stimulation of DNA synthesis in human endothelial cells. Am J Physiol 272:H1470–H1479

Ethier MF, Chander V, Dobson JG (1993) Adenosine stimulates proliferation of human endothelial cells in culture. Am J Physiol 265:H131–H138

Feoktistov I, Biaggioni I (1993) Characterization of adenosine receptors in human erythroleukemia cells. Further evidence for heterogeneity of adenosine A_2 receptors. Mol Pharmacol 43:909–914

Feoktistov I, Biaggioni I (1995) Adenosine A_{2B} receptors evoke interleukin-8 secretion in human mast cells. An enprofylline-sensitive mechanism with implications for asthma. J Clin Invest 96:1979–1986

Feoktistov I, Goldstein AE, Ryzhov S, Zeng D, Belardinelli L, Voyno-Yasenetskaya T, Biaggioni I (2002) Differential expression of adenosine receptors in human endothelial cells: role of A_{2B} receptors in angiogenic factor regulation. Circ Res 90:531–538

Feoktistov I, Ryzhov S, Goldstein AE, Biaggioni I (2003) Mast cell-mediated stimulation of angiogenesis: cooperative interaction between A_{2B} and A_3 adenosine receptors. Circ Res 92:485–492

Feoktistov I, Ryzhov S, Zhong H, Goldstein AE, Matafonov A, Zeng D, Biaggioni I (2004) Hypoxia modulates adenosine receptors in human endothelial and smooth muscle cells toward an A_{2B} angiogenic phenotype. Hypertension 44:649–654

Fischer S, Sharma HS, Karliczek GF, Schaper W (1995) Expression of vascular permeability factor/vascular endothelial growth factor in pig cerebral microvascular endothelial cells and its upregulation by adenosine. Brain Res Mol Brain Res 28:141–148

Fischer S, Knoll R, Renz D, Karliczek GF, Schaper W (1997) Role of adenosine in the hypoxic induction of vascular endothelial growth factor in porcine brain derived microvascular endothelial cells. Endothelium 5:155–165

Fredholm BB, Irenius E, Kull B, Schulte G (2001) Comparison of the potency of adenosine as an agonist at human adenosine receptors expressed in Chinese hamster ovary cells. Biochem Pharmacol 61:443–448

Gallus S, Tavani A, Negri E, La Vecchia C (2002) Does coffee protect against liver cirrhosis? Ann Epidemiol 12:202–205

Grant MB, Tarnuzzer RW, Caballero S, Ozeck MJ, Davis MI, Spoerri PE, Feoktistov I, Biaggioni I, Shryock JC, Belardinelli L (1999) Adenosine receptor activation induces vascular endothelial growth factor in human retinal endothelial cells. Circ Res 85:699–706

Grant MB, Davis MI, Caballero S, Feoktistov I, Biaggioni I, Belardinelli L (2001) Proliferation, migration, and ERK activation in human retinal endothelial cells through A_{2B} adenosine receptor stimulation. Investig Ophthalmol Vis Sci 42:2068–2073

Gu JW, Brady AL, Anand V, Moore MC, Kelly WC, Adair TH (1999) Adenosine upregulates VEGF expression in cultured myocardial vascular smooth muscle cells. Am J Physiol 277:H595–H602

Gu JW, Ito BR, Sartin A, Frascogna N, Moore M, Adair TH (2000) Inhibition of adenosine kinase induces expression of VEGF mRNA and protein in myocardial myoblasts. Am J Physiol 279:H2116–H2123

Hashimoto E, Kage K, Ogita T, Nakaoka T, Matsuoka R, Kira Y (1994) Adenosine as an endogenous mediator of hypoxia for induction of vascular endothelial growth factor mRNA in U-937 cells. Biochem Biophys Res Commun 204:318–324

Hudlicka O, Wright AJ, Ziada AM (1986) Angiogenesis in the heart and skeletal muscle. Can J Cardiol 2:120–123

Jen SC, Rovainen CM (1994) An adenosine agonist increases blood flow and density of capillary branches in the optic tectum of *Xenopus laevis* tadpoles. Microcirculation 1:59–66

Kalk P, Eggert B, Relle K, Godes M, Heiden S, Sharkovska Y, Fischer Y, Ziegler D, Bielenberg GW, Hocher B (2007) The adenosine A_1 receptor antagonist SLV320 reduces myocardial fibrosis in rats with 5/6 nephrectomy without affecting blood pressure. Br J Pharmacol 151:1025–1032

Khoa ND, Montesinos MC, Williams AJ, Kelly M, Cronstein BN (2003) Th1 cytokines regulate adenosine receptors and their downstream signaling elements in human microvascular endothelial cells. J Immunol 171:3991–3998

Klatsky AL, Armstrong MA (1992) Alcohol, smoking, coffee, and cirrhosis. Am J Epidemiol 136:1248–1257

Klatsky AL, Armstrong MA, Friedman GD (1993) Coffee, tea, and mortality. Ann Epidemiol 3:375–381

Klatsky AL, Morton C, Udaltsova N, Friedman GD (2006) Coffee, cirrhosis, and transaminase enzymes. Arch Intern Med 166:1190–1195

Kong T, Westerman KA, Faigle M, Eltzschig HK, Colgan SP (2006) HIF-dependent induction of adenosine A_{2B} receptor in hypoxia. FASEB J 20:2242–2250

Leibovich SJ, Chen JF, Pinhal-Enfield G, Belem PC, Elson G, Rosania A, Ramanathan M, Montesinos C, Jacobson M, Schwarzschild MA, Fink JS, Cronstein B (2002) Synergistic up-regulation of vascular endothelial growth factor expression in murine macrophages by adenosine A_{2A} receptor agonists and endotoxin. Am J Pathol 160:2231–2244

Li S, Huang S, Peng SB (2005) Overexpression of G protein-coupled receptors in cancer cells: involvement in tumor progression. Int J Oncol 27:1329–1339

Lutty GA, Mathews MK, Merges C, McLeod DS (1998) Adenosine stimulates canine retinal microvascular endothelial cell migration and tube formation. Curr Eye Res 17:594–607

Mall G, Schikora I, Mattfeldt T, Bodle R (1987) Dipyridamole-induced neoformation of capillaries in the rat heart. Quantitative stereological study on papillary muscles. Lab Invest 57:86–93

Man J, Dytoc MT (2004) Use of imiquimod cream 5% in the treatment of localized morphea. J Cutan Med Surg 8:166–169

Mattfeldt T, Mall G (1983) Dipyridamole-induced capillary endothelial cell proliferation in the rat heart: a morphometric investigation. Cardiovasc Res 17:229–237

Meininger CJ, Granger HJ (1990) Mechanisms leading to adenosine-stimulated proliferation of microvascular endothelial cells. Am J Physiol 258:H198–H206

Meininger CJ, Schelling ME, Granger HJ (1988) Adenosine and hypoxia stimulate proliferation and migration of endothelial cells. Am J Physiol 255:H554–H562

Merighi S, Benini A, Mirandola P, Gessi S, Varani K, Leung E, Maclennan S, Baraldi PG, Borea PA (2005) A_3 adenosine receptors modulate hypoxia-inducible factor-1alpha expression in human a375 melanoma cells. Neoplasia 7:894–903

Merighi S, Benini A, Mirandola P, Gessi S, Varani K, Simioni C, Leung E, Maclennan S, Baraldi PG, Borea PA (2007) Caffeine inhibits adenosine-induced accumulation of hypoxia-inducible factor-1α, vascular endothelial growth factor, and interleukin-8 expression in hypoxic human colon cancer cells. Mol Pharmacol 72:395–406

Mino RP, Spoerri PE, Caballero S, Player D, Belardinelli L, Biaggioni I, Grant MB (2001) Adenosine receptor antagonists and retinal neovascularization in vivo. Invest Ophthalmol Vis Sci 42:3320–3324

Mohsenin A, Burdick MD, Molina JG, Keane MP, Blackburn MR (2007a) Enhanced CXCL1 production and angiogenesis in adenosine-mediated lung disease. FASEB J 21:1026–1036

Mohsenin A, Mi T, Xia Y, Kellems RE, Chen JF, Blackburn MR (2007b) Genetic removal of the A_{2A} adenosine receptor enhances pulmonary inflammation, mucin production, and angiogenesis in adenosine deaminase-deficient mice. Am J Physiol 293:L753–L761

Montesinos MC, Gadangi P, Longaker M, Sung J, Levine J, Nilsen D, Reibman J, Li M, Jiang CK, Hirschhorn R, Recht PA, Ostad E, Levin RI, Cronstein BN (1997) Wound healing is accelerated by agonists of adenosine A_2 (Gαs-linked) receptors. J Exp Med 186:1615–1620

Montesinos MC, Desai A, Chen JF, Yee H, Schwarzschild MA, Fink JS, Cronstein BN (2002) Adenosine promotes wound healing and mediates angiogenesis in response to tissue injury via occupancy of A_{2A} receptors. Am J Pathol 160:2009–2018

Montesinos MC, Shaw JP, Yee H, Shamamian P, Cronstein BN (2004) Adenosine A_{2A} receptor activation promotes wound neovascularization by stimulating angiogenesis and vasculogenesis. Am J Pathol 164:1887–1892

Namazi H (2006) Imiquimod: a potential weapon against Dupuytren contracture. Med Hypotheses 66:991–992

Ohta A, Gorelik E, Prasad SJ, Ronchese F, Lukashev D, Wong MK, Huang X, Caldwell S, Liu K, Smith P, Chen JF, Jackson EK, Apasov S, Abrams S, Sitkovsky M (2006) A_{2A} adenosine receptor protects tumors from antitumor T cells. Proc Natl Acad Sci USA 103:13132–13137

Olah ME, Roudabush FL (2000) Down-regulation of vascular endothelial growth factor expression after A_{2A} adenosine receptor activation in PC12 pheochromocytoma cells. J Pharmacol Exp Ther 293:779–787

Panjehpour M, Castro M, Klotz KN (2005) Human breast cancer cell line MDA-MB-231 expresses endogenous A_{2B} adenosine receptors mediating a Ca^{2+} signal. Br J Pharmacol 145:211–218

Phelps PT, Anthes JC, Correll CC (2006) Characterization of adenosine receptors in the human bladder carcinoma T24 cell line. Eur J Pharmacol 536:28–37

Pinhal-Enfield G, Ramanathan M, Hasko G, Vogel SN, Salzman AL, Boons GJ, Leibovich SJ (2003) An angiogenic switch in macrophages involving synergy between Toll-like receptors 2, 4, 7, and 9 and adenosine A_{2A} receptors. Am J Pathol 163:711–721

Pueyo ME, Chen Y, D'Angelo G, Michel JB (1998) Regulation of vascular endothelial growth factor expression by cAMP in rat aortic smooth muscle cells. Exp Cell Res 238:354–358

Rodrigues S, De Wever O, Bruyneel E, Rooney RJ, Gespach C (2007) Opposing roles of netrin-1 and the dependence receptor DCC in cancer cell invasion, tumor growth and metastasis. Oncogene 26:5615–5625

Ruhl CE, Everhart JE, Ruhl CE, Everhart JE (2005) Coffee and tea consumption are associated with a lower incidence of chronic liver disease in the United States. Gastroenterology 129:1928–1936

Ryzhov S, McCaleb JL, Goldstein AE, Biaggioni I, Feoktistov I (2007) Role of adenosine receptors in the regulation of angiogenic factors and neovascularization in hypoxia. J Pharmacol Exp Ther 382:565–572

Ryzhov S, Novitskiy SV, Zaynagetdinov R, Goldstein AE, Biaggioni I, Carbone DC, Dikov MM, Feoktistov I (2008a) Host A2B adenosine receptors promote carcinoma growth. Neoplasia 10:987–995

Ryzhov S, Solenkova NV, Goldstein AE, Lamparter M, Fleenor T, Young PP, Greelish JP, Byrne JG, Vaughan DE, Biaggioni I, Hatzopoulos AK, Feoktistov I (2008b) Adenosine receptor-mediated adhesion of endothelial progenitors to cardiac microvascular endothelial cells. Circ Res 102:356–363

Schon M, Schon MP (2007) The antitumoral mode of action of imiquimod and other imidazoquinolines. Curr Med Chem 14:681–687

Schon MP, Schon M, Klotz KN (2006) The small antitumoral immune response modifier imiquimod interacts with adenosine receptor signaling in a TLR7- and TLR8-independent fashion. J Invest Dermatol 126:1338–1347

Sexl V, Mancusi G, Baumgartner-Parzer S, Schutz W, Freissmuth M (1995) Stimulation of human umbilical vein endothelial cell proliferation by A_2-adenosine and β_2-adrenoceptors. Br J Pharmacol 114:1577–1586

Sharp DS, Everhart JE, Benowitz NL (1999) Coffee, alcohol, and the liver. Ann Epidemiol 9:391–393

Sun LL, Xu LL, Nielsen TB, Rhee P, Burris D (1999) Cyclopentyladenosine improves cell proliferation, wound healing, and hair growth. J Surg Res 87:14–24

Sun CX, Zhong H, Mohsenin A, Morschl E, Chunn JL, Molina JG, Belardinelli L, Zeng D, Blackburn MR (2006) Role of A_{2B} adenosine receptor signaling in adenosine-dependent pulmonary inflammation and injury. J Clin Invest 116:2173–2182

Symons JD, Firoozmand E, Longhurst JC (1993) Repeated dipyridamole administration enhances collateral-dependent flow and regional function during exercise. A role for adenosine. Circ Res 73:503–513

Takagi H, King GL, Robinson GS, Ferrara N, Aiello LP (1996) Adenosine mediates hypoxic induction of vascular endothelial growth factor in retinal pericytes and endothelial cells. Invest Ophthalmol Vis Sci 37:2165–2176

Teuscher E, Weidlich V (1985) Adenosine nucleotides, adenosine and adenine as angiogenesis factors. Biomed Biochim Acta 44:493–495

Tornling G, Unge G, Ljungqvist A, Carlsson S (1978) Dipyridamole and capillary proliferation. A preliminary report. Acta Pathol Microbiol Scand A Pathol 86:82

Tornling G, Adolfsson J, Unge G, Ljungqvist A (1980a) Capillary neoformation in skeletal muscle of dipyridamole-treated rats. Arzneimittel-Forschung 30:791–792

Tornling G, Unge G, Adolfsson J, Ljungqvist A, Carlsson S (1980b) Proliferative activity of capillary wall cells in skeletal muscle of rats during long-term treatment with dipyridamole. Arzneimittel-Forschung 30:622–623

Tornling G (1982a) Capillary neoformation in the heart and skeletal muscle during dipyridamole: treatment and exercise. Acta Pathol Microbiol Immunol Scand 278(Suppl):1–63

Tornling G (1982b) Capillary neoformation in the heart of dipyridamole-treated rats. Acta Pathol Microbiol Immunol Scand A Pathol 90:269–271

Torry RJ, O'Brien DM, Connell PM, Tomanek RJ (1992) Dipyridamole-induced capillary growth in normal and hypertrophic hearts. Am J Physiol 262:H980–H986

Tverdal A, Skurtveit S (2003) Coffee intake and mortality from liver cirrhosis. Ann Epidemiol 13:419–423

Van Daele P, Van Coevorden A, Roger PP, Boeynaems JM (1992) Effects of adenine nucleotides on the proliferation of aortic endothelial cells. Circ Res 70:82–90

Victor-Vega C, Desai A, Montesinos MC, Cronstein BN (2002) Adenosine A_{2A} receptor agonists promote more rapid wound healing than recombinant human platelet-derived growth factor (becaplermin gel). Inflammation 26:19–24

Wakai A, Wang JH, Winter DC, Street JT, O'Sullivan RG, Redmond HP (2001) Adenosine inhibits neutrophil vascular endothelial growth factor release and transendothelial migration via A_{2B} receptor activation. Shock 15:297–301

Wakeno M, Minamino T, Seguchi O, Okazaki H, Tsukamoto O, Okada K, Hirata A, Fujita M, Asanuma H, Kim J, Komamura K, Takashima S, Mochizuki N, Kitakaze M (2006) Long-term stimulation of adenosine A_{2B} receptors begun after myocardial infarction prevents cardiac remodeling in rats. Circulation 114:1923–1932

Wothe D, Hohimer A, Morton M, Thornburg K, Giraud G, Davis L (2002) Increased coronary blood flow signals growth of coronary resistance vessels in near-term ovine fetuses. Am J Physiol 282:R295–R302

Zeng D, Maa T, Wang U, Feoktistov I, Biaggioni I, Belardinelli L (2003) Expression and function of A_{2B} adenosine receptors in the U87MG tumor cells. Drug Dev Res 58:405–411

Zhong H, Belardinelli L, Maa T, Zeng D (2005) Synergy between A_{2B} adenosine receptors and hypoxia in activating human lung fibroblasts. Am J Respir Cell Mol Biol 32:2–8

Zhou AM, Li WB, Li QJ, Liu HQ, Feng RF, Zhao HG (2004) A short cerebral ischemic preconditioning up-regulates adenosine receptors in the hippocampal CA1 region of rats. Neurosci Res 48:397–404

Ziada AM, Hudlicka O, Tyler KR, Wright AJ (1984) The effect of long-term vasodilatation on capillary growth and performance in rabbit heart and skeletal muscle. Cardiovasc Res 18:724–732

Adenosine Receptors and Cancer

P. Fishman, S. Bar-Yehuda, M. Synowitz, J.D. Powell, K.N. Klotz, S. Gessi, and P.A. Borea

Contents

Abstract The A_1, A_{2A}, A_{2B} and A_3 G-protein-coupled cell surface adenosine receptors (ARs) are found to be upregulated in various tumor cells. Activation of the receptors by specific ligands, agonists or antagonists, modulates tumor growth via a range of signaling pathways. The A_1AR was found to play a role in preventing the development of glioblastomas. This antitumor effect of the A_1AR is mediated via tumor-associated microglial cells. Activation of the A_{2A}AR results in inhibition of the immune response to tumors via suppression of T regulatory cell function and inhibition of natural killer cell cytotoxicity and tumor-specific CD4+/CD8+ activity. Therefore, it is suggested that pharmacological inhibition of A_{2A}AR activation

---

P. Fishman (✉)
Can-Fite BioPharma, 10 Bareket st., Kiryat Matalon, Petach Tikva, 49170, Israel
pnina@canfite.co.il

C.N. Wilson and S.J. Mustafa (eds.), *Adenosine Receptors in Health and Disease*, Handbook of Experimental Pharmacology 193, DOI 10.1007/978-3-540-89615-9_14,

by specific antagonists may enhance immunotherapeutics in cancer therapy. Activation of the A_{2B}AR plays a role in the development of tumors via upregulation of the expression levels of angiogenic factors in microvascular endothelial cells. In contrast, it was evident that activation of A_{2B}AR results in inhibition of ERK1/2 phosphorylation and MAP kinase activity, which are involved in tumor cell growth signals. Finally, A_3AR was found to be highly expressed in tumor cells and tissues while low expression levels were noted in normal cells or adjacent tissue. Receptor expression in the tumor tissues was directly correlated to disease severity. The high receptor expression in the tumors was attributed to overexpression of NF-κB, known to act as an A_3AR transcription factor. Interestingly, high A_3AR expression levels were found in peripheral blood mononuclear cells (PBMCs) derived from tumor-bearing animals and cancer patients, reflecting receptor status in the tumors. A_3AR agonists were found to induce tumor growth inhibition, both *in vitro* and *in vivo*, via modulation of the Wnt and the NF-κB signaling pathways. Taken together, A_3ARs that are abundantly expressed in tumor cells may be targeted by specific A_3AR agonists, leading to tumor growth inhibition. The unique characteristics of these A_3AR agonists make them attractive as drug candidates.

Keywords A_1 adenosine receptor · A_{2A} adenosine receptor · A_{2B} adenosine receptor · A_3 adenosine receptor · Expression · Tumor growth · Agonists · Antagonists

Abbreviations

A_1AR	A_1 adenosine receptor
A_{2A}AR	A_{2A} adenosine receptor
A_{2B}AR	A_{2B} adenosine receptor
A_3AR	A_3 adenosine receptor
APCs	Antigen-presenting cells
AR	Adenosine receptor
bFGF	Basic fibroblast growth factor
CCPA	2-Chloro-N^6-cyclopentyladenosine
CD39	Cluster of differentiation 39
CD73	Cluster of differentiation 73
GGAP	Cancer Genome Anatomy Project
CGS21680	2-*p*-(2-Carboxyethyl)phenethylamino-5′-*N*-ethylcarbox amidoadenosine 1680
CHO	Chinese hamster ovary cells
Cl–IB–MECA	2-Chloro-N^6-3-iodobenzyladenosine-5′-*N* -methyluronamide
CNS	Central nervous system
CPA	N^6-Cyclopentyladenosine
CTLA-4	Cytotoxic T lymphocyte-associated antigen 4
CTLs	Cytotoxic T lymphocytes

DPCPX	8-Cyclopentyl-1,3-dipropylxanthine
EGF	Epidermal growth factor
Epac	Exchange protein activated by cAMP
ER	Estrogen receptor
ERK	Extracellular signal-regulated kinase
G-CSF	Granulocyte colony stimulating factor
GPCR	G-protein-coupled receptor
GSK-3β	Glycogen synthase kinase 3β
HA	Hyaluronan
HCC	Hepatocellular carcinoma
HIF-1	Hypoxia-inducible factor 1
HMG1b	High mobility group 1b
HUGO	Human Genome Organization
IB–MECA	Methyl 1-[N^6-(3-iodobenzyl)-adenin-9-yl]-β-D-ribofuronamid
IKK	IκB kinase
IL	Interleukin
Lef/Tcf	Lymphoid enhancer factor/T-cell factor
MAP	Mitogen-activated protein
MMP	Metalloproteinase
MRS1191	3-Ethyl-5-benzyl-2-methyl-4-phenylethynyl-6-phenyl-1,4-(±)-dihydropyridine-3,5-dicarboxylate
MTT	1-(4,5-Dimethylthiazol-2-yl)-3,5-diphenylformazan thiazolyl
NECA	Adenosine-5′-*N*-ethyluronamide
NF-κB	Nuclear factor kappa B
NK	Natural killers
PAMPs	Pathogen-associated molecular patterns
PARP	Poly(ADP-ribose) polymerase
PBMCs	Peripheral blood mononuclear cells
PDTC	Pyrrolidine dithiocarbamate
PI3K	Phosphoinositide 3-kinase
PKA	Protein kinase A
PKB	Protein kinase B
PKB/Akt	Protein kinase B/Akt
PLC	Phospholipase C
PLD	Phospholipase D
TCR	T-cell receptor
TGF-β	Transforming growth factor β
thio-Cl–IB–MECA	2-Chloro-N^6-(3-iodobenzyl)-4′-thioadenosine-5′-*N*-methyluronamide
TNF-α	Tumor necrosis factor
VEGF	Vascular endothelial growth factor
Wt	Wild type

1 Introduction

During the last decade different approaches to treating cancer have been developed based mainly on specific targets that are mostly expressed in tumor but not in normal cells. Furthermore, it is now recognized that individualizing therapy for patients being treated with anticancer agents is an important goal, leading to the prediction of agents that will be efficacious. Adenosine is a purine nucleoside found within the interstitial fluid of tumors at concentrations that are able to modulate tumor growth by interacting with four G-protein-coupled adenosine receptor (AR) subtypes, designated A_1, A_{2A}, A_{2B} and A_3. Selective agonists and antagonists are now available for all four AR subtypes, enabling the examination of these ligands as immunomodulators and anticancer agents. Interestingly, AR levels in various tumor cells are upregulated, a finding which may suggest that the specific AR may serve as a biological marker and as a target for specific ligands leading to cell growth inhibition.

In this chapter, we will present the role played by each of the ARs in mediating tumor growth. Since immune cells such as lymphocytes, macrophages and natural killer (NK) cells were also found to express ARs, their ability to act as cytotoxic cells against tumor cells or to be involved in the antitumor process will be discussed as well. Based on these studies, possible drug candidates (anticancer agents that target ARs) will be presented.

2 A_1 Adenosine Receptor

The A_1AR is a G-protein-coupled receptor that mediates many of the physiological effects of adenosine in the brain. The binding of agonists to A_1AR induces inhibition of adenylate cyclase, leading to a decrease in intracellular cAMP levels or stimulation of phospholipase C (PLC). The A_1AR has a high affinity for adenosine and has been implicated in both pro- and anti-inflammatory aspects of disease processes. On the one hand, A_1AR signaling can promote neutrophil (Salmon and Cronstein 1990) and monocyte activation (Merrill et al. 1997; Salmon et al. 1993); on the other hand, A_1AR signaling is involved in anti-inflammatory and protective pathways in neuroinflammation and injury (Tsutsui et al. 2004), and in cardiac and renal injury (Liao et al. 2003; Lee et al. 2004a, b). Adenosine-mediated anti-inflammatory effects have been studied extensively in macrophages and macrophage cell lines. Adenosine inhibits the production of several proinflammatory cytokines (TNF-α, IL-6, and IL-8) by LPS-stimulated macrophages and enhances the release of the anti-inflammatory cytokine IL-10 (Hasko et al. 1996; Le Moine et al. 1996; Sajjadi et al. 1996). Recent studies suggest an anti-inflammatory role for chronic A_1AR activation by high levels of adenosine in the lung, a surprising and important finding in light of the fact that A_1AR antagonists are being investigated as a potential treatment for asthma (Sun et al. 2005). In the CNS, the A_1AR is highly expressed on microglia/macrophages and neurons (Johnston et al. 2001). In the latter, A_1AR is coupled to activation of K^+ channels (Trussell and Jackson 1985) and inhibition of Ca^{2+} channels (MacDonald et al. 1986), both of which are mechanisms that attenuate neuronal excitability, thereby reducing excitotoxicity, and so adenosine can act as a neuroprotective factor. Since A_1ARs are expressed throughout the brain (Dunwiddie 1985), adenosine has the potential to be involved in different brain pathologies. Although A_1ARs may play an important role in some physiological functions in the brain (e.g., sleep), A_1AR-deficient mice show no obvious abnormal behavior, levels of alertness, or appearance of focal neurological deficits, such as seizures (Synowitz et al. 2006). However, upon exposure to pathophysiological conditions like hypoxia, A_1AR-deficient mice show more neuronal damage and have a lower survival rate (Johansson et al. 2001). It was therefore concluded that, in the brain, A_1ARs are primarily important in mediating effects of adenosine during pathophysiological conditions (Gimenez-Llort et al. 2002; Johansson et al. 2001).

It has recently been reported that the deletion of functional ARs, specifically A_1AR, results in an increase in brain tumor growth, specifically glioblastoma tumor growth (Synowitz et al. 2006). This implies that adenosine acting via A_1AR impairs glioblastoma growth. In the context of glioblastoma, A_1ARs are prominently expressed by the tumor cells and those microglial cells associated with the glioblastoma tumor cells. In an experimental approach using an A_1AR-deficient mouse as a tumor host, the importance of the microglial cells for mediating the A_1AR anticancer effect is highlighted (Synowitz et al. 2006). In these studies, A_1AR-deficient mice and their wild-type littermate controls are inoculated with Gl261 tumor cells; thus, with this approach, the A_1AR is deleted in host cells but not tumor cells. In the control wild-type littermates the microglial cells accumulated at the tumor site, and this accumulation was even more pronounced in the A_1AR-deficient mice. However, tumor volume was significantly greater in A_1AR-deficient mice, suggesting that the microglial cells are the cellular candidates for inhibiting tumor growth. The importance of microglial A_1AR is further supported by a brain slice model where inhibition of tumor growth is only observed in the presence of microglial cells. To test the functional effect of A_1AR activity on glioblastoma growth, an organotypical brain slice model was employed where glioblastoma cells could be injected and ARs could be stimulated or inhibited (Synowitz et al. 2006). Brain slices (250 μm thick) were cultured for four days and 10^4 GFP-labeled Gl261 tumor cells were injected (suspended in 0.1 μL) into the tissue. The tumor size was evaluated by measuring the area occupied by the fluorescently labeled Gl261 cells. In these studies, adenosine and an A_1AR agonist, N^6-cyclopentyladenosine (CPA) significantly decreased tumor size. To determine if this effect of adenosine or activation of A_1ARs depends on the presence of microglia, endogenous microglia were selectively depleted from cultured organotypical brain slices by a 24 h treatment with clodronate-filled liposomes without affecting other cell types (e.g., neurons, oligodendrocytes, and astrocytes). As reported previously, activated microglia supported glioblastoma tumor growth, resulting in significantly smaller tumors in microglia-depleted slices compared with control slices. This serves as an internal control and thus confirms the observation that the presence of microglial cells per se is tumor promoting (Markovic et al. 2005). There was no significant change in the population of astrocytes or neural progenitor cells. The latter is of particular interest, since it was recently reported that neural progenitor cells are attracted to tumors or to gliomas and attenuate tumor growth (Glass et al. 2005). In these organotypical brain slice studies, tumor cells were injected three days after liposome treatment, and the size of the tumor bulk was evaluated with and without microglia. In theses studies, activation of A_1ARs with adenosine or CPA resulted in a larger tumor size in brain slices devoid of microglia. Moreover, as expected, the tumor size was greater in brain slices from A_1AR-deficient mice versus their littermate wild-type controls. Furthermore, in these studies, adenosine or CPA had no effect on tumor growth in brain slices from A_1AR-deficient mice. Taken together, the *in vivo* studies in A_1AR-deficient mice and *in vitro* studies in organotypical brain slices suggest that CPA and adenosine specifically act on A_1ARs on microglial cells to reduce tumor size.

The presence of ARs has been previously reported on astrocytoma cells (Prinz and Hanisch 1999) using an A_1AR-specific ligand. The presence of ARs on microglia is well established, and some functional implications of their activation have become apparent (Burnstock 2006; Farber and Kettenmann 2006). Cultured rat microglial cells express A_{2A}ARs, since the specific A_{2A}AR agonist CGS21680 triggers the expression of K^+ channels that are linked to microglial activation (Kust et al. 1999). In contrast, A_{2A}AR stimulation in rat microglia triggers the expression of nerve growth factor and its release, thereby exerting a neuroprotective effect (Heese et al. 1997). Moreover, cyclooxygenase-2 expression in rat microglia is induced by A_{2A}ARs, resulting in the release of prostaglandin (Fiebich et al. 1996). Hammarberg et al. provided evidence for functional A_3ARs in mouse microglial cells while A_1ARs were not detected in this study (Hammarberg et al. 2003). However, other studies, based on immunocytochemical data, indicate that microglial cells express A_1ARs and that the presence of tumor cells upregulates the expression of A_1ARs in microglia (Synowitz et al. 2006). Moreover, the results of these studies indicate that loss of A_1AR leads to an increase of tumor size associated with microglia, which may be due to infiltration and/or proliferation.

The potential source of extracellular adenosine in the brain is most likely ATP, which is released from presynaptic and postsynaptic terminals of neurons and also from glial cells (Fields and Burnstock 2006). In the extracellular space, adenosine is generated from ATP after dephosphorylation by specific ectoenzymes (e.g., cluster of differentiation 39 (CD39) and cluster of differentiation 73 (CD73)). These ectoenzymes represent a highly organized enzymatic cascade for the regulation of nucleotide-mediated signaling. They control the rate of nucleotide (ATP) degradation and nucleoside (adenosine) formation (Farber et al. 2008; Plesner 1995). Microglial cells express specific ectonucleotidase isoforms, CD39 and CD73, which are not expressed by any other cell type in the brain. Due to this specific expression, both molecules served as microglia-specific markers long before their functional importance was recognized (Braun et al. 2000; Schnitzer 1989; Schoen et al. 1992).

The role of adenosine in microglial proliferation remains controversial. One study reports that adenosine stimulates the proliferation of microglial cells through a mechanism that involves the simultaneous stimulation of A_1 and A_2 ARs (Gebicke-Haerter et al. 1996). By contrast, adenosine has been reported to inhibit the proliferation of microglial cells; i.e., phorbol 12-myristate 13-acetate-stimulated microglial proliferation is reduced following treatment with an A_1AR agonist (Si et al. 1996). Moreover, stimulation of the A_1AR can also cause microglial apoptosis (Ogata and Schubert 1996). Adenosine levels in the extracellular fluid are lower in human glioblastoma tissue than in control tissue, namely 1.5 and 3 μM, respectively. These values were obtained from human glioblastomas of high-grade malignancy and measured by brain microdialysis coupled to high-performance liquid chromatography (Bianchi et al. 2004). Whether this rather small difference causes the accumulation of microglia close to tumors is speculative.

Recent studies support the idea that ARs and specifically the A_1AR are good targets for drug development in several diseases that affect the CNS (Fredholm et al. 2005). A_1AR deficiency aggravates experimental allergic encephalomyelitis (Tsutsui et al. 2004), and it has been repeatedly shown that adenosine can protect tissues against the negative consequences of hypoxia or ischemia (Fredholm 1997), mainly by acting on the A_1AR. Hence, survival after a hypoxic challenge may be reduced if A_1ARs are absent or blocked (Johansson et al. 2001). The tissue-protective effect of A_1AR has been implicated in experimental paradigms using A_1AR-deficient mice. In a model of renal ischemia and reperfusion injury, A_1AR-deficient mice exhibited an increase in production of proinflammatory mediators and showed an increase in renal injury (Lee et al. 2004a, b). Similarly, in a model of experimental allergic encephalomyelitis, A_1AR deficiency led to increased neuroinflammation and demyelination and also augmented axonal injury. Both studies concluded that A_1AR serves anti-inflammatory functions that regulate subsequent tissue damage. Furthermore, metalloproteinase (MMP) 9 and MMP-12 are significantly elevated in A_1AR-deficient mice (Tsutsui et al. 2004). Indeed, MMPs play an important role in glioblastoma progression and, as was recently demonstrated, the expression of MMPs by microglia has an impact on tumor growth (Markovic et al. 2005). Matrix degradation by MMPs is an important prerequisite for glioblastoma invasion (Rao 2003). A_1AR activation on microglia/macrophages inhibits not only the production of cytokines like interleukin-1β but also matrix MMPs like MMP-12 (Tsutsui et al. 2004). MMP-12, also known as macrophage elastase, is an MMP that is produced by activated macrophages and preferentially degrades elastin (Werb and Gordon 1975). Hence, inhibition of microglial MMP-12 secretion via activation of A_1AR could explain the glioblastoma growth inhibition observed in the studies described above. Moreover, the lack of inhibition of MMP-12 by A_1ARs on microglia may explain why there is enhanced accumulation of microglia at the tumor sites in A_1AR-deficient mice along with their tumor-promoting effects (i.e., associated increased tumor size). Adenosine does not appear to directly regulate MMP-12 expression in microglia/macrophages since direct stimulation of cultured macrophages with AR agonists did not induce expression of MMP-12 (Sun et al. 2005). It is therefore likely that the removal of A_1AR signaling leads to enhanced production of mediators in the CNS, which then leads to enhanced MMP-12 production. A likely candidate for this is interleukin (IL)-13, since IL-13 has been shown to be involved in the production of MMP-12 in other model systems (Lanone et al. 2002).

The results from the studies described above suggest that the A_1AR plays an antitumorigenic role mediated by microglial cells in the development of glioblastomas. Further research into the mechanisms of how the pathways of A_1AR signaling modulate glioblastoma development may ultimately lead to treatments to reduce the progression of this disease.

3 A_{2A} Adenosine Receptor

3.1 *The* $A_{2A}AR$*: Protector of Host Tissue, Protector of Tumors*

The seminal observations of Ohta and Sitkovsky (2001) clearly established a role for the $A_{2A}AR$ in protecting host tissue from destruction by overexuberant immune responses. Considering that the tumor microenvironment contains relatively high levels of extracellular adenosine, data is emerging to support the hypothesis that tumor-derived adenosine is one mechanism by which tumors evade immune destruction (Blay et al. 1997; Ohta et al. 2006). In this section, we will discuss the role of adenosine in thwarting antitumor immunity and the potential pharmacologic interventions on the horizon that may serve to overcome this hurdle to immunotherapy.

3.2 *Tumors Evade the Immune System by Inhibiting Immune Cell Function*

The ability of the immune system to specifically recognize antigen makes it a potentially powerful tool in terms of developing modalities to treat cancer. However, in spite of many recent advances in understanding of and ability to identify tumor antigens, immunotherapy is clearly yet to live up to its full potential. In part, this is because tumors evade immune destruction by inhibiting tumor-specific immune cells (Pardoll 2002). For example, while a particular tumor may express a very unique and readily recognized tumor antigen, if this antigen is presented by resting or nonprofessional antigen-presenting cells (APCs), T-cell receptor (TCR) recognition will not lead to the destruction of the tumor but rather the inactivation of the tumor-specific T cell.

In this context, it is not the inability of T cells to recognize the tumor that is hampering cancer immunotherapy, but rather a lack of antigen-induced immune activation. That is, tumors readily express and T cells readily recognize tumor antigens (Overwijk and Restifo 2001). The problem is that T-cell recognition of the tumor does not lead to tumor destruction but rather to T-cell tolerance. In this regard, the tumor microenvironment is fraught with humors and cells that facilitate the ability of tumors to evade immune destruction (Drake et al. 2006). For example, the cytokines IL-10 and transforming growth factor β (TGF-β) in the tumor microenvironment can both directly inhibit T-cell function as well as promote the induction of regulatory T cells and tolerogenic APCs. Likewise, tumors can express coinhibitory ligands such as B7–H1 and B7–H4. These in turn engage molecules on the surfaces of T cells such as PD-1 that serve to inhibit T-cell function. In this context, it is becoming clear why tumor vaccines have failed to live up to their potential so far (Pardoll 2002). Vaccine regimens which have focused on trying to enhance tumor-specific T cells by utilizing viral vectors, DNA vaccines, cytokine-secreting cells and antigen-pulsed dendritic cells have all shown promise in animal models and even some clinical trials. Put simply, in spite of the ability of such approaches to generate activated tumor antigen-specific T cells, the efficacy of such cells is thwarted by the multiple immunologic checkpoints exploited by the tumor. With this in mind, current immunotherapeutic strategies are focused on blocking these checkpoints. In this regard, blocking antibodies against cytotoxic T lymphocyte-associated antigen 4 (CTLA-4 (a negative regulator of T-cell activation) has shown great promise in a number of animal models (Egen et al. 2002). Likewise, blocking anti-PD-1 antibodies are also currently being tested in order to enhance tumor immunotherapy (Blank and Mackensen 2007).

3.3 The A_{2A}AR Negatively Regulates Immune Responses

The ability of adenosine to inhibit immune function has been known for some time (Linden 2001). However, in light of the fact that there are four known AR subtypes, the critical, nonredundant role of the A_{2A}AR in mediating adenosine-induced anti-inflammatory responses was somewhat surprising. In a series of experiments, Sitkovsky's group demonstrated that normally nonlethal, self-limiting inflammation in wild-type (Wt) mice led to excessive inflammation and death in A_{2A}AR-null mice (Ohta and Sitkovsky 2001). These observations and additional studies led to a model whereby tissue damage resulting from inflammation leads to the release of extracellular adenosine, which then acts to quell the inflammatory response by acting on bone marrow-derived immune cells. Indeed, A_{2A}AR signaling on immune cells such as macrophages, T cells and dendritic cells has been shown to limit effector cell function (Erdmann et al. 2005; Huang et al. 1997; Khoa et al. 2001; Lappas et al. 2005; Naganuma et al. 2006; Panther et al. 2001; Schnurr et al. 2004). The existence of this negative feedback loop has led Sitkovsky to propose that, from an immunologic prospective, adenosine should be viewed as a metabokine that acts as an inhibitory second signal (Sitkovsky and Ohta 2005). For example, during an infection, pathogen-associated molecular patterns (PAMPs) along with host-derived uric acid, high mobility group (HMG1b) and hyaluronan (HA) would promote activating "danger signals" (Scheibner et al. 2006; Shi et al. 2003; Williams and Ireland 2008). As the inflammation progresses, the pathogen will be eliminated and the concentration of the potent immune-activating PAMPS will markedly decrease. In this setting, the inhibitory affects of adenosine released by damaged tissue will dominate to protect the tissue from further destruction by overacting immune responses.

Adenosine acting via the $A_{2A}AR$ has the ability to influence inflammation by inhibiting proinflammatory cytokine secretion, C2 activation, macrophage-mediated phagocytosis and superoxide production (Sullivan 2003). Likewise, $A_{2A}AR$ activation has profound effects on the adaptive immune response. $A_{2A}AR$ activation inhibits both CD4+ and CD8+ T-cell function (Erdmann et al. 2005; Lappas et al. 2005; Naganuma et al. 2006; Sevigny et al. 2007; Zarek et al. 2008). Interestingly, $A_{2A}AR$ activation on T cells seems to selectively inhibit proinflammatory cytokine expression while sparing anti-inflammatory cytokine expression (Naganuma et al. 2006). In addition, antigen activation in the presence of $A_{2A}AR$ agonists can promote T-cell tolerance in the form of anergy (Zarek et al. 2008). Likewise, $A_{2A}AR$ engagement can prevent the development of IL-17 producing cells and promote the development of Foxp3+ and LAG-3+ regulatory T-cells. Along these lines, it has been shown that adenosine acting via the $A_{2A}AR$ might partially mediate the suppressive function of regulatory T cells by engaging the $A_{2A}ARs$ on the suppressed cells (Deaglio et al. 2007). It was found that the ectoenzymes CD39 and CD73 appear to be more specific markers for $Foxp3^{+}$ regulatory cells than CD25 (Deaglio et al. 2007). Further data supporting the role of adenosine acting via the $A_{2A}AR$ in facilitating regulatory T-cell function has also been demonstrated in a colitis model of autoimmunity. In these studies, $CD45RB^{low}$ or CD25 + T cells derived from $A_{2A}AR$-null mice were unable to regulate $CD45RB^{high}$ cells and prevent disease (Naganuma et al. 2006). Furthermore, the $CD45RB^{high}$ cells from $A_{2A}AR$-null mice were not inhibited by regulatory T cells, even when they were derived from wild-type mice (Naganuma et al. 2006). Thus, with regard to the adaptive immune response, the $A_{2A}AR$ protects the host from excessive tissue destruction by not only acutely inhibiting T-cell function but also promoting the development of regulatory T cells.

3.4 Adenosine Protects Tumors from Immune Destruction

Tumors are very adept at usurping negative regulatory mechanisms of the immune system in order to evade antitumor responses. As mentioned above, the tumor microenvironment is replete with inhibitory cytokines, inhibitory ligands and regulatory T cells (Drake et al. 2006). Considering that $A_{2A}AR$ activation is a potent inhibitor of adaptive immune responses, it is not surprising that tumor-derived adenosine has been implicated in blocking antitumor immunity. Indeed, the tumor microenvironment has been shown to contain relatively high concentrations of adenosine (Blay et al. 1997). In part, this is due to the hypoxic nature of the tumor microenvironment (Lukashev et al. 2007). Hypoxia regulates the levels of adenosine by inhibiting enzymes involved in the destruction of adenosine and simultaneously increasing the activity of enzymes charged with the generation of adenosine.

Hoskin and colleagues were one of the first groups to propose that adenosine within the microenvironment of solid tumors might inhibit T-cell function (Hoskin et al. 1994). Their initial studies demonstrated that adenosine could inhibit natural killer (NK) cell function as well as the ability of cytotoxic T cells to adhere to tumor cell targets (MacKenzie et al. 1994; Williams et al. 1997). Subsequently, this group went on to formally demonstrate that the extracellular fluid of tumors contains concentrations of adenosine that are sufficient to inhibit lymphocyte activation (Blay et al. 1997). This observation has since been confirmed by others (Ohta et al. 2006). Note that the initial studies by the Hoskin's group did not implicate the $A_{2A}AR$ as playing a critical role in the inhibition of antitumor immune function. However, more recently it has been shown that adenosine can inhibit NK cell and IL-2/NKp46-activated NK cells specifically via the $A_{2A}AR$ (Raskovalova et al. 2006). These studies showed that $A_{2A}AR$-specific agonists inhibit the cytotoxicity of NK cells as well as their ability to elaborate cytokines. Interestingly, by employing various protein kinase A (PKA) inhibitors it was suggested that the ability of $A_{2A}AR$ activation to inhibit these functions is mediated downstream via PKA-I but not PKA-II. It has subsequently been shown that $A_{2A}AR$-specific agonists could also inhibit both tumor-specific CD4+ and CD8+ T cells (Raskovalova et al. 2007). In these studies, similar to the NK cell studies, $A_{2A}AR$-specific agonists inhibited the ability of human antimelanoma-specific cytotoxic T lymphocytes (CTLs) and human anti-melanoma-specific CD4+ T cells with regard to their ability to kill tumor cells and elaborate cytokines and chemokines in response to tumor cells. Biochemically, it was found that molecules that activated PKA-I but not PKA-II mimicked the affects of $A_{2A}AR$ activation on T-cell function. The $A_{2A}AR$-mediated inhibition, in turn, was blocked by Rp-8-Br-cAMPS, which antagonizes the binding of cAMP to the regulatory subunit of PKA-I. Alternatively, inhibitors of the PKA catalytic subunit did not mitigate the inhibitory affects of $A_{2A}AR$ activation.

As discussed, tumors evade host responses by acutely inhibiting immune function and promoting tolerance. Considering that $A_{2A}AR$ activation inhibits immune responses by suppressing immune activation and promoting tolerance, the following question arises: does tumor-derived adenosine play this role *in vivo*? Initial studies addressing this question suggest that the answer is yes (Ohta et al. 2006). $A_{2A}AR$-null mice have been shown to more readily reject melanoma and lymphoma tumor challenge. In addition, treating mice with $A_{2A}AR$ antagonists (including caffeine) led to increased tumor rejection by CD8+ T cells. These findings have been confirmed by another group that has also been able to demonstrate the ability of $A_{2A}AR$-null mice to more readily reject tumors and respond more robustly to tumor vaccines (Powell et al., unpublished data). In particular, the data from these studies suggest that genetic deletion of the $A_{2A}AR$ leads to more robust initial responses to vaccines. There are a number of important implications of these *in vivo* findings. First, the fact that genetic deletion of the $A_{2A}AR$ markedly enhances antitumor responses suggests that adenosine plays an important role in mediating tumor evasion of the immune system. Second, adenosine appears to block both the generation and effector phases of antitumor responses. Third, and perhaps most importantly, these findings support a role for pharmacologic inhibition of $A_{2A}AR$ activation as a means of enhancing immunotherapy.

3.5 $A_{2A}AR$ Antagonism as a Means of Enhancing Immunotherapy

Adenosine acting via the $A_{2A}AR$ has been shown to inhibit dendritic cell function, T-cell activation and differentiation, and T-cell effector function (Sitkovsky et al. 2004). Additionally, the $A_{2A}AR$ has been implicated in selectively enhancing anti-inflammatory cytokines, promoting the upregulation of PD-1 and CTLA-4, promoting the generation of LAG-3 and Foxp3+ regulatory T cells, and mediating the inhibition of regulatory T cells (Naganuma et al. 2006; Sevigny et al. 2007; Zarek et al. 2008). All of these immunosuppressive properties have also been identified as mechanisms by which tumors evade host responses. Initial *in vivo* studies demonstrating that genetically and pharmacologically inhibiting the $A_{2A}AR$ leads to robust antitumor responses suggest that adenosine is at least partially responsible for promoting these tumor defense mechanisms (Ohta et al. 2006). As such, the addition of $A_{2A}AR$ antagonists to cancer immunotherapeutic protocols represents an exciting approach to enhancing tumor immunotherapy. Interestingly, the safety of such compounds has already been shown in trials employing $A_{2A}AR$ antagonists for the treatment of Parkinson's disease (Jenner 2005).

Chemotherapy and radiation therapy result in the release of copious amounts of tumor antigen. However, this form of tissue destruction can also result in increases in extracellular adenosine. Therefore, the concomitant administration of $A_{2A}AR$ antagonists during chemotherapy or radiation therapy might actually lead to the expansion of tumor-specific T cells, while at the same time preventing the induction of tumor-specific regulatory T cells. In terms of combining $A_{2A}AR$ antagonists with tumor vaccines, we believe that there are two time points that are relevant. First, administration of antagonists during the perivaccination period might serve to enhance the generation of tumor-specific effector memory cells. This would be accomplished by both enhancing the activity of the antigen-presenting cells (e.g., dendritic cells), as well as blocking adenosine-mediated negative feedback on the T cells themselves. Second, the continued administration of $A_{2A}AR$ antagonists will enhance the effector function of these cells and potentially block the upregulation of regulatory T cells. Finally, perhaps the most effective use of $A_{2A}AR$ antagonists will be in combination with not only vaccines but also other checkpoint blockers. For example, blocking PD-1 engagement as well as the $A_{2A}AR$ will perhaps mitigate the ability of tumors to turn off tumor-specific effector T cells.

4 A_{2B} Adenosine Receptors

The A_{2B} adenosine receptor ($A_{2B}AR$) is found in many different cell types and requires higher concentrations of adenosine for activation than the A_1, A_{2A}, and A_3 AR subtypes (Fredholm et al. 2001). Thus, unlike the other AR subtypes, the $A_{2B}AR$ is not stimulated by physiological levels of adenosine, but may therefore play an important role in pathophysiological conditions associated with massive adenosine release. Such conditions occur in ischemia or in tumors where hypoxia is commonly observed (Illes et al. 2000; Merighi et al. 2003). Although potent and selective tools are scarce for the A_{2B}AR subtype, it has become increasingly clear in recent years that this AR subtype regulates a number of functions (e.g., vascular tone, cytokine release, and angiogenesis; Volpini et al. 2003). A_{2B}ARs may also play a role in cancer, based on a number of observations. Gaining an understanding of the exact mechanisms by which adenosine regulates the growth and proliferation of tumor cells via this AR subtype could potentially lead to a target for novel therapies or at least for cotherapies for cancer. In the following sections, potential mechanisms suggesting that A_{2B}AR might be involved in tumor development and progression are discussed.

One of the pivotal mechanisms for tumor growth is angiogenesis, a process that is highly regulated by an array of angiogenic factors and is triggered by adenosine under various circumstances that are associated with hypoxia. Although the A_3AR subtype is involved in the release of angiogenic factors, in some cases the A_{2B}AR also seems to be responsible for the release of a certain subset of cytokines (Feoktistov et al. 2003; Merighi et al. 2007). A_{2B}ARs are expressed in human microvascular endothelial cells, where they play a role in the regulation of the expression of angiogenic factors like vascular endothelial growth factor (VEGF), IL-8, and basic fibroblast growth factor (bFGF) (Feoktistov et al. 2002). Moreover, in HMC-1 cells derived from a highly malignant, undifferentiated human mastocytoma cancer, activation of A_{2B}ARs induces the release of IL-8 and VEGF, and the activation of A_3ARs induces angiopoietin 2 expression (Feoktistov et al. 2003). However, capillary formation induced by HMC-1 media was maximal when both HMC-1 A_{2B}ARs and A_3ARs were activated. Activation of A_{2B}ARs alone was less effective, suggesting a cooperation between A_{2B}ARs and A_3ARs on HMC-1 cells to produce angiogenesis. Furthermore, Merighi et al. demonstrated in HT29 human colon cancer cells that adenosine increases IL-8 expression via stimulation of A_{2B}ARs, while the stimulation of A_3AR caused an increase in VEGF (Merighi et al. 2007). In the glioblastoma cell line U87MG, a similar A_{2B}AR-mediated increase of IL-8 was observed (Zeng et al. 2003). In addition, it was shown that hypoxia caused an upregulation of A_{2B}ARs in these tumor cells. As these findings point to a crucial role for A_{2B}ARs in mediating the effects of adenosine on angiogenesis, blockade of A_{2B}ARs may limit tumor growth by limiting the oxygen supply.

There are numerous reports of a potential role of adenosine and ARs in breast cancer (Barry and Lind 2000; Madi et al. 2004; Panjehpour et al. 2005; Spychala et al. 2004). Although AR agonists acting through A_3ARs were shown to possess antitumor activity in breast cancer, it turned out (at least in some cases) that these effects were receptor independent (Chung et al. 2006; Lu et al. 2003). The very high concentrations of IB–MECA required for growth inhibition in some studies (Panjehpour and Karami-Tehrani 2004) may lend further support to the notion of A_3AR-independent effects.

A most striking observation was that the estrogen receptor-positive MCF-7 cells appeared to be devoid of any detectable amount of ARs, whereas the estrogen receptor-negative MDA-MB-231 cells express very high levels of A_{2B}ARs (Panjehpour et al. 2005). Both binding and functional experiments showed that other AR subtypes were not present in detectable levels in these tumor cells. Stimulation with the nonselective AR agonist 5′-(*N*-ethylcarboxamido)adenosine (NECA) resulted in the activation of adenylate cyclase, whereas 10 μM 2-*p*-(2-carboxyethyl)phenethylamino-5′-*N*-ethylcarboxamidoadenosine (CGS21680; which, at this concentration, activates all but the A_{2B}AR subtype) had no effect. Moreover, there was no A_1AR or A_3AR receptor-mediated inhibition of adenylate cyclase, confirming the exclusive presence of A_{2B}ARs as a functionally relevant AR subtype in MDA-MB-231 cells (Panjehpour et al. 2005).

In addition to the classical adenylate cyclase activation, A_{2B}ARs also mediate a Ca^{2+} signal (Feoktistov et al. 1994; Linden et al. 1999; Mirabet et al. 1997). A similar Ca^{2+} signal was detected in MDA-MB-231 cells, most likely as a result of the activation of G_q (Panjehpour et al. 2005). With the use of selective agonists and antagonists for A_1AR, A_{2A}AR, and A_3ARs, a pharmacological profile identical to the one found for the adenylate cyclase response was demonstrated for the Ca^{2+} signal in these cells, again suggesting an A_{2B}AR as the sole AR subtype in these cells.

The mitogen-activated protein (MAP) kinase pathways are critically important in the regulation of cell proliferation and differentiation (Raman et al. 2007). There are numerous extracellular signals feeding into these cascades, including input via GPCRs (Goldsmith and Dhanasekaran 2007). All four subtypes of ARs were shown to mediate extracellular signal-regulated kinase (ERK) 1/2 phosphorylation in transfected CHO cells (Graham et al. 2001; Schulte and Fredholm 2000). MAP kinase signaling and hence cell proliferation might be amenable to manipulation through specific ARs in tumor cells. Such a possibility seems to be particularly attractive in a situation where one AR subtype is highly expressed, as is the case for A_{2B}ARs in MDA-MB-231 cells. As mentioned above, A_{2B}ARs are stimulated only by pathophysiologically high concentrations of adenosine (Fredholm et al. 2001). Thus, selective blockade or stimulation of this AR subtype may not interfere with the numerous important physiological functions of adenosine mediated via other AR subtypes.

MDA-MB-231 cells show a very high basal ERK 1/2 phosphorylation, indicative of constitutively active growth signals (Bieber et al. 2008). This basal activity seems to be maximal, as stimulation of the MAP kinase pathway (e.g., with epidermal growth factor, EGF) does not cause a further increase in ERK phosphorylation. The nonselective AR agonist NECA, on the other hand, causes a time-dependent decrease in ERK 1/2 phosphorylation, whereas CGS 21680 shows no inhibitory effect. As described above, functional and binding studies suggest that only A_{2B}ARs are present in MDA-MB-231 cells. Therefore, it seems that this AR subtype is responsible for the unusual inhibitory signal on ERK 1/2 phosphorylation. Moreover, antagonists like 1,3-dipropyl-8-cyclopentylxanthine (DPCPX) block this response, confirming the identity of the AR subtype as the A_{2B}AR mediating the inhibition of ERK 1/2 phosphorylation (Bieber et al. 2008).

The exact pathway leading to A_{2B}AR-mediated inhibition is not fully understood at this point. Both the Ca^{2+} signal detected following A_{2B}AR stimulation in

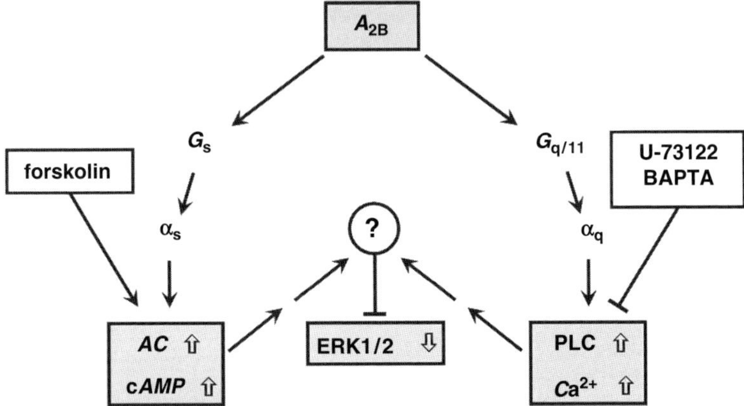

Fig. 1 Possible pathways leading to inhibition of extracellular signal regulated kinase (ERK)1/2 phosphorylation by A_{2B} adenosine receptors (A_{2B}ARs). Stimulation of adenylate cyclase (AC) via the G_s pathway results in inhibition of mitogen-activated protein kinase (MAPK) activity. Forskolin mimics this inhibition, confirming a role of cAMP. Alternatively, $G_{q/11}$ may be activated by A_{2B}AR stimulation, resulting in an increased activity of phospholipase C (PLC) and in intracellular Ca^{2+} signal. The PLC inhibitor U-73122 and the Ca^{2+} chelator BAPTA (applied as the cell-penetrating ester BAPTA-AM) both abolish A_{2B}AR-mediated inhibition of MAPK, providing evidence for a second pathway leading to the inhibition of ERK 1/2 phosphorylation. Both the G_s- and the $G_{q/11}$-mediated signals are linked to MAPK inhibition via currently unknown pathways

MDA-MB-231 cells (Panjehpour et al. 2005) and PLC activation are sufficient, as their blockade abolishes the inhibition of ERK 1/2 phosphorylation. On the other hand, forskolin stimulation mimics the effect of NECA, suggesting that cAMP may also play a role. Several inhibitors of PKA have no effect on NECA-induced inhibition of ERK 1/2 phosphorylation. Similarly without effect are activators of PKA and exchange protein activated by cAMP (Epac), making these effectors unlikely to be targets involved in mediating the inhibitory A_{2B}AR signal on MAP kinase activity. Figure 1 summarizes the current knowledge of potential pathways leading to A_{2B}AR-mediated inhibition of ERK 1/2 phosphorylation in MDA-MB-231 cells.

Although it was shown that A_{2B}ARs convey a stimulatory signal into MAP kinase pathways in transfected CHO cells (Schulte and Fredholm 2000), an inhibitory input was found in MDA-MB-231 cells. A few studies describe such an uncommon antiproliferative GPCR-mediated signal in glomerular mesangial cells (Haneda et al. 1996) and in vascular smooth muscle cells (Dubey et al. 2000). The high expression levels of A_{2B}ARs in an estrogen-negative breast cancer cell line together with a link to an antiproliferative signaling pathway make this AR subtype a potentially interesting target for tumor treatment, perhaps in combination with drugs interfering with downstream effectors in MAP kinase signaling pathways (Dhillon et al. 2007).

There is an increasing amount of data confirming that A_{2B}ARs play an important role in mediating the effects of adenosine on tumor growth and progression. The effects which are most interesting for a potential anticancer treatment based on A_{2B}ARs as a target are inhibition of angiogenesis and inhibition of ERK 1/2 phosphorylation. The dilemma is, however, that inhibition of angiogenesis requires the use of A_{2B}AR antagonists, whereas inhibition of growth signaling via the MAP kinase pathway might be achieved through treatment with A_{2B}AR agonists. The relative importance of these effects needs to be investigated using *in vivo* models before therapeutic suggestions can arise. It may eventually turn out that both agonists and antagonists will provide useful options for treatment in combination with other therapeutic measures if used at different stages of the disease and its treatment.

5 A_3 Adenosine Receptor

A_3AR belongs to the family of seven-transmembrane-domain GPCRs. The human A_3AR has been cloned and expressed and its adenosine agonist binding specificities characterized. The A_3AR was found to be most abundantly expressed in human lung and liver, with low amounts observed in the brain (Sajjadi and Firestein 1993). Low levels of expression were also observed in testes and heart. No expression was found in spleen or kidney. This expression profile differed from those for the A_1AR, A_{2A}AR and A_{2B}AR, which are expressed in variable levels in brain, heart, lung and kidney but not in liver tissues (Salvatore et al. 1993). Ligand structure–activity studies have identified selective agonists, partial agonists and antagonists for ARs (Cristalli et al. 2003; Muller 2003; Volpini et al. 2003; Zablocki et al. 2004). For the human and rat A_3AR, potent and selective agonists as well as selective A_3AR antagonists (e.g., PSB-10, PSB-11, MRE-3005F20 and MRS-1334) have been identified (Muller 2003). Site-directed mutagenesis and molecular modeling studies have also been performed that provide detailed information about the physical properties of ligand binding sites and the process of receptor activation (Gao et al. 2002; Muller 2003). Because of their selective tissue distribution and the development of specific A_3AR agonists and antagonists for them, A_3ARs have recently attracted considerable interest as novel drug targets.

Agonists to the A_3AR exert a differential effect on normal and tumor cells. In normal cells, the agonists induce the production of growth factors via induction of the NF-κB signaling pathway. In contrast, in tumor cells, the agonists induce apoptosis and tumor growth inhibition via deregulation of the NF-κB and the Wnt signaling pathways. This will be further detailed in Sect. 5.4.1 of this chapter.

Moreover, A_3AR agonists showed efficacy as cardioprotective, cerebroprotective, anti-inflammatory and immunosuppressive agents (Bar-Yehuda et al. 2007; Chen et al. 2006; Xu et al. 2006). For additional information on the pharmacology of the A_3AR and its role in disease, the reader is referred to Chap. 10, "A_3 Adenosine Receptor: Pharmacology and Role in Disease" (by Borea et al.), in this volume.

In this manuscript, the activity of A_3AR ligands as anticancer and chemoprotective agents will be presented. In addition, various aspects of A_3AR-targeted therapy, mainly in solid tumor malignancies such as melanoma, prostate, colon and hepatocellular carcinoma (HCC), will be discussed. Signal transduction pathways involved with A_3AR targeting utilizing highly selective A_3AR agonists and antagonists will be presented.

A significant part of the review is dedicated to the therapeutic effect of A_3AR agonists based on the concept that these compounds target mainly malignant cells that highly express A_3ARs without damaging normal body cells that barely express the receptor.

5.1 Overexpression of the A_3AR in Tumor Versus Normal Adjacent Tissues

Earlier studies revealed A_3AR expression in tumor cell lines including astrocytoma, HL-60 leukemia, B16–F10 and A378 melanoma, human Jurkat T-cell lymphoma, and murine pineal tumor cells, whereas low expression was described in most normal tissues (Auchampach et al. 1997; Gessi et al. 2002; Madi et al. 2003; Merighi et al. 2001; Suh et al. 2001; Trincavelli et al. 2002).

In more recent studies, a comparison between A_3AR expression in tumor vs. adjacent and relevant normal tissues supported the assumption that the receptor is upregulated in different types of malignancies. Recently, A_3AR in solid tumors was analyzed, leading to robust findings showing overexpression of the A_3AR in tumor tissues vs. low expression in the adjacent normal tissues. Furthermore, there is substantial evidence showing that A_3AR expression level is directly correlated to disease severity (Gessi et al. 2004; Madi et al. 2004).

In a comparative study, Morello et al. showed that primary thyroid cancer tissues express high levels of A_3ARs, as determined by immunohistochemistry analysis, whereas normal thyroid tissue samples do not express A_3ARs (Morello et al. 2007). Gessi et al. looked at the receptor binding values (K_d and B_{max}) of the A_3AR ligand [^{3}H]MRE 3008F20 in colon carcinoma tissue samples from 73 patients, and found an increased binding value in comparison to adjacent, remote and healthy colon mucosa (Gessi et al. 2004). Interestingly, they found that large adenomas showed increased binding versus small adenomas, which had affinity and density values that were very similar to those of the mucosa of healthy subjects. An additional important result of this study was that the high receptor binding values (K_d and B_{max}) were reflected in the peripheral blood lymphocytes and neutrophils of the patients with colon carcinoma. Upon tumor resection, the A_3AR binding value (K_d and B_{max}) returned to that of the healthy subjects, suggesting that the receptor may also serve as a biological marker (Gessi et al. 2004). Similar data were reported by Madi et al. showing higher A_3AR protein and mRNA expression levels in colon and breast carcinomas vs. adjacent non-neoplastic tissue or normal tissue (Madi et al. 2004). Further analysis revealed that the lymph node metastasis expressed even more A_3AR mRNA levels than the primary tumors, supporting the notion that A_3AR levels may reflect the status of tumor progression (Madi et al. 2004).

Madi et al. also reported that in human melanoma, colon, breast, small-cell lung, and pancreatic carcinoma tissues, A_3AR mRNA was upregulated compared to adjacent non-neoplastic tissue and normal tissue derived from healthy subjects (Madi et al. 2004). Moreover, computational analysis using different database sources supported the biological analysis that A_3AR is overexpressed in tumor tissues (Madi et al. 2004). A 2.3-fold increase in the expression of A_3AR in human colon adenoma versus normal colon tissue using microarray analysis (Princeton University database) was found. A search in the Cancer Genome Anatomy Project (CGAP); SAGE (website: http://cgap.nci.nih.gov/SAGE; Virtual Northern Legend) based on serial analysis of gene expression revealed that A_3AR was abundant in brain, kidney, lung, germ cells, placenta and retina, but that brain, lung, and pancreatic tumors expressed more A_3AR in the malignant than the normal non-cancerous tissues from the same organs of the same patients. A search of the Expression Viewer (Human Genome Organization (HUGO) Gene Nomenclature Committee/CleanEX) based on expressed sequence tags revealed that the relative expression of A_3AR was 1.6-fold higher in all of the cancer tissues compared with normal tissues (Madi et al. 2004).

In a recent study, Bar-Yehuda et al. showed that A_3AR mRNA expression is upregulated in HCC tissues in comparison to adjacent normal tissues (Bar-Yehuda et al. 2008). Remarkably, upregulation of A_3AR was also noted in peripheral blood mononuclear cells (PBMCs) derived from the HCC patients compared to healthy subjects. These results further show that A_3AR in PBMCs reflect receptor status in the remote tumor tissue (Bar-Yehuda et al. 2008). Moreover, the high expression level of the A_3AR was directly correlated to overexpression of NF-κB, a transcription factor for the A_3AR.

It is well established that G_i-protein-coupled receptors are internalized to early endosomes upon agonist binding (Bunemann et al. 1999; Claing et al. 2002). Early endosomes serve as the major site of receptor recycling, whereas the late endosomes are involved with the delivery of the internalized receptor to the lysosomes (Bunemann et al. 1999; Claing et al. 2002). Former studies have shown that chronic exposure of the A_3AR to the specific agonist methyl-1-[N^6-(3-iodobenzyl)-adenin-9-yl]-β-D-ribofuronamid (IB–MECA) resulted in receptor internalization/externalization in B16–F10 melanoma cells (Madi et al. 2003). It was also demonstrated that in experimental animal xenograft models of colon and prostate carcinoma, chronic treatment with IB–MECA (designated CF101) induced receptor downregulation shortly after agonist administration. Interestingly, 24 h after treatment there was no tachyphylaxis and the A_3AR was fully expressed, showing that the target is not downregulated upon chronic treatment with the agonist (Fishman et al. 2003, 2004).

The data showing a direct correlation in A_3AR expression between tumor tissue and PBMCs suggest that receptor expression in the PBMCs mirrors receptor status in the tumor tissue. It is possible that TNF-α upregulation induces an increase in the expression level and activity of NF-κB, a transcription factor for A_3ARs (Madi et al. 2004). This assumption is supported by the following finding. Upon treatment with 2-chloro-N^6-3-iodobenzyladenosine-5′-N-methyluronamide (Cl–IB–MECA; designated CF102), the expression levels of TNF-α and NF-κB were decreased, resulting in a downregulation of A_3AR expression in both PBMCs and the tumor tissue (Bar-Yehuda et al. 2008). Similar data were reported by Gessi et al., showing that A_3AR is upregulated in both colon carcinoma tissue and PBMCs of patients with colon carcinoma. This group further demonstrated that the expression levels of A_3AR were downregulated in the PBMCs upon tumor removal (Gessi et al. 2004).

Taken together, the findings described above that show A_3AR overexpression in different tumor cell types provide the rationale that this receptor may be utilized as a specific target to treat cancer.

5.2 *In Vitro Studies*

The A_3AR plays an important role in regulating normal and tumor cell growth. Cell response to a given A_3AR agonist is determined by a plethora of factors, including agonist concentration and affinity, receptor density, interaction between different ARs expressed on the cell surface, cell type, and the cell microenvironment.

5.2.1 Effect of Low-Concentration A_3AR Agonists on Tumor Cell Growth

The effects of A_3AR agonists, mainly IB–MECA and Cl–IB–MECA, on the proliferation of various tumor cells have been extensively tested. The rationale for using low concentrations of these two A_3AR agonists was based on their high affinity and selectivity at the A_3AR (approximately three orders of magnitude more than at the other ARs) (Fishman et al. 2007; Jeong et al. 2004; Joshi and Jacobson 2005). Moreover, Phase I clinical studies in healthy subjects, testing of IB–MECA (designated CF101) showed that the maximal tolerated dose of the drug was 5 mg kg^{-1}. At this dose, the plasma concentration was 40 ng ml^{-1}, which correlates with a concentration of 20 nM (van Troostenburg et al. 2004). This value correlates nicely with the affinity of IB–MECA to the mouse/rat/human A_3AR, exclusively activating this AR subtype, not any other AR subtype. Based on these data, IB–MECA and Cl–IB–MECA were tested both *in vitro* and *in vivo* at low concentrations and dosages, respectively. Remarkably, at this low concentration range these agonists induced a differential effect on tumor and normal cell proliferation.

Inhibition of the growth of tumor cells, including rat Nb2–11C and mouse Yac-1 lymphoma, K-562 leukemia, B16–F10 melanoma, MCA sarcoma, human LN-Cap and PC3 prostate carcinoma, MIA-PaCa pancreatic carcinoma and HCT-116 colon carcinoma, was found. The agonists induced a cytostatic effect towards the tumor cells, as manifested by a decrease in ^{3}[H]thymidine incorporation and cell cycle arrest at the G_0/G_1 phase (Bar-Yehuda et al. 2001; Fishman et al. 2000a, 2001, 2002a, b, 2003; Merimsky et al. 2003; Ohana et al. 2003). This effect was abolished by A_3AR antagonists (Madi et al. 2003), demonstrating that the response was A_3AR mediated. IB–MECA enhanced the cytotoxic effect of chemotherapy when tested in 1-(4,5-dimethylthiazol-2-yl)-3,5-diphenylformazan thiazolyl (MTT) and colony formation assays. A combined treatment of 5-flurouracil plus IB–MECA yielded higher growth inhibition of HCT-116 human colon carcinoma cells in comparison to the chemotherapy alone (Bar-Yehuda et al. 2005).

At the same time, Cl–IB–MECA stimulated the proliferation of bone marrow cells (Fishman et al. 2001). Interestingly, both IB–MECA and Cl–IB–MECA upregulated the production of granulocyte colony stimulating factor (G-CSF), known to act as a differentiation factor of neutrophils (Brandt et al. 1988). This novel activity mediated the stimulatory effect on bone marrow cell growth and prompted the examination of IB–MECA and Cl–IB–MECA as myeloprotective agents that prevent neutropenia upon treatment with chemotherapeutic agents (Bar-Yehuda et al. 2002, Fishman et al. 2000b, 2001, 2002b, 2003).

As opposed to the results of the studies described above, demonstrating an inhibition of tumor cell lines by A_3AR agonists, in a set of experiments conducted by Gessi et al., low-concentration (100 nM) Cl–IB–MECA stimulated the proliferation of some cancer cell lines such as Caco-2, DLD1, and HT29 human colon carcinoma cell line (Gessi et al. 2007). In addition, the same group showed that under hypoxic conditions, Cl–IB–MECA induced upregulation of hypoxia-inducible factor 1 (HIF-1) alpha and VEGF in HT-29 human colon carcinoma cells, A375 human melanoma cells, and A172 and U87MG glioblastoma cell lines. This effect could be blocked with the A_3AR antagonist (MRE3008F20) or by siRNA silencing (Merighi et al. 2005b, 2006, 2007). Moreover, Abbracchio et al. showed that Cl–IB–MECA modulates cytoskeleton reorganization, increases expression of Rho, and induces the intracellular distribution of the antiapoptotic protein Bcl–xL in ADF human astrocytoma cells (Abbracchio et al. 1997, 2001). Thus, A_3AR agonists can on the one hand induce the inhibition of tumor cell growth via cell cycle arrest, and on the other hand stimulate the proliferation of tumor cells, depending on cell type and culture conditions.

5.2.2 Effect of High-Concentration A_3AR Agonists on Tumor Cell Growth

The effect of high-concentration A_3AR agonists on tumor cell growth was an inhibitory one that was either A_3AR dependent or independent. Cl–IB–MECA at a concentration of 10 μM inhibited the growth of A375 human melanoma cells by inducing cell cycle arrest in the G_0/G_1 phase. This effect was blocked by an A_3AR antagonist, demonstrating the role of A_3AR activation in this response (Merighi et al. 2005a). Moreover, IB–MECA at high concentration (30–60 μM) produced cell growth inhibition in both ERα-positive MCF-7 cells and in ERα-negative MDAMB468 human breast carcinoma cells. In both cell types, the introduction of an A_3AR antagonist, MRS1220, blocked the effect of this A_3AR agonist (Panjehpour and Karami-Tehrani 2004, 2007).

The A_3AR agonist 2-chloro-N^6-(3-iodobenzyl)-4′-thioadenosine-5′-N-methyluronamide (thio-Cl–IB–MECA) has high affinity and specificity for the human A_3AR. The introduction of μM concentrations of this agonist to HL-60 human leukemia cell cultures resulted in apoptosis, as manifested by DNA fragmentation and poly(ADP-ribose) polymerase (PARP) cleavage (Lee et al. 2005).

Interestingly, an additional compound that inhibits the growth of tumor cells via A_3AR is cordycepin (3′-deoxyadenosine), an active ingredient of *Cordyceps sinensis*, a parasitic fungus used in traditional Chinese medicine (Nakamura et al. 2006). This molecule, at μM concentrations, induced a remarkable inhibitory effect on the growth of murine B16–BL6 melanoma and of Lewis lung carcinoma tumor cells. This inhibitory effect was abolished by the A_3AR antagonist 3-ethyl-5-benzyl-2-methyl-4-phenylethynyl-6-phenyl-1,4-(±)-dihydropyridine-3,5-dicarboxylate MRS1191 (Nakamura et al. 2006).

In contrast, IB–MECA and Cl–IB–MECA at μM concentrations inhibit the growth of various tumor cell lines (including NPA papillary thyroid carcinoma, HL-60 leukemia cells and U-937 lymphoma cells) in an A_3AR-independent mechanism (Kim et al. 2002; Morello et al. 2007). This inhibitory effect was characterized by apoptosis and was not abolished by antagonism or knockdown of the A_3AR. Based on these results, it was concluded that IB–MECA or Cl–IB–MECA at high concentrations can induce tumor cell death through receptor-independent mechanisms, perhaps via active transport into the cells through the nucleoside transporters (Kim et al. 2002; Merighi et al. 2002; Morello et al. 2007). Moreover, in MCF-7 human breast cancer cells, 100 μM of IB–MECA markedly reduced cell number and inhibited colony formation (Lu et al. 2003). These cancer cells do not express A_3ARs, overexpression of A_3AR did not lower the concentrations of IB–MECA needed to induce the inhibition of cell proliferation, and the introduction of MRS1191 (an A_3AR antagonist) did not abolish the IB–MECA inhibitory effect, suggesting that A_3AR was not involved in the cell growth inhibition of these human breast cancer cells. In these studies, an explanation for this inhibitory effect by IB–MECA may be related to its ability to reduce the expression level of estrogen receptor (ER) alpha, which plays a role in different signaling pathways leading to the transcription of genes responsible for G_1–S cell cycle progression (Lu et al. 2003). The effects of the various A_3AR agonists at low and high concentrations on tumor cell growth in *in vitro* studies are summarized in Table 1.

5.3 *In Vivo Studies*

In this part of the review, *in vivo* studies showing the efficacy of A_3AR agonists in various tumor-bearing animals will be presented, supporting the utilization of A_3AR as a target to treat cancer. In all experimental models, the A_3AR agonists were administered orally due to their stability and bioavailability profile. The dose used in these studies was calculated based on the affinity data, resulting in exclusive activation of the A_3AR. The studies included syngeneic, xenograft, orthotopic and metastatic experimental animal models utilizing IB–MECA and Cl–IB–MECA as the therapeutic agents.

Table 1 Effects of A_3AR agonists at low and high concentrations on tumor cell growth in *in vitro* studies

Drug	Low/high A_3AR agonist concentrations	Tumor cell type	A_3AR-related	Effect	Suggested mechanism of action	References
IB–MECA	Low (1–100 nM)	Murine NB2–11C lymphoma	Yes	Growth inhibition	Deregulation of the Wnt signaling pathway	Fishman et al. (2000a, 2002a, 2003) Madi et al. (2003)
		Murine B16–F10 melanoma			Cell cycle arrest at the G_0/G_1 phase	
		Human PC3 prostate carcinoma				
Cl–IB–MECA	Low (100 nM)	Human Caco2 colon carcinoma	Yes	Cell proliferation	Upregulation of HIF-1alpha and VEGF	Gessi et al. (2007)
		Human DLD1 colon carcinoma			Reorganization of cytoskeketon	Merighi et al. (2005b, 2006, 2007)
		Human HT29 colon carcinoma*			Increased expression of Rho	Abbracchio et al. (1997, 2001)
		Human A375 melanoma*			Induction of intracellular distribution of Bcl–xL	

		Human A172 and U87MG glioblastoma* Human ADF astrocytoma				
Thio-Cl–IB–MECA	Low (10nM)	Human HL-60 promyelocytic leukemia	Not determined	Growth inhibition	Deregulation of the Wnt signaling pathway	Lee et al. (2005)
IB–MECA	High (30–60 µM)	Human ERα-positive MCF-7 breast carcinoma Human ERα-negative MDAMB468 breast carcinoma	Yes	Growth inhibition	Inhibition of anchorage-dependent cell growth	Panjehpour and Karami-Tehrani (2004, 2007)
Cl–IB–MECA	High (10 µM)	Human A375 melanoma	Yes	Growth inhibition	Cell cycle arrest in the G_0/G_1 phase	Merighi et al. (2005a)
Thio-Cl–IB–MECA	High (25–50 µM)	Human HL-60 promyelocytic leukemia	Not determined	Growth inhibition	Downregulation of cyclin D1 and *c*-myc protein expression Cell cycle arrest in the G_0/G_1 phase Induction of apoptosis	Lee et al. (2005)

(continued)

Table 1 (continued)

Drug	Low/high A_3AR agonist concentrations	Tumor cell type	A_3AR-related?	Effect	Suggested mechanism of action	References
Cordycepin (3′-deoxyadenosine)	High (25–50 μM)	Mouse B16–BL6 melanoma Mouse Lewis lung carcinoma	Yes	Growth inhibition	Not determined	Nakamura et al. (2006)
IB–MECA	High (100 μM)	Human MCF-7 breast carcinoma	No	Growth inhibition	Downregulation estrogen receptor expression level Inhibition of colony formation	Lu et al. (2003)
Cl–IB–MECA	High (10, >30 μM)	Human HL-60 promyelocytic leukemia	No	Growth inhibition	Dephosphorylation of ERK1/2	Kim et al. (2002)
		Human MOLT-4 leukemia			Inhibition of cell growth blocking the G_1 cell cycle phase	
		Human NPA papillary thyroid carcinoma			Induction of apoptosis	

5.3.1 Melanoma

Oral administration of 10–100 μg kg^{-1} IB–MECA and Cl–IB–MECA once or twice daily inhibited the growth of primary B16–F10 murine melanoma tumors in syngeneic models (Madi et al. 2003). Moreover, in an artificial metastatic model, IB–MECA inhibited the development of B16–F10 murine melanoma lung metastases (Bar-Yehuda et al. 2001; Fishman et al. 2001, 2002b). The specificity of the response was demonstrated by the administration of an A_3AR antagonist that reversed the effect of the agonist (Madi et al. 2003).

Furthermore, IB–MECA or Cl–IB–MECA in combination with the chemotherapeutic agent cyclophosphamide induced an additive antitumor effect on the development of B16–F10 melanoma lung metastatic foci (Fishman et al. 2001, 2002b).

5.3.2 Colon Carcinoma

Oral administration of 10–100 μg kg^{-1} IB–MECA once or twice daily inhibited the growth of primary CT-26 colon tumors (Ohana et al. 2003). Furthermore, in xenograft models, IB–MECA inhibited the development of HCT-116 human colon carcinoma in nude mice (Ohana et al. 2003). In these studies, the combined treatment of IB–MECA and 5-fluorouracil resulted in an enhanced antitumor effect. IB–MECA was also efficacious in inhibiting liver metastases of CT-26 colon carcinoma cells inoculated in the spleen. (Bar-Yehuda et al. 2005; Fishman et al. 2002b, 2004; Ohana et al. 2003).

5.3.3 Prostate Carcinoma

IB–MECA inhibited the development of PC3 human prostate carcinoma in nude mice. Additionally, IB–MECA increased the cytotoxic index of Taxol in PC3 prostate carcinoma-bearing mice (Fishman et al. 2002b, 2003).

5.3.4 Hepatocellular Carcinoma

Recent studies showed that A_3AR is overexpressed in tumor tissues and in PBMCs of N1S1 HCC tumor-bearing Sprague–Dawley rats (Bar-Yehuda et al. 2008). For these studies, an orthotopic rat model was established in which a subxiphoid laparotomy was performed and N1S1 cells were injected into the right hepatic lobe. Treatments with Cl–IB–MECA at doses of 1, 50, 100, 500 and 1, 000 μg kg^{-1} three times daily were initiated on day 3 after tumor inoculation and continued until day 15. Cl–IB–MECA treatment exerted a bell-shaped, dose-dependent inhibitory effect on tumor growth with a maximal effect at a dose of 100 μg kg^{-1} (Bar-Yehuda et al. 2008).

5.3.5 Potentiation of Natural Killer Cell Activity

IB–MECA and Cl–IB–MECA also upregulate serum levels of IL-12 and potentiate NK cell activity (Harish et al. 2003). In mice, Cl–IB–MECA increased serum levels of IL-12 and potentiated the activity of NK cells (Harish et al. 2003). This effect of Cl–IB–MECA on NK cell activity was seen in adoptive transfer experiments utilizing melanoma-bearing mice where marked inhibition in the development of lung metastatic foci was observed in the mice engrafted with splenocytes derived from Cl–IB–MECA treated mice. Similar results were observed in HCT-116 human colon carcinoma-bearing nude mice treated with 10 $\mu g\ kg^{-1}$ IB–MECA (Ohana et al. 2003).

5.3.6 Chemoprotective Effect

IB–MECA and Cl–IB–MECA act also as chemoprotective agents. With cyclophosphamide treatment of B16–F10 melanoma-bearing mice or 5-fluorouracil treatment of HCT-116 human colon carcinoma-bearing nude mice, a marked decline in white blood cells and neutrophil counts occurs (Bar-Yehuda et al. 2002; Fishman et al. 2000b, 2001, 2002a, b, 2003). Administration of the A_3AR agonist restored the number of white blood cells and the percentage of neutrophils to their normal values. This was attributed to the ability of IB–MECA to induce the production of G-CSF (Bar-Yehuda et al. 2002; Fishman et al. 2000b, 2001, 2002a, b, 2003; Hofer et al. 2006, 2007)

Overall, the unique characteristics of the A_3AR agonists—they are orally bioavailable, exert their effects at low doses, enhance the effects of cytotoxic agents, and at the same time act as myeloprotective agents—together with their potential cardio- and neuroprotective activities suggest that this class of compounds may produce attractive clinical candidates as anticancer drugs.

5.4 Mechanisms of Action for the Anticancer Activity of the A_3AR

Adenosine receptors operate through distinct biochemical signaling mechanisms. The A_1 and A_3AR subtypes control most, if not all, of their cellular responses via pertussis toxin-sensitive G proteins of the G_i and G_o family. The A_3AR triggers G_i-protein activation, induces an intracellular signaling cascade that increases intracellular calcium concentrations, activates PLC and phospholipase D (PLD) as well as the production of intracellular second-messenger systems, which in turn, leads to related cellular responses such as cell proliferation or tumor cell apoptosis (Abbracchio et al. 1995; Murthy and Makhlouf 1995; Olah and Stiles 1995; Olah et al. 1995).

Activation of the A_3AR inhibits adenylate cyclase activity, thereby leading to a decrease in the level of the second messenger, cAMP. The latter modulates the level and activity of protein kinase A (PKA) that phosphorylates downstream elements of the MAPK and protein kinase B (PKB)/Akt (PKB/Akt) signaling pathways (Poulsen and Quinn 1998; Seino and Shibasaki 2005; Zhao et al. 2000). In addition, it was reported that PKA phosphorylates PKB/Akt directly, thereby mediating its activity (Fang et al. 2000). Both PKA and PKB/Akt regulate the NF-κB signaling pathway by phosphorylating and activating the downstream kinase IκB kinase (IKK), which phosphorylates IκB, thereby sorting it to degradation via the ubiquitin system. As a result, NF-κB is released from its complex with IκB and translocates to the nucleus to induce the transcription of genes such as cyclin D1 and *c*-Myc that control cell cycle progression (Karin and Ben-Neriah 2000; Li et al. 1999).

Taken together, since the activation of A_3AR induces the inhibition of adenylate cylase and reduces the level of intracellular cAMP, the downstream elements PKA and PKB/Akt are not activated and so do not phosphorylate IKK. This leads to the reduced activity and expression levels of the NF-κB, resulting in tumor cell cycle arrest and tumor growth inhibition.

5.4.1 Direct Effect of A_3AR Agonists on Tumor Cells: Deregulation of the NF-κB and Wnt Signaling Pathways

In melanoma, colon, prostate and hepatocellular carcinoma cell lines, treatment with IB–MECA or Cl–IB–MECA produced a decrease in PKA and PKB/Akt expression (Bar-Yehuda et al. 2008; Fishman et al. 2002a, b, 2003, 2004). As a result, the phosphorylation of IKK was inhibited, leading to the accumulation of IκB/NF-κB complex in the cytoplasm. This resulted in the downregulation of *c*-myc and cyclin D1 expression levels (Fig. 2) (Bar-Yehuda et al. 2008; Fishman et al. 2003, 2004).

Further studies showed that the Wnt signaling pathway is also involved in the anticancer activity mediated via the A_3AR. The rationale to investigate this pathway came from data showing that PKA and PKB/Akt phosphorylate and inactivate glycogen synthase kinase 3β (GSK-3β) (Cross et al. 1995; Fang et al. 2000). GSK-3β is a serine/threonine kinase that acts as a key element in the Wnt signaling pathway, which is known to play a pivotal role in dictating cell fate during embryogenesis and tumorigenesis (Peifer and Polakis 2000). GSK-3β phosphorylates the cytoplasmic protein β-catenin, which is sorted for degradation by the ubiquitin system. Upon phosphorylation, GSK-3β loses its ability to phosphorylate β-catenin, resulting in the accumulation of the latter in the cytoplasm and its subsequent translocation to the nucleus, where it associates with lymphoid enhancer factor/T-cell factor (Lef/Tcf) to induce the transcription of genes responsible for cell cycle progression, like *c*-myc and cyclin D1 (Fig. 2) (Ferkey and Kimelman 2000; Morin 1999; Novak and Dedhar 1999).

An inability of GSK-3β to phosphorylate β-catenin has been demonstrated in various malignancies, including colon carcinoma, melanoma and HCC (Bonvini et al. 1999; Cui et al. 2003; Robbins et al. 1996)

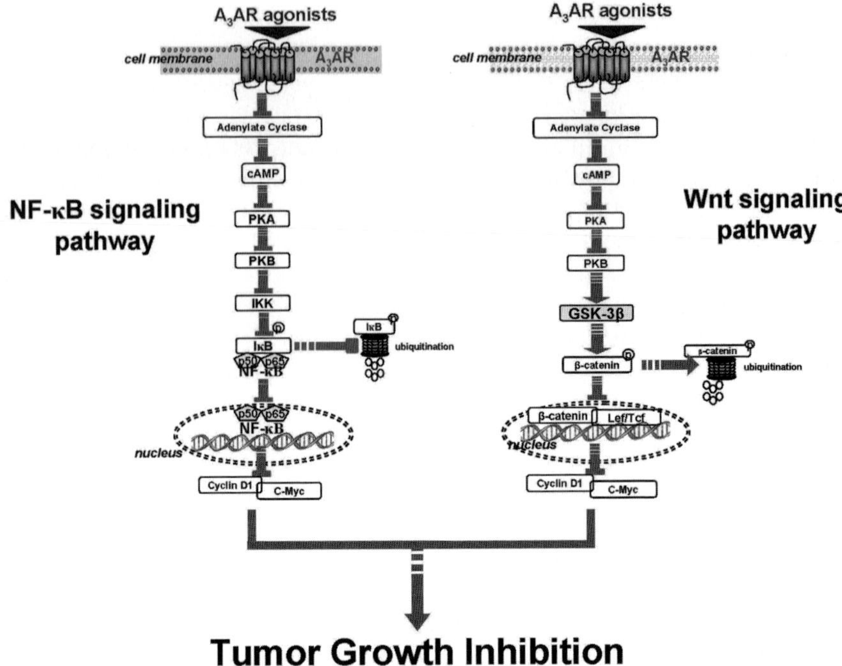

Fig. 2 Anticancer effect of A_3 adenosine receptor (A_3AR) agonists entails deregulation of the nuclear factor kappa B (NF-κB) and the Wnt signaling pathways. Activation of the A_3AR in tumor cells with specific agonists inhibits the activity of adenylate cyclase, inducing a decline in the level of cAMP, leading to decreased levels of protein kinase A (PKA) and its substrate protein kinase B (PKB)/Akt. Consequently, this leads to a downregulation in the expression levels of signal proteins that play a role in the NF-κB (IκB kinase (IKK) and IκB) and the Wnt (glycogen synthase kinase-3β (GSK-3β) and β-catenin) signaling pathways. As a result, the levels of *c*-Myc and cyclin D1, known to play a crucial role in cell cycle progression, are decreased. This chain of events leads to tumor growth inhibition

Treatment of B16–F10 melanoma, HCT-116 human colon carcinoma cells and PC-3 human prostate carcinoma cells *in vitro* with IB–MECA decreased PKA and PKB/Akt expression levels, resulting in the upregulation of GSK-3β and the subsequent phosphorylation and ubiquitination of β-catenin (Fishman et al. 2002a, 2003; Madi et al. 2003). In these studies, downregulation of cyclin D1 and *c*-myc expression levels, as well as tumor cell growth suppression, were observed (Fishman et al. 2002a, 2003; Madi et al. 2003). Moreover, the group of Lee et al. further reported that a highly specific A_3AR agonist, thio-Cl–IB–MECA, induced apoptosis of HL-60 promyelocytic leukemia cells and lung cancer cells via deregulation of the Wnt signaling pathway. The levels of β-catenin, phosphorylated forms of GSK3-β and Akt were downregulated upon treatment with thio-Cl–IB–MECA (10 nM) in a time-dependent manner (Kim et al. 2008; Lee et al. 2005).

Additional evidence to support the *in vitro* mechanistic pathways presented above came from the analysis of tumor tissues excised from melanoma, prostate, colon and HCC tumor-bearing animals treated with IB–MECA or Cl–IB–MECA (Bar-Yehuda et al. 2008; Fishman et al. 2003, 2004; Madi et al. 2003).

Both the NF-κB and Wnt signal transduction pathways were deregulated upon treatment with the A_3AR agonists, demonstrating a definitive molecular mechanism. Remarkably, Cl–IB–MECA induced marked apoptosis of tumor cells in the N1S1 HCC-bearing rats (Bar-Yehuda et al. 2008; Fishman et al. 2003, 2004; Madi et al. 2003).

In these studies, apoptosis of tumor cells was seen in the tunnel assay, and increases in the expression levels of the proapoptotic proteins Bad, BAX and capase 3 were observed as well (Bar-Yehuda et al. 2008; Fishman et al. 2003, 2004; Madi et al. 2003).

5.4.2 A_3AR Agonists as Myeloprotective Agents

Some chemotherapeutic agents are known to induce myelosuppression, as manifested by a decline in the number of white blood cells (especially neutrophils), making patients susceptible to infections and sepsis. G-CSF is a hematopoietic growth factor produced by endothelium, macrophages, and a number of other immune cells, and its synthesis is induced by activation of the transcription factor NF-κB. It stimulates the proliferation and differentiation of white blood cells. A recombinant form of G-CSF has become a standard supportive therapy for cancer patients to accelerate recovery from neutropenia after chemotherapy (Brandt et al. 1988; Rusthoven et al. 1998). In mice, IB–MECA induces G-CSF production and increases white blood cell and neutrophil counts in naïve and chemotherapy-treated animals (Bar-Yehuda et al. 2002). The myelostimulative effect of IB–MECA was also evidenced by high levels of G-CSF in bone marrow cells, splenocytes, and serum derived from IB–MECA-treated mice. Moreover, in splenocytes derived from IB–MECA-treated mice, increased expression levels of phosphoinositide 3-kinase (PI3K), known to play a role in the regulation of cell survival and proliferation (Gao et al. 2001), was noted. Consequently, the expression levels of PKB/Akt, IKK and NF-κB were enhanced, resulting in G-CSF upregulation (Fig. 3).

The role of the A_3AR and PI3K-NF-κB pathway in the production of G-CSF was further confirmed by treating the mice with pertussis toxin, a G_i-protein inactivator that interferes with the coupling of the receptor to the G_i protein. Splenocytes derived from mice that were treated with IB–MECA and pertussis toxin did not upregulate NF-κB levels. Moreover, the NF-κB inhibitor pyrrolidine dithiocarbamate (PDTC), known to suppress the release of IκB from the latent cytoplasmic form of NF-κB, counteracted the effect of IB–MECA and prevented the increase in NF-κB expression levels (Bar-Yehuda et al. 2002).

Taken together with the studies described in Sect. 5.4.1 above, these studies suggest that activation of the A_3AR by specific agonists induces differential effects on normal and tumor cells to produce modulations of definitive signal transduction

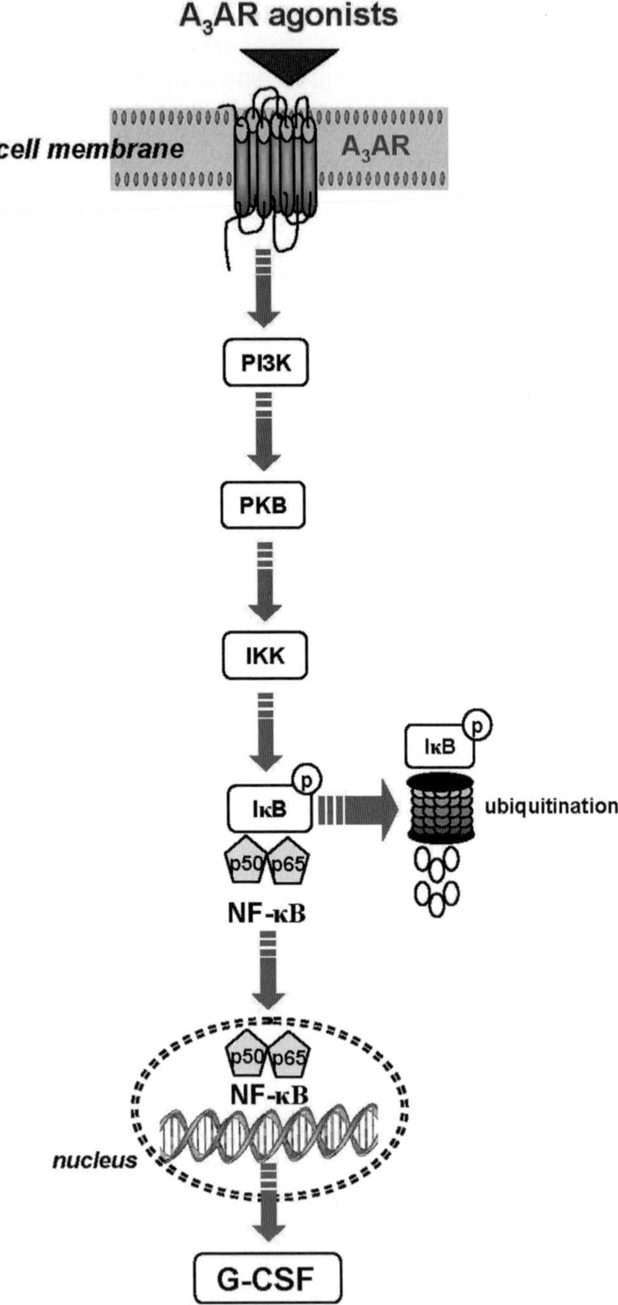

Fig. 3 Agonists for the A_3 adenosine receptor (A_3AR) induce granulocyte colony stimulating factor (G-CSF) production via nuclear factor kappa B (NF-κB). Activation of A_3AR in splenocytes induces upregulation of phosphoinositide 3-kinases (PI3K) and its downstream target protein kinase B (PKB)/Akt. The latter activates IκB kinase (IKK), which is responsible for the phosphorylation and ubiquitination of IκB. As a result, NF-κB translocates to the nucleus, where it induces the transcription of G-CSF

pathways that control cell growth regulatory mechanisms in the case of tumor cells and growth factor production in the case of normal hematopoietic cells (e.g., bone marrow cells and splenocytes).

6 Anticancer Activity of A_3AR Antagonists

A very interesting area of application of A_3AR ligands concerns cancer therapies. The possibility that the A_3AR plays an important role in the development of cancer has aroused considerable interest in recent years (Fishman et al. 2002b; Gessi et al. 2008; Merighi et al. 2003). The A_3AR subtype has been described in the regulation of the cell cycle, and both pro- and antiapoptotic effects have been reported, depending on the level of receptor activation (Gao et al. 2001; Gessi et al. 2007; Jacobson 1998; Merighi et al. 2005a; Yao et al. 1997). However, based on the studies presented above, it is important to note that A_3AR receptor activation appears to be involved in the inhibition of tumor growth both *in vitro* and *in vivo*.

Based on the relationships between tumors, hypoxia and adenosine concentrations, there are reports describing the potential utility of A_3AR antagonists for cancer treatment. Growing evidence from experimental and clinical studies points to the fundamental pathophysiological role of hypoxia in solid tumors. Hypoxia is the result of an imbalance between oxygen supply and consumption. Clinical investigations carried out over the last 15 years have clearly shown that the prevalence of hypoxic tissue areas is a characteristic pathophysiological feature of solid tumors. As the oxygen concentration decreases with increasing distance from the capillary, cell proliferation rates and drug concentrations both decrease. These two factors lead to resistance to anticancer drugs; firstly, because the majority of anticancer drugs are only effective against rapidly proliferating cells; secondly, because adequate levels of chemotherapy drugs have to reach the tumor cells from the blood vessels. Hypoxia inhibits enzymes that are involved in the breakdown of adenosine and increases the activities of those responsible for generating adenosine, thereby resulting in an increase in extracellular and intracellular adenosine. The elevated adenosine levels in response to hypoxia are not exclusive to tumor tissues, but, in this context, the increase in adenosine is localized to the tumor microenvironment, since the surrounding tissue is normally oxygenated (Blay et al. 1997). To survive under hypoxic conditions, tumor cells run numerous adaptive mechanisms, such as glycolysis, glucose uptake, and survival factor upregulation (Hockel and Vaupel 2001). Hypoxia-inducible factor (HIF) 1 is the most important factor involved in the cellular response to hypoxia (Semenza 2003). It is a heterodimer composed of an inducibly expressed HIF-1α subunit and a constitutively expressed HIF-1β subunit (Epstein et al. 2001). HIF-1α and HIF-1β mRNAs are constantly expressed under normoxic and hypoxic conditions (Wiener et al. 1996). However, during normoxia, HIF-1α is rapidly degraded by the ubiquitin proteasome system, whereas exposure to hypoxic conditions prevents its degradation (Minchenko et al. 2002; Semenza 2000). HIF-1α expression and activity are also regulated by the PI3K and MAPK signal transduction pathways (Semenza 2002; Zhong et al. 2000). A growing body of evidence indicates that HIF-1α contributes to tumor progression and metastasis (Hopfl et al. 2004; Welsh and Powis 2003). Immunohistochemical analyses have shown that HIF-1α is present in higher levels in human tumors than in normal tissues (Zhong et al. 1999), and the levels of HIF-1α activity in cells correlate with the tumorigenicity and angiogenesis in nude mice (Carmeliet et al. 1998). Tumor cells lacking HIF-1α expression are markedly impaired in their growth and vascularization (Jiang et al. 1997; Kung et al. 2000; Maxwell et al. 1997). Therefore, since HIF-1α expression and activity appear central to tumor growth and progression, HIF-1α inhibition becomes an appropriate approach to treating cancer (Kung et al. 2000; Ratcliffe et al. 2000; Semenza 2003). Hypoxia creates conditions that, on the one hand, are conducive to the accumulation of extracellular adenosine, and on the other hand stabilize hypoxia-inducible factors, such as HIF-1α (Fredholm 2003; Hockel and Vaupel 2001; Linden 2001; Minchenko et al. 2002; Semenza 2000; Sitkovsky et al. 2004). In particular, the correlation between AR stimulation and HIF-1α expression modulation in hypoxia has recently been investigated. It has been reported that adenosine increases HIF-1α protein accumulation in response to hypoxia in a dose- and time-dependent manner in human melanoma, glioblastoma and colon carcinoma through the involvement of the cell surface A_3AR (Merighi et al. 2005b, 2006, 2007). The signaling pathway involved in A_3AR-mediated accumulation of HIF-1α in hypoxia involves MAPKinase activity (Merighi et al. 2005b, 2006, 2007). It is well established that HIF-1α plays a major role in VEGF expression and angiogenesis. Furthermore, there is strong evidence that adenosine released from hypoxic tissues is an important player in driving the angiogenesis, by enhancing vascular growth through various mechanisms including the release of different factors, with VEGF being one of the most relevant (Adair 2005). A role for A_{2B}ARs in angiogenesis through an HIF-1α -independent intracellular pathway has been observed in human endothelial and smooth muscle cells (Feoktistov et al. 2004), but involvement of HIF-1α with the A_3AR has been demonstrated in different cancer cell lines (Merighi et al. 2005b, 2006, 2007). In particular, activation of the A_3AR subtype in glioblastoma and colon carcinoma cells stimulates VEGF expression in an HIF-1α-dependent manner (Merighi et al. 2006, 2007). In addition, A_3AR activation results in increased expression of another angiogenic factor, angiopoietin 2, in melanoma cells and HMC-1 cells derived from a highly malignant, undifferentiated human mastocytoma cancer (Feoktistov et al. 2003; Merighi et al. 2005b). This may be relevant because the effect of adenosine on new capillary formation is potentiated by the concomitant stimulation of A_{2B}ARs and A_3ARs acting on VEGF and angiopoietin 2 levels, respectively (Feoktistov et al. 2003). Recent studies indicate that pharmacologic inhibition of HIF-1α and particularly of HIF-regulated genes, which are important for cancer cell survival, may be more advantageous than HIF-gene-inactivation therapeutic approaches (Mabjeesh et al. 2003; Merighi et al. 2005b; Sitkovsky et al. 2004). In this regard, by blocking hypoxia-induced increases in HIF-1α, angiopoietin 2 and VEGF protein expression in the tumor microenvironment, A_3AR antagonists may represent a novel approach to the treatment of cancer.

7 Summary and Conclusions

Adenosine, the natural ligand of the four AR subtypes, affects all of these receptors under neoplastic conditions due to its mass accumulation in the tumor microenvironment. Its role in maintaining pro- and anticancer effects via each of its receptor subtypes was extensively reviewed in this chapter. Based on the studies presented in this review, it appears that all the AR subtypes are possible targets for the development of novel approaches to the treatment of cancer.

The antitumorigenic role of A_1AR in cancer was mainly studied in A_1AR-deficient mice, demonstrating that activation of the A_1AR on microglia inhibits the growth of glioblastomas.

Based on a number of reports, it has been suggested that the A_{2A}AR blocks antitumor immunity. In the tumor environment of hypoxia and high adenosine levels, activation of A_{2A}ARs leads to T-cell tolerance, inhibition of effector immune cells (including T cells, CTLs, NK cells, dendritic cells, and macrophages), an increase in regulatory T cells, and a decrease in proinflammatory cytokines, all of which thwart antitumor immunity and thus encourage tumor growth. Importantly, A_{2A}AR-null mice have been shown to more readily reject melanoma and lymphoma tumor challenge and to also respond to vaccines. Moreover, treating mice with A_{2A}AR antagonists (including caffeine) leads to increased tumor rejection by CD8+ T cells. For all these reasons, it was suggested that the addition of A_{2A}AR antagonists to cancer immunotherapeutic protocols may enhance tumor immunotherapy. Interestingly, the safety of such compounds has already been shown in trials employing A_{2A}AR antagonists for the treatment of Parkinson's disease.

The role of the A_{2B}AR in cancer is not clear. On the one hand, under conditions of hypoxia and high adenosine levels in the tumor microenvironment, activation of A_{2B}ARs leads to the release of angiogenic factors that promote tumor growth, suggesting that the use of A_{2B}AR antagonists may represent a novel approach to the treatment of cancer. On the other hand, the activation of A_{2B}ARs exclusively expressed on the surface of breast cancer cell line MDA-MB-231 cells exerts an inhibitory signal mediated via the inhibition of ERK 1/2 phosphorylation, suggesting that A_{2B}AR agonists may produce anticancer effects. The resolution of this dilemma will initially come from testing selective ligands for the A_{2B}AR in *in vitro* and *in vivo* studies in various cancer cell lines and tumor-bearing animals, and then, depending on the results of these studies, perhaps in humans with cancer.

The unique characteristics of the A_3ARs that are highly expressed in tumor cells suggest that this receptor subtype is an attractive target to combat cancer. Targeting the A_3AR with synthetic agonists results in cell cycle arrest and apoptosis towards different cancer cells both *in vitro* and *in vivo*. Preclinical and Phase I studies show that these agonists are safe and well tolerated in humans and thus may be considered possible therapeutic agents for certain neoplasmas such as HCC, where a significant apoptotic effect was demonstrated. However, by blocking hypoxia-induced increases in HIF-1α, angiopoietin 2 and VEGF protein expression in the tumor microenvironment, A_3AR antagonists may represent a novel approach for the treatment of cancer.

References

Abbracchio MP, Brambilla R, Ceruti S, Kim HO, von Lubitz DK, Jacobson KA, Cattabeni F (1995) G protein-dependent activation of phospholipase C by adenosine A_3 receptors in rat brain. Mol Pharmacol 48:1038–1045

Abbracchio MP, Rainaldi G, Giammarioli AM, Ceruti S, Brambilla R, Cattabeni F, Barbieri D, Franceschi C, Jacobson KA, Malorni W (1997) The A_3 adenosine receptor mediates cell spreading, reorganization of actin cytoskeleton, and distribution of Bcl-XL: studies in human astroglioma cells. Biochem Biophys Res Commun 241:297–304

Abbracchio MP, Camurri A, Ceruti S, Cattabeni F, Falzano L, Giammarioli AM, Jacobson KA, Trincavelli L, Martini C, Malorni W, Fiorentini C (2001) The A_3 adenosine receptor induces cytoskeleton rearrangement in human astrocytoma cells via a specific action on Rho proteins. Ann N Y Acad Sci 939:63–73

Adair TH (2005) Growth regulation of the vascular system: an emerging role for adenosine. Am J Physiol Regul Integr Comp Physiol 289:R283–R296

Auchampach JA, Xiaowei J, Tina CW, George H, Caughey GH, Linden J (1997) Canine mast cell adenosine receptors: cloning and expression of the A_3 receptor and evidence that degranulation is mediated by the A_{2B} receptor. Mol Pharmacol 52:846–860

Bar-Yehuda S, Barer F, Volfsson L, Fishman P (2001) Resistance of muscles to tumor metastasis: a role for A_3 adenosine receptor agonists. Neoplasia 3:125–131

Bar-Yehuda S, Madi L, Barak D, Mittelman M, Ardon E, Ochaion A, Cohn S, Fishman P (2002) Agonists to the A_3 adenosine receptor induce G-CSF production via NF-kappaB activation: a new class of myeloprotective agents. Exp Hematol 30:1390–1398

Bar-Yehuda S, Madi L, Silberman D, Slosman G, Shkapenuk M, Fishman P (2005) CF101, an agonist to the A_3 adenosine receptor enhances the chemotherapeutic effect of 5-flurouracil in a colon carcinoma murine model. Neoplasia 7:85–90

Bar-Yehuda S, Silverman MH, Kerns WD, Ochaion A, Cohen S, Fishman P (2007) The anti-inflammatory effect of A_3 adenosine receptor agonists: a novel targeted therapy for rheumatoid arthritis. Expert Opin Invest Drugs 16:1601–1613

Bar-Yehuda S, Stemmer SM, Madi L, Castel D, Ochaion A, Cohen S, Barer F, Zabutti A, Perez-Liz G, Del Valle L, Fishman P (2008) The A_3 adenosine receptor agonist CF102 induces apoptosis of hepatocellular carcinoma via de-regulation of the Wnt and NF-κB signal transduction pathways. Int J Oncol 33:287–295

Barry CP, Lind SE (2000) Adenosine-mediated killing of cultured epithelial cancer cells. Cancer Res 60:1887–1894

Bianchi L, De Micheli E, Bricolo A, Ballini C, Fattori M, Venturi C, Pedata F, Tipton KF, Della Corte L (2004) Extracellular levels of amino acids and choline in human high grade gliomas: an intraoperative microdialysis study. Neurochem Res 29:325–334

Bieber D, Lorenz K, Yadav R, Klotz K-N (2008) A_{2B} adenosine receptors mediate an inhibition of ERK 1/2 phosphorylation in the breast cancer cell line MDA-MB-231. Naunyn–Schmiedeberg's Arch Pharmacol 377(Suppl 1):19

Blank C, Mackensen A (2007) Contribution of the PD-L1/PD-1 pathway to T-cell exhaustion: an update on implications for chronic infections and tumor evasion. Cancer Immunol Immunother 56:739–745

Blay J, White TD, Hoskin DW (1997) The extracellular fluid of solid carcinomas contains immunosuppressive concentrations of adenosine. Cancer Res 57:2602–2605

Bonvini P, Hwang SG, el-Gamil M, Robbins P, Neckers L, Trepel J (1999) Melanoma cell lines contain a proteasome-sensitive, nuclear cytoskeleton-associated pool of beta-catenin. Ann N Y Acad Sci 886:208–211

Brandt SJ, Peters WP, Atwater SK, Kurtzberg J, Borowitz MJ, Jones RB, Shpall EJ, Bast RC Jr, Gilbert CJ, Oette DH (1988) Effect of recombinant human granulocyte-macrophage colony-stimulating factor on hematopoietic reconstitution after high-dose chemotherapy and autologous bone marrow transplantation. N Engl J Med 318:869–876

Braun N, Sévigny J, Robson SC, Enjyoji K, Guckelberger O, Hammer K, Di Virgilio F, Zimmermann H (2000) Assignment of ecto-nucleoside triphosphate diphosphohydrolase-1/cd39 expression to microglia and vasculature of the brain. Eur J Neurosci 12:4357–4366

Bunemann M, Lee KB, Pals-Rylaarsdam R, Roseberry AG, Hosey MM (1999) Desensitization of G-protein-coupled receptors in the cardiovascular system. Annu Rev Physiol 61:169–192

Burnstock G (2006) Purinergic signaling: an overview. Novartis Found Symp 276:26–48

Carmeliet P, Dor Y, Herbert JM (1998) HIF-1alpha in hypoxia-mediated apoptosis, cell proliferation and tumour angiogenesis. Nature 394:485–490

Chen GJ, Harvey BK, Shen H, Chou J, Victor A, Wang Y (2006) Activation of adenosine A_3 receptors reduces ischemic brain injury in rodents. J Neurosci Res 84:1848–1855

Chung H, Jung J-Y, Cho S-D, Hong K-A, Kim H-J, Shin D-H, Kim H, Kim HO, Shin D H, Lee HW, Jeong LS, Kong G (2006) The antitumor effect of LJ-529, a novel agonist to A_3 adenosine receptor, in both estrogen receptor-positive and estrogen receptor-negative human breast cancers. Mol Cancer Ther 5:685–692

Claing A, Laporte SA, Caron MG, Lefkowitz RJ (2002) Endocytosis of G protein-coupled receptors: roles of G protein-coupled receptor kinases and beta-arrestin proteins. Prog Neurobiol 66:61–79

Cristalli G, Lambertucci C, Taffi S, Vittori S, Volpini R (2003) Medicinal chemistry of adenosine A_{2A} receptor agonists. Curr Top Med Chem 3:387–401

Cross DA, Alessi DR, Cohen P, Andjelkovich M, Hemmings BA (1995) Inhibition of glycogen synthase kinase-3 by insulin mediated by protein kinase B. Nature 378:785–789

Cui J, Zhou X, Liu Y, Tang Z, Romeih M (2003) Wnt signaling in hepatocellular carcinoma: analysis of mutation and expression of beta-catenin, T-cell factor-4 and glycogen synthase kinase 3-beta genes. J Gastroenterol Hepatol 18:280–287

Deaglio S, Dwyer KM, Gao W, Friedman D, Usheva A, Erat A, Chen JF, Enjyoji K, Linden J, Oukka M, Kuchroo VK, Strom TB, Robson SC (2007) Adenosine generation catalyzed by CD39 and CD73 expressed on regulatory T-cells mediates immune suppression. J Exp Med 204:1257–1265

Dhillon AS, Hagan S, Rath O, Kolch W (2007) MAP kinase signalling pathways in cancer. Oncogene 26:3279–3290

Drake CG, Jaffee E, Pardoll DM (2006) Mechanisms of immune evasion by tumors. Adv Immunol 90:51–81

Dubey RK, Gillespie DG, Shue H, Jackson EK (2000) A2B receptors mediate antimitogenesis in vascular smooth muscle cells. Hypertension 35:267–272

Dunwiddie TV (1985) The physiological role of adenosine in the central nervous system. Int Rev Neurobiol 27:63–139

Egen JG, Kuhns MS, Allison JP (2002) CTLA-4: new insights into its biological function and use in tumor immunotherapy. Nat Immunol 3:611–618

Epstein AC, Gleadle JM, McNeill LA, Hewitson KS, O'Rourke J, Mole DR, Mukherji M, Metzen E, Wilson MI, Dhanda A, Tian YM, Masson N, Hamilton DL, Jaakkola P, Barstead R, Hodgkin J, Maxwell PH, Pugh CW, Schofield CJ, Ratcliffe PJ (2001) *C. elegans* EGL-9 and mammalian homologs define a family of dioxygenases that regulate HIF by prolyl hydroxylation. Cell 107:43–54

Erdmann AA, Gao ZG, Jung U, Foley J, Borenstein T, Jacobson KA, Fowler DH (2005) Activation of Th1 and Tc1 cell adenosine A_{2A} receptors directly inhibits IL-2 secretion *in vitro* and IL-2-driven expansion *in vivo*. Blood 105:4707–4714

Fang X, Yu SX, Lu Y, Bast RC Jr, Woodgett JR, Mills GB (2000) Phosphorylation and inactivation of glycogen synthase kinase 3 by protein kinase A. Proc Natl Acad Sci USA 24:11960–11965

Farber K, Kettenmann H (2006) Purinergic signaling and microglia. Pflugers Arch 452:615–621

Färber K, Markworth S, Pannasch U, Nolte C, Prinz V, Kronenberg G, Gertz K, Endres M, Bechmann I, Enjyoji K, Robson SC, Kettenmann H (2008) The ectonucleotidase cd39/ENTPDase1 modulates purinergic-mediated microglial migration. Glia 56:331–341

Feoktistov I, Murray JJ, Biaggioni I (1994) Positive modulation of intracellular Ca^{2+} levels by adenosine A2B receptors, prostacyclin, and prostaglandin E1 via a cholera toxin-sensitive mechanism in human erythroleukemia cells. Mol Pharmacol 45:1160–1167

Feoktistov I, Goldstein AE, Ryzhov S, Zeng D, Belardinelli L, Voyno-Yasenetskaya T, Biaggioni I (2002) Differential expression of adenosine receptors in human endothelial cells: role of A_{2B} receptors in angiogenic factor regulation. Circ Res 90:531–538

Feoktistov I, Ryzhov S, Goldstein AE, Biaggioni I (2003) Mast cell-mediated stimulation of angiogenesis: cooperative interaction between A_{2B} and A_3 adenosine receptors. Circ Res 92:485–492

Feoktistov I, Ryzhov S, Zhong H, Goldstein AE, Matafonov A, Zeng D, Biaggioni I (2004) Hypoxia modulates adenosine receptors in human endothelial and smooth muscle cells toward an A_{2B} angiogenic phenotype. Hypertension 44:649–654

Ferkey DM, Kimelman D (2000) GSK-3: new thoughts on an old enzyme. Dev Biol 225:471–479

Fiebich BL, Biber K, Lieb K, van Calker D, Berger M, Bauer J, Gebicke-Haerter PJ (1996) Cyclooxygenase-2 expression in rat microglia is induced by adenosine A_{2a}-receptors. Glia 18:152–160

Fields RD, Burnstock G (2006) Purinergic signalling in neuron–glia interactions. Nat Rev Neurosci 7:423–436

Fishman P, Bar-Yehuda S, Ohana G, Pathak S, Wasserman L, Barer F, Multani AF (2000a) Adenosine acts as an inhibitor of lymphoma cell growth: a major role for the A_3 adenosine receptor. Eur J Cancer 36:1452–1458

Fishman P, Bar-Yehuda S, Farbstein T, Barer F, Ohana G (2000b) Adenosine acts as a chemoprotective agent by stimulating G-CSF production: a role for A_1 and A_3 adenosine receptors. J Cell Physiol 183:393–398

Fishman P, Bar-Yehuda S, Barer F, Madi L, Multani AF, Pathak S (2001) The A_3 adenosine receptor as a new target for cancer therapy and chemoprotection. Exp Cell Res 269:230–236

Fishman P, Madi L, Bar-Yehuda S, Barer F, Del Valle L, Khalili K (2002a) Evidence for involvement of Wnt signaling pathway in IB-MECA mediated suppression of melanoma cells. Oncogene 21:4060–4064

Fishman P, Bar-Yehuda S, Madi L, Cohn I (2002b) A_3 adenosine receptor as a target for cancer therapy. Anticancer Drugs 13:1–8

Fishman P, Bar-Yehuda S, Rath-Wolfson L, Ardon E, Barrer F, Ochaion A, Madi L (2003) Targeting the A_3 adenosine receptor for cancer therapy: inhibition of prostate carcinoma cell growth by A_3AR agonist. Anticancer Res 23:2077–2083

Fishman P, Bar-Yehuda S, Ohana G, Ochaion A, Engelberg A, Barer F, Madi L (2004) An agonist to the A_3 adenosine receptor inhibits colon carcinoma growth in mice via modulation of GSK-3β and NF-κB. Oncogene 23:2465–2471

Fishman P, Jacobson KA, Ochaion A, Cohen S, Bar-Yehuda S (2007) The anti-cancer effect of A_3 adenosine receptor agonists: a novel targeted therapy. Immunol Endocr Metab Agents Med Chem 7:298–303

Fredholm BB (1997) Adenosine and neuroprotection. Int Rev Neurobiol 40:259–280

Fredholm BB (2003) Adenosine receptors as targets for drug development. Drug News Perspect 16:283–289

Fredholm BB, Irenius E, Kull B, Schulte G (2001) Comparison of the potency of adenosine as an agonist at human adenosine receptors expressed in Chinese hamster ovary cells. Biochem Pharmacol 61:443–448

Fredholm BB, Chen JF, Masino SA, Vaugeois JM (2005) Actions of adenosine at its receptors in the CNS: insights from knockouts and drugs. Annu Rev Pharmacol Toxicol 45:385–412

Gao Z, Li BS, Day YJ, Linden J (2001) A_3 adenosine receptor activation triggers phosphorylation of protein kinase B and protects rat basophilic leukemia 2H3 mast cells from apoptosis. Mol Pharmacol 59:76–82

Gao ZG, Kim SK, Biadatti T, Chen W, Lee K, Barak D, Kim SG, Johnson CR, Jacobson KA (2002) Structural determinants of A(3) adenosine receptor activation: nucleoside ligands at the agonist/antagonist boundary. J Med Chem 45:4471–4484

Gebicke-Haerter PJ, Christoffel F, Timmer J, Northoff H, Berger M, Van Calker D (1996) Both adenosine A_1- and A_2-receptors are required to stimulate microglial proliferation. Neurochem Int 29:37–42

Gessi S, Varani K, Merighi S, Cattabriga E, Iannotta V, Leung E, Baraldi PG, Borea PA (2002) A(3) adenosine receptors in human neutrophils and promyelocytic HL60 cells: a pharmacological and biochemical study. Mol Pharmacol 61:415–424

Gessi S, Cattabriga E, Avitabile A, Gafa' R, Lanza G, Cavazzini L, Bianchi N, Gambari R, Feo C, Liboni A, Gullini S, Leung E, Mac-Lennan S, Borea PA (2004) Elevated expression of A_3 adenosine receptors in human colorectal cancer is reflected in peripheral blood cells. Clin Cancer Res 10:5895–5901

Gessi S, Merighi S, Varani K, Cattabriga E, Benini A, Mirandola P, Leung E, Mac Lennan S, Feo C, Baraldi S, Borea PA (2007) Adenosine receptors in colon carcinoma tissues and colon tumoral cell lines: focus on the A_3 adenosine subtype. J Cell Physiol 211:826–836

Gessi S, Merighi S, Varani K, Leung E, Mac Lennan S, Borea PA (2008) The A_3 adenosine receptor: an enigmatic player in cell biology. Pharmacol Ther 117:123–140

Giménez-Llort L, Fernández-Teruel A, Escorihuela RM, Fredholm BB, Tobeña A, Pekny M, Johansson B (2002) Mice lacking the adenosine A_1 receptor are anxious and aggressive, but are normal learners with reduced muscle strength and survival rate. Eur J Neurosci 16:547–550

Glass R, Synowitz M, Kronenberg G, Walzlein JH, Markovic DS, Wang LP, Gast D, Kiwit J, Kempermann G, Kettenmann H (2005) Glioblastoma-induced attraction of endogenous neural precursor cells is associated with improved survival. J Neurosci 25:2637–2646

Goldsmith ZG, Dhanasekaran DN (2007) G Protein regulation of MAPK networks. Oncogene 26:3122–3142

Graham S, Combes P, Crumiere M, Klotz K-N, Dickenson JM (2001) Regulation of P42/P44 mitogen-activated protein kinase by the human adenosine A_3 receptor in transfected CHO cells. Eur J Pharmacol 420:19–26

Hammarberg C Schulte G, Fredholm BB (2003) Evidence for functional adenosine A_3 receptors in microglia cells. J Neurochem 86:1051–1054

Haneda M, Araki S-I, Sugimoto T, Togawa M, Koya D, Kikkawa R (1996) Differential inhibition of mesangial MAP kinase cascade by cyclic nucleotides. Kidney Int 50:384–391

Harish A, Hohana G, Fishman P, Arnon O, Bar-Yehuda S (2003) A_3 adenosine receptor agonist potentiates natural killer cell activity. Int J Oncol 23:1245–1249

Haskó G, Szabó C, Németh ZH, Kvetan V, Pastores SM, Vizi ES (1996) Adenosine receptor agonists differentially regulate IL-10, TNF-alpha, and nitric oxide production in RAW 264.7 macrophages and in endotoxemic mice. J Immunol 157:4634–4640

Heese K, Fiebich BL, Bauer J, Otten U (1997) Nerve growth factor (NGF) expression in rat microglia is induced by adenosine A_{2a}-receptors. Neurosci Lett 231:83–86

Hockel M, Vaupel P (2001) Tumor hypoxia: definitions and current clinical, biologic, and molecular aspects. J Natl Cancer Inst 93:266–276

Hofer M, Pospísil M, Vacek A, Holá J, Znojil V, Weiterová L, Streitová D (2006) Effects of adenosine A(3) receptor agonist on bone marrow granulocytic system in 5-fluorouracil-treated mice. Eur J Pharmacol 538:163–167

Hofer M, Pospísil M, Znojil V, Holá J, Vacek A, Streitová D (2007) Adenosine A(3) receptor agonist acts as a homeostatic regulator of bone marrow hematopoiesis. Biomed Pharmacother 61:356–359

Hopfl G, Ogunshola O, Gassmann M (2004) HIFs and tumors: causes and consequences. Am J Physiol Regul Integr Comp Physiol 286:R608–R623

Hoskin DW, Reynolds T, Blay J (1994) Adenosine as a possible inhibitor of killer T-cell activation in the microenvironment of solid tumours. Int J Cancer 59:854–855

Huang S, Apasov S, Koshiba M, Sitkovsky M (1997) Role of A_{2a} extracellular adenosine receptor-mediated signaling in adenosine-mediated inhibition of T-cell activation and expansion. Blood 90:1600–1610

Illes P, Klotz K-N, Lohse MJ (2000) Signaling by extracellular nucleotides and nucleosides. Naunyn–Schmiedeberg's Arch Pharmacol 362:295–298

Jacobson, KA (1998) Adenosine A_3 receptors: novel ligands and paradoxical effects. Trends Pharmacol Sci 19:184–191

Jenner P (2005) Istradefylline, a novel adenosine A_{2A} receptor antagonist, for the treatment of Parkinson's disease. Expert Opin Invest Drugs 14:729–738

Jeong LS, Kim MJ, Kim HO, Gao ZG, Kim SK, Jacobson KA, Chun MW (2004) Design and synthesis of 3′-ureidoadenosine-5′-uronamides: effects of the 3′-ureido group on binding to the A_3 adenosine receptor. Bioorg Med Chem Lett 14:4851–4854

Jiang BH, Agani F, Passaniti A, Semenza GL (1997) V-SRC induces expression of hypoxia-inducible factor 1 (HIF-1) and transcription of genes encoding vascular endothelial growth factor and enolase 1: involvement of HIF-1 in tumor progression. Cancer Res 57:5328–5335

Johansson B, Halldner L, Dunwiddie TV, Masino SA, Poelchen W, Giménez-Llort L, Escorihuela RM, Fernández-Teruel A, Wiesenfeld-Hallin Z, Xu XJ, Hårdemark A, Betsholtz C, Herlenius E, Fredholm BB (2001) Hyperalgesia, anxiety, and decreased hypoxic neuroprotection in mice lacking the adenosine A_1 receptor. Proc Natl Acad Sci USA 98:9407–9412

Johnston JB, Silva C, Gonzalez G, Holden J, Warren KG, Metz LM, Power C (2001) Diminished adenosine A_1 receptor expression on macrophages in brain and blood of patients with multiple sclerosis. Ann Neurol 49:650–658

Joshi BV, Jacobson KA (2005) Purine derivatives as ligands for A_3 adenosine receptors. Curr Top Med Chem 5:1275–1295

Karin M, Ben-Neriah Y (2000) Phosphorylation meets ubiquitination: the control of NF-κB activity. Annu Rev Immunol 18:621–663

Kim SG, Ravi G, Hoffmann C, Jung YJ, Kim M, Chen A, Jacobson KA (2002) p53-Independent induction of Fas and apoptosis in leukemic cells by an adenosine derivative, Cl-IB-MECA. Biochem Pharmacol 2002 63:871–880

Kim SJ, Min HY, Chung HJ, Park EJ, Hong JY, Kang YJ, Shin DH, Jeong LS, Lee SK (2008) Inhibition of cell proliferation through cell cycle arrest and apoptosis by thio-Cl-IB-MECA, a novel A(3) adenosine receptor agonist, in human lung cancer cells. Cancer Lett 264:309–315

Khoa ND, Montesinos MC, Reiss AB, Delano D, Awadallah N, Cronstein BN (2001) Inflammatory cytokines regulate function and expression of adenosine A(2A) receptors in human monocytic THP-1 cells. J Immunol 167:4026–4032

Kung AL, Wang S, Klco JM, Kaelin WG, Livingston DM (2000) Suppression of tumor growth through disruption of hypoxia-inducible transcription. Nat Med 6:1335–1340

Küst BM, Biber K, van Calker D, Gebicke-Haerter PJ (1999) Regulation of K^+ channel mRNA expression by stimulation of adenosine A_{2a}-receptors in cultured rat microglia. Glia 25:120–130

Lanone S, Zheng T, Zhu Z, Liu W, Lee CG, Ma B, Chen Q, Homer RJ, Wang J, Rabach LA, Rabach ME, Shipley JM, Shapiro SD, Senior RM, Elias JA (2002) Overlapping and enzyme-specific contributions of matrix metalloproteinases-9 and -12 in IL-13-induced inflammation and remodeling. J Clin Invest 110:463–474

Lappas CM, Rieger JM, Linden J (2005) A_{2A} adenosine receptor induction inhibits IFN-gamma production in murine CD4+ T-cells. J Immunol 174:1073–1080

Le Moine O, Stordeur P, Schandené L, Marchant A, de Groote D, Goldman M, Devière J (1996) Adenosine enhances IL-10 secretion by human monocytes. J Immunol 156:4408–4414

Lee HT, Gallos G, Nasr SH, Emala CW (2004a) A_1 adenosine receptor activation inhibits inflammation, necrosis, and apoptosis after renal ischemia-reperfusion injury in mice. J Am Soc Nephrol 15:102–111

Lee HT, Xu H, Nasr SH, Schnermann J, Emala CW (2004b) A_1 adenosine receptor knockout mice exhibit increased renal injury following ischemia and reperfusion. Am J Physiol Renal Physiol 286:298–306

Lee EJ, Min HY, Chung HJ, Park EJ, Shin DH, Jeong LS, Lee SK (2005) A novel adenosine analog, thio-Cl-IB-MECA, induces G_0/G_1 cell cycle arrest and apoptosis in human promyelocytic leukemia HL-60 cells. Biochem Pharmacol 70:918–924

Li ZW, Chu W, Hu Y, Delhase M, Deerinck T, Ellisman M, Johnson R, Karin M (1999) The IKKbeta subunit of IkappaB kinase (IKK) is essential for nuclear factor kappaB activation and prevention of apoptosis. J Exp Med 189:1839–1845

Liao Y, Takashima S, Asano Y, Asakura M, Ogai A, Shintani Y, Minamino T, Asanuma H, Sanada S, Kim J, Ogita H, Tomoike H, Hori M, Kitakaze M (2003) Activation of adenosine A_1 receptor attenuates cardiac hypertrophy and prevents heart failure in murine left ventricular pressure-overload model. Circ Res 93:759–766

Linden J (2001) Molecular approach to adenosine receptors: receptor-mediated mechanisms of tissue protection. Annu Rev Pharmacol Toxicol 41:775–787

Linden J, Thai T, Figler H, Jin X, Robeva AS (1999) Characterization of human A_{2B} adenosine receptors: radioligand binding, western blotting, and coupling to Gq in human embryonic kidney 293 cells and HMC-1 mast cells. Mol Pharmacol 56:705–713

Lu J, Pierron A, Ravid K (2003) An adenosine analogue, IB-MECA, down-regulates estrogen receptor alpha and suppresses human breast cancer cell proliferation. Cancer Res 63:6413–6423

Lukashev D, Ohta A, Sitkovsky M (2007) Hypoxia-dependent anti-inflammatory pathways in protection of cancerous tissues. Cancer Metastasis Rev 26:273–279

Mabjeesh NJ, Escuin D, LaVallee TM, Pribluda VS, Swartz GM, Johnson MS, Willard MT, Zhong H, Simons JW, Giannakakou P (2003) 2ME2 inhibits tumor growth and angiogenesis by disrupting microtubules and dysregulating HIF. Cancer Cell 3:363–375

MacDonald RL, Skerritt JH, Werz MA (1986) Adenosine agonists reduce voltage-dependent calcium conductance of mouse sensory neurones in cell culture. J Physiol 370:75–90

MacKenzie WM, Hoskin DW, Blay J (1994) Adenosine inhibits the adhesion of anti-CD3-activated killer lymphocytes to adenocarcinoma cells through an A_3 receptor. Cancer Res 54:3521–3516

Madi L, Bar-Yehuda S, Barer F, Ardon E, Ochaion A, Fishman P (2003) A3 adenosine receptor activation in melanoma cells: association between receptor fate and tumor growth inhibition. J Biol Chem 278:42121–42130

Madi L, Ochaion A, Rath-Wolfson L, Bar-Yehuda S, Erlanger A, Ohana G, Harish A, Merimski O, Barer F, Fishman P (2004) The A_3 adenosine receptor is highly expressed in tumor versus normal cells: potential target for tumor growth inhibition. Clin Cancer Res 10:4472–4479

Markovic DS, Glass R, Synowitz M, Rooijen N, Kettenmann H (2005) Microglia stimulate the invasiveness of glioma cells by increasing the activity of metalloprotease-2. J Neuropathol Exp Neurol 64:754–762

Maxwell PH, Dachs GU, Gleadle JM, Nicholls LG, Harris AL, Stratford IJ, Hankinson O, Pugh CW, Ratcliffe PJ (1997) Hypoxia-inducible factor-1 modulates gene expression in solid tumors and influences both angiogenesis and tumor growth. Proc Natl Acad Sci USA 94:8104–8109

Merighi S, Varani K, Gessi S, Cattabriga E, Iannotta V, Ulouglu C, Leung E, Borea PA (2001) Pharmacological and biochemical characterization of adenosine receptors in the human malignant melanoma A375 cell line. Br J Pharmacol 134:1215–1226

Merighi S, Mirandola P, Milani D, Varani K, Gessi S, Klotz KN, Leung E, Baraldi PG, Borea PA (2002) Adenosine receptors as mediators of both cell proliferation and cell death of cultured human melanoma cells. J Invest Dermatol 119:923–933

Merighi, S, Mirandola P, Varani K, Gessi S, Leung E, Baraldi PG, Tabrizi MA, Borea PA (2003) A glance at adenosine receptors: novel target for antitumor therapy. Pharmacol Ther 100:31–48

Merighi S, Benini A, Mirandola P, Gessi S, Varani K, Leung E, Maclennan S, Borea PA (2005a) A_3 adenosine receptor activation inhibits cell proliferation via phosphatidylinositol 3-kinase/Akt-dependent inhibition of the extracellular signal-regulated kinase 1/2 phosphorylation in A375 human melanoma cells. J Biol Chem 280:19516–19526

Merighi S, Benini A, Mirandola P, Gessi S, Varani K, Leung E, MacLennan S, Baraldi PG, Borea PA (2005b) A_3 adenosine receptors modulate hypoxia-inducible factor-1alpha expression in human A375 melanoma cells. Neoplasia 7:894–903

Merighi S, Benini A, Mirandola P, Gessi S, Varani K, Leung E, Maclennan S, Borea PA (2006) Adenosine modulates vascular endothelial growth factor expression via hypoxia-inducible factor-1 in human glioblastoma cells. Biochem Pharmacol 72:19–31

Merighi S, Benini A, Mirandola P, Gessi S, Varani K, Simioni C, Leung E, Maclennan S, Borea PA (2007) Caffeine inhibits adenosine-induced accumulation of hypoxia-inducible factor-1α, vascular endothelial growth factor and interleukin-8 expression in hypoxic human colon cancer cells. Mol Pharmacol 72:395–406

Merimsky O, Madi L, Bar-Yehuda S, Fishman P (2003) Modulation of the A_3 adenosine receptor by low agonist concentration induced anti-tumor and myelostimulation effects. Drug Dev Res 58:386–389

Merrill JT, Shen C, Schreibman D, Coffey D, Zakharenko O, Fisher R, Lahita RG, Salmon J, Cronstein BN (1997) Adenosine A_1 receptor promotion of multinucleated giant cell formation by human monocytes: a mechanism for methotrexate-induced nodulosis in rheumatoid arthritis. Arthritis Rheum 40:1308–1315

Minchenko A, Leshchinsky I, Opentanova I, Sang N, Srinivas V, Armstead V, Caro J (2002) Hypoxia-inducible factor-1-mediated expression of the 6-phosphofructo-2-kinase/fructose-2,6-bisphosphatase-3 (PFKFB3) gene. Its possible role in the Warburg effect. J Biol Chem 277:6183–6187

Mirabet M, Mallol J, Lluis C, Franco R (1997) Calcium mobilization in Jurkat cells via A_{2b} adenosine receptors. Br J Pharmacol 122:1075–1082

Morello S, Petrella A, Festa M, Popolo A, Monaco M, Vuttariello E, Chiappetta G, Parente L, Pinto A (2007) Cl-IB-MECA inhibits human thyroid cancer cell proliferation independently of A_3 adenosine receptor activation. Cancer Biol Ther 7:278–284

Morin PJ (1999) Beta-catenin signaling and cancer. Bioessays 21:1021–1030

Muller CE (2003) Medicinal chemistry of adenosine A_3 receptor ligands. Curr Top Med Chem 3:445–462

Murthy KS, Makhlouf GM (1995) Adenosine A_1 receptor-mediated activation of phospholipase C-beta 3 in intestinal muscle: dual requirement for alpha and beta gamma subunits of Gi3. Mol Pharmacol 47:1172–1179

Naganuma M, Wiznerowicz EB, Lappas CM, Linden J, Worthington MT, Ernst PB (2006) Cutting edge: critical role for A_{2A} adenosine receptors in the T cell-mediated regulation of colitis. J Immunol 177:2765–2769

Nakamura K, Yoshikawa N, Yamaguchi Y, Kagota S, Shinozuka K, Kunitomo M (2006) Antitumor effect of cordycepin (3′-deoxyadenosine) on mouse melanoma and lung carcinoma cells involves adenosine A_3 receptor stimulation. Anticancer Res 26:43–47

Novak A, Dedhar S (1999) Signaling through beta-catenin and Lef/Tcf. Cell Mol Life Sci 56:523–537

Ogata T, Schubert P (1996) Programmed cell death in rat microglia is controlled by extracellular adenosine. Neurosci Lett 218:91–94

Ohana G, Bar-Yehuda S, Arich A, Madi L, Dreznick Z, Silberman D, Slosman G, Volfsson-Rath L, Pnina Fishman (2003) Inhibition of primary colon carcinoma growth and liver metastasis by the A3 adenosine receptor agonist CF101. Br J Cancer 89:1552–1558

Ohta A, Sitkovsky M (2001) Role of G-protein-coupled adenosine receptors in downregulation of inflammation and protection from tissue damage. Nature 414:916–920

Ohta A, Gorelik E, Prasad SJ, Ronchese F, Lukashev D, Wong MK, Huang X, Caldwell S, Liu K, Smith P, Chen JF, Jackson EK, Apasov S, Abrams S, Sitkovsky M (2006) A_{2A} adenosine receptor protects tumors from antitumor T-cells. Proc Natl Acad Sci USA 103:13132–13137

Olah ME, Stiles GL (1995) Adenosine receptor subtypes: characterization and therapeutic regulation. Annu Rev Pharmacol Toxicol 35:581–606

Olah ME, Ren H, Stiles GL (1995) Adenosine receptors: protein and gene structure. Arch Pharmacodyn 329:135–150

Overwijk WW, Restifo NP (2001) Creating therapeutic cancer vaccines: notes from the battlefield. Trends Immunol 22:5–7

Panjehpour M, Karami-Tehrani F (2004) An adenosine analog (IB-MECA) inhibits anchorage-dependent cell growth of various human breast cancer cell lines. Int J Biochem Cell Biol 36:1502–1509

Panjehpour M, Karami-Tehrani F (2007) Adenosine modulates cell growth in the human breast cancer cells via adenosine receptors. Oncol Res 16:575–585

Panjehpour M, Castro M, Klotz K-N (2005) Human breast cancer cell line MDA-MB-231 expresses endogenous A_{2B} adenosine receptors mediating a Ca^{2+} signal. Br J Pharmacol 145:211–218

Panther E, Idzko M, Herouy Y, Rheinen H, Gebicke-Haerter PJ, Mrowietz U, Dichmann S, Norgauer J (2001) Expression and function of adenosine receptors in human dendritic cells. FASEB J 15:1963–1970

Pardoll DM (2002) Spinning molecular immunology into successful immunotherapy. Nat Rev Immunol 2:227–238

Peifer M, Polakis P (2000) Wnt signaling in oncogenesis and embryogenesis: a look outside the nucleus. Science 287:1606–1609

Plesner L (1995) Ecto-ATPases: identities and functions. Int Rev Cytol 158:141–214

Poulsen SA, Quinn RJ (1998) Adenosine receptors: new opportunities for future drugs. Bioorg Med Chem 6:619–641

Prinz M, Hanisch UK (1999) Murine microglial cells produce and respond to interleukin-18. J Neurochem 72:2215–2218

Raman M, Chen W, Cobb MH (2007) Differential regulation and properties of MAPKs. Oncogene 26:3100–3112

Rao JS (2003) Molecular mechanisms of glioma invasiveness: the role of proteases. Nat Rev Cancer 3:489–501

Raskovalova T, Lokshin A, Huang X, Jackson EK, Gorelik E (2006) Adenosine-mediated inhibition of cytotoxic activity and cytokine production by IL-2/NKp46-activated NK cells: involvement of protein kinase A isozyme I (PKA I). Immunol Res 36:91–99

Raskovalova T, Lokshin A, Huang X, Su Y, Mandic M, Zarour HM, Jackson EK, Gorelik E (2007) Inhibition of cytokine production and cytotoxic activity of human antimelanoma specific CD8+ and CD4+ T lymphocytes by adenosine-protein kinase A type I signaling. Cancer Res 67:5949–5956

Ratcliffe PJ, Pugh CW, Maxwell PH (2000) Targeting tumors through the HIF system. Nat Med 6:1315–1316

Robbins PF, El-Gamil M, Li YF, Kawakami Y, Loftus D, Appella E, Rosenberg SA (1996) A mutated beta-catenin gene encodes a melanoma-specific antigen recognized by tumor infiltrating lymphocytes. J Exp Med 183:1185–1192

Rusthoven J, Bramwell V, Stephenson B (1998) Use of granulocyte colony-stimulating factor (G-CSF) in patients receiving myelosuppressive chemotherapy for the treatment of cancer. Provincial systemic treatment disease site group. Cancer Prev Control 2:179–190

Sajjadi FG, Firestein GS (1993) cDNA cloning and sequence analysis of the human A_3 adenosine receptor. Biochim Biophys Acta 1179:105–107

Sajjadi FG, Takabayashi K, Foster AC, Domingo RC, Firestein GS (1996) Inhibition of TNF-alpha expression by adenosine: role of A_3 adenosine receptors. J Immunol 156:3435–3442

Salmon JE, Cronstein BN (1990) Fc gamma receptor-mediated functions in neutrophils are modulated by adenosine receptor occupancy. A_1 receptors are stimulatory and A_2 receptors are inhibitory. J Immunol 145:2235–2240

Salmon JE, Brogle N, Brownlie C, Edberg JC, Kimberly RP, Chen BX, Erlanger BF (1993) Human mononuclear phagocytes express adenosine A_1 receptors. A novel mechanism for differential regulation of Fc gamma receptor function. J Immunol 151:2775–2785

Salvatore CA, Jacobson MA, Taylor HE, Linden J, Johnson RG (1993) Molecular cloning an characterization of the human A_3 adenosine receptor. Proc Natl Acad Sci USA 90:10365–10369

Scheibner KA, Lutz MA, Boodoo S, Fenton MJ, Powell JD, Horton MR (2006) Hyaluronan fragments act as an endogenous danger signal by engaging TLR2. J Immunol 177:1272–1281

Schnitzer J (1989) Enzyme-histochemical demonstration of microglial cells in the adult and postnatal rabbit retina. J Comp Neurol 282:249–263

Schnurr M, Toy T, Shin A, Hartmann G, Rothenfusser S, Soellner J, Davis ID, Cebon J, Maraskovsky E (2004) Role of adenosine receptors in regulating chemotaxis and cytokine production of plasmacytoid dendritic cells. Blood 103:1391–1397

Schoen SW, Graeber MB, Kreutzberg GW (1992) 5′-Nucleotidase immunoreactivity of perineuronal microglia responding to rat facial nerve axotomy. Glia 6:314–317

Schulte G, Fredholm BB (2000) Human adenosine A_1, A_{2A}, A_{2B}, and A_3 receptors expressed in Chinese hamster ovary cells all mediate the phosphorylation of extracellular-regulated kinase 1/2. Mol Pharmacol 58:477–482

Seino S, Shibasaki T (2005) PKA-dependent and PKA-independent pathways for cAMP-regulated exocytosis. Physiol Rev 85:1303–1342

Semenza GL (2000) HIF-1: mediator of physiological and pathophysiological responses to hypoxia. J Appl Physiol 88:1474–1480

Semenza GL (2002) Signal transduction to hypoxia-inducible factor 1. Biochem Pharmacol 64:993–998

Semenza GL (2003) Targeting HIF-1 for cancer therapy. Nat Rev Cancer 3:721–732

Sevigny CP, Li L, Awad AS, Huang L, McDuffie M, Linden J, Lobo PI, Okusa MD (2007) Activation of adenosine 2A receptors attenuates allograft rejection and alloantigen recognition. J Immunol 178:4240–4249

Shi Y, Evans JE, Rock KL (2003) Molecular identification of a danger signal that alerts the immune system to dying cells. Nature 425:516–521

Si QS, Nakamura Y, Schubert P, Rudolphi K, Kataoka K (1996) Adenosine and propentofylline inhibit the proliferation of cultured microglial cells. Exp Neurol 137:345–349

Sitkovsky MV, Ohta A (2005) The 'danger' sensors that STOP the immune response: the A_2 adenosine receptors? Trends Immunol 26:299–304

Sitkovsky MV, Lukashev D, Apasov S, Kojima H, Koshiba M, Caldwell C, Ohta A, Thiel M (2004) Physiological control of immune response and inflammatory tissue damage by hypoxia-inducible factors and adenosine A_{2A} receptors. Annu Rev Immunol 22:657–682

Spychala J, Lazarowski E, Ostapkowiecz A, Ayscue LH, Jin A, Mitchell BS (2004) Role of estrogen receptor in the regulation of ecto-5′-nucleotidase and adenosine in breast cancer. Clin Cancer Res 10:708–717

Suh BC, Kim TD, Lee JU, Seong JK, Kim KT (2001) Pharmacological characterization of adenosine receptors in PGT-beta mouse pineal gland tumour cells. Br J Pharmacol 134:132–142

Sullivan GW (2003) Adenosine A_{2A} receptor agonists as anti-inflammatory agents. Curr Opin Invest Drugs 4:1313–1319

Sun CX, Young HW, Molina JG, Volmer JB, Schnermann J, Blackburn MR (2005) A protective role for the A_1 adenosine receptor in adenosine-dependent pulmonary injury. J Clin Invest 115:35–43

Synowitz M, Glass R, Färber K, Markovic D, Kronenberg G, Herrmann K, Schnermann J, Nolte C, van Rooijen N, Kiwit J, Kettenmann H (2006) A_1 adenosine receptors in microglia control glioblastoma–host interaction. Cancer Res 66:8550–8557

Trincavelli ML, Tuscano D, Marroni M, Falleni A, Gremigni V, Ceruti S, Abbracchio MP, Jacobson KA, Cattabeni F, Martini C (2002) A_3 adenosine receptors in human astrocytoma cells: agonist-mediated desensitization, internalization, and down-regulation. Mol Pharmacol 62:1373–1384

Trussell LO, Jackson MB (1985) Adenosine-activated potassium conductance in cultured striatal neurons. Proc Natl Acad Sci USA 82:4857–4861

Tsutsui S, Schnermann J, Noorbakhsh F, Henry S, Yong VW, Winston BW, Warren K, Power C (2004) A_1 adenosine receptor upregulation and activation attenuates neuroinflammation and demyelination in a model of multiple sclerosis. J Neurosci 24:1521–1529

van Troostenburg AR, Clark EV, Carey WDH, Warrington SJ, Kerns WD, Cohn I, Silverman MH, Bar-Yehuda S, Fong KLL, Fishman P (2004) Tolerability, pharmacokinetics, and concentration-dependent hemodynamic effects of oral CF101, an A_3 adenosine receptor agonist, in healthy young men. Int J Clin Pharmacol Ther 42:534–542

Volpini R, Costanzi S, Vittori S, Cristalli G, Klotz KN (2003) Medicinal chemistry and pharmacology of A_{2B} adenosine receptors. Curr Top Med Chem 3:427–443

Welsh SJ, Powis G (2003) Hypoxia inducible factor as a cancer drug target. Curr Cancer Drug Targets 3:391–405

Werb Z, Gordon S (1975) Elastase secretion by stimulated macrophages. Characterization and regulation. J Exp Med 142:361–377

Wiener CM, Booth G, Semenza GL (1996) In vivo expression of mRNAs encoding hypoxia-inducible factor 1. Biochem Biophys Res Commun 225:485–488

Williams JH, Ireland HE (2008) Sensing danger-Hsp72 and HMGB1 as candidate signals. J Leukoc Biol 83:489–492

Williams BA, Manzer A, Blay J, Hoskin DW (1997) Adenosine acts through a novel extracellular receptor to inhibit granule exocytosis by natural killer cells. Biochem Biophys Res Commun 231:264–269

Xu Z, Jang Y, Mueller RA, Norfleet EA (2006) IB-MECA and cardioprotection. Cardiovasc Drug Rev 24:227–238

Yao Y, Sei Y, Abbracchio MP, Jiang JL, Kim YC, Jacobson KA (1997) Adenosine A_3 receptor agonists protect HL-60 and U-937 cells from apoptosis induced by A_3 antagonists. Biochem Biophys Res Commun 232:317–322

Zablocki JA, Wu L, Shryock J, Belardinelli L (2004) Partial A(1) adenosine receptor agonists from a molecular perspective and their potential use as chronic ventricular rate control agents during atrial fibrillation (AF). Curr Top Med Chem 4:839–854

Zarek PE, Huang CT, Lutz ER, Kowalski J, Horton MR, Linden J, Drake CG, Powell JD (2008) A_{2A} receptor signaling promotes peripheral tolerance by inducing T-cell anergy and the generation of adaptive regulatory T-cells. Blood 111:251–259

Zeng D, Maa T, Wang U, Feoktistov I, Biaggioni I, Belardinelli L (2003) Expression and function of A_2 adenosine receptors in the U87MG tumor cells. Drug Dev Res 58:405–411

Zhao Z, Makaritsis K, Francis CE, Gavras H, Ravid K (2000) A role for the A_3 adenosine receptor in determining tissue levels of cAMP and blood pressure: studies in knock-out mice. Biochim Biophys Acta 1500:280–290

Zhong H, De Marzo AM, Laughner E, Lim M, Hilton DA, Zagzag D, Buechler P, Isaacs WB, Semenza GL, Simons JW (1999) Overexpression of hypoxia-inducible factor 1alpha in common human cancers and their metastases. Cancer Res 59:5830–5835

Zhong H, Chiles K, Feldser D, Laughner E, Hanrahan C, Georgescu MM, Simons JW, Semenza GL (2000) Modulation of hypoxia-inducible factor 1alpha expression by the epidermal growth factor/phosphatidylinositol 3-kinase/PTEN/AKT/FRAP pathway in human prostate cancer cells: implications for tumor angiogenesis and therapeutics. Cancer Res 60:1541–1545

Adenosine Receptors and the Kidney

Volker Vallon and Hartmut Osswald

Contents

V. Vallon (✉)
Departments of Medicine and Pharmacology, University of California San Diego and VA San Diego Healthcare System, 3350 La Jolla Village Dr (9151), San Diego, CA 92161, USA
vvallon@ucsd.edu

C.N. Wilson and S.J. Mustafa (eds.), *Adenosine Receptors in Health and Disease*, Handbook of Experimental Pharmacology 193, DOI 10.1007/978-3-540-89615-9_15,

Abstract The autacoid, adenosine, is present in the normoxic kidney and generated in the cytosol as well as at extracellular sites. The rate of adenosine formation is enhanced when the rate of ATP hydrolysis prevails over the rate of ATP synthesis during increased tubular transport work or during oxygen deficiency. Extracellular adenosine acts on adenosine receptor subtypes (A_1, A_{2A}, A_{2B}, and A_3) in the cell membranes to affect vascular and tubular functions. Adenosine lowers glomerular filtration rate by constricting afferent arterioles, especially in superficial nephrons, and thus lowers the salt load and transport work of the kidney consistent with the concept of metabolic control of organ function. In contrast, it leads to vasodilation in the deep cortex and the semihypoxic medulla, and exerts differential effects on NaCl transport along the tubular and collecting duct system. These vascular and tubular effects point to a prominent role of adenosine and its receptors in the intrarenal metabolic regulation of kidney function, and, together with its role in inflammatory processes, form the basis for potential therapeutic approaches in radiocontrast media-induced acute renal failure, ischemia reperfusion injury, and in patients with cardiorenal failure.

Keywords Adenosine receptors · Kidney · Tubuloglomerular feedback · Renin · Fluid and electrolyte transport · Metabolic control · Acute renal failure · Acute kidney injury · Radiocontrast media · Ischemia reperfusion injury · Heart failure

Abbreviations

AA	Afferent arteriole
ADO	Adenosine
ARF	Acute renal failure
A_XAR	Adenosine receptor subtype x
B	Bowman's capsule
BG9719	1,3-Dipropyl-8-[2-(5,6-epoxynorbornyl)] xanthine
BG9928	1,3-Dipropyl-8-[1-(4-propionate)-bicyclo-[2,2,2]octyl)]xanthine
BM	Basement membrane
BS	Bowman's space
cAMP	Cyclic adenosine monophosphate
CD39	Ecto-nucleoside triphosphate diphosphohydrolase-1
CD73	Ecto-5′-nucleotidase
CGS21680	2-[*p*-(2-Carboxyethyl)phenethylamino]-5′-*N*-ethylcarboxamido adenosine
CVT-124	*S*-Enantiomer of 1,3-dipropyl-8-[2-(5,6-epoxynorbornyl)] 1xanthine
DMPX	3,7-Dimethyl-1-propargylxanthine
DPCPX	1,3-Dipropyl-8-cyclopentylxanthine
DPSPX	1,3-Dipropyl-8-sulfophenylxanthine
DWH 146e	4-(3-(6-Amino-9-(5-ethylcarbamoyl-3,4-dihydroxytetrahydrofuran-2-yl)-9*H*-purin-2-yl)prop-2-ynyl)cyclohexanecarboxylic acid methyl ester

EA	Efferent arteriole
EGM	Extraglomerular mesangium
ENTPDase	Ectonucleoside triphosphate diphosphohydrolase
FK-453	(+)-(*R*)-[(*E*)-3-(2-Phenylpyrazolo[1,5-*a*]pyridin-3-yl)acryloyl]-2-piperidine ethanol
FK-838	6-Oxo-3-(2-phenylpyrazolo[1,5-*a*]pyridin-3-yl)-1(6*H*)-pyridazinebutanoic 'acid
GFR	Glomerular filtration rate
HSP27	Heat-shock protein 27
IMCD	Inner medullary collecting duct
KW-3902	8-(Noradamantan-3-yl)-1,3 dipropylxanthine
MBF	Medullary blood flow
MC	Mesangium cells
mTAL	Medullary thick ascending limb
NHE	$Na^+ - H^+$ exchanger
NKCC2	$Na^+ - K^+ - 2Cl^-$ cotransporter
NO	Nitric oxide
NY_2HA	New York Heart Association
pO_2	Partial oxygen pressure
PT	Proximal tubule
SNGFR	Single nephron glomerular filtration rate
TAL	Thick ascending limb
TGF	Tubuloglomerular feedback
T_{Na}	Transport of sodium
VSMC	Vascular smooth muscle cells

1 Introduction

Adenosine is a tissue hormone that is locally generated in many organs and that binds to cell surface receptors to mediate various aspects of organ function. Many of these effects revolve around a role of adenosine in metabolic control of organ function, including local matching of blood flow with energy consumption. According to this concept, the interstitial concentration of adenosine rises when cells are in negative energy balance. Adenosine locally activates vasodilatory adenosine A_2 receptor (A_2AR) and adjusts blood flow to meet demand. The role of adenosine in the kidney is analogous but, as a consequence of the specific renal structural organization and function, more complicated than its role in other organs. We will first describe the differential effects of adenosine on the renal cortical and medullary vascular structures, and its role in tubuloglomerular feedback (TGF), the regulation of renin secretion and in transport processes in the tubular and collecting duct system. These issues are subsequently discussed with regard to a potential role of adenosine receptors as new potential targets in the treatment of patients with radiocontrast media-induced acute renal failure, ischemia-reperfusion injury, and in patients with acute decompensated heart failure or cardiorenal failure. Please see recent reviews on the expression of adenosine receptors in the kidney and the role of adenosine in kidney function in general (Vallon et al. 2006), and in acute renal failure (Osswald and Vallon 2009) and fluid retention in particular (Welch 2002; Modlinger and Welch 2003; Vallon et al. 2008).

2 Vascular Effects of Adenosine in Kidney Cortex and Medulla

In contrast to other organs, blood flow into the cortex of the kidney generates, via the formation of an ultrafiltrate, the metabolic burden for tubular electrolyte transport and thus the demand for energy. Hence, to recover from negative energy balance in the kidney, a mechanism is required that lowers glomerular filtration rate (GFR) or the ratio between glomerular filtration rate and cortical renal blood flow. In comparison, blood flow in the renal medulla is nutritive. It derives from the postglomerular circulation of deep nephrons, and due to the way the kidney concentrates the urine, blood flow and O_2 supply are low in this area, although active NaCl reabsorption in the medullary thick ascending limb is essential for function. With regard to metabolic control, this requires a *vasodilator* to prevent hypoxic injury in the renal medulla. As outlined in the following, adenosine is a vasodilator in the renal medulla but induces cortical vasoconstriction and lowers GFR.

2.1 Activation of A_1AR Lowers Glomerular Filtration Rate

Healthy volunteers responded to an intravenous infusion or direct application of adenosine into the renal artery with a reduction in GFR of 15–25% while blood pressure and renal blood flow were unchanged (Edlund and Sollevi 1993; Edlund et al. 1994; Balakrishnan et al. 1996). Adenosine infusion into the renal artery of rats or dogs reduced single-nephron GFR (SNGFR) in superficial nephrons to a larger extent than whole-kidney GFR, indicating that deep-cortical vasodilation (see below) counteracts superficial vasoconstriction (Osswald et al. 1978a, b; Haas and Osswald 1981). Adenosine lowers SNGFR in superficial nephrons due to afferent arteriolar vasoconstriction (Osswald et al. 1978b; Haas and Osswald 1981) (Fig. 1). Direct videometric assessment of pre- and postglomerular arteries using the "split-hydronephrotic" rat kidney technique revealed adenosine-induced constriction of afferent arterioles via high-affinity A_1AR and dilation via activation of both high-affinity $A_{2A}AR$ and low-affinity $A_{2B}AR$ (Tang et al. 1999). Whereas activation of A_1AR led to the constriction of mainly afferent arterioles near the glomerulus, A_2AR activation lead to the dilation of mainly postglomerular arteries (Holz and Steinhausen 1987; Dietrich and Steinhausen 1993; Gabriels et al. 2000). A_1AR-mediated afferent arteriolar constriction involves a pertussis toxin-sensitive G_i protein and subsequent activation of phospholipase C, presumably through $\beta\gamma$

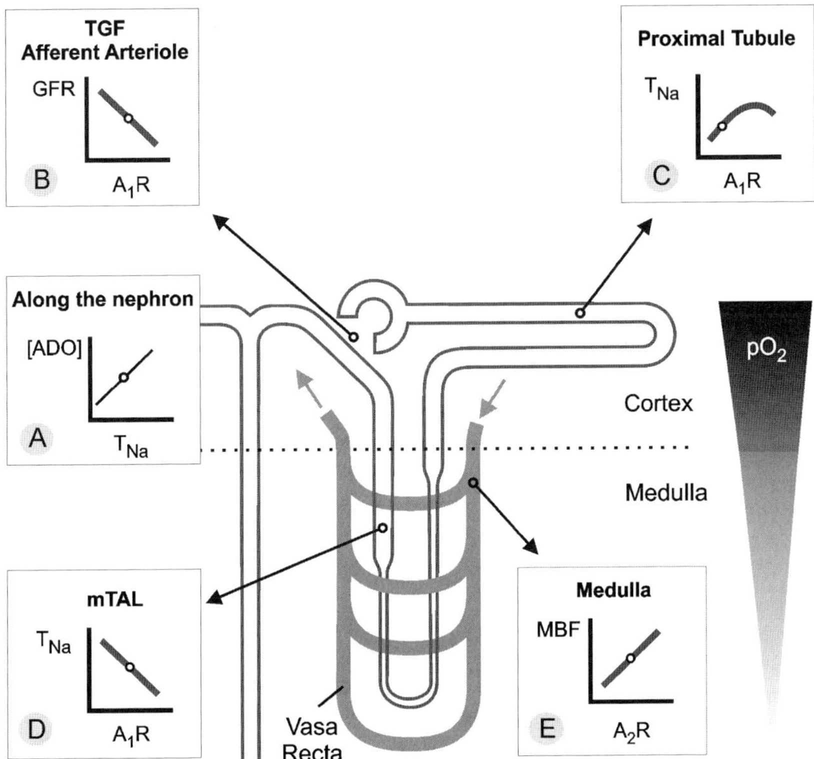

Fig. 1 a–e Control of renal hemodynamics and transport by adenosine (*ADO*). The *line plots* illustrate the relationships between the given parameters. *Small circles* on these lines indicate ambient physiological conditions. In general, the medulla is at greater risk for hypoxic damage than the cortex due to a lower partial oxygen pressure (pO_2). **a** In every nephron segment, an increase in reabsorption or transport of sodium (T_{Na}) increases extracellular ADO. **b** ADO via A_1AR mediates tubuloglomerular feedback (*TGF*) and constricts the afferent arteriole to lower GFR. **c** In the proximal tubule, ADO via A_1AR stimulates T_{Na} and thus lowers the Na^+ load to segments residing in the semihypoxic medulla. **d** In contrast, ADO via A_1AR inhibits T_{Na} in the medulla, including medullary thick ascending limb (*mTAL*). **e** In addition, ADO via A_2AR enhances medullary blood flow (*MBF*), which increases O_2 delivery and further limits O_2-consuming transport in the medulla (adapted from Vallon et al. 2006)

subunits released from $G_{\alpha i}$ (Hansen et al. 2003b). A_{2A}AR-mediated renal vasodilation may involve activation of ATP-regulated potassium channels (Tang et al. 1999) and endothelial nitric oxide synthase (Hansen et al. 2005).

Oral application of the A_1AR antagonist (+)-(*R*)-[(*E*)-3-(2-phenylpyrazolo [1,5-*a*]pyridin-3-yl)acryloyl]-2-piperidine ethanol (FK-453) to healthy male subjects increased GFR by ~20% without significantly altering effective renal plasma flow or mean arterial blood pressure (Balakrishnan et al. 1993), providing evidence that *endogenous* adenosine elicits a tonic suppression of GFR through the activation of A_1AR. Consistent with a prominent role of adenosine in the regulation of afferent arteriolar tone, autoregulation of renal blood flow and glomerular filtration rate (i.e., their constancy in spite of changes in renal perfusion pressure) is dependent upon the activation of A_1AR (Hashimoto et al. 2006).

2.2 *Factors Modulating Adenosine-Induced Cortical Vasoconstriction*

Suppression of the renin–angiotensin system by dietary salt or pharmacological means reduces or blocks the renal vasoconstrictive action of adenosine (Osswald et al. 1975, 1982; Spielman and Osswald 1979; Arend et al. 1985; Macias-Nunez et al. 1985; Dietrich et al. 1991; Dietrich and Steinhausen 1993). In contrast, activation of the renin–angiotensin system potentiates adenosine-induced vasoconstriction and lowering of GFR (Osswald et al. 1975, 1978a, 1982). Further studies identified a mutual dependency and cooperation of adenosine and angiotensin II in producing afferent arteriolar constriction (Weihprecht et al. 1994; Traynor et al. 1998; Hansen et al. 2003a). Adenosine enhances angiotensin II-induced constriction of afferent arterioles by receptor-dependent and -independent pathways. The latter involves adenosine uptake and intracellular effects that increase the calcium sensitivity by phosphorylating the myosin light chain (Lai et al. 2006; Patzak et al. 2007). Moreover, inhibiting the synthesis of vasodilators like nitric oxide (NO) (Barrett and Droppleman 1993; Pflueger et al. 1999b) or prostaglandins (Spielman and Osswald 1978; Pflueger et al. 1999a) increases the sensitivity of the kidney to adenosine-induced vasoconstriction. The outlined interactions can be of clinical relevance.

2.3 *Activation of A_2AR Induces Medullary Vasodilation*

Intrarenal adenosine infusion in rats initially induces vasoconstriction in all cortical zones; this is followed by persistent superficial cortical vasoconstriction but *deep cortical* vasodilation (Macias-Nunez et al. 1983; Miyamoto et al. 1988). While A_1AR-mediated afferent arteriolar constriction dominates in *superficial* nephrons, *deep cortical* glomeruli, which supply the blood flow to the renal medulla, can respond to adenosine with A_2AR-mediated vasodilation (Inscho et al. 1991; Weihprecht et al. 1992; Inscho 1996; Yaoita et al. 1999; Nishiyama et al. 2001). In accordance, renal interstitial infusion in rats of the A_2AR agonist 2-[*p*-(2-carboxyethyl)phenethylamino]-5′-*N*-ethylcarboxamido adenosine (CGS-21680) increased medullary blood flow (Agmon et al. 1993), whereas intramedullary infusion of the selective A_2AR antagonist 3,7-dimethyl-1-propargylxanthine (DMPX) (but not the A_1AR antagonist 1,3-dipropyl-8-cyclopentylxanthine (DPCPX)) decreased medullary blood flow (Zou et al. 1999). This indicates that *endogenous* adenosine dilates medullary vessels and sustains medullary blood flow via activation of A_2AR (Fig. 1).

2.4 *Adenosine is a Mediator of Tubuloglomerular Feedback via Activation of A_1AR*

The mammalian kidney has a rather high GFR (~180 l per day in humans). About 99% of the filtered fluid and NaCl are subsequently reabsorbed along the tubular and collecting duct system, such that urinary excretion closely matches intake. As a result, GFR is a significant determinant of renal transport work, and GFR and reabsorption have to be closely coordinated to avoid renal loss or retention of fluid and NaCl. The tubuloglomerular feedback (TGF) is a mechanism that helps to coordinate GFR with the tubular transport activity or capacity. In this mechanism, specialized tubular cells, the macula densa, sense the tubular NaCl load at the end of the thick ascending limb (TAL; where about 85% of the filtered NaCl has been reabsorbed), and induce a change in afferent arteriolar tone such that an inverse relationship is established between the tubular NaCl load and SNGFR of the same nephron. This way, the TGF stabilizes the NaCl load to further distal segments, where the fine regulation of NaCl and fluid balance takes place under *systemic* neurohumoral control.

The TGF response, in other words an inverse change in SNGFR or glomerular capillary pressure in response to changes in the NaCl concentration at the macula densa, is inhibited by unselective adenosine receptor blockers like theophylline or 1,3-dipropyl-8-sulfophenylxanthine (DPSPX) (Schnermann et al. 1977; Osswald et al. 1980; Franco et al. 1989), as well as by selective A_1AR antagonists like DPCPX, 8-(noradamantan-3-yl)-1,3-dipropylxanthine (KW-3902, rolofylline), CVT-124 (the *S*-enantiomer of the highly selective racemic A_1AR antagonist 1,3-dipropyl-8-[2-(5,6-epoxynorbornyl)] xanthine), or 6-oxo-3-(2-phenylpyrazolo [1,5-*a*]pyridin-3-yl)-1(6*H*)-pyridazinebutanoic acid (FK838) (Franco et al. 1989; Schnermann et al. 1990; Kawabata et al. 1998; Wilcox et al. 1999; Thomson et al. 2000; Ren et al. 2002a). Mice with gene knockout for A_1AR lack the TGF response (Brown et al. 2001; Sun et al. 2001; Vallon et al. 2004), and have an impaired ability to stabilize the Na^+ delivery to the distal tubule (Vallon et al. 2004). Most importantly, an intact TGF response requires local concentrations of adenosine to fluctuate depending on the NaCl concentration in the tubular fluid at the macula densa, indicating that adenosine serves as a *mediator* of TGF (Thomson et al. 2000).

In 1980, Osswald and colleagues proposed that adenosine may be a mediator of TGF. Figure 2 illustrates a current model. Changes in luminal concentrations of Na^+, K^+, and Cl^- alter NaCl uptake by macula densa cells via the furosemide-sensitive Na–K–2Cl cotransporter in the luminal membrane. This triggers basolateral ATP release (Bell et al. 2003; Komlosi et al. 2004) as well as transport-dependent hydrolysis by basolateral Na^+–K^+-ATPase (Lorenz et al. 2006) of ATP to AMP. Plasma membrane-bound ectonucleoside triphosphate diphosphohydrolase 1 (CD39) converts ATP and ADP to AMP (Oppermann et al. 2008) and ecto-5′-nucleotidase (CD73) converts extracellular AMP to adenosine (Thomson et al. 2000; Castrop et al. 2004; Ren et al. 2004; Huang et al. 2006). Part of the extracellular adenosine involved in the TGF response is generated independent of

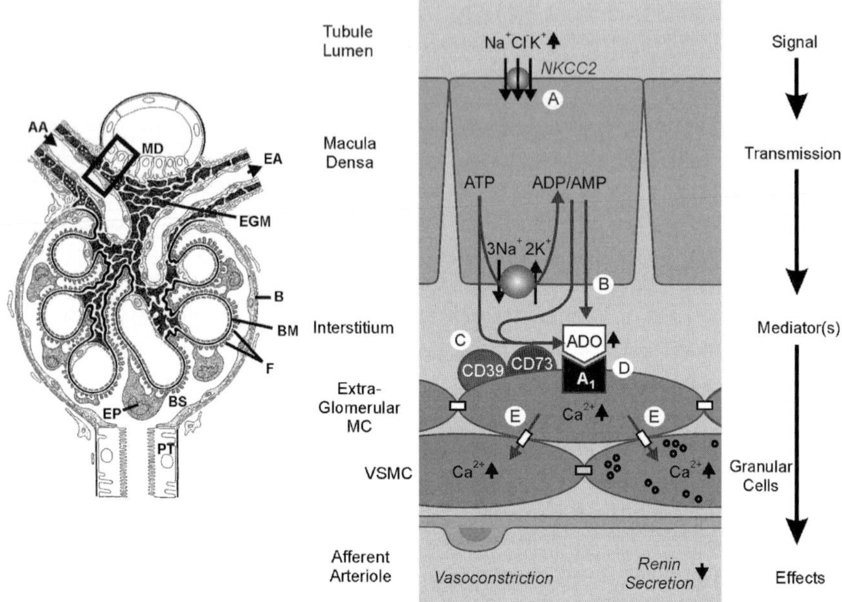

Fig. 2 a–e Adenosine is a mediator of the tubuloglomerular feedback: a proposed mechanism. *Left panel*: schematic drawing illustrating the macula densa (*MD*) segment at the vascular pole with the afferent arteriole (*AA*) entering and the efferent arteriole (*EA*) leaving the glomerulus; extraglomerular mesangium (*EGM*); glomerular basement membrane (*BM*); epithelial podocytes (*EP*) with foot processes (*F*); Bowman's capsule (*B*) and space (*BS*), respectively; proximal tubule (*PT*). (Adapted from Kriz, Nonnenmacher and Kaissling). *Right panel*: schematic enlargement of area in rectangle. An increase in concentration-dependent uptake of Na^+, K^+ and Cl^- via the furosemide-sensitive $Na^+ - K^+ - 2Cl^-$ cotransporter (*NKCC2*) **a** leads to transport-related, intra- and/or extracellular generation of adenosine (*ADO*) **b**, **c**. Extracellular ADO activates A_1AR, triggering an increase in cytosolic Ca^{2+} in extraglomerular mesangium cells (*MC*) **d**. The intensive coupling between extraglomerular MC, granular renin-containing cells, and vascular smooth muscle cells (*VSMC*) of the afferent arteriole by gap junctions allows propagation of the increased Ca^{2+} signal **e**, resulting in afferent arteriolar vasoconstriction and inhibition of renin release (adapted from Vallon et al. 2006)

ecto-5′-nucleotidase and may reflect direct adenosine release from macula densa cells (Huang et al. 2006). Extracellular adenosine binds to A_1AR at the surface of extraglomerular mesangial cells (Olivera et al. 1989; Weaver and Reppert 1992; Toya et al. 1993; Smith et al. 2001) and increases cytosolic Ca^{2+} concentrations (Olivera et al. 1989). Gap junctions between extraglomerular mesangial cells and smooth muscle cells of glomerular arterioles can transmit intracellular Ca^{2+} transients to these target structures, inducing afferent arteriolar constriction (Iijima et al. 1991; Ren et al. 2002b). Potential candidates for the formation of gap junctions in the juxtaglomerular apparatus include connexins 37, 40, and 43 (Wagner et al. 2007; Takenaka et al. 2008a, b).

3 Activation of A_1AR Inhibits Renin Secretion

Tagawa and Vander reported in 1970 that adenosine infusion into the renal artery of salt-depleted dogs inhibited the renal secretion of renin into the venous blood (Tagawa and Vander 1970). This was confirmed in various species including humans (Osswald et al. 1978b; Edlund et al. 1994). Most notably, a single application of the A_1AR antagonist FK-453 increased plasma renin concentrations in humans (Balakrishnan et al. 1993), indicating a tonic inhibition of renin secretion by A_1AR activation. In accordance, knockout mice for A_1AR have increased renal mRNA expression and content of renin (Schweda et al. 2003) as well as greater plasma renin activity (Brown et al. 2001; Rieg et al. 2007) compared with wild-type mice.

Jackson and coworkers proposed an extracellular cyclic adenosine monophosphate (cAMP)-adenosine pathway in the control of renin release: the increase in intracellular cAMP in renin-secreting cells causes efflux of cAMP, the latter being converted to adenosine in the extracellular space. The generated adenosine, by acting on A_1AR on the renin-secreting cells, then acts as a negative-feedback control on renin release (Jackson and Raghvendra 2004). In addition, *high* NaCl concentrations in the tubular lumen enhance adenosine generation in a macula densa-dependent way, and the adenosine generated inhibits renin release via activation of A_1AR (Itoh et al. 1985; Weihprecht et al. 1990; Lorenz et al. 1993; Kim et al. 2006) (Fig. 2). In contrast to A_1AR stimulation, activation of A_2AR can increase renin secretion (Churchill and Churchill 1985; Churchill and Bidani 1987). The latter may have contributed to the observation that the unselective adenosine receptor antagonist caffeine reduced plasma renin concentration in mice lacking A_1AR (Rieg et al. 2007).

4 Differential Effects of Adenosine on Fluid and Electrolyte Transport

In addition to its effects on renal blood flow, GFR, and renin release, adenosine induces *direct* effects on fluid and electrolyte transport along the tubular and collecting duct system.

4.1 Activation of A_1AR Increases Reabsorption in the Proximal Tubule

Endogenously formed adenosine can *stimulate* proximal tubular reabsorption of fluid, Na^+, HCO_3^-, and phosphate by activation of A_1AR (Takeda et al. 1993; Cai et al. 1994, 1995; Tang and Zhou 2003). Importantly, systemic application of selective A_1AR antagonists (such as CVT-124, DPCPX, KW-3902, or FK-453) elicits diuresis and natriuresis predominantly by inhibiting reabsorption in the proximal tubule in rats and humans (Mizumoto and Karasawa 1993; Balakrishnan et al. 1993;

van-Buren et al. 1993; Knight et al. 1993b; Wilcox et al. 1999; Miracle et al. 2007), indicating a tonic stimulation of proximal tubular reabsorption via A_1AR activation (Fig. 1). As a consequence, selective A_1AR antagonists are being developed as eukaliuretic natriuretics in Na^+-retaining states such as heart failure (see below). A_1AR-mediated increases in proximal tubular reabsorption may involve increases of intracellular Ca^{2+} (Di Sole et al. 2003), reductions of intracellular cAMP levels (Kost Jr et al. 2000), and activation of the Na^+–H^+ exchanger (NHE3) (Di Sole et al. 2003).

Similar to selective A_1AR blockade, systemic application or consumption of the unselective adenosine receptor antagonist theophylline or caffeine induces natriuretic and diuretic responses. These responses to theophylline and caffeine are absent in mice lacking A_1AR, strongly suggesting that A_1AR blockade mediates the natriuresis and diuresis in response to these compounds (Rieg et al. 2005).

4.2 Activation of A_1AR Inhibits Reabsorption in Medullary Thick Ascending Limb

In contrast to the proximal tubule, adenosine via activation of A_1AR inhibits NaCl reabsorption in medullary TAL (Torikai 1987; Burnatowska-Hledin and Spielman 1991; Beach and Good 1992). Medullary TAL is a site of adenosine release, and adenosine release in this segment is transport dependent (Beach et al. 1991; Baudouin-Legros et al. 1995) and enhances significantly during hypoxic conditions (Beach et al. 1991). Studies using pharmacological inhibition (Zou et al. 1999) or gene knockout (Vallon et al. 2004) are consistent with a tonic inhibition of Na^+ reabsorption in medullary TAL by A_1AR activation (Fig. 1). This is relevant since the renal medulla has a low partial oxygen pressure (Brezis and Rosen 1995). The described inhibitory effects of adenosine on transport work together with its A_2AR-mediated renal medullary vasodilation (see above) may serve to maintain metabolic balance in the renal medulla.

4.3 Effects of Adenosine on Transport in Distal Convolution and Cortical Collecting Duct

In general, natriuretics that act proximal to the aldosterone-sensitive distal nephron stimulate K^+ secretion in the latter segment and thus increase renal K^+ excretion. The natriuretic but eukaliuretic effect of A_1AR inhibitors suggests an additional site of action in the aldosterone-sensitive distal nephron, but the exact site of action and the involved mechanisms are unclear.

A_1AR activation can stimulate Mg^{2+} and Ca^{2+} uptake in the cortical collecting duct in vitro (Hoenderop et al. 1998, 1999; Kang et al. 2001), but the clinical relevance (e.g., during pharmacological inhibition of A_1AR) is not known.

4.4 Activation of A_1AR Counteracts Vasopressin Effects in Inner Medullary Collecting Duct

Extracellular adenosine feedback can inhibit vasopressin-induced cAMP-mediated stimulation of Na^+ and fluid reabsorption in the inner medullary collecting duct (IMCD) (Yagil 1990; Yagil et al. 1994; Rieg et al. 2008) and decrease vasopressin-stimulated electrogenic Cl^- secretion through the activation of A_1AR (Moyer et al. 1995). Vasopressin-induced adenosine may derive from the extracellular cAMP–adenosine pathway (Jackson et al. 2003) or follow the cellular release and breakdown of ATP (Vallon 2008). Studies on water transport in knockout mice indicate efficient compensation by other pathways in the absence of A_1AR, including upregulation of ATP-sensitive $P2Y_2$ receptors (Rieg et al. 2008).

5 Adenosine and Metabolic Control of Kidney Function

The above outlined functions of adenosine can be integrated into the concept of metabolic control of renal function (Fig. 1). Adenosine-induced vasoconstriction via A_1AR activation is predominant in the outer cortex by increasing the resistance of afferent arterioles, which lowers GFR and thus renal transport work. Under physiological conditions, adenosine-induced afferent arteriolar constriction primarily derives from tonic activation of the TGF, for which adenosine acts as a mediator. Adenosine via A_1AR tonically stimulates NaCl reabsorption in the cortical proximal tubule, which is a tubular segment with a relatively high basal oxygen supply, thereby limiting the NaCl load to downstream medullary segments. In the deep cortex and medulla, adenosine induces vasodilation via A_2AR activation, which is associated with an increase of medullary blood flow and thus increased medullary oxygenation. Moreover, adenosine inhibits NaCl reabsorption in medullary TAL and IMCD (i.e., nephron segments with relatively low oxygen delivery). In addition, the A_2AR-mediated rise in medullary blood flow lowers medullary transport activity by washing out the high osmolality in the medullary interstitium (Zou et al. 1999). In accordance, interstitial infusion of adenosine in rat kidney decreased partial pressure of O_2 in the cortex but increased it in the medulla, consistent with an important regulatory and protective role of adenosine in renal medullary O_2 balance (Dinour and Brezis 1991).

6 Adenosine and Acute Renal Failure

The renal effects of adenosine fit into the concepts of acute renal failure (ARF) in as much as adenosine is an *intrarenal* metabolite that accumulates in the kidney during renal ischemia and that can lower GFR. In addition, ischemia or nephrotoxins can

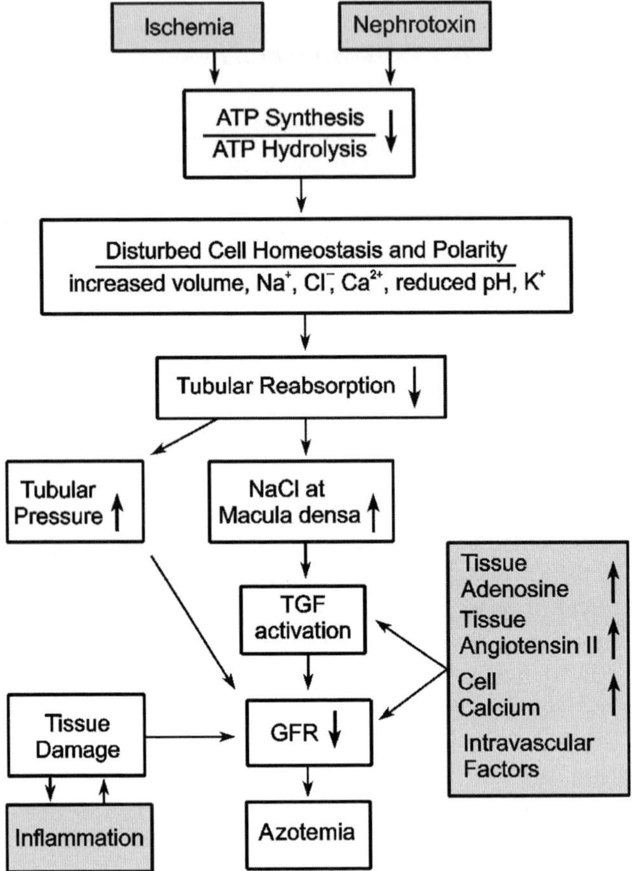

Fig. 3 Schematic illustration of intrarenal mechanisms in acute renal failure. See text for further explanation (adapted from Osswald and Vallon 2009)

inhibit renal transport activity, with the resulting increase in the NaCl concentration at the macula densa further lowering GFR (Fig. 3). Moreover, experimental models of ARF can be associated with increased expression of A_1AR in glomeruli, which may contribute to depressed GFR (Smith et al. 2000). Thus, inhibition of adenosine vasoconstrictor actions in the kidney could be beneficial in conditions of ARF. On the other hand, the ARF-associated reduction in GFR and thus in tubular NaCl load may, to some extent, protect the tubular system—especially the medulla—from hypoxic injury, and the body from excess NaCl loss. Moreover, adenosine can induce direct cytoprotective effects in renal cells. Therefore, inhibition of adenosine receptors in ARF could be a two-sided sword. In the following we discuss the role of adenosine in ARF induced by radiocontrast media and ischemia-reperfusion, respectively.

6.1 Radiocontrast Media-Induced Acute Renal Failure: Theophylline and A_1AR Antagonists Induce Protective Effects

Application of radiocontrast media to humans can lead to an impairment of renal function, including a fall in GFR. Concomitant volume and NaCl depletion increases the severity and can result in ARF. Unselective or A_1AR-selective antagonists can prevent renal impairment induced by radiocontrast media, as shown in dogs (Arend et al. 1987), rats (Erley et al. 1997), mice (Lee et al. 2006), and, most importantly, in humans (Erley et al. 1994; Katholi et al. 1995; Kolonko et al. 1998; Kapoor et al. 2002; Huber et al. 2002, 2003). In accordance, mice lacking A_1AR preserved kidney function better, and had lesser renal cortical vacuolization and enhanced survival 24 h after radiocontrast media treatment compared with wild-type mice (Lee et al. 2006). In comparison, dipyridamole, which increases extracellular adenosine concentrations, augmented the severity of renal impairment in response to radiocontrast media in dogs (Arend et al. 1987) and humans (Katholi et al. 1995). Two studies indicated that the unselective adenosine receptor antagonist theophylline is as effective as saline hydration at preventing ARF in response to contrast media, but the benefits of the two maneuvers are not additive (Abizaid et al. 1999; Erley et al. 1999). Thus, use of theophylline can be beneficial in patients where sufficient hydration may be impossible or in patients with a concomitant decrease in renal blood flow (e.g., congestive heart failure or chronic renal insufficiency (Erley et al. 1999; Huber et al. 2002)). A recent meta-analysis of clinical trials concluded that theophylline may reduce the incidence of radiocontrast media-induced nephropathy, and recommended a large, well-designed trial to more adequately assess the role of theophylline in this condition (Bagshaw and Ghali 2005). Notably, unselective or A_1AR-selective antagonists can also prevent renal impairment in response to other nephrotoxic substances (Table 1).

6.2 Ischemia-Reperfusion Injury

Ischemia-reperfusion injury plays a major role in delayed graft function and long-term changes after kidney transplantation. It has become evident that the cellular and molecular mechanisms that operate during ischemia and reperfusion resemble an acute inflammatory response (Gueler et al. 2004). To what extent the acute cellular alterations persist and affect organ function later on remains unclear.

In the kidney, extracellular adenosine derives to a large extent from the extracellular breakdown of ATP and ADP to AMP and adenosine via ectonucleoside triphosphate diphosphohydrolases (ENTPDases) and CD73 (Grenz et al. 2007a, b). Using knockout mouse models for these ectoenzymes, Grenz et al. showed that CD39-dependent nucleotide phosphohydrolysis as well as CD73-dependent adenosine formation serve to protect against renal ischemia-reperfusion injury and to

Table 1 Adenosine receptor antagonists improve renal function in various models of nephrotoxic acute renal failure (ARF)

Models of ARF	Species	Adenosine antagonist	References
Glycerol	Rat	Theophylline	Bidani and Churchill (1983); Bidani et al. (1987)
Injection		8-Phenyl-theophylline	Bowmer et al. (1986); Yates et al. (1987) Kellett et al. (1989); Panjehshahin et al. (1992)
		DPCPX	
		FK-453	Ishikawa et al. (1993)
		KW-3902	Suzuki et al. (1992)
Uranyl nitrate	Rat	Theophylline	Osswald et al. (1979)
Cisplatin	Rat	Theophylline	Heidemann et al. (1989)
	Human	DPCPX	Knight et al. (1991)
		KW-3902	Nagashima et al. (1995)
		Theophylline	Benoehr et al. (2005)
Contrast media	Dog	Theophylline	Arend et al. (1987)
	Human	Theophylline	Erley et al. (1994); Katholi et al. (1995); Kolonko et al. (1998); Kapoor et al. (2002); Huber et al. (2002, 2003)
	Rat	DPCPX, KW3902	Erley et al. (1997); Yao et al. (2001)
Endotoxin	Rat	DPCPX	Knight et al. (1993a)
Amphotericin B	Rat	Theophylline	Heidemann et al. (1983)
	Dog	Theophylline	Gerkens et al. (1983)
Gentamicin	Rat	KW-3902	Yao et al. (1994)

8-Cyclopentyl-1,3-dipropylxanthine (DPCPX), (+)-(*R*)-[(*E*)-3-(2-phenylpyrazolo[1,5-*a*]pyridin-3-yl)acryloyl]-2-piperidine ethanol (FK-453) and 8-(noradamantan-3-yl)-1,3 dipropylxanthine (KW-3902) are A_1AR-selective antagonists. Adapted from Vallon et al. (2006)

increase the ischemia tolerance of the kidney. In addition, the authors presented evidence that treatment with apyrase or soluble 5′-nucleotidase to increase extracellular adenosine concentrations could serve as potential novel pharmacological approaches to renal diseases precipitated by limited oxygen availability (Grenz et al. 2007a, b).

6.2.1 Theophylline Induces Protective Effects

Different animal studies assessed the effect of a *single* application of the unselective adenosine receptor antagonist theophylline in ischemia-reperfusion injury. Animals were pretreated with theophylline or it was given at day 5 after the renal ischemic/hypoxemic event. Pretreatment with a single dose of theophylline in rats attenuated the reduction in renal blood flow and GFR observed during the initiation phase of postischemic ARF as determined 1 h after releasing a 30 or 45 min occlusion of the renal artery (Lin et al. 1986). Similar results were obtained with theophylline in the rabbit (Gouyon and Guignard 1988). In rats subjected to 60 min occlusion of the left renal artery, theophylline given i.v. 20 min before the release of the renal artery clamp in doses which antagonize the renal actions of adenosine in vivo improved the recovery of renal function after ischemic injury by increasing urinary flow rate, GFR (measured by inulin clearance), and histology, as assessed by morphometric quantification of tubular damage, tubular obstruction and pathologic alteration of glomeruli at 3 h after initiating reperfusion (Osswald et al. 1979; Helmlinger 1979). In contrast, pretreating rats prior to renal artery occlusion for 30 min with dipyridamole, which increases extracellular adenosine concentrations, intensified the fall in renal blood flow and GFR determined about 1 h after releasing the clamp, and this impairment was blocked by theophylline (Lin et al. 1987).

Notably, single-dose pretreatment of rats with theophylline during a 30 min renal artery occlusion lead to increased renal blood flow and GFR during the maintenance phase of ARF after five days, indicating that the effects of theophylline in the acute phase affected the outcome in the maintenance phase (Lin et al. 1988). Similarly, a single dose of theophylline, given early after birth in asphyxiated full-term infants, has beneficial effects in reducing the renal involvement and fall in GFR as determined over the first five days (Bakr 2005). Finally, acute theophylline treatment given at five days after ischemia acutely increases renal blood flow and GFR in previously untreated rats, indicating that adenosine contributes to the suppression of renal blood flow and GFR in the maintenance phase of ischemia-reperfusion injury (Lin et al. 1988). These data provide strong evidence that pretreatment with theophylline can exert beneficial effects in the initiation and maintenance phase of ischemia-reperfusion injury.

6.2.2 Adenosine Induces Protective Effects via A_1AR and A_2AR

Similar to theophylline, systemic intravenous infusion of adenosine (1.75 mg kg^{-1} min^{-1} $\times$10 min, intravenously) 2 min before a 45 min ischemic insult protected renal function against ischemia and reperfusion injury, as indicated by lower blood urea nitrogen and creatinine and improved renal morphology after 24 h of reperfusion. The effects of adenosine were proposed to be mediated by A_1AR (Lee and Emala 2000), involve $G_{i/o}$ proteins and protein kinase C activation (Lee and Emala 2001a), and include a reduction in inflammation, necrosis, and apoptosis (Lee et al. 2004a). Direct cytoprotective effects of endogenous A_1AR activation in renal proximal tubules involve modulation of heat-shock protein (HSP)27 due to A_1AR-mediated enhancement of p38 and AP2 mitogen-activated protein kinase activities (Lee et al. 2007). In comparison, mice lacking A_1AR exhibited significantly higher plasma creatinines and worsened renal histology compared with wild-type mice at 24 h after renal ischemia for 30 min (Lee et al. 2004b). Similarly, wild-type mice pretreated with an A_1AR antagonist or agonist demonstrated worsened or improved renal function, respectively, after ischemia-reperfusion that was associated with increased or reduced markers of renal inflammation, respectively (Lee et al. 2004b) (Fig. 3). More recent work indicated that A_1AR activation produces not only acute but also delayed renal protection; i.e., pretreatment with a selective A_1AR agonist 24 h before renal ischemia was also protective against renal ischemia-reperfusion injury. Furthermore, the study showed that acute protection from A_1AR activation is dependent on protein kinase C and Akt activation, whereas the delayed protection is dependent on Akt activation and induction of HSP27 (Joo et al. 2007).

Continuous application in the reperfusion period of 4-(3-(6-amino-9-(5-ethylcarbamoyl-3,4-dihydroxytetrahydrofuran-2-yl)-9*H*-purin-2-yl)prop-2-ynyl) cyclohexanecarboxylic acid methyl ester (DWH-146e), a selective A_{2A}AR agonist, protected kidneys from ischemia-reperfusion injury, as evidenced by a lower rise in serum creatinine and blood urea nitrogen following 24 and 48 h of reperfusion. Histological examination revealed widespread tubular epithelial necrosis and vascular congestion in the outer medulla of vehicle-treated rats. These lesions were significantly reduced in DWH-146e-treated animals (Okusa et al. 1999). Similarly, systemic adenosine given after 45 min of renal ischemia but before reperfusion protected renal function, as indicated by lower rises in creatinine and less histologically evident renal tubular damage. Pharmacological maneuvers indicated that these effects of adenosine were mediated by A_{2A}AR activation (Lee and Emala 2001b). Whereas A_{2A}AR activation could improve medullary hypoxia, other studies suggested that protection from renal ischemia-reperfusion injury by A_2AR agonists or endogenous adenosine requires activation of A_2AR expressed on bone marrow-derived cells (Day et al. 2003). Activation of A_{2A}AR on macrophages was also shown to inhibit inflammation in a rat model of glomerulonephritis (Garcia et al. 2008). Moreover, activation of A_{2B}AR in the renal vasculature contributes to the increased ischemia tolerance produced by the procedure of renal ischemic preconditioning (Grenz et al. 2008).

Finally, A_3AR stimulation in rats deteriorated renal ischemia-reperfusion injury, whereas inhibition of A_3AR protected renal function as efficiently as preconditioning (Lee and Emala 2000). In accordance, mice lacking A_3AR presented significant renal protection, functionally and morphologically, from ischemic or myoglobinuric renal failure (Lee et al. 2003). The mechanisms of these A_3AR-mediated effects are not understood at present.

In summary, beneficial effects on GFR and renal morphology beyond 3–24 h of reperfusion after ischemia can be induced by (1) pretreatment with the unselective adenosine receptor antagonist theophylline, (2) pretreatment or treatment immediately before reperfusion with adenosine, (3) pretreatment with A_1AR agonists, (4) treatment immediately before or during reperfusion with A_{2A}AR agonists, (5) treatment with A_{2B}AR agonists, and (6) deficiency of A_3AR. In comparison, the outcome is worsened by (1) pretreatment with A_1AR antagonists or deficiency of A_1AR, (2) pretreatment with A_{2B}AR antagonists or deficiency of A_{2B}AR, and (3) pretreatment with A_3AR agonists. The findings appear contradictory because theophylline can inhibit both A_1AR and A_{2A}AR, and possibly acts as an agonist at A_3AR (Ezeamuzie 2001). Further studies are necessary to resolve this issue, which may relate to the nature of adenosine being a double-edged sword in ARF, and the situation being further complicated by the role of adenosine in inflammatory responses.

7 A_1AR Antagonists in the Treatment of Cardiorenal Failure

Concomitant renal dysfunction is one of the strongest risk factors for mortality in ambulatory heart failure patients (Dries et al. 2000; Hillege et al. 2000; Mahon et al. 2002). In patients hospitalized for decompensated heart failure, worsening of renal function further predicts an adverse outcome (Forman et al. 2004). Intravenous loop diuretics are the mainstay of therapy for patients with both systemic volume overload and acute pulmonary edema decompensated heart failure. Treatment, however, may be complicated by diuretic resistance and/or worsening of renal function, indicating the need for alternative approaches.

Volume overload heart failure in dogs increases myocardial adenosine release (Newman et al. 1984), and circulating levels of adenosine can be increased in patients with chronic heart failure (∼200 vs 60 nM) (Funaya et al. 1997) (Fig. 4). Whether this increases circulating adenosine to an extent that affects afferent arteriolar tone and thus GFR is unclear. Nonetheless, the renal vasculature in heart failure patients can be sensitized to the GFR-lowering effects of adenosine by the associated activation of the renin–angiotensin system and/or impairment of the local formation of NO (endothelial dysfunction) or prostaglandins (see above and Fig. 4). In addition, impaired renal perfusion and hypoxia enhance adenosine formation within the kidney (Nishiyama et al. 1999). As a consequence, the normally homeostatic adenosine system may become maladaptive and overshoots with regard to the downregulation of GFR in patients with heart failure. Fluid retention is further potentiated by stimulation of NaCl and fluid reabsorption in the proximal tubule, a mechanism also mediated by A_1AR activation (see above and Fig. 4). Based on this concept, pharmacological blockade of A_1AR could improve kidney function and fluid retention in heart failure. Since adenosine (through the activation of A_1AR) mediates TGF, the expected TGF-induced reduction in GFR in response to inhibition of proximal reabsorption by A_1AR antagonists should be blunted. In accordance, a study in rats showed that A_1AR antagonism with KW-3902 prevented the GFR-lowering effect of the proximal diuretic benzolamide, a carbonic anhydrase inhibitor (Miracle et al. 2007).

7.1 Animal Studies

Lucas et al. used a pig model of systolic dysfunction and induction of chronic heart failure by pacer-induced tachycardia. They observed that *acute* application of the selective A_1AR antagonist 1,3-dipropyl-8-[2-(5,6-epoxynorbornyl)xanthine (BG9719) (CVT-124) increased creatinine clearance and urinary flow rate and sodium excretion. This was associated with lower pulmonary capillary wedge pressure and pulmonary vascular resistance in the absence of significant changes in mean arterial blood pressure, heart rate or cardiac output compared with vehicle control (Lucas Jr et al. 2002). Similar effects were described by Jackson et al. in aged, lean SHHF/Mcc-fa(cp) rats, a rodent model of hypertensive dilated cardiomyopathy

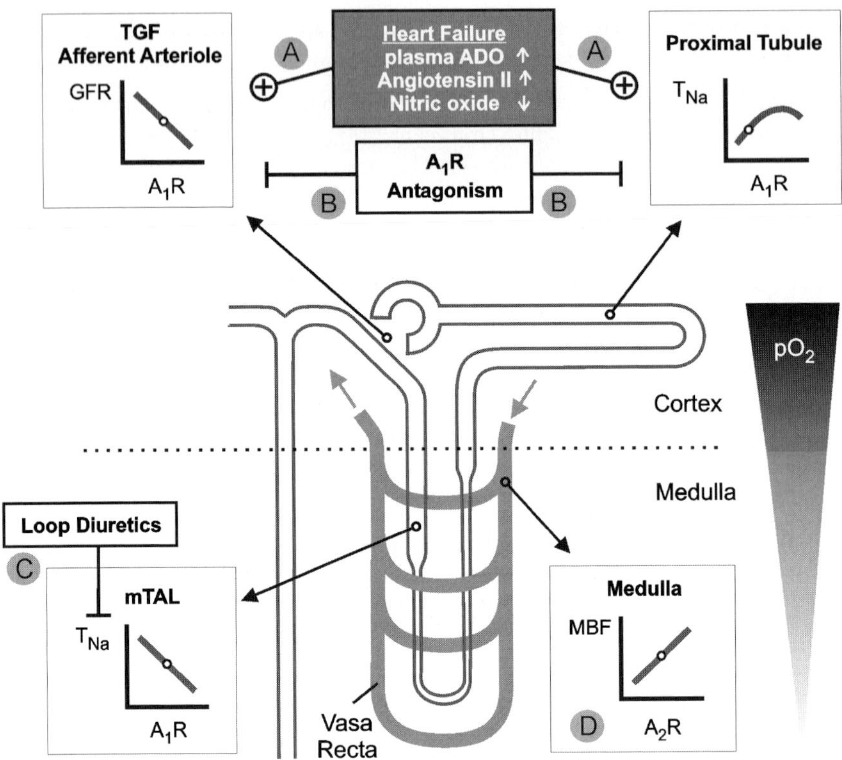

Fig. 4 a–d Basis for a therapeutic effect of A_1AR antagonism in heart failure. The basic effects of adenosine on renal functions are outlined in the legend to Fig. 1. **a** Heart failure can be associated with increased plasma concentrations of adenosine (*ADO*) and angiotensin II, and endothelial dysfunction can impair nitric oxide (*NO*) formation, all of which can enhance the A_1AR-mediated lowering of GFR and may, in addition, stimulate proximal reabsorption. **b** A_1AR antagonism induces natriuresis and diuresis by inhibiting proximal reabsorption and preserving or increasing GFR. **c** A_1AR antagonism can enhance sodium transport (T_{Na}) in semihypoxic medullary thick ascending limb (*mTAL*). This is prevented by coadministration of loop diuretics, and diuresis and natriuresis are potentiated. **d** A_2AR-mediated medullary vasodilation is preserved (adapted from Vallon et al. 2008)

in response to the same compound (Jackson et al. 2001). The rats were pretreated for 72 h before experiments with the loop diuretic furosemide to mimic the clinical setting of chronic diuretic therapy, and were given 1% NaCl as drinking water to reduce dehydration/sodium depletion. *Acute* application of BG9719 increased GFR and urinary fluid and sodium excretion. In comparison, acute application of furosemide decreased renal blood flow and GFR and increased fractional potassium excretion. Neither drug altered afterload or left ventricular systolic function ($+d\text{P}/dt$ (max)); however, furosemide, but not BG9719, decreased preload and attenuated diastolic function (decreased $-d\text{P}/dt$ (max), increased tau). Thus, in the setting of left ventricular dysfunction, chronic salt loading and prior loop diuretic treatment, selective A_1AR antagonists are effective diuretic/natriuretic agents that do not induce potassium loss and have a favorable renal hemodynamic/cardiac performance profile (Jackson et al. 2001).

7.2 Human Studies

Gottlieb et al. compared the *acute* effects of furosemide and BG9719 on renal function in 12 patients categorized as New York Heart Association (NYHA) functional classes II, III or IV (Gottlieb et al. 2000). Both BG9719 and furosemide increased sodium excretion compared with placebo. However, only furosemide lowered GFR. Subsequently, Gottlieb et al. compared BG9719 and furosemide in 63 patients categorized as NYHA functional classes II, III or IV, which despite receiving standard therapy, including furosemide (at least 80 mg daily) and angiotensin-converting enzyme inhibitors, remained edematous (Gottlieb et al. 2002). Patients received 7 h infusions of placebo or BG9719 to yield serum concentrations of 0.1, 0.75, or 2.5 $\mu g\ ml^{-1}$. BG9719 tripled urine output without lowering GFR or inducing kaliuresis. In comparison, furosemide increased urine output eightfold and increased potassium excretion while reducing GFR. Notably, when BG9719 was given with furosemide, GFR remained unaltered compared with placebo and sodium excretion increased further. These results indicate that A_1AR antagonism can preserve renal function while simultaneously promoting natriuresis during *acute* treatment of heart failure (Gottlieb et al. 2002).

Similar results were more recently reported in studies using the A_1AR antagonist KW-3902 in patients with congestive heart failure and impaired renal function (Dittrich et al. 2007; Givertz et al. 2007). Dittrich et al. assessed baseline GFR and renal plasma flow 3 h before and over 8 h following the intravenous administration of furosemide along with KW-3902 (30 mg) or placebo. After a washout period of 3–8 days (median six days), the crossover portion of the study was performed. KW-3902 increased GFR by 32% and renal plasma flow by 48% compared with placebo. Notably, subjects who initially received KW-3902 had a statistically significant 10 ml min^{-1} increase in GFR when they returned for the crossover phase compared with the previous baseline. Thus, the increase in GFR persisted for several days longer than predicted by pharmacokinetics. These findings suggest that KW-3902 reset the complex network that determines kidney function in these patients, and provided first evidence for potential longer-term benefits of using A_1AR antagonists (Dittrich et al. 2007). Greenberg et al. assessed the effects of the selective A_1AR antagonist 1,3-dipropyl-8-[1-(4-propionate)-bicyclo-[2,2,2]octyl)]xanthine (BG9928) given orally for ten days to 50 patients with heart failure and left ventricular systolic dysfunction who were receiving standard therapy (Greenberg et al. 2007). BG9928 (3–225 mg per day) increased sodium excretion without causing kaliuresis or reducing GFR. Notably, these effects were maintained over the ten-day period. BG9928 at doses of 15, 75, or 225 mg also reduced body weight at the end of the study compared with placebo (Greenberg et al. 2007).

In summary, the above described acute and short-term studies employing A_1AR antagonists in patients with heart failure yielded promising results. Since A_1AR blockade may increase transport in the semihypoxic medullary TAL, combining A_1AR antagonists with furosemide may potentiate natriuresis while helping to prevent transport-induced medullary hypoxia (Fig. 4). Whereas the presented animal and human studies were *acute* or *short-term* treatments, it remains to be determined whether longer-term application of A_1AR antagonism has beneficial effects. These studies should also reveal whether a clinically relevant effect of A_1AR blockade on renin release occurs. Consideration should also be given to the evidence that A_1AR activation is potentially important for protection in response to ischemia of the kidney (see above) and the heart (Cohen and Downey 2008). Apart from these issues, A_1AR blockade is unique in inducing natriuresis without potassium loss and lowering renal vascular resistance independent of all other organs. With regard to preserving renal function, this is an advantage over all vasodilator heart failure therapies that have been tried so far.

Acknowledgment The work from our laboratories was supported by the Deutsche Forschungsgemeinschaft (DFG VA 118/2-1, DFG OS 42/1–42/7), the Department of Veterans Affairs, the National Institutes of Health (DK56248, DK28602, GM66232, P30DK079337), and the American Heart Association (GiA 655232Y).

References

Abizaid AS, Clark CE, Mintz GS, Dosa S, Popma JJ, Pichard AD, Satler LF, Harvey M, Kent KM, Leon MB (1999) Effects of dopamine and aminophylline on contrast-induced acute renal failure after coronary angioplasty in patients with preexisting renal insufficiency. Am J Cardiol 83:260–263, A5

Agmon Y, Dinour D, Brezis M (1993) Disparate effects of adenosine A_1- and A_2-receptor agonists on intrarenal blood flow. Am J Physiol 265:F802–F806

Arend LJ, Thompson CI, Spielman WS (1985) Dipyridamole decreases glomerular filtration in the sodium-depleted dog. Evidence for mediation by intrarenal adenosine. Circ Res 56:242–251

Arend LJ, Bakris GL, Burnett JC Jr, Megerian C, Spielman WS (1987) Role for intrarenal adenosine in the renal hemodynamic response to contrast media. J Lab Clin Med 110:406–411

Bagshaw SM, Ghali WA (2005) Theophylline for prevention of contrast-induced nephropathy: a systematic review and meta-analysis. Arch Intern Med 165:1087–1093

Bakr AF (2005) Prophylactic theophylline to prevent renal dysfunction in newborns exposed to perinatal asphyxia: a study in a developing country. Pediatr Nephrol 20:1249–1252

Balakrishnan VS, Coles GA, Williams JD (1993) A potential role for endogenous adenosine in control of human glomerular and tubular function. Am J Physiol 265:F504–F510

Balakrishnan VS, Coles GA, Williams JD (1996) Effects of intravenous adenosine on renal function in healthy human subjects. Am J Physiol 271:F374–F381

Barrett RJ, Droppleman DA (1993) Interactions of adenosine A_1 receptor-mediated renal vasoconstriction with endogenous nitric oxide and ANG II. Am J Physiol 265:F651–F659

Baudouin-Legros M, Badou A, Paulais M, Hammet M, Teulon J (1995) Hypertonic NaCl enhances adenosine release and hormonal cAMP production in mouse thick ascending limb. Am J Physiol 269:F103–F109

Beach RE, Good DW (1992) Effects of adenosine on ion transport in rat medullary thick ascending limb. Am J Physiol 263:F482–F487

Beach RE, Watts BA, Good DW, Benedict CR, DuBose TD Jr (1991) Effects of graded oxygen tension on adenosine release by renal medullary and thick ascending limb suspensions. Kidney Int 39:836–842

Bell PD, Lapointe JY, Sabirov R, Hayashi S, Peti-Peterdi J, Manabe K, Kovacs G, Okada Y (2003) Macula densa cell signaling involves ATP release through a maxi anion channel. Proc Natl Acad Sci USA 100:4322–4327

Benoehr P, Krueth P, Bokemeyer C, Grenz A, Osswald H, Hartmann JT (2005) Nephroprotection by theophylline in patients with cisplatin chemotherapy: a randomized, single-blinded, placebo-controlled trial. J Am Soc Nephrol 16:452–458

Bidani AK, Churchill PC (1983) Aminophylline ameliorates glycerol-induced acute renal failure in rats. Can J Physiol Pharmacol 61:567–571

Bidani AK, Churchill PC, Packer W (1987) Theophylline-induced protection in myoglobinuric acute renal failure: further characterization. Can J Physiol Pharmacol 65:42–45

Bowmer CJ, Collis MG, Yates MS (1986) Effect of the adenosine antagonist 8-phenyltheophylline on glycerol-induced acute renal failure in the rat. Br J Pharmacol 88:205–212

Brezis M, Rosen S (1995) Hypoxia of the renal medulla: its implications for disease. N Engl J Med 332:647–655

Brown R, Ollerstam A, Johansson B, Skott O, Gebre-Medhin S, Fredholm B, Persson AE (2001) Abolished tubuloglomerular feedback and increased plasma renin in adenosine A_1 receptor-deficient mice. Am J Physiol Regul Integr Comp Physiol 281:R1362–R1367

Burnatowska-Hledin MA, Spielman WS (1991) Effects of adenosine on cAMP production and cytosolic Ca^{2+} in cultured rabbit medullary thick limb cells. Am J Physiol 260:C143–C150

Cai H, Batuman V, Puschett DB, Puschett JB (1994) Effect of KW-3902, a novel adenosine A_1 receptor antagonist, on sodium-dependent phosphate and glucose transport by the rat renal proximal tubular cell. Life Sci 55:839–845

Cai H, Puschett DB, Guan S, Batuman V, Puschett JB (1995) Phosphate transport inhibition by KW-3902, an adenosine A_1 receptor antagonist, is mediated by cyclic adenosine monophosphate. Am J Kidney Dis 26:825–830

Castrop H, Huang Y, Hashimoto S, Mizel D, Hansen P, Theilig F, Bachmann S, Deng C, Briggs J, Schnermann J (2004) Impairment of tubuloglomerular feedback regulation of GFR in ecto-5′-nucleotidase/CD73-deficient mice. J Clin Invest 114:634–642

Churchill PC, Bidani A (1987) Renal effects of selective adenosine receptor agonists in anesthetized rats. Am J Physiol 252:F299–F303

Churchill PC, Churchill MC (1985) A_1 and A_2 adenosine receptor activation inhibits and stimulates renin secretion of rat renal cortical slices. J Pharmacol Exp Ther 232:589–594

Cohen MV, Downey JM (2008) Adenosine: trigger and mediator of cardioprotection. Basic Res Cardiol 103:203–215

Day YJ, Huang L, McDuffie MJ, Rosin DL, Ye H, Chen JF, Schwarzschild MA, Fink JS, Linden J, Okusa MD (2003) Renal protection from ischemia mediated by A_{2A} adenosine receptors on bone marrow-derived cells. J Clin Invest 112:883–891

Dietrich MS, Steinhausen M (1993) Differential reactivity of cortical and juxtaglomerullary glomeruli to adenosine-1 and adenosine-2 receptor stimulation and angiotensin converting-enzyme inhibition. Microvasc Res 45:122–133

Dietrich MS, Endlich K, Parekh N, Steinhausen M (1991) Interaction between adenosine and angiotensin II in renal microcirculation. Microvasc Res 41:275–288

Dinour D, Brezis M (1991) Effects of adenosine on intrarenal oxygenation. Am J Physiol 261:F787–F791

Di Sole F, Cerull R, Petzke S, Casavola V, Burckhardt G, Helmle-Kolb C (2003) Bimodal acute effects of A_1 adenosine receptor activation on Na^+/H^+ exchanger 3 in opossum kidney cells. J Am Soc Nephrol 14:1720–1730

Dittrich HC, Gupta DK, Hack TC, Dowling T, Callahan J, Thomson S (2007) The effect of KW-3902, an adenosine A_1 receptor antagonist, on renal function and renal plasma flow in ambulatory patients with heart failure and renal impairment. J Card Fail 13:609–617

Dries DL, Exner DV, Domanski MJ, Greenberg B, Stevenson LW (2000) The prognostic implications of renal insufficiency in asymptomatic and symptomatic patients with left ventricular systolic dysfunction. J Am Coll Cardiol 35:681–689

Edlund A, Sollevi A (1993) Renal effects of i.v. adenosine infusion in humans. Clin Physiol 13:361–371

Edlund A, Ohlsen H, Sollevi A (1994) Renal effects of local infusion of adenosine in man. Clin Sci Colch 87:143–149

Erley CM, Duda SH, Schlepckow S, Koehler J, Huppert PE, Strohmaier WL, Bohle A, Risler T, Osswald H (1994) Adenosine antagonist theophylline prevents the reduction of glomerular filtration rate after contrast media application. Kidney Int 45:1425–1431

Erley CM, Heyne N, Burgert K, Langanke J, Risler T, Osswald H (1997) Prevention of radiocontrast-induced nephropathy by adenosine antagonists in rats with chronic nitric oxide deficiency. J Am Soc Nephrol 8:1125–1132

Erley CM, Duda SH, Rehfuss D, Scholtes B, Bock J, Muller C, Osswald H, Risler T (1999) Prevention of radiocontrast-media-induced nephropathy in patients with pre-existing renal insufficiency by hydration in combination with the adenosine antagonist theophylline. Nephrol Dial Transplant 14:1146–1149

Ezeamuzie CI (2001) Involvement of A(3) receptors in the potentiation by adenosine of the inhibitory effect of theophylline on human eosinophil degranulation: possible novel mechanism of the antiinflammatory action of theophylline. Biochem Pharmacol 61:1551–1559

Forman DE, Butler J, Wang Y, Abraham WT, O'Connor CM, Gottlieb SS, Loh E, Massie BM, Rich MW, Stevenson LW, Young JB, Krumholz HM (2004) Incidence, predictors at admission, and impact of worsening renal function among patients hospitalized with heart failure. J Am Coll Cardiol 43:61–67

Franco M, Bell PD, Navar LG (1989) Effect of adenosine A_1 analogue on tubuloglomerular feedback mechanism. Am J Physiol 257:F231–F236

Funaya H, Kitakaze M, Node K, Minamino T, Komamura K, Hori M (1997) Plasma adenosine levels increase in patients with chronic heart failure. Circulation 95:1363–1365

Gabriels G, Endlich K, Rahn KH, Schlatter E, Steinhausen M (2000) In vivo effects of diadenosine polyphosphates on rat renal microcirculation. Kidney Int 57:2476–2484

Garcia GE, Truong LD, Li P, Zhang P, Du J, Chen JF, Feng L (2008) Adenosine A_{2A} receptor activation and macrophage-mediated experimental glomerulonephritis. FASEB J 22:445–454

Gerkens JF, Heidemann HT, Jackson EK, Branch RA (1983) Effect of aminophylline on amphotericin B nephrotoxicity in the dog. J Pharmacol Exp Ther 224:609–613

Givertz MM, Massie BM, Fields TK, Pearson LL, Dittrich HC (2007) The effects of KW-3902, an adenosine A_1-receptor antagonist on diuresis and renal function in patients with acute decompensated heart failure and renal impairment or diuretic resistance. J Am Coll Cardiol 50:1551–1560

Gottlieb SS, Skettino SL, Wolff A, Beckman E, Fisher ML, Freudenberger R, Gladwell T, Marshall J, Cines M, Bennett D, Liittschwager EB (2000) Effects of BG9719 (CVT-124), an A_1-adenosine receptor antagonist, and furosemide on glomerular filtration rate and natriuresis in patients with congestive heart failure. J Am Coll Cardiol 35:56–59

Gottlieb SS, Brater DC, Thomas I, Havranek E, Bourge R, Goldman S, Dyer F, Gomez M, Bennett D, Ticho B, Beckman E, Abraham WT (2002) BG9719 (CVT-124), an A_1 adenosine receptor antagonist, protects against the decline in renal function observed with diuretic therapy. Circulation 105:1348–1353

Gouyon JB, Guignard JP (1988) Theophylline prevents the hypoxemia-induced renal hemodynamic changes in rabbits. Kidney Int 33:1078–1083

Greenberg B, Thomas I, Banish D, Goldman S, Havranek E, Massie BM, Zhu Y, Ticho B, Abraham WT (2007) Effects of multiple oral doses of an A_1 adenosine antagonist, BG9928, in patients with heart failure: results of a placebo-controlled, dose-escalation study. J Am Coll Cardiol 50:600–606

Grenz A, Zhang H, Eckle T, Mittelbronn M, Wehrmann M, Kohle C, Kloor D, Thompson LF, Osswald H, Eltzschig HK (2007a) Protective role of ecto-5′-nucleotidase (CD73) in renal ischemia. J Am Soc Nephrol 18:833–845

Grenz A, Zhang H, Hermes M, Eckle T, Klingel K, Huang DY, Muller CE, Robson SC, Osswald H, Eltzschig HK (2007b) Contribution of E-NTPDase1 (CD39) to renal protection from ischemia-reperfusion injury. FASEB J 21:2863–2873

Grenz A, Osswald H, Eckle T, Yang D, Zhang H, Tran ZV, Klingel K, Ravid K, Eltzschig HK (2008) The reno-vascular A_{2B} adenosine receptor protects the kidney from ischemia. PLoS Med 5:e137

Gueler F, Gwinner W, Schwarz A, Haller H (2004) Long-term effects of acute ischemia and reperfusion injury. Kidney Int 66:523–527

Haas JA, Osswald H (1981) Adenosine induced fall in glomerular capillary pressure. Effect of ureteral obstruction and aortic constriction in the Munich–Wistar rat kidney. Naunyn–Schmiedeberg's Arch Pharmacol 317:86–89

Hansen PB, Hashimoto S, Briggs J, Schnermann J (2003a) Attenuated renovascular constrictor responses to angiotensin II in adenosine 1 receptor knockout mice. Am J Physiol Regul Integr Comp Physiol 285:R44–R49

Hansen PB, Castrop H, Briggs J, Schnermann J (2003b) Adenosine induces vasoconstriction through Gi-dependent activation of phospholipase C in isolated perfused afferent arterioles of mice. J Am Soc Nephrol 14:2457–2465

Hansen PB, Hashimoto S, Oppermann M, Huang Y, Briggs JP, Schnermann J (2005) Vasoconstrictor and vasodilator effects of adenosine in the mouse kidney due to preferential activation of A_1 or A_2 adenosine receptors. J Pharmacol Exp Ther 315:1150–1157

Hashimoto S, Huang Y, Briggs J, Schnermann J (2006) Reduced autoregulatory effectiveness in adenosine 1 receptor-deficient mice. Am J Physiol Renal Physiol 290:F888–F891

Heidemann HT, Gerkens JF, Jackson EK, Branch RA (1983) Effect of aminophylline on renal vasoconstriction produced by amphotericin B in the rat. Naunyn–Schmiedeberg's Arch Pharmacol 324:148–152

Heidemann HT, Muller S, Mertins L, Stepan G, Hoffmann K, Ohnhaus EE (1989) Effect of aminophylline on cisplatin nephrotoxicity in the rat. Br J Pharmacol 97:313–318

Helmlinger J (1979) Das experimentell erzeugte akute ischämische Nierenversagen bei der Ratte. Dissertation, University of Aachen, Germany

Hillege HL, Girbes AR, de Kam PJ, Boomsma F, de Zeeuw D, Charlesworth A, Hampton JR, van Veldhuisen DJ (2000) Renal function, neurohormonal activation, and survival in patients with chronic heart failure. Circulation 102:203–210

Hoenderop JG, Hartog A, Willems PH, Bindels RJ (1998) Adenosine-stimulated Ca^{2+} reabsorption is mediated by apical A_1 receptors in rabbit cortical collecting system. Am J Physiol 274:F736–F743

Hoenderop JGJ, De Pont JJHH, Bindels RJM, Willems PHGM (1999) Hormone-stimulated Ca^{2+} reabsorption in rabbit kidney cortical collecting system is cAMP-independent and involves a phorbol ester-insensitive PKC isotype. Kidney Int 55:225–233

Holz FG, Steinhausen M (1987) Renovascular effects of adenosine receptor agonists. Renal Physiol 10:272–282

Huang DY, Vallon V, Zimmermann H, Koszalka P, Schrader J, Osswald H (2006) Ecto-5′-nucleotidase (cd73)-dependent and -independent generation of adenosine participates in the mediation of tubuloglomerular feedback in vivo. Am J Physiol Renal Physiol 291:F282–F288

Huber W, Ilgmann K, Page M, Hennig M, Schweigart U, Jeschke B, Lutilsky L, Weiss W, Salmhofer H, Classen M (2002) Effect of theophylline on contrast material-nephropathy in patients with chronic renal insufficiency: controlled, randomized, double-blinded study. Radiology 223:772–779

Huber W, Schipek C, Ilgmann K, Page M, Hennig M, Wacker A, Schweigart U, Lutilsky L, Valina C, Seyfarth M, Schomig A, Classen M (2003) Effectiveness of theophylline prophylaxis of renal impairment after coronary angiography in patients with chronic renal insufficiency. Am J Cardiol 91:1157–1162

Iijima K, Moore LC, Goligorsky MS (1991) Syncytial organization of cultured rat mesangial cells. Am J Physiol 260:F848–F855

Inscho EW (1996) Purinoceptor-mediated regulation of the renal microvasculature. J Auton Pharmacol 16:385–388

Inscho EW, Carmines PK, Navar LG (1991) Juxtamedullary afferent arteriolar responses to P1 and P2 purinergic stimulation. Hypertension 17:1033–1037

Ishikawa I, Shikura N, Takada K (1993) Amelioration of glycerol-induced acute renal failure in rats by an adenosine A_1 receptor antagonist (FR-113453). Renal Fail 15:1–5

Itoh S, Carretero OA, Murray RD (1985) Possible role of adenosine in the macula densa mechanism of renin release in rabbits. J Clin Invest 76:1412–1417

Jackson EK, Raghvendra DK (2004) The extracellular cyclic AMP-adenosine pathway in renal physiology. Annu Rev Physiol 66:571–599

Jackson EK, Kost CK Jr, Herzer WA, Smits GJ, Tofovic SP (2001) A(1) receptor blockade induces natriuresis with a favorable renal hemodynamic profile in SHHF/Mcc-fa(cp) rats chronically treated with salt and furosemide. J Pharmacol Exp Ther 299:978–987

Jackson EK, Mi Z, Zhu C, Dubey RK (2003) Adenosine biosynthesis in the collecting duct. J Pharmacol Exp Ther 307:888–896

Joo JD, Kim M, Horst P, Kim J, D'Agati VD, Emala CW Sr, Lee HT (2007) Acute and delayed renal protection against renal ischemia and reperfusion injury with A_1 adenosine receptors. Am J Physiol Renal Physiol 293:F1847–F1857

Kang HS, Kerstan D, Dai LJ, Ritchie G, Quamme GA (2001) Adenosine modulates Mg(2+) uptake in distal convoluted tubule cells via A(1) and A(2) purinoceptors. Am J Physiol Renal Physiol 281:F1141–F1147

Kapoor A, Kumar S, Gulati S, Gambhir S, Sethi RS, Sinha N (2002) The role of theophylline in contrast-induced nephropathy: a case–control study. Nephrol Dial Transplant 17:1936–1941

Katholi RE, Taylor GJ, McCann WP, Woods WT Jr, Womack KA, McCoy CD, Katholi CR, Moses HW, Mishkel GJ, Lucore CL, et al. (1995) Nephrotoxicity from contrast media: attenuation with theophylline. Radiology 195:17–22

Kawabata M, Ogawa T, Takabatake T (1998) Control of rat glomerular microcirculation by juxtaglomerular adenosine A_1 receptors. Kidney Int 67(Suppl):S228–S230

Kellett R, Bowmer CJ, Collis MG, Yates MS (1989) Amelioration of glycerol-induced acute renal failure in the rat with 8-cyclopentyl-1,3-dipropylxanthine. Br J Pharmacol 98:1066–1074

Kim SM, Mizel D, Huang YG, Briggs JP, Schnermann J (2006) Adenosine as a mediator of macula densa-dependent inhibition of renin secretion. Am J Physiol Renal Physiol 290:F1016–F1023

Knight RJ, Collis MG, Yates MS, Bowmer CJ (1991) Amelioration of cisplatin-induced acute renal failure with 8-cyclopentyl-1,3-dipropylxanthine. Br J Pharmacol 104:1062–1068

Knight RJ, Bowmer CJ, Yates MS (1993a) Effect of the selective A_1 adenosine antagonist 8-cyclopentyl-1,3-dipropylxanthine on acute renal dysfunction induced by *Escherichia coli* endotoxin in rats. J Pharm Pharmacol 45:979–984

Knight RJ, Bowmer CJ, Yates MS (1993b) The diuretic action of 8-cyclopentyl-1,3-dipropylxanthine, a selective A_1 adenosine receptor antagonist. Br J Pharmacol 109(1): 271–277

Kolonko A, Wiecek A, Kokot F (1998) The nonselective adenosine antagonist theophylline does prevent renal dysfunction induced by radiographic contrast agents. J Nephrol 11:151–156

Komlosi P, Peti-Peterdi J, Fuson AL, Fintha A, Rosivall L, Bell PD (2004) Macula densa basolateral ATP release is regulated by luminal [NaCl] and dietary salt intake. Am J Physiol Renal Physiol 286:F1054–F1058

Kost CK Jr, Herzer WA, Rominski BR, Mi Z, Jackson EK (2000) Diuretic response to adenosine A(1) receptor blockade in normotensive and spontaneously hypertensive rats: role of pertussis toxin-sensitive G-proteins. J Pharmacol Exp Ther 292:752–760

Lai EY, Martinka P, Fahling M, Mrowka R, Steege A, Gericke A, Sendeski M, Persson PB, Persson AE, Patzak A (2006) Adenosine restores angiotensin II-induced contractions by receptor-independent enhancement of calcium sensitivity in renal arterioles. Circ Res 99:1117–1124

Lee HT, Emala CW (2000) Protective effects of renal ischemic preconditioning and adenosine pretreatment: role of A(1) and A(3) receptors. Am J Physiol 278:F380–F387

Lee HT, Emala CW (2001a) Protein kinase C and G(i/o) proteins are involved in adenosine- and ischemic preconditioning-mediated renal protection. J Am Soc Nephrol 12:233–240

Lee HT, Emala CW (2001b) Systemic adenosine given after ischemia protects renal function via A(2a) adenosine receptor activation. Am J Kidney Dis 38:610–618

Lee HT, Ota-Setlik A, Xu H, D'Agati VD, Jacobson MA, Emala CW (2003) A_3 adenosine receptor knockout mice are protected against ischemia- and myoglobinuria-induced renal failure. Am J Physiol Renal Physiol 284:F267–F273

Lee HT, Gallos G, Nasr SH, Emala CW (2004a) A_1 adenosine receptor activation inhibits inflammation, necrosis, and apoptosis after renal ischemia-reperfusion injury in mice. J Am Soc Nephrol 15:102–111

Lee HT, Xu H, Nasr SH, Schnermann J, Emala CW (2004b) A_1 adenosine receptor knockout mice exhibit increased renal injury following ischemia and reperfusion. Am J Physiol Renal Physiol 286:F298–F306

Lee HT, Jan M, Bae SC, Joo JD, Goubaeva FR, Yang J, Kim M (2006) A_1 adenosine receptor knock-out mice are protected against acute radiocontrast nephropathy in vivo. Am J Physiol Renal Physiol 290:F1367–F1375

Lee HT, Kim M, Jan M, Penn RB, Emala CW (2007) Renal tubule necrosis and apoptosis modulation by A_1 adenosine receptor expression. Kidney Int 71:1249–1261

Lin JJ, Churchill PC, Bidani AK (1986) Effect of theophylline on the initiation phase of postischemic acute renal failure in rats. J Lab Clin Med 108:150–154

Lin JJ, Churchill PC, Bidani AK (1987) The effect of dipyridamole on the initiation phase of postischemic acute renal failure in rats. Can J Physiol Pharmacol 65:1491–1495

Lin JJ, Churchill PC, Bidani AK (1988) Theophylline in rats during maintenance phase of postischemic acute renal failure. Kidney Int 33:24–28

Lorenz JN, Weihprecht H, He XR, Skott O, Briggs JP, Schnermann J (1993) Effects of adenosine and angiotensin on macula densa-stimulated renin secretion. Am J Physiol 265:F187–F194

Lorenz JN, Dostanic-Larson I, Shull GE, Lingrel JB (2006) Ouabain inhibits tubuloglomerular feedback in mutant mice with ouabain-sensitive alpha1 Na,K-ATPase. J Am Soc Nephrol 17:2457–2463

Lucas DG Jr, Hendrick JW, Sample JA, Mukherjee R, Escobar GP, Smits GJ, Crawford FA Jr, Spinale FG (2002) Cardiorenal effects of adenosine subtype 1 (A1) receptor inhibition in an experimental model of heart failure. J Am Coll Surg 194:603–609

Macias-Nunez JF, Fiksen-Olsen MJ, Romero JC, Knox FG (1983) Intrarenal blood flow distribution during adenosine-mediated vasoconstriction. Am J Physiol 244:H138–H141

Macias-Nunez JF, Garcia Iglesias C, Santos JC, Sanz E, Lopez-Novoa JM (1985) Influence of plasma renin content, intrarenal angiotensin II, captopril, and calcium channel blockers on the vasoconstriction and renin release promoted by adenosine in the kidney. J Lab Clin Med 106:562–567

Mahon NG, Blackstone EH, Francis GS, Starling RC, III, Young JB, Lauer MS (2002) The prognostic value of estimated creatinine clearance alongside functional capacity in ambulatory patients with chronic congestive heart failure. J Am Coll Cardiol 40:1106–1113

Miracle CM, Rieg T, Blantz RC, Vallon V, Thomson SC (2007) Combined effects of carbonic anhydrase inhibitor and adenosine A_1 receptor antagonist on hemodynamic and tubular function in the kidney. Kidney Blood Press Res 30:388–399

Miyamoto M, Yagil Y, Larson T, Robertson C, Jamison RL (1988) Effects of intrarenal adenosine on renal function and medullary blood flow in the rat. Am J Physiol 255:F1230–F1234

Mizumoto H, Karasawa A (1993) Renal tubular site of action of KW-3902, a novel adenosine A-1-receptor antagonist, in anesthetized rats. Jpn J Pharmacol 31(3):251–253

Modlinger PS, Welch WJ (2003) Adenosine A_1 receptor antagonists and the kidney. Curr Opin Nephrol Hypertens 12:497–502

Moyer BD, McCoy DE, Lee B, Kizer N, Stanton BA (1995) Adenosine inhibits arginine vasopressin-stimulated chloride secretion in a mouse IMCD cell line (mIMCD-K2). Am J Physiol 269:F884–F891

Nagashima K, Kusaka H, Karasawa A (1995) Protective effects of KW-3902, an adenosine A_1-receptor antagonist, against cisplatin-induced acute renal failure in rats. Jpn J Pharmacol 67:349–357

Newman WH, Grossman SJ, Frankis MB, Webb JG (1984) Increased myocardial adenosine release in heart failure. J Mol Cell Cardiol 16:577–580

Nishiyama A, Miyatake A, Aki Y, Fukui T, Rahman M, Kimura S, Abe Y (1999) Adenosine A(1) receptor antagonist KW-3902 prevents hypoxia-induced renal vasoconstriction. J Pharmacol Exp Ther 291:988–993

Nishiyama A, Inscho EW, Navar LG (2001) Interactions of adenosine A_1 and A_{2a} receptors on renal microvascular reactivity. Am J Physiol Renal Physiol 280:F406–F414

Okusa MD, Linden J, Macdonald T, Huang L (1999) Selective A_{2A} adenosine receptor activation reduces ischemia-reperfusion injury in rat kidney. Am J Physiol 277:F404–F412

Olivera A, Lamas S, Rodriguez-Puyol D, Lopez-Novoa JM (1989) Adenosine induces mesangial cell contraction by an A_1-type receptor. Kidney Int 35:1300–1305

Oppermann M, Friedman DJ, Faulhaber-Walter R, Mizel D, Castrop H, Enjyoji K, Robson SC, Schnermann JB (2008) Tubuloglomerular feedback and renin secretion in NTPDase1/CD39-deficient mice. Am J Physiol Renal Physiol 294:F965–F970

Osswald H, Vallon V (2009) Adenosine and tubuloglomerular feedback in the pathophysiology of acute renal failure. In: Ronco C, Bellomo R, Kellum JA (eds) Critical Care Nephrology, 2nd edn, Saunders Elsevier, Philadelphia, pp 128–134

Osswald H, Schmitz HJ, Heidenreich O (1975) Adenosine response of the rat kidney after saline loading, sodium restriction and hemorrhagia. Pflügers Arch 357:323–333

Osswald H, Schmitz HJ, Kemper R (1978a) Renal action of adenosine: effect on renin secretion in the rat. Naunyn–Schmiedeberg's Arch Pharmacol 303:95–99

Osswald H, Spielman WS, Knox FG (1978b) Mechanism of adenosine-mediated decreases in glomerular filtration rate in dogs. Circ Res 43:465–469

Osswald H, Helmlinger J, Jendralski A, Abrar B (1979) Improvement of renal function by theophylline in acute renal failure of the rat. Naunyn–Schmiedebergs Arch Pharmacol 307(Suppl):R 47

Osswald H, Nabakowski G, Hermes H (1980) Adenosine as a possible mediator of metabolic control of glomerular filtration rate. Int J Biochem 12:263–267

Osswald H, Hermes H, Nabakowski G (1982) Role of adenosine in signal transmission of tubuloglomerular feedback. Kidney Int 22(Suppl 12):S136–S142

Panjehshahin MR, Munsey TS, Collis MG, Bowmer CJ, Yates MS (1992) Further characterization of the protective effect of 8-cyclopentyl-1,3-dipropylxanthine on glycerol-induced acute renal failure in the rat. J Pharm Pharmacol 44:109–113

Patzak A, Lai EY, Fahling M, Sendeski M, Martinka P, Persson PB, Persson AE (2007) Adenosine enhances long term the contractile response to angiotensin II in afferent arterioles. Am J Physiol Regul Integr Comp Physiol 293:R2232–R2242

Pflueger AC, Gross JM, Knox FG (1999a) Adenosine-induced renal vasoconstriction in diabetes mellitus rats: role of prostaglandins. Am J Physiol 277:R1410–R1417

Pflueger AC, Osswald H, Knox FG (1999b) Adenosine-induced renal vasoconstriction in diabetes mellitus rats: role of nitric oxide. Am J Physiol 276:F340–F346

Ren Y, Arima S, Carretero OA, Ito S (2002a) Possible role of adenosine in macula densa control of glomerular hemodynamics. Kidney Int 61:169–176

Ren Y, Carretero OA, Garvin JL (2002b) Role of mesangial cells and gap junctions in tubuloglomerular feedback. Kidney Int 62:525–531

Ren Y, Garvin JL, Liu R, Carretero OA (2004) Role of macula densa adenosine triphosphate (ATP) in tubuloglomerular feedback. Kidney Int 66:1479–1485

Rieg T, Steigele H, Schnermann J, Richter K, Osswald H, Vallon V (2005) Requirement of intact adenosine A_1 receptors for the diuretic and natriuretic action of the methylxanthines theophylline and caffeine. J Pharmacol Exp Ther 313:403–409

Rieg T, Schnermann J, Vallon V (2007) Adenosine A_1 receptors determine effects of caffeine on total fluid intake but not caffeine appetite. Eur J Pharmacol 555:174–177

Rieg T, Pothula K, Schroth J, Satriano J, Osswald H, Schnermann J, Insel PA, Bundey RA, Vallon V (2008) Vasopressin regulation of inner medullary collecting ducts and compensatory changes in mice lacking adenosine A_1 receptors. Am J Physiol Renal Physiol 294:F638–F644

Schnermann J, Osswald H, Hermle M (1977) Inhibitory effect of methylxanthines on feedback control of glomerular filtration rate in the rat kidney. Pflügers Arch 369:39–48

Schnermann J, Weihprecht H, Briggs JP (1990) Inhibition of tubuloglomerular feedback during adenosine1 receptor blockade. Am J Physiol 258:F553–F561

Schweda F, Wagner C, Kramer BK, Schnermann J, Kurtz A (2003) Preserved macula densa-dependent renin secretion in A_1 adenosine receptor knockout mice. Am J Physiol Renal Physiol 284:F770–F777

Smith JA, Whitaker EM, Bowmer CJ, Yates MS (2000) Differential expression of renal adenosine A(1) receptors induced by acute renal failure. Biochem Pharmacol 59:727–732

Smith JA, Sivaprasadarao A, Munsey TS, Bowmer CJ, Yates MS (2001) Immunolocalisation of adenosine A(1) receptors in the rat kidney. Biochem Pharmacol 61:237–244

Spielman WS, Osswald H (1978) Characterization of the postocclusive response of renal blood flow in the cat. Am J Physiol 235:F286–F290

Spielman WS, Osswald H (1979) Blockade of postocclusive renal vasoconstriction by an angiotensin II antagonist: evidence for an angiotensin–adenosine interaction. Am J Physiol 237:F463–F467

Sun D, Samuelson LC, Yang T, Huang Y, Paliege A, Saunders T, Briggs J, Schnermann J (2001) Mediation of tubuloglomerular feedback by adenosine: evidence from mice lacking adenosine 1 receptors. Proc Natl Acad Sci USA 98:9983–9988

Suzuki F, Shimada J, Mizumoto H, Karasawa A, Kubo K, Nonaka H, Ishii A, Kawakita T (1992) Adenosine A_1 antagonists. 2. Structure–activity relationships on diuretic activities and protective effects against acute renal failure. J Med Chem 35:3066–3075

Tagawa H, Vander AJ (1970) Effects of adenosine compounds on renal function and renin secretion in dogs. Circ Res 26:327–338

Takeda M, Yoshitomi K, Imai M (1993) Regulation of Na(+)–3HCO_3– cotransport in rabbit proximal convoluted tubule via adenosine A_1 receptor. Am J Physiol 265:F511–F519

Takenaka T, Inoue T, Kanno Y, Okada H, Hill CE, Suzuki H (2008a) Connexins 37 and 40 transduce purinergic signals mediating renal autoregulation. Am J Physiol Regul Integr Comp Physiol 294:R1–R11

Takenaka T, Inoue T, Kanno Y, Okada H, Meaney KR, Hill CE, Suzuki H (2008b) Expression and role of connexins in the rat renal vasculature. Kidney Int 73:415–422

Tang L, Parker M, Fei Q, Loutzenhiser R (1999) Afferent arteriolar adenosine A_{2a} receptors are coupled to KATP in in vitro perfused hydronephrotic rat kidney. Am J Physiol 277:F926–F933

Tang Y, Zhou L (2003) Characterization of adenosine A_1 receptors in human proximal tubule epithelial (HK-2) cells. Receptor Channel 9:67–75

Thomson S, Bao D, Deng A, Vallon V (2000) Adenosine formed by 5′-nucleotidase mediates tubuloglomerular feedback. J Clin Invest 106:289–298

Torikai S (1987) Effect of phenylisopropyladenosine on vasopressin-dependent cyclic AMP generation in defined nephron segments from rat. Renal Physiol 10:33–39

Toya Y, Umemura S, Iwamoto T, Hirawa N, Kihara M, Takagi N, Ishii M (1993) Identification and characterization of adenosine A_1 receptor–cAMP system in human glomeruli. Kidney Int 43:928–932

Traynor T, Yang T, Huang YG, Arend L, Oliverio MI, Coffman T, Briggs JP, Schnermann J (1998) Inhibition of adenosine-1 receptor-mediated preglomerular vasoconstriction in AT1A receptor-deficient mice. Am J Physiol 275:F922–F927

Vallon V (2008) P2 receptors in the regulation of renal transport mechanisms. Am J Physiol Renal Physiol 294:F10–F27

Vallon V, Richter K, Huang DY, Rieg T, Schnermann J (2004) Functional consequences at the single-nephron level of the lack of adenosine A_1 receptors and tubuloglomerular feedback in mice. Pflugers Arch 448:214–221

Vallon V, Muhlbauer B, Osswald H (2006) Adenosine and kidney function. Physiol Rev 86: 901–940

Vallon V, Miracle C, Thomson S (2008) Adenosine and kidney function: potential implications in patients with heart failure. Eur J Heart Fail 10:176–187

van-Buren M, Bijlsma JA, Boer P, van-Rijn HJ, Koomans HA (1993) Natriuretic and hypotensive effect of adenosine-1 blockade in essential hypertension. Hypertension 22:728–734

Wagner C, de Wit C, Kurtz L, Grunberger C, Kurtz A, Schweda F (2007) Connexin40 is essential for the pressure control of renin synthesis and secretion. Circ Res 100:556–563

Weaver DR, Reppert SM (1992) Adenosine receptor gene expression in rat kidney. Am J Physiol 263:F991–F995

Weihprecht H, Lorenz JN, Schnermann J, Skott O, Briggs JP (1990) Effect of adenosine1-receptor blockade on renin release from rabbit isolated perfused juxtaglomerular apparatus. J Clin Invest 85:1622–1628

Weihprecht H, Lorenz JN, Briggs JP, Schnermann J (1992) Vasomotor effects of purinergic agonists in isolated rabbit afferent arterioles. Am J Physiol 263:F1026–F1033

Weihprecht H, Lorenz JN, Briggs JP, Schnermann J (1994) Synergistic effects of angiotensin and adenosine in the renal microvasculature. Am J Physiol 266:F227–F239

Welch WJ (2002) Adenosine A_1 receptor antagonists in the kidney: effects in fluid-retaining disorders. Curr Opin Pharmacol 2:165–170

Wilcox CS, Welch WJ, Schreiner GF, Belardinelli L (1999) Natriuretic and diuretic actions of a highly selective adenosine A_1 receptor antagonist. J Am Soc Nephrol 10:714–720

Yagil Y (1990) Interaction of adenosine with vasopressin in the inner medullary collecting duct. Am J Physiol 259:F679–F687

Yagil C, Katni G, Yagil Y (1994) The effects of adenosine on transepithelial resistance and sodium uptake in the inner medullary collecting duct. Pflügers Arch 427:225–232

Yao K, Kusaka H, Sato K, Karasawa A (1994) Protective effects of KW-3902, a novel adenosine A_1-receptor antagonist, against gentamicin-induced acute renal failure in rats. Jpn J Pharmacol 65:167–170

Yao K, Heyne N, Erley CM, Risler T, Osswald H (2001) The selective adenosine A_1 receptor antagonist KW-3902 prevents radiocontrast media-induced nephropathy in rats with chronic nitric oxide deficiency. Eur J Pharmacol 414:99–104

Yaoita H, Ito O, Arima S, Endo Y, Takeuchi K, Omata K, Ito S (1999) Effect of adenosine on isolated afferent arterioles. Nippon Jinzo Gakkai Shi 41:697–703

Yates MS, Bowmer CJ, Kellett R, Collis MG (1987) Effect of 8-phenyltheophylline, enprofylline and hydrochlorothiazide on glycerol-induced acute renal failure in the rat. J Pharm Pharmacol 39:803–808

Zou AP, Nithipatikom K, Li PL, Cowley AWJr (1999) Role of renal medullary adenosine in the control of blood flow and sodium excretion. Am J Physiol 45:R790–R798

Adenosine Receptors and the Central Nervous System

Ana M. Sebastião and Joaquim A. Ribeiro

Contents

Abstract The adenosine receptors (ARs) in the nervous system act as a kind of "go-between" to regulate the release of neurotransmitters (this includes all known neurotransmitters) and the action of neuromodulators (e.g., neuropeptides, neurotrophic factors). Receptor–receptor interactions and AR–transporter interplay

J.A. Ribeiro (✉)
Institute of Pharmacology and Neurosciences, Faculty of Medicine and Unit of Neurosciences, Institute of Molecular Medicine, University of Lisbon, 1649-028 Lisboa, Portugal
jaribeiro@fm.ul.pt

C.N. Wilson and S.J. Mustafa (eds.), *Adenosine Receptors in Health and Disease*, Handbook of Experimental Pharmacology 193, DOI 10.1007/978-3-540-89615-9_16,

occur as part of the adenosine's attempt to control synaptic transmission. A_{2A}ARs are more abundant in the striatum and A_1ARs in the hippocampus, but both receptors interfere with the efficiency and plasticity-regulated synaptic transmission in most brain areas. The omnipresence of adenosine and A_{2A} and A_1 ARs in all nervous system cells (neurons and glia), together with the intensive release of adenosine following insults, makes adenosine a kind of "maestro" of the tripartite synapse in the homeostatic coordination of the brain function. Under physiological conditions, both A_{2A} and A_1 ARs play an important role in sleep and arousal, cognition, memory and learning, whereas under pathological conditions (e.g., Parkinson's disease, Alzheimer's disease, amyotrophic lateral sclerosis, stroke, epilepsy, drug addiction, pain, schizophrenia, depression), ARs operate a time/circumstance window where in some circumstances A_1AR agonists may predominate as early neuroprotectors, and in other circumstances A_{2A}AR antagonists may alter the outcomes of some of the pathological deficiencies. In some circumstances, and depending on the therapeutic window, the use of A_{2A}AR agonists may be initially beneficial; however, at later time points, the use of A_{2A}AR antagonists proved beneficial in several pathologies. Since selective ligands for A_1 and A_{2A} ARs are now entering clinical trials, the time has come to determine the role of these receptors in neurological and psychiatric diseases and identify therapies that will alter the outcomes of these diseases, therefore providing a hopeful future for the patients who suffer from these diseases.

Keywords Adenosine receptors · A1 adenosine receptor · A2A adenosine receptor · Central nervous system · Receptor cross-talk · G protein coupled receptors · Neurotrophic factor receptors · Ionotropic receptors · Receptor dimmers · Caffeine · Drug addiction · Neurodegenerative diseases · Pain · Ischemia · Hypoxia · Adenosine levels

Abbreviations

AC	Adenylate cyclase
ACh	Acetylcholine
ADO	Adenosine
ADP	Adenosine 5′-diphosphate
AK	Adenosine kinase
AMP	Adenosine 5′-monophosphate
AMPA	α-Amino-3-hydroxy-5-methyl-4-isoxazolepropionic acid
AOPCP	α, β-Methylene ADP
Ap5A	Diadenosine pentaphosphate
AR	Adenosine receptor
ATP	Adenosine 5′-triphosphate
BDNF	Brain-derived neurotrophic factor
BRET	Bioluminescence resonance energy transfer
CamK	Calmodulin-dependent kinase

cAMP	Cyclic adenosine 5′-monophosphate
CB	Cannabinoid
CGRP	Calcitonin gene-related peptide
DA	Dopamine
DARPP	Dopamine- and cAMP-regulated phosphoprotein
DPCPX	1,3-Dipropyl-8-cyclopentylxanthine
ENT	Equilibrative nucleoside transporter
ERK	Extracellular signal-regulated kinase
GABA	γ-Aminobutyric acid
GAT	GABA transporter
GLU	Glutamate
GDNF	Glial cell line-derived neurotrophic factor
GPCRs	G-protein-coupled receptors
HEK cells	Human embryonic kidney cells
HFS	High-frequency stimulation
IL-6	Interleukin 6
IP3	Inositol triphosphate
i.v.	Intravenous
KO	Knockout
LFS	Low-frequency stimulation
LTD	Long-term depression
LTP	Long-term potentiation
MAPK	Mitogen-activated protein kinase
mGluR	Metabotropic glutamate receptor (mGlu1–8 refer to mGluR subtypes)
NAc	Nucleus accumbens
nAChR	Nicotinic acetylcholine receptor
NBTI	Nitrobenzylthioinosine
NGF	Nerve growth factor
NMDA	*N*-Methyl-D-aspartate
NT	Neurotransmitter
NTR	Neurotransmitter receptor
NPY	Neuropeptide Y
NR	NMDA receptor subunit
NT-3	Neurotrophin 3
NTDase	Ecto-5′-nucleotidase
NTPDase	Ectonucleoside triphosphate diphosphohydrolase
PDE	Phosphodiesterase
PKA	Protein kinase A
PKC	Protein kinase C
PLC	Phospholipase C
PTX	Pertussis toxin
REM	Rapid eye movement
Trk receptors	Tropomyosin-related kinase receptors
VIP	Vasoactive intestinal peptide

1 Introduction

Before we go into the scope of this review, we would like to stress that we are starting with some areas of high general complexity: (a) the *nervous system*, the most complex biological system in the human body; (b) *adenosine*, which is ubiquitously present in all cells, with receptors distributed throughout all brain areas; any imbalance of such a widespread system is expected to lead to neurological diseases; (c) *caffeine*, an antagonist of all subtypes of ARs and the most widely consumed psychostimulant drug; moreover, chronic or acute intake of caffeine may affect ARs in different and even opposite ways; (d) finally, ARs or adenosine-related molecules are potential therapeutic targets for neurologic diseases, but this role can be *multifactorial*, with different receptors involved, different time windows of action, age-related changes, etc.

Many publications are now appearing that are devoted to research in animal models or humans that is directed towards many nervous system pathologies and towards novel therapeutic approaches based on adenosine and ARs. We have therefore chosen to focus the present review on the insights gained from recent studies related to the subtle way that ARs regulate other receptors and transporters for neurotransmitters and neuromodulators, and on the pathophysiological implications of this regulation. We believe that further advances in the therapeutic potential of adenosine-related drugs require a deeper understanding of the above mentioned complexities in the context of fine-tuning modulation by adenosine. This might be accomplished at the receptor–receptor level or through several receptors in sequence and/or in parallel and/or via the transducing system cascade. What occurs inside the cells with the transducing system's variability and crosstalk may also occur at the extracellular membrane level via receptor–receptor interaction and formation of heteromers.

The AR field in neuroscience started with an apparent paradox reported by Sattin and Rall (1970): the ability of adenosine to increase cyclic adenosine 5′-monophosphate (cAMP) in the brain was prevented by theophylline, which at the time was known only as a phosphodiesterase (PDE) inhibitor. This was the starting point for the hypothesis that adenosine was acting through a membrane receptor antagonized by theophylline, but several years elapsed before the birth of the first nomenclature of purinergic receptors, proposed by Burnstock (1976). The field for the identification of different subtypes of ARs was then opened, and a further breakthrough was attained by the end of the 1970s by van Calker et al. (1979), who first proposed ARs in brain cells as A_1 (inhibitory) and A_2 (stimulatory). At that time, as for most receptors, the AR classification relied upon pharmacological criteria and transducing pathways and the ability to stimulate or inhibit adenylate cyclase. AR cloning possibilities had to wait until the beginning of the 1990s. The first AR to be cloned was the A_1AR from brain tissue (Mahan et al. 1991). All four G-protein-coupled ARs (A_1, A_{2A} and A_{2B}, and A_3) have been cloned. Major advances have been made in the pharmacological tools available for all of them, as reviewed in great detail in the first five chapters of this book.

Animal research in the last two to three decades firmly established that ARs are involved in several pathophysiological conditions, and that manipulation of their degree of activation might prove therapeutically useful. Therefore, the time is now ripe for studies in humans. In fact, the number of adenosine-related research reports in humans is increasing. From the summary in Table 1, it is clear that the highest incidence of adenosine-related research in humans is related to sleep and Parkinson's disease. This is certainly due to the great advances made in basic research in these fields, which have allowed the clear identification of the role of A_{2A}ARs in Parkinson's disease as well as that of adenosine in sleep and epilepsy. The identification of caffeine and theophylline as AR antagonists, together with the empirical knowledge at the time that xanthine-rich beverages such as coffee and tea affect sleep also boosted the interest in adenosine-related research into human sleep. Objective-oriented adenosine-related research in epileptic humans is still scarce, but one retrospective (Miura and Kimura 2000) and one case report study (Bahls et al. 1991) clearly identified an increased risk of seizures in patients taking theophylline as a bronchodilator. Interestingly, and highlighting the frequent gap between basic and clinical research, neither of those two reports mentioned the putative scientific grounds for the increased risk of seizures induced by theophylline: its ability to antagonize ARs. At the time, adenosine had already been recognized as an anticonvulsant, with the pioneering report being published as early as 1984 (Barraco et al. 1984) and the first review highlighting the subject appearing in the 1980s (Chin 1989).

In this review, we will pay particular attention to the implications of AR function in neuropathophysiological conditions, but before we do this, we will briefly provide an overview of the state of the art on how adenosine acts as a neuromodulator, the distribution of ARs in the brain, and their ability to interact with other receptors to harmoniously fine-tune neuronal activity.

2 Adenosine as a Ubiquitous Neuromodulator

While ATP may function as a neurotransmitter in some brain areas (Burnstock 2007; Edwards et al. 1992), adenosine is neither stored nor released as a classical neurotransmitter since it does not accumulate in synaptic vesicles, and is released from the cytoplasm into the extracellular space through a nucleoside transporter. The adenosine transporters also mediate adenosine reuptake, the direction of the transport being dependent upon the concentration gradient on both sides of the membrane (Gu et al. 1995). Since it is not exocytotically released, adenosine behaves as an extracellular signaling molecule that influences synaptic transmission without itself being a neurotransmitter. Using G-protein-coupled mechanisms, that not only lead to changes in second messenger levels but also to the modulation of ion channels (such as calcium and potassium channels), adenosine modulates neuronal activity—presynaptically by inhibiting or facilitating transmitter release, postsynaptically by affecting the actions of other neurotransmitters, and nonsynaptically by

Table 1 AR research relating to the human central nervous system

Target	Comment/reference
Cognition	Caffeine facilitates information processing and motor output in healthy subjects. Dixit et al. (2006)
	Caffeine appears to reduce cognitive decline in women without dementia. Ritchie et al. (2007)
Sleep	Decrease in sleep efficiency and of total sleep in healthy subjects by preingestion (16 h) of caffeine. Landolt et al. (1995a, b)
	Attenuation of sleep propensity in healthy subjects. Landolt et al. (2004)
	Insomnia patients with greater sensitivity to awakening caffeine actions. Salín-Pascual et al. (2006)
	Involvement of adenosine in individual variations in sleep deprivation sensitivity. Rétey et al. (2006)
	Variations in A_{2A} receptor gene associated with objective and subjective responses to caffeine in relation to sleep. Rétey et al. (2007)
	Prolonged wakefulness induces A_1 receptor upregulation in cortical and subcortical brain regions. Elmenhorst et al. (2007)
Epilepsy	An increased risk of seizures after theophylline or caffeine intake. Bahls et al. (1991), Kaufman and Sachdeo (2003), Miura and Kimura (2000), Mortelmans et al. (2008)
	Increase (Angelatou et al. 1993) or *decrease* (Glass et al. 1996) in A_1 receptor density in different post-mortem brain areas of epileptic subjects
Parkinson's disease	Decrease in A_{2A} mRNA levels in post-mortem caudate and putamen, and increase in A_{2A} mRNA in the substantia nigra in Parkinson's disease patients. Hurley et al. (2000)
	Polymorphism of A_{2A} receptors did not confer susceptibility to Parkinson's disease in a Chinese population sample. Hong et al. (2005)
	Caffeine improved the "total akinesia" type of gait freezing in Parkinson's disease patients. Kitagawa et al. (2007)
	Caffeine administered before levodopa may improve its pharmacokinetics in some parkinsonian patients. Deleu et al. (2006)
	Significant association between higher caffeine intake and lower incidence of Parkinson's disease. Ross et al. (2000)
Alzheimer's disease	Caffeine intake inversely associated with risk of Alzheimer's disease. Maia and de Mendonça (2002), Ritchie et al. (2007)
	Co-localization of A_1 receptor and β-amyloid or tau and increase in A_{2A} receptor expression in post-mortem cerebral cortex and hippocampus of Alzheimer's disease patients. Angulo et al. (2003)
	Upregulation of A_1 and A_{2A} receptor expression and function in post-mortem cerebral cortex of Alzheimer's disease patients. Albasanz et al. (2008)
Pick's disease	Upregulation of A_1 and A_{2A} receptor expression and function in post-mortem frontal lobe of Pick's disease patients. Albasanz et al. (2006, 2007).
Pain	Beneficial effects of adenosine (i.v.) in 2 patients with neuropathic pain. Sollevi et al. (1995)
	Intrathecal adenosine reduces allodynia in patients with neuropathic pain, but has a side effect of backache. Eisenach et al. (2003)

(continued)

Table 1 (continued)

Target	Comment/reference
	Reduction of secondary hyperalgesia by adenosine in human models of cutaneous inflammatory pain. Sjölund et al. (1999)
	Local opioid receptor stimulation in the spinal cord of humans induces the release of adenosine. Eisenach et al. (2004)
	Theophylline improves esophageal chest pain (a randomized, placebo-controlled study), possibly by altering adenosine-mediated nociception. Rao et al. (2007)
Anxiety	A_{2A} receptor gene polymorphism associated with increases in anxiety in healthy volunteers. Alsene et al. (2003)
	A_{2A} receptor gene polymorphism associated with increases in anxiety response to amphetamine in healthy volunteers. Hohoff et al. (2005)
Panic disorder	Panic disorder patients with increased sensitivity to one cup of coffee. Boulenger et al. (1984)
	Anxiogenic and panic-inducing effects of caffeine in a double-blind study. Klein et al. (1991)
	Adenosine A_1 receptor supersensitivity, a probable compensatory process. DeMet et al. (1989)
Panic disorder and anxiety	A_{2A} receptor gene polymorphism associated with anxiety or panic disorder in Occidental but not Asiatic populations. Alsene et al. (2003), Deckert et al. (1998), Hamilton et al. (2004), Lam et al. (2005)
Schizophrenia	Increase in A_{2A} receptor density in post-mortem striatum of schizophrenic patients, a consequence of typical antipsychotic medication. Deckert et al. (2003)
	Polymorphism of A_{2A} receptors is not related to the pathogenesis of schizophrenia in a Chinese population sample. Hong et al. (2005)
Bipolar disorders	Typical, but not atypical, antipsychotics induce an upregulation of A_{2A} receptors assessed in platelets of patients with bipolar disorders. Martini et al. (2006)
Dependence behavior	Polymorphism of the A_{2A} receptor gene related to caffeine consumption in healthy volunteers. Cornelis et al. (2007)
Ventilation dyspnea and apnea	Adenosine produces hyperventilation and dyspnea in humans resulting from a direct activation of the carotid body. Uematsu et al. (2000), Watt et al. (1987)
	Caffeine and theophylline are effective in the treatment of apnea in premature and newborn infants. Aranda and Turmen (1979), Bairam et al. (1987), Uauy et al. (1975)
Miscellaneous (A_2/D_2 interaction)	A_2/D_2 interaction in post-mortem human striatal brain sections. Díaz-Cabiale et al. (2001)
	A_2/D_2 dimers in human platelets. Martini et al. (2006)
Miscellaneous (receptor localization)	Distribution of A_1 and A_{2A} receptors in post-mortem human brain. James et al. (1992), Svenningsson et al. (1997)
	Mapping A_{2A} receptors in the human brain by PET. Ishiwata et al. (2005), Mishina et al. (2007)
	Mapping A_1 receptors in the human brain by PET. Elmenhorst et al. (2007), Fukumitsu et al. (2003, 2005)

hyperpolarizing or depolarizing neurones. Curiously, the discovery of the presynaptic inhibitory action of adenosine, which we now know occurs through A_1AR, had the same paradoxical starting point as the identification of theophylline as a putative AR antagonist; that is, adenosine was used at the neuromuscular junction in an attempt to increase cAMP in motor nerve terminals and, in contrast with the expected excitatory effect, adenosine markedly inhibited neurotransmitter release (Ginsborg and Hirst 1971). ATP, released together with acetylcholine (ACh) (Silinsky 1975), mimicked the adenosine effect (Ribeiro and Walker 1973), decreasing both the synchronous and the asynchronous release of acetylcholine, with the maximum effect being about 50%. This presynaptic inhibitory action of ATP results from its extracellular hydrolysis into adenosine (Ribeiro and Sebastião 1987). Interestingly, high-affinity ARs positively coupled to adenylate cyclase do enhance neurotransmitter release, and this was also first identified in cholinergic synapses almost simultaneously in the central nervous system (Brown et al. 1990) and at the cholinergic nerve terminals of the neuromuscular junction (Correia-de-Sá et al. 1991). In the latter case, due to the reduced complexity of the model, it was possible to clearly demonstrate for the first time that both A_1 and A_{2A}ARs are present at the same nerve terminal (Correia-de-Sá et al. 1991). Adenosine research at cholinergic motor nerve endings to some degree anticipated and inspired the studies at central excitatory glutamatergic synapses, where adenosine decreases both synchronous and asynchronous transmitter release (Lupica et al. 1992; Prince and Stevens 1992).

The past few years have brought new insights into our understanding of the role of the tripartite synapse and gliotransmission in neurological diseases (Halassa et al. 2007). Adenosine and ATP are also among the most relevant players in neuron–glia communication (Fields and Burnstock 2006). ATP has a dual role since it acts upon its own receptors, mostly of the P2Y subtype, which are abundant in astrocytes and are relevant to calcium signaling; ATP is also a substrate of ectonucleotidases leading to adenosine formation, which then operates its own receptors. The adenosine system is critically involved in modulating glial cell functions, namely glycogen metabolism, glutamate transporters, astrogliosis and astrocyte swelling (Daré et al. 2007). ARs on oligodendrocytes regulate white matter development and myelinization (Daré et al. 2007; Fields and Burnstock 2006).

3 Manipulation of Endogenous Levels of Adenosine and its Neuromodulation

By using highly selective A_1AR antagonists such as 1,3-dipropyl-8-cyclopentylxanthine (DPCPX), it is possible to unmask a tonic A_1AR-mediated adenosinergic tonus. A similar strategy can be used towards endogenous A_{2A}AR activation, using selective A_{2A}AR antagonists such as SCH-58621 or ZM-241385. Simultaneous removal of tonic adenosinergic influences over all receptor types has been achieved through the use of adenosine deaminase, an enzyme used in isolated preparations that does not penetrate cell membranes and that deaminates adenosine into inosine, usually an inactive ligand for ARs. In studies with this enzyme, an appropriate control with an adenosine deaminase inhibitor together with adenosine deaminase (Sebastião and Ribeiro 1985) will allow actions of the enzyme not related to its enzymatic activity to be determined. The ability of inosine to influence the biological signal in the assay has to be directly tested since inosine can activate ARs, namely those of the A_3AR subtype, in some circumstances (Jin et al. 1997). Whenever aiming to evaluate endogenous adenosine actions in a mirror-like way (i.e., by using selective AR antagonists or extracellular adenosine removal), one must also bear in mind that irreversible AR-mediated actions will not be apparent. In these cases, further receptor activation by exogenously added ligands may be the only way to demonstrate a role for adenosine, providing that the system is not already fully saturated with endogenous adenosine actions.

Inhibition of the cascade of ectoenzymes that metabolize ATP to adenosine can be used whenever attempting to discriminate between ATP- and AR-mediated actions. Due to the multiplicity and redundancy of the ectoenzymes involved in the process (Zimmermann 2006), it has been difficult or almost impossible to fully inhibit extracellular ATP breakdown into adenosine. Ectonucleoside triphosphate diphosphohydrolase (NTPDase; EC 3.6.1.5), previously identified as ecto-ATPase, ecto-ATPDase or CD39, is the ectonucleotidase mainly responsible for the sequential hydrolysis of β- and γ-phosphates of tri- and diphosphonucleosides. Its inhibition by ARL-67156 has proven useful to enhance extracellular ATP actions and/or to avoid its breakdown into adenosine (Sperlágh et al. 2007). For a discussion on the therapeutic potential of NTPDase inhibition, see Gendron et al. (2002). Inhibition of adenosine formation from released adenine nucleotides can also be partially achieved through the use of an ecto-5′-nucleotidase inhibitor, α, β-methylene ADP (AOPCP). By comparing the action of adenosine deaminase with that of AOPCP, it was possible to estimate that the formation of adenosine from adenine nucleotides and the release of adenosine as such contribute in nearly equal amounts to the pool of endogenous adenosine that presynaptically inhibits acetylcholine release from motor nerve terminals at the neuromuscular junction (Ribeiro and Sebastião 1987), as well as in the hippocampus (Cunha et al. 1994c). However, this relationship may differ in other brain areas and according to neuronal activity and the regional distribution of ecto-5′-nucleotidase. For example, in cholinergic nerve terminals of the hippocampus, there is significant activity from ecto-5′-nucleotidase, whereas this is not the case in cerebral cortical cholinergic nerve terminals (Cunha et al. 1992). High-frequency neuronal firing favors ATP release (Cunha et al. 1996a) as well as activation of adenosine A_{2A}AR, and this has been shown to occur both in a peripheral nervous system preparation, the neuromuscular junction (Correia-de-Sá et al. 1996), and in a central nervous system preparation, the hippocampus (Almeida et al. 2003). At the neuromuscular junction, however, ecto-AMP deaminase shunts the pathway for adenosine formation, thus reducing its ability to activate A_{2A}AR (Magalhães-Cardoso et al. 2003).

An enhancement of extracellular adenosine levels can be achieved by inhibiting intracellular enzymes that are responsible for keeping intracellular adenosine concentrations low, such as the adenosine kinase (AK) that phosphorylates adenosine into AMP. Inhibition of this enzyme selectively amplifies extracellular adenosine concentrations at cell and tissue sites where adenosine release occurs. AK can be inhibited with iodotubercidin, which markedly enhances extracellular adenosine levels and causes an inhibition of synaptic transmission (Diógenes et al. 2004) at sites where A_1ARs are operative (e.g., hippocampus). The therapeutic antiepileptic potential of AK inhibition or of its underexpression in implanted cells has been highlighted recently (Li et al. 2007).

Manipulation of adenosine transporter activity with inhibitors such as dipyridamole or nitrobenzylthioinosine (NBTI) has also proven to be a useful approach, but one has to keep in mind that, due to the equilibrative nature of adenosine transporters in the brain cells, adenosine transport inhibition can either enhance or reduce extracellular adenosine levels according to the gradient of concentration of adenosine across the cell membrane, as well as according to the proportion of extracellular adenosine that is formed from the catabolism of released adenine nucleotides.

A still less explored way to increase extracellular adenosine levels is deep brain stimulation, and a very interesting report on this subject appeared recently (Bekar et al. 2008). Deep brain stimulation is used empirically to treat tremor and other movement disorders (Yu and Neimat 2008) as well as psychiatric diseases, including obsessive–compulsive disorders and depression (Larson 2008). As Bekar et al. (2008) clearly showed, deep brain stimulation is associated with a marked increase in the release of ATP from thalamic nuclei, resulting in accumulation of its catabolic product, adenosine. ATP, which is released in a nonexocytotic way, probably from astrocytes or other glial cells, is therefore crucial in adenosine accumulation following deep brain stimulation, leading to A_1AR-mediated inhibition of synaptic transmission in the thalamus (Bekar et al. 2008), in a way that is probably similar to the inhibition of synaptic transmission induced by ATP in the hippocampus, which requires localized extracellular catabolism by ectonucleotidases and channeling to A_1ARs (Cunha et al. 1998). Infusion of A_1AR agonists directly into the thalamus reduces tremor, whereas A_1AR-null mice show involuntary movements and seizures at stimulation intensities below the therapeutic level, suggesting that depression of synaptic transmission in the thalamus controls the spread of excitability and reduces side effects of deep brain stimulation (Bekar et al. 2008). Depression of synaptic transmission due to deep brain stimulation mimics in several aspects the depression of synaptic transmission caused by hypoxia, which is neuroprotective and can also be reversed by A_1AR antagonists (Sebastião et al. 2001). Hypoxia is, indeed, another highly efficient way to increase extracellular adenosine levels (Fowler 1993; Frenguelli et al. 2003), but in this case adenosine is mostly released as such (Latini and Pedata 2001), while deep brain stimulation appears to predominantly induce ATP release (Bekar et al. 2008).

A schematic representation of the different pathways involved in the control of extracellular adenosine concentrations, as well as the relevance of neuronal firing frequency to A_1AR vs. A_2AR activation by extracellularly formed adenosine, is depicted in Fig. 1.

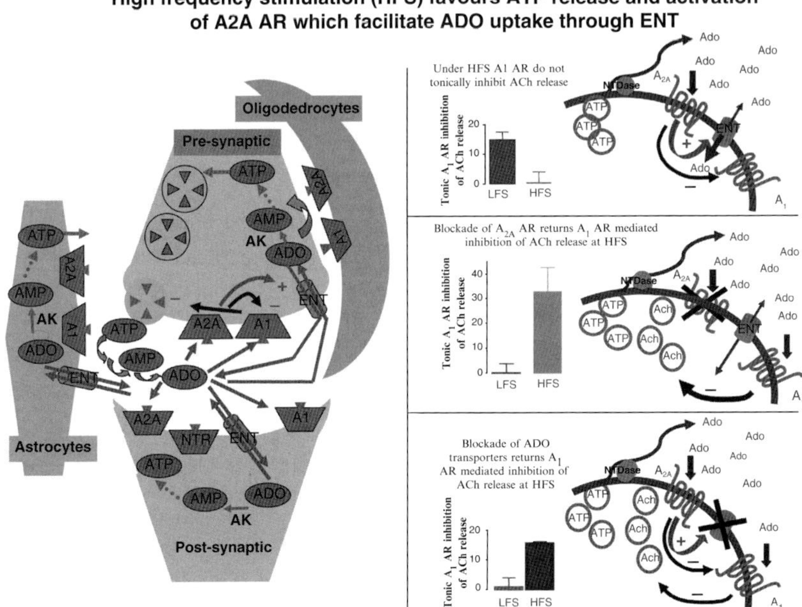

Fig. 1 Main pathways that control extracellular adenosine (*ADO*) concentrations, and their relationships to the activation of A_1 or A_{2A} adenosine receptors (*ARs*) under low or high neuronal firing rates induced experimentally by low- or high-frequency stimulation (*LFS* or *HFS*). Adenosine can be formed from extracellular catabolism of ATP by a cascade of ectoenzymes, the ecto-5′-nucleotidases (*NTDase*), or can be released as such through an equilibrative nucleoside transporter (*ENT*). Intracellularly the key enzyme influencing ADO concentration is adenosine kinase (*AK*), which is present in most cell types, including neurons and glia. The intracellular pathways for ADO metabolism into ATP are not depicted in oligodendrocytes for the sake of clarity. A_1 and A_{2A} receptors are present pre- and postsynaptically as well as in astrocytes and glia. At nerve terminals, A_1ARs decrease neurotransmitter (*NT*) release, thus reducing the availability to activate postsynaptic NT receptors (*NTR*). A_{2A} receptors have been shown to inhibit (*minus symbol*) A_1 receptor functiong in nerve terminals. A_1/A_{2A} receptor interactions might also occur in other cell types, namely in astrocytes (*see text*), but they are not represented for the sake of clarity. A_{2A}ARs are preferentially activated at high-frequency neuronal firing, which favors ATP release and adenosine formed from extracellular ATP catabolism. A_{2A}ARs enhance (*plus symbol*) adenosine transport through ENT, which in the case of HFS is in the inward direction, decreasing the availability of ADO for A_1ARs, the main consequence of which is a lower tonic inhibition of neurotransmitter release. Data shown in the *left panels* are adapted from data published by Pinto-Duarte et al. (2005), who reported the influence of the firing rate upon the tonic inhibition of acetylcholine (*ACh*) release from the CA3 area of hippocampal slices, and how it is related to the ability of A_{2A} receptors to enhance ENT activity at hippocampal nerve terminals. See text for further references

4 Distribution of ARs in the Central Nervous System and the Effect of Aging

Neuromodulation by adenosine is exerted through the activation of high-affinity A_1 and A_{2A} ARs, which are probably of physiological importance, and of low-affinity A_{2B}ARs, which may be relevant in pathological conditions. The A_3AR is a high-affinity receptor in humans, but has a low density in most tissues. These four ARs are also known as P1 purinoceptors, from the P1 (adenosine-sensitive)/P2 (ATP-sensitive) nomenclature (Burnstock 1976). They belong to the G-protein-coupled receptor (GPCR) family and all have been cloned and characterized from several mammalian species including humans (Fredholm et al. 2001).

The adenosine A_1AR is highly expressed in brain cortex, cerebellum, hippocampus, and dorsal horn of spinal cord (Ribeiro et al. 2003). The A_{2A}AR is highly expressed in the striatopallidal γ-aminobutyric acid (GABA)ergic neurones and olfactory bulb, and for a long time it was assumed that this receptor was circumscribed to these brain areas. The first evidence that the A_{2A}AR could influence neuronal communication outside the striatum or olfactory bulb was reported in 1992 using hippocampal slices (Sebastião and Ribeiro 1992). This was followed by evidence that A_{2A}AR mRNA and protein are expressed in the hippocampus (Cunha et al. 1994a). The initial scepticism was broken (Sebastião and Ribeiro 1996), and it is now widely recognized that A_{2A}ARs are expressed in several brain regions albeit in lower levels than in the striatum. A_{2B}ARs are expressed in low levels in the brain (Dixon et al. 1996), and the level of expression for the A_3AR is apparently moderate in the human cerebellum and hippocampus and low in most other areas of the brain (Fredholm et al. 2001) (Fig. 2).

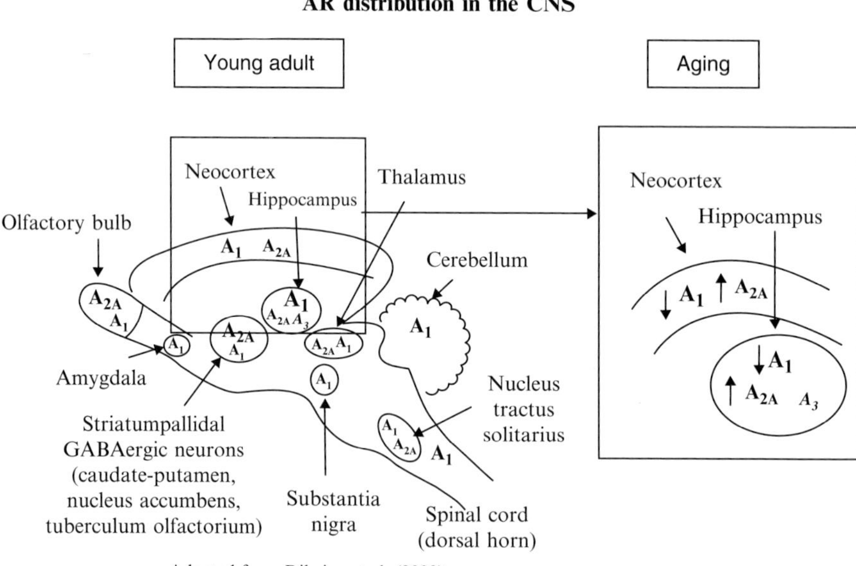

Fig. 2 Schematic representation of the distribution of adenosine receptors (*ARs*) in the different brain areas. The *inset* illustrates the reported changes in AR density in the forebrain (hippocampus and cortex). In aged rats, the density and functioning of A_{2A}ARs is increased (*upward arrow*) in the hippocampus and cortex, whereas the density and functioning of A_1ARs is decreased (*downward arrow*). No information, so far, is available for age-related changes in A_3AR density upon aging. See text for references

The relative densities of A_1 and A_{2A} ARs in subregions of the same brain area may differ. For instance, with respect to the modulation of acetylcholine in the hippocampus, there is a preponderance of A_1AR-mediated modulation by endogenous adenosine in both the CA1 and CA3 areas, but the CA3 has a relatively higher influence of A_{2A}ARs than the CA1 (Cunha et al. 1994b). Whenever two receptors coexist, one may ask about their relative importance (i.e., the hierarchy of one receptor with respect, to the other). This may change with neuronal activity, age, and even with other molecules that are in the vicinity of the site of action and that may be relevant for the production or inactivation of the ligand. High-frequency neuronal firing favours ATP release (Cunha et al. 1996a), and adenosine formed from released adenine nucleotides seems to prefer A_{2A}AR activation (Cunha et al. 1996b), which may be due to the geographical distribution of ecto-5-nucleotidadases and A_{2A}ARs. A_{2A}AR activation activates adenosine transport, which in the case of high neuronal activity and ATP release is in the inward direction (Fig. 1). This induces a decrease in extracellular adenosine levels and a reduced ability of A_1AR s to be activated by endogenous extracellular adenosine (Pinto-Duarte et al. 2005). By themselves, A_{2A}ARs are able to attenuate A_1AR activation (Cunha et al. 1994a), which may further contribute to a decreased activity of A_1AR s under high-frequency neuronal firing. The ability of A_1ARs to inhibit synaptic transmission is attenuated by protein kinase C (PKC) activation (Sebastião and Ribeiro 1990), and a similar mechanism appears to be involved in the A_{2A}AR-mediated attenuation of A_1AR responses (Lopes et al. 1999a).

Aging also decreases the ability of A_1ARs to inhibit neuronal activity (Sebastão et al. 2000a). This may be a function of an age-related decrease in the density of A_1ARs in the brain, which has been shown in both mice (Pagonopoulou and Angelatou 1992) and humans (Meyer et al. 2007). Low A_1AR receptor density and function, however, can be compensated for by higher levels of extracellular adenosine, which keep tonic inhibition high in aged animals (Bauman et al. 1992). While comparing changes in A_1AR density in the cerebral cortex, hippocampus and striatum, it was concluded that the most affected area was the cerebral cortex, followed by the hippocampus (Fig. 2), whereas the density of A_1ARs in the striatum was little affected by aging in rats (Cunha et al. 1995). A_1AR density in the cerebellum is also poorly affected by aging (Pagonopoulou and Angelatou 1992).

In contrast to A_1ARs, there is a significant increase in the density of A_{2A}ARs in the cortex (Cunha et al. 1995) and hippocampus (Diogenes et al. 2007) of aged rats, which correlates with their enhanced ability to facilitate glutamatergic synaptic transmission (Rebola et al. 2003) and acetylcholine release (Lopes et al. 1999b) in the hippocampus (Fig. 2). In the striatum there is a tendency for a decrease in A_{2A}AR density in aged rats (Cunha et al. 1995), and within the striatum, age may influence the A_{2A}ARs in glutamatergic, dopaminergic or GABAergic nerve terminals in different ways (Corsi et al. 1999, 2000). Taken together, these findings clearly show that there are age-related shifts in the A_1AR inhibitory/A_{2A}AR excitatory balance, and that this shift may be different in different areas of the brain, with the trend for the forebrain being towards an increase in A_{2A}AR-mediated influences and a decrease in A_1AR density. Due to the A_{2A}/A_1 AR interactions (see Sect. 5.1.5), an increase in $A_{2A}AR$ density may itself reduce A_1AR tonus. Due to the influence of $A_{2A}ARs$ on other receptors (see Sect. 5), the change in the $A_{2A}AR$ influence upon aging may markedly affect the action of other modulators. Indeed, the nonmonotonous age-related changes in the ability of brain-derived neurotrophic factor (BDNF) to influence synaptic transmission in the hippocampus are related to both a decrease in the density of tropomyosin-related kinase receptors (Trk) for BDNF (TrkB receptors) and an increase in the density of $A_{2A}ARs$, which allow TrkB receptor-mediated actions in the aged hippocampus (Diogenes et al. 2007).

5 Adenosine as a Modulator of Other Neuromodulators

Besides its direct pre- and postsynaptic actions on neurones, adenosine is rich in nuances of priming, triggering and braking the action of several neurotransmitters and neuromodulators. Because adenosine acts in such a subtle fashion, it was proposed as a fine tuner. In this way, adenosine is a partner in a very sophisticated interplay between its own receptors and receptors for other neurotransmitters and/or neuromodulators. Several possibilities exist for this interplay, either at the transducing system level (Sebastião and Ribeiro 2000) or as a consequence of receptor–receptor heteromerization (Ferré et al. 2007a), greatly expanding the number of possible receptor combinations to modulate cell signalling.

5.1 Interactions with G-Protein-Coupled Receptors

Besides the well known A_{2A}/D_2 dopamine interaction in the stritaum (Ferré et al. 1991), which has been explored intensively due to the implication of this receptor interaction for Parkinson's disease and other basal ganglia dysfunctions (Fuxe et al. 2007; Morelli et al. 2007), adenosine, mostly through activation of $A_{2A}ARs$, is also able to influence the functioning of other GPCRs (Fig. 3a). A brief overview of the influence of adenosine on these receptors will follow.

5.1.1 Dopamine Receptors

A first hint at the ability of $A_{2A}ARs$ to interact with dopamine D_2 receptors came from binding studies showing that activation of $A_{2A}ARs$ decreases the affinity of dopamine D_2 receptors in rat striatal membranes (Ferré et al. 1991). The possibility that this $A_{2A} - D_2$ receptor interaction is crucial to the behavioral effects of adenosine agonists and antagonists (like caffeine) was immediately highlighted (Ferré et al. 1991) and soon tested (Svenningsson et al. 1995). The functional consequences of $A_{2A}AR$ and D_2 receptor agonists upon dopamine and GABA release (Ferré et al. 1994; Mayfield et al. 1996) in the basal ganglia became evident soon thereafter.

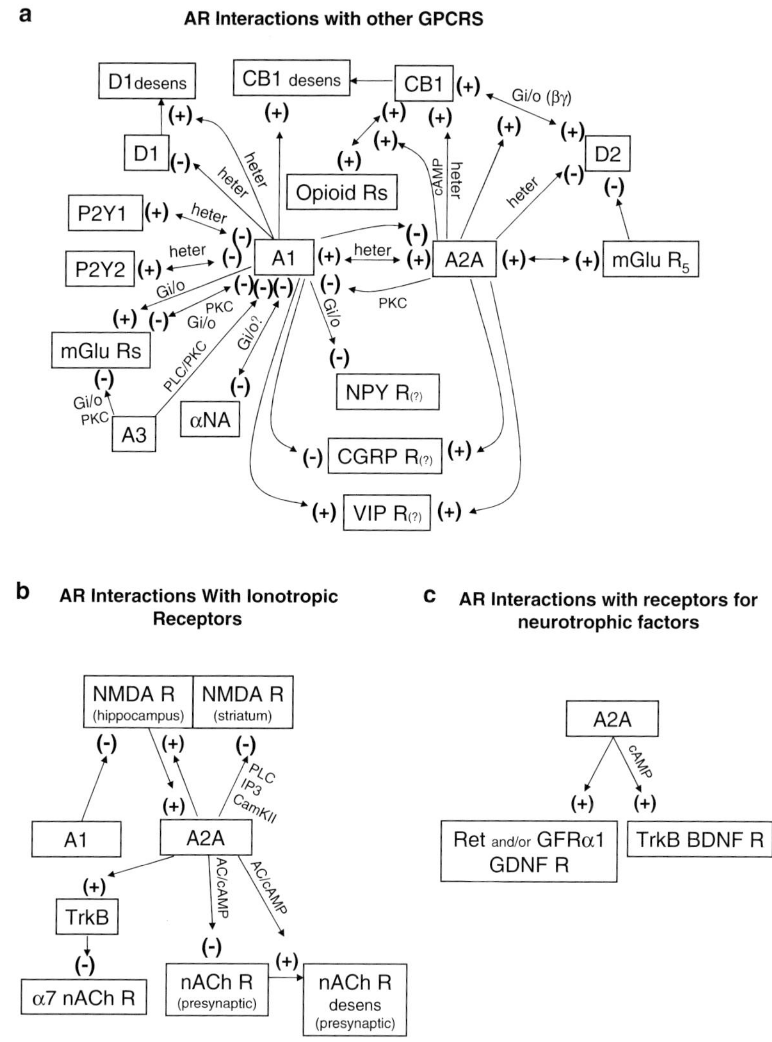

Fig. 3 a–c Interactions between adenosine receptors and receptors for other neurotransmitters. The known interactions with other G-protein-coupled receptors (GPCRs) **a**, with ionotropic receptors **b**, and with receptors for neurotrophic factors **c** are illustrated, where a *plus symbol* represents a facilitation or triggering of the action, or synergy between receptors, or the facilitation of desensitization (*desens*), and a *minus symbol* represents an inhibition, or an occlusion of the action, or less than addictive effects. Whenever the mechanisms involved in the interaction have been evaluated, they are indicated close to the *arrow*. G-protein sharing is indicated by the name of the G protein close to the *arrow*. Whenever receptor heteromerization (*heter*) has been shown to occur, it is also indicated close to the *arrow*. An absence of knowledge about the receptor subtype is indicated by a *question mark* close to the receptor name. See text for references. Other abbreviations: αNA, α receptor for noradrenaline; *BDNF*, brain-derived neurotrophic factor; *CB*, cannabinoid; CB_1: cannabinoid receptor type 1; *CGRP*, calcitonin gene-related peptide; *D*, dopamine receptor; $GFR\alpha 1$ *and Ret*: neurotrophic factors for GDNF; *GDNF*, glial cell line-derived neurotrophic factor; *mGluRs*, metabotropic glutamate receptor; *nAChR*: nicotinic acetylcholine receptor; *NMDAR*: *N*-methyl-D-aspartate receptor; *NPY*, neuropeptide Y; *VIP*, vasoactive intestinal peptide; *P2Y*, ATP receptor; *TrkB*, tropomyosin-related kinase receptor type B

Since this time, interest in the adenosine/D_2 interaction has continued to increase, extending to psychiatric and neurologic fields such as drug addiction, schizophrenia and Parkinson's disease, and has been the subject of many reviews by groups that have been involved in this subject since its origin (Ferré et al. 2007b). For more information on A_{2A}ARs and Parkinson's disease, please refer to Chap. 18, "Adenosine A_{2A} Receptors and Parkinson's Disease" (by Morelli et al.), in this volume.

A_1ARs and D_1 receptors also interact in the basal ganglia (Ferré et al. 1996), an interaction that has implications for the control of GABA release at the substantia nigra (Florán et al. 2002) and nucleus accumbens (Mayfield et al. 1999), as well as dopamine release in the striatum (O'Neill et al. 2007). Furthermore, A_1AR activation has been shown to facilitate D_1 receptor desensitization (Le Crom et al. 2002). D_1/A_1 receptor heteromerization may play a role in D_1 receptor desensitization mechanisms and be a molecular basis for the antagonistic modulation of A_1AR over D_1 receptor signaling (Ginés et al. 2000).

5.1.2 Neuropeptides

By activating A_{2A}ARs, adenosine tonically potentiates a facilitatory action of the neuropeptide calcitonin gene-related peptide (CGRP) on neurotransmitter release from motor nerve terminals (Correia-de-Sá and Ribeiro 1994a). The ability of CGRP to facilitate synaptic transmission in the CA1 area of the hippocampus is also under tight control by adenosine, with tonic A_1AR activation by endogenous adenosine "braking" the action of CGRP, and the A_{2A}ARs triggering this action (Sebastião et al. 2000b). This A_1AR-mediated inhibition of the action of CGRP, together with the A_1AR-induced inhibition of CGRP release (Carruthers et al. 2001), can be related to pain inhibition by adenosine (see Sect. 7). Indeed, CGRP is a potent vasodilator released from activated trigeminal sensory nerves that dilates intracranial blood vessels and transmits vascular nociception, and is implicated in the genesis of vascular pain such as migraine. Hence, inhibition of trigeminal CGRP release and CGRP receptor blockade have been proposed as promising antimigraine strategies (Goadsby 2008).

The facilitatory action of vasoactive intestinal peptide (VIP) on ACh release from motor nerve endings is prevented by A_{2A}AR blockade or by the removal of extracellular adenosine with adenosine deaminase, indicating that the activation of these A_{2A}ARs, attained with high-frequency motor neuron firing, is necessary to trigger the facilitatory action of VIP (Correia-de-Sá et al. 2001). VIP enhances synaptic transmission at the CA1 area of the hippocampus by enhancing GABA release from GABAergic neurones that make synapses with other interneurones, therefore reducing GABAergic inhibition into pyramidal glutamatergic neurones (Cunha-Reis et al. 2004, 2005). This action of VIP is dependent on both A_1 and A_{2A} AR activation by endogenous adenosine (Cunha-Reis et al. 2007, 2008). Interestingly, the finding that VIP-induced modulation of GABA release from hippocampal nerve terminals is under the control of adenosine A_1ARs constituted the first evidence of a role of A_1 receptors in hippocampal GABAergic terminals. This is an example of a situation where A_1ARs per se may not affect neurotransmitter release, just like GABA in the hippocampus (Lambert and Teyler 1991; Yoon and Rothman 1991), but instead influence the actions of other modulators of GABA release.

Neuropeptide Y (NPY) agonists inhibit presynaptic calcium influx through N- and P/Q-type calcium channels and inhibit glutamate release at the CA3–CA1 synapses of rat hippocampus, an action that is fully occluded by coactivation of adenosine A_1ARs (Qian et al. 1997). Interestingly, the inhibitory action of the $GABA_B$ agonist baclofen was not fully occluded by AR activation, indicating partially shared pathways between G-protein-coupled NPY, adenosine and GABA receptors. In PC12 cells, exocytosis of NPY-containing vesicles is facilitated by A_{2A}AR activation (Mori et al. 2004), but this does not occur in nerve endings from the rat mesenteric artery, where ARs affect noradrenaline but not NPY release (Donoso et al. 2006).

In cultured primary hippocampal neurones, agonists of delta-opioid receptors and of cannabinoid (CB) receptors of the CB_1 subtype act synergistically to activate protein kinase A (PKA) signaling through Gi-β/γ dimers, and this synergy requires A_{2A}AR activation (Yao et al. 2003). CB1 agonists also act synergistically with μ opiod receptors in primary nucleus accumbens/striatal neurones, and again this synergy requires adenosine A_{2A}ARs (Yao et al. 2006). Interestingly, A_{2A}AR blockade eliminates heroin-seeking behavior in addicted rats (Yao et al. 2006), suggesting that A_{2A}AR antagonists may be effective therapeutic agents in the management of abstinent heroin addicts (see Sect. 9).

5.1.3 Metabotropic Glutamate Receptors

Activation of metabotropic glutamate receptors (mGluR) with $1S,3R$-ACPD potentiates cAMP responses mediated by several receptors that are positively coupled to adenylate cyclase, namely A_2ARs and VIP and β-adrenergic receptors (Alexander et al. 1992; Winder and Conn 1993). mGluRs also influence A_1AR functioning in neurones, and this seems to involve PKC activity. In fact, PKC activity is required for the attenuation of the inhibitory effect of A_1AR activation on synaptic transmission at the hippocampus by agonists of group I mGluRs (mGlu1, mGlu5) which are coupled to phospholipase C, as well as by agonists of group III mGluRs (mGlu4, mGlu6, mGlu7, mGlu8), which are usually negatively coupled to cAMP (de Mendonça and Ribeiro 1997a). Agonists of group I mGluRs also attenuate $GABA_B$-mediated inhibition of synaptic transmission, a process that involves PKC activity (Shahraki and Stone 2003). In addition, activation of PKC by phorbol esters or activation of PKC-coupled mGluRs suppresses the inhibitory action of A_1AR agonists on glutamate release from cerebrocortical synaptosomes (Budd and Nichols 1995).

The inhibitory effects of an A_1AR agonist and of an agonist of group II mGluRs (mGlu2, mGlu 3) on glutamate release or cAMP formation was less than additive (Di Iorio et al. 1996), suggesting that the presynaptic A_1 and group II mGluRs are reciprocally occlusive, probably by sharing a pertussis toxin (PTX)-sensitive, PKC-regulated G protein (Zhang and Schmidt 1999).

Activation of A_3AR leads, through a PKC-dependent process, to a marked attenuation of the presynaptic inhibitory functions of cAMP-coupled mGluRs (groups II and III) at the CA1 area of the hippocampus (Macek et al. 1998). Again, the action of PKC and probably also that of A_3ARs on mGluRs might result from an inhibition of the coupling of mGluRs with G proteins, because PKC activation inhibits the increased [^{35}S]GTPγS binding induced by mGluR agonists (Macek et al. 1998). Thus, the actions of A_1 or A_3 ARs and those of mGluRs in neurones are mutually occlusive, through a process probably involving the crosstalk of transducing systems or the sharing of G proteins, as proposed several years ago to explain the mutual occlusion between presynaptic adenosine A_1 and α_2-adrenergic receptors (Limberger et al. 1988).

In contrast, in astrocytes, activation of A_1AR enhances the intracellular calcium response induced by mGluRs (Ogata et al. 1994), a process that involves a PTX-sensitive G protein (Cormier et al. 2001; Tom and Roberts 1999). Adenosine-induced calcium response in astrocytes requires A_1/A_2 AR cooperation, and is synergistic with mGluR response, leading to enhancement of cAMP levels (Ogata et al. 1996).

With respect to the interaction between $A_{2A}AR$ and mGluR, $A_{2A}AR$ agonists act synergistically with group I mGluR agonists to modulate dopamine D_2 receptors in the rat striatum, decreasing the affinity state of these receptors (Ferré et al. 1999). Furthermore, $A_{2A}ARs$ act synergistically with mGlu5 receptors to increase dopamine- and cAMP-regulated phosphoprotein of 32 kDa (DARPP-32) phosphorylation, so that blockade of one of the receptors is enough to prevent phosphorylation induced by activation of the other receptor (Nishi et al. 2003). $A_{2A}ARs$ and mGlu5 receptors are also co-localized presynaptically, namely at striatal glutamatergic terminals, where they facilitate glutamate release in a synergistic manner (Pintor et al. 2000; Rodrigues et al. 2005). Prevention of mGlu5 receptors and $A_{2A}AR$ synergy at the pre- and the postsynaptic level will, therefore, eventually lead to decreased glutamate release with consequent reduced excitotoxicity, together with a facilitation of D_2 dopaminergic receptor functioning, and this is the rational for the use of antagonists of these receptors as antiparkinsonian drugs (see also Chap. 18, "Adenosine A_{2A} Receptors and Parkinson's Disease," by Morelli et al., in this volume). Indeed simultaneous blockade of A_{2A} and mGlu5 receptors showed high efficacy in reversing parkinsonian deficits in rodents (Coccurello et al. 2004; Kachroo et al. 2005). Combined antagonism of mGlu5 receptors and $A_{2A}ARs$ also efficiently reduced alcohol self-administration and alcohol-seeking in rats (Adams et al. 2008), further reinforcing the importance of the mGlu5 and $A_{2A}AR$ interaction in the mesolimbic and basal ganglia areas.

5.1.4 Cannabinoid Receptors

The high density of adenosine $A_{2A}ARs$ in the basal ganglia, together with the profound motor-depressant effects of cannabinoids (CBs), prompted interest in investigating a putative crosstalk between $A_{2A}ARs$ and CB_1 receptors in this brain area. CB_1 receptor signaling in a human neuroblastoma cell line is dependent on $A_{2A}AR$ activation, and blockade of $A_{2A}ARs$ counteracts the motor-depressant effects produced by CB_1 receptor activation in vivo (Carriba et al. 2007). Interestingly, the motor-depressant effects produced by CB_1 receptor activation are attenuated by genetic inactivation of the DARPP-32 (Andersson et al. 2005), which is abundantly expressed in the medium spiny neurons of the striatum and is crucially involved in the striatal actions of cAMP-coupled receptors (Greengard 2001), as in the case of $A_{2A}ARs$. Molecular interactions between striatal $A_{2A}ARs$ and CB_1 receptors at the striatum may also exist since CB_1 and $A_{2A}ARs$ form heteromeric complexes once transfected to human embryonic kidney cells, HEK-293T (Carriba et al. 2007).

$A_{2A}AR$ activation is required for the synergistic actions between CB_1 receptors and μ opioid receptors in nucleus accumbens (NAc)/striatal neurons (Yao et al. 2006), as well as for the synergistic actions that occur between CB_1 agonists and D_2 agonists (Yao et al. 2003). Synergy between G_i-coupled receptors, such as CB_1 and D_2 receptors, with respect to the facilitation of cAMP-mediated signaling involves $\beta-\gamma$ dimers of G_i proteins, and these are also required for the interplay with A_2ARs (Yao et al. 2003). The implications of these multiple interactions as they pertain to therapies for drug addiction will be discussed below (see Sect. 9.4).

A_1ARs also appear to be involved in motor impairment by CBs, and this may occur at the cerebellum, since the incoordination induced by CB_1 agonists is attenuated by intracerebellar injection of an A_1AR-selective antagonist (DeSanty and Dar 2001). A reciprocal ability to heterologously desensitize CB_1 and A_1AR responses through prolonged agonist exposure has also been reported (Kouznetsova et al. 2002; Selley et al. 2004).

5.1.5 A_1, A_{2A} and A_3ARs

The existence of $A_1 - A_{2A}$ AR heteromers has been demonstrated and complicates the overall picture for adenosine as a neuromodulator and the role of ARs in neurotransmission. Co-immunoprecipitation and bioluminescence resonance energy transfer (BRET) techniques have shown the existence of $A_1 - A_{2A}$ AR heteromers in co-transfected HEK cells, as well as the existence of an intermolecular crosstalk, and radioligand-binding techniques have allowed the identification of an intramembrane receptor–receptor interaction in the $A_1 - A_{2A}$ receptor heteromer (Ciruela et al. 2006). According to Ferré et al. (2007a), the A_1–A_{2A} receptor heteromer may provide a "concentration-dependent switch" mechanism by which low and high concentrations of synaptic adenosine produce the opposite effects on glutamate release. Thus, a weak input might cause stimulation of the receptor with the highest affinity in the A_1/A_{2A} heteromer, while a strong input might cause additional stimulation of the other receptor, with crosstalk between both receptors that may allow a response that is different from the summation of both of them. However, as discussed in point 3 above, other factors such as the topographical arrangement of ectoenzymes, transporters and receptors as well as the neuronal firing frequency may also influence the A_1 versus $A_{2A}AR$-mediated actions at each synapse where both receptors co-localize.

With respect to crosstalk between A_1 and A_{2A} ARs, this was clearly documented with data obtained from experiments at the hippocampus, where activation of A_{2A}ARs attenuates the ability of A_1 AR agonists to inhibit excitability and synaptic transmission (Cunha et al. 1994a; O'Kane and Stone 1998). An A_{2A}AR-mediated decrease in A_1AR binding was also shown to occur in hippocampal (Lopes et al. 1999a) and striatal (Dixon et al. 1997) synaptosomes. A_{2A}AR-induced inhibition of A_1AR binding does not occur in membrane fragments, which indicates that the cross talk between A_1 and A_{2A} receptors involves a diffusible second messenger. The A_{2A}/A_1AR crosstalk might be related to PKC, rather than to the classical A_{2A}AR second messenger, the adenylate cyclase-cAMP–PKA pathway, because the interactions between A_{2A} and A_1ARs are prevented by PKC inhibitors but not by PKA inhibitors (Dixon et al. 1997; Lopes et al. 1999a). PKC activators, such as phorbol esters, mimic the ability of A_{2A} receptor agonists to decrease A_1AR binding (Lopes et al. 1999a). Thus, with respect to their ability to inhibit A_1AR-mediated responses, A_{2A}ARs appear to behave similarly to the phospholipase C-coupled metabotropic glutamate and muscarinic receptors (Worley et al. 1987); i.e., they operate through a phosphoinositide-PKC-dependent pathway. Activation of PKC inhibits presynaptic A_1ARs on motor nerve terminals without affecting the affinity of competitive receptor antagonists (Sebastião and Ribeiro 1990), suggesting that the target of PKC is not the receptor ligand-binding domain, but probably a locus related to G-protein coupling, the G protein itself, or both.

Besides the A_{2A}/A_1 AR interaction, which can be observed using either BRET, radioligand binding, or functional studies with selective agonists for both receptors, there are other ways through which A_{2A}AR activation can also induce a decrease in A_1AR tone. A_{2A}ARs enhance adenosine transport through equilibrative nucleoside transporter (ENT)s, with a consequent reduction in the availability of endogenous adenosine to tonically activate A_1ARs (Pinto-Duarte et al. 2005). As occurs with A_{2A}AR-mediated inhibition of A_1AR binding (Lopes et al. 1999a), the A_{2A}AR-induced enhancement of ENT activity is lost upon the inhibition of PKC but not PKA activity, suggesting the involvement of the phospholipase C (PLC) pathway rather the adenylate cyclase/cAMP one (Pinto-Duarte et al. 2005).

While evaluating the evoked release of acetylcholine at different frequencies of stimulation from hippocampal slices, it becomes clear that the A_{2A}AR-mediated enhancement of ENT activity plays a pivotal role in adjusting adenosine neuromodulation to different physiological needs (Pinto-Duarte et al. 2005). Thus, at high-frequency neuronal firing, there is a predominant release of ATP and a predominant formation of adenosine from released ATP (Cunha et al. 1996b). Therefore, the extracelluar adenosine concentration exceeds the intracellular one and the gradient of adenosine concentration across the plasma membrane will direct ENT to take up adenosine. Since A_{2A}ARs are concomitantly activated, the A_{2A}AR-induced enhancement of ENT activity leads to an enhancement of the removal of adenosine from the synaptic cleft, leading to a reduced tonic A_1 – AR-mediated inhibition of hippocampal acetylcholine release at high-frequency firing rates (Pinto-Duarte et al. 2005). This A_{2A}AR-mediated inhibition of tonic inhibitory adenosinergic tone may add to the A_{2A}AR inhibition of A_1AR activation (see above), thus efficiently reinforcing the enhanced firing rate of cholinergic afferents into the hippocampus, which are known to play a key role in the control of cognitive processes such as attention and memory (Hasselmo and Giocomo 2006).

Other interactions of A_1 and A_2 ARs include the influence of A_1ARs on A_{2A}AR activity, where desensitization of striatal A_1ARs is accompanied by a time-dependent amplification of $A_2 -$ AR-mediated stimulation of adenylate cyclase (Abbracchio et al. 1992). Moreover, presynaptic interactions between A_1 and A_{2A} ARs were clearly observed at motor nerve terminals where A_1AR inhibitory responses are enhanced in the presence of A_2AR antagonists, and A_{2A}AR excitatory responses are increased in the presence of A_1AR antagonists (Correia de Sá et al. 1996). However, in contrast to what occurs in neurones, positive cooperation between A_1 and A_2 ARs, which also requires concomitant activation of metabotropic glutamate receptors (groups I and II), was observed in cultured astrocytes (Ogata et al. 1996).

With respect to A_3ARs and the interaction of A_3ARs with other ARs, A_3AR activation attenuates the synaptic inhibitory actions of adenosine in the CA1 area of the hippocampus (Dunwiddie et al. 1997). Because adenosine A_3ARs might couple to phospholipase C, and phospholipase C-coupled receptors are able to inhibit $A_1 -$ AR-mediated responses (see above), it is possible that this $A_3 - A_1$ AR-mediated interaction involves this transduction pathway, in a similar manner to that described in relation to the $A_3 -$ AR-mediated inhibition of metabotropic receptor functioning (Macek et al. 1998).

5.1.6 P2 Purinoceptors

Although ATP and adenosine operate distinct families of receptors and although they may play very distinct roles in the CNS—adenosine being exclusively a neuromodulator and ATP behaving as a neurotransmitter, neuromodulator, or co-modulator—interactions between receptors for these two "family related" molecules have been reported. $P2Y_1$ receptors and A_1ARs can form heteromeric complexes and display a high degree of colocalization in the brain (Yoshioka et al. 2002). $P2Y_1$ receptors and A_1ARs are colocalized at glutamatergic synapses and surrounding astrocytes, and $P2Y_1$ receptor stimulation impairs the A_1AR coupling to the G protein probably by inducing heterologous desensitization (Tonazzini et al. 2008), whereas the stimulation of A_1ARs increases the functional responsiveness of $P2Y_1$ receptors (Tonazzini et al. 2007). Similar findings were found in relation to the crosstalk between A_1ARs and $P2Y_2$ receptors, where oligomerization of A_1ARs and $P2Y_2$ receptors generates a complex in which the simultaneous activation of the two receptors induces a structural alteration that interferes with signaling via $G_{i/o}$ but enhances signaling via $G_{q/11}$ (Suzuki et al. 2006).

The presynaptic facilitatory dinucleotide receptor is also under the control of ARs colocalized at the same nerve terminals. Thus, the apparent affinity of diadenosine pentaphosphate (Ap5A) for its receptor in hippocampal nerve terminals is increased up to the low nanomolar range by coactivation of A_1 or A_{2A} ARs, whereas it is decreased towards the high micromolar range when A_3ARs are coactivated (Díaz-Hernández et al. 2002). P2 purinoceptor activation by endogenous ATP may also inhibit dinucleotide receptor functioning (Díaz-Hernández et al. 2000).

5.2 *Interaction with Ionotropic Receptors*

ARs can interact with ionotropic receptors (Fig. 3b), with putative implications for neuroprotection, plasticity and learning, as it is the case for AMPA and NMDA glutamate receptors as well as nicotinic acetylcholine receptors (nAChRs). A brief overview of the published data follows.

5.2.1 Modulation of NMDA and AMPA Receptors by A_1 and A_2 ARs

In isolated rat hippocampal neurones (de Mendonça et al. 1995), as well as in bipolar retinal cells (Costenla et al. 1999), A_1AR activation inhibits *N*-methyl-D-aspartate (NMDA) receptor-mediated currents. Interestingly, the inhibitory postsynaptic action of A_1AR agonists is observed at very low concentrations, compatible with a tonic inhibitory action of adenosine. Accordingly, selective A_1AR antagonism enhances the NMDA component of excitatory postsynaptic currents in CA1 hippocampal neurones, probably due to the recruitment of previously silent NMDA receptors at synapses (Klishin et al. 1995). Through a postsynaptic action, endogenous adenosine also inhibits voltage- and NMDA receptor-sensitive dendritic spikes in the CA1 area of the hippocampus (Li and Henry 2000). Because of the important role played by NMDA receptors in synaptic plasticity phenomena, as well as in neuronal injury after prolonged stimulation or depolarizing conditions, it is conceivable that the ability of A_1ARs to inhibit NMDA receptor-mediated currents together with the well-known A_1AR-mediated inhibition of glutamate release are the basis for the A1–AR-mediated inhibition of synaptic plasticity phenomena such as long-term potentiation (LTP) and long-term depression (LTD) at CA3/CA1 excitatory synapses of the hippocampus (de Mendonça and Ribeiro 1997b). These two A_1AR-mediated actions also contribute to A_1–AR-mediated neuroprotective actions during hypoxia (Sebastião et al. 2001) and to stopping epileptiform firing in CA1 pyramidal cells (Li and Henry 2000).

On medium spiny neurones at the striatum, A_{2A}AR activation inhibits (rather than facilitates) the conductance of NMDA receptor channels by a mechanism involving the phospholipase C/inositol (1,4,5)-triphosphate/calmodulin and calmodulin kinase II pathway (Wirkner et al. 2000). In Mg^{2+}-free conditions, and therefore in conditions where NMDA receptors are not blocked, A_{2A}AR activation postsynaptically inhibits the NMDA receptors in a subpopulation of striatal neurones; however, if the NMDA receptors are blocked by Mg^{2+}, the predominant A_{2A}AR-mediated action is a presynaptic inhibition of GABA release (Wirkner et al. 2004). Whether the A_{2A}AR-mediated inhibition of NMDA receptors in the striatum explains the unexpected protective influence of $A_{2A}AR$ agonists towards NMDA-induced excitotoxicity (Popoli et al. 2004; Tebano et al. 2004) remains to be evaluated.

Interactions between $A_{2A}ARs$ and ionotropic glutamate receptors with implications for synaptic plasticity have been reported. LTP of synaptic transmission between CA3 and CA1 hippocampal areas of the hippocampus involves a postsynaptic facilitation of α-amino-3-hydroxy-5-methyl-4-isoxazolepropionic acid (AMPA) currents, a well-known process that requires previous activation of NMDA receptors and involves both pre- (enhanced glutamate release) and post- (depolarization-induced relief of NMDA receptor blockade by Mg^{2+}) synaptic mechanisms. Interestingly, A_2AR activation induces a form of LTP in the CA1 area that is NMDA receptor independent (Kessey and Mogul 1997). In contrast, $A_{2A}ARs$ localized postsynaptically at synapses between mossy fibers and CA3 pyramidal cells are essential for a form of LTP of NMDA currents, sparing AMPA currents (Rebola et al. 2008). Considering that CA3/CA1 LTP is predominantly NMDA receptor dependent, and that LTP at mossy fibers/CA3 synapses is predominantly presynaptic and NMDA receptor independent, it appears that $A_{2A}ARs$ are particularly devoted to unmasking nonpredominant forms of plasticity and therefore fine-tuning the networking and information flow within the hippocampus.

NMDA receptor activation suppresses neuronal sensitivity to adenosine in the hippocampus, and this interaction appears to result from an increase in the excitatory action of adenosine $A_{2A}ARs$ rather than a depression of A_1AR function (Nikbakht and Stone 2001).

Direct actions of purines upon NMDA receptor subunits (NR) may also occur. Thus, ATP, probably by directly binding to the glutamate-binding pocket of the NR2B subunit and not to ARs or ATP purinoceptors, can inhibit NMDA receptors and attenuate NMDA-mediated neurotoxicity (Ortinau et al. 2003).

5.2.2 Nicotinic Acetylcholine Receptors

Endogenous adenosine, by activating $A_{2A}ARs$ coupled to the adenylate cyclase/cAMP transduction pathway, tonically downregulates nAChR-mediated control of [3H]-ACh release at either the skeletal neuromuscular junction (Correia-de-Sá and Ribeiro 1994b) or myenteric plexus (Duarte-Araújo et al. 2004). Furthermore, at the skeletal neuromuscular junction, A_{2A} ARs enhance nicotinic receptor desensitization due to prolonged agonist exposure (Correia-de-Sá and Ribeiro 1994b).

The homopentameric α-7 subtype of nAChR is particularly relevant to brain functioning due to its high calcium permeability. By supplying calcium signals, these receptors influence several calcium-dependent events, including transmitter release and plasticity (Gray et al. 1996; Ji et al. 2001), and so several pathways must converge on their regulation. Adenosine, through $A_{2A}AR$ and BDNF, through TrkB receptors, exert double control over α-7-nicotinic currents at GABAergic interneurons in the hippocampus, since blockade of $A_{2A}ARs$ abolishes the BDNF-induced current inhibition (Fernandes et al. 2008). Since postsynaptic α7 nAChR-mediated inputs to GABAergic interneurons regulate inhibition within the hippocampus, A_{2A}AR, by allowing the inhibition of cholinergic currents by BDNF, may temporarily relieve GABAergic inhibition and therefore facilitate plasticity phenomena.

5.3 *Interaction with Receptors for Neurotrophic Factors*

Trk receptors belong to a third class of membrane receptors which, by themselves, possess catalytic activity involving autophosphorylation in tyrosine residues as a consequence of ligand binding, triggering a subsequent chain of phosphorylations that leads to the activation of several casacades involved in the regulation of cell death, survival and differentiation. Examples of this class of receptors are the receptors for neurotrophins, such as TrkA for nerve growth factor (NGF), TrkB for BDNF, TrkC for neurotrophin 3 (NT-3), and receptors for other neurotrophic factors, such as GFRα1 and Ret for GDNF. In spite of the structural differences between the GPCRs and receptor kinases, ARs, in particular A_{2A}AR, can tightly interact with receptors for neurotrophic factors, namely with receptors for BDNF and GDNF (Fig. 3c), which may have several implications for neurodegenerative diseases, as discussed below.

It has been known for several years that presynaptic depolarization (Boulanger and Poo 1999a)—which is known to increase extracellular adenosine levels, as well as enhancement of intracellular cAMP (Boulanger and Poo 1999b)—the most frequent A_{2A}AR transducing pathway, trigger synaptic actions of BDNF. On the other hand, A_{2A}ARs are known to transactivate TrkB receptors in the absence of the neurotrophin (Lee and Chao 2001). This transactivation requires long-term incubation with GPCR agonists and receptor internalization (Rajagopal et al. 2004), and it is not yet clear whether it operates the same mechanism as the more recently identified ability of A_{2A}ARs to trigger synaptic and promote survival actions of neurotrophic factors. Indeed, it has recently been recognized that adenosine A_{2A}AR activation is a crucial prerequisite for the functioning of neurotrophic receptors at synapses. This has been shown for the facilitatory actions of BDNF on synaptic transmission (Diógenes et al. 2004; Tebano et al. 2008) and on LTP (Fontinha et al. 2008) at the CA1 area of the hippocampus, as well as for the action of GDNF at striatal dopaminergic nerve endings (Gomes et al. 2006). A_{2A}ARs and TrkB BDNF receptors can coexist in the same nerve ending since the facilitatory action of A_{2A}ARs upon TrkB-mediated BDNF action is also visible at the neuromuscular junction (Pousinha et al. 2006), a single nerve ending synapse model.

The ability of BDNF to facilitate synaptic transmission is dependent on the age of the animals (Diógenes et al. 2007), and this may be related to the degree of activation of A_{2A}ARs by endogenous adenosine at different ages. Thus, in infant animals (i.e., immediately after weaning), in order to trigger a BDNF facilitatory action it is necessary to increase the extracellular levels of adenosine, either by inhibiting AK, through a brief depolarization (Diógenes et al. 2004; Pousinha et al. 2006), or by inducing high-frequency neuronal firing, such as those inducing LTP

(Fontinha et al. 2008); in all cases the actions of BDNF are lost by blocking A_{2A}ARs with selective antagonists. In adult animals, BDNF per se can facilitate synaptic transmission through TrkB receptor activation, but this effect is also fully lost with blockade of A_{2A}ARs (Diógenes et al. 2007) or in A_{2A}AR knockout (KO) mice (Tebano et al. 2008). Nicotinic α7 cholinergic currents in GABAergic hippocampal neurons are inhibited by BDNF, and this also requires coactivation of adenosine A_{2A}ARs (Fernandes et al. 2008). Inhibition of GABA transporters (GAT) of the predominant neural subtype, GAT1, by BDNF does not fully depend upon coactivation of A_{2A}ARs since it is not abolished by A_{2A}AR blockade; however, A_{2A}AR activation can facilitate this BDNF action (Vaz et al. 2008).

Whether the ability of A_{2A}ARs to protect retinal neurones against glutamate-induced excitotoxicity (Ferreira and Paes-de-Carvalho 2001) is due to its ability to facilitate actions of neurotrophic factors, as has been shown to occur in relation to A_{2A}AR-mediated neuroprotection of motor neurones (Wiese et al. 2007), requires further investigation. It is worth noting that while Wiese et al. (2007) reported a TrkB-mediated enhancement in the survival of injured facial motor neurons in vivo, TrkB receptor activation by BDNF may render spinal cord-cultured motor neurons more vulnerable to insult (Mojsilovic-Petrovic et al. 2006). Interestingly enough, in both cases, activation of A_{2A}ARs by endogenous adenosine was required, since A_{2A}AR antagonism prevented both the favorable (Wiese et al. 2007) and the deleterious (Mojsilovic-Petrovic et al. 2006) TrkB-mediated actions.

Activation of A_{2A}ARs enhances NGF-induced neurite outgrowth in PC12 cells and rescues NGF-induced neurite outgrowth impaired by blockade of the mitogen-activated protein kinase (MAPK) cascade, an action that requires PKA activation (Cheng et al. 2002). Furthermore, activation of A_{2A}ARs through Trk-dependent and phosphatidylinositol 3-kinase/Akt mechanisms promoted PC12 cell survival after NGF withdrawal (Lee and Chao 2001). A similar A_{2A}AR-mediated neuroprotection mechanism has been shown to occur in hippocampal neurones after BDNF withdrawal (Lee and Chao 2001). In contrast to A_{2A} receptors, which usually promote the actions of neurotrophic factors, A_{2A}ARs inhibit neurite outgrowth of cultured dorsal root ganglion neurons in both the absence and the presence of NGF (Thevananther et al. 2001).

Besides interactions at the neurotrophin receptor level, AR activation may also induce the release of neurotrophic factors. Thus, the expression and/or release of NGF are enhanced by the activation of A_{2A}ARs in microglia (Heese et al. 1997) and by the activation of A_1ARs in astrocytes (Ciccarelli et al. 1999). A_{2B}ARs in astrocytes are also able to enhance GDNF expression (Yamagata et al. 2007). A_{2A}ARs are required for normal BDNF levels in the whole hippocampus (Tebano et al. 2008).

Interactions among purinergic, growth factor and cytokine signaling are also highly relevant in nonpathologic brain functioning, namely in the regulation of neuronal and glial maturation as well as development. In neuronal-dependent glial maturation, both ATP purinoceptors and adenosine ARs are involved (Fields and Burnstock 2006). The extracellular adenosine levels attained during high-frequency neuronal firing are sufficient to stimulate ARs in oligodendrocyte ancestor cells, inhibiting their proliferation and stimulating their differentiation into myelinating oligodendrocytes (Stevens et al. 2002), but unfortunately the nature of the AR involved was not identified in this work. In premyelinating Schwann cells, A_{2A}ARs activate phosphorylation of extracellular signal-regulated kinases (ERKs), namely ERK1/2, and inhibit Schwann cell proliferation without arresting differentiation (Stevens et al. 2004).

Decreases in the levels and/or actions of neurotrophic factors have been implicated in the pathophysiological mechanisms of many diseases of the nervous system, such as Alzheimer's disease, Parkinson's disease, Huntington's disease, diabetic neuropathies, amyotrophic lateral sclerosis, and even depression, making the use of naturally occurring neurotrophic factors a very promising approach to the treatment of these disorders (Schulte-Herbrüggen et al. 2007). However, the pharmacological administration of neurotrophic factors in vivo has not been easy so far because these molecules are unable to cross the blood–brain barrier, making invasive application strategies like intracerebroventricular infusion necessary. The evidence that A_{2A}ARs trigger or facilitate actions of neurotrophins upon synaptic strength and neuronal survival has led to a new therapeutic strategy: the use of adenosine A_{2A}AR agonists that cross the blood–brain barrier to potentiate neurotrophic actions in the brain.

However, we should particularly mention epilepsy, where neurotrophic factors have been considered both harmful, being causal mediators in the development of acquired epileptic syndromes, and also eventually useful in treating epilepsy-associated damage (Scharfman and Hen 2007; Simonato et al. 2006). On top of this controversy, we can add discrepant findings of both anticonvulsive (Huber et al. 2002) and proconvulsive (Zeraati et al. 2006) A_{2A}AR-mediated actions, with the proconvulsive actions being the more expected due to the usually excitatory nature of these receptors.

Finally, the crosstalk between A_{2A}ARs and receptors for neurotrophins also points to the need for caution about therapies with A_{2A}AR antagonists in neurodegenerative diseases, as has been proposed for Parkinson's disease to ameliorate L-DOPA-induced dyskinesias. For more information on A_{2A}ARs and Parkinson's disease, please refer to Chap. 18, "A_{2A} Adenosine Receptors and Parkinson's Disease" (by Morelli et al.), in this volume. Indeed, the identification of postsynaptic A_{2A}/D_2 receptor interactions in the striatum, together with the findings that A_{2A}AR antagonists are neuroprotective in Parkinson's disease models (Chase et al. 2003) and increase dopamine synthesis from L-DOPA (Golembiowska and Dziubina 2004), led to the proposed use of A_{2A}AR antagonists in Parkinson's disease. On the other hand, neurotrophic factors, in particular GDNF, may be a potential therapeutic approach in the management of Parkinson's disease (Love et al. 2005; Patel et al. 2005). Enhancing GDNF actions via A_{2A}AR agonists (Gomes et al. 2006) may also be of high therapeutic interest. In any case, the finding that the actions of GDNF on dopamine release in the striatum are prevented by A_{2A}AR antagonism (Gomes et al. 2006) points to the need for further studies on the consequences of long-term therapy with A_{2A} receptor blockers in neurodegenerative diseases where neurotrophic factors may play a beneficial role. One issue that should be explored in the future is the optimal time window for combined beneficial effects of neurotrophic factors and A_{2A}AR agonists/antagonists. Perhaps A_{2A}AR antagonists may be advantageous in the late stages of neurodegenerative diseases; however, in the early stages, where an enhancement of neurotrophic factors is highly desirable, A_{2A}AR antagonists should be avoided and A_2AR agonists should perhaps be considered, in order to allow neurotrophic influences. A schematic representation of what has been reported so far on the interactions of ARs and neurotrophin receptors and on neurotrophin release, as well as the implications of these interactions at the hippocampus and striatum in relation to Alzheimer's and Parkinson's diseases, is illustrated in Fig. 4.

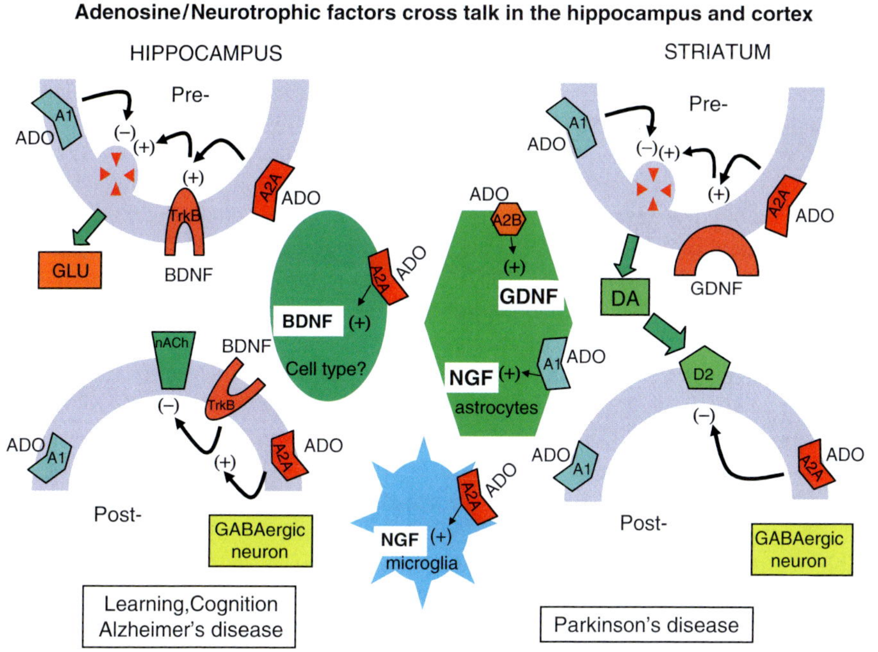

Fig. 4 Schematic representation of what has been reported regarding the interaction between adenosine receptors (*ARs*) and neurotrophic factor receptors, as well as the influence of *ARs* on neurotrophic factor synthesis or release, focusing on brain areas with implications for learning, cognition and Alzheimer's disease (hippocampus) or Parkinson's disease (striatum). A *plus symbol* denotes facilitation and a *minus symbol* denotes inhibition of receptor functioning or neurotrophic factor synthesis or release. The positive influence of A_{2A}ARs upon brain-derived neurotrophic factor (*BDNF*) levels was studied in slices, so the cell type cannot be identified. See text for references. Other abbreviations: *ADO*, adenosine; *D*, dopamine receptor; *DA*, dopamine; *GDNF*, glial cell line-derived neurotrophic factor; *GLU*, glutamate; *nAChR*, nicotinic acetylcholine receptor; *NGF*, nerve growth factor; *TrkB*, tropomyosin-related kinase receptor type B

6 Hypoxia and Ischemia

6.1 Adenosine and Control of Synaptic Transmission During Hypoxia

A very intimate relationship between hypoxia/ischemia and adenosine is well established. This relationship has been the subject of extensive reviews (de Mendonça et al. 2000), and the implications of this relationship for neuroprotection are discussed in Chap. 17, "Adenosine Receptors and Neurological Disease: Neuroprotection and Neurodegeneration" (by Stone et al.), in this volume. Thus, we will focus on hypoxia and the synaptic actions of adenosine.

High amounts of adenosine, and perhaps surprisingly, of ATP are released into the synaptic cleft during a hypoxic/ischemic insult (Frenguelli et al. 2007), leading to A_1AR activation and profound inhibition of synaptic transmission (Fowler 1993). This A_1AR-mediated inhibition promotes recovery after the insult, since blockade of A_1ARs reduces inhibition of synaptic transmission but also impairs recovery after reoxygenation (Sebastião et al. 2001). Similar observations have been made using A_1AR KO mice (Johansson et al. 2001) or through focal deletion of the presynaptic A_1ARs (Arrigoni et al. 2005). The facilitation of recovery of synaptic transmission after a hypoxic insult involves both presynaptic inhibition of glutamate release and subsequent reduction of NMDA receptor activation during the hypoxic episode (Sebastião et al. 2001). In other words, the neuroprotection induced by adenosine operates two well-known synaptic actions of A_1ARs that also occur under normoxic conditions and are of particular relevance in the case of hypoxia: a decrease of neurotransmitter release via inhibition of presynaptic calcium entry through the blocking of calcium channels (Ribeiro et al. 1979), and postsynaptically inhibiting calcium entry via inhibition of NMDA receptors (de Mendonça et al. 1995).

The mammalian brain can adapt to injurious insults such as cerebral ischemia to promote cell survival in the face of subsequent injury, a phenomenon known as ischemic preconditioning (Gidday 2006). Adenosine, through A_1AR, is responsible for the protective actions of ischemic preconditioning in the hippocampus; A_{2A}ARs are not involved in this process, whereas A_3AR activation is harmful to ischemic preconditioning, impairing recovery (Pugliese et al. 2003). However, hypoxia leads to a rapid (<90 min) homologous desensitization of A_1AR-mediated inhibition of synaptic transmission that is likely due to an internalization of A_1ARs in nerve terminals (Coelho et al. 2006). This may alleviate A_1AR-mediated functional disconnection of GABAergic neurones (Congar et al. 1995; Lucchi et al. 1996), allowing sequential time windows for a protective role of adenosine and GABA during hypoxia (Sebastião et al. 1996). Changes in the activity of adenosine-producing enzymes also occur during hypoxic/ischemic episodes. This is the case for AK, which is downregulated (Pignataro et al. 2008), and for the enzyme chain-hydrolyzing extracellular ATP, which is upregulated (Braun et al. 1998), with both processes leading to more intense extracellular adenosine production and contributing to its neuroprotection.

6.2 *Adenosine and Control of Ventilation*

The partial pressure of oxygen in the blood is sensed by the carotid body located at the bifurcation of the carotid artery. Low levels of oxygen in the arterial blood activate the carotid body and ventilation is subsequently enhanced. Since the early 1980s, when the first description of the excitatory effects of adenosine on carotid body chemoreceptor activity appeared (McQueen and Ribeiro 1981), this nucleoside has emerged as a key molecule in the regulation of chemosensory activity and ventilation (Lahiri et al. 2007). Adenosine enhances carotid body chemosensory activity either in vivo (McQueen and Ribeiro 1986) or in vitro (Runold et al. 1990), as well as ventilation (Monteiro and Ribeiro 1987). This action of adenosine is mediated by A_{2B}ARs in sensory terminals and A_{2A}ARs at carotid body cells, which are activated by its endogenous release as a consequence of a decrease in the partial pressure of oxygen around those cells (Conde et al. 2006; McQueen and Ribeiro 1986). A_{2A}AR mRNA, but not A_1AR mRNA, is expressed in type I carotid body cells, and these receptors modulate Ca^{2+} homeostasis during hypoxia (Kobayashi et al. 2000a).

Upon denervation of the carotid bodies, AR agonists depress ventilation by activating A_{2A}ARs in the CNS (Koos and Chau 1998). Adenosine also modulates cardiorespiratory control through presynaptic actions in the nucleus tractus solitarius, where it modulates transmitter release (Spyer and Thomas 2000).

In humans, intravenous (i.v.) injection of adenosine produces hyperventilation and dyspnea resulting from direct activation of the carotid body (Watt et al. 1987). However, some secondary effects, including heat sensation, flushed face, dyspnea and chest discomfort in humans, have been reported after i.v. adenosine infusion (Uematsu et al. 2000). Adenosine enhances the ventilatory response to hypoxia but not to hypercapnia (Maxwell et al. 1986), which argues against a major contribution from the central chemosensory centers, where adenosine increases the sensitivity to hypercapnia (Phillis 2004), suggesting a major role for peripheral sensors in the ventilatory response to adenosine in humans (Lahiri et al. 2007).

The usefulness of the carotid body in maintaining oxygen homeostasis is magnified by its plasticity, which to a large extent is due to changes in gene expression (Lahiri et al. 2007). The contribution of purines to the control of carotid body activity may also be developmentally regulated. For example, the A_{2A}AR and D_2 dopaminergic receptors are differentially expressed in glomus cells during development, with greater relative expression of mRNA message for the A_{2A}AR found in earlier stages, and for the D_2 receptors in the adult animal (Gauda et al. 2000). A_{2B}ARs in the carotid body are slightly downregulated within 24 h exposure to moderate (10% O_2) hypoxia (Ganfornina et al. 2005), whereas A_{2A}ARs are upregulated by chronic hypoxia, at least in PC12 cells (Seta and Millhorn 2004). As occurs in brief hypoxic/ischemic episodes (see Sect. 6.1), chronic hypoxia decreases the expression of AK, adenosine deaminase and the adenosine transporter, while it increases the expression of ecto-5′-nucleotidase (Kobayashi et al. 2000b). All of these hypoxia-induced changes in the expression of ARs and the enzymes involved in the control of extracellular adenosine levels may contribute to a protective adaptation to hypoxia.

6.2.1 Adenosine and Respiration in the Newborn

The inhibitory effect of CNS A_{2A}AR activation on respiratory drive is more evident early in life, and is mediated via GABAergic inputs to the inspiratory timing neural circuitry (Mayer et al. 2006). Blockade of these receptors is probably the mechanism by which xanthine therapy alleviates apnea in prematures (Aranda and Turmen 1979; Bairam et al. 1987; Uauy et al. 1975). Indeed, blockade of A_{2A}ARs blunts the respiratory roll-off response to hypoxia in newborn lambs (Koos et al. 2005). Xanthine therapy in the newborn may, however, increase the risk of seizures (see Table 1).

7 Role of ARs in Pain

Pain can have multiple causes and origins, and therefore the ability of adenosine to influence pain also has multiple sites of action and diverse mechanisms. Activation of A_1AR in the spinal cord produces antinociceptive properties in acute nociceptive, inflammatory and neuropathic pain tests (Sawynok 2007; Sawynok and Liu 2003). In humans, the first evidence for antinociceptive actions of adenosine was detected during adenosine infusion (i.v.), which had beneficial effects in two patients with neuropathic pain (Sollevi et al. 1995). A few years later, the same group showed that adenosine can also reduce secondary hyperalgesia in two human models of cutaneous inflammatory pain (Sjölund et al. 1999). Although peripheral A_2AR activation can exacerbate pain responses (Sawynok 1998), its anti-inflammatory action may also contribute to decreasing inflammatory pain. As a consequence, A_1AR agonists have entered clinical trials for neuropathic pain, whereas A_{2A}AR agonists are entering clinical trials as anti-inflammatory agents (Gao and Jacobson 2007). There is also growing interest in the use of allosteric enhancers of A_1AR activation due to the putative tissue selectivity of A_1ARs. Allosteric modulation of adenosine A_1ARs reduces allodynia, and this has been shown to occur not only after intrathecal injection but also after systemic administration (Pan et al. 2001).

The pain-relieving effect of activating A_1ARs at the level of the spinal cord is related to their ability to presynaptically inhibit excitatory transmission to neurons of the substantia gelatinosa (Lao et al. 2001). The inhibition of NMDA receptors by adenosine (see Sect. 5.2.1) probably also occurs at the level of the spinal cord (DeLander and Wahl 1988) and contributes to a reduction of central sensitization and plasticity mechanisms involved in chronic pain. In contrast, adenosine A_{2A}AR activation sensitizes peripheral afferent fibers that project to the spinal cord, enhancing nociception (Hussey et al. 2007). Accordingly, mice lacking the A_{2A}AR have reduced responses to thermal nociceptive stimuli (Ledent et al. 1997), whereas mice lacking the A_1AR show increased nociceptive response (Wu et al. 2005)

The peripheral administration of adenosine in humans produces pain responses resembling those generated under ischemic conditions (Sawynok 1998). This pain-initiating effects of adenosine are augmented by substance P (Gaspardone et al.

1994) and nicotine (Sylvén et al. 1990), and are usually a limiting factor in the use of adenosine-related compounds for the control of chronic pain. Activation of A_3ARs produces pain due to the release of histamine and 5-hydroxytryptamine from mast cells and subsequent actions on the sensory nerve terminal (Sawynok 1998). However, in spite of these algogenic consequences of peripheral administration of adenosine, its net action on pain processing is inhibitory, since enhancers of extracellular adenosine levels have antinociceptive action (see below).

Due to the simultaneous A_1AR-mediated antinociceptive and A_{2A}AR-mediated anti-inflammatory actions of adenosine, there has been increasing interest in the development of drugs that, by influencing extracellular adenosine levels, could have analgesic actions. Successful examples include inhibitors of AK (see Sect. 3), whose spinally-mediated antinociceptive properties were noted over a decade ago (Keil and DeLander 1992). Most likely due to the anti-inflammatory actions of adenosine, AK inhibitors administered orally are more effective at reducing inflammatory pain than neuropathic or acute pain (Jarvis et al. 2002). By comparing the antinociceptive and anti-inflammatory properties of AK inhibitors administered at the ipsilateral or contralateral sides of the injury, it was concluded that much of the anti-inflammatory action is locally mediated, whereas the antinociceptive action is systemically mediated, exerted predominantly at the level of the spinal dorsal horn (Poon and Sawynok 1999). Indeed, AK inhibitors are able to reduce the increased *c-Fos* expression in the spinal dorsal horn induced by peripheral injection of an inflammatory (carrageenan) substance (Poon and Sawynok 1999).

Antidepressants are widely used in the treatment of neuropathic pain, but their analgesic efficacy seems to occur irrespective of mood-altering effects, and may involve an increase in extracellular adenosine levels. This has been shown after either acute (Esser and Sawynok 2000) or chronic (Esser et al. 2001) amitriptyline administration in rat models of neuropathic pain. Similarly, endogenous adenosine seems to be involved in the antiallodynic action of amitriptyline in a rat model of painful diabetic neuropathy (Ulugol et al. 2002). As pointed out by Esser and Sawynok (2000), the manipulation of endogenous adenosine by amitriptyline, while important, is unlikely to be the sole mechanism underlying its ability to reduce pain, but the attenuation of its effect by modest doses of caffeine (within those levels easily attained in humans after two cups of strong coffee) raise the possibility that dietary caffeine consumption may influence the efficacy of amitriptyline in alleviating neuropathic pain in humans.

Increases in adenosine levels may contribute to the analgesic action of opioids. An increase in adenosine levels in the cerebrospinal fluid has been detected in humans following intrathecal administration of morphine (Eisenach et al. 2004). It is of interest that in neuropathic rats the release of adenosine induced by morphine is reduced (Sandner-Kiesling et al. 2001), which may explain a decreased efficacy and potency of opioids in the treatment of neuropathic pain. Moreover, modifications in the expression of several types of opioid receptors were recently detected in mice lacking the A_{2A}AR gene (Bailey et al. 2002), suggestive of a functional interplay between A_{2A}AR and opioid receptors with respect to pain modulation.

A critical review of the applications of adenosine and ATP in pain control, summarizing most of the human studies, suggests a high potential for adenosine compounds to alleviate pain (Hayashida et al. 2005). This review suggests that the doses, the routes and the timing of administration together with the tissue penetration of the drugs must be taken into consideration, and that there is a need for more basic research to clarify several points. Caffeine, via its antagonistic actions on ARs, can modulate pain; however, as recently discussed (Shapiro 2007), the type of effect (e.g., generation or alleviation of headache) depends on the site of action as well as the dosage and timing of exposure. Both A_{2A} and A_{2B} ARs are probably involved in the interaction between paracetamol and caffeine in pain control. Blockade of A_{2B}ARs causes an enhancement of the action of paracetamol in tail immersion and hot-plate tests in mice, and blockade of A_{2A}ARs produces an antinociceptive effect, even in the absence of paracetamol (Godfrey et al. 2006). Moreover, theophylline ameliorates chest pain in patients with a hypersensitive esophagus, possibly by altering adenosine-mediated nociception (Rao et al. 2007).

As a potent vasodilator, CGRP, which is released by the trigeminocerebrovascular system, plays a key role in the pathophysiology of migraine headache; antagonism of CGRP has been suggested as a promising new approach for the treatment of this condition (Goadsby 2008). Another approach to blocking the trigeminovascular system and CGRP to treat migraine headache may include the use of A_1AR agonists. Activation of A_1ARs inhibits trigeminovascular activation by acting on the trigeminal nucleus and by inhibiting the release of CGRP in the cranial circulation, with this second action being attributable to activation of A_1ARs on peripheral terminals of the trigeminal nerve (Goadsby et al. 2002). Tonic activation of A_1ARs may also prevent the facilitatory actions of CGRP, as has been shown to occur in the hippocampus (Sebastião et al. 2000b). Interestingly, A_{2A}AR activation facilitates the actions of CGRP (Correia-de-Sá and Ribeiro 1994a; Sebastião et al. 2000a, b), but the relevance of these observations to an approach for the treatment of migraine headache (i.e., with A_2AR antagonists) remains to be established.

8 Caffeine and ARs

Ever since the delights of tea were first discovered by Emperor Shen Nung in 2737 BC, methylxanthines, including caffeine, have been widely consumed by humans all over the world. The broad caffeine intake associated with common beverages, together with the impact of xanthines on biomedical research, prompted many studies that have focused on specific caffeine effects rather than using it as a tool to antagonize ARs (Daly 2007; Ferré 2008). Indeed, as a pharmacological tool, caffeine is no longer very useful, because its affinity for ARs is low and its selectivity towards the different ARs is also very poor. It is interesting to note that the first proposal for the existence of an A_3AR was based upon pharmacological characteristics, namely high affinity for agonists and xanthine sensitivity (Ribeiro and Sebastião 1986). Cloning and cellular expression of the rat A_3AR (Zhou et al. 1992) challenged these criteria, since the rat A_3 receptor is xanthine insensitive and has low agonist affinity. Cloning and expression of the human A_3AR (Salvatore et al. 1993) reversed the situation again, since the human A_3AR is xanthine sensitive and a high-affinity receptor for A_3AR ligands. For more information on the affinity of the human A_3AR for A_3AR ligands, the reader is referred to Chap. 5, "Medicinal Chemistry of the A_3 Adenosine Receptor: Agonists, Antagonists, and Receptor Engineering" (Jacobson et al.), in this volume.

However, xanthines such as caffeine have other biological actions besides AR antagonism. They inhibit PDEs (PDE4, PDE1, PDE5), promote calcium release from intracellular stores, and interfere with $GABA_A$ receptors (Daly 2007). Caffeine analogs can be developed to target any of these mechanisms rather than ARs, and this may be explored therapeutically (Daly 2007), but in the case of caffeine, the effects seen at the low doses taken in during normal human consumption are mostly due to AR antagonism (Fredholm et al. 1999). Due to its ability to antagonize ARs, to cross the blood–brain barrier, and also due to the low risk of intake, caffeine has therapeutic potential in central nervous system dysfunctions (e.g., Alzheimer's disease and Parkinson's disease). Adverse effects of caffeine may include anxiety, hypertension, drug interactions, and withdrawal symptoms (Daly 2007). In human volunteers, caffeine improves cognition; however, it also affects sleep (see Table 1). Moreover, a relationship between adenosine A_{2A}ARs and genetic variability in caffeine metabolism associated with habitual caffeine consumption has been proposed (Cornelis et al. 2007), which provides a biological basis for caffeine consumption. In this study, persons with the *ADORA2A TT* genotype were significantly more likely to consume less caffeine than carriers of the *C* allele.

The therapeutic or adverse effects of caffeine are quite different depending on whether it is administered chronically or acutely. For example, chronic caffeine intake, which increases plasma concentrations of adenosine (Conlay et al. 1997), may be neuroprotective. This contrasts with the consequences of acutely antagonizing A_1ARs (de Mendonça et al. 2000). Chronic AR antagonism with caffeine may also influence cognition and motor activity in a way that resembles the acute effects of AR agonists (Jacobson et al. 1996). Such opposite actions of chronic versus acute treatment not only have important implications for the development of xanthine-based compounds as therapeutic agents but also constitute a frequent confounding parameter in research. Upregulation of A_1ARs after chronic AR antagonism with xanthines does occur, but A_{2A}AR levels apparently do not change; in addition, there are changes in the levels of receptors for neurotransmitters with chronic administration of xanthines, namely a marked decrease in β-adrenergic receptors and an increase in 5-HT and $GABA_A$ receptors (Jacobson et al. 1996). The increased expression of A_1ARs in response to chronic antagonism of ARs by caffeine, as compared with A_{2A}ARs, may lead to a shift in the A_1/A_{2A} AR balance after prolonged caffeine intake (Ferré 2008). Moreover, chronic caffeine treatment leads to modifications in the function of the A_1R–A_{2A}R heteromer and this may, in part, be the scientific basis for the strong tolerance to the psychomotor effects of chronic caffeine (Ciruela et al. 2006).

Furthermore, alteration of astrocytogenesis via $A_{2A}AR$ blockade during brain development raises the possibility that postnatal caffeine treatment could have long-term negative consequences on brain function, and should perhaps be avoided in breast-feeding mothers (Desfrere et al. 2007)

8.1 Influence on Brain Function and Dysfunction

8.1.1 Sleep

One of the main reasons for drinking a cup of strong coffee is to repel sleep. Most studies on ARs and sleep regulation in humans rely upon consequences of caffeine ingestion by human volunteers (see Table 1), and it is now widely accepted that caffeine prolongs wakefulness by interfering with the key role of adenosine upon sleep homeostasis (Landolt 2008). In an innovative review of the role of adenosine upon sleep regulation, Porkka-Heiskanen et al. (2002) proposed adenosine as a sleeping factor and hypothesized that adenosine functions in a similar way to neuroprotection against energy depletion. In the critical arousal area (basal forebrain), extracellular adenosine levels start to rise in response to prolonged neuronal activity during wakeful periods. This increase leads to a decrease in neuronal activity, and sleep is induced before the energy balance in the whole brain is affected. Microdialysis measurements performed in freely moving cats showed an increase in the concentrations of adenosine during spontaneous wakefulness, and adenosine transport inhibitors mimicked the sleep-wakefulness profile that occurs after prolonged wakefulness (Porkka-Heiskanen et al. 1997). In contrast, AR antagonists like caffeine increase wakefulness (see Table 1). Prolonged wakefulness induces signs of energy depletion in the brain, which induces an increase in sleep (Benington and Heller 1995). Molecular imaging provided evidence for an A_1 receptor upregulation in cortical and subcortical brain regions after prolonged wakefulness in humans (Elmenhorst et al. 2007). Adenosinergic mechanisms contribute to individual differences associated with sleep deprivation sensitivity in humans (Rétey et al. 2006); furthermore, a genetic variation in the adenosine $A_{2A}AR$ gene may contribute to individual sensitivity to the effects of caffeine on sleep (Rétey et al. 2007, see Table 1).

It is well documented that A_1ARs are involved in sleep regulation through the inhibition of ascending cholinergic neurons of the basal forebrain (Basheer et al. 2004). However, more recent studies, which include experiments with A_{2A} and A_1 AR KO mice, indicate that $A_{2A}ARs$ (most probably localized in the ventrolateral preoptic area of the hypothalamus) also play a crucial role in the sleep-promoting effects of adenosine and the arousal-enhancing effects of caffeine (Huang et al. 2005). These studies suggest that $A_{2A}AR$ antagonists may represent a novel approach as potential treatments for narcolepsy and other sleep-related disorders (Ferré et al. 2007b). Adenosine $A_{2A}ARs$ in the pontine reticular formation promote acetylcholine release, rapid eye movement (REM) and non-REM sleep in mice. This effect on non-REM sleep is probably due to $A_{2A}AR$-induced enhancement of GABAergic inhibition of arousal promoting neurons (Coleman et al. 2006). In addition to its effect in the basal forebrain, adenosine exerts its sleep-promoting effect in the lateral hypothalamus by A_1AR-mediated inhibition of hypocretin/orexin neurons (Liu and Gao 2007; Thakkar et al. 2008).

In conclusion, the two high-affinity ARs, the A_1 and the A_{2A} ARs, affect multiple mechanisms in several brain areas involved in the regulation of sleep and arousal. Therefore, the influences of caffeine upon sleep, felt by many humans and recently also documented in controlled studies in healthy volunteers (see Table 1), can be attributed to both A_1 and A_{2A} AR blockade. As discussed above (see Sect. 8), chronic caffeine consumption may alter AR function and the A_1/A_{2A} AR balance and consequently the influence of both ARs upon sleep.

8.1.2 Epilepsy

There are several clinical reports on caffeine or theophylline intake and seizure susceptibility (Kaufman and Sachdeo 2003; Mortelmans et al. 2008), but surprisingly, no mention is made of the main cause of seizure induction by these drugs, AR antagonism.

Indeed, after the initial observation that adenosine has an anticonvulsant action (Barraco et al. 1984), the therapeutic potential of adenosine-related compounds in epilepsy was immediately pointed out (Dragunow et al. 1985), and it is now widely accepted that adenosine is an endogenous anticonvulsant, an action mediated by inhibitory A_1ARs that restrain excessive neuronal activity. Other ARs are, however, involved in seizure control, though their roles are most frequently related to exacerbating seizures. An influence of A_3 and A_2 ARs in $GABA_A$ receptor stability has been suggested recently (Roseti et al. 2008), based on the observation that A_3 or A_{2B} AR antagonists reduce rundown of $GABA_A$ currents. A_{2A}ARs, by promoting neuronal excitability, may also increase seizure susceptibility. Indeed, A_{2A}AR KO mice are less sensitive to pentylenetetrazol-induced seizures (El Yacoubi et al. 2008).

It has been shown that A_1AR activation by locally released adenosine is an efficient way to keep an epileptic focus localized (Fedele et al. 2006). Therefore, attention is now focused on the development of biocompatible materials for adenosine-releasing intrahippocampal implants (Wilz et al. 2008). In line with the evidence for the antiepileptic role of A_1ARs, A_1AR KO mice are more susceptible to seizures and develop lethal status epilepticus after experimental traumatic brain injury (Kochanek et al. 2006). There are, however, limitations on the use of A_1AR agonists as anticonvulsant drugs due to their pronounced peripheral side effects, like cardiac asystole as well as central side effects like sedation (Dunwiddie 1999). One possibility would be the use of partial agonists, which are more likely to display tissue selectivity. An N^6,C8-disubstituted adenosine derivative with low efficacy towards A_1AR activation in whole brain membranes but with high efficacy as an inhibitor of hippocampal synaptic transmission was identified (Lorenzen et al. 1997). Another approach that has been more intensely explored is with the use of compounds that increase the extracellular concentrations of adenosine. This has been attempted with AK inhibitors, which showed beneficial effects in animal models of epilepsy and an improved preclinical therapeutic index over direct-acting AR agonists (McGaraughty et al. 2005). An even more refined approach would be local reconstitution of the inhibitory adenosinergic tone by intracerebral implantation of cells engineered to release adenosine, and this has been done using AK-deficient cells (Güttinger et al. 2005). The reverse also holds true, since transgenic mice overexpressing AK in the brain have increased seizure susceptibility (Fedele et al. 2005). Furthermore, intrahippocampal implants of AK-deficient stem cell-derived neural precursors suppress kindling epileptogenesis (Li et al. 2007). The above evidence suggests that adenosine-augmenting cell and gene therapies may lead to improved treatment options for patients suffering from intractable epilepsy (Boison 2007).

AK is mostly expressed in astrocytes (Studer et al. 2006), and overexpression of AK after seizures, with consequent reduced adenosine inhibitory tone, contributes to seizure aggravation (Fedele et al. 2005). However, release of interleukin-6 (IL-6) from astrocytes induces an upregulation of A_1ARs in both astrocytes (Biber et al. 2001) and neurons (Biber et al. 2007). This leads to an amplification of A_1AR function, enhances the responses to readily released adenosine, enables neuronal rescue from glutamate-induced death, and protects animals from chemically induced convulsing seizures (Biber et al. 2007). Indeed, IL-6 KO mice are more susceptible to seizures and lack the well-known seizure-induced upregulation of A_1ARs (Biber et al. 2007).

Seizure-induced release of neurotrophic factors, such as BDNF, may have beneficial and aggravating actions upon epilepsy, with the beneficial ones being mostly related to promotion of cell survival and the deleterious ones being related to excessive cell proliferation and neuronal sprouting (Simonato et al. 2006). Adenosine, through A_{2A}AR activation, triggers and facilitates BDNF actions in neurons (Diógenes et al. 2004; Fontinha et al. 2008, see Sect. 5.3 above), but the relevance of this interplay for epilepsy remains to be explored. This may be of particular relevance whenever designing therapies that lead to enhanced extracellular adenosine levels, since besides A_1ARs, A_{2A}ARs can also be activated.

8.1.3 Cognition, Learning, and Memory

Endogenous adenosine, through A_1ARs, inhibits long-term synaptic plasticity phenomena such as LTP (de Mendonça and Ribeiro 1994), LTD, and depotentiation (de Mendonça et al. 1997c). In accordance, A_1AR antagonists have been proposed for the treatment of memory disorders (Stone et al. 1995). Cognitive effects of caffeine are mostly due to its ability to antagonize adenosine A_1ARs in the hippocampus and cortex, the brain areas mostly involved in cognition, but as already discussed in detail (see Fredholm et al. 1999), positive actions of caffeine on information processing and performance may also be attributed to improvements in behavioral routines, arousal enhancement and sensorimotor gating. This interpretation was supported by the observation that the AR antagonist theophylline enhances spatial memory performance only during the light period, which is the time of sleepiness in rats (Hauber and Bareiss 2001). Independently of the processes caffeine or theophylline use to improve cognition, there is little doubt that the beneficial effects most of us feel after a few cups of coffee or tea are due to the actions of these psychoactive substances upon ARs. Recent evidence that blockade of A_1 receptors improves cognition came from a study using a mixed A_1/A_{2A} receptor antagonist, ASP5854 (Mihara et al. 2007). This orally active drug could reverse scopolamine-induced memory deficits in rats, whereas a specific adenosine A_{2A}AR antagonist, KW-6002, did not. Reduced A_{2A}AR activation may also be relevant for cognitive improvements, since A_{2A}AR KO mice have improved spatial recognition memory (Wang et al. 2006). Accordingly, overexpression of A_{2A}ARs leads to memory deficits (Giménez-Llort et al. 2007).

8.1.4 Alzheimer's Disease

There is the possibility that chronic intake of caffeine during one's lifetime might protect from cognitive decline associated with aging. Elderly women who drank relatively large amounts of coffee over their lifetimes gave better performances in memory and other cognitive tests than nondrinkers (Johnson-Kozlow et al. 2002). A case–control study was specifically designed to evaluate whether chronic intake of caffeine might be related to a lower risk of Alzheimer's disease (Maia and de Mendonça 2002), the most common form of dementia. Levels of caffeine consumption in the 20 years that preceded the diagnosis in patients were compared with those taken by age- and sex-matched controls with no signs of cognitive impairment. Data analysis showed that caffeine intake was inversely associated with the risk of Alzheimer's disease and that this association was not explained by several possible confounding variables related to habits and medical disorders (Maia and de Mendonça 2002). This was confirmed in a larger-scale study (4,197 women and 2,820 men) with similar objectives, showing that the psychostimulant properties of caffeine appear to reduce cognitive decline in aged women without dementia (Ritchie et al. 2007).

Long-term protective effects of dietary caffeine intake were also shown in a controlled longitudinal study involving a transgenic murine model of Alzheimer's disease. Caffeine was added to the drinking water of mice between four and nine months of age, with behavioral testing done during the final six weeks of treatment; the results revealed that moderate daily intake of caffeine may delay or reduce the risk of cognitive impairment in these mice (Arendash et al. 2006). Amnesia can be induced experimentally in mice by central administration of β-amyloid peptides, a process that involves cholinergic dysfunction (Maurice et al. 1996). Acute i.v. administration of caffeine or A_{2A}AR antagonists afforded protection against β-amyloid-induced amnesia (Dall'Igna et al. 2007). These acute effects of A_{2A}AR blockade are somewhat unexpected, because A_{2A}ARs are known to facilitate cholinergic function (namely in the hippocampus; Cunha et al. 1994b), and therefore either adenosine A_{2A}AR agonists or A_1AR antagonists (to prevent A_1AR-mediated inhibition of acetylcholine release) were expected to be cognitive enhancers. Indeed, the most widely used drugs in Alzheimer's disease are directed towards an increase in cholinergic function by inhibiting acetylcholinesterase (Doody et al. 2001). These apparent discrepancies point towards the need for more basic research to understand the biological basis and the potential benefits of the emerging adenosine-based therapies for Alzheimer's disease.

8.1.5 Anxiety

The inhibitory action of A_1ARs on the nervous system, together with the identification of crosstalk mechanisms between benzodiazepines and ARs (Boulenger et al. 1982) and transporters (Bender et al. 1980), soon suggested that adenosine could mediate the anxiolytic action of several centrally active drugs (Phillis and Wu 1982). The possibility that drugs that facilitate A_1AR-mediated actions could be effective for anxiety was supported by the observations that A_1AR agonists have anxiolytic actions in rodents (Florio et al. 1998; Jain et al. 1995). Accordingly, A_1AR KO mice showed increased anxiety-related behavior (Johansson et al. 2001), but this also holds true for A_{2A}AR KO mice (Ledent et al. 1997). A_1 and A_{2A}ARs are involved in benzodiazepine withdrawal signs. In mice, these signs of withdrawal are manifested by increased seizure susceptibility, and agonists of A_1ARs (Listos et al. 2005) or A_{2A}ARs (Listos et al. 2008) attenuate them. The potential of A_1AR agonists to reduce the anxiogenic effects during ethanol withdrawal has also been suggested (Prediger et al. 2006).

It is of interest that patients suffering from panic disorder, a serious form of anxiety disorder, appear to be particularly sensitive to small amounts of caffeine (Boulenger et al. 1984). Caffeine is well known to promote anxious behavior in humans and animal models, and can precipitate panic attacks (Klein et al. 1991). It is, however, worth noting that chronic and acute caffeine consumption may lead to quite different consequences with respect to the function of ARs (see above; Boulenger et al. 1983; Jacobson et al. 1996). The short-term anxiety-like effect of caffeine in mice may not be related solely to the blockade of A_1 and A_{2A}ARs, since it is not shared by selective antagonists of each receptor (El Yacoubi et al. 2000). In contrast, anxiolytic effects of a xanthine derivative have been reported, but this is most probably related to agonist activity at serotonin receptors (Daly 2007).

A significant association between self-reported anxiety after caffeine administration and two linked polymorphisms of the A_{2A}AR gene has been reported (Alsene et al. 2003). Furthermore, evidence for a susceptibility locus for panic disorder, either within the A_{2A}AR gene or in a nearby region of chromosome 22, was reported (Deckert et al. 1998, Hamilton et al. 2004). However, this positive association between A_{2A}AR gene polymorphism and panic disorder may not occur in the Asian population (Lam et al. 2005), suggesting an ethnicity-dependent association.

8.1.6 Depression

A_{2A}AR KO mice and wild-type mice injected with A_{2A}AR antagonists were found to be less sensitive to "depressant" challenges than controls (El Yacoubi et al. 2001), suggesting that blockade of adenosine A_{2A}ARs might be an interesting target for the development of antidepressant agents. This antidepressant-like effect of selective A_{2A}AR antagonists is probably linked to an interaction with dopaminergic transmission, possibly in the frontal cortex, since administration of the dopamine D_2 receptor antagonist haloperidol prevented antidepressant-like effects elicited by selective A_{2A}AR antagonists in the forced swim test (putatively involving cortex), whereas it had no effect on stimulant motor effects of selective A_{2A}AR antagonists (putatively linked to ventral striatum) (El Yacoubi et al. 2003). Depression is frequently associated with loss of motivation and psychomotor slowing. In this context, it is interesting to note that A_{2A}ARs in the nucleus accumbens appear to regulate effort-related processes and action that could be related to modulation of the ventral striatopallidal pathway (Mingote et al. 2008).

Besides A_{2A}ARs, A_1ARs are also probably involved in the antidepressant-like effect of adenosine (Kaster et al. 2004), which may be of consequence for interactions with the opioid system (Kaster et al. 2007).

It is worth noting that that deep brain stimulation, now widely used by neurosurgeons to treat tremor and other movement disorders, as well as in a number of psychiatric diseases, including obsessive–compulsive disorders and depression, produces its effects by inducing the release of ATP, which is subsequently converted extracellularly to adenosine (Bekar et al. 2008).

Results from clinical and basic studies have demonstrated that stress and depression decrease BDNF expression and neurogenesis, leading to the neurotrophic hypothesis of depression (Castrén et al. 2007; Kozisek et al. 2008). How adenosine A_{2A}AR-dependent facilitation of BDNF actions on hippocampal synapses (see Sect. 5.3), namely enhancement of synaptic transmission (Diógenes et al. 2004) and enhancement of synaptic plasticity (Fontinha et al. 2008), may contribute to these antidepressive actions of adenosine remains to be established.

8.1.7 Schizophrenia

No study, so far, has directly evaluated the influence of caffeine in schizophrenia, but there is growing evidence that adenosine dysfunction may contribute to the neurobiological and clinical features of schizophrenia (Lara et al. 2006). Indeed, adenosine, via activation of A_1 and A_{2A}ARs, is uniquely positioned to influence glutamatergic and dopaminergic neurotransmission, the two neurotransmitter systems that are most affected by the disease. It is possible that an adenosine inhibitory deficit may emerge, resulting in reduced control of dopamine activity and increased vulnerability to excitotoxic glutamate action in the mature brain. Interactions between A_{2A}ARs and D_2 receptors allow further opportunity for mutual modulation between the adenosine and dopamine systems (Fuxe et al. 2007). These mechanisms could provide a rationale for an antipsychotic-like profile for AR agonists, in particular A_{2A}AR agonists to promote a reduction in D_2 receptor signaling (Fuxe et al. 2007) and A_1AR agonists to promote a reduction in dopamine release (Lara et al.

2006). Indeed, dypiridamole, a well-known inhibitor of adenosine transporters and therefore an enhancer of extracellular adenosine levels, may be of some therapeutic interest in schizophrenic patients (Akhondzadeh et al. 2000).

Reduced NMDA receptor function may contribute to the cognitive and negative symptoms of schizophrenia (Ross et al. 2006). The relationships between adenosine and NMDA receptor function are complex and may operate in opposite ways. Thus, NMDA receptor activation induces adenosine release (Hoehn and White 1989; Schotanus et al. 2006), and therefore NMDA receptor hypofunction may induce a decrease in adenosine-mediated actions. On the other hand, NMDA receptor activation suppresses neuronal sensitivity to adenosine (Nikbakht and Stone 2001). In addition, both A_1 and A_{2A}ARs can influence NMDA receptor functioning, with both receptors being able to inhibit NMDA currents in different brain areas (see Sect. 5.2.1 above).

8.1.8 Huntington's Disease

The role played by ARs in Hungtington's disease was recently reviewed and discussed (Popoli et al. 2007) and is a topic in another chapter in this volume, Chap. 17, "Adenosine Receptors and Neurological Disease: Neuroprotection and Neurodegeneration" (by Stone et al.). Therefore, only a few considerations will be mentioned in this section. The complexity inherent to a genetically based, slowly progressing neurodegenerative disease; the different experimental models, which are very frequently nonchronic or subchronic models; as well as changes in receptor levels due to cell loss or to prolonged drug administration give an apparent contradictory picture of the AR involvement in this disease. The pre- versus postsynaptic localization of ARs, in particular A_{2A}ARs, which have highly distinct roles in striatal function according to their synaptic localization, may also contribute to conflicting neuroprotective/neurotoxic consequences of AR manipulation (Blum et al. 2003). Indeed, A_1AR agonists (Blum et al. 2002), A_{2A}AR agonists (Popoli et al. 2007), as well as A_{2A}AR antagonists (Domenici et al. 2007) are all able to influence diverse symptoms in experimental models of Huntington's disease.

Another aspect that applies to all neurodegenerative diseases, and that may be particularly relevant in the case of Huntington's disease, is related to the loss of neurotrophic support. Huntington's disease is caused by a mutation in a protein named huntingtin that, in its mutated form, is neurotoxic. It happens that wild-type huntingtin upregulates transcription of BDNF (Zuccato et al. 2001), and decreased BDNF levels may be an initial cause of neuronal death in this disease. A_{2A}AR activation can facilitate or even trigger BDNF actions in the brain (Diógenes et al. 2004, 2007; Fontinha et al. 2008), pointing to the possibility that A_{2A}AR activation, at least in the early stages of the disease, may rescue striatal neurons from death due to diminished trophic support by BDNF. It is worth noting that A_{2A}ARs have a dual action in Huntington's disease (Popoli et al. 2007). The ability of A_{2A}ARs to facilitate the actions of BDNF, which is clearly deficient in this neurodegenerative disease (Zuccato and Cattaneo 2007), is most probably some of the positive influence that A_{2A}ARs have on the disease.

8.1.9 Parkinson's Disease

A significant association between higher caffeine intake and lower incidence of Parkinson's disease was reported some years ago (Ross et al. 2000). Moreover, the beneficial effects of caffeine in Parkinson's disease patients have also been reported (Kitagawa et al. 2007, see Table 1). Furthermore, caffeine administered before levodopa may improve its pharmacokinetics in some patients with Parkinson's disease (Deleu et al. 2006).

Caffeine has well known stimulatory actions upon locomotion due to the antagonism of A_{2A} and A_1ARs in the striatum (Ferré 2008), and in most animal models of Parkinson's disease, antagonizing A_{2A}ARs attenuates some disease symptoms. Since a full chapter in this volume is devoted to ARs and Parkinson's disease, Chap. 18, "Adenosine A_{2A} Receptors and Parkinson's Disease" (by Morelli et al.), and since a recent sequence of reviews were published as proceedings of a meeting on the topic (Chen et al. 2007; Fredholm et al. 2007; Morelli et al. 2007; Schiffmann et al. 2007), we will only highlight a point that is focused upon less, which concerns interactions between adenosine and neurotrophic factors. The putative role of the neurotrophic factor GDNF in slowing or halting disease progression through the facilitation of neuronal survival (Peterson and Nutt 2008) and the facilitatory action of A_{2A}ARs upon the actions of GDNF in striatal dopaminergic nerve endings (Gomes et al. 2006) indicate the need for great caution when blocking A_{2A}ARs in the early phases of Parkinson's disease. Indeed, if the actions of GDNF in dopaminergic neurons depend upon coactivation of A_{2A}ARs (Gomes et al. 2006), it is highly probable that blockade of A_{2A}ARs will be deleterious during the time window when it is possible to rescue neurons with trophic support.

Another relevant consideration is related to the recent finding (Bekar et al. 2008) that deep brain stimulation, a procedure now used to reduce tremor in Parkinson's disease patients, involves the release of considerable amounts of ATP, with its subsequent extracellular metabolism to adenosine. Activation of A_1ARs by adenosine during this procedure is an essential step in reducing tremor and controlling spread of excitability, thereby reducing the side effects of deep brain stimulation. However, since A_{2A}ARs are highly expressed in thalamic areas, it could be expected that A_{2A}ARs are also activated during deep brain stimulation. Thus, in the late stages of the disease, where it is desirable to prevent A_{2A}AR-mediated inhibition of dopamine D_2 receptor function, the use of an A_{2A}AR antagonist in combination with deep brain stimulation may be beneficial.

9 Drug Addiction and Substances of Abuse

It is currently believed that molecular adaptations of the corticoaccumbens glutamatergic synapses are involved in compulsive drug seeking and relapse. The high density of A_{2A}ARs that pre- and postsynaptically regulate glutamatergic transmission in this brain area lead to the proposal that A_{2A}AR-related compounds could

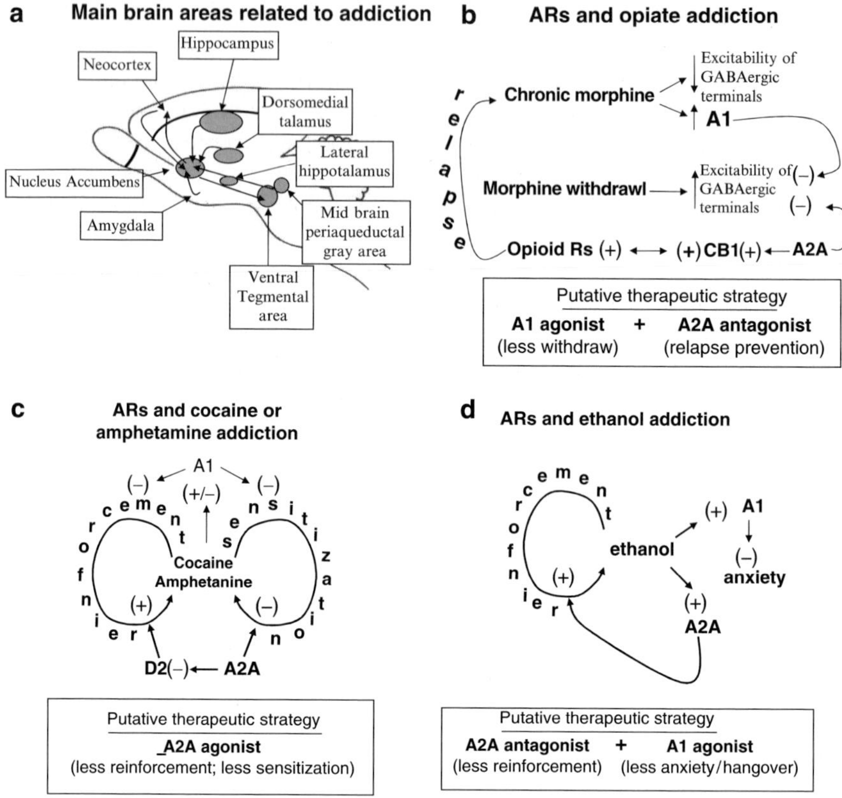

Fig. 5 a–d Brain areas mostly involved in addiction **a**, and the role of A_1 and A_{2A} adenosine receptors (ARs) in addictive behavior **b–d**. The putative therapeutic stratagy based on the ARs is indicated *below each panel*. A *plus symbol* denotes facilitation and a *minus symbol* denotes inhibition. See text for references

become new therapeutic agents for drug addiction (Ferré et al. 2007b). Other brain areas involved in reinforcement, motivational and withdrawal consequences of drug use and abuse are the limbic areas, such as the hippocampus and amygdala (Fig. 5a). Accordingly, there is a growing body of evidence suggesting that adenosine is involved in drug addiction and withdrawal, that both A_1 and A_2 ARs may be involved (Hack and Christie 2003), and that a considerable degree of compensation may occur.

9.1 Opioids

Caffeine combined with the opioid antagonist naloxone produces a characteristic quasi-morphine withdrawal syndrome in opiate-naive animals that is almost completely abolished in A_{2A}AR KO mice and has intermediate intensity in heterozygous animals, suggesting an involvement of A_{2A}ARs in the withdrawal syndrome (Bilbao et al. 2006). These observations are in agreement with previous data that adenosine reduces morphine withdrawal in an acute model, while caffeine aggravates it (Capasso and Loizzo 2001).

Chronic treatment with opioids induces adaptations in neurons that lead to tolerance and dependence. Endogenous adenosine, through A_1AR activation, reduces the hyperexcitability of GABAergic terminals of the midbrain periaqueductal gray area (Fig. 5b) that occurs during withdrawal from chronic morphine treatment (Hack et al. 2003). Chronic morphine treatment significantly increased the number of A_1ARs (Kaplan et al. 1994) and adenosine transporters (Kaplan and Leite-Morris 1997) as well as the adenosine sensitivity in the nucleus accumbens (Brundege and Williams 2002). Surprisingly, chronic blockade of opioid receptors also causes upregulation of A_1ARs (Bailey et al. 2003), suggesting an adaptative mechanism in the purinergic system with chronic opioid receptor manipulation. Interestingly, A_{2A}AR levels in the striatum appear to be unaffected by chronic morphine (Kaplan et al. 1994) or chronic opioid antagonism (Bailey et al., 2003).

Both A_1 and A_{2A}AR agonists attenuate opiate withdrawal symptoms (Fig. 5b), but the specific symptoms affected by each AR are different, and the corresponding AR antagonists exacerbate those symptoms (Kaplan and Sears 1996), suggesting that AR agonists rather than AR antagonists may be useful as therapeutics for opioid withdrawal. In line with this idea is the observation that AK inhibitors attenuate opiate withdrawal symptoms (Kaplan and Coyle 1998). Adenosine also seems to act as a regulator of regional cerebral blood flow in both morphine-dependent rats and morphine withdrawal in rats (Khorasani et al., 2006).

Relapse is the most serious limitation of effective medical treatment of opiate addiction. In this respect, A_{2A}AR antagonists may prove useful since A_{2A}AR antagonists administered either directly into the nucleus accumbens or indirectly by intraperitoneal injection eliminate heroin-induced reinstatement in rats that are trained to self-administer heroin, a model of human craving and relapse (Yao et al. 2006). The mechanism wherein A_{2A}AR antagonists block heroin reinstatement most likely involves opiate receptors and their synergy with other GPCRs, namely crosstalk between CB_1 receptors and A_{2A}AR signaling, as well as $\beta-\gamma$ dimers (see Sect. 5.1.4 and Fig. 5b).

9.2 Cocaine

Activation of A_{2A}ARs is required to develop the addictive effects to cocaine, since the lack of A_{2A}ARs diminishes the reinforcing efficacy of cocaine (Soria et al. 2006). On the other hand, A_{2A}AR activation protects against cocaine sensitization (Filip et al. 2006), which suggests a therapeutic potential of A_{2A}AR agonists in the treatment of cocaine dependence (Fig. 5c). This is not unexpected, since A_{2A}ARs inhibit D_2 receptor functioning, and these receptors are highly involved in brain-reinforcing circuits. In line with this idea are the observations that A_{2A}AR agonists inhibit cocaine self-administration in rats (Knapp et al. 2001), and that a nonselective AR antagonist reinstates cocaine-seeking behavior and maintains self-administration in baboons (Weerts and Griffiths 2003). Interestingly, in high-risk situations, prophylactic activation of A_{2A}AR activation may prove beneficial, since A_{2A}AR agonists inhibit the initiation of cocaine self-administration in rats (Knapp et al. 2001). However, the ability of caffeine to prevent the extinction of cocaine-seeking behavior (Kuzmin et al. 1999) or even to reinstate extinguished cocaine self-administration (Green and Schenk 2002) may be related to its blocking effects on A_1ARs, rather than A_{2A}ARs. Moreover, in the nucleus accumbens, sorting and recycling of A_1ARs is dysregulated as a consequence of repeated cocaine administration, so that the amount of A_1AR protein and mRNA is upregulated but the number of membrane receptors, their coupling to G proteins, and their ability to form dimers with D_1 receptors is downregulated (Toda et al. 2003). Furthermore, adenosine uptake in the nucleus accumbens seems to be augmented after cocaine withdrawal (Manzoni et al. 1998)

9.3 Amphetamine

Daily treatment with amphetamine markedly enhances locomotor responses, and this enhancement remains after washout, a process known as sensitization. No sensitization to amphethamines occurs either in conditional A_{2A}AR KO mice or in the presence of A_{2A}AR activation (Bastia et al. 2005), indicating that A_{2A} receptors reduce sensitization (Fig. 5c). Also, selective A_1AR agonists may have some attenuating influence on the development of amphetamine dependence (Poleszak and Malec 2003).

9.4 Cannabinoids

Several studies have reported crosstalk between ARs and CB receptors, as mentioned above (see Sect. 5.1.4). In this section, only the studies specifically addressing the influence of ARs upon CB addiction or tolerance will be mentioned. Crosstolerance between A_1AR and CB_1 receptor agonists has been reported in motor incoordination induced by CBs (DeSanty and Dar 2001). A significant reduction in tetrahydrocannabinol-induced rewarding and aversive effects was found in mice lacking A_{2A}ARs, indicating a specific involvement of A_{2A}ARs in the addiction-related properties of CBs (Soria et al. 2004). Somatic manifestations of tetrahydrocannabinol withdrawal were also significantly attenuated in A_{2A}AR KO mice; however, antinociception, hypolocomotion and hypothermia induced by acute tetrahydrocannabinol administration were not affected (Soria et al. 2004).

9.5 Ethanol

The anxiolytic properties of ethanol are generally accepted to be an important motivational factor in its consumption and the development of alcohol dependence. The anxiolytic-like effect induced by ethanol in mice involves the activation of A_1ARs but not A_{2A}ARs (Prediger et al. 2004). The anxiety-like behavior observed during acute ethanol withdrawal (hangover) in mice is attenuated by nonanxiolytic doses of A_1AR agonists (Prediger et al. 2006). Tolerance to ethanol-induced motor incoordination is prevented by A_1AR and dopamine D_1 receptor antagonists, but not by A_{2A}AR antagonists (Batista et al. 2005). However, the reinforcing properties of ethanol are partially mediated via an A_2AR activation of cAMP/PKA signaling in the nucleus accumbens, indicating that administration of an A_{2A}AR antagonist may decrease ethanol reward and consumption (Fig. 5d). Indeed, A_{2A}AR antagonism produces a robust and behaviorally selective reduction of ethanol reinforcement (Thorsell et al. 2007).

10 Concluding Remarks

Several years ago, we (Sebastião and Ribeiro 2000) pointed out that "*In addition to its direct pre- and post-synaptic actions on neurones, adenosine is rich in nuances of priming, triggering and inhibiting the action of several neurotransmitters and neuromodulators* (...) *The harmonic way adenosine builds its influence at synapses to control neuronal communication is operated through fine-tuning, 'synchronizing' or 'desynchronizing' receptor activation...*". In a recent review, Uhlhaas and Singer (2006) considered that abnormal neural synchronization is central to and the underlying basis for several neurological diseases such as epilepsy, schizophrenia, autism, Alzheimer's disease, and Parkinson's disease. These authors highlighted the role of GABAergic neurons and their pivotal role in the primary generation of high-frequency oscillations and local synchronization, the role of glutamatergic connections in controlling their strength, duration, and long-range synchronization, and the role of cholinergic modulation in the fast state-dependent facilitation of high-frequency oscillations and the associated response synchronization. As reviewed in the present work, adenosine is a molecule involved in brain homeostasis that has recently been proposed to be crucial to the effects of deep brain stimulation (Bekar et al. 2008), which mainly aims to affect neuronal synchronization and therefore influence several psychiatric and neurodegenerative diseases. This review suggests that adenosine is a sort of "universal modulator" or a "maestro;" the main molecule involved in coordinating and controlling the synchronization of the release and actions of many synaptic mediators. It also suggests that targeting approaches that increase adenosine levels to provide this synchronization, or targeting ARs with novel safe, selective, and effective therapeutics that are currently in (or are poised to enter) clinical trials, will enhance our understanding of the role of this important endogenous "universal modulator" signaling molecule and its receptors in cognition, neurodegenerative diseases, psychiatric diseases, and drug addiction.

Acknowledgements The work in the authors' laboratory is supported by research grants from Fundação para a Ciência e Tecnologia (FCT), Gulbenkian Foundation and the European Union (COST B30).

References

Abbracchio MP, Fogliatto G, Paoletti AM, Rovati GE, Cattabeni F (1992) Prolonged in vitro exposure of rat brain slices to adenosine analogues: selective desensitization of A_1 but not A_2 receptors. Eur J Pharmacol 227:317–324

Adams CL, Cowen MS, Short JL, Lawrence AJ (2008) Combined antagonism of glutamate mGlu5 and adenosine A_{2A} receptors interact to regulate alcohol-seeking in rats. Int J Neuropsychopharmacol 11:229–241

Akhondzadeh S, Shasavand E, Jamilian H, Shabestari O, Kamalipour A (2000) Dipyridamole in the treatment of schizophrenia: adenosine–dopamine receptor interactions. J Clin Pharm Ther 25:131–137

Albasanz JL, Rodríguez A, Ferrer I, Martín M (2006) Adenosine A_{2A} receptors are up-regulated in Pick's disease frontal cortex. Brain Pathol 16:249–255

Albasanz JL, Rodríguez A, Ferrer I, Martín M (2007) Up-regulation of adenosine A_1 receptors in frontal cortex from Pick's disease cases. Eur J Neurosci 26:3501–3508

Albasanz JL, Perez S, Barrachina M, Ferrer I, Martín M (2008) Up-regulation of adenosine receptors in the frontal cortex in Alzheimer's disease. Brain Pathol 18:211–219

Alexander SP, Curtis AR, Hill SJ, Kendall DA (1992) Activation of a metabotropic excitatory aminoacid receptor potentiates A_{2b} adenosine receptor-stimulated cyclic AMP accumulation. Neurosci Lett 146:231–233

Almeida T, Rodrigues RJ, de Mendonça A, Ribeiro JA, Cunha RA (2003) Purinergic P2 receptors trigger adenosine release leading to adenosine A_{2A} receptor activation and facilitation of long-term potentiation in rat hippocampal slices. Neuroscience 122:111–121

Alsene K, Deckert J, Sand P, de Wit H (2003) Association between A_{2a} receptor gene polymorphisms and caffeine-induced anxiety. Neuropsychopharmacology 28:1694–1702

Andersson M, Usiello A, Borgkvist A, Pozzi L, Dominguez C, Fienberg AA, Svenningsson P, Fredholm BB, Borrelli E, Greengard P, Fisone G (2005) Cannabinoid action depends on phosphorylation of dopamine- and cAMP-regulated phosphoprotein of 32 kDa at the protein kinase A site in striatal projection neurons. J Neurosci 25:8432–8438

Angelatou F, Pagonopoulou O, Maraziotis T, Olivier A, Villemeure JG, Avoli M, Kostopoulos G (1993) Upregulation of A_1 adenosine receptors in human temporal lobe epilepsy: a quantitative autoradiographic study. Neurosci Lett 163:11–14

Angulo E, Casadó V, Mallol J, Canela EI, Viñals F, Ferrer I, Lluis C, Franco R (2003) A_1 adenosine receptors accumulate in neurodegenerative structures in Alzheimer disease and mediate both amyloid precursor protein processing and tau phosphorylation and translocation. Brain Pathol 13:440–451

Aranda JV, Turmen T (1979) Methylxanthines in apnea of prematurity. Clin Perinatol 6:87–108

Arendash GW, Schleif W, Rezai-Zadeh K, Jackson EK, Zacharia LC, Cracchiolo JR, Shippy D, Tan J (2006) Caffeine protects Alzheimer's mice against cognitive impairment and reduces brain beta-amyloid production. Neuroscience 142:941–952

Arrigoni E, Crocker AJ, Saper CB, Greene RW, Scammell TE (2005) Deletion of presynaptic adenosine A_1 receptors impairs the recovery of synaptic transmission after hypoxia. Neuroscience 132:575–580

Bahls FH, Ma KK, Bird TD (1991) Theophylline-associated seizures with "therapeutic" or low toxic serum concentrations: risk factors for serious outcome in adults. Neurology 41: 1309–1312

Bailey A, Ledent C, Kelly M, Hourani SM, Kitchen I (2002) Changes in spinal delta and kappa opioid systems in mice deficient in the A_{2A} receptor gene. J Neurosci 22:9210–9220

Bailey A, Hawkins RM, Hourani SM, Kitchen I (2003) Quantitative autoradiography of adenosine receptors in brains of chronic naltrexone-treated mice. Br J Pharmacol 139:1187–1195

Bairam A, Boutroy MJ, Badonnel Y, Vert P (1987) Theophylline versus caffeine: comparative effects in treatment of idiopathic apnea in the preterm infant. J Pediatr 110:636–639

Barraco RA, Swanson TH, Phillis JW, Berman RF (1984) Anticonvulsant effects of adenosine analogues on amygdaloid-kindled seizures in rats. Neurosci Lett 46:317–322

Basheer R, Strecker RE, Thakkar MM. McCarley RW (2004) Adenosine and sleep-wake regulation. Prog Neurobiol 73:379–396

Bastia E, Xu YH, Scibelli AC, Day YJ, Linden J, Chen JF, Schwarzschild MA (2005) A crucial role for forebrain adenosine A(2A) receptors in amphetamine sensitization. Neuropsychopharmacology 30:891–900

Batista LC, Prediger RD, Morato GS, Takahashi RN (2005) Blockade of adenosine and dopamine receptors inhibits the development of rapid tolerance to ethanol in mice. Psychopharmacology 181:714–721

Bauman LA, Mahle CD, Boissard CG, Gribkoff VK (1992) Age-dependence of effects of A_1 adenosine receptor antagonism in rat hippocampal slices. J Neurophysiol 68:629–638

Bekar L, Libionka W, Tian GF, Xu Q, Torres A, Wang X, Lovatt D, Williams E, Takano T, Schnermann J, Bakos R, Nedergaard M (2008) Adenosine is crucial for deep brain stimulation-mediated attenuation of tremor. Nat Med 14:17–19

Bender AS, Phillis JW, Wu PH (1980) Diazepam and flurazepam inhibit adenosine uptake by rat brain synaptosomes. J Pharm Pharmacol 32:293–294

Benington JH, Heller HC (1995) Restoration of brain energy metabolism as the function of sleep. Prog Neurobiol 45:347–360

Biber K, Lubrich B, Fiebich BL, Boddeke HW, van Calker D (2001) Interleukin-6 enhances expression of adenosine A(1) receptor mRNA and signaling in cultured rat cortical astrocytes and brain slices. Neuropsychopharmacology 24:86–96

Biber K, Pinto-Duarte A, Wittendorp MC, Dolga AM, Fernandes CC, Von Frijtag Drabbe Künzel J, Keijser JN, de Vries R, Ijzerman AP, Ribeiro JA, Eisel U, Sebastião AM, Boddeke HW (2007) Interleukin-6 upregulates neuronal adenosine A(1) receptors: implications for neuromodulation and neuroprotection. Neuropsychopharmacology 33:2237–2250

Bilbao A, Cippitelli A, Martín AB, Granado N, Ortiz O, Bezard E, Chen JF, Navarro M, Rodríguez de Fonseca F, Moratalla R (2006) Absence of quasi-morphine withdrawal syndrome in adenosine A_{2A} receptor knockout mice. Psychopharmacology 185:160–168

Blum D, Gall D, Galas MC, d'Alcantara P, Bantubungi K, Schiffmann SN (2002) The adenosine A_1 receptor agonist adenosine amine congener exerts a neuroprotective effect against the development of striatal lesions and motor impairments in the 3-nitropropionic acid model of neurotoxicity. J Neurosci 22:9122–9133

Blum D, Galas MC, Pintor A, Brouillet E, Ledent C, Muller CE, Bantubungi K, Galluzzo M, Gall D, Cuvelier L, Rolland AS, Popoli P, Schiffmann SN (2003) A dual role of adenosine A_{2A} receptors in 3-nitropropionic acid-induced striatal lesions: implications for the neuroprotective potential of A_{2A} antagonists. J Neurosci 23:5361–5369

Boison D (2007) Adenosine-based cell therapy approaches for pharmacoresistant epilepsies. Neurodegener Dis 4:28–433

Boulanger L, Poo M (1999a) Presynaptic depolarization facilitates neurotrophin-induced synaptic potentiation. Nat Neurosci 2:346–351

Boulanger L, Poo M (1999b) Gating of BDNF-induced synaptic potentiation by cAMP. Science 284:1982–1984

Boulenger JP, Patel J, Marangos PJ (1982) Effects of caffeine and theophylline on adenosine and benzodiazepine receptors in human brain. Neurosci Lett 30:161–166

Boulenger JP, Patel J, Post RM, Parma AM, Marangos PJ (1983) Chronic caffeine consumption increases the number of brain adenosine receptors. Life Sci 32:1135–1142

Boulenger JP, Uhde TW, Wolff EA 3rd, Post RM (1984) Increased sensitivity to caffeine in patients with panic disorders. Preliminary evidence. Arch Gen Psychiat 41:1067–1071

Braun N, Zhu Y, Krieglstein J, Culmsee C, Zimmermann H (1998) Upregulation of the enzyme chain hydrolyzing extracellular ATP after transient forebrain ischemia in the rat. J Neurosci 18:4891–4900

Brown SJ, James S, Reddington M, Richardson PJ (1990) Both A_1 and A_{2a} purine receptors regulate striatal acetylcholine release. J Neurochem 55:31–38

Brundege JM, Williams JT (2002) Increase in adenosine sensitivity in the nucleus accumbens following chronic morphine treatment. J Neurophysiol 87:1369–1375

Budd DC, Nichols DG (1995) Protein kinase C-mediated suppression of the presynaptic adenosine A_1 receptor by a facilitatory metabotropic glutamate receptor. J Neurochem 65:615–621

Burnstock G (1976) Purinergic receptors. J Theor Biol 62:491–503

Burnstock G (2007) Physiology and pathophysiology of purinergic neurotransmission. Physiol Rev 87:659–797

Capasso A, Loizzo A (2001) Purinoreceptors are involved in the control of acute morphine withdrawal. Life Sci 69:2179–2188

Carriba P, Ortiz O, Patkar K, Justinova Z, Stroik J, Themann A, Müller C, Woods AS, Hope BT, Ciruela F, Casadó V, Canela EI, Lluis C, Goldberg SR, Moratalla R, Franco R, Ferré S (2007) Striatal adenosine A_{2A} and cannabinoid CB1 receptors form functional heteromeric complexes that mediate the motor effects of cannabinoids. Neuropsychopharmacology 32:2249–2259

Carruthers AM, Sellers LA, Jenkins DW, Jarvie EM, Feniuk W, Humphrey PP (2001) Adenosine A(1) receptor-mediated inhibition of protein kinase A-induced calcitonin gene-related peptide release from rat trigeminal neurons. Mol Pharmacol 59:1533–1541

Castrén E, Võikar V, Rantamäki T (2007) Role of neurotrophic factors in depression. Curr Opin Pharmacol 7:18–21

Chase TN, Bibbiani F, Bara-Jimenez W, Dimitrova T, Oh-Lee JD (2003) Translating A_{2A} antagonist KW6002 from animal models to Parkinsonian patients. Neurology 61:S107–S111

Chen JF, Sonsalla PK, Pedata F, Melani A, Domenici MR, Popoli P, Geiger J, Lopes LV, de Mendonça A (2007) Adenosine A_{2A} receptors and brain injury: broad spectrum of neuroprotection, multifaceted actions and "fine tuning" modulation. Prog Neurobiol 83:310–331

Cheng HC, Shih HM, Chern Y (2002) Essential role of cAMP-response element-binding protein activation by A_{2A} adenosine receptors in rescuing the nerve growth factor-induced neurite outgrowth impaired by blockage of the MAPK cascade. J Biol Chem 277:33930–33942

Chin JH (1989) Adenosine receptors in brain: neuromodulation and role in epilepsy. Ann Neurol 26:695–698

Ciccarelli R, Di Iorio P, Bruno V, Battaglia G, D'Alimonte I, D'Onofrio M, Nicoletti F, Caciagli F (1999) Activation of A(1) adenosine or mGlu3 metabotropic glutamate receptors enhances the release of nerve growth factor and S-100beta protein from cultured astrocytes. Glia 27:275–281

Ciruela F, Casadó V, Rodrigues RJ, Luján R, Burgueño J, Canals M, Borycz J, Rebola N, Goldberg SR, Mallol J, Cortés A, Canela EI, López-Giménez JF, Milligan G, Lluis C, Cunha RA, Ferré S, Franco R (2006) Presynaptic control of striatal glutamatergic neurotransmission by adenosine A_1–A_{2A} receptor heteromers. J Neurosci 26:2080–2087

Coccurello R, Breysse N, Amalric M (2004) Simultaneous blockade of adenosine A_{2A} and metabotropic glutamate mGlu5 receptors increase their efficacy in reversing Parkinsonian deficits in rats. Neuropsychopharmacology 29:1451–1461

Coelho JE, Rebola N, Fragata I, Ribeiro JA, de Mendonça A, Cunha RA (2006) Hypoxia-induced desensitization and internalization of adenosine A_1 receptors in the rat hippocampus. Neuroscience 138:1195–1203

Coleman CG, Baghdoyan HA, Lydic R (2006) Dialysis delivery of an adenosine A_{2A} agonist into the pontine reticular formation of C57BL/6J mouse increases pontine acetylcholine release and sleep. J Neurochem 96:1750–1759

Conde SV, Obeso A, Vicario I, Rigual R, Rocher A, Gonzalez C (2006) Caffeine inhibition of rat carotid body chemoreceptors is mediated by A_{2A} and A_{2B} adenosine receptors. J Neurochem 98:616–628

Congar P, Khazipov R, Ben-Ari Y (1995) Direct demonstration of functional disconnection by anoxia of inhibitory interneurons from excitatory inputs in rat hippocampus. J Neurophysiol 73:421–426

Conlay LA, Conant JA, deBros F, Wurtman R (1997) Caffeine alters plasma adenosine levels. Nature 389:136

Cormier RJ, Mennerick S, Melbostad H, Zorumski CF (2001) Basal levels of adenosine modulate mGluR5 on rat hippocampal astrocytes. Glia 33:24–35

Cornelis MC, El-Sohemy A, Campos H (2007) Genetic polymorphism of the adenosine A_{2A} receptor is associated with habitual caffeine consumption. Am J Clin Nutr 86:240–244

Correia-de-Sá P, Ribeiro JA (1994a) Potentiation by tonic A_{2a}-adenosine receptor activation of CGRP-facilitated [3H]-ACh release from rat motor nerve endings. Br J Pharmacol 111: 582–588

Correia-de-Sá P, Ribeiro JA (1994b) Tonic adenosine A_{2A} receptor activation modulates nicotinic autoreceptor function at the rat neuromuscular junction. Eur J Pharmacol 271:349–355

Correia-de-Sá, P, Sebastião AM, Ribeiro JA (1991) Inhibitory and excitatory effects of adenosine receptor agonists on evoked transmitter release from phrenic nerve ending of the rat. Br J Pharmacol 103:1614–1620

Correia-de-Sá P, Timóteo MA, Ribeiro JA (1996) Presynaptic A_1 inhibitory/A_{2A} facilitatory adenosine receptor activation balance depends on motor nerve stimulation paradigm at the rat hemidiaphragm. J Neurophysiol 76:3910–3919

Correia-de-Sá P, Timóteo MA, Ribeiro JA (2001) Synergism between A(2A)-adenosine receptor activation and vasoactive intestinal peptide to facilitate [3H]-acetylcholine release from the rat motor nerve terminals. Neurosci Lett 2001 309:101–114

Corsi C, Melani A, Bianchi L, Pepeu G, Pedata F (1999) Effect of adenosine A_{2A} receptor stimulation on GABA release from the striatum of young and aged rats in vivo. Neuroreport 10:3933–3937

Corsi C, Melani A, Bianchi L, Pedata F (2000) Striatal A_{2A} adenosine receptor antagonism differentially modifies striatal glutamate outflow in vivo in young and aged rats. Neuroreport 11:2591–2595

Costenla AR, De Mendonça A, Sebastião A, Ribeiro JA (1999) An adenosine analogue inhibits NMDA receptor-mediated responses in bipolar cells of the rat retina. Exp Eye Res 68:367–370

Cunha RA, Sebastião AM, Ribeiro JA (1992) Ecto-5′-nucleotidase is associated with cholinergic nerve terminals in the hippocampus but not in the cerebral cortex of the rat. J Neurochem 59:657–666

Cunha RA, Johansson B, van der Ploeg I, Sebastião AM, Ribeiro JA, Fredholm BB (1994a) Evidence for functionally important adenosine A_{2a} receptors in the rat hippocampus. Brain Res 649:208–216

Cunha RA, Milusheva E, Vizi ES, Ribeiro JA, Sebastião AM (1994b) Excitatory and inhibitory effects of A_1 and A_{2A} adenosine receptor activation on the electrically evoked [3H]acetylcholine release from different areas of the rat hippocampus. J Neurochem 63:207–214

Cunha RA, Ribeiro JA, Sebastião AM (1994c) Purinergic modulation of the evoked release of [3H]acetylcholine from the hippocampus and cerebral cortex of the rat: role of the ectonucleotidases. Eur J Neurosci 6:33–42

Cunha RA, Constantino MC, Sebastião AM, Ribeiro JA (1995) Modification of A_1 and A_{2a} adenosine receptor binding in aged striatum, hippocampus and cortex of the rat. Neuroreport 6:1583–1588

Cunha RA, Vizi ES, Ribeiro JA, Sebastião AM (1996a) Preferential release of ATP and its extracellular catabolism as a source of adenosine upon high- but not low-frequency stimulation of rat hippocampal slices. J Neurochem 67:2180–2187

Cunha RA, Correia-de-Sá P, Sebastião AM, Ribeiro JA (1996b) Preferential activation of excitatory adenosine receptors at rat hippocampal and neuromuscular synapses by adenosine formed from released adenine nucleotides. Br J Pharmacol 119:253–260

Cunha RA, Sebastião AM, Ribeiro JA (1998) Inhibition by ATP of hippocampal synaptic transmission requires localized extracellular catabolism by ecto-nucleotidases into adenosine and channeling to adenosine A_1 receptors. J Neurosci 18:1987–1995

Cunha-Reis D, Sebastião AM, Wirkner K, Illes P, Ribeiro JA (2004) VIP enhances both pre- and post-synaptic GABAergic transmission to hippocampal interneurones leading to increased excitatory synaptic transmission to CA1 pyramidal cells. Br J Pharmacol 143:733–744

Cunha-Reis D, Ribeiro JA, Sebastião AM (2005) VIP enhances synaptic transmission to hippocampal CA1 pyramidal cells through activation of both VPAC1 and VPAC2 receptors. Brain Res 1049:52–60

Cunha-Reis D, Fontinha BM, Ribeiro JA, Sebastião AM (2007) Tonic adenosine A_1 and A_{2A} receptor activation is required for the excitatory action of VIP on synaptic transmission in the CA1 area of the hippocampus. Neuropharmacology 52:313–320

Cunha-Reis D, Ribeiro JA, Sebastião AM (2008) A_1 and A_{2A} receptor activation by endogenous adenosine is required for VIP enhancement of K^+-evoked [3H]-GABA release from rat hippocampal nerve terminals. Neurosci Lett 430:207–212

Dall'Igna OP, Fett P, Gomes MW, Souza DO, Cunha RA, Lara DR (2007) Caffeine and adenosine A(2a) receptor antagonists prevent beta-amyloid (25–35)-induced cognitive deficits in mice. Exp Neurol 203:241–245

Daly JW (2007) Caffeine analogs: biomedical impact. Cell Mol Life Sci 64:2153–2169

Daré E, Schulte G, Karovic O, Hammarberg C, Fredholm BB (2007) Modulation of glial cell functions by adenosine receptors. Physiol Behav 92:15–20

de Mendonça A, Ribeiro JA (1994) Endogenous adenosine modulates long-term potentiation in the hippocampus. Neuroscience 62:385–390

de Mendonça A, Ribeiro JA (1997a) Influence of metabotropic glutamate receptors agonists on the inhibitory effects of adenosine A_1 receptor activation in the rat hippocampus. Br J Pharmacol 121:1541–1548

de Mendonça JA, Ribeiro JA (1997b) Adenosine and neuronal plasticity. Life Sci 60:245–251

de Mendonça A, Sebastião AM, Ribeiro JA (1995) Inhibition of NMDA receptor-mediated currents in isolated rat hippocampal neurones by adenosine A_1 receptor activation. Neuroreport 6: 1097–1100

de Mendonça A, Almeida T, Bashir ZI, Ribeiro JA (1997c) Endogenous adenosine attenuates long-term depression and depotentiation in the CA1 region of the rat hippocampus. Neuropharmacology 36:161–167

de Mendonça A, Sebastião AM, Ribeiro JA (2000) Adenosine: does it have a neuroprotective role after all? Brain Res Rev 33:258–274

Deckert J, Nöthen MM, Franke P, Delmo C, Fritze J, Knapp M, Maier W, Beckmann H, Propping P (1998) Systematic mutation screening and association study of the A_1 and A_{2a} adenosine receptor genes in panic disorder suggest a contribution of the A_{2a} gene to the development of disease. Mol Psychiatr 3:81–85

Deckert J, Brenner M, Durany N, Zöchling R, Paulus W, Ransmayr G, Tatschner T, Danielczyk W, Jellinger K, Riederer P (2003) Up-regulation of striatal adenosine A(2A) receptors in schizophrenia. Neuroreport 14:313–316

DeLander GE, Wahl JJ (1988) Behavior induced by putative nociceptive neurotransmitters is inhibited by adenosine or adenosine analogs coadministered intrathecally. J Pharmacol Exp Ther 246:565–570

Deleu D, Jacob P, Chand P, Sarre S, Colwell A (2006) Effects of caffeine on levodopa pharmacokinetics and pharmacodynamics in Parkinson disease. Neurology 67:897–899

DeMet E, Stein MK, Tran C, Chicz-DeMet A, Sangdahl C, Nelson J (1989) Caffeine taste test for panic disorder: adenosine receptor supersensitivity. Psychiatr Res 30:231–242

DeSanty KP, Dar MS (2001) Involvement of the cerebellar adenosine A(1) receptor in cannabinoid-induced motor incoordination in the acute and tolerant state in mice. Brain Res 905:178–187

Desfrere L, Olivier P, Schwendimann L, Verney C, Gressens P (2007) Transient inhibition of astrocytogenesis in developing mouse brain following postnatal caffeine exposure. Pediatr Res 62:604–609

Di Iorio P, Battaglia G, Ciccarelli R, Ballerini P, Giuliani P, Poli A, Nicoletti F, Caciagli (1996) F Interaction between A_1 adenosine and class II metabotropic glutamate receptors in the regulation of purine and glutamate release from rat hippocampal slices. J Neurochem 67:302–309

Díaz-Cabiale Z, Hurd Y, Guidolin D, Finnman UB, Zoli M, Agnati LF, Vanderhaeghen JJ, Fuxe K, Ferré S (2001) Adenosine A_{2A} agonist CGS 21680 decreases the affinity of dopamine D2 receptors for dopamine in human striatum. Neuroreport 12:1831–1834

Díaz-Hernández M, Pintor J, Miras-Portugal MT (2000) Modulation of the dinucleotide receptor present in rat midbrain synaptosomes by adenosine and ATP. Br J Pharmacol 130:434–440

Díaz-Hernández M, Pereira MF, Pintor J, Cunha RA, Ribeiro JA, Miras-Portugal MT (2002) Modulation of the rat hippocampal dinucleotide receptor by adenosine receptor activation. J Pharmacol Exp Ther 301:441–450

Diógenes MJ, Fernandes CC, Sebastião AM, Ribeiro JA (2004) Activation of adenosine A_{2A} receptor facilitates brain-derived neurotrophic factor modulation of synaptic transmission in hippocampal slices. J Neurosci 24:2905–2913

Diógenes MJ, Assaife-Lopes N, Pinto-Duarte A, Ribeiro JA, Sebastião AM (2007) Influence of age on BDNF modulation of hippocampal synaptic transmission: interplay with adenosine A_{2A} receptors. Hippocampus 17:577–585

Dixit A, Vaney N, Tandon OP (2006) Evaluation of cognitive brain functions in caffeine users: a P3 evoked potential study. Indian J Physiol Pharmacol 50:175–180

Dixon AK, Gubitz AK, Sirinathsinghji DJ, Richardson PJ, Freeman TC (1996) Tissue distribution of adenosine receptor mRNAs in the rat. Br J Pharmacol 118:1461–1468

Dixon AK, Widdowson L, Richardson PJ (1997) Desensitisation of the adenosine A_1 receptor by the A_{2A} receptor in rat striatum. J Neurochem 69:315–321

Domenici MR, Scattoni ML, Martire A, Lastoria G, Potenza RL, Borioni A, Venerosi A, Calamandrei G, Popoli P (2007) Behavioral and electrophysiological effects of the adenosine A_{2A} receptor antagonist SCH 58261 in R6/2 Huntington's disease mice. Neurobiol Dis 28:197–205

Donoso MV, Aedo F, Huidobro-Toro JP (2006) The role of adenosine A_{2A} and A_3 receptors on the differential modulation of norepinephrine and neuropeptide Y release from peripheral sympathetic nerve terminals. J Neurochem 96:1680–1695

Doody RS, Geldmacher DS, Gordon B, Perdomo CA, Pratt RD (2001) Donepezil Study Group Open-label, multicenter, phase 3 extension study of the safety and efficacy of donepezil in patients with Alzheimer disease. Arch Neurol 58:427–433

Dragunow M, Goddard GV, Laverty R (1985) Is adenosine an endogenous anticonvulsant? Epilepsia 26:480–487

Duarte-Araújo M, Timóteo MA, Correia-de-Sá P (2004) Adenosine activating A(2A)-receptors coupled to adenylate cyclase/cyclic AMP pathway downregulates nicotinic autoreceptor function at the rat myenteric nerve terminals. Neurochem Int 45:641–651

Dunwiddie TV (1999) Adenosine and suppression of seizures. Adv Neurol 79:1001–1010

Dunwiddie TV, Diao L, Kim HO, Jiang JL, Jacobson KA (1997) Activation of hippocampal adenosine A_3 receptors produces a desensitization of A_1 receptor-mediated responses in rat hippocampus. J Neurosci 17:607–614

Edwards FA, Gibb AJ, Colquhoun D (1992) ATP receptor-mediated synaptic currents in the central nervous system. Nature 359:144–147

Eisenach JC, Rauck RL, Curry R (2003) Intrathecal, but not intravenous adenosine reduces allodynia in patients with neuropathic pain. Pain 105:65–70

Eisenach JC, Hood DD, Curry R, Sawynok J, Yaksh TL, Li X (2004) Intrathecal but not intravenous opioids release adenosine from the spinal cord. J Pain 5:64–68

El Yacoubi M, Ledent C, Parmentier M, Costentin J, Vaugeois JM (2000) The anxiogenic-like effect of caffeine in two experimental procedures measuring anxiety in the mouse is not shared by selective A(2A) adenosine receptor antagonists. Psychopharmacology 148:153–163

El Yacoubi M, Ledent C, Parmentier M, Bertorelli R, Ongini E, Costentin J, Vaugeois JM (2001) Adenosine A_{2A} receptor antagonists are potential antidepressants: evidence based on pharmacology and A_{2A} receptor knockout mice. Br J Pharmacol 134:68–77

El Yacoubi M, Costentin J, Vaugeois JM (2003) Adenosine A_{2A} receptors and depression. Neurology 61:S72–S73

El Yacoubi M, Ledent C, Parmentier M, Costentin J, Vaugeois JM (2008) Evidence for the involvement of the adenosine A(2A) receptor in the lowered susceptibility to pentylenetetrazol-induced seizures produced in mice by long-term treatment with caffeine. Neuropharmacology 55:35–40

Elmenhorst D, Meyer PT, Winz OH, Matusch A, Ermert J, Coenen HH, Basheer R, Haas HL, Zilles K, Bauer A (2007) Sleep deprivation increases A_1 adenosine receptor binding in the human brain: a positron emission tomography study. J Neurosci 27:2410–2415

Esser MJ, Sawynok J (2000) Caffeine blockade of the thermal antihyperalgesic effect of acute amitriptyline in a rat model of neuropathic pain. Eur J Pharmacol 399:131–139

Esser MJ, Chase T, Allen GV, Sawynok J (2001) Chronic administration of amitriptyline and caffeine in a rat model of neuropathic pain: multiple interactions. Eur J Pharmacol 430:211–218

Fedele DE, Gouder N, Güttinger M, Gabernet L, Scheurer L, Rülicke T, Crestani F, Boison D (2005) Astrogliosis in epilepsy leads to overexpression of adenosine kinase, resulting in seizure aggravation. Brain 128:2383–2395

Fedele DE, Li T, Lan JQ, Fredholm BB, Boison D (2006) Adenosine A_1 receptors are crucial in keeping an epileptic focus localized. Exp Neurol 200:184–190

Fernandes CC, Pinto-Duarte A, Ribeiro JA and Sebastião AM (2008) Postsynaptic action of brain-derived neurotrophic factor attenuates alpha7 nicotinic acetylcholine receptor-mediated responses in hippocampal interneurons. J Neurosci 28:5611–5618

Ferré S (2008) An update on the mechanisms of the psychostimulant effects of caffeine. J Neurochem 105:1067–1079

Ferré S, von Euler G, Johansson B, Fredholm BB, Fuxe K (1991) Stimulation of high-affinity adenosine A_2 receptors decreases the affinity of dopamine D2 receptors in rat striatal membranes. Proc Natl Acad Sci USA 88:7238–7241

Ferré S, O'Connor WT, Snaprud P, Ungerstedt U, Fuxe K (1994) Antagonistic interaction between adenosine A_{2A} receptors and dopamine D2 receptors in the ventral striopallidal system. Implications for the treatment of schizophrenia. Neuroscience 63:765–773

Ferré S, Popoli P, Tinner-Staines B, Fuxe K (1996) Adenosine A_1 receptor-dopamine D1 receptor interaction in the rat limbic system: modulation of dopamine D1 receptor antagonist binding sites. Neurosci Lett 208:109–112

Ferré S, Popoli P, Rimondini R, Reggio R, Kehr J, Fuxe K (1999) Adenosine A_{2A} and group I metabotropic glutamate receptors synergistically modulate the binding characteristics of dopamine D2 receptors in the rat striatum. Neuropharmacology 38:129–140

Ferré S, Ciruela F, Woods AS, Lluis C, Franco R (2007a) Functional relevance of neurotransmitter receptor heteromers in the central nervous system. Trends Neurosci 30:440–446

Ferré S, Diamond I, Goldberg SR, Yao L, Hourani SM, Huang ZL, Urade Y, Kitchen I (2007b) Adenosine A_{2A} receptors in ventral striatum, hypothalamus and nociceptive circuitry implications for drug addiction, sleep and pain. Prog Neurobiol 83:332–347

Ferreira JM, Paes-de-Carvalho R (2001) Long-term activation of adenosine A(2a) receptors blocks glutamate excitotoxicity in cultures of avian retinal neurons. Brain Res 900:169–176

Fields RD, Burnstock G (2006) Purinergic signalling in neuron–glia interactions. Nat Rev Neurosci 7:423–436

Filip M, Frankowska M, Zaniewska M, Przegaliski E, Muller CE, Agnati L, Franco R, Roberts DC, Fuxe K (2006) Involvement of adenosine A_{2A} and dopamine receptors in the locomotor and sensitizing effects of cocaine. Brain Res 1077:67–80

Florán B, Barajas C, Florán L, Erlij D, Aceves J (2002) Adenosine A_1 receptors control dopamine D1-dependent [(3)H]GABA release in slices of substantia nigra pars reticulata and motor behavior in the rat. Neuroscience 115:743–751

Florio C, Prezioso A, Papaioannou A, Vertua R (1998) Adenosine A_1 receptors modulate anxiety in CD1 mice. Psychopharmacology 136:311–319

Fontinha BM, Diógenes MJ, Ribeiro JA, Sebastião AM (2008) Enhancement of long-term potentiation by brain-derived neurotrophic factor requires adenosine A(2A) receptor activation by endogenous adenosine. Neuropharmacology 54:924–933

Fowler JC (1993) Changes in extracellular adenosine levels and population spike amplitude during graded hypoxia in the rat hippocampal slice. Naunyn–Schmiedeberg's Arch Pharmacol 347: 73–78

Fredholm BB, Bättig K, Holmén J, Nehlig A, Zvartau EE (1999) Actions of caffeine in the brain with special reference to factors that contribute to its widespread use. Pharmacol Rev 51: 83–133

Fredholm BB, IJzerman AP, Jacobson KA, Klotz KN, Linden J (2001) International Union of Pharmacology. XXV. Nomenclature and classification of adenosine receptors. Pharmacol Rev 53:527–552

Fredholm BB, Chern Y, Franco R, Sitkovsky M (2007) Aspects of the general biology of adenosine A_{2A} signaling. Prog Neurobiol 83:263–276

Frenguelli BG, Llaudet E, Dale N (2003) High-resolution real-time recording with microelectrode biosensors reveals novel aspects of adenosine release during hypoxia in rat hippocampal slices. J Neurochem 86:1506–1515

Frenguelli BG, Wigmore G, Llaudet E, Dale N (2007) Temporal and mechanistic dissociation of ATP and adenosine release during ischaemia in the mammalian hippocampus. J Neurochem 101:1400–1413

Fukumitsu N, Ishii K, Kimura Y, Oda K, Sasaki T, Mori Y, Ishiwata K (2003) Imaging of adenosine A_1 receptors in the human brain by positron emission tomography with [11C]MPDX. Ann Nucl Med 17:511–515

Fukumitsu N, Ishii K, Kimura Y, Oda K, Sasaki T, Mori Y, Ishiwata K (2005) Adenosine A_1 receptor mapping of the human brain by PET with 8-dicyclopropylmethyl-1–11C-methyl-3-propylxanthine. J Nucl Med 46:32–37

Fuxe K, Ferré S, Genedani S, Franco R, Agnati LF (2007) Adenosine receptor-dopamine receptor interactions in the basal ganglia and their relevance for brain function. Physiol Behav 92: 210–217

Ganfornina MD, Pérez-García MT, Gutiérrez G, Miguel-Velado E, López-López JR, Marín A, Sánchez D, González C (2005) Comparative gene expression profile of mouse carotid body and adrenal medulla under physiological hypoxia. J Physiol 566:491–503

Gao ZG, Jacobson KA (2007) Emerging adenosine receptor agonists. Expert Opin Emerg Drugs 12:479–492

Gaspardone A, Crea F, Tomai F, Iamele M Crossman DC, Pappagallo M, Versaci F, Chiariello L, Gioffre' PA (1994) Substance P potentiates the algogenic effects of intraarterial infusion of adenosine. J Am Coll Cardiol 24:477–482

Gauda EB, Northington FJ, Linden J, Rosin DL (2000) Differential expression of a(2a), A(1)-adenosine and D(2)-dopamine receptor genes in rat peripheral arterial chemoreceptors during postnatal development. Brain Res 872:1–10

Gendron FP, Benrezzak O, Krugh BW, Kong Q, Weisman GA, Beaudoin AR (2002) Purine signaling and potential new therapeutic approach: possible outcomes of NTPDase inhibition. Curr Drug Targets 3:229–245

Gidday JM (2006) Cerebral preconditioning and ischemic tolerance. Nat Rev Neurosci 7:437–448

Giménez-Llort L, Schiffmann SN, Shmidt T, Canela L, Camón L, Wassholm M, Canals M, Terasmaa A, Fernández-Teruel A, Tobeña A, Popova E, Ferré S, Agnati L, Ciruela F, Martínez E, Scheel-Kruger J, Lluis C, Franco R, Fuxe K, Bader M (2007) Working memory deficits in transgenic rats overexpressing human adenosine A_{2A} receptors in the brain. Neurobiol Learn Mem 87:42–56

Ginés S, Hillion J, Torvinen M, Le Crom S, Casadó V, Canela EI, Rondin S, Lew JY, Watson S, Zoli M, Agnati LF, Verniera P, Lluis C, Ferré S, Fuxe K, Franco R (2000) Dopamine D1 and adenosine A_1 receptors form functionally interacting heteromeric complexes. Proc Natl Acad Sci USA 97:8606–8611

Ginsborg BL, Hirst GD (1971) Cyclic AMP, transmitter release and the effect of adenosine on neuromuscular transmission. Nat New Biol 232:63–64

Glass M, Faull RL, Bullock JY, Jansen K, Mee EW, Walker EB, Synek BJ, Dragunow M (1996) Loss of A_1 adenosine receptors in human temporal lobe epilepsy. Brain Res 710:56–68

Goadsby PJ (2008) Calcitonin gene-related peptide (CGRP) antagonists and migraine: is this a new era? Neurology 70:1300–1301

Goadsby PJ, Hoskin KL, Storer RJ, Edvinsson L, Connor HE (2002) Adenosine A_1 receptor agonists inhibit trigeminovascular nociceptive transmission. Brain 125:1392–1401

Godfrey L, Yan L, Clarke GD, Ledent C, Kitchen I, Hourani SM (2006) Modulation of paracetamol antinociception by caffeine and by selective adenosine A_2 receptor antagonists in mice. Eur J Pharmacol 531:80–86

Golembiowska K, Dziubina A (2004). Striatal adenosine A(2A) receptor blockade increases extracellular dopamine release following L-DOPA administration in intact and dopamine-denervated rats. Neuropharmacology 47:414–426

Gomes CA, Vaz SH, Ribeiro JA, Sebastião AM (2006) Glial cell line-derived neurotrophic factor (GDNF) enhances dopamine release from striatal nerve endings in an adenosine A_{2A} receptor-dependent manner. Brain Res 1113:129–136

Gray R, Rajan AS, Radcliffe KA, Yakehiro M, Dani JA (1996) Hippocampal synaptic transmission enhanced by low concentrations of nicotine. Nature 383:713–716

Green TA, Schenk S (2002) Dopaminergic mechanism for caffeine-produced cocaine seeking in rats. Neuropsychopharmacology 26:422–430

Greengard P (2001) The neurobiology of slow synaptic transmission. Science 294:1024–1030

Griffiths TL, Christie JM, Parsons ST, Holgate ST (1997) The effect of dipyridamole and theophylline on hypercapnic ventilatory responses: the role of adenosine. Eur Respir J 10:156–160

Gu JG, Foga IO, Parkinson FE, Geiger JD (1995) Involvement of bidirectional adenosine transporters in the release of L-[3H]adenosine from rat brain synaptosomal preparations. J Neurochem 64:2105–2110

Güttinger M, Fedele D, Koch P, Padrun V, Pralong WF, Brüstle O, Boison D (2005) Suppression of kindled seizures by paracrine adenosine release from stem cell-derived brain implants. Epilepsia 46:1162–1169

Hack SP, Christie MJ (2003) Adaptations in adenosine signaling in drug dependence: therapeutic implications. Crit Rev Neurobiol 15:235–274

Hack SP, Vaughan CW, Christie MJ (2003) Modulation of GABA release during morphine withdrawal in midbrain neurons in vitro. Neuropharmacology 45:575–584

Halassa MM, Fellin T, Haydon PG (2007) The tripartite synapse: roles for gliotransmission in health and disease. Trends Mol Med 13:54–63

Hamilton SP, Slager SL, De Leon AB, Heiman GA, Klein DF, Hodge SE, Weissman MM, Fyer AJ, Knowles JA (2004) Evidence for genetic linkage between a polymorphism in the adenosine 2A receptor and panic disorder. Neuropsychopharmacology 29:558–565

Hasselmo ME, Giocomo LM (2006) Cholinergic modulation of cortical function. J Mol Neurosci 30:133–135

Hauber W, Bareiss A (2001) Facilitative effects of an adenosine A_1/A_2 receptor blockade on spatial memory performance of rats: selective enhancement of reference memory retention during the light period. Behav Brain Res 118:43–52

Hayashida M, Fukuda K, Fukunaga A (2005) Clinical application of adenosine and ATP for pain control. J Anesth 19:225–235

Heese K, Fiebich BL, Bauer J, Otten U (1997) Nerve growth factor (NGF) expression in rat microglia is induced by adenosine A_{2a}-receptors. Neurosci Lett 231:83–86

Hoehn K, White TD (1989) Evoked release of endogenous adenosine from rat cortical slices by K^+ and glutamate. Brain Res 478:149–151

Hohoff C, McDonald JM, Baune BT, Cook EH, Deckert J, de Wit H (2005) Interindividual variation in anxiety response to amphetamine: possible role for adenosine A_{2A} receptor gene variants. Am J Med Genet B Neuropsychiatr Genet 139:42–44

Hong CJ, Liu HC, Liu TY, Liao DL, Tsai SJ (2005) Association studies of the adenosine A_{2a} receptor (1976T > C) genetic polymorphism in Parkinson's disease and schizophrenia. J Neural Transm 112:1503–1510

Huang ZL, Qu WM, Eguchi N, Chen JF, Schwarzschild MA, Fredholm BB, Urade Y, Hayaishi O (2005) Adenosine A_{2A}, but not A_1, receptors mediate the arousal effect of caffeine. Nat Neurosci 8:858–859

Huber A, Güttinger M, Möhler H, Boison D (2002) Seizure suppression by adenosine A(2A) receptor activation in a rat model of audiogenic brainstem epilepsy. Neurosci Lett 329:289–292

Hurley MJ, Mash DC, Jenner P (2000) Adenosine A(2A) receptor mRNA expression in Parkinson's disease. Neurosci Lett 291:54–58

Hussey MJ, Clarke GD, Ledent C, Hourani SM, Kitchen I (2007) Reduced response to the formalin test and lowered spinal NMDA glutamate receptor binding in adenosine A_{2A} receptor knockout mice. Pain 129:287–294

Ishiwata K, Mishina M, Kimura Y, Oda K, Sasaki T, Ishii K (2005) First visualization of adenosine A(2A) receptors in the human brain by positron emission tomography with [11C]TMSX. Synapse 55:133–136

Jacobson KA, von Lubitz DKJE, Daly JW, Fredholm BB (1996) Adenosine receptor ligands: differences with acute versus chronic treatment. Trends Pharmacol Sci 17:108–113

Jain N, Kemp N, Adeyemo O, Buchanan P, Stone TW (1995) Anxiolytic activity of adenosine receptor activation in mice. Br J Pharmacol 116:2127–2133

James S, Xuereb JH, Askalan R, Richardson PJ (1992) Adenosine receptors in post-mortem human brain. Br J Pharmacol 105:238–244

Jarvis MF, Yu H, McGaraughty S, Wismer CT, Mikusa J, Zhu C, Chu K, Kohlhaas K, Cowart M, Lee CH, Stewart AO, Cox BF, Polakowski J, Kowaluk EA (2002) Analgesic and anti-inflammatory effects of A-286501, a novel orally active adenosine kinase inhibitor. Pain 96:107–118

Ji D, Lape R, Dani JA (2001) Timing and location of nicotinic activity enhances or depresses hippocampal synaptic plasticity. Neuron 31:131–141

Jin X, Shepherd RK, Duling BR, Linden J (1997) Inosine binds to A_3 adenosine receptors and stimulates mast cell degranulation. J Clin Invest 100:2849–2857

Johansson B, Halldner L, Dunwiddie TV, Masino SA, Poelchen W, Giménez-Llort L, Escorihuela RM, Fernández-Teruel A, Wiesenfeld-Hallin Z, Xu XJ, Hårdemark A, Betsholtz C, Herlenius E, Fredholm BB (2001) Hyperalgesia, anxiety, and decreased hypoxic neuroprotection in mice lacking the adenosine A1 receptor. Proc Natl Acad Sci USA 98:9407–9412

Johnson-Kozlow M, Kritz-Silverstein D, Barrett-Connor E, Morton D (2002) Coffee consumption and cognitive function among older adults. Am J Epidemiol 156:842–850

Kachroo A, Orlando LR, Grandy DK, Chen JF, Young AB, Schwarzschild MA (2005) Interactions between metabotropic glutamate 5 and adenosine A_{2A} receptors in normal and parkinsonian mice. J Neurosci 25:10414–10419

Kaplan GB, Coyle TS (1998) Adenosine kinase inhibitors attenuate opiate withdrawal via adenosine receptor activation. Eur J Pharmacol 362:1–8

Kaplan GB, Leite-Morris KA (1997) Up-regulation of adenosine transporter-binding sites in striatum and hypothalamus of opiate tolerant mice. Brain Res 763:215–220

Kaplan GB, Sears MT (1996) Adenosine receptor agonists attenuate and adenosine receptor antagonists exacerbate opiate withdrawal signs. Psychopharmacology 123:64–70

Kaplan GB, KA Leite-Morris, MT Sears (1994) Alterations in adenosine A receptors in morphine dependence. Brain Res 657:347–350

Kaster MP, Rosa AO, Rosso MM, Goulart EC, Santos AR, Rodrigues AL (2004) Adenosine administration produces an antidepressant-like effect in mice: evidence for the involvement of A_1 and A_{2A} receptors. Neurosci Lett 355:21–24

Kaster MP, Budni J, Santos AR, Rodrigues AL (2007) Pharmacological evidence for the involvement of the opioid system in the antidepressant-like effect of adenosine in the mouse forced swimming test. Eur J Pharmacol 576:91–98

Kaufman KR, Sachdeo RC (2003) Caffeinated beverages and decreased seizure control. Seizure 12:519–521

Keil GJ 2nd, DeLander GE (1992) Spinally-mediated antinociception is induced in mice by an adenosine kinase-, but not by an adenosine deaminase, inhibitor. Life Sci 51:PL171–PL176

Kessey K, Mogul DJ (1997) NMDA-Independent LTP by adenosine A_2 receptor-mediated postsynaptic AMPA potentiation in hippocampus. J Neurophysiol 78:1965–1972

Khorasani MZ, Hajizadeh S, Fathollahi Y, Semnanian S (2006) Interaction of adenosine and naloxone on regional cerebral blood flow in morphine-dependent rats. Brain Res 1084:61–66

Kitagawa M, Houzen H, Tashiro K (2007) Effects of caffeine on the freezing of gait in Parkinson's disease. Mov Disord 22:710–712

Klein E, Zohar J, Geraci MF, Murphy DL, Uhde TW (1991) Anxiogenic effects of m-CPP in patients with panic disorder: comparison to caffeine's anxiogenic effects. Biol Psychiat 30: 973–984

Klishin A, Tsintsadze T, Lozovaya N, Krishtal O (1995) Latent N-methyl-D-aspartate receptors in the recurrent excitatory pathway between hippocampal CA1 pyramidal neurons: Ca(2+)-dependent activation by blocking A_1 adenosine receptors. Proc Natl Acad Sci USA 92: 12431–12435

Knapp CM, Foye MM, Cottam N, Ciraulo DA, Kornetsky C (2001) Adenosine agonists CGS 21680 and NECA inhibit the initiation of cocaine self-administration. Pharmacol Biochem Behav 68:797–803

Kobayashi S, Conforti L, Millhorn DE (2000a) Gene expression and function of adenosine A(2A) receptor in the rat carotid body. Am J Physiol Lung Cell Mol Physiol 279:L273–L282

Kobayashi S, Zimmermann H, Millhorn DE (2000b) Chronic hypoxia enhances adenosine release in rat PC12 cells by altering adenosine metabolism and membrane transport. J Neurochem 74:621–632

Kochanek PM, Vagni VA, Janesko KL, Washington CB, Crumrine PK, Garman RH, Jenkins LW, Clark RS, Homanics GE, Dixon CE, Schnermann J, Jackson EK (2006) Adenosine A_1 receptor knockout mice develop lethal status epilepticus after experimental traumatic brain injury. J Cereb Blood Flow Metab 26:565–575

Koos BJ, Chau A (1998) Fetal cardiovascular and breathing responses to an adenosine A_{2a} receptor agonist in sheep. Am J Physiol 274:R152–R159

Koos BJ, Kawasaki Y, Kim YH, Bohorquez F (2005) Adenosine A_{2A}-receptor blockade abolishes the roll-off respiratory response to hypoxia in awake lambs. Am J Physiol Regul Integr Comp Physiol 288:R1185–R1194

Kouznetsova M, Kelley B, Shen M, Thayer SA (2002) Desensitization of cannabinoid-mediated presynaptic inhibition of neurotransmission between rat hippocampal neurons in culture. Mol Pharmacol 61:477–485

Kozisek ME, Middlemas D, Bylund DB (2008) Brain-derived neurotrophic factor and its receptor tropomyosin-related kinase B in the mechanism of action of antidepressant therapies. Pharmacol Ther 117:30–51

Kuzmin A, Johansson B, Zvartau EE, Fredholm BB (1999) Caffeine, acting on adenosine A(1) receptors, prevents the extinction of cocaine-seeking behavior in mice. J Pharmacol Exp Ther 290:535–542

Lahiri S, Mitchell CH, Reigada D, Roy A, Cherniack NS (2007) Purines, the carotid body and respiration. Respir Physiol Neurobiol 157:123–129

Lam P, Hong CJ, Tsai SJ (2005) Association study of A_{2a} adenosine receptor genetic polymorphism in panic disorder. Neurosci Lett 378:98–101

Lambert NA, Teyler TJ (1991) Adenosine depresses excitatory but not fast inhibitory synaptic transmission in area CA1 of the rat hippocampus. Neurosci Lett 122:50–52

Landolt HP (2008) Sleep homeostasis: a role for adenosine in humans? Biochem Pharmacol 75:2070–2079

Landolt HP, Werth E, Borbély AA, Dijk DJ (1995a) Caffeine intake (200 mg) in the morning affects human sleep and EEG power spectra at night. Brain Res 675:67–74

Landolt HP, Dijk DJ, Gaus SE, Borbély AA (1995b) Caffeine reduces low-frequency delta activity in the human sleep EEG. Neuropsychopharmacology 12:229–238

Landolt HP, Rétey JV, Tönz K, Gottselig JM, Khatami R, Buckelmüller I, Achermann P (2004) Caffeine attenuates waking and sleep electroencephalographic markers of sleep homeostasis in humans. Neuropsychopharmacology 29:1933–1939

Lao LJ, Kumamoto E, Luo C, Furue H, Yoshimura M (2001) Adenosine inhibits excitatory transmission to substantia gelatinosa neurons of the adult rat spinal cord through the activation of presynaptic A(1) adenosine receptor. Pain 94:315–324

Lara DR, Dall'Igna OP, Ghisolfi ES, Brunstein MG (2006) Involvement of adenosine in the neurobiology of schizophrenia and its therapeutic implications. Prog Neuropsychopharmacol Biol Psychiat 30:617–629

Larson PS (2008) Deep brain stimulation for psychiatric disorders. Neurotherapeutics 5:50–58

Latini S, Pedata F (2001) Adenosine in the central nervous system: release mechanisms and extracellular concentrations. J Neurochem 79:463–484

Le Crom S, Prou D, Vernier P (2002) Autocrine activation of adenosine A_1 receptors blocks D1A but not D1B dopamine receptor desensitization. J Neurochem 82:1549–1552

Ledent C, Vaugeois JM, Schiffmann SN, Pedrazzini T, El Yacoubi M, Vanderhaeghen JJ, Costentin J, Heath JK, Vassart G, Parmentier M (1997) Aggressiveness, hypoalgesia and high blood pressure in mice lacking the adenosine A_{2a} receptor. Nature 388:674–678

Lee FS, Chao MV (2001) Activation of Trk neurotrophin receptors in the absence of neurotrophins. Proc Natl Acad Sci USA 98:3555–3560

Li H, Henry JL (2000) Adenosine receptor blockade reveals *N*-methyl-D-aspartate receptor- and voltage-sensitive dendritic spikes in rat hippocampal CA1 pyramidal cells in vitro. Neuroscience 100:21–31

Li T, Steinbeck JA, Lusardi T, Koch P, Lan JQ, Wilz A Segschneider M, Simon RP, Brustle O, Boison D (2007) Suppression of kindling epileptogenesis by adenosine releasing stem cell derived brain implants. Brain 130:1276–1288

Limberger N, Späth L, Starke K (1988) Presynaptic alpha 2-adrenoceptor, opioid kappa-receptor and adenosine A_1-receptor interaction on noradrenaline release in rabbit brain cortex. Naunyn–Schmiedeberg's Arch Pharmacol 338:53–61

Listos J, Malec D, Fidecka S (2005) Influence of adenosine receptor agonists on benzodiazepine withdrawal signs in mice. Eur J Pharmacol 523:71–78

Listos J, Talarek S, Fidecka S (2008) Adenosine receptor agonists attenuate the development of diazepam withdrawal-induced sensitization in mice. Eur J Pharmacol 588:72–77

Liu ZW, Gao XB (2007) Adenosine inhibits activity of hypocretin/orexin neurons by the A_1 receptor in the lateral hypothalamus: a possible sleep-promoting effect. J Neurophysiol 97:837–848

Lopes LV, Cunha RA, Ribeiro JA (1999a) Cross talk between A(1) and A(2A) adenosine receptors in the hippocampus and cortex of young adult and old rats. J Neurophysiol 82:3196–3203

Lopes LV, Cunha RA, Ribeiro JA (1999b) Increase in the number, G protein coupling, and efficiency of facilitatory adenosine A_{2A} receptors in the limbic cortex, but not striatum, of aged rats. J Neurochem 73:1733–1738

Lorenzen A, Sebastião AM, Sellink A, Vogt H, Schwabe U, Ribeiro JA, IJzerman AP (1997) Biological activities of N6,C8-disubstituted adenosine derivatives as partial agonists at rat brain adenosine A_1 receptors. Eur J Pharmacol 334:299–307

Love S, Plaha P, Patel NK, Hotton GR, Brooks DJ, Gill SS (2005) Glial cell line-derived neurotrophic factor induces neuronal sprouting in human brain. Nat Med 11:703–704

Lucchi R, Latini S, de Mendonça A, Sebastião AM, Ribeiro JA (1996) Adenosine by activating A1 receptors prevents GABAA-mediated actions during hypoxia in the rat hippocampus. Brain Res 732:261–266

Lupica CR, Proctor WR, Dunwiddie TV (1992) Presynaptic inhibition of excitatory synaptic transmission by adenosine in rat hippocampus: analysis of unitary EPSP variance measured by whole-cell recording. J Neurosci 12:3753–3764

Macek TA, Schaffhauser H, Conn PJ (1998) Protein kinase C and A_3 adenosine receptor activation inhibit presynaptic metabotropic glutamate receptor (mGluR) function and uncouple mGluRs from GTP-binding proteins. J Neurosci 18:6138–6146

Magalhães-Cardoso MT, Pereira MF, Oliveira L, Ribeiro JA, Cunha RA, Correia-de-Sá P (2003) Ecto-AMP deaminase blunts the ATP-derived adenosine A_{2A} receptor facilitation of acetylcholine release at rat motor nerve endings. J Physiol 549:399–408

Mahan LC, McVittie LD, Smyk-Randall EM, Nakata H, Monsma FJ Jr, Gerfen CR, Sibley DR (1991) Cloning and expression of an A_1 adenosine receptor from rat brain. Mol Pharmacol 40:1–7

Maia L, de Mendonça A (2002) Does caffeine intake protect from Alzheimer's disease? Eur J Neurol 9:1–6

Manzoni O, Pujalte D, Williams J, Bockaert J (1998) Decreased presynaptic sensitivity to adenosine after cocaine withdrawal. J Neurosci 18:7996–8002

Martini C, Tuscano D, Trincavelli ML, Cerrai E, Bianchi M, Ciapparelli A, Alessio L, Novelli L, Catena M, Lucacchini A, Cassano GB, Dell'Osso L (2006) Upregulation of A_{2A} adenosine receptors in platelets from patients affected by bipolar disorders under treatment with typical antipsychotics. J Psychiat Res 40:81–88

Maurice T, Lockhart BP, Privat A (1996) Amnesia induced in mice by centrally administered beta-amyloid peptides involves cholinergic dysfunction. Brain Res 706:181–193

Maxwell Dl, Fuller RW, Nolop KB, Dixon CM, Hughes JM (1986) Effects of adenosine on ventilatory responses to hypoxia and hypercapnia in humans. J Appl Physiol 61:1762–1766

Mayer CA, Haxhiu MA, Martin RJ, Wilson CG (2006) Adenosine A_{2A} receptors mediate GABAergic inhibition of respiration in immature rats. J Appl Physiol 100:91–97

Mayfield DR, Larson G, Orona RA, Zahniser NR (1996) Opposing actions of adenosine A_{2a} and dopamine D2 receptor activation on GABA release in the basal ganglia: evidence for an A_{2a}/D2 receptor interaction in globus pallidus. Synapse 22:132–138

Mayfield RD, Jones BA, Miller HA, Simosky JK, Larson GA, Zahniser NR (1999) Modulation of endogenous GABA release by an antagonistic adenosine A_1/dopamine D1 receptor interaction in rat brain limbic regions but not basal ganglia. Synapse 33:274–281

McGaraughty S, Cowart M, Jarvis MF, Berman RF (2005) Anticonvulsant and antinociceptive actions of novel adenosine kinase inhibitors. Curr Top Med Chem 5:43–58

McQueen DS, Ribeiro JA (1981) Effect of adenosine on carotid chemoreceptor activity in the cat. Br J Pharmacol 74:129–136

McQueen DS, Ribeiro JA (1986) Pharmacological characterization of the receptor involved in chemoexcitation induced by adenosine. Br J Pharmacol 88:615–620

Meyer PT, Elmenhorst D, Boy C, Winz O, Matusch A, Zilles K, Bauer A (2007) Effect of aging on cerebral A_1 adenosine receptors: a [18F]CPFPX PET study in humans. Neurobiol Aging 28:1914–1924

Mihara T, Mihara K, Yarimizu J, Mitani Y, Matsuda R, Yamamoto H, Aoki S, Akahane A, Iwashita A, Matsuoka N (2007) Pharmacological characterization of a novel, potent adenosine A_1 and A_{2A} receptor dual antagonist, 5-[5-amino-3-(4-fluorophenyl)pyrazin-2-yl]-1-isopropylpyridine-2(1*H*)-one (ASP5854), in models of Parkinson's disease and cognition. J Pharmacol Exp Ther 323:708–719

Mingote S, Font L, Farrar AM, Vontell R, Worden LT, Stopper CM, Port RG, Sink KS, Bunce JG, Chrobak JJ, Salamone JD (2008) Nucleus accumbens adenosine A_{2A} receptors regulate exertion of effort by acting on the ventral striatopallidal pathway. J Neurosci 28:9037–9046

Mishina M, Ishiwata K, Kimura Y, Naganawa M, Oda K, Kobayashi S, Katayama Y, Ishii K (2007) Evaluation of distribution of adenosine A_{2A} receptors in normal human brain measured with [11C]TMSX PET. Synapse 61:778–784

Miura T, Kimura K (2000) Theophylline-induced convulsions in children with epilepsy. Pediatrics 105:920

Mojsilovic-Petrovic J, Jeong GB, Crocker A, Arneja A, David S, Russell DS, Kalb RG (2006) Protecting motor neurons from toxic insult by antagonism of adenosine A_{2a} and Trk receptors. J Neurosci 26:9250–9263

Monteiro EC, Ribeiro JA (1987) Ventilatory effects of adenosine mediated by carotid body chemoreceptors in the rat. Naunyn–Schmiedeberg's Arch Pharmacol 335:143–148

Morelli M, Di Paolo T, Wardas J, Calon F, Xiao D, Schwarzschild MA (2007) Role of adenosine A_{2A} receptors in parkinsonian motor impairment and L-DOPA-induced motor complications. Prog Neurobiol 83:293–309

Mori Y, Higuchi M, Masuyama N, Gotoh Y (2004) Adenosine A_{2A} receptor facilitates calcium-dependent protein secretion through the activation of protein kinase A and phosphatidylinositol-3 kinase in PC12 cells. Cell Struct Funct 29:101–110

Mortelmans LJ, Van Loo M, De Cauwer HG, Merlevede K (2008) Seizures and hyponatremia after excessive intake of diet coke. Eur J Emerg Med 15:51

Nikbakht MR, Stone TW (2001) Suppression of presynaptic responses to adenosine by activation of NMDA receptors. Eur J Pharmacol 427:13–25

Nishi A, Liu F, Matsuyama S, Hamada M, Higashi H, Nairn AC, Greengard P (2003) Metabotropic mGlu5 receptors regulate adenosine A_{2A} receptor signaling. Proc Natl Acad Sci USA 100:1322–1327

Ogata T, Nakamura Y, Tsuji K, Shibata T, Kataoka K, Schubert P (1994) Adenosine enhances intracellular Ca^{2+} mobilization in conjunction with metabotropic glutamate receptor activation by t-ACPD in cultured hippocampal astrocytes. Neurosci Lett 170:5–8

Ogata T, Nakamura Y, Schubert P (1996) Potentiated cAMP rise in metabotropically stimulated rat cultured astrocytes by a Ca^{2+}-related A_1/A_2 adenosine receptor cooperation. Eur J Neurosci 8:1124–1131

O'Kane EM Stone TW (1998) Interactions between adenosine A_1 and A_2 receptor-mediated responses in the rat hippocampus in vitro. Eur J Pharmacol 362:17–25

O'Neill C, Nolan BJ, Macari A, O'Boyle KM, O'Connor JJ (2007) Adenosine A_1 receptor-mediated inhibition of dopamine release from rat striatal slices is modulated by D1 dopamine receptors. Eur J Neurosci 26:3421–428

Ortinau S, Laube B, Zimmermann H (2003) ATP inhibits NMDA receptors after heterologous expression and in cultured hippocampal neurons and attenuates NMDA-mediated neurotoxicity. J Neurosci 23:4996–5003

Pagonopoulou O, Angelatou F (1992) Reduction of A_1 adenosine receptors in cortex, hippocampus and cerebellum in ageing mouse brain. Neuroreport 3:735–737

Pan, H.L, Xu Z, Leung E, Eisenach JC (2001) Allosteric adenosine modulation to reduce allodynia. Anesthesiology 95:416–420

Patel NK, Bunnage M, Plaha P, Svendsen CN, Heywood P, Gill SS (2005) Intraputamenal infusion of glial cell line-derived neurotrophic factor in PD: a two-year outcome study. Ann Neurol 57:298–302

Peterson AL, Nutt JG (2008) Treatment of Parkinson's disease with trophic factors. Neurotherapeutics 5:270–280

Phillis JW (2004) Adenosine and adenine nucleotides as regulators of cerebral blood flow: roles of acidosis, cell swelling, and KATP channels. Crit Rev Neurobiol 16:237–270

Phillis JW, Wu PH (1982) Adenosine mediates sedative action of various centrally active drugs. Med Hypotheses 9:361–367

Pignataro G, Maysami S, Studer FE, Wilz A, Simon RP, Boison D (2008) Downregulation of hippocampal adenosine kinase after focal ischemia as potential endogenous neuroprotective mechanism. J Cereb Blood Flow Metab 28:17–23

Pinto-Duarte A, Coelho JE, Cunha RA, Ribeiro JA, Sebastião AM (2005) Adenosine A_{2A} receptors control the extracellular levels of adenosine through modulation of nucleoside transporters activity in the rat hippocampus. J Neurochem 93:595–604

Pintor A, Pèzzola A, Reggio R, Quarta D, Popoli P (2000) The mGlu5 receptor agonist CHPG stimulates striatal glutamate release: possible involvement of A_{2A} receptors. Neuroreport 11:3611–3614

Poleszak E, Malec D (2003) Effects of adenosine receptor agonists and antagonists in amphetamine-induced conditioned place preference test in rats. Pol J Pharmacol 55:319–26

Poon A, Sawynok J (1999) Antinociceptive and anti-inflammatory properties of an adenosine kinase inhibitor and an adenosine deaminase inhibitor. Eur J Pharmacol 384:123–138

Popoli P, Minghetti L, Tebano MT, Pintor A, Domenici MR, Massotti M (2004) Adenosine A_{2A} receptor antagonism and neuroprotection: mechanisms, lights, and shadows. Crit Rev Neurobiol 16:99–106

Popoli P, Blum D, Martire A, Ledent C, Ceruti S, Abbracchio MP (2007) Functions, dysfunctions and possible therapeutic relevance of adenosine A_{2A} receptors in Huntington's disease. Prog Neurobiol 81:331–348

Porkka-Heiskanen T, Strecker RE, Thakkar M, Bjorkum AA, Greene RW, McCarley RW (1997) Adenosine: a mediator of the sleep-inducing effects of prolonged wakefulness. Science 276:1265–1268

Porkka-Heiskanen T, Alanko L, Kalinchuk A, Stenberg D (2002) Adenosine and sleep. Sleep Med Rev 6:321–332

Pousinha PA, Diogenes MJ, Ribeiro JA, Sebastião AM (2006) Triggering of BDNF facilitatory action on neuromuscular transmission by adenosine A_{2A} receptors. Neurosci Lett 404:143–147

Prediger RD, Batista LC, Takahashi RN (2004) Adenosine A_1 receptors modulate the anxiolytic-like effect of ethanol in the elevated plus-maze in mice. Eur J Pharmacol 499:147–154

Prediger RD, da Silva GE, Batista LC, Bittencourt AL, Takahashi RN (2006) Activation of adenosine A_1 receptors reduces anxiety-like behavior during acute ethanol withdrawal (hangover) in mice. Neuropsychopharmacology 31:2210–2220

Prince DA, Stevens CF (1992) Adenosine decreases neurotransmitter release at central synapses. Proc Natl Acad Sci USA 89:8586–8590

Pugliese AM, Latini S, Corradetti R, Pedata F (2003) Brief, repeated, oxygen-glucose deprivation episodes protect neurotransmission from a longer ischemic episode in the in vitro hippocampus: role of adenosine receptors. Br J Pharmacol 140(2):305–14

Qian J, Colmers WF, Saggau P (1997) Inhibition of synaptic transmission by neuropeptide Y in rat hippocampal area CA1: modulation of presynaptic Ca^{2+} entry. J Neurosci 17:8169–8177

Rajagopal R, Chen ZY, Lee FS, Chao MV (2004) Transactivation of Trk neurotrophin receptors by G-protein-coupled receptor ligands occurs on intracellular membranes. J Neurosci 24: 6650–6658

Rao SS, Mudipalli RS, Remes-Troche JM, Utech CL, Zimmerman B (2007) Theophylline improves esophageal chest pain: a randomized, placebo-controlled study. Am J Gastroenterol 102:930–938

Rebola N, Sebastião AM, de Mendonca A, Oliveira CR, Ribeiro JA, Cunha RA (2003) Enhanced adenosine A_{2A} receptor facilitation of synaptic transmission in the hippocampus of aged rats. J Neurophysiol 90:1295–303

Rebola N, Lujan R, Cunha RA, Mulle C (2008) Adenosine A_{2A} receptors are essential for long-term potentiation of NMDA-EPSCs at hippocampal mossy fiber synapses. Neuron 57:121–134

Rétey JV, Adam M, Gottselig JM, Khatami R, Dürr R, Achermann P, Landolt HP J (2006) Adenosinergic mechanisms contribute to individual differences in sleep deprivation-induced changes in neurobehavioral function and brain rhythmic activity. Neurosci 26:10472–10479

Rétey JV, Adam M, Khatami R, Luhmann UF, Jung HH, Berger W, Landolt HP (2007) A genetic variation in the adenosine A_{2A} receptor gene (ADORA2A) contributes to individual sensitivity to caffeine effects on sleep. Clin Pharmacol Ther 81:692–698

Ribeiro JA, Sebastião AM (1986) Adenosine receptors and calcium: basis for proposing a third (A_3) adenosine receptor. Prog Neurobiol 26:179–209

Ribeiro JA, Sebastião AM (1987) On the role, inactivation and origin of endogenous adenosine at the frog neuromuscular junction. J Physiol 384:571–585

Ribeiro JA, Walker J (1973) Action of adenosine triphosphate on endplate potentials recorded from muscle fibres of the rat-diaphragm and frog sartorius. Br J Pharmacol 49:724–725

Ribeiro JA, Sá-Almeida AM, Namorado JM (1979) Adenosine and adenosine triphosphate decrease ^{45}Ca uptake by synaptosomes stimulated by potassium. Biochem Pharmacol 28: 1297–1300

Ribeiro JA, Sebastião A.M, de Mendonça A (2003) Adenosine receptors in the nervous system: pathophysiological implications. Prog Neurobiol 68:377–392

Ritchie K, Carrière I, de Mendonca A, Portet F, Dartigues JF, Rouaud O, Barberger-Gateau P, Ancelin ML (2007) The neuroprotective effects of caffeine: a prospective population study (the Three City Study). Neurology 69:536–545

Rodrigues RJ, Alfaro TM, Rebola N, Oliveira CR, Cunha RA (2005) Co-localization and functional interaction between adenosine A(2A) and metabotropic group 5 receptors in glutamatergic nerve terminals of the rat striatum. J Neurochem 92:433–441
Roseti C, Martinello K, Fucile S, Piccari V, Mascia A, Di Gennaro G, Quarato PP, Manfredi M, Esposito V, Cantore G, Arcella A, Simonato M, Fredholm BB, Limatola C, Miledi R, Eusebi F (2008) Adenosine receptor antagonists alter the stability of human epileptic GABAA receptors. Proc Natl Acad Sci USA 105:15118–15123
Ross GW, Abbott RD, Petrovitch H, Morens DM, Grandinetti A, Tung KH, Tanner CM, Masaki KH, Blanchette PL, Curb JD, Popper JS, White LR (2000) Association of coffee and caffeine intake with the risk of Parkinson disease. JAMA 283:2674–2679
Ross CA, Margolis RL, Reading SA, Pletnikov M, Coyle JT (2006) Neurobiology of schizophrenia. Neuron 52:139–153
Runold M, Cherniack NS, Prabhakar NR (1990) Effect of adenosine on isolated and superfused cat carotid body activity. Neurosci Lett 113:111–114
Salín-Pascual RJ, Valencia-Flores M, Campos RM, Castaño A, Shiromani PJ (2006) Caffeine challenge in insomniac patients after total sleep deprivation. Sleep Med 7:141–145
Salvatore CA, Jacobson MA, Taylor HE, Linden J, Johnson RG (1993) Molecular cloning and characterization of the human A_3 adenosine receptor. Proc Natl Acad Sci USA 90: 10365–10369
Sandner-Kiesling A, Li X, Eisenach JC (2001) Morphine-induced spinal release of adenosine is reduced in neuropathic rats. Anesthesiology 95:1455–1459
Sattin A, Rall TW (1970) The effect of adenosine and adenine nucleotides on the cyclic adenosine 3′, 5′-phosphate content of guinea pig cerebral cortex slices. Mol Pharmacol 6:13–23
Sawynok J (1998) Adenosine receptor activation and nociception. Eur J Pharmacol 347:1–11
Sawynok J (2007) Adenosine and ATP receptors. Handb Exp Pharmacol 177:309–328
Sawynok J, Liu XJ (2003) Adenosine in the spinal cord and periphery: release and regulation of pain. Prog Neurobiol 69:313–340
Scharfman HE, Hen R (2007) Is more neurogenesis always better? Science 315:336–338
Schiffmann SN, Fisone G, Moresco R, Cunha RA, Ferré S (2007) Adenosine A_{2A} receptors and basal ganglia physiology. Prog Neurobiol 83:277–292
Schotanus SM, Fredholm BB, Chergui K (2006) NMDA depresses glutamatergic synaptic transmission in the striatum through the activation of adenosine A_1 receptors: evidence from knockout mice. Neuropharmacology 51:272–282
Schulte-Herbrüggen O, Braun A, Rochlitzer S, Jockers-Scherübl MC, Hellweg R (2007) Neurotrophic factors: a tool for therapeutic strategies in neurological, neuropsychiatric and neuroimmunological diseases? Curr Med Chem 14:2318–2329
Sebastião AM, Ribeiro JA (1985) Enhancement of transmission at the frog neuromuscular junction by adenosine deaminase: evidence for an inhibitory role of endogenous adenosine on neuromuscular transmission. Neurosci Lett 62:267–270
Sebastião AM, Ribeiro JA (1990) Interactions between adenosine and phorbol esters or lithium at the frog neuromuscular junction. Br J Pharmacol 100:55–62
Sebastião AM, Ribeiro JA (1992) Evidence for the presence of excitatory A_2 adenosine receptors in the rat hippocampus. Neurosci Lett 138:41–44
Sebastião AM, Ribeiro JA (1996) Adenosine A_2 receptor-mediated excitatory actions on the nervous system. Prog Neurobiol 48:167–189
Sebastião AM, Ribeiro JA (2000) Fine-tuning neuromodulation by adenosine. Trends Pharmacol Sci 21:341–346
Sebastião AM, Lucchi R, Latini S, De Mendonça A, Ribeiro JA (1996) Functional negative interaction between adenosine (A_1) and GABA (GABAA) during hypoxia in the rat hippocampus. In: Okada Y (ed) The role of adenosine in the nervous system. Elsevier, Amesterdam, pp 177–184
Sebastião AM, Cunha RA, de Mendonça A, Ribeiro JA (2000a) Modification of adenosine modulation of synaptic transmission in the hippocampus of aged rats. Br J Pharmacol 131:1629–1634
Sebastião AM, Macedo MP, Ribeiro JA (2000b) Tonic activation of A(2A) adenosine receptors unmasks, and of A(1) receptors prevents, a facilitatory action of calcitonin gene-related peptide in the rat hippocampus. Br J Pharmacol 129:374–380

Sebastião AM, de Mendonça A, Moreira T, Ribeiro JA (2001) Activation of synaptic NMDA receptors by action potential-dependent release of transmitter during hypoxia impairs recovery of synaptic transmission on reoxygenation. J Neurosci 21:8564–8571

Selley DE, Cassidy MP, Martin BR, Sim-Selley LJ (2004) Long-term administration of Delta9-tetrahydrocannabinol desensitizes CB_1-, adenosine A_1-, and $GABA_B$-mediated inhibition of adenylyl cyclase in mouse cerebellum. Mol Pharmacol 66:1275–1284

Seta KA, Millhorn DE (2004) Functional genomics approach to hypoxia signaling. J Appl Physiol 96:765–773

Shahraki A, Stone TW (2003) Interactions between adenosine and metabotropic glutamate receptors in the rat hippocampal slice. Br J Pharmacol 138:1059–1068

Shapiro RE (2007) Caffeine and headaches. Neurol Sci 28(Suppl 2):S179–S183

Silinsky EM (1975) On the association between transmitter secretion and the release of adenine nucleotides from mammalian motor nerve terminals. J Physiol 247:145–162

Simonato M, Tongiorgi E, Kokaia M (2006) Angels and demons: neurotrophic factors and epilepsy. Trends Pharmacol Sci 27:631–638

Sjölund KF, Segerdahl M, Sollevi (1999) Adenosine reduces secondary hyperalgesia in two human models of cutaneous inflammatory pain. Anesth Analg 88:605–610

Sollevi A, Belfrage M, Lundeberg T, Segerdahl M, Hansson P (1995) Systemic adenosine infusion: a new treatment modality to alleviate neuropathic pain. Pain 61:155–158

Soria G, Castañé A, Berrendero F, Ledent C, Parmentier M, Maldonado R, Valverde O (2004) Adenosine A_{2A} receptors are involved in physical dependence and place conditioning induced by THC. Eur J Neurosci 20:2203–2213

Soria G, Castañé A, Ledent C, Parmentier M, Maldonado R, Valverde O (2006) The lack of A_{2A} adenosine receptors diminishes the reinforcing efficacy of cocaine. Neuropsychopharmacology 31:978–987

Sperlágh B, Zsilla G, Baranyi M, Illes P, Vizi ES (2007) Purinergic modulation of glutamate release under ischemic-like conditions in the hippocampus. Neuroscience 149:99–111

Spyer KM, Thomas T (2000) A role for adenosine in modulating cardiorespiratory responses: a mini-review. Brain Res Bull 53:121–124

Stevens B, Porta S, Haak LL, Gallo V, Fields RD (2002) Adenosine: a neuron-glial transmitter promoting myelination in the CNS in response to action potentials. Neuron 36:855–868

Stevens B, Ishibashi T, Chen JF, Fields RD (2004) Adenosine: an activity-dependent axonal signal regulating MAP kinase and proliferation in developing Schwann cells. Neuron Glia Biol 1: 23–34

Stone TW, Collis MG, Williams M, Miller LP, Karasawa A, Hillaire-Buys D (1995) Adenosine: some therapeutic applications and prospects. In: Cuello AC, Collier B (eds) Pharmacological sciences: perspectives for research and therapy in the late 1990s. Birkhäuser, Basel, pp 303–309

Studer FE, Fedele DE, Marowsky A, Schwerdel C, Wernli K, Vogt K, Fritschy J-M, Boison D (2006) Shift of adenosine kinase expression from neurons to astrocytes during postnatal development suggests dual functionality of the enzyme. Neuroscience 142:125–137

Suzuki T, Namba K, Tsuga H, Nakata H (2006) Regulation of pharmacology by hetero-oligomerization between A_1 adenosine receptor and $P2Y_2$ receptor. Biochem Biophys Res Commun 351:559–565

Svenningsson P, Nomikos GG, Fredholm BB (1995) Biphasic changes in locomotor behavior and in expression of mRNA for NGFI-A and NGFI-B in rat striatum following acute caffeine administration. J Neurosci 15:7612–24

Svenningsson P, Hall H, Sedvall G, Fredholm BB (1997) Distribution of adenosine receptors in the postmortem human brain: an extended autoradiographic study. Synapse 27:322–335

Sylvén C, Beermann B, Kaijser L, Jonzon B (1990) Nicotine enhances angina pectoris-like pain and atrioventricular blockade provoked by intravenous bolus of adenosine in healthy volunteers. J Cardiovasc Pharmacol 16:962–965

Tebano MT, Pintor A, Frank C, Domenici MR, Martire A, Pepponi R, Potenza RL, Grieco R, Popoli P (2004) Adenosine A_{2A} receptor blockade differentially influences excitotoxic mechanisms at pre- and postsynaptic sites in the rat striatum. J Neurosci Res 77:100–107

Tebano MT, Martire A, Potenza RL, Grò C, Pepponi R, Armida M, Domenici MR, Schwarzschild MA, Chen JF, Popoli P (2008) Adenosine A(2A) receptors are required for normal BDNF levels and BDNF-induced potentiation of synaptic transmission in the mouse hippocampus. J Neurochem 104:279–286

Thakkar MM, Engemann SC, Walsh KM, Sahota PK (2008) Adenosine and the homeostatic control of sleep: effects of A_1 receptor blockade in the perifornical lateral hypothalamus on sleep-wakefulness. Neuroscience 153:875–880

Thevananther S, Rivera A, Rivkees SA (2001) A_1 adenosine receptor activation inhibits neurite process formation by Rho kinase-mediated pathways. Neuroreport 12:3057–3063

Thorsell A, Johnson J, Heilig M (2007) Effect of the adenosine A_{2a} receptor antagonist 3,7-dimethyl-propargylxanthine on anxiety-like and depression-like behavior and alcohol consumption in Wistar Rats. Alcohol Clin Exp Res 31:1302–1307

Toda S, Alguacil LF, Kalivas PW (2003) Repeated cocaine administration changes the function and subcellular distribution of adenosine A_1 receptor in the rat nucleus accumbens. J Neurochem 87:1478–1484

Tom NJ, Roberts PJ (1999) Group 1 mGlu receptors elevate $[Ca^{2+}]$ in rat cultured cortical type 2 astrocytes: $[Ca^{2+}]$ synergy with adenosine A_1 receptors. Neuropharmacology 38:1511–1517

Tonazzini I, Trincavelli ML, Storm-Mathisen J, Martini C, Bergersen LH (2007) Co-localization and functional cross-talk between A_1 and $P2Y_1$ purine receptors in rat hippocampus. Eur J Neurosci 26:890–902

Tonazzini I, Trincavelli ML, Montali M, Martini C (2008) Regulation of A_1 adenosine receptor functioning induced by $P2Y_1$ purinergic receptor activation in human astroglial cells. J Neurosci Res 86:2857–2866

Uauy R, Shapiro DL, Smith B, Warshaw JB (1975) Treatment of severe apnea in prematures with orally administered theophylline. Pediatrics 55:595–598

Uematsu T, Kozawa O, Matsuno H, Niwa M, Yoshikoshi H, Oh-uchi M, Kohno K, Nagashima S, Kanamaru M (2000) Pharmacokinetics and tolerability of intravenous infusion of adenosine (SUNY4001) in healthy volunteers. Br J Clin Pharmacol 50:177–181

Uhlhaas PJ, Singer W (2006) Neural synchrony in brain disorders: relevance for cognitive dysfunctions and pathophysiology. Neuron 52:155–168

Ulugol A, Karadag HC, Tamer M, Firat Z, Aslantas A, Dokmeci I (2002) Involvement of adenosine in the anti-allodynic effect of amitriptyline in streptozotocin-induced diabetic rats. Neurosci Lett 328:129–132

van Calker D, Müller M, Hamprecht B (1979) Adenosine regulates via two different types of receptors, the accumulation of cyclic AMP in cultured brain cells. J Neurochem 33:999–1005

Vaz S, Cristóvão-Ferreira S, Ribeiro JA, Sebastiao AM (2008) Brain-derived neurotrophic factor inhibits GABA uptake by the rat hippocampal nerve terminals. Brain Res 1219:19–25

Wang JH, Ma YY, van den Buuse M (2006) Improved spatial recognition memory in mice lacking adenosine A_{2A} receptors. Exp Neurol 199:438–445

Watt AH, Reid PG, Stephens MR, Routledge PA (1987) Adenosine-induced respiratory stimulation in man depends on site of infusion. Evidence for an action on the carotid body? Br J Clin Pharmacol 23:486–490

Weerts EM, Griffiths RR (2003) The adenosine receptor antagonist CGS15943 reinstates cocaine-seeking behavior and maintains self-administration in baboons. Psychopharmacology 168: 155–163

Wiese S, Jablonka S, Holtmann B, Orel N, Rajagopal R, Chao MV, Sendtner M (2007) Adenosine receptor A_{2A}-R contributes to motoneuron survival by transactivating the tyrosine kinase receptor TrkB. Proc Natl Acad Sci USA 104(43):17210–17215

Wilz A, Pritchard EM, Li T, Lan JQ, Kaplan DL, Boison D (2008) Silk polymer based adenosine release: therapeutic potential for epilepsy. Biomaterials 29:3609–3616

Winder DG, Conn PJ (1993) Activation of metabotropic glutamate receptors increases cAMP accumulation in hippocampus by potentiating responses to endogenous adenosine. J Neurosci 13:38–44

Wirkner K, Assmann H, Köles L, Gerevich Z, Franke H, Nörenberg W, Boehm R, Illes P (2000) Inhibition by adenosine A_{2A} receptors of NMDA but not AMPA currents in striatal neurons. Br J Pharmacol 130:259–269

Wirkner K, Gerevich Z, Krause T, Günther A, Köles L, Schneider D, Nörenberg W, Illes P (2004) Adenosine A2A receptor-induced inhibition of NMDA and $GABA_A$ receptor-mediated synaptic currents in a subpopulation of rat striatal neurons. Neuropharmacology 46:994–1007

Worley PF, Baraban JM, McCarren M, Snyder SH, Alger BE (1987) Cholinergic phosphatidylinositol modulation of inhibitory, G protein-linked neurotransmitter actions: electrophysiological studies in rat hippocampus. Proc Natl Acad Sci USA 84:3467–3471

Wu WP, Hao JX, Halldner L, Lövdahl C, DeLander GE, Wiesenfeld-Hallin Z, Fredholm BB, Xu XJ (2005) Increased nociceptive response in mice lacking the adenosine A_1 receptor. Pain 113:395–404

Yamagata K, Hakata K, Maeda A, Mochizuki C, Matsufuji H, Chino M, Yamori Y (2007) Adenosine induces expression of glial cell line-derived neurotrophic factor (GDNF) in primary rat astrocytes. Neurosci Res 59:467–474

Yao L, Fan P, Jiang Z, Mailliard WS, Gordon AS, Diamond I (2003) Addicting drugs utilize a synergistic molecular mechanism in common requiring adenosine and Gi-beta gamma dimers. Proc Natl Acad Sci USA 2003 100:14379–14384

Yao L, McFarland K, Fan P, Jiang Z, Ueda T, Diamond I (2006) Adenosine A_{2a} blockade prevents synergy between mu-opiate and cannabinoid CB_1 receptors and eliminates heroin-seeking behavior in addicted rats. Proc Natl Acad Sci USA 103:7877–7882

Yoon K-W, Rothman SM (1991) Adenosine inhibits excitatory but not inhibitory synaptic transmission in the hippocampus. J Neurosci 11:1375–1380

Yoshioka K, Hosoda R, Kuroda Y, Nakata H (2002) Hetero-oligomerization of adenosine A1 receptors with $P2Y_1$ receptors in rat brains. FEBS Lett 531:299–303

Yu H, Neimat JS (2008) The treatment of movement disorders by deep brain stimulation. Neurotherapeutics 5:26–36

Zeraati M, Mirnajafi-Zadeh J, Fathollahi Y, Namvar S, Rezvani ME (2006) Adenosine A_1 and A_{2A} receptors of hippocampal CA_1 region have opposite effects on piriform cortex kindled seizures in rats. Seizure 15:41–48

Zhang C, Schmidt JT (1999) Adenosine A_1 and class II metabotropic glutamate receptors mediate shared presynaptic inhibition of retinotectal transmission. J Neurophysiol 82:2947–2955

Zhou QY, Li C, Olah ME, Johnson RA, Stiles GL, Civelli O (1992) Molecular cloning and characterization of an adenosine receptor: the A_3 adenosine receptor. Proc Natl Acad Sci USA 89:7432–7436

Zimmermann H (2006) Ectonucleotidases in the nervous system. Novartis Found Symp 276: 113–128

Zuccato C, Cattaneo E (2007) Role of brain-derived neurotrophic factor in Huntington's disease. Prog Neurobiol 81:294–330

Zuccato C, Ciammola A, Rigamonti D, Leavitt BR, Goffredo D, Conti L, MacDonald ME, Friedlander RM, Silani V, Hayden MR, Timmusk T, Sipione S, Cattaneo E (2001) Loss of huntingtin-mediated *BDNF* gene transcription in Huntington's disease. Science 293:493–498

Adenosine Receptors and Neurological Disease: Neuroprotection and Neurodegeneration

Trevor W. Stone, Stefania Ceruti, and Mariapia P. Abbracchio

Contents

T.W. Stone (✉)
Institute of Biomedical and Life Sciences, University of Glasgow, Glasgow G12 8QQ, UK
T.W.Stone@bio.gla.ac.uk

C.N. Wilson and S.J. Mustafa (eds.), *Adenosine Receptors in Health and Disease*,
Handbook of Experimental Pharmacology 193, DOI 10.1007/978-3-540-89615-9_17,

Abstract Adenosine receptors modulate neuronal and synaptic function in a range of ways that may make them relevant to the occurrence, development and treatment of brain ischemic damage and degenerative disorders. A_1 adenosine receptors tend to suppress neural activity by a predominantly presynaptic action, while A_{2A} adenosine receptors are more likely to promote transmitter release and postsynaptic depolarization. A variety of interactions have also been described in which adenosine A_1 or A_2 adenosine receptors can modify cellular responses to conventional neurotransmitters or receptor agonists such as glutamate, NMDA, nitric oxide and P2 purine receptors. Part of the role of adenosine receptors seems to be in the regulation of inflammatory processes that often occur in the aftermath of a major insult or disease process. All of the adenosine receptors can modulate the release of cytokines such as interleukins and tumor necrosis factor-α from immune-competent leukocytes and glia. When examined directly as modifiers of brain damage, A_1 adenosine receptor (AR) agonists, A_{2A}AR agonists and antagonists, as well as A_3AR antagonists, can protect against a range of insults, both in vitro and in vivo. Intriguingly, acute and chronic treatments with these ligands can often produce diametrically opposite effects on damage outcome, probably resulting from adaptational changes in receptor number or properties. In some cases molecular approaches have identified the involvement of ERK and GSK-3β pathways in the protection from damage. Much evidence argues for a role of adenosine receptors in neurological disease. Receptor densities are altered in patients with Alzheimer's disease, while many studies have demonstrated effects of adenosine and its antagonists on synaptic plasticity in vitro, or on learning adequacy in vivo. The combined effects of adenosine on neuronal viability and inflammatory processes have also led to considerations of their roles in Lesch–Nyhan syndrome, Creutzfeldt–Jakob disease, Huntington's disease and multiple sclerosis, as well as the brain damage associated with stroke. In addition to the potential pathological relevance of adenosine receptors, there are earnest attempts in progress to generate ligands that will target adenosine receptors as therapeutic agents to treat some of these disorders.

Keywords Neuroprotection · Neurodegeneration · Ischaemia · Alzheimer's disease · β-amyloid · Huntington's disease · Parkinson's disease · Neurotoxicity · Aging · Stroke · Lesch-Nyhan syndrome · Multiple sclerosis · Creutzfeldt-Jacob syndrome · Prion disease · Acute administration · Chronic administration · Receptor up-regulation · Receptor down-regulation

Abbreviations

ADAC	Adenosine amine congener
AMP	Adenosine monophosphate
AR	Adenosine receptor
BDNF	Brain-derived neurotrophic factor
BIIP20	*S*-(−)-8-(3-Oxocyclopentyl)-1,3-dipropyl-7*H*-purine-2,6-dione
cAMP	Cyclic adenosine monophosphate
CCPA	2-Chloro-N^6-cyclopentyladenosine
CGS15943	5-Amino-9-chloro-2-(2-furyl)-1,2,4-triazolo[1,5-*c*]quinazoline
CGS21680	2-[4-(2-Carboxyethyl)-phenylethylamino]-5′*N*-ethyl-carbox amido-adenosine
CHA	N^6-Cyclohexyladenosine
CJD	Creutzfeldt–Jakob disease
Cl-IB-MECA	2-Chloro-N^6-(3-iodobenzyl)adenosine-5′-*N*-methyluronamide
CNS	Central nervous system
CP66,713	4-Amino-1-phenyl[1,2,4]-triazolo[4,3-*a*]quinoxaline
CPA	Cyclopentyl adenosine
8-CPT	8-Cyclopentyltheophylline
CREB	Cyclic AMP responsive element binding protein
CSC	8-(3-Chloro styryl)caffeine
DMPX	3,7-Dimethyl-1-propargylxanthine
DPCPX	8-Cyclopentyl-1,3-dipropylxanthine
EAE	Allergic encephalomyelitis
ERK1/2	Extracellular signal-regulated kinases 1 and 2
GABA	Gamma-aminobutyric acid
HD	Huntington's disease
HGPRT	Hypoxanthine-guanine phosphoribosyltransferase
IB–MECA	N^6-(3-Iodobenzyl)adenosine-5′-*N*-methyluronamide
IL	Interleukin
KFM19	*RS*-(−)-8-(3-oxocyclopentyl)-1,3-dipropyl-7*H*-purine-2,6-dione
LNS	Lesch–Nyhan syndrome
MAP-2	Microtubule-associated protein 2
MAPK	Mitogen-activated protein kinases
MCAo	Middle cerebral artery occlusion
MPTP	1-Methyl-4-phenyl-1,2,3,6-tetrahydropyridine
MRS2179	N^6-Methyl-2′-deoxyadenosine-3′, 5′-bisphosphate
MRS1706	*N*-(4-Acetylphenyl)-2-[4-(2,3,6,7-tetrahydro-2,6-dioxo-1,3-dipropyl-1*H*-purin-8-yl)-phenoxy]acetamide
MS	Multiple sclerosis
NBTI	Nitrobenzylthioinosine
NECA	5′-*N*-Ethylcarboxamidoadenosine
NGF	Nerve growth factor
NMDA	*N*-Methyl-D-aspartate
3-NP	3-Nitro-propionic acid

PKC	Protein kinase C
PLC	Phospholipase C
R-PIA	*R*-Phenylisopropyladenosine
SAH	*S*-Adenosylhomocysteine
SCH58261	7-(2-Phenylethyl)-5-amino-2-(2-furyl)-pyrazolo-[4,3-*e*]-1,2,4-triazolo[1,5-*c*]-pyrimidine
TNF-α	Tumor necrosis factor alpha
Trk	Tropomyosin-related kinase
ZM241385	4-(2-[7-Amino-2-(2-furyl)(1,2,4)-triazolo(2,3-*a*)-(1,3,5)triazin-5-yl-amino]ethyl)phenol

1 Introduction

As will be evident elsewhere in this volume, adenosine receptors are essentially ubiquitous, with almost all cell types expressing functional forms of at least one of the four known subtypes (A_1, A_{2A}, A_{2B}, A_3). Each of these subtypes has been associated with a range of actions, some of which may become over- or underexpressed, over- or underactive. Such a change in activity could lead to abnormalities of tissue function, which may be severe enough to lead to overt disease. In this chapter, the evidence for a possible contribution of adenosine receptors to the processes of neurodegeneration and neurological disorders involving neurodegeneration will be addressed, together with the potential for developing adenosine receptor ligands as therapeutic agents to modify those disorders.

2 Relevant General Features of Adenosine Receptor Actions

2.1 A_1 Adenosine Receptors

A_1 adenosine receptors occur throughout the central nervous system (CNS), with a high density in the hippocampus and neocortex. The widespread distribution of these receptors is seen in almost all mammalian species examined, including humans (Fastbom et al. 1986, 1987a, b). All cell types in the CNS possess these receptors, including both neurons and microglia (Goodman and Snyder 1982; Lee and Reddington 1986; Rivkees et al. 1995; Fiebich et al. 1996b; Svenningsson et al. 1997; Ochiishi et al. 1999a, b), with neuronal receptors existing on presynaptic terminals and postsynaptic membranes (Ochiishi et al. 1999a, b). Probably the most prominent consequence of activating the A_1 adenosine receptor (AR) is the inhibition of neurotransmitter release from synaptic terminals, an action that has been linked to the reduction of calcium influx in response to action potential invasion of the terminals (Wu and Saggau 1997). A_1ARs are able to suppress the release of a variety of neurotransmitters, including glutamate (Corradetti et al. 1984; Fastbom and Fredholm 1985; Andine et al. 1990; Butcher et al. 1990), acetylcholine (Spignoli et al. 1984; Brown et al. 1990) and dopamine (Michaelis et al. 1979; Chowdhury and Fillenz 1991). There is a significant degree of specificity in this action, however, since it seems to result primarily in a suppression of release of excitatory transmitters such as the major excitatory transmitter glutamate (Corradetti et al. 1984; Héron et al. 1993; Poli et al. 1991), rather than inhibitory transmitters such as gamma-aminobutyric acid (GABA). While a depression of GABA release can be demonstrated using A_1AR agonists, the potency of these compounds and the amount of release inhibition that can be produced are far less than those that have been reported on glutamate release (Hollins and Stone 1980). This difference may be fundamentally important to understanding the relevance of adenosine receptors in neurodegeneration and neuroprotection, since the brain damage which follows strokes and traumatic (mechanical) injuries to the brain (Corsi et al. 1999a) has been attributed to a massive release of glutamate, and it is a suppression of this that may contribute to the neuroprotective efficacy of adenosine A_1 (and A_{2A}) AR. The much smaller effect on GABA release means that the risk of reinstating a degree of hyperexcitability, as a result of blocking inhibitory transmission, is greatly reduced.

Activation of A_1AR reduces calcium influx, or inhibits calcium availability, as demonstrated in neuronal and cardiac tissues (Dolphin and Prestwich 1985; Fredholm and Dunwiddie 1988; Rudolphi et al. 1992; Scholz and Miller 1992). This may be related to the frequently observed ability of A_1AR to modulate the potassium conductances of several types, including the ATP-sensitive potassium channels in heart and hippocampal neurons (Trussel and Jackson 1985; Regenold and Illes 1990; Hosseinzadeh and Stone 1998). There appear to be neuronal chloride conductances which are also sensitive to purines, resulting in an increased chloride influx which should contribute to neuronal inhibition in most areas of the brain (Mager et al. 1990; Schubert et al. 1991).

2.2 *A_{2A} Adenosine Receptors*

A population of A_{2A}ARs is usually distinguished from A_{2B}ARs on the basis of the higher affinity of A_{2A}ARs for the agonist ligand 2-[4-(2-carboxyethyl)-phenylethylamino]-5′*N*-ethyl-carboxamido-adenosine(CGS21680).CGS21680 sho ws an approximately 140-fold selectivity for A_{2A}ARs relative to A_1ARs, (Bridges et al. 1988; Hutchison et al. 1989; Merkel et al. 1992). The A_{2A}ARs occur predominantly on neurons in the striatum, especially the GABAergic striatopallidal projection neurons and on cholinergic interneurons (Jarvis and Williams 1989; Schiffmann et al. 1991; Cunha et al. 1994; Kurokawa et al. 1994; Latini et al. 1996; Ongini and Fredholm 1996; Moreau and Huber 1999). They are also found in the nucleus accumbens and olfactory tubercle, and the hippocampus and cerebral cortex (Cunha et al. 1994; Dixon et al. 1996), although in the last two areas there are significant pharmacological differences between the nominally A_{2A} sites and those classically described in striatum (Cunha et al. 1996). A broadly similar distribution exists in human brain, since, although they were initially reported to exist primarily in striatal regions (Martinez-Mir et al. 1991), subsequent work has shown their presence more widely throughout the CNS (Svenningsson et al. 1997).

There is abundant evidence from a number of biochemical and electrophysiological investigations that the activation of A_{2A}AR promotes the release of neurotransmitters, including glutamate (Sebastiao and Ribeiro 1992; Cunha et al. 1994), an effect probably produced by increasing presynaptic calcium influx (Goncalves et al. 1997). Administration of the A_{2A}AR agonist CGS21680 in vivo does not itself alter the extracellular levels of glutamate in the CNS, but in the rat it can increase the efflux of glutamate triggered by ischemia (Fredholm and Dunwiddie 1988; O'Regan et al. 1992). Consistent with this, the AR antagonist 5-amino-9-chloro-2-(2-furyl)-1,2,4-triazolo[1,5-*c*]quinazoline (CGS15943) can depress glutamate release, possibly by blocking the enhancing effect of endogenous adenosine at A_{2A}AR (Fredholm and Dunwiddie 1988). The facilitation of release by A_{2A}AR agonists has also been demonstrated for other transmitters such as GABA. Hence it is possible that neuroprotection by A_{2A}AR agonists may result, at least in part, from increased extracellular levels of GABA causing generalized inhibition of cell activity, calcium influx and damage (Mayfield et al. 1993; Kurokawa et al. 1994).

2.3 A_{2B} Adenosine Receptors

The low-affinity A_{2B}AR was cloned in the early 1990s, and has long remained the least known adenosine receptor subtype. The A_{2B} receptor positively couples to both adenylyl cyclase and phospholipase C (PLC), the latter occurring through G_q proteins and representing the most important pathway responsible for A_{2B}-mediated effects (Linden et al. 1999). The A_{2B}AR is expressed at low levels in almost all tissues including brain and spinal cord, and its low affinity for the natural ligand suggests that it could be mainly recruited under pathological conditions.

In the CNS, A_{2B}ARs have been suggested to mediate the outgrowth of dorsal spinal cord axons (Corset et al. 2000) and to interact with inflammatory cytokines in the induction of long-term brain responses to trauma and ischemia, such as reactive astrogliosis. A complex interaction between A_{2B}AR and tumor necrosis factor alpha (TNF-α) has been reported, depending upon specific pathophysiological conditions. In particular, prolonged treatment of human astrocytes with the proinflammatory cytokine TNF-α increased the functional responsiveness of A_{2B}AR, which, in turn, synergized with the cytokine in inducing the morphological signs of chronic reactive gliosis (Trincavelli et al. 2004). Conversely, short-term exposure of astrocytes to TNF-α caused the phosphorylation of A_{2B}AR and impairment in their coupling to G_s proteins, with consequent decreases of cyclic adenosine monophosphate (cAMP) production. TNF-α-mediated downregulation of A_{2B}AR was demonstrated to occur via protein kinase C (PKC) intracellular kinase. This event likely represents a defense mechanism to counteract excessive A_{2B} receptor activation under acute damage conditions characterized by massive release of both cytokines and adenosine, such as those occurring during trauma or ischemia (Trincavelli et al. 2008).

A_{2B}ARs have been also suggested to inhibit taurine release from pituicytes, the astroglial cells of the neurohypophyses. In whole rat neurohypophyses preloaded with [^{3}H]taurine, taurine efflux elicited by hypotonic shocks was about 30–50% smaller in the presence of 10 mM adenosine or 1 mM NECA (5′-*N*-ethylcarboxamidoadenosine). The A_{2B}AR antagonists MRS1706 {*N*-(4-acetyl-phenyl)-2-[4-(2,3,6,7-tetrahydro-2,6-dioxo-1,3-dipropyl-1*H*-purin-8-yl)-phenoxy] acetamide} or alloxazine partially reversed the inhibition of release by NECA, while neither agonists of the adenosine A_1, A_{2A} or A_3 ARs nor the A_1AR antagonist DPCPX (8-cyclopentyl-1,3-dipropylxanthine) had any effect (Pierson et al. 2007). Based on evidence implicating taurine not only in cell osmoregulation but also in olfactory, auditory and visual development, as well as in long-term potentiation in the striatum (Warskulat et al. 2007), if confirmed by further studies, this observation may unveil entirely new pathophysiological roles for this as-yet neglected adenosine receptor subtype.

2.4 A_3 Adenosine Receptors

The A_3ARs (Zhou et al. 1992) have been less well studied than the A_1 and A_{2A} AR populations recognized earlier. A_3 sites exist primarily in peripheral tissues, but they are believed to occur on neuronal and glial cells membranes in most species examined, including human (Jacobson 1998), although at least one group has reported failing to find either the A_3 receptor protein or its mRNA in the CNS (Rivkees et al. 2000). This report was accompanied by claims that several of the purportedly selective ligands used in the functional study of A_3 receptor effects actually have significant activity at A_1AR that could complicate the interpretation of results, and could possible account entirely for the supposed actions attributed to A_3AR.

2.5 Receptor Interactions

There is good evidence for interactions between receptors for adenosine and other neuroactive compounds. For instance, activation of *N*-methyl-D-aspartic acid (NMDA) receptors can inhibit the actions of A_1AR agonists on presynaptic terminals (Bartrup and Stone 1990; Bartrup et al. 1991; Nikbakht and Stone 2001). In a situation in which the levels of glutamate increase significantly, therefore, there is a real danger that the protective activity of endogenous adenosine could be compromised by NMDA receptors. The direction of receptor interactions is reversed at postsynaptic sites. On hippocampal and striatal neurons, for example, adenosine can depress the activation of NMDA receptors. This action can be produced by A_1 or A_{2A}AR (de Mendonça et al. 1995; Norenberg et al. 1997, 1998; Wirkner et al. 2000,

2004; Gerevich et al. 2002). The relevance of these interactions remains unclear, as do the circumstances under which one or the other would be dominant. Thus, if an increase in the ambient levels of glutamate occurs prior to any elevation of adenosine levels, then a loss of AR-mediated protection would be expected, leading to enhanced cell damage. If, however, any increase in adenosine levels precedes a change in glutamate, the purine could limit the release of the amino acid, and block the activation of NMDA receptors by the lower amounts of glutamate present.

There may be a significant contribution to A_{2A}AR antagonist neuroprotection by the modulation of responses to other neuroactive agents via influences directly on the receptors. Simpson et al. (1992) and later Cunha et al. (1994) reported the ability of A_{2A}AR to antagonize the activation of A_1AR, a proposal subsequently confirmed and supported by other groups (Dixon et al. 1997; O'Kane and Stone 1998; Latini et al. 1999). The A_{2A} agonist CGS21680 inhibits neuronal responses to the A_1 ligand CPA (O'Kane and Stone 1998), an action that may be related to its ability to induce a low-affinity site for the highly selective A_1 receptor agonist 2-chloro-N^6-cyclopentyladenosine (CCPA) (Dixon et al. 1997). This interaction may occur between the membrane receptors themselves, or via an intermediate, diffusible messenger. Both A_1 and A_{2A}ARs suppress the electrophysiological effects of glutamate or NMDA applied directly to neurons (de Mendonça et al. 1995; Norenberg et al. 1997, 1998; Gerevich et al. 2002; Wirkner et al. 2000, 2004), while CGS21680 reduces the increased postsynaptic influx of calcium induced by quinolinic acid, and the A_{2A}AR antagonist SCH58261 increases it (Popoli et al. 2002), although another antagonist 4-(2-[7-amino-2-(2-furyl)(1,2,4)-triazolo(2,3-*a*)-(1,3,5)triazin-5-yl-amino]-ethyl)phenol (ZM241385) appears not to do so (Tebano et al. 2004). One implication of this interaction is that, when the extracellular level of adenosine reaches levels sufficient to activate A_{2A}AR, as it can do after kainate administration or ischemia, it may inhibit A_1 receptor function. This phenomenon may explain the curious observation that neuroprotection by ZM241385 is lessened by DPCPX (Jones et al. 1998a,b). The protection by ZM241385 could be due to its blockade of A_{2A}AR, thus "releasing" A_1AR from tonic suppression by A_{2A}AR. If the heightened activation of A_1AR were then responsible for the neuroprotection, it would be prevented by DPCPX, as observed (Jones et al. 1998a, b).

As in the case of the A_{2A} agonists noted above, there is evidence that agonists at A_3AR, administered acutely, may reduce responses to A_1AR agonists, and thus decrease the protective activity of endogenous adenosine levels (von Lubitz et al. 1999a). On the contrary, a chronic activation of A_3AR exerts protective effects, as detailed below (see Sect. 3.4). However, it is not known whether these effects are mediated by an opposite activity on the A_1AR subtype.

Finally, there is evidence that some AR subtypes can physically interact with other neurotransmitter receptors, leading to the generation of receptor heteromers characterized by unique pharmacological properties. Yoshioka et al. (2001) co-expressed A_1AR and $P2Y_1$ receptor for ADP in HEK293 cells. These receptors co-immunoprecipitated in western blots of whole cell membrane lysates. Coexpressing the $P2Y_1$ receptor did not alter surface expression of the A_1 receptor, but it did inhibit the binding of radiolabeled A_1AR agonists and antagonists in membrane preparations. This change was not seen in a mixture of membranes from cells expressing each receptor individually. Additionally, the binding of an A_1AR agonist was displaced by the $P2Y_1$ agonist ADPβS and the $P2Y_1$ antagonist N^6-methyl-2′-deoxyadenosine-3′, 5′-bisphosphate (MRS2179) in cotransfected cells, but not in cells expressing the A_1 receptor only. Globally, these data indicate formation of a functional heteromeric complex where A_1ARs physically interact with $P2Y_1$ receptors (Abbracchio et al. 2006).

A_1ARs couple to G_i, mediating depression of intracellular cAMP levels, whereas $P2Y_1$ receptors interact with $G_{q/11}$ and have no effect on cAMP. ADPβS inhibited cAMP production in co-transfected cells only, an effect that was antagonized by the A_1 antagonist DPCPX, but not by MRS2179, and was abolished by pertussis toxin. Thus, ADPβS appears to have acted via the A_1AR ligand-binding site; i.e., the $P2Y_1/A_1$ dimer has novel pharmacological properties compared with the parent receptors. Interestingly, although ADPβS induced inositol phosphate synthesis, the A_1 agonist cyclopentyl adenosine (CPA) did not. Thus, dimerization did not lead to a complete change in pharmacological properties in this case.

Using confocal laser microscopy to study the subcellular distribution of the $P2Y_1$ and A_1AR, Yoshioka et al. (2001) showed that both were expressed mainly near the plasma membrane of HEK293 cells. Furthermore, there was a strong overlap in their distribution in individual cells. This was confirmed in a subsequent study using the biophysical technique of bioluminescence resonance energy transfer (Yoshioka et al. 2002b). In the absence of agonists, the receptors showed a homogeneous colocalization across the cells. Addition of ADPβS and CPA together, but not alone, induced an increase in the bioluminescence resonance energy transfer ratio over 10 min. Thus, although the receptors have a constitutive association, their coactivation increased the association. This association was also seen with native receptors in central neurons. Using confocal laser microscopy and double immunofluorescence, Yoshioka et al. (2002a) demonstrated that the $P2Y_1$ and A_1AR colocalized in neurons of the rat cortex, hippocampus, and cerebellum. A direct association was then shown by their coimmunoprecipitation in membrane extracts from these regions.

The structural requirements for the receptor–receptor interaction are not known at present. The physiological roles of the $P2Y_1/A_1$ dimer also remain to be determined, although Nakata et al. (2003) have pointed out that its pharmacological properties resemble those of a presynaptic receptor that mediates inhibition of neurotransmitter release in some tissues. Finally, Yoshioka et al. (2001) reported that the rat $P2Y_2$ receptor also coimmunoprecipitated with the A_1 receptor when they were coexpressed in HEK293 cells. Thus, the formation of oligomers by A_1AR receptors is likely to be widespread and to greatly increase the diversity of purinergic signaling.

In a similar way, $A_{2A}ARs$ have been demonstrated to dimerize with D_2 receptors, an interaction which involved peculiar peptide residues (Canals et al. 2003). The formation of A_{2A}/D_2 receptor heteromers in the plasma membrane contributes to explain the early observation of agonist affinity loss at the D_2 receptor after activation of the A_{2A}AR (Ferré et al. 1991) and provides a molecular explanation to the functional interaction between adenosine and dopamine in basal ganglia.

2.6 *Anti-inflammatory Effects*

One line of argument that tissue protection by purines is more dependent on modulation of the immune system than on neurotransmitter release or activity is that protection against damage is shown in a range of tissues besides the CNS. Adenosine antagonizes the release and actions of several proinflammatory cytokines such as TNF-α and complement (Lappin and Whaley 1984; Cronstein et al. 1992; LeVraux et al. 1993; Barnes et al. 1995; Ritchie et al. 1997). A_{2A}ARs specifically inhibit the production of IL-12 by human monocytes but increase the generation of IL-10 (Link et al. 2000). This ability to modulate the relative release of several cytokines could be a significant factor in determining the overall immune profile that occurs in response to different primary activating stimuli in different inflammatory situations. Adenosine suppresses phagocytosis, free radical generation and cell adherence by white blood cells activated by immune stimulation (Cronstein et al. 1985, 1987, 1990, 1992; Burkey and Webster 1993; Cronstein 1994). There is now clear evidence that A_{2A}AR play a major role in this form of cellular regulation (Dianzani et al. 1994; Hannon et al. 1998), probably acting via the activation of a serine/threonine protein phosphatase (Revan et al. 1996). Most strikingly, adenosine receptors protect the heart against damage occasioned by ischemia (Zhao et al. 1993; Matherne et al. 1997). Indeed, all anti-inflammatory actions of adenosine have been demonstrated in the myocardium, including suppression of TNF-α production (Meldrum et al. 1997; Wagner et al. 1998a, b; Cain et al. 1998) and regulation of neutrophil adherence to myocytes (Bullough et al. 1995). There is, however, some confusion as to the nature of the ARs involved. Human neutrophils possess A_1 and A_{2A}ARs (Varani et al. 1998) and Cronstein et al. (1992) have demonstrated that both receptors are able to modulate several aspects of the immune response, including chemotaxis. Lozza et al. (1997) have suggested that A_1 and A_{2A}AR agonists are both able to protect the heart against ischemia/reperfusion injury, but there are reports that A_1 agonists but not A_{2A}AR agonists provide cardiac protection (Casati et al. 1997), whereas other groups have claimed the opposite (Cargnoni et al. 1999). The former claim is more consistent with evidence that resistance to myocardical ischemia is correlated with the level of expression of A_1AR. In most cases, the two populations of receptor exhibit opposing actions, suggesting that their joint presence could be the basis of a control system in which low concentrations of adenosine, via A_1AR, are normally able to enhance the sensitivity of white blood cells to immune stimuli but, at the higher concentrations likely to occur at the time of an established immune response, A_{2A}AR can restrain the extent of cellular activity (Cronstein et al. 1992).

The regulation of cytokines by A_3AR is quite selective. Production of several cytokines, including some such as IL-1β and IL-6, which are also proinflammatory, can be modified by A_3AR activation (Ramakers et al. 2006). A_3AR may also suppress the oxidative burst that accompanies the response of defensive leucocytes to immune activation. They can reduce superoxide generation in human eosinophils (Ezeamuzie and Philips 1999), for example, although there is apparently no similar suppression of oxidative activity in human neutrophils (Hannon et al. 1998). The former action could be secondary to an increase in the level of antioxidant enzymes, including superoxide dismutase, which has been shown to be produced by A_3AR agonists in endothelial cells (Maggirwar et al. 1994).

3 Role of Adenosine Receptors in Brain Cell Survival and in Neurodegenerative Diseases

3.1 A_1 Adenosine Receptors and Neuroprotection

Although much of the interest in the therapeutic value of purine receptor ligands has centered on protection following strokes, there remains the possibility that overactivation of glutamate receptors may contribute to chronic neurodegenerative disorders such as Alzheimer's disease and Huntington's disease. This possibility is the rationale for studying the protective effects of agents against excitotoxins, which are frequently used as a model of stroke and neurodegenerative disease. The most commonly used excitotoxins are kainic acid and quinolinic acid, a tryptophan metabolite for which the evidence for a role in some degenerative disorders is substantial (see Stone 1993, 2001; Stone and Darlington 2002 for reviews). Not only do they produce a controllable degree and extent of injury, but the mechanisms of damage have much in common with natural causes. Thus, even the damage produced by kainic acid probably involves a presynaptic action of kainate, which induces the release of endogenous compounds such as glutamate and aspartate (Kohler et al. 1978; Ferkany et al. 1982; Lehmann et al. 1983; Jacobson and Hamberger 1985; Connick and Stone 1986; Virgili et al. 1986; Okazaki and Nadler 1988). Whether this secondary release is the primary cause of cell death or only a contributory (perhaps permissive) component is irrelevant, since the essential issue is that the inhibition of their release by an agent such as an adenosine A_1AR agonist will have the same net protective activity.

Both A_1AR agonists and $A_{2A}AR$ agonists and antagonists (see also below) can protect against kainic acid-induced damage (MacGregor and Stone 1993; Jones et al. 1998a, b). *R*-Phenylisopropyladenosine (R-PIA) protected against kainic acid neurotoxicity in several regions of the CNS in addition to the hippocampus (MacGregor and Stone 1993; MacGregor et al. 1993, 1996), the involvement of A_1AR being further confirmed by showing that protection could be prevented by the simultaneous administration of an A_1AR antagonist such as DPCPX. In addition to the use of DPCPX to confirm the involvement of A_1AR, there have been several studies showing that DPCPX and other selective A_1AR blockers increase the amount of neuronal damage after ischemia or the administration of excitotoxins (Rudolphi et al. 1987; von Lubitz et al. 1994a; Phillis 1995). R-PIA prevented the kainate-induced damage in areas such as the basolateral amygdala, pyriform cortex and rhinal fissure. An observation that has not been pursued, but which may be of considerable pathological and therapeutic importance, was that some areas of the brain, especially those located in more caudal regions such as the entorhinal cortex, the posteromedial cortical amygdaloid nucleus and the amygdalopyriform transition, were not protected by A_1AR activation. It is still uncertain why R-PIA showed such regionally selective protection. There may be fewer A_1ARs in the resistant areas, or a greater susceptibility to damage which the adenosine agonist was unable to overcome at the doses used. The protection afforded by A_1ARs does not necessarily require the use of a selective exogenous agonist, since compounds which raise the concentrations of endogenous adenosine, either by inhibiting transporter function or adenosine metabolism, can also produce protection (Parkinson et al. 1994; Pazzagli et al. 1994).

There is an even greater contribution of presynaptic release in the neuronal damage caused by quinolinic acid, although its major action seems to be the activation of NMDA receptors and the generation of reactive oxygen species (Stone and Darlington 2002; Stone 2001).

A number of other studies have demonstrated protection by adenosine analogs against damage produced by toxins or excitotoxins (Arvin et al. 1989; Connick and Stone 1989; Finn et al. 1991). One especially interesting report showed that protection could be produced by the adenosine A_1 receptor agonist N^6-cyclohexyladenosine (CHA) against the selective dopaminergic neurotoxin 1-methyl-4-phenyl-1,2,3,6-tetrahydropyridine (MPTP) (Lau and Mouradian 1993). This protection raises the real possibility that a selective A_1AR agonist could be useful in Parkinson's disease, where a proportion of cases may be caused by the exposure of patients to exogenous toxins with molecular structures and a propensity to generate oxidative stress similar to those of MPTP. However, the mechanism of protection against MPTP remains unclear, although antagonists at NMDA receptors can also block MPTP damage, raising the possibility that glutamate receptors may play a critical role comparable to that exhibited by them in stroke-induced damage, and against which A_1AR agonists are also effective.

3.2 A_{2A} Adenosine Receptors and Neuroprotection

At variance from the clearcut neuroprotective role exerted by A_1AR, contrasting data have been reported so far on the beneficial/detrimental roles mediated by A_{2A}AR on brain cells.

As with the A_1AR, very early studies indicated that agonists at A_{2A}AR can produce protection of the CNS against several insults, including ischemia (Phillis

1995; Sheardown and Knutsen 1996), and excitotoxins such as kainate (Sperk 1994; Jones et al. 1998a, b). However, protection by CGS21680 was largely prevented by 8-(*p*-sulfophenyl)-theophylline (8PST), a nonselective xanthine antagonist that blocks both A_1 and A_2ARs. Since this antagonist does not penetrate the blood–brain barrier, it was suggested that the protective activity of CGS21680 was generated via sites on the systemic rather than central side of the barrier. The effect was believed to be primarily exerted on the vascular system, modifying blood flow to the potentially damaged regions of brain, or on white blood cells of the immune system, reducing their penetration to and activation by the early neuronal damage. This conclusion was supported by findings that administering CGS21680 directly into the hippocampus did not induce protection.

On the other hand, neuroprotection by an antagonist at A_{2A}AR was first reported by Gao and Phillis (1994). They found that the nonselective A_2AR antagonist CGS15943 protected the gerbil brain against ischemic damage, an observation later supported using the more selective compounds 8-(3-chlorostyryl)caffeine (CSC) and 4-amino-1-phenyl[1,2,4]-triazolo[4,3-*a*]quinoxaline (CP66,713) (Phillis 1995). Many of the earlier studies examined protection against global cerebral ischemia, but protection has also been demonstrated against focal ischemic damage (Ongini et al. 1997). More recent work has involved a range of different receptor ligands and ischemic models (von Lubitz et al. 1995b; Sheardown and Knutsen 1996; Monopoli et al. 1998). In addition, protection by A_{2A}AR antagonists occurs against excitotoxins such as kainic acid, glutamate and quinolinic acid (Jones et al. 1998a, b). The ability of A_{2A}AR antagonists to protect the CNS has received strong support from the generation of transgenic mice lacking these receptors. These knockout animals exhibit a significantly lower level of brain injury following excitotoxins or ischemia (Bona et al. 1997; Chen et al. 1999).

Interestingly, a possible mechanism at the basis of the neuroprotective effects of A_{2A}AR antagonists may reside in blockade of A_{2A}AR-mediated glutamate release by astrocytes.

Adenosine causes a two- to threefold increase in glutamate release from cultured hippocampal astrocytes (Nishizaki et al. 2002; Nishizaki 2004). Such an effect is mimicked by the A_{2A}AR agonist CGS21680 and inhibited by the A_{2A}AR antagonist 3,7-dimethyl-1-propargylxanthine (DMPX), but not by the A_1AR antagonist 8-cyclopentyltheophylline (8-CPT) (Li et al. 2001; Nishizaki et al. 2002). These observations suggest that adenosine stimulates vesicular glutamate release from astrocytes via A_{2A}AR. This agrees with recent findings demonstrating that the A_{2A} receptor antagonist ZM241385 (5 nM via probe) completely prevents the increase in extracellular glutamate outflow induced by dihydrokainic acid, a blocker of glial glutamate uptake (Pintor et al. 2004).

More recently, however, the equation A_{2A} receptor blockade = neuroprotection has appeared too simplistic (in this respect, see Popoli et al. 2007). First, it is now definitely clear that, besides mediating “bad” responses (for example, stimulation of glutamate outflow and excessive glial activation), A_{2A}ARs also promote “good” responses (such as trophic and anti-inflammatory effects). This implies that blockade of A_{2A}AR can result in either protoxic or neuroprotective effects according to the mechanisms involved in a given experimental model.

Confirmation that A_{2A}AR activation could be neuroprotective came with the development of more selective compounds. Thus, ZM241385 is highly selective for A_{2A}AR, with an approximately 80-fold greater affinity at A_{2A}AR compared with A_{2B}AR. It has an affinity for A_{2A}AR that is around 1,000 times greater than for A_1AR (Palmer et al. 1995). When examined for its ability to protect the CNS against kainic acid, ZM241385 was as effective as the agonist ligand CGS21680. Indeed, the agonist and antagonist together produced a synergistic protection leading to the complete protection of hippocampal neurones (Jones et al. 1998a, b).

To explain these puzzling results, several hypotheses have been invoked, including different degrees of presynaptic versus postsynaptic A_{2A} receptor blockade. The question of presynaptic versus postsynaptic sites of action of A_{2A}AR has been explored by Blum et al. (2003a), with the conclusion that the overall response will depend on the balance of involvement of the former, at which A_{2A}AR activation appears to be deleterious, whereas A_{2A}AR stimulation is protective at postsynaptic sites. In line with this hypothesis, the increase in intracellular calcium levels induced by quinolinic acid in striatal neurons (an effect mediated by postsynaptic NMDA receptors) is significantly potentiated by the A_{2A}AR antagonist 7-(2-phenylethyl)-5-amino-2-(2-furyl)-pyrazolo-[4,3-*e*]-1,2,4-triazolo[1,5-*c*]-pyrimidine (SCH58261) and prevented by the A_{2A}AR agonist CGS21680 (Popoli et al. 2002). In agreement, CGS-21680 was reported to reduce NMDA currents in striatal neurons (Norenberg et al. 1997; Wirkner et al. 2000). Moreover, ZM241385 potentiated NMDA-induced effects in rat corticostriatal slices (Tebano et al. 2004), and the A_{2A}AR antagonist CSC potentiated NMDA-induced toxicity in the hippocampus (Robledo et al. 1999). Thus, as far as NMDA-dependent toxicity is concerned, it seems that A_{2A}AR activation, rather than its blockade, can exert neuroprotective effects.

However, the activity of A_{2A}AR antagonists on the "postsynaptic side" of excitotoxicity appears to be far more problematic. At variance from protective receptors on postsynaptic neuronal cells, postsynaptic A_{2A}ARs localized on microglial inflammatory cells might play a detrimental role. In addition, A_{2A}ARs expressed by bone marrow-derived cells have been proposed as potential contributors to striatal damage induced by mitochondrial dysfunctions in Huntington's and Parkinson's disease (Huang et al. 2006), as was previously suggested in the ischemic context (Yu et al. 2004). In accordance with these findings, in an established in vitro model of reactive astrogliosis, blockade of A_{2A}ARs abolished growth-factor mediated astrocytic activation, an event that may potentially contribute to inflammation and neuronal damage in neurodegenerative diseases (Brambilla et al. 2003).

Finally, A_{2A}ARs can mediate neuroprotection by potentiating brain-derived neurotrophic factor (BDNF) survival signaling pathways. The first link between BDNF and adenosine was provided in 2001, with the demonstration that the activation of tropomyosin-related kinase (Trk)A receptors in PC12 cells and TrkB in hippocampal neurons could be obtained in the absence of neurotrophins by treatment with adenosine (Lee and Chao 2001). These effects were reproduced by using the adenosine agonist CGS21680 and were counteracted with the antagonist ZM241385, indicating that this transactivation by adenosine involves the A_{2A}AR subtype. At hippocampal synapses, presynaptic activity-dependent release of adenosine, through the activation of $A_{2A}AR$, facilitates BDNF modulation of synaptic transmission (for a review, see: Popoli et al. 2007). A similar positive interaction has more recently been confirmed to occur at the neuromuscular junction, which possesses both adenosine $A_{2A}AR$ and BDNF TrkB receptors. The following sequence of events in what concerns cooperativity between $A_{2A}AR$ and TrkB receptors has been suggested: $A_{2A}ARs$ activate the PKA pathway, which promotes the action of BDNF through TrkB receptors coupled to PLCγ, leading to the enhancement of neuromuscular transmission (Pousinha et al. 2006; see also below). Preliminary data indicate that $A_{2A}ARs$ also regulate BDNF levels in the striatum. The importance of $A_{2A}AR$ in regulating BDNF has recently been strengthened by the demonstration that both BDNF levels and functions are significantly reduced in the brains of $A_{2A}AR$ knockout (KO) mice (Popoli et al. 2007).

The possible detrimental/beneficial effects elicited by $A_{2A}AR$ activation or blockade on different brain cell populations are summarized in Fig. 1.

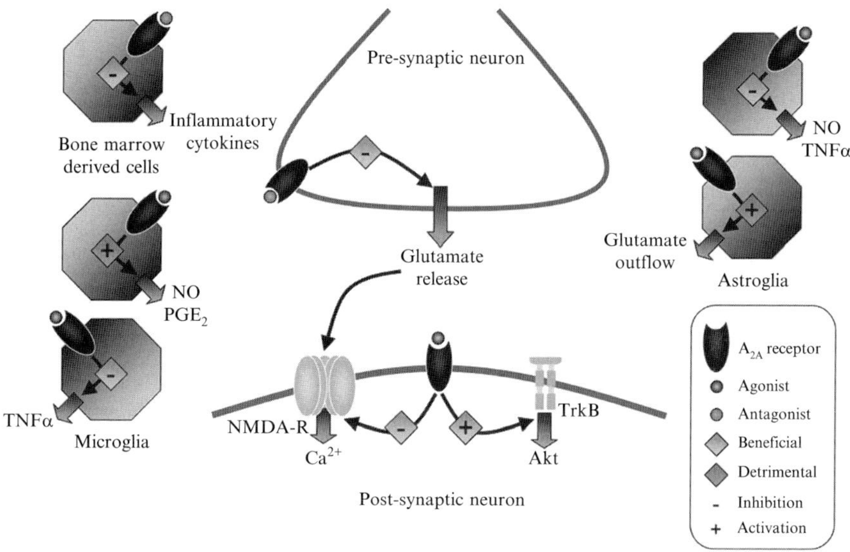

Fig. 1 Schematic representation of the possible effects elicited by A_{2A} adenosine receptor (AR) activation or blockade on different brain cell populations. In the presynaptic neurons, $A_{2A}AR$ blockade may exert beneficial effects through the inhibition of glutamate release. In the postsynaptic neurons, adenosine $A_{2A}ARs$ inhibit *N*-methyl-D-aspartate (NMDA) receptor currents and activate tropomyosin-related kinase (Trk)B receptors, both being potentially beneficial effects. The picture is further complicated by the different effects elicited by the stimulation or blockade of $A_{2A}ARs$ expressed on non-neuronal cells. In astrocytes, $A_{2A}AR$ stimulation can induce both deleterious effects by an increase in glutamate outflow (for a more detailed description of the effects elicited by $A_{2A}ARs$ on glial-mediated modulation of glutamate outflow, see the main text), and beneficial effects through an inhibition of nitric oxide (NO) and tumor necrosis factor alpha (TNF-α) release. This latter beneficial effect has been observed also in microglial cells, although the stimulation of $A_{2A}ARs$ can also induce potentially deleterious effects on this cell population (see also Saura et al. 2005). Finally, in bone marrow-derived cells, it seems to be the blockade of $A_{2A}ARs$ that, through the reduction of cytokine release, can induce beneficial effects. Reproduced and modified from Popoli et al. (2007) with permission from Elsevier

3.3 A_{2B} Adenosine Receptors and Neuroprotection

Far less is known about the role of A_{2B}ARs in neuroprotection compared to that of A_1 and of A_{2A}ARs. As already mentioned, expression of A_{2B}ARs on glial cells and their lower affinity for adenosine suggests a role under emergency conditions (when adenosine levels are massively increased) in mediating long-term inflammatory changes. In line with this hypothesis, A_{2B}ARs were found to synergize with the proinflammatory cytokine TNF-α in mediating the induction of reactive astrogliosis (see Trincavelli et al. 2004).

3.4 A_3 Adenosine Receptors and Neuroprotection

A dual, biphasic role of A_3AR in neuroprotection has been described in several experimental models, both in vivo and in vitro. In fact, von Lubitz et al. clearly demonstrated that an acute administration of the A_3AR selective agonist N^6-(3-iodobenzyl)adenosine-5′-*N*-methyluronamide (IB–MECA) to gerbils dramatically worsened the outcome of a subsequent ischemic episode, whereas chronic stimulation of this receptor subtype protected the animals from stroke, probably through the induction of preconditioning (von Lubitz et al. 1999b; see also below). The protective action of A_3AR agonists against ischemic damage has been recently confirmed by Chen et al. (2006), who also showed that neuroprotection was completely lost in A_3 knockout mice, thus demonstrating the specific involvement of this receptor subtype. Similar results have been obtained in in vitro models. In fact, in non-neuronal cells, low concentrations of the A_3AR agonist 2-chloro-N^6-(3-iodobenzyl)adenosine-5′-*N*-methyluronamide (Cl–IB–MECA; 10 nM or 1 μM) protected against the cell death induced by selective antagonists at this receptor subtype (Yao et al. 1997). Thus, it is suggested that there is a tonic low level of A_3AR activation, possibly induced by the release of endogenous adenosine, which results in cell protection. Protection by Cl–IB–MECA against cell death has been also demonstrated in primary cortical cultures subjected to oxygen–glucose deprivation (Chen et al. 2006). Opposite toxic effects can be achieved when concentrations of agonists$\geqslant$10 μM are used. This has been proven true in several non-neuronal cell lines, with induction of apoptosis and Bak expression (Yao et al. 1997), but also in rat cerebellar granule cells (Sei et al. 1997) and in astrocytic cultures (Abbracchio et al. 1998; Di Iorio et al. 2002), where the reduction of the Bcl-2 expression and the activation of the proapoptotic enzyme caspase 3 by Cl–IB–MECA have been demonstrated (Appel et al. 2001).

3.5 Adenosine Receptors and Therapeutic Possibilities

These various findings have aroused great interest in the search for new drugs that could be used to slow or prevent the neuronal damage that characterizes neurodegenerative conditions such as Alzheimer's disease, Parkinson's disease, and Huntington's disease. That interest is attributable not only to the efficacy of the compounds available, but also to the fact that they should be relatively free of major side effects. Whereas the use of A_1AR agonists would lead to a suppression of transmitter release at many sites within the central and peripheral nervous system, whatever the physiopathological state of those sites, A_{2A}AR antagonists should only produce effects when the receptors are being activated by endogenous adenosine. In practice, this means that A_{2A}AR antagonists have little effect on heart rate, blood pressure, or other vital signs under normal conditions. During ischemia in the brain, however, the levels of adenosine may rise to levels at which A_{2A}AR are activated. Stimulation of A_{2A}AR increases the release of the excitotoxic amino acid glutamate (O'Regan et al. 1992; Popoli et al. 1995), which would tend to cause or facilitate the occurrence of damage. Under these circumstances, A_{2A}AR antagonists should reduce the enhanced release of glutamate and thus decrease the extent of neuronal damage. Their beneficial activity would therefore be restricted to those areas of the brain experiencing ischemia, with little or no effect on other areas of the brain or peripheral tissues.

A particularly exciting aspect of A_{2A}AR protection is that it may contribute to the long-term benefits of treating patients with Parkinson's disease with A_{2A} receptor antagonists. It is clear that A_{2A}ARs potently modulate cell sensitivity to dopamine receptors, accounting for the beneficial effects of adenosine antagonists in this disease (Mally and Stone 1994, 1996, 1998). This phenomenon has led to clinical trials with A_{2A}AR antagonists in Parkinson's disease with promising, though as-yet unpublished, results. In lower primates, A_{2A}AR antagonists are certainly effective against toxin-induced models of the disorder (Kanda et al. 1998; Grondin et al. 1999). The occurrence of protection has been supported strongly by the demonstration that MPTP was able to produce little damage in transgenic mice engineered to be deficient in A_{2A}AR (Ongini et al. 2001).

As noted earlier, the protective effect of A_{2A}AR antagonists may be the result of their removal of the A_{2A}AR suppression of A_1AR (Jones et al. 1998a, b; Pedata et al. 2001). Since there is clear evidence that A_1AR activation is protective, this interaction would explain both the protection by A_{2A}AR antagonists and the blockade of that protection by A_1AR antagonists (Jones et al. 1998a, b). A_1AR activation suppresses excitatory transmitter amino acid release, as does blockade of A_{2A}AR by CGS15943 (Simpson et al. 1992), whereas blockade of A_1AR or activation of A_{2A}AR enhances the release.

3.6 *Molecular Basis of Neuroprotection*

There has to date been little progress in identifying the molecular basis of the neuroprotective activity of adenosine receptors, quite apart from identifying the relative importance of neurons and glia in neuroprotection. Staurosporine is a well-recognized activator of apoptosis, and in many cells, including astrocytes, this activity is accompanied by caspase 3, p38 mitogen-activated protein kinase (MAPK) and glycogen synthase kinase 3β (GSK3β) activation (D'Alimonte et al. 2007). The induction of apoptosis can be prevented by CCPA at A_1 receptor-selective concentrations that are blocked by DPCPX. In addition, these authors noted that CCPA induced the phosphorylative activation of Akt and thus activation of phosphatidylinositol 3-kinase (PI3K), leading to the proposal that this action caused inhibition of the staurosporine effects. The same group has reported a similarly protective action of CCPA against astrocyte apoptosis induced by the quasi-ischemic procedure of oxygen/glucose deprivation (Ciccarelli et al. 2007). Abnormalities in both the p38 and GSK3β pathways have been implicated in the neuronal damage following acute (stroke) and chronic (Alzheimer's disease) neurodegenerative conditions, so that the modulation of adenosine A_1 receptor function may have a more fundamental and direct relevance to cell protection in these cases than merely a global influence on cell excitability or transmitter release. Protection was again accompanied by activation of PI3K. Pharmacological modifiers of apoptosis led to the overall conclusion that A_1 receptor activation protects by activating the PI3K and extracellular signal-regulated kinase 1 and 2 (ERK1/2) MAPK pathways.

Some of the mechanisms at the basis of A_{2A}AR-mediated neuroprotection have already been described above (see Sect. 3.2). In addition, part of the neuroprotective effects of A_{2A}AR may stem from the reduction of nitric oxide production. Saura et al. (2005) reported that CGS21680 potentiated the lipolysaccharide-induced increase of NOS expression and NO production in mixed neuron/glial cultures, whereas ZM241385 blocked the effect. Similarly, Fiebich et al. (1996b) have shown that the activation of A_{2A}AR can induce the expression of COX-2, a key proinflammatory molecule giving rise to eicosanoids and, indirectly, to increased oxidative stress. The A_{2A}AR antagonists could therefore suppress this expression as part of their neuroprotective mechanism (in this respect, see also Sect. 3.6).

The mechanisms underlying the protective effects exerted by low doses of A_3AR agonists have not been clearly understood. In an in vivo model of ischemia, protection by IB–MECA appears to be associated with preservation of cytoskeletal proteins (such as microtubule-associated protein) and increased deposition of glial fibrillary acidic protein in injured areas (von Lubitz et al. 1999b). This would accord with studies in vitro, using glial cultures, in which Cl–IB–MECA induced a number of cytoskeletal changes with the formation of actin filaments (the so-called "stress fibers") accompanied by alterations of cell morphology, such as the emission of long and thick processes in parallel with the alteration of cytoskeletal-associated RhoGTPases (Abbracchio et al. 1997). These changes resulted in a significant reduction of spontaneous apoptosis in culture (Abbracchio et al. 1998), suggesting that astrocytes exposed to nanomolar concentrations of A_3AR agonists are more resistant to cell death, probably due to increased adherence to the culture substrate. Therefore, it can be envisaged that neuroprotection observed in vivo could (at least in part) be due to the beneficial effects of A_3AR agonists on astrocytes, which might in turn help neurons to survive the ischemic episode.

3.7 Trophic Activity

It is possible that an important feature of adenosine receptor activation or blockade contributing to the regulation of neuronal and glial function and viability is the ability of these receptors to directly influence the growth and development of nerve and glial cells. Much of this work has been performed and reviewed by Rathbone et al. (1992, 1999). Both A_1 and A_{2A}ARs can promote neuritogenesis in neuroblastoma cells, the A_{2A}AR acting via PKA (Canals et al. 2005). Trophic effects are exerted by A_{2A}ARs via a positive synergistic interaction with BDNF prosurvival pathways (see above). This interaction occurred through activation of PI3K/Akt via a Trk-dependent mechanism, resulting in increased cell survival after nerve growth factor or brain-derived neurotrophic factor withdrawal.

The ability of adenosine to mediate trophic effects via activation of its receptors presents yet another factor to be considered in using adenosine ligands therapeutically, since antagonists might inhibit a valuable degree of structural reorganization and recovery following a brain insult or limit the degree of damage produced in a degenerative disorder.

4 Aging and Alzheimer's Disease

4.1 Changes of Adenosine Receptors with Aging

Since ARs can modify the neural release and actions of acetylcholine, one of the neurotransmitters most intimately associated with the loss of cortical afferent neurons arising from the nucleus basalis of Meynert, and therefore the transmitter most commonly linked to the development of Alzheimer's disease, they have attracted some attention in relation to dementias.

It is interesting to compare the range of studies that have examined adenosine receptors during the normal aging process with those that have concentrated selectively on changes found in the brains of patients with dementias. Animal studies to date have centered largely on A_1AR presence and distribution in view of their ability to inhibit transmitter release. Reports of alterations with ageing have been confusing, no doubt (at least in part) due to the differing choices of species, brain region, methodology and ligands employed. Some groups have reported clear decreases in A_1AR binding in limited regions of animal brain (Araki et al. 1993; Cunha et al. 1995), whereas others have found more generalized losses (Pagonopoulou and Angelatou 1992) or no change (Virus et al. 1984; Hara et al. 1992; Fredholm et al. 1998) with aging. In one of the earliest of these studies, the loss of a low-affinity subtype of A_1AR was described (Corradetti et al. 1984), although binding was examined using an agonist ligand, and the pharmacological tools to explore the nature of the receptor in more detail were not yet available. A later study using gerbils classified as "middle aged" (16 months old), and which may not therefore have direct relevance for neurodegeneration in the elderly, found significant reductions of A_1AR density in the hippocampus compared with young animals (1 month old), whereas increased binding was found in the neocortex (Araki et al. 1993). When changes in the presence of A_1AR were studied using quantitative autoradiography in the brains of young, old, and senescent rats (3, 24, or 30 months), the density of receptors diminished with age, although the dynamics of that reduction were very different in the various brain regions examined. Thus, while a gradual decline in receptor numbers was seen in hippocampus, cortical sites were lost only after 24 months of age (Meerlo et al. 2004). Fredholm et al. (1998) noted that, while they could find no change in receptor binding, mRNA for the A_1AR was decreased in aging rats, a finding which emphasizes the importance for interpretation of examining the receptor message as well as the protein and, ideally, a measure of receptor function.

Results with A_{2A}ARs have been more consistent, usually indicating a reduction in receptor binding in regions of high density such as striatum (Fredholm et al. 1998). Although these changes were statistically significant, the limited magnitude of the change (20% decrease between 6 and 99 weeks of age) leaves open the question of the functional meaning of that change in the light of the innate adaptive plasticity of the brain.

No data on the possible changes of A_3AR with age are available at the moment.

4.2 Alterations of Adenosine Receptors in Alzheimer's Patients

The examination of human brain tissue from patients who died with a confirmed diagnosis of Alzheimer's disease seems to consistently show a loss of A_1AR (Jansen et al. 1990; Kalaria et al. 1990; Ulas et al. 1993; Deckert et al. 1998), especially and most clearly in the hippocampus, a region of the brain most intimately involved in the processes of learning and memory.

Jansen et al. (1990) described a decrease in receptor densities for several neuroactive compounds in post-mortem tissue from Alzheimer's disease patients. Losses were found in receptors for most of these, including adenosine A_1ARs, which were reduced by 46% in the dentate gyrus. An autoradiographic study using DPCPX as a ligand also reported marked decreases in A_1AR binding in the outer layers of the dentate gyrus, probably reflecting the loss of perforant path input (Jaarsma et al. 1991). The surprising observation was made that the CA1 and CA3 regions showed no loss of A_1AR, despite clear cellular degeneration and reduced numbers of NMDA receptors. Although the difference was attributed to a dendritic location of the A_1AR, the recognized association of A_1AR with presynaptic terminals leaves open the question of whether the perforant path is far more profoundly affected by degeneration than intrinsic hippocampal fibers.

Ulas et al. (1993) also found a similar decrease in A_1 receptor binding in the hippocampus and parahippocampal gyrus of Alzheimer individuals and age-matched controls, with a loss of binding density, though not affinity, in the dentate gyrus (molecular layer). However, decreases were also seen in the CA1 stratum oriens and outer layers of the para-hippocampal gyrus, with subnormal levels of antagonist binding in the CA3 region. Coupling to G proteins was similar in the control and patient populations, indicating a normal transduction pathway for the remaining receptors.

Striatal A_1ARs are also decreased in patients with Alzheimer's disease. Quantitative autoradiography in the post-mortem striatum indicated a reduction of A_1-binding sites in Alzheimer's disease patients compared with matched controls. No comparable change of another presynaptic site, that for kappa opiate receptors, was noted, but the loss of A_1AR showed a strong correlation with the decreased activity of choline acetyltransferase measured in the same tissue samples (Ikeda et al. 1993). In contrast, the levels of A_1AR and A_2ARs appear to be increased in the frontal cortex, in parallel with an increased functional activity of these receptors (Albasanz et al. 2008).

In a fascinating analysis of post-mortem neocortical and hippocampal tissue from patients with Alzheimer's disease, Angulo et al. (2003) reported a significant colocalization of A_1AR with β-amyloid in senile plaques. They also showed that, in human neuroblastoma cells, activation of A_1AR activated PKC, p21 Ras and ERK1/2, leading to increased formation of soluble β-amyloid fragments, raising the possibility that agonists at A_1AR might be valuable drugs in the treatment of established or late-stage Alzheimer's disease.

4.3 Adenosine Receptors and Cognition

In considering both the possible role of adenosine receptors in the symptomatology of Alzheimer's disease, and the potential value of adenosine ligands in treatment, it is clearly important to consider not only histologically or functionally defined neuronal damage but also the reflection of that damage at the behavioral level, especially for cognition.

There is increasing epidemiological evidence for a role of adenosine receptors in cognitive decline with aging. Much of this evidence relates to the use of coffee, which, in several recent studies, has been concluded to produce a protective effect against the cognitive decline in Alzheimer's disease (van Gelder et al. 2007; Quintana et al. 2007). It is clear, however, that the variations in methodology between studies rather confuse attempts to compare results. It is also clear that the relationship between coffee and cognition is not simple, with major questions remaining, such as the role of caffeine versus other constituents of the brew, and the existence of an optimal coffee intake, above and below which cognitive decline may be enhanced.

Despite this caveat, studies specifically focused on caffeine have reached similar conclusions. The risk of developing Alzheimer's disease, for example, is inversely related to caffeine consumption (Maia and de Mendonça 2002). Both caffeine and ZM241385 prevent the neuronal toxicity caused by β-amyloid peptide in vitro or in vivo (Dall'lgna et al. 2003, 2007). Caffeine was also effective in the Swedish mutation transgenic mouse model of Alzheimer's disease, in which cognitive deficits are associated with the induced overexpression of β-amyloid in the brain. Caffeine was able to reduce the β-amyloid load and behavioral indications of cognitive impairment in these mice (Arendash et al. 2006). Associated proteins such as presenilin 1 and β-secretase were also reduced. Confirmation that these effects were likely to be a direct result of actions on the neurons rather than glia or peripheral mechanisms was obtained by showing a similar reduction of β-amyloid formation in neuronal cultures with the same mutation.

Psychological studies have investigated the effects of caffeine on a range of behaviors in human subjects, including vigilance and aspects of learning, as well as in a variety of modified states, including subject age, frequency of caffeine use, level of tolerance or withdrawal, and state of sleep deprivation. However, the relevant doses and their molecular mechanism of action often remain unproven. In a representative study, Riedel et al. (1995) noted that, in healthy subjects, 250 mg of caffeine reduced the scopolamine-induced performance deficit in memory tasks. The provocative conclusion was drawn that any cognition-enhancing drug being considered for therapeutic use should be shown to be at least as active as this dose of caffeine: an amount equivalent to only three cups of coffee. Results of the many studies on caffeine are, however, often confusing. In one study of almost 1,000 people, it was reported that the consumption of (caffeinated) coffee was associated with improved cognitive performance in women, especially those aged over 80 years, but not men. A possible attribution of this finding to caffeine was made on the basis that decaffeinated coffee seemed to have no influence on cognitive function (Johnson-Kozlow et al. 2002).

It seems likely that the effects of adenosine antagonists, especially the nonselective ones such as caffeine, may have quite subtle effects on learning. Angelucci et al. (2002) suggested that this was due to an effect to improve memory retention, with less or no effect on memory acquisition, while Hauber and Bareiss (2001) showed an improvement by theophylline of spatial reference memory when acquisition was achieved under light conditions, but not in the dark.

Whereas most studies have found that agonists at A_1ARs tend to impair learning and memory function (Normile and Barraco 1991; Zarrindast and Shafaghi 1994; Corodimas and Tomita 2001), there are occasional reports of learning facilitation or improvement after the acute (Hooper et al. 1996) or chronic (von Lubitz et al. 1993) administration of an agonist. Antagonists have clear ability to enhance cognition and to reverse induced cognitive deficits. One of the earliest studies on animal learning used the compound *RS*-(−)-8-(3-oxocyclopentyl)-1,3-dipropyl-7*H*-purine-2,6-dione (KFM19), an A_1AR antagonist, which showed cognition-enhancing properties in a rat model (Schingnitz et al. 1991). In a more recent study of olfactory discrimination and social memory in rats, Prediger et al. (2005) demonstrated that deficits in the behaviors of both 12- and 18-month-old animals could be prevented by caffeine or ZM241385. Interestingly, A_1AR blockade by DPCPX was ineffective. Similarly, DPCPX was reported not to affect the acquisition of a shock-induced avoidance task, even though caffeine, or a selective

A_{2A}AR antagonist, did so (Kopf et al. 1999). It is important to note, however, that knockout studies have not been consistent with many of the pharmacological studies using antagonists. Thus, mice lacking A_1AR exhibited normal learning of spatial tasks in the water maze (Gimenez-Llort et al. 2002).

Maemoto et al. (2004) have also shown recently that a new A_1AR-selective antagonist, FR194921 (2-(1-methyl-4-piperidinyl)-6-(2-phenylpyrazolo[1,5-*a*]pyridin-3-yl)-3(2*H*)-pyridazinone) was able to reverse scopolamine-induced deficits on a passive avoidance test, with little effect on behavioral paradigms related to anxiety and depression. Pitsikas and Borsini (1997) obtained similar results using the A_1AR antagonist *S*-(−)-8-(3-oxocyclopentyl)-1,3-dipropyl-7*H*-purine-2,6-dione (BIIP20). A number of detailed structure–activity studies have attempted to define the molecular requirements of A_1AR antagonism that are needed for cognition enhancement (Suzuki et al. 1993).

Few studies have been performed using A_{2A} or A_3 AR ligands in human subjects. Most recently, a specific relationship between A_{2A}AR and Alzheimer's disease was reported by Scatena et al. (2007). The administration of SCH58261 to mice in which β-amyloid (25–35) peptide was delivered into the cerebral ventricles was found able to prevent the subsequent neuronal loss, raising the possibility of reversing the neuronal loss in Alzheimer's disease that is attributable to β-amyloid accumulation. This is particularly interesting in relation to the report, described above, of the colocalization of A_1AR with β -amyloid in senile plaques and the ability of A_1AR to increase the formation of soluble β-amyloid fragments. Since there are several strands of evidence that A_{2A}AR can inhibit the activation of A_1AR (Cunha et al. 1994; Dixon et al. 1997; O'Kane and Stone 1998), it is possible that the effect of SCH58261 is the result of removing an A_{2A}AR-mediated suppression of A_1AR, unmasking the protective efficacy of A_1ARs.

Part of the difficulty in accounting for the detailed mechanism of A_{2A}AR antagonist protection lies in the fact that modulation of A_{2A}AR results in a plethora of actions, many of which are in functional opposition to each other. Thus, while A_{2A}AR agonists promote glutamate release (Sebastiao and Ribeiro 1996) and antagonists should therefore have a valuable action in suppressing excessive release (see also Sect. 3.2), the opposite applies to the inflammatory cytokines. Activation of A_{2A}AR inhibits both the initial calcium influx and the subsequent release of TNF-α induced by various stimuli, including the neurotoxic HIV protein Tat (Fotheringham et al. 2004). Blockade of A_{2A}AR should therefore increase the release of TNF-α and, presumably, related proinflammatory cytokines, thus potentially increasing cell damage. Perhaps the net neuroprotective effects of A_{2A}AR blockade are a complex result of pro- and anti-inflammatory activities, at least some of which may be time dependent, determined by whether the antagonists are present acutely or chronically.

The A_3AR agonist IB–MECA appeared to have little effect alone on measures of learning using simple tests such as spontaneous alternation and passive avoidance. However, this compound did prevent the deficits in these behaviors induced by scopolamine or dizocilpine (Rubaj et al. 2003).

4.4 The Enigma of Propentofylline

The xanthine derivative propentofylline has been the subject of research for almost 20 years, yet in many respects it remains an enigma. It is also an enigma that requires decoding, since its activity may have significance for understanding the role of adenosine receptors in health and disease. Propentofylline is a weak antagonist at adenosine receptors. Its main actions seem to be an inhibition of adenosine uptake into cells, resulting in increased extracellular concentrations, and an inhibition of cyclic AMP phosphodiesterases. But at the level of the behaving animal, its overall effect is to promote cognitive function. Propentofylline has been shown to protect against cerebral ischemia in gerbils (Dux et al. 1990).

Even in humans, this compound is an effective cognition enhancer (Noble and Wagstaff 1997), and has been found to improve cognitive function in patients with vascular dementias (Mielke et al. 1996a, b). In animal models of Alzheimer's disease, it has been shown to prevent the cognitive impairment caused by intracerebral administration of β-amyloid (1–40) (Yamada et al. 1998). This effect was attributed to the promotion of nerve growth factor (NGF) production, which raises further questions about the relationship between this hypothesis and the activation of adenosine receptors. One possibility is that raised extracellular adenosine levels activate A_{2A}AR, and these in turn, as shown by Heese et al. (1997), then promote the generation of NGF and other neurotrophins. The balance of activation of adenosine receptors could be tipped from A_1AR to A_{2A}AR activation by virtue of the inhibitory effect of propentofylline on phosphodiesterase (Schubert et al. 1997). This hypothesis would be consistent with the fact that propentofylline is able to suppress TNF-α production (Meiners et al. 2004), an action that could be mediated partly via the activation (direct or indirect) of A_{2A}AR.

There may also be more direct influences of propentofylline on microglial cells which regulate their degree of inflammatory activity (McRae et al. 1994; Schubert et al. 1996; Rudolphi and Schubert 1997), though the extent to which adenosine receptors might be involved in this also remains unclear. Some of these effects are almost certainly mediated via changes in calcium dynamics within neurons and glia (McLarnon et al. 2005).

4.5 Adenosine, Homocysteinuria and Alzheimer's Disease

Homocysteinuria has been widely linked to vascular abnormalities leading, directly or indirectly, to the compromise of neuronal function and cognitive dysfunction seen in vascular dementia and Alzheimer's disease, and there have been suggestions that a deficiency of adenosine may contribute to the neurological manifestations of increased homocysteine levels. One of the consequences of raised extracellular homocysteine is a parallel reduction of adenosine concentrations, possibly resulting from the formation of *S*-adenosylhomocysteine (SAH). A strong negative correlation between plasma levels of the two compounds has been recorded in Alzheimer's disease patients (Selley 2004). It is possible, therefore, that a raised homocysteine level could induce a fall of adenosine concentrations to the extent that activation of protective receptors, including A_1 and A_{2A}ARs, is compromised.

4.6 Genetic Studies

In an attempt to assess the possible relevance to Alzheimer's disease of mutations in the A_{2A} receptor gene, Liu et al. (2005) have examined 174 patients and 141 controls for the presence of the 1976 T>C polymorphism. No significant differences were noted in the genotype distribution or allelic frequency of this molecule, implying that a change of A_{2A}AR function characterized by this mutation was not likely to be a major contributor to the Alzheimer's disease susceptibility. However, the numbers of patients are not high for this type of study, and there may be alternative polymorphisms that are more relevant.

5 Creutzfeldt–Jakob Disease

Creutzfeldt–Jakob disease (CJD) is one of the prion diseases, characterized by the presence of protease-resistant prion protein within the brain parenchyma, leading to neuronal degeneration, motor impairment and ultimately death. CJD is often considered to be the human equivalent of scrapie, a disease primarily of sheep and related animals, and bovine spongiform encephalopathy (BSE) in cattle. The involvement of transmitters and other endogenous neural molecules in the development of prion-induced brain damage has received rather little attention, other than a degree of focus on glutamate and its receptor subtypes. However, Rodriguez et al. (2006) have examined the levels of adenosine A_1AR in the neocortex of 12 patients with CJD and six age-matched controls. Elevated numbers of A_1AR were identified in the patient group, together with increased receptor activity in cyclic AMP assays but normal levels of mRNA, suggesting increased receptor efficacy together with a possible decrease in the rate of receptor turnover.

When similar measurements were made in mice expressing bovine BSE prion protein, a similar increase in A_1AR number in the brain occurred in parallel with the appearance of prion protein and the development of motor symptoms (Rodriguez et al. 2006). A simplistic interpretation of these data would be consistent with an up-regulation of A_1AR function as a protective adaptation to the potentially injurious prion protein. However, it will be important to assess how the changes in A_1ARs compare with changes in other purine receptors, purine transporters and purine metabolic enzymes, in addition to other ARs and other neuroactive substances, before a significant role of A_1ARs can be considered in isolation.

6 Lesch–Nyhan Syndrome

Lesch–Nyhan syndrome (LNS) is the result of an X-linked deficiency of hypoxanthine–guanine phosphoribosyltransferase (HGPRT). The lack of this major purine salvage enzyme results in high levels of hypoxanthine and uric acid, the latter producing a range of consequences in peripheral tissues, such as gouty arthritis and nephrolithiasis. In some cases, especially those with a complete absence of enzyme activity, there is also involvement of the CNS, with mental retardation and self-mutilation. Since the realization that the latter behavior could be induced by the administration of high doses of caffeine, the question has arisen of whether the various behavioral symptoms are due to a lack of adenosine or its receptors. To date, in spite of the increased de novo synthesis of purines, there is little evidence for any abnormality of adenosine levels or function, but it has been found that hypoxanthine can inhibit adenosine uptake. Levels of hypoxanthine comparable with those found in LNS patients suppress the equilibrative nucleoside transporters in human leucocytes, whether they are sensitive or not to nitrobenzylthioinosine (NBTI) (Torres et al. 2004; Prior et al. 2006). An examination of ARs in a mouse HGPRT knockout model of LNS has revealed an increase in the expression of A_1AR and a decrease of A_{2A}AR in the brain (Bertelli et al. 2006). What remains unclear is whether these receptor changes are induced by the alterations of adenosine uptake, and whether either of these phenomena can account for any of the behavioral symptoms in mouse models or human patients.

7 Multiple Sclerosis

Multiple sclerosis (MS) is an autoimmune disorder that results in damage to areas of the CNS. It has been widely considered that the primary site of damage is the oligodendrocyte and myelin sheath surrounding central axons, but more recent work is beginning to indicate a significant involvement of neuronal damage, produced either directly by autoantibodies or occurring secondary to the loss of myelin.

The various adenosine receptors are effective modulators of cytokine release from immune-competent cells (Haskò and Cronstein 2004; Bours et al. 2006; Haskò et al. 2007). Adenosine levels in the blood of MS patients are lower than in controls (Mayne et al. 1999), raising the possibility that this could contribute to the induction of an autoimmune attack. The actions of adenosine on blood mononuclear cells also differ between patients and controls. Both groups of cells release similar amounts of the proinflammatory cytokines TNF-α and IL-6 in the resting state, but when activated, the increased production of TNF-α is reduced by A_1AR activation in controls but not patients with MS. Conversely, A_1ARs inhibit IL-6 but not TNF-α release in patients (Mayne et al. 1999). Both results are consistent with an apparently lower A_1AR density in cells from MS patients (Johnston et al. 2001). These data are also reflected in transgenic mice lacking A_1AR, which show a marked propensity to develop experimental allergic encephalomyelitis (EAE), a condition widely recognized as the murine equivalent of MS (Tsutsui et al. 2004). The signs and symptoms of EAE develop in parallel with increased production of proinflammatory cytokines, consistent with the inhibitory activity of A_1AR activation in monocytes from control humans.

Some of this work may be translated into therapeutic application, since methylthioadenosine has now been shown to not only suppress proinflammatory cytokine production by human white blood cells but also prevent and reverse EAE in animals (Moreno et al. 2006). These effects were attributed to an interference with the activation of the nuclear transcription factor NF-kB, and the involvement of AR activation of blockade was left open. Nevertheless, the potential implications of this activity of methylthioadenosine on MS treatment will no doubt encourage much further work on its molecular basis.

8 Huntington's Disease

Huntington's disease (HD) is an inherited neurodegenerative disease caused by loss of neurons in the striatum—especially medium spiny neurons containing GABA and enkephalin—and cortex. These changes result in motor abnormalities such as chorea, with the development of mental and psychological deterioration. The molecular origin of the degeneration has been ascribed to the production of an abnormal form of the protein huntingtin, in which an extended polyglutamine sequence (CAG triplets at the gene level) occurs.

Among the earliest proposals for the mechanism of neurodegeneration in HD was that excessive stimulation of glutamate receptors could be responsible for neuronal damage and death (Coyle and Schwarcz 1976; Lipton and Rosenberg 1994). Indeed, a large number of studies have demonstrated that overactivation of NMDA receptors in particular can produce many of the symptoms of HD in animals. The most effective agonist in this regard is quinolinic acid (Stone 2001), an endogenous metabolite of tryptophan that, unlike glutamate itself, is a selective agonist at the NMDA receptors (Stone and Perkins 1981; Stone and Darlington 2002). Administration of quinolinic acid into the striatum produces chronic neurodegeneration, which reproduces many of the electrophysiological, histological, motor and other behavioral symptoms of human HD (Beal et al. 1986, 1991; Ferrante et al. 1993; Popoli et al. 1994, 2002). Since ARs are important regulators of glutamate-mediated neurotransmission, there have been many suggestions that adenosine may be relevant to understanding HD, either as a key to the underlying cellular actions of huntingtin, and thus the molecular basis of the disorder, or as a means to treat the development or progress of the condition.

A second major hypothesis is that mutant huntingtin induces changes in mitochondrial function, and it is this that represents the primary cellular abnormality ("gain of function" hypothesis). A number of other potentially pathogenetic events have been attributed to mutant huntingtin. For example, proteolytic cleavage of mutant huntingtin generates fragments that aggregate into the nucleus and cytoplasm, thus contributing to early neuropathology. Accumulation of proteolytic huntingtin fragments and their aggregation may also trigger a cascade of damaging processes, leading to increasing dysfunctions in neurons through oxidative injury, transcriptional dysregulation, glutamate receptor excitotoxicity and apoptotic signals (Popoli et al. 2007 and references therein). In addition to this toxicity, there may also be a "loss of function" effect due to the loss of some beneficial actions exerted by normal huntingtin, which has been shown to be antiapoptotic, essential for normal embryonic development, and stimulatory on the production of BDNF into the cortex and its delivery to the striatal targets (see Popoli et al. 2007 and references therein).

Animal models of HD have become widely used based on each of these defects, namely the intrastriatal application of quinolinic acid or the administration (intrastriatal or systemic) of the mitochondrial toxin 3-nitro-propionic acid (3-NP). In addition, there are several transgenic models involving the induced expression of mutant hungtingtin, the R6/2 model being the most commonly used (Mangiarini et al. 1996).

8.1 Adenosine Receptors in HD

As noted above, A_1AR activation suppresses glutamate release from neurons. In line with the excitotoxic hypothesis, Blum et al. (2002) have reported that an A_1AR agonist, referred to as an adenosine amine congener (ADAC), was able to prevent the neuronal degeneration and motor sequelae of 3-NP administration to mice. Since no protection was apparent in cell cultures, the results were interpreted to indicate an action on presynaptic sites, presumably those at which the release of glutamate could be inhibited. Conversely, the A_1AR antagonist DPCPX exacerbates damage induced by a similar mitochondrial poison, malonate (Alfinito et al. 2003).

The ARs that have become of greatest interest in HD are the A_{2A}ARs. Activation of these promotes the release of glutamate, depending on the age of animals and the presence of a depolarizing stimulus (Corsi et al. 1999a, 2000), and increased numbers or functional activity of A_{2A}ARs could cause or contribute to an excitotoxic process (Domenici et al. 2007). The administration of CGS21680 itself increases extracellular glutamate levels (Popoli et al. 1995).

Consistent with this, A_{2A}AR antagonists have been shown to reduce the toxic consequences of quinolinic acid administration, an effect correlated with a reduction of glutamate release triggered by quinolinic acid (Reggio et al. 1999; Popoli et al. 2002; Scattoni et al. 2007). A similar phenomenon has been described in R6/2 mice, in which SCH58261 reduced the motor abnormalities and loss of brain tissue (Chou et al. 2005) and glutamate release in the striatum of R6/2 mice (Gianfriddo et al. 2004).

While this provides comforting support of the concept that quinolinic acid administration provides an acceptable model of HD, it is important to establish whether glutamate release is elevated in mutant mice or HD patients and, if so, the mechanism involved. The most obvious possibility, that of raised quinolinic acid levels, has been supported directly by evidence from Guidetti et al. (2004), who measured increased amounts in patients at an early stage of HD. As to the mechanism, the presence of mutant huntingtin has been shown to reduce the uptake of glutamate by astrocytes (Behrens et al. 2002), a result that could cause increased activation of glutamate receptors, contributing to excitotoxicity.

Neuroprotection has also been demonstrated for the 3-NP model of HD. The blockade of A_{2A}AR by CSC-protected mice treated with 3-NP against neuronal loss (Fink et al. 2004). Similarly, less cell death was seen when 3-NP was administered to A_{2A}AR knockout mice compared with wild-type controls. Consistent with this, another inhibitor of mitochondrial complex II, malonate, produced a degeneration of striatal neurons that was also prevented by DMPX (Alfinito et al. 2003).

However, as already noted above (Fig. 1), the role of A_{2A}ARs in HD is far more complex. Activation of A_{2A}ARs has also been reported to mediate beneficial effects. The A_{2A}AR agonist CGS21680 enhances the neurotrophic activity of growth factors such as BDNF, a key factor promoting the viability of striatal neurons, by facilitating TrkB receptor function (Lee and Chao 2001). Agonists at A_{2A}AR are also associated with a normalization of cyclic AMP response element binding protein (CREB) in transgenic animals (Chiang et al. 2005).

Moreover, CGS21680 reduces the incidence of abnormal extracellular macromolecular deposits that are present in HD brains in a similar way to β-amyloid deposits in Alzheimer's disease and Lewy bodies in Parkinson's disease. In R6/2 mice, ubiquitinated deposits have indeed been demonstrated in striatal cells, both in vivo and in cell cultures, which appear to depend on the expression of mutant huntingtin protein (Chou et al. 2005). These deposits are reduced by CGS21680. In the same study, it was also noted that CGS21680 corrected the abnormally high levels of blood glucose and 5′-adenosine monophosphate (AMP)-activated protein kinase activity in the mutant mice, strongly suggesting a more fundamental role of A_{2A}AR than had hitherto been suspected in the regulation of cellular biochemistry.

There is a significant depletion of A_{2A}AR in the striatum of patients with HD and in transgenic mice (Blum et al. 2003b) or rats (Bauer et al. 2005) expressing mutant huntingtin. On the other hand, the density of A_{2A}AR on blood platelets is increased in human HD patients (Varani et al. 2003), showing a significant correlation with both CAG repeat length (Maglione et al. 2006) and anticipation of symptoms between generations (Maglione et al. 2005). In a total of 126 HD gene-positive individuals, A_{2A}AR B_{max} values were found to be robustly increased at all HD stages as well as in 32 presymptomatic subjects (Varani et al. 2007). The same abnormality is present also in other neurological diseases characterized by an extended polyglutamine sequence (polyQ), but not in non-polyQ inherited disorders (Varani et al. 2007). The same peripheral cells exhibited altered membrane fluidity, a finding that may explain the observed change in receptor density. Authors argue that the observed alteration in lymphocytes reflects the presence of the mutant protein and suggest that the measurement of the A_{2A}AR binding activity might be of potential interest for a peripheral assessment of chemicals capable of interfering with the immediate toxic effects of the mutation.

There is clear evidence for increased activity of $A_{2A}AR$ associated with HD. Striatal neurons expressing mutant huntingtin were found to show increased $A_{2A}AR$ activation of adenylate cyclase (Varani et al. 2001), and a similar result was observed subsequently in blood cells of HD patients (Varani et al. 2003). The ability of $A_{2A}AR$ stimulation to raise cyclic AMP levels is also increased in R6/2 mice (Chou et al. 2005; Tarditi et al. 2006), and in animals treated with 3-NP (Blum et al. 2003a). In the Tarditi et al. study, an increase in both the number of $A_{2A}ARs$ as well as their activation of adenylate cyclase was reported, which was apparent within a few days of birth of R6/2 mice. Both of these parameters then fell to the values seen in wild-type animals. The mRNA for $A_{2A}AR$, in contrast, showed no change until 21 days postnatally, after which it decreased substantially. Two conclusions may be drawn from this work. Firstly, the mismatch between $A_{2A}AR$ protein and mRNA could indicate changes in factors that affect translation or transcription of the $A_{2A}AR$, or which regulate receptor activity. Secondly, a loss rather than an increase of $A_{2A}AR$ seems to be associated with older mice in which motor symptoms of HD are beginning to occur. A further intriguing observation in this study was that in the young mice, $A_{2A}AR$ function was not prevented by ZM241385, whereas sensitivity to this antagonist was established in the older animals after 21 days of age. Whether this also implies differences in the regulation of receptor function, or different variant structures of the receptor protein at different ages, remains to be explored.

In conclusion, available data on the potential exploitation of $A_{2A}AR$ ligands in HD are controversial and reflect the complexity of $A_{2A}AR$ regulation in this disease (for further comments, see Popoli et al. 2007). The complex mutual relationship between AR activities mediating detrimental or beneficial effects (see also Sect. 3) makes it difficult to establish whether targeting $A_{2A}AR$ would really be of interest to treat HD. Further basic research is needed to solve several specific questions, in particular: (1) neuronal versus non-neuronal receptor localization, and (2), for receptors expressed in neurons, pre- versus postsynaptic sites (see Fig. 1).

9 Cerebral Ischemia and Reperfusion: Stroke

9.1 Role of A_1 Adenosine Receptors

One of the earliest reports of neuroprotection against ischemia was that the nonselective agonist 2-chloroadenosine would prevent hippocampal damage in rats (Evans et al. 1987). Similar results were obtained subsequently using A_1AR-selective agonists (von Lubitz et al. 1989; Phillis and O'Regan 1993; von Lubitz et al. 1995a), with suggestions that the protection could involve an inhibition of leukocyte adherence and extravasation (Grisham et al. 1989).

The finding that theophylline could increase the release of glutamate produced by ischemia certainly suggests that endogenous adenosine is exerting an inhibitory action on glutamate release (Héron et al. 1993), although this could have been due to A_1 or A_{2A} AR blockade. The simultaneous measurement of purine and glutamate release into the extracellular space of brain, together with the neuronal damage and behavioral consequences of an ischemic episode, revealed a significant relationship between these parameters, with a lower extracellular glutamate being associated with less cell damage (Melani et al. 1999). It is interesting that several nonpurine compounds that can depress the release of excitatory amino acids are also protective against ischemic damage (Ochoa et al. 1992; Graham et al. 1993). Conversely, A_1AR blockade exacerbates ischemic damage (Phillis 1995).

On the other hand, it has been argued that the release of endogenous glutamate is not actually related to ischemic-induced brain damage. Systemic administration of R-PIA, CHA or an adenosine uptake inhibitor did not prevent the increase of glutamate levels in brain during ischemia (Héron et al. 1993, 1994; Cantor et al. 1992; Kano et al. 1994), although other groups have reported a decreased release using CPA (Simpson et al. 1992). The differences may depend on the pharmacokinetics of the agonists used or the model used for inducing damage.

Although little is known of the signaling pathways that underlie ischemic damage or adenosine-mediated protection in vivo, some clues may be gleaned from in vitro work. Di Capua et al. (2003), for example, found that A_1AR agonism protected primary rat neurons against "chemical ischemia" (produced by iodoacetate) via the activation of protein kinase C-epsilon. The activity of A_1ARs themselves may change under ischemic conditions. Adenosine A_1ARs are desensitized and internalized by a period of hypoxia in brain slices (Coelho et al. 2006). A period of ischemia in vivo followed by reperfusion has been said to result in no change in the number of A_1ARs or their inhibitory efficacy on presynaptic transmitter release (Shen et al. 2002), although Lai et al. (2005) have reported an increase in A_1AR expression in the cerebral cortex following ischemia in Wistar rats.

9.2 *Role of A_2 Adenosine Receptors*

The activation of A_{2A}AR can protect neurons against ischemia-induced damage. One of the best-tested A_{2A}AR agonists is ATL-146e, which prevents ischemic damage in the spinal cord (Cassada et al. 2001a; Reece et al. 2006) as well as damage induced by mechanical trauma (Reece et al. 2004; Okonkwo et al. 2006). This protection afforded by ATL-146e was accompanied by the normalization of several molecular markers, such as those for apoptosis (Cassada et al. 2001b), microtubule-associated protein 2 (MAP-2) and TNF-α levels (Reece et al. 2004). However, the protection is not completely prevented by ZM241385, implying that there are relevant sites of action other than A_{2A}AR. A period of ischemia of the spinal cord does, however, induce a highly significant increase in A_{2A}AR number, a finding that may contribute to the protective effect of A_{2A}AR agonists. There is also a greater inhibition of TNF-α levels in postischemic spinal cord, as well as reduced platelet adhesion to endothelial cells (Cassada et al. 2002), consistent with an important role of A_{2A}AR on blood cells

On the other hand, blockade of A_{2A}AR is also neuroprotective against ischemic damage caused by transient or permanent arterial occlusion (Gao and Phillis 1994; Phillis 1995; Monopoli et al. 1998; Pedata et al. 2005). Confirmation of the detrimental influence of A_{2A}AR has come from an examination of A_{2A}AR-deficient transgenic mice (Chen et al. 1999). These animals showed substantial resistance to ischemia-induced brain damage compared with their normal littermates.

An interesting observation reported by Corsi et al. (1999a, b) is that the agonist CGS-21680 only increased the spontaneous efflux of glutamate and GABA in young (not old) rats, although it enhanced potassium-evoked release similarly in both groups of animals. This may have implications for the utility of A_{2A}AR agonist and antagonist ligands in treating older patients after cerebral ischemia, since chronic treatment might show fewer side effects attributable to increased basal release of glutamate, while retaining neuroprotective activity against the depolarization-induced release occurring during and immediately after cerebral ischemia or trauma. The reason for the increased damage may depend, at least partly, on the increased release of glutamate and related amino acids that these compounds produced during cerebral ischemia (O'Regan et al. 1992)

It is interesting to note that, while most of the work in this area has employed adult rodents, there is some evidence that the reverse situation occurs for young animals. Thus, in neonatal rats, Aden et al. (2003) found that it was activation of A_{2A}AR that protected against a period of hypoxia and ischemia, with A_{2A}AR knockout mice showing greater brain damage than wild-type controls.

The release of proinflammatory cytokines such as TNF-α from macrophages is suppressed by activation of A_{2A}AR (Kreckler et al. 2006). Work by Chen and colleagues (Yu et al. 2004), however, has revealed a fascinating insight into the sites through which protection is mediated. By generating populations of rats lacking A_{2A}AR generally and replacing bone marrow tissue selectively with cells reconstituted to contain A_{2A}AR, they have been able to comment directly on the roles of receptors intrinsic to the CNS relative to those in the blood. The results showed that the presence of A_{2A}ARs on blood cells alone was sufficient to reverse the protective effect of generalized A_{2A}AR knockout, while wild-type mice given A_{2A}AR knockout bone marrow cells were protected against ischemic damage. This illuminating study strongly suggests that the A_{2A}ARs relevant to protection against ischemic damage are those on blood cells. This may also imply that the mechanism of A_{2A}AR antagonist protection is more strongly dependent on, for example, the release of inflammatory cytokines, than had previously been thought.

Although the A_{2B}AR has received relatively little attention with respect to neuroprotection, its activation has a number of consequences that could well contribute significantly to the phenomenon. For example, there is evidence that its activation of p38 MAPK leads to the increased expression of IL-6 in macrophages (Fiebich et al. 1996a, 2005). Since IL-6 is a cytokine that has been reported to protect neurons against a range of insults (Bensadoun et al. 2001; Carlson et al. 1999), its production, either in central glia or peripheral cells, may result in some protective efficacy.

Brain inflammation induced in rats by a chronic intraventricular infusion of LPS was associated with a loss of neuronal A_{2B}AR. This loss was prevented by a nitro derivative of the anti-inflammatory drug flurbiprofen, while the parent compound was inactive (Rosi et al. 2003). The authors' conclusion was that an NO-releasing anti-inflammatory compound might be an effective inhibitor of brain inflammation in conditions such as Alzheimer's disease, and that changes in the density of A_{2B}AR might be involved. It is becoming increasingly clear that much more work is required to expand our knowledge of the effects of A_{2B}AR activation or loss on the overall profile of pro- and anti-inflammatory cytokines in the brain and elsewhere, especially in relation to the net effects on neurotransmission, β-amyloid production, and neuronal or glial cell viability.

9.3 *Role of A_3 Adenosine Receptors*

As already mentioned above (see Sect. 3.4), A_3AR activation can protect isolated cells from hypoxia-induced death (Chen et al. 2006), and it reduced infarct size in rats subjected to middle cerebral artery occlusion (MCAo). Conversely, animals lacking A_3AR exhibit substantially increased infarct volumes, suggesting that the activation of these receptors by endogenous adenosine normally acts as a physiological brake on those processes causing damage (Chen et al. 2006; Fedorova et al. 2003).

The chronic administration of an A_3AR agonist such as IB–MECA affords protection against a subsequent period of cerebral ischemia (von Lubitz et al. 1999b, 2001).

At least part of the protective activity of A_3AR agonists may involve modulation of immune-competent cells and the inflammatory reaction to cellular damage. Agonists have been shown to inhibit the generation of several proinflammatory cytokines from cells, including interleukin (IL) 10, IL-12, interferon-γ and TNF-α (Haskò et al. 1998; McWhinney et al. 1996). The latter action is sufficiently robust to have been developed as a screen for new agonist compounds (Knutsen et al. 1998). Indeed, it has been suggested that activation of A_3AR may be responsible for the reported inhibition by adenosine of TNF-α secretion in the human U937 macrophage cell line (Sajjadi et al. 1996).

The opposite effects obtained on the outcome of brain ischema upon acute or chronic treatment with selective A_3AR agonists are discussed below (see Sect. 9.5).

9.4 *Time Course of Protection Induced by Adenosine Receptor Ligands*

One of the valuable features of neuroprotection by A_1AR activation is that it can be demonstrated for a period of several hours following the occurrence of a vascular or toxic insult. This is a major consideration for any drug intended for clinical use as a neuroprotectant following an acute incident such as a stroke, since the expansion of damage from a limited central region into a more extensive penumbral area occurs over a period of hours or days, and it is essential to limit the degree of that expansion if patient recovery is to be optimized. Most authorities consider that there is a window of opportunity for neuroprotection of up to several hours after the occurrence of stroke. Several A_1AR-selective agonists such as R-PIA certainly exhibit protection, even when administered up to 2 h after excitotoxic insults, indicating that the neuronal network and intracellular signaling processes that contribute to damage continue to operate over this time frame (Miller et al. 1994). Against ischemia-induced damage, cyclohexyladenosine (CHA) remains protective when administered up to at least 30 min following cerebral ischemia (von Lubitz et al. 1989), and ADAC similarly has a window of efficacy of several hours after cerebral ischemia in gerbils (von Lubitz et al. 1996). This latter compound is of special interest since it seems to possess fewer of the cardiovascular side effects associated with some other A_1AR agonists (Bischofberger et al. 1997), and its efficacy is still apparent when administered chronically in very low doses (von Lubitz et al. 1999a). The importance of this finding is that many other purine receptor ligands produce opposite effects when used chronically rather than in a single acute dose paradigm. Since most patients needing neuroprotection may be taking the drugs for prolonged periods of time, this could be a highly significant advantage of ADAC and related compounds.

The timing of acute administration of A_3AR agonists is also important. Treatment prior to ischemia increased infarct size, while postischemic administration reduced damage, probably as a result of altered dynamics of receptor activation, on neurons, glia and blood components (von Lubitz et al. 2001).

9.5 Acute Versus Chronic Administration

Despite the evidence for a neuroprotective action of adenosine and A_1AR agonists, caffeine—a nonselective antagonist at both A_1 and A_2 ARs—was also found to protect against ischemic damage in the CNS after its chronic administration (Rudolphi et al. 1989; Sutherland et al. 1991). Single, acute injections of more selective A_1AR antagonists, including DPCPX, were also found to exacerbate ischemic damage (Phillis 1995; von Lubitz et al. 1994a), while their chronic administration reduced damage and produced neuroprotection (von Lubitz et al. 1994a). This dichotomy of response probably indicates compensatory changes of receptor density that follow the prolonged presence of any receptor ligand. However, such changes may be limited in extent, or restricted to certain cell subtypes, since no significant changes in A_1AR binding were detected after chronic administration of antagonists (Traversa et al. 1994). However, others have reported that chronic administration of the AR antagonists caffeine and theophylline increase A_1ARs in cerebral cortex (Murray 1982; Szot et al. 1987) and the hippocampal CA1 region (Rudolphi et al. 1989)

Chronic administration of low doses of ADAC generated the opposite result, with marked protection of the brain. The reasons for this difference from other A_1AR agonists is not entirely clear, although the authors point out the substantial difference in molecular structure between ADAC and other compounds, with the implication that it may yield a different spectrum or time course of action on a range of cellular targets whose balance determines the overall production of neuronal damage or protection (von Lubitz et al. 1999a).

The effects of acute and chronic treatment with $A_{2A}AR$ ligands show less disparity than in the case of the A_1AR ligands described above. Overall, the qualitative effects of agonists and antagonists are similar whether they are administered acutely or chronically. This assertion would be consistent with evidence that receptor numbers and affinities change little in vivo (von Lubitz et al. 1995b) or in vitro (Abbracchio et al. 1992) in the continued presence of $A_{2A}AR$ ligands.

As mentioned above (see Sects. 3.4 and 3.5), the acute administration of agonists at A_3ARs, and their application to neurons in cell culture, does appear to induce neuronal death (Sei et al. 1997). In addition, an A_3AR agonist can potentiate the degree of CNS damage following cerebral ischemia (von Lubitz et al. 1994b). On the other hand, the maintained presence or chronic intermittent administration of A_3AR agonists produces protection, probably as a result of compensatory adaptations in the number or sensitivity of receptors. Thus, acute administration of the selective agonist ligand IB–MECA significantly enhanced the extent of brain damage following ischemia in gerbils. Chronic administration of the same compound, however, resulted in a highly significant reduction in the ischemic damage (von Lubitz et al. 1994b; 1999b; Chen et al. 2006).

9.6 *Therapeutic Implications of Preconditioning*

Relatively short periods of hypoxia, hypoglycemia or ischemia can result in protection of tissues against a subsequent and more severe insult. This is the phenomenon of preconditioning. Neuronal preconditioning has been demonstrated using both in vivo and in vitro preparations (Schurr et al. 1986; Khaspekov et al. 1998). One factor contributing to this is a change in the number of A_1ARs, which increases after the preconditioning period, probably as an adaptive protective development against further ischemia (Zhou et al. 2004). Adenosine is known to be involved in preconditioning, mainly through its opening of K_{ATP} channels (Yao and Gross 1994). In many models, even those in which it is induced by an anesthetic agent such as isoflurane (Liu et al. 2006), preconditioning can be prevented almost completely by A_1AR blockers (Hiraide et al. 2001; Nakamura et al. 2002; Yoshida et al. 2004; Pugliese et al. 2003), although $A_{2A}ARs$ seem to contribute little to the phenomenon. The lack of involvement of $A_{2A}ARs$ is also surprising given the foregoing discussion on the clear neuroprotective activity of $A_{2A}AR$ antagonists (Phillis 1995; Jones et al. 1998a, b), although it is likely that the differences between in vitro slice preparations and in vivo studies are largely responsible for this difference. Interestingly, A_3AR antagonists can enhance neuronal recovery after simulated ischemia in vitro, consistent with the work quoted above that acute activation of A_3AR worsens ischemic damage in vivo (Pugliese et al. 2007).

10 Prospects for Adenosine Receptor-Based Therapeutics

In summary, there is an increasingly acceptable rationale, at the cellular, biochemical and behavioral levels, for believing that changes in AR function might contribute to the symptoms and possibly progression of neurodegenerative disorders (Ribeiro et al. 2002), and that ligands acting at the various ARs may have a potential role in the therapeutic treatment of some of those disorders (Muller 1997, 2000; Mally and Stone 1998; Broadley 2000; Press et al. 2007; Baraldi et al. 2008). Of course, no receptor population is likely to function in isolation in the CNS. The activation or blockade of other neurotransmitter receptors may have significant effects on the number or efficacy of ARs.

For example, von Lubitz et al. (1995a) tested combinations of ligands acting at NMDA and A_1AR, using either acute or chronic treatments. The results revealed changes of animal responses with combined treatments that suggested important interactions between NMDA and A_1AR contributing to the changes of seizure generation and motor impairment. There were parallel changes of A_1AR density which indicated that the interactions were occurring at a deeper level of cellular function than merely a degree of nonspecificity in the ligand efficacy at the different receptors.

If ARs are indeed significant contributors to neuroprotection by responding to altered endogenous levels of the purine, or if they are used as targets for therapeutic agents that act directly upon them, it will be necessary to obtain information on the manner in which those receptors behave throughout the period of insult and subsequently. A range of factors, such as acidity, oxygen levels, cytokines, peptides, growth factors, and undoubtedly many more, could act to modify receptor responsiveness in a fashion that reduces or enhances the expected efficacy of agonists or antagonists. Examples of this include the report that tissue oxidation reduces the affinity of A_1AR antagonists, but not agonists, although the density of binding sites was decreased for both (Oliveira et al. 1995). Changes in the balance of agonist to antagonist activity could be produced in this way, which could significantly alter the anticipated response to ligands.

The ability of AR antagonists to reverse cognitive dysfunction has been taken to indicate that they may have a standalone place in the treatment of dementias. However, given the undoubted existence of this and many other receptor interaction phenomena, the generalized loss of neurons that can occur with aging or disease, and the complexity of neuronal interactions that underlie cognitive performance, it is likely that future attention will shift to compounds that retain specificity of action but act at a defined number of different sites (Van der Schyf et al. 2006). In this context, however, it seems likely that blockade of ARs would be one of the more valuable sites to include in the profile of optimum targets.

For the treatment of ischemic damage, however, there is clearly a potential use for A_1AR agonists and A_{2A}AR antagonists. An alternative approach to using conventional agonists for stroke-induced brain damage could be to inhibit adenosine kinase. This enzyme is a major route for the removal of adenosine, converting it to AMP. Consequently, the overexpression of the kinase results in an exacerbation of ischemic damage (Pignataro et al. 2007), whereas inhibition has been found to raise extracellular adenosine levels and produce protection against damage (Jiang et al. 1997).

Whichever strategic approach is used, and whichever receptor subtype is selected, it seems likely that ARs will in the future represent a valuable series of targets for protection of the brain against a range of insults. However, we will have to solve some pending issues concerning the opposite (beneficial versus detrimental) effects exerted by some AR subtypes depending on their cellular and/or pre- versus postsynaptic localization. This especially applies to the A_{2A}AR subtype (see Fig. 1). Blockade of A_{2A}AR can result either in protoxic or neuroprotective effects according to the mechanisms involved in a given experimental model and, in some cases, to the disease stage. In this respect, it is envisaged that notable advances will be achieved by the availability of transgenic mice bearing selective defects of A_{2A}AR on specific cell populations (Yu et al. 2004). The use of these mice will help in addressing the therapeutical use of A_{2A}AR ligands in not only HD but also all other neurodegenerative diseases characterized by a dysfunction of the adenosinergic system.

References

Abbracchio MP, Fogliatto G, Paoletti AM, Rovati GE, Cattabeni F (1992) Prolonged invitro exposure of rat-brain slices to adenosine-analogs: selective desensitization of adenosine-A(1) but not adenosine-A(2) receptors. Eur J Pharmacol 227:317–324

Abbracchio MP, Rainaldi G, Giammarioli AM, Ceruti S, Brambilla R, Cattabeni F, Barbieri D, Franceschi C, Jacobson KA, Malorni W (1997) The A_3 adenosine receptor mediates cell spreading, reorganisation of actin cytoskeleton and distribution of Bcl-x(L): studies in human astroglioma cells. Biochem Biophys Res Commun 241:297–304

Abbracchio MP, Ceruti S, Brambilla R, Barbieri D, Camurri A, Franceschi C, Giammarioli AM, Jacobson KA, Cattabeni F, Malorni W (1998) Adenosine A_3 receptors and viability of astrocytes. Drug Dev Res 45:379–386

Abbracchio MP, Burnstock G, Boeynaems JM, Barnard EA, Boyer JL, Kennedy C, Knight GE, Fumagalli M, Gachet C, Jacobson KA, Weisman GA (2006) International Union of Pharmacology LVIII: update on the P2Y G protein-coupled nucleotide receptors: from molecular mechanisms and pathophysiology to therapy. Pharmacol Rev 58:281–341

Aden U, Halldner L, Lagercrantz H, Dalmau I, Ledent C, Fredholm BB (2003) Aggravated brain damage after hypoxic ischemia in immature adenosine A(2A) knockout mice. Stroke 34: 739–744

Albasanz JL, Perez S, Barrachina M, Ferrer I, Martin M (2008) Up-regulation of adenosine receptors in the frontal cortex in Alzheimer's disease. Brain Pathol 18:211–219

Alfinito PD, Wang SP, Manzino L, Rijhsinghani S, Zeevalk GD, Sonsalla PK (2003) Adenosinergic protection of dopaminergic and GABAergic neurons against mitochondrial inhibition through receptors located in the substantia nigra and striatum, respectively. J Neurosci 23:10982–10987

Andiné P, Rudolphi KA, Fredholm BB, Hagberg H (1990) Effect of propentofylline (HWA285) in extracellular purines and excitatory amino acids in CA1 of rat hippocampus during transient ischaemia. Br J Pharmacol 100:814–818

Angelucci ME, Cesário C, Hiroi RH, Rosalen PL, Da Cunha C (2002) Effects of caffeine on learning and memory in rats tested in the Morris water maze. Braz J Med Biol Res 35: 1201–1208

Angulo E, Casado V, Mallol J, Canela EI, Vinals F, Ferrer I, Lluis C, Franco R (2003) A(1) adenosine receptors accumulate in neurodegenerative structures in Alzheimer disease and mediate both amyloid precursor protein processing and tau phosphorylation and translocation. Brain Pathol 13:440–451

Appel E, Kazimirsky G, Ashkenazi E, Kim SG, Jacobson KA, Brodie C (2001) Roles of BCL-2 an caspase 3 in the adenosine A_3 receptor-induced apoptosis. J Mol Neurosci 17:285–292

Araki T, Kato H, Kanai Y, Kogure K (1993) Selective changes of neurotransmitter receptors in middle-aged gerbil brain. Neurochem Intern 23:541–548

Arendash GW, Schleif W, Rezai-Zadeh K, Jackson EK, Zacharia LC, Cracchiolo JR, Shippy D, Tan J (2006) Caffeine protects Alzheimer's mice against cognitive impairment and reduces brain β-amyloid production. Neuroscience 142:941–952

Arvin B, Neville LF, Pan J, Roberts PJ (1989) 2-Chloroadenosine attenuates kainic acid-induced toxicity within the rat striatum: relationship to release of glutamate and Ca^{2+} influx. Br J Pharmacol 98:225–235

Baraldi PG, Tabrizi MA, Gessi S, Borea PA (2008) Adenosine receptor antagonists: translating medicinal chemistry and pharmacology into clinical utility. Chem Rev 108:238–263

Barnes CR, Mandell GL, Carper HT, Luong S, Sullivan GW (1995) Adenosine modulation of TNFα-induced neutrophil activation. Biochem Pharmacol 50:1851–1857

Bartrup JT, Stone TW (1990) Activation of NMDA receptor-coupled channels suppresses the inhibitory action of adenosine on hippocampal slices. Brain Res 530:330–334

Bartrup JT, Addae JI, Stone TW (1991) Interaction between adenosine and excitatory agonists in rat hippocampal slices. Brain Res 564:323–327

Bauer A, Zilles K, Matusch A, Holzmann C, Riess O, von Horsten S (2005) Regional and subtype selective changes of neurotransmitter receptor density in a rat transgenic for the Huntington's disease mutation. J Neurochem 94:639–650

Beal MF, Kowall NW, Ellison DW, Mazurek MF, Swartz KJ, Martin JB (1986) Replication of the neurochemical characteristics of Huntington's disease by quinolinic acid. Nature 321:168–171

Beal MF, Ferrante RJ, Swartz KJ, Kowall NW (1991) Chronic quinolinic acid lesions in rats closely resemble Huntington's disease. J Neurosci 11:1649–1659

Behrens PF, Franz P, Woodman B, Lindenberg KS, Landwehrmeyer GB (2002) Impaired glutamate transport and glutamate–glutamine cycling: downstream effects of the Huntington mutation. Brain 125:1908–1922

Bensadoun JC, de Almeida LP, Dréano M, Aebischer P, Déglon N (2001) Neuroprotective effect of interleukin-6 and IL6/IL6R chimera in the quinolinic acid rat model of Huntington's syndrome. Eur J Neurosci 14:1753–1761

Bertelli M, Cecchin S, Lapucci C, Jacomelli G, Jinnah HA, Pandolfo M, Micheli V (2006) Study of the adenosinergic system in the brain of HPRT knockout mouse (Lesch–Nyhan disease). Clin Chim Acta 373:104–107

Bischofberger N, Jacobson KA, von Lubitz DKJE (1997) Adenosine A(1) receptor agonists as clinically viable agents for treatment of ischemic brain disorders. Ann NY Acad Sci 825:23–29

Blum D, Gall D, Galas MC, d'Alcantara P, Bantubungi K, Schiffmann, SN (2002) The adenosine A(1) receptor agonist adenosine amine congener exerts a neuroprotective effect against the development of striatal lesions and motor impairments in the 3-nitropropionic acid model of neurotoxicity. J Neurosci 22:9122–9133

Blum D, Galas MC, Pintor A, Brouillet E, Ledent C, Muller CE, Bantubungi K, Galluzzo M, Gall D, Cuvelier L, Rolland AS, Popoli P, Schiffmann SN (2003a) A dual role of adenosine A(2A) receptors in 3-nitropropionic acid-induced striatal lesions: implications for the neuroprotective potential of A(2)A antagonists. J Neurosci 23:5361–5369

Blum D, Hourez R, Galas MC, Popoli P, Schiffmann SN (2003b) Adenosine receptors and Huntington's disease: implications for pathogenesis and therapeutics. Lancet Neurol 2:366–374

Bona E, Adén U, Gilland E, Fredholm BB, Hagberg H (1997) Neonatal cerebral hypoxia-ischaemia: the effect of adenosine receptor antagonists. Neuropharmacology 9:1327–1338

Bours MJ, Swennen EL, Di Virgilio F, Cronstein BN, Dagnelie PC (2006) Adenosine 5′-triphosphate and adenosine as endogenous signaling molecules in immunity and inflammation. Pharmacol Ther 112:358–404

Brambilla R, Cottini L, Fumagalli M, Ceruti S, Abbracchio MP (2003) Blockade of A_{2A} adenosine receptors prevents basic fibroblast growth factor-induced reactive astrogliosis in rat striatal primary astrocytes. Glia 43:190–194

Bridges AJ, Bruns RF, Ortwine DF, Priebe SR, Szotek DL, Trivedi BK (1988) N^6-[2-(3,5-Dimethoxyphenyl)-2-(2-methylphenyl)-ethyl]adenosine and its uronamide derivatives. Novel adenosine agonists and antagonists with both high affinity and high selectivity for the adenosine A_2 receptor. J Med Chem 31:1282–1285

Broadley KJ (2000) Drugs modulating adenosine receptors as potential therapeutic agents for cardiovascular diseases. Exp Opin Ther Pat 10:1669–1692

Brown SJ, James S, Reddington M, Richardson PJ (1990) Both A_1 and A_{2A} purine receptors regulate striatal acetylcholine release. J Neurochem 55:31–38

Bullough DA, Magill MJ, Firestein GS, Mullane KM (1995) Adenosine activates A_2 receptors to inhibit neutrophil adhesion and injury to isolated cardiac myocytes. J Immunol 155:2579–2586

Burkey TH, Webster RD (1993) Adenosine inhibits fMLP-stimulated adherence and superoxide anion generation by human neutrophils at an early step in signal transduction. Biochem Biophys Acta 1175:312–318

Butcher SP, Bullock R, Graham DI, McCulloch J (1990) Correlation between amino acid release and neuropathologic outcome in rat brain following middle cerebral artery occlusion. Stroke 21:1727–1733

Cain BS, Meldrum DR, Dinarello CA, Meng X, Banerjee A, Harken AH (1998) Adenosine reduces cardiac TNF-α production and human myocardial injury following ischaemia-reperfusion. J Surg Res 76:117–123

Canals M, Marcellino D, Fanelli F, Ciruela F, de Benedetti P, Goldberg SR, Neve K, Fuxe K, Agnati LF, Woods AS, Ferré S, Lluis C, Bouvier M, Franco R (2003) Adenosine A_{2A}-dopamine D_2 receptor–receptor heteromerization: qualitative and quantitative assessment by fluorescence and bioluminescence energy transfer. J Biol Chem 278:46741–46749

Canals M, Angulo E, Casado V, Canela EI, Mallol J, Vinals F, Staines W, Tinner B, Hillion J, Agnati L, Fuxe K, Ferre S, Lluis C, Franco R (2005) Molecular mechanisms involved in the adenosine A(1) and A(2A) receptor-induced neuronal differentiation in neuroblastoma cells and striatal primary cultures. J Neurochem 92:337–348

Cantor SL, Zornow MH, Miller LP, Yaksh TL (1992) The effect of cyclohexyladenosine on the periischemic increases of hippocampal glutamate and glycine in the rabbit. J Neurochem 59:1884–1892

Cargnoni A, Ceconi C, Boraso A, Bernocchi P, Monopoli A, Curello S, Ferrari R (1999) Role of A_{2A} receptors in the modulation of myocardial reperfusion damage. J Cardiovasc Pharmacol 33:883–893

Carlson NG, Wieggel WA, Chen J, Bacchi A, Rogers SW, Gahring LC (1999) Inflammatory cytokines IL-1α, IL-1β, IL-6, and TNF-α impart neuroprotection to an excitotoxin through distinct pathways. J Immunol 163:3963–3968

Casati C, Forlani A, Lozza G, Monopoli A (1997) Hemodynamic changes do not mediate the cardioprotection induced by the A_1 adenosine receptor agonist CCPA in the rabbit. Pharmacol Res 35:51–55

Cassada DC, Tribble CG, Kaza AK, Fiser SM, Long SM, Linden J, Rieger JM, Kron IL, Kern JA (2001a) Adenosine analogue reduces spinal cord reperfusion injury in a time-dependent fashion. Surgery 130:230–235

Cassada DC, Tribble CG, Laubach VE, Nguyen BN, Rieger JM, Linden J, Kaza AK, Long SM, Kron IL, Kern JA (2001b) An adenosine A(2A) agonist, ATL-146e, reduces paralysis and apoptosis during rabbit spinal cord reperfusion. J Vasc Surg 34:482–488

Cassada DC, Tribble CG, Long SM, Laubach VE, Kaza AK, Linden J, Nguyen BN, Rieger JM, Fiser SM, Kron IL, Kern JA (2002) Adenosine A2(A) analogue ATL-146e reduces systemic tumor necrosing factor-α and spinal cord capillary platelet-endothelial cell adhesion molecule-1 expression after spinal cord ischemia. J Vasc Surg 35, 994–998

Chen JF, Huang Z, Ma J, Zhu J, Moratalla R, Standaert D, Moskowitz MA, Fink JS, Schwarzschild MA (1999) A_{2A} adenosine receptor deficiency attenuates brain injury induced by transient focal ischaemia in mice. J Neurosci 19:9192–9200

Chen G-J, Harvey BK, Shen H, Chou J, Victor A, Wang Y (2006) Activation of adenosine A_3 receptors reduces ischemic brain injury in rodents. J Neurosci Res 84:1848–1855

Chiang MC, Lee YC, Huang CL, Chern YJ (2005) cAMP-response element-binding protein contributes to suppression of the A(2A) adenosine receptor promoter by mutant huntingtin with expanded polyglutamine residues. J Biol Chem 280:14331–14340

Chou SY, Lee YC, Chen HM, Chiang MC, Lai HL, Chang HH, Wu YC, Sun CN, Chien CL, Lin YS, Wang SC, Tung YY, Chang C, Chern YJ (2005) CGS21680 attenuates symptoms of Huntington's disease in a transgenic mouse model. J Neurochem 93:310–320

Chowdhury M, Fillenz M (1991) Presynaptic adenosine A_2 and NMDA receptors regulate dopamine synthesis in rat striatal synaptosomes. J Neurochem 56:1783–1788

Ciccarelli R, D'Alimonte I, Ballerini P, D'Auro M, Nargi E, Buccella S, Di Iorio P, Bruno V, Nicoletti F, Caciagli F (2007) Molecular signalling mediating the protective effect of A(1) adenosine and mGlu3 metabotropic glutamate receptor activation against apoptosis by oxygen/glucose deprivation in cultured astrocytes. Mol Pharmacol 71:1369–1380

Coelho JE, Rebola N, Fragata I, Ribeiro JA, De Mendonca A, Cunha RA (2006) Hypoxia-induced desensitization and internalization of adenosine A(1) receptors in the rat hippocampus. Neuroscience 138:1195–1203

Connick JH, Stone TW (1986) The effects of kainate, β-kainate and quinolinic acids on the release of endogenous amino acids from rat brain slices. Biochem Pharmacol 35:3631–3635

Connick JH, Stone TW (1989) Quinolinic acid neurotoxicity: protection by intracerebral phenylisopropyl adenosine (PIA) and potentiation by hypotension. Neurosci Lett 101:191–196

Corodimas KP, Tomita H (2001) Adenosine A_1 receptor activation selectively impairs the acquisition of contextual fear conditioning in rats. Behav Neurosci 115(6):1283–1290

Corradetti R, Lo Conte G, Moroni F, Passani MB, Pepeu G (1984) Adenosine decreases aspartate and glutamate release from rat hippocampal slices. Eur J Pharmacol 140:19–26

Corset V, Nguyen-Ba-Charvet KT, Forcet C, Moyse E, Chetodal, A, Mehlen P (2000) Netrin-1-mediated axon outgrowth and cAMP production requires interaction with adenosine A_{2B} receptor. Nature 407:747–750

Corsi C, Melani A, Bianchi L, Pepeu G, Pedata F (1999a) Striatal A_{2A} adenosine receptors differentially regulate spontaneous and K^+-evoked glutamate release in vivo in young and aged rats. Neuroreport 10:687–691

Corsi C, Melani A, Bianchi L, Pepeu G, Pedata F (1999b) Effect of adenosine A_{2A} stimulation on GABA release from the striatum of young and aged rats in vivo. Neuroreport 10:3933–3937

Corsi C, Melani A, Bianchi L, Pedata F (2000) Striatal A_{2A} adenosine receptor antagonism differentially modifies striatal glutamate outflow in vivo in young and aged rats. Neuroreport 11:2591–2595

Coyle JT, Schwarcz R (1976) Lesion of striatal neurones with kainic acid provides a model for Huntington's chorea. Nature 263:244–246

Cronstein BN (1994) Adenosine, an endogenous anti-inflammatory agent. J Appl Physiol 76:5–13

Cronstein BN, Rosenstein ED, Kramer SB, Weissmann G, Hirschhorn R (1985) Adenosine: a physiologic modulator of superoxide anion generation by human neutrophils. Adenosine acts via an A_2 receptor on human neutrophils. J Immunol 135:1366–1371

Cronstein BN, Kubersky SM, Weissmann G, Hirschhorn R (1987) Engagement of adenosine receptors inhibits hydrogen peroxide-release by activated human neutrophils. Clin Immunol Immunopathol 42:76–85

Cronstein BN, Daguma L, Nichols D, Hutchison AJ, Williams M (1990) The adenosine/neutrophil paradox resolved: human neutrophils possess both A_1 and A_2 receptors that both promote chemotaxis and inhibit O_2 generation respectively. J Clin Invest 85:1150–1157

Cronstein BN, Levin RI, Philips M, Hirschhorn R, Abramson SB, Weissmann G (1992) Neutrophil adherence to endothelium is enhanced via adenosine A_1 receptors and inhibited via adenosine A_2 receptors. J Immunol 148:2201–2206

Cunha RA, Johansson B, van der Ploeg I, Sebastião AM, Ribeiro JA, Fredholm BB (1994) Evidence for functionally important adenosine A_{2a} receptors in the rat hippocampus. Brain Res 649:208–216

Cunha RA, Constantino MC, Sebastião AM, Ribeiro JA (1995) Modification of A_1 and A_{2a} adenosine receptor binding in aged striatum, hippocampus and cortex of the rat. Neuroreport 6:1583–1588

Cunha RA, Johansson B, Constantino MD, Sebastião AM, Fredholm BB (1996) Evidence for high affinity binding sites for the adenosine A_{2A} receptor agonist [3H]CGS21680 in the rat hippocampus and cerebral cortex that are different from striatal A_{2A} receptors. Naunyn–Schmiedeberg's Arch Pharmacol 353:261–271

D'Alimonte I, Ballerini P, Nargi E, Buccella S, Giuliani P, Di Iorio P, Caciagli, F (2007) Staurosporine-induced apoptosis in astrocytes is prevented by A(1) adenosine receptor activation. Neurosci Lett 418:66–71

Dall'lgna OP, Porciuncula LO, Souza DO, Cunha RA, Lara DR (2003) Neuroprotection by caffeine and adenosine A(2A) receptor blockade of β-amyloid neurotoxicity. Br J Pharmacol 138: 1207–1209

Dall'Igna OP, Fett P, Gomes MW, Souza DO, Cunha RA, Lara DR (2007) Caffeine and adenosine A(2a) receptor antagonists prevent β-amyloid (25–35)-induced cognitive deficits in mice. Exp Neurol 203:241–245

Deckert J, Abel F, Kunig G, Hartmann J, Senitz D, Maier H, Ransmayr G, Riederer P (1998) Loss of human hippocampal adenosine A(1) receptors in dementia: evidence for lack of specificity. Neurosci Lett 244:1–4

De Mendonca A, Sebastiao AM, Ribeiro JA (1995) Inhibition of NMDA receptor-mediated currents in isolated rat hippocampal-neurons by adenosine A(1) receptor activation. Neuroreport 6:1097–1100

Dianzani C, Brunelleschi S, Viano I, Fantozzi R (1994) Adenosine modulation of primed human neutrophils. Eur J Pharmacol 263:223–226

Di-Capua N, Sperling O, Zoref-Shani E (2003) Protein kinase C-epsilon is involved in the adenosine-activated signal transduction pathway conferring protection against ischemia-reperffision injury in primary rat neuronal cultures. J Neurochem 84:409–412

Di Iorio P, Kleywegt S, Ciccarelli R, Traversa U, Andrew CM, Crocker CE, Werstiuk ES, Rathbone MP (2002) Mechanisms of apoptosis induced by purine nucleosides in astrocytes. Glia 38: 179–190

Dixon AK, Gubitz AK, Sirinathsinghji DJ, Richardson PJ, Freeman TC (1996) Tissue distribution of adenosine receptor mRNAs in the rat. Br J Pharmacol 118:1461–1468

Dixon AK, Widdowson L, Richardson PJ (1997) Desensitisation of the adenosine A_1 receptor by the A_{2A} receptor in the rat striatum. J Neurochem 69:315–321

Dolphin AC, Prestwich SA (1985) Pertussis toxin reverses adenosine inhibition of neuronal glutamate release. Nature 316:148–150

Domenici MR, Scattoni ML, Martire A, Lastoria G, Potenza RL, Borioni A, Venerosi A, Calamandrei G, Popoli P (2007) Behavioral and electrophysiological effects of the adenosine A_{2A} receptor antagonist SCH58261 in R6/2 Huntington's disease mice. Neurobiol Disease 28:197–205

Dux E, Fastbom J, Ungerstedt U, Rudolphi K, Fredholm BB (1990) Protective effect of adenosine and a novel xanthine derivative propentofylline on the cell damage after bilateral carotid occlusion in the gerbil hippocampus. Brain Res 516:248–256

Evans MC, Swan JH, Meldrum BS (1987) An adenosine analog, 2-chloroadenosine, protects against long-term development of ischemic cell loss in the rat hippocampus. Neurosci Lett 83:287–292

Ezeamuzie CI, Philips E (1999) Adenosine A_3 receptors on human eosinophils mediate inhibition of degranulation and superoxide anion release. Br J Pharmacol 127:188–194

Fastbom J, Fredholm BB (1985) Inhibition of (3H)glutamate release from rat hippocampal slices by L-PIA. Acta Physiol Scand 125:121–123

Fastbom J, Pazos A, Probst A, Palacios JM (1986) Adenosine A_1-receptors in human brain: characterisation and autoradiographic visualisation. Neurosci Lett 65:127–132

Fastbom J, Pazos A, Palacios JM (1987a) The distribution of adenosine A_1 receptors and 5′-nucleotidase in the brain of some commonly used experimental animals. Neuroscience 22:813–826

Fastbom J, Pazos A, Probst A, Palacios JM (1987b) Adenosine A_1 receptors in the human brain: a quantitative autoradiographic study. Neuroscience 22:827–839

Fedorova IM, Jacobson MA, Basile A, Jacobson KA (2003) Behavioral characterization of mice lacking the A(3) adenosine receptor: sensitivity to hypoxic neurodegeneration. Cell Mol Neurobiol 23:431–447

Ferkany JW, Zaczek R, Coyle JT (1982) Kainic acid stimulates excitatory amino acid neurotransmitter release at presynaptic receptors. Nature 298:757–759

Ferrante RJ, Kowall NW, Cipolloni PB, Storey E, Beal MF (1993) Excitotoxin lesions in primates as a model for Huntington's disease: histopathologic and neurochemical characterization. Exp Neurol 119:46–71

Ferré S, von Euler G, Johansson B, Fredholm BB, Fuxe K (1991) Stimulation of high-affinity adenosine A_2 receptors decreases the affinity of dopamine D_2 receptors in rat striatal membranes. Proc Natl Acad Sci U S A 88:7238–7241

Fiebich BL, Biber K, Gyufko K, Berger R, Bauer J, vanCalker D (1996a) Adenosine A(2b) receptors mediate an increase in interleukin (IL)-6 mRNA and IL-6 protein synthesis in human astroglioma cells. J Neurochem 66:1426–1431

Fiebich BL, Biber K, Lieb K, vanCalker D, Berger M, Bauer J, GebickeHaerter PJ (1996b) Cyclooxygenase-2 expression in rat microglia is induced by adenosine A(2a)-receptors. Glia 18:152–160

Fiebich BL, Akundi RS, Biber K, Hamke M, Schmidt C, Butcher RD, van Calker D, Willmroth F (2005) IL-6 expression induced by adenosine A(2b) receptor stimulation in U373MG cells depends on p38 mitogen activated kinase and protein kinase C. Neurochem Intern 46:501–512

Fink JS, Kalda A, Ryu H, Stack EC, Schwarzschild MA, Chen JF, Ferrante RJ (2004) Genetic and pharmacological inactivation of the adenosine A(2A) receptor attenuates 3-nitropropionic acid-induced striatal damage. J Neurochem 88:538–544

Finn SF, Swartz KJ, Beal MF (1991) 2-Chloroadenosine attenuates NMDA, kainate and quisqualate toxicity. Neurosci Lett 126:191–194

Fotheringham J, Mayne M, Holden C, Nath A, Geiger JD (2004) Adenosine receptors control HIV-1 Tat-induced inflammatory responses through protein phosphatase. Virology 327:186–195

Fredholm BB, Dunwiddie TV (1988) How does adenosine inhibit transmitter release? Trends Pharmacol Sci 9:130–134

Fredholm BB, Johansson B, Lindstrom K, Wahlstrom G (1998) Age-dependent changes in adenosine receptors are not modified by life-long intermittent alcohol administration. Brain Res 791:177–185

Gao Y, Phillis JW (1994) CGS 15943, an adenosine A_2 receptor antagonist, reduces cerebral ischemic injury in the Mongolian gerbil. Life Sci 55:PL61–PL65

Gerevich Z, Wirkner K, Illes P (2002) Adenosine A(2A) receptors inhibit the *N*-methyl-D-aspartate component of excitatory synaptic currents in rat striatal neurons. Eur J Pharmacol 451:161–164

Gianfriddo M, Melani A, Turchi D, Giovannini MG, Pedata F (2004) Adenosine and glutamate extracellular concentrations and mitogen-activated protein kinases in the striatum of Huntington transgenic mice. Selective antagonism of adenosine A(2A) receptors reduces transmitter outflow. Neurobiol Dis 17:77–88

Giménez-Llort L, Fernández-Teruel A, Escorihuela RM, Fredholm BB, Tobeña A, Pekny M, Johansson B (2002) Mice lacking the adenosine A_1 receptor are anxious and aggressive, but are normal learners with reduced muscle strength and survival rate. Eur J Neurosci 16:547–550

Goncalves ML, Cunha RA, Ribeiro JA (1997) Adenosine A_{2A} receptors facilitate $^{45}Ca^{2+}$ uptake through class A calcium channels in rat hippocampal CA3 but not CA1 synaptosomes. Neurosci Lett 238:73–77

Goodman RR, Snyder SH (1982) Autoradiographic localisation of adenosine receptors in rat brain using [3H]cyclohexyladenosine. J Neurosci 2:1230–1241

Graham SH, Chen J, Sharp FR, Simon RP (1993) Limiting ischaemic injury by inhibition of excitatory amino acid release. J Cereb Blood Flow Metab 13:88–97

Grisham MB, Hernandez LA, Granger DN (1989) Adenosine inhibits ischemia-reperfusion-induced leukocyte adherence and extravasation. Am J Physiol 257:H1334–H1339

Grondin R, Bédard PJ, Hadj Tahar A, Grégoire L, Mori A, Kase H (1999) Antiparkinsonian effect of a new selective adenosine A_{2A} receptor antagonist in MPTP-treated monkeys. Neurology 52:1673–1677

Guidetti P, Luthi-Carter RE, Augood SJ, Schwarcz R (2004) Neostriatal and cortical quinolinate levels are increased in early grade Huntington's disease. Neurobiol Dis 17:455–461

Hannon JP, Bray-French KM, Phillips RM, Fozard JR (1998) Further pharmacological characterization of the adenosine receptor subtype mediating inhibition of oxidative burst in human isolated neutrophils. Drug Dev Res 43:214–224

Hara H, Onodera H, Kato H, Kogure K (1992) Effects of aging on signal transmission and transduction systems in the gerbil brain: morphological and autoradiographic study. Neuroscience 46:475–488

Haskó G, Cronstein BN (2004) Adenosine: an endogenous regulator of innate immunity. Trends Immunol 25(1):33–39

Haskó G, Németh ZH, Vizi ES, Salzman AL, Szabó C (1998) An agonist of adenosine A_3 receptors decreases interleukin-12 and interferon-gamma production and prevents lethality in endotoxemic mice. Eur J Pharmacol 358:261–268

Haskó G, Pacher P, Deitch EA, Vizi ES (2007) Shaping of monocyte and macrophage function by adenosine receptors. Pharmacol Ther 113:264–275

Hauber W, Bareiss A (2001) Facilitative effects of an adenosine A_1/A_2 receptor blockade on spatial memory performance of rats: selective enhancement of reference memory retention during the light period. Behav Brain Res 118(1):43–52

Heese K, Fiebich BL, Bauer J, Otten U (1997) Nerve growth factor (NGF) expression in rat microglia is induced by adenosine A(2a)-receptors. Neurosci Lett 231:83–86

Héron A, Lasbennes F, Seylez J (1993) Adenosine modulation of amino acid release in rat hippocampus during ischaemia and veratridine depolarisation. Brain Res 608:27–32

Héron A, Lekieffre D, Le Peillet E, Lasbennes F, Seylaz J, Plotkine M, Boulu RG (1994) Effects of an A_1 adenosine receptor agonist on the neurochemical, behavioural and histological consequences of ischaemia. Brain Res 641:217–224

Hiraide T, Katsura K, Muramatsu H, Asano G, Katayama Y (2001) Adenosine receptor antagonists cancelled the ischemic tolerance phenomenon in gerbil. Brain Res 910:94–98

Hollins C, Stone TW (1980) Adenosine inhibition of GABA release from slices of rat cerebral cortex. Br J Pharmacol 69:107–112

Hooper N, Fraser C, Stone TW (1996) Effects of purine analogues on spontaneous alternation in mice. Psychopharmacology 123(3):250–257

Hosseinzadeh H, Stone TW (1998) Tolbutamide blocks postsynaptic but not presynaptic effects of adenosine on hippocampal CA1 neurons. J Neural Transm 105:161–172

Huang QY, Wei C, Yu LQ, Coelho JE, Shen HY, Kalda A, Linden J, Chen JF (2006) Adenosine A_{2A} receptors in bone marrow-derived cells but not in forebrain neurons are important contributors to 3-nitropropionic acid-induced striatal damage as revealed by cell-type-selective inactivation. J Neurosci 26:11371–11378

Hutchison AJ, Webb RL, Oei HH, Ghai GR, Zimmerman MB, Williams M (1989) CGS21680, an A_2 selective adenosine receptor agonist with preferential hypotensive activity. J Pharmacol Exp Ther 251:47–55

Ikeda M, Mackay KB, Dewar D, Mcculloch J (1993) Differential alterations in adenosine-A(1) and kappa-1-opioid receptors in the striatum in Alzheimer's-disease. Brain Res 616:211–217

Jaarsma D, Sebens JB, Korf J (1991) Reduction of adenosine-A_1-receptors in the perforant pathway terminal zone in Alzheimer hippocampus. Neurosci Lett 121:111–114

Jacobson KA (1998) Adenosine A_3 receptors: novel ligands and paradoxical effects. Trends Pharmacol Sci 19:184–191

Jacobson I, Hamberger A (1985) Kainic acid-induced changes of extracellular amino acid levels, evoked potentials and EEG activity in the rabbit olfactory bulb. Brain Res 348:289–296

Jansen KLR, Faull RLM, Dragunow M, Synek BL (1990) Alzheimer's disease: changes in hippocampal *N*-methyl-D-aspartate, quisqualate, neurotensin, adenosine, benzodiazepine, serotonin and opioid receptors—an autoradiographic study. Neuroscience 39:613–627

Jarvis MF, Williams M (1989) Direct autoradiographic localisation of adenosine A_2 receptors in the brain using the A_2-selective agonist, [3H]CGS21680. Eur J Pharmacol 168:243–246

Jiang N, Kowaluk EA, Lee CH, Mazdiyasni H, Chopp M (1997) Adenosine kinase inhibition protects brain against transient focal ischemia in rats. Eur J Pharmacol 320:131–137

Johnson-Kozlow M, Kritz-Silverstein D, Barrett-Connor E, Morton D (2002) Coffee consumption and cognitive function among older adults. Am J Epidemiol 156:842–850

Johnston JB, Silva C, Gonzalez G, Holden J, Warren KG, Metz LM, Power C (2001) Diminished adenosine A_1 receptor expression on macrophages in brain and blood of patients with multiple sclerosis. Ann Neurol 49:650–658

Jones PA, Smith RA, Stone TW (1998a) Protection against intrahippocampal kainate excitotoxicity by intracerebral administration of an adenosine A_{2A} receptor antagonist. Brain Res 800: 328–335

Jones PA, Smith RA, Stone TW (1998b) Protection against kainate-induced excitotoxicity by adenosine A_{2A} receptor agonists and antagonists. Neuroscience 85:229–237

Kalaria RN, Sromek S, Wilcox BJ, Unnerstall JR (1990) Hippocampal adenosine-A_1-receptors are decreased in Alzheimer's-disease. Neurosci Lett 118:257–260

Kanda T, Jackson MJ, Smith LA, Pearce RK, Nakamura J, Kase H, Kuwana Y, Jenner P (1998) Adenosine A_{2A} antagonist: a novel antiparkinsonian agent that does not provoke dyskinesia in parkinsonian monkeys. Ann Neurol 43:507–513

Kano T, Katayama Y, Kawamata T, Hirota H, Tsubokawa T (1994) Propentofylline administered by microdialysis attenuates ischaemia-induced hippocampal damage but not excitatory amino acid release in gerbils. Brain Res 641:149–154

Khaspekov L, Shamloo M, Victorov I, Wieloch T (1998) Sublethal in vitro glucose-oxygen deprivation protects cultured hippocampal neurons against a subsequent severe insult. Neuroreport 9:1273–1276

Knutsen LJS, Sheardown MJ, Roberts SM, Mogensen JP, Olsen UB, Thomsen C, Bowler AN (1998) Adenosine A_1 and A_3 selective N-alkoxypurines as novel cytokine modulators and neuroprotectants. Drug Dev Res 45:214–221

Kohler C, Schwarcz R, Fuxe K (1978) Perforant path transections protect hippocampal granule cells from kainate lesion. Neurosci Lett 10:241–246

Kopf SR, Melani A, Pedata F, Pepeu G (1999) Adenosine and memory storage: effect of A(1) and A(2) receptor antagonists. Psychopharmacology 146:214–219

Kreckler LM, Wan TC, Ge ZD, Auchampach JA (2006) Adenosine inhibits tumor necrosis factor-alpha release from mouse peritoneal macrophages via A(2A) and A(2B) but not the A(3) adenosine receptor. J Pharmacol Exp Ther 317:172–180

Kurokawa M, Kirk IP, Kirkpatrick KA, Kase H, Richardson PJ (1994) Inhibition by KF17837 of adenosine A_{2A} receptor-mediated modulation of striatal GABA and acetylcholine release. Br J Pharmacol 113:43–48

Lai DM, Tu YK, Liu IM, Cheng JT (2005) Increase of adenosine A_1 receptor gene expression in cerebral ischemia of Wistar rats. Neurosci Lett 387:59–61

Lappin D, Whaley K (1984) Adenosine A_2 receptors on human monocytes modulate C2 production. Clin Exp Immunol 57:454–460

Latini S, Pazzagli M, Pepeu G, Pedata F (1996) A_2 adenosine receptors: their presence and neuromodulatory role in the CNS. Gen Pharmacol 27:925–933

Latini S, Bordoni F, Corradetti R, Pepeu G, Pedata F (1999) Effect of A_{2A} adenosine receptor stimulation and antagonism on synaptic depression induced by in vitro ischaemia in rat hippocampal slices. Br J Pharmacol 128:1035–1044

Lau Y-S, Mouradian MM (1993) Protection against acute MPT-induced dopamine depletion in mice by adenosine. J Neurochem 60:768–771

Lee FS, Chao MV (2001) Activation of Trk neurotrophin receptors in the absence of neurotrophins. Proc Natl Acad Sci USA 98:3555–3560

Lee KS, Reddington M (1986) Autoradiographic evidence for multiple CNS binding sites for adenosine derivative. Neuroscience 19:535–549

Lehmann A, Isacsson H, Hamberger A (1983) Effects of in vivo administration of kainic acid on the extracellular amino acid pool in the rabbit hippocampus. J Neurochem 40:1314–1320

LeVraux V, Chen YL, Masson I, De Sousa M, Giroud JP, Florentin I, Chauvelot-Moachon L (1993) Inhibition of human monocyte TNF production by adenosine receptor agonists. Life Sci 52:1917–1924

Li XX, Nomura T, Aihara H, Nishizaki T (2001) Adenosine enhances glial glutamate efflux via A_{2A} adenosine receptors. Life Sci 68:1343–1350

Linden J, Thai T, Figler H, Jin X, Robeva AS (1999) Characterization of human A_{2B} adenosine receptors: radioligand binding, western blotting, and coupling to Gq in human embryonic kidney 293 cells and HMC-1 mast cells. Mol Pharmacol 56:705–713

Link AA, Kino T, Worth JA, McGuire JL, Crane ML, Chrousos GP, Wilder RL, Elenkov IJ (2000) Ligand activation of the adenosine A_{2A} receptors inhibits IL-12 production by human monocytes. J Immunol 164:436–442

Lipton SA, Rosenberg PA (1994) Excitatory amino acids as a final common pathway for neurologic disorders. N Engl J Med 330:613–622

Liu HC, Hon CJ, Liu TY, Tsai, SJ (2005) Association analysis of adenosine A_{2a} receptor 1976T > C polymorphisms and Alzheimer's disease. Eur Neurol 53:99–100

Liu YH, Xiong L, Chen SY, Wang Q (2006) Isoflurane tolerance against focal cerebral ischemia is attenuated by adenosine A(1) receptor antagonists. Can J Anaes 53:194–201

Lozza G, Conti A, Ongini E, Monopoli A (1997) Cardioprotective effects of adenosine A_1 and A_{2A} receptor agonists in the isolated heart. Pharmacol Res 35:57–64

MacGregor DG, Stone TW (1993) Inhibition by the adenosine analogue, (R)-N^6-phenylisopropyladenosine, of kainic acid neurotoxicity in rat hippocampus after systemic administration. Br J Pharmacol 109:316–321

MacGregor DG, Miller WJ, Stone TW (1993) Mediation of the neuroprotective action of *R*-phenylisopropyladenosine through a centrally located adenosine A_1 receptor. Br J Pharmacol 110:470–476

MacGregor DG, Jones PA, Maxwell WL, Graham DI, Stone TW (1996) Prevention by a purine analogue of kainate-induced neuropathology in rat hippocampus. Brain Res 725:115–120

Maemoto T, Tada M, Mihara T, Ueyama N, Matsuoka H, Harada K, Yamaji T, Shirakawa K, Kuroda S, Akahane A, Iwashita A, Matsuoka N, Mutoh S (2004) Pharmacological characterization of FR194921, a new potent, selective, and orally active antagonist for central adenosine A(1) receptors. J Pharmacol Sci 96:42–52

Mager R, Ferroni S, Schubert P (1990) Adenosine modulates a voltage-dependent chloride conductance in cultured hippocampal neurons. Brain Res 532:58–62

Maggirwar SB, Dhanraj DN, Somani SM, Ramkumar V (1994) Adenosine acts as an endogenous activator of the cellular antioxidant defense system. Biochem Biophys Res Commun 201: 508–515

Maglione V, Giallonardo P, Cannella M, Martino T, Frati L, Squitieri F (2005) Adenosine A(2A) receptor dysfunction correlates with age at onset anticipation in blood platelets of subjects with Huntington's disease. Am J Med Genet B Neuropsychiatr Genet 139B:101–105

Maglione V, Cannella M, Martino T, De Blasi A, Frati L, Squitieri F (2006) The platelet maximum number of A_{2A}-receptor binding sites (Bmax) linearly correlates with age at onset and CAG repeat expansion in Huntington's disease patients with predominant chorea. Neurosci Lett 393:27–30

Maia L, de Mendonça A (2002) Does caffeine intake protect from Alzheimer's disease? Eur J Neurol 9:377–382

Mally J, Stone TW (1994) The effect of theophylline on parkinsonian symptoms. J Pharm Pharmacol 46:515–517

Mally J, Stone TW (1996) Potential role of adenosine antagonist therapy in pathological tremor disorders. Pharmacol Ther 72:243–250

Mally J, Stone TW (1998) Potential of adenosine A(2A) receptor antagonists in the treatment of movement disorders. CNS Drugs 10:311–320

Mangiarini L, Sathasivam K, Seller M, Cozens B, Harper A, Hetherington C, Lawton M, Trottier T, Lehrach H, Davies SW, Bates GP (1996) Exon 1 of the HD gene with an expanded CAG repeat is sufficient to cause a progressive neurological phenotype in transgenic mice. Cell 87:493–506

Martinez-Mir MI, Probst A, Palacios JM (1991) Adenosine A_2 receptors: selective localisation in the human basal ganglia and alterations with disease. Neuroscience 42:697–706

Matherne GP, Linden J, Byford AM, Gauthier NS, Headrick JP (1997) Transgenic A_1 adenosine receptor overexpression increases myocardial resistance to ischaemia. Proc Nat Acad Sci USA 94:6541–6546

Mayfield RD, Suzuki F, Zahniser NR (1993) Adenosine A_{2A} receptor modulation of electrically evoked endogenous GABA release from rat globus pallidus. J Neurochem 60:2334–2337

Mayne M, Shepel PN, Jiang Y, Geiger JD, Power C (1999) Dysregulation of adenosine A(1) receptor-mediated cytokine expression in peripheral blood mononuclear cells from multiple sclerosis patients. Ann Neurol 45:633–639

McLarnon JG, Choi HB, Lue LF, Walker DG, Kim SU (2005) Perturbations in calcium-mediated signal transduction in microglia from Alzheimer's disease patients. J Neurosci Res 81:426–435

McRae A, Rudolphi KA, Schubert P (1994) Propentofylline depresses amyloid and Alzheimer's CSF microglial antigens after ischaemia. Neuroreport 5:1193–1196

McWhinney CD, Dudley MW, Bowlin TL, Peet NP, Schook L, Bradshaw M, De M, Borcherding DR, Edwards CK 3rd (1996) Activation of adenosine A_3 receptors on macrophages inhibits TNFα. Eur J Pharmacol 310:209–216

Meerlo P, Roman V, Farkas E, Keijser JN, Nyakas C, Luiten PGM (2004) Ageing-related decline in adenosine A_1 receptor binding in the rat brain: an autoradiographic study. J Neurosci Res 78:742–748

Meiners I, Hauschildt S, Nieber K, Munch G (2004) Pentoxyphylline and propentophylline are inhibitors of TNF-ot release in monocytes activated by advanced glycation endproducts. J Neural Transm 111:441–447

Melani A, Pantoni L, Corsi C, Bianchi L, Monopoli A, Bertorelli R, Pepeu G, Pedata F (1999) Striatal outflow of adenosine, excitatory amino acids, GABA, and taurine in awake, freely moving rats after middle cerebral artery occlusion: correlation with neurological deficit and histopathological damage. Stroke 30:2448–2454

Meldrum DR, Cain BS, Cleveland JC Jr, Meng X, Ayala A, Banerjee A, Harken AH (1997) Adenosine decreases post-ischemic cardiac TNF-α production: anti-inflammatory implications for preconditioning and transplantation. Immunology 92:472–477

Merkel LA, Lappe RW, Rivera LM, Cox BF, Perrone MH (1992) Demonstration of vasorelaxant activity with an A_1-selective adenosine agonist in porcine coronary artery: involvement of potassium channels. J Pharmacol Exp Ther 260:437–443

Michaelis ML, Michaelis EK, Myers SL (1979) Adenosine modulation of synaptosomal dopamine release. Life Sci 24:2083–2092

Mielke R, Kessler J, Szelies B, Herholz K, Wienhard K, Heiss WD (1996a) Vascular dementia: perfusional and metabolic disturbances and effects of therapy. J Neural Transm 47:183–191

Mielke R, Kittner B, Ghaemi M, Kessler J, Szelies B, Herholz K, Heiss WD (1996b) Propentofylline improves regional cerebral glucose metabolism and neuropsychologic performance in vascular dementia. J Neurol Sci 141:59–64

Miller WJ, Macgregor DG, Stone TW (1994) Time-course of purine protection against kainate-induced increase in hippocampal [H-3] PK11195 binding. Brain Res Bull 34:133–136

Monopoli A, Lozza G, Forlani A, Mattavelli A, Ongini E (1998) Blockade of adenosine A_{2A} receptors by SCH58261 results in neuroprotective effects in cerebral ischaemia in rats. Neuroreport 9:3955–3959

Moreau JL, Huber G (1999) Central adenosine A_{2A} receptors: an overview. Brain Res Rev 31: 65–82

Moreno MB, Hevia H, Santamaria M, Sepulcre J, Munoz J, Garcia-Trevijano ER, Berasain C, Corrales FJ, Avila MA, Villoslada P (2006) Methylthioadenosine reverses brain autoimmune disease. Ann Neurol 60:323–334

Muller CE (1997) A(1)-adenosine receptor antagonists. Exp Opin Ther Pat 7:419–440

Muller CE (2000) A(2A) adenosine receptor antagonists—future drugs for Parkinson's disease? Drugs Future 25:1043–1052

Murray TF (1982) Up-regulation of rat cortical adenosine receptors following chronic administration of theophylline. Eur J Pharmacol 82:113–114

Nakamura M, Nakakimura K, Matsumoto M, Sakabe T (2002) Rapid tolerance to focal cerebral ischemia in rats is attenuated by adenosine A(1) receptor antagonist. J Cereb Blood Flow Metab 22:161–170

Nakata H, Yoshioka K, Saitoh O (2003) Hetero-oligomerization between adenosine A_1 and $P2Y_1$ receptors in living cells: formation of ATP-sensitive adenosine receptors. Drug Dev Res 58:340–349

Nikbakht M-R, Stone TW (2001) Activation of NMDA receptors suppresses the presynaptic effects of adenosine. Br J Pharmacol 133:155P

Nishizaki T (2004) ATP- and adenosine-mediated signaling in the central nervous system: adenosine stimulates glutamate release from astrocytes via A_{2A} adenosine receptors J Pharmacol Sci 94:100–102

Nishizaki T, Nagai K, Nomura T, Tada H, Kanno T, Tozaki H, Li XX, Kondoh T, Kodama N, Takahashi E, Sakai N, Tanaka K, Saito N (2002) A new neuromodulatory pathway with a glial contribution mediated via A(2a) adenosine receptors. Glia 39:133–47

Noble S, Wagstaff AJ (1997) Propentofylline. CNS Drugs 8:257–264

Norenberg W, Wirkner K, Illes P (1997) Effect of adenosine and some of its structural analogues on the conductance of NMDA receptor channels in a subset of rat neostriatal neurones. Br J Pharmacol 122:71–80

Norenberg W, Wirkner K, Assmann H, Richter M, Illes P (1998) Adenosine A(2A) receptors inhibit the conductance of NMDA receptor channels in rat neostriatal neurons. Aminoacids 14:33–39

Normile HJ, Barraco RA (1991) N^6-Cyclopentyladenosine impairs passive avoidance retention by selective action at A_1 receptors. Brain Res Bull 27:101–104

O'Kane EM, Stone TW (1998) Interactions between A_1 and A_2 adenosine receptor-mediated responses in the rat hippocampus in vitro. Eur J Pharmacol 362:17–25

O'Regan MH, Simpson RE, Perkins LM, Phillis JW (1992) The selective A_2 adenosine receptor agonist CGS21680 enhances excitatory amino acid release from the ischaemic rat cerebral cortex. Neurosci Lett 138:169–172

Ochiishi T, Chen L, Yukawa A, Saitoh Y, Sekino Y, Arai T, Nakata H, Miyamoto H (1999a) Cellular localisation of adenosine A_1 receptors in rat forebrain: immunohistochemical analysis using adenosine A_1 receptor-specific monoclonal antibody. J Comp Neurol 411:301–316

Ochiishi T, Saitoh Y, Yukawa A, Saji M, Ren Y, Shirao T, Miyamoto H, Nakata H, Sekino Y (1999b) High levels of adenosine A_1 receptor-like immunoreactivity in the CA2/CA3a region of the adult rat hippocampus. Neuroscience 93:955–967

Ochoa CM, Jackson TA, Aaron CS, Lahti RA, Strain GM, Von Voigtlander PF (1992) Antagonism of kainic acid lesions in the mouse hippocampus by U-54494A and U-50488H. Life Sci 51:1135–1143

Okazaki MM, Nadler JV (1988) Protective effects of mossy fibre lesions against kainic acid induced seizures and neuronal degeneration. Neuroscience 26:763–781

Okonkwo DO, Reece TB, Laurent JJ, Hawkins AS, Ellman PI, Linden J, Kron IL, Tribble CG, Stone JR, Kern JA (2006) A comparison of adenosine A(2A) agonism and methylprednisolone in attenuating neuronal damage and improving functional outcome after experimental traumatic spinal cord injury in rabbits. J Neurosurg 4:64–70

Oliveira JC, Constantino MD, Sebastiao AM, Ribeiro JA (1995) Ascorbate/Fe^{3+}-induced peroxidation and inhibition of the binding of A_1 adenosine receptor ligands in rat-brain membranes. Neurochem Int 26:263–268

Ongini E, Fredholm BB (1996) Pharmacology of adenosine A_{2A} receptors. Trends Pharmacol Sci 17:364–372

Ongini E, Adami M, Ferri C, Bertorelli R (1997) Adenosine A_{2A} receptors and neuroprotection. Ann NY Acad Sci 825:30–48

Ongini E, Monopoli A, Impagnatiello F, Fredduzzi S, Schwarzschild M, Chen JF (2001) Dual actions of A_{2A} adenosine receptor antagonists on motor dysfunction and neurodegenerative processes. Drug Dev Res 52:379–386

Pagonopoulou O, Angelatou F (1992) Reduction of A_1 adenosine receptors in cortex, hippocampus and cerebellum in ageing mouse brain. Neuroreport 3:735–737

Palmer TM, Poucher SM, Jacobson KA, Stiles GL (1995) ^{125}I-4-[7-amino-2-{2-furyl} {1,2,4}triazolo{2,3-a} {1,3,5}triazin-5-yl-amino]ethyl)phenol, a high affinity antagonist radioligand selective for the A_{2A} adenosine receptor. Mol Pharmacol 48:970–974

Parkinson FE, Rudolphi KA, Fredholm BB (1994) Propentofylline: a nucleoside transport inhibitor with neuroprotective effects in cerebral ischaemia. Gen Pharmacol 25:1053–1058

Pazzagli M, Corsi C, Latini S, Pedata F, Pepeu G (1994) In vivo regulation of extracellular adenosine levels in the cerebral cortex by NMDA and muscarinic receptors. Eur J Pharmacol 254:277–282

Pedata F, Corsi C, Melani A, Bordoni F, Latini S (2001) Adenosine extracellular brain concentrations and role of A_{2A} receptors in ischemia. Ann N Y Acad Sci 939:74–84

Pedata F, Gianfriddo M, Turchi D, Melani A (2005) The protective effect of adenosine A(2A) receptor antagonism in cerebral ischemia. Neurol Res 27:169–174

Phillis JW (1995) The effects of selective A_1 and A_{2A} adenosine receptor antagonists on cerebral ischemic injury in the gerbil. Brain Res 705:79–84

Phillis JW, O'Regan MH (1993) Prevention of ischemic brain injury by adenosine receptor activation. Drug Dev Res 28:390–394

Pierson PM, Peteri-Brunbäck B, Pisani DF, Abbracchio MP, Mienville JM, Rosso L (2007) A(2b) receptor mediates adenosine inhibition of taurine efflux from pituicytes. Biol Cell 99:445–454

Pignataro G, Simon RP, Boison D (2007) Transgenic overexpression of adenosine kinase aggravates cell death in ischemia. J Cereb Blood Flow Metab 27:1–5

Pintor A, Galluzzo M, Grieco R, Pèzzola A, Reggio R, Popoli P (2004) Adenosine A_{2A} receptor antagonists prevent the increase in striatal glutamate levels induced by glutamate uptake inhibitors. J Neurochem 89:152–156

Pitsikas N, Borsini F (1997) The adenosine A_1 receptor antagonist BIIP 20 counteracts scopolamine-induced behavioral deficits in the passive avoidance task in the rat. Eur J Pharmacol 328:19–22

Poli A, Lucchi R, Vibio M, Barnabei O (1991) Adenosine and glutamate modulate each other's release from rat hippocampal synaptosomes. J Neurochem 57:298–306

Popoli P, Pèzzola A, Domenici MR, Sagratella S, Diana G, Caporali MG, Bronzetti E, Vega J, Scotti de Carolis A (1994) Behavioral and electrophysiological correlates of the quinolinic acid rat model of Huntington's disease in rats. Brain Res Bull 35:329–335

Popoli P, Betto P, Reggio R, Ricciarello G (1995) Adenosine A(2A) receptor stimulation enhances striatal extracellular glutamate levels in rats. Eur J Pharmacol 287:215–217

Popoli P, Pintor A, Domenici MR, Frank C, Tebano MT, Pèzzola A, Scarchilli L, Quarta D, Reggio R, Malchiodi-Albedi F, Falchi M, Massotti M (2002) Blockade of striatal adenosine A_{2A} receptor reduces, through a presynaptic mechanism, quinolinic acid-induced excitotoxicity: possible relevance to neuroprotective interventions in neurodegenerative diseases of the striatum. J Neurosci 22:1967–1975

Popoli P, Blum D, Martire A, Ledent C, Ceruti S, Abbracchio MP (2007) Functions, dysfunctions and possible therapeutic relevance of adenosine A_{2A} receptors in Huntington's disease. Prog Neurobiol 81:331–348

Pousinha PA, Diogenes JM, Ribeiro AJ, Sebastiao AM (2006) Triggering of BDNF facilitatory action on neuromuscular transmission by adenosine A_{2A} receptors. Neurosci Lett 404:143–147

Prediger RDS, Batista LC, Takahashi RN (2005) Caffeine reverses age-related deficits in olfactory discrimination and social recognition memory in rats: involvement of adenosine A(1) and A(2A) receptors. Neurobiol Ageing 26:957–964

Press NJ, Gessi S, Borea PA, Polosa R (2007) Therapeutic potential of adenosine receptor antagonists and agonists. Exp Opin Ther Pat 17:979–991

Prior C, Torres RJ, Puig JG (2006) Hypoxanthine effect on equilibrative and concentrative adenosine transport in human lymphocytes. Implications in the pathogenesis of Lesch–Nyhan syndrome. Nucleosides Nucleotides Nucl Acids 25:1065–1069

Pugliese AM, Latini S, Corradetti R, Pedata F (2003) Brief, repeated, oxygen-glucose deprivation episodes protect neurotransmission from a longer ischemic episode in the in vitro hippocampus: role of adenosine receptors. Br J Pharmacol 140:305–314

Pugliese AM, Coppi E, Volpini R, Cristalli G, Corradetti R, Jeong LS, Jacobson KA, Pedata F (2007) Role of adenosine A(3) receptors on CA1 hippocampal transmission during oxygen-glucose deprivation episodes of different duration. Biochem Pharmacol 74:768–779

Quintana JLB, Allam MF, Del Castillo AS, Navajas RFC (2007) Alzheimer's disease and coffee: a review. Neurol Res 29:91–95

Ramakers BP, Riksen NP, Rongen GA, van der Hoeven JG, Smits P, Pickkers P (2006) The effect of adenosine receptor agonists on cytokine release by human mononuclear cells depends on the specific Toll-like receptor subtype used for stimulation. Cytokine 35:95–99

Rathbone MP, Middlemiss PJ, Kim JK, Gysbers JW, Deforge SP, Smith RW, Hughes DW (1992) Adenosine and its nucleotides stimulate proliferation of chick astrocytes and human astrocytoma cells. Neurosci Res 13:1–17

Rathbone MP, Middlemiss PJ, Gysbers JW, Andrew C, Herman MAR, Reed JK, Ciccarelli R, Di Iorio P, Caciagli F (1999) Trophic effects of purines in neurons and glial cells. Prog Neurobiol 59:663–690

Reece TB, Okonkwo DO, Ellman PI, Warren PS, Smith RL, Hawkins AS, Linden J, Kron IL, Tribble CG, Kern JA (2004) The evolution of ischemic spinal cord injury in function, cytoarchitecture, and inflammation and the effects of adenosine A(2A) receptor activation. J Thorac Cardiovasc Surg 128:925–932

Reece TB, Kron IL, Okonkwo DO, Laurent JJ, Tache-Leon C, Maxey TS, Ellman PI, Linden J, Tribble CG, Kern JA (2006) Functional and cytoarchitectural spinal cord protection by ATL-146e after ischemia/reperfusion is mediated by adenosine receptor agonism. J Vasc Surg 44:392–397

Regenold JT, Illes P (1990) Inhibitory adenosine A_1-receptors on rat locus coeruleus neurones. An intracellular electrophysiological study. Naunyn–Schmied Arch Pharmacol 341:225–231

Reggio R, Pezzola A, Popoli P (1999) The intrastratial injection of an adenosine A(2) receptor antagonist prevents frontal cortex EEG abnormalities in a rat model of Huntington's disease. Brain Res 831:315–318

Revan S, Montesinos MC, Naime D, Landau S, Cronstein BN (1996) Adenosine A_2 receptor occupancy regulates stimulated neutrophil function via activation of a serine/threonine protein phosphatase. J Biol Chem 271:17114–17118

Ribeiro JA, Sebastiao AM, de Mendonça A (2002) Adenosine receptors in the nervous system: pathophysiological implications. Prog Neurobiol 68:377–392

Riedel W, Hogervorst E, Leboux R, Verhey F, Vanpraag H, Jolles J (1995) Caffeine attenuates scopolamine-induced memory impairment in humans. Psychopharmacology 122:158–168

Ritchie PK, Spangelo BL, Krzymowski DK, Rossiter TB, Kurth E, Judd AM (1997) Adenosine increases interleukin-6 release and decreases TNF release from rat adrenal zona glomerulosa cells, ovarian cells, anterior pituitary cells and peritoneal macrophages. Cytokine 9:187–198

Rivkees SA, Price SL, Zhou FC (1995) Immunohistochemical detection of A_1 adenosine receptors in rat brain with emphasis on localisation in the hippocampal formation, cerebral cortex, cerebellum and basal ganglia. Brain Res 677:193–203

Rivkees SA, Thevananther S, Hao H (2000) Are A_3 adenosine receptors expressed in the brain? Neuroreport 11:1025–1030

Robledo P, Ursu G, Mahy N (1999) Effects of adenosine and gamma-aminobutyric acid A receptor antagonists on *N*-methyl-D-aspartate induced neurotoxicity in the rat hippocampus. Hippocampus 9:527–533

Rodriguez A, Martin M, Albasanz JL, Barrachina M, Espinosa JC, Torres JM, Ferrer I (2006) Adenosine A(1) receptor protein levels and activity is increased in the cerebral cortex in Creutzfeldt–Jakob disease and in bovine spongiform encephalopathy-infected bovine-PrP mice. J Neuropathol Exp Neurol 65:964–975

Rosi S, McGann K, Hauss-Wegrzyniak B, Wenk GL (2003) The influence of brain inflammation upon neuronal adenosine A(2B) receptors. J Neurochem 86:220–227

Rubaj A, Zgodzinski W, Sieklucka-Dziuba M (2003) The influence of adenosine A_3 receptor agonist: IB-MECA, on scopolamine- and MK-801-induced memory impairment. Behav Brain Res 141:11–17

Rudolphi KA, Schubert P (1997) Modulation of neuronal and glial cell function by adenosine and neuroprotection in vascular dementia. Behav Brain Res 83:123–128

Rudolphi KA, Keil M, Hinze HJ (1987) Effect of theophylline on ischemically induced hippocampal damage in Mongolian gerbils: a behavioural and histopathological study. J Cereb Blood Flow Metab 7:74–81

Rudolphi KA, Keil M, Fastbom J, Fredholm BB (1989) Ischemic damage in gerbil hippocampus is reduced following upregulation of adenosine (A1) receptors by caffeine treatment. Neurosci Lett 103:275–280

Rudolphi KA, Schubert P, Parkinson FE, Fredholm BB (1992) Adenosine and brain ischaemia. Cerebrovasc Brain Metab Rev 4:346–369

Sajjadi FG, Takabayashi K, Foster AC, Domingo RC, Firestein GS (1996) Inhibition of TNF-α expression by adenosine: role of A_3 adenosine receptors. J Immunol 156:3435–3442

Saura J, Angulo E, Ejarque A, Casado V, Tusell JM, Moratalla R, Chen JF, Schwarzschild MA, Lluis C, Franco R, Serratosa J (2005) Adenosine A(2A) receptor stimulation potentiates nitric oxide release by activated microglia. J Neurochem 95:919–929

Scatena R, Martorana GE, Bottoni P, Botta G, Pastore P, Giardina B (2007) An update on pharmacological approaches to neurodegenerative diseases. Exp Opin Invest Drugs 16:59–72

Scattoni ML, Valanzano A, Pezzola A, De March Z, Fusco FR, Popoli P, Calamandrei G (2007) Adenosine A(2A) receptor blockade before striatal excitotoxic lesions prevents long term behavioural disturbances in the quinolinic rat model of Huntington's disease. Behav Brain Res 176:216–221

Schiffmann SN, Jacobs O, Vanderhaegen JJ (1991) Striatal restricted adenosine A_2 receptor (RDC8) is expressed by enkephalin but not by substance P neurons: an in situ hybridisation histochemistry study. J Neurochem 57:1062–1067

Schingnitz G, Kufnermuhl U, Ensinger H, Lehr E, Kuhn FJ (1991) Selective A_1-antagonists for treatment of cognitive deficits. Nucleosides Nucleotides 10:1067–1076

Scholz KP, Miller RJ (1992) Inhibition of quantal transmitter release in the absence of calcium influx by a G protein-linked adenosine receptor at hippocampal synapses. Neuron 8:1139–1150

Schubert P, Ferroni S, Mager R (1991) Pharmacological blockade of chloride pumps on chloride channels reduces the adenosine-mediated depression of stimulus train-evoked calcium fluxes in rat hippocampal slices. Neurosci Lett 124:174–177

Schubert P, Ogata T, Ferroni S, McRae A, Nakamura Y, Rudolphi K (1996) Modulation of glial cell signaling by adenosine and pharmacological reinforcement: a neuroprotective strategy? Mol Chem Neuropathol 28:185–190

Schubert P, Ogata T, Rudolphi K, Marchini C, McRae A, Ferroni S (1997) Support of homeostatic glial cell signaling: a novel therapeutic approach by propentofylline. Ann NY Acad Sci 826:337–347

Schurr A, Reid KH, Tseng MT, West C, Rigor BM (1986) Adaptation of adult brain-tissue to anoxia and hypoxia invitro. Brain Res 374:244–248

Sebastiao AM, Ribeiro JA (1992) Evidence for the presence of excitatory A_2 adenosine receptors in the rat hippocampus. Neurosci Lett 138:41–44

Sebastiao AM, Ribeiro JA (1996) Adenosine A(2) receptor-mediated excitatory actions on the nervous system. Progr Neurobiol 48:167–189

Sei Y, von Lubitz DKJE, Abbracchio MP, Ji X-D, Jacobson KA (1997) Adenosine A_3 receptor agonist-induced neurotoxicity in rat cerebellar granule neurons. Drug Dev Res 40:267–273

Selley ML (2004) Increased homocysteine and decreased adenosine formation in Alzheimer's disease. Neurol Res 26:554–557

Sheardown MJ, Knutsen LJS (1996) Unexpected neuroprotection observed with the adenosine A_{2A} receptor agonist CGS21680. Drug Dev Res 39:108–114

Shen H, Zhang L, Yuen D, Logan R, Jung BP, Zhang G, Eubanks JH (2002) Expression and function of A_1 adenosine receptors in the rat hippocampus following transient forebrain ischemia. Neuroscience 114:547–556

Simpson RE, O'Regan MH, Perkins LM, Phillis JW (1992) Excitatory transmitter amino acid release from the ischaemic rat cerebral cortex: effects of adenosine recptor agonists and antagonists. J Neurochem 58:1683–1690

Sperk G (1994) Kainic acid seizures in the rat. Prog Neurobiol 42:1–32

Spignoli G, Pedata F, Pepeu G (1984) A_1 and A_2 adenosine receptors modulate acetylcholine release from brain slices. Eur J Pharmacol 97:341–342

Stone TW (1993) The neuropharmacology of quinolinic acid and kynurenic acid. Pharmacol Rev 45:309–379

Stone TW (2001) Kynurenines in the CNS: from endogenous obscurity to clinical relevance. Prog Neurobiol 64:185–218

Stone TW, Darlington LG (2002) Endogenous kynurenines as targets for drug discovery and development. Nat Rev Drug Discov 1:609–620

Stone TW, Perkins MN (1981) Quinolinic acid: a potent endogenous excitant at amino acid receptors in the CNS. Eur J Pharmacol 72:411–412

Sutherland GR, Peeling J, Lesiuk HJ, Brownstone RM, Rydzy M, Saunders JK, Geiger JD (1991) The effects of caffeine on ischemic neuronal injury as determined by magnetic-resonance-imaging and histopathology. Neuroscience 42:171–182

Suzuki F, Shimada J, Shiozaki S, Ichikawa S, Ishii A, Nakamura J, Nonaka H, Kobayashi H, Fuse E (1993) Adenosine A_1 antagonists. 3. Structure–activity relationships on amelioration against scopolamine- or N^6-((R)-phenylisopropyl)adenosine-induced cognitive disturbance. J Med Chem 36:2508–2518

Svenningsson P, Hall H, Sedvall G, Fredholm BB (1997) Distribution of adenosine receptors in the postmortem human brain: an extended autoradiographic study. Synapse 27:322–335

Szot P, Sanders RC, Murray TF (1987) Theophylline-induced up-regulation of adenosine-A_1-receptors associated with reduced sensitivity to convulsants. Neuropharmacology 26: 1173–1180

Tarditi A, Camurri A, Varani K, Borea PA, Woodman B, Bates G, Cattaneo E, Abbracchio MP (2006) Early and transient alteration of adenosine A(2A) receptor signaling in a mouse model of Huntington disease. Neurobiol Dis 23:44–53

Tebano MT, Pintor A, Frank C, Domenici MR, Martire A, Pepponi R, Potenza RL, Grieco R, Popoli P (2004) Adenosine A(2A) receptor blockade differentially influences excitotoxic mechanisms at pre- and postsynaptic sites in the rat striatum. J Neurosci Res 77:100–107

Torres RJ, DeAntonio I, Prior C, Puig JG (2004) Effects of hypoxanthine on adenosine transport in human lymphocytes. Implications in the pathogenesis of Lesch–Nyhan syndrome. Nucleosides Nucleotides Nucl Acids 23:1177–1179

Traversa U, Rosati AM, Chiara F, Rodolfo V (1994) Effects of chronic administration of adenosine antagonists on adenosine A_1 and A_{2A} receptors in mouse brain. In vivo 8:1073–1078

Trincavelli ML, Marroni M, Tuscano D, Ceruti S, Mazzola A, Mitro N, Abbracchio MP, Martini C (2004) Regulation of A_{2B} adenosine receptor functioning by tumour necrosis factor a in human astroglial cells. J Neurochem 91:1180–1190

Trincavelli ML, Tonazzini I, Montali M, Abbracchio MP, Martini C (2008) Short-term TNF-α treatment induced A_{2B} adenosine receptor desensitization in human astroglial cells. J Cell Biochem 104:150–161

Trussel LD, Jackson MB (1985) Adenosine activated potassium conductance in cultured striatal neurones. Proc Nat Acad Sci USA 82:4857–4861

Tsutsui S, Schnermann J, Noorbakhsh F, Henry S, Yong VW, Winston BW, Warren K, Power C (2004) A_1 adenosine receptor upregulation and activation attenuates neuroinflammation and demyelination in a model of multiple sclerosis. J Neurosci 24:1521–1529

Ulas J, Brunner LC, Nguyen L, Cotman CW (1993) Reduced density of adenosine-A_1-receptors and preserved coupling of adenosine-A_1-receptors to G-proteins in Alzheimer hippocampus: a quantitative autoradiographic study. Neuroscience 52:843–854

Van der Schyf CJ, Gal S, Geldenhuys WJ, Youdim MBH (2006) Multifunctional neuroprotective drugs targeting monoamine oxidase inhibition, iron chelation, adenosine receptors, and cholinergic and glutamatergic action for neurodegenerative diseases. Expert Opin Invest Drugs 15:873- 886

van Gelder BM, Buijsse B, Tijhuis M, Kalmijn S, Giampaoli S, Nissinen A, Kromhout D (2007) Coffee consumption is inversely associated with cognitive decline in elderly European men: the FINE Study. Eur J Clin Nutr 61:226–232

Varani K, Gessi S, Dionisotti S, Ongini E, Borea PA (1998) [3H]-SCH58261 labelling of functional A_{2A} receptors in human neutrophil membranes. Br J Pharmacol 123:1723–1731

Varani K, Rigamonti D, Sipione S, Camurri A, Borea PA, Cattabeni F, Abbracchio MP, Cattaneo E (2001) Aberrant amplification of A(2A) receptor signaling in striatal cells expressing mutant huntingtin. FASEB J 15:1245–1247

Varani K, Abbracchio MP, Cannella M, Cislaghi G, Giallonardo P, Mariotti C, Cattabriga E, Cattabeni F, Borea PA, Squitieri F, Cattaneo E (2003) Aberrant A_{2A} receptor function in peripheral blood cells in Huntington's disease. FASEB J 17:2148–2150

Varani K, Bachoud-Lévi AC, Mariotti C, Tarditi A, Abbracchio MP, Gasperi V, Borea PA, Dolbeau G, Gellera C, Solari A, Rosser A, Naji J, Handley O, Maccarrone M, Peschanski M, DiDonato S, Cattaneo E (2007) Biological abnormalities of peripheral A(2A) receptors in a large representation of polyglutamine disorders and Huntington's disease stages. Neurobiol Dis 27:36–43

Virgili M, Poli A, Contestabile A, Migani P, Barnabei O (1986) Synaptosomal release of newly synthesised or recently accumulated amino acids; differential effects of kainic acid on naturally occurring excitatory amino acids and on [D-3H]-aspartate. Neurochem Int 9:29–33

Virus RM, Baglajewski T, Radulovacki M (1984) [3H]N^6-(L-Phenylisopropyl) adenosine binding in brains from young and old rats. Neurobiol Aging 5:61–62

von Lubitz DKEJ, Dambrosia JM, Redmond DJ (1989) Protective effect of cyclohexyladenosine in treatment of cerebral ischaemia in gerbils. Neuroscience 30:451–462

von Lubitz DK, Paul IA, Bartus RT, Jacobson KA (1993) Effects of chronic administration of adenosine A_1 receptor agonist and antagonist on spatial learning and memory. Eur J Pharmacol 249:271–280

von Lubitz DK, Lin RC, Melman N, Ji XD, Carter MF, Jacobson KA (1994a) Chronic administration of selective adenosine A_1 receptor agonist or antagonist in cerebral ischaemia. Eur J Pharmacol 256:161–167

von Lubitz DK, Lin RC, Popik P, Carter MF, Jacobson KA (1994b) Adenosine A_3 receptor stimulation and cerebral ischaemia. Eur J Pharmacol 263:59–67

von Lubitz DKJE, Kim J, Beenhakker M, Carter MF, Lin RCS, Meshulam Y, Daly JW, Shi D, Zhou LM, Jacobson KA (1995a) Chronic NMDA receptor stimulation: therapeutic implications of its effect on adenosine A(1) receptors. Eur J Pharmacol 283:185–192

von Lubitz DKJE, Lin RC-S, Jacobson KA (1995b) Cerebral ischaemia in gerbils: effects of acute and chronic treatment with adenosine A_{2A} receptor agonist and antagonist. Eur J Pharmacol 287:295–302

von Lubitz DK, Lin RC, Paul IA, Beenhakker M, Boyd M, Bischofberger N, Jacobson KA (1996) Postischemic administration of adenosine amine congener (ADAC): analysis of recovery in gerbils. Eur J Pharmacol 316:171–179 von Lubitz DJKE, Lin RCS, Bischofberger N, Beenhakker M, Boyd M, Lipartowska R, Jacobson KA (1999a) Protection against ischemic damage by adenosine amine congener, a potent and selective adenosine A_1 receptor agonist. Eur J Pharmacol 369:313–317

von Lubitz DKJE, Lin RC-S, Boyd M, Bischofberger N, Jacobson KA (1999b) Chronic administration of adenosine A_3 receptor agonist and cerebral ischaemia: neuronal and glial effects. Eur J Pharmacol 367:157–163

Von Lubitz DKJE, Simpson KL, Lin RCS (2001) Right thing at a wrong time? Adenosine A(3) receptors and cerebroprotection in stroke. Neuroprotective Agents 939:85–96

Wagner DR, Combes A, McTiernan C, Sanders VJ, Lemster B, Feldman AM (1998a) Adenosine inhibits lipopolysaccharide-induced cardiac expression of TNFα. Circ Res 82:47–56

Wagner DR, McTiernan C, Sanders VJ, Feldman AM (1998b) Adenosine inhibits lipopolysaccharide-induced secretion of TNF-α in the failing human heart. Circulation 97:521–524

Warskulat U, Heller-Stilb B, Oermann E, Zilles K, Haas H, Lang F, Häussinger D (2007) Phenotype of the taurine transporter knockout mouse. Methods Enzymol 428:439–58

Wirkner K, Assmann H, Koles l, Gerevich Z, Franke H, Norenberg, Boehm R, Illes P (2000) Inhibition by adenosine A(2A) receptors of NMDA but not AMPA currents in rat neostriatal neurons. Br J Pharmacol 130:259–269

Wirkner K, Gerevich Z, Krause T, Gunther A, Koles L, Schneider D, Norenberg W, Illes P (2004) Adenosine A(2A) receptor-induced inhibition of NMDA and GABA(A) receptor-mediated synaptic currents in a subpopulation of rat striatal neurons. Neuropharmacology 46:994–1007

Wu LG, Saggau P (1997) Presynaptic inhibition of elicited neurotransmitter release. Trends Neurosci 20:204–212

Yamada K, Tanaka T, Senzaki K, Kameyama T, Nabeshima T (1998) Propentofylline improves learning and memory deficits in rats induced by β-amyloid protein-(1–40). Eur J Pharmacol 349:15–22

Yao ZH, Gross GJ (1994) The ATP-dependent potassium channel - an endogenous cardioprotective mechanism. J Cardiovasc Pharmacol 24: S28–S34

Yao Y, Sei Y, Abbracchio MP, Jiang JL, Kim YC, Jacobson KA (1997) Adenosine A_3 receptor agonists protect HL-60 and U-937 cells from apoptosis induced by A_3 antagonists. Biochem Biophys Res Commun 232:317–22

Yoshida M, Nakakimura K, Cui YJ, Matsumoto M, Sakabe T (2004) Adenosine a, receptor antagonist and mitochondrial ATP-sensitive potassium channel blocker attenuate the tolerance to focal cerebral ischemia in rats. J Cereb Blood Flow Metab 24:771–779

Yoshioka K, Saitoh O, and Nakata H (2001) Heteromeric association creates a P2Y-like adenosine receptor. Proc Natl Acad Sci USA 98:7617–7622

Yoshioka K, Hosada R, Kuroda Y, Nakata H (2002a) Hetero-oligomerization of adenosine A_1 receptors with $P2Y_1$ receptors in rat brains. FEBS Lett 531:299–303

Yoshioka K, Saitoh O, Nakata H (2002b) Agonist-promoted heteromeric oligomerization between adenosine A_1 and $P2Y_1$ receptors in living cells. FEBS Lett 523:147–151

Yu LQ, Huang Z, Mariani J, Wang Y, Moskowitz M, Chen JF (2004) Selective inactivation or reconstitution of adenosine A_{2A} receptors in bone marrow cells reveals their significant contribution to the development of ischemic brain injury. Nat Med 10:1081–1087

Zarrindast MR, Shafaghi B (1994) Effects of adenosine receptor agonists and antagonists on acquisition of passive avoidance learning. Eur J Pharmacol 256:233–239

Zhao ZQ, McGee S, Nakanishi K, Toombs CF, Johnston WE, Ashar MS, Vinten-Johansen J (1993) Receptor-mediated cardioprotective effects of endogenous adenosine are exerted primarily during reperfusion after coronary occlusion in the rabbit. Circulation 88:709–719

Zhou AM, Li WB, Li QJ, Liu HQ, Feng RF, Zhao HG (2004) A short cerebral ischemic preconditioning up-regulates adenosine receptors in the hippocampal CA1 region of rats. Neurosci Res 48:397–404

Zhou QY, Li C, Olah ME, Johnson RA, Stiles GL, Civelli O (1992) Molecular cloning and characterisation of an adenosine receptor: the A3 receptor. Proc Nat Acad Sci USA 89:7432–7436

Adenosine A_{2A} Receptors and Parkinson's Disease

Micaela Morelli, Anna R. Carta, and Peter Jenner

Contents

Abstract The drug treatment of Parkinson's disease (PD) is accompanied by a loss of drug efficacy, the onset of motor complications, lack of effect on non-motor symptoms, and a failure to modify disease progression. As a consequence, novel approaches to therapy are sought, and adenosine A_{2A} receptors (A_{2A}ARs) provide

---

M. Morelli (✉)
Department of Toxicology and Center of Excellence for Neurobiology of Addiction, University of Cagliari, via Ospedale 72, 09124 Cagliari, Italy
morelli@unica.it, acarta@unica.it

C.N. Wilson and S.J. Mustafa (eds.), *Adenosine Receptors in Health and Disease*, Handbook of Experimental Pharmacology 193, DOI 10.1007/978-3-540-89615-9_18,

a viable target. A_{2A}ARs are highly localized to the basal ganglia and specifically to the indirect output pathway, which is highly important in the control of voluntary movement. A_{2A}AR antagonists can modulate γ-aminobutyric acid (GABA) and glutamate release in basal ganglia and other key neurotransmitters that modulate motor activity. In both rodent and primate models of PD, A_{2A}AR antagonists produce alterations in motor behavior, either alone or in combination with dopaminergic drugs, which suggest that they will be effective in the symptomatic treatment of PD. In clinical trials, the A_{2A}AR antagonist istradefylline reduces "off" time in patients with PD receiving optimal dopaminergic therapy. However, these effects have proven difficult to demonstrate on a consistent basis, and further clinical trials are required to establish the clinical utility of this drug class. Based on preclinical studies, A_{2A}AR antagonists may also be neuroprotective and have utility in the treatment of neuropsychiatric disorders. We are only now starting to explore the range of potential uses of A_{2A}AR antagonists in central nervous system disorders, and their full utility is still to be uncovered.

Keywords A_{2A} antagonist · Clinical trial · Dyskinesia · Motor dysfunction · Basal ganglia · MPTP · 6-OHDA · Neuroprotection

Abbreviations

AIMs	Abnormal involuntary movements
A_{2A}AR	Adenosine A_{2A} receptor
AUC	Area under the curve
AMPA	Alpha-amino-3-hydroxy-5-methyl-4-isoxazolepropionic acid
BG	Basal ganglia
COMT	Catechol-*O*-methyl transferase
CPu	Caudate-putamen
CGI	Clinical global impression
DA	Dopamine
DYN	Dynorphin
ENK	Enkephalin
GABA	γ-Aminobutyric acid
GAD67	Glutamic acid decarboxylase
GP	Globus pallidus
GPe	Globus pallidus, external segment
GPi	Globus pallidus, internal segment
5-HT	5-Hydroxytryptamine
LOCF	Last observation carried forward
KO	Knockout
L-DOPA	3,4-Dihydroxy-L-phenylalanine
LTP/LDP	Long-term potentiation/long-term depression
mGlu5	Metabotropic glutamate subtype 5

MAO B	Monoamine oxidase B
6-OHDA	6-Hydroxydopamine
MPTP	1-Methyl-4-phenyl-1,2,3,6-tetrahydropyridine
PD	Parkinson's disease
STN	Subthalamus
SNr	Substantia nigra pars reticulata
TJM	Tremulous jaw movement
UPDRS	Unified Parkinson's disease rating scale

1 Introduction

Increasing life expectancy will inevitably lead to an increase in the incidence of neurodegenerative illnesses, such as Parkinson's disease (PD), constituting an increasing social and economic burden (Dorsey et al. 2007). At the same time, the dopaminergic therapies currently used to treat the motor symptoms of PD, while effective in the initial stages of the illness, become inadequate as the disease progresses, do not reverse non-motor symptomatology, and become associated with adverse effects that prove difficult to manage (Fahn and Janlovic 2007; Jankovic 2006). In this situation, drug treatments that act beyond the damaged dopaminergic system, for example adenosine A_{2A} receptor (A_{2A}AR) antagonists, are becoming important targets for the treatment of PD since they may be effective in both the early and late stages of PD and avoid the unwanted side effects currently associated with chronic dopaminergic treatment.

2 Parkinson's Disease

PD affects 1 in 500 of the general population and 1 in 100 of those individuals aged 60 or over. The incidence of the illness is age related and this remains the only clearly established predisposing factor (Weintraub et al. 2008a). It is characterized by akinesia, rigidity, tremor and postural abnormalities, but increasingly there is awareness that it is a much broader illness that induces a range of non-motor symptoms such as sweating, falling, speech and swallowing difficulties, and neuropsychiatric components such as depression, anxiety and cognitive decline (Chaudhuri et al. 2005). Many of these features can precede the onset of motor symptoms and they, and others, are being actively investigated as early diagnostic features of those individuals that are likely to go on to develop clinical PD (Berg 2006; Siderowf and Stern 2006). The motor symptoms of PD are due primarily to the degeneration of the dopaminergic nigrostriatal pathway, with the mesolimbic/mesocortical dopaminergic pathways remaining relatively intact. However, pathology is widespread, with cell loss also occurring in many other brain areas, such as the locus coeruleus, raphe nuclei, dorsal motor nucleus of the vagus and the ventral forebrain, leading to changes in a range of neurotransmitters, including noradrenaline, 5-hydroxytryptamine (5-HT) and acetylcholine (Agid 1991; Jellinger 2002). Precisely how these contribute to the symptomatology of PD is not known, but they may be the origin of the non-motor features of the illness. Recently, the suggestion was made that PD is a progressive pathological disorder that starts in the periphery and then affects the brain, sweeping from the brainstem through to the cortex and only leading to a diagnosis of PD when the pathological process starts to affect the basal ganglia (BG) (Braak et al. 2006a, b; Braak and Del 2008). Although this is controversial, it implies that treatment strategies should be more broadly based and that pathological change in the BG may be a later feature of PD than previously thought.

PD can be induced by gene defects in rare familial cases, but the bulk of the PD population is considered to have idiopathic disease (Gasser 2007; Hardy et al. 2006). In all probability, it is not a single disorder but a syndrome with multiple causes and with clear differences between, for example, young-onset PD and late-onset illness, and between tremor-dominant and akinetic manifestations. The usual description of PD is that it is due to a combination of genetic and environmental factors that can interact to varying degrees and at different levels (McCulloch et al. 2008). The pathogenic process responsible for neuronal loss in PD remains unknown, but contributing factors are oxidative and nitrative stress, mitochondrial dysfunction, excitotoxicity and altered proteolysis (Jenner and Olanow 2006; Litvan et al. 2007a, b). Cells are presumed to die by apoptosis, but this has not been conclusively demonstrated. There are, however, two key features of PD that probably provide the major clues to the underlying mechanisms. First, pathological change is always accompanied by the appearance of cytoplasmic inclusions, termed Lewy bodies, in surviving neurons (Wakabayashi et al. 2007), and second, there is a reactive microgliosis and to some extent astrocytosis that leads to inflammatory change and that may contribute to the progression of pathology in PD (McGeer and McGeer 2008).

The primary effect of dopaminergic loss in the striatum in PD leads to a disruption of the parallel processing loops between the motor cortex, basal ganglia, thalamus and back to pre-motor and motor cortex that are responsible for the integration of motor, sensory and cognitive information that controls voluntary movement (Obeso et al. 2000, 2004). Dopamine plays three important roles in the striatum that are lost in PD. It controls the activity of the corticostriatal glutamatergic input, it determines the activity of the GABAergic medium spiny neurons that make up the major striatal output pathways—the direct and indirect pathways (see below), and it plays a key role in motor programming through the maintenance of long-term potentiation or long-term depression (LTP/LDP)-type processes (Calabresi et al. 2006, 2007). All of these are key to how dopaminergic therapy reverses the motor symptoms of PD and to how non-dopaminergic drugs, such as adenosine antagonists, can also alter basal ganglia function in PD.

3 Treatment of PD and Limitations of Therapy

The current therapy for PD is based on dopaminergic replacement therapy using 3,4-dihydroxy-L-phenylalanine (L-DOPA) and dopamine agonists, notably ropinirole and pramipexole (Horstink et al. 2006a, b; Weintraub et al. 2008b). These lead to almost complete reversal of motor symptoms in the early stages of the disease, but the dopamine agonists do not possess as great an efficacy as L-DOPA. This may be related to their more selective effects on dopamine receptor subtypes, largely D_2/D_3 receptors, and to the fact that L-DOPA stimulates all dopamine receptor populations and also enhances noradrenergic and serotoninergic transmission and can alter glutamate release among a range of actions. Adjuncts to dopaminergic therapy are the other major drug types used in PD. These are the catechol-*O*-methyl transferase (COMT) inhibitors entacapone and tolcapone, which prevent the metabolism of L-DOPA to 3-*O*-methyl-DOPA, as well as the monoamine oxidase B (MAO B) inhibitors selegiline and rasagiline, which prevent the breakdown of endogenous dopamine and dopamine derived from L-DOPA. Otherwise, the only other drugs routinely used to treat PD are anticholinergics, which are particularly effective against tremor, or the weak NMDA antagonist amantadine, which has some mild symptomatic actions but is usually employed to suppress dyskinesia (see below).

However, the symptomatic treatment of PD becomes more complex with disease progression and with chronic drug treatment (Fabbrini et al. 2007; Jankovic 2005; Jankovic and Stacy 2007; Stacy and Galbreath 2008; Stocchi 2003). Dopaminergic drugs show a shortening of duration of effect (wearing-off), and the clinical response becomes unpredictable and subject to rapid oscillations, with patients switching rapidly between mobility and immobility (on–off). This can be treated by using a longer-acting dopamine agonist drug or by adding a COMT inhibitor or MAO B inhibitor to therapy, but this is only a short-term measure. A significant proportion of PD patients develop involuntary movements or dyskinesia (chorea, dystonia, athetosis), particularly when treated with L-DOPA. Once established, these are evoked by every dose of dopaminergic medication that is administered. Treatment is usually by dose reduction, but this worsens PD; or by the addition of amantadine, but this is poorly tolerated by many patients; or by the use of continuous drug infusions (subcutaneous apomorphine or intraduodenal L-DOPA); or by referral for deep brain stimulation, employing electrode placement in the subthalamic nucleus (Guridi et al. 2008).

Dopaminergic medications induce a range of acute side effects that further complicate current treatment. These include acute effects such as nausea and vomiting and more prolonged changes in cardiovascular function and in hormonal status. Probably most worrying, however, are the neuropsychiatric complications of dopaminergic treatment usually seen after longer periods of treatment in more advanced patients with PD. Psychosis induced by dopaminergic medication, particularly in elderly patients showing cognitive decline, can become treatment limiting. More recently, dopaminergic dysregulation syndromes, such as compulsive gambling and hypersexuality, have been identified as affecting significant numbers of individuals (Stamey and Jankovic 2008; Stocchi 2005) and leading to legal action that may limit the use of this drug class. All of this leads to the conclusion that new approaches to treatment are required. While dopaminergic medication is highly effective against the motor symptoms of PD, it has little effect on the non-motor components of PD, which are largely non-dopaminergic in origin. Cognitive decline in PD and the high incidence of anxiety and depression require particular attention (Weintraub et al. 2008c). These have become a major problem in treating PD, and novel therapeutic approaches are required.

All current treatment of PD is orientated towards symptomatic therapy. There are no proven treatments that alter the rate of progression of PD. A key objective is to find disease-modifying treatments that stop or slow disease progression. However, neuroprotection is proving a difficult issue, with drugs that look highly effective in preclinical models of PD turning out to be ineffective in clinical trials (Ahlskog 2007; Hung and Schwarzschild 2007; Kieburtz and Ravina 2007; LeWitt 2006; Schapira 2008; Stocchi and Olanow 2003). This has occurred with MAO B inhibitors, glutamate antagonists, inhibitors of apoptotic mechanisms, enhancers of mitochondrial function, trophic factors, and dopamine agonists, amongst others. The reasons for this are not entirely clear, but it may relate to the inappropriateness of the animal models or to the multiple causes of PD and the use of patient populations with different pathogenic mechanisms underlying the origin of their disease.

New approaches to neuroprotection are needed, and clues may be gained by looking at factors that are thought to reduce the risk of developing PD in the human population. Some of the more robust, although still controversial, include cigarette smoking, the use of nonsteroidal anti-inflammatory drugs, antihypertensive agents (notably calcium channel blockers), and caffeine (Becker et al. 2008; Bornebroek et al. 2007; Esposito et al. 2007; Hu et al. 2007; Powers et al. 2008; Ritz et al. 2007). The ability of caffeine to reduce risk may be highly relevant to the potential therapeutic effects of A_{2A}AR antagonists in the treatment of PD.

4 Basal Ganglia Organization

4.1 Localization of A_{2A}ARs in Basal Ganglia

The BG comprise a group of tightly interconnected forebrain nuclei, intercalated among the cerebral cortex, thalamus and brainstem, and mainly involved in motor control and sensorimotor integration. Within the last decade, a number of dedicated studies have extensively shown how dopamine and adenosine interact to modulate motor function at this level (Fuxe et al. 2007; Schwarzschild et al. 2006; Schiffmann et al. 2007).

Adenosine binds at least four different G-protein-coupled receptors, namely A_1, A_{2A}, A_{2B}, A_3 (Fredholm et al. 1994). In contrast to the widespread distribution of A_1, A_{2B} and A_3 adenosine receptors in the brain, A_{2A}ARs are more selectively distributed, being abundantly expressed in the BG, and reaching the highest levels of expression in the caudate-putamen (CPu) (Rosin et al. 1998; Schiffmann et al. 1991). This selective distribution of A_{2A}ARs, involving a potentially low incidence of side effects, first led to the consideration of A_{2A}AR antagonists among the most promising non-dopaminergic agents for the treatment of PD motor symptoms.

The CPu is mainly composed of medium spiny GABAergic neurons, which are equally divided into two neuronal populations: striatonigral neurons, which connect the CPu with the substantia nigra pars reticulata (SNr) or globus pallidus (GP) internal segment (GPi), otherwise called the entopeduncular nucleus in rodents, and striatopallidal neurons, which connect the CPu with the GP or GPe (globus pallidus external segment) in primates (Fig. 1). Within this system, A_{2A}ARs are restricted to GABAergic neurons projecting to the GP which also selectively express the D_2 dopamine receptor and the peptide enkephalin (ENK) (Fink et al. 1992; Schiffmann et al. 1991). Conversely, striatonigral neurons, which selectively express the D_1 dopamine receptor and the peptide dynorphin (DYN), do not contain appreciable levels of A_{2A}AR. At the molecular level, G_s-coupled A_{2A}ARs activate adenylate cyclase, resulting in stimulation of neuronal activity, and opposing the dopamine-mediated inactivation of adenylate cyclase through the G_i-coupled D_2 receptor (Fredholm 1995). Recent studies have demonstrated that in striatopallidal neurons the A_{2A}AR can form heteromers with the D_2 receptor to attenuate coupling to the signaling pathway of the latter, offering a molecular mechanism of interaction which has compelling implications for PD treatment (Fuxe et al. 2005; Hillion et al. 2002).

The second most abundant neuronal population within the CPu are the large cholinergic aspiny interneurons, which represent about 5% of the entire population (Gerfen 1992). Striatal cholinergic nerve terminals express A_{2A}ARs, which, by modulating the release of acethylcholine in the rat CPu (Fig. 1), represent a novel interesting target for tremor control in PD models (see later).

4.2 *Function of A_{2A}ARs in Basal Ganglia*

In an intact CPu, adenosine via A_{2A}ARs excites striatopallidal neurons, opposing the inhibitory effect exerted by dopamine (Fig. 1). In PD, lack of dopamine generates an imbalance in the activity of striatal output pathways. Striatonigral neurons become hypoactive, whereas striatopallidal neurons, losing the inhibitory effect of dopamine while undergoing the stimulatory influence of adenosine, become hyperactive, boosting their inhibitory influence on GP neurons. Such imbalanced activity leads to a markedly increased inhibitory output from SNr/GPi to thalamocortical neurons, which produces hypokinetic symptoms in PD. Many authors have suggested that the positive effects of A_{2A}AR antagonists in PD rely on the blockade of A_{2A}ARs on striatopallidal neurons, which should dampen their excessive activity and restore some balance between striatonigral and striatopallidal neurons, consequently relieving thalamocortical activity. This mechanism offers a rationale

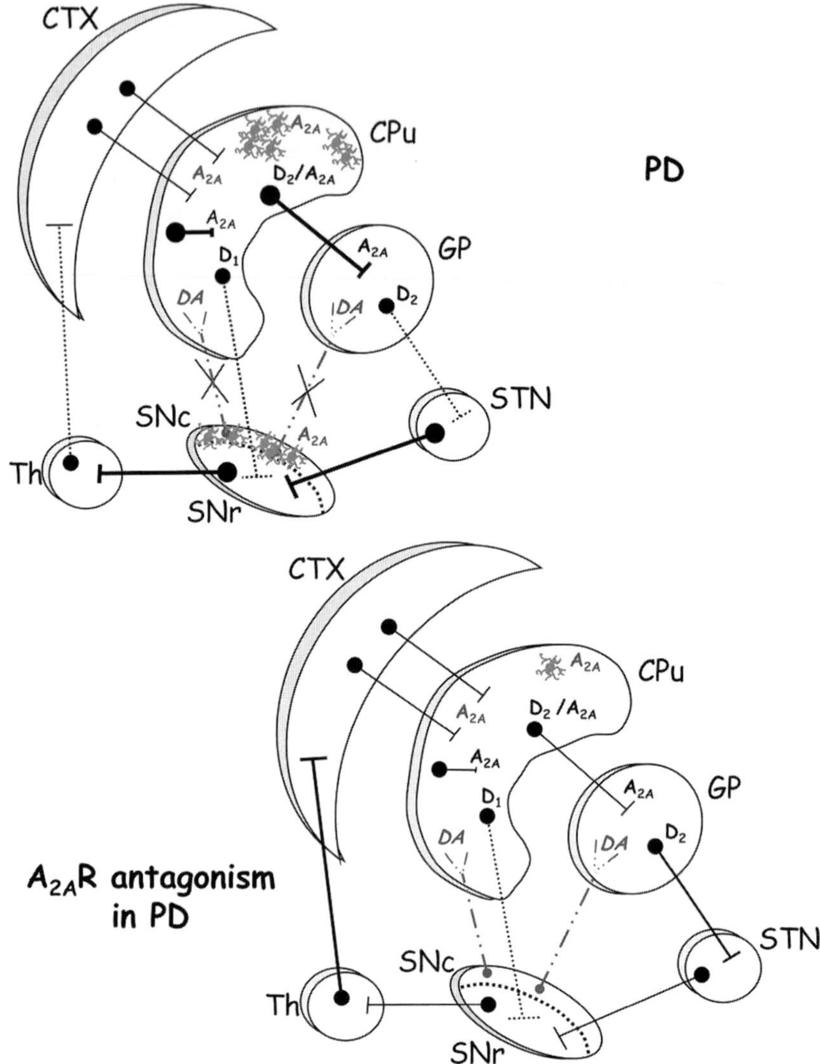

Fig. 1 Proposed mechanisms of adenosine A_{2A} receptor ($A_{2A}AR$) antagonist activity in Parkinson's disease (*PD*). Mechanisms of symptomatic effects are drawn in *black*, whereas mechanisms of neuroprotection are drawn in *gray*. In PD, lack of dopamine (*DA*) induces hypoactivity of striatonigral D_1-containing neurons and hyperactivity of striatopallidal D_2-containing neurons, resulting in subthalamus (*STN*) and substantia nigra pars reticulata (*SNr*) hyperactivity. Acetylcholine (*Ach*) interneurons in the caudate-putamen (*CPu*) are also hyperactive. The final outcome is depressed activity of thalamocortical (*Th*) projections, which produces characteristic symptoms of akinesia. $A_{2A}AR$ blockade in striatopallidal neurons, and likely in the globus pallidus (*GP*), relieves their hyperactivity, restoring balance between the output pathways. As a consequence, SNr and Th-cortical neurons become normoactive, relieving the akinesia. Moreover, $A_{2A}AR$ blockade in Ach interneurons restores Ach tone, which may contribute to counteracting tremor. In the parkinsonian state, glial proliferation is present in both the CPu and the substantia nigra pars compacta (*SNc*). As neuroprotective agents, $A_{2A}AR$ antagonists attenuate dopaminergic cell degeneration through a mechanism that may involve $A_{2A}ARs$ located presynaptically or alternatively $A_{2A}ARs$ in glial cells

for the use of A_{2A}R antagonists as a monotherapy in PD, as well as for the synergistic effect observed upon the concurrent administration of A_{2A}AR antagonists with L-DOPA or dopaminergic agonists, which restore dopamine receptor stimulation (Jenner 2003; Morelli 2003).

Of great interest is the neuronal colocalization and synergistic interaction observed between striatal A_{2A} receptor and metabotropic glutamate subtype 5 (mGlu5), glutamate receptor, which itself represents one of the most promising targets for treatment of PD symptoms (Ferré et al. 2002; Rodrigues et al. 2005). A potentiation of motor activity has been reported upon combined administration of A_{2A} and mGlu5 receptor antagonists, together with a synergistic interaction at the level of signal transduction pathways (Coccurello et al. 2004; Ferré et al. 2002; Kachroo et al. 2005; Nishi et al. 2003). The recent discovery of A_{2A}–mGlu5 heteromers in CPu has further strengthened the rational for studying antiparkinsonian strategies that simultaneously block A_{2A}ARs and mGlu5 receptors (Ferré et al. 2002).

4.3 Role of Globus Pallidus A_{2A} Adenosine Receptors

An important function of A_{2A}ARs located outside the CPu, particularly in the GP, has been evidenced by the positive effects displayed by A_{2A}AR antagonists when administered in association with dopaminergic therapies. In recent years, several works have led to a reconsideration of the role played by the GP in BG circuits, with this nucleus now placed at a critical functional position to modulate the excitability of afferent (CPu and STN) and efferent (SNr) nuclei (Obeso et al. 2006). The infusion of GABA agonists directly into the GP has been found to severely hamper motor function, whereas the antagonism of pallidal GABAergic transmission results in beneficial motor effects (Hauber 1998). The GP receives a direct dopaminergic innervation, being enriched in D_2 dopamine receptors. In the parkinsonian state, in which the GP discharge rate and oscillatory activity are altered, intrapallidal dopaminergic antagonists produce akinesia, whereas dopamine stops this symptom (Galvan et al. 2001; Hauber and Lutz 1999), suggesting that dopamine depletion either directly or indirectly disrupts the modulatory function of GP within the BG. A_{2A}ARs are highly expressed in the GP, mainly in the neuropil, where they can regulate pallidal extracellular GABA concentration and, thereafter, GP activity (Rosin et al. 1998; 2003). While stimulation of pallidal A_{2A}ARs enhances striatopallidal GABA outflow, their blockade reduces it (Ochi et al. 2004; Shindou et al. 2003). Recently, it was reported that while intrapallidal infusion of A_{2A}AR antagonists in 6-hydroxydopamine (6-OHDA)-lesioned rats does not elicit any motor response per se, it does potentiate motor activity induced by L-DOPA or dopaminergic agonists, suggesting that the beneficial effect exerted by these compounds in PD might also rely on the blockade of pallidal A_{2A}ARs (Simola et al. 2006; 2008). It might be hypothesized that in PD, the blockade of pallidal A_{2A}ARs, by reducing extracellular GABA, may contribute to restoring GP activity and in turn subthalamic nucleus activity, leading to a more balanced activation of direct and indirect pathways and, when associated with dopaminergic agonists, an enhancement of their motor-stimulating effects.

5 Motor-Behavioral Effects of A_{2A}AR Antagonists in Animal Models of Parkinson's Disease

5.1 Effects of Acute A_{2A}AR Antagonism on Motor Deficits

The highly enriched distribution of adenosine A_{2A}ARs in striatopallidal neurons, and their ability to form functional heteromeric complexes with dopamine D_2 and metabotropic glutamate mGlu5 receptors, mean that A_{2A}AR antagonists are of particular interest for the modulation of motor behavior, whilst at the same time they display a low predisposition to induce non-motor side effects.

Research performed to evaluate the effects produced by AR ligands on motor behavior in experimental rodents has provided the first evidence that adenosine is implicated in the modulation of movement. The critical role of A_{2A}AR in the regulation of motor behavior was first highlighted by data showing inhibition of motor behavior by the A_{2A}AR agonist 2-*p*-[(2-carboxyethyl)-phenethylamino]-5′-*N*-ethylcarboxamidoadenosine (CGS-21680), while the A_{2A}AR antagonist 7-(2-phenylethyl)-5-amino-2-(2-furyl)-pyrazolo-[4,3-*e*]-1,2,4-triazolo[1,5-*c*]pyrimidine (SCH-58261) was found to stimulate motor activity (Morelli et al. 1994; Pollack and Fink 1996; Pinna et al. 1996).

A large number of A_{2A}AR antagonists have been demonstrated to affect motor behavior by reversing catalepsy in rodents (reducing its duration and severity), hence accounting for an improvement in parkinsonian motor deficit by these drugs. Moreover, combined administration of the A_{2A}AR antagonists with L-DOPA has been shown to potentiate the L-DOPA-induced anticataleptic effect, indicating the existence of a synergistic interaction between L-DOPA and A_{2A}AR antagonists (Kanda et al. 1994; Shiozaki et al. 1999; Wardas et al. 2001).

In line with results obtained in the catalepsy protocol, A_{2A}AR antagonists showed motor-facilitatory activity in animals rendered parkinsonian by the administration of dopaminergic neurotoxins, such as 6-OHDA and 1-methyl-4-phenyl-1,2,3,6-tetrahydropyridine (MPTP), which induce the degeneration of nigrostriatal dopaminergic neurons, resulting in models of parkinsonian-like disabilities (akinesia, bradykinesia, etc.) in the animals treated.

Acute administration of the A_{2A}AR agonist CGS 21680 to unilaterally 6-OHDA-lesioned rats has been shown to significantly reduce the turning behavior induced by L-DOPA and either D_1 or D_2 dopamine receptor agonists (Morelli et al. 1994). Conversely, the A_{2A} receptor antagonist SCH 58261, when administered acutely to 6-OHDA-lesioned rats, has been demonstrated to significantly potentiate turning behavior induced by L-DOPA and either D_1 or D_2 dopamine receptor agonists (Pinna et al. 1996). An increase in the turning behavior stimulated by L-DOPA or apomorphine was observed following acute A_{2A}AR blockade by 1,3-dipropyl-7-methyl-8-(3,4-dimethoxystyryl)xanthine (KF-17837), 2-butyl-9-methyl-8-(2*H*-1,2,3-triazol-2-yl)-9*H*-purin-6-ylamine (ST-1535) or (*E*)-1,3-diethyl-8-(3,4-dimethoxystyryl)-7-methyl-3,7-dihydro-1*H*-purine-2,6-dione (KW-6002) (Koga et al. 2000; Rose et al. 2007; Tronci et al. 2007).

Besides turning behavior, subtle aspects of PD symptomatology develop in rats as a consequence of dopamine neuron degeneration, such as forelimb akinesia, gait impairment and sensory-motor integration deficits that are considered analogous to the PD symptoms seen in humans. Acute administration of the A_{2A}AR antagonists SCH-58261 and ST-1535, in a similar manner to L-DOPA although with a lower intensity, counteracted the lesion-induced impairments to the initiation time of the stepping test, to adjusting steps, and to vibrissae-evoked forelimb placing (Pinna et al. 2007). These results suggest that A_{2A}AR antagonists might ameliorate parkinsonian symptoms in PD patients, even when used as a monotherapy.

Most importantly, the efficacy of A_{2A}AR antagonists in MPTP-treated nonhuman primates, provided the impetus for experimentating with these compounds in clinical trials. Acute administration of the A_{2A}AR antagonist KW-6002 counteracted motor impairments and increased locomotor activity in primates previously treated with MPTP (Kanda et al. 1998a, b). Furthermore, a synergistic interaction between A_{2A}AR antagonists and L-DOPA, as well as dopaminergic agonists, in decreasing motor impairment has been observed in MPTP-treated common marmosets (Kanda et al. 2000; Rose et al. 2007).

The crucial role of CPu in the effects of A_{2A}AR antagonists has been confirmed by data indicating that the intrastriatal infusion of the A_{2A}AR antagonist MSX-3 significantly counteracted catalepsy produced by D_1 or D_2 receptor antagonists (Hauber et al. 2001). However, further to the well-documented role of CPu in mediating motor facilitation produced by A_{2A}AR antagonists, extrastriatal circuits may also be involved in this effect (see Sects. 4.3 and 5.5).

5.2 *Efficacy of A_{2A}AR Antagonists in Relieving Parkinsonian Tremor and Muscular Rigidity*

To date, tremor and rigidity are devoid of adequate pharmacological treatments, and so preclinical evidence showing that A_{2A}AR antagonists may be effective in relieving rigidity as well as resting tremor, one of the first symptoms manifested in individuals affected by PD, has greatly increased the attention directed towards A_{2A}AR antagonist compounds.

Promising effects of A_{2A}AR antagonists have been observed in rat models of parkinsonian-like muscular rigidity. Haloperidol and reserpine induce a muscular stiffness that displays electromyographic and mechanographic features that partly overlap with those of parkinsonian muscular rigidity. Both effects are attenuated by the administration of the $A_{2A}AR$ antagonist SCH-58261, suggesting the existence of a potential beneficial effect of $A_{2A}AR$ blockade on parkinsonian-like muscular rigidity (Wardas et al. 2001).

Blockade of $A_{2A}ARs$ effectively counteracts tremulous jaw movements (TJM), a valuable model for the screening of new antitremorigenic agents in rats. Administration of either the $A_{2A}AR$ antagonist SCH-58261 or ST-1535 has been demonstrated to significantly suppress tacrine-induced TJM and, in line with this finding, antagonism of $A_{2A}AR$ by KF-17837 has been reported to relieve TJM elicited by haloperidol, suggesting a beneficial use of these drugs as specific agents against this parkinsonian symptom (Correa et al. 2004; Mally and Stone 1996; Simola et al. 2004). In addition, intracranial infusion of $A_{2A}AR$ antagonists revealed a critical role of the ventrolateral portion of the CPu in counteracting TJM (Simola et al. 2004). Interestingly, a specific increase in $A_{2A}AR$ mRNA expression in this striatal portion was detected following dopamine denervation in the 6-OHDA model of PD (Pinna et al. 2002).

In order to explain the antitremorigenic effect, it should be noted that striatal cholinergic nerve terminals express $A_{2A}ARs$, and $A_{2A}AR$ antagonists can reduce the evoked release of acethylcoline in rat CPu (Kurokawa et al. 1996), whereas increased acetylcholine transmission, particularly in the ventrolateral portion of CPu, is believed to play an important role in the genesis of TJM in rats (Salamone et al. 1998).

5.3 Effects of Chronic $A_{2A}AR$ Antagonism on Motor Complications and Dyskinesia

In line with data obtained following acute administration, long-term treatment with $A_{2A}AR$ antagonists has been shown to significantly counteract motor disabilities in rodent and nonhuman primate PD models (Kanda et al. 1998b; Pinna et al. 2001). Moreover, chronic $A_{2A}AR$ antagonism has been shown not to induce tolerance to motor-stimulant effects in both rats and primates (Halldner et al. 2000; Jenner 2003; Pinna et al. 2001). Lack of tolerance to motor-stimulant effects of $A_{2A}AR$ antagonists is of particular significance in PD, in which the motor-improving properties of therapeutic agents are required to persist during the chronic regimen.

A major finding emerging from studies on chronic $A_{2A}AR$ antagonists is represented by the results reported on motor fluctuations ("wearing off") and dyskinesia in experimental animals treated with $A_{2A}AR$ antagonists and L-DOPA (Koga et al. 2000). The wearing off of L-DOPA that is observed in humans is mimicked in 6-OHDA-lesioned rats, where the duration of rotational behavior elicited by L-DOPA is progressively reduced during chronic administration. Combined administration of the $A_{2A}AR$ antagonist KW-6002 prevented the shortening of rotational behavior, reflecting a potential beneficial influence of $A_{2A}AR$ blockade on L-DOPA wearing off (Koga et al. 2000). At the same time, sensitization of rotational behavior and development of abnormal involuntary movements (AIMs) is thought to mimic dyskinetic effects elicited by L-DOPA. In this paradigm, interesting results concerning the modulation of dyskinesia by A_{2A}AR blockade have been obtained by comparing the rotational behavior elicited by long-term administration of a higher dose of L-DOPA to an equipotent combination of a lower dose of L-DOPA plus the A_{2A}AR antagonist ST-1535 (Rose et al. 2007; Tronci et al. 2007). Although both L-DOPA (high dose) and L-DOPA (lower dose) plus ST-1535 produced a comparable degree of rotations on the first administration, sensitization of rotational behavior and AIMs were observed only in response to chronic L-DOPA alone, not to chronic L-DOPA plus ST-1535, suggesting that the association between the two drugs represents a treatment with low dyskinetic potential (Tronci et al. 2007). These results have been strengthened by studies showing that genetic deletion of the A_{2A}AR prevents the sensitization of rotational behavior stimulated by L-DOPA in 6-OHDA-lesioned A_{2A}AR knockout (KO) mice (Fredduzzi et al. 2002).

Results obtained in MPTP-treated primates confirm and further extend those deriving from 6-OHDA-lesioned rats. First, A_{2A}AR antagonists do not induce dyskinesia per se, since administration of KW-6002 to parkinsonian primates relieved motor disability without stimulating abnormal movements (Grondin et al. 1999; Kanda et al. 1998, 2000). Second, in MPTP-treated marmosets previously exposed to chronic L-DOPA in order to develop dyskinesia, motor stimulation induced by KW-6002 was not associated with an exacerbation of dyskinetic movements (Kanda et al. 1998). Furthermore, no sign of apomorphine-induced dyskinesia was observed in parkinsonian cynomolgus monkeys chronically treated with a combination of apomorphine and KW-6002 (Bibbiani et al. 2003). Interestingly, when KW-6002 (but not apomorphine) administration was interrupted, primates previously treated with KW-6002 displayed apomorphine-induced dyskinesia only 10–12 days after KW-6002 discontinuation, thus accounting for a potential preventive effect of A_{2A}AR blockade on the development of dyskinesia (Bibbiani et al. 2003; Morelli 2003). It should be noted, however, that while A_{2A}AR antagonists associated with a low nondyskinetic dosage of L-DOPA may achieve satisfactory results in motor stimulation, whilst at the same time limiting the severity of L-DOPA-induced dyskinesia, no study has yet demonstrated the ability of A_{2A}AR antagonists to revert an already established dyskinesia in animal models.

In this regard, in MPTP-treated common marmosets previously rendered dyskinetic by chronic L-DOPA, it has been shown that the relief of motor impairment produced by an optimal dose of L-DOPA presenting a high dyskinetic potential was adequately mimicked by a combination of KW-6002 plus a suboptimal dose of L-DOPA, which, in contrast, was associated with weak induction of dyskinesia (Bibbiani et al. 2003).

Taken together, data obtained from several preclinical studies indicate the existence of beneficial effects of chronic A_{2A}AR antagonists on PD motor disability and on motor complications produced by long-term L-DOPA. These effects are of considerable interest in light of the fact that motor complications are one of the intrinsic limitations of L-DOPA therapy, and are often insensitive to pharmacological manipulation.

5.4 *Effects of Acute and Chronic $A_{2A}AR$ Antagonism on Biochemical Parameters*

The study of the effects of $A_{2A}AR$ antagonists on behavioral parameters in both rat and primate models has been paralleled by the analysis of the influence of $A_{2A}AR$ blockade on the biochemical modifications induced by chronic L-DOPA in 6-OHDA lesioned rats in the basal ganglia. Prolonged administration of L-DOPA, according to a regimen capable of inducing a sensitized (dyskinetic-like) rotational response and AIMs, has been shown to modify the expression of the neuropeptides ENK and DYN as well as of the enzyme glutamic acid decarboxylase (GAD67) in the basal ganglia of 6-OHDA-lesioned rats (Carta et al. 2002; Cenci et al. 1998). Although a direct relationship between these biochemical changes and L-DOPA-induced dyskinesia onset has not been unequivocally demonstrated, they have nevertheless been postulated to reflect a more general aberrant functionality of BG produced by long-term L-DOPA, which is thought to underlie the dyskinesia elicited by this drug.

Interestingly, combined administration of low doses of L-DOPA with the $A_{2A}AR$ antagonists SCH-58261 or ST-1535, which (as reported above) induce the same degree of contralateral rotation upon the first administration, did not induce the modifications in the striatal levels of ENK, DYN and GAD67 mRNAs produced by chronic higher doses of L-DOPA in 6-OHDA-lesioned rats (Carta et al. 2002; Tronci et al. 2007).

Moreover, beneficial effects of $A_{2A}AR$ blockade on the regulation of the phosphorylation state of the α-amino-3-hydroxy-5-methyl-4-isoxazolepropionic acid (AMPA) type of glutamate receptor by L-DOPA have been described. Hyperphosphorylation of the striatal AMPA receptor consequent to chronic administration of L-DOPA to 6-OHDA-lesioned rats is in fact prevented by combined administration with KW-6002 (Chase et al. 2003).

In addition to the postulated $A_{2A}AR$ regulatory effects on neuronal responsiveness following prolonged dopaminergic stimuli, it should be considered that $A_{2A}AR$ antagonists, by potentiating the motor effects of L-DOPA or dopamine agonist drugs, allow the use of dopaminomimetic compounds at low nondyskinetic doses. Therefore, the sparing of these agents produced by combined administration with $A_{2A}AR$ antagonists might contribute towards reducing, or at least delaying, the onset of neuroplastic modifications in BG.

5.5 *Biochemical Changes in Extrastriatal Basal Ganglia Areas*

In the context described above, the increase in GAD67 mRNA in the GP of 6-OHDA-lesioned rats treated subchronically with L-DOPA (full effective dose) but not with SCH-58261 plus L-DOPA (threshold dose) is particularly important, indicating that chronic L-DOPA—but not an equally effective combination of SCH-58261 plus L-DOPA—elicits abnormal modifications of GP neuronal activity (Carta et al. 2003).

Moreover, subchronic studies have shown that, while a fully effective dose of L-DOPA reduces the 6-OHDA lesion-induced increase in GAD67 mRNA in SNr, it simultaneously reduces GAD67 mRNA values to below the levels present on the intact side, producing an excessive inhibition of SNr efferent neurons (Carta et al. 2003). In contrast, the combined subchronic administration of SCH-58261 plus L-DOPA reduces GAD67 mRNA to a lesser extent, decreasing GAD67 mRNA to levels similar to those present on the intact nonlesioned side (Carta et al. 2003). Excessive inhibition of SNr in rodents and GP internal segment in primates, together with an altered firing pattern, is correlated with the onset of dyskinetic movements after L-DOPA (Boraud et al. 2001; Papa et al. 1999). Thus, the ability of subchronic SCH-58261 plus L-DOPA to produce a decrease in GAD67 mRNA values to levels similar to those present in nonlesioned SNr may correlate with the presence of contralateral turning (index of therapeutic response) and to the failure to produce sensitization in contralateral turning (index of dyskinetic movements).

The results of those studies underline the importance of the role played by the indirect CPu-GP-STN-SNr pathway in eliciting the therapeutic response of A_{2A}AR receptor antagonists, and its involvement in abnormal motor responses produced by subchronic L-DOPA.

6 Clinical Actions of Adenosine A_{2A}AR Antagonists

The anatomic localization of A_{2A}ARs and their biochemical and pharmacological properties suggest that modulation of striatal GABAergic output will modify motor function in PD, and that this should occur with no risk of the development or expression of dyskinesia (Kase et al. 2003). The activity of A_{2A}ARs in functional models of PD also points to actions of the A_{2A}AR antagonists as monotherapy and as adjuncts to L-DOPA and dopamine agonists. Only one A_{2A}AR antagonist has undergone detailed clinical evaluation so far: istradefylline (KW-6002).

In healthy subjects, istradefylline (40, 60, 80 and 160 mg per day for 14 days) showed dose-proportional increases in the area under the curve (AUC) and a C_{max} with a half-life ($t_{1/2}$) of 67–95 h, suggesting that once-daily dosing should be effective (Rao et al. 2005a). Similar studies in patients with PD showed that istradefylline (60 and 80 mg per day for 14 days) also exhibits a dose-proportional pharmacokinetic profile (Rao et al. 2005b). The occupation of striatal A_{2A}ARs by istradefylline was shown using ^{11}C-istradefylline as a ligand for PET investigations in healthy subjects (Brooks et al. 2008). These studies showed >90% occupation of A_{2A}ARs at doses of istradefylline exceeding 5 mg, while this decreased proportionally at lower doses. From these studies, it was concluded that 20 or 40 mg per day istradefylline would provide consistent A_{2A}AR occupation, and that this would be an appropriate dosage for subsequent clinical investigations.

Some early clinical efficacy studies to establish proof of concept in patients with PD took place prior to the completion of the PET A_{2A}AR imaging investigations, and so these studies utilized higher doses. These involved studies of the effects of istradefylline (40 or 80 mg per day over four weeks) alone or in combination with subsequent steady-state intravenous infusions of L-DOPA using an optimal or low infusion rate (Bara-Jimenez et al. 2003). Perhaps surprisingly, istradefylline alone had no effect on motor disability. This finding contrasts with the mild symptomatic effects of istradephylline seen in MPTP-treated primates, but is more consistent with the absence of significant rotation in 6-OHDA-lesioned rats. The data suggest that the drug would not be effective as monotherapy in the treatment of PD, but there is only one recent report on the efficacy of istradefylline as sustained monotherapy, which was inconclusive (Fernandez et al. 2008).

The results of the effects of istradefylline in conjunction with L-DOPA infusions gave the first indication of the clinical actions of the effect of A_{2A}AR receptor occupation. Istradefylline in conjunction with an optimal L-DOPA infusion had no effect on the severity of motor deficits (Bara-Jimenez et al. 2003). However, when combined with a low dose of L-DOPA, istradefylline (80 mg per day) potentiated the improvement in motor function by 36% while dyskinesia was unchanged. All primary motor symptoms of PD were improved by the addition of istradefylline. Istradefylline also increased the duration of efficacy of L-DOPA by 76%, as judged by the length of time patients remained mobile ("on" time) following cessation of L-DOPA infusion.

These findings are interesting, as they strongly support the results of the preclinical investigations in 6-OHDA-lesioned rats and in MPTP-treated primates, which showed that istradefylline potentiated the effects of low-threshold doses of L-DOPA but that little effect was seen when combined with high effective doses of the drug. The implication is that the optimal clinical effects would therefore be observed under similar conditions, but, as will be seen, the major clinical trials were undertaken in patients receiving optimal administration of dopaminergic therapy for regulatory reasons related to the need to demonstrate efficacy as a decrease in the length of time patients were immobile during the waking day ("off" time) in a group not adequately controlled by currently available medication.

In a 12-week exploratory study of safety and efficacy in advanced PD patients receiving L-DOPA therapy and other dopaminergic agents with both motor fluctuations and peak dose dyskinesia, istradefylline (up to 20 or 40 mg per day) reduced off time by 1.2 h during the waking day in the later stages of the study, as assessed using a home diary, although no change in the unified Parkinson's disease rating scale (UPDRS) scores for motor function or clinical global impression (CGI) of improvement in parkinsonian symptoms was found (Hauser et al. 2003; Hauser and Schwarzschild 2005). This is similar to the reductions produced by the COMT inhibitor entacapone when added to L-DOPA therapy. No overall increase in dyskinesia was observed, but perhaps surprisingly based on the preclinical findings, there was an increase in the amount of on time during which dyskinesia occurred. The overall success of this study then paved the way for a series of longer-term clinical investigations in larger patient populations.

These studies have largely confirmed the effects seen in the initial investigations with istradefylline. In a double-blind multicenter study, in PD patients with prominent end-of-dose wearing off, istradefylline (40 mg per day) reduced off time during the waking day by 1.2 h compared to placebo (LeWitt et al. 2004, 2008; Stacy et al. 2004). There was no increase in dyskinesia that was disabling to the patient, but on time with dyskinesia was increased as a result of an increase in mild dyskinesia that was not troublesome to the patient and did not impair mobility. This was not unexpected on the basis of the earlier clinical studies, but it does conflict with the preclinical data on dyskinesia in MPTP-treated primates, although these studies were largely carried out using low doses of L-DOPA. In another study of istradefylline in PD patients with motor complications using 20 or 60 mg per day istradefylline versus placebo, almost identical findings were obtained except that the decreases in off time were 0.64 and 0.72 h, respectively, for the 20 and 60 mg per day doses, respectively (LeWitt et al. 2004; Stacy et al. 2004, 2008). A long-term open-label efficacy study lasting 52 weeks in advanced-stage PD patients who had previously completed a double-blind placebo-controlled investigation showed that the efficacy of the drug in reducing off time in doses of between 20 and 60 mg per day was maintained in patients who were already taking the drug at the start of the study (Mark et al. 2005). In those patients from the placebo arm of the previous double-blind study who started istradefylline, or those who had been off the drug for more than two weeks and were restarted on the drug, off time was reduced after two weeks and then maintained. The findings of these studies have more or less set the scene for the clinical effects of this A_{2A}AR antagonist in advanced PD patient populations.

However, problems have recently been encountered relating to the efficacy of istradefylline in other Phase III clinical studies, which are probably due to the problem of large and maintained placebo effects in PD and the modest duration of the decrease in off time seen throughout the clinical development. In patients with advanced PD exhibiting motor fluctuations, as defined by an average of at least 3 h off time, 20 mg per day istradefylline reduced the off times at two and four weeks but not at eight or twelve weeks (Hauser et al. 2006, 2008; Shulman et al. 2006; Trugman et al. 2006), although the effect was significant at the end-point (determined by the last observation carried forward, LOCF), with a 0.73 h reduction in off time. An analysis of secondary end-points showed a reduction in UPDRS Part 3 for motor symptoms at four weeks, a trend at two and eight weeks, and no effect at twelve weeks. Similarly, in patients with PD showing motor complications that were not adequately controlled by L-DOPA, istradefylline (10, 20 or 40 mg per day) did not decrease off time compared to a larger than expected placebo effect, although a trend for the improvement in response to increase with increasing istradefylline dosage (a dose-ordered response) was observed between the istradefylline-treated groups (Guttman et al. 2006; Pourcher et al. 2006). The results from these studies have led the FDA to issue a nonapprovable letter for the use of istradefylline in late-stage PD.

Since istradefylline is the only A_{2A}AR antagonist with results from clinical trials for PD reported to date, it is difficult to know whether the profile seen with this drug is typical of this class of drugs, or whether the design of the clinical trials in line with regulatory end-points will provide further insights into the efficacy of this class of drugs for PD. A number of other A_{2A}AR antagonists are in clinical trials at this time, such as V2006 and SCH-58261, and the results of these studies are eagerly awaited. Based on its preclinical profile, istradefylline would have been expected to have some modest symptomatic effects as a monotherapy, but this needs further investigation. Moreover, based on preclinical investigations, istradefylline should produce an additive effect with L-DOPA, but perhaps the necessity of undertaking the clinical studies in patients on optimal dopaminergic medication has masked its ability to potentiate the effects of low-threshold doses of L-DOPA, an effect that was clearly demonstrated in preclinical studies. Thus, the design of clinical trials for istradefylline with this in mind may have provided a different outcome.

7 Future Directions

7.1 Effects on Cognition

Clinical evidence demonstrates the occurrence of cognitive impairments irrespective of motor disability in parkinsonian patients, including both overt dementia during later stages of the disease and less marked deficits displayed by the majority of subjects during the early stages. PD-associated cognitive symptoms involve abnormalities in visuospatial performance and memory deficits, with both short- and long-term memory being affected. Alterations in organization, planning, regulation of goal-directed behaviors and information retrieval and attention are widely observed in PD patients and are key events triggering the manifestations of PD-associated cognitive decline (Appollonio et al. 1994).

L-DOPA has been found to exert contradictory effects, if any, on cognitive deficits in PD, improving several symptoms whilst worsening others. Thus, the development of new therapeutic options currently constitutes an important requirement in the treatment of cognitive decline observed in PD, and A_{2A}AR antagonists may represent a valid option. Several data obtained in experimental animals have evidenced how counteracting A_{2A}AR-mediated signaling by drugs or genetic deletion of the gene encoding for the A_{2A}AR may significantly improve cognitive functions, whereas working memory deficits have been demonstrated in rats overexpressing the A_{2A}AR (Giménez-Llort et al. 2007; Wang et al. 2006). Moreover, studies employing the A_{2A}AR antagonists KW-6002 and SCH-412348 have revealed how A_{2A}AR blockade exerts beneficial effects on cognition-related functions other than memory, enhancing both motivation and attention, facilitating reward-related behaviors, increasing motor readiness, and speeding up motor-preparatory responses (O'Neill and Brown 2006; Takahashi et al. 2008).

Several authors have hypothesized how a defective functionality of the frontostriatal dopaminergic circuit connecting the CPu to the frontal cortex contributes towards cognitive deficits associated with PD (Gao and Goldman-Rakic 2003; Kulisevsky et al. 2000). A_{2A}ARs are particularly abundant in the CPu, and are also (although to a lesser extent) expressed in the frontal cortex (Rosin et al. 2003).

Hence, by facilitating dopamine receptor-mediated effects, A_{2A}AR antagonists may boost neurotransmission at the level of the frontostriatal circuit, eventually exerting a positive influence on parkinsonian cognitive deficits. Moreover, in addition to the modulation of dopaminergic transmission by A_{2A}ARs, cholinergic system functioning may also be affected. Interestingly, the A_{2A}AR antagonist SCH-58261 has been found to increase acetylcholine release in rat frontal cortex (Acquas et al. 2002). The latter finding may be potentially relevant to the treatment of cognitive deficits in PD, suggesting the potential ability of A_{2A}AR antagonism to modify hypofunctionality of the frontal cortex cholinergic system, implicated to some extent in cognitive decline in PD, an effect which may contribute towards improving this specific symptom of PD. These results do not exclude a potential role of adenosine A_1 receptor in contrasting cognitive decline in PD (Mihara et al. 2007).

7.2 Neuroprotective Potential

One of the major limitations of the current pharmacological treatment of PD is represented by its substantial ineffectiveness in counteracting the degeneration of dopaminergic neurons, which underlies this condition. In this regard, it has recently been emphasized that the blockade of adenosine A_{2A}ARs may potentially represent a valuable approach in counteracting neuronal death in PD (Chen et al. 2007).

Neuroprotective effects have been obtained in different PD animal models by drug administration or in A_{2A}AR KO mice. In the MPTP mouse model, blockade of A_{2A}ARs by either SCH-58261 or KW-6002 or deletion of the gene encoding for the A_{2A}AR has been shown to substantially reduce both the demise of dopaminergic nigral neurons and the fall in striatal dopamine concentration elicited by MPTP administration (Chen et al. 2001; Ikeda et al. 2002; Pierri et al. 2005).

Despite the fact that neuroprotection elicited by A_{2A}AR antagonists in PD animal models is clearly manifested, the neuronal mechanisms underlying this effect have not yet been ascertained, although they would seem to differ from those mediating the motor-stimulating effects of these agents.

An abnormal increase in glutamate outflow may be implicated in triggering the demise of dopaminergic neurons observed in PD, and so an involvement of glutamate in A_{2A}AR blockade-mediated neuroprotection has been suggested, since A_{2A}ARs located presynaptically on glutamatergic terminals control glutamate release in a negative way (Cunha 2001; Popoli et al. 2002). It should nevertheless be taken into account that mechanisms other than that regulating glutamate release may be involved in the neuroprotection mediated by A_{2A}AR blockade, in view of the modulation by A_{2A}ARs of a large number of brain functions. The ability of A_{2A}ARs to modulate the activity of non-neuronal cell types (e.g., microglia or astroglia) is of particular interest to this regard, in view of the crucial role played by glia-mediated neuroinflammation in PD. Therefore, interference with glial-released neurotoxic factors might confer protective properties on these agents as well, leading to the compelling possibility that a unique broad mechanism might subserve A_{2A}AR-mediated neuroprotection in diverse neurodegenerative pathologies (Kust et al. 1999; Nishizaki et al. 2002).

To date no clinical studies have been carried out to investigate potential neuroprotective effects on the dopaminergic system following the administration of A_{2A}AR antagonists. However, epidemiological studies have demonstrated how the incidence of idiopathic PD negatively correlates with caffeine intake, being significantly lower in individuals that regularly consume caffeine throughout their lifetime (Ascherio et al. 2001).

Therefore, direct evidence of neuroprotection mediated by A_{2A}AR antagonists in experimental animals, as well as data from epidemiological studies, provide new insights into the study of the antiparkinsonian potential of these drugs. It can therefore be postulated that A_{2A}AR antagonists may not only relieve motor deficits in established PD but may also potentially prevent the the pathology from progressing by arresting the degeneration of dopaminergic mesencephalic neurons.

8 Conclusions

Although the neuroprotective and symptomatic effects of A_{2A}AR antagonists on parkinsonian neuronal demise appear to be most promising, it should be noted that (i) by acting on A_{2A} ARs to produce vasodilation, adenosine affects oxygen supply:demand, (ii) by acting on A_{2A}ARs on inflammatory cells, adenosine produces anti-inflammatory responses, and (iii) by acting on A_{2A}ARs on endothelial cells, adenosine decreases endothelial permeability. Therefore, blockade of A_{2A}ARs may produce adverse effects in regions other than the brain, such as the heart, kidney, lung and inflammatory responses in general. For more information on A_{2A}ARs in other organs, please refer to other chapters in this volume, such as those focusing on adenosine receptors and the kidney (Chap. 15), heart (Chaps. 6 and 7), asthma (Chap. 11), and inflammation (Chap. 8). As a consequence, more detailed studies should be undertaken in the future in both experimental animals and humans to clarify whether (and under which specific conditions) A_{2A}AR antagonists may be used as safe and effective agents in the treatment of PD.

References

Acquas E, Tanda G, Di Chiara G (2002) Differential effects of caffeine on dopamine and acetylcholine transmission in brain areas of drug-naive and caffeine-pretreated rats. Neuropsychopharmacology 27:182–193

Agid Y (1991) Parkinson's disease: pathophysiology. Lancet 337:1321–1324

Ahlskog JE (2007) I can't get no satisfaction: still no neuroprotection for Parkinson disease. Neurology 69:1476–1477

Appollonio I, Grafman J, Clark K, Nichelli P, Zeffiro T, Hallett M (1994) Implicit and explicit memory in patients with Parkinson's disease with and without dementia. Arch Neurol 51:359–367

Ascherio A, Zhang SM, Hernán MA, Kawachi I, Colditz GA, Speizer FE, Willett WC (2001) Prospective study of caffeine consumption and risk of Parkinson's disease in men and women. Ann Neurol 50:56–63

Bara-Jimenez W, Sherzai A, Dimitrova T, Favit A, Bibbiani F, Gillespie M, Morris MJ, Mouradian MM, Chase TN (2003) Adenosine A(2A) receptor antagonist treatment of Parkinson's disease. Neurology 61:293–296

Becker C, Jick SS, Meier CR (2008) Use of antihypertensives and the risk of Parkinson disease. Neurology 70:1438–1444

Berg D (2006) Marker for a preclinical diagnosis of Parkinson's disease as a basis for neuroprotection. J Neural Transm Suppl:123–132

Bibbiani F, Oh JD, Petzer JP, Castagnoli N Jr, Chen JF, Schwarzschild MA (2003) A_{2A} antagonist prevents dopamine agonist-induced motor complications in animal models of Parkinson's disease. Exp Neurol 184:285–294

Boraud T, Bezard E, Bioulac B, Gross CE (2001) Dopamine agonist-induced dyskinesias are correlated to both firing pattern and frequency alterations of pallidal neurones in the MPTP-treated monkey. Brain 124:546–557

Bornebroek M, de Lau LM, Haag MD, Koudstaal PJ, Hofman A, Stricker BH, Breteler MM (2007) Nonsteroidal anti-inflammatory drugs and the risk of Parkinson disease. Neuroepidemiology 28:193–196

Braak H, Bohl JR, Muller CM, Rub U, de Vos RA, Del TK (2006a) Stanley Fahn Lecture 2005: the staging procedure for the inclusion body pathology associated with sporadic Parkinson's disease reconsidered. Mov Disord 21:2042–2051

Braak H, Muller CM, Rub U, Ackermann H, Bratzke H, de Vos RA, Del TK (2006b) Pathology associated with sporadic Parkinson's disease: where does it end? J Neural Transm Suppl 89–97

Braak H, Del TK (2008) Invited article: Nervous system pathology in sporadic Parkinson disease. Neurology 70:1916–1925

Brooks DJ, Doder M, Osman S, Luthra SK, Gunn R, Hirani E, Hume S, Kase H, Kilborn J, Martindill S, Mori A (2008) Positron emission tomography analysis of [11C]KW-6002 binding to human and rat adenosine A_{2A} receptors in the brain. Synapse 62:671–681

Calabresi P, Picconi B, Parnetti L, Di FM (2006) A convergent model for cognitive dysfunctions in Parkinson's disease: the critical dopamine–acetylcholine synaptic balance. Lancet Neurol 5:974–983

Calabresi P, Picconi B, Tozzi A, Di FM (2007) Dopamine-mediated regulation of corticostriatal synaptic plasticity. Trends Neurosci 30:211–219

Carta AR, Pinna A, Cauli O, Morelli M (2002) Differential regulation of GAD67, enkephalin and dynorphin mRNAs by chronic-intermittent L-dopa and A_{2A} receptor blockade plus L-dopa in dopamine-denervated rats. Synapse 44:166–174

Carta AR, Tabrizi MA, Baraldi PG, Pinna A, Pala P, Morelli M (2003) Blockade of A_{2A} receptors plus L-DOPA after nigrostriatal lesion results in GAD67 mRNA changes different from L-DOPA alone in the rat globus pallidus and substantia nigra reticulata. Exp Neurol 184:679–87

Cenci MA, Lee CS, Bjorklund A (1998) L-DOPA-induced dyskinesia in the rat is associated with striatal overexpression of prodynorphin- and glutamic acid decarboxylase mRNA. Eur J Neurosci 10:2694–706

Chase TN, Bibbiani F, Bara-Jimenez W, Dimitrova T, Oh-Lee JD (2003) Translating A_{2A} antagonist KW6002 from animal models to parkinsonian patients. Neurology 61:S107–11

Chaudhuri KR, Yates L, Martinez-Martin P (2005) The non-motor symptom complex of Parkinson's disease: a comprehensive assessment is essential. Curr Neurol Neurosci Rep 5:275–283

Chen JF, Xu K, Petzer JP, Staal R, Xu YH, Beilstein M, Sonsalla PK, Castagnoli K, Castagnoli N Jr, Schwarzschild MA (2001) Neuroprotection by caffeine and A(2A) adenosine receptor inactivation in a model of Parkinson's disease. J Neurosci 21:RC143

Chen JF, Sonsalla PK, Pedata F, Melani A, Domenici MR, Popoli P, Geiger J, Lopes LV, de Mendonça A (2007) Adenosine A_{2A} receptors and brain injury: broad spectrum of neuroprotection, multifaceted actions and fine tuning modulation. Prog Neurobiol 83:310–331

Coccurello R, Breysse N, Amalric M (2004) Simultaneous blockade of adenosine A_{2A} and metabotropic glutamate mGlu5 receptors increase their efficacy in reversing Parkinsonian deficits in rats. Neuropsychopharmacology 29:1451–61

Correa M, Wisniecki A, Betz A, Dobson DR, O'Neil MF, O'Neil MJ, Salamone JD (2004) The adenosine A_{2A} antagonist KF17837 reverses the locomotor suppression and tremulous jaw movements induced by haloperidol in rats: possible relevance to parkinsonism. Behav Brain Res 148:47–54

Cunha RA (2001) Adenosine as a neuromodulator and as a homeostatic regulator in the nervous system: different roles, different sources and different receptors. Neurochem Int 38:107–125

Dorsey ER, Constantinescu R, Thompson JP, Biglan KM, Holloway RG, Kieburtz K, Marshall FJ, Ravina BM, Schifitto G, Siderowf A, Tanner CM (2007) Projected number of people with Parkinson disease in the most populous nations, 2005 through 2030. Neurology 68:384–386

Esposito E, Di MV, Benigno A, Pierucci M, Crescimanno G, Di GG (2007) Non-steroidal anti-inflammatory drugs in Parkinson's disease. Exp Neurol 205:295–312

Fabbrini G, Brotchie JM, Grandas F, Nomoto M, Goetz CG (2007) Levodopa-induced dyskinesias. Mov Disord 22:1379–1389

Fahn S, Jankovic J (2007) Principles and practice of movement disorders. Elsevier, Philadelphia, pp 1–652

Fernandez HH, 6002-US-051 Clinical Investigator Group (2008) The safety and efficacy of istradefylline, an adenosine A_{2A} antagonist, as monotherapy in Parkinson's Disease: results of the KW-6002-US-051 trial. Presented at the 12th International Congress of Parkinson's Disease and Movement Disorders, Chicago. Mov Disord 23(Suppl 1):S87

Ferré S, Karcz-Kubicha M, Hope BT, Popoli P, Burgueño J, Gutiérrez MA, Casadó V, Fuxe K, Goldberg SR, Lluis C, Franco R, Ciruela F (2002) Synergistic interaction between adenosine A_{2A} and glutamate mGlu5 receptors: implications for striatal neuronal function. Proc Natl Acad Sci USA 99:11940–11945

Fink JS, Weaver DR, Rivkees SA, Peterfreund RA, Pollack AE, Adler EM, Reppert SM (1992) Molecular cloning of the rat A_2 adenosine receptor: selective co-expression with D_2 dopamine receptors in rat striatum. Brain Res Mol Brain Res 14:186–195

Fredduzzi S, Moratalla R, Monopoli A, Cuellar B, Xu K, Ongini E (2002) Persistent behavioral sensitization to chronic L-DOPA requires A_{2A} adenosine receptors. J Neurosci 22:1054–1062

Fredholm BB (1995) Purinoceptors in the nervous system. Pharmacol Toxicol 76:228–239

Fredholm BB, Abbracchio MP, Burnstock G, Daly JW, Harden TK, Jacobson KA, Leff P, Williams M (1994) Nomenclature and classification of purinoceptors. Pharmacol Rev 46:143–156

Fuxe K, Ferré S, Canals M, Torvinen M, Terasmaa A, Marcellino D, Goldberg SR, Staines W, Jacobsen KX, Lluis C, Woods AS, Agnati LF, Franco R (2005) Adenosine A_{2A} and dopamine D_2 heteromeric receptor complexes and their function. J Mol Neurosci 26:209–220

Fuxe K, Marcellino D, Genedani S, Agnati L (2007) Adenosine A(2A) receptors, dopamine D(2) receptors and their interactions in Parkinson's disease. Mov Disord 22:1990–1917

Galvan A, Floran B, Erlij D, Aceves J (2001) Intrapallidal dopamine restores motor deficits induced by 6-hydroxydopamine in the rat. J Neural Transm 108:153–166

Gao WJ, Goldman-Rakic PS (2003) Selective modulation of excitatory and inhibitory microcircuits by dopamine. Proc Natl Acad Sci USA 100:2836–2841

Gasser T (2007) Update on the genetics of Parkinson's disease. Mov Disord 22(Suppl 17):S343–S350

Gerfen CR (1992) The neostriatal mosaic: multiple levels of compartmental organization. Trends Neurosci 15:133–139

Giménez-Llort L, Schiffmann SN, Shmidt T, Canela L, Camón L, Wassholm M, Canals M, Terasmaa A, Fernández-Teruel A, Tobeña A, Popova E, Ferré S, Agnati L, Ciruela F, Martínez E, Scheel-Kruger J, Lluis C, Franco R, Fuxe K, Bader M (2007) Working memory deficits in transgenic rats overexpressing human adenosine A_{2A} receptors in the brain. Neurobiol Learn Mem 87:42–56

Grondin R, Bedard PJ, Hadj Tahar A, Gregoire L, Mori A, Kase H (1999) Antiparkinsonian effect of a new selective adenosine A_{2A} receptor antagonist in MPTP-treated monkeys. Neurology 52:1673–1677

Guridi J, Obeso JA, Rodriguez-Oroz MC, Lozano AA, Manrique M (2008) L-Dopa-induced dyskinesia and stereotactic surgery for Parkinson's disease. Neurosurgery 62:311–23

Guttman M, 6002-US-018 Clinical Investigator Group (2006) Efficacy of istradephylline in Parkinson's disease patients treated with levodopa with motor response complications: results of the KW-6002-US-018 study. Presented at the 10th International Congress of Parkinson's Disease and Movement Disorders, Kyoto. Mov Disord 21(Suppl. 15):S585

Halldner L, Lozza G, Lindström K, Fredholm BB (2000) Lack of tolerance to motor stimulant effects of a selective adenosine A(2A) receptor antagonist. Eur J Pharmacol 406:345–354

Hardy J, Cai H, Cookson MR, Gwinn-Hardy K, Singleton A (2006) Genetics of Parkinson's disease and parkinsonism. Ann Neurol 60:389–398

Hauber W (1998) Involvement of basal ganglia transmitter systems in movement initiation. Prog Neurobiol 56:507–540

Hauber W, Lutz S (1999) Dopamine D_1 or D_2 receptor blockade in the globus pallidus produces akinesia in the rat. Behav Brain Res 106:143–150

Hauber W, Neuscheler P, Nagel J, Muller CE (2001) Catalepsy induced by a blockade of dopamine D_1 or D_2 receptors was reversed by a concomitant blockade of adenosine A(2A) receptors in the caudate-putamen of rats. Eur J Neurosci 14:1287–93

Hauser RA, Hubble JP, Truong DD (2003) Randomized trial of the adenosine A(2A) receptor antagonist istradefylline in advanced PD. Neurology 61:297–303

Hauser RA, Schwarzschild MA (2005). Adenosine A_{2A} receptor antagonists for Parkinson's disease: rationale, therapeutic potential and clinical experience. Drugs Aging 22:471–482

Hauser R, 6002-US-013 Clinical Investigator Group (2006) Effects of istradephylline (KW-6002) in levodopa treated Parkinson's disease patients with motor response complications: secondary efficacy results of the KW-6002-US-013 study. Presented at the 10th International Congress of Parkinson's Disease and Movement Disorders, Kyoto. Mov Disord 21(Suppl 15):S510

Hauser RA, Shulman LM, Trugman JM, Roberts J, Mori A, Ballerini R, Sussman NM (2008) Study of istradefylline in patients with Parkinson's disease on levodopa with motor fluctuations. Mov Disord 23:2177–2185

Hillion J, Canals M, Torvinen M, Casado V, Scott R, Terasmaa A, Hansson A, Watson S, Olah ME, Mallol J, Canela EI, Zoli M, Agnati LF, Ibanez CF, Lluis C, Franco R, Ferre S, Fuxe K (2002) Coaggregation, cointernalization, and codesensitization of adenosine A_{2A} receptors and dopamine D_2 receptors. J Biol Chem 277:18091–18097

Horstink M, Tolosa E, Bonuccelli U, Deuschl G, Friedman A, Kanovsky P, Larsen JP, Lees A, Oertel W, Poewe W, Rascol O, Sampaio C (2006a) Review of the therapeutic management of Parkinson's disease. Report of a joint task force of the European Federation of Neurological Societies and the Movement Disorder Society–European Section. Part I: early (uncomplicated) Parkinson's disease. Eur J Neurol 13:1170–1185

Horstink M, Tolosa E, Bonuccelli U, Deuschl G, Friedman A, Kanovsky P, Larsen JP, Lees A, Oertel W, Poewe W, Rascol O, Sampaio C (2006b) Review of the therapeutic management of Parkinson's disease. Report of a joint task force of the European Federation of Neurological Societies (EFNS) and the Movement Disorder Society–European Section (MDS-ES). Part II: late (complicated) Parkinson's disease. Eur J Neurol 13:1186–1202

Hu G, Bidel S, Jousilahti P, Antikainen R, Tuomilehto J (2007) Coffee and tea consumption and the risk of Parkinson's disease. Mov Disord 22:2242–2248

Hung AY, Schwarzschild MA (2007) Clinical trials for neuroprotection in Parkinson's disease: overcoming angst and futility? Curr Opin Neurol 20:477–483

Ikeda K, Kurokawa M, Aoyama S, Kuwana Y (2002) Neuroprotection by adenosine A_{2A} receptor blockade in experimental models of Parkinson's disease. J Neurochem 80:262–270

Jankovic J (2005) Motor fluctuations and dyskinesias in Parkinson's disease: clinical manifestations. Mov Disord 20(Suppl 11):S11–S16

Jankovic J (2006) An update on the treatment of Parkinson's disease. Mt Sinai J Med 73:682–689

Jankovic J, Stacy M (2007) Medical management of levodopa-associated motor complications in patients with Parkinson's disease. CNS Drugs 21:677–692

Jellinger KA (2002) Recent developments in the pathology of Parkinson's disease. J Neural Transm Suppl 347–376

Jenner P (2003) A_{2A} antagonists as novel non-dopaminergic therapy for motor dysfunction in PD. Neurology 61:S32–S38

Jenner P, Olanow CW (2006) The pathogenesis of cell death in Parkinson's disease. Neurology 66:S24–S36

Kachroo A, Orlando LR, Grandy DK, Chen JF, Young AB, Schwarzschild MA (2005) Interactions between metabotropic glutamate 5 and adenosine A_{2A} receptors in normal and parkinsonian mice. J Neurosci 25:10414–19

Kanda T, Shiozaki S, Shimada J, Suzuki F, Nakamura J (1994) KF17837: a novel selective adenosine A_{2A} receptor antagonist with anticataleptic activity. Eur J Pharmacol 256:263–268

Kanda T, Tashiro T, Kuwana Y, Jenner P. (1998a) Adenosine A_{2A} receptors modify motor function in MPTP-treated common marmosets. Neuroreport 9:2857–2860

Kanda T, Jackson MJ, Smith LA, Pearce RK, Nakamura J, Kase H, Kuwana Y, Jenner P (1998b) Adenosine A_{2A} antagonist: a novel antiparkinsonian agent that does not provoke dyskinesia in parkinsonian monkeys. Ann Neurol 43:507–513

Kanda T, Jackson MJ, Smith LA, Pearce RK, Nakamura J, Kase H (2000) Combined use of the adenosine A(2A) antagonist KW-6002 with L-DOPA or with selective D_1 or D_2 dopamine agonists increases antiparkinsonian activity but not dyskinesia in MPTP-treated monkeys. Exp Neurol 162:321–327

Kase H, Aoyama S, Ichimura M, Ikeda K, Ishii A, Kanda T, Koga K, Koike N, Kurokawa M, Kuwana Y, Mori A, Nakamura J, Nonaka H, Ochi M, Saki M, Shimada J, Shindou T, Shiozaki S, Suzuki F, Takeda M, Yanagawa K, Richardson PJ, Jenner P, Bedard P, Borrelli E, Hauser RA, Chase TN (2003) Progress in pursuit of therapeutic A_{2A} antagonists: the adenosine A_{2A} receptor selective antagonist KW6002: research and development toward a novel nondopaminergic therapy for Parkinson's disease. Neurology 61:S97–S100

Kieburtz K, Ravina B (2007) Why hasn't neuroprotection worked in Parkinson's disease? Nat Clin Pract Neurol 3:240–241

Koga K, Kurokawa M, Ochi M, Nakamura J, Kuwana Y (2000) Adenosine A(2A) receptor antagonists KF17837 and KW-6002 potentiate rotation induced by dopaminergic drugs in hemi-Parkinsonian rats. Eur J Pharmacol 408:249–255

Kulisevsky J, García-Sánchez C, Berthier ML, Barbanoj M, Pascual-Sedano B, Gironell A, Estévez-González A (2000) Chronic effects of dopaminergic replacement on cognitive function in Parkinson's disease: a two-year follow-up study of previously untreated patients. Mov Disord 15(4):613–626

Kurokawa M, Koga K, Kase H, Nakamura J, Kuwana Y (1996) Adenosine A_{2A} receptor-mediated modulation of striatal acetylcholine release in vivo. J Neurochem 66:1882–1888

Küst BM, Biber K, van Calker D, Gebicke-Haerter PJ (1999) Regulation of K^+ channel mRNA expression by stimulation of adenosine A_{2A}-receptors in cultured rat microglia. Glia 25:120–130

LeWitt PA (2006) Neuroprotection for Parkinson's disease. J Neural Transm Suppl:113–122

LeWitt PA, 6002-US-005/6002-US-006 Clinical Investigator Group (2004) OFF time reduction from adjunctive use of istradephylline (KW-6002) in levodopa-treated patients with advanced Parkinson's disease. Presented at 8th International Congress of Parkinson's Disease and Movement Disorders, Rome. Mov Disord 19(Suppl 9):S222

LeWitt PA, Guttman M, Tetrud JW, Tuite PJ, Mori A, Chaikin P, Sussman NM (2008) Adenosine A_{2A} receptor antagonist istradefylline (KW-6002) reduces OFF time in Parkinson's disease: a double-blind, randomized, multicenter clinical trial (6002-US-005). Ann Neurol 63:295–302

Litvan I, Chesselet MF, Gasser T, Di Monte DA, Parker D Jr, Hagg T, Hardy J, Jenner P, Myers RH, Price D, Hallett M, Langston WJ, Lang AE, Halliday G, Rocca W, Duyckaerts C, Dickson DW, Ben-Shlomo Y, Goetz CG, Melamed E (2007a) The etiopathogenesis of Parkinson disease and suggestions for future research. Part I. J Neuropathol Exp Neurol 66:251–257

Litvan I, Halliday G, Hallett M, Goetz CG, Rocca W, Duyckaerts C, Ben-Shlomo Y, Dickson DW, Lang AE, Chesselet MF, Langston WJ, Di Monte DA, Gasser T, Hagg T, Hardy J, Jenner P, Melamed E, Myers RH, Parker D Jr, Price DL (2007b) The etiopathogenesis of Parkinson disease and suggestions for future research. Part II. J Neuropathol Exp Neurol 66:329–336

Mally J, Stone TW (1996) Potential role of adenosine antagonist therapy in pathological tremor disorders. Pharmacol Ther 72:243–250

Mark MH, 6002-US-007 Clinical Investigator Group (2005) Long-term efficacy of istradephylline in patients with advanced Parkinson's disease. Presented at 9th International Congress of Parkinson's disease and Movement Disorders, New Orleans. Mov Disord 20(Suppl 10):S93

McCulloch CC, Kay DM, Factor SA, Samii A, Nutt JG, Higgins DS, Griffith A, Roberts JW, Leis BC, Montimurro JS, Zabetian CP, Payami H (2008) Exploring gene–environment interactions in Parkinson's disease. Hum Genet 123:257–265

McGeer PL, McGeer EG (2008) Glial reactions in Parkinson's disease. Mov Disord 23:474–483

Mihara T, Mihara K, Yarimizu J, Mitani Y, Matsuda R, Yamamoto H, Aoki S, Akahane A, Iwashita A, Matsuoka N (2007) Pharmacological characterization of a novel, potent adenosine A_1 and A_{2A} receptor dual antagonist, 5-[5-amino-3-(4-fluorophenyl)pyrazin-2-yl]-1-isopropylpyridine-2(1H)-one (ASP5854), in models of Parkinson's disease and cognition.J Pharmacol Exp Ther 323:708–719

Morelli M (2003) Adenosine A_{2A} antagonists: potential preventive and palliative treatment for Parkinson's disease. Exp Neurol 184:20–23

Morelli M, Fenu S, Pinna A, Di Chiara G (1994) Adenosine A_2 receptors interact negatively with dopamine D_1 and D_2 receptors in unilaterally 6-hydroxydopamine-lesioned rats. Eur J Pharmacol 251:21–25

Nishi A, Liu F, Matsuyama S, Hamada M, Higashi H, Nairn AC, Greengard P (2003) Metabotropic mGlu5 receptors regulate adenosine A_{2A} receptor signaling. Proc Natl Acad Sci USA 100:1322–1327

Nishizaki T, Nagai K, Nomura T, Tada H, Kanno T, Tozaki H, Li XX, Kondoh T, Kodama N, Takahashi E, Sakai N, Tanaka K, Saito N (2002) A new neuromodulatory pathway with a glial contribution mediated via A(2a) adenosine receptors. Glia 39:133–147

Obeso JA, Rodriguez-Oroz MC, Rodriguez M, Lanciego JL, Artieda J, Gonzalo N, Olanow CW (2000) Pathophysiology of the basal ganglia in Parkinson's disease. Trends Neurosci 23:S8–S19

Obeso JA, Rodriguez-Oroz M, Marin C, Alonso F, Zamarbide I, Lanciego JL, Rodriguez-Diaz M (2004) The origin of motor fluctuations in Parkinson's disease: importance of dopaminergic innervation and basal ganglia circuits. Neurology 62:S17–S30

Obeso JA, Rodriguez-Oroz MC, Javier Blesa F, Guridi J (2006) The globus pallidus pars externa and Parkinson's disease. Ready for prime time? Exp Neurol 202:1–7

Ochi M, Shiozaki S, Kase H (2004) Adenosine A(2A) receptor-mediated modulation of GABA and glutamate release in the output regions of the basal ganglia in a rodent model of Parkinson's disease. Neuroscience 127:223–231

O'Neill M, Brown VJ (2006) The effect of the adenosine A(2A) antagonist KW-6002 on motor and motivational processes in the rat. Psychopharmacology 184:46–55

Papa SM, Desimone R, Fiorani M, Oldfield EH (1999) Internal globus pallidus discharge is nearly suppressed during levodopa-induced dyskinesias. Ann Neurol 46:732–738

Pierri M, Vaudano E, Sager T, Englund U (2005) KW-6002 protects from MPTP induced dopaminergic toxicity in the mouse. Neuropharmacology 48:517–524

Pinna A, di Chiara G, Wardas J, Morelli M (1996) Blockade of A_{2A} adenosine receptors positively modulates turning behaviour and c-Fos expression induced by D_1 agonists in dopamine-denervated rats. Eur J Neurosci 8:1176–1181

Pinna A, Fenu S, Morelli M (2001) Motor stimulants effects of the adenosine A_{2A} receptor antagonist SCH 58261 do not develop tolerance after repeated treatments in 6-hydroxydopamine lesioned rats. Synapse 39:233–239

Pinna A, Corsi C, Carta AR, Valentini V, Pedata F, Morelli M (2002) Modification of adenosine extracellular levels and adenosine A_{2A} receptor mRNA by dopamine denervation. Eur J Pharmacol 446:75–82

Pinna A, Pontis S, Morelli M (2007) Adenosine A_{2A} receptor antagonists improve deficits in initiation of movement and sensory motor integration in the unilateral 6-hydroxydopamine rat model of Parkinson's disease. Synapse 61:606–614

Pollack AE, Fink JS (1996) Synergistic interaction between an adenosine antagonist and a D_1 dopamine agonist on rotational behavior and striatal c-Fos induction in 6-hydroxydopamine-lesioned rats. Brain Res 743:124–130

Popoli P, Pintor A, Domenici MR, Frank C, Tebano MT, Pèzzola A, Scarchilli L, Quarta D, Reggio R, Malchiodi-Albedi F, Falchi M, Massotti M (2002) Blockade of striatal adenosine A_{2A} receptor reduces, through a presynaptic mechanism, quinolinic acid-induced excitotoxicity: possible relevance to neuroprotective interventions in neurodegenerative diseases of the striatum. J Neurosci 22:1967–1975

Pourcher E, 6002-US-018 Clinical Investigator Group (2006) Safety and tolerability of istradephylline (KW-6002) in Parkinson's disease with motor response complications: results of the KW-6002-US-018 study. Presented at the 10th International Congress of Parkinson's Disease and Movement Disorders, Kyoto. Mov Disord 21(Suppl 15):S508

Powers KM, Kay DM, Factor SA, Zabetian CP, Higgins DS, Samii A, Nutt JG, Griffith A, Leis B, Roberts JW, Martinez ED, Montimurro JS, Checkoway H, Payami H (2008) Combined effects of smoking, coffee, and NSAIDs on Parkinson's disease risk. Mov Disord 23:88–95

Rao N, Uchimura T, Mori A (2005a) Evaluation of safety, tolerability, and multiple dose pharmacokinetics of istradephylline in healthy subjects. Clin Pharmacol Ther 83(Suppl):PIII-89

Rao N, Uchimura T, Mori A (2005b) Evaluation of safety, tolerability, and multiple dose pharmacokinetics of istradephylline in Parkinson's disease patients. Clin Pharmacol Ther 83(Suppl):PIII-88

Ritz B, Ascherio A, Checkoway H, Marder KS, Nelson LM, Rocca WA, Ross GW, Strickland D, Van Den Eeden SK, Gorell J (2007) Pooled analysis of tobacco use and risk of Parkinson disease. Arch Neurol 64:990–997

Rodrigues RJ, Alfaro TM, Rebola N, Oliveira CR, Cunha RA (2005) Co-localization and functional interaction between adenosine A(2A) and metabotropic group 5 receptors in glutamatergic nerve terminals of the rat striatum. J Neurochem 92:433–441

Rose S, Ramsay Croft N, Jenner P. The novel adenosine A_{2A} antagonist (2007) ST1535 potentiates the effects of a threshold dose of L-dopa in unilaterally 6-OHDA-lesioned rats. Brain Res 1133:110–114

Rosin DL, Robeva A, Woodard RL, Guyenet PG, Linden J (1998) Immunohistochemical localization of adenosine A_{2A} receptors in the rat central nervous system. J Comp Neurol 401:163–186

Rosin DL, Hettinger BD, Lee A, Linden J (2003) Anatomy of adenosine A_{2A} receptors in brain: morphological substrates for integration of striatal function. Neurology 61:S12–S18

Salamone JD, Mayorga AJ, Trevitt JT, Cousins MS, Conlan A, Nawab A (1998) Tremulous jaw movements in rats: a model of parkinsonian tremor. Prog Neurobiol 56:591–611

Schapira AH (2008) Progress in neuroprotection in Parkinson's disease. Eur J Neurol 15(Suppl 1):5–13

Schiffmann SN, Jacobs O, Vanderhaeghen JJ (1991) Striatal restricted adenosine A_2 receptor (RDC8) is expressed by enkephalin but not by substance P neurons: an in situ hybridization histochemistry study. J Neurochem 57:1062–1067

Schiffmann SN, Fisone G, Moresco R, Cunha RA, Ferré S (2007) Adenosine A_{2A} receptors and basal ganglia physiology. Prog Neurobiol 83:277–292

Schwarzschild MA, Agnati L, Fuxe K, Chen JF, Morelli M (2006) Targeting adenosine A_{2A} receptors in Parkinson's disease. Trends Neurosci 29:647–654

Shindou T, Richardson PJ, Mori A, Kase H, Ichimura M (2003) Adenosine modulates the striatal GABAergic inputs to the globus pallidus via adenosine A_{2A} receptors in rats. Neurosci Lett 352:167–170

Shiozaki S, Ichikawa S, Nakamura J, Kitamura S, Yamada K, Kuwana Y (1999) Actions of adenosine A_{2A} receptor antagonist KW-6002 on drug-induced catalepsy and hypokinesia caused by reserpine or MPTP. Psychopharmacology 147:90–95

Shulman LM, 6002-US-013 Clinical Investigator Group (2006) The safety profile of istradephylline (KW-6002) in Parkinson's disease with motor response complications on levodopa/carbidopa: results of KW-6002-US-013 study. Presented at 10th International Congress of Parkinson's Disease and Movement Disorders, Kyoto. Mov Disord 21(Suppl 15):S488

Siderowf A, Stern MB (2006) Preclinical diagnosis of Parkinson's disease: are we there yet? Curr Neurol Neurosci Rep 6:295–301

Simola N, Fenu S, Baraldi PG, Tabrizi MA, Morelli M (2004) Blockade of adenosine A_{2A} receptors antagonizes parkinsonian tremor in the rat tacrine model by an action on specific striatal regions. Exp Neurol 189:182–188

Simola N, Fenu S, Baraldi PG, Tabrizi MA, Morelli M (2006) Involvement of globus pallidus in the antiparkinsonian effects of adenosine A(2A) receptor antagonists. Exp Neurol 202:255–257

Simola N, Fenu S, Baraldi PG, Tabrizi MA, Morelli M (2008) Blockade of globus pallidus adenosine A(2A) receptors displays antiparkinsonian activity in 6-hydroxydopamine-lesioned rats treated with D(1) or D(2) dopamine receptor agonists. Synapse 62:345–351

Stacy M, Galbreath A (2008) Optimizing long-term therapy for Parkinson disease: levodopa, dopamine agonists, and treatment-associated dyskinesia. Clin Neuropharmacol 31:51–56

Stacy M, 6002-US-005/6002-US-006 Clinical Investigator Group (2004) Istradephylline (KW-6002) as adjunctive therapy in patients with advanced Parkinson's disease: a positive safety profile with supporting efficacy. Presented at 8th International Congress on Parkinson's Disease and Movement Disorders, Rome. Mov Disord 19(Suppl 9):S215

Stacy M, Silver D, Mendis T, Sutton J, Mori A, Chaikin P, Sussman NM (2008) A 12-week, placebo-controlled study (6002-US-006) of istradefylline in Parkinson disease. Neurology 70:2233–2240

Stamey W, Jankovic J (2008) Impulse control disorders and pathological gambling in patients with Parkinson disease. Neurologist 14:89–99

Stocchi F (2003) Prevention and treatment of motor fluctuations. Parkinsonism Relat Disord 9(Suppl 2):S73–S81

Stocchi F (2005) Pathological gambling in Parkinson's disease. Lancet Neurol 4:590–592

Stocchi F, Olanow CW (2003) Neuroprotection in Parkinson's disease: clinical trials. Ann Neurol 53(Suppl 3):S87–S97

Takahashi RN, Pamplona FA, Prediger RD (2008) Adenosine receptor antagonists for cognitive dysfunction: a review of animal studies. Front Biosci 13:2614–2632

Tronci E, Simola N, Borsini F, Schintu N, Frau L, Carminati P, Morelli M (2007) Characterization of the antiparkinsonian effects of the new adenosine A_{2A} receptor antagonist ST1535: acute and subchronic studies in rats. Eur J Pharmacol 566:94–102

Trugman JM, 6002-US-013 Clinical Investigator Group (2006) Efficacy of istradephylline (KW-6002) in levodopa-treated Parkinson's disease patients with motor response complications: primary efficacy results of the KW-6002-US-013 study. Presented at 10th International Congress of Parkinson's Disease and Movement Disorders, Kyoto. Mov Disord 21(Suppl 15):S513

Wakabayashi K, Tanji K, Mori F, Takahashi H (2007) The Lewy body in Parkinson's disease: molecules implicated in the formation and degradation of alpha-synuclein aggregates. Neuropathology 27:494–506

Wang JH, Ma YY, van den Buuse M (2006) Improved spatial recognition memory in mice lacking adenosine A_{2A} receptors. Exp Neurol 199:438–445

Wardas J, Konieczny J, Lorenc-Koci E (2001) SCH 58261, an A(2A) adenosine receptor antagonist, counteracts parkinsonian-like muscle rigidity in rats. Synapse 41:160–171

Weintraub D, Comella CL, Horn S (2008a) Parkinson's disease—Part 1: Pathophysiology, symptoms, burden, diagnosis, and assessment. Am J Manag Care 14:S40–S48

Weintraub D, Comella CL, Horn S (2008b) Parkinson's disease—Part 2: Treatment of motor symptoms. Am J Manag Care 14:S49–S58

Weintraub D, Comella CL, Horn S (2008c) Parkinson's disease—Part 3: Neuropsychiatric symptoms. Am J Manag Care 14:S59–S69

Adenosine Receptor Ligands and PET Imaging of the CNS

Andreas Bauer and Kiichi Ishiwata

Contents

Abstract Advances in radiotracer chemistry have resulted in the development of novel molecular imaging probes for adenosine receptors (ARs). With the availability of these molecules, the function of ARs in human pathophysiology as well as the safety and efficacy of approaches to the different AR targets can now be determined. Molecular imaging is a rapidly growing field of research that allows the identification of molecular targets and functional processes in vivo. It is therefore gaining increasing interest as a tool in drug development because it permits the process of evaluating promising therapeutic targets to be stratified. Further, molecular imaging has the potential to evolve into a useful diagnostic tool, particularly for neurological and psychiatric disorders. This chapter focuses on currently available AR

A. Bauer, M.D. (✉)
Institute of Neuroscience and Biophysics (INB-3), Research Center Jülich, 52425 Jülich, Germany
an.bauer@fz-juelich.de

C.N. Wilson and S.J. Mustafa (eds.), *Adenosine Receptors in Health and Disease*, Handbook of Experimental Pharmacology 193, DOI 10.1007/978-3-540-89615-9_19,

ligands that are suitable for molecular neuroimaging and describes first applications in healthy subjects and patients using positron emission tomography (PET).

Keywords Adenosine receptors · Brain disorders · Drug development · Molecular imaging · Positron emission tomography · Radioligands · Radiosynthesis

Abbreviations

AMP	Adenosine monophosphate
AR	Adenosine receptor
A_1AR	A_1 adenosine receptor
$A_{2A}AR$	A_{2A} adenosine receptor
$A_{2B}AR$	A_{2B} adenosine receptor
A_3AR	A_3 adenosine receptor
AD	Alzheimer's disease
BS–DMPX	(*E*)-8-(3-Bromostyryl)-3,7-dimethyl-1-propargylxanthine
Bq	Becquerel
CNS	Central nervous system
CPFPX	8-Cyclopentyl-3-(3-fluoropropyl)-1-propylxanthine
CSC	(*E*)-8-Chlorostyryl-1,3,7-trimethylxanthine (8-chlorostyrylcaffeine)
D_2R	Dopamine D_2 receptor
DMPX	3,7-Dimethyl-1-propylxanthine
DPCPX	8-Cyclopentyl-1,3-dipropylxanthine
ED_{50}	50% Efficient dose
EPDX	2-Ethyl-8-dicyclopropylmethyl-3-propylxanthine
FDG	2-Deoxy-2-fluoro-D-glucose
[^{18}F]FE@SUPPY	5-(2-[^{18}F]fluoroethyl)-2,4-diethyl-3-(ethylsulfanylcarbonyl)-6-phenylpyridine-5-carboxylate
FR194921	2-(1-Methyl-4-piperidinyl)-6-(2-phenylpyrazolo [1,5-*a*]pyridin-3-yl)-3(2*H*)-pyridazinone
IS–DMPX	(*E*)-3,7-Dimethyl-8-(3-iodostyryl)-1-propargylxanthine
keV	Kiloelectron volt
KF15372	8-Dicyclopropylmethyl-1,3-dipropylxanthine
MPDX	8-Dicyclopropylmethyl-1-methyl-3-propylxanthine
KF17837	(*E*)-8-(3,4-Dimethoxystyryl)-1,3-dipropyl-7-methylxanthine
KF18446 (TMSX)	(*E*)-8-(3,4,5-Trimethoxystyryl)-1,3,7-trimethylxanthine
KF19631	(*E*)-1,3-Diallyl-7-methyl-8-(3,4,5-trimethoxystyryl) xanthine

KF21213	(*E*)-8-(2,3-Dimethyl-4-methoxystyryl)-1,3,7-trimethylxanthine
KF21652	3-[1-(6,7-Dimethoxyquinazolin-4-yl)piperidin-4-yl]-1,6-dimethyl-2, 4(*1H*, *3H*)-quinazolinedione
KW-6002 (istradefylline)	(*E*)1,3-Diethyl-8-(3,4-dimethoxystyryl)-7-methylxanthine
PET	Positron emission tomography
PD	Parkinson's disease
SCH442416	5-Amino-7-(3-(4-methoxyphenyl)propyl)-2-(2-furyl)-pyrazolo[4,3-*e*]-1,2,4-triazolo[1,5-*c*]pyrimidine
SCH 58261	7-(2-Phenylethyl)-5-amino-2-(2-furyl)-pyrazolo [4,3-*e*]-1,2,4-triazolo[1,5-*c*]pyrimidine
SPECT	Single-photon emission computed tomography
SUV	Standard uptake value
Sv	Sievert

1 Introduction

Adenosine contributes to many physiological processes, particularly in excitable tissues such as the heart and brain. In the brain, adenosine acts as a neuromodulator and seems to have an inhibitory net effect on neuronal tissue (Dunwiddie and Masino 2001). It participates in the autoregulation of cerebral blood flow (Berne et al. 1981; Dirnagl et al. 1994), functions as a retrograde synaptic messenger (Brundege and Dunwiddie 1997), and is involved in the induction and maintenance of sleep and the regulation of arousal (Elmenhorst et al. 2007b; Porkka-Heiskanen 1999; Portas et al. 1997). Given the broad range of adenosine involvement in physiological and pathophysiological processes, numerous agonists and antagonists of adenosine receptors (ARs) are presently under evaluation in order to explore their therapeutic and diagnostic potential.

Molecular imaging is a means to get access to these processes in vivo in the human body. It will, therefore, aid in stratifying the process of evaluating promising therapeutic compounds from bench to market, and it has also the potential to evolve into a useful diagnostic tool of adenosine-related diseases, particularly, neurodegenerative disorders [e.g., Parkinson's disease (PD) and Alzheimer's disease (AD)], and brain pathologies including epilepsy, ischemia, and sleep disorders (Jacobson and Gao 2006). This chapter will primarily focus on AR-related ligands suitable for molecular neuroimaging, and their research and clinical applications using positron emission tomography (PET).

Adenosine exerts its physiological actions through four subtypes of G-protein-coupled receptor ARs (A_1, A_{2A}, A_{2B}, and A_3) (Fredholm et al. 1997, 2001; Olah and Stiles 2000). The A_1 adenosine receptor (A_1AR) is densely and heterogeneously expressed in the brain. High densities occur in thalamus and basal ganglia, as well as in neocortical and allocortical regions. A_1AR density is low in cerebellum, midbrain, and brain stem (Chaudhuri et al. 1998; Deckert et al. 1998; Fastbom et al. 1986; Glass et al. 1996; Schindler et al. 2001; Svenningsson et al. 1997). Pre- and postsynaptic A_1ARs mediate the depressant, sedative, and anticonvulsant effects of cerebral adenosine. A_1ARs are involved in the pathology of seizure disorders (Franklin et al. 1989; Moraidis and Bingmann 1994) and are reduced in cerebral inflammatory diseases (Johnston et al. 2001). In AD there are reports of regional losses of A_1AR binding sites (Deckert et al. 1998; Jaarsma et al. 1991; Schubert et al. 2001; Ulas et al. 1993) and local increases of A_1AR immunoreactivity (Albasanz et al. 2008; Angulo et al. 2003), which could reflect a specific regional and stage-related pattern of cerebral A_1AR involvement in AD. Therefore, evidence is accumulating that cerebral A_1ARs are potential targets for diagnostic imaging and therapeutic interventions in these diseases (Abbracchio and Cattabeni 1999; Fukumitsu et al. 2008; Ribeiro et al. 2003; Schubert et al. 1997).

The interaction and coexpression of A_{2A} adenosine receptors (A_{2A}ARs) and D_2 dopamine receptors (D_2Rs) in medium-sized cells of the striatum have drawn attention to the therapeutic potentials of A_{2A}AR antagonists. Treatment with these compounds alleviates symptoms in PD and seems to decelerate the neurodegenerative process (Xu et al. 2005). Given the importance of A_1ARs and A_{2A}ARs in brain physiology and pathology, they were the first AR subtypes to be successfully visualized in the human brain in vivo (Bauer et al. 2003; Fukumitsu et al. 2003, 2005; Ishiwata et al. 2005a; Mishina et al. 2007).

Adenosine A_{2B} receptors (A_{2B}ARs) and A_3 receptors (A_3ARs) seem to be primarily activated under pathological conditions, such as ischemia and various types of cancer. For both AR subtypes, there is currently no radiotracer that has successfully been applied in the human brain.

Molecular imaging methods, such as PET and single-photon emission computed tomography (SPECT), are characterized by a high sensitivity that allows the visualization of receptors of neurotransmitters and neuromodulators (e.g., adenosine; i.e., ARs) in vivo with excellent temporal and reasonable spatial resolution, respectively. PET is based on the imaging of radiopharmaceuticals labeled with positron-emitting radionuclides such as ^{11}C, ^{15}O, and ^{18}F, and on measuring the annihilation radiation using a coincidence technique. Two 511 keV γ-rays are emitted at $\sim$180° as a result of the collision between a positron emitted from a radionuclide and a nearby electron. The two 511 keV γ-rays are detected by external coincidence circuits. Importantly, the nanomolar amount of mass for the radionuclide that is injected intravenously is too small to affect the steady state of the biochemical process under investigation. Therefore, the advantage of PET is its ability to measure low-density binding sites without perturbing the biochemistry of the system. Besides, PET can determine the pharmacokinetics of labeled drugs and assess the effects of drugs on metabolism in vivo in a quantitative manner. Because only very low amounts of the radiolabeled drug have to be administered (far below toxicity levels) human studies can be carried out even before the drug is entered in Phase I clinical trials. Such studies can provide cost-effective predictive toxicology data and information on the metabolism and mode of action of drugs. Especially valuable is the contribution of PET to bridge the gap between molecular biology/pathophysiology and the design of new drugs. Regarding ARs, there are several reports of successful visualizations of A_1ARs, A_{2A}ARs (in humans and different animal species), and recently A_3ARs (in the rat) using PET, which clearly demonstrate the feasibility of these powerful modalities to further enhance the role of radiotracer studies in drug-effect monitoring. However, so far, all of these applications are of an experimental nature and have not yet reached the arena of clinical diagnostic use.

This chapter provides an overview of the current status regarding the development of both PET radioligands for mapping ARs and new lead compounds for potential PET radioligands. It also summarizes preclinical and clinical results that have so far been obtained by molecular imaging of ARs.

2 Development of PET Radioligands

For the last two decades, ARs have been extensively studied biologically and pharmacologically, and advancements in the synthesis and screening of a large number of compounds have resulted in the identification of selective ligands with high affinity and high specific binding for each receptor subtype. Since 1995, several PET ligands with xanthine-type structures, which are expected to penetrate the blood–brain barrier, have been proposed for mapping A_1ARs (Furuta et al. 1996; Holschbach et al. 1998; Ishiwata et al. 1995; Noguchi et al. 1997) and A_{2A}ARs (Hirani et al. 2001; Ishiwata et al. 1996, 2000a, b, d, 2003a; Marian et al. 1999; Noguchi et al. 1998; Stone-Elander et al. 1997; Wang et al. 2000) in the central nervous system (CNS). Later, nonxanthine-type ligands were also developed (Matsuya et al. 2005; Todde et al. 2000). Among them, at least five PET ligands for A_1AR and A_{2A}AR subtypes have been applied to clinical studies (Fig. 1) (Bauer

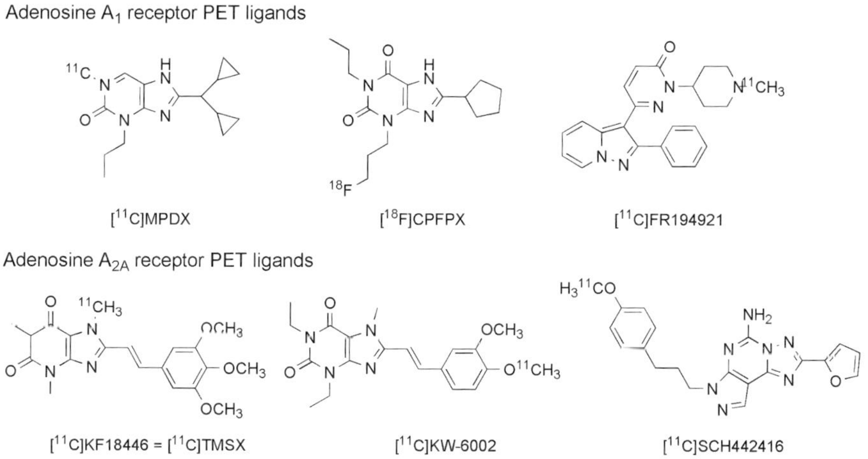

Adenosine A_1 receptor PET ligands

$[^{11}C]$MPDX $[^{18}F]$CPFPX $[^{11}C]$FR194921

Adenosine A_{2A} receptor PET ligands

$[^{11}C]$KF18446 = $[^{11}C]$TMSX $[^{11}C]$KW-6002 $[^{11}C]$SCH442416

Fig. 1 Representative PET ligands for mapping adenosine A_1 receptors (A_1ARs) and adenosine A_{2A} receptors (A_{2A}ARs). All ligands except for $[^{11}C]$FR194921 have been used clinically, but only preliminary results have been published for $[^{11}C]$SCH442416

et al. 2003; Fukumitsu et al. 2003; Hunter 2006; Ishiwata et al. 2005a). On the other hand, PET ligands for the A_3AR subtype (Wadsak et al. 2008) and the adenosine uptake site (Ishiwata et al. 2001; Mathews et al. 2005) are limited, and no PET ligand for the A_{2B}AR subtype has been reported until now. Early works on the development of PET ligands have been described (Holschbach and Olsson 2002; Ishiwata et al. 2002c; Suzuki and Ishiwata 1998), and recent advances in the development of PET ligands and medicinal chemistry, including candidates for this purpose, have been reviewed (Ishiwata et al. 2008).

2.1 Adenosine A_1 Receptor Ligands

In Table 1, in vitro and in vivo properties of A_1AR PET ligands are summarized. Xanthine derivatives such as 8-dicyclopropylmethyl-1,3-dipropylxanthine (KF15372) (Shimada et al. 1991; Suzuki et al. 1992) and 8-cyclopentyl-1,3-dipropylxanthine (DPCPX) (Bruns et al. 1987; Lohse et al. 1987) are selected as leading compounds for PET ligands. [^{3}H]DPCPX has been used in vitro as a radioligand with high affinity and selectivity for the A_1AR (Deckert et al. 1998; Jaarsma et al. 1991; Svenningsson et al. 1997; Ulas et al. 1993). Both compounds have two propyl groups, each of which can potentially be labeled with ^{11}C (half-life of 20.4 min). Ishiwata et al. prepared [^{11}C]KF15372 and its [^{11}C]ethyl and [^{11}C]methyl derivatives (2-[^{11}C]ethyl-8-dicyclopropylmethyl-3-propylxanthine ([^{11}C]EPDX) and 8-dicyclopropylmethyl-1-[^{11}C]methyl-3-propylxanthine ([^{11}C]MPDX), respectively) (Furuta et al. 1996; Ishiwata et al. 1995; Noguchi et al. 1997). [^{11}C]MPDX (Fig. 1) showed a slightly lower affinity for A_1ARs than [^{11}C]KF15372; however, [^{11}C]MPDX was selected for further investigations among the three ligands because of a high radiochemical yield and easy penetration through the blood–brain barrier. Later, Holschbach et al. examined a series of DPCPX analogs and found several candidates containing fluorine or iodine (Holschbach et al. 1998). The selected ligand was [^{18}F]8-cyclopentyl-3-(3-fluoropropyl)-1-propylxanthine ([^{18}F]CPFPX) (Fig. 1) (^{18}F, half-life of 110 min), in which a [^{18}F]fluoropropyl group was incorporated into DPCPX instead of ^{11}C labeling a propyl group (Holschbach et al. 2002). This substitution greatly enhanced the affinity and selectivity for A_1ARs. Radioiodine-labeled ligands may be used for PET (^{124}I, half-life of 4.18 days) and SPECT (^{123}I, half-life of 13.3 h). Recently, nonxanthine-type pyrazolpyridine compounds were proposed for A_1AR ligands (Kuroda et al. 2001; Maemoto et al. 2004), and Matsuya et al. prepared [^{11}C]2-(1-methyl-4-piperidinyl)-6-(2-phenylpyrazolo [1,5-*a*]pyridin-3-yl)-3(2*H*)-pyridazinone ([^{11}C]FR194921) (Fig. 1) (Matsuya et al. 2005).

Among five ligands, [^{18}F]CPFPX shows the highest affinity and selectivity in vitro as well as high uptake and specific binding in vivo (Table 1). In mice, the brain uptake was rapid and remained constant for 40 min after injection, followed by a gradual decrease because of high affinity, suggesting that a long PET scan covering

Table 1 PET ligands for the A_1 adenosine receptor (A_1AR)

	In vitro studies				In vivo studies		
	Affinity (K_i nM)		Selectivity		Uptake[a] (cerebral cortex) (SUV)	Specific binding[b] (%)	Reference
	A_1	A_{2A}	A_{2A}/A_1	Reference			
DPCPX	6.4	590	92	Shimada et al. (1991); Suzuki et al. (1992)			
[^{11}C]KF15372	3.0	430	143	Shimada et al. (1991); Suzuki et al. (1992)	0.43 (m, 15 min)	57 (m, 15 min)[d]	Noguchi et al. (1997)
[^{11}C]EPDX	1.7	>100	>59	Noguchi et al. (1997)	0.66 (m, 15 min)	47 (m, 15 min)[d]	Noguchi et al. (1997)
[^{11}C]MPDX	4.2	>100	>24	Noguchi et al. (1997)	0.54 (m, 15 min)	43 (m, 15 min)[d]	Noguchi et al. (1997)
						61–64 (r, 15 min)[e, c]	Wang et al. (2003)
[^{18}F]CPFPX	0.183			Holschbach et al. (1998)			
	0.63–1.37 (K_d)	812–940 (K_d)	>700	Holschbach et al. (2002)	0.88 (m, 40 min)	70–80 (m, 10–40 min)[e,f]	Holschbach et al. (2002)
[^{11}C]FR194921	2.91	>100	>34	Matsuya et al. (2005); Maemoto et al. (2004)	0.3 (r, 30 min)	50 (r, 30 min)[e]	Matsuya et al. (2005)

[a]Uptake was normalized as the standardized uptake value [SUV, (tissue activity/total injected activity) × (gram body weight/gram tissue weight)], assuming the body weights of rats and mice were 300 g and 35 g, respectively. In the parentheses, "r" and "m" express the uptakes in the brain of rats and mice, respectively, which were killed at the indicated time after injection of the tracer. The tissue uptake was measured by the tissue dissection method, except in one case (marked by [c]), where it was measured by ex vivo autoradiography

[b]The reduced percentages of the uptake by blockade with injection of selective appropriate adenosine A_1 receptor ligand together with the tracer[d] or before[e] or after[f] injection of the tracer

the pseudoequilibrium state of the ligand–receptor binding may be preferable. The other ligands showed reasonable brain uptake and specific binding due to the affinity in vitro and the liphophilicity.

Xanthine derivatives are unstable in relation to peripheral metabolism. Percentages of the unchanged form in rodent plasma were <30% for both [^{11}C]MPDX and [^{18}F]CPFPX 30 min postinjection, whereas [^{11}C]FR194921 was much more stable (87% at 60 min) (Bier et al. 2006; Matsuya et al. 2005; Noguchi et al. 1997). The metabolic pathway of [^{18}F]CPFPX was extensively investigated (Bier et al. 2006), and Matusch et al. (2006) identified that cytochrome P-450 1A2 catalyzed the metabolism of it. Later [^{11}C]MPDX was confirmed to be much more stable in human plasma (75% was unchanged at 60 min) (Fukumitsu et al. 2005), while [^{18}F]CPFPX was metabolized faster in humans (Bauer et al. 2003).

However, the evaluation of PET ligands at a single or a limited number of time points after injection, as shown in Table 1, was not adequate when comparing several ligands. Dynamic PET studies in monkeys or cats were carried out for [^{11}C]KF15372 (Wakabayashi et al. 2000), [^{11}C]MPDX (Ishiwata et al. 2002a; Shimada et al. 2002) and [^{11}C]FR194921 (Matsuya et al. 2005). Although [^{11}C]KF15372 and [^{11}C]FR194921 have similar affinities in vitro, the brain kinetics were considerably different in monkeys. [^{11}C]KF15372 accumulated and reached a maximum at 10 min followed by a gradual decrease, while [^{11}C]FR194921 accumulated over 60 min. In the time frame of a PET scan using a ^{11}C-labeled tracer (60–90 min), [^{11}C]KF15372 showed preferable brain kinetics for quantitative evaluation of the ligand–receptor binding, while the affinity of [^{11}C]FR194921 may be too high. Compared with [^{11}C]KF15372, [^{11}C]MPDX showed a faster brain clearance in monkeys and cats, but quantitative evaluation of A_1 ARs in the cat brain was nevertheless successfully performed by PET.

The other radioligands labeled with positron emitters are 5′-*N*-(2-[^{18}F]fluoroethyl)-carboxamidoadenosine and 5′-(methyl[^{75}Se]seleno)-N^6-cyclopentyladenosine (^{75}Se, half-life of 7.1 h) (Lehel et al. 2000; Blum et al. 2004). Although the biological evaluation of these tracers has not been reported, they may be suitable ligands for peripheral organs but not for the CNS, if available for PET studies; however, 5′-*N*-(2-[^{18}F]fluoroethyl)-carboxamidoadenosine may not be a selective ligand for A_1 ARs (Lehel et al. 2000).

2.2 Adenosine A_{2A} Receptor Ligands

Considering 3,7-dimethyl-1-propylxanthine (DMPX) as a lead for A_{2A}AR-selective antagonists (Seale et al. 1988), Shimada et al. have discovered that xanthines with the styryl group in the 8 position have selective A_{2A}AR antagonistic properties (Nonaka et al. 1994; Shimada et al. 1992). Later, Müller et al. also introduced brominated and chlorinated styryl groups in the 8 position of DMPX to produce A_{2A}AR-selectivity (Müller et al. 1997, 1998). The representative compound (*E*)-8-(3,4-dimethoxystyryl)-1,3-dipropyl-7-methylxanthine (KF17837) has been used for pharmacological and neurochemical studies as a selective antagonist for

A_{2A}ARs (Correa et al. 2004; Hayaishi 1999; Koga et al. 2000). So far KF17837 and seven other derivatives have been labeled with ^{11}C, and these radiotracers were investigated as potential PET ligands (Ishiwata et al. 1996, 2000a, b; Noguchi et al. 1998; Stone-Elander et al. 1997; Wang et al. 2000) (Table 2). [^{11}C](*E*)-8-(3-Bromostyryl)-3,7-dimethyl-1-propargylxanthine ([^{11}C]BS-DMPX) and [^{11}C](*E*)-3,7-dimethyl-8-(3-iodostyryl)-1-propargylxanthine ([^{11}C]IS-DMPX) (Ishiwata et al. 2000d) can potentially be labeled with radiolabeled bromines (^{75}Br, $t_{1/2} = 1.7$ h or ^{76}Br, $t_{1/2} = 16.1$ h) and iodines (^{124}I, half-life of 4.18 days, and ^{123}I, half-life of 13.3 h), respectively, for PET or SPECT. Most of these studies were done by Ishiwata et al. in collaboration with Kyowa Hakko Kogyo Co., Ltd. (Tokyo, Japan). Later, Kyowa Hakko Kogyo chose the selective A_{2A}AR antagonist (*E*)-1,3-diethyl-8-(3,4-dimethoxystyryl)-7-methylxanthine (KW-6002), known as istradefylline, for clinical evaluation as an antiPD agent (Bara-Jimenez et al. 2003; Hauser et al. 2003) after an experimental study of [^{11}C]KW-6002 (Fig. 1) (Hirani et al. 2001). It was noted that photoisomerization occurred in the styryl group at the 8 positions of xanthine-type A_2AR-selective ligands such as (*E*)-8-(3,4,5-trimethoxystyryl)-1,3,7-trimethylxanthine ([^{11}C]KF18446), later designated [^{11}C]TMSX) (Fig. 1) (Ishiwata et al. 2003b; Nonaka et al. 1993). Consequently, all procedures in PET studies were carried out under dim light until injection and also during plasma metabolite analysis.

Besides xanthine derivatives, a number of nonxanthine heterocycles have also been synthesized as A_{2A}AR antagonists. 7-(2-Phenylethyl)-5-amino-2-(2-furyl)-pyrazolo[4,3-*e*]-1,2,4-triazolo[1,5-*c*]pyrimidine (SCH 58261) is a representative ligand with a high and selective affinity for the A_2AR (Zocchi et al. 1996a, b); however, it does not have an appropriate synthon for labeling with positron emitters. Todde et al. used 5-amino-7-(3-(4-methoxyphenyl)propyl)-2-(2-furyl)-pyrazolo [4,3-*e*]-1,2,4-triazolo[1,5-*c*]pyrimidine (SCH442416) with its 4-methoxyphenylpropyl group, and prepared [^{11}C]SCH442416 (Fig. 1) by *O*-[^{11}C]methylation (Todde et al. 2000).

Table 2 summarizes the in vitro and in vivo properties of A_{2A}AR PET ligands. The highest affinity for A_{2A}ARs was found in SCH442416, followed by KF17837, KW-6002, and (*E*)-8-(2,3-dimethyl-4-methoxystyryl)-1,3,7-trimethylxanthine (KF21213). SCH442416, KF21213 and IS–DMPX showed superior A_{2A}AR selectivity. (*E*)-1,3-Diallyl-7-methyl-8-(3,4,5-trimethoxystyryl)xanthine (KF19631), TMSX, (*E*)-8-chlorostyryl)-1,3,7-trimethylxanthine (8-chlorostyrylcaffeine, CSC), and BS–DMPX showed moderate selectivity, but their affinities for the A_1ARs were too low to bind in vivo. In evaluation studies in rodents, all radioligands showed A_2AR-selective uptake in the striatum where the expression of A_{2A}ARs is high; however, specific binding was also observed in the cerebral cortex as well as cerebellum to a certain extent for most radioligands except for [^{11}C]KF21213. Thus, the highest A_{2A}AR selectivity in vivo was observed in [^{11}C]KF21213, followed by [^{11}C]SCH442416 and [^{11}C]TMSX, when evaluated based on the uptake ratio of receptor-rich striatum to receptor-poor cerebellum.

Compared with A_1AR receptor ligands, a slow peripheral degradation of two xanthine compounds was confirmed in the metabolite analysis in plasma;

Table 2 PET ligands for the A_{2A} adenosine receptor ($A_{2A}AR$)

	In vitro studies				In vivo studies			
	Affinity (K_i)		Selectivity		Striatal uptake[a]	Selectivity[b]		
	A_1	A_{2A}	A_1/A_{2A}	References	(SUV)	(Str/Cer)	Specific binding[c] (%)	References
DMPX	12,000	8,600	1.4	Shimada et al. (1991); Suzuki et al. (1992)				
[^{11}C]KF17837	62	1.0	62	Nonaka et al. (1994)	0.82(m, 15 min)	2.0 (m, 60 min) 1.2 (r, 15 min)[d]	43 (m, 15)[e]	Noguchi et al. (1998)
[^{11}C]KF19631	860	3.5	250	Ishiwata et al. (2002c)	0.33 (m, 15 min)	1.2 (m, 60 min) 1.2 (r, 15 min)[d]	31 (m, 15)[e]	Ishiwata et al. (2000a)
[^{11}C]KF18446 = [^{11}C]TMSX	1,600	5.9	270	Ishiwata et al. (2000a)	1.54 (m, 15 min) 1.68 (r, 15 min)	2.8 (m, 15 min)	72 (m, 15)[e]	Ishiwata et al. (2000a)
[^{11}C]CSC	28,000	54	520	Nonaka et al. (1994)	ND	ND	ND	
[^{11}C]BS–DMPX	2,300	7.7	300	Ishiwata et al. (2000d)	0.90 (m, 15 min)	1.2 (m, 60 min)	51 (m, 15)[e]	Ishiwata et al. (2000d)
[^{11}C]IS–DMPX	>10,000	8.9	>1,100	Ishiwata et al. (2000d)	0.70 (m, 15 min)	1.2 (m, 60 min)	17[ns] (m, 15)[e]	Ishiwata et al. (2000d)
[^{11}C]KF21213	>10,000	3.0	>3,300	Wang et al. (2000)	1.40 (m, 15 min)	10.5 (m, 60 min)	69 (m, 15)[e]	Wang et al. (2000)

[^{11}C]KW-6002	150	2.2	68	Nonaka et al. (1994)	2.9 (r, 15 min)	1.1 (r, 75 min)	88 (r, 75)[f]	Hirani et al. (2001)
SCH-58261	121	2.3	53	Zocchi et al. (1996a)				
[^{11}C]SCH442416	1,800	0.50	3,630	Todde et al. (2000)	1.15 (r, 15 min)	4.6 (r, 15 min)	ND	Todde et al. (2000)
					ca. 1 (r, 15 min)	4.97 (r, 15 min)	31–86 (r, 15)[f]	Moresco et al. (2005)

[a]Uptake was normalized as the standardized uptake value [SUV, (tissue activity/total injected activity) × (gram body weight/gram tissue weight)], assuming the body weights of rats and mice were 300 g and 35 g, respectively. In the parentheses, "r" and "m" express the uptakes in rat and mouse brain, respectively, which were killed at the indicated time after injection of the tracer. The tissue uptake was measured by the tissue dissection method, except in some cases (marked by [d]), where it was measured by ex vivo autoradiography

[b]Selectivity was determined as the uptake ratio of striatum to cerebellum (Str/Cer). This concept is based on the finding that the striatum is rich in A_{2A}ARs, while the expression of A_{2A}ARs is low or negligible in the cerebellum

[c]Reduced percentages of uptake by the blockade with injection of selective appropriate A_{2A}AR ligand together with the tracer[e] or before[f] injection of the tracer

ND, not determined; *ns*, no significance (control versus blocked animals)

percentages of the unchanged form were 81% for [^{11}C]TMSX at 30 min in mice (Ishiwata et al. 2000a) and 66% for [^{11}C]KW-6002 at 45 min in rats (Hirani et al. 2001). [^{11}C]SCH442416 was slightly unstable: 40% was unchanged at 30 min in rats (Todde et al. 2000). Later, [^{11}C]TMSX was confirmed to be much more stable in human plasma (>90% of the unchanged form at 60 min) (Mishina et al. 2007).

Dynamic PET studies in monkeys were carried out for [^{11}C]KF17837, [^{11}C]TMSX and [^{11}C]SCH442416. The striatal uptake of [^{11}C]TMSX was approximately tenfold higher at 5–10 min compared with [^{11}C]KF17837, and the uptake ratios of striatum to cortex and striatum to cerebellum for [^{11}C]TMSX were slightly higher than those for [^{11}C]KF17837 (Ishiwata et al. 2000a). A slightly lower affinity of [^{11}C]TMSX resulted in a faster clearance of the radioligand from the striatum compared to [^{11}C]KF17837. Because it exhibited the highest affinity among the three ligands, [^{11}C]SCH442416 showed more preferable brain kinetics for quantitative evaluating the ligand–receptor binding (Moresco et al. 2005). Although [^{11}C]KF21213 showed the most preferable properties in rodents, in a preliminary PET study using monkeys [^{11}C]TMSX showed better brain kinetics than [^{11}C]KF21213 (Ishiwata et al. 2005b).

Most studies of radioligands have focused on ARs in the CNS. On the other hand, Ishiwata et al. demonstrated that xanthine-type ligands can be applicable to studies on peripheral A_{2A}ARs (Ishiwata et al. 1997, 2003a, 2004). In rodents, specific binding of [^{11}C]TMSX was observed in the muscle and heart. Swimming exercise caused fluctuations in [^{11}C]TMSX-receptor binding in these tissues, and the specific binding of [^{11}C]TMSX to these tissues was also preliminarily demonstrated clinically (Ishiwata et al. 2004). Furthermore, the [^{11}C]TMSX-receptor binding in the cardiac and skeletal muscles was greater in endurance-trained men than in untrained men (Mizuno et al. 2005).

2.3 Adenosine A_3 Receptor Ligands

Recently, Wadsak et al. (2008) reported on the synthesis of 5-(2-[^{18}F]fluoroethyl)-2,4-diethyl-3-(ethylsulfanylcarbonyl)-6-phenylpyridine-5-carboxylate ([^{18}F]FE@SUPPY) for the A_3AR and a preliminary evaluation. The tracer was taken up in the rat brain at intermediate levels and bound to rat brain slices in vitro; however, further in vivo studies are essential for the evaluation of its specificity and selectivity.

2.4 Ligands for the Adenosine Uptake Site

[1-Methyl-^{11}C]-3-[1-(6,7-dimethoxyquinazolin-4-yl)piperidin-4-yl]-1,6-dimethyl-2,4($1H,3H$)-quinazolinedione ([^{11}C]KF21652), with a K_{i} value of 13 nM, was prepared by N-[^{11}C]methylation (Ishiwata et al. 2001). The brain uptake of [^{11}C]KF21562 was very low in vivo, probably because of its relatively high lipophilicity (log$P$ 3.6), although in vitro autoradiography showed specific binding to adenosine uptake sites to a certain extent (less than 25% of total binding). Peripherally, only the liver showed carrier-saturable uptake. The compound is not a suitable PET ligand.

Another potential labeled tracer for adenosine uptake sites is [^{11}C]adenosine monophosphate ([^{11}C]AMP) (Mathews et al. 2005). In mice, this tracer was not incorporated in the brain, and the highest uptake was observed in the lung, blood, and heart. The lung uptake was significantly reduced to about 40% by blocking with dipyridamole, a ligand for adenosine uptake sites. The putative value of this ligand needs to be investigated further.

2.5 *Radiosynthesis*

All ^{11}C-labeled ligands except for [^{11}C]AMP have been synthesized by N- or O-alkylation with [^{11}C]methyl iodide or [^{11}C]alkyl iodide. Practically speaking, the production of [^{11}C]methyl iodide is much easier than those of [^{11}C]ethyl iodide and [^{11}C]propyl iodide, which were used for the syntheses of [^{11}C]EPDX and [^{11}C]KF15372, respectively, and usually achieved high radiochemical yields of the ligands (Noguchi et al. 1997). [^{11}C]Methyl triflate is a highly reactive alternative to [^{11}C]methyl iodide (Kawamura and Ishiwata 2004). [^{11}C]AMP was produced by reacting [^{11}C]formaldehyde with the corresponding amino-imidazolyl-carboxamide, giving a low radiochemical yield (Mathews et al. 2005).

On the other hand, ^{18}F-labeled ligands were prepared by nucleophilic, cryptate-mediated substitution using ^{18}F anion. In general, ^{18}F-labeled ligands have practical advantages: the specific activity is usually higher than that of ^{11}C-labeled ligands, fluorine-18 provides slightly better resolution of the images, and its longer half-life is more suitable for clinical purposes than that of ^{11}C-labeled tracers. On the other hand, ^{11}C-labeled ligands provide reduced radiation doses for human subjects compared to ^{18}F-labeled ligands. Also, the shorter half-life of ^{11}C can allow successive PET measurements experimentally (Nariai et al. 2003) and clinically (Ishiwata et al. 2005a) on the same day.

3 Experimental Studies

Several studies using experimental animals have demonstrated the usefulness of AR ligands and PET. In the rat model, in which monocular enucleation was performed in order to destroy the anterior visual input, a loss of A_1 ARs was detected by ex vivo autoradiography using [^{11}C]MPDX (Kiyosawa et al. 2001). The decrease in presynaptic A_1ARs in the superior colliculus following enucleation was coupled with an upregulation of postsynaptic benzodiazepine receptors (Wang et al. 2003). In an occlusion and reperfusion model of the cat, [^{11}C]MPDX PET was more sensitive to the detection of severe cerebral ischemic insult than [^{11}C]flumazenil PET when measuring central benzodiazepine receptors (Nariai et al. 2003).

In a glioma-bearing rat model, Bauer et al. found that the binding of [^{18}F]CPFPX was increased in the zone surrounding tumors (136–146% as compared to control brain tissue) due to the upregulation of A_1ARs in activated astrocytes (Bauer et al. 2005; Dehnhardt et al. 2007). Furthermore, in a preliminary study, the same group demonstrated A_1AR occupancy by caffeine in the rat brain by [^{18}F]CPFPX PET (Meyer et al. 2003).

In a Huntington's disease model, induced by intrastriatal injection of quinolinic acid and consecutive degeneration of striatopallidal γ-aminobutyric acid/enkephalin neurons, degeneration of A_{2A}ARs in the lesioned striatum was detected to a similar extent as degeneration of D_2Rs using PET and ex vivo and in vitro autoradiography with [^{11}C]TMSX (Ishiwata et al. 2002b). Another A_{2A}AR ligand, [^{11}C]SCH442416, was applied to the same rat model of Huntington's disease (Moresco et al. 2005), demonstrating that the striatal uptake of [^{11}C]SCH442416 was reduced on the quinolinic acid-lesioned side. Furthermore, an ex vivo autoradiography study showed that [^{11}C]TMSX, but not [^{11}C]raclopride for D_2Rs, was incorporated into the globus pallidus to a lesser extent (the striatum-to-globus pallidus uptake ratio was approximately 0.6), and showed a remarkably reduced uptake in both the striatum and globus pallidus for the lesioned side in the rat model of Huntington's disease (Ishiwata et al. 2000c). These findings suggest that [^{11}C]TMSX is a candidate tracer for imaging the pallidal terminals of striatal projection neurons.

4 Clinical Studies

A large number of selective AR agonists and antagonists have been discovered, and some of them have been taken to the next level and evaluated in Phase I, II, and III clinical trials. So far, no compound has received regulatory approval. The same is true of adenosine and AR-based ligands used as PET tracers, which are under evaluation for diagnostic purposes or as markers to evaluate the efficacy of therapeutics.

4.1 Adenosine A_1 Receptor Imaging

To date, two PET ligands have successfully been applied for the visualization of A_1ARs in the human brain, [^{18}F]CPFPX (Bauer et al. 2003) and [^{11}C]MPDX (Fukumitsu et al. 2003, 2005). A direct comparison of postmortem brain material using autoradiography demonstrated a close correlation between the regional [^{18}F]CPFPX binding potential and the cerebral [^{3}H]CPFPX distribution (Bauer et al. 2003). Consistent with results from [^{3}H]CPFPX autoradiography, high A_1AR densities were found in the putamen and mediodorsal thalamus using [^{18}F]CPFPX PET. Neocortical areas showed regional differences in [^{18}F]CPFPX binding, with high accumulation in temporal > occipital > parietal > frontal lobes and a lower level of binding in the sensorimotor cortex. Ligand accumulation was low in the cerebellum, midbrain, and brain stem (Bauer et al. 2003; Meyer et al. 2004). The specificity of [^{18}F]CPFPX binding was established in a displacement study using cold CPFPX (Meyer et al. 2006).

The clinical applicability of [^{18}F]CPFPX was assured by test–retest (Elmenhorst et al. 2007a) and dosimetrical studies (Herzog et al. 2008), respectively. The dosimetrical studies showed that an injection of 3×10^8 Bq [^{18}F]CPFPX resulted in an effective dose of 5.3×10^{-3} Sv, which is comparable to other ^{18}F-labeled ligands and thus suitable for clinical applications. Test–retest evaluations were performed in order to study the physiological intrasubject variability of [^{18}F]CPFPX binding. This factor is extremely important for the definition of the normal range of cerebral receptor binding and thus highly accountable for the discriminative power of the method as a diagnostic tool. Elmenhorst et al. (2007a) demonstrated that test–retest variability was low (5.9–13.2% on average) and therefore highly suitable for diagnostic purposes. They also showed that noninvasive quantification (i.e., without the need to take blood samples during the PET scan) is even superior to invasive measurements, which greatly improves the clinical applicability of [^{18}F]CPFPX PET. A series of horizontal planes of the cerebral [^{18}F]CPFPX distribution as well as a three-dimensional reconstruction of the neocortical surface of the brain of a healthy subject are depicted in Fig. 2.

The spatial distribution of [^{11}C]MPDX differed significantly from the regional cerebral blood flow measured by PET using [^{15}O]H_2O and the regional cerebral metabolism of glucose evaluated using 2-deoxy-2-[^{18}F]fluoro-D-glucose ([^{18}F]FDG), and was in good agreement with autoradiographic data from other highly specific A_1AR ligands (Fukumitsu et al. 2003). Moreover, this A_1AR radiotracer showed a better metabolic stability than [^{18}F]CPFPX but had a lower affinity to A_1ARs (4.2 nM in comparison to 0.183 nM).

For both tracers, quantitative methods have been developed to measure the A_1AR binding potential in vivo in the human brain (Kimura et al. 2004; Meyer et al. 2005a, b). For clinical applications, noninvasive but fully quantitative methods with significantly shortened scan durations and without blood sampling have been developed (Naganawa et al. 2008; Meyer et al. 2005b).

With respect to the use of AR-based PET tracers in humans to define the role of ARs in neuropathology, only a limited number of clinical studies have been published so far. Boy et al. (2008) reported lower cortical and subcortical A_1AR binding in patients suffering from liver cirrhosis and hepatic encephalopathy in comparison to controls. They concluded that regional cerebral adenosinergic neuromodulation is heterogeneously altered in cirrhotic patients, and that the decrease in cerebral A_1AR binding may further aggravate neurotransmitter imbalance at the synaptic cleft in hepatic encephalopathy.

In a recent study utilizing an A_1AR-based PET tracer, Fukumitsu et al. (2008) reported on changes in A_1ARs in the brains of patients with AD. They applied two

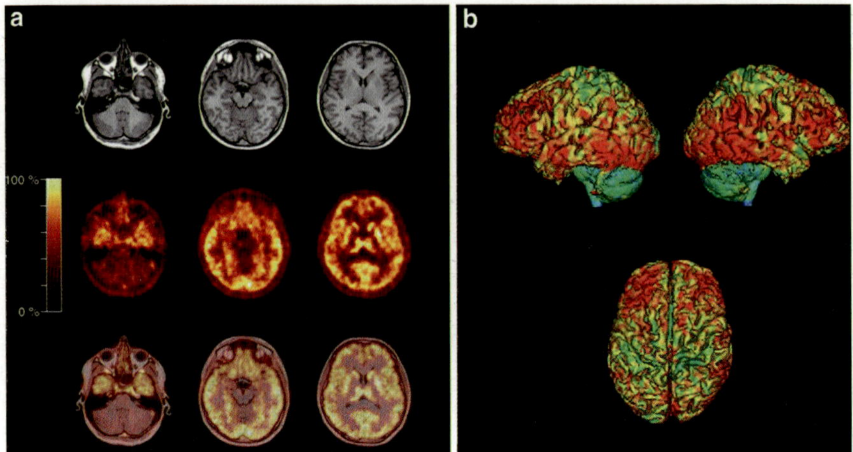

Fig. 2 a–b Distribution of adenosine A_1 receptors (A_1ARs) in the human brain. **a** Serial horizontal MRI (*upper line*) and coregistered PET images (*middle line*) from a healthy subject. Summed data from 5 to 60 min after intravenous injection of [^{18}F]CPFPX are depicted. The fusion images (*bottom line*) show high ligand binding in neocortical areas as well as thalamus and basal ganglia (as indicated by *bright yellow colors*); low binding is found in the cerebellum (depicted by *dark orange colors*). **b** Three-dimensional reconstruction of the brain surface generated from serial planes from the same PET scan as in **a**. Note that A_1ARs are ubiquitously but not homogeneously distributed in the neocortex. There are clusters with high A_1AR binding in prefrontal and temporoparietal cortices (high binding is indicated by *red* and *orange colors*, low binding is indicated by *green* and *cyan colors*)

PET scans with [^{11}C]MPDX and [^{18}F]FDG to the same patients to directly compare A_1ARs and glucose metabolism reflecting neural activity in the brain. There was significantly reduced binding of [^{11}C]MPDX in patients with AD in the temporal and medial temporal cortices and in the thalamus. Thus, the regional pattern of A_1AR changes in AD was different from the well known and previously reported hypometabolic brain regions (temporoparietal cortex and posterior cingulate gyrus), where [^{18}F]FDG uptake was typically decreased in AD. This pilot study was the first study to show with the use of a PET tracer for A_1ARs that A_1ARs are reduced in AD. It clearly demonstrates that A_1AR PET ligands could become valuable tools for the investigation of neurodegenerative disorders like AD.

An interesting example of the scientific potential of A_1AR imaging in neuroscience has been published in a study on the effect of sleep deprivation for 24 h on healthy subjects, which shows promise for clinical applications in sleep disorders (Elmenhorst et al. 2007b). It is currently hypothesized that adenosine is involved in the induction of sleep after prolonged wakefulness. This effect is partially reversed by the application of caffeine, which is a nonselective blocker of ARs. Elmenhorst et al. (2007b) report that the A_1AR is upregulated after 24 h of sleep deprivation in a region-specific pattern in a broad spectrum of brain regions, with a maximum increase in the orbitofrontal cortex. There were no changes in the control group, who had regular sleep. Thus, the study provides in vivo evidence for an A_1AR contribution to the homeostatic regulation of sleep in humans. Molecular imaging using A_1AR ligands therefore shows significant potential for sleep research and, in the long run, sleep medicine.

These findings are also of importance regarding the role of caffeine as a neurostimulant and nonselective antagonist of adenosine effects at A_1ARs and $A_{2A}ARs$. Throughout the world, caffeine is the most widely used pharmacological agent; it is present in beverages such as coffee, tea, and soft drinks. As a stimulant, caffeine promotes wakefulness and reduces sleep and sleep propensity (Fredholm et al. 1999; Landolt 2008; Schwierin et al. 1996; Virus et al. 1990; Yanik and Radulovacki 1987). Molecular imaging using adenosine tracers has great potential to provide insights into the regional and temporal modes of caffeine action in the human brain. In vivo A_1AR occupancy by caffeine has so far only been demonstrated in the rat brain by [^{18}F]CPFPX PET (Meyer et al. 2003).

4.2 *Adenosine A_{2A} Receptor Imaging*

With regard to adenosine $A_{2A}AR$ imaging, the most promising clinical application is currently PD. Striatopallidal $A_{2A}ARs$ have been implicated in the modulation of motor functions because they partially antagonize the functions of striatal D_2Rs. Since $A_{2A}ARs$ show a highly enriched distribution in basal ganglia cells and are able to form functional heteromeric complexes with D_2Rs and metabotropic glutamate mGluR5 receptors, $A_{2A}ARs$ are of particular interest with regard to the nondopaminergic modulation of motor behavior (Ferré and Fuxe 1992; Fuxe et al. 1993; Marino et al. 2003). Additional evidence for an adenosinergic contribution to PD comes from epidemiological studies showing that chronic consumption of caffeine, a nonselective AR antagonist, is able to reduce the risk of developing PD (Ascherio et al. 2001; Ross et al. 2000). Given the relevance of $A_{2A}ARs$ in PD, an important advance was made by Ishiwata et al. (2005a), who were able to introduce [^{11}C]TMSX, allowing $A_{2A}ARs$ to be imaged in the living human brain for the first time. The specificity of [^{11}C]TMSX PET was confirmed by theophylline challenge (Ishiwata et al. 2005a), and the cerebral distribution pattern was consistent with previous autoradiographic findings in human postmortem brain. The binding potential was largest in the anterior (1.25) and posterior putamen (1.20), followed by the head of caudate nucleus (1.05) and thalamus (1.03). Low ligand binding was found in the cerebral cortex, particularly in the frontal lobe (0.46). Interestingly, the binding of [^{11}C]TMSX was relatively large in the thalamus in comparison with previous reports for other mammals (Mishina et al. 2007). For clinical purposes, the authors developed a modeling method (Naganawa et al. 2007) and proposed recently a noninvasive method for in vivo receptor quantification (Naganawa et al. 2008). A preliminary application of [^{11}C]TMSX to patients suffering from PD was presented at an international meeting (Mishina et al. 2006). Figure 3 depicts [^{11}C]TMSX PET images of a unilaterally affected patient with early-stage PD and a healthy control subject. [^{11}C]TMSX binding was reduced in the left putamen, which

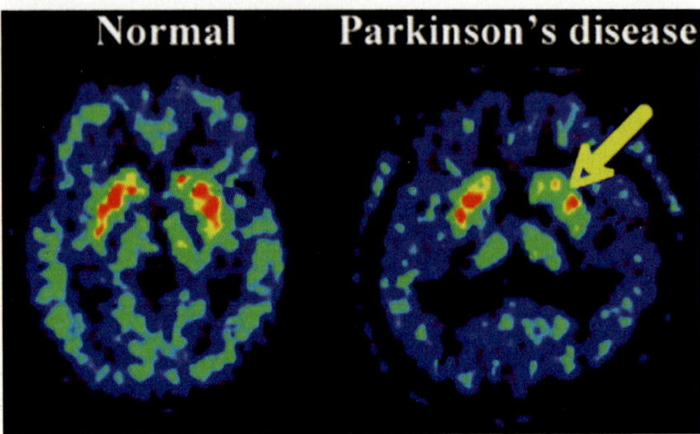

Fig. 3 Distribution of adenosine A_{2A} receptors (A_{2A}ARs) in the human brain: a normal subject (*left*) and a patient with Parkinson's disease (PD) (*right*). The binding potential of [^{11}C]TMSX (Naganawa et al. 2007) in a patient with early-stage PD (*right*) was lower in the putamen of the left hemisphere (*arrow*), which was consistent with more severe clinical symptoms on the right body side. In contrast, the binding of [^{11}C]raclopride to dopamine D_2 receptors (D_2Rs) was slightly increased in the left putamen (Mishina et al. 2006). See text for comments on the findings of this PET study in humans

is contralateral to the primarily affected body side, while binding of [^{11}C]raclopride to D_2Rs was slightly increased. Upregulation of D_2Rs most likely reflects a postsynaptic compensation to impaired presynaptic dopamine release. Simultaneous downregulation of A_{2A}ARs and upregulation of D_2Rs is therefore likely to reflect an imbalance of adenosinergic and dopaminergic transmission at the postsynaptic site as a consequence of PD pathophysiology. This study suggests that PET imaging with A_{2A}AR-selective radiotracer PET ligands may be used to monitor the natural history and progression of PD in both animal models of PD and humans with PD, and may serve as guide for therapy with A_{2A}AR antagonists in patients with PD. Moreover, PET imaging with A_{2A}AR-selective radiotracer PET ligands may be used to stratify patients for recruitment into clinical trials (i.e., patients with early versus later stages of PD), in order to determine the safety and efficacy of A_{2A}AR antagonists in this patient population.

The above mentioned development of the selective A_{2A}AR antagonist istradefylline (KW-6002) as a nondopaminergic drug for PD (Kase et al. 2003) is another good example of the usefulness of PET imaging in the process of drug development. In a study of healthy subjects, seven groups received doses of cold istradefylline ranging from 0 to 40 mg per day for 14 days (Brooks et al. 2008). Thereafter, ^{11}C-labeled istradefylline ([^{11}C]KW-6002) and PET were applied in order to determine the binding potential of [^{11}C]KW-6002. Estimates of the striatal binding potential were used to derive saturation kinetics in the presence of cold KW-6002, assuming that nonspecific binding was constant across subjects and the binding potential was proportional to the concentration of available A_{2A}AR binding sites.

Brain [^{11}C]KW-6002 uptake was well characterized by a two-tissue compartmental model with a blood volume term, and the 50% efficient dose (ED_{50}) of cold KW-6002 was 0.5 mg in the striatum. The study revealed that over 90% receptor occupancy was achieved with daily oral doses of greater than 5 mg.

5 Conclusion

Both basic neuroscience and clinical research have established substantial evidence for an important role of adenosine and its receptors in the pathophysiology of the brain. Molecular in vivo imaging of ARs in the human brain is therefore an attractive means to study the role of adenosine, its receptor subtypes and their alterations under disease conditions in patients suffering from neurologic and psychiatric disorders, sleep disorders, and perhaps drug addiction. The first two high-affinity and subtype-selective AR ligands dedicated for use in PET, [^{18}F]CPFPX and [^{11}C]MPDX permit quantitative measurements of A_1ARs in the living human brain. The clinically important A_{2A}AR has been made accessible through the use of [^{11}C]TMSX and [^{11}C]KW-6002, a radiolabeled drug. Reports on human applications are currently focused on A_1ARs and A_{2A}ARs, reflecting current understanding of their specific implications in cerebral neuropathology and their potential as neuroprotective targets. Regarding A_{2B}ARs and A_3ARs, their relatively low densities and their disease-specific appearance make it more challenging to assess them in vivo. However, given that it is now clear that adenosine plays a greater role in the pathophysiology of neurological and psychiatric disorders than previously thought, and the systematic and intensive search that is now underway for ligands with high affinity and selectivity, the molecular imaging of ARs will become an increasingly important tool in clinically oriented research.

References

Abbracchio MP, Cattabeni F (1999) Brain adenosine receptors as targets for therapeutic intervention in neurodegenerative diseases. Ann N Y Acad Sci 890:79–92

Albasanz JL, Perez S, Barrachina M, Ferrer I, Martín M (2008) Up-regulation of adenosine receptors in the frontal cortex in Alzheimer's disease. Brain Pathol 18:211–219

Angulo E, Casadó V, Mallol J, Canela EI, Viñals F, Ferrer I, Lluis C, Franco R (2003) A_1 adenosine receptors accumulate in neurodegenerative structures in Alzheimer disease and mediate both amyloid precursor protein processing and tau phosphorylation and translocation. Brain Pathol 13:440–451

Ascherio A, Zhang SM, Hernán MA, Kawachi I, Colditz GA, Speizer FE, Willett WC (2001) Prospective study of caffeine consumption and risk of Parkinson's disease in men and women. Ann Neurol 50:56–63

Bara-Jimenez W, Sherzai A, Dimitrova T, Favit A, Bibbiani F, Gillespie M, Morris MJ, Mouradian MM, Chase TN (2003) Adenosine A_{2A} receptor antagonist treatment of Parkinson's disease. Neurology 61:293–296

Bauer A, Holschbach MH, Meyer PT, Boy C, Herzog H, Olsson RA, Coenen HH, Zilles K (2003) In vivo imaging of adenosine A_1 receptors in the human brain with [^{18}F]CPFPX and positron emission tomography. Neuroimage 19:1760–1769

Bauer A, Langen KJ, Bidmon H, Holschbach MH, Weber S, Olsson RA, Coenen HH, Zilles K (2005) ^{18}F-CPFPX PET identifies changes in cerebral A_1 adenosine receptor density caused by glioma invasion. J Nucl Med 46:450–454

Berne RM, Winn HR, Rubio R (1981) The local regulation of cerebral blood flow. Prog Cardiovasc Dis 24:243–260

Bier D, Holschbach MH, Wutz W, Olsson RA, Coenen HH (2006) Metabolism of the A_1 adenosine receptor positron emission tomography ligand [^{18}F]8-cyclopentyl-3-(3-fluoropropyl)-1-propylxanthine ([^{18}F]CPFPX) in rodents and humans. Drug Metab Dispos 34:570–576

Blum T, Ermert J, Wutz W, Bier D, Coenen HH (2004) First no-carrier-added radioselenation of an adenosine-A_1 receptor ligand. J Label Compd Radiopharm 47:415–427

Boy C, Meyer PT, Kircheis G, Holschbach MH, Herzog H, Elmenhorst D, Kaiser HJ, Coenen HH, Häussinger D, Zilles K, Bauer A (2008) Cerebral A_1 adenosine receptors (A_1AR) in liver cirrhosis. Eur J Nucl Med Mol Imaging 35:589–597

Brooks DJ, Doder M, Osman S, Luthra SK, Hirani E, Hume S, Kase H, Kilborn J, Martindill S, Mori A (2008) Positron emission tomography analysis of [^{11}C]KW-6002 binding to human and rat adenosine A_{2A} receptors in the brain. Synapse 62:671–681

Brundege JM, Dunwiddie TV (1997) Role of adenosine as a modulator of synaptic activity in the central nervous system. Adv Pharmacol 39:353–391

Bruns RF, Fergus JH, Badger EW, Bristol JA, Santay LA, Hartman JD, Hays SJ, Huang CC (1987) Binding of the A_1-selective adenosine antagonist 8-cyclopentyl-1,3-dipropylxanthine to rat brain membranes. Naunyn–Schmiedeberg's Arch Pharmacol 335:59–63

Chaudhuri A, Cohen RZ, Larocque S (1998) Distribution of adenosine A_1 receptors in primary visual cortex of developing and adult monkeys. Exp Brain Res 123:351–354

Correa M, Wisniecki A, Betz A, Dobson DR, O'Neill MF, O'Neill MJ, Salamone JD (2004) The adenosine A_{2A} antagonist KF17837 reverses the locomotor suppression and tremulous jaw movements induced by haloperidol in rats: possible relevance to parkinsonism. Behav Brain Res 148:47–54

Deckert J, Abel F, Kunig G, Hartmann J, Senitz D, Maier H, Ransmayr G, Riederer P (1998) Loss of human hippocampal adenosine A_1 receptors in dementia: evidence for lack of specificity. Neurosci Lett 244:1–4

Dehnhardt M, Palm C, Vieten A, Bauer A, Pietrzyk U (2007) Quantifying the A_1AR distribution in peritumoral zones around experimental F98 and C6 brain tumours. J Neurooncol 85:49–63

Dirnagl U, Niwa K, Lindauer U, Villringer A (1994) Coupling of cerebral blood flow to neuronal activation: role of adenosine and nitric oxide. Am J Physiol 267:H296–H301

Dunwiddie TV, Masino SA (2001) The role and regulation of adenosine in the central nervous system. Annu Rev Neurosci 24:31–55

Elmenhorst D, Meyer PT, Matusch A, Winz OH, Zilles K, Bauer A (2007a) Test-retest stability of cerebral A_1 adenosine receptor quantification using [^{18}F]CPFPX and PET. Eur J Nucl Med Mol Imaging 34:1061–1070

Elmenhorst D, Meyer PT, Winz OH, Matusch A, Ermert J, Coenen HH, Basheer R, Haas HL, Zilles K, Bauer A (2007b) Sleep deprivation increases A_1 adenosine receptor binding in the human brain: a [^{18}F]CPFPX PET study. J Neurosci 27:2410–2415

Fastbom J, Pazos A, Probst A, Palacios JM (1986) Adenosine A_1-receptors in human brain: characterization and autoradiographic visualization. Neurosci Lett 65:127–132

Ferré S, Fuxe K (1992) Dopamine denervation leads to an increase in the intramembrane interaction between adenosine A_2 and dopamine D_2 receptors in the neostriatum. Brain Res 594:124–130

Franklin PH, Zhang G, Tripp ED, Murray TF (1989) Adenosine A_1 receptor activation mediates suppression of (−)bicuculline methiodide-induced seizures in rat prepiriform cortex. J Pharmacol Exp Ther 251:1229–1236

Fredholm BB, Abbracchio MP, Burnstock G, Dubyak GR, Harden TK, Jacobson KA, Schwabe U, Williams M (1997) Towards a revised nomenclature for P1 and P2 receptors. Trends Pharmacol Sci 18:79–82

Fredholm BB, Bättig K, Holmén J, Nehlig A, Zvartau EE (1999) Actions of caffeine in the brain with special reference to factors that contribute to its widespread use. Pharmacol Rev 51: 83–133

Fredholm BB, IJzerman AP, Jacobson KA, Klotz KN, Linden J (2001) International union of pharmacology. XXV. Nomenclature and classification of adenosine receptors. Pharmacol Rev 53:527–552

Fukumitsu N, Ishii K, Kimura Y, Oda K, Sasaki T, Mori Y, Ishiwata K (2003) Imaging of adenosine A_1 receptors in the human brain by positron emission tomography with [^{11}C]MPDX. Ann Nucl Med 17:511–515

Fukumitsu N, Ishii K, Kimura Y, Oda K, Sasaki T, Mori Y, Ishiwata K (2005) Adenosine A_1 receptor mapping of the human brain by PET with 8-dicyclopropylmethyl-1–^{11}C-methyl-3-propylxanthine. J Nucl Med 46:32–37

Fukumitsu N, Ishii K, Kimura Y, Oda K, Hashimoto M, Suzuki M, Ishiwata K (2008) Adenosine A_1 receptors using 8-dicyclopropylmethyl-1-[^{11}C]methyl-3-propylxanthine PET in Alzheimer's disease. Ann Nucl Med 22:841–847

Furuta R, Ishiwata K, Kiyosawa M, Ishii S, Saito N, Shimada J, Endo K, Suzuki F, Senda M (1996) Carbon-11-labeled KF15372: a potential central nervous system adenosine A_1 receptor ligand. J Nucl Med 37:1203–1207

Fuxe K, Ferré S, Snaprud P, von Euler G, Jahansson B, Ferdholm B (1993) Antagonistic A_{2A}/D_2 receptor interaction in the striatum as a basis for adenosine/dopamine interactions in the central nervous system. Drug Dev Res 28:374–380

Glass M, Faull RL, Dragunow M (1996) Localization of the adenosine uptake site in the human brain: a comparison with the distribution of adenosine A_1 receptors. Brain Res 710:79–91

Hauser RA, Hubble JP, Truong DD (2003) Randomized trial of the adenosine A_{2A} receptor antagonist istradefylline in advanced PD. Neurology 61:297–303

Hayaishi O (1999) Prostaglandin D_2 and sleep: a molecular genetic approach. J Sleep Res 8(Suppl 1):60–64

Herzog H, Elmenhorst D, Winz O, Bauer A (2008) Biodistribution and radiation dosimetry of the A_1 adenosine receptor ligand ^{18}F-CPFPX determined from human whole-body PET. Eur J Nucl Med Mol Imaging 35:1499–1506

Hirani E, Gillies J, Karasawa A, Shimada J, Kase H, Opacka-Juffry J, Osman S, Luthra SK, Hume SP, Brooks DJ (2001) Evaluation of [4-*O*-methyl-^{11}C]KW-6002 as a potential PET ligand for mapping central adenosine A_{2A} receptors in rats. Synapse 42:164–176

Holschbach MH, Olsson RA (2002) Applications of adenosine receptor ligands in medical imaging by positron emission tomography. Curr Pharm Des 8:2345–2352

Holschbach MH, Fein T, Krummeich C, Lewis RG, Wutz W, Schwabe U, Unterlugauer D, Olsson RA (1998) A_1 adenosine receptor antagonists as ligands for positron emission tomography (PET) and single-photon emission tomography (SPET). J Med Chem 41:555–563

Holschbach MH, Olsson RA, Bier D, Wutz W, Sihver W, Schuller M, Palm B, Coenen HH (2002) Synthesis and evaluation of no-carrier-added 8-cyclopentyl-3-(3-[^{18}F]fluoropropyl)-1-propylxanthine ([^{18}F]CPFPX): a potent and selective A_1-adenosine receptor antagonist for in vivo imaging. J Med Chem 45:5150–5156

Hunter JC (2006) SCH 420814: a novel adenosine A_{2A} antagonist. Exploring Parkinson's disease and beyond. In: Targeting Adenosine A_{2A} Receptors in Parkinson's Disease and Other CNS Disorders (meeting), Boston, MA, 17–19 May 2006

Ishiwata K, Furuta R, Shimada J, Ishii S, Endo K, Suzuki F, Senda M (1995) Synthesis and preliminary evaluation of [^{11}C]KF15372, a selective adenosine A_1 antagonist. Appl Radiat Isotopes 46:1009–1013

Ishiwata K, Noguchi J, Toyama H, Sakiyama Y, Koike N, Ishii S, Oda K, Endo K, Suzuki F, Senda M (1996) Synthesis and preliminary evaluation of [^{11}C]KF17837, a selective adenosine A_{2A} antagonist. Appl Radiat Isotopes 47:507–511

Ishiwata K, Sakiyama Y, Sakiyama T, Shimada J, Toyama H, Oda K, Suzuki F, Senda M (1997) Myocardial adenosine A_{2a} receptor imaging of rabbit by PET with [^{11}C]KF17837. Ann Nucl Med 11:219–225

Ishiwata K, Noguchi J, Wakabayashi S, Shimada J, Ogi N, Nariai T, Tanaka A, Endo K, Suzuki F, Senda M (2000a) ^{11}C-labeled KF18446: a potential central nervous system adenosine A_{2a} receptor ligand. J Nucl Med 41:345–354

Ishiwata K, Ogi N, Shimada J, Nonaka H, Tanaka A, Suzuki F, Senda M (2000b) Further characterization of a CNS adenosine A_{2A} receptor ligand [^{11}C]KF18446 with in vitro autoradiography and in vivo tissue uptake. Ann Nucl Med 14:81–89

Ishiwata K, Ogi N, Shimada J, Wang W, Ishii K, Tanaka A, Suzuki F, Senda M (2000c) Search for PET probes for imaging the globus pallidus studied with rat brain ex vivo autoradiography. Ann Nucl Med 14:461–466

Ishiwata K, Shimada J, Wang WF, Harakawa H, Ishii S, Kiyosawa M, Suzuki F, Senda M (2000d) Evaluation of iodinated and brominated [^{11}C]styrylxanthine derivatives as in vivo radioligands mapping adenosine A_{2A} receptor in the central nervous system. Ann Nucl Med 14:247–253

Ishiwata K, Takai H, Nonaka H, Ishii S, Simada J, Senda M (2001) Synthesis and preliminary evaluation of a carbon-11-labeled adenosine transporter blocker [^{11}C]KF21562. Nucl Med Biol 28:281–285

Ishiwata K, Nariai T, Kimura Y, Oda K, Kawamura K, Ishii K, Senda M, Wakabayashi S, Shimada J (2002a) Preclinical studies on [^{11}C]MPDX for mapping adenosine A_1 receptors by positron emission tomography. Ann Nucl Med 16:377–382

Ishiwata K, Ogi N, Hayakawa N, Oda K, Nagaoka T, Toyama H, Suzuki F, Endo K, Tanaka A, Senda M (2002b) Adenosine A_{2A} receptor imaging with [^{11}C]KF18446 PET in the rat brain after quinolinic acid lesion: comparison with the dopamine receptor imaging. Ann Nucl Med 16:467–475

Ishiwata K, Shimada J, Ishii K, Suzuki F (2002c) Assessment of adenosine A_{2A} receptors with PET as a new diagnostic tool for neurological disorders. Drugs Future 27:569–576

Ishiwata K, Kawamura K, Kimura Y, Oda K, Ishii K (2003a) Potential of an adenosine A_{2A} receptor antagonist [^{11}C]TMSX for myocardial imaging by positron emission tomography: a first human study. Ann Nucl Med 17:457–462

Ishiwata K, Wang WF, Kimura Y, Kawamura K, Ishii K (2003b) Preclinical studies on [^{11}C]TMSX for mapping adenosine A_{2A} receptors by positron emission tomography. Ann Nucl Med 17:205–211

Ishiwata K, Mizuno M, Kimura Y, Kawamura K, Oda K, Sasaki T, Nakamura Y, Muraoka I, Ishii K (2004) Potential of [^{11}C]TMSX for the evaluation of adenosine A_{2A} receptors in the skeletal muscle by positron emission tomography. Nucl Med Biol 31:949–956

Ishiwata K, Mishina M, Kimura Y, Oda K, Sasaki T, Ishii K (2005a) First visualization of adenosine A_{2A} receptors in the human brain by positron emission tomography with [^{11}C]TMSX. Synapse 55:133–136

Ishiwata K, Tsukada H, Kimura Y, Kawamura K, Harada N, Hendrikse NH (2005b) In vivo evaluation of [^{11}C]TMSX and [^{11}C]KF21213 for mapping adenosine A_{2A} receptors: brain kinetics in the conscious monkey and P-glycoprotein modulation in the mouse brain. In: 22nd Int Symp on Cerebral Blood Flow, Metabolism, and Function/7th Int Conf on Quantification of Brain Function with PET, Amsterdam, Netherlands, 7–11 June 2005

Ishiwata K, Kimura Y, de Vries EFJ, Elsinga PH (2008) PET tracers for mapping adenosine receptors as probes for diagnosis of CNS disorders. CNS Agents Med Chem 7:57–77

Jaarsma D, Sebens JB, Korf J (1991) Reduction of adenosine A_1-receptors in the perforant pathway terminal zone in Alzheimer hippocampus. Neurosci Lett 121:111–114

Jacobson KA, Gao ZG (2006) Adenosine receptors as therapeutic targets. Nat Rev Drug Discov 5:247–264

Johnston JB, Silva C, Gonzalez G, Holden J, Warren KG, Metz LM, Power C (2001) Diminished adenosine A_1 receptor expression on macrophages in brain and blood of patients with multiple sclerosis. Ann Neurol 49:650–658

Kase H, Aoyama S, Ichimura M, Ikeda K, Ishii A, Kanda T, Koga K, Koike N, Kurokawa M, Kuwana Y, Mori A, Nakamura J, Nonaka H, Ochi M, Saki M, Shimada J, Shindou T, Shiozaki S, Suzuki F, Takeda M, Yanagawa K, Richardson PJ, Jenner P, Bedard P, Borrelli E, Hauser RA, Chase TN; KW-6002 US-001 Study Group (2003) Progress in pursuit of therapeutic A_{2A} antagonists: the adenosine A_{2A} receptor selective antagonist KW6002: research and development toward a novel nondopaminergic therapy for Parkinson's disease. Neurology 61(Suppl 6):S97–S100

Kawamura K, Ishiwata K (2004) Improved synthesis of [^{11}C]SA4503, [^{11}C]MPDX and [^{11}C]TMSX by use of [^{11}C]methyl triflate. Ann Nucl Med 18:165–168

Kimura Y, Ishii K, Fukumitsu N, Oda K, Sasaki T, Kawamura K, Ishiwata K (2004) Quantitative analysis of adenosine A_1 receptors in human brain using positron emission tomography and [1-methyl-^{11}C]8-dicyclopropylmethyl-1-methyl-3-propylxanthine. Nucl Med Biol 31:975–981

Kiyosawa M, Ishiwata K, Noguchi J, Endo K, Wang WF, Suzuki F, Senda M (2001) Neuroreceptor bindings and synaptic activity in visual system of monocularly enucleated rat. Jpn J Ophthalmol 45:264–269

Koga K, Kurokawa M, Ochi M, Nakamura J, Kuwana Y (2000) Adenosine A_{2A} receptor antagonists KF17837 and KW-6002 potentiate rotation induced by dopaminergic drugs in hemi-Parkinsonian rats. Eur J Pharmacol 408:249–255

Kuroda S, Takamura F, Tenda Y, Itani H, Tomishima Y, Akahane A, Sakane K (2001) Design, synthesis and biological evaluation of a novel series of potent, orally active adenosine A_1 receptor antagonists with high blood–brain barrier permeability. Chem Pharm Bull (Tokyo) 49:988–998

Landolt HP (2008) Sleep homeostasis: a role for adenosine in humans? Biochem Pharmacol 75:2070–2079

Lehel SZ, Horvath G, Boros I, Mikecz P, Marian T, Szentmiklosi AJ, Tron L (2000) Synthesis of 5′-*N*-(2-[^{18}F]Fluoroethyl)-carboxamidoadenosine: a promising tracer for investigation of adenosine receptor system by PET technique. J Label Compd Radiopharm 43:807–815

Lohse MJ, Klotz KN, Lindenborn-Fotinos J, Reddington M, Schwabe U, Olsson RA (1987) 8-Cyclopentyl-1,3-dipropylxanthine (DPCPX): a selective high affinity antagonist radioligand for A_1 adenosine receptors. Naunyn–Schmiedeberg's Arch Pharmacol 336:204–210

Maemoto T, Tada M, Mihara T, Ueyama N, Matsuoka H, Harada K, Yajima T, Shirakawa K, Kuroda S, Akihane A, Iwashita A, Matsuoka N, Mutoh S (2004) Pharmacological characterization of FR194921, a new potent, selective, and orally active antagonist for central adenosine A_1 receptors. J Pharmacol Sci 96:42–52

Marian T, Boros I, Lengyel Z, Balkay L, Horvath G, Emri M, Sarkadi E, Szentmiklosi AJ, Fekete I, Tron L (1999) Preparation and primary evaluation of [^{11}C]CSC as a possible tracer for mapping adenosine A_{2A} receptors by PET. Appl Radiat Isotopes 50:887–893

Marino MJ, Valenti O, Conn PJ (2003) Glutamate receptors and Parkinson's disease: opportunities for intervention. Drugs Aging 20:377–397

Mathews WB, Nakamoto Y, Abraham EH, Scheffel U, Hilton J, Ravert HT, Tatsumi M, Rauseo PA, Traughber BJ, Salikhova AY, Dannals RF, Wahl RL (2005) Synthesis and biodistribution of [^{11}C]adenosine 5′-monophosphate ([^{11}C]AMP). Mol Imaging Biol 7:203–208

Matsuya T, Takamatsu H, Murakami Y, Noda A, Ichise R, Awaga Y, Nishimura S (2005) Synthesis and evaluation of [^{11}C]FR194921 as a nonxanthine-type PET tracer for adenosine A_1 receptors in the brain. Nucl Med Biol 32:837–844

Matusch A, Meyer PT, Bier D, Holschbach MH, Woitalla D, Elmenhorst D, Winz OH, Zilles K, Bauer A (2006) Metabolism of the A_1 adenosine receptor PET ligand [^{18}F]CPFPX by CYP1A2: implications for bolus/infusion PET studies. Nucl Med Biol 33:891–898

Meyer PT, Bier D, Holschbach MH, Cremer M, Tellmann L, Bauer A (2003) Image of the month. In vivo imaging of rat brain A_1 adenosine receptor occupancy by caffeine. Eur J Nucl Med Mol Imaging 30:1440

Meyer PT, Bier D, Holschbach MH, Boy C, Olsson RA, Coenen HH, Zilles K, Bauer A (2004) Quantification of cerebral A_1 adenosine receptors in humans using [^{18}F]CPFPX and PET. J Cereb Blood Flow Metab 24:323–333

Meyer PT, Elmenhorst D, Bier D, Holschbach MH, Matusch A, Coenen HH, Zilles K, Bauer A (2005a) Quantification of cerebral A_1 adenosine receptors in humans using [^{18}F]CPFPX and PET: an equilibrium approach. Neuroimage 24:1192–1204

Meyer PT, Elmenhorst D, Zilles K, Bauer A (2005b) Simplified quantification of cerebral A_1 adenosine receptors using [^{18}F]CPFPX and PET: analyses based on venous blood sampling. Synapse 55:212–223

Meyer PT, Elmenhorst D, Matusch A, Winz O, Zilles K, Bauer A (2006) A_1 adenosine receptor PET using [^{18}F]CPFPX: displacement studies in humans. Neuroimage 32:1100–1105

Mishina M, Ishii K, Kitamura S, Kimura Y, Naganawa M, Hashimoto M, Suzuki M, Oda K, Kobayashi S, Katayama, Y Ishiwata K (2006) Distribution of adenosine A_{2A} receptors in de novo Parkinson's disease using TMSX PET: a preliminary study (Poster 44). In: Targeting Adenosine A_{2A} Receptors in Parkinson's Disease and Other CNS Disorders (meeting), Boston, MA, 17–19 May 2006

Mishina M, Ishiwata K, Kimura Y, Naganawa M, Oda K, Kobayashi S, Kitamura S, Katayama Y, Ishii K (2007) Evaluation of distribution of adenosine A_{2A} receptors in normal human brain measured with [^{11}C]TMSX PET. Synapse 61:778–784

Mizuno M, Kimura K, Tokizawa K, Ishii K, Oda K, Sasaki T, Nakamura Y, Muraoka I, Ishiwata K (2005) Greater adenosine A_{2A} receptor densities in cardiac and skeletal muscle in endurance trained men: a [^{11}C]TMSX PET study. Nucl Med Biol 32:831–836

Moraidis I, Bingmann D (1994) Epileptogenic actions of xanthines in relation to their affinities for adenosine A_1 receptors in CA3 neurons of hippocampal slices (guinea pig). Brain Res 640:140–145

Moresco RM, Todde S, Belloli S, Simonelli P, Panzacchi A, Rigamonti M, Galli-Kienle M, Fazio F (2005) In vivo imaging of adenosine A_{2A} receptors in rat and primate brain using [^{11}C]SCH442416. Eur J Nucl Med Mol Imaging 32:405–413

Müller CE, Geis U, Hipp J, Schobert U, Frobenius W, Pawlowski M, Suzuki F, Sandoval-Ramírez J (1997) Synthesis and structure-activity relationship of 3,7-dimethyl-1-propargylxanthine derivatives, A_{2A}-selective adenosine receptor antagonists. J Med Chem 40:4396–4405

Müller CE, Sandoval-Ramírez J, Schobert U, Geis U, Frobenius W, Klotz KN (1998) 8-(Sulfosytryl)xanthines; water-soluble A_{2A}-selective adenosine receptor antagonists. Bioorg Med Chem 6:707–719

Naganawa M, Kimura Y, Mishina M, Manabe Y, Chihara K, Oda K, Ishii K, Ishiwata K (2007) Quantification of adenosine A_{2A} receptors in the human brain using [^{11}C]TMSX and positron emission tomography. Eur J Nucl Med Mol Imaging 34:679–687

Naganawa M, Kimura Y, Yano J, Mishina M, Yanagisawa M, Ishii K, Oda K, Ishiwata K (2008) Robust estimation of the arterial input function for Logan plots using an intersectional searching algorithm and clustering in positron emission tomography for neuroreceptor imaging. Neuroimage 40:26–34

Nariai T, Shimada Y, Ishiwata K, Nagaoka T, Shimada J, Kuroiwa T, Ono K, Ohno K, Hirakawa K, Senda M (2003) PET imaging of adenosine A_1 receptors with ^{11}C-MPDX as an indicator of severe cerebral ischemic insult. J Nucl Med 44:1839–1844

Noguchi J, Ishiwata K, Furuta R, Simada J, Kiyosawa M, Ishii S, Endo K, Suzuki F, Senda M (1997) Evaluation of carbon-11 labeled KF15372 and its ethyl and methyl derivatives as a potential CNS adenosine A_1 receptor ligand. Nucl Med Biol 24:53–59

Noguchi J, Ishiwata K, Wakabayashi S, Nariai T, Shumiya S, Ishii S, Toyama H, Endo K, Suzuki F, Senda M (1998) Evaluation of carbon-11-labeled KF17837: a potential CNS adenosine A_{2a} receptor ligand. J Nucl Med 39:498–503

Nonaka Y, Shimada J, Nonaka H, Koike N, Aoki N, Kobayashi H, Kase H, Yamaguchi K, Suzuki F (1993) Photoisomerization of a potent and selective adenosine A_2 antagonist, (E)-1,3-dipropyl-8-(3,4-dimethoxystyryl)-7-methylxanthine. J Med Chem 36:3731–3733

Nonaka H, Ichimura M, Takeda M, Nonaka Y, Shimada J, Suzuki F, Yamakuchi K, Kase A (1994) KF17837 ((E)-8-(3,4-dimethoxystyryl)-1,3-dipropyl-7-methylxanthine) a potent and selective adenosine A_2 receptor antagonist. Eur J Pharmcol 267:335–341

Olah ME, Stiles GL (2000) The role of receptor structure in determining adenosine receptor activity. Pharmacol Ther 85:55–75

Porkka-Heiskanen T (1999) Adenosine in sleep and wakefulness. Ann Med 31:125–129

Portas CM, Thakkar M, Rainnie DG, Greene RW, McCarley RW (1997) Role of adenosine in behavioral state modulation: a microdialysis study in the freely moving cat. Neuroscience 79:225–235

Ribeiro JA, Sebastiao AM, de Mendonca A (2003) Participation of adenosine receptors in neuroprotection. Drug News Perspect 16:80–86

Ross GW, Abbott RD, Petrovitch H, Morens DM, Grandinetti A, Tung KH, Tanner CM, Masaki KH, Blanchette PL, Curb JD, Popper JS, White LR (2000) Association of coffee and caffeine intake with the risk of Parkinson disease. JAMA 283:2674–2679

Schindler M, Harris CA, Hayes B, Papotti M, Humphrey PP (2001) Immunohistochemical localization of adenosine A_1 receptors in human brain regions. Neurosci Lett 297:211–215

Schubert P, Ogata T, Marchini C, Ferroni S, Rudolphi K (1997) Protective mechanisms of adenosine in neurons and glial cells. Ann N Y Acad Sci 825:1–10

Schubert P, Ogata T, Marchini C, Ferroni S (2001) Glia-related pathomechanisms in Alzheimer's disease: a therapeutic target? Mech Ageing Dev 123:47–57

Schwierin B, Borbély AA, Tobler I (1996) Effects of N^6-cyclopentyladenosine and caffeine on sleep regulation in the rat. Eur J Pharmacol 300:163–171

Seale TW, Abla KA, Shamim MT, Carney JM, Daly JW (1988) 3,7-Dimethyl-1-propargylxanthine: a potent and selective in vivo antagonist of adenosine analogs. Life Sci 43:1671–1684

Shimada J, Suzuki F, Nonaka H, Karasawa A, Mizumoto H, Ohno T, Kubo K, Ishii A (1991) 8-(Dicyclopropylmethyl)-1,3-dipropylxanthine: a potent and selective adenosine A_1 antagonist with renal protective and diuretic activities. J Med Chem 34:466–469

Shimada J, Suzuki J, Nonaka H, Ishii A, Ichikawa S (1992) (*E*)-1,3-Dialkyl-7-methyl-8-(3,4,5-trimethoxystyryl)xanthine: potent and selective adenosine A_2 antagonists. J Med Chem 35:2342–2345

Shimada Y, Ishiwata K, Kiyosawa M, Nariai T, Oda K, Toyama H, Suzuki F, Ono K, Senda M (2002) Mapping adenosine A_1 receptors in the cat brain by positron emission tomography with [^{11}C]MPDX. Nucl Med Biol 29:29–37

Stone-Elander S, Thorell JO, Eriksson L, Fredholm BB, Ingvar M (1997) In vivo biodistribution of [N-^{11}C-methyl]KF 17837 using 3-D-PET: evaluation as a ligand for the study of adenosine A_{2A} receptors. Nucl Med Biol 24:187–191

Suzuki F, Ishiwata K (1998) Selective adenosine antagonists for mapping central nervous system adenosine receptors with positron emission tomography: Carbon-11 labeled KF15372 (A_1) and KF17837 (A_{2a}). Drug Dev Res 45:312–323

Suzuki F, Shimada J, Mizumoto H, Karasawa A, Kubo K, Nonaka H, Ishii A, Kawakita T (1992) Adenosine A_1 antagonists. 2. Structure–activity relationships on diuretic activities and protective effects against acute renal failure. J Med Chem 35:3066–3075

Svenningsson P, Hall H, Sedvall G, Fredholm BB (1997) Distribution of adenosine receptors in the postmortem human brain: an extended autoradiographic study. Synapse 27:322–335

Todde S, Moresco RM, Simonelli P, Baraldi PG, Cacciari B, Spalluto G, Varani K, Monopoli A, Matarrese M, Carpinelli A, Magni F, Kienle MG, Fazio F (2000) Design, radiosynthesis, and biodistribution of a new potent and selective ligand for in vivo imaging of the adenosine A_{2A} receptor system using positron emission tomography. J Med Chem 43:4359–4362

Ulas J, Brunner LC, Nguyen L, Cotman CW (1993) Reduced density of adenosine A_1 receptors and preserved coupling of adenosine A_1 receptors to G proteins in Alzheimer hippocampus: a quantitative autoradiographic study. Neuroscience 52:843–854

Virus RM, Ticho S, Pilditch M, Radulovacki M (1990) A comparison of the effects of caffeine, 8-cyclopentyltheophylline, and alloxazine on sleep in rats. Possible roles of central nervous system adenosine receptors. Neuropsychopharmacology 3:243–249

Wadsak W, Mien LK, Shanab K, Ettlinger DE, Haeusler D, Sindelar K, Lanzenberger RR, Spreitzer H, Viernstein H, Keppler BK, Dudczak R, Kletter K, Mitterhauser M (2008) Preparation and first evaluation of [^{18}F]FE@SUPPY: a new PET tracer for the adenosine A_3 receptor. Nucl Med Biol 35:61–66

Wakabayashi S, Nariai T, Ishiwata K, Nagaoka T, Hirakawa K, Oda K, Sakiyama Y, Shumiya S, Toyama H, Suzuki F, Senda M (2000) A PET study of adenosine A_1 receptor in anesthetized monkey brain. Nucl Med Biol 27:401–406

Wang WF, Ishiwata K, Nonaka H, Ishii S, Kiyosawa M, Shimada J, Suzuki F, Senda M (2000) Carbon-11-labeled KF21213: a highly selective ligand for mapping CNS adenosine A_{2A} receptors with positron emission tomography. Nucl Med Biol 27:541–546

Wang WF, Ishiwata K, Kiyosawa M, Shimada J, Senda M, Mochizuki M (2003) Adenosine A_1 and benzodiazepine receptors and glucose metabolism in the visual structures of rats monocularly deprived by enucleation or eyelid suture at a sensitive period. Jpn J Ophthalmol 47:182–190

Xu K, Bastia E, Schwarzschild M (2005) Therapeutic potential of adenosine A_{2A} receptor antagonists in Parkinson's disease. Pharmacol Ther 105:267–310

Yanik G, Radulovacki M (1987) REM sleep deprivation up-regulates adenosine A_1 receptors. Brain Res 402:362–364

Zocchi C, Ongini E, Conti A, Monopoli A, Negretti A, Baraldi PG, Dionisotti S (1996a) The non-xanthine heterocyclic compound SCH 58261 is a new potent and selective A_{2A} adenosine receptor antagonist. J Pharmacol Exp Ther 276:398–404

Zocchi C, Ongini E, Ferrara S, Baraldi PG, Dionisotti S (1996b) Binding of the radioligand [^{3}H]-SCH 58261, a new non-xanthine A_{2A} adenosine receptor antagonist, to rat striatal membranes. Br J Pharmacol 117:1381–1386

Index

D

Printing: Krips bv, Meppel, The Netherlands
Binding: Stürtz, Würzburg, Germany